Mechanical behavior of materials-IV.

MECHANICAL BEHAVIOUR OF MATERIALS — IV

(in two volumes)

Volume 1

INTERNATIONAL SERIES ON THE
STRENGTH AND FRACTURE OF MATERIALS AND STRUCTURES

Other Titles in the Series

BUNSELL *et al.*	Advances in Composite Materials, 2 Volumes
EASTERLING	Mechanisms of Deformation & Fracture
FRANCOIS	Advances in Fracture Research, 6 Volumes
GARRETT & MARRIOTT	Engineering Applications of Fracture Analysis
GIFKINS	Strength of Metals & Alloys (ICSMA 6), 3 Volumes
HAASEN *et al.*	Strength of Metals & Alloys (ICSMA 5), 3 Volumes
MILLER & SMITH	Mechanical Behaviour of Materials (ICM 3), 3 Volumes
PIGGOTT	Load Bearing Fibre Composites
OSGOOD	Fatigue Design, 2nd Edition
RADON	Fracture & Fatigue: Elasto-Plasticity, Thin Sheet & Micromechanisms
SIH	Progress in Fracture Mechanics
SMITH	Fracture Mechanics: Current Status, Future Prospects
TAPLIN	Advances in Research on the Strength & Fracture of Materials, 6 Volumes

Related Pergamon Journals
(free specimen copy gladly sent on request)

Acta Metallurgica
Canadian Metallurgical Quarterly
Computers & Structures
Corrosion Science
Engineering Fracture Mechanics
Fatigue of Engineering Materials & Structures
International Journal of Solids & Structures
Materials Research Bulletin
Metals Forum
Physics of Metals & Metallography
Scripta Metallurgica

U.K.	Pergamon Press Ltd., Headington Hill Hall, Oxford OX3 0BW, England
U.S.A.	Pergamon Press Inc., Maxwell House, Fairview Park, Elmsford, New York 10523, U.S.A.
CANADA	Pergamon Press Canada Ltd., Suite 104, 150 Consumers Road, Willowdale, Ontario M2J 1P9, Canada
AUSTRALIA	Pergamon Press (Aust.) Pty. Ltd., P.O. Box 544, Potts Point, N.S.W. 2011, Australia
FRANCE	Pergamon Press SARL, 24 rue des Ecoles, 75240 Paris, Cedex 05, France
FEDERAL REPUBLIC OF GERMANY	Pergamon Press GmbH, Hammerweg 6, D-6242 Kronberg-Taunus, Federal Republic of Germany

First edition 1984

Library of Congress Cataloging in Publication Data

Main entry under title:
Mechanical behaviour of materials-IV.
(International series on the strength and fracture of materials and structures)
Papers submitted for the Fourth International Conference on Mechanical Behaviour of Materials. Cf. Pref.
Includes indexes.
1. Materials—Mechanical properties—Congresses.
I. Carlsson, Janne, 1932- . II. Ohlson, N. G.
III. International Conference on Mechanical Behavior of Materials (4th : 1983 : Stockholm, Sweden)
IV. Series. V. Title: Mechanical behaviour of materials-4.
TA417.6.M42 1983 620.1'1 83-12166

British Library Cataloguing in Publication Data

Mechanical behaviour of materials-IV.—(International series on the strength and fracture of materials and structures)
1. Materials—Congresses
I. Carlsson. J. II. Ohlson, N. G. III. Series
620.1'1292 TA401.3

ISBN 0-08-029340-9

In order to make this volume available as economically and as rapidly as possible the authors' typescripts have been reproduced in their original forms. This method unfortunately has its typographical limitations but it is hoped that they in no way distract the reader.

Printed in Great Britain by A. Wheaton & Co. Ltd., Exeter

MECHANICAL BEHAVIOUR OF MATERIALS — IV

Proceedings of the Fourth International Conference,
Stockholm, Sweden, 15–19 August 1983

Edited by

J. CARLSSON

and

N. G. OHLSON

Royal Institute of Technology, Stockholm, Sweden

Volume 1

PERGAMON PRESS

OXFORD · NEW YORK · TORONTO · SYDNEY · PARIS · FRANKFURT

CONTENTS OF VOLUME 1

THEME I: High Temperature Deformation and Failure

CONTENTS OF VOLUME 2

INVITED PAPERS

Due to the good response to our call for papers it was found necessary to restrict the length of normally contributed papers to seven pages. We are pleased that most authors have gladly complied with this restriction.

It was the policy of the organizing committee to have the deadline for the contributions as close as possible to the conference in order to get fresh, interesting material presented there. The consequence of this was that papers requiring corrections and changes could not be placed under the proper heading in the proceedings but had to be placed under a common heading "Additional papers" at the end of Volume 2. We are sorry for this since we realize it will cause some inconvenience to the readers.

Finally, on behalf of the local organizing committee, we would like to thank all those who contributed to the preparations of the conference. In particular, we would like to express our gratitude to Mrs. Ingunn Wester, Chief Assistant of the Organizing Committee, who worked faithfully for the ICM 4 from the moment at which preparations began, and to Mrs. Anna Brising who was responsible for the registration of participants.

Financial assistance from the following organizations is gratefully acknowledged.

ASEA
ASEA-ATOM
The Oskarshamn Nuclear Plant Energy Group, Ltd
Saab-Scania, Linköping
Sandvik
SKF
Sweden's Nuclear Inspectorate
The Swedish Board for Technical Development (STU)
The Swedish Ironmasters' Association
Swedish Plant Inspectorate
Uddeholm
Volvo

In conclusion, we hope that the Fourth International Conference on Mechanical Behaviour of Materials has contributed to the progress of material science and mechanics as successfully as its predecessors in the same series. Certainly this was promoted by the high scientific quality of the papers contributed and the great interest shown during the conference by the participants and their diligent participation in the scientific sessions as well as in the social programme. We are grateful to all contributors and participants for this.

Janne Carlsson	N.G. Ohlson
Chairman	Secretary

PREFACE

The First International Conference on Mechanical Behaviour of Materials was held in Kyoto, Japan in 1971. This was followed by conferences in Boston, Mass., in 1975 and in Cambridge, UK, in 1979.

The purpose of these conferences is to bring together engineers, continuum mechanics specialists and material scientists in order to discuss on a broad basis current engineering problems. For this reason each conference has been concentrated to a limited number of topics selected by the local organizing committee.

The Swedish community of specialists in strength of materials, solid mechanics and materials science consider it an honour to have been trusted to arrange the Fourth International Conference on Mechanical Behaviour of Materials. Once Stockholm had been chosen as the locality for the conference it was quite natural to give the responsibility for the arrangement to the Secretariates of the Materials Center and the Division of Strength of Materials and Solid Mechanics at the Royal Institute of Technology.

The themes of the conference were chosen with due regard both to important international trends in materials science and to the qualifications and interests of the Swedish scientific community and industry. The themes selected were:

1. High temperature deformation and failure

2. Mechanical properties of surface coated materials

3. Design with fibre reinforced (composite) materials

4. Sheet forming of high-strength materials

5. Materials aspects of fracture mechanics in engineering practice

The distribution of papers between the themes has turned out to be somewhat uneven. There has also been a broadening of the individual themes compared to what was originally planned.

MATERIALS ASPECTS OF CRACK TIP YIELDING AND SUBCRITICAL CRACK GROWTH IN ENGINEERING ALLOYS

J. Lankford

Department of Materials Sciences, Southwest Research Institute, 6220 Culebra Road, San Antonio, Texas 78284, USA

ABSTRACT

The micromechanics of crack opening, and plastic yielding within the highly deformed near-tip region, are described. Recent high resolution experimental observations are compared with the results of continuum analyses, and deviations due to material and environmental influences noted. The role of detailed crack yielding information in the modeling and understanding of subcritical crack advance is discussed.

KEYWORDS

Crack tip yielding; crack tip opening; fatigue crack growth mechanisms; small fatigue cracks; local strain measurement.

INTRODUCTION

Efforts to predict the lifetimes of flawed structures have led to a continuing need for reliable, detailed knowledge of crack tip yielding parameters. For the case of a monotonically loaded crack, this modeling requirement has been met with considerable success by recent finite element and J-integral type calculations. However, subcritical crack growth involving cyclic loading, aggressive environments, and microstructural/metallurgical influences can introduce significant uncertainties into what often are considered to be reasonably safe, but nevertheless critical, calculations.

At least two distinct types of analysis are usually required in order to describe crack growth rates in engineering materials. The first kind of analysis is a micromechanical one, based upon an assumed or, at best, inferred, micromechanism of crack tip extension (sliding off, crack renucleation, process zone damage accumulation, etc.). Such models invariably are derived in terms of material response parameters related to crack tip yielding (stress, strain, crack tip opening displacement, plastic zone size, crack opening mode). Values for the latter parameters must likewise be determined through some kind of analytical procedure, usually one based on continuum considerations.

The crack tip factors which normally figure in subcritical crack growth models are sketched in Fig. 1. This representation summarizes the following widely-held, continuum-based assumptions for Mode I, plane strain cracks subject to a uniaxial stress σ:

1. Outside the elastic-plastic boundary, elastic stress and strain scale as $r^{-1/2}$, where r is the radial distance from the crack tip.

2. The elastic-plastic boundary is shaped approximately as shown, and is defined by $(K_{max}/\sigma_y)^2$, where K_{max} is the maximum value of the stress intensity, and σ_y is the yield stress.

3. Over the bulk of the plastic zone, strain scales as $r^{-(1/1+n)}$, where n is the strain hardening coefficient.

4. Crack tip blunting is accommodated in local intense slip bands in which strain is proportional to r^{-1}.

5. The crack tip blunts, with the crack tip opening displacement (CTOD) being proportional to $K^2/\sigma_y E$, where E is Young's modulus; during fatigue, the crack blunts to the same extent on each cycle at a given cyclic stress intensity.

6. The crack opens in Mode I.

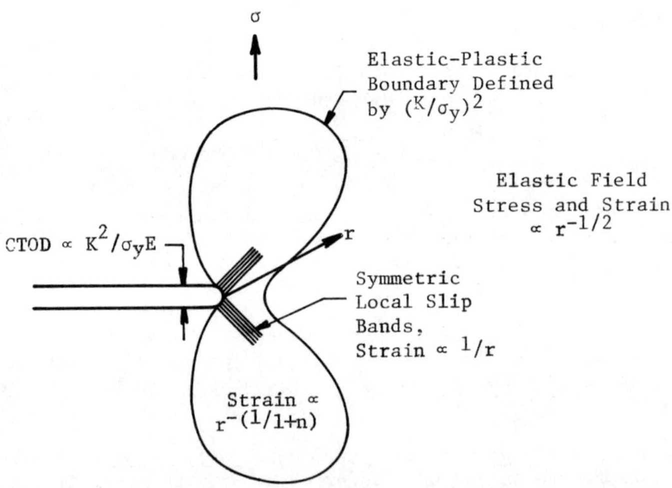

Fig. 1. General concept of crack tip yielding for a Mode I, plane strain crack, based on continuum analyses.

The principal objective of this paper is to explore the validity of this general picture of crack tip yielding. In particular, it will be shown that by using appropriate experimental techniques, it is possible to characterize the details of the deformation imposed by crack tip stress fields. Further, it is critical that this be done, since cyclic loading, environment, and microstructural factors will be found to significantly modify certain details of continuum predictions; these modifications can strongly influence the micromechanical modeling of crack extension. On the other hand, existing analyses can be used to provide useful insight as to which crack tip deformation factors actually control fatigue crack growth; this will be demonstrated for the case of small fatigue cracks.

CRACK TIP YIELDING AND OPENING

Much of what follows will be based upon the results of recent experiments involving new techniques for characterizing extremely localized plasticity. In particular, operation of the SEM in the selected area electron channeling mode permits the accurate mapping of crack tip plastic zones (Davidson, 1975), and the estimation of the crack tip strain gradient from the elastic-plastic boundary to within approximately 50 μm of the tip (Davidson and Lankford, 1980). To characterize crack tip opening, mode, displacement, and strain nearer to, in fact within ~1 μm of, the tip, the technique of stereoimaging strain analysis has been developed (Williams, Davidson, and Lankford, 1980). The latter involves obtaining high magnification pairs of photographs, representing the crack tip in the loaded and unloaded conditions; by viewing the photographs in a stereoviewer, crack tip opening mode and load can be determined. Utilization of quantitative stereophotogrammetry permits quantification of crack tip displacements (Fig. 2; note the mixed Modes I and II opening of the crack tip) and, by spatial differentiation, crack tip strains. The photographs derive from a special servo-controlled hydraulic load cycling stage (Davidson and Nagy, 1978) situated within the SEM specimen chamber.

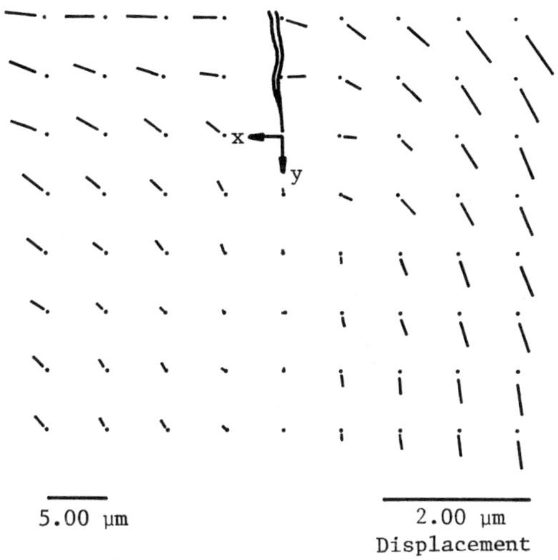

Fig. 2. Displacements due to the loading portion of the cycle near a fatigue crack grown in 7075-T651 in vacuum at $\Delta K=6.2$ $MNm^{-3/2}$. Note the asymmetry of material motion about the crack plane, resulting in considerable shear, even though the loading was Mode I along the x-axis. Displacement magnitude is enhanced relative to the spacing of the measurements.

Specimens are essentially macroscopic, i.e., 2-mm thick by 25-mm wide. Since the imposed stress intensities are relatively low, small scale yielding obtains, and it is thought that the state of stress at the crack tip is basically plane strain. Although it would be most desirable to be able to measure the crack tip parameters of interest inside the specimen, one is constrained to operate at the surface. It is felt that the accuracy, direct interpretation, and insight afforded by the measurements offset the drawback of their surface origin, especially insofar as regards relative changes in crack tip yielding parameters due to material and environmental differences. The experiments to be discussed here have involved precipitation-hardened aluminum alloys, whose properties are given in Table 1.

TABLE 1 Material Properties

Alloy	$\sigma_y (MN/m^2)$	Monotonic Strain Hardening Coefficient	Cyclic Strain Hardening Coefficient	Preparation Technique
2024-T4	325	0.20	0.09	Ingot cast-rolled
6061-T6	280	0.04	0.10	Ingot cast-rolled
7075-T651	508	0.11	0.10	Ingot cast-rolled
MA-87	503	0.05	0.075	Powder Metallurgy-extruded
IN 9052	634	--	--	Mechanically Alloyed-extruded

Plastic Zone Size and Shape

The extent and shape of a crack tip strain field can be determined theoretically by applying the appropriate yield criterion in the crack tip stress field equations. Figure 3 shows plastic zone size and shape according to Von Mises and Tresca yield criteria, between which most low work hardening engineering alloys probably fall. The sizes of the zones are given in terms of the non-dimensional parameter

$$\alpha = r_p / \left(\frac{K_{max}}{\sigma_y}\right)^2 \tag{1}$$

where r_p is the plastic zone radius corresponding to a strain of 0.2%, K_{max} is the maximum value of the stress intensity at the crack tip, and σ_y is the 0.2% yield stress. More accurate finite element calculations (Rice, McMeeking, Parks, and Sorensen, 1979) change these size and shape relationships only slightly, and indicate that strain hardening is a relatively minor influence.

For present purposes, it should be observed that for materials which have similar strain hardening parameters and yield criteria, plastic zone size and shape would be expected, from fracture mechanics theory, to be governed principally by yield strength and state of stress. For a specimen thickness such that the same state of stress prevails for various K_{max} and σ_y, the plastic zone shape should be essentially constant, and zone size should vary with $(K_{max}/\sigma_y)^2$. It is not known exactly how fatigue loading would modify these expectations, but it generally is thought that for zero-tension

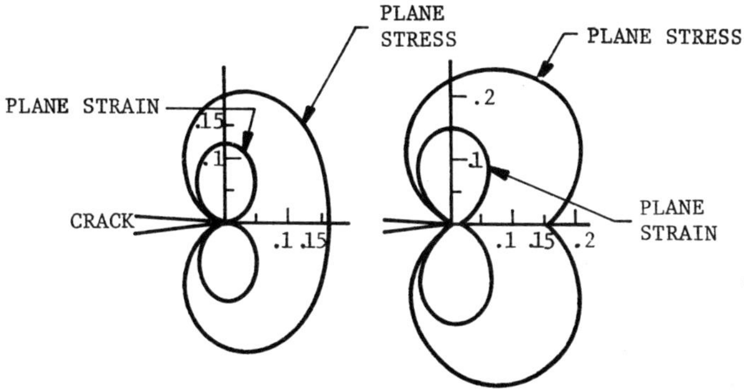

(a) Von Mises (b) Tresca

Fig. 3. Normalized plastic zone dimension $\alpha = r_p/(K/\sigma_y)^2$, based on theory.

loading, cyclic influences would be significant only for the strain distribution in the local crack tip region.

Relative shapes and sizes of the plastic zones for 2024-T4, 6061-T6, and 7075-T651 aluminum, normalized according to Eq. (1), are shown in Fig. 4. Note that the modulus of the axes is the same for each material; thus, a marked lack of normalization is apparent. For example, the maximum values of α for the three alloys are approximately 0.2, 0.37, and 0.76 for 6061-T6, 2024-T4, and 7075-T651, respectively (Davidson and Lankford, 1980). In addition, these maxima do not correspond to the same angular orientation, since the fundamental shapes of the isostrain contours are distinctly different for the three materials.

It would be anticipated that by taking account of yield strength and stress intensity through fracture mechanics theory, plastic zone sizes and shapes for the three alloys would be found to be equivalent. The alloys are all precipitation hardened, with low tensile and cyclic strain hardening coefficients, and presumably, they all obey the same yield criteria. Specimen geometries were identical, with a plastic zone size to thickness ratio of ~250 μm/2500 μm = 0.1, suggesting a state of near plane strain within the plastic zone. This is in agreement with a conventional calculation (Broek, 1974) of the minimum thickness required for plane strain under these conditions, i.e.,

$$t_{min} = 2.5 \left(\frac{K}{\sigma_y}\right)^2 \qquad (2)$$

For 6061-T6, characterized by the minimum σ_y, the required value of t_{min} is 2.65 mm, slightly greater than the actual specimen thickness of 2.5 mm; the other two alloys have t_{min} requirements less than the specimen thickness. It therefore seems likely that plane strain conditions for the three alloys were satisfied. Nevertheless, despite all of these factors which would imply equivalent plastic zones for the three alloys, they exhibit fundamentally different plastic zone sizes and shapes when normalized for yield stress and stress intensity.

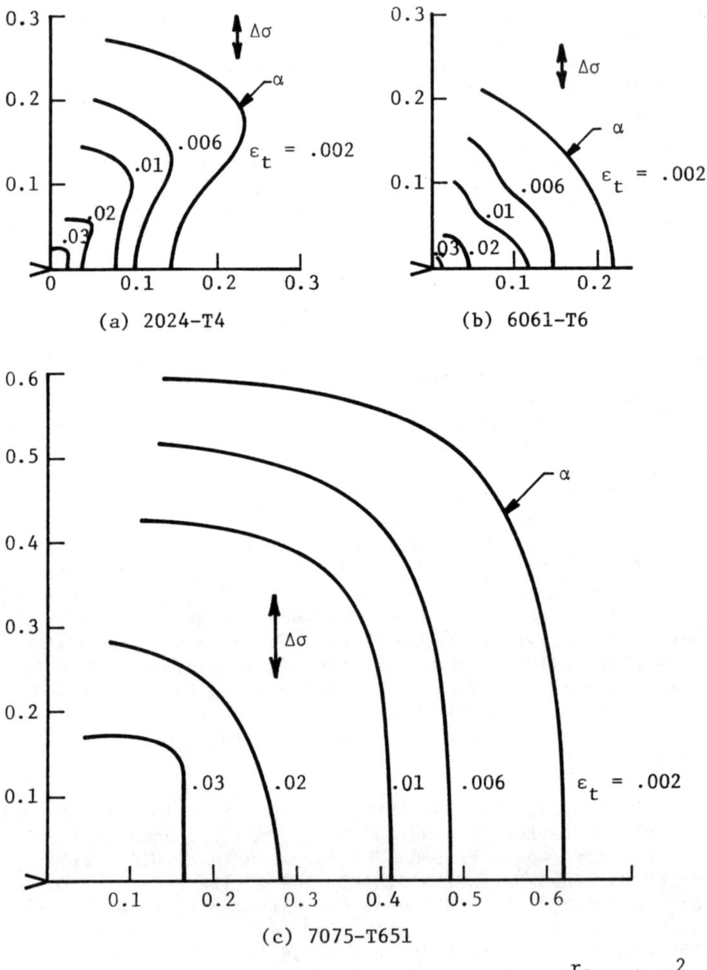

Fig. 4. Normalized plastic zone dimension $\alpha = r_p/(K/\sigma_y)^2$, and equivalent tensile strain distribution within the plastic zone, based on selected area electron channeling measurements.

It is interesting that in the classic work by Hahn, Hoagland, and Rosenfield (1972) on Fe-3Si steel, dislocation etching also showed that the outer bounds of the fatigue plastic zone, even for interior sections, considerably exceeded predictions of continuum theory; α_x (normal to the crack) was found to average ~.25. More recently, Tanaka, Hojo, and Nakai (1982) performed similar experiments on silicon steel, and measured values of α_x on the order of 0.4 for interior, plain strain sections. It is thought that cyclic microyielding within the nominal elastic region surrounding the continuum plastic zone may explain both the latter and the aluminum results described above. Thus, it would appear that both cyclic loading, and rather simple alteration of material properties and microstructure (the three aluminum alloys), can cause significant deviation from continuum predictions of plastic zone sizes and shapes.

The experiments described here were performed in air, although continuum theories in principal pertain only to inert environments. Considerable evidence (Verkin and Grinberg, 1979; Ranganathan, Petit, and Bouchet, 1979), involving several classes of material, suggests that plastic zone sizes in vacuum or inert gases exceed those formed in air. However, rather than pursue further the issue of plastic zone size, it is more useful in the present context to consider the effect of environment, and other factors, upon the plastic strain distribution within the plastic boundary.

Crack Tip Strain

Continuum analyses do not treat the effect of environment upon crack tip strain, and the analyses usually encounter particular difficulty very near the crack tip. Experiments show that the actual near-tip strain fields may differ markedly in inert and aggressive environments, as demonstrated in Fig. 5. Shown are the maximum shear strain distributions within ~25 μm of the crack tip for 7075-T651 cycled at $\Delta K = 6\ MNm^{-3/2}$ in air and dry nitrogen. For both environments, strains tend to peak ahead of the crack. However, the maximum strain value for moist air is only about 2%, versus approximately 6.5% in dry nitrogen. Furthermore, average strains are higher throughout the plastic zone for the inert case.

Figure 6 represents a slightly larger scale (the region within ~50 μm of the tip) view of the local strain field, for $\Delta K = 8\ MNm^{-3/2}$ in dry nitrogen. The maximum strain amplitude is now nearly 8%. It is evident that the strain distribution drops off fairly smoothly ahead of the crack, and theoretically, should fit one of two equations:

$$\Delta\gamma_{max} = \Delta\gamma_o - m\ \ln(r+A) \quad (3)$$

and

$$\Delta\gamma_{max} = \frac{B}{(A+r)^b} \quad (4)$$

where $\Delta\gamma_o$, m, A, B, and b are constants. It is found (Davidson and Lankford, 1983a), however, that Eq. (3) fits experimental data (Figure 7) much better than Eq. (4). Equation (3) has the general form derived by Rice (1968) for a moving crack, while Eq. (4) represents the class of function derived by Hutchinson (1968) and Rice and Rosengren (1968) for unidirectionally loaded, static cracks in work hardening materials. Both equations have been modified here by inclusion of the parameter A, in order to accommodate the finite strain observed at r=0.

This result is further supported by selected area electron channeling data, which measure equivalent tensile strains ε_t accumulated further out in the plastic zone (Davidson and Lankford, 1980). Figure 8 shows ε_t data for 7075-T651 in air along a radial path at 45° to the crack plane; it is clear that the data agree better with the logarithmic type relationship (Eq. (3), ε_t substituted for $\Delta\gamma_{max}$) than with the appropriate power law function. It is interesting that Eq. (3) seems to hold in both inert and aggressive environments.

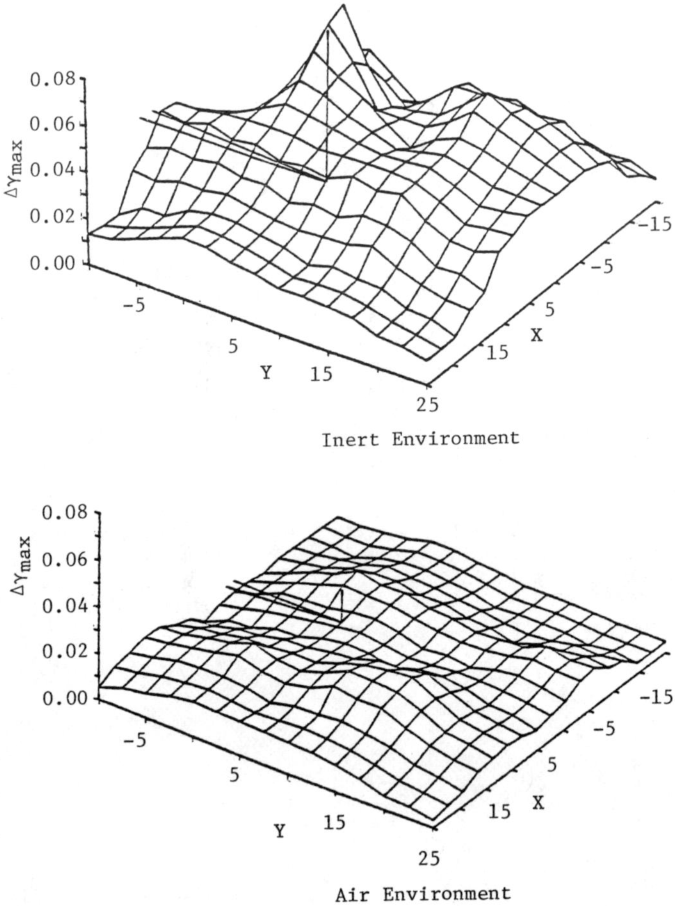

Fig. 5. Distribution of maximum shear strain range for 7075-T651 in moist and dry environments at $\Delta K = 6\ MNm^{-3/2}$. Crack is shown on the plane of zero strain; X and Y are in micrometers. Scale on strain axis is the same for both plots.

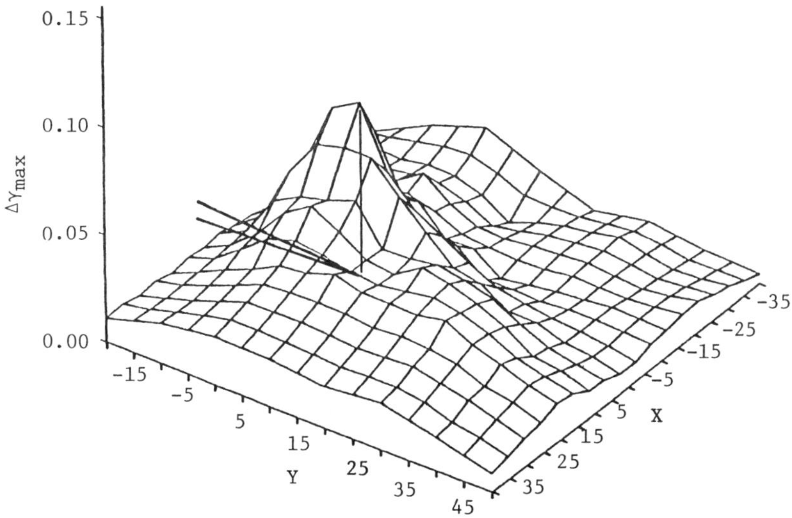

Fig. 6. Distribution of maximum shear strain for 7075-T651 in dry nitrogen at ΔK = 8 $MNm^{-3/2}$.

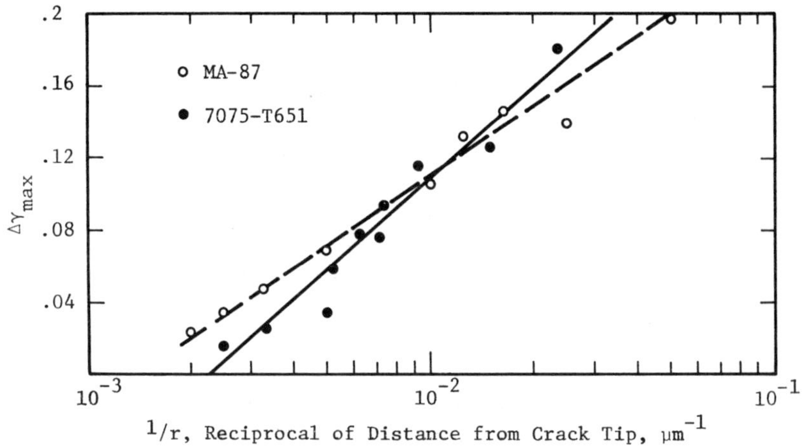

Fig. 7. Maximum cyclic shear strain versus reciprocal of distance from the crack tip, ΔK = 10 $MNm^{-3/2}$, R = .2, dry nitrogen environment. Lines represent best fit logarithmic relationship (Eq. (3)).

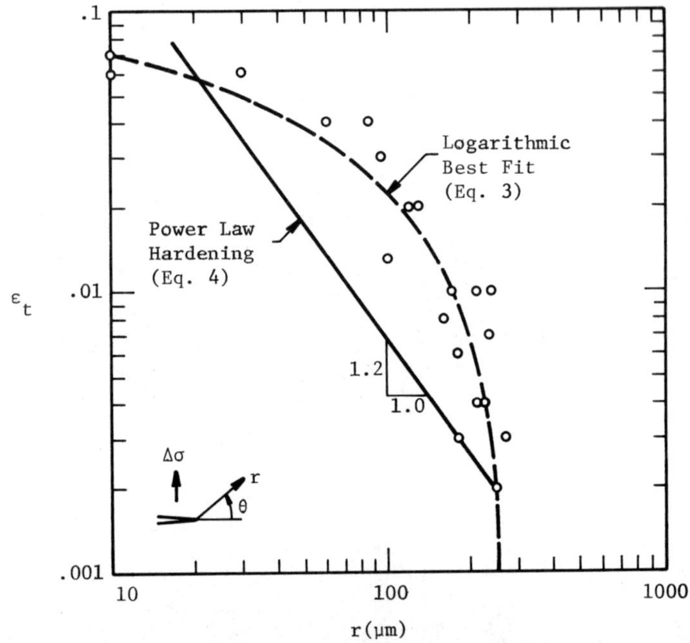

Fig. 8. Equivalent tensile strain versus distance from the crack tip for 7075-T651, $\Delta K = 9.4\ MNm^{-3/2}$, $\theta = 45°$.

As in the case of plastic zone size, alloying and/or microstructure can significantly alter the local plastic strain distribution. This is shown rather dramatically in Fig. 9 for ingot metallurgy, powder metallurgy, and mechanically-alloyed aluminum alloys (Davidson and Lankford, 1983b), all cycled in a dry nitrogen environment at $\Delta K = 10\ MNm^{-3/2}$ and a low R value. As will be discussed shortly, crack tip strains are not identical on each load cycle, and the data of Fig. 9 represent typical maximum values of the maximum shear strain distributions. Although the materials have very similar material properties (Table 1), and the stress intensity is sufficiently low to satisfy small scale, plane strain yielding, the strain fields are remarkably different. The maximum cyclic shear strain amplitudes, for example, are 0.539, 0.332, and 0.165 for the IN-9052, the MA-87, and the 7075-T651, respectively. On the other hand, despite these differences, the radial distribution of crack tip strain is best described for all three materials by a logarithmic relationship.

In analyzing the strains near the crack tip, it has been found useful to utilize the effective total strain range, defined as

$$\Delta\varepsilon_p = \frac{2}{\sqrt{3}}(\Delta\varepsilon_1{}^2 + \Delta\varepsilon_1\Delta\varepsilon_2 + \Delta\varepsilon_2{}^2)^{1/2} - 0.0065 \quad (5)$$

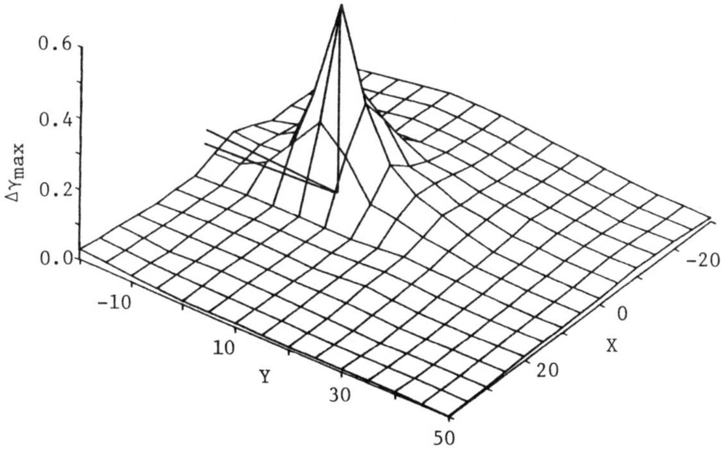

(a) IN-9052, R = 0.1

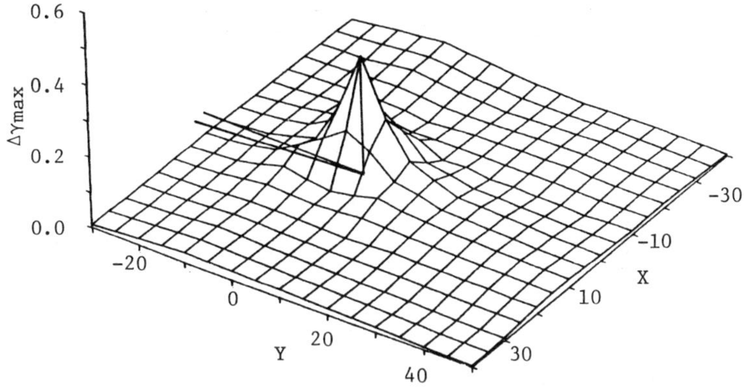

(b) MA-87, R = 0.26

Fig. 9. Effect of alloying/processing upon maximum cyclic shear strain distribution, $\Delta K = 10\ MNm^{-3/2}$, dry nitrogen environment, X and Y are in micrometers. Crack is shown schematically on the plane of zero strain.

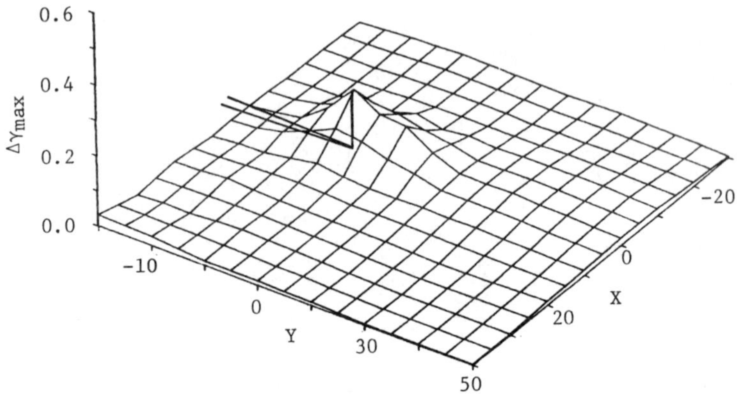

(c) 7075-T6, R = 0.15

Fig. 9 (Continued). Effect of alloying/processing upon maximum cyclic shear strain distribution, $\Delta K = 10\ MNm^{-3/2}$, dry nitrogen environment, X and Y are in micrometers. Crack is shown schematically on the plane of zero strain.

where $\Delta\varepsilon_1$ and $\Delta\varepsilon_2$ are the principal strain increments and 0.0065 is approximately twice the strain at yield; at the tip itself (R=0), this strain is denoted by $\Delta\varepsilon_p(0)$. The dependence of $\Delta\varepsilon_p(0)$ upon ΔK is shown in Fig. 10 for 7075-T651 cycled in air. Several interesting features are associated with this plot.

First, the crack tip strain increases with ΔK, rising (on average) from ~0.03 at $\Delta K = 6\ MNm^{-3/2}$ to approximately 0.25 for $\Delta K = 15\ MNm^{-3/2}$. Secondly, it should be noted that this is precisely the information required by micromechanical crack growth models based on low cycle fatigue failure or cumulative damage within the crack tip process zone. No current continuum analysis predicts $\Delta\varepsilon_p(0)$ as a function of ΔK.

The third interesting feature of Fig. 10 is that the correlation between crack tip strain and cyclic stress intensity shows a surprising amount of scatter; for other alloys and environments, even more scatter has been observed. However, it is possible to rectify this situation if, instead of ΔK, one plots the CTOD corresponding to each value of crack tip strain. As shown in Fig. 11, the resulting correlation is considerably better than in the preceding case.

Crack Tip Opening Displacement

Among the most significant observations to date regarding the opening of fatigue cracks is the fact that they do not open to the same extent on every cycle. Regardless of the procedure used to calculate crack opening

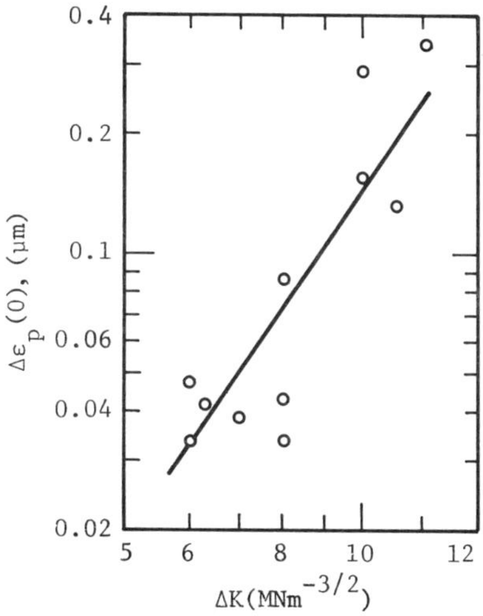

Fig. 10. Crack tip plastic strain increment vs ΔK, for 7075-T651 in dry nitrogen.

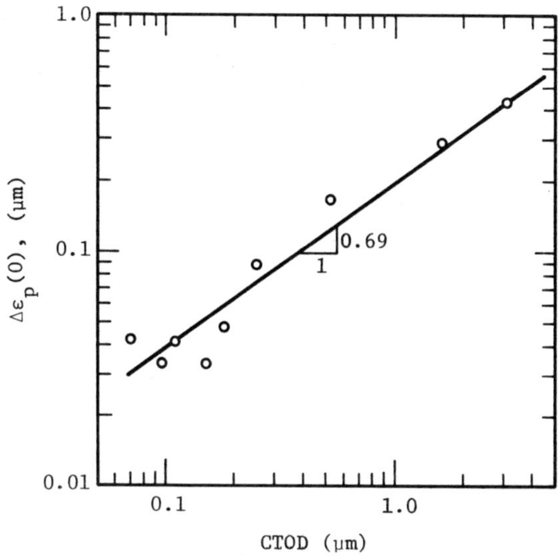

Fig. 11. Crack tip plastic strain range vs crack tip opening displacement 1 μm behind crack tip, for 7075-T651 in dry nitrogen.

displacement (COD), it is invariably found that this parameter scales as $(\Delta K)^2/\sigma_y E$, where E is Young's modulus. Such a computation is shown in Fig. 12, based on the finite element COD calculations by Dean and Hutchinson (1980) for small scale, plane strain yielding in a homogeneous, isotropic, power hardening material with $\Delta K = 6\ MNm^{-3/2}$. The solid and dashed lines represent curves which describe experimental results of the accuracy indicated, for 7075-T651; data points are omitted in the interest of clarity.

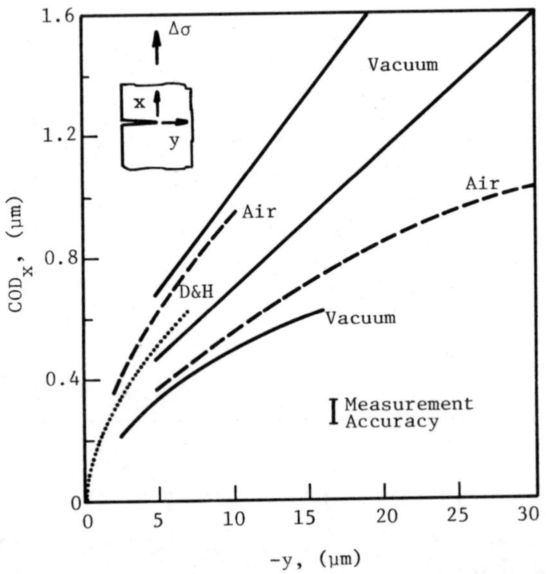

Fig. 12. Crack opening displacement (COD_x) versus distance behind crack tip for 7075-T651, $\Delta K = 6\ MNm^{-3/2}$. The computed crack opening from Dean and Hutchinson (1980) is shown as D & H. Measurement accuracy is determined from the sensitivity of the stereoimaging measurement technique.

The main point here is that the experimental curves bracket the theoretical prediction, but encompass a large range of variability. In an inert environment, for example, the COD 15 μm behind the crack tip may range from (at least) as low as .55 μm to (at least) as high as 1.3 μm. The measurements shown were obtained at various times during crack growth. Subsequent dynamic observation of cracks growing over many cycles within the SEM made it clear that fatigue cracks in all of the alloys studied undergo a gradual blunting/sudden sharpening sequence of events in achieving an increment of extension. The gradual blunting corresponds to the spread in crack opening displacement shown in Fig. 12.

Similar behavior is found for other materials, as indicated (Fig. 13) in the case of MA-87. At the higher ΔK shown (~10 $MNm^{-3/2}$), the experimental results no longer bracket the Dean and Hutchinson prediction, but exceed it, with the inert environment crack openings clearly being higher than those

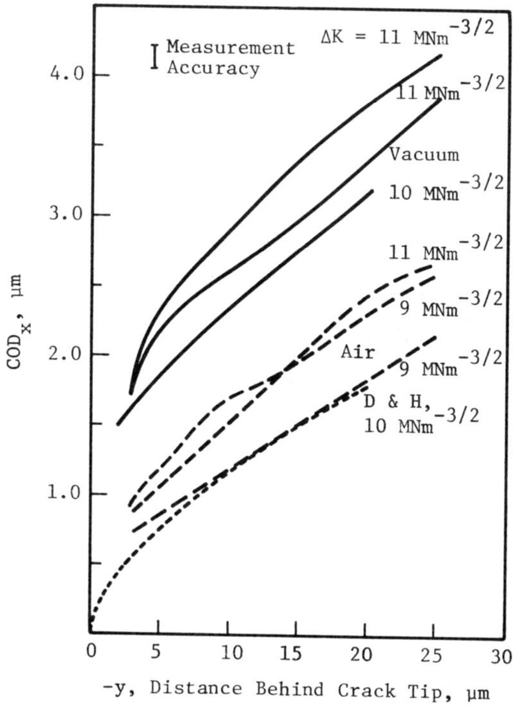

Fig. 13. Crack opening measurements for MA-87, $\Delta K \approx 9\text{-}11$ MNm$^{-3/2}$, vacuum environment. Computed COD_x from Dean and Hutchinson (1980) for $\Delta K = 10$ MNm$^{-3/2}$ is shown as D & H.

observed in moist air. This is quite interesting, since the theoretical curve should correspond to inert conditions, but nevertheless correlates best with the results obtained in air.

The observed variability in COD, in terms of sequential blunting over numerous cycles, explains the scatter in crack tip strain at a given ΔK. Apparently there is no unique value for either CTOD or $\Delta\varepsilon_p(0)$, but rather a maximum value of each which is achieved prior to resharpening of the crack tip. As will be discussed later, this "resharpening" actually corresponds to the incremental extension of the crack tip over a small, cumulatively damaged region. Further advance requires the reattainment of the critical crack tip conditions via another gradual blunting sequence.

The fact that the cyclic CTOD is not constant has actually been predicted by finite element analysis, although numerical comparison is premature. Two years ago, at ICF-5, Gortemaker, de Pater, and Spiering (1981) presented an analysis of plane strain cracks in elastic-perfectly plastic and strain hardening materials loaded in tension over two cycles at R=0. It was found that on application of the second cycle, the crack opening at maximum load was considerably larger than that calculated for the first cycle; likewise, tensile plastic strains ahead of the crack tip on the second cycle exceeded those reached on the first.

Crack Tip Opening Mode

Yielding on symmetrically oriented, inclined slip planes, as modelled by Tomkins (1968), has been observed for certain situations. Indeed, the author has documented such behavior for Paris-regime fatigue crack growth in 4340 steel (Lankford and Kusenberger, 1972) and 304 stainless steel (Lankford and Barbee, 1974), and shown that within such bands, the strains do indeed go as r^{-1}. However, the description of fatigue crack tip yielding in aluminum alloys by a such discrete slip line model seems to be inappropriate, especially at low cyclic stress intensities. Instead, slip lines at the tip are not evident even at electron optical magnifications, despite the high strains which obtain, and the local plasticity appears "spread out" around the tip, as Rice (1967) suggested would probably be the case for plane strain plastic yielding. Within this region, as noted earlier, a log r, rather than r^{-1}, strain relationship is found.

Furthermore, local yielding is often not symmetric. A composite sketch of the average crack opening behavior at $\Delta K = 7.5$ MNm$^{-3/2}$ in vacuum is shown, in Fig. 14, for 7075-T651 and MA-87 (Lankford and Davidson, 1983a). The data actually were obtained by stereoimaging, but for purposes of discussion, the profiles indicate the displacement of the two segments of a hypothetical reference scratch located 1 μm behind the tip.

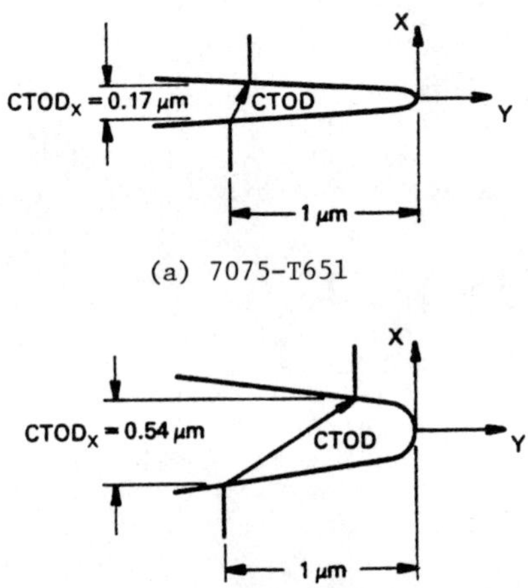

(a) 7075-T651

(b) MA-87

Fig. 14. Scaled schematic of crack tip opening in aluminum alloys, $\Delta K = 7.5$ MNm$^{-3/2}$, vacuum environment; crack tip opening displacement shows relative motion of a hypothetical reference scratch 1 μm behind the crack tip.

It is immediately apparent that the crack tip openings are markedly different, despite the almost identical material properties of the two alloys

(Table I). While the ingot alloy opens mostly in Mode I, crack opening in the powder metallurgy material is mixed, with the Mode II component actually exceeding the Mode I opening. In addition, the total crack opening (CTOD, Fig. 14) for the MA-87 exceeds that for the 7075-T651 by a factor of about 4.5 times.

The large Mode II crack opening component of MA-87 is not predicted by current continuum theories, and is thought to be a real effect caused by intrinsic material response. It is not considered to be a surface or stress state phenomenon, for several reasons. First, no macroscopic shear lip is present, and the fracture surface is uniformly flat. Further, the shear direction switches from one side of the tip to the other, as the latter crosses grain boundaries. Finally, recent fractographic study (Lankford and Davidson, 1983b) has shown that the entire fracture surface is covered with striations plus spherical wear particles produced by in-plane sliding. By stereoimaging the crack tip at various stages of the loading cycle, it was determined that the crack tips in MA-87 begin sliding in Mode II at approximately 0.1 K_{max}, but do not open in Mode I until $\Delta K \approx 0.5\ K_{max}$.

At present, it is not at all clear why fatigue cracks in MA-87 choose to open in mixed Modes I and II, while 7075-T651 cracks open predominantly in Mode I. Neither is it apparent why the CTOD for MA-87 is apparently greater than that for 7075-T651. Aside from grain size, ~50 μm for 7075-T651 versus ~5 μm for MA-87, the principal differences between the two alloys lie in their microstructures. MA-87 is overaged and has a low dispersoid population, while 7075-T651 is peak-aged, with a high dispersoid content.

Regardless of the origin of the Mode II opening in MA-87, its presence is remarkable, considering the relatively low driving force responsible for it. Analysis of the present crack profiles shows that on average, for both alloys in air and vacuum, a given crack segment is oriented approximately 25° from normal to the stress axis. Under such circumstances (Patel and Pandey, 1981) the Mode I stress intensity normal to the local crack plane is .82 of the Mode I stress intensity (K_I) for the average crack plane. Correspondingly, the local Mode II stress intensity is only .38 K_I.

CRACK GROWTH MODELING

In order to demonstrate the relevance of detailed local crack tip yielding information, two particular situations involving the micromechanical modeling of crack advance will be considered. The first is the ongoing problem of how to conceptualize and quantify fatigue crack growth in conventional laboratory specimens containing large cracks. The second situation involves the extension of such results to practical cases in which the crack in question may be "small."

Growth of Large Fatigue Cracks

An interesting review of the problem of predicting fatigue crack growth rates through micromechanical concepts was presented a few years ago by Irving and McCartney (1977). It was shown that there really are only three types of such laws, based respectively on: (1) instantaneous values of cyclic CTOD, which yield a Paris-type law with a ΔK exponent n=2; (2) damage or strain accumulation at the crack tip, yielding n=4; (3) energy balance criteria and the notion of a "process zone," which yield n=2-4, depending on the size of the process zone. A close survey of experimental data has shown that for an inert environment, $n \simeq 4$, suggesting damage or work-of-fracture as the most appropriate bases for fatigue crack growth models.

This kind of rationale, and supporting experimental observations, have led to the recent development of a series of models (Chakraborty, 1979; Lanteigne and Bailon, 1981; Antolovich, Saxena, and Chanani, 1975; Weertman, 1973; Majumdar and Morrow, 1974) based upon strain (damage) accumulation within the plastic zone as a criterion for local, periodic crack tip failure. Fundamental to such low cycle fatigue analyses is recognition of the fact that at low ΔK levels, each stress cycle may not correspond to an increment of crack growth. Examples of such behavior are shown in Figs. 15 and 16, which indicate that for both 7075-T651 and MA-87, striations produced by incremental crack advance in vacuum, for $\Delta K \lesssim 15\ \mathrm{MNm}^{-3/2}$, significantly exceed the average rate of crack growth da/dN. Extensive evidence is presented elsewhere (Lankford and Davidson, 1983) to support the assertion that within this region, an increment of crack advance equals one striation.

Under these circumstances, one is led to a crack growth relationship of the form

$$da/dN = \frac{\Delta a}{\Delta N_c} \tag{6}$$

where Δa is an increment of crack advance, and ΔN_c is the number of cycles required for crack growth. Since da/dN is known to be a function of ΔK, i.e.,

$$da/dN = A(\Delta K)^n \tag{7}$$

where n is a constant, the trick is to determine the ΔK-dependence of Δa and ΔN_c. In the low cycle fatigue analyses cited above, this feat is accomplished by appealing to continuum analyses of crack tip yielding, and to the results of low cycle fatigue tests of bulk specimens.

In particular, it is usually necessary to assume the validity of the Manson-Coffin version of the low cycle fatigue law

$$\Delta N_c^{\ \beta} \Delta\varepsilon_p = \varepsilon_c \tag{8}$$

where β is the Manson-Coffin exponent for a particular material, and ε_c is the ΔK-independent, critical cumulative strain required to fracture the element Δa. By writing down an expression for $\Delta\varepsilon_p$, and summing it over an assumed distance ahead of the crack, one then obtains ΔN_c as required by Eq. (6). The crack growth increment is obtained by assuming a microscopic failure process. Problems arise in the following:

1. Should one really use the Manson-Coffin exponent, which relates to the initiation and growth of a crack in a bulk specimen, rather than the failure of a microscopic crack tip element?

2. What is the strain distribution near the crack tip? Does it fall off as $1/r$, or log r, or is it constant within some local "process zone?"

3. Over what region should the plastic strain be summed, i.e., over the whole plastic zone, the reverse plastic zone, the "process zone," a fraction of the CTOD?

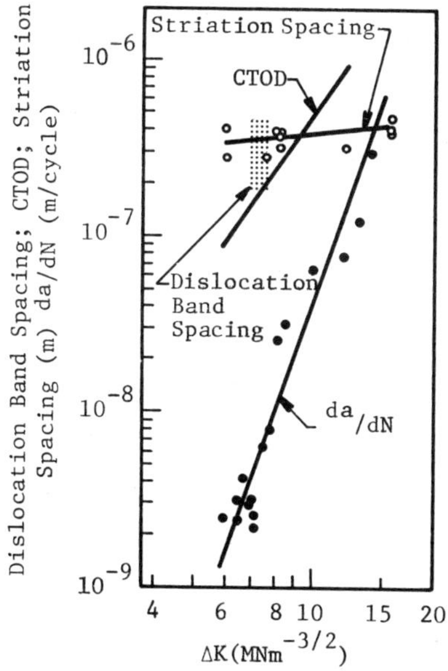

Fig. 15. Crack tip parameters versus ΔK for 7075-T651 in vacuum.

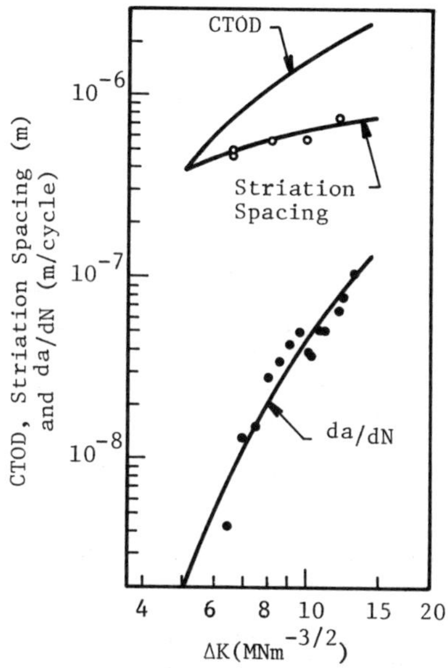

Fig. 16. Crack tip parameters versus ΔK for MA-87 in vacuum.

4. What are the sizes of these zones, and how large is the CTOD?

5. Is Δa constant, based on some microstructural dimension, or is it related to some crack yielding parameter (CTOD, plastic zone size)?

Rather than review the choices made in each of these instances for each of the low cycle fatigue models cited above, suffice it to say that the choices all are based on continuum estimates for the relevant parameters, the validity of Manson-Coffin at crack tips is assumed, and CTOD and crack tip strain are assumed to be repeatable on each cycle.

On the other hand, the results presented in the foregoing indicate that strains and crack tip openings are environment and material dependent, in ways not predicted by continuum analysis. Further, strain does not accumulate identically each cycle, but rather as shown in Fig. 17. The increment of crack extension is equal to one striation spacing. By taking account of the incremental nature of the crack yielding process, using the appropriate stereoimaging information for $\Delta\varepsilon_p$ during a blunt-sharp sequence, and measuring Δa fractographically, it is possible to derive a physically realistic crack growth relationship (Davidson and Lankford, 1983a). By fitting this equation to experimental da/dN results, relating the measured Δa to the experimentally determined maximum CTOD, through ΔK, and assuming that ε_c is ΔK-independent, it is possible to establish β and ε_c for the crack tip element of width Δa.

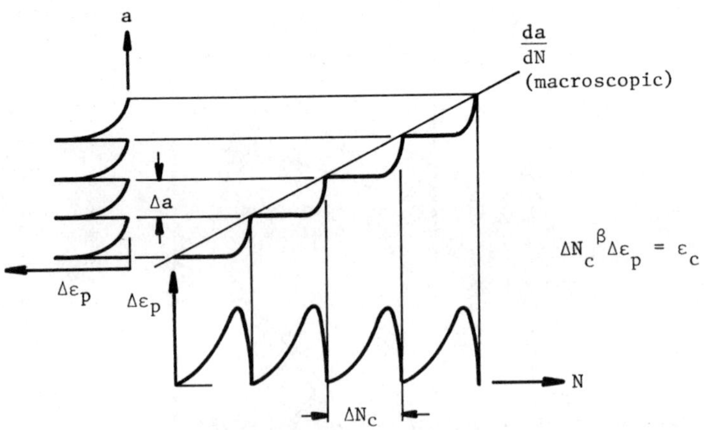

Fig. 17. Diagram of the relationships embodied in the low cycle damage law, showing schematically how $\Delta\varepsilon_p$ and Δa change microscopically with N, leading to the macroscopic growth da/dN. Crack growth is microscopically intermittent: cycling increases $\Delta\varepsilon_p$, while a remains constant. Upon reaching the critical condition given by the equation, the crack extends a distance Δa, which results in a decrease in the crack tip strain.

The net result for a particular cyclic stress intensity is summarized in Table 2. Considering 7075-T651 in air and vacuum, it is clear that the effect of an aggressive specie is to lower the maximum strain at the crack tip, increase the low cycle fatigue exponent, and reduce both the number of cycles and the critical strain required to fail an element Δa. The crack growth increment also is reduced, even though the maximum CTOD is slightly increased.

TABLE 2 Crack Extension Parameters, $\Delta K = 6\ MNm^{-3/2}$

Material	Environment	$\Delta\varepsilon_p(0)^{max}$	ΔN_c	ε_c	β	Δa (μm)	$CTOD^{max}$ (μm)
7075-T651	Air	.018	41	.280	.74	.15	.15
7075-T651	Vacuum	.030	220	.430	.49	.35	.10
MA-87	Vacuum	.150	100	.336	.172	.46	.75

Similarly, a change in alloy has a profound effect on crack tip failure, despite the fact that the material properties are nominally equivalent. Crack tip strains in MA-87 are five times higher in vacuum for MA-87 than in 7075-T651, but the low cycle fatigue exponent of the former is lower by a factor of nearly 3. For both materials, the crack growth increments are practically identical, despite a seven-fold difference in maximum CTOD.

It is worth noting that the β-values determined for the local low cycle fatigue specimen (Δa) differ from that obtained by cycling laboratory specimens to failure (Davidson and Lankford, 1980; 1983a). Further, it is found that the most reasonable values for ε_c correspond to summing strains only when the crack tip is within Δa of a material element, suggesting that analyses which sum strains over larger fractions of the plastic zone grossly overestimate the cumulative strain required to cause an increment of crack growth.

These crack tip micromechanics are not predictable by continuum analysis. Unfortunately, they are nevertheless very important, since they provide guidance for purposes such as alloy design, in terms of the specific parameters which one hopes to modify.

GROWTH OF SMALL FATIGUE CRACKS

In the preceding section, it was shown that reliance upon continuum concepts to predict behavior within the process zone very close to a fatigue crack tip may lead to unrealistic comprehension of what really is taking place there. This is not to suggest, however, that such analyses may not provide considerable insight regarding cyclic crack extension under the appropriate circumstances, especially when reinforced, as necessary, by direct observational techniques. Recent studies of fatigue microcracks provide an example of such interaction.

Nearly a decade ago, Pearson (1975) demonstrated that "small" fatigue cracks in aluminum alloys grow at anomalously fast rates, compared to "large" cracks in linear elastic fracture mechanics (LEFM) specimens. In this context, "small" is defined relative either to some microstructural element, to some localized plastic zone in which the crack is immersed, or to the specimen. Typical growth rate data for small cracks in 7075-T651 are shown in Fig. 18 (Lankford, 1982). In this case, the half-penny surface cracks are on the order of, or smaller than, the grain size. The key points about this figure are that the small cracks: (1) grow rapidly well below the LEFM threshold; (2) are retarded at a characteristic length; (3) experience a degree of retardation which varies from a slight reduction in the rate of growth, to total arrest.

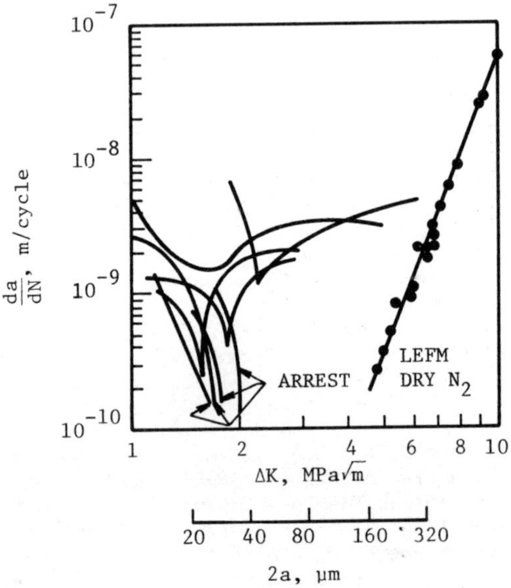

Fig. 18. Growth of fatigue microcracks in 7075-T651 in dry nitrogen compared with large crack (LEFM) results. The surface length of the half-penny microcracks is 2a.

Theories which have been suggested to explain the rapid growth of small cracks have involved crack closure (Newman, 1982), surface microplasticity (Morris, James, and Inman, 1981), plastic zone size (Morris, 1980), and excess crack tip strain (Chan and Lankford, 1983), while the retardation phenomenon has been postulated as due to grain boundary blockage of slip (Morris, James, and Inman, 1981). For the situation shown in Fig. 18, these possibilities can be assessed as follows.

Using the SEM load cycling stage, crack tip opening loads were estimated with a sensitivity of 5%, for both large and small cracks, and found to be essentially identical. In both cases, the opening load was .5-.55 of the maximum cyclic load, which would appear to confine crack closure differences, in this particular case, to a secondary role.

The possibility of cyclic surface microplasticity contributing to microcrack tip opening was investigated by subjecting numerous grains to interrogation by electron channeling, during tens of thousands of fatigue cycles. No degradation in channeling patterns could be detected, which implied that the cumulative deformation due to any fatigue microstrain did not exceed ~0.3% equivalent tensile strain. Morris, James, and Inman (1981) used a variation of stereoimaging to obtain data on a similar alloy; their results suggested that strains on just such a level (0.002-0.004) were indeed achieved under these conditions. This would imply the occurrence of cyclic microstrains on a much lower strain level; 100x10-6 might be taken as a conservative maximum. However, a rough continuum estimate can be used to show that this is inconsequential.

Given a small crack immersed in a plastic field, Tomkins (1979) has shown that the extra CTOD contribution δ which is added to that due to the crack tip plastic zone itself can be estimated by a modification of the BCS continuum model, i.e.,

$$\delta \simeq \frac{\pi\sigma_{max}\Delta\varepsilon_p^{\ m} a}{\sigma_y(1+n)} \tag{9}$$

where σ_{max} is the maximum applied cyclic stress, and $\Delta\varepsilon_p^{\ m}$ is the plastic strain in the matrix around the crack. If we suppose that in the present instance $\Delta\varepsilon_p^{\ m} \approx 100\times10^{-6}$, then δ is ~0.006 μm. However, for typical microcracks, with ΔK as low as ~3 $MNm^{-3/2}$, the measured CTOD is on the order of 0.3 μm, so that δ is only about 2% of the total cyclic CTOD — hardly enough to account for the enhanced rate of growth.

Plastic zone sizes for aluminum alloy microcracks were suggested by Morris (1980) to be independent of ΔK, and equal instead to the size of the grain in which their tips lay. This idea has been evaluated in the present case by electron channeling; it turns out (Lankford and Davidson, 1983c) that grains containing the tips of microcracks have plastic zones of approximately the size which would be predicted on the basis of experiments carried out on large cracks in the same material (Davidson and Lankford, 1980). Furthermore, the microcrack tip plastic zones do not fill up entire grains. Thus, the plastic zones of small cracks in 7075-T651 do not seem to be anomalously large.

However, the ratio of the plastic zone size to the crack size <u>is</u> large — in fact, it was found that r_p/a averaged ≥ 0.5. This is more than an order of magnitude greater than the ratio generally considered allowable for LEFM, through ΔK, to offer a valid description of the state of the crack tip. The significance of this has been established by the excellent, recent finite element study by Chiang and Miller (1982) of stresses and strains around very short cracks — so short that they are effectively immersed in their own plastic zones. The calculations demonstrate that because of the latter fact, the plastic strain at the tip of a short (small) crack is significantly higher than for an equivalent large crack.

Finally, the location of crack tips during retardation has been investigated fractographically (Lankford, 1982) and dynamically, using the SEM loading stage (Lankford and Davidson, 1983d). It has been clearly established that during these periods, the crack tips are located at, or very near, the next grain boundary.

The concepts of a larger plastic strain range associated with the tips of a small crack, grain boundary blockage of crack tip slip, and the same low cycle fatigue crack tip failure process established for large cracks, have been combined (Chan and Lankford, 1983) into a crack growth law for small cracks, i.e.,

$$\frac{da}{dN} = \frac{\Delta a\ C}{\Delta\varepsilon_p(0)}\ \Delta K^{n'}\ [1-(1 - {}^{\tau_B}/\tau_A)(\frac{D-2X}{D})^m] \qquad (10)$$

where C and n′ are experimentally measured constants describing the ΔK-dependence of the strain at microcrack tips far from grain boundaries, m also is an experimentally-established constant related to the rate at which strain drops off as a boundary is approached, τ_A and τ_B are the resolved shear stresses of particular slip systems in adjoining grains, D is the average grain size, and X is the distance from a crack tip to the next grain boundary.

Figure 19 demonstrates the behavior of the model in simulating the approach of a small crack to a grain boundary, whose "strength" is reflected in the factor

$$k(\phi) = 1 - {}^{\tau_B}/\tau_A \qquad (11)$$

Available experimental data for the constants, and an estimate for C, were used to generate the figure, which closely matches the experimental results of Fig. 18. The chosen value for C corresponds to a maximum microcrack tip cyclic strain on the order of 0.04 for $\Delta K \simeq 2\ MNm^{-3/2}$, far above what would be derived by extrapolation of the large crack data in Fig. 10, but perfectly compatible with the observed microcrack tip opening displacements of 0.1 to 0.2 μm (compare with $\Delta\varepsilon_p(0)$ versus CTOD data, Fig. 11).

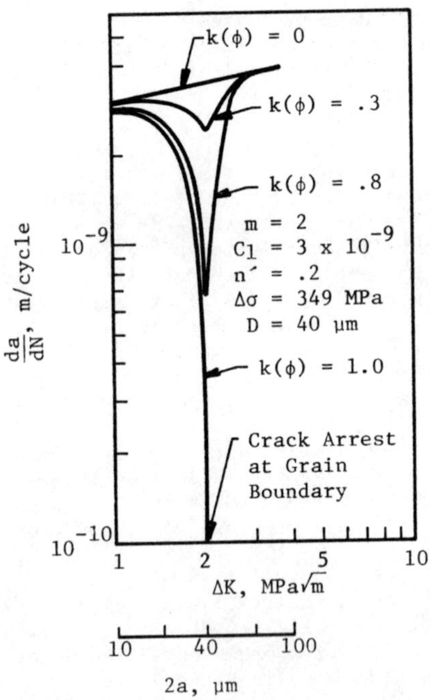

Fig. 19. Predicted behavior of fatigue microcracks in 7075-T651, based on crack tip strain model.

Stereoimaging strain analysis experiments are in progress to check the predictions of this model, and to refine the experimental factors involved in it. It is evident that the development of the approach required considerable effort to observe plastic yielding events on a very small scale. Moreover, the modeling was helped greatly by the several continuum analyses invoked to decide upon the key, controlling element in the problem; the usefulness of these analyses was not compromised by their inability to accurately quantify cyclic plastic strains at the tip of a small crack.

CONCLUSIONS AND IMPLICATIONS

From the foregoing presentation, it appears that in a given situation, any of the items in the original list of "conventional" crack tip yielding assumptions may be wrong. The true state of affairs requires careful observation, which is rendered difficult because of the highly localized scale. While further analytical effort to describe crack tip yielding certainly is warranted, extreme care must be exercised in extrapolating the results into the crack tip process zone. It will be very difficult to incorporate cyclic effects into these analyses, although some progress has been made, and even more difficult to include microstructural and environmental effects. Thus, there will exist a continuing need for high resolution experimental characterization of the local crack tip micromechanics.

The type of information obtained by such characterization is critical, in that it has practical significance for both alloy design (optimization) and lifetime prediction. As noted earlier, both of these activities utilize analytical constructs whose parameters reflect assumed crack tip mechanics. However, since these parameters are usually chosen to fit macroscopic (da/dN, etc.) experimental data, they <u>can</u> simply amount to adjustable constants, <u>unless</u> some effort is made to verify that the postulated mechanisms are really operating. In the absence of such knowledge, alloy or design adjustments aimed at altering a hypothetical micromechanism may be quite pointless.

Throughout this paper, the growth of subcritical cracks subject to cyclic stresses has been emphasized. It is relevant to consider the extent to which the "anomalous" behavior described here is actually a consequence of fatigue, versus material microstructure. If due primarily to the latter, the same types of crack yielding phenomena would be expected to hold for other types of slow fracture situations, such as stress corrosion cracking (SCC). Here, too, micromechanical models are heavily dependent upon conceptualizations and calculations of highly localized plastic yielding at crack tips. In fact, in this case, local strain <u>rates</u> begin to enter as parameters in the crack extension models, suggesting the need for the characterization of the detailed dynamics of very localized yielding. Efforts to perform such experiments are just beginning, and if anything, our knowledge of the micromechanics involved in SCC is even less certain than those related to fatigue.

It should be noted that the stereoimaging information presented here represents, to the author's knowledge, the most localized crack tip displacement/strain data obtained to date, i.e., within 1-2 μm of the tip. However, the even more local region corresponding to an increment of crack growth has not yet been adequately characterized. In particular, Δa has usually been obtained indirectly, through striation measurements (except for IN 9052, in which Δa is so large that it is readily discernible during dynamic cycling in the SEM). This is alright insofar the magnitude of Δa is concerned, but it does not reveal how the growth increment takes place. The latter, seemingly "brittle" event, could proceed by slipoff, micro "cleavage", or some other means, and remains to be ascertained.

Finally, phenomena such as mixed mode crack opening in nominal single mode loading deserve careful consideration. Under certain circumstances, Mode I-loaded cracks open only elastically in Mode I, but exhibit extensive Mode II sliding (Lankford, 1983). It is the latter opening which accounts for crack growth, but which current continuum models fail to predict. Further experimental microstructure-micromechanics analysis will be required in order to provide a physical rationale for this effect. Aside from implications regarding CTOD and crack tip strain, crack sliding can have a profound effect upon corrosion processes such as oxide wedging debris layers (Vasudevan and Suresh, 1982). These layers have a tendency to thicken, and thereby further reduce the effective stress intensity, if cyclic crack tip sliding can occur.

ACKNOWLEDGEMENTS

The author is grateful to the Army Research Office for support of the small crack research, and during the preparation of this paper. The work on large cracks was supported by the Air Force Office of Scientific Research (effect of alloying on plastic zone size and strain distribution) and by the Office of Naval Research (effect of environment on local crack tip yielding). The writer would especially like to acknowledge his continuous professional collaboration with Dr. David L. Davidson during the course of the studies on which this paper is based.

REFERENCES

Antolovich, S. D., A. Saxena, and G. R. Chanani (1975). Eng. Fract. Mech. 7, 649-652.
Broek, D. (1974). Elementary Engineering Fracture Mechanics, Noordhoff International Publishing Co., The Netherlands.
Chakrabortty, S. B. (1979). Fat. Eng. Mat. Struct. 2, 331-344.
Chan, K. S., and J. Lankford (1983). Scripta Met. (in press).
Chiang, W. T., and K. J. Miller (1982). Fat. Eng. Mat. Struct. 5, 249-263.
Davidson, D. L. (1975). Int. J. Fract. 11, 1047-1048.
Davidson, D. L., and A. Nagy (1978). J. Phys. E 11, 207-210.
Davidson, D. L., and J. Lankford (1980). Fat. Eng. Mat. Struct. 3, 289-303.
Davidson, D. L., and J. Lankford (1983a). Fat. Eng. Mat. Struct. (in press).
Davidson, D. L., and J. Lankford (1983b). High Performance Aluminum Powder Metallurgy Alloys, TMS-AIME (in press).
Dean, R. H., and J. W. Hutchinson (1980). ASTM STP 700, 383-405.
Gortemaker, P.C.M., C. de Pater, and R.M.E.J. Spiering (1981). Advances in Fracture Research, ICF5, Pergamon Press, 1, 151-160.
Hahn, G. T., R. G. Hoagland, and A. R. Rosenfield (1972). Met. Trans. 3, 1189-1202.
Hutchinson, J. W. (1968). J. Mech. Phys. Sol. 16, 337-347.
Irving, P. E., and L. N. McCartney (1977). Met. Sci. 11, 351-361.
Lankford, J., and F. N. Kusenberger (1972). Phil. Mag. 26, 1485-1487.
Lankford, J., and J. G. Barbee (1974). J. Mat. Sci. 9, 1908-1910.
Lankford, J., and D. L. Davidson (1978). Int. J. Fract. 14, R87-R90.
Lankford, J. (1982). Fat. Eng. Mat. Sci. 5, 233-248.
Lankford, J. (1983). Fat. Eng. Mat. Sci. 5 (in press).
Lankford, J., and D. L. Davidson (1983a). Acta Met. (in press).

Lankford, J., and D. L. Davidson (1983b). Met. Trans. (in press).
Lankford, J., and D. L. Davidson (1983c). Met. Trans. (submitted).
Lankford, J., and D. L. Davidson (1983d). Crack Length Measurement II, 79-93.
Lanteigne, J., and J.-P. Bailon (1981). Met. Trans. 12A, 459-466.
Majumdar, S., and J. Morrow (1974). ASTM STP 559, 159-182.
Morris, W. L. (1980). Met. Trans. 11A, 1117-1123.
Morris, W. L., M. R. James, and R. V. Inman (1981). Mechanisms by Which Humidity Alters Ductility, ONR Technical Report, Contract No. N00014-79-C-0334.
Newman, J. C. (1983). Behavior of Short Cracks in Airframe Components, Proceeding AGARD Meeting, Toronto, Canada, September, 1982 (in press).
Patel, A. B., and R. K. Pandey (1981). Fat. Eng. Mat. Struct. 4, 65-77.
Pearson, S. (1975). Eng. Fract. Mech. 7, 235-247.
Ranganathan, N., J. Petit, and B. Bouchet (1979). Eng. Fract. Mech. 11, 775-789.
Rice, J. R. (1967). ASTM STP 415, 247-309.
Rice, J. R. (1968). Fracture Vol. II, ed. H. Liebowitz, Academic Press, N.Y., 191-311.
Rice, J. R., and G. F. Rosengren (1968). J. Mech. Phys. Sol. 16, 1-12.
Rice, J. R., R. M. McMeeking, D. M. Parks, and E. P. Sorensen (1979). Comp. Meth. App. Mech. Eng. 17/18, 411-442.
Tanaka, K., M. Hojo, and Y. Nakai (1982). Mat. Sci. Eng. 55, 85-96.
Tomkins, B. (1968). Phil. Mag. 18, 1041-1066.
Tomkins, B. (1979). Met. Sci. 13, 387-395.
Vasudevan, A. K., and S. Suresh (1982). Met. Trans. 13A, 2271-2280.
Verkin, B. I., and N. M. Grinberg (1979). Mat. Sci. Eng. 41, 149-181.
Weertman, J. (1973). Int. J. Fract. 9, 125-131.
Williams, D. R., D. L. Davidson, and J. Lankford (1980). Exp. Mech. 20, 134-139.

MODERN MANUFACTURING AND PERFORMANCE CONCEPTS OF COMPOSITE MATERIALS

G. Lubin

George Lubin Inc., 5373D Lakefront Blvd., Delray Beach, FL 33445, USA

ABSTRACT

The successful use of composites as modern manufacturing materials depends upon several factors which should be optimized for these materials to be integrated into practical products. The design phase has reached that stage where data and analytical methods are available for almost any type of structure. What is still lacking is long time performance and aging data, which has been difficult to obtain due to the extremely fast growth of this industry. The first "modern composites" have only been developed since the early 1940's, fourty years ago. Much accelerated aging data has been developed but its accuracy depends on numerous assumptions and is seldom precise. The preferred method is to obtain real life aging performance of these materials. However, the constant improvements in resins, fibers and sizings left the industry with numerous parts in service, but little aging time for any specific formulation.

The author was fortunate in locating aircraft parts which have been in service for up to 20 years without any change in formulation, and which went through many stress and exposure cycles. Complete data on their original mechanical strength was also available.

With the cooperation of the U.S. Navy and Grumman Aerospace Corp.,these parts were obtained, cut into specimens and tested. The data was then compared with the original strength figures, and percent of strength retention determined. It was found that maximum strength retention was in parts protected with high quality surface paint.

The effectiveness of various coatings for weathering protection will also be discussed.

The manufacturing and producibility phases of composite production have advanced to a point where automation is now standard procedure. However, this applies mainly to mass produced parts of the type used in automotive industry. A large number of parts in the aircraft industry are manufactured in smaller quantities and expensive automation may in many cases not be cost effective. In this presentation several phases of composite molding will be discussed showing how effective cost savings can be obtained by modifying bagging processes, which have not changed much since their conception 40 years ago.

The use of mold release agents and their properties has not been covered to any depth in literature. New studies of low cost molds with long lasting release surfaces show that very extensive cost and time reductions are available for the industry.

Studies by FREKOTE INC. and Grumman show the results of tests on various release agents, their durability and transfer characteristics. New test methods and mold designs developed by the author will be described.

The last part of the presentation is devoted to the listing of some of the latest techniques of composite machining and fabrication.

KEYWORDS

Real Life Aging; protective coatings; performance history; strength retention; protective coatings; low cost molds; mold release agents; mold release ratings; part removal processe.

COMPARISON BETWEEN EXPERIMENTAL AND THEORETICAL FORMING LIMIT DIAGRAMS

A. Melander

Swedish Institute for Metals Research, Drottning Kristinas väg 48, S-114 28 Stockholm, Sweden

ABSTRACT

A detailed comparison is presented between experimental and theoretical forming limit diagrams. Deep drawing quality steel, austenitic stainless steel, copper-nickel and brass were studied. Forming limit diagrams were obtained both for straight and non-linear strain paths. The theory assumes that necking is initiated by a heterogeneous distribution of inclusions in the sheet and surface roughness caused by the grain structure. The theory utilizes work hardening data and strain rate sensitivities determined in uniaxial tension and biaxial straining. The flow surface is assumed to be smooth and its shape is characterized by the $\bar{r}$ value. Void growth at inclusions is considered to cause softening in the material. A satisfactory agreement was obtained between theoretical and experimental forming limit diagrams.

KEYWORDS

Forming limit diagram; work hardening; strain rate sensitivity; plastic anisotropy; flow surface; inclusion; void; surface roughness.

INTRODUCTION

Formability is a complex engineering concept which cannot easily be defined by a single measure. Instead it must be broken down into a number of components. One essential component is the ability of the material to distribute strains during forming. The formability is usually decreased if large strain differences are created in the sheet. The ultimate formability is controlled by sharp necking or fracture in the sheet. It is usually characterized by the forming limit diagram (FLD).The position of the FLD depends on intrinsic material properties but is also related to the forming process. If the FLD is known and the ability of the material to distribute strains is characterized a fair picture of the formability has been created.

The present paper is devoted to the FLD. For a wide range of materials a detailed comparison is made between measurements of FLD and material properties on one hand and theoretical calculations of the FLD on the other hand. The influence of a number of material properties on the FLD is considered such as the work hardening rate, the strain rate sensitivity, plastic anisotropy, void growth at inclusions and surface roughness caused by the grain structure. The effect of non-linear deformation paths is also investigated.

FACTORS INFLUENCING THE FLD

During the last decade significant progress has been made in the understanding of how different factors influence the FLD. This includes both theoretical and experimental findings.

The FLD is usually determined experimentally by measuring the strain just outside the sharp neck. The results are sensitive to the exact proceedure of measurement which often differs from one laboratory to the next. Important factors in this respect are size of measuring circles in particular in relation to sheet thickness and exact choise of circles for evaluation. Different forming techniques like hydraulic bulge testing, forming with a cylindrical or spherical punch or tensile testing produce different strain distributions and strain paths which might influence the FLD. Small strain gradients and straight deformation paths are desirable if one wishes to characterize the influence of intrinsic material properties on the FLD. Such experimental conditions are, however, often not at hand.

Several theoretical techniques can be used to model the FLD. Conditions for strain localization might be considered in a homogeneous material. Alternatively the development of an initial heterogeneity might be followed up to sharp necking. Several types of heterogeneities might be considered. It, however, seems as if surface roughness and distributions of inclusions in the sheet are presently considered to be the most likely sources of necking in commercial sheet products.

Surface roughness might act as a heterogeneity since it creates a locally varying sheet thickness. Initially during forming the roughness increases due to intrinsic properties of the grain structure. Eventually the roughness growth is accelerated and a macroscopic sharp neck is created at a location where the sheet thickness was previously smaller than in the surrounding material (Yamaguchi and Mellor, 1976, Wilson, Roberts and Rodrigues, 1981a; Melander and Thuvander, 1982a). Several factors might influence the growth of surface roughness such as grain size, texture and local inhomogeneities of those (Wilson, Roberts and Rodrigues, 1981b).

Inclusions might act as a heterogeneity if voids exist at the inclusions during forming. The voids have a softening effect on the material which accelerates necking if a larger density of inclusions is present locally in the sheet (Needleman and Triantafyllidis, 1978; Chu and Needleman, 1980; Melander, 1981; Melander, 1982; Schmitt and Jalinier, 1982). Significant local variations in the density of inclusions are known to exist in commercial materials.

The flow properties of the material have a strong effect on the shape and position of the FLD.

The shape of the flow surface is known to influence the FLD strongly in particular for positive minor strains i.e. for biaxial stretching. If the flow surface contains a sharp vertex localized necking might occur both on the left and right hand sides of the FLD at realistic levels of strain even if no initial heterogeneity is present (Hutchinson and Neale; 1978). With a smooth yield surface this is not the case. If a heterogeneity is present in the sheet the local curvature of the yield surface influences the FLD significantly. A smaller curvature of the yield surface around biaxial stretching usually has a depressing effect on the right hand side of the FLD but a rather small influence on the left hand side. This has been shown by considering materials with flow surfaces according to Hill (1950,1979) with different anisotropy (Marciniak and Kuczynski, 1967; Parmar and Mellor, 1978). The anisotropy is often characterized by the $\bar{r}$-value. A high $\bar{r}$-value gives a small curvature of the flow surface around biaxial stretching (Hill, 1950) and a low FLD for positive minor strains. An accurate description of the shape of the flow surface is critical to the theoretical description of the FLD. Much remains to be done in this field both experimentally and theoretically.

The work hardening properties of the material are known to influence the FLD significantly. A larger work hardening rate usually produces a higher FLD. Sharp necking usually occurs at large effective strains which means that the work hardening behaviour must be known at such levels of strain. This means that the strain range studied with tensile tests is often too small. Other techniques like compression or hydraulic bulge testing which can be evaluated up to large strains should be utilized. The mentioned methods represent different deformation modes, uniaxial tension respectively biaxial straining, and can thus provide information on the anisotropy of work hardening rates. Such an anisotropy would imply that the flow surface does not expand homogeneously during deformation but its shape changes. In most theories of the FLD it is assumed that the flow surface expands homogeneously during forming (Marciniak and Kuczynski, 1967). Kinematic theories have, however, also been used where the flow surface has a constant shape and size but is translated in stress space during deformation (Tvergaard, 1978).

The strain rate sensitivity of the material is another factor which influences the FLD. The strain rate might influence several material properties but the influence on flow stress and work hardening rate is usually considered most important to the formability (Melander and Thuvander, 1982b). A more positive strain rate sensitivity usually produces a higher FLD.

Void generation and growth at particles might initiate necking as was mentioned previously. It also influences the flow properties of the material in the respect that an increasing void fraction decreases the flow stress and thus the work hardening ability. The effect of void growth is usually to lower the FLD (Melander, 1981) but more complicated situations might arise if void nucleation occurs during forming (Chu and Needleman, 1980).

THEORY OF THE FORMING LIMIT DIAGRAM

The present theory takes into account two initiating factors for necking namely a heterogeneous distribution of inclusions and surface roughness. The theory was originally developed only for inclusion heterogeneities (Melander 1981) but was later on extended to take the surface roughness into account (Melander and Thuvander; 1982a). The reader is referred to these papers for a detailed description. The present paper only gives a short summary.

The theory considers two separate volumes of the sheet, Fig 1, one local volume where the neck eventually develops (superscript n) and the surrounding material (superscript 0) which deforms according to the externally prescribed strains. The local volume of the sheet has a higher volume fraction of particles f_p^n than the surrounding material f_p^o and an imposed thickness reduction of f_{th} compared to the thickness of the surrounding material. The neck is inclined by an angle ψ to the minor strain direction 2.

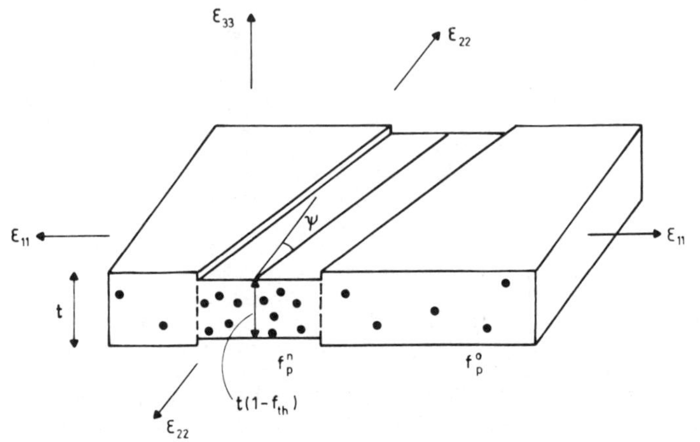

Fig. 1 Schematic representation of the geometry used in the model of the FLD.

During deformation voids are created and grow at the particles. The distribution of voids make the material softer than the matrix itself. The softening effect of voids increases with the void volume fraction. This means that softening proceeds more rapidly in the particle rich region of the sheet than in the surrounding material even if the strain levels are identical. Softening promotes strain localization and leads to sharp necking in the particle rich region of the sheet.

Cold rolled sheet often has a small initial surface roughness. During sheet forming the roughness increases due to different deformation behaviour of grains with different orientations. This is considered in the model by prescribing a small initial thickness reduction of the sheet which increases at a prescribed rate which is typical to surface roughness. The prescribed thickness reduction of the sheet produces a reduced load carrying section which accelerates necking.

It is assumed that necking occurs in a region of the sheet where an increased population of inclusions and a thickness reduction due to surface roughness are present simultaneously.

The deformation in the sheet is restricted by two conditions: Equilibrium of forces exists in the normal direction of the plane between the necking region and the surrounding material. The sheet is assumed to deform at the same rate in the direction of the neck both in- and outside the neck. The latter condition and thus the whole theory is expected to be more realistic on the right hand than on the left hand side of the FLD.

The constitutive behaviour of the material is described with flow theory assuming normal anisotropy, $\bar{r}$, with a smooth yield surface according to Hill (1950). The influence of voids on the flow properties of the material is modelled with a flow rule based on a regular distribution of voids (Gurson, 1977). In this model softening due to voids depends both on the volume fraction of voids and the magnitude of the hydrostatic pressure. The flow of matrix material, excluding voids, in described by

$$\sigma_o = K \cdot \varepsilon_e^n \cdot (\dot{\varepsilon}_e/\dot{\varepsilon}_{eo})^M \qquad (1)$$

where ε_e is the effective strain and $\dot{\varepsilon}_e$ the effective strain rate. K, n and M are parameters which depend both on the deformation mode and on the effective stress or strain. $\dot{\varepsilon}_{eo}$ is the imposed strain rate at the beginning of testing. The model thus assumes homogeneous expansion of the flow surface during deformation. The rate of expansion might, however, be different for different modes of deformation.

The calculation of the FLD requires data representing the strain hardening properties, the strain rate sensitivity and the anisotropy parameter of the material. The average volume fraction of particles f_p^o in the material must be known. The volume fraction of particles in the necking region f_p^n is used as a fitting parameter. The initial surface roughness and the development of roughness with strain must be known. The constant of proportionality between the used measure of surface roughness and the prescribed thickness reduction is considered a fitting parameter. The theory thus incorporates two fitting parameters.

The calculation of the forming limit is performed for a range of imposed strain ratios from uniaxial tension to equibiaxial tension. For each strain ratio the development of effective strain in the neck is studied until the strain goes to infinity. The strain components ε_{11}^o and ε_{22}^o in the sheet surrounding the neck at the moment of infinite strain in the neck define the theoretical forming limit. The forming limit is calculated for a range of initial inclinations ψ_o of the particle rich region. The value of ψ_o which leads to final necking at the smallest amount of externally imposed effective strain defines the forming limit which is introduced into the FLD. On the right hand side of the FLD necking always occurs for neck orientations parallel to the minor strain direction, ψ=0. On the left hand side of the FLD a range of inclinations must be tested.

VOID GROWTH DURING METAL FORMING

Within the present theory void growth at particles is a central phenomenon in the necking process. In Melander (1981) void growth during metal forming was studied in detail for tough-pitch copper. In that material voids grow at oxide inclusions from the beginning of metal forming. It was found that the homogenous void growth model described experimental results well.

Void growth at inclusions is considered in the investigated materials. The inclusions are assumed to be stiff during metal forming and void nucleation is assumed to have already occurred during cold rolling. This means that voids exist in the as-recieved material and can grow from the beginning of metal forming. The volume fraction of voids can be written

$$f_v = f_p(\exp(\varepsilon_e + \varepsilon_o) - 1) \qquad (2)$$

where f_p is the inclusions fraction and ε_o is used to describe the fraction of voids in the as-recieved material. In Melander (1981) ε_o was found to be 0.2 for tough pitch copper and this value is used in the present work. The volume fraction of inclusions outside the neck is set to the average volume fraction of inclusions, which was measured with optical microscopy.

GROWTH OF SURFACE ROUGHNESS

The development of surface roughness during sheet forming has been studied in the literature both for steel (Fukuda, Yamaguchi, Takakura and Sakano, 1974) and brass (Wilson, Roberts and Rodrigues, 1981a). In the case of brass the influence of grain size and sheet thickness was investigated. The surface roughness was evaluated both as an average value and as an extremum value of the surface profile. In both cases it was found that the roughness from a small initial value increased linearly with strain until it finally started to accelerate. The growth of surface roughness for different strain paths could be well represented as a function of a single parameter namely the effective strain, ε_e. The roughness was found to be proportional to the grain size.

These results are incorporated into the present model by imposing a relative reduction in sheet thickness between the neck and the surrounding material of

$$f_{th} = f_{th}^{o} + S \cdot \varepsilon_e^{o} \cdot \frac{d_o}{t_o} \tag{3}$$

where S is a constant, ε_e^o the effective strain in the material outside the neck and d_o/t_o the ratio of grain intercept length to sheet thickness.

This means that the reduction in sheet thickness in the neck during forming is caused both by the microstructurally imposed growth of surface roughness and by macroscopic strain localization. The macroscopic strain localization process is partly driven by growth of surface roughness and partly by void growth at inclusions.

DETERMINATION OF MATERIAL PARAMETERS

Results on five materials will be presented in this paper - a deep drawing quality steel (SS1147), an austenitic stainless steel (SS2343), copper 10% nickel, and two 70/30 brasses with grain sizes of 8 respectively 82 µm. The chemical compositions of the materials are presented in Table 1. All materials are cold rolled to a sheet thickness of 0.7 mm.

TABLE 1 Chemical composition of investigated materials in weight percent.

Material	Cu	Fe	Zn	Ni	Cr	Mn	Si	S	C	P	Mo
Deep drawing steel	.024	bal.	-	.010	.007	.188	.024	.013	.074	.011	-
Stainless steel	.12	bal.	-	11.8	16.7	1.44	.55	.007	.037	.028	2.61
Copper-Nickel	bal.	1.20	.05	10.6	-	1.09	.003	-	-	-	-
70/30-8 Brass	70.1	.014	bal.	.004	-	-	.003	-	-	.003	-
70/30-82 Brass	69.8	.017	bal.	.003	-	-	.003	.001	-	.003	-

Tensile testing was performed at an initial strain rate of $\dot{\varepsilon}_{to} \approx 10^{-3}$ per second. Mechanical properties in the rolling direction of the sheet are presented in Table 2 where the proof stress $R_{p0.2}$, the ultimate tensile strength, R_m, the elongation to maximum load, δ_m, and the total elongation, δ_{tot}, are listed. The r-value was evaluated in three tensile directions and Table 2 gives the values of $\bar{r}$ and Δr.

TABLE 2 Mechanical properties of investigated materials

Material	$R_{p0.2}$ MPa	R_m MPa	δ_m	δ_{tot}	$\bar{r}$	$\Delta\bar{r}$	ε_o	f_p^o
Deep drawing steel	179	315	.24	.41	1.73	.93	.2	.0008
Stainless steel	289	618	.52	.58	.98	-.16	.2	.002
Coppel-nickel	173	341	.31	.37	.80	-.09	.2	.001
70/30-8 Brass	197	398	.46	.51	.83	-.08	.2	.0004
70/30-82 Brass	106	337	.65	.71	1.04	-.63	.2	.0004

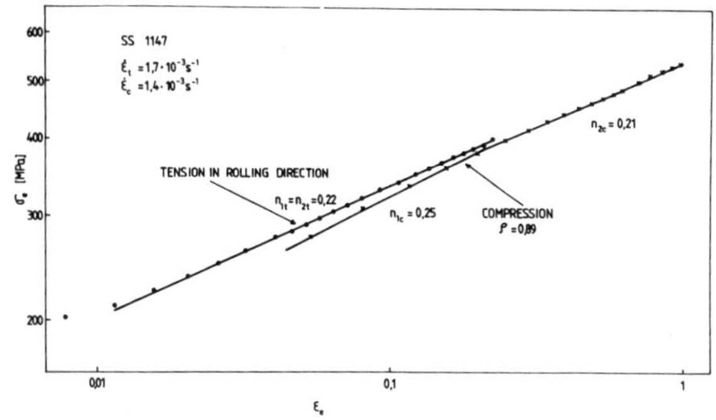

Fig 2a Flow curves in uniaxial tension and compression for the deep drawing quality steel SS1147.

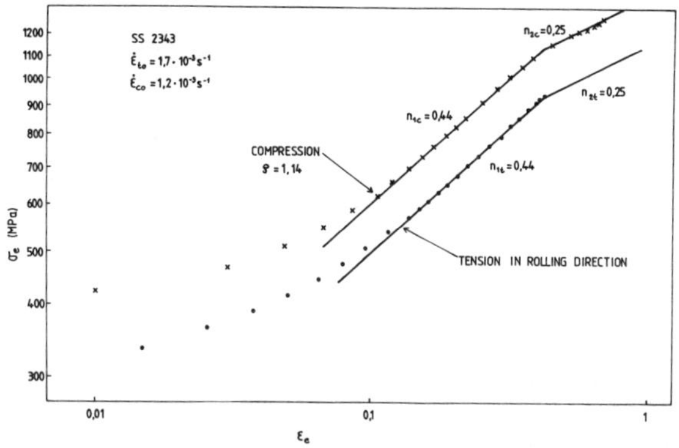

Fig 2b Flow curves in uniaxial tension and compression for austenitic stainless steel SS2343.

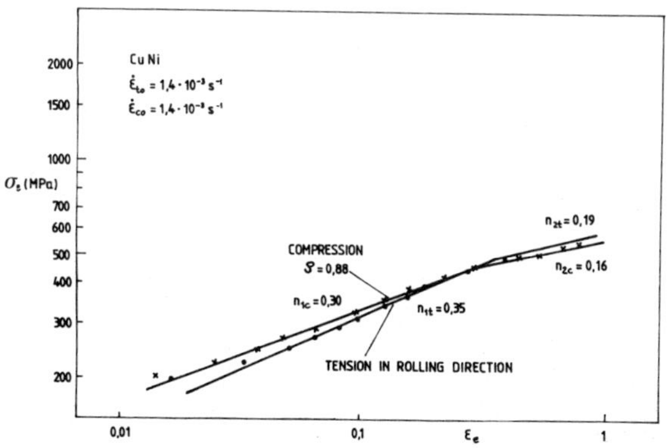

Fig 2c Flow curves in uniaxial tension and compression for copper-10% nickel.

Flow curves for tensile testing in the rolling direction are presented in Figs 2a to 2c. The effective strains and stresses are calculated with Hill's (1950) anisotropic theory of plasticity assuming normal anisotropy. In the case of tensile testing the effective stresses and strains are identical to the tensile components of stresses and strains, $\sigma_e = \sigma_{11}$ and $\varepsilon_e = \varepsilon_{11}$.

In order to determine the anisotropy of the work hardening behaviour for different deformation modes and to analyse the work hardning rate for large strains compression tests were performed. The compression specimens were prepared by gluing together approximately nine sheets of material. The specimens were compressed in the normal direction of the sheets. This resulted in a deformation mode with $\rho=\varepsilon_{22}/\varepsilon_{11}$ close to unity i.e. equibiaxial testing. The initial strain rate $\dot{\varepsilon}_{oc}$ was approximately $10^{-3}s^{-1}$ or approximately the same as in the tensile test. The effective stresses and strains in compression can be written

$$\varepsilon_e = -\sqrt{\frac{b(b + 2\rho + b\rho^2)}{b^2 - 1}}\, \frac{1}{1+\rho} \cdot \varepsilon_{33} \tag{4a}$$

$$\sigma_e = \sqrt{2(1 - \frac{1}{b})}\, \sigma_{33} \tag{4b}$$

where $b = (1+r)/r$. Index 3 denotes the normal direction of the sheet. The r-value was set to the mean value $\bar{r}$ of Table 2.

The flow parameters were evaluated in terms of the Ludwik-Hollomon model, $\sigma_e=K\varepsilon_e^n$. Both in tension and compression the present materials exhibit a multiple n behaviour. Since the most important part of the flow curve in this application is at high strains where necking occurs the part of the curve lying below about 5% effective strain is disregarded. For each material two strain hardening exponents n_1 and n_2 have been evaluated. The strain hardening exponent is n_1 up to an effective strain of ε_e^d. Above ε_e^d the strain hardening exponent is n_2. The present materials all exhibit a decrease in the strain hardening exponent above ε_e^d during compression. The same behaviour is assumed to occur during tension although necking prevents the observation of the change from n_1 to n_2. Accordingly n_2 and ε_e^d from tensile testing have been estimated assuming proportionality to the corresponding parameters from compression testing. The estimated values are indicated by parenthesis in Table 3. Indices t and c in Table 3 indicate tensile and compression testing respectively. The constant K (Table 3) of the Ludwik-Hollomon expression was evaluated below ε_e^d both for tension and compression.

TABLE 3 Parameters used to calculate the FLD. Parenthesis indicates estimated values

	Tension				Compression					
Material	n_{1t}	ε_{et}^d	n_{2t}	K_t	n_{1c}	ε_{ec}^d	n_{2c}	K_c	$M_o \cdot 10^3$	$M_1 \cdot 10^6$
Deep drawing steel	.22	-	-	552	.25	.21	.21	577	10.5	-
Stainless steel	.44	(.41)	(.25)	1370	.44	.41	.25	1673	3.4	2.3
Copper-nickel	.35	(.35)	(.19)	707	.30	.30	.16	658	-5.0	20.
70/30-8 Brass	.42	(.48)	(.13)	880	.42	.48	.13	895	-12.6	23.6
70/3-82 Brass	.63	(.44)	(.43)	873	.53	.38	.37	804	-6.2	16.3

If Hill's (1950) anisotropic theory of plasticity was fully satisfactory the flow curves for tension and compression would coincide. The differences both in stress levels and strain hardening rates which can be observed in Figs 2a to 2c thus indicate that Hill's theory cannot exactly describe the anisotropy of the materials.

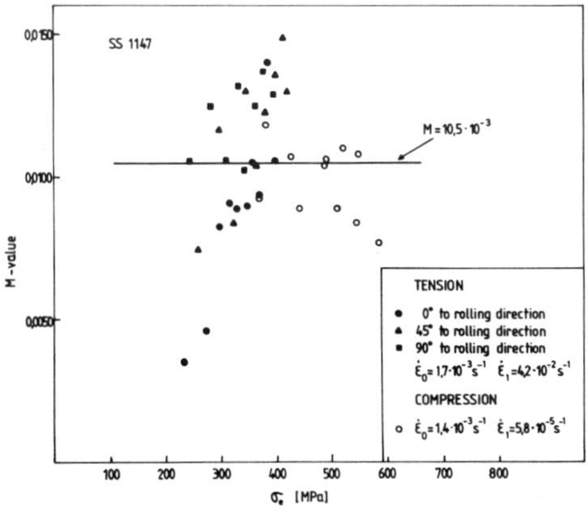

Fig 3a Instantaneous strain rate sensitivity of deep drawing quality steel.

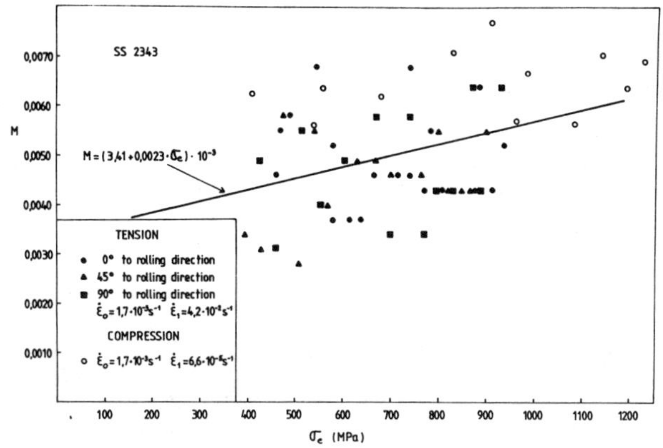

Fig 3b Instantaneous strain rate sensitivity of austenitic stainless steel.

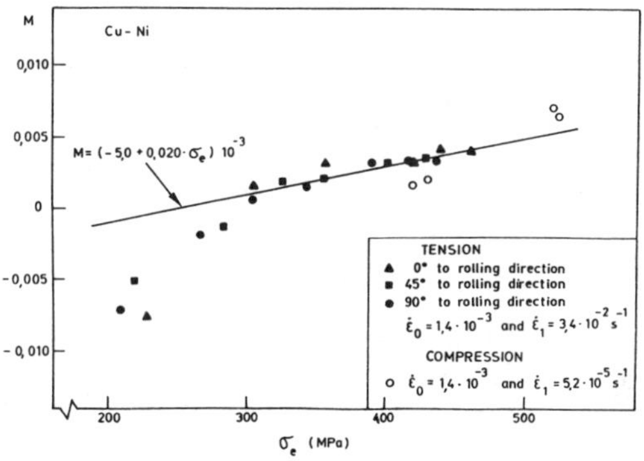

Fig 3c Instantaneous strain rate sensitivity of copper-10% nickel.

In order to get a satisfactory description of the flow curves in tension, compression and other strain paths it is assumed that all parameters in Ludwik-Hollomon's expression (Table 3) vary linearly with ρ between uniaxial tension ($\rho= -1/b$) and equibiaxial tension ($\rho=1$).

The strain rate sensitivity of the present materials was determined by measuring the change in effective stress when changing the strain rate instantaneously in a tensile or compression test.

$$M = \partial \ln \sigma_e / \partial \ln \dot{\varepsilon}_e \tag{5}$$

For each material tensile tests were performed on specimens oriented in three different sheet directions, and a compression test was performed in the normal direction of the sheet. In these tests the strain rate was changed repeatedly by a factor of approximately $\dot{\varepsilon}_{e1}/\dot{\varepsilon}_{eo}=25$.The M-values are presented in Figs 3a to 3c. A single straight line has been fitted to all experimental points for each material. The strain rate sensitivity is represented by

$$M = M_o + M_1 \cdot \sigma_e \tag{6}$$

The coefficients M_o and M_1 are given in Table 3.

The parameters describing the work hardening properties and the strain rate sensitivity of the two brasses of this paper are included in Table 3 and will not be represented in diagrams.

EXPERIMENTAL DETERMINATION OF THE FORMING LIMIT DIAGRAM

The FLDs of the present paper were determined with hydraulic bulge testing for positive minor strains and with tensile testing for negative minor strains.

Hydraulic bulge testing was performed with masks of elliptical shape with aspect ratios 1.0, 0.9, 0.8, 0.63 and 0.4. The major axis is 100 mm. The minor axis of the ellipse is parallel to the rolling direction of the sheet. In the case of conventional FLDs determined with closely linear deformation paths (Figs 4 and 5) notched tensile specimens were used on the left hand side of the FLD. The specimens have external dimensions of 200 by 230 mm. The gauge width is 68 mm and the root radius of the notches are 10, 20, 30 and 66 mm respectively. The tensile specimen without notches has a gauge width of 12 mm and gauge length of 60 mm. In the experiments on non-linear deformation paths, Figs 6, rectangular tensile specimens were used on the left hand side of the FLD. The height to width ratios of the deforming part of the specimens were 2, 0.8, 0.56, 0.32 and 0.2 (Schedin and Melander, 1982). The tensile direction of the specimens is parallel to the rolling direction of the sheet. Tensile testing and hydraulic bulge testing were performed at strain rates of the order of 10^{-3} per second. Double tests were used for all specimen geometries. A circular grid was imposed on the sheets with a circle diameter of 2 mm. Testing continued until a crack was visible and at that moment the test was interrupted. For each specimen the true major and minor strains (rolling and transverse direction of the sheet respectively) were measured adjacent to the crack outside the visible sharp neck. The experimental results are presented by circles in Figs 4a to c, Figs 5a to b and Figs 6a to b. The data points lie within a scatter band which reflects the strain gradients just outside the sharp neck.

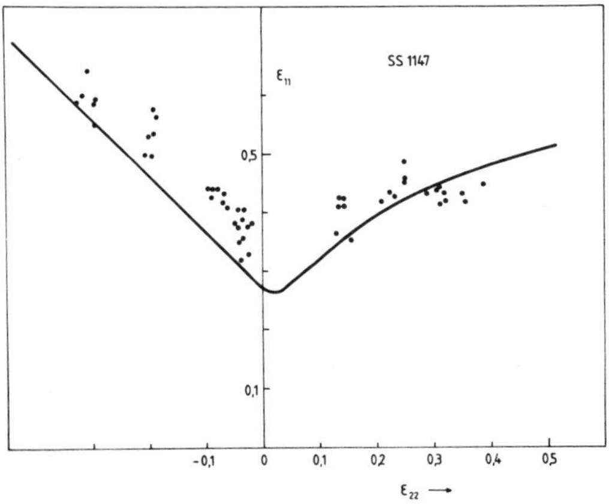

Fig 4a FLD of deep drawing quality steel. The fitting parameters were S=0 and f_p^n/f_p^o = 2.5. f_{th}^o= 0.

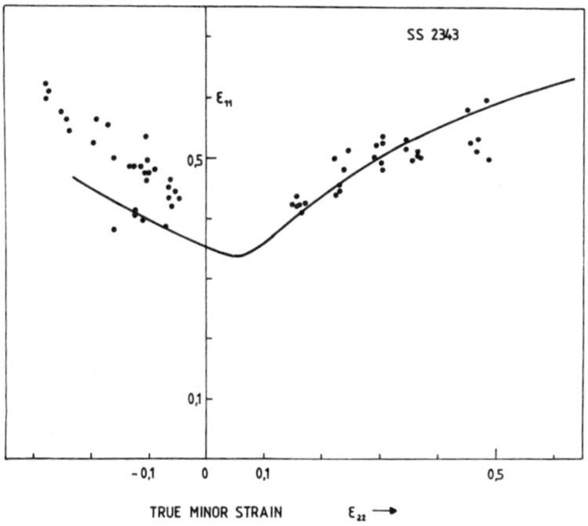

Fig 4b FLD of austenitic stainless steel. The fitting parameters were S=0 and f_p^n/f_p^o = 2.3. f_{th}^o =0.

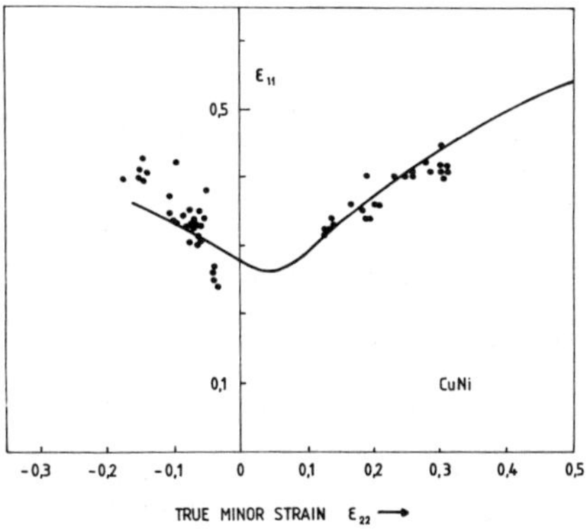

Fig 4c FLD of copper-10% nickel. The fitting parameters were S=0 and f_p^n/f_p^o= 5. f_{th}^o=0.

THE FLD OF DEEP DRAWING STEEL, AUSTENITIC STAINLESS STEEL AND COPPER-NICKEL

In this section the experimental FLDs of the deep drawing quality steel, SS1147, the austenitic stainless steel, SS2343, and copper-10%nickel are compared to theoretical calculations. This work was previously reported by Melander, Schedin, Karlsson and Steninger (1982), Melander (1981) and

Melander (1982). The three materials represent a wide range of work hardening properties, strain rate sensitivities, plastic anisotropy and inclusion contents. The theory is based on the material parameters of Tables 2 and 3. We make the simplifying approximation that the surface roughness can be neglected as an initiating factor for necking in comparison to inhomogeneous distributions of inclusions. This means that only one fitting parameter remains to be determined namely the locally higher volume fraction of inclusions in the necking region f_p^n. Fitting is performed on the right hand side of the FLD since the theory is considered to be more satisfactory on this side than on the left hand side of the FLD.

The theoretical curve for the deep drawing quality steel of Fig 4a was obtained with a fitting parameter of $f_p^n/f_p^o = 2.5$. This does not appear to be an unrealistic scatter in the density of inclusions. The material has an average volume fraction of inclusions of $f_p^o = 0.0008$.

The theoretical FLD of Fig 4b for the austenitic stainless steel was calculated with a fitting parameter of $f_p^n/f_p^o = 2.3$ which is quite close to the value for the deep drawing steel. The particle fraction outside the neck was $f_p^o = 0.002$.

Fig 4c illustrates the theoretical FLD of copper-nickel. The fitting parameter f_p^n/f_p^o was 5 for the average volume fraction of inclusions $f_p^o = 0.001$.

Figs 4 show a very satisfactory agreement between theory and experiment on the right hand side of the FLDs. The theory underestimates the experimental FLD for negative minor strains. It was found that the fitting parameter f_p^n/f_p^o took slightly different values in the three studied materials. It has in general been found that the value of f_p^n/f_p^o decreases with increasing average fraction of particles. (Melander, Schedin, Karlsson and Steninger, 1982). This can partly be explained by considering the scatter in particle density in a random distribution of particles. In a certain volume of material the relative scatter in the number of particles is larger if the average number of particles is smaller or the particle fraction is smaller (Melander, 1981).

With the present theory it can be illustrated how a variation in individual material parameters influences the FLD. The reader is referred to (Melander, 1981 ; Melander, Schedin, Karlsson and Steninger, 1982) for such results.

INFLUENCE OF GRAIN SIZE ON THE FORMING LIMIT DIAGRAM

In the present section it is illustrated how the grain size influences the forming limit diagram of 70/30 brass (Melander and Thuvander; 1982a). The two studied materials have the same sheet thickness of 0.7 mm but grain sizes of 8 respectively 82 μm as determined with the intercept method. Table 3 shows that the fine grained material has a significantly smaller work hardening capability than the coarse grained material. At a constant stress level the strain rate sensitivity M increases with increasing grain size.

In the theoretical analysis of the FLDs of the two brasses both a heterogeneity in the distribution of inclusions and surface roughness are assumed to cause necking. This means that two fitting parameters must be determined namely f_p^n/f_p^o and the constant of proportionality S which describes growth of surface roughness or rather sheet thickness reduction in eq.(3). It is assumed that the surface roughness of the as cold rolled sheet can be neglected, $f_{th}^o=0$. The fitting of f_p^n/f_p^o and S was performed by neglecting in

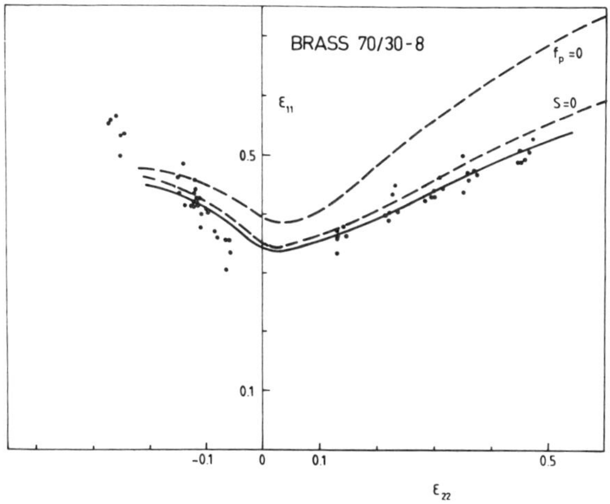

Fig 5a FLD of 70/30-8 brass. The full drawn line corresponds to S=0.5 and $f_p^n/f_p^o=9$. For the broken lines one of these parameters was set to zero. ($d_o/t_o=0.012$ and $f_{th}^o=0$).

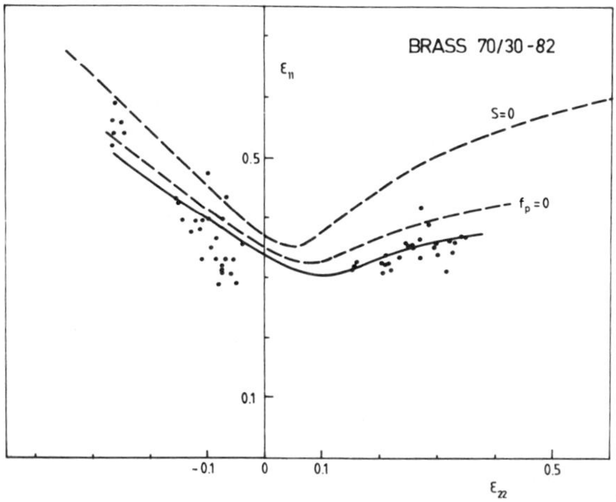

Fig 5b FLD of 70/30-82 brass. The full drawn line corresponds to S=0.5 and $f_p^n/f_p^o=9$. For the broken lines one of these parameters was set to zero. ($d_o/t_o=0.12$ and $f_{th}^o=0$).

a first stage the effect of the grain size in the fine grained material, S=0, and neglecting the effect of inclusions in the coarse grained material, $f_p^o=f_p^n=0$. The parameters thus obtained were in a second stage varied slightly so that both inclusions and surface roughness could be considered simultaneously. The following parameter values were obtained: $f_p^n/f_p^o=9$ and S = 0.5.

The theory thus predicts that the inclusion fraction is nine times larger inside than outside the neck. This large scatter in the density of inclusions is probably caused by the very low volume fraction of inclusions in the present materials of $f_p^o=0.0004$. The fitted constant of proportionality S= 0.5 for the imposed growth of surface roughness or thickness reduction is close to the value of approximately 0.4 which was determined experimentally (Wilson, Roberts and Rodrigues; 1981a) for the maximum peak to valley depth of the surface profile of brass.

The solid lines of Figs 5a and b show a very satisfactory agreement with experimental data on the right hand side of the FLDs. The agreement is not as satisfactory for negative minor strains.The respective roles of inclusion heterogeneities and surface roughness in the neck initiation porcess are illustrated in Figs 5 by neglecting either the surface roughness,S=0, or the inclusion contents, $f_p^o=f_p^n=0$.The lower broken line for the fine grained brass in Fig 5a shows that surface roughness has a small influence on the FLD if the grain size is small enough.If void growth at inclusions is neglected as for the upper broken line of Fig 5a the FLD is drastically increased.Heterogeneities in the distribution of inclusions is thus the dominating process of neck initiation for the 70/30-8 brass.The broken lines of Fig 5b for the coarse grained brass,70/30-82,show that surface roughness has a significantly larger influence on the necking process than inclusion heterogeneities if the grain size is as large as 82 µm.

INFLUENCE OF NON-LINEAR STRAIN PATHS ON THE FORMING LIMIT DIAGRAM

The FLDs presented so far were obtained with approximately linear strain paths. It is, however, known that strain paths in engineering sheet forming industry are often curved. It is also known that the shape and position of the FLD is very sensitive to non-linearities in the strain paths (Müschenborn and Sonne, 1975; Rasmussen, 1982).

In this section FLDs are presented which were determined with strain paths composed of linear segments. Two types of initial straining were used namely uniaxial tension and biaxial stretching. These operations were performed on so large specimens that FLDs could be determined subsequently with hydraulic bulge testing and tensile testing with rectangular specimens.

The experiments were performed on the deep drawing quality steel of Table 1 (Schedin; 1982). Tensile and compression testing was done on the preformed specimens. The flow curves often deviated significantly from the ones for as received material during an initial period of strain but subsequently the differences decreased. In the theoretical description of the FLD the work hardening properties of the material are given by Table 3 assuming one set of parameters during preforming and another set for the final deformation path. The combination of parameters is determined by the instantaneous strain path $\rho_i=d\varepsilon_{22}^o/d\varepsilon_{11}^o$. The anomalous work hardening behaviour initially during the second deformation sequence is thus neglected. No significant differences were observed in the strain rate sensitivity between as received and preformed materials at the same effective stress level.

The theoretical calculation of the FLD proceeds as for straight strain paths with the only exception that the instantaneous value of the imposed strain path ρ_i is considered. We make the same assumptions as in the previous calculation, Fig 4a, for the deep drawing quality steel namely that the influence of surface roughness can be neglected and that the inclusion fraction is by a factor of 2.5 higher in the necking region than outside.

Fig 6a shows the FLD after uniaxial preforming of the deep drawing quality steel. The experimental results can be compared to Fig4a which was obtained for straight strain paths. It was mentioned previously that different types of samples were used to determine the left hand side of the FLD of as received respectively preformed materials. It has, however, been shown that these techniques produce identical FLDs within experimental scatter. The experimental FLD for non-linear strain paths in Fig 6a is displaced towards negative minor strains compared to Fig 4a. This is important from a practical forming point of view since the FLD is shifted upwards for final states of

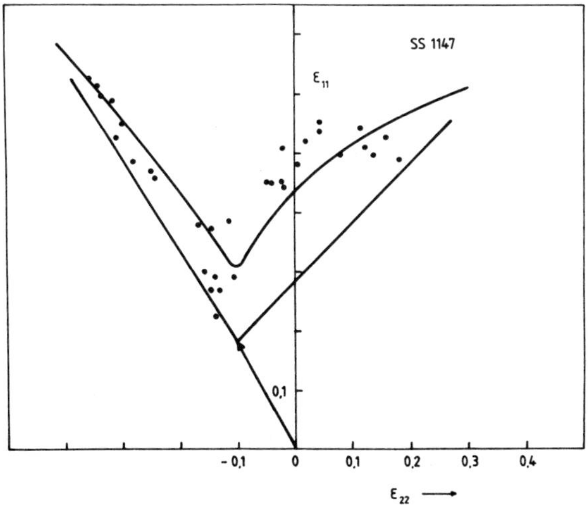

Fig 6a FLD after uniaxial preforming for deep drawing quality steel. The fitting parameters were S=0 and f_p^n/f_p^o=2.5. f_{th}^o=0.

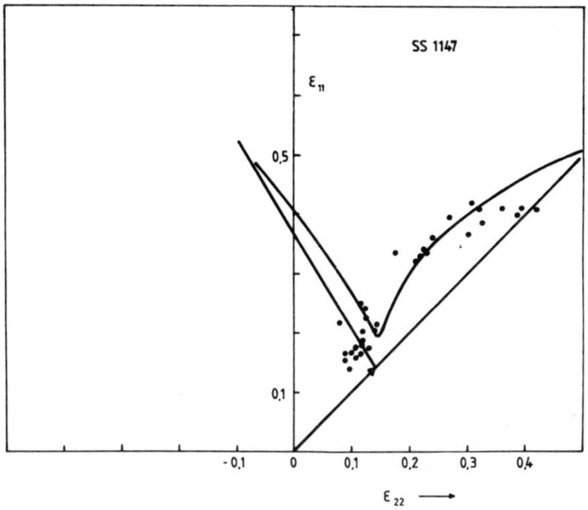

Fig 6b FLD after equibiaxial preforming for deep drawing quality steel. The fitting parameters were S=0 and f_p^n/f_p^o=2.5. f_{th}^o=0.

deformation of plane strain and biaxial stretching if uniaxial preforming is used.

The theoretical FLD of Fig 6a shows a very satisfactory agreement with the experimental results. In fact the agreement is better for the preformed than for the as received material.

The FLD after equibiaxial preforming is shown in Fig 6b. The FLD is displaced downwards and towards positive minor strains compared to the as received material. No improvement of formability relative to as received material was observed. The agreement between theory and experiment is satisfactory.

CONCLUSIONS

A satisfactory agreement between theoretical and experimental forming limit diagrams, in particular for positive minor strains, was obtained for a range of materials (deep drawing steel, austenitic stainless steel, copper-nickel and brass) both for straight and non-linear strain paths if the following factors were considered theoretically:

- neck initiation is caused by surface roughness, originating from the grain structure, and heterogeneous distributions of inclusions.
- the flow surface is smooth with a shape characterized by normal anisotropy measured by the $\bar{r}$-value.
- work hardening rates which were determined in uniaxial tension and biaxial straining.
- instantaneous strain rate sensitivity which was determined in uniaxial tension and biaxial straining.
- void growth at inclusions which causes softening in the material.

ACKNOWLEDGEMENTS

The present research was financed by the General Research Programme of the Swedish Institute for Metals Research, Jernkontoret, Svenskt Stål AB, Rautaruukki OY, Avesta Jernverks AB, Fagersta AB, Nyby-Uddeholm AB, Saab-Scania, Volvo, Gränges Metallverkan and Outokumpu OY which is gratefully acknowledged. The reported experiments were performed by the following colleagues of the author - A.Thuvander, E.Schedin and S.Karlsson. The paper is published with permission by the Swedish Institute for Metals Research.

REFERENCES

Chu, C-C. and Needleman, A. (1980) J.Engr.Mater.Tech. 102, 249-256.

Fukuda, M., Yamaguchi, K., Takakura, N. and Sakano, Y. (1974) J.Japan Soc. Techn.Plasticity 15, 994-1002.

Gurson, A.L. (1977) J.Engr.Mater. Tech 99, 2.

Hill, R. (1950) The mathematical theory of plasticity, Publ. by Oxford university Press.

Hill, R. (1979) Math.Proc.Camb. Phil.Soc. 85, 179-191.

Hutchinson, J.W. and Neale, K.W. (1978) in Mechanics of sheet metal forming, Ed.by D.P.Koistinen and N-M.Wang, Publ. by Plenum Press, New York, p. 111-126.

Marciniak, Z. and Kuczynski, K. (1967) Int.J.Mech Sci 9, 609-620.

Melander, A. (1981), A new model of the forming limit diagram applied to experiments on four copper base alloys, Swedish Institute for Metals Research, Report IM-1618, to appear in Mater.Sci.Eng.

Melander, A. (1982), Proc.12th Biennial Congress of IDDRG, S.Margherita Ligure, Italy, Publ.by Associazione Italiana di Metallurgia, vol. 2, p. 17-30.

Melander, A. and Thuvander, A.(1982a) Influence of surface roughness and void growth at inclusions on the forming limit diagram of brass, Swedish Institute for Metals Research, Report IM-1694.

Melander, A. and Thuvander, A.(1982b) Res.Mech. 5, 129-149.

Melander, A., Schedin, E., Karlsson, S. and Steninger, J.(1982) A theoretical and experimental study of the forming limit diagram of deep drawing steels, dual phase steels, austenitic and ferritic stainless steels and titanium, Swedish Institute for Metals Research, Report IM-1676. Available 1985.

Müschenborn, W. and Sonne, H-M.(1975) Arch.Eisenhüttenwes. 46, 597-602.
Needleman, A. and Triantafyllidis, N.(1978) J.Engr.Mater.Tech. 100, 164-169.
Parmar, A. .and Mellor, P.B.(1978) Int.J.Mech. Sci 20, 385-391.
Rasmussen, S.N.(1982) Proc 12th Biennial Congress of IDDRG, S.Margherita Ligure, Italy, Publ by Associazione Italiana di Metallurgia,vol 2, p. 83-94.
Schedin, E. (1982) private communication.
Schedin, E. and Melander, A. (1982) Determination of the forming limit diagram with the rectangular sheet tension test, Swedish Institute for Metals Research, Report IM-1682.
Schmitt, J.H. and Jalinier, J.M.(1982) Acta Met. 30, 1789-1798 and 1799-1809.
Tvergaard, V.(1978) Int.J.Mech.Sci 20, 651-658.
Wilson,D.V., Roberts,W.T. and Rodrigues,P.M.B.(1981a)Met.Trans 12A, 1595-1602.
Wilson,D.V., Roberts,W.R. and Rodrigues,P.M.B.(1981b)Met.Trans 12A, 1603-1611.
Yamaguchi, K. and Mellor, P.B. (1976) Int.J.Mech.Sci 18, 85-90.

LIMITS TO FORMABILITY IN RATE-SENSITIVE METAL SHEETS

A. Needleman* and V. Tvergaard**

*Division of Engineering, Brown University, Providence, RI 02912, USA
**Department of Solid Mechanics, The Technical University of Denmark, Lyngby, Denmark

ABSTRACT

Localized necking in biaxially stretched, high-strength metal sheets is discussed, with focus on the effect of material rate-sensitivity and on the implications of various models of multi-axial material response. Results of plane-stress analyses are presented, comparing isotropic hardening and kinematic hardening materials, under proportional or non-proportional straining. Also the influence of progressive rupturing on the micro-scale, at second phase particles, is analysed. Limitations of the plane stress theory, due to the actual three dimensional stress state in the necking region, are pointed out.

1. INTRODUCTION

Failure modes which serve to limit ductility in sheet forming processes include wrinkling, puckering, fracture and local necking. Here we will focus on the local necking mode which is associated with the formation of a narrow zone of intense thinning. A fruitful concept in correlating observations on local necking is the forming limit diagram, Keller (1968), which exhibits the dependence of the imposed strain at local necking (the limit strain) on the imposed strain ratio. Much effort has been directed toward the experimental determination of forming limit diagrams, see e.g. Hecker (1978).

There has also been considerable interest in predicting forming limit diagrams analytically. When there is a line of zero extension in the plane of the sheet, the onset of local necking is predicted by the bifurcation analysis of Hill (1952). This analysis predicts that local necking will not occur from positive biaxial stretching states. However, observations show that sheets subject to biaxial tension fail by a process of local necking. In order to account for the occurrence of local necking in biaxial tension, Marciniak and Kuczynski (1967) postulated the presence of an initial thickness reduction and demonstrated that this initial inhomogeneity could precipitate local necking by instigating a drift of the strain state in the neck toward plane strain, while the material outside the neck continues to deform in biaxial tension.

Azrin and Backofen (1970) carried out experiments which did show the drift of the strain state in the neck toward plane strain predicted by the M-K model. However, the magnitude of the assumed initial thickness reduction required to fit the predictions to the data was much larger than actually measured in the specimens. It was also found that the shape of the observed forming limit diagram, for most of the materials tested, was different from that given by the analysis.

Both the bifurcation analysis of Hill (1952) and the imperfection analysis of Marciniak and Kuczynski (1967) were based on classical smooth yield surface plasticity theory. Stören and Rice (1975) showed that a simple model of a solid with a vertex on its yield surface does predict a bifurcation corresponding to local necking in biaxial tension. The forming limit curves given by this analysis were found to be in better qualitative accord with observations than those given by the M-K analysis. Although physical models of polycrystalline aggregates based on the concept of single crystal slip do predict the development of a vertex on the yield surface at the current loading point, Hill (1967), Hutchinson (1970), direct experimental evidence for vertex formation is ambiguous and contradictory, e.g. Hecker (1976), Michno and Findley (1977). However, there is evidence for the development of a region of high curvature on the yield surface at the current stress point, Hecker (1976). Tvergaard (1978) carried out an analysis of sheet necking for a material that obeys a kinematic hardening rule, which can be regarded as a model of a solid with a yield surface that remains smooth but that develops a high curvature at the current loading point. The bifurcation predictions based on kinematic hardening theory are essentially the same as those obtained using isotropic hardening. The imperfection magnitude required to obtain limit strains in accord with observed values is much less than needed when isotropic hardening is employed.

In essence the vertex model (and the kinematic hardening model) attribute the onset of local necking to the destabilizing effect of an inherent feature of plastic flow in metals - the discrete crystallographic nature of slip. Needleman and Triantafyllidis (1978) employed the basic model of Marciniak and Kuczynski (1967) in conjunction with a constitutive relation for porous ductile solids due to Gurson (1977a,b). This model attributes local necking to the progressive weakening induced by incipient ductile rupture. Both this theory and the vertex (or "rounded vertex" kinematic hardening theory) give predicted forming limit diagram shapes in better accord with observation than those given by classical isotropic hardening theory (Needleman and Rice (1978)).

Investigations have been carried out examining the influence of various constitutive properties on local necking predictions, for example the effects of material strain rate sensitivity and anisotropy have been documented (see e.g. Marciniak et al. (1973), Hutchinson and Neale (1978), Bassani, Hutchinson and Neale (1979)). In this paper we will focus primarily on rate effects in localized necking.

We begin by presenting a formulation of the governing equations for analyzing local necking of elastic-viscoplastic sheets within the context of plane stress theory. We then consider the prediction of the onset of local necking in proportional straining histories for several material models. One model is the usual rate dependent generalization of isotropic hardening, another is an analogous generalization of kinematic hardening. We also analyze the onset of sheet necking using the rate sensitive generalization of the Gurson (1977a,b) model of a porous plastic solid introduced in Pan et al. (1981).

Although the idealized proportional loading considered in this type of analysis can be closely approximated in tests (Azrin and Backofen (1975)), the deformation histories encountered in other circumstances can differ considerably from proportional loading. There is experimental evidence that limit strains can depend on the loading path (e.g. Laukonis and Ghosh (1978)) and several analytical investigations of the effect of deformation path on local necking have recently been carried out, Chu (1982) and Rasmussen (1981). Here we will illustrate the effect of loading path on the onset of local necking for both the rate dependent kinematic and isotropic hardening material models.

We close with a discussion of three dimensional effects on sheet necking. Sheets, although thin, are inherently three dimensional objects and there are significant effects that need to be considered in a three dimensional context; one example being sheet thickness effects.

2. ANALYSIS OF LOCAL NECKING

The Lagrangian formulation of localized necking in thin sheets presented here is a generalization of that given by Tvergaard (1980) to accommodate rate dependent as well as rate independent plasticity. The sheet is subjected quasi-statically to applied displacement rates that, in a uniform sheet, would give rise to homogeneous deformations. As sketched in Fig. 1 the neck is regarded as a narrow band in an otherwise homogeneously deformed sheet. Across the neck jumps in stress and strain, consistent with the requirements of equilibrium and compatibility, are permitted. Plane stress theory is employed so that stresses, strains and displacements in the theory are thickness averages and only the in-plane stresses are taken to be non-zero.

In the initial (reference) configuration the position of a material point in the plane of the sheet is denoted by $\mathbf{x}$. In the current configuration its position is $\bar{\mathbf{x}}$. Both $\mathbf{x}$ and $\bar{\mathbf{x}}$ are referred to a fixed Cartesian frame. Although it is possible to refer all field quantities to the fixed Cartesian frame we find that the numerical implementation of the analysis is simplified by employing a convected coordinate formulation of the governing equations. The convected coordinate net can be visualized as being inscribed on the body in the initial configuration and deforming with the material. Thus even though the coordinate net is taken to be initially Cartesian, the fact that the coordinate net deforms with the material necessitates the use of general tensors. We will use standard tensor notation with subscripts indicating covariant quantities, superscripts indicating contravariant quantities and the summation convention employed for repeated diagonal indices. Greek indices range from 1 to 2. Latin indices, when employed, range from 1 to 3, with the 3-direction normal to the plane of the sheet.

The in-plane base vectors of the convected net in the reference configuration are denoted by $\mathbf{g}_\alpha$. In the current (deformed) configuration these become $\bar{\mathbf{g}}_\alpha$, where, expressed in terms of the planar deformation gradient $\mathbf{F}$,

$$\bar{\mathbf{g}}_\alpha = F^{\beta}_{\cdot\alpha}\, \mathbf{g}_\beta \quad , \quad F^{\beta}_{\cdot\alpha} = \frac{\partial \bar{x}^\beta}{\partial x^\alpha} \tag{2.1}$$

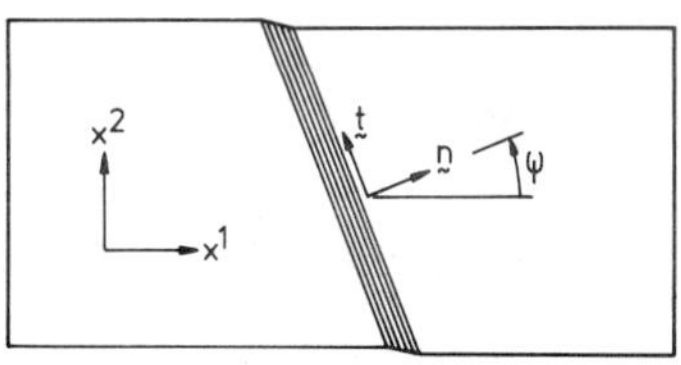

Fig. 1. Narrow band in biaxially stretched sheet.

The covariant components of the metric tensor on the deformed net are $\bar{g}_{\alpha\beta}(\bar{g}_{\alpha\beta} = \bar{g}_\alpha \cdot \bar{g}_\beta)$ and the contravariant components are denoted by $\bar{g}^{\alpha\beta}(\bar{g}^{\alpha\beta}\bar{g}_{\alpha\beta} = \delta^\alpha_\beta$, where δ^α_β is the Kronecker delta).

In-plane compatibility is expressed by the requirement that the in-plane displacements be continuous over the neck interface. This condition is satisfied if the tangential displacement derivatives are continuous across the neck. The neck is taken to be homogeneously deformed and compatibility implies (see Hill (1962)) that the deformation gradient inside the neck takes the form

$$(F_{\alpha\beta})_b = (F_{\alpha\beta})_o + s_\alpha n_\beta \tag{2.2}$$

Here, $(\ )_b$ denotes quantities within the band, $(\ )_o$ denotes corresponding quantities outside the band and $\underset{\sim}{n}$ is the normal to the line of the local neck in the initial configuration.

Equilibrium across the band is expressed by $h_b T^\alpha_b = h_o T^\alpha_o$, where h_b and h_o , respectively, denote the initial thickness of the sheet inside and outside the band and $\underset{\sim}{T}$ is the nominal traction vector, i.e., the force per unit reference area. The rate form of this equilibrium relation can be written as

$$h_b n_\alpha \left[F^\beta_{\cdot\gamma}\dot{\tau}^{\alpha\gamma} + \dot{F}^\beta_{\cdot\gamma}\tau^{\alpha\gamma} \right]_b = h_o n_\alpha \left[F^\beta_{\cdot\gamma}\dot{\tau}^{\alpha\gamma} + \dot{F}^\beta_{\cdot\gamma}\tau^{\alpha\gamma} \right]_o \tag{2.3}$$

where $(\dot{\ })$ denotes $\partial(\ )/\partial t$ at a fixed material point. The contravariant components of Kirchhoff stress $\tau^{\alpha\gamma}$ appearing in (2.3) are related to the corresponding Cauchy (or true) stress components $\sigma^{\alpha\gamma}$ by

$$\tau^{\alpha\gamma} = \det(F^\rho_{\cdot\omega})\sigma^{\alpha\gamma} \tag{2.4}$$

with $\det(\ )$ denoting the determinant of a matrix.

To proceed further the mechanical response of the sheet material needs to be specified. The constitutive relations to be considered here can be written in the general form

$$\dot{\tau}^{\alpha\beta} = C^{\alpha\beta\gamma\delta}\dot{\eta}_{\gamma\delta} + \dot{P}^{\alpha\beta} \tag{2.5}$$

where $\underset{\sim}{C}$ is the tensor of instantaneous moduli, $\dot{\underset{\sim}{P}}$ is an initial stress type term that arises for rate dependent solids and $\dot{\eta}_{\gamma\delta}$ are the covariant Lagrangian strain rate components

$$\dot{\eta}_{\gamma\delta} = \frac{1}{2}(F^\rho_{\cdot\gamma}\dot{F}_{\rho\delta} + \dot{F}^\rho_{\cdot\gamma}F_{\rho\delta}) \tag{2.6}$$

Substituting (2.5) and (2.6) into the rate statement of equilibrium (2.3) and using the rate form of (2.2) for $\underset{\sim}{F}$ we obtain

$$\left[h_b n_\alpha (F^\beta_{\cdot\gamma}C^{\alpha\gamma\mu\nu}F^\rho_{\cdot\mu} + \tau^{\alpha\nu}g^{\beta\rho})_b n_\nu \right]\dot{s}_\rho = h_o n_\alpha (F^\beta_{\cdot\gamma}C^{\alpha\gamma\mu\nu}F^\rho_{\cdot\mu} + \tau^{\alpha\nu}g^{\beta\rho})_o \overset{\bullet o}{F}_{\rho\nu}$$
$$- h_b n_\alpha (F^\beta_{\cdot\gamma}C^{\alpha\gamma\mu\nu}F^\rho_{\cdot\mu} + \tau^{\alpha\nu}g^{\beta\rho})_b \overset{\bullet o}{F}_{\rho\nu} + h_o n_\alpha (F^\beta_{\cdot\gamma}\dot{P}^{\alpha\gamma})_o - h_b n_\alpha (F^\beta_{\cdot\gamma}\dot{P}^{\alpha\gamma})_b \tag{2.7}$$

Equation (2.7) is a system of two equations for the two unknowns $\dot{s}_\rho$. Given the initial conditions, the initial sheet thickness and material properties inside and outside the neck and given the imposed deformation history outside the neck, these equations can be solved in an incremental fashion to obtain the deformation history inside the neck. Localization is said to occur when the ratio of some scalar measure of the amount of straining inside the neck to the corresponding value outside the neck becomes unbounded.

For an initially uniform sheet the right hand side of (2.7) vanishes identically and the deformation state remains homogeneous unless the determinant of the coefficient matrix vanishes. This is the condition governing a local necking bifurcation.

The analysis of local necking presented here is a precise two dimensional analog of the three dimensional theory of shear band localization (see e.g. Rice (1977) and Needleman and Rice (1978)). Consistent with this analogy the compatibility and equilibrium conditions have been interpreted as jump conditions across a narrow band in the presentation.

3. PROPORTIONAL STRAINING

Most analytical studies of local necking have been carried out for cases in which the x_1 and x_2 axes in Fig. 1 remain principal strain directions throughout the loading history and with the ratio, ρ, of the principal logarithmic strains, $\varepsilon_1 = \ell n(F^1_{\cdot 1})$ and $\varepsilon_2 = \ell n(F^2_{\cdot 2})$, being constant

$$\rho = \varepsilon_2/\varepsilon_1 = \dot{\varepsilon}_2/\dot{\varepsilon}_1 \tag{3.1}$$

Such studies have proved useful in elucidating the effect of various constitutive features on the onset of local necking. Studies carried out within this framework have shown that a relatively small amount of strain rate hardening results in a substantial delay in the onset of local necking, Marciniak, Kuczynski and Pokora (1973), Ghosh (1978), Hutchinson and Neale (1978). The amount of this delay also depends on the imposed strain ratio so that, particularly for high hardening materials, the shape of the forming limit diagram varies with the strain rate sensitivity of the material.

The rate sensitive constitutive relations that have been employed in previous analyses of local necking have been strain rate dependent generalizations of the simplest (J_2) flow and deformation theories of plasticity. In this section we present some results on local necking using a rate dependent generalization of Prager-Ziegler kinematic hardening to accommodate rate sensitive material response. As discussed by Rice (1970) an explicit vertex is, on very general grounds, ruled out for rate sensitive solids. The kinematic hardening solid is a simple but consistent model of a rate sensitive solid with a flow potential surface (there is no yield surface in the rate dependent formulation we employ) of relatively high curvature at the current loading point. For comparison purposes we also consider a rate sensitive isotropic hardening solid.

The Lagrangian strain rate (see (2.6)) is written as the sum of an elastic part $\dot{\eta}^e_{ij}$ and a plastic part $\dot{\eta}^p_{ij}$. The elastic part is given by

$$\dot{\eta}^e_{ij} = M_{ijk\ell} \overset{\Delta}{\tau}{}^{k\ell} \tag{3.2}$$

where $\underset{\sim}{M}$ is the elastic compliance tensor and $\overset{\Delta}{\tau}{}^{k\ell}$ are the contravariant components of the Jaumann stress rate on the deformed base vectors.

The flow potential surfaces in stress space are circular cylinders centered about a deviatoric stress $\underset{\sim}{\alpha}$ giving a flow rule of the form

$$\dot{\eta}^p_{ij} = \frac{3}{2} \dot{\bar{\varepsilon}} \frac{s_{ij} - \alpha_{ij}}{\bar{\sigma}} \tag{3.3}$$

Here, $\dot{\bar{\varepsilon}}$, the plastic strain rate magnitude, is a function of the accumulated plastic strain $\bar{\varepsilon}$ and the equivalent stress $\bar{\sigma}$ and

$$s^{ij} = \tau^{ij} - \frac{1}{3}\tau^k_k \bar{g}^{ij} \quad , \quad \bar{\sigma}^2 = \frac{3}{2}(s^{ij} - \alpha^{ij})(s_{ij} - \alpha_{ij}) \tag{3.4}$$

Combining (3.2) and (3.3), and solving for $\overset{\Delta}{\tau}{}^{ij}$ gives

$$\overset{\Delta}{\tau}{}^{ij} = L^{ijk\ell}\dot{\eta}_{k\ell} - \dot{\bar{\varepsilon}}\,Q^{ij} \quad , \quad Q^{ij} = \frac{3}{2}L^{ijk\ell}\left[\frac{s_{k\ell} - \alpha_{k\ell}}{\bar{\sigma}}\right] \tag{3.5}$$

Using the standard relation between contravariant components of the Jaumann and convected stress rates leads to

$$\dot{\tau}^{ij} = \left[L^{ijk\ell} - \tau^{ik}\bar{g}^{j\ell} - \tau^{j\ell}\bar{g}^{ik}\right]\dot{\eta}_{k\ell} - \dot{\bar{\varepsilon}}\,Q^{ij} = \hat{L}^{ijk\ell}\dot{\eta}_{k\ell} - \dot{\bar{\varepsilon}}\,Q^{ij} \tag{3.6}$$

For the kinematic hardening solid the evolution equation for $\underset{\sim}{\alpha}$ is taken to be the finite strain generalization of Ziegler's hardening rule used by Tvergaard (1978), here specified on the form

$$\dot{\alpha}^{ij} = \frac{2}{3}b\,\bar{g}^{ik}\bar{g}^{j\ell}\dot{\eta}^{p}_{k\ell} - \bar{g}^{ik}\alpha^{j\ell}\dot{\eta}_{k\ell} - \bar{g}^{j\ell}\alpha^{ik}\dot{\eta}_{k\ell} \tag{3.7}$$

The constitutive description is completed by specifying the dependence of $\dot{\bar{\varepsilon}}$ on the accumulated plastic strain $\bar{\varepsilon}$ and equivalent stress $\bar{\sigma}$. For the isotropic hardening solid we take

$$\dot{\bar{\varepsilon}} = \dot{\varepsilon}_o\left[\frac{\bar{\sigma}}{g(\bar{\varepsilon})}\right]^{1/m} \quad , \quad g(\bar{\varepsilon}) = \sigma_o(\bar{\varepsilon}/\varepsilon_o + 1)^N \tag{3.8}$$

and for the kinematic hardening solid

$$\dot{\bar{\varepsilon}} = \dot{\varepsilon}_o(\bar{\sigma}/\sigma_o)^{1/m} \quad , \quad b = \frac{\bar{\sigma}N}{\varepsilon_o}[\bar{\varepsilon}/\varepsilon_o + 1]^{N-1} \tag{3.9}$$

In (3.8) and (3.9) N is the strain hardening exponent and m is the strain rate hardening exponent; $\dot{\varepsilon}_o$ is a reference strain rate, ε_o is a reference strain and σ_o is $E\varepsilon_o$, E being Young's modulus. In uniaxial tension, with the small effect of material elasticity neglected, (3.9) gives

$$\sigma = \sigma_o\left[\left(\frac{\dot{\varepsilon}}{\dot{\varepsilon}_o}\right)^m + \int_0^t\left(\frac{\dot{\varepsilon}}{\dot{\varepsilon}_o}\right)^m N\left[\frac{\varepsilon}{\varepsilon_o} + 1\right]^{N-1}\frac{\dot{\varepsilon}}{\varepsilon_o}dt\right] \tag{3.10}$$

For constant plastic strain rate (3.10) reduces to the power hardening relation

$$\sigma = \sigma_o\left(\frac{\dot{\varepsilon}}{\dot{\varepsilon}_o}\right)^m\left[\frac{\varepsilon}{\varepsilon_o} + 1\right]^N \tag{3.11}$$

which also represents the uniaxial response of the isotropic hardening solid; (3.10) and (3.11) also coincide in the rate independent limit $m \to 0$. The strain rate history dependence embodied in (3.10) has, as will be illustrated subsequently, a significant effect on necking behaviour. It should be noted here, in connexion with (3.9), that there are other natural ways of incorporating rate-sensitivity into kinematic hardening.

Employing $\dot{\tau}^{33} = 0$ to solve for $\dot{\eta}_{33}$ gives the in-plane constitutive relations needed for the sheet necking analysis. The plane stress moduli and initial stress term in (2.5) are given in terms of their three dimensional counterparts by

$$C^{\alpha\beta\gamma\delta} = \hat{L}^{\alpha\beta\gamma\delta} - \hat{L}^{\alpha\beta 33}\hat{L}^{33\gamma\delta}/\hat{L}^{3333} \quad , \quad \dot{P}^{\alpha\beta} = \dot{\bar{\varepsilon}}[Q^{\alpha\beta} - \hat{L}^{\alpha\beta 33}Q^{33}/\hat{L}^{3333}] \tag{3.12}$$

Note that for the rate dependent plastic solid the instantaneous moduli are the elastic moduli. The plasticity enters only through the initial stress term $\underset{\sim}{P}$. This has significant theoretical and numerical implications. Firstly, for the rate dependent solid a local necking bifurcation is ruled out at achievable strains. Thus, in carrying out a local necking analysis for the rate dependent solid it is always necessary to include some initial inhomogeneity, even in

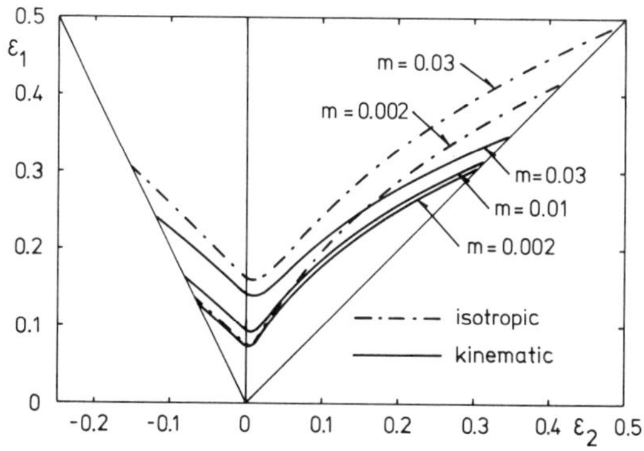

Fig. 2. Forming limit curves for various rate hardening exponents and proportional straining, with $h_b/h_o = 0.99$.

the regime $\rho < 0$ where a local necking bifurcation occurs for an isotropic hardening rate independent solid. Secondly, the elastic moduli put a severe restriction on allowable step size in a numerical integration scheme. Pierce et al. (1983) have introduced a "rate tangent modulus" procedure that greatly facilitates the integration of this type of constitutive relation. The algebraic details will not be repeated here but we note that the procedure is based on a Taylor series expansion of the initial stress term about the current known state. This procedure permits time steps nearly an order of magnitude larger than can be employed with (3.6).

Forming limit curves obtained for imposed proportional straining at a constant strain rate $\dot{\varepsilon}_o$ (the forming limit curves are nearly independent of the magnitude of the imposed rate) are depicted in Fig. 2 for a relatively low hardening material, $N = 0.1$. In each case we use $\dot{\varepsilon}_o = 0.005[\sec^{-1}]$, $\varepsilon_o = \sigma_o/E = 0.005$ and $\nu = 0.3$ (the results are not sensitive to the values of these parameters) and the initial imperfection is specified as $h_b/h_o = 0.99$. The two dashed curves are based on isotropic hardening with $m = 0.002$ and $m = 0.03$, while the three solid curves are for kinematic hardening with $m = 0.002$, $m = 0.01$ and $m = 0.03$. When the material behaviour is nearly rate independent $(m = 0.002)$ the limit strains in plane strain tension for the two material models nearly coincide. With $m = 0.03$, due to the effect of strain rate history, the limit strain in plane strain tension for the kinematic hardening solid falls well below the corresponding value for the isotropic hardening solid. We note that in plane strain tension the local necking predictions of the rate dependent flow and deformation theories considered by Hutchinson and Neale (1978) coincide.

The stabilizing effect associated with material strain rate dependence comes from the increased stiffness with increasing strain rate. Local necking brings with it a rapidly increasing strain rate in the incipient neck and it is the response of the material to this changing strain rate that influences the onset of localization. Hence it is the response of the material to a changing strain rate that governs necking behaviour. The results in Fig. 2 illustrate that identical response at constant (plastic) strain rate does not imply identical necking behaviour.

4. NON-PROPORTIONAL LOADING PATHS

Experimental evidence indicates that abrupt changes in loading path can have a marked effect on the observed forming limit, e.g. Laukonis and Ghosh (1978), and there has been recent interest in analyzing this effect within the M-K framework. Rasmussen (1981,1982) has carried out analyses employing rate independent smooth yield surface plasticity theory while Chu (1982) has employed the J_2 corner theory of Christoffersen and Hutchinson (1979).

Here we will present results illustrating the effect of path changes on forming limit diagrams for the rate dependent kinematic hardening solid. All material parameters are as specified in Section 3 and we focus on the case $m = 0.03$. We confine attention to bilinear strain paths of the form

$$\dot{\varepsilon}_2 = \begin{cases} \rho_1 \dot{\varepsilon}_1 , & \text{for} \quad \varepsilon_1 \leq \varepsilon_s \\ \rho_2 \dot{\varepsilon}_1 , & \text{for} \quad \varepsilon_1 > \varepsilon_s \end{cases} \tag{4.1}$$

Forming limit diagrams have been calculated for two histories, one with $\rho_1 = -0.5$ (nearly uniaxial tension), $\rho_2 = 1.0$ (equal biaxial tension) and the other with $\rho_1 = 1.0$, $\rho_2 = 0.0$ (plane strain tension). Several values of ε_s have been specified and the resulting forming limit diagrams are displayed in Fig. 3. For comparison purposes the forming limit diagram for proportional straining from Fig. 2 is reproduced in the figure. A prestrain in uniaxial tension followed by equal biaxial tension gives a forming limit diagram higher than for proportional stretching, while prestrain in equal biaxial tension followed by plane strain tension results in a lowered forming limit diagram.

The trends illustrated in Fig. 3 are consistent with a number of experimental observations discussed by Rasmussen (1981,1982), who compared the results of his M-K analysis with two proposed empirical rules, a surface strain rule and an effective strain rule (see Chu (1982) for an additional discussion of the effective strain rule). For the case $\rho_1 = -0.5$, $\rho_2 = 1.0$ Rasmussen (1982) found quite good agreement between the M-K results and the effective strain based rule, with the M-K based forming limit diagram falling above the one given by this empirical rule. The surface strain rule led to forming limit curves significantly above the calculated ones. However, M-K results obtained for isotropic hardening in the present investigation (not shown in Fig. 3) lie below the predictions of the empirical effective strain rule. Thus, the effective strain rule predicts a greater rate of increase of limit strain with tensile prestrain than is exhibited by the present calculations.

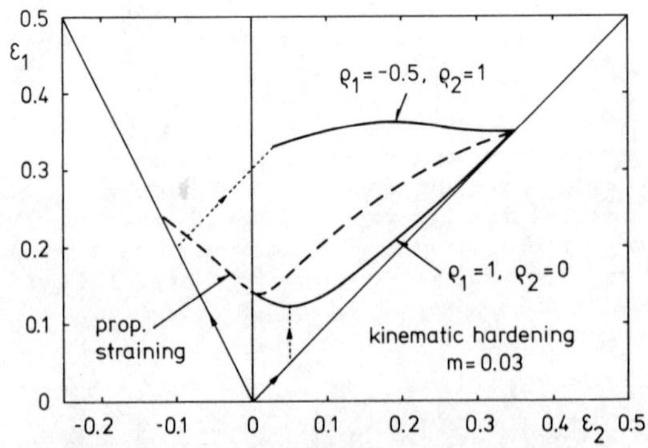

Fig. 3. Forming limit curves for non-proportional straining, with $h_b/h_o = 0.99$.

The increase in limit strain due to a uniaxial prestrain found for the kinematic hardening solid in Fig. 3 is somewhat smaller than that found for isotropic hardening. Also in this case, the effective strain rule predicts limit strains above those calculated by the M-K analysis. For example, in Fig. 3 with a tensile prestrain of 0.15 the M-K analysis gives a limit strain of 0.36 , while using the effective strain rule gives a limit strain of 0.43 .

5. RATE-SENSITIVITY WITH VOID NUCLEATION

The porosity arising from the cracking or decohesion of second phase particles in a ductile metal results in a dilatancy and pressure sensitivity of the macroscopic plastic deformations. A set of constitutive relations that, approximately, account for this porosity have been suggested by Gurson (1977a,b). These relations, formulated for rate-independent plasticity, have been used by Needleman and Triantafyllidis (1978) to study the influence of void growth on necking instabilities in biaxially stretched sheets, and the effect of void nucleation has been studied by Chu and Needleman (1980). In the following, the influence of a rate sensitive, void-nucleating material on sheet necking will be analysed, based on a rate dependent version of the Gurson model introduced by Pan et al. (1981).

The approximate yield condition proposed by Gurson (1977a,b) takes the form

$$\Phi(\sigma^{ij}, \sigma_M, f) = \frac{\sigma_e^2}{\sigma_M^2} + 2f q_1 \cosh\left\{\frac{\sigma_k^k}{2\sigma_M}\right\} - (1 + q_1^2 f^2) = 0 \tag{5.1}$$

where σ^{ij} is the average macroscopic Cauchy stress tensor, σ_M is an equivalent tensile flow stress representing the microscopic stress-state in the matrix material, and f is the current void volume fraction. In terms of the stress deviator $s^{ij} = \sigma^{ij} - \bar{g}^{ij}\sigma_k^k/3$ the macroscopic effective Mises stress is $\sigma_e = (3s_{ij}s^{ij}/2)^{\frac{1}{2}}$. The expression (5.1), for $q_1 = 1$, was derived by Gurson (1977a) based on a rigid-perfectly plastic solution for spherically symmetric deformations around a single spherical void; but considerably improved agreement with numerical studies of periodically distributed circular cylindrical voids or spherical voids in elastic-plastic solids is obtained using $q_1 = 1.5$ (Tvergaard, 1981,1982a).

The rate-sensitivity is introduced by characterizing the matrix material as visco-plastic, with the microscopic effective plastic strain rate $\dot{\varepsilon}_M^P$ represented by a power law relation of the form (3.8). Here, the function $g(\varepsilon_M^P)$ is given by

$$\varepsilon_M^P = \frac{\sigma_o}{E}\left\{\frac{g(\varepsilon_M^P)}{\sigma_o}\right\}^{1/N} - \frac{g(\varepsilon_M^P)}{E} \quad , \quad g(0) = \sigma_o \tag{5.2}$$

with the strain hardening exponent being N , Young's modulus E and the reference stress σ_o . The yield function (5.1) is here used as a plastic potential, such that the plastic part of the macroscopic strain rate is given by the expression

$$\dot{\eta}_{ij}^P = \Lambda \frac{\partial \Phi}{\partial \sigma^{ij}} \tag{5.3}$$

Assuming an equivalent plastic work expression, $\sigma^{ij}\dot{\eta}_{ij}^P = (1 - f)\sigma_M\dot{\varepsilon}_M^P$, the parameter Λ is given by

$$\Lambda = (1 - f)\sigma_M\dot{\varepsilon}_M^P\left[\sigma^{ij}\frac{\partial \Phi}{\partial \sigma^{ij}}\right]^{-1} \tag{5.4}$$

The total strain rate $\dot{\eta}_{ij}$ is taken to be the sum of an elastic part and the plastic part (5.3). Then, the expression for the Jaumann rate of the Cauchy stress can be written in the form (3.6).

The rate of increase of the void volume fraction is taken to be

$$\dot{f} = (1-f)\bar{g}^{ij}\dot{\eta}^P_{ij} + \mathcal{B}\left(\dot{\sigma}_M + (\sigma^k_k)^{\bullet}/3\right) + \mathcal{D}\,\dot{\varepsilon}^P_M \tag{5.5}$$

where the first term on the right hand side gives the contribution from void growth, the second term models nucleation controlled by the maximum normal stress on the particle matrix interface, as suggested by Needleman and Rice (1978), and the last term models plastic strain controlled nucleation. The parameters $\mathcal{B}$ and $\mathcal{D}$ are chosen so that void nucleation follows a normal distribution (Chu and Needleman, 1980), with the volume fraction of void nucleating particles denoted by f_N, the mean stress and the mean strain for nucleation denoted by σ_N and ε_N, respectively, and the corresponding standard deviation denoted by s. Furthermore, the consistency condition, $\dot{\Phi} = 0$, gives

$$\frac{(\sigma_e^2)^{\bullet}}{\sigma_M^2} + \frac{f}{\sigma_M}q_1 \sinh\left\{\frac{\sigma^k_k}{2\sigma_M}\right\}(\sigma^k_k)^{\bullet} + \frac{\partial\Phi}{\partial\sigma_M}\dot{\sigma}_M + \frac{\partial\Phi}{\partial f}\dot{f} = 0 \tag{5.6}$$

For known strain-rates and macroscopic stress-rates the values of $\dot{f}$ and $\dot{\sigma}_M$ are found from (5.5) and (5.6).

The concept of elastic unloading is not directly incorporated in this visco-plastic model; but since we consider small values of the rate-hardening exponent m, the plastic strain rates are negligible if σ_M is only slightly smaller than $g(\varepsilon^P_M)$. Thus, $g(\varepsilon^P_M)$ acts effectively as a flow strength. In the limiting case $m \to 0$ all relevant Φ-surfaces coincide, thus representing the yield surface of time-independent plasticity.

The plane-stress constitutive relationship required for the incremental M-K-analysis is obtained in a straightforward manner, as in (3.12). Also, the details of a forward gradient method (Peirce et al., 1982) used to increase the stable step size are omitted.

For $f \equiv 0$ throughout the process the present material model reduces to rate-sensitive J_2 flow theory, and the sheet necking results of Hutchinson and Neale (1978) are directly reproduced. Here, we shall investigate cases with no initial voids $f_I = 0$, but with plastic strain controlled nucleation specified by the volume fraction $f_N = 0.04$ of particles, nucleating around the mean plastic strain $\varepsilon_N = 0.5$ with standard deviation $s = 0.1$. An additional

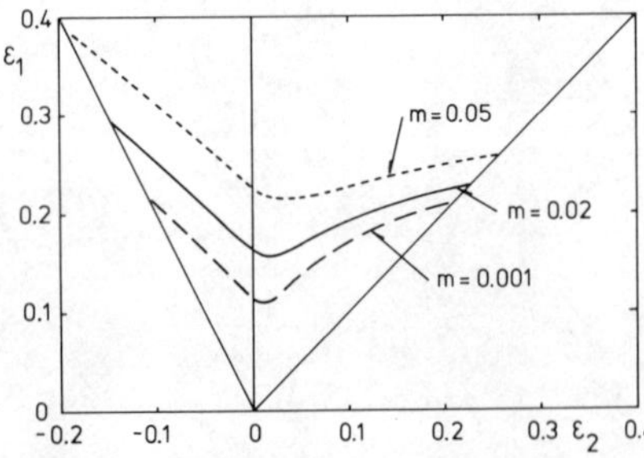

Fig. 4. Forming limit curves for various rate hardening exponents. Stress controlled nucleation inside the band, $\Delta f_N = 0.01$, in addition to strain controlled nucleation everywhere, $f_N = 0.04$.

particle concentration $\Delta f_N = 0.01$ inside the band is assumed to nucleate according to the stress criterion, with mean stress $\sigma_N = 2.1\sigma_o$ and standard deviation $s = 0.4\sigma_o$. This imperfection, Δf_N, could represent inclusion colonies or relatively large inclusions of low strength in a structural steel, while the remaining nucleation could represent smaller voids that nucleate from carbides or other precipitates at large strains. Furthermore, the material parameters $\sigma_o/E = 0.001$, $\nu = 0.3$, $N = 0.2$ and $q_1 = 1.5$ are used, and a uniform initial plate thickness is assumed.

Fig. 4 shows forming limit curves for different values, 0.05, 0.02 and 0.001, of the rate-hardening exponent m, with proportional straining, $\rho = \varepsilon_2/\varepsilon_1$, prescribed outside the band. The critical angle of inclination of the band, relative to the x^2-axis, is determined as that giving the minimum localization strain, and is zero for positive ρ. Comparison with rate-independent plasticity results shows that localization is only slightly delayed with $m = 0.001$; but the results in Fig. 4 display considerable necking retardation over the whole range of ρ values, for rate sensitivities that are still relatively small.

All results in the present figures are obtained for a constant prescribed rate of the major principal strain outside the band, $\dot{\varepsilon}_1 = \dot{\varepsilon}_o$, where $\dot{\varepsilon}_o$ is the reference strain-rate of (3.8). Hutchinson and Neale (1978) found the interesting feature of their material models that the localization strains are independent of the strain-rate for given material parameters, and this is also found to be essentially true here. However, the stress level during the process depends on the strain rate, and therefore when nucleation is stress controlled the localization strain will be strain rate dependent.

The necking development, leading to the forming limit curves of Fig. 4, is illustrated in Fig. 5 by the strain ε_1^b and the void volume fraction f^b inside the band, for the special case of $\rho = 1$. Since relatively large stresses are reached at an early stage, the stress controlled nucleation inside the band takes place at small strains, and simultaneously the strains in the neck grow somewhat larger than those outside. Subsequently the behaviour is mainly controlled by void growth, since the plastic strain ε_M^P remains well below the mean value ε_N for strain controlled nucleation until shortly before localization. It is seen that the necking delay due to a higher degree of rate-sensitivity also results in a higher void volume fraction prior to localization.

The strong sensitivity to the magnitude of the initial imperfection Δf_N is illustrated in Fig. 6, for $m = 0.02$. The material is otherwise identical with that of Fig. 4, and only positive values of ρ are considered, since the imperfection sensitivity is most pronounced at $\rho \approx 1$. For the smaller

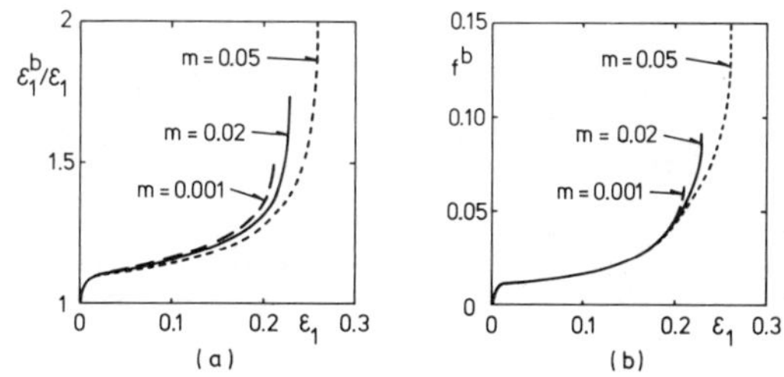

Fig. 5. Growth of maximum principal strain ε_1^b and void volume fraction f^b inside the band, for $\rho = 1$, $\Delta f_N = 0.01$ and $f_N = 0.04$.

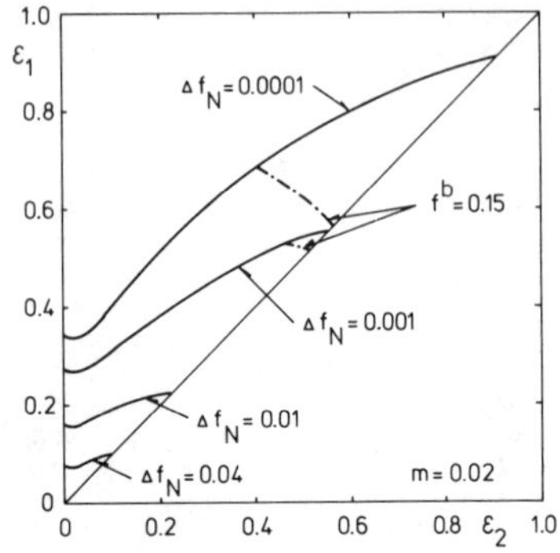

Fig. 6. Forming limit curves for various initial inhomogeneities Δf_N , in addition to the uniformly distributed particle concentration $f_N = 0.04$.

imperfections, $\Delta f_N = 0.0001$ and $\Delta f_N = 0.001$, the figure also shows curves at which the void volume fraction f^b inside the band reaches the value 0.15 . Experimental studies of ductile fracture by Brown and Embury (1973) indicate that for a number of materials a representative void volume fraction at coalescence is of the order of 0.15 . Therefore, the curves $f^b = 0.15$ indicate a competing failure mechanism, so that ductile fracture occurs before noticeable necking for values of ρ close to unity, whereas sheet formability is limited by localization for the remaining range of stretching ratios. Such "fracture curves" have also been discussed by Chu and Needleman (1980).

6. THREE DIMENSIONAL EFFECTS

The assumption of a plane stress state in sheet necking analyses results in a simple, efficient method for predicting forming limit curves, as has been shown in the previous sections. However, the influence of the actual three dimensional stress state in the necking region should be kept in mind when evaluating M-K-model predictions.

Full three dimensional bifurcation analyses are available for the special cases of $\rho = 0$ and $\rho = 1$. The analysis of the plane strain case by Hill and Hutchinson (1975) shows that for a sufficiently long wavelength relative to the plate thickness bifurcation coincides with the plane stress prediction (except, possibly, for very high hardening materials), whereas bifurcation into shorter wavelength modes is somewhat delayed. Similar results were found by Needleman and Tvergaard (1977) for the case of axisymmetric deformations in a circular plate, which is representative of equal biaxial stretching. In practice, the length to thickness ratio of sheets subjected to biaxial stretching is usually quite large, so that M-K-model bifurcation predictions are sufficiently accurate.

Even though bifurcation is associated with a long wavelength mode, the plane stress theory is no longer appropriate when the neck begins to develop. The length of the well developed localized necking region is comparable with the plate thickness, and the rapid thickness changes give rise to significant out-of-plane shear stresses. The resulting shear stiffening, neglected in the plane stress analysis, must delay the actual three dimensional localization process. This is true for a neck initiated at a bifurcation point, as well as for a neck growing from a small initial imperfection. Quantitative estimates of the delay require comparison with full three dimensional analyses, which are rather scarce. However, axisymmetric circular plate analyses by Needleman and Tvergaard (1977) for J_2 flow theory and deformation theory, and by Tvergaard (1978) for kinematic hardening, show that localization strains predicted by the plane stress analysis for $\rho = 1$ are about 15 - 35% smaller than those obtained by the three dimensional analyses. These results correspond to thickness imperfections, with the better agreement found for the smallest imperfection considered, $h_b/h_o = 0.99$. The delay is smaller for a long-wavelength imperfection, relative to the plate thickness, than for a short-wavelength imperfection, as has been discussed by Hutchinson, Neale and Needleman (1978). Thus, also the effect of a material inhomogeneity, such as a local concentration of void nucleating particles, will be much delayed if the band containing the inhomogeneity is narrow relative to the plate thickness.

The actual failure mechanism inside the neck is a question not considered by the M-K-analyses. In the context of plane stress theory, bifurcation of a uniformly stretched sheet into a localized necking mode coincides with loss of ellipticity of the equations governing incremental equilibrium. But usually, the equations governing the actual three dimensional problem retain ellipticity well beyond the point of necking. Final fracture is often associated with the formation of localized shear bands, when the three dimensional incremental equations loose ellipticity inside the neck. These shear bands may be triggered at an earlier stage by the formation of short wave surface instabilities, Hutchinson and Tvergaard (1980).

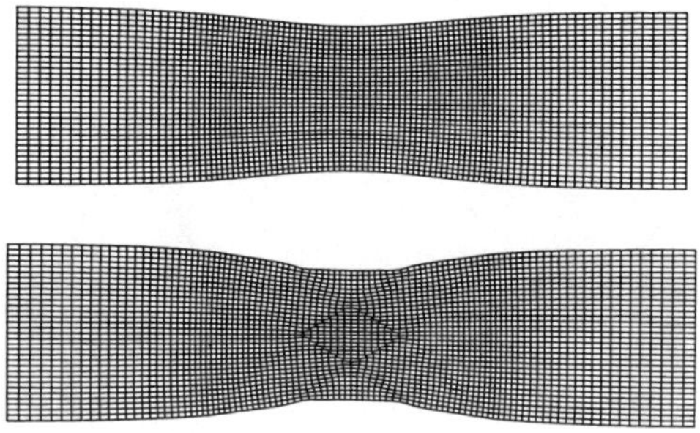

Fig. 7. Neck development in plane strain, analysed by J_2 corner theory. Shear bands have formed at the last stage.

Some insight in the failure mechanism is obtained from a numerical investigation of the plane strain case (Tvergaard, Needleman and Lo, 1981), based on J_2 corner theory. The plate analysed has an 0.375% initial thickness inhomogeneity (not visible on a figure), and a uniform rectangular grid is specified in the necking region. Flow localization is characterized by strain-rate reversal outside the neck, and Fig. 7 shows the deformed grid at two stages beyond this point. In the first stage the sheet exhibits a typical diffuse necking pattern; but slightly later the critical strain for shear band development is reached, and further deformations concentrate in the narrow shear bands that can be seen in the second deformed mesh.

Shear band formation is also predicted by the Gurson model, accounting for the nucleation and growth of micro-voids. Based on this material model Tvergaard (1982b) has analysed the full development of the final shear fracture inside shear bands, by coalescence of voids in a void sheet mechanism.

Plane stress, long wavelength analyses of sheet necking are very useful in evaluating the effects of various material properties and inhomogeneities on necking. However, a description of the final failure mechanism inside a neck requires three dimensional considerations. Viewing plane stress theory from the perspective of the three dimensional theory clearly exhibits the physical assumptions underlying this type of analysis.

ACKNOWLEDGEMENT

The work of A.N. was supported by the Materials Research Laboratory at Brown University funded by the U.S. National Science Foundation.

REFERENCES

Azrin, M. and W.A. Backofen (1970). Met. Trans., 1, 2857-2865.

Bassani, J.L., J.W. Hutchinson and K.W. Neale (1979). On the prediction of necking in anisotropic sheets. Metal Forming Plasticity (ed. H. Lippmann), 1-13, Springer-Verlag.

Brown, L.M., and J.D. Embury (1973). Proc. 3rd. Int. Conf. on Strength of Metals and Alloys, 164-169, Inst. of Metals, London.

Christoffersen, J., and J.W. Hutchinson (1979). J. Mech. Phys. Solids, 27, 465-487.

Chu, C.C. (1982). Int. J. Solids Structures, 18, 205-215.

Chu, C.C., and A. Needleman (1980). J. Engr. Mat. Tech., 102, 249-256.

Ghosh, A.K. (1978). Plastic flow properties in relation to localized necking in sheets. Mechanics of Sheet Metal Forming (ed. D.P. Koistinen and N.-M. Wang), 287-312, Plenum Publ. Corp.

Gurson, A.L. (1977a). J. Engr. Mat. Tech., 99, 2-15.

Gurson, A.L. (1977b). Proc. Int. Conf. Fracture (ed. D.M.R. Taplin), 2A, 357-364, Pergamon Press, Oxford.

Hecker, S.S. (1976). Experimental studies of yield phenomena in biaxially loaded metals. Constitutive Equations in Plasticity: Computational and Engineering Aspects, 1-33, ASME.

Hecker, S.S. (1978). Experimental studies of sheet formability. Formability: Analysis, Modeling and Experimentation (ed. S.S. Hecker, A.K. Ghosh and H.L. Gegel), 150-182, AIME.

Hill, R. (1952). J. Mech. Phys. Solids, 1, 19-30.

Hill, R. (1962). J. Mech. Phys. Solids, 10, 1-16.

Hill, R. (1967). J. Mech. Phys. Solids, 15, 79-95.

Hill, R., and J.W. Hutchinson (1975). J. Mech. Phys. Solids, 23, 239-264.

Hutchinson, J.W. (1970). Proc. Roy. Soc. Lond., A318, 247-272.

Hutchinson, J.W., and K.W. Neale (1978). Sheet necking - III. Strain rate effects. Mechanics of Sheet Metal Forming (ed. D.P. Koistinen and N.-M. Wang), 269-285, Plenum Publ. Corp.

Hutchinson, J.W., K.W. Neale and A. Needleman (1978). Sheet necking - I. Validity of the plane stress assumptions of the long-wavelength approximation. Mechanics of Sheet Metal Forming (ed. D.P. Koistinen and N.-M. Wang), 111-126, Plenum Publ. Corp.

Hutchinson, J.W., and V. Tvergaard (1980). Int. J. Mech. Sci., 22, 339-354.

Keeler, S.P. (1968). Understanding sheet metal formability. Machinery., 74, Nos. 6-11.

Laukonis, J.V., and A.K. Ghosh (1978). Met. Trans., 9A, 1849-1856.

Marciniak, Z., and K. Kuczynski (1967). Int. J. Mech. Sci., 9, 609-620.

Marciniak, Z., K. Kuczynski and T. Pokora (1973). Int. J. Mech. Sci., 15, 789-805.

Michno, M.J., Jr. and W.N. Findley (1976). Int. J. Non-Linear Mech., 11, 59-82.

Needleman, A., and J.R. Rice (1978). Limits to ductility set by plastic flow localization. Mechanics of Sheet Metal Forming (ed. D.P. Koistinen and N.-M. Wang), 237-267, Plenum Publ. Corp.

Needleman, A., and N. Triantafyllidis (1978). J. Engr. Mat. Tech., 100, 164-169.

Needleman, A., and V. Tvergaard (1977). J. Mech. Phys. Solids, 25, 159-183.

Pan, J., M. Saje, and A. Needleman (1981). Localization of deformation in rate sensitive porous solids. Materials Research Laboratory, MRL E-136, Brown University.

Peirce, D.P., C.F. Shih and A. Needleman (1983). A tangent modulus method for rate dependent solids. Brown University Report DAAG-81-K-0121/1.

Rasmussen, S.N. (1981). Annals CIRP, 30, 179-184.

Rasmussen, S.N. (1982). Proc. 12th Biennial Congress Int. Deep Drawing Research Group, Genoa, Italy, 2, 83-93.

Rice, J.R. (1970). J. Appl. Mech., 37, 728-737.

Rice, J.R. (1977). The localization of plastic deformation. Proc. 14th Int. Congr. Theor. and Appl. Mech. (ed. W.T. Koiter), 207-220.

Stören, S., and J.R. Rice (1975). J. Mech. Phys. Solids, 23, 421-441.

Tvergaard, V. (1978). Int. J. Mech. Sci., 20, 651-658.

Tvergaard, V. (1980). ZAMM, 60, T26-T34.

Tvergaard, V. (1981). Int. J. Fracture, 17, 389-407.

Tvergaard, V. (1982a). Int. J. Fracture, 18, 237-252.

Tvergaard, V. (1982b). J. Mech. Phys. Solids, 30, 399-425.

Tvergaard, V., A. Needleman, and K.K. Lo (1981). J. Mech. Phys. Solids, 29, 115-142.

CONSTRAINED GRAIN BOUNDARY CAVITATION IN A CREEPING BODY CONTAINING A MACROSCOPIC CRACK

H. Riedel

Max-Planck-Institut fuer Eisenforschung, 400 Düsseldorf, Federal Republic of Germany

ABSTRACT

This paper extends previous models of constrained diffusive grain boundary cavitation under creep conditions to multiaxial states of stress. Constitutive equations for cavitating materials are given. In the limiting case where the stress on cavitating grain facets is completely relaxed, the macroscopic behavior of the material is still nonlinear viscous (but compressible) allowing for the use of the C*-integral in creep crack growth testing. An approximate method to determine C* in cavitating material is proposed. The transition to unconstrained diffusive cavity growth at higher stresses, in principle, offsets the nonlinear viscous description but the effect of cavitation on the strain rate is expected to be small in this range. Therefore the departure from viscous behavior is small and C* can be applied. At very high stresses, voids grow primarily by creep flow of the matrix. In this range, a purely viscous description becomes increasingly inappropriate.

KEYWORDS

Creep cavities; creep crack growth; fracture mechanics; damage mechanics.

INTRODUCTION

Creep crack growth in the more ductile materials has been described successfully by the C*-integral (e.g. Koterazawa and Mori, 1977, Landes and Begley, 1976, Nikbin and others, 1976, Ohji and others, 1974, Taira and others, 1979, and Wagner and Riedel, 1981). Theoretically, C* is expected to be the appropriate load parameter if the bulk of the specimen, with the possible exception of a small zone near the crack tip, deforms predominantly by (linear or nonlinear) viscous creep. If this prerequisite is not satisfies, C* cannot work. For example, in the more brittle materials, often elastic deformation predominates so that the stress intensity factor K_I is the appropriate load parameter (Ohji and others, 1980, Riedel and Rice, 1980, Wagner and Riedel, 1981). Other limitations to C* set by instantaneous plasticity, transient creep, crack tip blunting and cavity formation have been explored by Riedel (1983). The effect of cavitation will be discussed further in the present paper.

If cavitation is confined to a small fracture process zone near the crack tip ("small-scale cavitation") while the bulk of the specimen undergoes viscous creep, it is clear that C* is still the right macroscopic load parameter. Since cavitation and its effects on the stress fields are explored in greater detail here, also situations with wide-spread cavitation in the whole specimen can be analyzed. Of course, this is only possible under certain special assumptions. Here we assume that void nucleation occurs readily so that cavitating grain facets are present from the beginning of the test. A rather rigorous treatment is possible if the voids are assumed to grow by the constrained diffusive mechanism in the limiting case of completely creep-controlled growth (Hutchinson, 1982). In this case, C* is still a valid parameter. It is then argued that the transition to unconstrained diffusive growth at higher stresses does not necessarily render the C*-approach useless, although it is no longer exact.

CONSTRAINED DIFFUSIVE CAVITY GROWTH IN AXISYMMETRIC STRESSS FIELDS

Creep rupture often occurs by the initiation and coalescence of cavities on grain boundaries. These cavities are probably nucleated by vacancy condensation at grain boundary particles where the local tensile stress is elevated and the nucleation barrier is reduced (Raj and Ashby, 1975, Raj, 1978). Void nucleation is not further discussed here. It is rather assumed that nucleation occurs readily and growth is the rate-controlling process. The voids can grow by diffusive transport of atoms (or vacancies) in the grain boundary and by creep deformation of the grains. Both aspects have been considered in the literature, e.g., by Needleman and Rice (1980).

An important point has been overlooked in many of these papers, namely that the stress acting on the cavitating grain boundary, σ_b, may be very different from the macroscopically applied stress. This is the idea of the constrained diffusive cavity growth model (Dyson, 1976, 1979). Since cavitation usually occurs on relatively isolated, favorably oriented grain boundaries, the excess volume of the voids cannot be accomodated locally on the cavitating boundary unless creep flow of the surrounding material accomodates the excess volume. Thus, although the single voids may grow by diffusive transport of atoms in the grain boundary, the rate of the whole process may be controlled by creep flow of the matrix. If creep deformation can proceed only slowly at a given stress level, the diffusive plating of atoms on the grain boundary creates back stresses so that the actual stress on the boundary, σ_b, is reduced.

To quantify this idea, consider an isolated circular grain facet of diameter d subjected to axisymmetric loading. The facet is oriented normal to the remote maximum principal stress σ_I^∞, while the remote transverse stresses are $\sigma_{II}^\infty=\sigma_{III}^\infty$ and the remote equivalent tensile stress is therefore $\sigma_e^\infty=\sigma_I^\infty-\sigma_{II}^\infty$. Assume that the normal stress σ_b transmitted by the cavitating grain facet is uniformly distributed over the facet. This assumption has also been made by Rice (1981) who considered a similar model for uniaxial tension. Then the volume growth rate of a void in this facet by the usual diffusive growth mechanism is given by Needleman and Rice (1980) as

$$\dot{V} = (8\pi\ \Omega\ \delta D_b/kT)\ [\ \sigma_b - 2\ (1-\omega)\ \gamma\ \sin\psi\ /R]\ /\ [-2\ \ln\omega - (3-\omega)(1-\omega)]. \quad (1)$$

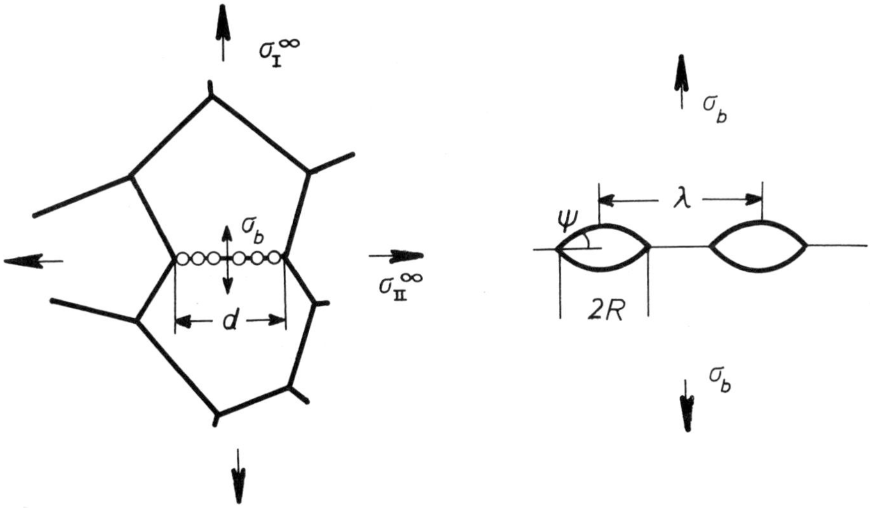

Fig. 1. Cavitating grain boundary facet under remote axisymmetric loading. Voids on cavitating facet.

The superposed dot denotes the time derivative, Ω is atomic volume, δD_b is the grain boundary diffusion coefficient times grain boundary thickness, γ is surface energy, ψ is the void tip angle, k is Boltzmann's constant, T is absolute temperature, $\omega=(2R/\lambda)^2$ is the voided area fraction of the grain boundary, 2R is the diameter of the axisymmetric void in the grain boundary and λ is the void spacing. Figure 1 shows the geometry.

The total volume growth rate of all voids in the facet, $\dot{V}(d/\lambda)^2(\pi/4)$, must be compatible with the volume that is accomodated by creep flow. This requirement determines σ_b. The grain facet with uniformly distributed tractions σ_b can be considered as a penny-shaped crack with constant tractions on its faces. The crack is embedded in a creeping material described by the power-law viscous relation

$$\dot{\varepsilon}_{ij} = (3/2)\ B\ \sigma_e^{n-1}\sigma'_{ij} \tag{2}$$

where $\dot{\varepsilon}$ and σ are local strain rate and stress, respectively. The equivalent stress is defined as

$$\sigma_e = (3\ \sigma'_{ij}\ \sigma'_{ij}/\ 2)^{1/2}. \tag{3}$$

The prime denotes the deviator and the summation convention for repeated indexes is employed; B and n are material parameters measured, for example, in uniaxial creep tests where $\dot{\varepsilon}= B\sigma^n$. This uniaxial form of eq. (2) is known as Norton's creep law.

He and Hutchinson (1981) have calculated the volume growth rate of a penny-shaped crack in power-law elastic (or, by analogy, viscous) material, however, for traction-free crack faces only ($\sigma_b = 0$). They develop a perturbation solution for small $\sigma_I^\infty/\sigma_e^\infty$, which turns out to be very accurate up to at least $\sigma_I^\infty/\sigma_e^\infty=2$. (For n=1, the solution is exact for arbitrary stress ratios). Since the approximate solution follows from equations that are linearized in σ_I^∞, it can easily be extended to the case where tractions σ_b act on the crack faces just by replacing σ_I^∞ by $\sigma_I^\infty - \sigma_b$. This gives the volume growth rate of the whole facet by creep:

$$\dot{V}_{creep} = \dot{\varepsilon}_e^\infty d^3 (1+3/n)^{-1/2} (\sigma_I^\infty - \sigma_b)/\sigma_e^\infty. \quad (4)$$

Equating this to the total volume growth rate by diffusive void growth of all voids in the facet gives the desired equation for the determination of σ_b,

$$\Sigma_e^{n-1} (\sigma_I^\infty - \sigma_b) = \sigma_b / [-2 \ln \omega - (3-\omega)(1-\omega)]. \quad (5)$$

Here Σ_e is the applied equivalent stress in non-dimensional form, viz.,

$$\Sigma_e = \sigma_e^\infty \{B\ d\ \lambda^2\ kT / [2\ \pi^2 (1+3/n)^{1/2}\ \Omega\ \delta D_b]\}^{1/(n-1)}, \quad (6)$$

and the sintering stress in eq. (1) associated with the surface energy γ has been neglected. This simplifies the following discussion but may not be justified in all cases, particularly at low applied stress levels.

In order to determine the two unknowns, σ_b and ω, from eqs. (1 and 5) we use the fact that the void volume is related to the radius R by

$$V = (4\pi/3)\ h(\psi)\ R^3 \quad \text{with} \quad h(\psi) = [1/(1 + \cos\psi) - .5 \cos\psi]/\sin\psi$$
$$V = (\pi/4)\ h(\psi)\ \lambda^3\ \omega^{1/2}\ \omega. \quad (7)$$

The dimensionless function $h(\psi)$ accounts for the lenticular void shape which consists of spherical caps. A typical value is $h(70^\circ)=0.61$. In terms of the voided area fraction $\omega = (2R/\lambda)^2$, eq. (1) becomes

$$\mu\sqrt{\omega}\,\dot{\omega} = \sigma_b / [-2 \ln \omega - (3-\omega)(1-\omega)], \quad (8)$$

where the sintering stress has again been neglected and the abbreviation

$$\mu = h(\psi)\ \lambda^3\ kT / (32\Omega\ \delta D_b) \quad (9)$$

has the dimension of (stress·time). Now σ_b can be eliminated from eqs. (5 and 8). This leads to a differential equation of first order in time for ω

$$[-2 \ln\omega - (3-\omega)(1-\omega) + \Sigma_e^{1-n}]\ \omega^{1/2}\ \dot{\omega} = \sigma_I^\infty/\mu, \quad (10)$$

which is readily integrated for time-independent Σ_e and σ_I^∞ to give

$$\sigma_I^\infty t/\mu = \omega^{3/2} [(2/3)\ \Sigma_e^{1-n} - (4/3) \ln\omega - 10/9 + 8\omega/5 - 2\omega^2/7]. \quad (11)$$

<u>The time to void coalescence</u> follows from eq. (11) above by setting $\omega = 1$:

$$t(\omega=1) = (\mu/\sigma_I^\infty)\ (0.667\Sigma_e^{1-n} + 0.203) \quad (12)$$
$$= \frac{\pi^2}{24} h(\psi) \left[\frac{(1+3/n)^{1/2}\ \lambda/d}{B\ (\sigma_e^\infty)^{n-1}\ \sigma_I^\infty} + 0.015 \frac{\lambda^3\ kT}{\sigma_I^\infty\ \Omega\ \delta D_b} \right].$$

For small stresses, the time to void coalescence is determined by the first term in brackets, i.e., creep deformation of the surrounding material is rate controlling and $t \propto 1/\sigma^n$. For large stresses, the second term in brackets dominates. In this case, creep can occur so rapidly that accomodating the excess volume of the voids is no problem and diffusive transport of atoms in the grain boundary between the voids becomes rate controlling. In this limit, the time to coalescence varies as $t \propto 1/\sigma$.

<u>The time to relax the stress on the facet by cavitation.</u> For the macroscopic mechanical properties, the stress σ_b carried by the cavitating facets is an important quantity. In the first instant after load is applied and cavitation commences, σ_b is equal to the applied tensile stress. After long times, the stress on the facet is completely relaxed and the full applied stress contributes to the opening of the facet. From eq. (5) it follows that the stress on the facet is reduced to some fraction α of its initial value σ_I^∞ when ω attains the value defined by

$$\alpha/(1-\alpha)\ \Sigma_e^{1-n} = -2\ \ln\omega - (3-\omega)\ (1-\omega) \tag{13}$$

Eqs. (11 and 13) together define the time within which the tractions on the grain facet are reduced to the fraction α of the remotely applied stress:

$$t(\alpha) = \frac{h(\psi)}{32}\left[\frac{\lambda^3\ kT}{\Omega\ \delta D_b}\right]^{n/(n-1)}\left[\frac{Bd/\lambda}{2\pi^2(1+3/n)^{1/2}}\right]^{1/(n-1)}\frac{\sigma_e^\infty}{\sigma_I^\infty}F(\Sigma_e,\alpha). \tag{14}$$

The dimensionless function $F(\Sigma_e,\alpha)$ can be evaluated numerically by varying the parameter ω from 0 to 1 and calculating Σ_e from eq. (13) and $F(\Sigma_e)$ from eqs. (11 and 14) for each ω. The result is shown in Fig. 2 for n=5. It appears that for small and large stresses the time to relax the stress is short. For intermediate stresses, the relaxation time has a maximum. The

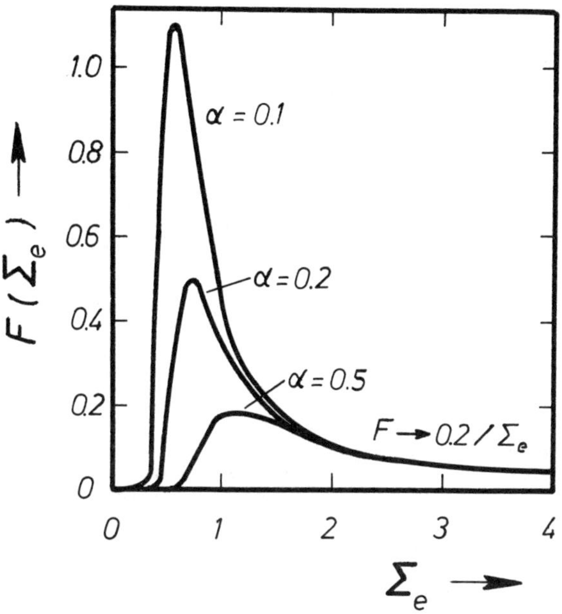

Fig. 2. $F(\Sigma_e)$ = normalized time for stress relaxation on a cavitating grain boundary facet; Σ_e = normalized remote stress. n = 5.

reason for this behavior is that at high stresses, cavitation proceeds quickly and the stress is relaxed because the voided area fraction becomes large. At low stresses, void growth by diffusion is fast compared to the possibility of accomodating the excess volume on a larger scale. Hence the stress on the grain facet is quickly reduced by plating atoms on the boundary between the voids although the voided area fraction is still small.

According to Fig. 2, the maximum of $F(\Sigma_e)$ occurs at $\Sigma_e = 1.1$ if $\alpha = 1/2$. Eq. (6) gives this stress in physical coordinates. As an example we take the material parameters for AISI 304 stainless steel at 650°C from Frost and Ashby (1977): $B=2\ 10^{-23}$ MPa^{-n}/s, $n = 7.5$, $\lambda^3 kT/\Omega\delta D_B = 10^9$ MPa s, and $\lambda=3\mu m$, $d=30\mu m$. Then the stress at which $F(\Sigma_e)$ has its maximum follows from eq. (6) as $\sigma_e^\infty = 160$ MPa. At this stress, the time for stress relaxation on a facet by cavitation is $t(\alpha=1/2) = 6h$ if $\sigma_e^\infty/\sigma_I^\infty=1$. At other stress levels the relaxation time is shorter.

The relaxation time $t(\alpha)$ is used to estimate the time after which cavitating grain facets can be considered as being effectively traction free ($\sigma_b=0$). For this limiting case, Hutchinson has formulated the constitutive equations of creeping solids with cavitating facets (cf. the next section). These equations have the favorable property for fracture mechanics applications to allow for the use of C*.

STRESS FIELDS IN CRACKED BODIES FOR FULLY RELAXED STRESSES ON CAVITATING FACETS

Hutchinson's constitutive model. For times larger than $t(\alpha)$ (eq. 14), cavitating grain facets can be considered to behave effectively like traction-free microcracks. Hutchinson (1982) considers a power-law viscous solid described locally by eq. (2) containing penny-shaped microcracks that are aligned normal to the maximum principal tensile stress σ_I. On a scale large compared with the microcrack spacing, the constitutive material law of the cavitating solid is

$$\dot{\varepsilon}_{ij} = (3/2)\ B\ \sigma_e^{n-1}\ \{\sigma'_{ij} + \rho\ [(n-1)\ \sigma'_{ij}\ (\sigma_I/\sigma_e)^2 + (4/3)\sigma_I\ m_{ij}]/(n+1)\}, \quad (15)$$

where the components of the tensor m_{ij} in the principal axes of stress are 0 exept in the maximum principal stress direction where $m_I=1$. The parameter ρ is related to the number density, N, of cavitating facets and to their diameter d by

$$\rho = 0.5\ d^3\ N\ (n+1)\ (1+3/n)^{-1/2}. \quad (16)$$

Obviously, the material law, eq. (15) describes compressible nonlinear viscous behavior, i.e., the strain rate is a unique function of stress. For fracture mechanics this has the important consequence that the C*-integral is path-independent for two-dimensional crack problems. Hence, C* is the appropriate load parameter to describe creep crack growth in solids with traction-free, cavitating grain facets. Of course, this is only true as long as ρ is constant. The importance of continuous void nucleation on an increasing number of grain facets remains to be discussed.

Experimental determination of C*. The general form to determine C* from load/displacement records is

$$C^* = -(\partial/\partial a) \int P d\dot{\Delta}\ , \quad (17)$$

where P is load per unit specimen thickness and $\dot{\Delta}$ is the load line deflection rate. The differentiation with respect to crack length, a, can be done in principle by testing specimens with neighboring crack lengths. An experimental procedure based an eq. (17) has been described by Landes and Begley (1976). It is more convenient, however, to determine C* from the current load and deflection rate of a single specimen. This is particularly simple if the strain rate is a homogenous function of arbitrary degree n of the stress components. Eq. (15) has the desired property of being homogeneous of degree n. Then the stress field can be shown to be a proportional field, $\sigma_{ij}(r,P) = P\bar{\sigma}_{ij}(r)$, where r represents the spatial coordinates and $\bar{\sigma}_{ij}$ is a function of r but not of P. Together with the material law, eq. (15), and with the line-integral definition of C* (Landes and Begley, 1976), this implies that C* can be determined from either one of the formulas

$$C^* = a\ B\ \sigma_{net}^{n+1}\ g_1(a/W,n) = g_2(a/W,n)\ \sigma_{net}\ \dot{\Delta} \tag{18}$$

where σ_{net} is the load divided by the area of the uncracked ligament and W is the specimen width. The dimensionless functions g_1 und g_2 must generally be determined numerically for any given specimen geometry. For incompressible, power-law viscous materials, g_1 and g_2 are related to the functions h_1, h_3 and η tabulated by Kumar, German and Shih (1981) through

$$g_1 = h_1\ (W/a-1)/(1.455\,\eta\,)^{n+1}, \qquad g_2 = (W/a-1)\ h_1/(1.455\ \eta\, h_3). \tag{19}$$

Musicco (1983) gives plots of $g_2(a/W,n)$ for incompressible materials.

An approximation for the function g_2. For cavitating viscous solids, the geometrical functions have not been calculated so far, but it is anticipated that g_2 is not strongly sensitive to details of the material law. This conjecture is supported by the following argument. A general (compressible or incompressible) viscous material law can be written in terms of the potential $\Phi(\sigma_{ij})$ as

$$\dot{\varepsilon}_{ij} = \partial\Phi\,/\partial\sigma_{ij}. \tag{20}$$

Following He and Hutchinson (1981) we start a perturbation solution for the stress field from the trivial solution where only stresses transverse to the crack are applied so that the crack is not opened and the transverse stress is uniform. This trivial state is denoted by the subscript o. Now small forces normal to the crack plane are superimposed giving rise to small increments of stress, $\delta\sigma_{ij}$, and strain rate, $\delta\dot{\varepsilon}_{ij}$. Then, from eq. (20)

$$\dot{\varepsilon}_{ij,o} + \delta\dot{\varepsilon}_{ij} = (\partial\Phi/\partial\sigma_{ij})_o + (\partial^2\Phi/\,\partial\sigma_{ij}\ \partial\sigma_{kl})_o\ \delta\sigma_{kl}$$

$$\delta\dot{\varepsilon}_{ij} = (\partial^2\Phi/\partial\sigma_{ij}\ \partial\sigma_{kl})_o\ \delta\sigma_{kl}. \tag{21}$$

This is a linear viscous (or by analogy, linear elastic) material law for the incremental quantities. The moduli $(\partial^2\Phi/\partial\sigma_{ij}\ \partial\sigma_{kl})_o$ have the anisotropy of the trivial stress state. For the further discussion, it is necessary that the moduli are uniform. In our case, this means that the cavitating facets are uniformly aligned in the trivial state.

The solution of the linear problem with uniform, transversely isotropic moduli, generally requires numerical techniques for arbitrary specimen geometries. For plane-strain cracks and penny-shaped cracks under axisymmetric loading in infinite bodies, however, sufficient analytic information is available for our present purposes. He and Hutchinson (1981) show that for elliptic cracks in an infinite, linear, transversely isotropic solid, the C*-integral can be expressed in terms of the remotely applied tensile stress $\delta\sigma_{22}^{\infty}$ and the opening rate in the center of the crack, $\Delta\dot{u}$. For penny-shaped cracks there results

$$C^* = (1/2)\,\Delta\dot{u}\;\delta\sigma_{22}^{\infty}, \tag{22a}$$

and for plane-strain cracks is

$$C^* = (\pi/4)\,\Delta\dot{u}\;\delta\sigma_{22}^{\infty}. \tag{22b}$$

(He and Hutchinson's argument refers to the J-integral in linear elastic material, but the elastic/viscous analogy allows a direct application to C*). The result is independent of the moduli since the derivation makes no use of them. Thus, for the linearized problem, the C*-integral if written in the form of eq. (22) has the desired property of being independent of the material law. Now, for incompressible, power-law materials, He and Hutchinson (1981) show that the linearized solution has a surprisingly wide range of approximate validity. For uniaxial tension, for example, the results are still rather accurate. There is no obvious reason why this favorable behavior should be lost for compressible materials.

In summary, the second form of eq. (18) with the geometrical function g_2 for incompressible, power-law material can be used to determine C* approximately even for cavitating materials which do not necessarily obey a power-law relation between strain rate and stress.

<u>Crack-tip fields.</u> If r and θ are polar coordinates centered at the crack tip, the asymptotic stress field in materials described by eq. (15) has the form

$$\sigma_{ij} = [C^*/(I_n B r)]^{1/(n+1)}\;\tilde{\sigma}_{ij}(\theta). \tag{23}$$

This can be verified by insertion into the constitutive equations. The dimensionless factor I_n and the dimensionless angular functions $\tilde{\sigma}_{ij}(\theta)$ have been given by Hutchinson (1968) and Rice and Rosengren (1968) for incompressible material, while Hutchinson (1982) gives the results for a cavitating material obeying eq. (15). He finds that the angular functions $\tilde{\sigma}_{ij}(\theta)$ are not changed substantially by the occurrence if cavitation. (They are normalized such that the maximum of $\tilde{\sigma}_e(\theta)$ is unity). The normalizing factor I_n increases with ρ according to

$$I_n(\rho) = I_n(0) + \rho I_n', \tag{24}$$

where ρ as a measure of cavitation has been defined in eq. (16); $I_n(0)$ is the value for non-cavitating material (with $I_1(0)$=6.28, $I_3(0)$=5.51, $I_5(0)$=5.01) and I_n'=9.17, 13.3 and 12.4 for n=1,3 and 5, respectively. This increase of I_n with ρ implies an only moderate reduction of the crack-tip stress field by the occurence of cavitation. For example, if n=5 and ρ =1, the stress level near the crack tip is reduced by only 20%, although ρ=1 corresponds to a spacing between the cavitating facets which is less than twice their diameter. Of course, the model based on non-interacting microcracks becomes increasingly inaccurate at such high crack densities. Further, Hutchinson (1982) observes that the product of stress and strain rate remains relatively unaffected by cavitation.

<u>The relation of crack-tip fields to the far-field C* in "small-scale cavitation".</u> From the approximate constancy of the dissipation rate

$\dot{\varepsilon}_{ij}\,\sigma_{ij}$, Hutchinson (1982) concludes that C* should be approximately path-independent even in cases where cavitation is still confined to the vicinity of the crack tip ("small-scale cavitation"). This conjecture has important consequences for fracture mechanics under creep conditions. The ligament of a small test specimen might be completely cavitated, while creep crack growth in a large structure at the same C*-level might occur under small-scale cavitation conditions. In order to transfer the test data to the structure it is necessary that the same far-field C*-value (measured, e.g. according to eq. (18)) guarantees at least approximately identical conditions of stress near the crack tips. This is analogous to elastic-plastic fracture mechanics where the critical J-integral is measured using small, fully yielded specimens and the result is applied to small-scale yielding situations in large structures. This approach works since the J-integral happens to be approximately path-independent for small-scale yielding so that the far-field J (= $K_I^2(1-\ ^2)/E$) creates nearly the same conditions near the crack tip as does the same J-value in a fully yielded specimen.

<u>Creep crack growth rates for strain controlled local failure.</u> Riedel (1981, 1983) studied creep crack growth by the coalescence of grain boundary cavities. The crack-tip stress and strain-rate fields (eq. 23) were assumed to be those of an incompressible viscous material. This inconsistency of considering crack growth by cavity coalescence and, on the other hand, neglecting the effect of the cavities on the stress distribution, can now be partly removed. The analysis that has been done for the incompressible case can be directly transferred to compressible, cavitating material, simply by using eq. (23) with the appropriate constants for compressible material. If cavity growth is controlled by creep of the surrounding matrix, the extension of the macroscopic crack is controlled by a critical strain criterion. Then the crack growth rate is found to be of the order

$$\dot{a} \simeq (Bx_c)^{1/(n+1)}\ C^{*n/(n+1)}/\varepsilon_c \qquad (25)$$

for both, compressible and incompressible materials. The factor α is of order unity, and ε_c is the critical strain to be attained over a microstructural length x_c ahead of the crack tip. The fact that cavitation reduces the stress level slightly modifies the factor α. The dependence on C* which is confirmed by numerous experimental studies (e.g. Koterazawa and Mori, 1977, Ohji and others, 1980, Saxena, 1980, Taira and others, 1979, Wagner and Riedel, 1981) is not affected, nor is the mild dependence on temperature (through the creep coefficient B raised to a small power 1/(n+1)) which is also found experimentally.

DISCUSSION

It has been shown that cavitation does not necessarily invalidate the macroscopic C*-approach to creep crack growth. Under small-scale cavitation conditions when cavitation is confined to a small process zone near the crack tip while the bulk of the specimen creeps without substantial cavitation, C* is clearly a valid parameter for analogous reasons why K_I is the appropriate load parameter under small-scale yielding conditions. This should be kept in mind during the following discussion. But even if cavitation spreads across the whole ligament of the specimen, C* remains useful provided that the creep-constrained void growth mechanism predominates and the number density of cavitating grain facets is sufficiently constant in a certain vicinity of the crack tip. It is not expected that the constancy of the density is a very stringent requirement since the stress distribution is surprisingly insensitive even to a high density of cavitating facets. Recall that the stress level near the crack tip is reduced by as little as 20% if cavitation occurs on about every second tensile grain boundary facet.

At relatively high stresses, the voids grow by the unconstrained diffusive mechanism. In this case, the cavitating facets transmit about the same tractions as do non-cavitating boundaries and the contribution of the increasing void volume to the total strain rate is negligible. This implies that cavitation has an even smaller effect on the macroscopic stress distribution than in the limit of completely constrained void growth.

Thus unconstrained diffusive void growth on isolated grain facets is mechanically intermediate between the cases of no cavitation and fully creep-constrained cavitation. In both of these limits, C* is applicable and the stress fields do not differ substantially. It can be concluded that C* works approximately in the intermediate case of diffusive void growth as well. Furthermore, the characteristic time $t(\alpha)$ given in eq. (14) turns out to be rather short compared to typical creep test durations in many practical cases so that indeed creep-constrained cavitation predominates.

The situation becomes different if the defects (single cavities or cavitated grain facets) must be considered as being more or less spherical, as distinct from crack-like, and if they grow by creep deformation of the surrounding material. For cavities, this happens at rather high stresses, when

$$\sigma > [\Omega\,\delta D_b/(kT\ B\ R^3)]^{1/(n-1)} \tag{26}$$

(Beere and Speight, 1978, Edward and Ashby, 1979, Needleman and Rice, 1980). In this case the volumetric growth rate of the voids is proportional to their current volume and thus depends on the previous history. The compressible part of the macroscopic strain rate is therefore not a unique function of the current stress only, but also of the volume fraction of voids. A creeping material with voids that possess a finite volume does not behave purely nonlinear viscous on a scale large compared with the void spacing, and the C*-integral is not directly applicable. To be definite, consider a power-law viscous material containing well separated spherical (or nearly spherical) voids. Budiansky and others (1981) give an approximation for the growth rate of an isolated spherical void under highly triaxial remote stressing:

$$\dot{V} = \dot{\varepsilon}_e^{\infty}\, V\, \Lambda \tag{27}$$

with

$$\Lambda = (3/2)\ \{(3/2n)\ \sigma_m^{\infty}/\sigma_e^{\infty} + (n-1)(n+0.432)/n^2\}^n$$

for $\Lambda>0$. The mean stress $\sigma_m^{\infty} = \sigma_{kk}^{\infty}/3$ must be greater than about twice the equivalent stress for the high-triaxiality approximation to work. Using this result in a procedure like that employed by Gurson (1977) for time-independent cavitating material, gives the macroscopic strain rate

$$\dot{\varepsilon}_{ij} = (3/2)\ B\ \sigma_e^{n-1}\ \sigma'_{ij} + (\delta_{ij}/3)\ \Lambda\ N\ V\ B\ \sigma_e^n, \tag{28}$$

where NV is the volume fraction of voids and V obeys the differential equation in time, eq. (27). This makes a notable difference to the constitutive equation (15) for solids with crack-like defects, where ρ is a constant.

It appears that under creep conditions the constrained diffusive void growth mechanism is the most important one since the stresses are usually not high enough to satisfy the condition (26) for void growth by creep flow.

CONCLUSIONS

The preceding discussion supports the conclusion that extensive grain boundary cavitation does not necessarily prevent the application of C* to creep crack growth provided that none of several other possible factors limits the validity of C*. Numerical analyses should be done to substantiate the conjecture that the C*-integral is approximately path-independent under small-scale cavitation conditions.

As long as no accurate finite element analyses for cavitating materials are available, it is suggested to determine C* from

$$C^* = g_2(a/W,n)\ \dot{\Delta}\ \sigma_{net}$$

with the same function g_2 which has been calculated and tabulated for non-cavitating material.

REFERENCES

Beere, W. and M.V. Speight (1978). Creep Cavitation by Vacancy Diffusion in Plastically Deforming Solid. Metal Sci., 12, 172-176.

Budiansky, B., J.W. Hutchinson and S. Slutsky (1981). Void Growth and Collapse in Viscous Solids. In: H.G. Hopkins and M.J. Sewell (Eds.), Mechanics of Solids, The Rodney Hill 60th Anniversary Volume, Pergamon Press, Oxford, 13-45.

Dyson, B.F. (1976). Constraints on Diffusional Cavity Growth Rates. Metal Sci., 10, 349-353.

Dyson, B.F. (1979). Constrained Cavity Growth, its Use in Quantifying Recent Creep Fracture Results. Can. Metall. Q., 18, 31-38.

Edward, G.H. and M.F. Ashby (1979). Intergranular Fracture during Power-Law Creep. Acta Metall., 27, 1505-1518.

Frost, H.J. and M.F. Ashby (1977). Deformation-Mechanism Maps for Pure Iron, Two Austenitic Stainless Steels, and a Low-Alloy Ferritic Steel. In: R.I. Jaffee and B.A. Wilcox (Eds.), Fundamental Alloy Design, Plenum Press, New York and London, 27-58.

Gurson, A.L. (1977). Continuum Theory of Ductile Rupture by Void Nucleation and Growth: Part I - Yield Criteria and Flow Rules for Porous Ductile Media. J. Eng. Mater. Technology, 99, 2-15.

He, M.Y. and J.W. Hutchinson (1981). The Penny-Shaped Crack and the Plane Strain Crack in an Inifinite Body of Power-Law Material, J. Appl. Mech., 48, 830-840.

Hutchinson, J.W. (1968). Singular Behaviour at the End of a Tensile Crack in a Hardening Material. J. Mech. Phys. Solids, 16, 13-31.

Hutchinson, J.W. (1982). Constitutive Behavior and Crack Tip Fields for Materials Undergoing Creep-Constrained Grain Boundary Cavitation, Division of Applied Sciences, Harvard Univ., Report Mech-34.

Koterazawa, R. and T. Mori (1977). Applicability of Fracture Mechanics Parameters to Crack Propagation under Creep Condition. J. Eng. Mater. Technology, 99, 298-305.

Kumar, V., M.D. German and C.F. Shih (1981). An Engineering Approach for Elastic-Plastic Fracture Analysis, EPRI-Report NP-1931, Palo Alto.

Landes, J.D. and J.A. Begley (1976). A Fracture Mechanics Approach to Creep

Crack Growth. In: Mechanics of Crack Growth, ASTM STP 590, Philadelphia, 128-148.
Musicco, G.G. (1983). Third Advanced Seminar on Fracture Mechanics, Ispra, Oct 1981, L.H. Larsson, Ed., Applied Science Publishers, London, in press.
Needleman, A., and J.R. Rice (1980). Plastic Creep Flow Effects in the Diffusive Cavitation of Grain Boundaries, Acta Metall., 28, 1315-1332.
Nikbin, K.M., G.A. Webster and C.E. Turner (1976). Relevance of Nonlinear Fracture Mechanics to Creep Cracking. In: Cracks and Fracture, ASTM STP 601, Philadelphia, 47-62.
Ohji, K., K. Ogura and S. Kubo (1974). Application of J-Integral to Creep Crack Problem. Preprint of Japan. Soc. Mech. Engrs., No. 740-11, 207 (in Japanese).
Ohji, K., K. Ogura and S. Kubo (1980a). The Stress Field and Modified J-Integral near a Crack Tip under the Condition of Confined Creep Deformation. J. Soc. Mater. Sci. Japan, 29, No. 320, 465-471.
Ohji, K., K. Ogura and S. Kubo (1980b). The Application of Modified J-Integral to Creep Crack Growth. In: Engineering Aspects of Creep, Vol. 2, The Institution of Mechanical Engineers, London, 9-16.
Raj, R. (1978). Nucleation of Cavities at Second-Phase Particles in Grain Boundaries. Acta Metall., 26, 995-1006.
Raj, R. and M.F. Ashby (1975). Intergranular Fracture at Elevated Temperature. Acta Metall., 23, 653-666.
Rice, J.R. (1981). Constraints on the Diffusive Cavitation of Isolated Grain Boundary Facets in Creeping Polycrystals. Acta Metall., 29, 675-681.
Rice, J.R., and G.F. Rosengren, (1968). Plane Strain Deformation near a Crack Tip in a Power-Law Hardening Material. J. Mech. Phys. Solids, 16, 1-12.
Riedel, H. (1981). The Extension of a Macroscopic Crack at Elevated Temperature by the Growth and Coalescence of Microvoids. In: A.R.S. Ponter and D.R. Hayhurst (Eds.), Creep in Structures, Springer Verlag, Berlin Heidelberg New York, 504-519.
Riedel, H. (1983). The Use and the Limitations of C* in Creep Crack Growth Testing. Proceedings of the International Symposium on Fracture Mechanics, Beijing, China.
Riedel, H., and J.R. Rice (1980). Tensile Cracks in Creeping Solids. In: P.C. Paris (Ed.), Fracture Mechanics: Twelfth Conference, ASTM STP 700, Philadelphia, 112-130.
Saxena, A. (1980). Evaluation of C* for the Characterization of Creep Crack Growth Behavior in 304 Stainless Steel. In: P.C. Paris (Ed.), Fracture Mechanics: The Twelfth Conference, ASTM STP 700, Philadelphia, 13-151.
Taira, S., R. Ohtani and T. Kitamura (1979). Application of J-Integral to High Temperature Crack Propagation. J. Eng. Mater. Technology, 101, 154-161.
Wagner, W. and H. Riedel (1981). Creep Crack Growth in Nimonic 80A and in a 1%Cr1/2%Mo Steel. In: 13. Sitzung des AK Bruchvorgaenge, Deutscher Verband fuer Materialpruefung, Berlin, 142-150 (in German).

HIGH TEMPERATURE DEFORMATION AND FAILURE OF WELDED JOINTS

J. A. Williams

Central Electricity Generating Board, Marchwood Engineering Laboratories, Welding Section, Marchwood, Southampton, UK

ABSTRACT

Welds, by their method of manufacture, are inhomogeneous in both a microstructural and a property sense. The differences are briefly reviewed and the simplified models used to characterise weld behaviour are outlined. The analytical or computed predictions for deformation and failure, based on these models, are compared with experimental data from a number of sources. The results suggest that the current analysis methods are adequate for characterising steady state deformation of welds. However, the results of failure analysis, although promising, have not been verified to the same degree.

KEYWORDS

Creep; welds; high temperature deformation; fracture.

INTRODUCTION

There are many methods available for joining components. For high temperature plant, the most commonly used process is fusion welding and this is considered here. However, the basic concepts discussed are similar for other welding procedures, such as diffusion bonding and friction welding, although there will be some differences in detail.

The outstanding feature of a weld, vis a vis parent material, is the local microstructural difference found in the weld vicinity. This can occur in several ways and for the purposes of this paper is referred to as macroscopic or microscopic inhomogeneity.

Macroscopic inhomogeneity concerns the case of weldments fabricated between two similar materials using weld metal of a different composition or, in the extreme case, of two different materials welded with a weld metal of yet another composition. An example of the latter is given by dissimilar welds, where AISI type 316 austenitic steel and 2¼Cr1Mo ferritic alloy are welded with a nickel based weld metal. A more usual case is found from power plant where $\frac{1}{2}$Cr$\frac{1}{2}$Mo$\frac{1}{4}$V ferritic pipework is fabricated with 2$\frac{1}{4}$Cr1Mo weld metal, although many other instances can be found.

Microscopic inhomogeneity results directly from the sequence followed during welding and can be considered in terms of the local generation and dissipation of heat as the weld metal is deposited. Heat is dissipated through the parent material and the structures produced within the heat affected zone, HAZ, will be a function of the peak temperature reached, the rate of cooling and the material's grain growth kinetics and transformation characteristics, (Alberry and Jones, 1977). For a single weld bead on $\frac{1}{2}$Cr$\frac{1}{2}$Mo$\frac{1}{4}$V plate, typical structures are shown schematically in Fig.1a and the complicating factor of laying additional beads on the previous weld metal is shown in Fig.1b. It is clear that in a real weld, the HAZ consists of a repetitive distribution of different microstructures where the spacing and extent of each region is controlled by welding details such as weld bead overlap and heat input etc. (Alberry and Jones, 1979). Similar structural differences will occur within the weld metal, see Fig.1b.

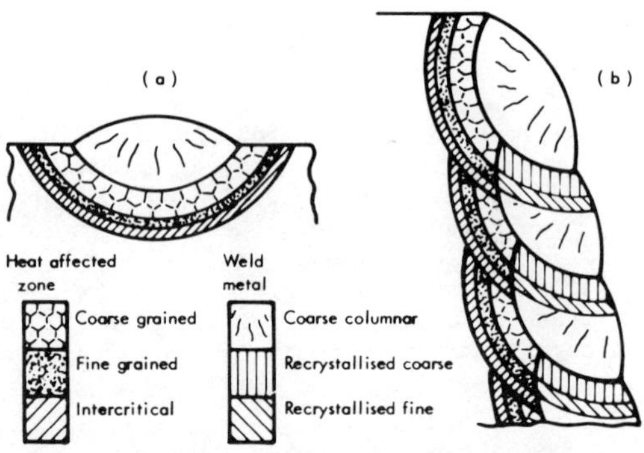

Fig.1. Schematic representation of microstructures in single and multiple bead weldments (after Coleman, 1978).

These concepts have been verified and have led to the development of welding procedures which can successfully produce specific microstructural distributions (Alberry and Jones, 1979; Coleman, 1982). The most detailed investigation of a single $\frac{1}{2}$Cr$\frac{1}{2}$Mo$\frac{1}{4}$V:2$\frac{1}{4}$Cr1Mo weld has been reported by Middleton and Cane (1977) where in the HAZ region, up to 13 separate structures were identified.

Further modification of the structures can occur by tempering during the latter stages of welding, during any subsequent post weld heat treatment, PWHT, and during service. The tempering characteristics of each constituent may also differ; a further complicating factor.

The problem is therefore to determine how these microstructural variations, in particular their extent, geometry, distribution and any property differences, affect the creep and failure behaviour of complete weldments. As shown, the structure in real weldments is complex, but progress can be made as the structures are definable and repetitive and also their distribution can be controlled. In addition, not all the structures identified have markedly different mechanical and high temperature properties. Furthermore, operating experience with high temperature weldments can identify the more important parameters.

This paper reviews published research in this field and illustrates how these factors affect the high temperature deformation and failure processes. Recently published analyses, which represent these structural variations as a small number of individual zones of different creep properties, are used to illustrate the key features and compared with experimental results on real welds. An additional factor, the effect of welding residual stress, is also briefly considered. Such secondary stresses are particularly important as certain welds, for example those in austenitic materials, may not be subjected to PWHT whereas in other welds PWHT may be inadequate.

PERFORMANCE OF HIGH TEMPERATURE WELDMENTS

The main service cracking modes found during service for $\frac{1}{2}$Cr$\frac{1}{2}$Mo$\frac{1}{4}$V:2$\frac{1}{4}$Cr1Mo welds are briefly described. In addition, general statements are made concerning the property differences which can be found between the different material structures found in welds. More complete reviews have been published elsewhere (Cane, 1981; Price and Williams, 1982).

Failure Modes

Design codes generally consider only the behaviour of the homogeneous component and do not specifically examine the weld properties or take account of microstructural distributions. However, many service failures are associated with welds.

High temperature creep failure in $\frac{1}{2}$Cr$\frac{1}{2}$Mo$\frac{1}{4}$V:2$\frac{1}{4}$Cr1Mo welds can be considered in four categories (Price and Williams, 1982). The first, stress relief cracking, is associated with the relaxation of residual welding stresses and occurs in a circumferential mode within the coarse grained bainitic, low ductility, layer adjacent to the fusion boundary. Another form of cracking referred to as transverse weld metal cracking (Wolstenholme, 1976) arises in 2$\frac{1}{4}$Cr1Mo weld metal and occurs in the coarse columnar regions within the weld metal. The mechanism of cracking is similar to that of stress relief cracking and is a direct consequence of inadequate PWHT. This leads to low ductility in the weld metal and high residual welding stresses which can relax in service. The third form, first reported in Germany, is associated with welds adjacent to section changes and/or subjected to system loading, (Schuller and co-workers, 1974). The cracks are circumferential and occur in the partially transformed and overtempered region at the low temperature edge of the HAZ and adjacent to the parent material. In these premature failure modes, the cracks are restricted to specific microstructural regions.

The last and most nebulous form of failure is that occurring at the component design life. This is difficult to specify as the majority of weld failures to date have been premature. The final failure mode under these conditions can only be inferred from analysis.

The local pattern of creep damage build-up is similar for all failure modes. Creep strain accumulates, leading to nucleation and growth of cavities which link up to form macroscopic cracks which cause component failure. With the exception of certain types of dissimilar welds, the basic failure mechanisms are identical to those obtained from uniaxial tests on homogeneous materials in the laboratory. However, in welds, the size and distribution of the damaged regions will be controlled by the distribution of microstructures within the weld.

Properties Of Individual Constituents Within A Weld

For the ferritic $\frac{1}{2}$Cr$\frac{1}{2}$Mo$\frac{1}{4}$V:2$\frac{1}{4}$Cr1Mo combination, Middleton and Cane (1977) identified numerous different structures within the weld region. However, only four are structures worthy of special note: the parent material, the weld metal and two HAZ regions, coarse grained bainite and the partially transformed region. For these regions, the properties are as follows:

1. Coarse grained bainite is the strongest in creep, but the least ductile.

2. The parent material has good creep strength at design temperatures with a medium to high ductility.

3. The weld metal tends to be weaker than the parent material and generally has lower ductility.

4. The partially transformed overtempered region has the lowest strength and the highest ductility.

These regions have been identified in Fig. 1a, b. However, the weld metal is usually considered to deform uniformly as specific differences in weld metal properties have not been investigated. With very few exceptions, these regions can be used to characterise the properties of ferritic weldments.

A similar picture can be presented for austenitic weldments. Within the limited data base available in the literature, the HAZ regions tend to be slightly stronger than the parent under creep conditions (Sandstrom and Ivarsson, 1979; Roode and co-workers, 1980). The weld metal is more variable, being weaker or stronger than the parent, depending on the particular austenitic weld metal used.

Based on these simple microstructural models for characterising welded structures, deformation behaviour is reviewed in the following text to identify the effects of structural variations with the objective of quantifying creep in welds.

CREEP DEFORMATION OF INHOMOGENEOUS STRUCTURES

For the purposes of analysis, the individual structures present in the weld were assumed to have the following characteristics:

1. They obey Norton's secondary creep law under uniaxial loading.

i.e. $$\dot{\varepsilon} = A'\sigma^n \qquad (1)$$

where $\dot{\varepsilon}$ = secondary or minimum creep rate,
σ = applied stress and
A' and n are material, stress and temperature dependent constants.

2. They obey the multiaxial equivalent of Norton's Law, based on the von Mises equivalent stress, σ^*, where:-

$$\dot{\varepsilon}_{ij} = 3/2\ A'(\sigma^*)^{n-1}\ S_{ij} \quad (2)$$

using the normal nomenclature.

3. As steady state conditions are assumed, the relative creep rate of two materials, A and B, is given by:-

$$\dot{\varepsilon}_A/\dot{\varepsilon}_B = [A/B]\ \sigma^{(n_A - n_B)} \quad (3)$$

where the suffices,A and B, refer to the material considered.

If $n_A = n_B$ and the stress is constant, then $\dot{\varepsilon}_A/\dot{\varepsilon}_B = A/B$, allowing the relative creep properties of any two materials to be represented by a simple parameter. This approach was originally used by Browne and co-workers (1981).

As illustrated in previous sections, the problem is complex and to identify the controlling features, simple models have been used. Initially, the effect of a single interface between two materials, A and B, of different creep properties subjected to plane strain loading is considered, (Nicol, 1982). The results are then extended to cover the effect of two interfaces, that is a material combination, A:B:A, (Nichol and Williams, 1983) and are compared with the results of other workers for similar geometries. Finally, more realistic pressure loaded tube and pipe welds are considered. (Walters and Cockroft, 1972; Browne, et al., 1981).

A Simple Two Material Model Subjected To Plane Strain Loading Under Creep Conditions

Nicol (1982) examined the steady state creep deformation of a sandwich structure, A:B:A, subjected to plane strain loading, see Fig. 2. The zone width is z, the specimen thickness is h and as above, $n_A = n_B$. The analysis was carried out for a range of B/A values from 0.01 to 1000, that is, hard to soft zone materials, z/h from 0.5 to infinity and n from 1 to 10 allowing a full parametric study of the effect of these variables on the steady state deformation rates.

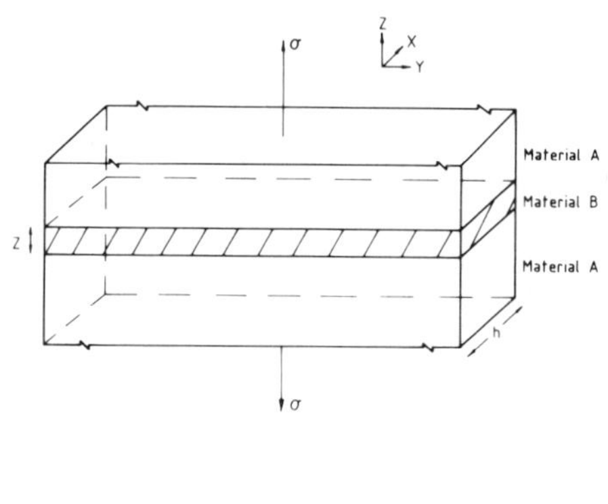

Fig. 2. Cross weld geometry (after Nicol,1982).

For the single interface case, i.e. $z/h = \infty$, the strain rate distribution along the specimen is shown in Fig. 3 for a range of B/A ratios from 1 to 10 and $n = 4$. In all cases, the strain rate remote from the interface was given directly by the applied stress and the uniaxial creep properties of the individual materials. However, within a half to one specimen thickness either side of the interface, the strain rate is modified so that it changes smoothly from the high remote value on the soft side to the lower remote value on the hard side.

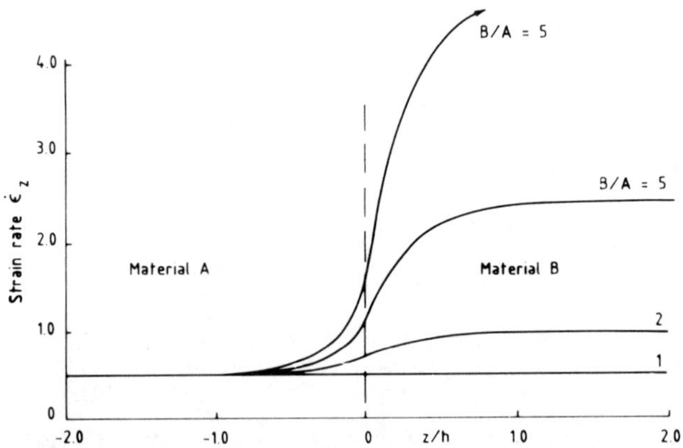

Fig. 3. The effect of the ratio, B/A, on the creep strain rate distribution across the interface (Nicol and Williams, 1983).

If a second interface is introduced, i.e. $z/h < \infty$, interaction occurs once the value of z/h is less than unity. The effect of increasing the value of n is shown in Fig. 4 for a very soft zone, $B/A = 200$, at various zone width ratios, z/h. It is clear that increases in the value of n, even from 2 to 8, reduce the rate of change of strain rate near the interface. This is because when $n \to \infty$, plastic solutions will be applicable.

Finally, Figs. 5a and b, for hard and soft zones respectively, show the effect of reducing the zone width ratio from ∞ to 0.5. The key features are:-

1. For a soft zone, decreasing the zone width relative to the specimen thickness, increases the constraint and reduces the strain rate below that of a homogeneous sample of the zone material subjected to plane strain loading with the maximum strain rate being recorded at the zone centre, $\dot{\varepsilon}_{zcl}$.

2. For a hard zone, decreasing the zone width ratio, z/h, causes the strain rate in the hard zone to be <u>increased</u> over that found from uniaxial data for a homogeneous sample. The maximum strain rate is at the zone interface, $\dot{\varepsilon}_{zi}$.

It may be inferred from these results that the soft zones should have a small effect on weld behaviour as their strain rates are reduced and if narrow, only slightly increase the overall strain rate. On the other hand, hard narrow zones have increased relative strain rates as the zone width is reduced.

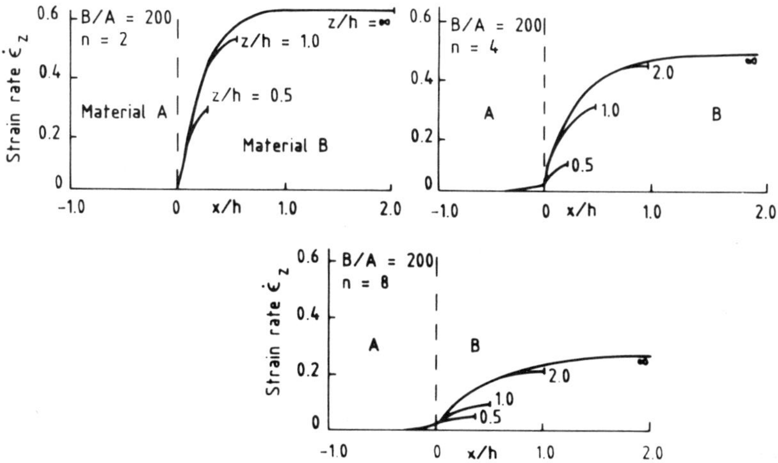

Fig. 4. The effect of creep rate index, n, on the creep strain rate distribution for a soft zone composite (Nicol, 1982).

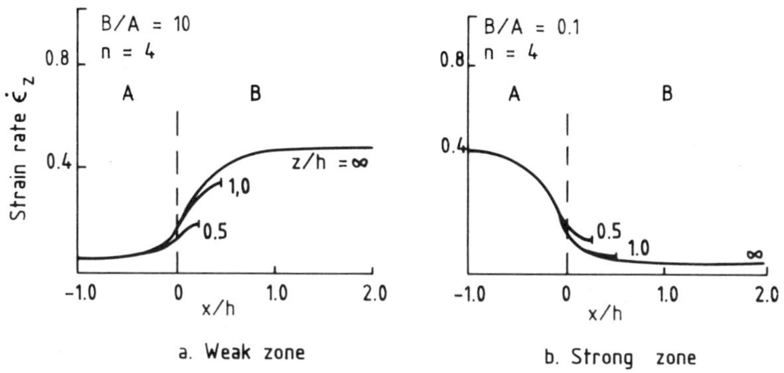

Fig. 5. The distribution of creep strain rate across the zone as a function of z/h (after Nicol, 1982).

If the strain rates at the zone centre and interface positions are normalised to that for a large homogeneous specimen of the zone material, $\dot{\varepsilon}_{z\infty}$, then approximately linear relationships are obtained between the logarithms of the normalised strain rates and B/A, (Nicol and Williams, 1983) for restricted ranges of B/A etc., Fig. 6, and:-

$\dot{\varepsilon}_{zi}/\dot{\varepsilon}_{z\infty} = 0.970\ [1/\sqrt{B/A}]$ for the interface, and (4)

$\dot{\varepsilon}_{zc1}/\dot{\varepsilon}_{z\infty} = [B/A]^c$ for the zone centre, where c is a (5)
function of z/h and n.

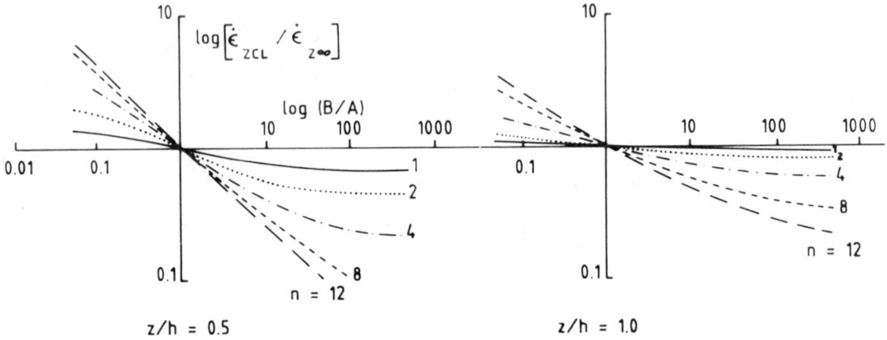

Fig. 6. The logarithm of the normalised strain rate at the zone centre as a function of B/A and z/h (Nicol, 1982).

Relationships (4) and (5) can be rewritten in terms of the uniaxial data for the zone material, for example

$$\dot{\varepsilon}_{zi} = \dot{\varepsilon}_{z\infty}\ [0.970/\sqrt{B/A}] = B\sigma^{n}\ [0.970/\sqrt{B/A}] \qquad (6)$$

and similar relationships are obtained for the zone centre strain rate. Conceptually, this approach can be used to derive an apparent reference stress for such a composite, i.e. $\dot{\varepsilon}_{zi} = B\sigma_R^n$, where σ_R is the reference stress and is given by:-

$$\sigma_R = \sigma[0.970\sqrt{A/B}]^{1/n} \qquad (7)$$

Reference stresses obtained in this way are numerically similar to those obtained using a limit load approach (Williams, 1981).

DATA AVAILABLE FROM THE LITERATURE

The analysis discussed so far relates to a simplified metal structure. In this section the limited experimental data for more realistic configurations are reviewed.

Uniaxial Experiments

Deformation of austenitic cross weld specimens has been studied using finite element and experimental methods by Ettiene, van Rossum and Roode, (1980) and Roode, Ettiene and van Rossum, (1980). Their analysis incorporated a weld metal, two HAZ regions and parent material. Creep data was primarily represented by the steady state Norton's Law or by Blackburn's formulation for combined primary and steady state creep. The weld metal was weaker in creep than the parent whereas both HAZ regions were stronger though to different degrees. Two specific geometries were considered, a wide plate under plane strain and a smaller specimen under plane stress loading.

The plane strain problem analysed is examined first as it is closer to the Nicol model. Examination of the published strain data, Fig. 7 taken from Ettiene and co-workers (1980), found that the principles outlined above are sound, namely that the strain rate in the hard zone is increased over that expected from uniaxial data for a homogeneous sample of the zone material subjected to the same load. This suggests that stress redistribution is occurring. Ettiene and co-workers obtained very good agreement between their finite element and experimental results for the smaller plane stress geometry. However, subtle differences concerned with plasticity in these analyses, and considered in the original paper, make any further comparisons with the Nicol model difficult.

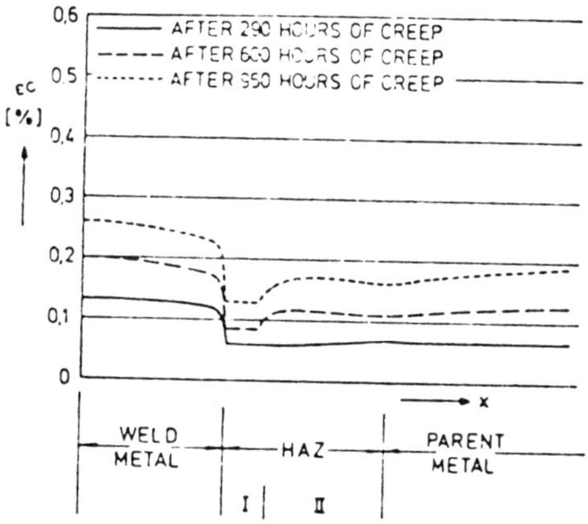

Fig. 7. Longitudinal creep strain at the centre of the analysed section of thick plate. Nominal stress = 165 MNm^{-2}, Temperature = $600^{o}C$ (Roode, Ettiene and van Rossum, 1980).

Few other workers have examined the steady state deformation of cross weld samples loaded normal to the fusion boundary. However, Manjoine (1982) considered the case of a stress applied parallel to the weld axis, and found that marked stress redistribution occurred even in this quite different geometry.

Pressure Loading Of Tube And Pipe Welds Under Creep Conditions

Internal pressure is a more usual loading system in practice, although laboratory verification is difficult and expensive. Early in the 1970's, finite element analysis using steady state constitutive relationships, established the basis of models for pressurised tube and pipe welds (Walters and Cockroft, 1972; Walters, 1976). Simplistically, on increasing the pressure, the tube or pipe section loads elastically and then with time, creep and stress redistribution occur. In practice, a weld metal which is less creep resistant than the parent deforms at a faster rate predominantly in the hoop direction and sheds load on to the harder adjacent material. This was referred to as "offloading". A weld comprising a less resistant weld metal and a more resistant HAZ relative to the parent offloads stress on to the adjacent HAZ. These stresses are also predominantly in the hoop direction and Fig. 8 shows a typical example from the work of Browne, et al. (1981). Stress redistribution also occurs in the axial direction but to a more limited extent.

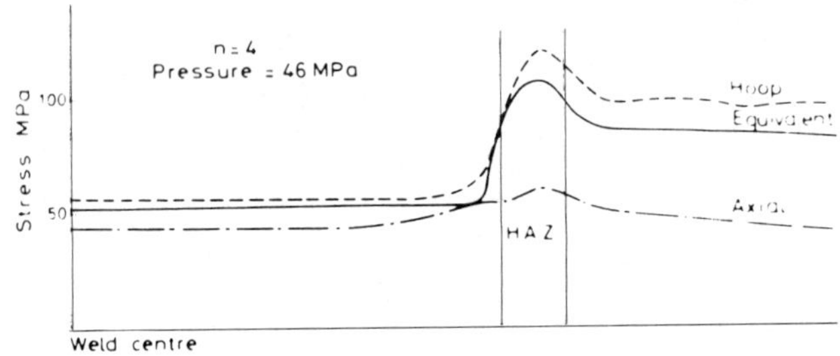

Fig.8. Calculated stress distribution along the outer surface of a weld (Browne and co-workers, 1981).

These analyses can be examined using the results from a major research programme carried out within the CEGB into the creep behaviour of welded pressurised pipe. It involved uniaxial data collection for individual weldment structures and tube and full sized vessel tests, coupled with a comprehensive stress analysis programme based on the original finite element work of Walters and Cockroft (1972). The main objectives of the work have been documented by Rowley and Coleman (1973). The ferritic $\frac{1}{2}Cr\frac{1}{2}Mo\frac{1}{4}V$ steel was chosen as the parent material and two major variables were studied. The first involved the weld metal:parent material creep strength ratio by including mild steel, $1Cr\frac{1}{2}Mo$, $2\frac{1}{4}Cr1Mo$ and $\frac{1}{2}Cr\frac{1}{2}Mo\frac{1}{4}V$ weld metals in the programme. The second involved the effect of PWHT on welds fabricated with $2\frac{1}{4}Cr1Mo$ weld metal. In addition, other full size pressure vessel programmes at MEL have examined other factors which can be encountered in practice. The subsequent discussion relies heavily on these programmes, as well as the results of research work in Sweden.

In the CEGB work, the finite element studies used steady state creep relationships of the Norton's Law form and the relative creep property differences were related to the parent material properties, see equation (3). A range of weld metal:parent material creep strain rate ratios, A_{wm}/A_{pm}, from unity to 1690 were examined. It was found that the steady state stresses generated at the weld metal outer surface centre were independent of whether a two or three material model were used. However, near the HAZ region, a three material model should be more appropriate. The detailed results are published by Browne and co-workers (1981) and Coleman, Parker and Walters (1981) so only the general principles will be given here.

The analysed results are completely consistent with the original work in the 1970's, with the weaker weld metals showing the characteristic stress offloading illustrated in Fig. 8 for an outer surface. The effect of differences in the weld metal:parent material creep properties can be best illustrated by examining the steady state creep stresses and is shown for the weld metal centre line and the HAZ positions in Fig. 9. It should be noted that in both cases, the stresses are normalised by the steady state creep stresses expected in the pipe remote from the weld. The key characteristics are:-

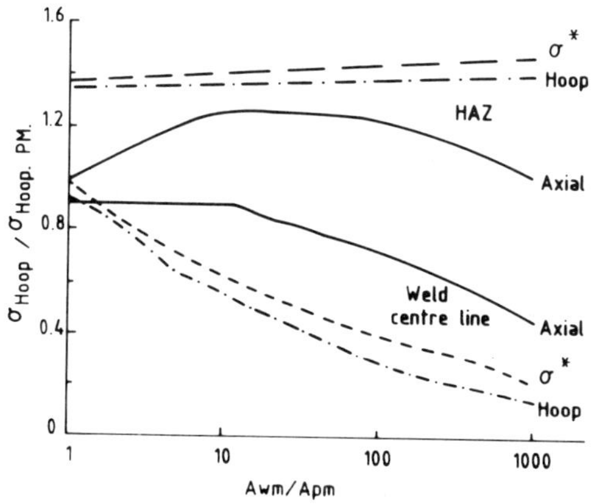

Fig.9. Normalised steady state stresses at HAZ and weld metal centre line as a function of weld metal:parent material creep strain rate ratio (from Coleman, Parker and Walters, 1981).

1. At the weld metal outer surface centre line position, the axial stress is relatively unaffected when $A_{wm}/A_{pm} \lesssim 20$, although both the equivalent and hoop stress decrease continuously with increasing creep property mismatch.

2. At the outer surface HAZ position, both the equivalent and hoop stresses increase continuously, albeit slowly, with increasing weld metal:parent material creep property mismatch; the axial stress peaks at A_{wm}/A_{pm} ratios of 10 to 100.

As stress in itself is not a measurable quantity, strain or strain rate data are required to verify any predictions of the analysis. Predictions of strain were obtained directly from mesh displacements once steady state stresses were established. As no primary creep was incorporated within the analysis, strain rate, rather than total strain was used. This method has the advantage that the gauge length over which any strain monitoring system operated could be incorporated into the strain predictions. An alternative method for determining a strain rate at a point in the structure is to incorporate the steady state stresses in the multiaxial creep laws.

The calculated creep strain rates, Table 1, suggest that the hoop strain rate across the outer surface of the weld metal is relatively unaffected by the weld metal strength, with the pipe controlling the deformation rate of the weld (Coleman, Parker and Walters, 1981). Experimentally, the gauge length for strain measurement is set by the strain gauge type used and the results, Table 2, indicate that the strain rate in this region is essentially constant. However, for extreme combinations of creep strength ratio, the results showed more variation, particularly close to the interface. This was further examined by calculating strain rates at specific points with the results

TABLE 1 Hoop and axial steady state creep rates calculated from the finite element analysis at positions across the weld (Coleman, Parker and Walters, 1981).

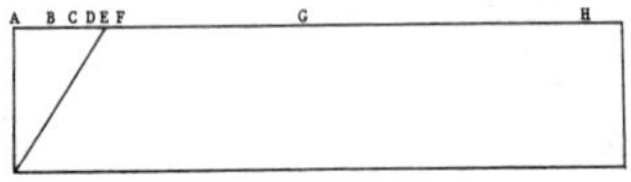

Material		Weld Metal				HAZ	Parent		
Distance from weld centreline mm		A	B	C	D	E	F	G	H
"A" ratios WM/PM/HAZ	Direct'n	0-10	14.5	18	22	28	53	140	235
1:1:0.25	Hoop	11	11	11	11	11	11	11	12
	Axial	-1	-1	-1	-1		-1	0	0
3:1:0.25	Hoop	12	12	12	12	13	12	12	12
	Axial	2	2	2	3		0.7	0	0
10:1:0.25	Hoop	13	14	14	14	14	13	12	12
	Axial	5	6	8	9		3	0	-1
15:1:0.25	Hoop	14	14	14	14	15	14	12	12
	Axial	7	9	11	13		4	-1	-1
1690:1:0.25	Hoop	12	13	16	22	37	31	19	17
	Axial	58	61	190	340		100	-6	-1

All creep rates x 10^{-8} h^{-1}, with constant n = 4.

shown in Fig. 10. This compares the hoop strain rate normalised by the pipe strain rate remote from the weld and directly illustrates the near constancy of strain rate at least up to weld metal:parent material ratios of 100.

Strain predictions have also been verified in a separate series of full size vessel tests (Coleman, Parker and Walters, 1981) and are supported by other unpublished test data. The results can be summarised as follows:-

1. Initial deformation on loading was elastic for both pipe section and welds and the pressurisation strains were compatible with those calculated from the Lamé equations.

2. For a wide range of weld metal:parent material strength ratios, the steady state hoop strain rate is virtually independent of the weld metal, being primarily controlled by the parent pipe. Tables 1 and 2, from the work of Coleman, Parker and Walters (1981), illustrate the agreement. Unpublished data from other vessel programmes for both thin, D/d $\sim$ 1.05, and thicker, D/d $\sim$ 1.5, pipe welds are consistent with these results.

3. When strain rather than strain rate is examined, the softer welds deform initially more than the pipe due to the stress redistribution. However, once steady state conditions are established, the hoop strain rates are similar.

4. When the weld metal is very undermatched, the weld metal hoop stress off-loads to below the level of the axial stress. Under these circumstances, the predominant strain direction will be axial.

These results were obtained for conventional "J" preparations where the weld width approximated to the weld thickness. However, Walters (1976) has illustrated that changing the volume of the weld has a relatively small

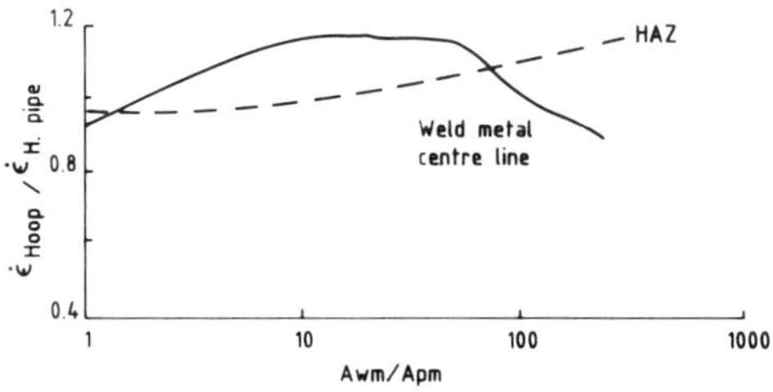

Fig. 10. Calculated hoop strain rates at weld metal centre line and HAZ positions.

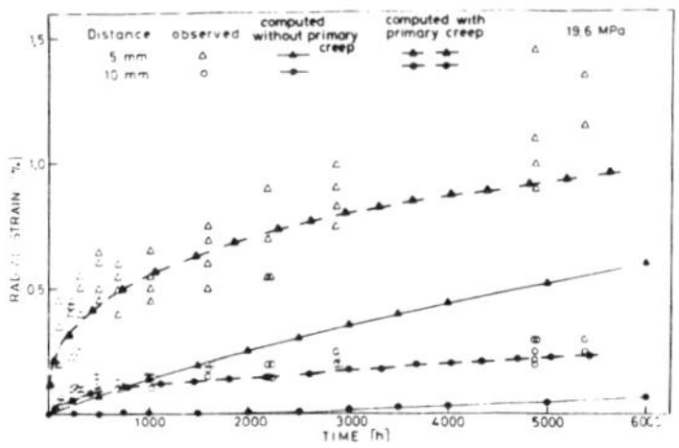

Fig. 11. Hoop strains at 5 and 10 mm from the centre line of austenitic tube welds. (Ivarsson and Sandstrom, 1981).

effect except at extremes of weld width or angle. However, experimental verification of these effects is not, as yet, available.

A similar programme on pressurised austenitic tube welds has been reported from Sweden, (Sandstrom and Ivarsson, 1979; Ivarsson and Sandstrom, 1981). The primary objective was similar to the CEGB objectives, but there were differences between the programmes. Sandstrom and Ivarsson based their analysis on thin walled tubes, $D/d < 1.1$, and in addition, used finite difference rather than finite element methods. Conceptually their results are in very good agreement with the basic offloading and redistribution model. They also extend the range of validity of the models by including austenitic materials and overmatched weld metals and, in one analysis, examined the effect of sharp and of diffuse HAZ regions.

To verify the model, the measured and calculated strains were compared through primary and secondary creep. It was found that although the strain rate was characterised by the analysis, the strain measured was very much greater than the calculated value. However, once primary creep was incorporated into the analysis, the agreement was very good, see Fig. 11. This was due to the low yield strength and the large primary creep shown by austenitics at high stresses.

ADDITIONAL FACTORS INFLUENCING WELD BEHAVIOUR

The analyses reported and discussed in the previous sections have examined specific geometries. However, there are additional features of welds which need to be considered.

Constraint Factors

In general, pressure circuits often terminate with a seal such as an end cap

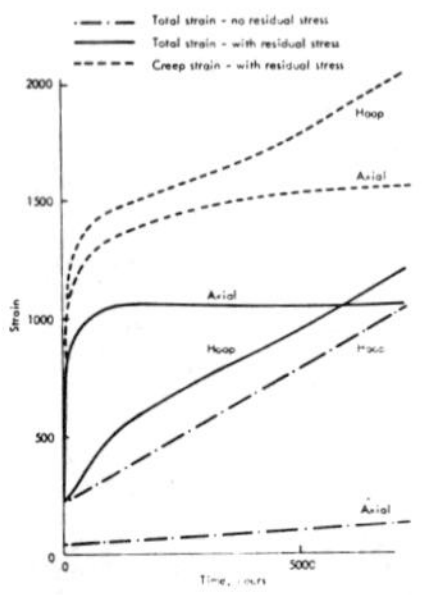

Fig. 12. Effect of residual stress on the calculated creep strain at the centre line of a weaker weld metal pipe weld (Hepworth, 1981).

which, in itself, applies a constraint to any adjacent weld and pipe sections. This constraint will affect the creep in the weldment locally. As part of the CEGB's programme on weld behaviour, secondary creep finite element analyses have been carried out for several end cap designs and, in one case, has been compared with experimental data. The preliminary results are promising and suggest that when the stresses produced by end cap constraint are incorporated into the stress analysis, the original concepts are equally applicable to this case (Coleman and Walters, private communication).

Welds Not Subjected To Post Weld Heat Treatment

Examination of strain data from as-welded and PWHT welds by Coleman and Parker (1981) showed that for as-welded joints, the hoop strain rate at the weld metal outer surface was appreciably higher than the parent pipe value and that the axial strains and strain rates exceeded those in PWHT joints. The materials were $\frac{1}{2}$Cr$\frac{1}{2}$Mo$\frac{1}{4}$V pipe welded with 2$\frac{1}{4}$Cr1Mo weld metal. However, after extended testing equivalent to times sufficient to relax the residual stress to low values, the creep rates reverted to those found for PWHT joints. These findings have recently been confirmed on other thick section vessel welds between $\frac{1}{2}$Cr$\frac{1}{2}$Mo$\frac{1}{4}$V and 2$\frac{1}{4}$Cr1Mo,(Coleman, 1982) and thinner section welds subjected to inadequate or no PWHT, (Williams and Hepworth, 1982, private communication).

Hepworth (1981) carried out a finite element analysis of a thin walled, $\sim$ 30 mm, large diameter, $\sim$ 500 mm, pipe to model the behaviour of as-welded joints. He imposed possible as-welded residual stress distributions on the finite element mesh and, using creep data typical of $\frac{1}{2}$Cr$\frac{1}{2}$Mo$\frac{1}{4}$V parent and 2$\frac{1}{4}$Cr1Mo weld metals, generated the stresses and strains accumulated during creep under internal pressure loading. A typical result for outer surface strains is shown in Fig. 12 for an undermatched weld metal. In this figure, the total strain refers to that strain which would be measured experimentally, whereas the creep strain also includes the strain calculated to occur from conversion of elastic to creep strains. The work identified several main features:-

1. The total hoop strain at the weld metal centre line is much greater than that found in a PWHT weld and a plain pipe.

2. The axial strain rate is also affected in a similar manner, but is also influenced by the under or overmatching capability of the weld metal.

3. The effect is transient and the weld behaviour reverts to that of a PWHT weld once the accumulated creep strain, to a first approximation, exceeds the elastic strains generated during welding, which constitute the residual stress field.

The Hepworth analysis is in good agreement with the experimental data for primary and secondary creep. However, it should be emphasised that if cavitation occurs during this early transient stage, the later results may be invalid.

Dissimilar Joints

These welds, usually between austenitic and ferritic materials, are very complex and their detailed deformation behaviour is not fully understood. They have special features which are not generally found in the welds considered so far. They are constructed from materials which have different coefficients of thermal expansion, rendering them sensitive to the effects of thermal cycling. Furthermore, they are constructed from materials of completely different creep strengths. In addition, element transfer can occur across the interface during service leading to local precipitation, carbon depletion and alloy transfer.

The mechanisms of stress redistribution and the effect of secondary stress such as residual or thermal stresses in causing transient behaviour must be applicable to these welds although the detailed structures may differ. However, the integration of these processes into a comprehensive model has not yet been reported.

Summary

It is clear that although welds are very complex structurally, simplifying assumptions can be used so that a realistic analysis can be carried out. Such assumptions have been used by various workers and now, for a range of different geometries and material combinations, a realistic and validated model can be produced which characterises steady state creep deformation and, if necessary, primary creep.

CREEP FAILURE OF WELDED JOINTS

In the previous section, primary and steady state creep were considered. Practically, however, the rupture behaviour is more important and, failure criteria need to be established so that the work can be incorporated directly into design codes.

Failure Criteria

At present two approaches are used to define the life of welded joints. Firstly, the steady state creep stresses can be used with uniaxial data to identify the life of the component parts of the weld; with the implicit assumption that local failure corresponds to weld failure. This approach was originally put forward by Walters and Cockroft (1972) and has been used with some success by Sandstrom and Ivarsson (1979); Ivarsson and Sandstrom (1980) and Browne and co-workers (1981). It is referred to here as a stress controlled approach and the important factor is the form of the stress used in the calculations.

An alternative approach, primarily put forward by Manjoine in 1974, concerns the material ductility and defines a strain limit for the constitutent parts of a component. The strain limit is a function of the uniaxial ductility, as measured in the laboratory, and the complex stresses present in the material. There are advantages and disadvantages with both methods but, in fact, once the key features of each are identified, they should be complementary.

Creep failure under uniaxial loading occurs by nucleation and growth of cavities, leading to microscopic cracking, macroscopic cracking and final failure. Similar sequences and features are found in welds, albeit limited in extent by the metallurgical structures present and, mechanistically the models should be identical. However, complex stresses are generated during creep deformation of welds and it is essential that any mechanistic failure model incorporates any effects of stress state on creep damage and rupture. This has been examined in some detail for several materials.

Laboratory scale work indicates that multiaxial creep rupture is a function of σ_1, the maximum principal stress, and σ^*, the von Mises equivalent stress; see for example, Dyson and Maclean (1977), Needham and Gladman (1980) and Cane (1982a). The exact form of the function appears to be influenced by the applied stress state, the absolute stress magnitude, the temperature, material and the heat treatment condition. The available data for ferritic steel has been recently reviewed by Browne, Lonsdale and Flewitt (1982) but are very limited.

From a study of cavity nucleation and growth kinetics, combined with the criterion that failure occurs at a constant critical cavitated area, then the multiaxial rupture life, t_{RM}, is given by:-

$$t_{RM} \propto [\sigma_1^{\nu} \sigma^{*(m-\nu)}]^{-1} \qquad (10)$$

where ν is the principal stress exponent of rupture and m is the slope of the uniaxial stress rupture plot. Based on this, a representative rupture stress, σ_R, for use with uniaxial data, is given by:-

$$\sigma_R = \sigma_1^{\nu/m} \sigma^{*(m-\nu)/m} \qquad (11)$$

(Cane, 1982a; Browne, Lonsdale and Flewitt, 1982).

The values of σ_1 and σ^* are the steady state values and tertiary creep is implicitly incorporated through the use of multiaxial rupture data to define ν. Using such an approach for the constituent parts of the weld, Browne and co-workers (1981) predicted the creep life of welded tube from uniaxial rupture data. In practice, as steady state stresses are used and tertiary creep is not explicitly included, the approach should be conservative. In addition, the analysis should allow the position and orientation of the cracks causing failure to be identified.

For thick ferritic butt welded tube, the analysis showed that the offloaded hoop stresses should nucleate and grow short cracks, orientated parallel to the tube axis, within the HAZ and this has been shown experimentally by Coleman and co-workers (1981). Final failure should be associated with this region. If, however, the tube or pipe is welded with a severely undermatched weld metal, the hoop stress can be offloaded to below the axial stress leading to the axial strain predominating and failure by circumferential cracking within the weld metal. (Coleman, Parker and Walters, 1981; Coleman and Walters, 1983, private communication).

An alternative method for defining the representative stress was suggested by Ivarsson and Sandstrom (1980), based on the concept that for their thin walled tube welds, the equivalent strain rates of the weld and parent material were identical. The stress characterising the weld behaviour is therefore that stress sufficient to give equal equivalent strain rates in the weld and parent metals. A comparison of the stresses generated by this simple method and the more detailed finite different method is shown in Table 3. This table also illustrates the effect of assuming a sharp or a diffuse HAZ region.

The previous approaches exclude tertiary creep and in most cases, should yield a conservative result. The major problem in incorporating tertiary creep is the lack of reliable constitutive relationships. The earliest forms were modifications of the Garafalo-McVetty relationship as given by Wilshire and others who added extra terms with increasing powers of time to the primary and secondary creep representation. More recent work by Evans, Parker and Wilshire (1982) has rationalised the approach for certain materials if constant stress creep data are available. Other relationships are the Sandstrom and Kandyr (1979) form:-

$$\dot{\varepsilon} = A \, \text{Exp} \, [\beta t] \tag{12}$$

and the very recent use of the Kachanov relationships by Cane (1982a):-

$$\dot{\varepsilon} = A \, \sigma^n \, \left[\frac{1}{1-t/tR}\right]^{\frac{\lambda-1}{\lambda}} \tag{13}$$

where $\lambda = \varepsilon_R/\varepsilon_{sm}$ and ε_R = rupture strain and ε_{sm} = Monkman-Grant constant. These are being incorporated into analytical methods and investigated by a series of benchmark experiments but the results are not available at present.

A failure criterion based on strain accumulation has many advantages for existing components in service, as strain can be measured on plant. There is no reason to believe that finite element analyses incorporating tertiary creep will not be successful and thus it is essential to define the limiting strain for failure of the material under a given stress state.

If there is no effect of stress multiaxiality on failure ductility, the conventional uniaxial ductility can be used in its various forms. This implicitly assumes that the final failure position is most likely to be the point of highest strain rate within that region, irrespective of the form of failure strain used. In detail, the rate of exhaustion of ductility of each region, $\varepsilon/\varepsilon_f$ should be considered.

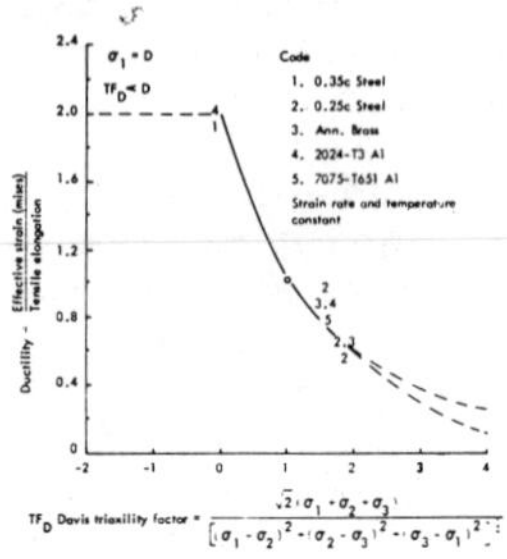

Fig. 13. The effect of stress state on ductility (after Manjoine, 1982).

Over the last ten years, Manjoine has rigorously examined such strain limits and in 1974, he considered high and low ductility failures. He examined uniaxial ductility as a function of strain rate, material and stress and combined with other reported multiaxial data, he suggested an empirical relationship between the multiaxial:uniaxial failure ductility ratio, $\varepsilon_f^*/\varepsilon_o$ and the Davis Triaxiality Factor (DTF):-

$$\varepsilon_f^*/\varepsilon_o = 2^{(1-DTF)} \qquad (14)$$

where DTF = $(\sigma_1+\sigma_2+\sigma_3)/\sigma^*$. His results are shown in Figure 13. As the DTF increases, the multiaxial ductility is reduced. Such a relationship has been suggested previously but was not quantified, even empirically. Immediately, the equivalence between multiaxial stress and strain failure criteria is illustrated, each being controlled, to a first approximation, by a function of σ_1 and σ^*.

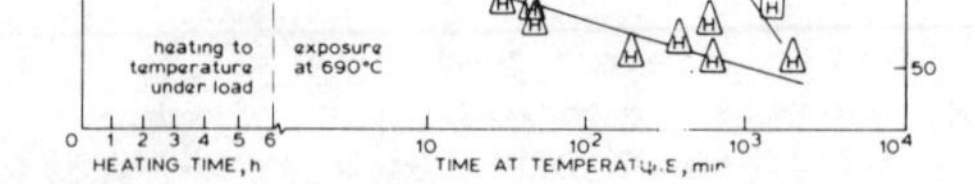

Fig. 14. Rupture data on $\frac{1}{2}Cr\frac{1}{2}Mo\frac{1}{4}V:2\frac{1}{4}Cr1Mo$ cross weld geometries (Myers and Clark, 1981).

COMPARISON OF PREDICTION AND EXPERIMENT

Cross Weld Geometries

Cross weld specimens subjected to stress rupture testing have been used for many years to examine the creep performance of welds but in more recent years, this specimen type has received more scrutiny. Two factors can be obtained from such tests in addition to the rupture life. These are the zone where failure occurs and the position, within this zone, of the maximum creep damage. Unfortunately, as shown by Williams (1982) the behaviour of a cross weld specimen is a function of specimen size and zone width ratio and in general, the majority of specimens tested have investigated either weld metal or parent material behaviour. Thus, not all the available data can be used to examine the effect of soft and hard zones.

Myers and Clark (1981) examined the creep failure of large cross weld specimens subjected to load during heating to 690°C followed by creep to failure at 690°C. They obtained results for a range of various HAZ structures generated by different welding procedures and showed conclusively that the final failure position could be defined from a consideration of the rupture behaviour of constituent parts of the weldment. As the stress rupture curves had different stress indices, the failure mode could be changed by varying the applied load, see Fig. 14. The more limited results of Etienne and co-workers (1979) and Endo and Sakon (1980) support this approach.

More recent work has compared the rupture behaviour of $\frac{1}{2}$Cr$\frac{1}{2}$Mo$\frac{1}{4}$V furnace heat treated samples containing a homogeneous distribution of coarse grained bainite with thermally cycled samples containing a central band of identical structure at the gauge length centre. (Clarke and Alberry, 1980; Clark,1981). It was shown that although the coarse grained bainitic regions were identical, the rupture life of the two types of sample differed by an amount outside the expected errors,with the homogeneous sample having the longer life. This effect was attributed to strain incompatibility associated with the hard central zone in agreement with the later model of Nicol (1982). In addition, Clark (1981), using samples with different lengths of coarse grained bainite in the central sections examined the creep damage accumulation in the central zone. The creep damage accumulated near, but not at, the zone interface and also, as the zone width was reduced, the damage remained at the same distance from the interface but moved,relatively, towards the zone centre. Although these samples contained more complex zone structures than the original single zone models, the work supports the strain redistribution, creep damage and strain exhaustion models. (Nicol, 1982; Nicol and Willians, 1982).

Tube and Pipe Butt Welds

As shown in the previous sections, stress and strain redistribution is predicted to be a maximum in the hoop direction and stress is offloaded on to the HAZ region. The results produced so far by both the UK and Swedish research groups support the general approach of rupture being predicted by the steady state stress levels, using a representative rupture stress, with the relevant uniaxial data. Table 4 compares the predicted and actual rupture lives (Browne, et al. 1981) for welded ferritic tubes using different failure criteria. Similar agreement was found by Sandstrom and Ivarsson (1981) for welded austenitic tubes. With very few exceptions, the results are conservative though there is obviously benefit to be gained from more accurate prediction.

The predicted microdamage and cracking for ferritic thick wall pipe and tube welds fabricated with weld metals of different strength, is again as expected. Initially short axially oriented cracks occur in the HAZ but not at the interface, but later grow toward the interface. In addition, for very weak weld metals, due to the large amount of hoop offloading, the predominant strain direction can become axial resulting in circumferential

TABLE 2 The measured steady state strain rates at the centre line outer surface position for four weld metals (Coleman, Parker and Walters, 1981).

Weld Metal	Experimental Measurements					Calculated Stresses		
	Uniaxial Creep Consts.		Pressure Vessel Strain Rates $\times 10^{-8}$ h^{-1}		$\dot{\varepsilon}_H/\dot{\varepsilon}_A$	σ_H/σ_A	σ_H MNm^{-2}	σ_A MNm^{-2}
	A	n	Hoop	Axial				
Mild Steel	5.4×10^{-12}	4	6	70	0.086	0.60	14.1	23.5
1CrMo	4.5×10^{-14}	4	8	5	1.60	1.17	44.2	37.8
2CrMo	1.6×10^{-14}	4	5	4	1.25	1.08	49.0	45.4
½CrMoV	3.2×10^{-15}	4	8	0	-	2	95.0	47.5

TABLE 3 Comparison of simplified and finite difference method for determining rupture life (Ivarsson and Sandstrom, 1980).

Weld metal	Internal pressure (MPa)	Rupture times (h) Unaffected parent tube	Smooth transition	Sharp transition	Estimated
Swcc	16	12000	3600	3200	3100
	24.5	430	260	270	250
Thomas (3)	17	7200	2600	2000	1300
	21	1400	-	190	130
	24.5	430	34	-	24

TABLE 4 Comparison of experimental rupture lives with predictions based on various stress criteria (Browne and co-workers, 1981).

Weld Metal	Internal Pressure MPa	Observed Rupture Life h.	Life Prediction, h							
			Maximum Equivalent Stress			Mean Diameter Hoop Stress			Maximum Rupture Stress HAZ	Maximum Principal Stress HAZ
			Weld Metal	HAZ	Parent	σ_{mdh}	$1.25\ \sigma_{mdh}$	$1.5\ \sigma_{mdh}$		
Mild Steel	45.3	14,691	23,000	5,000	40,000	31,000	5,200	1,500	8,000	17,000
1CrMo	45.3	18,774	50,000	14,000	40,000	31,000	5,200	1,500	17,000	28,000
2¼CrMo	45.3	22,039	50,000	18,000	40,000	31,000	5,200	1,500	18,000	26,000
½CrMoV	45.3	43,871	40,000	21,000	40,000	31,000	5,2001	1,500	24,000	24,000
Mild Steel	51.5	7,752	18,000	2,000	18,000	11,000	1,900	300	3,500	16,000
1CrMo	51.5	7,712	36,000	7,500	18,000	11,000	1,900	300	11,000	13,000
2¼CrMo	51.5	9,285	28,000	10,000	18,000	11,000	1,900	300	12,500	15,000
½CrMoV	51.5	16,511	18,000	12,000	18,000	11,000	1,900	300	14,500	18,000

cracks in the weld metal. Finally, if the loading system is modified by adding primary end loading or bending stresses from end caps, the circumferential weld metal cracking mode can be emphasised.

The more complex case of welds containing residual stress has not, as yet, been incorporated within a generalised failure assessment route.

CONCLUDING REMARKS

Although conventional welds are very complex structurally, the results of the simplistic models for a range of different geometries have confirmed conclusively that the weld structures can be considered as a series of zones or regions of finite size and have emphasised the general applicability of stress redistribution. The analyses for primary and steady state creep have been validated for both ferritic and austenitic welds in the post weld heat treated state. In addition, the effect of residual stress on the deformation processes has been predicted and verified practically.

Failure predictions, based on a representative rupture stress have been shown to be successful as a conservative method of assessing the weldment life, although more accurate methods will be required to incorporate tertiary creep and, possibly, limiting strain as a failure criterion. However, the analyses methods, at present, do successfully predict both the position and orientation of the cracking phenomena for the available experiments.

ACKNOWLEDGEMENTS

The author wishes to thank colleagues for useful discussions. This paper is published by permission of the Central Electricity Generating Board.

REFERENCES

Alberry, P.J. and Jones, W.K.C. (1977). "Diagram for the prediction of weld heat affected zone microstructure". Metals Technol., 4, 360.

Alberry, P.J. and Jones, W.K.C. (1979). "A computer model for the prediction of heat affected zone microstructures in multipass weldments". CEGB Report R/M/R282.

Alberry, P.J. and Lambert,J.A. (1982). "The welding metallurgy of SA508 Class 2 heat affected zones", 1982, Int.Conf. on Welding Technology for Energy Applications, Gatlinburg, Tenn., USA, AWS-ASM.

Browne, R.J., Lonsdale, D. and Flewitt, P.E.J. (1982). "Multiaxial stress rupture testing and compendium of data for creep resisting steels". J.Eng.Mats. and Technol. 104, 291.

Browne, R.J., Cane, B.J., Parker, J.D. and Walters, D.J. (1981). "Creep Failure Analysis of butt welded tubes". Conf. "Creep and Fracture of Engineering Materials and Structures", Swansea, 1981. Pineridge Press, Swansea, UK.

Cane, B.J. (1981). "A Collaborative Programme on the Correlation of Test Data for High Temperature Design of Welded Steam pipes:- Creep Data". CEGB Report RD/L/2101/N81.

Cane, B.J. (1982a). "Damage Accumulation and Failure Criteria in "Creep Brittle" Ferritic Weldment Structures". Int.Conf. on Welding Technology for Energy Applications, Gatlinburg, Tenn., USA, AWS-ASM.

Cane, B.J. (1982b). "Remaining Creep Life Estimation by Strain Assessment on Plant". J.Press.Vess. and Piping, 10, 11.

Clark, J.N. (1981). "The effect of a mismatch in creep properties on creep damage distributions", Scripta Met., 15, 1285.

Clark, J.N. and Alberry, P.J. (1980). "An observation of a creep rate mismatch interaction". Scripta Met., 14, 595.

Coleman, M.C. (1978). "The structure of weldments and its relevance to high temperature failure"., Conf. "Weldments: Physical Metallurgy and Failure Phenomena", 5th Bolton Landing Conf., Ed. Christoffel, R.J., Nippes, E.F. and Solomon, H.D., General Electric Co., USA.

Coleman, M.C. (1982). "Creep deformation and crack growth in a low alloy steel welded pressure vessel containing defects", Int.Conf. on "Welding Technology for Energy Applications", Gatlinburg, Tenn., USA., AWS-ASM.

Coleman, M.C. and Parker, J.D. (1981). "Deformation in 2CrMo-½CrMoV pressure vessel weldments at elevated temperature", Conf. "Creep and Fracture of Engineering Materials and Structures", Swansea, 1981. Pineridge Press, Swansea, UK.

Coleman, M.C., Parker, J.D. and Walters, D.J. (1981). "The behaviour of ferritic weldments in thick section ½CrMoV pipe at elevated temperatures", CEGB Report RD/M/1204/R81.

Dyson, B.F. and McLean, D. (1977). "Creep of Nimonic 80A in Torsion and Tension", Metal Sci.J., Feb. 37.

Endo, T. and Sakon, T. (1980). "A study on the Limitation of Inelastic Strain in Welded Joints at Elevated Temperatures", Conf. Engineering Aspects of Creep", 1980., Sheffield, Inst.Mech. Engrs., London, UK.

Ettiene, C.F., van Rossum, O. and Roode, F. (1980). "Creep of welded joints in AISI 316", Conf. "Engineering Aspects of Creep", Sheffield 1980, Inst. of Mech. Engrs., London, UK.

Evans, R.W., Parker, J.D. and Wilshire, B. (1982). "An extrapolation technique for long term creep strain and creep life prediction with special reference to ½CrMoV steels". Recent Advances in Creep and Fracture of Engineering Materials and Structures, Ed.Wilshire,B. and Owen,D.R.J., Pineridge Press, Swansea, UK.

Hepworth, J.K. (1981). "The effect of residual stress on the creep deformation of welded pipe". CEGB Report RD/M/1202/R81.

Ivarsson,B.G. and Sandstrom, R. (1980). "Calculated Methods of determining the Rupture Life of Butt-welded Tubes"., Conf. "Engineering Aspects of Creep", Sheffield 1980, Inst.Mech.Engrs., London UK.

Ivarsson, B.G. and Sandstrom, R. (1981). "Creep of Butt-welded Tube"., Conf. "Creep and Fracture of Engineering Materials and Structures", Swansea, 1981, Pineridge Press, Swansea, UK.

Manjoine, M.J. (1974). "Elevated temperature mechanics of metals". Conf. "Mechanical Behaviour of Materials", Kyoto, Japan.

Manjoine, M.J. (1982). "Creep rupture behaviour of weldments", Amer.W.J., Res.Supp., Feb., 50s.

Middleton,C.J. and Cane, B.J. (1977). "Characterisation of nature and distribution of constituents of multipass butt welds", CEGB Report RD/L/N24.

Myers, J. and Clark, J.N. (1981). "Influence of welding procedure on stress relief cracking in CrMoV steels, Part I. Use of Cross Weld Specimens", Metals Technology, 8, 382.

Needham, N.G. and Gladman, T. (1980). "The effect of stress state on the processes controlling creep fracture in a 2¼Cr1Mo steel", Conf. "Engineering Aspects of Creep", Sheffield, 1980. Inst.Mech.Engrs, London, UK.

Nicol, D.A.C. (1982). "Creep behaviour of butt welded joints". CEGB Report TPRD/M/1257/N82.

Nicol, D.A.C. and Williams, J.A. (1983). "The Creep Behaviour of Cross Weld Specimens under Uniaxial Loading", CEGB Report TPRD/M/1302/R83.

Price, A.T. and Williams, J.A. (1982). "The influence of welding on the creep properties of steel", Recent Advances in Creep and Fracture of Engineering Materials and Structures, Ed.Wilshire,B. and Owen, D,R.J.

Roode, F., Ettiene, C.F. and van Rossum, O. (1980). "Stress and Strain analyses for creep and plasticity in welded joints in AISI 316", Conf. Engineering Aspects of Creep", Sheffield 1980., Inst.Mech.Engrs., London, UK.

Rowley, T. and Coleman, M.C. (1973). "Collaborative Programme on the Correlation of Test Data for the design of Welded Steam Pipes", CEGB Report RD/M/N710.

Sandstrom, R. and Ivarsson, B. (1979). "Creep in Butt-welded Tubes of Cold Worked AISI 316 with Internal Pressure"., Conf. ICM3, Cambridge, UK. Pergamon Press, Oxford, UK.

Sandstrom, R. and Kondyr, A. (1979). "Model for tertiary creep in Mo and CrMo steels"., Conf. ICM3, Cambridge, UK., Pergamon Press, Oxford, UK.

Schüller, H.J., Hagn, L. and Woitscheck, A. (1974). "Cracking in the weld region of shaped components in hot steam lines". Der Maschineschaden, 47, 1.

Walters, D.J. (1976). "The stress analysis of cylindrical pipe welds under creep conditions", CEGB Report RD/B/N3716.

Walters, D.J. and Cockcroft, R.D.H. (1972). "A stress analysis and failure criterion for high temperature butt welds", Symposium, Toronto, Canada, International Institute of Welding.

Williams, J.A. (1982). "A simplified approach to the effect of specimen size on the creep rupture of cross weld samples", J.Eng.Mats. and Technol., 104, 36.

Wolstenholme, D.A. (1976). "Transverse cracking on reheating of a 2CrMo submerged arc weld metal", CEGB Report R/M/N868, CEGB, UK.

SIMPLIFIED DESIGN PROCEDURES FOR COMPOSITE MATERIALS

S. W. Tsai

Air Force Wright Aeronautical Laboratories, Wright-Patterson Air Force Base, Ohio 45433, USA

ABSTRACT

Designing with composite materials can be simplified considerably by making the laminates balanced and symmetric. By taking advantage of invariants of the coordinate transformation of strain and stiffness modulus, further simplifications can be made. Only one off-axis component of the stiffness modulus need to be calculated. The remaining three components are derived from symmetry and invariants. The failure envelope in strain space is preferred over that in stress space. By using the rigid body rotation of the envelope in the Mohr's circle space, the first-ply-failure envelope can be readily determined. The strength analysis of composite materials becomes essentially the same as that of the conventional material.

KEYWORDS

Composite materials; design methodology; laminate stiffness; laminate strength.

INTRODUCTION

We have found that in training engineers to design with composite materials it is easier to introduce the simplest composite materials first, and then gradually evolve with higher degree of complexity as time goes on. The simplest composite construction is the orthotropic, symmetric laminate. This class of laminates consists of 0 and 90 degrees, and balanced plies; i.e., the positive and negative off-axis ply orientations appear in pairs. It is the purpose of this paper to show how easy it is to design with these simple laminates. Many important features of composite materials are illustrated using formulas and diagrams which are no more complicated than those for the conventional materials.

We will further limit ourselves to the in-plane loading of this class of laminates. We will start with a set of basic ply data. The notations used here follow essentially those in our book (1). In our sample calculations, we will use T-300/5208, a popular graphite-epoxy composite. Figure 1 shows the definition of terms used for the longitudinal properties. Analogous definitions apply to the transverse and shear properties.

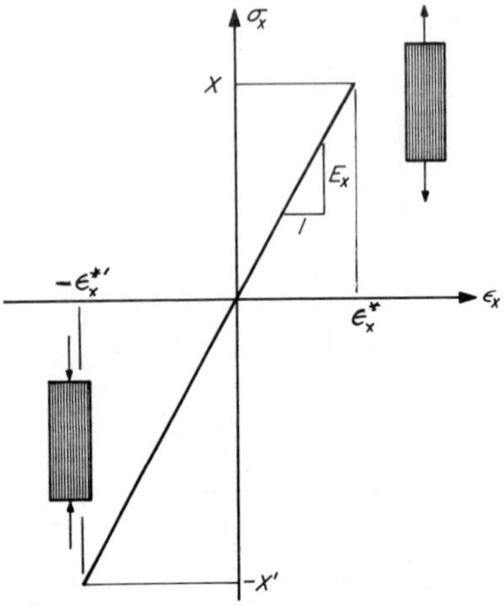

Figure 1. Longitudinal Properties

Table 1. On-Axis Stress-Strain Relations

	ϵ_x	ϵ_y	ϵ_s
σ_x	Q_{xx}	Q_{xy}	
σ_y	Q_{yx}	Q_{yy}	
σ_s			Q_{ss}

Table 2. Stiffness Modulus Transformation Equations

	Q_{xx}	Q_{yy}	Q_{xy}	Q_{ss}
Q_{11}	m^4	n^4	$2m^2n^2$	$4m^2n^2$
Q_{22}	n^4	m^4	$2m^2n^2$	$4m^2n^2$
Q_{12}	m^2n^2	m^2n^2	m^4+n^4	$-4m^2n^2$
Q_{66}	m^2n^2	m^2n^2	$-2m^2n^2$	$(m^2-n^2)^2$
Q_{16}	m^3n	$-mn^3$	mn^3-m^3n	$2(mn^3-m^3n)$
Q_{26}	mn^3	$-m^3n$	m^3n-mn^3	$2(m^3n-mn^3)$

$m = \cos\theta,\ n = \sin\theta$

$$
\begin{aligned}
E_x &= \text{longitudinal stiffness} = 181 \text{ GPa} \\
E_y &= \text{transverse stiffness} = 10.3 \text{ GPa} \\
\nu_x &= \text{longitudinal Poisson's ratio} = .28 \\
\nu_y &= \text{transverse Poisson's ratio} = \nu_x E_y / E_x = .016 \\
E_s &= \text{longitudinal shear modulus} = 7.17 \text{ GPa} \\
h_o &= \text{unit ply thickness} = 125 \times 10^{-6} \text{m}
\end{aligned} \tag{1}
$$

$$
\begin{aligned}
X &= \text{longitudinal tensile strength} = 1500 \text{ MPa} \\
X' &= \text{longitudinal compressive strength} = 1500 \text{ MPa} \\
Y &= \text{transverse tensile strength} = 40 \text{ MPa} \\
Y' &= \text{transverse compressive strength} = 246 \text{ MPa} \\
S &= \text{longitudinal shear strength} = 68 \text{ MPa}
\end{aligned} \tag{2}
$$

Ply Stiffness

From the elastic constants in Equation (1) above, we can calculate the stiffness modulus Q_{ij} of this unidirectional composite:

$$
\begin{aligned}
Q_{xx} &= \mu E_x = 181.8 \text{ GPa} \\
Q_{yy} &= \mu E_y = 10.34 \text{ GPa} \\
Q_{xy} &= \nu_x Q_{yy} = \nu_y Q_{xx} = 2.89 \text{ GPa} \\
Q_{ss} &= E_s = 7.17 \text{ GPa}
\end{aligned} \tag{3}
$$

where $\mu = (1-\nu_x\nu_y)^{-1}$. These moduli are coefficients of the stress-strain relations shown as components in Table 1, where matrix multiplication is implied. From these on-axis stiffness modulus, we can find the off-axis modulus through simple geometric relations, which can be found in standard textbooks. These relations are shown in Table 2.

The angle is the ply orientation as defined in Figure 2, where the laminate axes are 1-2; and the lamina axes, x-y. The off-axis stress strain relations are different from those in Table 1 and are shown in Table 3.

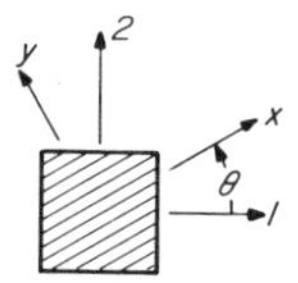

Figure 2. On- and Off-Axis Coordinate Systems

Table 3. Off-Axis Stress-Strain Relations

	ϵ_1	ϵ_2	ϵ_6
σ_1	Q_{11}	Q_{12}	Q_{16}
σ_2	Q_{21}	Q_{22}	Q_{26}
σ_6	Q_{61}	Q_{62}	Q_{66}

For every transformation relations, there are a number of invariants. By combining rows in Table 2, we can easily show that

$$I_1 = Q_{11} + Q_{22} + 2Q_{12} = Q_{xx} + Q_{yy} + 2Q_{xy} \tag{4}$$

$$I_2 = Q_{66} - Q_{12} = Q_{ss} - Q_{xy} \tag{5}$$

$$I_3 = I_1 + I_2 = Q_{11} + Q_{22} + Q_{12} + Q_{66} = Q_{xx} + Q_{yy} + Q_{xy} + Q_{ss} \tag{6}$$

where for T300/5208, I_1 = 197.8 GPa, I_2 = 4.28 GPa and I_3 = 202.2 GPa. These invariants impose limits on the stiffness moduli and are illustrated in Figure 3 by using Table 2. A simpler method for the calculation of the off-axis moduli will be discussed shortly. While each modulus can vary in

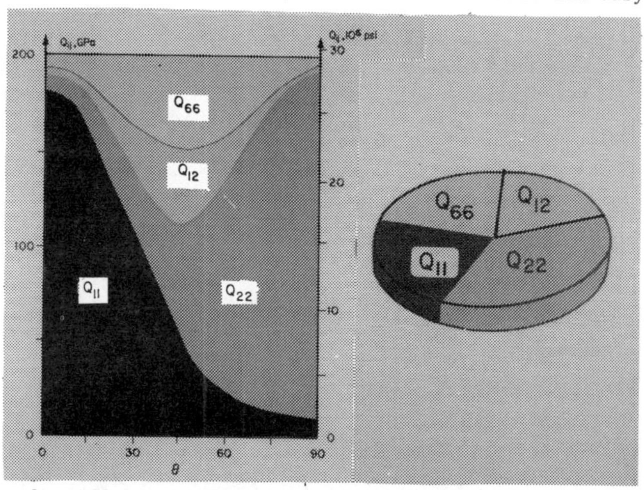

Figure 3. Off-Axis Stiffness Modulus of T300/5208.

values but the sum is constrained to I_3, with a value of 202 GPa as stated in Equation 6. As we design laminates consisting of the same unidirectional plies with varying ply orientations, the laminate stiffness is constrained by the same invariants. This is helpful to understand the capability of composite laminates, and, at the same time, simplifies the determination of the laminate stiffness.

The off-axis stiffness is very easy to determine. We take the first equation in Table 2.

$$Q_{11} = m^4 Q_{xx} + n^4 Q_{yy} + 2m^2n^2 \; (Q_{xy} + 2Q_{ss}) \tag{7}$$

Take the values for T300/5208 in Equation (1), we can calculate the following off-axis moduli shown in Table 4. By symmetry, we can write down Q_{22} without calculation. We can also calculate the sum of Q_{11} and Q_{22}.

Table 4. Off-Axis Stiffness Modulus

θ	0	15	30	45	60	75	90
Q_{11}	181.8	160.5	109.3	56.6	23.4	12.0	10.3
Q_{22}	10.3	12.0	23.4	56.6	109.3	160.5	181.8
$Q_{11} + Q_{22}$	192.1	172.5	132.7	113.2	132.7	172.5	192.1

Note this sum is symmetric with respect to 45-degree ply orientation. Taking advantage of the invariants in Equations (4), (5) and (6), we can directly calculate the remaining off-axis moduli using the values in Table 4.

$$\begin{aligned} 2Q_{12} &= [I_1 - (Q_{11} + Q_{22})] = 198 - (Q_{11} + Q_{22}) \\ 2Q_{66} &= I_1 + 2I_2 - (Q_{11} + Q_{22}) = 206.6 - (Q_{11} + Q_{22}) \end{aligned} \tag{8}$$

A graphical construction of the values in Table 4 and those in Equation (8) is shown in Figure 4. This is very similar to Figure 3 except the scales

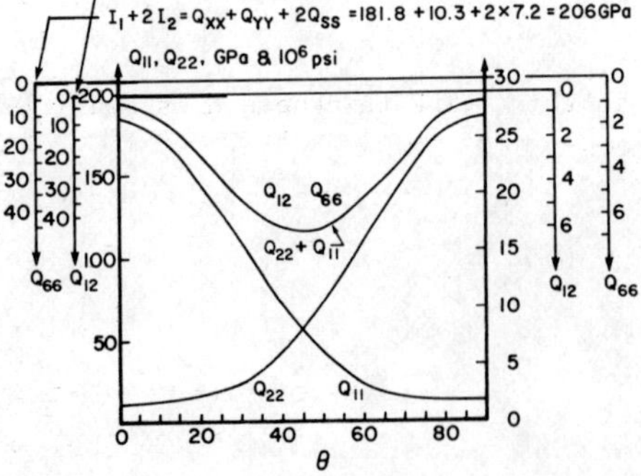

Figure 4. Off-Axis Stiffness of T300/5208.

of Q_{12} and Q_{66} are shown, with their origins at I_1(=198) and $I_1 + 2I_2$(=206.6) GPa on the vertical axis, respectively. The scales are twice as large due to the factor of two in Equation (8).

The off-axis stiffness modulus can be read directly from this figure. Figure 5 shows the values for θ = 30 degrees in matrix form.

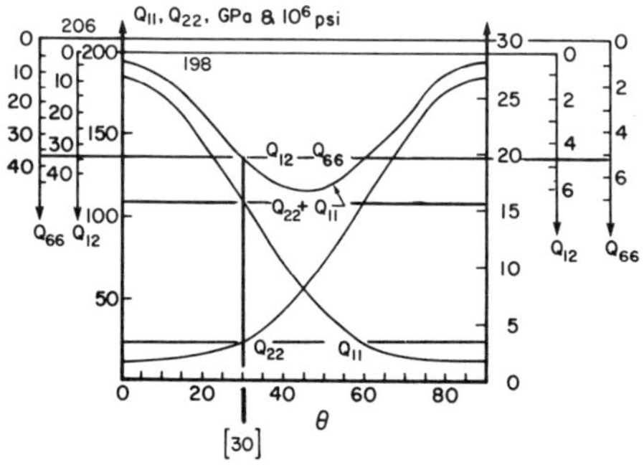

Figure 5. Stiffness of [30] T300/5208.

$$Q_{ij} = \begin{bmatrix} 109 & 32 & Q_{16} \\ 32 & 23 & Q_{26} \\ Q_{61} & Q_{62} & 37 \end{bmatrix}, \text{GPa} = \begin{bmatrix} 16 & 4.7 & Q_{16} \\ 4.7 & 3.4 & Q_{26} \\ Q_{61} & Q_{62} & 5.3 \end{bmatrix}, 10^6 \text{ psi} \tag{9}$$

We did not calculate the Q_{16} and Q_{26} components because we do not need them for our balanced laminates.

Laminated Plate Theory

As stated earlier, we will limit our discussion to balanced (therefore orthotropic), symmetric laminates. The same assumptions of the classical laminated plate theory remain valid; i.e.,

- The laminate is thin. The thickness h is small in comparison with the length and width of the laminate
- The strains across the thickness of the laminate are constant. The lamina and laminate strains are equal.

$$\varepsilon_1 = \varepsilon_1^o, \ \varepsilon_2 = \varepsilon_2^o, \ \varepsilon_6 = \varepsilon_6^o \tag{10}$$

The super zero refers to the in-plane quantities. The average stresses across the laminate are

$$\bar{\sigma}_1 = \frac{1}{h}\int_{-h/2}^{h/2} \sigma_1 dz,\ \bar{\sigma}_2 = \frac{1}{h}\int_{-h/2}^{h/2} \sigma_2 dz,\ \bar{\sigma}_6 = \frac{1}{h}\int_{-h/2}^{h/2} \sigma_6 dz \tag{11}$$

where the over bar represents average quantities. Substituting the off-axis stress-strain relations in Table 3 into the first of Equation (11),

$$\bar{\sigma}_1 = \frac{1}{h}\int (Q_{11}\varepsilon_1 + Q_{12}\varepsilon_2 + Q_{16}\varepsilon_6) dz \tag{12}$$

Using Equation (10), where the strains are not z-dependent, and can then be taken out of the integral signs, we have

$$\begin{aligned} \bar{\sigma}_1 &= \frac{1}{h}[\int Q_{11}dz]\varepsilon_1^o + \frac{1}{h}[\int Q_{12}dz]\varepsilon_2^o + \frac{1}{h}[\int Q_{16}dz]\varepsilon_6^o \\ &= \bar{Q}_{11}\varepsilon_1^o + \bar{Q}_{12}\varepsilon_2^o + \bar{Q}_{16}dz \end{aligned} \tag{13}$$

where the barred quantities refer to the laminate stiffness which are simply the average of the ply stiffness. The two other analogous stress-strain relations for the laminate are similarly derived. They are:

$$\begin{aligned} \bar{\sigma}_2 &= \bar{Q}_{21}\varepsilon_1^o + \bar{Q}_{22}\varepsilon_2^o + \bar{Q}_{26}\varepsilon_6^o \\ \bar{\sigma}_6 &= \bar{Q}_{61}\varepsilon_1^o + \bar{Q}_{62}\varepsilon_2^o + \bar{Q}_{66}\varepsilon_6^o \end{aligned} \tag{14}$$

The laminate stiffness is simply the linear average of the ply stiffness; i.e.

$$\begin{aligned} \bar{Q}_{11} &= \frac{1}{h}\int Q_{11}dz = \frac{1}{h}\sum_{t=1}^{m} Q_{11}^{(t)}(z_{t+1} - z_t) = \frac{1}{h}\sum_{t=1}^{m} Q_{11}^{(t)} h_{(t)} \\ &= \sum_{t=1}^{m} v_{(t)} Q_{11}^{(t)} \end{aligned} \tag{15}$$

where m = number of ply groups or ply orientations

$h_{(t)}$ = thickness of the t-th ply group or ply orientation

$v_{(t)}$ = volume fraction of the t-th ply group

Equation (15) is simply the rule-of-mixture relation between the stiffness of the ply and the laminate. The other five stiffness moduli follow the same formula.

From the transformation equations in Table 2, we can see that Q_{16} and Q_{26} changes signs when the ply orientation changes from positive to negative. Thus for balanced laminates; i.e., equal pairs of plus and minus off-axis plies, the laminate becomes orthotropic;

$$\bar{Q}_{16} = \bar{Q}_{26} = 0. \tag{16}$$

We have only four laminate stiffness moduli and the resulting stress-strain relations are identical to those for the conventional material except the number of moduli are increased from two to four. But the equations have the same linear relationship. The Q_{ij} of the unit ply are the key stiffness parameters. They provide the window or bridge between the ply and laminate. Engineering constants, on the other hand, do not provide the window. We cannot say

$$\bar{E}_1 = \Sigma\, v_{(t)} E_1^{(t)} \tag{17}$$

This is incorrect. We must use Equation (15).

Laminate Stiffness

Since the stiffness moduli of the orthotropic laminate are based on rule-of-mixtures relations, as seen in Equation (15), we can readily determine the laminate stiffness. We only need to determine $\bar{Q}_{11}$ and $\bar{Q}_{22}$; we use the invariant relations to determine $\bar{Q}_{12}$ and $\bar{Q}_{66}$.

$$\begin{aligned} \bar{Q}_{12} &= \frac{1}{2}[I_1 - (\bar{Q}_{11} + \bar{Q}_{22})] = 98.9 - \frac{1}{2}(\bar{Q}_{11} + \bar{Q}_{22}) \\ \bar{Q}_{66} &= I_2 + \bar{Q}_{12} = 4.28 + \bar{Q}_{12} \end{aligned} \tag{18}$$

where we claim that

$$I_1 = \bar{I}_1,\ I_2 = \bar{I}_2,\ I_3 = \bar{I}_3,$$

that is, the laminate and plies have the same invariants. This can be easily shown by using Equation (15) that:

$$I_1 = Q_{11} + Q_{22} + 2Q_{12} = \Sigma v_{(t)} Q_{11}^{(t)} + \Sigma v_{(t)} Q_{22}^{(t)} + 2\Sigma v_{(t)} Q_{12}^{(t)} = \Sigma v_{(t)} I_1 = \bar{I}_1 \tag{19}$$

where $\Sigma v_{(t)} = 1$.

As a numerical example, we will determine the stiffness modulus of cross-ply laminate $[0_7/90_2]$. We first find $\bar{Q}_{11}$ using Equation (15) and data in Table 4.

$$\bar{Q}_{11} = \frac{1}{9}(7 \times 181.8 + 2 \times 10.3) = 144 \text{ GPa}$$

$$\bar{Q}_{22} = \frac{1}{9}(2 \times 181.8 + 7 \times 10.3) = 48 \text{ GPa}$$

From Equation (18)

$$\bar{Q}_{12} = 98.9 - 96.05 = 2.9 \text{ GPa};\ \bar{Q}_{66} = 4.28 + 2.9 = 7.2 \text{ GPa}$$

We can obtain the same values above graphically using straight lines and the lever rule of the phase diagram. This is shown in Figure 6.

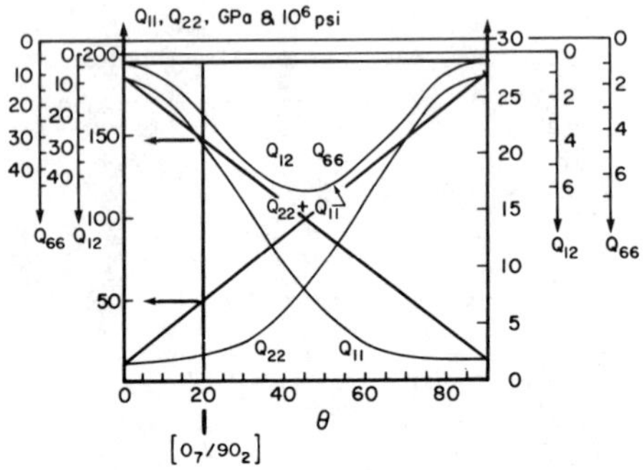

Figure 6. Stiffness of $[0_7/90_2]$ T300/5208.

Applying Equation (3) for orthotropic materials to our orthotropic laminate, we can determine the following effective in-plane engineering constants:

$$\nu^o_{12} = \bar{Q}_{12}/\bar{Q}_{11} = 3/144$$

$$\nu^o_{21} = \bar{Q}_{12}/\bar{Q}_{22} = 3/48$$

$$\mu^o = (1 - \nu^o_{21}\nu^o_{12})^{-1} = (.999)^{-1} \quad (20)$$

$$E^o_1 = \bar{Q}_{11}/\mu^o = 144 \text{ GPa}$$

$$E^o_2 = \bar{Q}_{22}/\mu^o = 48 \text{ GPa}$$

$$E^o_6 = \bar{Q}_{66} = 7.2 \text{ GPa}$$

For $[0/\pm 60]$, we simply use the values from Table 4 and Equations 15 and 8:

$$\bar{Q}_{11} = \bar{Q}_{22} = 76 \text{ GPa}, \bar{Q}_{12} = 99 - 76 = 23 \text{ GPa}, \bar{Q}_{66} = 4.3 + 23 = 27.3 \text{ GPa} \quad (21)$$

This is a quasi-isotropic laminate which is a very common laminate. The values above can also be determined graphically from Figure 6.

$$\nu^o_{12} = \nu^o_{21} = 23/76 = .30, \mu^o = (1 - \nu^o_{12}\nu^o_{21})^{-1} = (.91)^{-1} \quad (22)$$

$$E^o_1 = E^o_2 = 76 \times .91 = 69 \text{ GPa}, E^o_6 = 27.3 \text{ GPa}$$

The stiffness moduli of balanced laminates can be determined by either numerical or graphical method. Instant estimate of the laminate stiffness is achieved without lengthy calculation. The resulting $\bar{Q}_{ij}$ can be used to calculate laminate stress from laminate strain, like those in Table 1 for the ply stress from ply strain. The effective engineering constants can be used to calculate the laminate strain from laminate stress. This is shown in Table 5.

Table 5. Stress-Strain Relation for Balanced Laminates.

	$\bar{\sigma}_1$	$\bar{\sigma}_2$	$\bar{\sigma}_6$
ϵ_1^o	$\frac{1}{E_1^o}$	$-\frac{\nu_{12}^o}{E_2^o}$	
ϵ_2^o	$-\frac{\nu_{21}^o}{E_1^o}$	$\frac{1}{E_2^o}$	
ϵ_6^o			$\frac{1}{E_6^o}$

The procedure described here can be extended to hybrids. The rule-of-mixtures relations and the invariants shall be applied to each of the composite materials to obtain the stiffness of the sublaminates. The final laminate stiffness is again a rule-of-mixtures relation between the sub-laminates stiffness.

Failure Criteria

Failure of unidirectional plies can be based on maximum stress, maximum strain, or quadratic form in stress and strain space. Strain space representation is preferred over the stress space because in a laminate the strain is assumed to be constant across the thickness, as stated in Equation (10). The strain failure envelope for each ply orientation is fixed in strain space, independent of plies with different orientations in the laminate. The failure envelope in stress space, on the other hand, changes with each particular laminate, the [0/90] and $[0_2/90]$ laminates have the same failure envelopes in strain space, but different ones in stress space. That is why strain space is preferred. The maximum strain criterion is easy to construct graphically. As a minimum, five strengths under simple states of stress must be measured for a unidirectional ply. Figure 1 shows the longitudinal strengths in tension and compression. Assuming linear stress-strain relation to failure, we have from the strength data in Equation (2):

$$\text{Longitudinal tensile strain} = \varepsilon_x^* = \frac{X}{E_x} = \frac{1500}{181} \cong .8\%$$

$$\text{Longitudinal compressive strain} = \varepsilon_x^{*\prime} = \frac{X'}{E_x} = \frac{1500}{181} \cong .8\% \qquad (23)$$

Similarly, we can calculate

$$\text{Transverse tensile strain} = \varepsilon_y^* = \frac{Y}{E_y} = \frac{40}{10.3} \cong .4\%$$

$$\text{Transverse compressive strain} = \varepsilon_y^{*\prime} = \frac{Y'}{E_y} = \frac{246}{10.3} \cong 2.4\% \qquad (24)$$

$$\text{Longitudinal shear strain} = \varepsilon_s^* = \frac{S}{E_s} = \frac{68}{7.17} \cong 1.0\%$$

These ultimate strains are shown as solid dots in Figure 7.

In the principal strain or the zero shear strain plane, the maximum strain failure envelope is simply a rectangle drawn through the maximum longitudinal and transverse strains. The shear strain sets the r-axis value, defined by Eq. (25).

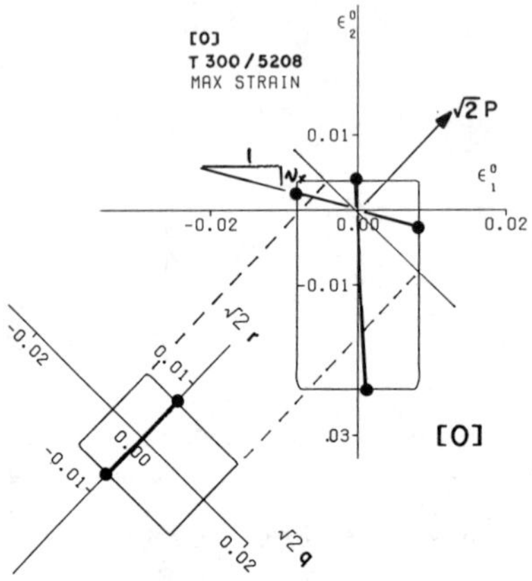

Figure 7. Maximum Strain Failure Envelope for [0] T300/5208.

This is the failure envelope of a 0-degree T300/5208 unidirectional composite. The uniaxial stress appears biaxial in strain space because of the Poisson contraction. The longitudinal and transverse loading lines in strain space have slopes equal to the longitudinal and transverse Poisson's ratio, respectively. The transverse loading is nearly vertical or follows that of the ε_2-axis because the transverse Poisson's ratio is .016 or nearly zero (see Equation (1)). These loading lines are also shown in Figure 7 going from origins to the ultimate strains shown as solid dots.

The 90-degree envelope can be determined by interchanging the x and y directions. Graphically, this can be achieved by turning over Figure 7 for the 0-degree. Figure 8 represents failure envelopes of $[0_j/90_k]$ laminates for any values of j and k. Thus, the positions and shapes of the envelopes in strain space are invariant. That is one of the chief advantages of depicting failure envelopes in strain space.

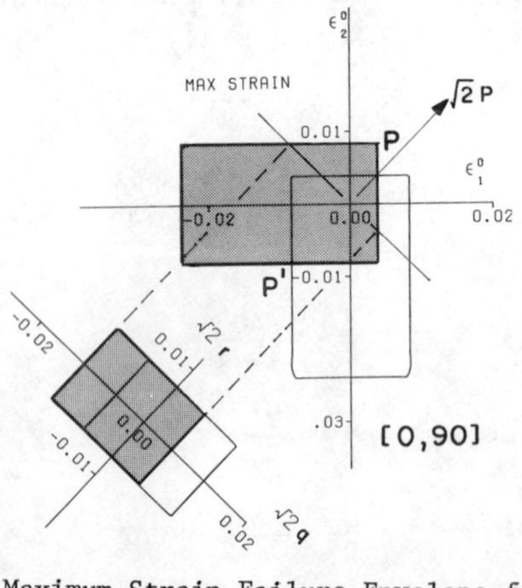

Figure 8. Maximum Strain Failure Envelope for $[0_j/90_k]$ T300/5208.

The turning over from the 0 to 90 degree envelope is actually achieved through a rigid body rotation about the p-axis in Figure 8, where $p = (\varepsilon_1 + \varepsilon_2)/2$. The hydrostatic tensile and compressive points are designated as P and P'. These two points are invariant; i.e., failure envelopes for all ply orientations must pass through these points. If we further define

$$q = (\varepsilon_1 - \varepsilon_2)/2, \; r = \varepsilon_6/2 \tag{25}$$

We go back to the classical Mohr's circle for strain components where in the q-r plane strain components vary in accordance to a rigid body rotation with twice the angular displacement from that of the reference coordinate axes. Thus failure envelopes will undergo rigid body rotation when viewed in the q-r plane. This is illustrated in Chapter 7 of Reference 1 for the quadratic failure envelope. But the same rigid body rotation in the q-r plane applies to all failure criteria.

It is straight forward to determine the failure enveloped in the principal strain plane for a plus or minus 45-degree ply. This is done in Figure 9.

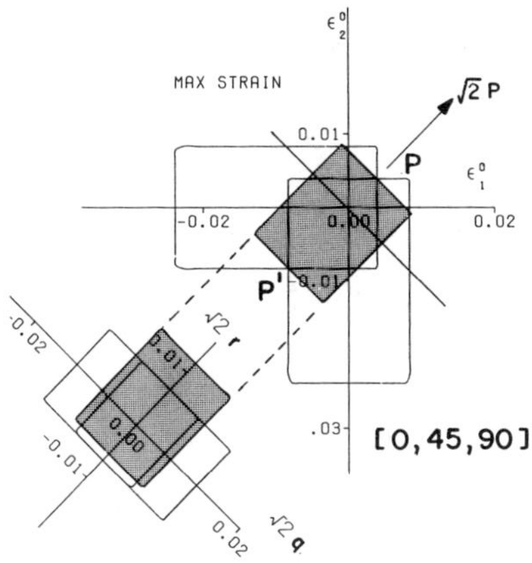

Figure 9. Maximum Strain Failure Envelope for $[0_j/90_k/45_f]$ T300/5208.

Note first the symmetry in the measured positive and negative shear strains. The failure lines through these maximum shear strains must be parallel to the p-axis. This is the result of the rigid body rotation in the q-r plane cited in the last paragraph. The other failure lines must pass through the hydrostatic points P and P' in Figure 9. This is required by the invariance of hydrostatic straining. From symmetry and the rigid body rotation consideration, the failures of the 45-degree ply must be perpendicular to the p-axis, or parallel to the q-axis. Thus the failure envelope for the 45-degree ply can be readily constructed.

The inner envelope in Figure 9 is called the first-ply failure envelope. If the shear strain is increased beyond the failure line of transverse tensile failure, the 45-degree ply will not be the limiting strain for the first-ply failure.

The low transverse tensile failure is the limiting factor in the first-ply failure envelope. If laminate strength is to be improved, this strain is more critical than the other strains in Equations (23) and (24). For this material, improvements over the longitudinal tensile have no impact on the first-ply failure envelopes.

We have studied many composite materials and found in most cases, the 0 and 90 degree plies control the first-ply failure envelopes. Other ply orientations either do not control or when they control they do not differ much from the 0 or 90 degree plies. So it is reasonable to assume that the quick construction of the failure envelopes in Figure 9 are adequate to determine the first-ply failure envelopes for all ply orientations.

For any applied in-plane stresses, we must first use the stress-strain relations in Table 5 to determine the resulting in-plane strains. The strain then must be converted to the principal components:

$$\varepsilon_I = I + R$$

$$\varepsilon_{II} = I - R \tag{26}$$

where $I = p = \frac{1}{2}(\varepsilon_1 + \varepsilon_2)$

$$R^2 = q^2 + r^2 = \frac{1}{4}[(\varepsilon_1 - \varepsilon_2)^2 + \varepsilon_6^2]$$

The principal strains can be used directly to determine if first-ply failure has occurred. These strains are functions of the applied in-plane stresses and the effective in-plane stiffness modulus. The resulting strains will change when the ply orientations and the thickness of the laminate change. This is the basic difference between the conventional and composite materials. When we design with conventional materials, thickness change is the only variable. The laminate stiffness remains constant.

For the hybrid laminate, the failure envelopes for different materials can be superposed provided the scale is the same. The laminate strains are calculated the same way, i.e., by Table 5 using the effective engineering constants of the hybrid.

SUMMARY

A simplified design procedure has been outlined to show how the ply properties can be passed onto the laminate properties. For the elastic properties, the window of this passage is through the off-axis stiffness moduli. The resulting laminate stiffness and the stress-strain relations are shown in Table 5 for balanced, symmetric laminates. Only simple arithmetics or graphical methods using Figure 6 is involved. Failure envelopes can also be readily constructed using the maximum strain criterion. The simplified procedure is straight forward and leads to design methods which are nearly identical to those for the conventional material. The simplified procedure is a stepping stone for beginners to go to more complex design of composite materials.

REFERENCES

Tsai, S. W., and H. T. Hahn (1980). Introduction to Composite Materials. Technomic Publishing Company, Lancaster, Pennsylvania, USA.

MECHANICAL PROPERTIES OF STRUCTURAL MATERIALS AT LOW TEMPERATURES WITH REFERENCE TO THE TYPE OF STRESSED STATE AND MODE OF LOADING

G. S. Pisarenko, A. A. Lebedev and V. A. Strizhalo

Institute for Problems of Strength, Academy of Sciences of the Ukrainian SSR, 2 Timiryazev Street, 252014, Kiev - 14, USSR

ABSTRACT

Considered in this paper are the results of experimental studies of the regularities in deformation and failure of structural steels and alloys at cryogenic temperatures under conditions of static and cyclic (low-cycle) loading. Special attention is devoted to the description of effects associated with the phenomenon of intermittent yielding of the material at temperatures close to the absolute zero and to the analysis of strain-hardening laws in relation to the effect of temperature and the type of stressed state. Data obtained in the recent years at the Institute for Problems of Strength by the authors are given in this paper.

KEYWORDS

Low temperatures, stress-strain diagrams, intermittent yielding, strain hardening, linear and plane stressed state, cyclic creep, loading trajectories, principal stresses, yield surfaces.

INTRODUCTION

The problems of enhancing the reliability of modern machinery strength analysis and the improvement of advanced technology require a comprehensive study of physico-mechanical properties of structural materials under varying conditions of mechanical and temperature loading. Due to extensive use of cryogenic engineering, more and more attention is recently devoted to the investigation of elasto-plastic deformation process and to strength of materials of various grades at low temperatures. The analysis and systematization of published experimental data show that the degree and the character of the cooling effect on the mechanical properties of metals and alloys are determined by their nature and structural state (Krasowsky,1980; Novikov and co-workers, 1979; Strizhalo, 1978).

In materials with fcc-lattice (austenitic steel, aluminium alloys) the yield strength changes slightly with temperature decrease, while the ultimate plasticity somewhat rises, and the character of strain hardening changes.

A characteristic feature in materials with bcc-lattice (carbon steel and many alloyed steels) is a considerable rise of yield strength and an essential drop of plasticity (embrittlement) with the decrease of temperature. The intensity of strain hardening of bcc-metals with temperature decrease stays either constant or decreases.

Metals with hcp-lattice (titanium alloys) by the temperature dependence of their mechanical properties occupy a midposition between metals with bcc- and fcc-lattices and are characterized by a marked increase of the yield strength and the intensity of the strain hardening with the temperature decrease.

In cooling down to 77K the value of uniform deformation of the above metals rises and then slightly drops.

In a number of materials with a low stability of solid solution (e.g. in chromium-nickel steels) the low-temperature deformation is accompanied by phase transition which causes qualitative changes in the nature of strain hardening.

In cryogenic machine-building non-ferrous metal alloys and chromium-nickel steels, which have no cold brittleness threshold in the entire low-temperature range, are used as structural materials. Therefore, major attention in this paper is focused on the description of regularities in deformation and failure of such alloys, as well as on the estimation of the effect of cryogenic temperatures on their strength and plasticity with the account of the mode of loading and the type of the stressed state.

EXPERIMENTAL TECHNIQUE

The size and shape of test samples were chosen depending on the test conditions and mode.

Static tests were conducted on flat specimens cut out of a sheet with the gauge length measuring 2 x 5 x 30 mm and on tubular specimens made from round billets, their outer diameter being 26 mm, with 1 mm thick wall and 80 mm gauge length.

In cyclic loading specimens were subjected to repeated tension with the loading frequency of 1...2 c/min, the maximum load being constant. Under such conditions the processes of cyclic creep develop in the metals and fracture occurs after a small number of cycles typical of the low-cycle fatigue region.

The tests at temperatures of 77 and 4.2K were conducted under conditions of linear stressed state in a special cryostat in the medium of liquid coolants: nitrogen and helium.

To conduct tests under conditions of complex stressed state, an automated test system "CHT-YBM" was developed at the Institute for Problems of Strength of the Academy of Sciences of the Ukr.SSR which included experimental set-ups providing stress-controlled (model CHT-5) and strain-controlled (model CHT-10) loading and a digital control computer. Under conditions of complex stressed state the tubular specimens were cooled by a flow-through type cooler placed into the inside of the specimen which ensured cooling of the gauge length down to any temperature within the range from 293 to 93 K. Strains were measured by electro-mechanical extensometers on the basis of 20-25 mm (Pisarenko and Lebedev, 1976).

UNIAXIAL TENSION TEST RESULTS

The temperature decrease caused an increase of the static strength in all the alloys investigated. The plasticity of titanium alloys decreases, whereas that in the aluminium and copper alloys increases, as a rule, Table 1.

TABLE 1 Mechanical Properties of the Alloys

Material	T /K/	σ_u /MPa/	δ / % /	n	$\Delta\sigma$ /MPa/
Titanium	293	405	27,4	-	-
alloy	77	858	40,0	-	-
VT1-0	4.2	990	25.7	105	110
Titanium	293	780	13.2	-	-
alloy	77	1300	14.3	-	-
VT5-1	4.2	1415	3.4	5	220
Titanium	293	840	13.3	-	-
alloy	77	1280	12.7	-	-
AT2-2	4.2	1440	9.0	10	235
Titanium	293	990	11.7	-	-
alloy	77	1545	11.7	-	-
VT6S	4.2	1590	2.7	2	250
Aluminium	293	352	9.7	-	-
alloy	77	465	9.7	-	-
1201	4.2	553	15.8	75	40
Copper	293	231	34.0	-	-
alloy	77	380	44.0	-	-
BrKh08	4.2	468	55.5	250	10

With the temperature decrease to 4.2K a change in the character of metal deformation is observed, this being quite well illustrated by data given in Fig. 1. Thus, at temperatures 293K and 77K in the region of plastic deformation the stress-strain curves are smooth and plastic deformations are distributed practically uniformly throughout the macrovolume of the specimen, while at 4.2K strain accumulation in the process of the stress increase occurs due to intermittent yielding of the material in the local zones and it is of a discrete character. Moreover, with the increase of alloy strength the jumps in the stress-strain curves become higher, and their number decreases (Fig. 1, and Table 1).

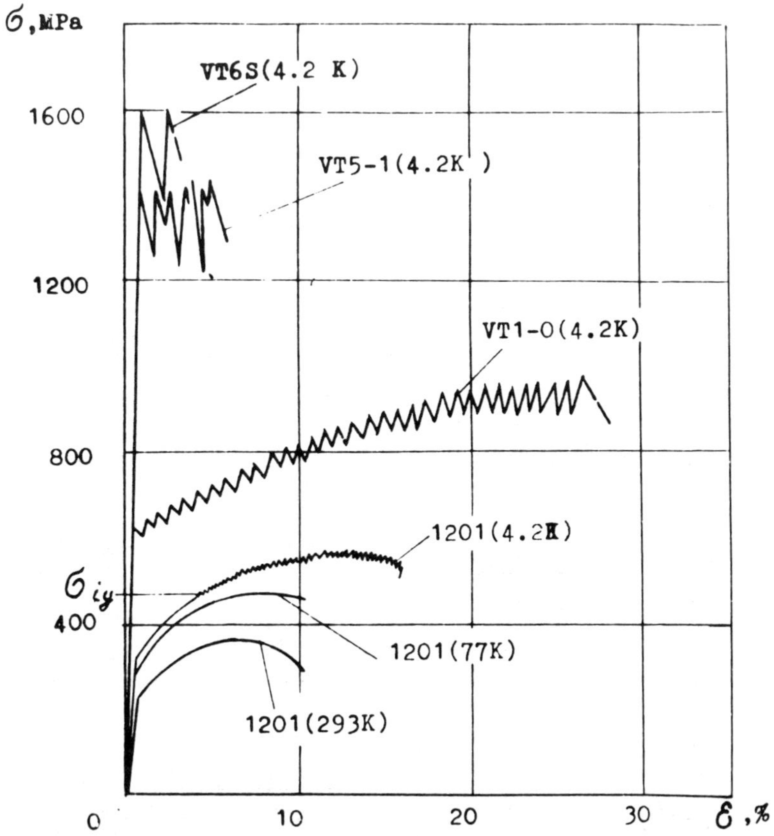

Fig. 1. Stress-strain curves for the alloys.

It should also be noted that the stress-strain curves of less strong aluminium and copper alloys at the temperature of 4.2K exhibit sections of both intermittent and monotonic deformation. The boundary between these sections is determined by stress σ_{iy}.

The cyclic strength of the alloys investigated also increases throughout the entire low-cycle fatigue region with the temperature decrease, Fig. 2, and at the liquid helium temperature the accumulation of plastic strains in the process of low - cycle loading occurs, as in the case of the static loading, due to intermittent yielding of the material, Fig. 3

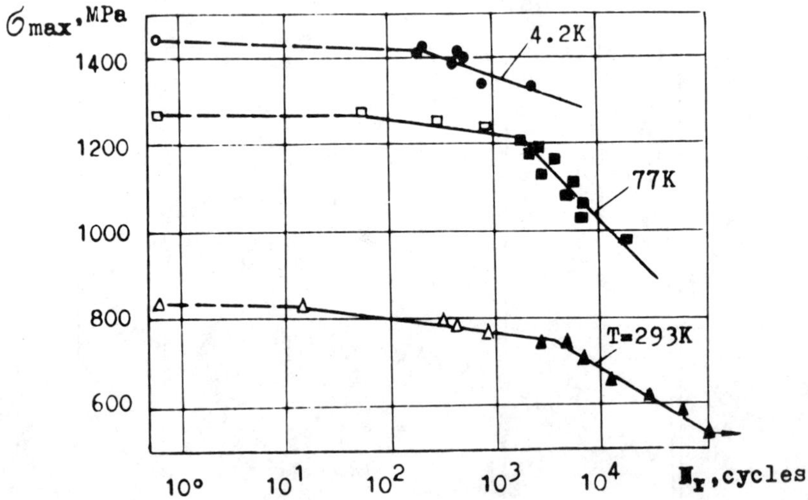

Fig. 2. Low-cycle fatigue curves for AT2-2 titanium alloy.

Thus, no qualitative changes are observed in the processes of non-ferrous alloys deformation as the temperature decreases from 293 to 77 K under conditions of linear stressed state both at static and cyclic loading. Such changes take place only during the transition to the region of quite low temperatures close to the absolute zero, they being associated with the localization of plastic deformations zones.

Therefore, we shall dwell on the phenomenon of the intermittent yielding the realization of which in the materials investigated at liquid helium temperature is indicative of the qualitative changes in the deformation mechanism. A series of approaches are known which elucidate this phenomenon. In a general case specific features of the intermittent yielding depend on the temperature level, loading conditions and the rigidity of the test machine. It is related to deformation twinning, martensite transformation, to the formation of rough slip bands and dynamic strain aging. But the most general explanation of this phenomenon can be given from the standpoint of the adiabatic strain according to which strain jumps occur due to a sharp temperature rise in local zones of the specimen during its loading. The lower the heat capacity and heat conduction of the material, the higher the temperature jumps (at temperatures close to liquid helium ones the heat capacity and heat conduction of metals drop as compared to normal conditions).

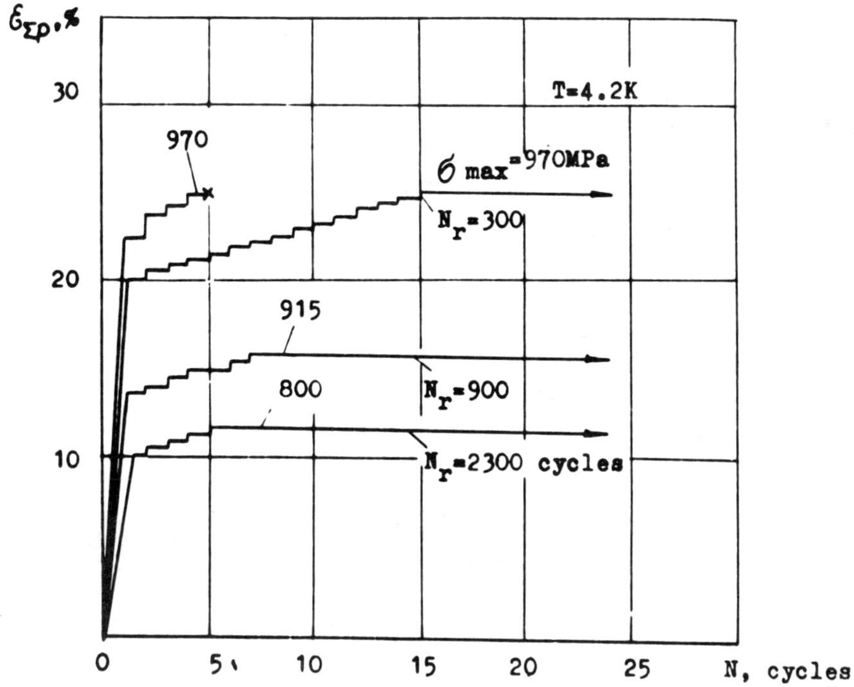

Fig. 3. Initial portions of cyclic creep curves for VT1-0 titanium alloy at 4.2K.

With the rise of temperature in a local volume (in the process of conventional slipping at external loading) the yield stress of the material within this volume decreases and rapid deformation begins. As shown in Fig. 1, each subsequent act of the intermittent yielding during the process of static loading is characterized by a greater magnitude of the stress jump, $\Delta\sigma$ (the maximum values of $\Delta\sigma$ for the material are listed in Table 1). Each subsequent deformation act involves new sections of the specimen until the ultimate strength is reached. Our investigations have shown that the total number of the intermittent yielding acts and the magnitude of stress jumps at static loading essentially depend on the alloys strength and plasticity, Fig. 1, Table 1. Thus, for a comparatively low-strength and high-plasticity copper BrKh08 the number of stress jumps at low temperatures can reach 250, and for a high-strength but low-plasticity titanium alloy VT6S it is only 2; naturally, the magnitude of the stress jump in a higher-strength alloy is also higher.

At cyclic loading and at the temperature of 4.2K the strain accumulated to rupture is realized, mainly, in the first cycle with the stress increase from 0 to σ_{max} and partially in several subsequent loading cycles. In cyclic loading the number of strain jumps drops sharply as compared with the static one; and for the alloy VT1-0, for example, it is not more than 20, Fig. 3. Here, a fewer number of jumps corresponds to a lower stress in a cycle. The intermittent yielding occurs in this case primarily at the initial stage of the test and is practically not observed during further loading.

Quantitative estimation and comparison of the effect of low temperatures on mechanical characteristics of alloys can be carried out by means of temperature effect coefficients $K_{\sigma_u} = \frac{\sigma_{u2} - \sigma_{u1}}{\sigma_{u1}}$ and $K_\delta = \frac{\delta_2 - \delta_1}{\delta_1}$ (here σ_u is the ultimate strength; δ is the residual elongation, subscript 1 denotes the temperature of 293 K, subscript 2 corresponds to low temperatures). The values of these coefficients for the alloys investigated are given in Table 2. Thus, for the alloy VT1-0 these coefficients at the temperature of 4.2K are + 1.44 and -0.06, respectively (sign "+" indicates the increase in the value of the characteristic and sign "-" shows its decrease). It signifies that in low-temperature region the ultimate strength increases by 144% and the residual elongation drops by 6% as compared to the room temperature conditions.

TABLE 2 Temperature Effect Coefficients for the Alloys Investigated

Material	T,K	K_{σ_u}	K_δ
Titanium alloy	77	1.12	0.46
VT1-0	4.2	1.44	-0.06
Titanium alloy	77	0.66	0.08
VT5-1	4.2	0.81	-0.74
Titanium alloy	77	0.52	-0.04
AT2-2	4.2	0.71	-0.32
Titanium alloy	77	0.56	0
VT6S	4.2	0.61	-0.77
Copper alloy	77	0.32	0
BrKh08	4.2	0.57	0.63
Aluminium alloy	77	0.64	0.26
1201	4.2	1.03	0.63

The comparison of data listed in Table 2 shows that with the decrease in temperature to 4.2K the strength of all the alloys increases, while the plasticity of the titanium alloy decreases and that of the aluminium and copper alloys increases.

COMPLEX STRESSED-STATE TEST RESULTS

The analysis of the experimental data obtained for the range of temperatures from 293 to 93K in testing tubular specimens by the test system CHT-YBM has shown that the general mechanisms of the low-temperature deformation of materials sufficiently plastic both at room and at low temperatures remain, under conditions of complex stressed state, qualitatively the same as at room temperature. The effect of the type of the stressed state is strongly pronounced in materials which lose in their plastic properties with temperature decrease.

For the isotropic materials of this group (copper, chromium-nickel steels of stable structure) the onset of the yielding at low temperatures is described to satisfaction by the von Mises criterion; under conditions of simple (proportional) loading the hypothesis of the similarity of stress and strain diviators is satisfied, as well as the hypothesis of the invariance of the generalized curve to the type of the stressed state in the coordinates "stress intensity σ_i - strain intensity ε_i", which signifies that the theory of small elasto-plastic deformations at low temperatures is applicable to the class of materials considered.

Investigations conducted on tubular specimens from aluminium alloys (D16, D16T, AMg3, AMg6, AMcS) and titanium alloys (VT5-1kt, VT6) have shown that their elasticity, strength and plasticity characteristics are anisotropic. It is interesting to note that the anisotropy of alloys decreases with temperature reduction. The generalized stress-strain curves, $\sigma_i(\varepsilon_i)$, corresponding to various ratios of principal stresses do not coincide either at room temperature or at low temperatures.

With the aim of studying the vectorial properties of anisotropic materials, the loading and deformation trajectories of alloys were investigated in the planes $S_1\ S_2$ and $Э_1^p\ Э_2^p$ of five-dimensional deviatoric spaces $S_{(s)}$ and $Э^p_{(s)}$ /Iljushin, 1963/. In the case of the tubular specimens loaded with axial force and internal pressure

$$S_1 = \sqrt{\frac{2}{3}}\left(\sigma_z - \frac{1}{2}\sigma_\theta\right), \quad Э_1^p = \sqrt{\frac{3}{2}}\,\varepsilon_z^p,$$

$$S_2 = \frac{\sqrt{2}}{2}\cdot\sigma_\theta, \quad Э_2^p = \frac{\sqrt{2}}{2}\left(\varepsilon_\theta^p - \varepsilon_r^p\right), \qquad (1)$$

where S_1, S_2 are the equivalent stresses, $Э_1^p$, $Э_2^p$ are the equivalent strains, σ_z, σ_θ are axial and tangential stresses, ε_z, ε_θ, ε_r are axial, tangential and radial deformations of the specimen.

It was found that in the region of small deformations of initially anisotropic media, a simple deformation process corresponds to a simple loading process.But the stress and strain deviators are not proportional. The angle between the loading and deformation trajectories depends on the orientation of the loading trajectory in the space of stresses.

A great number of investigations was carried out aimed at experimental studies of the fundamental laws of the yielding theory, in particular, of the law of hardening which describes the evolution of the yield surface in the process of the plastic deformation.

The analysis of the experimental data has shown that in ray-trajectory loading the transformation of the yield surface of alloys in the plane S_1 S_2 corresponds to the law of the isotropic-kinematic hardening.

$$f = h\sqrt{\frac{(S_1-a_1)^2}{r_1^2}+\frac{(S_2-a_2)^2}{r_2^2}} + (1-h)\left[C_1(S_1-a_1)+C_2(S_2-a_2)\right] - \rho = 0 \qquad (2)$$

where a_j and ρ are the kinematic and isotropic components of the hardening, respectively; $h < 1$ is a parameter; r_1, r_2, C_1 and C_2 are the yield stress functions determined in the direction of the main axis of the anisotropy.

The processing of the experimental data by the method of least squares has shown the coefficient of isotropic expansion is linearly dependent on the plastic strain value and is independent of the preloading trajectory orientation in the space of stresses. The translation vector $\vec{a}$ is collinear with the stress vector $\vec{S}^{(k)}$ corresponding to the preloading, and the dependence of its modulus upon the plastic strain level with the account of the preloading trajectory orientation can be well described by the power function

$$a^{(k)} = A^{(\infty)} \frac{S_T^{(k)}}{S_T^{(\infty)}} (Э^P)^n ,$$

where $S_T^{(\infty)}$ and $S_T^{(k)}$ are the yield stress at uniaxial tension ($K = \infty$) and at the ratio of the principal stresses K; $A^{(\infty)}$, n are the constants determined at the uniaxial tension.

The vectors of plastic strain increment in repeated simple loading practically coincide with the directions of the normals to the subsequent yield surfaces described by function (2) both at room and at low temperatures, Fig. 4.

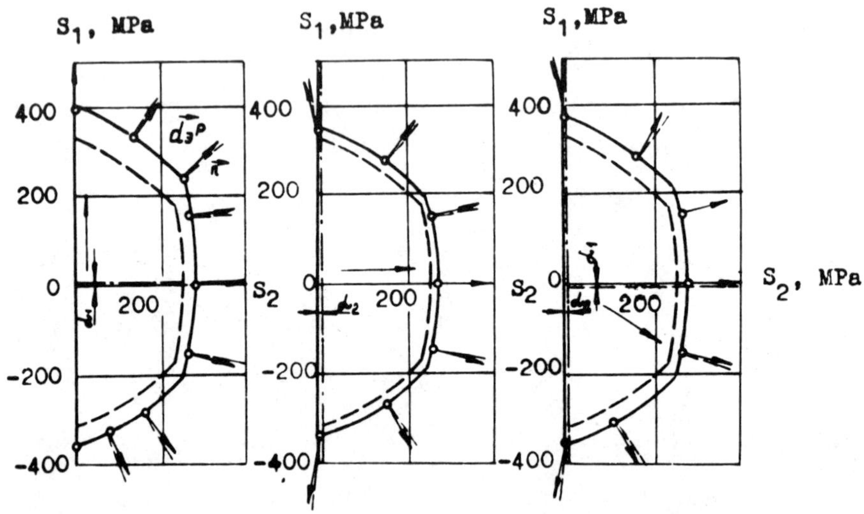

Fig. 4. Yield surfaces of D16T aluminium alloy after preloading with different trajectories.

An essentially important result of these investigations is the establishing of the invariance of parameters of the subsequent yield surfaces to the temperature of prestraining for a wide range of embrittling materials and for the medium-carbon steels, in particular. This permits a conclusion that under conditions of non-isothermic loading the hardening of the above materials does not depend on the thermal loading history but is determined by the plastic strain level and the current temperature value. The dependence of the parameter of subsequent yield surfaces upon the temperature of the repeated deformation is given in Fig. 5 from which it can be seen that the value of isotropic hardening component (surface radius R) increases considerably with temperature reduction, while the value of isotropic component characterized by the surface shift within a wide temperature range does not change. A noticeable decrease is observed only at $T \geqslant 673K$ caused by the effects associated with creep and structural changes in the material as a result of recovery and recrystallization, and at $T \leqslant 77K$ when, with the approach to the visco-brittle transition region, a change in the surface shape is observed.

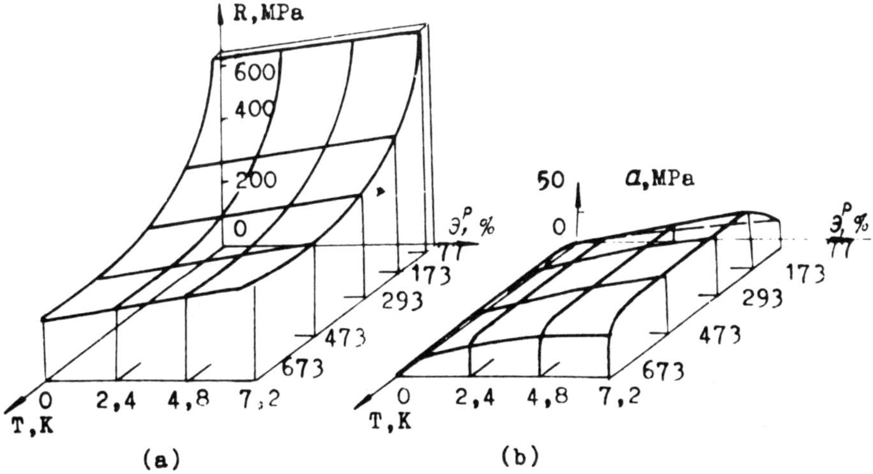

Fig. 5. Temperature and plastic deformation level dependence of the radius /a/ and of the magnitude of subsequent yield surfaces shift /b/ for steel 45.

Thus, the hardening of the materials considered under conditions of non-isothermic loading can be described by means of two thermomechanical hardening surfaces $\Phi_1(a, Э^P, T)=0$ and $\Phi_2(R, Э^P, T)=0$, Fig. 5, determined under isothermic conditions.

REFERENCES

Iljushin A.A. (1963). Fundamentals of general mathematical theory. Izd.Akad.Nauk SSSR, Moscow.p.271.

Krasowsky, A.J. (1980). Brittleness of materials at low temperatures. Naukova Dumka, Kiev. p.340.

Novikov, N.V., A.A.Maistrenko, and A.P.Uljanenko (1979). Structural strength at low temperatures. Naukova Dumka, Kiev. p.232.

Pisarenko, G.S. and A.A.Lebedev (1976). Deformation and strength of materials in a complex stressed state. Naukova Dumka, Kiev. p.416.

Strizhalo, V.A.(1978). Cyclic strength and creep of metals under low-cycle loading at low and high temperatures. Naukova Dumka, Kiev. p.238.

FACTORS AFFECTING PLASTIC INSTABILITY AND SHEET FORMABILITY

S. S. Hecker

Materials Science and Technology Division, Los Alamos National Laboratory, Los Alamos, NM 87545, USA

ABSTRACT

The strong influence of geometry and materials properties on plastic instability and sheet formability is illustrated with several experimental observations. Geometry (shape) of the specimen or work piece and the evolution of shape with deformation is of great importance. Experiments on sheet and thin-walled tubes have demonstrated convincingly that sheets stretched over a punch are more stable than sheets stretched in-plane; which, in turn, are more stable than expanded thin-walled tubes. All can be explained by the specific interaction of specimen shape with loads and deformations. The materials parameters of greatest importance (Hecker, 1978a) are strain hardening, strain rate sensitivity, and plastic anisotropy. Several experiments are cited that demonstrate the importance of stress state, large strains, and path changes on the strain hardening response and on subsequent stability.

KEYWORDS

Plastic instability, sheet metal forming, strain hardening, punch stretching, multiaxial stresses, large strains.

INTRODUCTION

The prediction of the onset and development of plastic instability is a most difficult and important problem. It requires the proper treatment of large geometrical changes and an accurate description of strain hardening at large strains and under changing deformation paths. The problem of plastic instability in thin sheets and its relation to sheet formability has received much recent attention (see the proceedings edited by Hecker, Gosh and Gegel, 1978; Koistinen and Wang, 1978). The important role of geometry and materials properties on plastic instability is being recognized. In this paper, some of the unusual experimental observations in the sheet forming literature will be reviewed and explained on the basis of the effects of geometry and material properties on instability.

THE INFLUENCE OF GEOMETRY

Ghosh and Hecker (1975) demonstrated that the increased useful deformation (as defined by a forming limit curve) observed in punch (out-of-plane) stretching compared to in-plane stretching of sheet results from differences in deformation geometry. Figure 1a illustrates the different test techniques used by Hecker (1978a) for the results on commercially pure

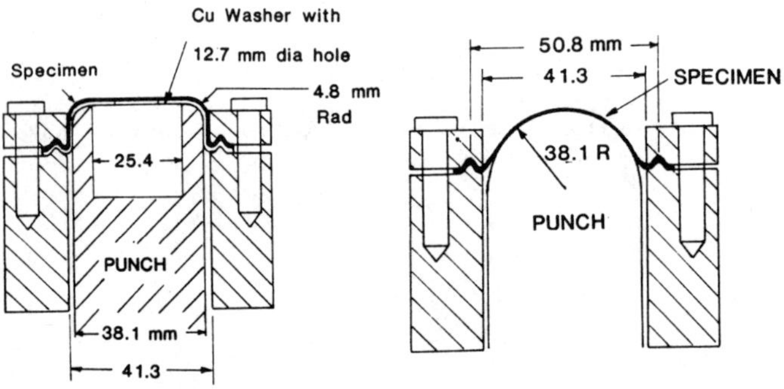

Fig. 1a. Schematic diagrams of punch apparatuses used for in-plane and punch-stretch experiments

annealed aluminum (Fig. 1b). For in-plane stretching, deformation is uniform and local necking should not occur in biaxial tension unless local inhomogeneities are admitted (Marciniak and Kuczynski, 1967; Marciniak, Kuczynski, and Pokora, 1973) or constitutive descriptions other than J_2 flow theory are used (Stören and Rice, 1975; Hutchinson and Neale, 1973; Tvergaard, 1978). Specifically, it has been demonstrated that deformation

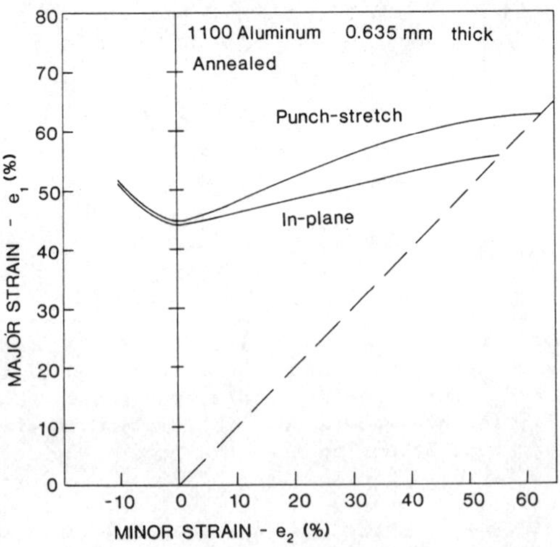

Fig. 1b. Forming limit curves for 1100 aluminum. A comparison of in-plane and punch stretching.

(in contrast to flow) theories of plasticity predict localized making in biaxial stretching at reasonable stress levels. Bifurcation analyses by Stören and Rice (1975) using J_2 deformation theory with a yield locus vertex and by Tvergaard (1978) using kinematic hardening and small initial inhomogeneities predict realistic critical strain levels at instability. In punch stretching, no restriction to local instability exists because a strain gradient develops immediately upon punch to sheet contact. In spite of this strain gradient, stability during punch stretching is prolonged because the geometry of the ever-expanding cup delays the attainment of localized plane strain required for local necking.

The geometry differences can also cause a significantly different sensitivity to some materials properties. In punch stretching, the deformation is always nonuniform and the location of the neck is, in large part, dictated by geometry. Hence, second-phase inclusions or particles that may trigger early fracture are detrimental only if they exist in the narrow band where necking eventually occurs. For in-plane stretching, the entire sheet is stretched uniformly and, hence, inclusinons in any portion of the sheet may trigger early fracture. Therefore, the in-plane techniques show a much greater sensitivity to imperfections (including material cleanliness) than punch stretching techniques or what is normally found in metal stampings.

The profound influence of geometry on plastic instability was even more convincingly demonstrated by Stout and Hecker (1983) in the comparison of sheet to thin-walled tubes. The onset of diffuse instability (maximum uniform deformation) was found to be very sensitive to geometry as demonstrated in Fig. 2. In biaxial, combined tension-internal pressure experiments, we found that the uniform strain in tubes for axial plane strain

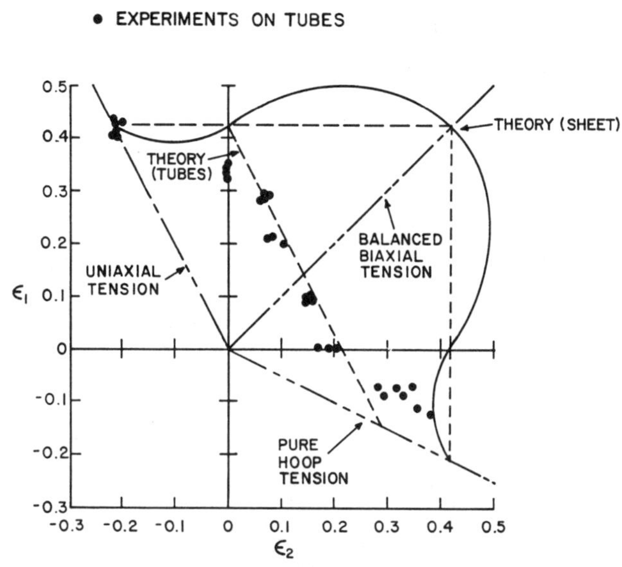

Fig. 2. A comparison diffuse instability strains for the 70-30 brass tubes and sheet. ε_1 and ε_2 refer to major and minor time strains, respectively

(no length change) is twice that for hoop plane strain (no diameter change). In the comparison of tubes and sheet, we found that the uniform strain in balanced biaxial tension in tubes is only one-third of that in sheet. The strain levels for local necking and fracture were similar for tubes and sheet (no geometry effect). However, the useful deformation beyond diffuse instability for tubes is very limited because localization occurs rapidly. In punch-stretched sheet the geometry is stable and localization occurs gradually, providing substantial post-uniform deformation.

PLANE STRAIN BEHAVIOR

The resistance to plastic instability during stretching of α-brass is considerably less than what one expects on the basis of its uniaxial tensile behavior (Ghosh, 1978; Wagoner, 1982). This is particularly true for plane-strain deformation. It is well recognized that brass typically exhibits zero strain rate sensitivity and, hence, unlike low-carbon steels, its stability does not benefit from a contribution of positive strain rate sensitivity (Hecker, 1978a). Ghosh (1978) and Wagoner (1982) explained the brass instability behavior on the basis of a stress-state dependence of work hardening. A limited number of experiments in plane-strain tension suggested a lower work hardening rate than in uniaxial tension. Hecker and Stout (1983) performed extensive stress-state tests on α-brass tubes and also found a definite stress state effect. As observed by Ghosh and Wagoner, plane-strain hardening is low; however, balanced biaxial hardening is similar to that in uniaxial tension (see Fig. 3).

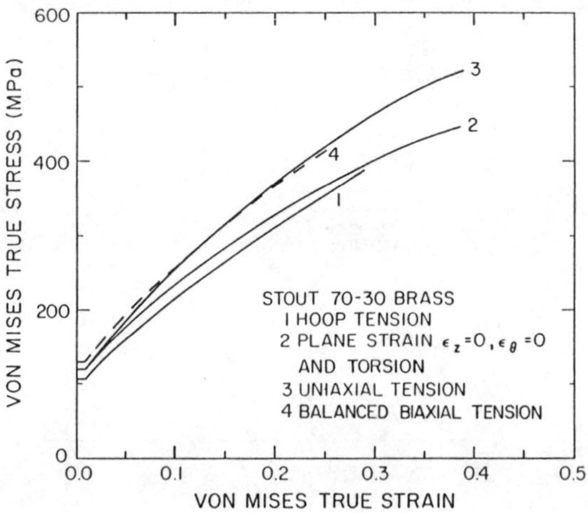

Fig. 3. Comparison of stress-strain curves for 70-30 brass thin-walled tubes (Hecker and Stout, 1983). Curve #2 represents the results for three different stress states; torsion, plane strain with no length change ($\varepsilon_z = 0$), and plane strain with no diameter change ($\varepsilon_\theta = 0$). Curve #1 represents uniaxial hoop tension.

The thin-walled tube tests on brass are limited to strains <0.4 because of the onset of plastic instability. As mentioned above, sheets are able to sustain larger useful deformation and, hence, it would be beneficial to assess strain hardening at larger strains. This is a most difficult task because of the large geometric changes that occur and the tendency for instabilities and other deformation inhomogeneities. Hecker and Stout (1983) compared the strain hardening in axisymmetric compression to torsion at strains up to 3.0. The results shown in Fig. 4 clearly demonstrate the lower work hardening rate (when compared on the basis of a von Mises criterion) in plane strain torsion compared to axisymmetric compression. The hardening response in tension up to instability is similar to that in compression. Hecker and Stout demonstrated that in brass most of the difference in hardening can be explained on the basis of textural evolution. A review of several other materials including other modes of deformation indicates that the lower hardening in plane strain deformation modes appears to be quite common (Hecker and Stout, 1983). The implications of this observation to sheet forming is obviously of great importance.

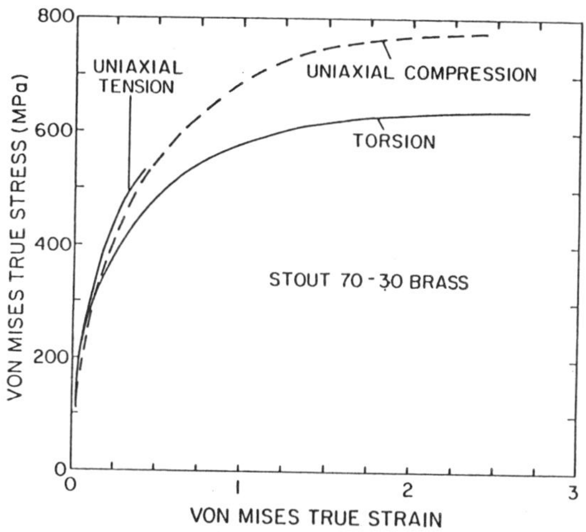

Fig. 4. Comparison of stress-strain curves for 70-30 brass for uniaxial tension, compression and torsion (Hecker and Stout, 1983). Tension and torsion were carried out on identical thin walled tubes. Compression was carried out on solid rod, remachined often to avoid barreling.

We should point out that the most thorough study comparing torsion and axisymmetric deformation was carried out by Jonas and co-workers. (1982) on copper. In conjunction with a new theoretical development on texture evolution (Canova, Kocks, Jonas, 1982), they found that texture explained approximately half of the hardening difference. The rest must be a result of differences in slip behavior for different stress states.

LARGE STRAIN BEHAVIOR

Comparison of forming limit curves of low-carbon steels with aluminum alloys shows that steels are much more resistant to local plastic instability (Hecker, 1978a). Much of this difference can be attributed to the positive strain rate sensitivity of steel compared to either the zero or negative sensitivity of many aluminum alloys. However, another important factor is the strain hardening at large plastic strains, beyond those typically attained in uniaxial tensile tests. These strain levels are important because in punch stretching (or in many metal stamping operations) much useful deformation can be obtained beyond the maximum uniform strain in tension. However, it is extremely difficult to measure strain hardening accurately at large strains. Hecker and Stout (1983) have measured strain hardening at large strains for a few materials using compression, torsion, and rolling followed by tension. From their work and a review of the literature they concluded that hardening persists at large strain for most materials. Several factors, however, favor a saturation in hardening. The most prominent are plane strain deformation modes and high purity (low solute content). Also, other microstructural features may favor inhomogeneities such as shear banding and also result in saturation.

The importance of strain hardening at large strains was demonstrated by Bird and Duncan (1981). They compared the forming limit curves (FLC) of 1008 steel, 1100 (commercially-pure) Al and 2036-T4 Al (the experimental curves as determined by Hecker (1978a) are shown in Fig. 5). The strain hardening

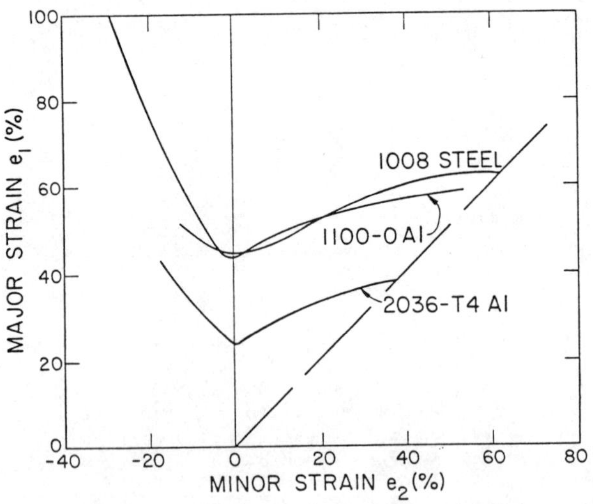

Fig. 5. Forming limit curves of three materials from Hecker (1978a, 1978b).

exponents (n-values from $\sigma = K\varepsilon^n$, where σ is the flow stress, ε the plastic strain, and K a strength constant) in uniaxial tension were between 0.21 and 0.26 for all three materials. The low FLC for 2036-T4 Al was explained by Hecker on the basis of its negative strain rate sensitivity (m-value from $\sigma = k'\dot{\varepsilon}^m$ of -0.005). However, steel has a positive m value of 0.012, whereas 1100 Al is rate insensitive, which suggests that the FCL for steel should be higher. Bird and Duncan (1981) showed that strain hardening at large strains provides the explanation. Figure 6 shows the strain hardening data derived by Bird and Duncan from hydraulic bulge tests on sheet. Strain hardening is represented by the normalized hardening slope, $\frac{1}{\sigma}\frac{d\sigma}{d\varepsilon}$. In uniaxial tension the maximum uniform strain attainable occurs when $\frac{1}{\sigma}\frac{d\sigma}{d\varepsilon} \cong 1$. At strains up to this level the hardening rates are similar. However, at larger strains the hardening rates begin to differ dramatically. The high hardening rate of 1100 Al apparently offsets the zero strain rate sensitivity when comparing the FLCs for 1100 Al and steel. Hardening for 2036-T4 Al is the lowest, contributing further to its poor FLC behavior.

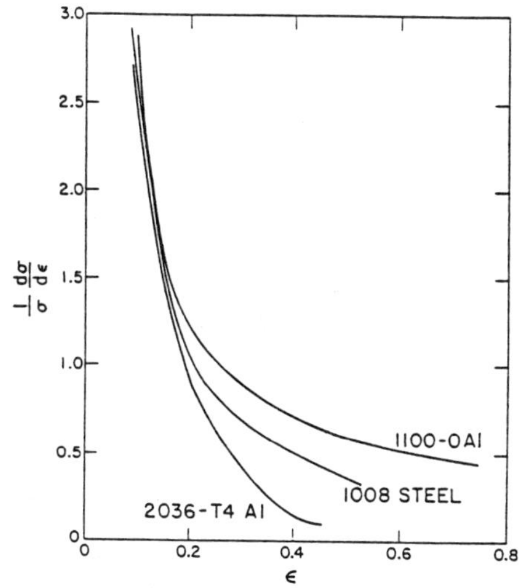

Fig. 6. Hardening rate normalized by stress as a function of strain for hydraulic bulge tests from Bird and Duncan (1981).

STRAIN PATH CHANGES

The importance of strain path changes on plastic instability and the forming limit curve have been discussed by several authors (Ghosh and Laukonis, 1976; Kleemola and Pelkkikangas, 1979). The effect on stability arises from transients in hardening and geometrical considerations. Most path changes result in some softening (and hence destabilizing) transients because of the generalized Bauschinger effect. Hardening or softening beyond the initial transient depends on the stress state influence. For instance, a change in path towards plane strain from axisymmetric deformation will most likely result in a decreased strain hardening rate (Kocks, 1983).

The combined effect of hardening transients and geometry can be illustrated very effectively by examining the ductility or (formability) of heavily cold worked metals. Hecker (1978b) demonstrated the effects of prior cold work (by rolling) on the FLCs of 1100 aluminum. The results (Fig. 7) show that increasing prior cold work drastically reduces subsequent uniaxial and plane-strain ductility, with little effect on balanced biaxial ductility.

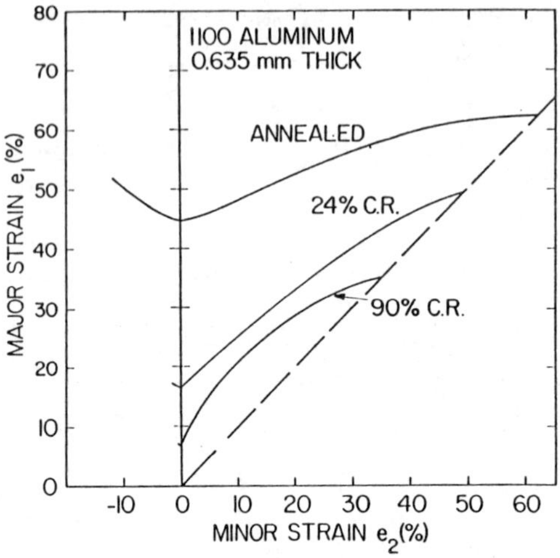

Fig. 7. Failure limit curves of 1100 aluminum for annealed and cold-rolled conditions. Failure is defined by local necking. Sheet thickness is 0.635 mm (Hecker, 1978b).

Experiments by Hecker and Stout (1983) established the large-strain hardening response of 1100 aluminum. The stress-strain curve determined from rolling followed by tension experiments exhibits continued hardening to very large strains. In Fig. 8 we see that after a prestrain of 90 percent ($\varepsilon = 2.3$),

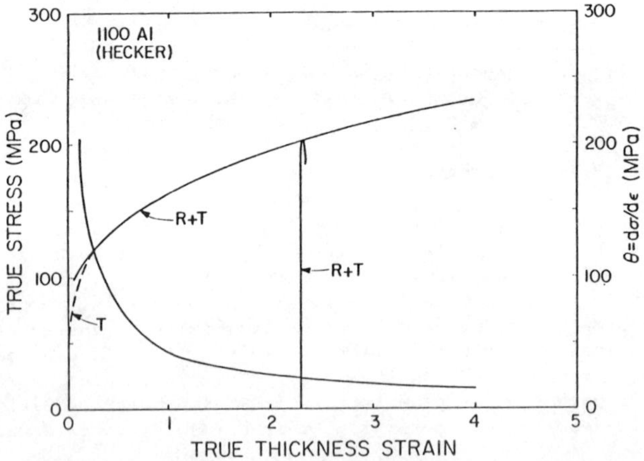

Fig. 8. Stress-strain curve and hardening rate (θ) for 1100 aluminum (Hecker and Stout, 1983). Dashed curve labelled T represents uniaxial tension on annealed material. The curve at true thickness strain of 2.3 represents tension following a prestrain of 2.3.

aluminum is still quite capable of hardening. However, after a prestrain of 90 percent a specimen pulled in uniaxial tension becomes unstable almost immediately as shown in Fig. 8. The lack of stability in tension is easily understood. For rate-insensitive material, tensile instability occurs when $d\sigma/d\varepsilon = \sigma$. Figure 8 shows the curve for σ and $d\sigma/d\varepsilon$ as a function of strain. They cross at a point where tensile instability is observed in an annealed specimen ($\varepsilon \cong 0.25$). During rolling the flow stress is increased much beyond the ultimate strength in uniaxial tension. However, the intrinsic hardening rate ($d\sigma/d\varepsilon$) decreases. For the prestrain of 2.3 chosen in Fig. 8, the flow stress is obviously much greater than the level that could be supported by the intrinsic hardening rate. Therefore, the tensile specimen should go unstable immediately upon yielding. In actuality, some stability is observed because of the generalized Bauschinger effect. Yielding occurs before the prestrain stress level is reached and the hardening rate is temporarily increased to help support the stress level. As demonstrated in Fig. 9, this effect is transient and leads to limited ductility.

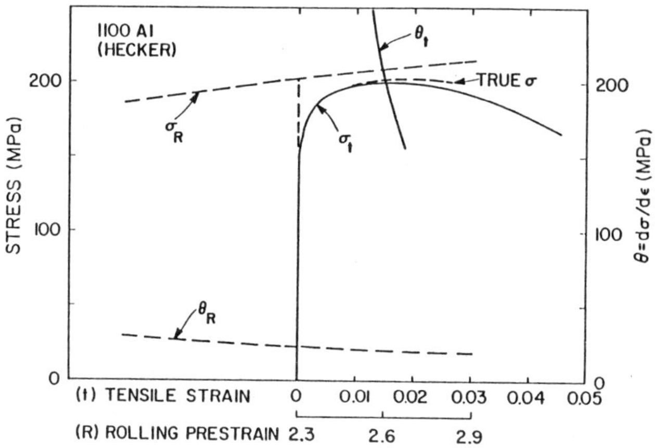

Fig. 9. Enlargement of Fig. 8 near the prestrain level of 2.3. Subscript R refers to rolling curve and t to subsequent tension. The solid σ_t curve represents the engineering stress-strain curve. True stress-strain behavior is labelled True σ.

In biaxial stretching, particularly in punch stretching, deformation is much more stable geometrically and, hence, the material is able to take better advantage of its intrinsic hardening and exhibit much greater ductility (see Fig. 7).

In summary, the importance of geometry and strain hardening (especially the influence of stress state, strain and path changes) on plastic instability have been demonstrated with the aid of several experimental observations on sheet and thin-walled tubes.

ACKNOWLEDGEMENT

The support of the Division of Materials Sciences, Office of Basic Energy Sciences, US Department of Energy is gratefully acknowledged.

REFERENCES

Bird, J. E., and J. L. Duncan (1981). Strain Hardening at High Strains in Aluminum Alloys and its Effect on Strain Localization. Met. Trans., 12A, 235-241.

Canova, G. R., U. F. Kocks, and J. J. Jonas (1982). To be published.

Ghosh, A. K., S, S. Hecker (1975). Failure in Thin Sheets Stretched Over Rigid Punches. Met. Trans., 6A, 1065-1074.

Ghosh, A. K., J. V. Laukonis (1976). Proceedings International Conference on Mechanical Behavior, Boston, p. 278.

Gosh, A. K. (1978). Plastic Flow Properties in Relation to Localized Necking in Sheets. (Eds.) D. P. Koistinen and N.-M. Wang, Mechanics of Sheet Metal Forming, Material Behavior and Deformation Analysis. Plenum Press, New York, p.287-312

Hecker, S. S. (1978a). Experimental Studies of Sheet Stretchability. Formability: Analysis, Modeling, and Experimentation, Eds. S. S. Hecker, A. K. Gosh, and H. L. Gegel. The Metallurgical Society of AIME, New York. pp. 150-182.

Hecker, S. S. (1978b). Sheet Stretching Experiments. Applications of Numerical Methods to Forming Processes, Eds. H. Armen and R. F. Jones, Jr., American Society of Mechanical Engineers, AMD-Vol. 28, New York, pp. 85-94.

Hecker, S. S., A. K, Gosh and H. L. Gegel Eds. (1978), Formability: Analysis, Modeling, and Experimentation, The Metallurgical Society of AIME, New York.

Hecker, S. S. and Stout, M. G. (1983). To appear in Proceedings of ASM Seminar on Deformation, Processing, and Structure, Ed. G. Krauss, American Society for Metals, Metals Park, Ohio.

Hutchinson, J. W. and Neale, K. W. (1978). Sheet Necking - II Time Independent Behavior. (Eds.) D. P. Koistinen, and N.-M. Wang, Mechanics of Sheet Metal Forming, Material Behavior and Deformation Analysis. Plenum Press, New York, p. 287.

Jonas, J. J., Canova, G. R., Shrivastava, S. C., and Christodoulou, N. (1982). (Eds.) E. H. Lee, and R. L. Mallett, Plasticity of Metals at Finite Strain: Theory, Computation and Experiment, Stanford University, Palo Alto, Ca. pp. 206-229.

Kocks, U. F. (1983). Private discussion.

Koistinen, D. P. and N.-M Wang (1978). Mechanics of Sheet Metal Forming, Material Behavior and Deformation Analysis, Plenum Press, New York.

Marciniak, Z. and K. Kuczynski (1967). Limit Strains in the Processes of Stretch-Forming Sheet Metal. Int. J. Mech. Sci., 9, 609.

Marciniak, Z. and K. Kuczynski and J.Pokora (1973). Influence of the Plastic Properties of a Material on the Forming Limit Diagram for Sheet Metal in Tension. Int. J. Mech. Sci., 15, 789.

Stören, S. and J. R. Rice (1975). Localized Necking in Thin Sheets. J. Mech. Phys. Solids, 23, 421.

Stout, M. G. and S. S. Hecker (1983). Role of Geometry in Plastic Instability and Fracture of Tubes and Sheet. to appear in Mech. of Materials.

Tvergaard, V. (1978). Effect of Kinematic Hardening on Localized Necking in Biaxially Stretched Sheets. Int. J. Mech. Sci., 20, 651.

Wagoner, R. H. (1982). Plastic Behavior of 70/30 Brass Sheet. Met. Trans., 13A, 1481-1500.

THEME I

HIGH TEMPERATURE DEFORMATION AND FAILURE

CONSTITUTIVE LAWS OF 316 STAINLESS STEEL IN HIGH TEMPERATURE CREEP BY BIAXIAL EXPERIMENTS

P. Delobelle, C. Oytana and A. Mermet

Laboratoire de Mécanique Appliquée LA 040004, Faculté des Sciences et des Techniques de Besançon, 25030 Besancon Cedex, France

ABSTRACT

The creep strain of a 316 low carbon stainless steel is examined through tensile and traction torsion tests. When time dependant strain is considered biaxial tests show that a single internal variable takes a part in the plastic state equation : internal stress (kinematical hardening). The coefficients of this equation are determined. It is observed that hardening is temperature dependant but depends very little on stress while recovery varies according to a power law, with $n_r = 8$.

KEYWORDS

Creep, plasticity, biaxial experiments, dip test, internal stresses, flow rules, 316 stainless steel.

NOMENCLATURE

Tensile creep test :

σ : tensile creep stress ; ε : strain, $\dot{\varepsilon}$ strain rate ; σ_i internal stress

Biaxial traction-torsion test :

$\sigma_{ij},\sigma_{zz},\sigma_{z\vartheta}$ stress components

$\varepsilon_{ij},\varepsilon_{zz},\varepsilon_{z\vartheta}$ corresponding strain components

$\dot{\varepsilon}_{ij},\dot{\varepsilon}_{zz},\dot{\varepsilon}_{z\vartheta}$ strain rate components

$\alpha_{ij},\alpha_{zz},\alpha_{z\vartheta}$ internal stress components

$\sigma'_{ij},\sigma'_{z\vartheta},\sigma'_{zz}$ deviatoric stress components

$\alpha'_{ij},\alpha'_{z\vartheta},\alpha'_{zz}$ deviatoric internal stress components

$\sigma^*_{ij},\sigma^*_{zz},\sigma^*_{z\vartheta}$ effective stress components $\sigma^*_{ij} = \sigma_{ij}-\alpha_{ij}$

$\overline{\sigma} = \sqrt{\frac{3}{2}}\,(\sigma'_{ij}\ \sigma'_{ij})^{\frac{1}{2}}$

$\sigma_{eq} = \sqrt{\frac{3}{2}}\,(\sigma^*_{ij}\ \sigma^*_{ij})^{\frac{1}{2}}$

$\overline{\dot{\varepsilon}} = \sqrt{\frac{2}{3}}\,(\dot{\varepsilon}_{ij}\ \dot{\varepsilon}_{ij})^{\frac{1}{2}}$ equivalent strain rate

$\overline{\alpha}_s^{st} = (\alpha_{zz}^{st^2} + 3\alpha_{z\vartheta}^{st^2})^{\frac{1}{2}}$ equivalent internal stress.

1 - INTRODUCTION

Many studies have been devoted to plasticity constitutive laws in which qualitative rheological models are first proposed from uniaxial tensile tests with complex stress/strain paths and then identified from some of these same experiments (see for instance Chaboche and Rousselier, 1981). The experimental determination of the nature of the internal variables representing structural modifications during the plastic flow is one of the modelisation difficulties when only uniaxial tests results are available. Due to obvious experimental problems, the available biaxial tests results about high temperature visco-plasticity are still few (Findley, Cho and Ding, 1979 ; Cho and Findley, 1980; Oytana, Delobelle, Mermet 1982). Such a study performed on a 316 L stainless steel is reported in the present paper.

2 - EXPERIMENTAL METHOD

The experiments are pure tensile tests or traction-torsion tests on tubes. The method used to establish the flow rules of high temperature plasticity is based on the work of Oytana, Delobelle and Mermet 1982. Therefore, only the basic ideas will be summarized here.

Let the plastic equation of state be written as :

(1) $f(\dot{\varepsilon}_{ij}, T, \sigma_{kl}, v_k) = 0.$

The v_k being the internal variables. A set of possible values of the variables $\dot{\varepsilon}_{ij}$, T, σ_{kl}, v_k is callen a plastic state. The v_k are measured through a biaxial generalisation of the tensile dip test technique first proposed by Ahlquist and Nix (1969) and explained in §3.1 ; then (1) is determined.
When (1) is established it is possible to study the kinetics of the v_k:

(2) $\dot{v}_k = \dot{v}_k\,(\dot{\varepsilon}_{ij}, T, \sigma_{kl}, v_j)$ the dot meaning a time derivative.

The studied alloy is a low carbon, controled nitrogen content 316 L stainless steel the composition of which is given in table I.

Table I

C	Mn	Si	S	P	N_i	C_r	M_o	N_2	B
0,028	1,88	0,3	0,001	0,028	12,46	17,31	2,44	0,077	12ppm

The tests are performed in the temperature range 500-700°C and for equivalent strain rate lying in the range 10^{-9} s^{-1} to 5×10^{-5} s^{-1}.

3 - EXPERIMENTAL PLASTIC STATE EQUATION

3.1 - Biaxial dip test

In a biaxial stress state experiment let σ_{zz}, $\sigma_{z\vartheta}$, $\dot{\varepsilon}_{zz}$, $\dot{\varepsilon}_{z\vartheta}$ and T be the external variables of a given plastic state. Starting from this state, a stress drop is applied, i.e. σ_{zz} is suddenly reduced to $\sigma_{zz}-\Delta\sigma_{zz}$, $\sigma_{z\vartheta}$ to $\sigma_{z\vartheta}-\Delta\sigma_{z\vartheta}$ and $\dot{\varepsilon}^t_{zz}$ and $\dot{\varepsilon}^t_{z\vartheta}$ are the resulting instantaneous creep rates. Before every new stress drop the sample is set back to the pre-fixed σ_{zz} and $\sigma_{z\vartheta}$ values and the sample allowed to creep until the strain rates components $\dot{\varepsilon}_{zz}$ and $\dot{\varepsilon}_{z\vartheta}$ have reached pre-fixed values in such a way to start from a given plastic state. In the examples reported on figure 1, $\dot{\varepsilon}^t_{zz}$ and $\dot{\varepsilon}^t_{z\vartheta}$ are the instantaneous creep rates following respectively a stress drop $\Delta\sigma_{zz}$ ($\sigma_{z\vartheta}$ constant) and $\Delta\sigma_{z\vartheta}$ (σ_{zz} constant). Several remarks must be done :

1 - For each case reported on figure 1, the values of $\Delta\sigma_{zz}$ and $\Delta\sigma_{z\vartheta}$ giving respectively $\dot{\varepsilon}^t_{zz} = 0$ and $\dot{\varepsilon}^t_{z\vartheta} = 0$ should be considered as the measures of the effective stress components σ^*_{zz} and $\sigma^*_{z\vartheta}$ actually responsible of the plastic flow just before the stress drop.

2 - The various examples of figure 1 have been obtained on the same sample at the same applied stress $\sigma_{zz} = 32\times10^7\,N/m^2$ and $\sigma_{z\vartheta} = 4\times10^7\,N/m^2$.

3 - The values of $\dot{\varepsilon}^t_{zz}$ and $\dot{\varepsilon}^t_{z\vartheta}$ are average rates on 30 secondes.

More generally one can consider the two creep rate components $\dot{\varepsilon}^t_{zz}$ and $\dot{\varepsilon}^t_{z\vartheta}$ following a biaxial stress drop $\Delta\sigma_{zz}$, $\Delta\sigma_{z\vartheta}$. Three examples are reported in fig.2.

They correspond to the same examples as those of figure 1 ; and (A) is the stress representative point of the considered plastic state. From A the stress drop $\Delta\sigma_{zz}$, $\Delta\sigma_{z\vartheta}$ brings the stress representative point nearer to the origin. At each of these points, a vector (instantaneous rate vector) which represents the direction of the instantaneous creep rate ($\dot{\varepsilon}^t_{zz}$, $\dot{\varepsilon}^t_{z\vartheta}$) is associated. Finally ellipses $(\sigma_{zz}-\alpha_{zz})^2 + \frac{3}{2}(\sigma_{z\vartheta}-\alpha_{z\vartheta})^2$ = constant, have been drawn. If the instantaneous rate vectors are normal to these ellipses no anisotropy variable has to be introduced in (1). Here this rule is not quite well obeyed at small strain rates (i.e. for those points close to the center of the ellipses α_{zz}, $\alpha_{z\vartheta}$) but it should be noticed that the accuracy on transients is low when the creep rates are very small and that a small error on α_{zz} and $\alpha_{z\vartheta}$ induces an important desagreement with the normality rule for points close to the ellipses center.

3.2 - Invariant properties

At this point the question arises to know if kinematical hardening is the only internal variable taking place in (1). If this is true in multiaxial state of stress, then one should find that the equivalent creep rate $\bar{\dot{\varepsilon}}$ depends only on the equivalent effective stress σ_{eq}. Fig.3 shows the results for the 316L stainless steel. The experimental points come from various tests : those labelled P and St correspond respectively to measurements performed during primary and steady creep while those labelled A_i or E_k are due to complex biaxial stress experiments ; $\bar{\dot{\varepsilon}}$ and σ_{eq} are calculated according to their definitions. The points "ins^t" come from tests performed at 650°C at tensile stresses of 16×10^7 N/m^2 and 12×10^7 N/m^2 (i.e. lower than the macroscopic yield stress) for which the creep rates and therefore $\dot{\varepsilon}$ just after loading could be appreciated and for which σ_{eq} has been calculated assuming that internal stresses were zero at the very begining of primary creep; it is noteworthy that these points fit the same curve as all of the others.

Thus, in a large range of different experiments, figure 3 demonstrates that, for a given temperature, $\bar{\dot{\varepsilon}}$ depends only on σ_{eq}. Therefore, the normality rule has to be applied to equipotential surfaces of the Von-Mises type shifted by kinematical hardening which can be written down as :

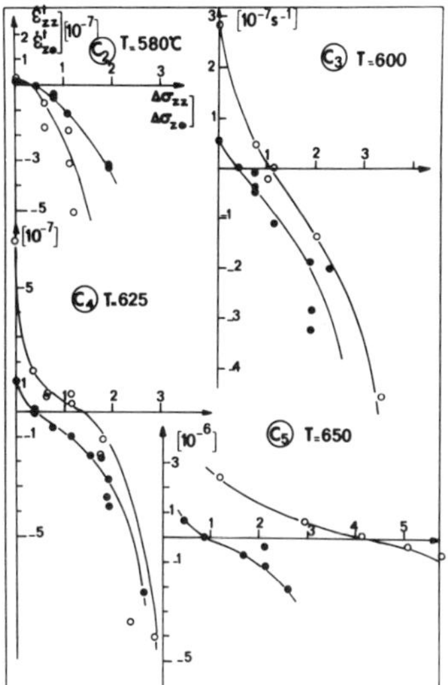

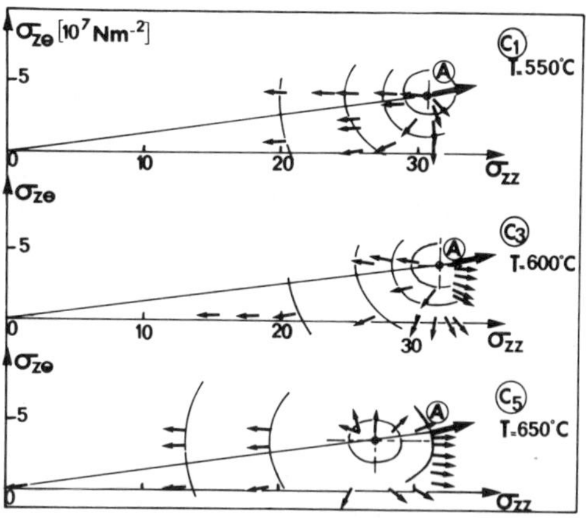

Fig.1 : Biaxial dip test determination of internal stress.
Fig.2 : Instantaneous creep rate vectors for three plastic states. Direction of the vectors is instantaneous strain rate after a stress drop. C_1,C_3,C_5 corresponds to the experiments of fig. 1.

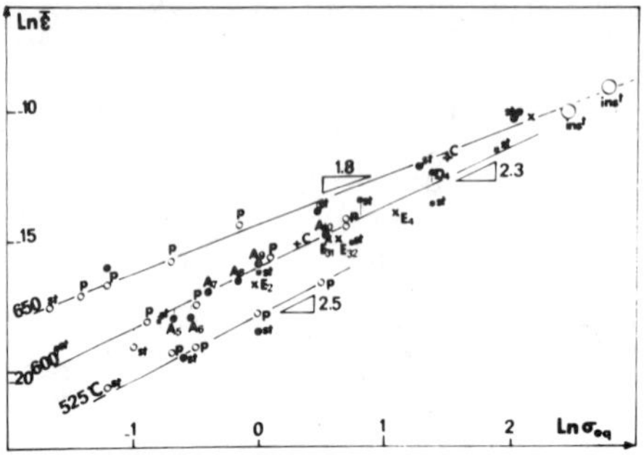

Fig.3 : Isothermal curves of equivalent strain rate versus equivalent effective stress for various states.

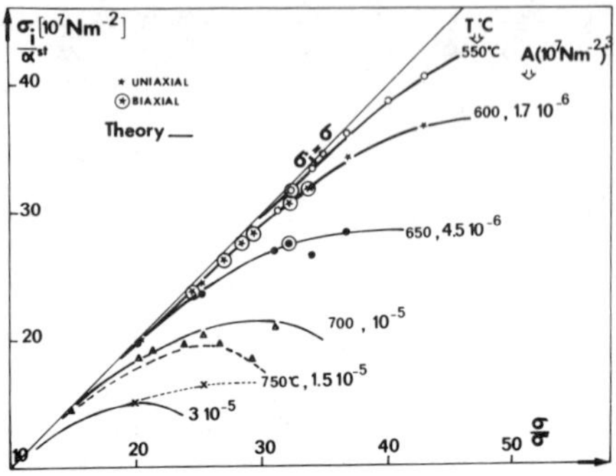

Fig. 4 : Evolution of internal stress versus applied stress ; σ_i,σ : tensile tests, $\bar{\alpha}^{st}$, $\bar{\sigma}$: traction-torsion tests.

(4) $$\dot{\varepsilon}_{ij} = \frac{\partial\Phi(\sigma_{eq})}{\partial\tau_{ij}} = \dot{\varepsilon}_o(T)\left(\frac{\sigma_{eq}}{\sigma^*_o}\right)^{n^*}\frac{\sigma'_{ij}-\alpha'_{ij}}{\sigma_{eq}}$$

An important experimental study with tensile tests and biaxial verifications led to the following results :

(5) $$\dot{\varepsilon}_o(t) = 2.5\times10^{-11}\,\frac{D_{cr}\,G\,b}{kT}$$ where

D_{cr} is the diffusion coefficient of Cr in the f.c.c. Stainless steel, n^* has been found to depend on T :

(6) $$n^* = \frac{3500}{T} - 1.8.$$

It is convenient for a separation of the diffusion activation energy effect from that of n^* and for unit reasons to introduce σ^*_o which has been calculated as $\sigma^*_o = 0.08\times10^7$ N/m^2.

Equation (4) completed by (5) and (6) and this σ^*_o value is the plastic equation of state of the 316L stainless steel considered.

4 - Kinetics of the internal variable

4.1 - Evolution of α_{ij}

The most physically based theory of creep comes from the Bailey-Orowan equation which explains the strain flow mechanism with the opposite effects of hardening and recovery.

Figure 4 shows the evolution of σ_i during steady state creep versus σ when purely tensile tests are performed together with $\bar{\alpha}^{st}$ versus the equivalent stress $\bar{\sigma}$. Pure tensile and biaxial experiments lead to the same curves which supports the validity of the method. The shape obtained qualitatively agrees with an hardening - recovery mechanism governing the kinematical hardening but a decrease of σ_i and $\bar{\alpha}^{st}$ is observed at very high stresses. During steady creep the following relationship is found :

(7) $\sigma_{eq} = \beta^{st}\,\bar{\sigma}$; β^{st} being a scalar (This result will be explained further

If (4) is used through classical invariant calculation to get $\bar{\dot{\varepsilon}}$, with (7) one obtains :

(8) $$\bar{\dot{\varepsilon}}^{st} = \dot{\varepsilon}_o(T)\,\frac{\beta^{n^*}\sigma^{n^*}}{\sigma_o^{n^*}}$$

. The power law is also obtained when the creep rate is given versus the applied stress :

(9) $$\bar{\dot{\varepsilon}}^{st} = K(T)\,\bar{\sigma}^{n}$$. Therefore :

$$(10)\qquad \overline{\alpha}^{st} = \left(1 - \left(\frac{K(T)}{\dot{\varepsilon}_o(T)}\right)^{\frac{1}{n^*}} \sigma_o \cdot \overline{\sigma}^{\frac{n-n^*}{n^*}}\right) \overline{\sigma} \ .$$

The theoretical curves of fig.4 have been obtained with (10) using the experimental values found in our experiments. This curve is fairly well fitted by the experimental values ; at least it foresees the decrease of σ_i and $\overline{\alpha}^{st}$ at the highest stresses.

Another important result about kinematical hardening is illustrated by fig.5.

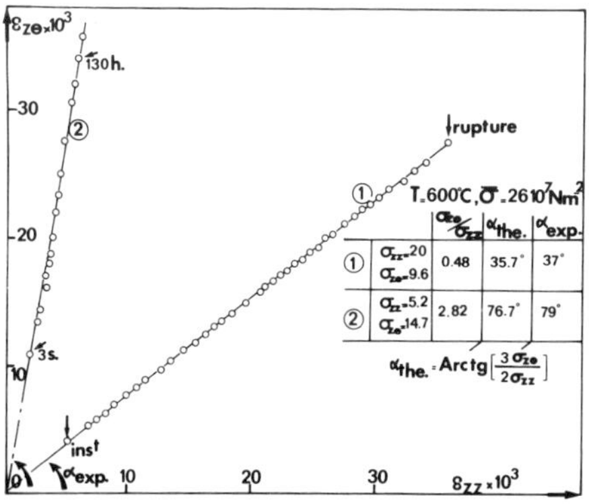

Fig. 5 : Direction of creep and instantaneous stresses as compared to the previsions of (11).

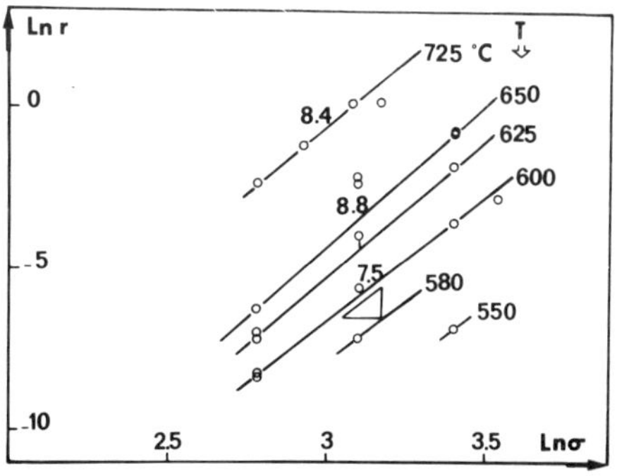

Fig.6: Values of recovery measurements (Mitra-McLean Method) versus applied stress.

If the experimental conditions are such that :

$$(11)\qquad \frac{\alpha_{ij}}{\sigma_{ij}} = 1 - \beta \ (\beta \text{ being a scalar}), \text{ then (7) leads to :}$$

$$(12)\qquad \frac{\dot{\varepsilon}_{z\vartheta}}{\dot{\varepsilon}_{zz}} = \frac{2}{3}\,\frac{\sigma_{z\vartheta}}{\sigma_{zz}} \ \text{(Oytana, Delobelle, Mermet, 1982)}$$

Fig.5 shows two examples of proportional loading ; it can be observed that in this case (12) and therefore (11) are obeyed; α_{the} is calculated from the applied stress using (12). This means that in proportional loading, in the present experiments one has : $\alpha_{z\vartheta}/\sigma_{z\vartheta} = \alpha_{zz}/\sigma_{zz} = 1-\beta$, a relationship corrobarated by direct measurements of α_{ij}. This relationship is also valid when steady creep is considered (see for instance fig.2 and Delobelle, Mermet, Oytana 1982) and this explains the use of (7) just above.

Another striking feature observed on every experiment of the type of those displayed on fig.6 is that (11) applies until rupture, therefore far into tertiary creep ; this indicates that, in proportional loadings, the damage parameter D behaves as a scalar and this property will have to be checked by any equation governing the damage tensor ($\overline{\overline{D}}$). When such relationships as those given by (11) and (12) exist during steady creep, Oytana, Delobelle and Mermet (1982) have shown that the Orowan law could be written as :

$$(13)\qquad \frac{d\alpha'_{ij}}{dt} = p\ (q\sigma'_{ij} - \alpha_{ij})\ \dot{\varepsilon} - R\ (\overline{\alpha})\ \alpha'_{ij}$$

In this case of the studied 316 stainless steel, the saturation value of α'_{ij} can be taken as σ'_{ij}, therefore q=1. One can try to determine p and R in different ways.

4.2 - <u>Adjustment to the monotonic tensile creep curves</u>. (Or to proportional loading curves). The initial part of a monotonic tensile creep curve is given by (4) and (13) in which R is supposed negligible as α is small :

$$(14) \qquad \frac{d\sigma_i}{dt} = p\ (\sigma - \sigma_i)\ \dot{\varepsilon}$$

If the simple case $p = p_o \sigma_{eq}^m$ is choosen (14) and (4) lead to :

$$(15) \qquad \varepsilon = \frac{1}{p_o m \sigma^m} \left(1 - \left(1 + \frac{2}{3} \frac{p_o (n^*+m).\sigma^{n^*+m}\ \dot{\varepsilon}_o(T)}{\sigma_o^{n^*}} t\right)^{n^*/n^*+m}\right)$$ if $m \neq 0$ and to

$$(16) \qquad \varepsilon = \frac{1}{p_o n^*} \mathrm{Ln}\left(1 + \frac{2}{3} p_o . n^* .\ \dot{\varepsilon}_o(T) \frac{\sigma^{n^*}}{\sigma_o^{n^*}} t\right) \quad \text{if } m = 0$$

Similar results are to be found in proportional loadings. The comparison of the theoretical curves with tensile creep tests has been done for m=0 as this value gives a good agreement through the p adjustment. For the identification of $R(\alpha)$ (13) is used in the stationary case for which $d\sigma_i/dt = 0$; the measurements of $\dot{\varepsilon}^{st}$ and the knowledge of hardening allow for a determination of $R\sigma_i$.

The method is not accurate because it is based on the determination of hardening with a condition which reduces the range of possible experiments to $\sigma_{zz} \leq 15.10^7 N/m^2$ and prohibits the use of an extensive part of the $\varepsilon(t)$ curves.

4.2 - The Mitra-Mc Lean method

A more direct method, first proposed by Mitra and Mc Lean (1967) is available for the determination of r (if (13) is written as $d\sigma_i/dt = h\ \dot{\varepsilon} - r$). Fig.6 shows the values of r measured in steady tensile creep tests versus the applied stress (due to the large values of σ_i very similar curves would be obtained if r was reported versus internal stress). r depends on σ (or σ_i) through a power law with $n_r = (\partial \mathrm{Ln}\ r / \partial \mathrm{Ln}\ \sigma)_T$ lying in the range 7.5 to 8.8.

These values are close to those found for $n = \partial \mathrm{Ln}\ \bar{\dot{\varepsilon}}^{st} / \partial \mathrm{Ln}\ \bar{\sigma}$. As $d\sigma_i/dt = 0$ in steady creep, this means that h depends very little on applied stress, which supports the result of 4.2 giving little variations of hardening with stresses. On the other hand, the r activation energy :

$H_r = -\left(\partial \mathrm{Ln}\ r / \partial \mathrm{Ln} \frac{1}{kT}\right)_\sigma$ (average value 4.04 ev) is different from :

$H_c = -\left(\partial \mathrm{Ln}\ \dot{\varepsilon}^{st} / \partial \frac{1}{kT}\right)_\sigma$, and therefore h must be temperature dependant.

In the same way, experimental methods exist allowing a direct measurement of h from stress increment tests, (Gasca-Neri, Barret and Nix 1971). However, because of the occurence of very important instantaneous strains at stresses above $15 \times 10^7 N/m^2$ it has been necessary to use the following sequency : a small stress drop $\delta\sigma$ is set on the sample during a tensile creep test for the time necessary to creep rate to recover a steady positive value, then $\delta\sigma$ is removed and the resulting transient strain measured ; h is then evaluated using $h = \lim_{\delta\sigma \to 0} \delta\sigma / \delta\varepsilon$. The h values thus obtained are such that the ratio $r/h = \dot{\varepsilon}^{st}$ (r being obtained from the Mitra-Mc Lean method) ; this strongly supports the reliability of the used methods. It has been so found for h an average activation energy $H_h = -1.18$ e.v. and a weak dependance of h on σ ($h \propto \sigma^p$ with p lying between 0 and -1) or on $\sigma - \sigma_i$. Let us notice too that h values are very large : about $3.10^{12} N/m^2$. Moreover $R(\bar{\alpha}) = r_o \exp - \frac{4,04}{kT} . (1-\beta)^{-n_r} \bar{\alpha}^{(n_r - 1)}$, n_r being of the order of 8.

CONCLUSION

It has been shown that, as far as the instantaneous plastic strain occuring chiefly for equivalent stresses above $15.10^7 N/m^2$ is disregarded, a single internal kinematical variable is representative of the part taken by the induced structural modifications in creep flow rules of 316L stainless steel.

In this case the viscoplastic state equation has been established and its coefficients experimentally determined, particularly, the internal variable has no longer been identified from a creep curve, but has been directly obtained from biaxial dip tests.

The equation governing the variations of the internal variable has been studied as a Bailey-Orowan one. This problem appears more difficult but quantitative results concerning the r_{ij} and h_{ij} dependances on T and applied stresses have been obtained. To ascertain them, it is now necessary to use these results in more complex paths such as non proportional loadings or unloadings. For higher stressings it is necessary to consider the very important plastic strain occuring at the loading. The experimental results and the corresponding modelisation will be reported in another paper.

AKNOWLEDGMENT. The stainless steel was kindly supplied by Creusot-Loire, D.R.U. and the present work received financiel Support from Electricité de France. The authors wish also to thank M. Balladon, M. HERITIER and M. RABBE from Creusot-Loire, MM. Masson and Rousselier from EDF for fruitful discussions.

REFERENCES

AHLQUIST, C.N., W.D. NIX (1969) A technique for measuring mean internal stress during high temperature creep. Scripta Met.3, 679-682

CHABOCHE J.L., G.ROUSSELIER (1981) On the plastic and viscoplastic constitutive equations based on the internal variables concept. 3rd International Seminar SMIRT 6, Paris 24-25 Août 1981

CHO, U.W., W.N. FINDLEY (1980). Creep and creep recovery of 304 stainless steel under combined stress. J. Appl. mech., 47, 755-761.

DELOBELLE,P., A.MERMET, C.OYTANA (1982b). Biaxial dip tests measurements of internal stresses during high temperature plastic flow. In R.C. Gifkins (Ed.), Strength of metals and alloys (ICSMA 6), vol.2, Pergamon Press, Oxford and New York, 575-580.

FINDLEY,W.N., U.W.CHO, J.L. DING (1979). Creep of metals and plastics under combined stresses, a review. J.Engng.Mat.Tech., 101, 365-368.

GASCA-NERI R., C.R. BARRETT, W.D. NIX (1971) Ande on the recovery work hardening theory of creep. Scripta Met. 5, 733-740.

MITRA S.K. and D. Mc Lean (1967) Cold work and recovery in creep at Ostensibly constant structure, Met. Sci. J. 1, 192-198.

OYTANA,C., P. DELOBELLE, A. MERMET (1982a). Constitutive equations study in biaxial stress experiments. J. Engn. Mat. Techn., 104, 1-11.

CONSTITUTIVE RELATIONSHIPS FOR CREEP-FATIGUE IN HIGH-TEMPERATURE MATERIALS

A. Berkovits* and S. Nadiv**

*Department of Aeronautical Engineering,
Technion - Israel Institute of Technology, Haifa, Israel
**Department of Materials Engineering,
Technion - Israel Institute of Technology, Haifa, Israel

ABSTRACT

Results of a continuing effort to correlate the response of high-temperature materials under cyclic straining with the behavior of such materials under monotonic creep and plasticity are reported. The mechanisms governing deformation in Udimet 700 superalloy and 316 stainless steel are shown to be the same in both the monotonic and cyclic cases. The direction of future research is indicated.

KEYWORDS

High temperature fatigue; creep-fatigue; strainrange partitioning; deformation mechanisms; superalloys.

INTRODUCTION

A major problem in predicting low-cycle fatigue at elevated temperatures in turbomachine applications has been estimating the effect on life, of deformation processes which are strongly rate dependent. These include creep while strain is increasing in tension or compression during the fatigue cycle, and stress relaxation while the strain is held constant. Success in estimating low-cycle fatigue life under such conditions has varied considerably. The work presented in this paper was undertaken in order to achieve an understanding of the basic mechanisms which are responsible for observed material response. Specific objectives included:

1. Analysis of the plasticity and creep damage mechanisms operating during fatigue loading, through metallographic and fractographic examination of specimens which have undergone thermomechanical strain-cycling test conditions;

2. Formulation of the relationships between the observed cyclic response and monotonic creep and plasticity materials response;

3. Finally development of an analytical model for the estimation of the strain-life materials response under arbitrary cyclic conditions, based on the results obtained.

One of the more successful phenomenological methods for predicting high-temperature, low-cycle fatigue life is the strainrange partitioning approach, developed at NASA by Manson, Halford, and Hirschberg (1971). The procedure involves the determination of the four basic life relationships, resulting from the four possible combinations of plastic or creep strainrange in the tensile or compressive halves of the fatigue cycle, for a given material, and their use in conjunction with the interaction damage rule to predict the cycle lives.

At present the strainrange partitioning approach is entirely empirical and, although qualitative understanding exists (Manson, 1983), there is as yet no quantitative explanation for the strainrange partitioning phenomenon as it is observed. Partitioned strain-life curves have not been correlated analytically with monotonic material response in the way that was achieved for room temperature fatigue through the Manson-Coffin Universal Slopes equation (Morrow, 1965). The relationship between monotonic and cyclic response in the high-temperature creep range needed to carry out such a correlation can be found when another fundamental variable, namely the strain-rate, is considered. Strain-rate has been shown to be the controlling variable in many situations involving monotonic creep (Berkovits, 1972, 1974a) and may be expected to have an important influence in cyclic problems (Berkovits, 1974b).

The present paper reports on a continuing research effort in high-temperature materials. The program is concerned with three materials, namely Mar-M-200 and Udimet 700 nickel superalloys, and 316 stainless steel. Tests are being conducted in the range of temperatures encountered by these material in jet engine applications. At these temperatures both time-dependent and time-independent plastic deformations occur under load. In the initial stages the mechanical properties of the materials were defined under monotonic and cyclic conditions, ranging from low strain-rate, time-dependent straining, to high strain-rate, time-independent straining, as well as combinations of the two extremes. Some of the tests were interrupted in order to provide samples for metallographic investigations of the micromechanical processes occurring. Also fractographic studies were conducted on failed specimens. Subsequently cyclic tests were conducted in strain and strain-rate ranges similar to those imposed during monotonic tests. Cyclic tests on Udimet 700 and 316SS have been completed, while tests on Mar-M-200 nickel alloy are in progress.

The paper discusses the approach to the problem, and presents results and conclusions drawn to this juncture. Thus the following sections will be devoted to discussions of strainrange partitioning micromechanisms, strain-range-life relationships, strain-rate dependency, and possible service prediction models.

MONOTONIC CHARACTERISTICS

Before cyclic response of a material can be understood fully, the mechanisms which determine monotonic behavior must be examined. By monotonic are meant nonreversed loadings, such as strength and creep tests leading to failure. Monotonic loading includes tests in which such loadings are preceded by a single loading in the opposite direction in order to induce residual strain. Extensive tests of this type were conducted on hollow axial specimens of Udimet 700 and of 316SS. Monotonic tests consisted of tensile and compressive tests conducted at constant strain-rate throughout. This type of strength test permits stress to develop at a rate compatible with the imposed strain-rate condition and the characteristic rates of deformation processes of the material. Data for U-700 under tension (solid lines) and compression (dashed lines) at 925C are presented in Fig. 1 as a material hodograph. Berkovits (1972, 1974a) showed that characteristics common to diverse mechanical responses may be clearly evaluated if material data are plotted in hodographic form, as inelastic strain-rate against inelastic strain, with stress and temperature as discrete parameters. The hodograph is particularly useful because it presents the governing materials parameters, strain-rate and strain, as continuous variables without introducing the extraneous time term. At low values of strain the characteristic curves in Fig. 1 tend to be horizontal. This shows that for a given strain-rate there is a corresponding stress at which flow can be initiated. At high strains the characteristic curves become relatively steep. In between these extremes of strain the characteristics pass through a minimum, which indicates a maximum stress for a given constant strain-rate, or conversely a minimum strain-rate for a given constant stress.

The dashed lines in Fig. 1 represent data obtained in compression. Although the general shape of the curves are similar to the tensile curves, it is significant that, at low inelastic strain, strain-rates are somewhat higher in compression than in tension, but a high strain compressive strain-rate is lower and more constant over a substantial strain region. This appears to be particularly so at high stresses. Similar response has been reported previously (Manson, 1983) for 316SS.

The data in Fig. 1 are time-dependent and are due to creep-type flow in the material. At lower temperatures or at higher strain-rates the material hodograph delineates the boundary between time-dependent and time-independent (plastic) deformation. The tensile hodograph for U-700 at 760C is shown in Fig. 2. At high strain rates the curves become vertical, indicating that the strain is insensitive to strain-rate. However as inelastic strain accumulates time-dependent effects can enter, after which

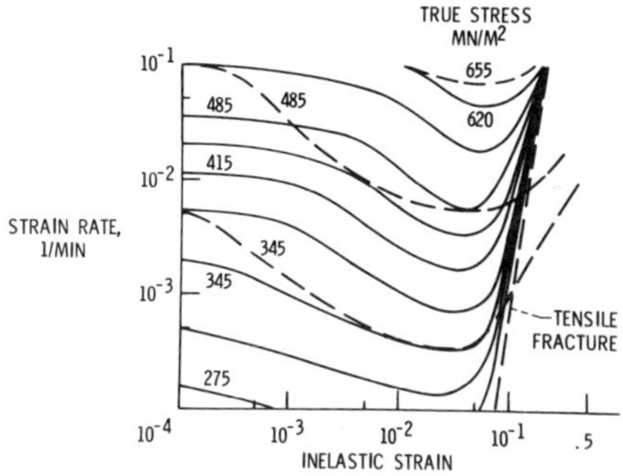

Fig. 1. Tensile and compressive hodograph for U-700 at 925C.

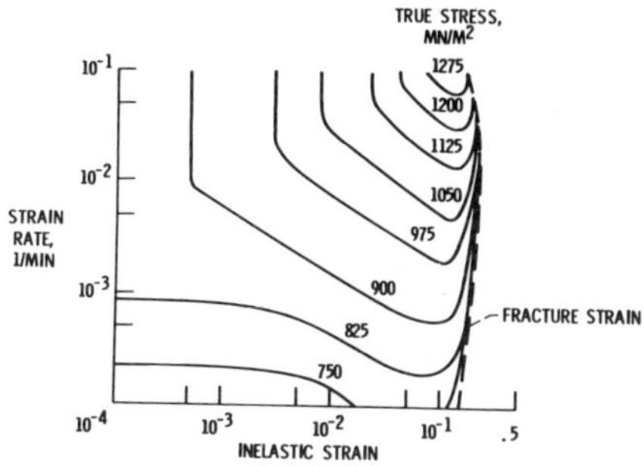

Fig. 2. Tensile hodograph for U-700 at 760C.

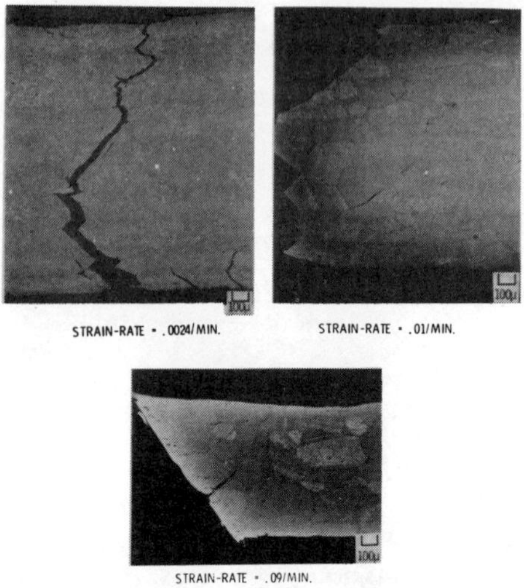

Fig. 3. Fracture of U-700 at 760C.

the inelastic strain consists of creep as well as plasticity. As a result, at high strain-rate the failure strain falls between the expected creep ductility and the true, time-independent ductility.

The two types of inelastic deformation discernible in the hodograph for 760° C are supported by the micrographs in Fig. 3. The specimens tested at moderate strain rates (0.0024/min and 0.01/min) indicate time-dependent failure, as at the higher temperature. The presence of many grain-boundary cracks and fracture lines which follow the grain-boundaries is evidence of the flow common to the creep phenomenon. On the other hand, the specimen tested at a strain rate of 0.09 per min. clearly indicates predominantly time-independent flow and fracture, as at low temperatures. The micrograph is similar to that of the undeformed material, except that the grains near the edges and the fracture line are plastically elongated. The 45° fracture surface is plainly transcrystalline. Similar effects were discerned when tensile tests were carried out following prior compressive straining.

MONOTONIC TENSION AFTER COMPRESSIVE STRAINING

It has become common usage in high-temperature materials fatigue research to partition the applied strainrange into four basis types, and to study the interactive responses of either plasticity or creep in either or both of the tensile and compressive portions of the strain cycle. Eight corresponding monotonic tests can be devised to demonstrate the effects of prior straining in the opposite direction. Four of these are of practical importance because they lead to failure:

Tensile Loading	Prior Compressive Loading	Notation
Plasticity	Plasticity	PP
Creep	Creep	CC
Plasticity	Creep	PC
Creep	Plasticity	CP

Tensile Plasticity - Compressive Plasticity (PP)

The effects of compression into the plastic zone upon subsequent tensile plasticity are well-documented. They are the same at elevated temperatures as they are in the sub-creep region. Inelastic deformation is generally along slip planes, and compressive permanent strains are largely washed out when the material is strained beyond yield in subsequent tension.

Tensile Creep - Compressive Creep (CC) and Tensile Plasticity - Compressive Creep (PC)

The role of compressive creep-strain on later straining in tension is unchanged, whether the tensile strain is time-dependent creep or time-independent plasticity. The grain-boundary sliding which constitutes creep is a rate process, and tends to maintain the internal equilibrium of the material structure. Moreover, in compression voids are not opened. Thus the material is essentially intact when loading direction is reversed and tensile strain is applied, and either tensile plasticity or tensile creep would be expected to be little affected by the prior compression creep.

In Fig. 4 the dashed curves indicate the modification which occurs in the U-700 tensile hodograph at 925C after compressive creep strain at the same temperature. The general shape of the hodograph is retained. However, at a given strain rate the stress developed is now significantly lower, as it would be in virgin material which had been strained on the order of the total compressive and tensile creep combined. Nevertheless the inelastic fracture strain is unaffected by prior compressive creep, as can be seen by comparing the prestrained stress-strain curves in Fig. 5 with the curve representing virgin material. Clearly prior compressive creep-strain at 925C had little effect on the fracture strain.

Tensile Creep - Compressive Plasticity (CP)

A different situation exists when the compressive prestrain occurs at 525C, at which temperature all inelastic strain is plastic for U-700. Plasticity in compression reduces subsequent creep elongation by 50%, as result of a severe embrittling effect. Microstructurally the plastic compression strain occurs along slip lines, causing microscopic protrusions at the edge of the material. Nearby grain-boundary sliding during subsequent tensile creep forms local microcracks at the surface, which quickly oxidize and propagate into the substrate.

Another embrittling mechanism which operates in this case is due to dislocation pile-up at the intersection of plastic slip lines with grain boundaries. Both dislocation pile-up and severe oxidation hasten the premature failure seen in Fig. 5. Final fracture is completely intercrystalline.

The four monotonic damage mechanisms described above are identical to those operating during the basic

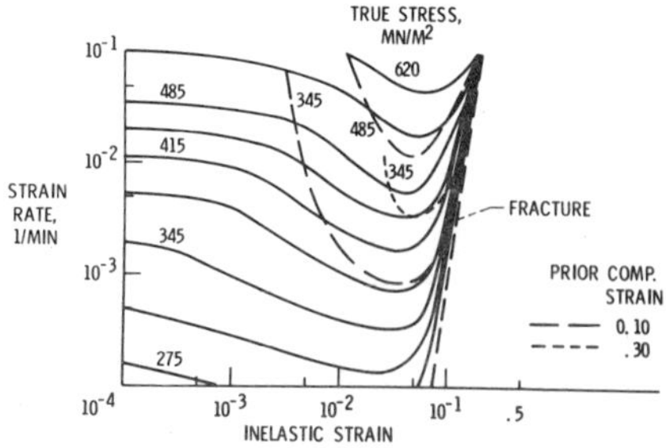

Fig. 4. Hodograph for U-700 in tension after compressive creep, 925C.

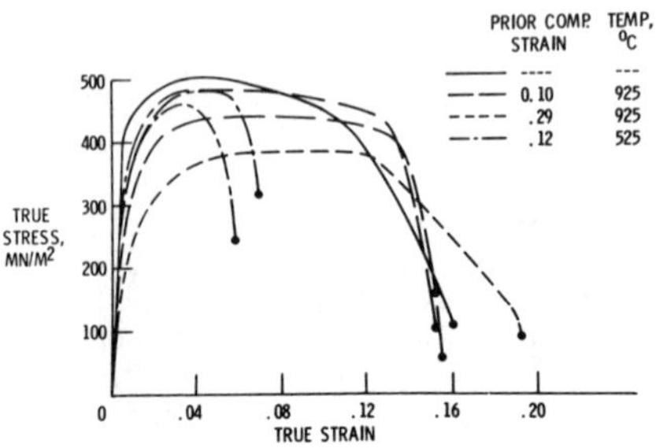

Fig. 5. Tensile response of U-700 at $\dot{\varepsilon}$=.01/min, 925C, after compressive strain.

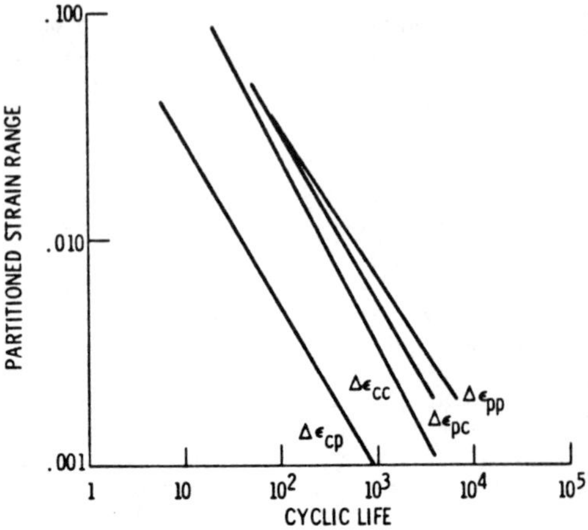

Fig. 6. Partitioned strain life of 316SS, 705C.

types of strainrange partitioning defined in high-temperature fatigue applications (Manson, 1983). It is reasonable to expect that cyclic and monotonic data can be correlated such that cyclic behavior will be predictable from monotonic materials response.

CORRELATION OF STRAINRANGE PARTITIONED LIFE WITH MONOTONIC BEHAVIOR

For many materials fatigue-life curves for the four distinct types of strainranges differ from one another, due to the distinct micromechanisms causing fatigue damage in each case. Strainrange partitioned life curves for 316SS at 705C are shown in Fig. 6. The $\Delta\varepsilon_{pp}$-line is essentially independent of time (i.e., frequency) and its position can be estimated by use of the Manson-Coffin Equation. Both the CC- and PC-lines exhibit lives comparable to those of the $\Delta\varepsilon_{pp}$-line at high strainrange, low cyclic life. However, at lower strainranges the PC- and CC-lines indicate progressively lower lives than corresponding PP lives.

This behavior is readily explained with the aid of the micromechanisms described earlier under monotonic loading. At high strainranges little or no strain-hardening takes place either in tension or compression. Furthermore the relatively high strain-rates applied do not permit significant oxidation to occur during creep. Thus similar lifetimes are obtained under PP-, CC-, and PC-straining for this material. Lower strainranges necessarily involve lower strain-rates during PC- and CC-straining. Under PC-straining sufficient surface oxidation occurs in the compressive creep phase of the cycle to cause brittle cracking when plastic reloading occurs in tension. Such cracking is clearly seen in Fig. 7. In CC-cycling, which was conducted under constant tensile and compressive load amplitudes, the compressive half-cycle is responsible for more oxide build-up than is the tensile half-cycle. The fact that the true compressive stress amplitude developed is lower than the true tensile stress amplitude on the order of magnitude of the strainrange only partially explains why the compressive creep rate is only 33 to 50% of the tensile creep rate. Whatever the cause, significant surface oxidation again occurs in compression, with commensurate reduction in fatigue life (Fig. 8).

FUTURE WORK

The results presented here from tests on dimet 700 and 316SS shed considerable light on many unifying aspects of elevated temperature phenomena. However, the picture is necessarily incomplete, due to the large number of parameters involved, and the limited nature of the testing. Extensive work is presently being conducted on directionally solidified MAR-M-200 nickel-base alloy. This alloy is notable for high oxidation-resistance as well as very strong, brittle grain-boundaries. Thus the role of oxidation and grain-boundary sliding in creep (monotonic or cyclic) is small. For example it has been shown that all four types of strainrange partitioned failure data fall on a single life curve for this material (Manson, 1983).

The participation of fewer micromechanisms in the deformation process make MAR-M-200 a good candidate material for study. Cross-section of a test specimen of directionally solidified MAR-M-200 (Hf-modified) is shown in the micrograph in Fig. 9. Besides the usual γ' particles in γ-phase matrix (left), columnar dendritic structure (right) shows γ-γ' eutectic nodules adjacent to a dendritic-arm boundary. Small carbides decorate the boundary. Monotonic tests are being conducted in tension, compression, and in tension after prior compression, over a range of constant strain rates. Cyclic strainrange partitioned tests are planned in the same range of strain-rates and temperatures. Mechanical response is being correlated in all tests with the aid of metallographic and fractographic investigation.

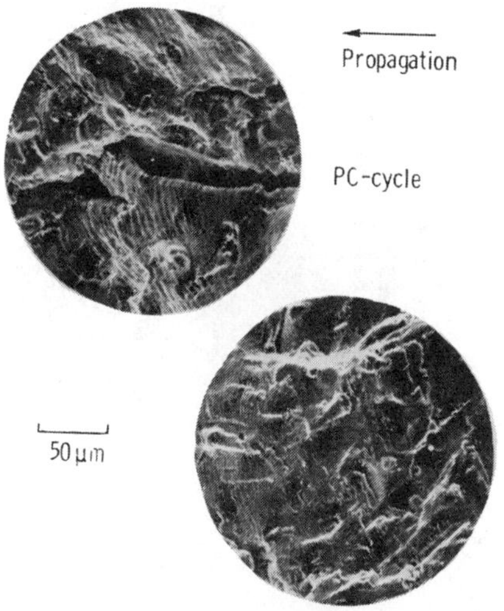

Fig. 7. Planar distortion in tensile creep of PC-cycle in 316SS, 705C.

Fig. 8. Cracking in CC-cycle in 316SS, 705C.

ACKNOWLEDGMENT

Some of the earlier work reported herein was performed while the author was a visiting scientist at NASA Lewis Research Center, Cleveland, Ohio.

REFERENCES

Berkovits, A. (1972). Hodographic approach to predicting inelastic strain at high temperature. NASA TN D-6973.

Berkovits, A. (1974a). ASME J. of Eng. Mats. and Tech., 96H, pp. 106-108.

Berkovits, A. (1974b). AIAA J. of Aircraft, 11, pp. 10-14.

Manson, S. S. (1983). Complexities of high temperature metal fatigue: some steps toward understanding. Proc. 25th Israel Annual Conf. on Aviation and Astronautics, Haifa.

Manson, S. S., G. R. Halford, and M. H. Hirschberg (1971). Creep-fatigue analysis by strainrange partitioning. Proc. 1st. Symp. on Design for Elevated Temperature Environment, San Francisco, pp. 12-24.

Morrow, J. (1965). ASTM STP378. pp. 45-87.

Oldgrieve, R. E. (1978). Fractographic evaluation of creep effects on strain-controlled fatigue-cracking of AISI 304LC and 316 stainless steel. NASA TM-78913.

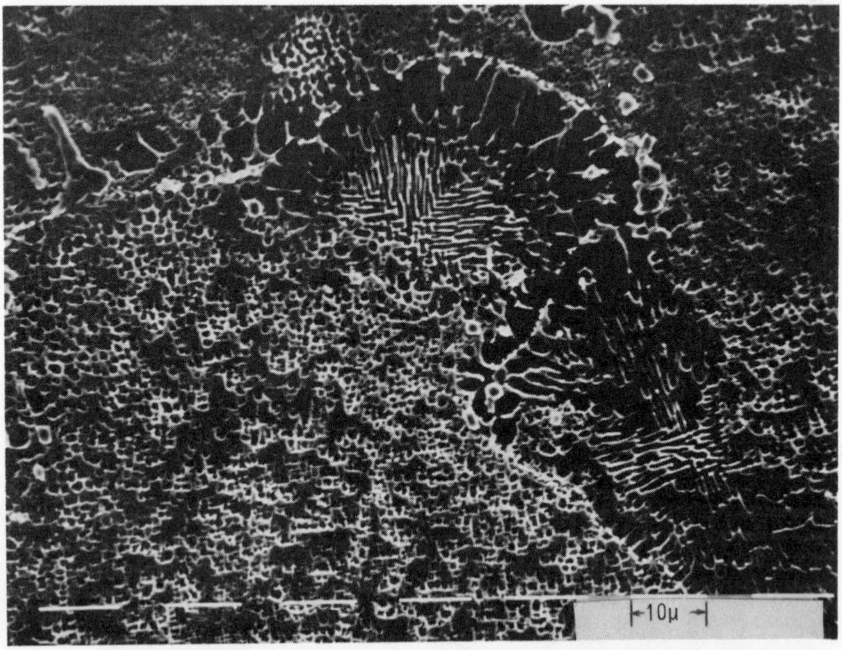

Fig. 9. Columnar dendritic structure of directionally-solidified MAR-M-200 (Hf-modified) alloy.

HIGH TEMPERATURE LOW CYCLE FATIGUE OF TWO AUSTENITIC STAINLESS STEELS — A COMPARISON BETWEEN MECHANICAL PROPERTIES AND MICROSTRUCTURE

J.-O. Nilsson

Steel Division Research and Development Centre, Sandvik AB, S-811 81 Sandviken, Sweden

ABSTRACT

Low cycle fatigue tests have been conducted at 600°C on AISI 316 and Sandvik 253MA.

These tests showed that Sandvik 253MA is more resistant to fatigue than AISI 316, the difference in number of cycles being approximately a factor of 2. On the basis of fractographic measurements it could be shown that the difference in fatigue resistance resulted from a longer initiation stage in 253MA combined with a lower crack propagation rate.

Transmission electron microscopy revealed that 253MA is a planar slip alloy whereas wavy slip predominated in 316, indicating that cross slip and climb occurred more easily in 316. The difference in fatigue resistance between the two alloys could be explained by inhibition of cross slip in 253MA.

KEY WORDS

High temperature fatigue, crack initiation, crack propagation, AISI 316, Sandvik 253MA.

INTRODUCTION

Thermal stresses which cause high temperature low cycle fatigue (e.g. Coffin, 1977) may arise in essentially two ways: either as a result of time-variations of temperature, or as a result of temperature gradients or a combination of both. Because of the high thermal expansion coefficient of austenite, the situation is more severe in austenitic steels than in ferritic steels. Furthermore, the incompatibility of thermal expansion coefficients may also cause problems when ferritic and austenitic steels are linked in such components as, for example, composite tubes. The materials considered in the present investigation are AISI 316 and Sandvik

253MA. The latter steel contains 21% Cr, 11% Ni with additions of Si, N and rare earth metals. AISI 316 is widely used as superheater tubes in power plants whereas 253MA is a candidate material for many high-temperature components in the petrochemical industry, for instance reactor tubes in cracking furnaces for styrene production. Both these cases represent situations where the influence of thermally induced stresses is significant. AISI 316 was used in the present study mainly because it has been the subject of many previous investigations and therefore serves as a good reference. Emphasis has been put on relations between microstructure and mechanical behaviour. It will be shown that differences between 316 and 253MA in terms of fatigue resistance can be related to dissimilar slip behaviour.

EXPERIMENTAL

The chemical compositions of the two alloys investigated are given in Table 1 below:

TABLE 1 Chemical Composition

Material	C	Si	Mn	P	S	Cr	Ni	Mo	Al	N	Ce
SANDVIK 253MA	.084	1.51	0.45	.017	<.003	20.85	10.97	0.03	.014	.21	.02
AISI 316	.043	0.57	1.67	.022	.003	16.57	12.24	2.62	.002	.056	-

The starting materials were in the form of extruded bar having a diameter of 20 mm. Both alloys were solution treated at 1100°C for 10 minutes and given an aging treatment at 600°C for 1000h. The grain sizes were 48 μm for 253MA and 63 μm for AISI 316 with the corresponding hardnesses (HV30) of 200 and 160. The aging treatment did not alter grain size and hardness significantly as compared to non-aged material.

Cylindrical specimens, having a gauge length of 15 mm were tested isothermally at 600°C in strain control using axial extensometry at a constant strain rate of $5 \cdot 10^{-3} s^{-1}$. Failure was defined as a 25% reduction of peak stress in tension compared to the maximum stress.

RESULTS

Mechanical Behaviour

The diagram shown in Fig. 1 summarizes the fatigue results obtained for the two alloys. From the diagram we may infer that 253MA is more fatigue resistant than 316, the difference in life-time being approximately a factor of 2 throughout the whole interval of strains investigated.

It may also be concluded from Fig. 1 that the aging treatment at 600°C for 1000h tends to decrease the fatigue life as compared to non-aged material. This effect is more pronounced for 253MA than for 316. Although the aged specimens resulted in a larger scatter than non-aged specimens, it seems clear that 253MA is more fatigue resistant than 316 even in aged condition.

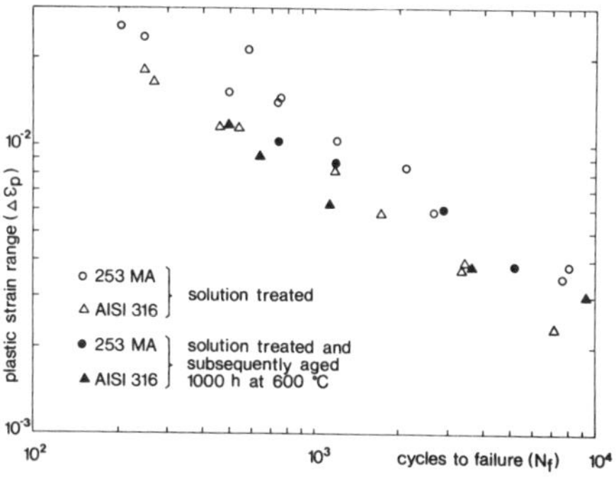

Fig. 1. Effect of plastic strain range on fatigue life of 316 and 253MA.

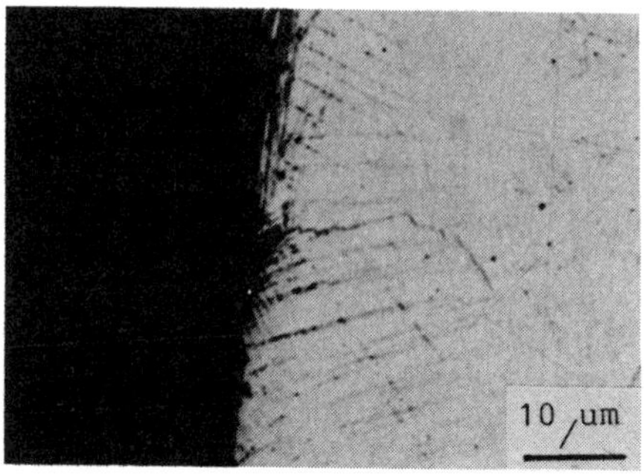

Fig. 2. Optical micrograph showing intrusion at specimen surface.

Optical Microscopy

Optical examinations showed that crack initiation always occurred at the specimen surface, probably through the formation of intrusions or extrusions. In fact, several observations of intrusions in association with subcracks have been made, an example of which is given in Fig. 2. Crack propagation paths were almost exlusively transgranular but isolated examples of intergranular crack paths were seen. It should be pointed out that the intergranular precipitation of large $M_{23}C_6$ particles in the aged materials did not alter the mode of crack propagation. However, examination of sectioned specimens from aged material showed subcracks which could have been nucleated at grain boundary surface intersections. An example of this is given in Fig. 3.

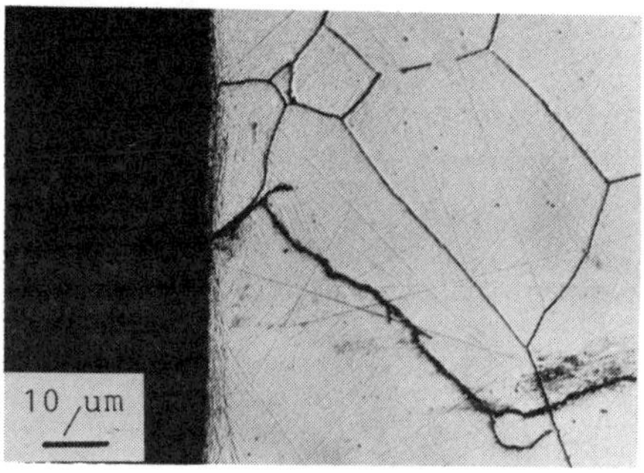

Fig. 3. Optical micrograph of subcrack which has nucleated in a grain boundary.

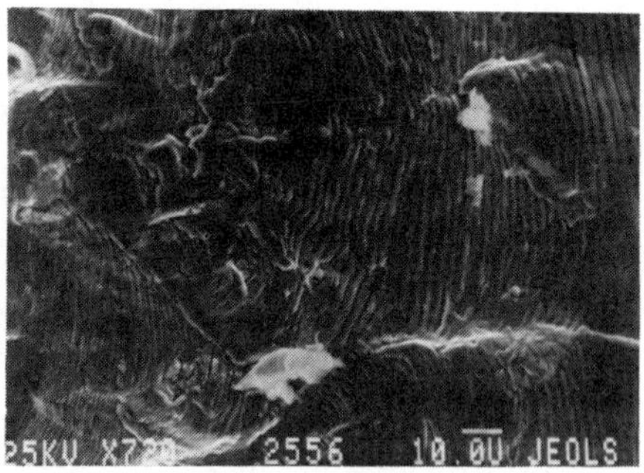

Fig. 4. Scanning electron micrograph of striations formed in 316 at $\Delta\varepsilon_p$ = 1.14%.

Scanning Electron Microscopy

Fatigue striations were clearly visible in the region corresponding to stage II crack propagation. Striation spacings were measured in non-aged material from SEM micrographs similar to that shown in Fig. 4. This particular micrograph is taken from 316 deformed at $\Delta\varepsilon_p$ = 1.14% to a crack length (distance from initiation point at the specimen edge) of $\sim$0.32 mm. The striation spacing determined in this case was 2.2 μm. It is evident from Fig. 4 that the striations were somewhat distorted, probably resulting from damage to the fracture surface in the compression part of fatigue cycle. However, this did not affect the measurements of striation spacings.

The results from striation measurements are summarized in Fig. 5, from which it may be seen that the relationship between striation spacing and crack length is approximately linear. Using the same approach as Wareing and Vaughan (1977) we can integrate the number of cycles N spent in Stage II, as follows:

$$N = \int_{a_i}^{a_f} \frac{da}{ka} = \frac{1}{k} \ln \frac{a_f}{a_i} \qquad (1)$$

where a is crack length and k is a constant of proportionality. It was noted that the striations became visible at $a_i \sim 0.1$ mm and were discernible throughout the whole test, i.e. to $a_f \sim 3$ mm corresponding to a 25% load reduction.

The number of cycles predicted from eq. (1) are compared with the observed number of cycles to failure in Fig. 6. Results of Kim and Laird (1978, 1978b) seem to prove that the extent of stage I is of minor importance in high strain fatigue, implying that the difference between observed and predicted lives, N_f-N, may be interpreted as the number of cycles spent in the nucleation stage. Assuming this to be true, two conclusions can be drawn from Fig. 6. Firstly, crack propagation rates are lower in 253MA since the predicted values lie to the right of those corresponding to 316. Secondly, the nucleation stage defined as N_f-N, occupies more cycles in 253MA than in 316.

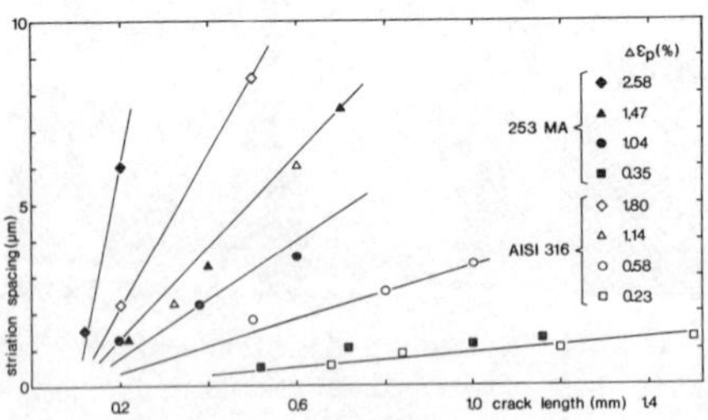

Fig. 5. Diagram showing linear relation between striation spacing and crack length.

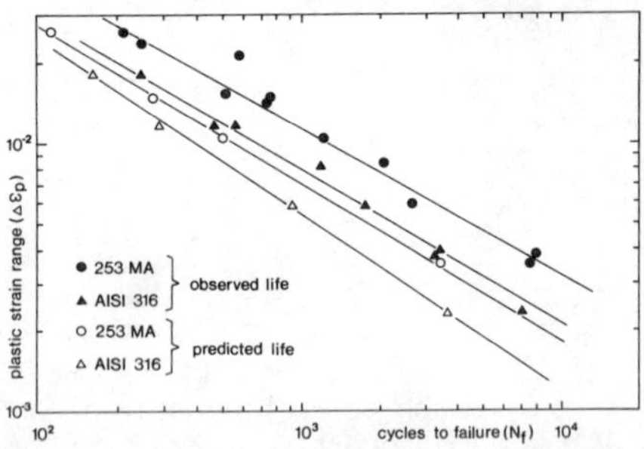

Fig. 6. Comparison between calculated life and observed life of 316 and 253MA.

Transmission Electron Microscopy

Both starting materials had a very low intragranular dislocation density. Grain boundaries in non-aged starting material were precipitate-free as opposed to grain boundaries in aged starting material which showed heavy precipitation of $M_{23}C_6$ even in coherent twin boundaries. The precipitation of $M_{23}C_6$ was particularly pronounced in high angle boundaries, an example of which is shown in Fig. 7. In deformed non-aged material rapid nucleation and growth of $M_{23}C_6$ occurred, particularly in high angle boundaries. As is shown in Fig. 8 intergranular precipitates were already visible after 30 minutes, in 316 cycled at $\Delta\varepsilon_p = 1.15\%$.

Regarding the dislocation structure in deformed materials, there are striking dissimilarities between the two alloys.Dislocation cells developed in 316 at all of the strain levels investigated. Examples of this are given in Fig. 8 and 9. Even at low strain levels, a tendency to form cells was observed, Fig. 10, although they were less regular than those produced at high strain. In contrast to 316, planar dislocation arrays formed in 253MA at all strain levels. At low strain, the dislocation density was so low that individual dislocations in the arrays could be resolved, Fig. 11. In this particular micrograph it is evident that the dislocations often appear in pairs. The symmetric contrast of each pair shows that they are dislocation dipoles (Forwood and Humble, 1970). At high strain the planar slip character still predominated, although individual dislocations were no longer resolved owing to the high dislocation density, Fig. 12. It should be noted in this context that although planar slip predominated in 253MA, dislocation cells were occassionally observed at $\Delta\varepsilon_p > 0.50\%$. Conversely, no evidence of planar slip was found in 316 even at low strain.

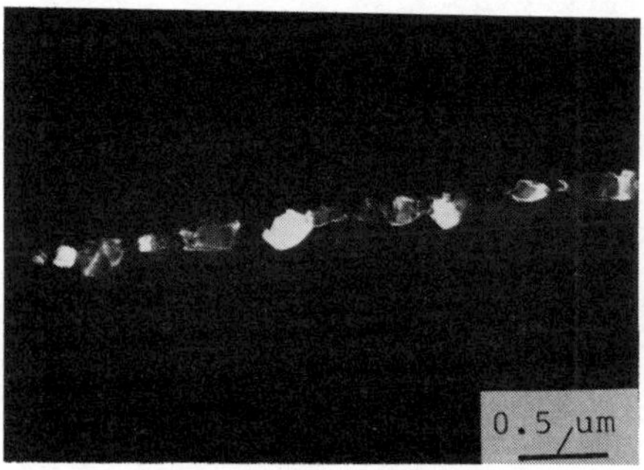

Fig. 7. Dark field image of intergranular precipitation of $M_{23}C_6$ in aged 253MA.

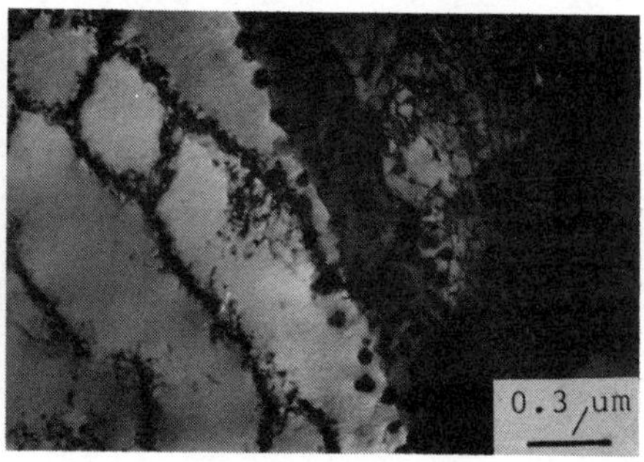

Fig. 8. TEM micrograph of intergranular precipitation of $M_{23}C_6$ in non-aged 316 cycled at $\Delta\varepsilon_p = 1.15\%$ for 30 minutes.

Fig. 9. Dislocation cells formed in 316 at $\Delta\varepsilon_p$ = 1.15%.

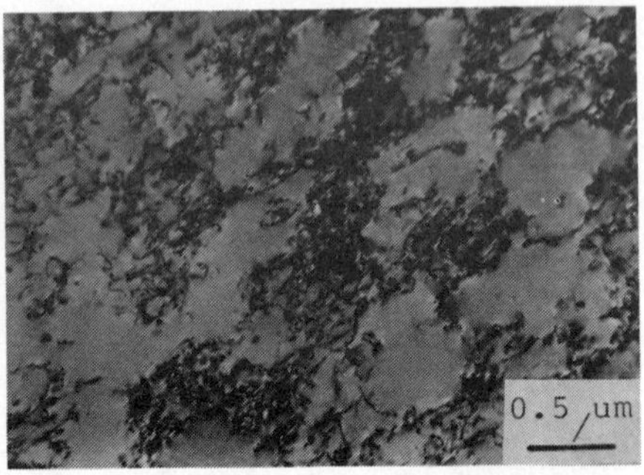

Fig. 10. Irregular dislocation cells formed in 316 at $\Delta\varepsilon_p$ = 0.37%.

Fig. 11. Planar dislocation arrays produced in 253MA at $\Delta\varepsilon_p$ = 0.22%. Dislocation dipoles are visible (arrowed).

Fig. 12. Micrograph showing that planar slip character is retained at high strain. $\Delta\varepsilon_p$ = 1.07%.

DISCUSSION

The results clearly show that there is a significant difference in fatigue resistance between the two alloys, 253MA having approximately twice the endurance of 316 in solution treated conditions. There are also pronounced differences in terms of intragranular slip behaviour. The most striking feature is the absence of cell formation in 253MA indicating that cross slip and climb occur less readily in this alloy than in 316. In a previous report (Nilsson, 1983a) the difficulty of cross-slip in 253MA was attributed to a low stacking fault energy (SFE) combined with an elastic interaction between interstitials and dislocations. A comparatively low SFE in 253MA is supported by experiments which show that N and Si reduce the SFE (Schramm and Reed, 1975), although the relative importance of these two species has not been assessed. Recent work on AISI 316N, containing 0.20% N, has revealed a transition to a planar slip mode and a concomitant improvement of fatigue resistance as compared to 316 (Nilsson, 1983b). The fatigue life was at low strain ($\Delta\varepsilon_p$ < 0.40%) increased by almost a factor of 2, i.e. approached the strength of 253MA. However, at high strain ($\Delta\varepsilon_p$ > 0.40%) the difference in fatigue life between 316 and

316N was somewhat reduced. Similar qualitative effects on the fatigue strength of 316 have recently been observed by Degallaix and co-workers (1982).

These results provide evidence that N contributes toward improved fatigue properties although it is apparent that the sole effect of N is insufficient to explain the performance of 253MA. Conceivable additional factors which should be investigated more closely are for instance the role of Si and Ni - both of which alter the SFE (Schramm and Reed, 1975) - and the influence of oxidation resistance on crack nucleation. It may be inferred from the results shown in Fig. 6 that the high fatigue resistance of 253MA at 600°C can be ascribed to a retarded crack nucleation in conjunction with a low crack propagation rate during stage II. This is consistent with results obtained by other workers who report both prolonged crack initiation stages (McEvily and Böttner, 1963; Laird and Feltner, 1967) and low crack propagation rates (McEvily and Böttner, 1963; Miller and co-workers, 1965; Saxena and Antolovich, 1975) in planar slip alloys. Although most of these studies were performed on the Cu-Al system the present results indicate that this principle may have a more general validity.

The advantageous effects of a planar slip mode seem to be explicable in terms of the concept of reversible plastic flow used by Grosskreutz (1972). Regarding crack initiation, reversible slip may facilitate slip step elimination at the surface (Margolin and co-workers, 1976), whereas irreversible slip often results in the formation of crack nuclei at grain boundaries through surface rumpling (Laird and Feltner, 1967). Similarly, the effect of planar slip on crack propagation may be understood using this concept, since Neumann (1974) has confirmed by direct observations of a moving crack tip that striations are formed as a result of 'imperfect slip reversal'.

Concerning the effect of aging to produce heavy intergranular precipitation, there appears to be a reduction in fatigue resistance as compared to non-aged material. This effect seems to be more pronounced in 253MA than in 316. The reason for this loss in fatigue strength has not been investigated in detail, but Fig. 3 suggests that the more brittle boundaries in aged material reduce the crack initiation stage. However, the effect of aging is less severe than under creep conditions, when a drastic loss in creep ductility occurs in these alloys at 600°C as a consequence of the intergranular precipitation of $M_{23}C_6$.

CONCLUSIONS

1. Sandvik 253MA exhibits a higher resistance to low cycle fatigue at 600°C than AISI 316, the difference in number of cycles to failure is approximately a factor of 2.

2. Aging these alloys for 1000h at 600°C produces heavy grain boundary precipitation of $M_{23}C_6$, and results in a somewhat reduced fatigue resistance in both alloys.

3. Planar slip prevails in 253MA as opposed to wavy slip in AISI 316.

4. The fatigue results are explicable in terms of the difficulty of dislocation cross-slip and climb in 253MA which promotes partially reversible plastic flow.

5. Measurement of striation spacings indicate that the comparatively high fatigue resistance of 253MA can be related to a long crack initiation stage in conjunction with a comparatively low crack propagation rate.

ACKNOWLEDGEMENTS

This paper is published by permission of Sandvik AB. Valuable discussions with Mr. T. Andersson, Dr. G.L. Dunlop and Prof. A.J. McEvily as well as technical assistance of Mr. G. Svensk is gratefully acknowledged.

REFERENCES

Coffin, L.F. (1977). Proceedings 4th ICF Meeting, Waterloo, Canada, Vol 1, 263-292

Degallaix, S., J.B. Vogt, and J. Foct (1982). Mem. Sci. Rev. Met., 9. 443.

Forwood, C.T., and P. Humble (1970). Aust. J. Phys., 23, 697-704.

Grosskreutz, J.C. (1972). Met. Trans., 3, 1255-1262.

Kim, W.H., and C. Laird (1978a) Acta Met., 26, 777-787.

Kim, W.H., and C. Laird (1978b) Acta Met., 26, 789-799.

Laird, C., and C.E. Feltner (1967). Trans. Met. Soc. AIME, 239, 1074-1083.

Margolin, H., Y. Mahajan, and Y. Saleh (1976). Scripta Met., 10, 1115-1118.

McEvily, A.J., and R.C. Böttner (1963). Acta Met., 11, 725-743.

Miller, G.A., D.H. Avery, and W.A. Backofen (1966). Trans. Met. Soc.AIME, 236, 1667-1673.

Neumann, P. (1974). Acta Met., 22, 1155-1165.

Nilsson, J.-O. (1983a). Scripta Met., to appear.

Nilsson, J.-O. (1983b). Sandvik internal report.

Saxena, A., and S.D. Antolovich (1975). Met. Trans. A, 6, 1809-1828.

Schramm, R.E., and R.P. Reed (1975). Met. Trans. A, 6, 1345-1351.

Wareing, J., and H.G. Vaughan (1977). Met. Sci., 11, 439-446.

CREEP MODELING OF TEXTURED ZIRCALOY UNDER BIAXIAL STRESSING

B. L. Adams* and K. L. Murty**

*Department of Materials Science and Engineering, University of Florida, Gainesville, Florida, USA

**Department of Nuclear Engineering, N. Carolina State University, Raleigh, N. Carolina, USA

ABSTRACT

Anisotropic biaxial creep behavior of textured Zircaloy tubing was modeled using a crystal-plastic uniform strain-rate upper-bound and a uniform stress lower-bound approach. Power-law steady-state creep is considered to occur on each crystallite glide system by fixing the slip rate to be proportional to the resolved shear stress raised to a power. Prismatic, basal, and pyramidal slip modes were considered. The crystallographic texture is characterized using the orientation distribution function determined from a set of three pole-figures. This method is contrasted with a Von-Mises-Hill phenomenological model in comparison with experimental data obtained at 673°K. The resulting creep-dissipative loci show the importance of the basal slip mode on creep in heavily cold-worked cladding, whereas prismatic slip is more important for the recrystallized materials.

KEYWORDS

Multiaxial deformation; creep; anisotropy; texture; crystallite orientation distribution function; slip; zirconium alloys.

INTRODUCTION

Zirconium alloys are commonly used in light-water reactors as tubing materials to clad nuclear fuel. Creep behavior of these tubes under multiaxial loading has been an important consideration in fuel-life predictions and core licensing criteria. The mechanical properties of hcp metals such a Zircaloy are very sensitive to fabrication variables and the resulting crystallographic texture. Previous efforts in correlating multiaxial creep behavior with texture parameters based solely on the basal-pole distribution have met with only limited success. For example, recrystallized Zircaloy cladding was shown to exhibit relatively (with respect to axial) higher hoop creep resistance in comparison with cold-worked stress-relieved (CW-SR) materials (Beauregard, Clevinger and Murty, 1977; Kallstrom, Andersson, and Hofvenstem, 1974); this result, however, is not readily predicted from considerations of the basal-pole distribution (Stehle,

Steinberg and Tenckhoff, 1977).

We have studied anisotropic multiaxial creep from both phenomenological and crystal-plastic viewpoints (Adams and Hirth, 1981: Murty and Adams, 1980). The phenomenological approach is derived from the Von-Mises-Hill theory of yielding in anisotropic metals (Hill, 1948). The crystal-plastic approach is based on the upper-bound model of Hutchinson (1976, 1977) for power-law steady-state creep. This latter methodology requires a complete description of the crystallite orientation distribution function (ODF).

In this paper both modeling approaches were compared with all known sources of biaxial creep data for Zircaloy at ~673°K; this includes the data of Murty (reported by Adams, 1979); Baty and Associates (1982); and Stehle, Steinberg and Tenckhoff (1977). Steady-state creep data are compared with predictions at constant rates of creep dissipation. It is shown that an acceptable degree of agreement with the data is possible with both the crystal-plastic bounding analysis and the phenomenological approach. An additional benefit of the crystal-plastic bounding analysis, however, is the emergence of important insights relating microstructure and texture to creep anisotropy. These new insights may be utilized to select processing parameters which optimize the cladding's technological performance in specific applications.

MATERIALS

The most commonly observed texture of Zircaloy tubing is a bimodal distribution of the basal poles concentrated in the plane normal to the tube axis (r-θ plane) with intensity maxima at ±20-45 deg from the radial direction (Fig. 1). The basal [0001] pole texture angle $\hat{\phi}$ (defined as the angle between the radial direc-

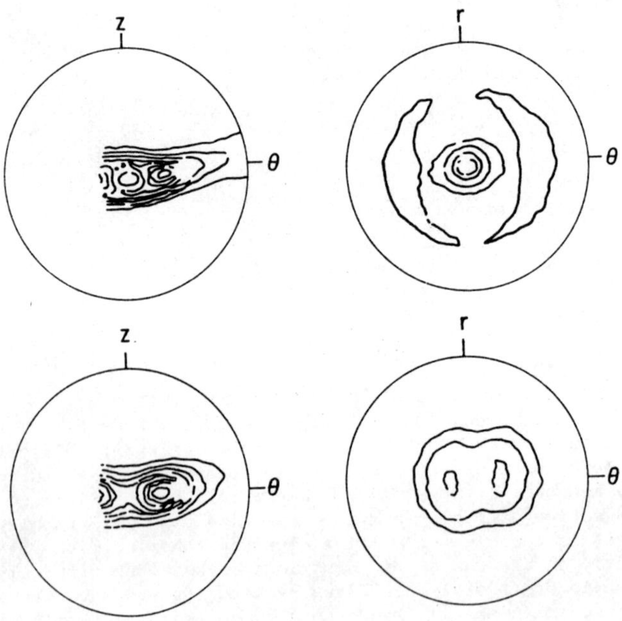

Fig. 1 Basal (left) and prism pole distributions for CW-SR (top) and S-2 claddings.

tion and the $\hat{c}$-axis of the average crystal in the r-θ plane) is sensitive to the strain path or cold work prior to final annealing. The final annealing temperature has at best a minor effect on the basal pole angle and increases the population of randomly oriented crystallites (Fig. 1). However, the final annealing temperature significantly affects the prism [10$\bar{1}$0] pole distribution as illustrated in Fig. 1 where the texture pole-figures for cold-worked stress-relieved (CW-SR) and recrystallized (S-2) tubing are depicted. The major texture modification due to increased amounts of partial recrystallization involves the rotation of the basal plane about the $\hat{c}$-axis. Thus there exists a relatively high population of prism poles along the axial direction in CW-SR while the population at ±30 deg from the axial direction is increased at complete recrystallization and relatively few first order (10$\bar{1}$0) prism poles are observed along the tube axial direction.

Zircaloy-4 is a dilute alloy of Zirconium containing ~1.5 weight percent tin and minor additions of Fe, Cr and Ni. The tubing was tested in the as-received condition. The data of Stehle, Steinberg, and Tenckhoff (1977) (designated as ⑤) is also for a fully recrystallized cladding, but the details of its processing history and crystallographic texture were not available.

VON-MISES-HILL FORMULATION

Hill (1948) proposed a modified Von-Mises Yield criterion for anisotropic material when the applied stress axes lie coincident with the axes of material anisotropy. Later modifications by Backofen (1972) and Duncombe, Friedrich and Guilinger (1971), in conjunction with the Prandtl-Reuss energy balance (see Lucas and Bement, 1975) allow expression of the principal creep-rates at constant structure (steady-state creep) in the form:

$$\begin{pmatrix} \dot{e}_r \\ \dot{e}_\theta \\ \dot{e}_z \end{pmatrix} = \frac{\dot{e}_g}{P(R+1)\sigma_g} \begin{pmatrix} (R+P) & -R & -P \\ -R & R(P+1) & -RP \\ -P & -RP & P(R+1) \end{pmatrix} \begin{pmatrix} \sigma_r \\ \sigma_\theta \\ \sigma_z \end{pmatrix}. \qquad (1)$$

Here r, θ, z refer to principal directions in the cladding, and R, P refer to the anisotropy parameters given by contractile strain (rate) ratios in the uniaxial (z) and uniaxial-hoop (θ) tests respectively. $\dot{e}_g$ and σ_g are the generalized strain (rate) and generalized stress. Equation (1) can be normalized to the uniaxial creep-rate in a cylindrical specimen loaded axially to a stress σ_g, given in its complete form as:

$$\sigma_g^2 = \frac{R(\sigma_r - \sigma_\theta)^2 + RP(\sigma_\theta - \sigma_z)^2 + P(\sigma_z - \sigma_r)^2}{P(R+1)} \qquad (2)$$

At a constant energy dissipation-rate ($\dot{W} = \sigma_{ij}\dot{e}_{ij} = \sigma_g\dot{e}_g$ = constant) equation (1) becomes a locus of creep dissipation which can be readily compared with the experimental data.

CRYSTAL PLASTIC FORMULATION

Hutchinson (1976, 1977) first adapted the slip models of crystal plasticity to power-law steady-state creep constitutive behavior of polycrystalline aggregates, including isotropically distributed hexagonal polycrystals. His uniform strain-rate upper-bound estimate is readily adapted to a full textural description of the metal. Three prominent slip modes have been allowed—{0001}<11$\bar{2}$0> basal, {10$\bar{1}$0}<11$\bar{2}$0> prismatic, and {11$\bar{2}$2}<$\bar{1}\bar{1}$23> pyramidal. The creep-rate in the crystallite is given as:

$$\dot{e}^{c}_{ij} = \sum_{k} \mu^{k}_{ij} \dot{\gamma}^{k} , \qquad (3)$$

where $\dot{\gamma}^k$ is the slip-rate in the kth slip system, and μ^k_{ij} is the geometric tensor connecting slip coordinates with crystallite coordinates. For power-law steady-state creep the slip-rate can be expressed in the form:

$$\dot{\gamma}^k = (A/\tau_o^{\ k}) \left| \tau^k/\tau_o^{\ k} \right|^{n-1} \tau^k \qquad (4)$$

where A is the reference strain-rate, and τ_o^k is the reference shear stress—always a positive quantity in this context. n is the stress exponent. The shear stress is also related to the crystallite stress using the fundamental tensor:

$$\tau^k = \sigma_{ij} \ \mu^k_{ij} \qquad (5)$$

If the stress is presumed known in equation (5) (i.e., $\sigma^c_{ij} = \sigma_{ij}$), a lower-bound estimate of the creep-rate, $\dot{e}^{lb}_{ij}$, can be calculated from the ODF (here represented as $f(\tilde{g})$, where $\tilde{g}$ denotes orientation) using the averaging procedure:

$$\dot{e}^{lb}_{ij} = \{\dot{e}^{c}_{ij}\} = \int_{\tilde{g}} f(\tilde{g}) \cdot \dot{e}^{c}_{ij} \, d\tilde{g} \ . \qquad (6)$$

In the lower-bound estimate stress equilibrium is artificially preserved at the expense of violating strain-rate compatibility.

The upper-bound estimate is computed by iteratively solving equation (3) for the stress, σ^c_{ij}, at a fixed (imposed) strain-rate $\dot{e}^c_{ij} = \dot{e}_{ij}$ (macroscopic strain-rate). Upper-bound stresses, σ^{ub}_{ij} are estimated from the textural average of crystallite stresses:

$$\sigma^{ub}_{ij} = \{\sigma^{c}_{ij}\} = \int_{\tilde{g}} f(\tilde{g}) \cdot \sigma^{c}_{ij} \, d\tilde{g} \ . \qquad (7)$$

Thus strain compatibility is preserved but stresses vary across crystallite boundaries; in this manner the first order distribution of internal stress is approximated. (Note that the upper-bound method substantially over-estimates the magnitude of the internal stress.)

Equations (3) through (7) allow calculation of the rate of creep dissipation, $\dot{W} = \sigma_{ij}\dot{e}_{ij}$. For biaxial stressing a locus of combined stresses can be plotted in stress space for which creep occurs at constant $\dot{W}$. One can readily demonstrate, for both the upper- and lower-bounds, that stress components scale with rate of creep dissipation according to:

$$\sigma^{o}_{ij} = \left[\frac{(A/\tau^{n}_{o})}{(A/\tau^{n}_{o})^{o}} \left(\frac{\dot{W}^{*}}{\dot{W}}\right)\right]^{\frac{1}{n+1}} \sigma_{ij} \tag{8}$$

where $\dot{W}^{*} = \sigma^{*}_{ij}\dot{e}^{*}_{ij}$. Since the constant factor A/τ^{n}_{o} in equation (4) is not known a priori; it is usually necessary, not only to scale $\dot{W}$, but also to scale A/τ^{n}_{o} in order to compare model predictions with experimental data. If we denote experimental values of the stress component as σ^{o}_{ij}, we can compute consistent values of the constant A/τ^{n}_{o} (here denoted as $(A/\tau^{n}_{o})^{o}$ by comparison with experimental values.

Upper- and lower-bound estimates were calculated for the tube textures considered in the experimental data, utilizing various choices for the reference shear stress τ_{o}. τ_{o} was assumed to be a function of slip-mode (i.e., τ^{basal}_{o} τ^{prism}_{o}, and $\tau^{pyramidal}_{o}$) and independent of resolved shear stress for a particular loading condition. As with the phenomenological formulation, the crystal-plastic method was found to be capable of predicting the loci of creep dissipation taken from all three data sets.

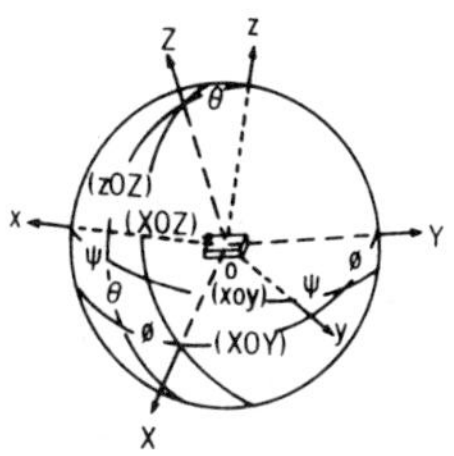

Fig. 2 Definition of Euler angles ψ,θ,ϕ connecting crystallite coordinates (O-XYZ) and fabrication coordinates (o-xyz).

THE CRYSTALLITE ORIENTATION DISTRIBUTION FUNCTION (ODF)

The ODF $f(\tilde{g}) = f(\psi,\theta,\phi)$ expresses the probability that a crystallite has an orientation (ψ,θ,ϕ) with respect to the specimen axes. The Euler angles are defined in Fig. 2. The $\hat{c}$-axis of the crystallite was chosen to be coincident with the Z-axis in Fig. 2, while X and Y are parallel to digonal Type-II $[10\bar{1}0]$ and Type-I $[11\bar{2}0]$ axes respectively. The series method of Bunge (1965) and Roe (1965) was utilized in conjunction with symmetry restrictions, enumerated by Morris and Heckler (1969) for hexagonal crystallite symmetry and orthrotropic specimen symmetry, to approximate the ODF from three pole-figures.

Figures 3 and 4 show the ODF for CW-SR and S-2 cladding. The necessary range of angles for representing the ODF is 0 to $\pi/2$ in θ and ψ, and 0 to $\pi/3$ for ϕ in hexagonal materials. Each separate plot represents the ODF at constant ϕ. As expected, the CW-SR ODF (Fig. 3) shows a very pronounced texture for orientations fixing the $\hat{c}$-axis on (or near) the transverse plane ($\theta \simeq \pi/2$). The times-random intensity was found to reach values as high as 720. This extreme sharpness of texture coupled with the truncation of the series expansion at sixteenth order lead to larger than usual negative regions in the ODF, approaching 30% of peak height. For less pronounced textures negative regions of 10% are typical. It would be desirable to increase the number of pole figures from 3 to 5 or more in order to improve the fit of the series expansion.

The degree of texture severity is markedly reduced in the S-2 cladding where the times-random intensity had a peak value of 8. The major difference, as compared with the CW-SR texture, is that now the peak intensity occured at $\phi = \pi/6$ ([10$\bar{1}$0] prism direction located at ~π/6 from the cladding axis). The predominant location of the crystallite $\hat{c}$-axis was again in the cladding transverse plane. Again, this is consistent with the features of the basal and prism pole figures shown in Fig. 1.

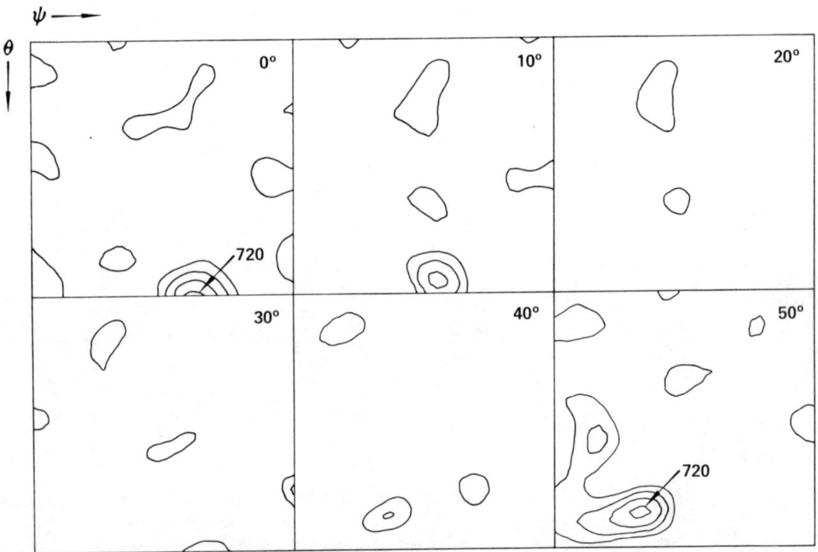

Fig. 3 Crystallite orientation distribution function (ODF) for CW-SR cladding.

RESULTS AND DISCUSSION

Figure 5 illustrates the measured creep dissipative loci for the CW-SR, S-2, and 5 data. These are fitted with selected values of R and P using equation (1). The data were normalized to creep dissipation rates of 3 X 10^{-2} MPa/hr.

The annealed cladding types (S-2 and 5) were found to exhibit near planar isotropy, but with substantial through thickness resistance to creep. In contrast, CW-SR cladding showed substantially higher creep resistance in the axial direction when compared with both types of annealed cladding. Each of the three type of cladding showed marked deviations from isotropic behavior.

The results of the crystal plastic modeling are illustrated in Figs. 6 and 7. In these figures Z represents the axial stress for creep under uniaxial-axial loading at fixed $\dot{W}$. Figure 6a,b illustrates the comparisons of experimental dat with upper- and lower-bound predictions for easy prism slip ($\tau_o^{prism} \lesssim 10\ \tau_o^{basal}$, $\tau_o^{pyramidal}$). Although minor differences may be apparent between the CW-SR and

S-2 textures, the shapes are very similar. (Upper-bound predictions were not available for $\sigma_z/\sigma_\theta < 1$, but the general shape is expected to be similar to the lower-bound results.) Both models predict substantial creep resistance for internal pressurization loading ($\sigma_z/\sigma_\theta = 0.5$). For easy basal slip ($\tau_o^{basal} \lesssim 10\ \tau_o^{prism}, \tau_o^{pyramidal}$), as illustrated in Fig. 7a,b, substantial creep resistance in the axial direction is predicted as compared with hoop direction. The effect is most pronounced for the lower-bound model, but the upper-bound model closely fits the experimental data for CW-SR creep. Adams and Hirth (1981)

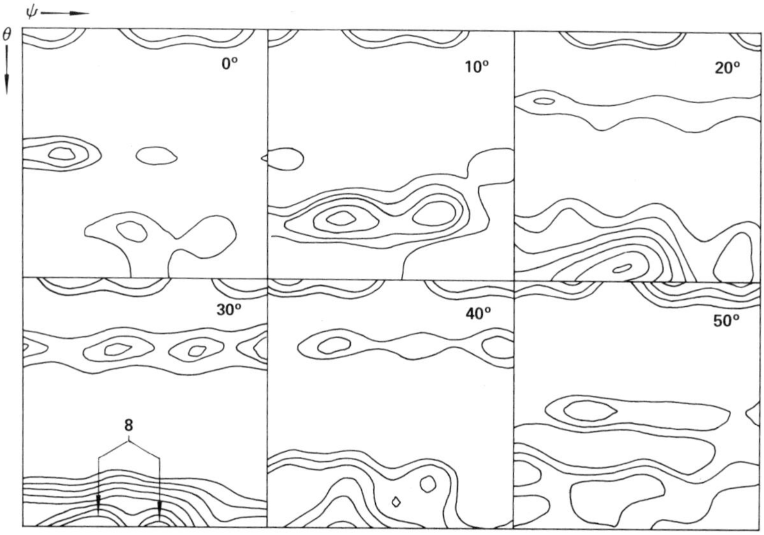

Fig. 4 Crystallite orientation distribution function (ODF) for S-2 cladding.

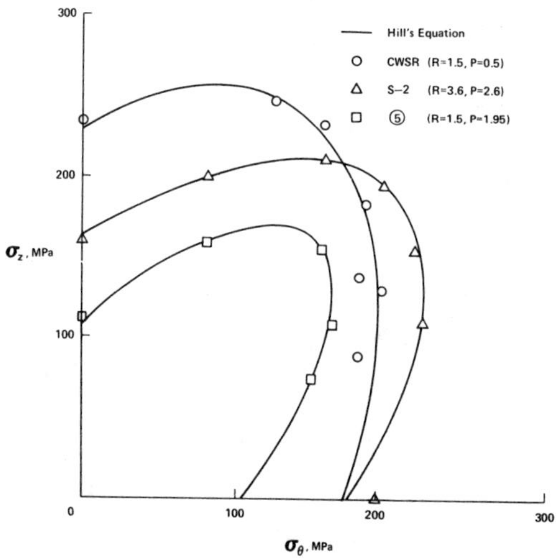

Fig. 5 Creep dissipative loci for CW-SR, S-2, and 5 cladding types at W = 3 X 10^{-2} MPa/hr. Solid lines indicate predictions of Hill's equation for selected values of R_1P.

previously suggested that this implies a relaxation of basal slip with respect to prismatic slip at creep temperatures of 673°K, presumably due to prior cold working.

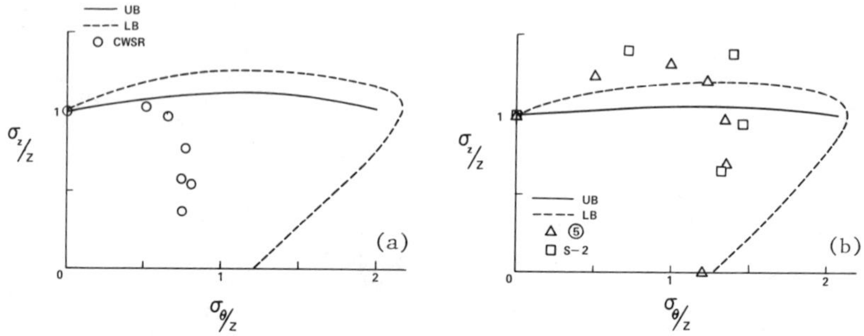

Fig. 6 Upper- and lower-bound model predictions for easy {10$\bar{1}$0}<11$\bar{2}$0> prism slip ($\tau_o^{prism} \lesssim 10\ \tau_o^{basal}, \tau_o^{pyramidal}$). (a) CW-SR texture, (b) S-2 texture

For easy pyramidal slip ($\tau_o^{pyramidal} \lesssim 10\ \tau_o^{prism}, \tau_o^{basal}$), we predict a nearly isotropic behavior for both upper- and lower-bound models. This is not surprising considering that pyramidal slip alone provides the five degrees of freedom required for a general deformation. It is apparent that combinations of prism slip and pyramidal slip (or prism slip and basal slip) could be selected which would more closely approximate the creep behavior of the annealed S-2 cladding.

Note that the crossover of the upper- and lower-bound predictions which is so evident in Fig. 7 does not imply that the two models are not 'upper' and 'lower' bounds on the creep stress at constant $\dot{W}$. The crossover is due to normalizing the axial stress under uniaxial-axial loading to the experimental value. Thus, in plots shown in Figures 6 and 7 we selected different values of $(A/\tau_o^n)^o$ to normalize upper- and lower-bound predictions. At equivalent values of $(A/\tau_o^n)^o$, the

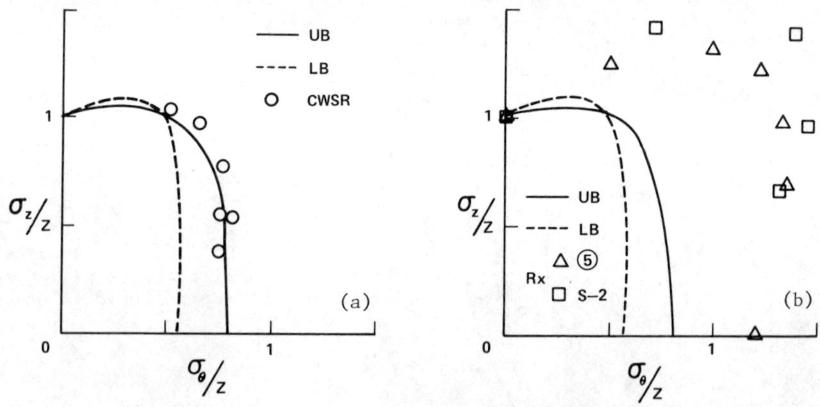

Fig. 7 Upper- and lower-bound model predictions for easy {0001}<11$\bar{2}$0> basal slip ($\tau_o^{basal} \lesssim 10\ \tau_o^{prism}, \tau_o^{pyramidal}$). (a) CW-SR texture, (b) S-2 texture.

two predictions would indeed constitute 'upper' and 'lower' bounds. Analyses along these lines are underway and will be published later.

ACKNOWLEDGEMENT

One of the authors (B. L. Adams) gratefully acknowledges the support of the National Science Foundation under grant DMR-8112970.

REFERENCES

Adams, B. L. (1979). An Upper-bound Analysis of Inelastic Deformation in Textured Hexagonal Polycrystals (Zircaloy). Ph.D. Dissertation, The Ohio State University.

Adams, B. L. and J. P. Hirth (1981). In M. F. Ashby, R. Bullough, C. S. Hartley and J. P. Hirth (Eds.), Dislocation Modelling of Physical Systems. Pergamon Press, Oxford. pp. 368-372.

Backofen, W. A. (1972). Deformation Processing. Addison-Wesley, Reading, Massachusetts.

Baty, D. L., W. A. Pavinich, M. R. Dietrich, G. S. Clevinger, and T. P. Papazoglou. In Proc. 6th Int. Conf. on Zirconium in the Nuclear Industry, Vancouver 1982. To be published, A.S.T.M.-STP.

Beauregard, R. J., G. S. Clevinger and K. L. Murty (1977). In Proc. SMiRT-IV, San Francisco. Paper C3/5.

Bunge, H. J. (1965). Z. Metallk., 56, 872.

Duncombe, E., C. M. Friedrich and W. H. Guilinger (1971). Nucl. Tech., 12, 194.

Hill, R. (1948). Proc. R. Soc. (London), 193A, 281.

Hutchinson, J. W. (1976). Proc. R. Soc. (London), 348A, 101-128.

Hutchinson, J. W. (1977). Met. Trans., 8A, 1465.

Kallstrom, K., T. Andersson and A. Hofvenstem (1974). In Proc. 3rd Int. Conf. Zirconium in the Nuclear Industry, A.S.T.M.-STP 551, pp. 160.

Lucas, G. E. and A. L. Bement (1975). J. Nucl. Mater., 55, 246.

Morris, P. R. and A. J. Heckler (1969). Trans. A.I.M.E., 245, 1877.

Murty, K. L. and B. L. Adams (1982). In Mechanical Testing for Deformation Model Development. A.S.T.M.-STP 765, pp. 382.

Roe, R. J. (1965). J. Appl. Phys., 36, 2024.

Stehle, H., E. Steinberg and E. Tenckhoff (1977). In Proc. 4th Int. Conf. Zirconium in Nuclear Industry, A.S.T.M.-STP 633, pp. 486.

THERMAL FATIGUE TEST BY DIRECT PASSAGE METHOD OF LARGE ELECTRIC CURRENT ON STAINLESS STEELS

K. Kamachi*, N. Tani, T. Ishida**, M. Kawano** and T. Kuhohori***

*Faculty of Science & Technology, Kinki University, Higashiosaka, 577, Japan
**Nagasaki Institutes of Science, Aba, 851-01, Nagasaki, Japan

ABSTRACT

In the study of thermal fatigue, direct passage method of large electric current was newly established. Thermal impact were imposed on stainless steels under tensile or compressive stress, and S-N curves were obtained. Thermal stresses were measured by X-ray method which has non-contactive and non-destructive nature.

KEYWORDS

Thermal fatigue; Thermal stresses; X-ray stress measurement; Inertial thermo nuclear fusion reactor; Stainless steel.

INTRODUCTION

In considering a number of equipments and constructions working at elevated temperature under high pressure, the composing materials are not allowed to deform by the constraints of their surroundings, therefor the increment of temperature produces compressive stress on the materials and later tensile stress would be generated by succeeding cooling.
When thermal impact cycles are imposed on a structure repeatedly, thermal fatigue is induced on the structure.
In the case of inertial thermo nuclear fusion reactor severe thermal impacts would be imposed on the wall at every implosion, and the mechanical behaviours of materials under such repeated thermal impacts are one of most important problems.
In order to study the fatigue strength by these repeated thermal impacts, a number of impact heating should be applied to specimen rapidly. The direct passage method of large electric current intermittently was developed in our laboratory.
By this method, comparatively rapid heating of specimen is feasible in wide range of specimen uniformly.

The strains by the thermal fatigue tests were plotted against repeated numbers of thermal impacts to get S-N curves, and metallographical observations were made as well as stress measurement at elevated temperature by X-ray stress measureing method.

EXPERIMENTAL PROCEDURE

Testing Device of Thermal Fatigue Test by Direct Passage Method of Large Electric Current

In order to study the thermal fatigue,impact heatings of specimen were carried out by direct passage of 1.3 KA electric current to the specimen intermittently. The testing device of the experiment is shown in Fig.1(a), and the block diagram of the circuit is in Fig.1(b). One end of the specimen is fixed to the electrode E-1 and the other end is fixed to the electrode E-2, which is slidable in the horizontal direction, and the both electrodes are cooled by water. The electrode E-1 is connected to a load cell and the electrode E-2 can be pulled or pressed through a connecting rod to give tensile or compressive stress to the specimen by a lever mechanism.
The temperature of the specimen is controlled by the passage time of electric current and by the interval of intermittent passage, which is regulated by an I.C. timer and a thyrister circuit.

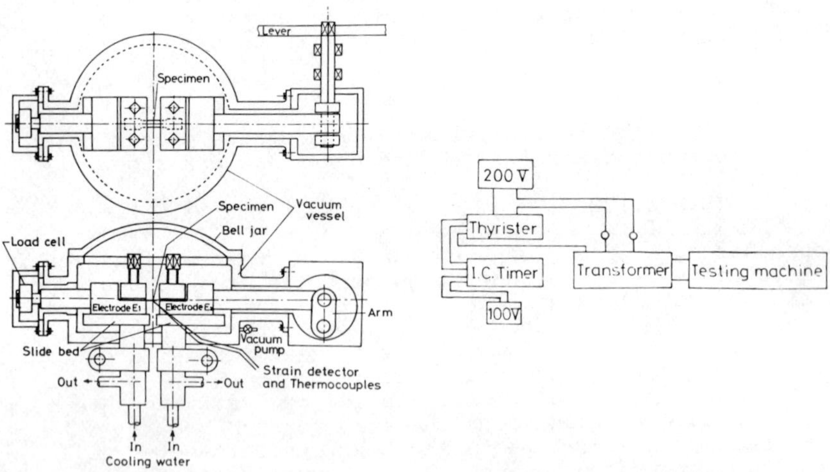

(a) The testing device (b) The block diagram

Fig.1 The testing device and the block diagram

Specimen

Specimens employed are 2 x 10^{-3}m thick plate of SUS 316 and SUS 329 having rectangular shape. The chemical compositions is as shown in TABLE 1.
SUS 316 is a stable austenite steel and 329 is a two phase steel

of mixed structure of b.c.c.(α) and f.c.c.(γ) phases.

TABLE 1 Chemical Compositions (Mass %)

Specimen	C	Si	Mn	P	S	Ni	Cr	Mo	Cu	Ti	Al
SUS 329	0.019	0.54	0.50	0.033	0.006	5.05	23.03	1.52	0.87	0.015	0.019
SUS 316	0.057	0.61	0.50	0.027	0.001	12.18	17.57	2.15	0.26	-	-

Measurement of Stresses at Elevated Temperature

Stresses of specimen were measured at elevated temperature by X-ray stress measuring method, using an apparatus that had been specially designed for high temperature use.
The calculation of stresses was made by the next equation.

$$\sigma = -\frac{E}{2(1+\nu)}\cdot\cot\theta\cdot\frac{\partial 2\theta_\psi}{\partial \sin^2\psi} = K\cdot M \qquad (1)$$

where, K is X-ray constant that is determind for the specimen by the X-ray measurement and M is the tangent of a stright line for $\Delta 2\theta$ plotted against $\sin^2\psi$ diagram with the incident angle of ψ. (S.Taira)

RESULTS AND DISCUSSIONS

Thermal Impact

One pulse of thermal impact by a passage of 1.3 KA electric current for 20/60, 30/60, 60/60 cycle increases the temperature of specimen up to about 800°K, 1000°K and 1500°K respectively, and so on. And the temperature decreases back to room temperature in 30 40 seconds as shown in Fig.2. Repetitions of intermittent thermal cycles give rise to some temperature versus time (Number of repetition) relation as shown in Fig.3.

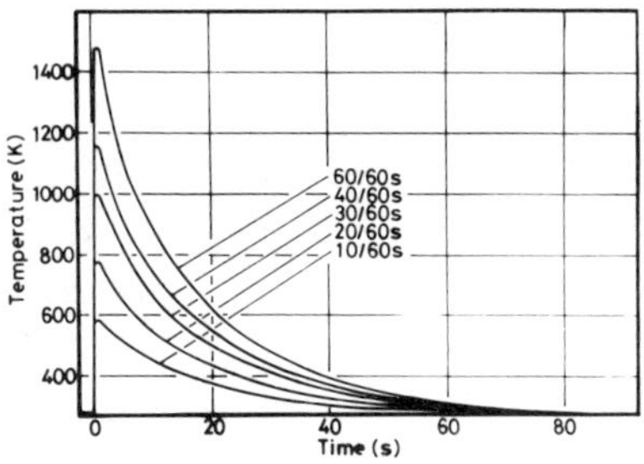

Fig.2 Time-Temperature curve by one pulse of thermal impact.

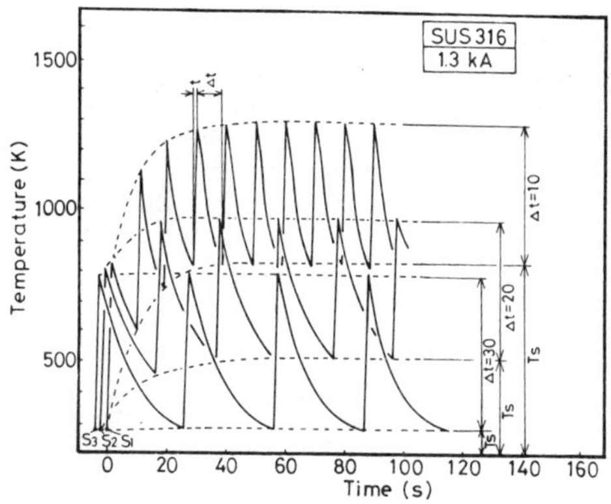

Fig.3 Temperature of specimen for passage of 1.3 KA. (Passage time:t, and interval: Δt)

As can be seen from Fig.3, the initial stage of the temperature variation is unstable, but it attains the steady state within several seconds. (Origins of curves in Fig.3

is shifted for convenience.)

Thermal Fatigue Test

Thermal impacts were given to the specimen fixed to the electrodes by the manner mentioned above under tensile, compressive or unloaded condition. Typical strain behaviours are shown in Fig.4, in which the ordinate is strain and the abscissa is repeated numbers of impact heating. Fig.4(a) is a curve for SUS 316,Fig.4 (b) is under the same condition but unstressed and Fig.4(c) is under compressive stress of -49 MPa. Fig.5 is the same sorts of curves for SUS 316 and SUS 329 (Each serration are omitted). Applied stresses against repeated numbers of thermal impact, S-N curve is shown in Fig.6.

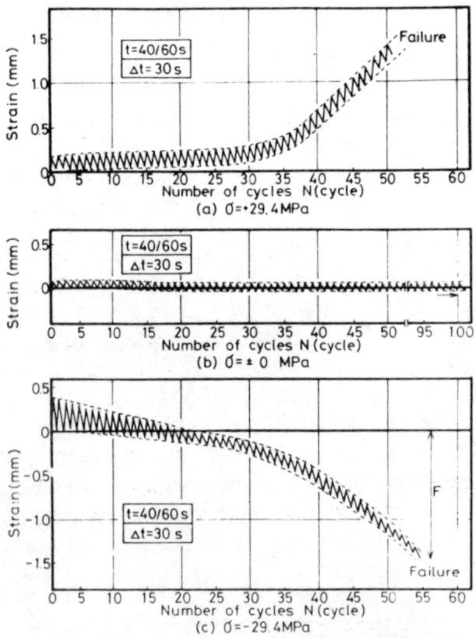

Fig.4 Typical curves of strain behaviour of SUS 316,(a)under tensile stress of 29.4 MPa, (b) under no loaded and (c) under compressive stress of -29.4 Mpa.

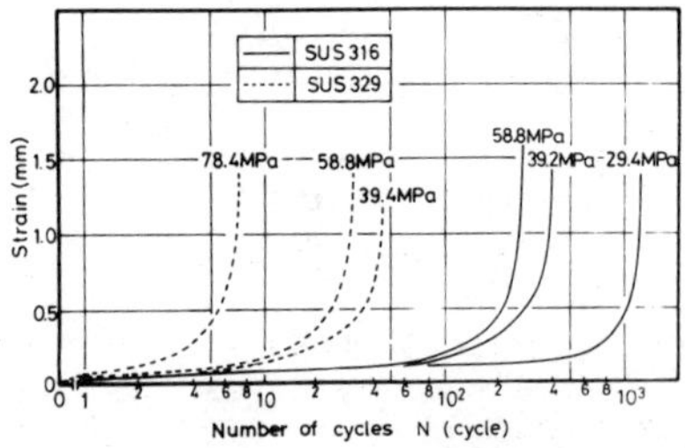

Fig.5 Strain behaviours of SUS 316 and SUS 329.

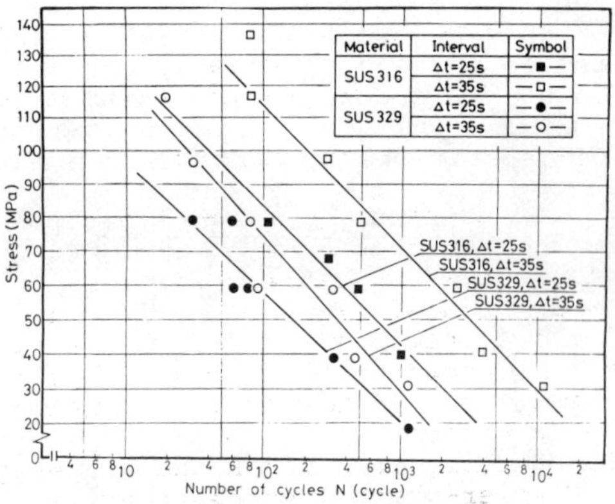

Fig.6 S-N curve of thermal fatigue tests.(t=30/60)

Enlarged views of serrations of the temperature-time and strain - time curves are shown in Fig.7 and Fig.8 respectively. As can be seen, the increasing ratio of the temperature and the strain behave differently for tension and compression. The difference of strain might be induced by the difference of mechanical properties at elevated temperature, and the difference of the increasing ratio of temperature might be due to the difference of electrical resistance owing to the atomic vibration mode around the lattice sites by a stressed state. (The details will be reported elswhere.)

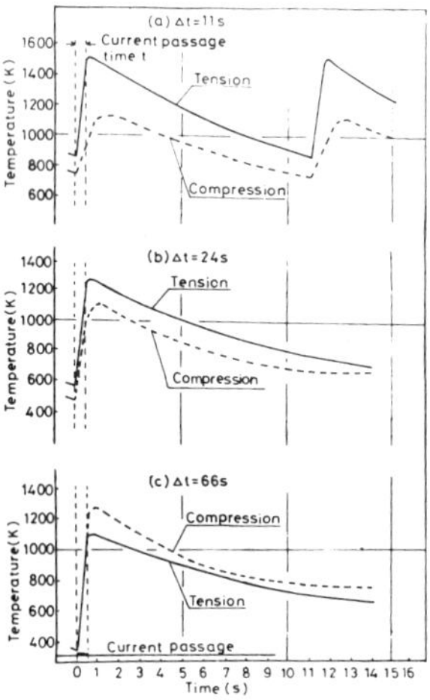

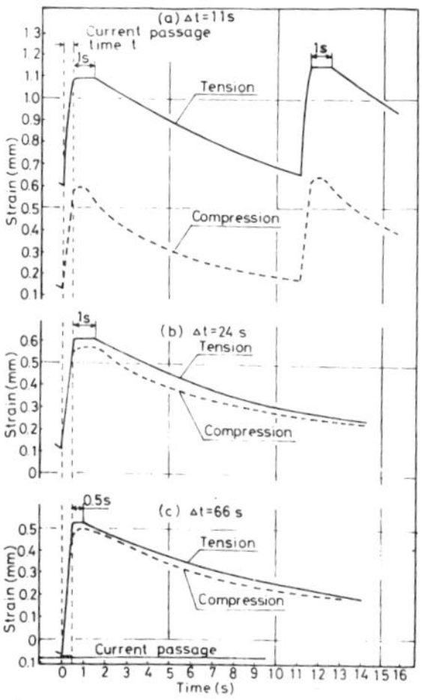

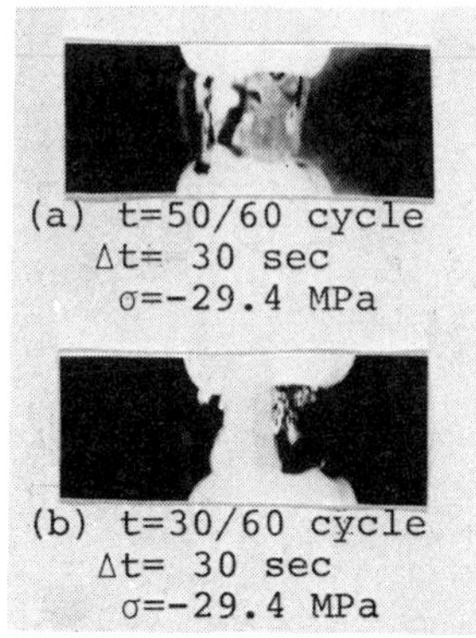

Fig.7(left) Enlarged view of a serration of temperature vs time relation and

Fig.8 (Right) One of a strain serration.

Photo.1 Fractured part of failed specimen under compression.

The envelopes of strain serrations shown in broken lines are corr-sponding to a creep curve, and under a tensile stress the fracture occurs accompanying a lateral contraction. However, under a com-pressive stress a bulgy deformation is induced during the heating process, a tensile stress is generated during the following cool-ing process with little elongation and fracture occurs.
The later fracture mode resembles to that of fatigue fracture and cracks are mostly trans-crystalline type. Photo. 1 shows the fractured parts of some specimen failed under compressive stress of -29.4 MPa and passage time of 50/60 cycle at every 30 seconds.

Stress Measurement and Inter-phasial Stresses

It can be seen from Fig.6 that the fatigue strength of SUS 329 is much lower than that of SUS 316, which is considered to be due to the effect of internal fatigue induced by the superposition of inter-phasial stresses. Therefor, the measurements of inter-phasial stresses are made by X-ray method. The measured stresses are plotted in Fig.9. Residual stresses are about 200 MPa in f.c.c. phase and -140 MPa in b.c.c. phase at room temperature respective-ly, and these stresses become zero at arround 600°K. At higher temperature than 600°K the tendency of these stresses is reversed. The inter-phasial stresses against thermal cycles is shown in Fig. 10 and the residual stresses both in b.c.c.and f.c.c. phases remain even after the repetition of thermal impacts of more than 1000 times. From these facts, it may be possible to conclude that two phase alloys is not preferable for the structure which would suffer from thermal fatigue.

Metallographic Observation

In the impact heating cycles, even though the total exposure time to high temperature is not so long, a remarkable grin growth is observed, especially in SUS 329.

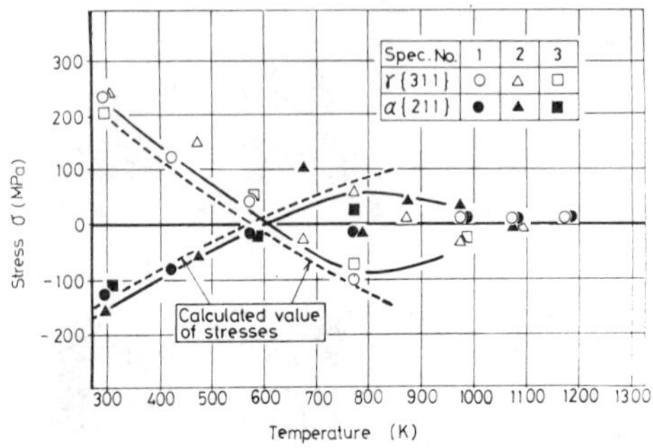

Fig.9 Measured stresses in f.c.c. and b.c.c. phase at elevated temperature.

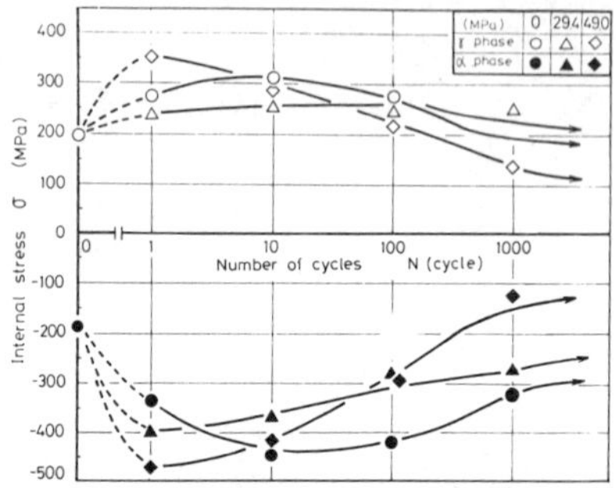

Fig.10 Residual stresses in f.c.c. and b.c.c. phase after repetition of thermal cycles.

Therefor, it can be considered that the grain growth is accelerated during stressing and heating process rather than the holding period at high temperature. Photo.2 shows the accelerated grain growth of SUS 329 and Fig.11 is the meam diameter of grains plotted against repeated numbers of impact heating.

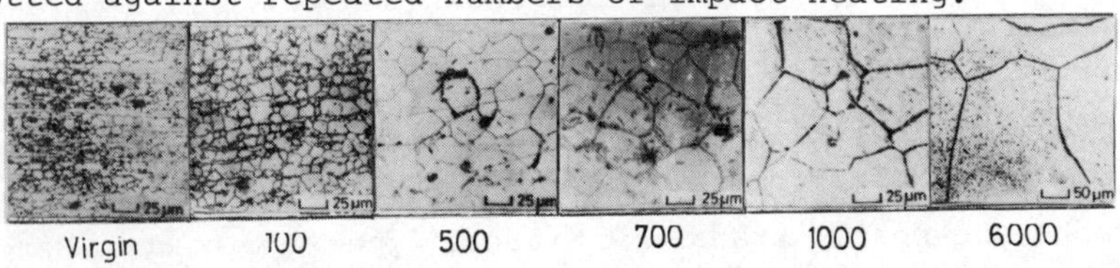

Phto.2 Growing process of crystal grains of SUS 329 by repetition of thermal impacts.

SUMARY

Thermal fatigue is a highly complicated problem, for the metallurgical effects, such as recovery, recrystallization,enhanced diffusion and grain growth, are superposed on the stressing process as well as the deterioration of the mechanical properties. Half width of X-ray diffraction lines decreases at one time and increases at another time, and their tendencies depend on the temperature range and stressing conditions.

The results of the presented investigation are summarized as follows:

(1) In the study of thermal fatigue, experimental procedures by direct passage method of large electric current intermittently have been established. And S-N curves are obtained.

(2) Thermal fatigue strength is much lower than obtained by the mechanicl tests.

(3) By this method, thermal fatigue strength can be estimated in short time.

(4) The strain behaviours of thermal fatigue process is dose to that of creep, it can be say " fatigue-creep", and the fracture

mode by compressive stress is similar to fatigue fracture and trans-crystalline cracks are ofen observed.

(5) In two phase alloys, internal fatigue accelerates the accumulation of fatigue damages owing to the inter-phasial stresses.

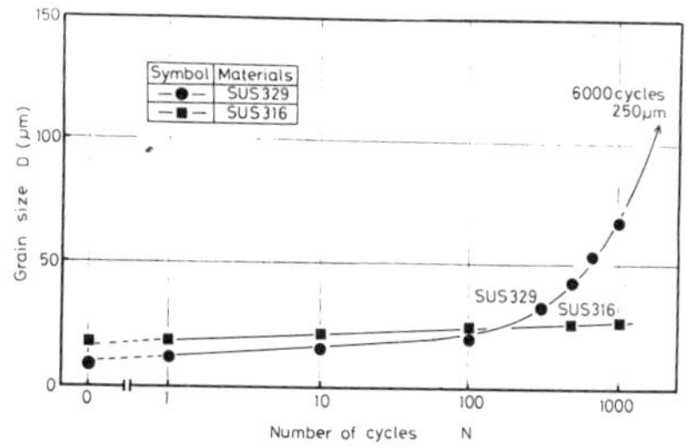

Fig. 11 Mean diameter of crystal grains against repeated numbers of thermal impacts.

REFERENCE

Taira,S. (1974). X-ray Studies on Mechanical Behaviour of Materials. Soc. Material. Sci. Japan.

THE PREDICTIVE CAPABILITY OF THREE HIGH TEMPERATURE LOW CYCLE FATIGUE MODELS IN THE ALLOY IN-738

M. Y. Nazmy and C. Wüthrich

Brown Boveri Research Center, CH-5405 Baden, Switzerland

ABSTRACT

In this investigation three models for high temperature low cycle fatigue were evaluated for their capability to correlate fatigue data and to predict fatigue behavior at 850°C of cast nickel-base IN 738 blade superalloy. These models are the strain range partitioning model, the frequency-separation model, and the model of Ostergren. The results from a series of strain-controlled fatigue tests were used to evaluate these models. A second set of results on the same material, extracted from literature, were predicted using the established relationships for the three models. The best correlation of the data as well as the best predictability of the second set of data were achieved by the strain range partitioning model.

KEYWORDS

Low cycle fatigue; life prediction; strain range partitioning; frequency separation; Ostergren model; nickel-base superalloys.

INTRODUCTION

A major thrust in the development and use of gas turbines is efficiency through better life-cycle management. Turbine blades, and rotors represent expensive and critical engine components. A reliable and accurate lifetime prediction of these components is essential for optimal use of gas turbines. A major difficulty in predicting the crack initiation life is accounting for the high temperature effects of creep-fatigue interaction, and environmental attack. The relative importance of these effects depends on the alloy system and on the loading conditions. Many models have been proposed for predicting the low cycle fatigue life at elevated temperature. Of these models the three prominent ones, that have been applied to other materials, are the subject of study of the present investigation. These models are the strain range partitioning model (Manson, and co-workers, 1971; Manson, 1973; Manson and co-workers, 1975), the frequency-separation model (Coffin, 1976), and the model of Ostergren (Ostergren, 1976). In many cases the experimental data were generated to test one of these models; there are few instances in the literature where the same core of results from one heat of material is used to evaluate the prediction capability of several models.

The purpose of this study is to evaluate the prediction capabilitiy of these models as applied to the alloy system IN 738 at 850°C. For making this evaluation, a sizable data base was used. These data came from a series of strain-controlled fatigue tests at 850°C using seven different types of strain wave shapes. Using the computed coefficients in the main equation of each model, literature data (Witt and co-workers, 1980) were predicted. The significance of using a criterion for crack initiation was also mentioned.

RESULTS AND DISCUSSION

Strain Range Partitioning Model

This model extends the Manson-Coffin law, which is valid at room temperature, to high temperature by considering the interaction of time-dependent inelastic strains and time-independent inelastic strains. At high temperature, both creep and plastic strains are reversed by a combination of creep and plastic strains during compressive loading. The manner in which these different types of inelastic strains are postulated to reverse themselves, is the basis of the strain range partitioning model (Manson, and co-workers, 1971). Using the interaction damage rule proposed by Manson (1973), the high temperature low cycle fatigue data of the present investigation were correlated. A full account of the results, and the experimental procedure are reported by Nazmy (1983). The coefficients in this equation were calculated using a least square fit computer program. In order to make a quantitative comparison of the data correlation capability as well as the predictive capacility of the various models, the standard deviation defined as follows

$$S = \sqrt{\frac{1}{n-1}\sum_{i=1}^{n}[\ln N^{i}_{calc} - \ln N^{i}_{obs}]^{2}} \qquad (1)$$

was used. N^{i}_{calc}, N^{i}_{obs} are the calculated and the observed numbers of cycles to crack initiation, respectively, and n is the number of data points. Figure 1, shows the data correlation using the strain range partitioning model with S = 0.38.

The calculated constants in the interaction damage equation (Manson, 1973) were used to predict the high temperature low cycle fatigue results at 850°C and on the same material reported by Witt and co-workers (1980). Figure 2, shows a plot of the observed vs. the predicted numbers of cycles to crack initiation for the data reported by Witt and co-workers (1980), with S = 0.69.

It is noteworthy to point out that the results of Witt, and co-workers (1980) were obtained from testing of hollow specimens, where crack propagation did not play a significant role. The base line data used in this investigation were obtained by Nazmy (1983) by testing solid specimens and adopting an experimentally defined crack intiation criterion. Hence, the use of this crack initiation criterion made it possible to predict data coming from other type of specimen geometry. This is not possible if one uses the final separation of the specimen as a failure criterion. The description of this experimentally defined crack initiation criterion as well as the significance of its use were reported by Nazmy, and Meixner (1983).

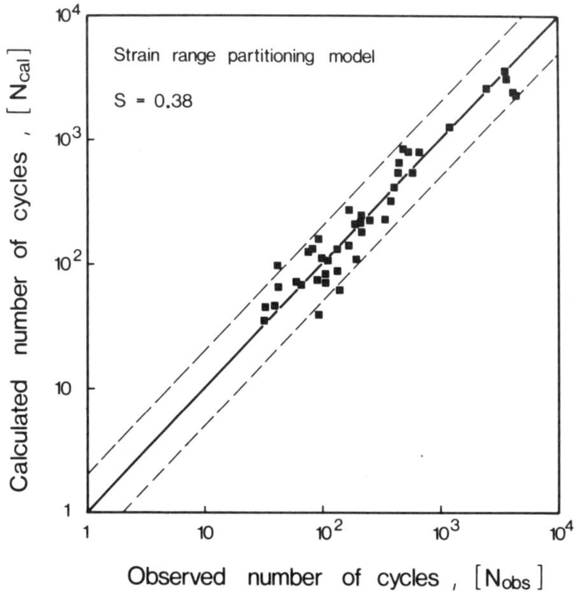

Fig. 1. High temperature low cycle fatigue data correlation.

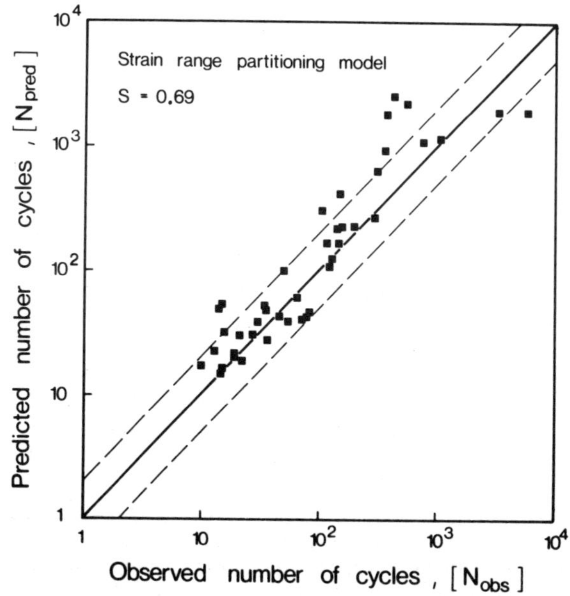

Fig. 2. Observed vs. predicted numbers of cycles to crack initiation for the results of Witt and co-workers (1980).

Frequency-Separation

The frequency-separation model (Coffin, 1976) postulates that the basic parameters necessary to predict the creep-fatigue life are the inelastic strain range, the tension-going frequency, and the hysteresis loop-time unbalance. The data of this investigation were correlated using this model and the result is shown in Fig. 3, with S = 0.41.

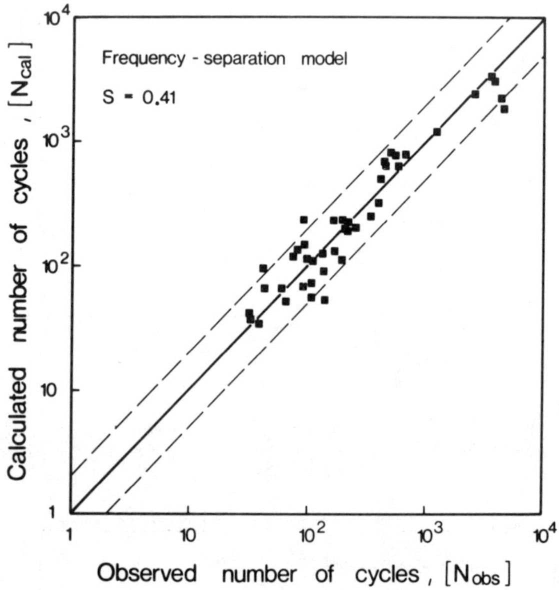

Fig. 3. High temperature low cycle fatigue data correlation.

The coefficients in the equation of this model were calculated using a least square fit computer program and then used to predict the set of data reported by Witt, and co-workers (1980). The result is shown in Fig. 4, with S = 0.76.

Ostergren Model

The measure of deformation used by this model (Ostergren, 1976) is the tensile hysteretic energy obsorbed by the specimen. This energy is approximated as the product of the width of the hysteresis loop, its height, and a shape factor. A frequency term is used in the equation to account for time-dependent effects. Figure 5, shows the correlation of the data of the present investigation using the model of Ostergren with S = 0.39.

Using the calculated coefficients in the equation of this model the results of Witt and co-workers (1980) were predicted. This is shown in Fig. 6, with S = 0.57.

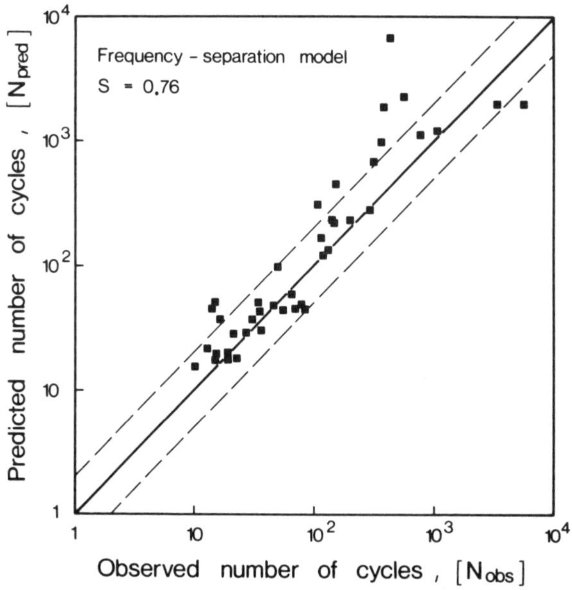

Fig. 4. Observed vs. predicted numbers of cycles to crack initiation for the results of Witt, and co-workers (1980).

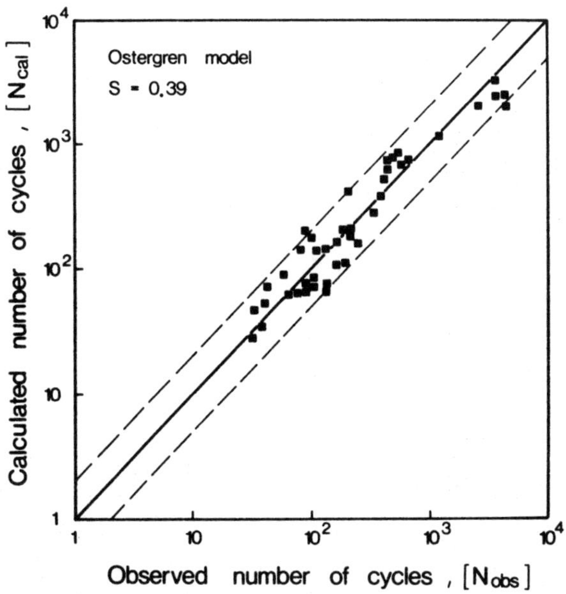

Fig.5. High temperature low cycle fatigue data correlation.

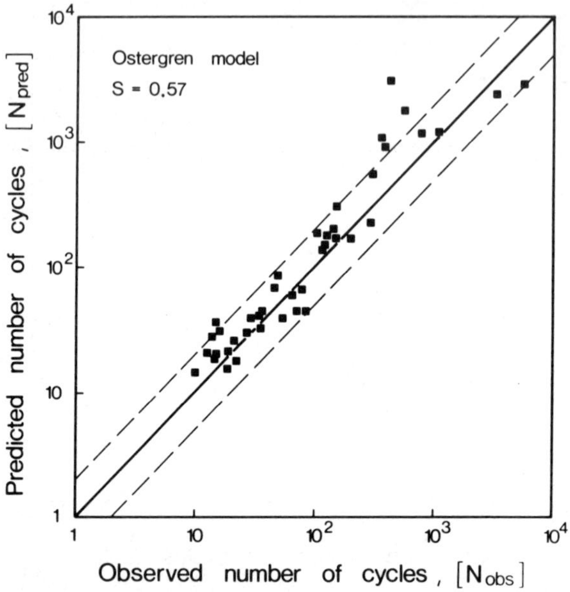

Fig. 6. Observed vs. predicted numbers of cycles to crack initiation for the results of Witt and co-workers (1980).

CONCLUSIONS

1. Using one set of high temperature low cycle fatigue data the strain range partitioning, frequency-separation, and Ostergren models were evaluated.

2. The strain range partitioning model was capable of correlating the data better than the other models. However, there were no significant differences between the predictability of these models.

3. Using the determined constants in the equations of these models, it was possible to predict a second set of high temperature low cycle fatigue data extracted from literature. The prediction of this literature data using the strain range partitioning model had the lowest standard deviation compared to the other two models.

4. Using a well defined crack initiation criterion it was possible to predict data coming from testing of hollow specimens using fatigue data obtained from testing of solid ones.

REFERENCES

Manson S.S., Halford G.R. and Hirschberg M.H., (1971), Creep Fatigue Analysis by Strainrange Partitioning, NASA TMX-67838.

Manson S.S. (1973, Fatigue at Elevated Temperatures, ASTM STP 520, pp. 744-775.

Manson S.S., Halford G.R., and Hirschberg M.H. (1975), "Strain-range Partitioning - A Tool for Characterizing High Temperature Low Cycle Fatigue", NASA TMX-71691.

Coffin L.F., Jr., (1976), ASME-MPC Symposium on Creep-Fatigue Interaction, MPC-3, pp. 349-364.

Witt A., König R., and Eckard C., (1980), Final Report, Project D8, COST 50-II.

Ostergren W.J. (1976), ASME-MPC Symposium on Creep-Fatigue Interaction, MPC-3, pp. 179-202.

Nazmy M.Y. (1983). Metall. Trans. 14A.

Nazmy M.Y. and Meixner W., (1983). Material und Technik, No. 1.

REHEAT CRACKING DURING STRESS RELIEF ANNEALING — A NUMERICAL APPROACH

L. Josefson

Division of Solid Mechanics, Chalmers University of Technology, S-412 96 Göteborg, Sweden

ABSTRACT

Reheat cracking during different Post Weld Heat Treatments (PWHT) of a butt-welded thick-walled pipe made of the low-alloy steel ASTM A508 cl.2 is numerically studied. The Finite Element Method (FEM) is applied. The formation of micro-cracks is modelled by use of a scalar damage field as proposed in the Kachanov theory for brittle fracture. It is found that a low accumulation of damage can be achieved by use of a high heating rate and a high holding temperature.

KEYWORDS

creep relaxation; reheat cracking; damage; PWHT; FEM; CDM.

INTRODUCTION

Intercrystalline cracks may form in the coarse-grained Heat Affected Zone (HAZ) of a weldment during PWHT of steels designed for use at elevated temperatures (Dhooge and others, 1978; Liljestrand and others,1978). Tensile tests performed on weld-simulated specimens (Ruge and Rabe, 1976; Dhooge and others, 1978) have shown that the HAZ may have a considerably lower ductility (often < 3-5%) than the unaffected base metal. Cracks in the HAZ are believed to be caused by a combination of low ductility and large accumulation of plastic strain (Dhooge and Vinckier, 1979; Myers and Clark, 1981). Several experimental methods have been proposed to quantify (phenomenologically) the susceptibility to reheat cracking, see Dhooge and others (1978). They range from tests on complete weldments to creep or relaxation tests (at a constant temperature or with a controlled heating rate) on specimens containing a thermally simulated HAZ. Hereby it has been possible to compare different combinations of steels and parameters of welding and PWHT with respect to the resulting sensitivity to reheat cracking. In some cases (Liljestrand and others, 1978; Nakao,1981) use of a high heating rate in the PWHT seems to reduce the susceptibility to reheat cracking.

In the following, the process of formation and growth of cracks (voids) in the HAZ is modelled by use of a scalar damage field according to the single-coupled Kachanov theory for brittle rupture, see Odqvist (1966). The stress redistribution during annealing of a multi-pass butt-welded pipe is studied numerically by use of FEM. The accumulation of damage (and of non-thermal strain) is calculated for different combinations of heating rate, holding temperature and holding time of the PWHT. The material studied is the low-alloy steel ASTM A508 cl.2.

CONSTITUTIVE EQUATIONS

The total strain increment is calculated as the sum of elastic, plastic and thermal strain increments. Small strains and small rotations are assumed. The plastic strain is taken as decomposed into two mutually independent parts: one time-independent part which (possibly) is accumulated at lower temperatures and is modelled by the von Mises yield function and the associated flow rule assuming ideal plasticity, and one time-dependent (creep) part which is accumulated at elevated temperatures and is modelled by the Norton law extended to multi-axial deformation (Odqvist, 1966). Values of the two material parameters needed in Norton's law were evaluated from reported relaxation (and creep) experiments on specimens with a simulated reheated HAZ, see Josefson (1982, 1983) for details.

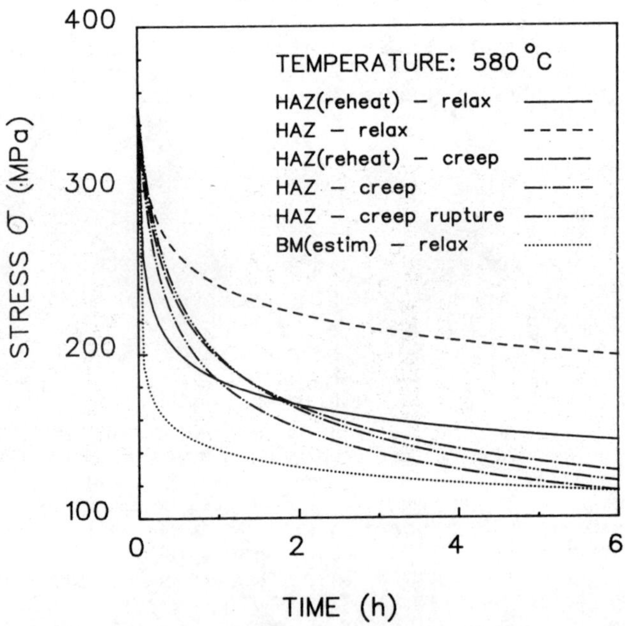

Fig. 1. Calculated uni-axial stress relaxation at 580 °C for HAZ of A508 cl.2 steel.

Figure 1 shows calculated curves for uni-axial stress relaxation at 580 ^{o}C (for the HAZ) where the material parameters have been evaluated from some relaxation and creep tests. The chosen initial stress 350 MPa corresponds to the yield stress of the HAZ at 580 ^{o}C. The relaxation tests used were reported by Detert, Banga and Bertram (1974) and were performed on both a simulated HAZ and a simulated reheated HAZ. The creep tests used were reported by Ruge and Rabe (1976) for a simulated HAZ and by Liljestrand and others (1978) for a simulated reheated HAZ. The creep rupture tests on a simulated HAZ reported by Detert, Banga and Bertram (1974) were evaluated by employing an average strain rate method as proposed by Bråthe and Josefson (1979). Included in Fig. 1 is also a curve for the stress relaxation of the base metal at 580 ^{o}C (extrapolated from the reported relaxation (Liljestrand and others, 1978) of the base metal and the HAZ at 620 ^{o}C). One finds from Fig. 1 that the initial part of the stress reduction can not be predicted by use of values of material parameters calculated from the creep tests available here.

DAMAGE LAW

The development of the scalar damage ω with time is formulated as (see Odqvist, 1966)

$$\dot{\omega} = C(\sigma_a/(1-\omega))^{\nu} \tag{1}$$

Here σ_a is a so called active stress whereas C and ν are material parameters determined from creep rupture tests on specimens with a simulated HAZ (van den Boom and Mulder, 1972). The damage ω is defined to be zero for undamaged material and unity for ruptured material. The specific values of C and ν as well as of the other material parameters were taken to be the same as in Josefson (1983). One may note that the damage ω does not enter (in the present formulation) in Norton's law for creep deformation. It has been observed that the formulas for rupture under uni-axial relaxation in the single-coupled theory (Bråthe, 1976) give a rupture time in good agreement with a corresponding reported experimental result (Detert, Banga and Bertram, 1974) for a simulated HAZ at 580 ^{o}C.

Biaxial creep rupture tests on low-alloy steels (Hayhurst, 1972) indicate that the stress σ_a in Eq. (1) might be taken as

$$\sigma_a = 0.25\ \sigma_{kk} + 0.75\ \sigma_e \tag{2}$$

in a multi-axial case. Here σ_{kk} is the first invariant of the stress tensor and σ_e the von Mises effective stress.

Figure 2 shows the influence of tri-axiality in a welding stress field on the damage accumulation during a PWHT. The reduction of the effective stress σ_e and the accumulation of the damage ω (for different active stresses σ_a) are calculated (for a certain material point in space) during heating from 500 ^{o}C to the holding temperature 650 ^{o}C and for a subsequent holding period. The sum of the elastic and plastic strains is held constant during this anisothermal relaxation. The initial stresses (σ_x=283, σ_y=176, σ_z=58 MPa) correspond to the situation at the highest stressed point in the HAZ of a multi-pass butt-welded pipe (Josefson, 1982). One finds that the damage accumulation is strongly influenced by the chosen active stress (maximum

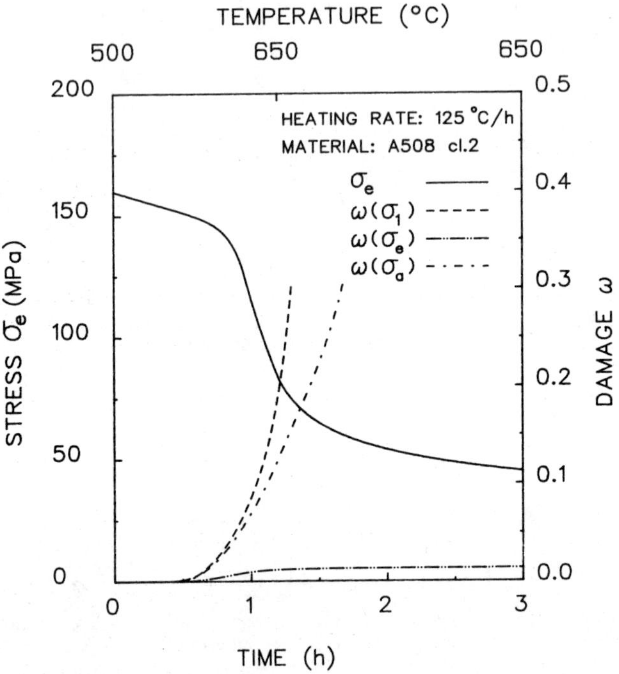

Fig. 2. Calculated stress reduction and damage accumulation during tri-axial anisothermal relaxation.

principal stress σ_1, σ_e, or the stress σ_a given in Eq. (2)). For the case $\sigma_a = \sigma_e$ rupture is seen not to occur. Similar calculations for a 2Cr1Mo steel resulted in a much lower damage accumulation than for A508 cl.2. Use of double-coupled theory (Bråthe, 1976) in the anisothermal relaxation problem above was found to result in a damage accumulation which is noticeably lower than the presented result in Fig. 2 for a single-coupled theory.

PWHT OF A MULTI-PASS BUTT-WELDED PIPE

The stress redistribution and damage development during PWHT have been studied for a butt-welded pipe made of the A508 cl.2 steel. The geometry of the pipe weld joint is shown in Fig. 3. The welding residual stress field calculated by Josefson (1982) for a C-Mn steel pipe was used here as the initial stress field. The FE-program for axi-symmetric thermal stress and damage analysis described in Josefson (1983) was employed. The damage ω is thus evaluated in each of the 6 gauss points of the triangular (20 dof) elements at the end of each subincrement of a time step. Gauss points where ω has reached the value unity are, within the program, attributed both a low value of the Young modulus and zero values of all stress components. During the annealing process the temperature in the kinematically unrestrained pipe was uniform in space thus simulating a furnace heat treatment.

TABLE 1 Calculated maximum effective stress σ_e and maximum damage ω (or number of gauss points where ω=1 has been attained) after holding period in A508 cl.2 steel pipe for different heating rates HR, holding temperatures T_{hold} and holding times t_{hold}.

T_{hold} [°C]	HR [°C/h]	t_{hold} [h] 0 σ_e [MPa]	ω -	0.5 σ_e [MPa]	ω -	1.5 σ_e [MPa]	ω -
	200	210	0.005	186	0.072	172	(11)
580	125	201	0.009	185	0.081	174	(10)
	50	191	0.025	183	0.142	180	(15)
	200	156	0.039	102	0.193	82	(2)
620	125	145	0.072	101	(2)	82	(3)
	50	125	(4)	118	(9)	90	(15)
	200	94	0.092	64	(1)	51	(2)
660	125	84	(2)	62	(2)	50	(6)
	50	93	(11)	85	(12)	66	(16)

Table 1 summarizes the calculated maximum values of σ_e and ω (or number of gauss points where ω is unity) in the pipe for some combinations of parameters in a PWHT. These maximum values for σ_e and ω were found to appear in the HAZ near the outer surface where both axial and hoop stresses are large. A large damage accumulation (and a stress increase in the adjacent unruptured areas) was observed after use of either a low heating rate or a holding temperature around 580 °C. The increased susceptibility at about 600 °C has also been reported in experiments (Liljestrand and others, 1978). A separate numerical study showed that a two-stage annealing (with holding at an intermediate temperature) does not reduce the calculated susceptibility to reheat cracking. For example, holding at 620 °C for 15 minutes (after use of HR=200 or 125 °C/h) followed by heating (HR=200 °C/h) to the new holding temperature 660 °C resulted in slightly lower σ_e-values and slightly higher ω-values than those presented in Table 1.

Figure 3 shows the cross-sectional areas of the HAZ where the calculated ω has reached unit values after holding at 660 °C for 1.5h (and after use of HR=50 °C/h). The total sum of elastic and plastic strains (accumulated during the PWHT) was found to be about 0.11% in these areas prior to rupture.

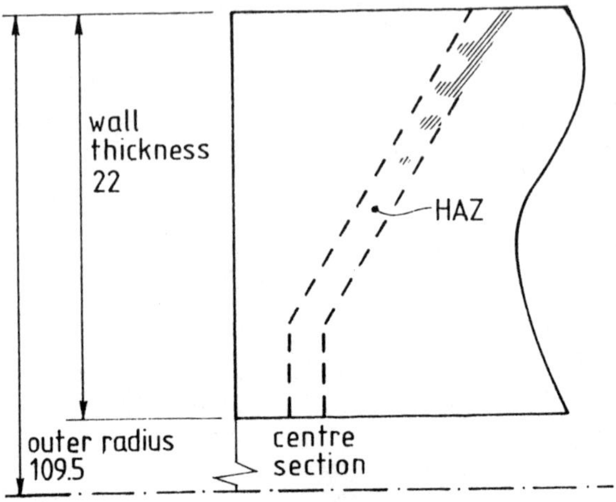

Fig. 3. Weld joint geometry (mm) used in analysis. Shaded area contains gauss points where ω=1 has been attained after holding at 660 ^{o}C for 1.5 h. Heating rate used was 50 ^{o}C/h.

DISCUSSION

Welding stress fields in thick-walled structures often include domains with tri-axial tensile stress states (high σ_{kk}/σ_e ratio) which are believed to increase the cracking susceptibility. The damage development would thus be strongly influenced by the σ_{kk} term in Eq. (2). It may be noted that the damage measure used here is related to the spatial variation of the density of micro-cracks only. A micro-crack development in certain preferred directions has been reported for alloys that show a similar damage behaviour, see Hayhurst (1972). Some possible methods to model a directional dependence for the damage have been summarized by Chaboche (1981).

ACKNOWLEDGEMENTS

This work was performed under the supervision of Professor Bengt Åkesson at the Division of Solid Mechanics, Chalmers University of Technology, Göteborg, Sweden. The Swedish Board for Technical Development (STU) financially supported the work.

REFERENCES

Bråthe, L. (1976). Rupture under relaxation. In Klason, C. and J. Kubat(Eds.) Proc. of the VII Int. Congr. on Rheology, Göteborg, Sweden, 616-617.

Bråthe, L. and L. Josefson (1979). Estimation of Norton-Bailey parameters from creep rupture data. Met. Sci., 13, 660-664.

Chaboche, J.-L. (1981). Continuous damage mechanics - a tool to describe phenomena before crack initiation. Nucl. Eng. Des., 64, 233-247.
Detert. K., R. Banga and W. Bertram (1974). Effect of stress-relieving on the mechanical properties and the structure of welding-simulation heat-treated specimens of high-temperature fine-grained structural steels. Arch. Eisenhuttenwes., 45, 245-255.
Dhooge, A. and A.G. Vinckier (1979). Reheat cracking in welded structures during stress relief heat treatments. J. Heat Treating, 1, 72-80.
Dhooge, A., R.E. Dolby, J. Sebille, R. Steinmetz and A.G. Vinckier (1978). A review of work related to reheat cracking in nuclear reactor pressure vessel steels. Int. J. Pressure Vessels and Piping, 6, 329-408.
Hayhurst, D.R. (1972). Creep rupture under multi-axial states of stress. J. Mech. Phys. Solids, 20, 381-390.
Josefson, B.L. (1982). Stress redistribution during annealing of a multi-pass butt-welded pipe. J. Pressure Vessel Technol. Trans. ASME. Accepted for publication.
Josefson, B.L. (1983). Local annealing of a multi-pass butt-welded pipe - Computation of stress redistributions and cracking susceptibilities. Fourth Nat. Congr. Pressure Vessels Piping Technol., Portland Ore, USA. Accepted for presentation.
Liljestrand, L.-G., M. Aucouturier, G. Östberg and P. Lindhagen (1978). Formation of microcracks during stress-relief annealing of a weldment in pressure vessel steel of type A508 Cl.2. J. Nucl. Mater., 75, 52-67.
Myers, J. and J.N. Clark (1981). Influence of welding procedure on stress-relief cracking in Cr-Mo-V steels: Part 1 use of cross-weld specimens. Met. Technol., 8, 382-388.
Nakao, Y. (1981). Fundamental study on the reheat crackings in steels. Proc. Int. Conf. on Welding Research in the 1980:s, Osaka, Japan, Japan Welding Research Institute, 143-148.
Odqvist, F.K.G. (1966). Mathematical Theory of Creep and Creep Rupture. Oxford University Press.
Ruge, J.L. and W. Rabe (1976). Study on the susceptibility to stress-relief cracking of low-alloy steel weldments by means of slow tensile tests and creep tests. Doc. X-803-76, Int. Inst. of Welding (IIW).
van den Boom, J.E. and J.P.F. Mulder (1972). Mechanical properties of strip clad weld heat affected zones determined for steels ASTM A508, cl.2, A533, gr. B, cl.1, A302, gr. B and A387, gr. D, in view of susceptibility to stress relief cracking. Weld. Res. Int., 2, 20-36.

INFLUENCE OF PRESTRESS ON THE YIELD SURFACE OF THE CAST NICKEL SUPERALLOY MAR-M002 AT ELEVATED TEMPERATURE

M. R. Winstone

Royal Aircraft Establishment, Pyestock, Farnborough, Hampshire GU14 0LS, UK

ABSTRACT

The effect of prestress on the yield surface of the cast nickel superalloy Mar-M002 has been determined at 750°C. A tension/torsion testing machine was used with an acoustic emission technique to detect yield. The initial yield surfaces indicated a ratio of shear yield stress to tensile yield stress of 0.7. The von Mises and Tresca yield criteria predict 0.58 and 0.5 respectively. The subsequent yield surfaces were in qualitative agreement with the kinematic hardening model but included an expansion of the yield surface in the direction of the prestress with some contraction in the reverse direction.

KEYWORDS

Yield strength; yield surface; tension tests; torsion tests; nickel alloys; superalloys; acoustic emission.

INTRODUCTION

During typical flight cycles, a gas turbine blade is subjected to a complex multiaxial stress cycle due to the combination of centrifugal, thermal and gas loading stresses. Cyclic deformation of the blade occurs due to time independent plastic strain and time dependent creep strain. The effects of plasticity on the redistribution of the cyclic stresses should be included in the life-prediction analysis of the blade. Before such calculations are possible the multiaxial deformation behaviour of the material must be known, both during the first loading cycle and during subsequent cycles when the effects of work hardening are significant. Few multiaxial data are available for gas turbine blade alloys at elevated temperatures (Dyson and McLean, 1977) so the biaxial tensile and creep behaviour of the nickel alloy Mar-M002 are being studied in an experimental programme using combined tension/torsion testing machines. The cast superalloy Mar-M002 was selected because of its important applications in British gas turbine engines. This paper is concerned with the determination of the initial yield surface of

Mar-M002 at elevated temperature and the effects of prestress on subsequent yield surfaces.

Several theoretical models have been proposed to describe the effect of work hardening on the subsequent yield surfaces. The isotropic hardening model assumes a uniform expansion of the yield surface, with no shift of its centre. The kinematic hardening (Prager, 1955) model assumes a rigid body translation of the yield surface in the direction of the applied stress. No distortion or rotation of the yield surface occurs and the model predicts a Bauschinger effect. The kinematic model has also been modified to include some cross-hardening (Hodge, 1957) i.e. hardening in directions other than the prestress direction.

None of these theories has been shown to correlate closely with published results. Tests on copper (Mair and Pugh, 1964); (Sudderar and Duffy, 1967), aluminium (Phillips, Liu and Justusson, 1972); (Shiratori, Ikegami and Yoshida, 1976), brass (Miastkowski, 1965) and steel (Michro and Findley, 1974); (Moreton, Moffat and Hornby, 1978) demonstrate that an empirical modification to the kinematic theory is necessary to fit experimental data but the form of this modification is dependent upon the experimental procedure employed. The yield surface can be determined by using a new test piece for each loading condition or loading a single specimen to yield under a series of different loading conditions. If more than one test piece is used the results are susceptible to material scatter and therefore most previous experimental work has been done using a single test piece to determine a complete yield surface. However, the results of this technique are sensitive to the definition of yield (Mair and Pugh, 1964); (Phillips, Liu and Justusson, 1972) and the ability of the experimental method to detect yield. If yield is not detected sufficiently early then the single specimen technique produces a distortion of the yield surface which can result in a reduced Bauschinger effect and some cross-hardening. In the present work an acoustic emission technique was used to detect yield, thus minimising post-yield strains. A transducer attached to the loading system monitored the first indications of dislocation motion and hence yield was defined before any measurable plastic strain was incurred. This technique has the additional advantage of allowing a remote sensing position to be used, therefore avoiding the problems inherent in extensometry at high temperatures.

PROCEDURE

Conventional investment castings of the nickel superalloy Mar-M002 were made in a vacuum induction furnace. The test bars had a columnar grain structure with crystals approximately 1 mm diameter growing radially towards the centre of the bar. Solid and tubular test pieces were machined from the castings with gauge lengths of 50 mm and external diameters of 8.25 mm. The tubular test pieces had a wall thickness of 0.95 mm. Combined tension/torsion tests were carried out on a servohydraulic testing machine capable of applying a maximum tensile load of 50 kN and a maximum torque of 200 Nm. The testing was undertaken using constant rates of extension and torsion. An acoustic emission transducer was attached to a water-cooled flange on the lower draw bar and the amplified signal was plotted, together with load and torque, on a strip chart recorder.

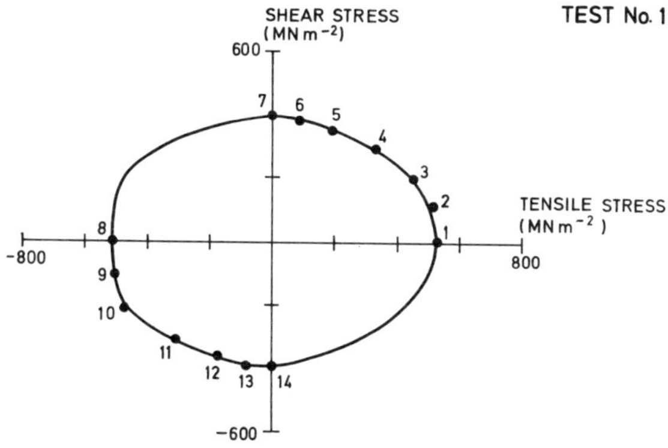

Fig.1. Initial yield surface of Mar-M002 at 750°C.

The sequence of a typical test to determine the initial yield surface at 750°C is shown in Fig. 1. The yield surface was obtained by probing into the plastic region under various combinations of tensile and torsional loads, both loads being reduced to zero as soon as "noise" was detected on the acoustic emission equipment. Repeating the test sequence or changing the order of the loading combinations (See Fig. 2) had no significant effect on the shape of the yield surface. Experimental points were only obtained in the first and third quadrants of the yield surface but the second and fourth quadrants have been completed by symmetry. This was due to limitations in the control signal generator which prevented tests with torque and tensile loads of opposite sign.

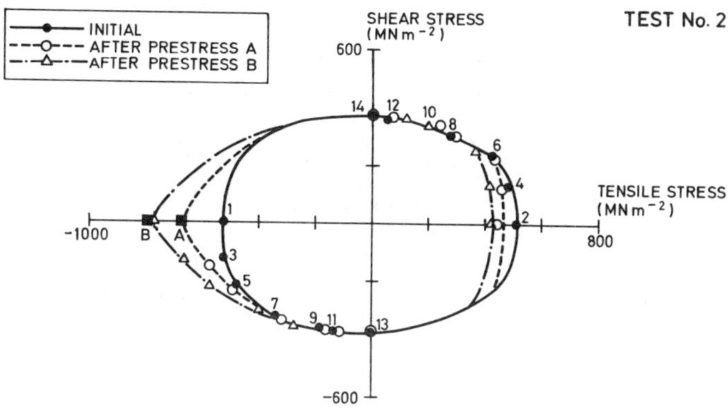

Fig. 2. Initial and subsequent yield surfaces, prestressed in compression at 750°C.

After obtaining the initial yield surface, each test piece was prestressed at 750°C to approximately 20% beyond the yield stress in tension, torsion, compression or combined tension/torsion. The new yield surface was then determined in the same manner before increasing the prestress by a further 20%.

All the yield surface tests used tubular test pieces. The solid test pieces were only used to check the acoustic emission system at higher loads. No acoustic emission signal was detected from the solid specimens at twice the tubular tensile/compression yield loads or 1.7 times the torsional yield load. It is therefore confirmed that the acoustic emission signal detected in the tubular samples did originate from the gauge length and was not a machine or gripping "noise".

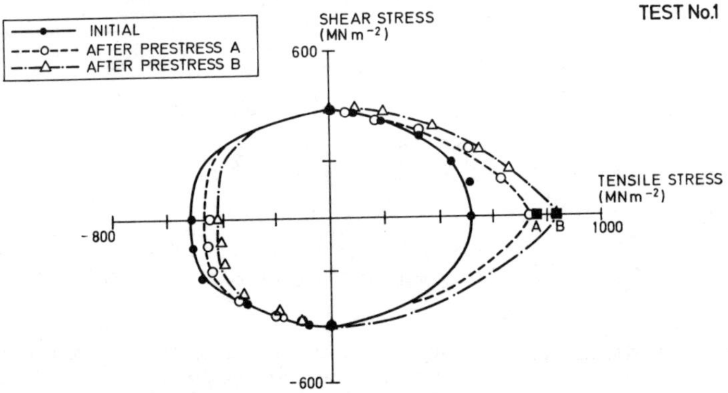

Fig. 3. Initial and subsequent yield surfaces, prestressed in tension at 750°C.

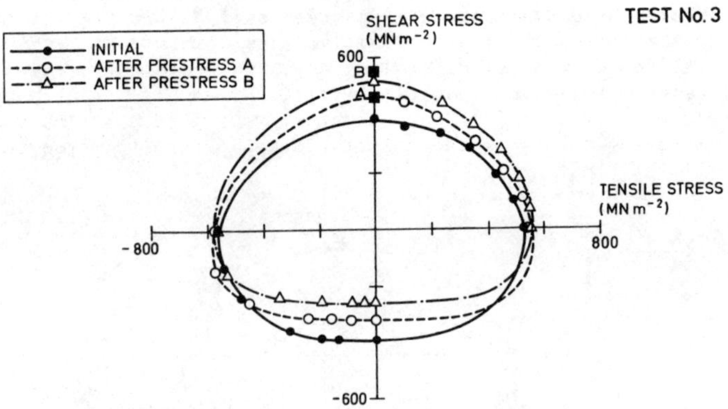

Fig. 4. Initial and subsequent yield surfaces, prestressed in torsion at 750°C.

RESULTS AND DISCUSSION

Figures 2-4 show the yield surfaces obtained at 750°C after compression, tension or torsion prestresses. The complete yield surface ellipses for the combined tension/torsion prestresses cannot be drawn (Fig. 5) because the subsequent yield surfaces are not symmetrical about the axial or torsional shear stress axes.

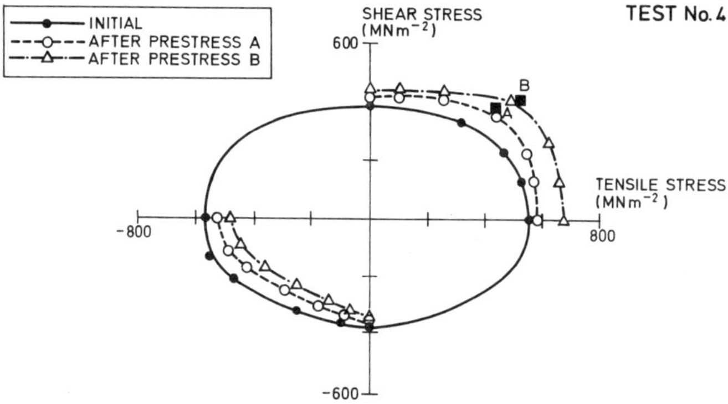

Fig. 5. Initial and subsequent yield surfaces, prestressed in combined tension/torsion at 750°C

Initial Yield Surfaces

All the initial yield surfaces are plotted together in Fig. 6. They lie within a tight scatter band despite the fact that the castings had a rather large grain size which can exagerate test piece scatter. The ratio of torsional shear yield stress to tensile yield stress is ≈ 0.7, surprisingly high when compared with the values of 0.58 and 0.5 expected from the von Mises and Tresca yield criteria respectively. Clearly neither of these criteria can accurately model the biaxial yield behaviour of Mar-M002 castings. Published data for other face centred cubic metals show reasonable agreement with the von Mises criterion and the discrepancy reported here may have resulted from the combined effects of the very coarse grain structure and the preferred orientation that resulted from the columnar grain growth.

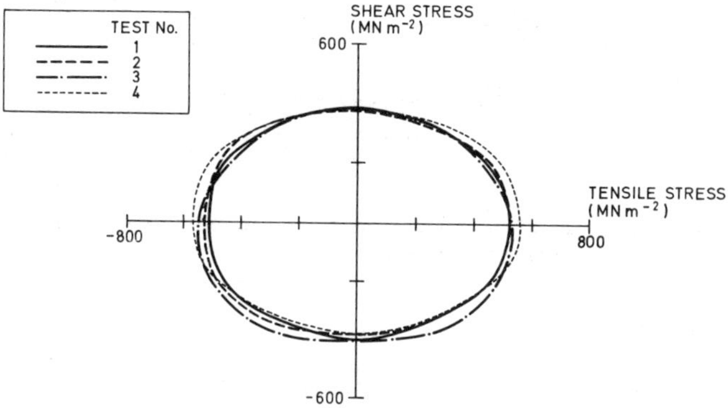

Fig. 6. A comparison of the initial yield surface determinations

It has been shown that the tensile yield stress of many metals decreases as the section thickness approaches the grain size (Miyazaki, Shibata and

Fujita, 1979). This effect arises because the thin section exerts less constraint on the deformation of the individual grains. It is to be expected therefore that the tensile yield stress of the tubular samples used in this work will be significantly lower than that of the solid samples. In torsion the individual grains of the thin tube may be under more constraint from surrounding material since the applied shear stress is in the plane of the tube wall and has no component perpendicular to the thin section. Directionally solidified (DS) castings are less prone to section size effects (Winstone and Northwood, 1980) and future tests on DS castings may show better agreement with the von Mises criterion. DS castings would also modify the preferred orientation of the test pieces from the present <100> radial texture to a 100 longitudinal structure. The texture effect may also have contributed to the high ratio of torsional to tensile yield stress.

Subsequent yield surfaces

All the subsequent yield surface plots display the same features. The yield surfaces are displaced in the direction of the prestress, with a significant Bauschinger effect and no cross-hardening. This is qualitatively consistent with the kinematic hardening model but considerable distortion of the yield surfaces also occurs, with an expansion towards the prestress direction and a contraction in the reverse direction. These observations are similar to the yield surface behaviour of aluminium (Phillips, Liu and Justusson, 1972); (Shiratori, Ikegami and Yoshida, 1976) and brass (Miastkowski and Szczepinski, 1965) but in these materials the prestress points were far outside the subsequent yield surface because significant recovery occurred between the prestressing and the determination of the yield surface. The prestress points for the Mar-M002 tests always lie on, or very close to, the subsequent yield surface. This demonstrates the great stability of deformation structures in Mar-M002. Indeed, one test piece was held at 750°C for 16 hours and the subsequent yield surface still passed through the prestress point.

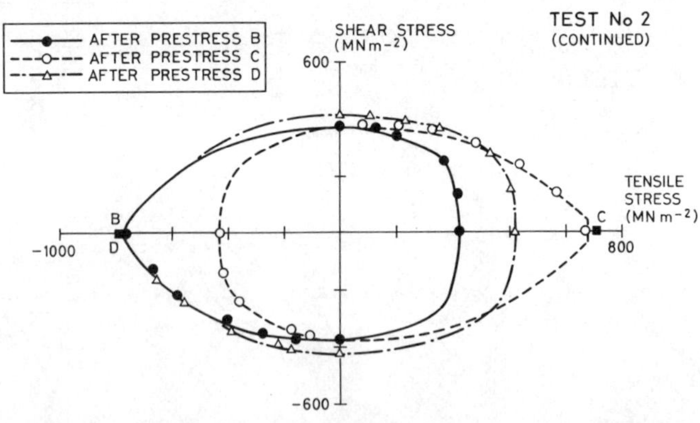

Fig. 7. Subsequent yield surfaces with alternating prestresses

The results presented so far refer to simple monotonic prestresses. Under service conditions turbine blades experience complex cyclic loads which introduce alternating tensile and compressive stresses. Under such conditions it is to be expected that "shakedown" yield surfaces will be established for both the maximum and minimum of the cycle. As a first step towards determining a cyclic stress yield surface, Fig. 7 shows the results obtained when Test 2 was continued beyond the compressive prestress B by applying a tensile prestress C followed by a further compressive prestress D. The total area of the yield surface has increased suggesting that cyclic deformation will introduce an isotropic component to the hardening. This expansion of the yield surface is a characteristic of cyclic work hardening which has been observed at the start of low cycle fatigue tests of nickel alloys (Harrison and Weaver, 1978).

CONCLUDING REMARKS

The acoustic emission technique can provide an accurate and sensitive method for determining yield surfaces. Yield surfaces of the cast nickel superalloy Mar-M002 have been determined at 750°C. The subsequent yield surfaces after prestressing were in qualitative agreement with a kinematic hardening model. Further tests are in progress to investigate the effect of an alternating tensile and compressive prestress on the yield surface and to extend the test conditions to higher temperatures where recovery and time dependent deformation will become significant.

REFERENCES

Dudderar, T. D., and J. Duffy (1967). Trans Am. Soc. mech. Engrs., E, 34, 200-209.

Dyson, B. F., and D. McLean (1977). Metal Sci. J., 11, 37-45.

Harrison. G. F.. and M. J. Weaver (1978). AGARD Conference Proceedings, 243. 6/1-6/19.

Hodge, P. G. (1957). J. Appl. Mech., 79, 482.

Mair, W, N., and H. Ll. D. Pugh (1964). J. Mech. Eng. Sci., 6, 150-163.

Miastkowski, J., and W. Szczepinski (1965). Int. J. Solids and Structures, 1, 189-194.

Michno, M. J., and W. N. Findley (1974). Trans. Am. Soc. mech. Engrs, H, 97, 25-32.

Miyazaki, S., K. Shibata and H. Fujita (1979). Acta Met., 27, 855-862.

Moreton, D. N., D. G. Moffat and R. P. Hornby (1978). J. Strain Analysis, 13, 185-191.

Phillips, A., C. S. Liu and J. W. Justusson (1972). Acta Mechanica, 14, 119-146.

Prager, W. (1955). Proc. Inst. Mech. Engrs. 169, 41-57.

Shiratori, E., K. Ikegami and F. Yoshida (1976). Bull. J.S.M.E., 19, 1122-1128.

Winstone. M. R.. and J. E. Northwood (1980). Solidification Technology in the Foundry and Casthouse, Paper 47, Metals Society Conference at University of Warwick.

FATIGUE CRACK PROPAGATION IN STAINLESS STEEL SUBJECTED TO REPEATED THERMAL SHOCK

M. Shimizu*, M. W. Brown and K. J. Miller****

*Department of Mechanical Engineering, Keio University, Yokohama, Japan
**Department of Mechanical Engineering, University of Sheffield, Sheffield, UK

ABSTRACT

Fatigue crack growth has been studied in an AISI 316 stainless steel subjected to two types of repeated thermal down-shock; (i) a thermal shock where a jet of air at room temperature is blown onto a hot plate heated to about 900°C and (ii) a thermal shock where liquid nitrogen is sprayed onto the centre of a disc-shaped specimen heated to 180°C. The effect of a mean mechanical stress on fatigue growth for the type (i) thermal shock is discussed in relation to fatigue-creep interactions. The correlation of crack growth behaviour with stress intensity factors is examined and compared with isothermal strain cycling fatigue data.

KEYWORDS

Thermal shock; crack growth; fatigue (materials); fatigue-creep interaction; stress intensity; low temperature; stainless steel.

INTRODUCTION

It has recently become important to design against fatigue fracture due to thermal shock. Several studies have been carried out on thermal shock fatigue and the propagation behaviour of cracks has been discussed mainly from the standpoint of fracture mechanics (Howes, 1973; Marsh, 1981; Mowbray et al, 1973; Newton, 1976; Skelton, 1979).

The aim of the present study was to help clarify the initiation and propagation behaviour of cracks when a thermal shock is repeatedly applied to AISI 316 stainless steel with respect to the following two cases. The first study was a thermal down-shock where the maximum temperature in the thermal cycle was in the range of temperature producing creep damage. The second case was a thermal down-shock where the minimum temperature was -196°C i.e., liquid nitrogen temperature. For the first study the effect of a mechanical mean stress on crack growth behaviour has been examined in relation to fatigue-creep interaction.

SPECIMEN MATERIAL AND TESTING METHODS

Solution annealed AISI 316 austenitic stainless steel was used to make specimens for both kinds of tests. The composition and tensile properties are given in Table 1 (material A) of a previous paper (Brown and Miller, 1982).

(i) Low-temperature thermal down-shock (LTDS):
Liquid nitrogen was sprayed for about 80 seconds through a nozzle of 4 mm diameter onto the centre of a disc-shaped test piece (Fig. 1a) which had been uniformly heated to a temperature of 180°C on a hot ceramic plate. The central part of the test piece ($r \leq 25$ mm) was machined to form a thin disc of 2 mm thickness so that a high, biaxial thermal stress was produced in this part by rapid cooling, owing to constraint by the surrounding thick plate of high thermal capacity.

(ii) High-temperature thermal down-shock (HTDS):
A 20 mm wide jet of air at room temperature was blown for about 1 second onto one edge of the plate specimen (Fig. 1b) which had been almost uniformly heated to a temperature of 860°C by a high-frequency induction current. The test machine was designed to apply a mechanical load to the test piece by hydraulic pressure, independent of the thermally induced strain.

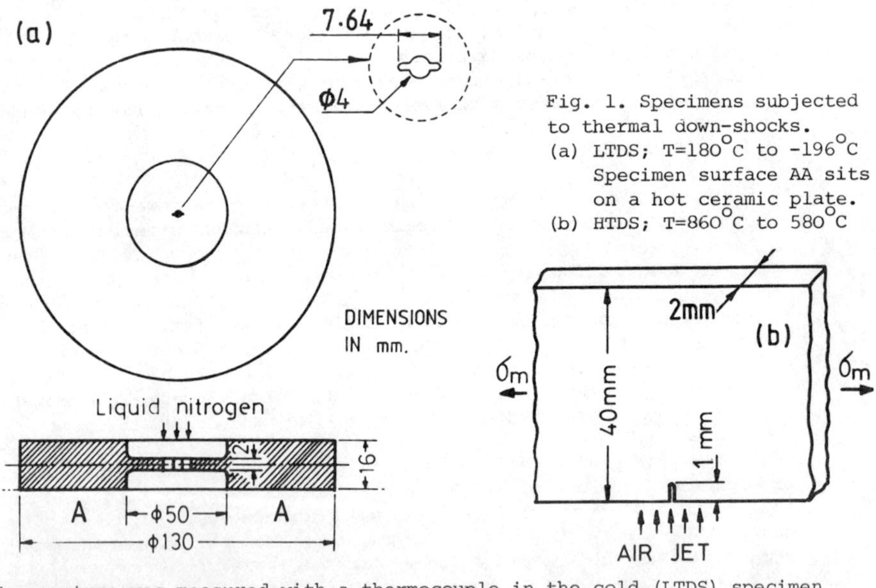

Fig. 1. Specimens subjected to thermal down-shocks.
(a) LTDS; T=180°C to -196°C Specimen surface AA sits on a hot ceramic plate.
(b) HTDS; T=860°C to 580°C

Temperature was measured with a thermocouple in the cold (LTDS) specimen and with a pyrometer having a 0.7 mm diameter target area on the hot (HTDS) specimen. Temperature distributions are shown in Fig. 2.

RESULTS AND DISCUSSION

Low-temperature thermal down-shock fatigue tests.

The crack growth curve is shown in Fig. 3a. Because of insufficient data for the early stage of growth, the exact number of cycles for crack initiation was not obtained. However, the extrapolated curve in Fig. 3a for early growth indicates that cracks appear to start immediately at the notch. Fig. 3b shows a SEM fractograph of the fracture surface taken less than 5 μm from the slit. The striations can be clearly recognised, showing that cracks are already in stage II growth at small depth.

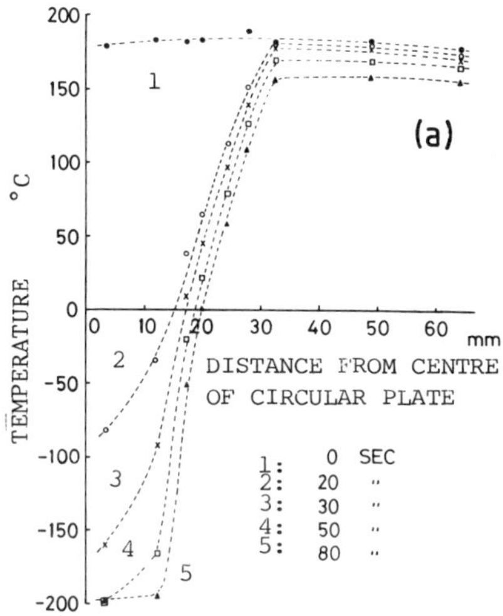

Fig. 2. Temperature distributions.
(a) LTDS specimen
(b) HTDS specimen

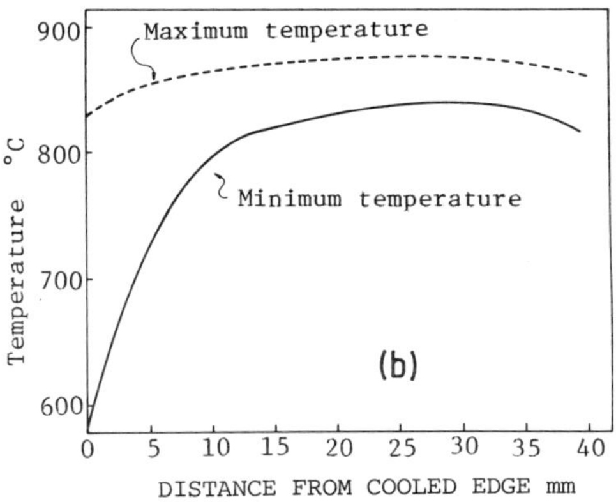

Fig. 4a shows a comparison of the striation spacings with the macroscopic crack growth rate da/dN determined from the slope of the crack growth curve. It is observed that both agree well, indicating that the crack extends each time a thermal shock is applied.

When examining the relationship between da/dN and the stress intensity factor (S.I.F.) K_I, the value of $K_I max$ in a cycle, was calculated by the method of superposition (Okamura 1976), using the results of a numerical calculation of the thermal stress transient and assuming elastic deformation. The relation between

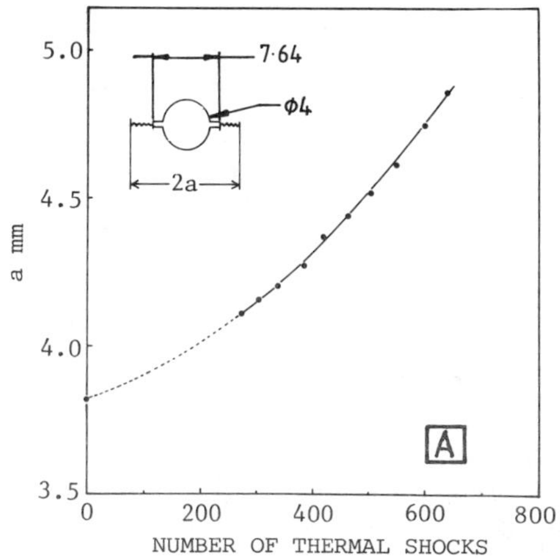

Fig. 3. Low-temperature down-shocks
(A) fatigue crack growth curve
(B) striations near the slit; arrow shows crack growth direction.

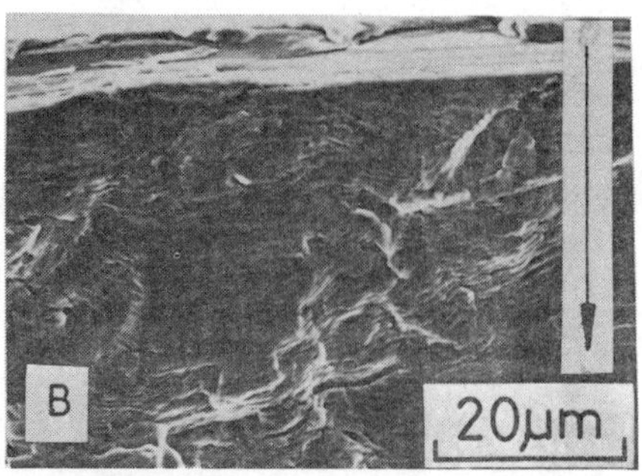

ΔK (equal to $Y\Delta\sigma\sqrt{\pi a}$) and da/dN was also plotted for a series of crack propagation tests under uniaxial and equibiaxial fully reversed cyclic loading at room temperature (Brown and Miller, 1982), as shown in Fig. 4b. It is evident that the thermal shock test result agrees with the upper bound of such a ΔK-da/dN curve obtained from mechanical fatigue tests at room temperature. It should be noted that room temperature is approximately equal to the mean temperature in a cycle, implying that thermal shock crack growth can be estimated from isothermal data at the average temperature of the cycle. This appears to be valid, even though yield stress was exceeded in thermal loading, because elastic

Fig. 4. Comparison of crack growth data in LTDS specimens with
(a) striation spacing
(b) isothermal strain-cycling

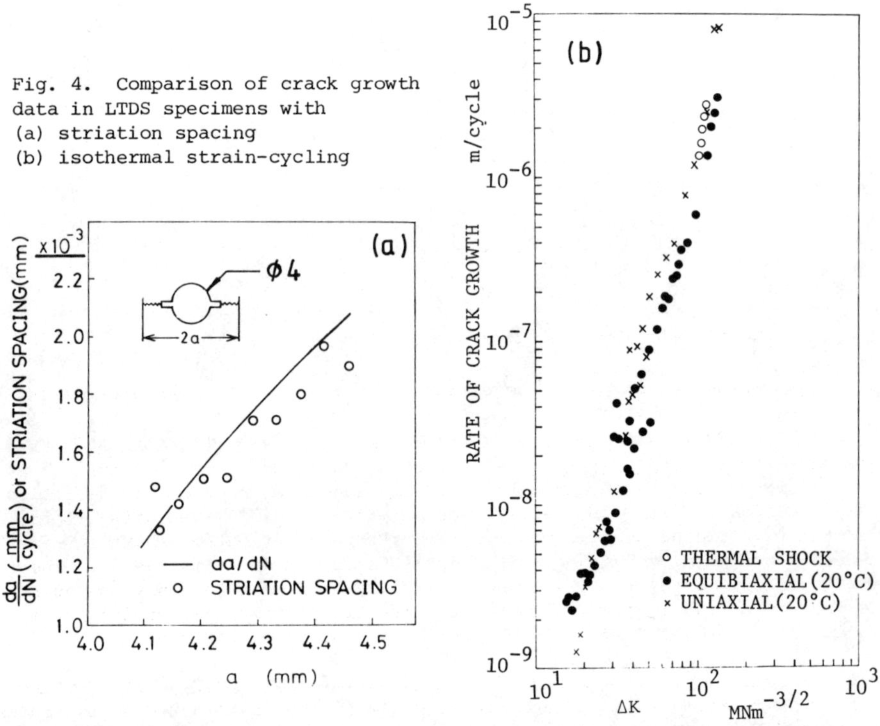

thermal stress analysis is based on strains due to thermal contraction on cooling, so that the value of K_I derived above is related to strain intensity at the crack tip. The strain intensity factor is effective in describing crack propagation once cyclic yield has been exceeded (Skelton, 1982).

High-temperature thermal down-shock fatigue tests.

Fig. 5 shows the fatigue crack growth curves for HTDS specimens, indicating variations in crack growth rate versus crack length for various mean stresses, σm. The following characteristics of crack growth behaviour can be observed.

(i) In a purely thermal shock condition (σm = O), the crack increases its growth rate in the early growth stage. After showing a maximum at a certain length however, the rate rapidly decreases and the crack does not penetrate beyond a certain length, i.e., it shows a tendency to arrest.

(ii) When a small mean stress ($\sigma m = 19\ N/mm^2$) is superimposed on a repeated thermal shock, the crack growth has the same tendency as the case of $\sigma m = 0$, but the maximum value of da/dN appears at a longer crack length and the crack penetrates more deeply.

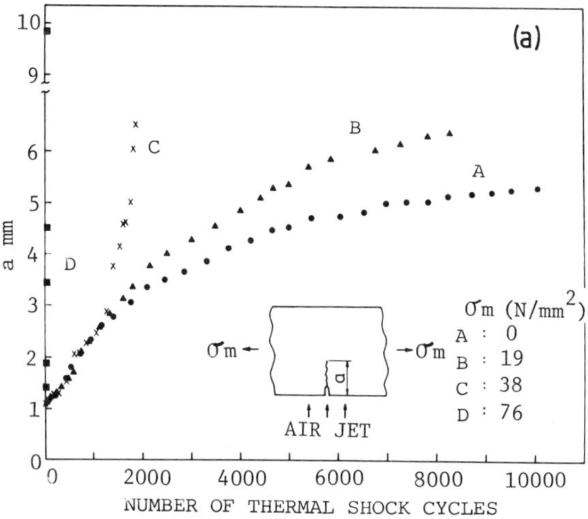

Fig. 5. The effect of mean stress on fatigue crack growth in high temperature specimens subjected to down-shock

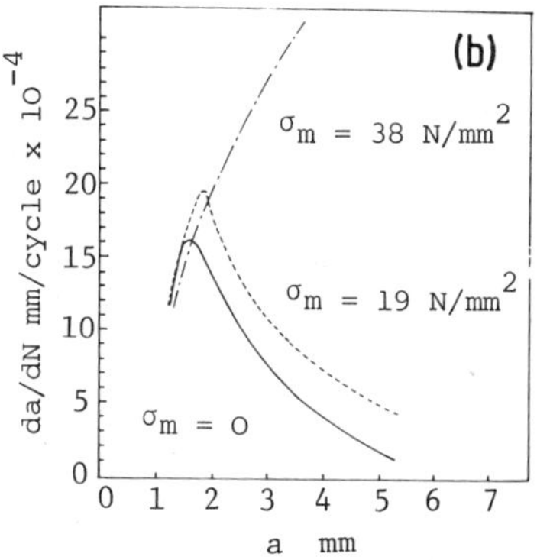

(iii) With a much larger mean stress ($\sigma m = 38$ and $76\ N/mm^2$), the crack growth rate increases steadily with length, leading to final fracture of the specimen.

This crack growth behaviour can be interpreted by considering both the stress intensity factor (SIF) variation with increasing crack length and the effect of creep damage on the crack extension. First, the dependence on SIF variation will be discussed. The SIF profile was numerically calculated with respect to the cracked plate having the minimum temperature distribution shown in Fig. 2b in accordance with Emery's method (Emery, Walker and Williams, 1969). The result is shown in Fig. 6. Although this result illustrates only the approximate nature of the K_I max profile in the cracked plate, it is useful for a qualitative discussion of crack growth in thermal shock fatigue.

When the results for $\sigma m = 0$ and $19\ N/mm^2$ are compared with Fig. 6, it is evident that the variations of da/dN and SIF with the crack length agree well, indicating that SIF decreases as the crack extends beyond a certain depth, giving a decreasing crack growth rate with eventual crack arrest. The slightly deeper penetration of the fatigue crack under a small mean stress ($\sigma m = 19\ N/mm^2$) may be ascribed to the increase of K_{Imax} by an amount $1.12\sigma m\sqrt{\pi a}$ due to the mean mechanical stress.

The effect of a large mean stress ($\sigma m = 38$ and $76\ N/mm^2$) may be explained by taking into consideration the effect of creep damage on crack growth behaviour. Fig. 7 shows the crack growth curve obtained from a creep test at a temperature equal to the maximum value in the HTDS test, compared with the HTDS result for the same steady stress ($\sigma m = 38\ N/mm^2$) used in the creep test. In this figure, the time scale of the horizontal axis for the creep test corresponds to the elapsed time in the thermal shock fatigue test.

For the creep test there is an initial incubation period of about 350 minutes in which damage accumulates, followed by rapid creep crack propagation. In the thermal shock test, the crack is able to extend to about 2.5 mm during the incubation period, but there is an abrupt change from fatigue to a creep mechanism after 350 minutes, demonstrated by the parallel cases in Fig. 7 and the marked increase in da/dN in Fig. 5b. For $\sigma m = 76\ N/mm^2$, the incubation period is short, so that fatigue plays no part in crack extension, see curve D, Fig. 5a.

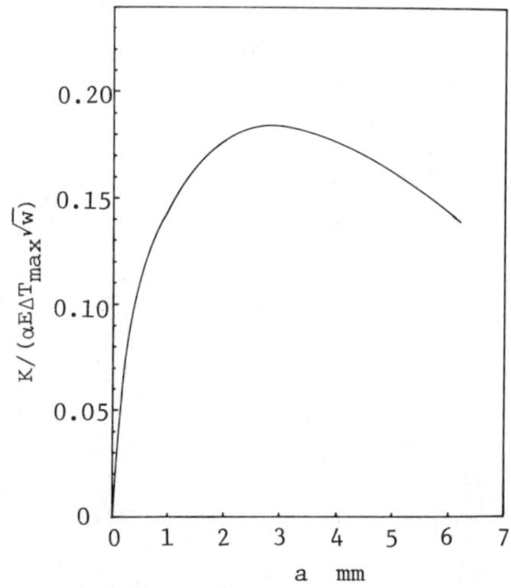

Fig. 6 SIF profile calculated with respect to Fig. 2b temperature distribution.

α: Coefficient of thermal expansion
E: Young's modulus
ΔTmax: Maximum temperature difference
W: Plate width

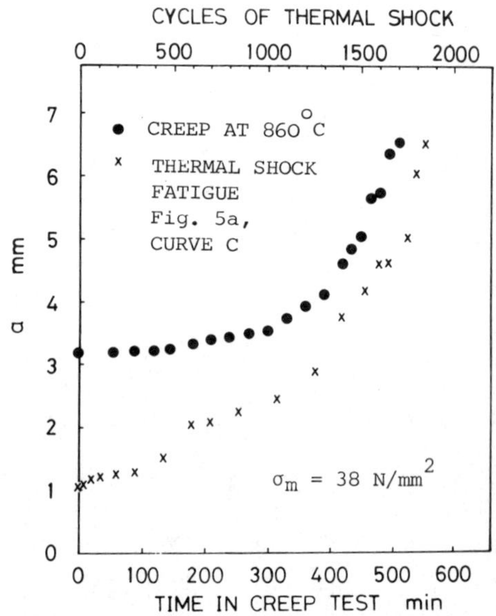

Fig. 7 Comparison of crack growth behaviour in creep with that in thermal shock fatigue with the same steady stress.

CONCLUSIONS

1. Thermal shock fatigue crack propagation rate may be related to the elastic stress intensity factor, and correlated with isothermal crack growth data for the mean temperature of the cycle.

2. As the crack extends from the cooled surface, its speed will reduce leading to arrest, since the stress intensity factor tends to zero with the reduction in temperature gradient.

3. At elevated temperatures where creep damage can accumulate, crack arrest may not occur if a mean stress can cause a change from fatigue to creep crack growth.

ACKNOWLEDGEMENTS

Financial support from the Fukuzawa memorial fund in Keio University for one of the authors (M. Shimizu) is gratefully acknowledged. The authors are also grateful to Oak Ridge National Laboratory, U.S.A. for supplying the material and the Nuclear Regulatory Commission, U.S.A. for financial support.

REFERENCES

Brown, M.W., and Miller, K.J. (1982). Presented at ASTM Symposium on Biaxial/Multiaxial Fatigue, San Francisco, December, 1982.

Emery, A.F., Walker, G.E. and Williams, J.A. (1969). J. Basic Engng, Trans. ASME, Series D. 91, 618-624.

Howes, M.A.H., (1973). Fatigue at Elevated Temperatures, Spec. Tech. Publ. 520, Amer. Soc. Testing Mater., Philadelphia, pp. 242-254.

Marsh, D.J. (1981). Fatigue Engng. Mater. Struct., 4, 179-195.

Mowbray, D.F., Woodford, D.A. and Brandt, D.E. (1973). Fatigue at Elevated Temperatures, Spec. Tech. Publ. 520, Amer. Soc. Testing Mater., Philadelphia, pp. 416-426.

Newton, J.D. (1976). J. Austral. Inst. Met., 21, 94-102.

Okamura, H. (1976). Introduction to Linear Fracture Mechanics, Baifukan, Tokyo (in Japanese).

Skelton, R.P. (1979). Proc. ICM 3, Cambridge, Pergamon, Vol. 2 pp 13-22.

Skelton, R.P. (1982). Low Cycle Fatigue and Life Prediction, Spec. Tech. Publ. 770, Amer. Soc. Testing Mater., Philadelphia, pp. 337-381.

A STUDY OF CREEP BEHAVIOR OF WELDED JOINT

K. Satoh* and M. Toyoda

Department of Welding Engineering, Osaka University, Suita, Osaka, Japan
***Nuclear Systems Division, Kawasaki Heavy Industries Ltd., Kawasaki, Japan**

ABSTRACT

The creep behaviors of welded joints which are subjected to transverse and longitudinal loads respectively have been discussed from bimetallic viewpoints. The rupture behavior of the joints is dominated by the combination of the rupture time ratio and the creep rate ratio of weld metal to base metal.

KEYWORDS

Creep behavior; welded joint; transverse load; longitudinal load; rupture time ratio of weld metal to base metal; creep rate ratio of weld metal to base metal.

INTRODUCTION

In the Very High Temperature Gas-cooled Reactor or Fast Breeder Reactor, it has become an important subject from the viewpoint of high temperature structure design to obtain information on creep behavior for structural materials and the welded joints. Most of creep properties of the welded joints are obtained from the transverse joint specimen where the loading is perpendicular to weld line (Arata and others, 1978). Moreover, special attention is not paid to correlation between specimen size and weld metal configuration for the specimen. Generally, creep properties of weld metal and base metal of which a welded joint is composed are different from each other. Therefore, the creep behavior of the joint may be affected by the creep properties and size of base metal and weld metal. Further, considering the actual component, the welded joints are subjected to load of various direction. For example, in the case of a cylindrical shell or a tube operating under internal pressure, the hoop stress acts parallel to the welding direction for the circumferential butt welds, whereas the stress acts perpendicular to the longitudinal butt welds. The creep behaviors of these welds may be simplified as the creep behaviors of the joints subjected to longitudinal and transverse loads, respectively. In this paper the effect of difference in creep properties between base metal and weld metal on creep behaviors of welded joints has been discussed for welded joints subjected to trasverse and longitudinal loads respectively as the most basic case.

CREEP BEHAVIOR OF WELDED JOINT

The creep properties of weld metal in a welded joint are obviously different from the creep properties of base metal as shown in Figs. 1 and 2 (Arata, Shimizu, Satoh, 1979). It appears from these data that secondary creep rate $\dot{\varepsilon}$ and creep rupture time t_F can be expressed for base metal and weld metal as follows.

$$\left.\begin{aligned} \dot{\varepsilon}_B &= K_B\sigma^n && \text{for base metal} \\ \dot{\varepsilon}_W &= K_W\sigma^n && \text{for weld metal} \\ t_F^B &= A_B\sigma^{-m} && \text{for base metal} \\ t_F^W &= A_W\sigma^{-m} && \text{for weld metal} \end{aligned}\right\} \quad (1)$$

Where, K, A, n and m are constants, and subscripts or superscripts B and W indicate base metal and weld metal, respectively.

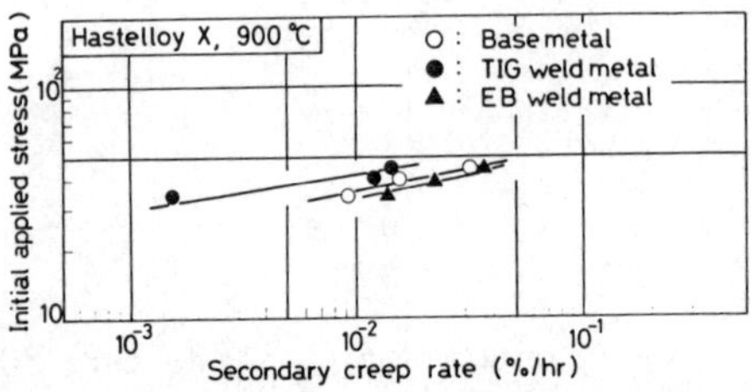

Fig.1 Secondary creep rate of Hastelloy X and the weld metals

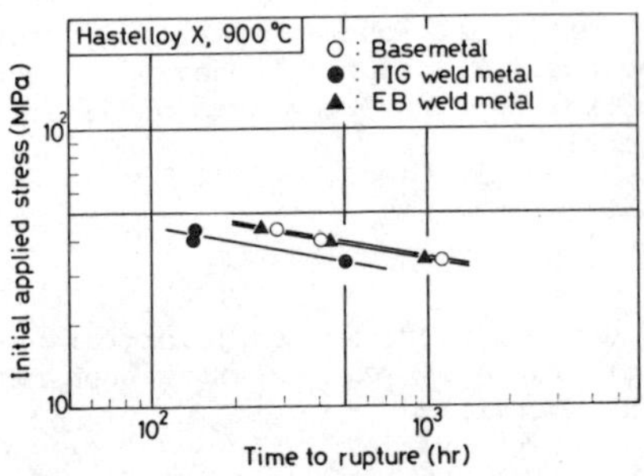

Fig.2 Rupture time of Hastelloy X and the weld metals

Welded Joint Subjected to Transverse Load

In the transversely loading state as shown in Fig. 3(a), the same stress σ acts on the base metal and the weld metal in the welded joint. Therefore, the rupture time of the welded joint t_F^{JT} may be equal to the lower value of the rupture time for weld metal t_F^W or the rupture time for base metal t_F^B. This hypothesis is descrived as follows.

$$\left.\begin{aligned} t_F^{JT} &= t_F^W \text{ or } t_F^{JT}/t_F^B < 1 && \text{for } A_W < A_B \text{ or } A_W/A_B < 1 \\ t_F^{JT} &= t_F^B \text{ or } t_F^{JT}/t_F^B = 1 && \text{for } A_W \geqq A_B \text{ or } A_W/A_B \geqq 1 \end{aligned}\right\} \quad (2)$$

Figure 4 shows the experimental results for the joints. In case of $A_W > A_B$, the rupture times t_F^{JT} of the joints were neary equal to those of the base

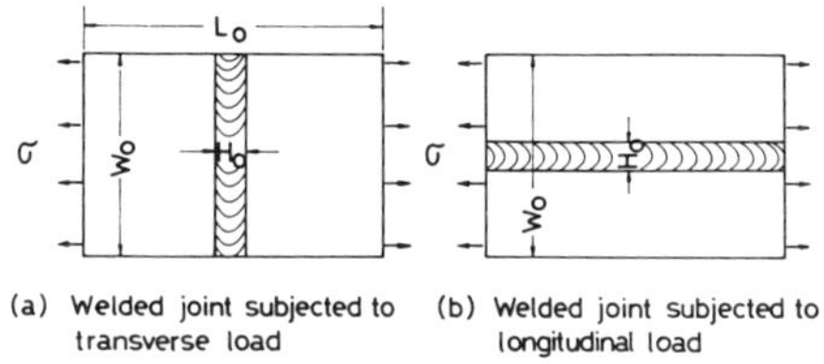

Fig.3 Relation between butt-welded joints and direction of loads

metal and all of the joints ruptured at base metal. In case of $A_W < A_B$, all of the joints ruptured at weld metal. However, the rupture times of the joints were longer than the time expected by Eq.(2), where K_W was larger than K_B. The rupture time t_F^{JT} of the joint is equal to the rupture time of weld metal at larger relative thickness Ho/Do, but it has a tendency to increase toward the rupture time of base metal at smaller Ho/Do. This phenomenon seems to be attributable to the multiaxial stress state of weld metal with restraint of base metal. As the results, Eq.(2) is an excellent method estimating the creep rupture time of the welded joint being transversely loaded, but in some cases it has conservativeness.

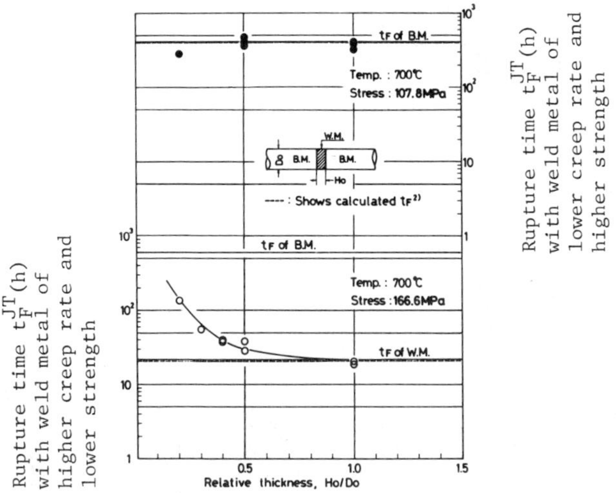

Fig.4 Comparison of experimental rupture time and calculated rupture time for the transversely loaded joint

Welded Joint Subjected to Longitudinal Load

In a welded joint subjected to longitudinal load as shown in Fig. 3(b), it may be assumed that any cross section perpendicular to weld line is always kept in a plane surface during deformation (Satoh, Toyoda, 1975). According to this assumption, the stresses in weld metal and base metal depend on the creep rates of base metal and weld metal. As shown in Fig. 5, the stress in weld metal decreases below the stress in base metal when $K_W/K_B > 1$, while the stress in weld metal increases beyond the stress in base metal when $K_W/K_B \leq 1$.

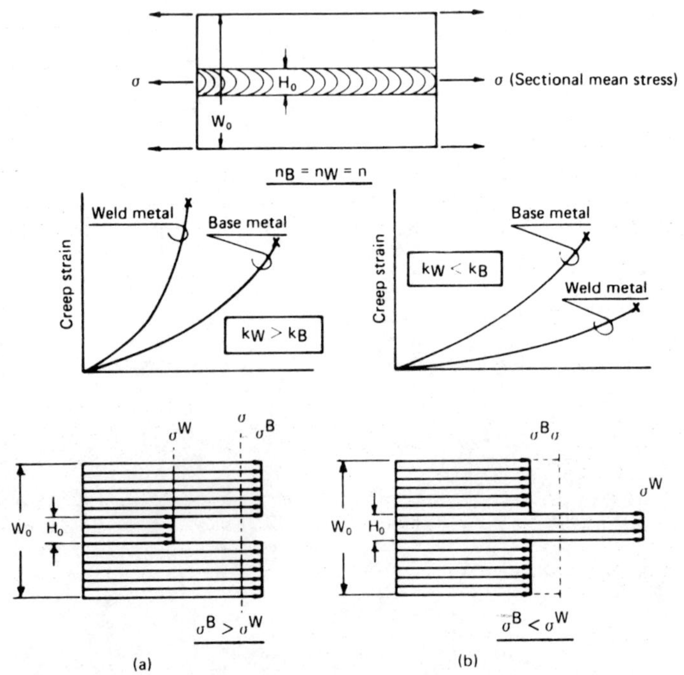

Fig.5 Stress distribution of welded joint subjected to longitudinal load in the case when the creep rate of weld metal is (a) higher and (b) lower than of base metal.

According to Eq.(1), the rupture time for base metal subjected to a stress of σ^B is $A_B(\sigma^B)^{-m}$ and that for weld metal subjected to a stress of σ^W is $A_W(\sigma^W)^{-m}$. Here, the failure occurring at weld metal or base metal is defined as the rupture of the joint. According to this assumption, the rupture time t_F^{JL} for the welded joint is equal to the lower one of $A_B(\sigma^B)^{-m}$ and $A_W(\sigma^W)^{-m}$. After mathematical transformation of Eq.(1) with above mentioned assumption and relation, the rupture time t_F^{JL} of the joint is derived as follows.

$$\left.\begin{aligned} \frac{t_F^{JL}}{t_F^B} &= \left(\frac{K_W}{K_B}\right)^{-\frac{m}{n}} \left\{\left(\frac{K_W}{K_B}\right)^{\frac{1}{n}} + \left[1-\left(\frac{K_W}{K_B}\right)^{\frac{1}{n}}\right]\frac{Ho}{Wo}\right\}^m \quad \text{for } \left(\frac{A_W}{A_B}\right)\left(\frac{K_W}{K_B}\right)^{\frac{m}{n}} \geqq 1 \\ \frac{t_F^{JL}}{t_F^B} &= \left(\frac{A_W}{A_B}\right) \left\{\left(\frac{K_W}{K_B}\right)^{\frac{1}{n}} + \left[1-\left(\frac{K_W}{K_B}\right)^{\frac{1}{n}}\right]\frac{Ho}{Wo}\right\}^m \quad \text{for } \left(\frac{A_W}{A_B}\right)\left(\frac{K_W}{K_B}\right)^{\frac{m}{n}} < 1 \end{aligned}\right\} \quad (3)$$

Where t_F^B is the rupture time for base metal subjected to sectional mean stress $\bar{\sigma}$. It is observed from Eq.(3) that the rupture time t_F^{JL} for the welded joint subjected to longitudinal load depends not only on the rupture time ratio A_W/A_B, but also on the creep rate ratio K_W/K_B and the width ratio Ho/Wo. Figures 6 and 7 show the comparison of the experimental rupture time and the rupture time calculated by Eq.(3). The calculated values agree well with the experimental results in both cases which are the soft joint ($K_W > K_B$, $A_W < A_B$) and the hard joint ($K_W < K_B$, $A_W > A_B$).

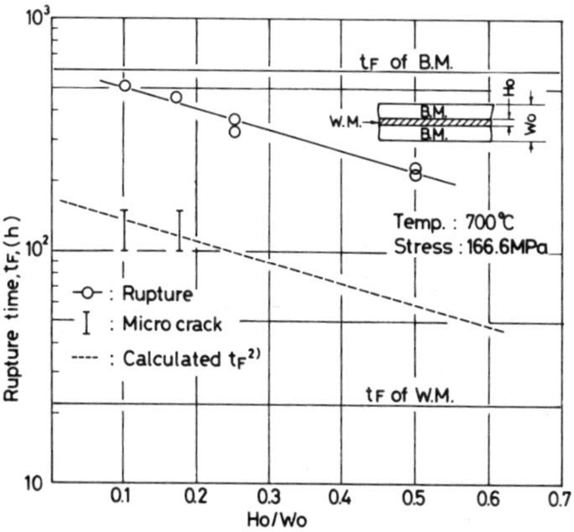

Fig.6 Comparison of experimental rupture time and calculated repture time for the longitudinally loaded joint (soft joint)

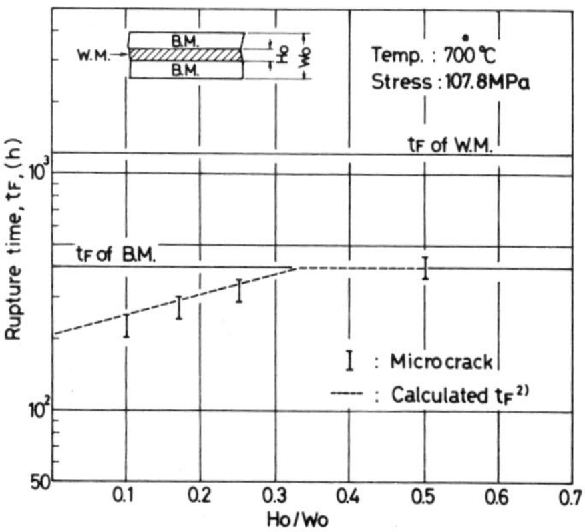

Fig.7 Comparison of experimental rupture time and calculated rupture time for the longitudinally loaded joint (hard joint)

CLASSIFICATION OF CREEP BEHAVIOR FOR WELDED JOINT

Now, we consider the creep rupture time for an infinite plate with a welded joint. Creep rupture strength of the joint depends on the creep rate ratio K_W/K_B of weld metal to base metal and on the rupture time ratio A_W/A_B of weld metal to base metal. In other words, creep behavior of the joint can be classified into six regions based on the ratios of creep rate and rupture time as shown in Fig. 8. And the inequalities of the four kinds of creep rupture times, that is, t_F^{JT}, t_F^{JL}, t_F^B and t_F^W, in each region are as follows.

Region I : $t_F^W > t_F^{JT} = t_F^{JL} = t_F^B$

Region II-1 : $t_F^W = t_F^{JT} < t_F^{JL} = t_F^B$

Region II-2 : $t_F^W = t_F^{JT} < t_F^{JL} < t_F^B$

Region III : $t_F^{JL} < t_F^W = t_F^{JT} < t_F^B$

Region IV-1 : $t_F^W > t_F^{JT} = t_F^{JL} = t_F^B$

Region IV-2 : $t_F^W > t_F^{JT} = t_F^B > t_F^{JL}$

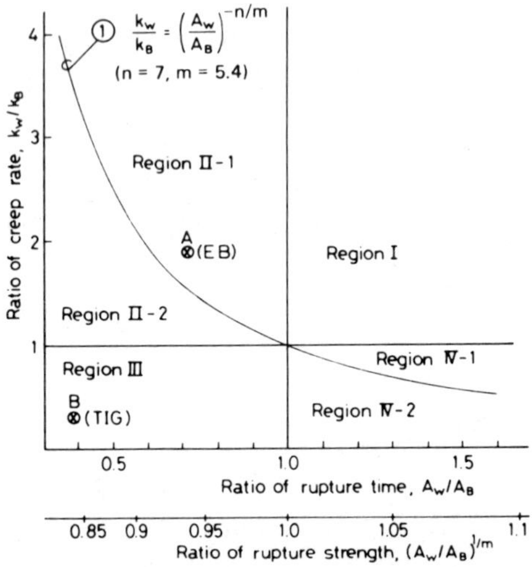

Fig.8 Classification diagram of creep behavior for welded joints.

The welded joints in the Regions I and IV-1 are ideal, because the creep rupture times of the joints are equal to that of base metal even if the joint is subjected to any load acting longitudinally or transversely. In a limited case of the longitudinal load, even the joints in the Region II-1 are good. The points A and B shown in Fig. 8 stand for EB welded joint and TIG welded joint for Hastelloy X, respectively. These points are derived from the data of Figs. 1 and 2. Hastelloy X TIG welded joint has remarkably low rupture time as compared with base metal in both cases, that is, longitudinal load and transverse load. It suggests that it is necessary to improve TIG weld metal. On the other hand, in case of Hastelloy X EB welded joint, the joint subjected to longitudinal load is estimated to have the same or longer rupture time than that of base metal. So, it is probable that there is no problem in applying EB welding to circumferential joints of tubes which are operating under internal pressure at high temperature. In order to ensure satisfactory creep behavior of a welded joint, it is necessary to select a combination of weld metal and base metal properly based on such classification as shown in Fig. 8.

CONCLUSION

The effect of differences in the creep properties of base metal and weld metal on the creep behavior of welded joints has been discussed. As the result, it has been clarified that the rupture time of the joint is dominated by the combination of the rupture time ratio and the creep rate ratio of weld metal to base metal. In order to ensure satisfactory creep behavior of a welded joint, it is necessary to make the combination of weld metal and base metal most suitable. Although there are many problems to be studied, based on the fundamental discussion on suitability of weld metal to base metal as described in this report, actualization for a method to evaluate creep behavior for welded joint will be a leading subject for a future study.

REFERENCE

Arata, Y., K. Terai, H. Nagai, S. Shimizu, T. Aota, K. Satoh, and Y. Ikemoto (1978). Fundamental studies of EB welding of heat-resistant superalloys for nuclear plants. Transaction of JWRI, 7, 73-83.

Arata, Y., S. Shimizu, and K. Satoh (1979). Creep properties of EB welded joint on Hastelloy X. IIW Doc. IV-258-79.

Satoh, K., and M. Toyada (1975). Joint strength of heavy plates with lower strength weld metal. Welding J., 54, 311S-319S.

EFFECT OF CYCLIC STRAIN ON CUMULATIVE FATIGUE DAMAGE OF TYPE 304 STAINLESS STEEL AT HIGH TEMPERATURE

H. L. Nguyen* and T. Bui-Quoc**

***Industrial Materials Research Institute, National Research Council, Montreal, Canada**
****Department of Mechanical Engineering, Ecole Polytechnique, Montreal, Canada**

ABSTRACT

Cumulative damage tests with controlled stresses using both increasing and decreasing sequences have been carried out on Type 304 stainless steel at 427°C. The variation of the maximum strain due to cyclic creep is investigated and its effect is taken into consideration in the evaluation of the cumulative damage effect.

KEY WORDS

Stress-controlled fatigue; cyclic creep; strain accumulation; cumulative damage.

INTRODUCTION

Under push-pull loading with controlled stresses in the low-cycle region at high temperature, the material is usually subjected to an important accumulation of cyclic strain due to the creep effect. This strain may have a significant influence on the fatigue behavior, either under constant amplitude loading or under cumulative damage loading. Some specific components in nuclear power plants are subjected to this type of loading.

Under cumulative damage tests, depending upon the phenomenon involved in the damage process, the sum of life fractions may be different; for example, with two increasing stresses, this sum is larger than unity for pure fatigue loading and is smaller than unity for creep loading (Bui-Quoc, 1982).

The purpose of the present paper is to study the effect of the accumulation of cyclic strain on the fatigue behavior of Type 304 stainless steel at 427°C. Constant amplitude tests lead to the basic fatigue diagram; then sequences of either two increasing stresses or two decreasing stresses are used to evaluate the cumulative damage effect in the region where the effect of cyclic creep is important. An expression is suggested for analysing the data, taking into account the combined effect of cyclic creep and fatigue.

EXPERIMENTAL WORK

Fatigue tests on Type 304 stainless steel with balanced controlled stresses at 427°C have been carried out using a close-loop servo-hydraulic testing machine. A triangular waveform with a constant stress rate of 184 MPa/s (this corresponds to frequencies ranging from 7 to 17 cycles/min) has been imposed on cylindrical specimens of 12.8 mm in diameter. During a test, the axial strain has been measured by means of an axial extensometer with 9.8 mm in gage length; the cyclic (nominal) stress has also been measured by means of strain gages mounted on columns in parallel with the specimen. The stress-strain hysteresis loops have been recorded periodically using an X-Y plotter.

Both sequences of increasing and decreasing stresses have been considered for evaluating the life fraction summation. In these two-step tests, the two stress levels of 207 and 276 MPa have been imposed on the specimens with various life fractions at the first level.

ANALYSIS OF RESULTS

Constant Amplitude Tests

The basic fatigue diagram obtained from constant amplitude tests is shown in Fig. 1 and typical stress-strain loops recorded in a test with σ = 276 MPa are reproduced in Fig. 2.

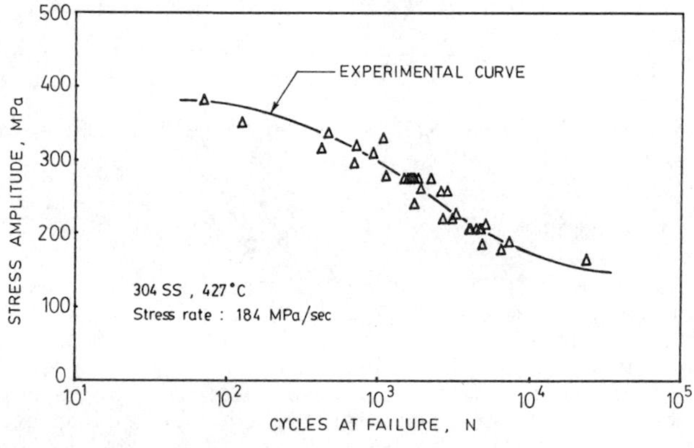

Fig. 1. Stress-controlled fatigue diagram of Type 304 stainless steel.

In general, the material exhibits an initial strain softening characterized by an extensive strain accumulation. Then, the loops are relatively stable during a large part of the specimen life, hence the total strain range is nearly constant. At a low stress, the fracture is achieved by the propagation of a fatigue crack with a limited plastic deformation. However, for a large stress amplitude, the specimen fails in a ductile manner and the fracture is characterized by necking at the working section; this behavior is reflected by a rapid shift of the stress-strain loops along the strain axis.

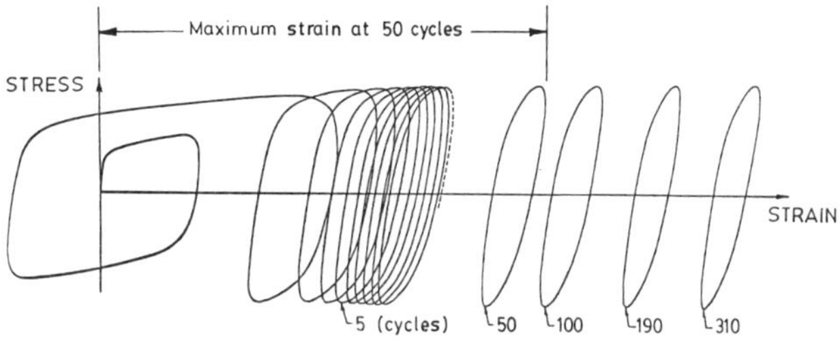

Fig. 2. Typical stress-strain hysteresis loops (σ = 276 MPa, Temp. = 427°C).

This stress-strain characteristic has also been observed for a similar material (Turner and Martin, 1980) and for a cast mild steel (Radon and coworkers, 1972).

A plot of the maximum strain in terms of the applied cycles is presented in Fig. 3. At a low stress, the rate of strain accumulation is nearly constant until the appearance of a large fatigue crack after which this rate increases rapidly. At a higher stress, the strain accumulation rate is more pronounced and it increases continuously throughout the specimen life. It has been noted that, the accumulated strain corresponding to failure increases with an increase of the stress amplitude.

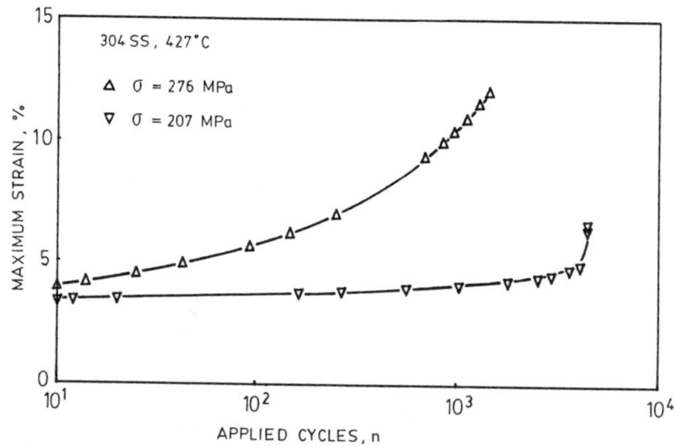

Fig. 3. Variation of the maximum strain with applied cycles in constant stress amplitude tests.

Cumulative Damage Tests

Increasing sequence. Fig. 4a shows a typical variation of the maximum cyclic strain in a two-step test with increasing stresses. After the stress changes, the rate of the strain accumulation is more pronounced than that obtained in the corresponding tests with constant amplitude.

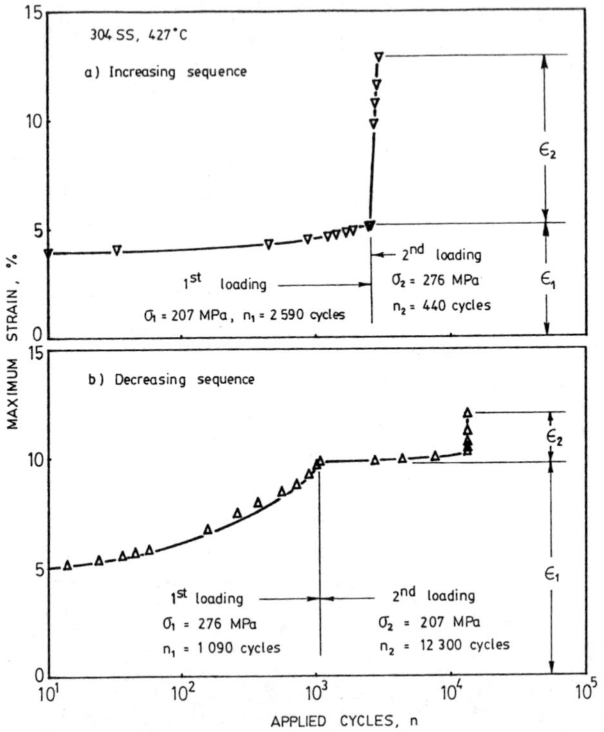

Fig. 4. Variation of maximum strain with applied cycles in two-step tests.

The experimental values of sums of cycle ratios range from 0.67 to 0.96; this is in contrast to the results usually obtained in similar tests under pure fatigue. In fact, stress-controlled tests with an increasing sequence performed by Rey (1958) on SAE 4130 alloy steel at room and high temperatures at a high frequency (80 Hz) indicate that the values of cycle ratio summations are larger than unity.

Decreasing sequence. A change of the stress amplitude results in a relatively lower rate of strain accumulation as shown in Fig. 4b. The rapid increase of the maximum strain in the last part of the curve is explained by the presence of large fatigue cracks near the end of the specimen life.

The sum of cycle ratios stands between 1.70 and 2.70. The results are opposite to those obtained in tests with a high frequency. In fact, with decreasing sequences used in stress-controlled tests on Superalloy IN 100 at 900 and 1000°C at 5 Hz (Chaboche and Policella, 1977), the experimental values of cycle ratio summations range from 0.40 to 0.80.

Figure 5 illustrates the relationship between the cycle ratios at the first and second stress levels obtained in cumulative damage experiments. The scatter of the experimental results in this diagram is quite large, in particular for the decreasing sequence. This situation is closely related to

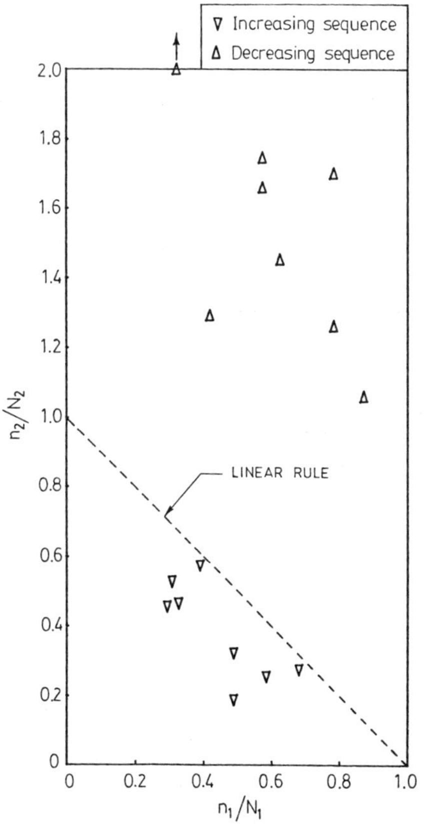

Fig. 5. Cumulative damage diagram based on cycle ratios.

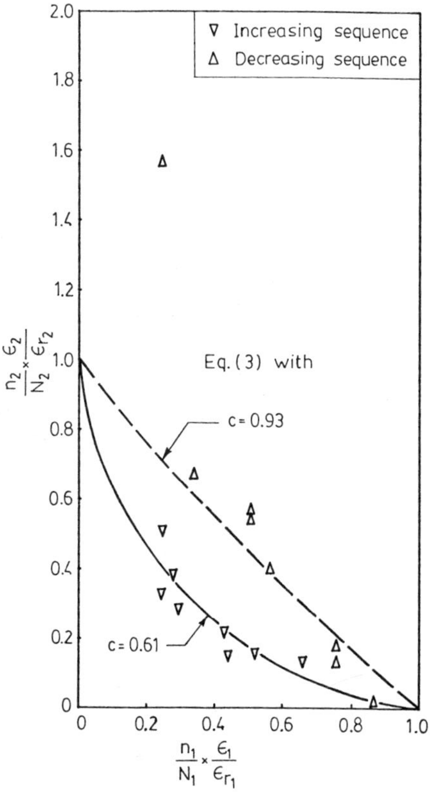

Fig. 6. Modified cumulative damage diagram.

the large variation of the cyclic strain accumulation at the second level. From this diagram, it is seen that the linear damage rule is not adequate for correlating the present experimental data.

It appears that, the accumulated strain, and to an extent the material behavior, is significantly influenced by the previous loading. Furthermore, considering the testing conditions, i.e. the relatively high temperature and low frequencies used in the present experimental work, the creep effect on the material fatigue behavior is not negligible.

A RULE FOR LIFE FRACTION SUMMATION

Because of the change in the rate of strain accumulation at the second stress level, the mechanisms involved in the damage evolution in the material are not due only to fatigue, but also due to creep. On the basis of a life fraction summation, the experimental values obtained are probably more closely associated with creep than with fatigue. In fact, for cumulative damage tests on a Cr-Mo-V steel at high temperatures, the experimental values of time ratio summations are smaller than unity for increasing sequences and vice versa (Goldhoff and Woodford, 1972).

In the study of the cumulative damage under pure creep loading, Voorhees and Freeman (1959) have emphasized that the exposure time t and the accumulated strain ε should be considered for evaluating the material behavior and they have proposed the following expression:

$$\sum_i \left(\frac{t_i}{t_{ri}} \frac{\varepsilon_i}{\varepsilon_{ri}}\right)^{0.5} = 1 \qquad (1)$$

where t_r and ε_r are time and accumulated strain at rupture respectively.

Adapting the above concept to the case of fatigue with the existence of cyclic creep, the time ratio t/t_r is replaced by the cycle ratio n/N. Then eq. (1) becomes:

$$\sum_i \left(\frac{n_i}{N_i} \frac{\varepsilon_i}{\varepsilon_{ri}}\right)^{c} = 1 \qquad (2)$$

where c is a constant. The residual cyclic life at the second stress level is thus expressed in terms of the parameters associated with the first level as follows:

$$\frac{n_2}{N_2} \frac{\varepsilon_2}{\varepsilon_{r2}} = \left[1 - \left(\frac{n_1}{N_1} \frac{\varepsilon_1}{\varepsilon_{r1}}\right)^{c}\right]^{1/c} \qquad (3)$$

In the above expression, the fatigue damage is represented by the cycle ratio and creep damage is accounted for by the strain ratio. With the modified parameter, the cumulative damage diagram takes a new form, as shown in Fig. 6. It is seen that the scatter band has been significantly reduced with respect to that associated with the cycle ratio only (Fig. 5). With the present experimental data obtained from a fixed temperature and a combination of two stress levels, it seems that constant c in eq. (3) depends upon, among several parameters, the sequence of loading, as shown in Fig. 6.

It should be emphasized that eq. (3) is suggested on the basis of a limited number of test results. There is an improvement in the data presentation; however, further investigations on the subject are necessary before the suggested relation becomes readily applicable for practical purposes.

CONCLUSION

Low frequency, completely reversed stress-controlled fatigue tests have been carried out on Type 304 stainless steel at 427°C using constant amplitude stress and sequences of either two increasing or two decreasing levels.

The presence of the accumulated strain due to cyclic creep leads to a sum of cycle ratios which deviates from the trend commonly observed in cumulative fatigue damage tests. A rule of summation, modified from the expression already proposed for evaluating the cumulative damage effect under pure creep, is suggested for analysing the experimental results in cases where the cyclic creep is present under loading with variable amplitude.

ACKNOWLEDGMENTS

This work was supported by the Natural Sciences and Engineering Research Council of Canada (Grant A-3952) and the Direction Générale de l'Enseignement Supérieur du Gouvernement du Québec (Grant CRP-295-72), Canada.

REFERENCES

Bui-Quoc, T. (1982). Recent developments of continuous damage approaches for the analysis of material behavior under fatigue-creep loading. ASME PVP-59, 211-226.

Chaboche, J. L., and H. Policella (1977). Caractérisation expérimentale des matériaux sous chargement cyclique à haute température. Rech. Aérosp., 5, 307-324.

Goldhoff, R. M., and D. A. Woodford (1972). The evaluation of creep damage in a Cr-Mo-V steel. Testing for Prediction of Material Performance in Structures and Components, ASTM STP 515, 89-106.

Radon, J. C., P. W. J. Oldroyd, and D. J. Burns (1972). Mean strain and hardness changes in uniaxial stress cycling. Proc. First Int. Conf. on Mechanical Behavior of Materials - ICM I, Vol. II, Society of Materials Science, Japan, 285-298.

Rey, W. K. (1958). Cumulative fatigue damage at elevated temperature. NACA TN 4284.

Turner, A. P. L, and T. J. Martin (1980). Cyclic creep of type 304 stainless steel during unbalanced tension-compression loading at elevated temperature. Met. Trans., 11A, 475-481.

Voorhees, H. R., and J. W. Freeman (1959). Notch sensitivity of aircraft structural and engine alloys. Wright Air Development Center Tech. Rep., Part II.

THE EFFECTS OF THE WAVY GRAIN BOUNDARIES ON THE CREEP FRACTURE BEHAVIOR IN UDIMET-700 ALLOY

Guixiong Guo, Huiwang Shen and Qigong Cai

Central Iron and Steel Research Institute, Beijing, China

ABSTRACT

The wavy grain boundaries (GBs) in U-700 alloy improve remarkably the creep rupture strain and the resistance to the creep fracture. By comparing the measured and predicted creep crack initiation life at the sharp notch root it is demonstrated that creep J*-integral is the effective parameter controlling the asymptotic field and the damage rate at the tip of the sharp notch (and so the crack). Based on this result a combined analysis using the linear damage model and the HRR singularity is carried out and an expression for creep crack growth rate is derived, which is in good agreement with experimental results of the alloy with smooth and wavy GBs. A microscopic mechanism of the creep damage in the alloy with wavy GBs is proposed from the point of view of the cavitation along the GBs.

KEYWORDS

Wavy grain boundary; fracture mechanics; creep J*-integral; creep damage; cavitation.

INTRODUCTION

Wavy GBs improve remarkably the resistance to creep fracture (Larson and Floreen, 1977) and have been used in some Nickel-based superalloys. Recently, the formation mechanism of wavy GBs and their effects on the creep and fatigue fracture behavior have become an important and interesting problem in the theory and application of superalloys. In this paper the effects of wavy GBs on the creep fracture behavior in U-700 alloy are discussed from the point of view of the fracture mechanics and the cavitation damage along the GBs.

EXPERIMENTAL PROCEDURE

The chemical compositions in wt pct of U-700 alloy tested are as follows: Ni (55.77), Cr (14.71), Co (16.57), Mo (5.21), Al (4.04), Ti (3.59), C (0.045), B (0.022), Mg (0.04). Two heat treatment were used in order to obtain smooth and wavy GBs with grain size of 0.27 mm:

Treatment A (smooth GBs): 1200°C/4h A.C. + 1080°C/4h A.C. + 845°C/24h A.C. + 760°C/16h A.C. ;

Treatment B (wavy GBs): 1200°C/4h $\xrightarrow{1.1°C/min}$ 1000°C A.C. + 850°C/16h A.C. .

6 mm diam specimens with 20 mm gage length were used for creep rupture test. The specimens for creep crack growth test had rectangular cross-section and a gage size of 3 mm x 12 mm x 15 mm. The test was carried out on Instron 1251. Temperature was maintained by a three zone resistance furnace and controlled within 800 ± 3°C over the gage length. The instant crack length was measured by a tool-microscope with accuracy of 0.005 mm after each unloading the specimen from the testing machine and then repeated the test.

EXPERIMENTAL RESULTS AND DISCUSSION

Heat Treatment for Wavy GBs

Wavy GBs are usually obtained by slow cooling after solution treatment. In the present work it was found that during aging at temperature range of $M_{23}C_6$-type carbide precipitation the amplitude of wavy GBs obtained after slow cooling treatment increases remarkably. The wavy GBs with amplitude of 2-10 μm and good generality were obtained by Treatment B (Fig. 1).

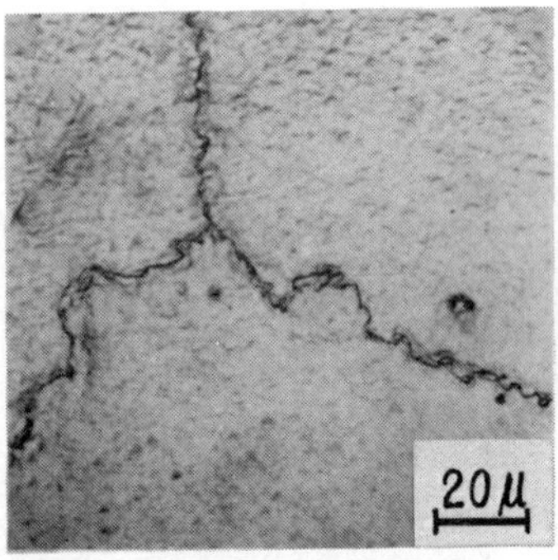

Fig. 1 Optical micrograph of the wavy GBs obtained by Treatment B.

Effects of the Wavy GBs

The results of the tests are summarized in Fig. 2-4. They shows that although the wavy GBs do not improve the stress rupture life of the alloy (Fig. 2), they improve remarkably the rupture strain (Fig. 3) and the resistance to the creep crack growth (Fig. 4). The later effect is a most important one

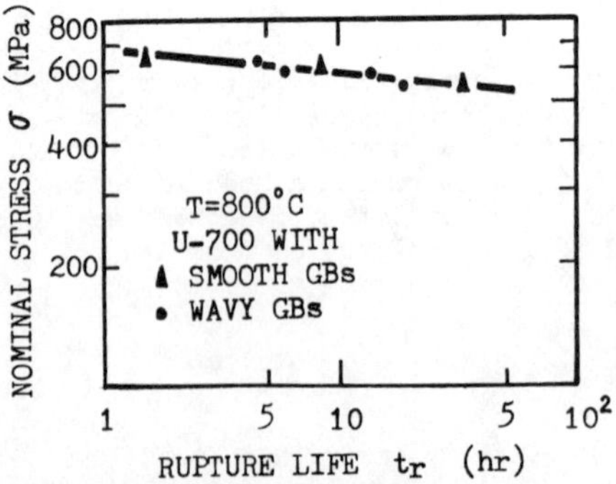

Fig. 2 The nominal stress versus the rupture life.

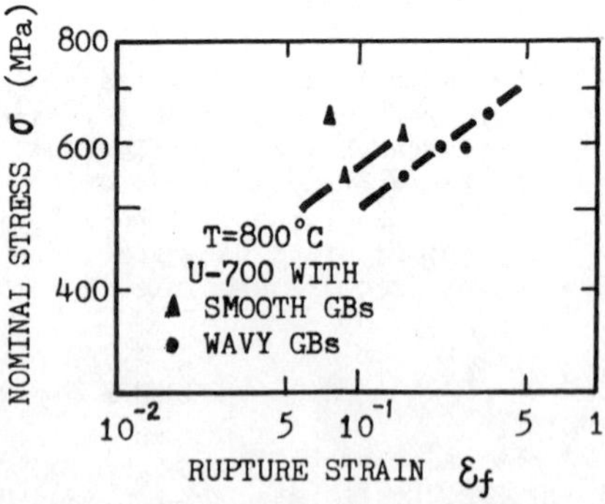

Fig. 3 The nominal stress versus the rupture strain.

since the premature failures and the life of the high temperature components are usually caused by the growth of some surface crack initiated owing to hot corrosion and thermal fatigue.

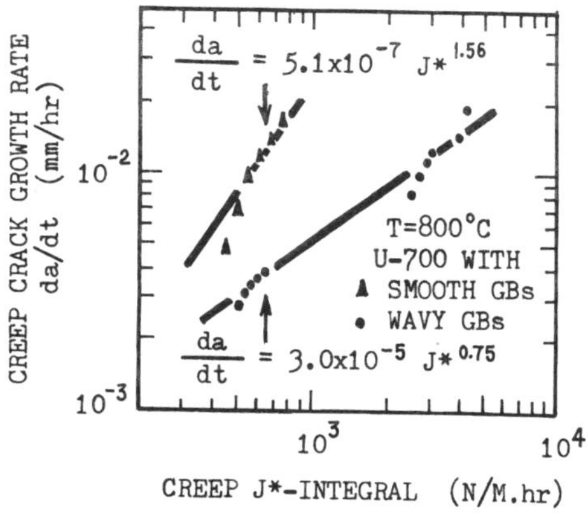

Fig. 4 The creep crack growth rate versus J*-integral.

Creep Fracture Mechanics Analysis

Analysis of the creep crack initiation life at a sharp notch root. In the following the appropriate parameter governing creep crack stress field is determined by comparison of experimental and analytical results on the creep crack initiation life at the sharp notch root. Since creep crack always initiates along GBs it is reasonable to define the crack initiation life as the time of formation of a crack with one grain size depth. In the analysis it is assumed that at a rather short time after loading the steady-state creep stress field dominates the region near the sharp notch root, so the C*-integral (in this paper named J*-integral) proposed by Landes and Begley (1976) is assumed to be an appropriate parameter for the analysis. Based on the similarity of the constitutive equations and their solutions between power law creep materials and pure hardening materials, the theory and the computing results for J-integral of pure hardening materials (Cai, 1976, 1977; Goldman and Hutchinson, 1975; Shih and Hutchinson, 1976) can be used for evaluating the creep J*-integral, thus we have:

$$J^* = 2\pi q a \int_0^{\dot{\varepsilon}} \sigma \, d\dot{\varepsilon} \qquad (1)$$

for path Γ_1 and

$$J^* = A\rho \int_0^{\dot{\varepsilon}_1} \sigma_1 \, d\dot{\varepsilon}_1 \qquad (2)$$

for path Γ_2. where σ, $\dot{\varepsilon}$ is remote nominal stress and steady-state creep rate, σ_1, $\dot{\varepsilon}_1$ the stress and creep rate at the root, a and ρ are defined as in Fig. 5, q the shape factor, A is a constant between 1 and 2 which is related to the distribution of deformation work rate density along the path Γ_2 and one can take A=2 to ensure the fracture process zone to contain some GBs. Using stress-strain rate relationship:

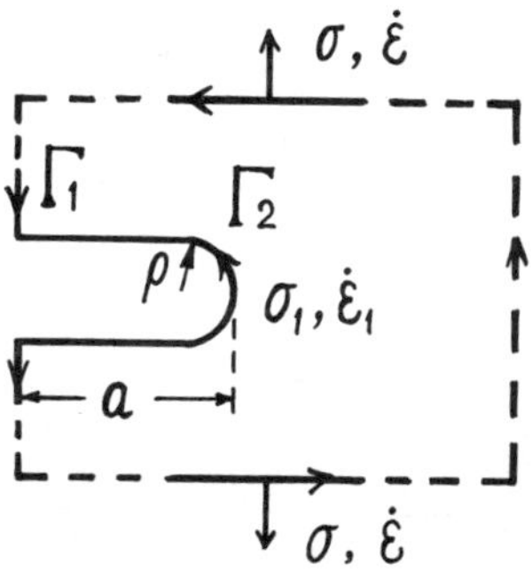

Fig. 5 The creep J*-integral analysis.

$$\frac{\dot{\varepsilon}_e}{\dot{\varepsilon}_0} = \alpha\left(\frac{\sigma_e}{\sigma_0}\right)^n \qquad (3)$$

and the path-independence of J-integral, from eqs. (1) and (2) we obtain:

$$\frac{\sigma_1}{\sigma} = \left(\frac{\pi q a}{\rho}\right)^{\frac{1}{n+1}} \qquad (4)$$

$$\frac{\dot{\varepsilon}_1}{\dot{\varepsilon}} = \left(\frac{\pi q a}{\rho}\right)^{\frac{n}{n+1}} \qquad (5)$$

Eqs. (4) and (5) represent the stress and creep rate concentration coefficient at the notch root under steady-state creep condition. It is assumed that the crack initiates when the accumulated creep strain at the root attains a critical value ε_f, the rupture strain measured by rupture test. Using this criterion and empirical relation obtained in the present work:

$$\varepsilon_f = B\,\dot{\varepsilon}^k t_r = B\,\dot{\varepsilon}_1^k t_i \tag{6}$$

and eqs. (3), (5), we obtain the creep crack initiation life at the root:

$$t_i = \frac{\varepsilon_f}{B(\alpha\dot{\varepsilon}_0)^k}\left(\frac{\sigma_0}{\sigma}\right)^{kn}\left(\frac{\rho}{\pi q a}\right)^{\frac{kn}{n+1}} \tag{7}$$

The numerical values of the parameters are obtained by rupture test and shown in Table 1. It is emphasized that ε_f varies with the stress level. To estimate t_i using eq. (7) one can calculate σ_1 from eq. (4) and find the value of ε_f corresponded to σ_1 in Fig. 3. The shape factor q is calculated by using the computing result of Kumar, German and Shih (1981) as following:

$$q = \frac{1}{2\pi}(C\sigma_0)^{n+1}\left(1-\frac{a}{b}\right)h_1\left(\frac{a}{b},n\right)\left(\frac{b}{P_0}\right)^{n+1}\left(1+\frac{1}{n}\right) \tag{8}$$

where C is 1 for plane stress and $2/\sqrt{3}$ for plane strain.

TABLE 1 Numerical Values of the Parameters

Parameter	α	n	B	k	σ_0 (MPa)	$\dot{\varepsilon}_0$ (h^{-1})	γ	kn/(n+1)
Smooth GBs	1	15.2	205	1.61	500	1.38×10^{-4}	1.56	1.51
Wavy GBs	2.03	15.2	1.76	0.78	500	1.38×10^{-4}	0.75	0.73

Comparison of the predicted result by eq. (7) and experimental results is shown in Fig. 6. In all specimens in Fig. 6 the crack did not go through out the surface of the plate. The specimens were then broken at room temperature and it was found that the creep crack initiated at the notch root had only one grain size in depth. This shows that predicted t_i by eq. (7) is in good agreement with the experimental results. The predicted t_i by using linear elastic concentration factor K_t (or by LEFM) also by elastic-plastic J-integral analysis all depart considerably from the experimental date (the predicted t_i by K_t is also shown in Fig. 6). This confirms that under the present test conditions only creep J*-integral governs the asymptotic field near the tip of the sharp notch (and so the crack).

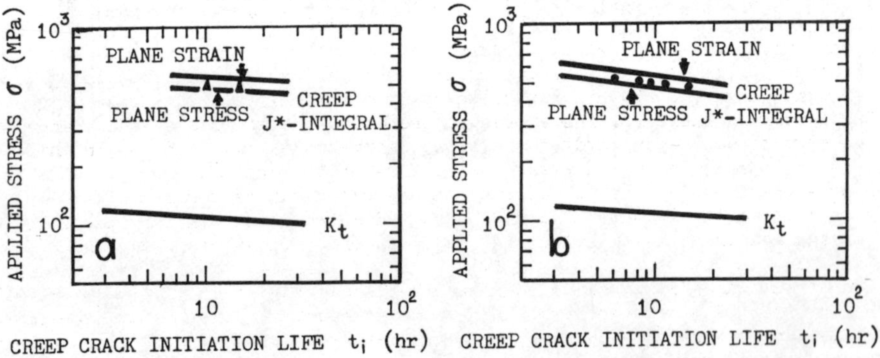

Fig. 6 The comparison of J* and K_t analysis with experimental results of U-700 alloy with (a) smooth GBs; (b) wavy GBs. ($a/\rho=5$).

Creep crack propagation analysis. The experiment shows that the creep crack propagation is stepwise. During the incubation period of each step the crack is stationary, so that it is assumed that the σ_{ij} - $\dot{\varepsilon}_{ij}$ field of the stationary crack governs the creep damage rate and fracture time in the near-tip region and the HRR-type singularity field of creep J*-integral is suitable for describing the creep crack propagation.

Appling the linear damage model (Kubo, Ohji and Ogura, 1979) to the analysis. the creep damage $\Delta\eta$ during the time interval Δt is defined as $\Delta t/t_r$ and the damage rate $\dot{\eta}$ is:

$$\dot{\eta} = \frac{\Delta\eta}{\Delta t} = \frac{1}{t_r} = \frac{B\dot{\varepsilon}^k}{\varepsilon_f} \qquad (9)$$

where the eq. (6) is used. The fracture criterion is that the fracture occurs when accumulated creep damage reaches unity. Thus when the crack propagates at the distance x_1 in front of the initial crack, we have (Fig. 7):

$$\eta_{x_1} = \sum_{t=0}^{t_{x_1}} \dot{\eta}_{x_1}(t)\,\Delta t = 1 \qquad (10)$$

where t_{x_1} is the time for crack tip to arrive at x_1. The crack propagation from the tip toward the point x_1 can be regarded as the movement of the point x_1 toward the crack tip, so substituting $\Delta t = -\Delta x/\dot{a}$ and eq. (9) into eq. (10), we obtain:

$$\sum_{x=0}^{x_1} \frac{B\,[\dot{\varepsilon}(x)]^k}{\varepsilon_f}\,\frac{\Delta x}{\dot{a}} = 1 \qquad (11)$$

where $\dot{\varepsilon}(x)$ is the steady- state creep rate at $x+\Delta x/2$. In first approximation we neglect the nonsingular damage outside x_0, the extent in which the HRR singular field dominates, and apply the HRR singularity to eq. (11), the creep crack growth rate is derived as:

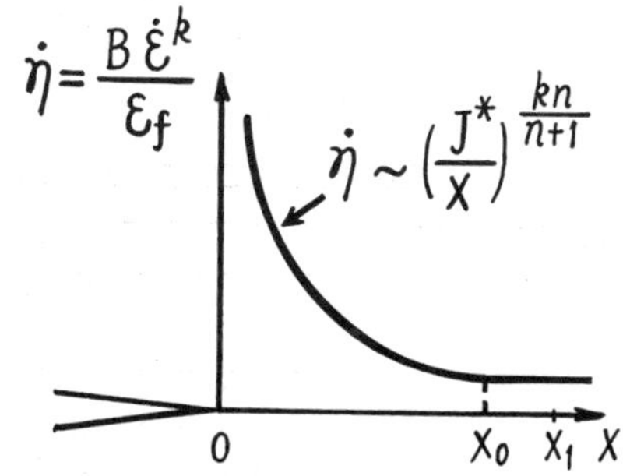

Fig. 7 The creep crack propagation analysis.

$$\frac{da}{dt} = B\,\alpha^{\frac{k}{n+1}}\,D(x_0)\,J^{*\frac{kn}{n+1}} \qquad (12)$$

with

$$D\,(x_0) = \left(\frac{\dot{\varepsilon}_0}{\sigma^n}\right)^{\frac{k}{n+1}}\left(\frac{1}{I_n}\right)^{\frac{kn}{n+1}}\left[\tilde{\dot{\varepsilon}}_{22}(\theta)\right]^k \sum_{x=0}^{x_0} \frac{\Delta x}{\varepsilon_f\,x^{\frac{kn}{n+1}}} \qquad (13)$$

Comparing the eq. (12) with the experimental relation (Fig. 4): $da/dt = AJ^{*\gamma}$, we have: $\gamma = kn/(n+1)$. The predicted exponent $kn/(n+1)$ by eq. (12) is in good agreement with the measured experimentally of the alloy with smooth and wavy GBs (shown in Table 1). The expression reflects the effects of the change of the microstructure on da/dt through the macro-parameters α, B, k. In this expression one finds that decrease of either α or B reduces the da/dt. α is a macro-parameter related to material strength and can be reduced effectively by γ' precipitation in superalloys. The wavy GBs treatment makes γ' particles coarsen and increases α value. But da/dt is proportional to $\alpha^{k/(n+1)}$ and $\alpha^{k/(n+1)}$ of the alloy with wavy GBs is only 1.03 times as large as that for smooth GBs, so its effect on da/dt is small. On the other hand, the alloy with wavy GBs makes B two order of magnitude lower than that of smooth GBs and has an excellent resistance to creep crack growth. The expression shows that in order to improve creep crack growth resistance, reducing B is a more effective method.

Mechanism of Creep Damage in the Alloy with Wavy GBs

The creep fracture of the alloy with wavy GBs is induced by initiation and coalescence of cavities along the GBs, so the creep damage must be related to the cavitation on the GBs. The cavities are nucleated by the incompatibility between the grain boundary sliding (GBS) and the grain deformation. Raj and Ashby (1971) discussed the sliding rate v by diffusion accommodation and their result shows that v is proportional to λ/h^2. where λ is the wave length and h the wave amplitude. One can take the amplitude of the smooth GBs as one-half of the $M_{23}C_6$ particle size i.e. 0.1-0.5 μm. Measured amplitude of the wavy GBs is 1-5 μm and is an order of magnitude larger than that of smooth GBs. Further, the wave length of the wavy GBs is an order of magnitude too, so the sliding rate of the wavy GBs is smaller for one order of magnitude.

When the sliding is accommodated by plasticity within the grains, the sliding rate of a segment with length Δl of the wavy GBs, to which a resolved shear stress $\tau\cos\theta$ is applied (Fig. 8), is given by (Eberhardt and Baudelet, 1980):

$$v=\frac{4}{3}\frac{D b^2 \Delta l}{h' k T \mu}\tau^2\cos^2\theta \qquad (14)$$

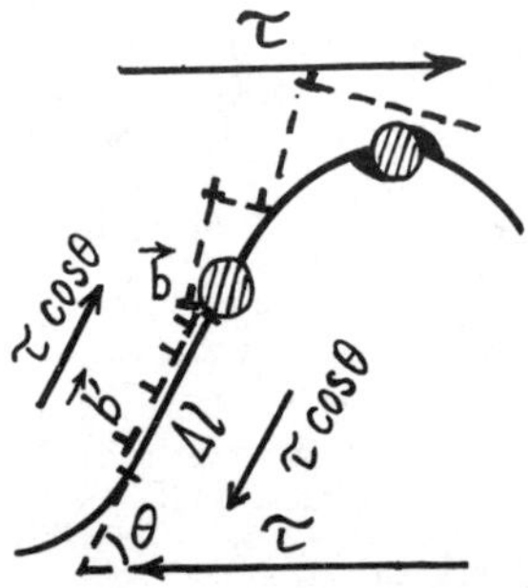

Fig. 8 GBS of the wavy GBs accommodated by plasticity within the grains.

where D is the diffusion coefficient of vacancies in the grains, μ the shear modulus, k Boltzmann's constant, T the absolute temperature, b the Burgers vector of the matrix dislocation, h' the dislocation climb distance. Eq. (14) shows that v is proportional to $\cos^2\theta$. At θ=0 i.e. at the peak or valley of the wavy GBs the sliding rate achieves maximum, so the cavities nucleate mostly at the peak or valley of the wavy GBs. The SEM fractograph reveals that the cavities on the wavy GBs are much less in quantity but much larger in size than those on the smooth GBs (Fig. 9). The spacing between two cavities on the wavy GBs is about 3-8 μm and closes to the wave length. One can image from this that the cavities were nucleated at the peak or valley of the wavy GBs. This is supported by the metallographic fracture profile (Fig. 10).

Eq. (14) also shows the sliding rate is inversely proportional to h', the

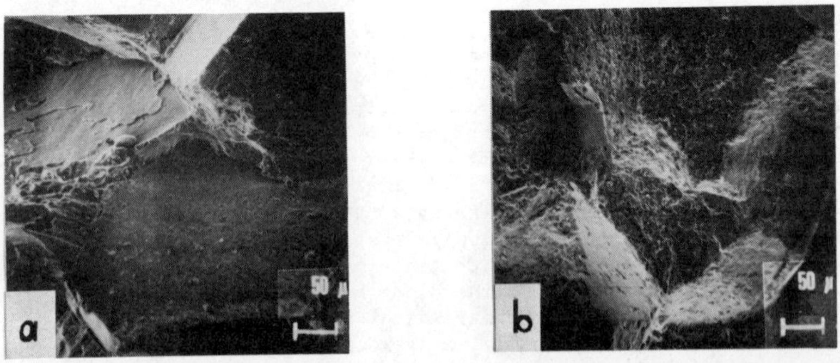

Fig. 9 The SEM fractograph of U-700 alloy with (a) smooth GBs, σ=617 MPa, t_r=8.3 h ; (b) wavy GBs, σ=591 MPa, t_r=13.7 h .

dissociated dislocation climb distance in adjacent grains. Only after the dislocation had climbed over the peak, the subsequent leading boundary dislocation dissociation can be proceeded to ensure the continuity of the accommodation. This means that h' is proportional to amplitude h and so the sliding rate of the wavy GBs with plastic deformation accommodation is an order of magnitude smaller too.

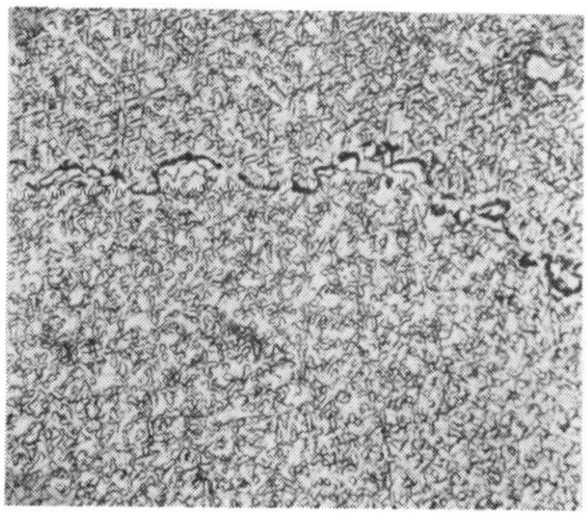

Fig. 10 Metallographic fracture profile of U-700 alloy with wavy GBs revealing that cavities are nucleated at the peak.

Under higher stress level although GBS is not the principal mechanism of creep deformation, but it remains the principal cause of the cavitation along the GBs. In this case the sliding rate of the wavy GBs is much smaller and their grain deformation rate is larger. These make the accommodation ease and inhibit effectively the cavitation along the wavy GBs.

Since the cavities are nucleated at the peak or valley and the wave length of the wavy GBs is an order of magnitude larger than that of smooth GBs, the cavities coalescence on the wavy GBs is more difficult and the damage rate and creep crack growth rate is remarkably lower.

CONCLUSION

Aging at 850-950°C after slow cooling treatment is important for obtaining wavy GBs with large amplitude and wave length in U-700 alloy; The wavy GBs improve remarkably the rupture strain and the resistance to the creep fracture; At 800°C in the stress range of 500-600 MPa the creep J*-integral governs the asymptotic field near the tip of the sharp notch (and so the crack); The derived creep crack initiation life at the notch root and the creep crack growth rate are in good agreement with the experimental results of the alloy with smooth and wavy GBs; The GBS rate of the wavy GBs is an order of magnitude lower and the cavities are nucleated at the peak or valley of wavy GBs, these inhibit the cavities nucleation and coalescence and are the causes for wavy GBs to improve remarkably the resistance to creep fracture.

REFERENCES

Cai Qigong (1976). Acta Metallurgica Sinica, 12, 45-67.
Cai Qigong (1977). Acta Metallurgica Sinica, 13, 246-262.
Eberhardt, A., and B. Baudelet (1980). Phil. Mag., 41, 843-869.
Goldman, N. L., and J.W. Hutchinson (1975). Int. J. Solids Structures, 11, 575-591.
Kubo, S., K. Ohji, and K. Ogura (1979). Eng. Fracture Mech., 11, 315-329.
Kumar, V., M. D. German, and C.F. Shih (1981). An Engineering Approach for Elastic-Plastic Fracture Analysis, EPRI NP-1931, Chap. 3.
Landes, J. D., and J. A. Begley (1976). ASTM STP 590 , 128-148.
Larson, J.M., and S. Floreen (1977). Met. Trans., 8A, 51-55.
Raj, R., and M. F. Ashby (1971). Met. Trans., 2, 1113-1127.
Shih, C. F., and J. W. Hutchinson (1976). J. Eng. Mater. & Technol., 98, 289-295.

THE NON-ISOTHERMAL FATIGUE TESTS AND ANALYSIS

K. Bohatec, C. Budac and J. Zdarek

Sigma Research Institute, Prague, Czechoslovakia

ABSTRACT

Piping and valves of fossil and future nuclear power stations with fast reactors are exposed to high temperature environment. Primary loading conditions include internal pressure, external forces and moments and non-stationary thermal gradients. In our paper we present methodical principles of the non-isothermal fatigue tests as well as test results of several structural materials. The calculation method for fatigue life evaluation under high temperature is suggested.

KEYWORDS

Non-isothermal fatigue, fatigue evaluation, experiment, life calculation, valves.

THE METHODICAL PRINCIPLES OF THE NON-ISOTHERMAL FATIGUE TESTS

In the Sigma Research Institute, for the non-isothermal fatigue tests, the same components, equipment and principles as for isothermal fatigue tests are employed /Zdarek, 1975; Budac, 1980/. Test specimen with test section diameter 6 mm, extensometer for diametral strain measurement and load cell are used; specimen is heated by direct electric current resistance. Cyclic heating and cooling is controlled by temperature regulator. Cooling is realized by free heat transfer to both upper and lower water cooled grips.

The magnitude of mechanical deformation in case of isothermal fatigue test is lineary proportional to output signal from diametral extensometer. Under non-isothermal fatigue test when the ΔT parameter value differes from zero, cyclic change of the specimen test section diameter in the range of $\Delta\varepsilon_{tep}$ /range of diametral thermal deformation/ due to thermal expansion occures. On imposing cyclic mechanical deformation on the specimen, the output signal from extensometer is proportional to the magnitude of total diametral deformation:

$$\Delta\varepsilon = \Delta\varepsilon_{tep} + \Delta\varepsilon_{mech}$$

The thermal deformation can be expressed as:

$$\Delta\varepsilon_{tep} = K_1 \Delta T$$

Mechanical component of the longitudinal deformation is controlled by analog computer according to equation:

$$\Delta\varepsilon_1 = /\Delta\varepsilon - K_1 \Delta T + K_2/.2 - \sigma/E \cdot /1 - \nu_{el}/$$

where K_2 is a constant used for symetrization of the thermal deformation signal, E is an arithmetic mean of Young´s modulus for maximum and minimum temperature of the cycle. Block diagram for electric control signal is on Fig. 1

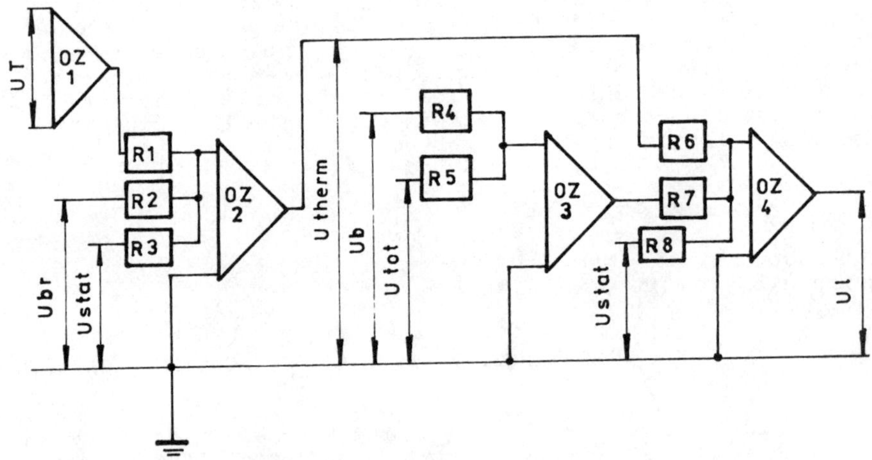

Fig. 1 Block diagram for the determination of mechanical component of the total longitudinal deformation.

The electric signal output U_1 is the mechanical component of the total longitudinal deformation. This voltage value is compared with the control voltage U_c from the external generator in closed loop. Therefore combined thermo-mechanical loading with non synchronized relation of the temperature and mechanical deformation is achived. However positive[1] and negative type of the thermal fatigue cycle require synchronized relation of both controlling signals. The synchronization is accomplished by temperature function generator, loading function being derived from the temperature time profile. The instanteneous value of the control temperature voltage is as follows:

$$U_{c,tep} = K_3 U_T + K_4$$

where K_3 and K_4 are arbitrary positive or negative constant values, i.e., that positive or negative loading cycle is available.

As a part of our research program, tests on selected heats of the austenitic steels similiar to AISI 304L and 316L were carried out. Results of the thermal fatigue tests with positive and negative type of cycle as well as results from isothermal fatigue tests are shown on Fig. 2 and 3. Also thermal fatigue tests with positive type of cycle were performed on CrMoV steel. These results compared with isothermal fatigue tests are on Fig. 4.

For evaluation of the thermal fatigue experiments it is necessary to consider following four parameters: the temperature range ΔT, magnitude of the maximum temperature T_{max}, range of mechanical deformation $\Delta \varepsilon$ and type of loading cycle. At present a simple 4^2 design is carried out, each of the parameters mentioned above being tested at two levels.

Based on available results and gained experience we can conclude that thermal fatigue from the point of view of experimentation and results interpretation is a very complex subject. Most of the structural materials tested have more significant fatigue damage for positive than for negative type of the thermal fatigue loading cycle. This effet is more pronounced for large plastic deformation.

---

1 The positive type of the loading cycle reaches maximum temperature at maximum tensile deformation, negative type of the cycle has maximum temperature at maximum compressive deformation.

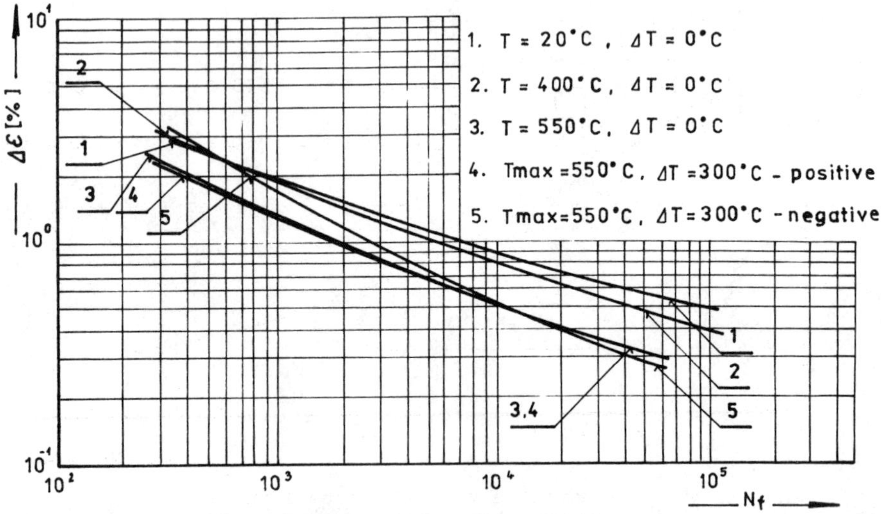

Fig. 2 Test results of the isothermal and non-isothermal fatigue tests on O3Cr18Ni11Mo3 steel.

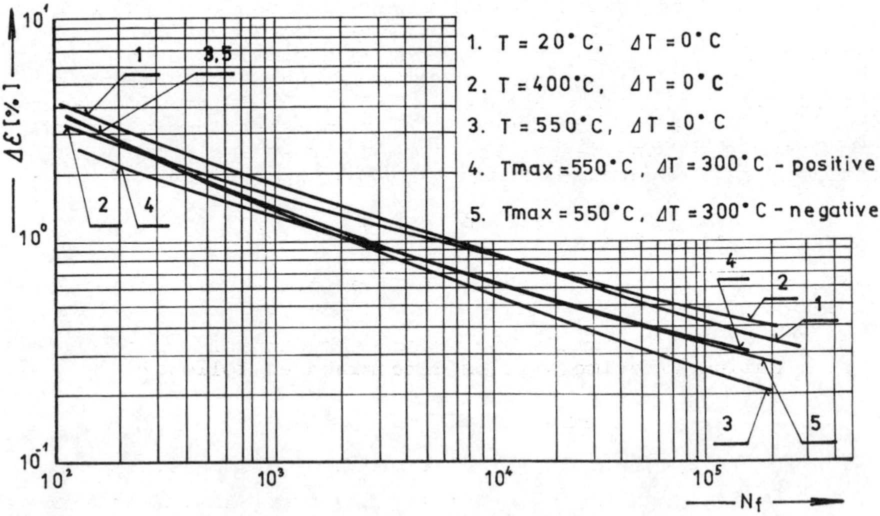

Fig. 3 Test results of the isothermal and non-isothermal fatigue tests on O3Cr18Ni11 steel.

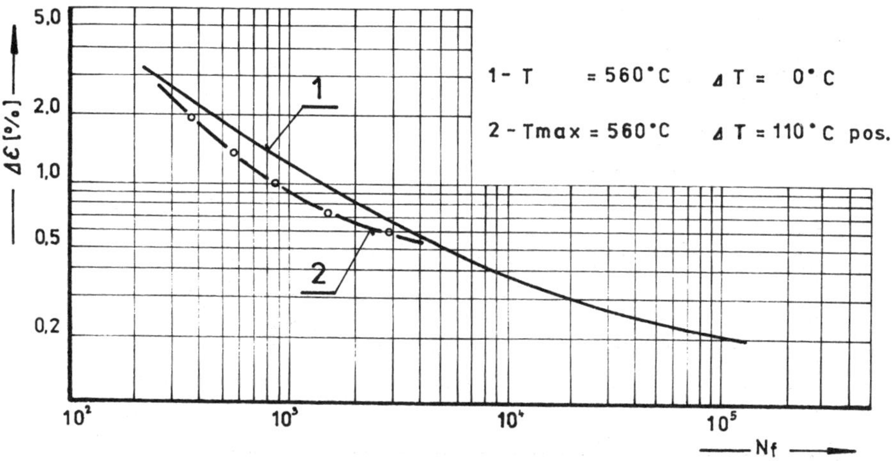

Fig. 4 Test results of the isothermal and non-isothermal fatigue tests on CrMoV steel.

THE HIGH TEMPERATURE FATIGUE STRENGTH CALCULATION

The valve bodies used in fossil power stations are subject to the combined thermo-mechanical loading. Stress analysis calculations of selected types of valves were carried out by simplified methods as well as by three dimensional finite element method /Zdarek, Hnatek, Pospisil, 1982/. The internal pressure, external forces and moments as well as non-stationery thermal gradients were considered. The cyclic stress amplitude is a maximum value determined from following expressions:

$$\sigma_a = \Delta\sigma_{ekv,ij}/2; \qquad \sigma_a = /\Delta\sigma_{ekv,ij}/2/.C$$

where constant C is a function of allowable stress values for maximum and minimum loading cycle temperature, $\Delta\sigma_{ekv,ij}$ is a stress intensity according to Tresca´s hypothesis. The total damage during lifetime will be determined as follows:

$$\sum_i N_i/[N] + (1{,}25\,\sigma_c/\sigma_\pi^t)^m \leq D$$

where σ_c is the calculated value of creep stress to rupture

σ_π^t is the allowable stress value for maximum temperature of the loading cycle

[N] is the allowable number of cycles

D is the total damage

m is the characteristic of material

The allowable number of cycles $[N]$ eventually the allowable stress amplitude $[\sigma_a]$ without creep effect are determined from design fatigue curves. The influence of creep is included in modified allowable values $[N^*]$, $[\sigma_a^*]$ by coefficients π_N and π_σ:

$$[N^*] = [N]/\pi_N; \quad [\sigma_a^*] = [\sigma_a]/\pi_\sigma$$

where

$$\pi_N = 1/[D-(1{,}25\,\sigma_c/\sigma_\pi^t)^m]; \quad \pi_\sigma = (\pi_N)^{-m}$$

The total allowable damage D is determined from Fig. 5.

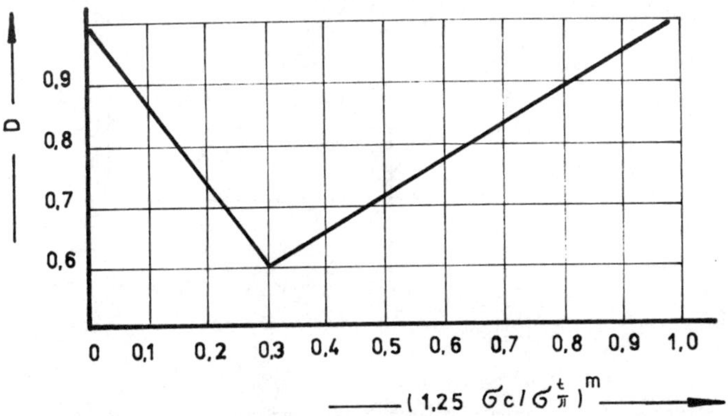

Fig. 5 The determination of allowable damage D

CONCLUSION

The presented paper summarizes methods and results of thermal fatigue tests as well as high temperature fatigue strength calculation methods. At present it is possible to conclude that thermal fatigue tests with possitive type of loading cycle represent a reasonable simulation of combined thermo-mechanical loading which is typical for service conditions. The temperature range magnitude ΔT, used in thermal fatigue tests on CrMoV steel, is higher than expected from service loading conditions and experimentally determined values of the fatigue strength are therefore more conservative. The comparison of the non--isothermal and isothermal fatigue test results shows that for CrMoV steel the thermal fatigue strength is lower for higher plastic deformation. For lower values of plastic deformation which are more comparable with service conditions the difference of fatigue strength values for non-isothermal and isothermal fatigue is relatively small.

The estimation of fatigue strength in critical section of a component requires determination of corresponding stress amplitude. The allowable number of cycles and allowable stress amplitude are derived from design fatigue curves. The design curves according to the GOST Code /1973/ have no allowance for the creep and therefore coefficients π_N and π_σ are introduced. The total allowable creep-and-fatigue damage D is determined from corresponding plot according to the calculated creep stress and properties of material.

At present the equations for high temperature application similar to the equations from the GOST Code /1973/ are being evaluated. It will be possible to determine allowable number of cycles or allowable stress amplitude for low carbon and low alloyed steels in the temperature range of 350°C and 500°C, and for Cr-Ni austenitic steels in temperature range of 450°C and 600°C. In equations strength and ductility characteristics of specified material for corresponding temperature will be used. The above mentioned calculation procedure together with experimental results can be included in computer program for high temperature fatigue strength evaluation.

REFERENCES

Budac, C. /1980/. Summary of the cyclic and fatigue characteristics of steels considered for nuclear stations with fast reactors. Sigma Research Institute, Internal Report, 1980.

GOST Code /1973/. Code for strength calculation of reactors, steam generators and piping components. Moscow, Metalurgija, 1973.

Zdarek, J., Hnatek, T., Pospisil, L. /1982/. State of stress and Life Prediction Calculations of the Main Valve Body for Primary Circuits. Cerpadla, potrubi, armatury, 2-3/1982.

Zdarek, J. /1975/. PhD thesis. CVUT-FJFI, Prague, 1975.

THERMAL FATIGUE BEHAVIOUR OF MAR M509 SUPERALLOY

F. Rezai-Aria*, L. Rémy*, C. Herman and B. Dambrine*****

*Centre des Matériaux de l'Ecole des Mines de Paris, ERA CNRS 767, 91003 Evry Cedex, France
**SNECMA, Centre d'Evry, B.P. 81, 91003 Evry Cedex, France
***SNECMA, Centre de Villaroche, 77550 Moissy Cramayel, France

ABSTRACT

The thermal fatigue behaviour of a cast cobalt based superalloy MAR M509 was investigated using wedge type specimens and flame heating. A standard cycle from 200 to 1100°C was used and the strain range applied to the thin edge was varied through modifications in specimen geometry. Finite element computations of the temperature stress-strain history of the thin edge showed significant variations in strain and stress ranges which actually correspond to thermal fatigue lives (to 1 mm crack) ranging from a few tens to thousand cycles. Mechanisms of crack initiation and early growth were actually found to be the same in thermal and isothermal fatigue. The relevance of the results to the engineering approach of thermal fatigue damage was briefly examined.

KEYWORDS

Thermal fatigue, crack initiation, crack growth, finite element analysis, constitutive equation, cobalt base superalloy, fatigue damage.

INTRODUCTION

Many critical components of high temperature machinery are subjected to rapid temperature changes during operation. The associated expansion and contraction across the section of the component is non uniform and produces a system of transient strains and stresses. The repetition of this transient may lead to the initiation and propagation of cracks. So thermal fatigue (TF) is a problem of major importance in jet engine components such as blades and guide vanes.

Therefore we have investigated the TF behaviour of a cast cobalt based superalloy MAR M509. This alloy is mainly used for guide vane applications. In this program we made a study of metallurgical damage under conventional low cycle fatigue (LCF) in air and under TF. LCF fatigue results were recently reported (Reuchet and Rémy 1979, 1983a, 1983b). TF cycling was carried out on wedge specimens. between 200 and 1100°C. Care was taken to use a similar casting structure for both LCF specimens and TF ones. The transient strain amplitude of the thin edge was changed through variations in specimen geometry. Therefore the present work will report the influence of wedge specimen geometry on life to initiation. This influence will be discussed using finite element computations of the stress-strain-temperature history. TF and isothermal LCF behaviour will be compared.

EXPERIMENTAL PROCEDURE

The alloy studied is a cast cobalt based superalloy MAR M509 which contains in wt pct 0.59C, 11Ni, 23.2Cr, 6.95W, 3.31Ta, 0.30Zr, 0.22Ti, 0.17Fe, 0.008B, 0.005P and 0.003S. A single casting microstructure was used for TF and LCF specimens. Castings were specially made in which two wedge type specimens or two cylindrical LCF specimens could be machined. Heat treatment and structure are given by Reuchet and Rémy (1979, 1983a). The section of wedge type TF specimens is shown in Fig. 1. Two specimens have a standard section with an edge radius of 0.25 mm (A) and 1 mm (B) respectively; the third one has a shorter bulk part with an edge radius of 1 mm (C). These variations in specimen geometry were designed to decrease the maximum temperature difference between the thin edge and the bulk of the specimen from A to C. This was expected to lead to a decrease of transient strain range and to an increase in fatigue life (Glenny 1967). The thin edge of the specimen is heated by the flame of a burner using a mixture of air and propane gas, for 60s and then it is cooled by high pressure air, for 20s. Maximum and a minimum temperatures were 1100 and 200°C respectively. Tests were periodically interrupted to inspect specimen surfaces. The evolution of the major crack is reported here though the length of all the cracks was monitored. Metallographic observations were carried out on specimen surface and lengthwise sections using optical and scanning electron microscopy (SEM). Experimental details are given elsewhere (Rémy and coworkers 1983).

COMPUTATION PROCEDURE

Computation of the transient temperature at every point of the TF specimens was achieved using a two dimension finite element analysis, the boundary conditions which were given by measurements of temperature at various locations of specimen surface and the temperature dependent values of specific heat and thermal conductivity. The present analysis of the stress-strain history used the viscoplastic constitutive equation of Chaboche (1977) which accounts properly for the Bauschinger effect. In uniaxial conditions, the viscoplastic strain rate $\dot{\varepsilon}_p$ is given as

$$\dot{\varepsilon}_p = \left\langle \frac{|\sigma - X| - R}{K} \right\rangle^n \text{sign}\,(\sigma - X) \qquad [1]$$

with $\dot{X} = c(a\,\dot{\varepsilon}_p - X|\dot{\varepsilon}_p|)$

where n, K, R, a, C are five temperature dependent coefficients. The parameter X refers to the position of the center of the elastic domain. All the coefficients were identified on isothermal stress-strain hysteresis loops for various plastic strain amplitudes and test frequencies. The identification was carried out at 600, 800, 900, 1000 and 1100°C. A parabolic interpolation was used for intermediate temperatures. A simple uniaxial analysis of the wedge specimen was made as the strain tensor has one major principal strain. Such an approximation was shown to be in goodagreement with a full axisymmetric finite element analysis of tapered disks by Mowbray and McConnelee (1976). In the present analysis a plane section of the wedge specimen is assumed to remain plane but to be free of rotating along an axis lying in its plane. Every center of gravity of elements is supposed to deform independently of neighbouring elements.

RESULTS AND DISCUSSION

Temperature-stress-strain history

The temperature of the first element of the edge of TF specimens is shown in Fig.2 for the various specimen geometries. The maximum and minimum temperatures increase slightly from specimen A to C. The temperature variation is slightly slower from specimen A to C as it was expected from geometry changes. The total mechanical strain is plotted as a function of temperature in Fig. 3 for the center of gravity of the same element. This curve refers to a stabilized loop and the time is indicated along the curves. From specimen A to C the variation of geometry gives rise to a major change in total strain range and to a slight shift of temperatures at which extremum strains occur (between 500 and 700°C). The influence of the edge radius on the stress-total strain hysteresis loops is shown in Fig. 4. The temperature is indicated along the loops. The loops exhibit small mean tensile strains and stresses. The temperature of maximum and minimum stress is shifted to lower temperatures when decreasing the edge radius especially for the peak tensile stress. A fourfold reduction of edge radius produces a 2.5 times reduction in inelastic strain (defined at zero stress) but much smaller variations in total strain and stress ranges.

Thermal fatigue results

The variation of major crack depth is plotted as a function of the number of cycles in Fig. 5 for the various geometries. Each set of symbols refers to the major crack in a single specimen; all the results were duplicated. Cracks nucleate rather early, however there are strong differences in crack development among specimen geometries. For specimen A, the crack growth rate is very high (30 μm/cycle) and nearly constant up to a depth of a few millimeters. Specimens B and C exhibit slow crack growth rates before reaching the linear regime, since they have a large edge radius. In the linear regime the crack growth rate of specimen B is still very high (15 μm/cycle) and that of specimen C much lower (1 to 2 μm/cycle). TF life was defined as the number of cycles to 1 mm crack since LCF life corresponds generally to the presence of a major fatigue crack about 1 mm. This life increases from 30-50 cycles to 200-300 cycles when increasing the edge radius from 0.25 to 1 mm (from A to B). It further increases to about 1000 cycles when reducing the bulk part of the specimen (from B to C).

Metallographic examinations

The LCF behaviour has been extensively reported elsewhere (Reuchet and Rémy 1983a). Fatigue life was found to be weakly temperature dependent for a given total strain amplitude up to 650°C about. However above 650°C there is a strong reduction of fatigue life within an order of magnitude or more. The high temperature behaviour is due to an early initiation of cracks along MC carbides which are preferentially oxidized at much higher rates

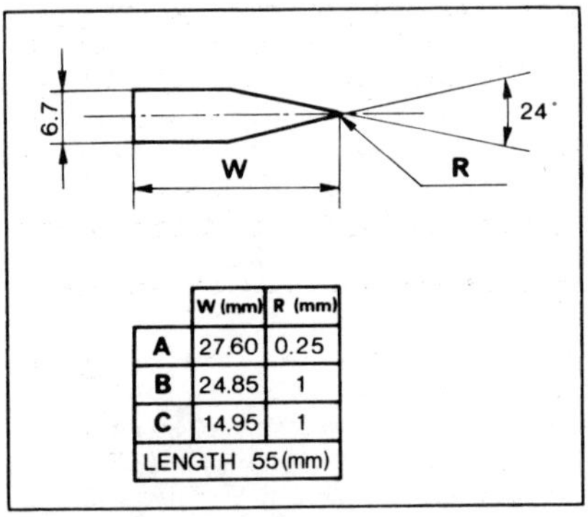

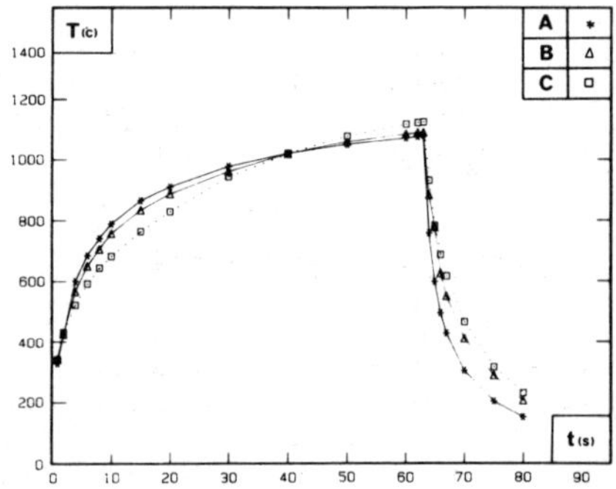

Fig. 1. Geometry of the thermal fatigue specimens.

Fig. 2. Variation of temperature (T) of the element at the edge as a function of time (t) for the various geometries.

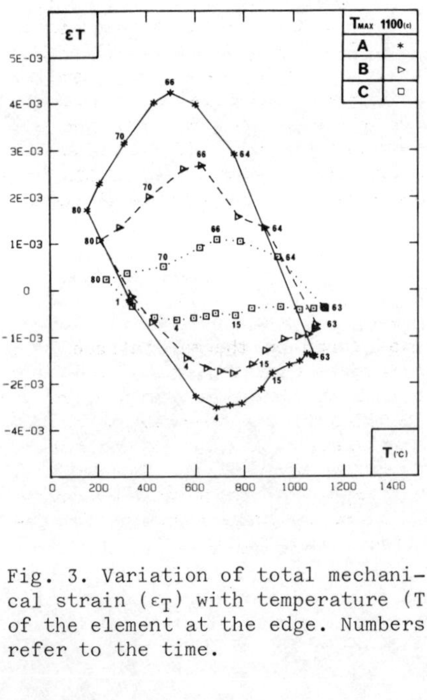

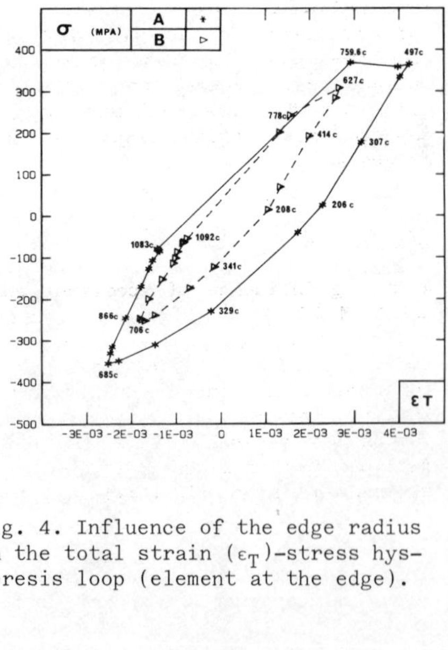

Fig. 3. Variation of total mechanical strain (ε_T) with temperature (T) of the element at the edge. Numbers refer to the time.

Fig. 4. Influence of the edge radius on the total strain (ε_T)-stress hysteresis loop (element at the edge).

than the surrounding matrix (Reuchet and Rémy 1983b). Fig. 6 shows this mechanism at surface of a specimen which was cycled at 900°C. Similar observations were made on TF specimens for the early development of cracks using optical and scanning electron microscopy. Fig. 7 shows a typical example of crack initiation at oxidized carbides (specimen A, after 10 cycles) on specimen surface. Crack propagation mechanisms were also found to be identical in TF and LCF specimens (Rémy and coworkers 1983). These observations are especially important with respect to the engineering approaches to TF life prediction, that identifies TF damage either to creep damage (and uses simple creep stress time to rupture curves, Spera 1969) or to isothermal LCF damage (and attempts to define an equivalent LCF temperature, Taira 1972, Halford and Manson 1976). These results give a physical basis to the latter point of view in the present material.

Comparison between thermal and isothermal fatigue behaviours

Thermal fatigue lives (defined to 1 mm crack) and LCF results under saw-tooth strain cycling at 600, 900 and 1100°C (strain rate $\dot{\varepsilon}_t \sim 7\ 10^{-4} s^{-1}$ i.e. for cycle periods between 10 and 40 s) are reported as a function of total strain amplitude in Fig. 8 and of stress amplitude in Fig. 9. TF data points do not seem to be connected by a smooth curve as LCF results do; a breakdown in slope is likely but should be confirmed by further tests. The usual engineering approach to TF is to define an equivalent isothermal temperature using a strain based criterion. In the present case TF life is always much shorter than isothermal LCF life, even at 1100°C, for same total strain amplitude. On the contrary the TF life - stress amplitude curve is rather similar to the LCF curve at 900°C (with some deviation in the long life range). Therefore a fatigue damage approach to TF, which has been shown to be physically based in this material, must take into account the stress range. This suggests that all the criteria which are only based on strain amplitude, are of poor capability for this material. Furthermore oxidation seems to have a major influence on crack initiation mechanisms and also on crack propagation in TF and LCF. Recent results on this material have actually shown a strong interaction between strain cycling and oxidation kinetics (Reuchet and Rémy 1983b). LCF life was reasonably predicted using a damage model based on the interaction of oxidation and fatigue crack growth; the model was able to account for differences in LCF crack growth rates in air and in vacuum (Réger and Rémy 1982). The application of this damage model to TF life prediction is currently under progress.

CONCLUSIONS

A strong influence of specimen geometry was found on thermal fatigue life of MARM 509 superalloy using wedge type specimens cycled between 200 and 1100°C. This is due to variations in stress and total strain ranges as shown by finite element computations. Metallographic observations showed that the same mechanisms of initiation and early crack growth operate in thermal fatigue and high temperature low cycle fatigue. However the comparison between Manson-Coffin plots of thermal fatigue and low cycle fatigue results suggested that the damage criterion should account for the stress amplitude and, hopefully, for the effect of oxidation.

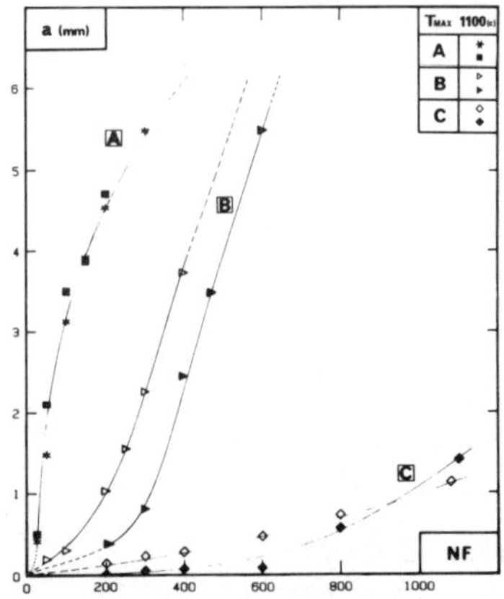

Fig. 5. Variation of major crack depth (a) with the number of thermal cycles (N).

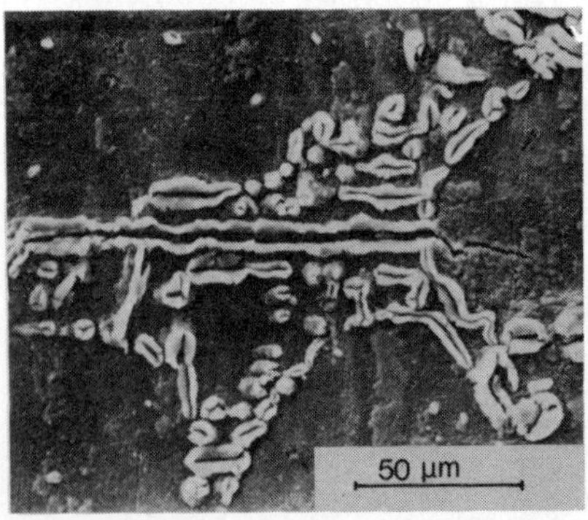

Fig. 6. Crack initiation at a preferentially oxidized MC carbide : LCF specimen cycled at 900°C (total strain)

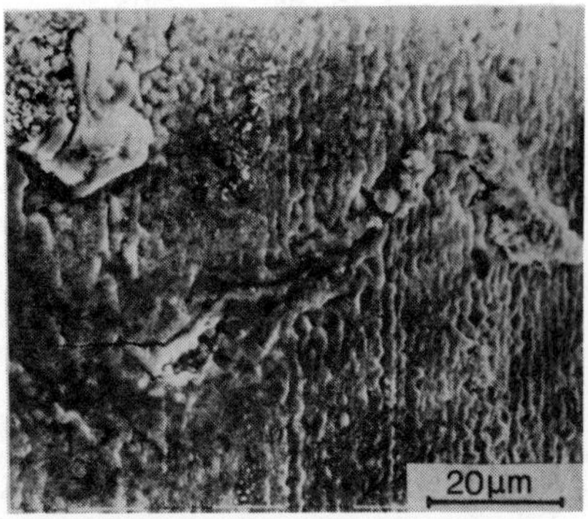

Fig. 7. Crack initiation at a preferentially oxidized MC carbide : TF specimen A after 10 thermal cycles (SEM).

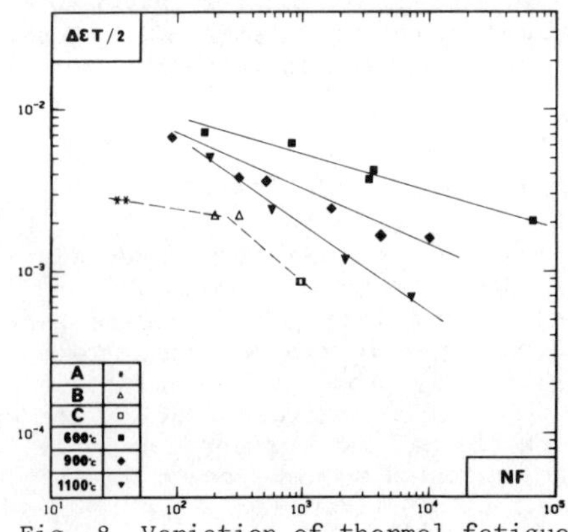

Fig. 8. Variation of thermal fatigue life and isothermal low cycle fatigue (N_f) with total strain range ($\Delta\varepsilon_T/2$).

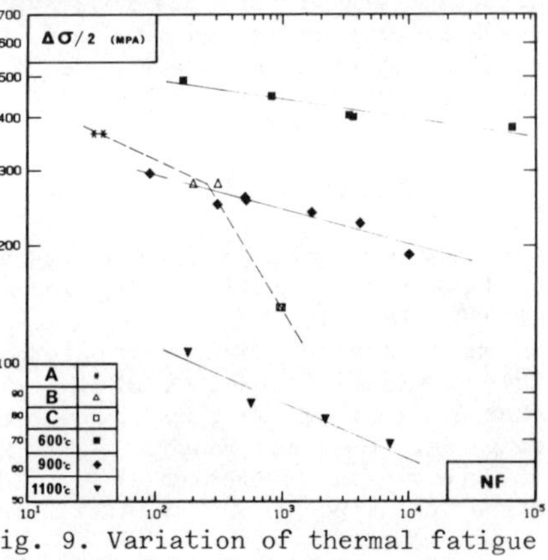

Fig. 9. Variation of thermal fatigue life and of isothermal low cycle fatigue life (N_f) with stress range ($\Delta\sigma/2$).

ACKNOWLEDGMENTS

This work is part of a cooperative project between the Centre des Matériaux and the SNECMA. Financial support from the DDSTI (Direction du Développement Scientifique et Technologique et de l'Innovation) under the auspices of COST 50 is gratefully acknowledged. Thanks are due to M. François for his assistance in TF experiments.

REFERENCES

Chaboche, J.L. (1977). Bulletin Acad. Polonaise des Sciences, 25, 33-42.

Glenny, E. (1967). The Influence of Specimen Geometry on Thermal Fatigue Behaviour. In Thermal and High Strain Fatigue, The Institute of Metals and The Iron and Steel Institute, London, 346-363.

Halford, G.R. and S.S. Manson (1976). Life Prediction of Thermal Mechanical Fatigue using Strain range Partitioning. In D.A. Spera and D.F. Mowbray (Eds), Thermal Fatigue of Materials and Components, ASTM STP 612, American Society for Testing and Materials, 239-254.

Mowbray, D.F. and J.E. McConnelee (1976). Non linear Analysis of a Tapered Disk Thermal Fatigue Specimen. In D.A. Spera and D.F.Mowbray (Eds), Thermal Fatigue of Materials and Components; ASTM STP 612, American Society for Testing and Materials, 10-29.

Rémy, L., F. Rezaï-Aria, M. François, C. Herman, B. Dambrine, Y. Honnorat (1983). Application of Isothermal Fatigue to the Study of Thermal Fatigue, Final Report F6, Cost 50 - Round III.

Reger, M. and L. Rémy (1982). Influence on Microcrack Propagation in High Temperature Low Cycle Fatigue in K.L. Maurer, F.E. Matzer (Eds), Fracture and the Role of Microstructure, E.M.A.S., London, 2, 531-538.

Reuchet, J. and L. Rémy (1979). Fatigue Eng. Mater. Struct., 2, 51-62.

Reuchet, J. and L. Rémy (1983a). Mater. Sci. Eng., 58, 19-32 and 33-42.

Reuchet, J. and L. Rémy (1983b). Metall. Trans. A., 14A, 141-149.

Spera, D.A. (1969). Calculation of Thermal-Fatigue Life based on Accumulated Creep Damage. NASA TND-5489, National Aeronautics and Space Administration, Washington, D.C.

Taira, S. (1973). Relationship between Thermal Fatigue and Low Cycle Fatigue at Elevated Temperature. In A.J. McEvily (Eds), Fatigue at Elevated Temperatures, ASTM STP 520, American Society for Testing and Materials, 80-101.

THERMAL FATIGUE OF A THERMALLY UNSTABLE ALLOY

G. Cailletaud, J. P. Culié and H. Kaczmarek

Office National d'Etudes et de Recherches Aérospatiales (ONERA), 92320 Châtillon, France

ABSTRACT

The problem of life prediction under thermal fatigue conditions is considered from different points of view :

- experimental : a thermal fatigue equipment ("two bar" type) is described,
- stress-strain behaviour : the abnormal behaviour under variable temperature is explained by some tests including isothermal cycling with temporary overheating periods under remoted load,
- failure prediction : creep-fatigue models are used to predict failure in both cases ; the result is fairly good if stress is chosen as the critical variable (creep fatigue cumulation methods) but becomes deficient if plastic strain range is used (for example Strain Range Partitioning). In the latter case, a ductility correction would be necessary to obtain reasonable predictions.

The material of the study is the Nickel Base alloy IN 100.

KEYWORDS

Fatigue (materials); thermal fatigue; thermal cycling tests; viscoplasticity; creep-fatigue models; life prediction.

INTRODUCTION

In this paper, we set the problem of life prediction in thermal fatigue. Two main types of model are available to describe creep-fatigue interaction, using either stress or plastic strain as critical variable. In the case where the cyclic hardening curve is unique, these two types of approaches are generally consistent for a given set of tests (one goes from a formulation to the other using the hardening exponent) but if a microstructural instability or a memory effect arises in the material, there is not a single hardening curve and the hypotheses are no longer equivalent.

Our purpose is to choose between stress and plastic strain, so we present some special discriminating tests :

- under thermal fatigue conditions, using a simple "two bar" system,

- under isothermal cycling, interrupted by temporary overheating periods with with remoted loading.

In these tests, we use the Nickel Base alloy IN 100, which microstructure is unstable under variable temperature in the 700°C - 1100°C range so that a very important hardening effect may arise in some conditions (Cailletaud, Chaboche, 1979).

A SIMPLE THERMAL FATIGUE TEST

In order to avoid a complex analysis (as in tapered discs or edge specimens), thermal fatigue data may be generated on an uniaxial specimen (i) with a servocontrolled fatigue machine, (ii) or simply using a constraining frame. We developed a system of the second type, as described by Manson (1973), in which the stresses induced by thermal cycling are similar to thermal stresses in structures (compressive at maximum temperature, tensile at minimum temperature). The figure 1a shows the central part of the system, and the figure 1b its schematic representation : the element tested is a tensile-compressive specimen attached to two plates connected by three springs which produce the loading by acting in opposition to the thermal strains. An additive system consists in three small pillars of adjustable length fixed on a plate and creating an unilateral contact during the cold period of the cycle, so that we obtain an adjustable maximum load in tension for a given temperature cycle. An induction heating is used, the springs are used as a dynamometer; acquisition and analysis of results are made by means of a mini-computer thus it is possible to obtain stress-strain curves at variable temperature as long as the temperature profile on the specimen and temperature evolution are known.

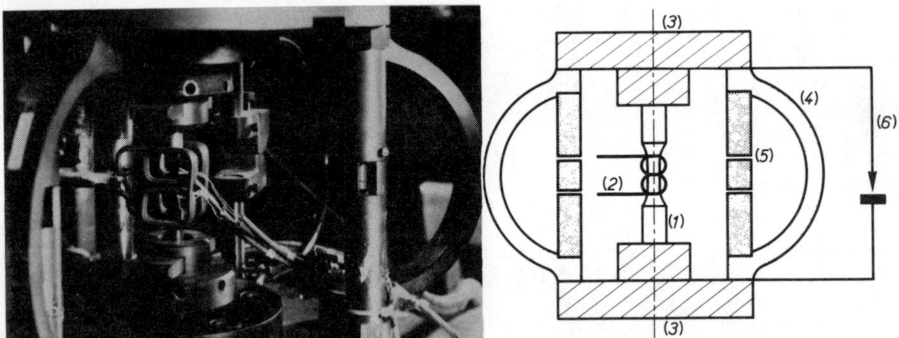

a b

(1) Specimen
(2) Induction heating
(3) Plate and anchoring system
(4) Springs
(5) Pillars
(6) Extensometer

Fig. 1 — a) The central part of the system ; b) Schematical representation.

Such a system simulates, in a very simplified form, a cooled turbine blade the specimen acting as the trailing (or leading) edge of the blade and the springs as the center of the blade.: a more detailed description of the rig is given elsewhere (Cailletaud, Culié, Kaczmarek, 1981), together with the experimental methodology and the formulation of the equations used for stress strain analysis. Two different temperature cycles are studied (725-1125°C and 900-1150°C, the "cold" period being respectively 1 and 10 mn long, the time at high temperature being 1 mn). In each of them, pillars of various length are introduced (see the element (5) on fig. 1). After a compressive ratchetting period, a stabilized stress-plastic strain loop is obtained, with a compressive mean strain and a tensile mean stress. The non-isothermal behaviour is illustrated (Fig. 2) by plotting on the cyclic hardening diagram : the plastic flow at T_{min} under tensile stress σ_{max} and the plastic flow at T_{max} under compressive stress σ_{min}.

For both thermal histories, the compressive flow seems to be normal, but during the tensile part of the cycle, at lower temperature, a slight hardening is noted for the 725-1125°C cycle, and a very large one for 900-1150°C.

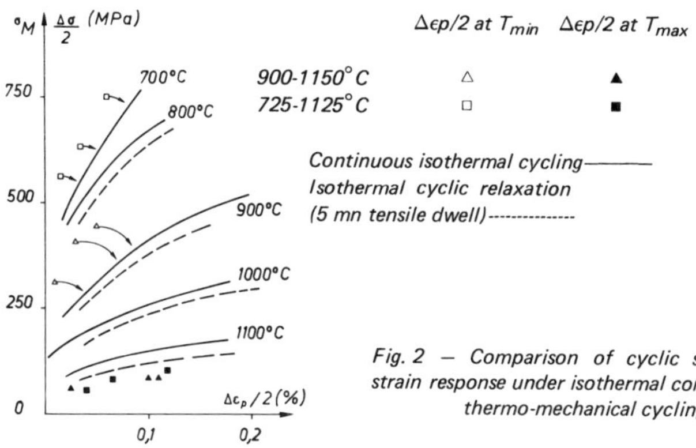

Fig. 2 — Comparison of cyclic stress-plastic strain response under isothermal conditions and thermo-mechanical cycling.

COMMENTS ON STRESS-STRAIN BEHAVIOUR

The mechanisms of strengthening by a second phase are now well known : part of the book by Sims and Hagel (1972) is devoted to this subject, Decker (1969) has reviewed the applications to Nickel base alloys. In the case of IN 100, it was observed (microphotographs are shown by Culie and coll., 1982), that an overheating period induces the dissolution of a part of the γ' precipitates and that a new generation of very fine precipitates appears during the cooling and produces an additional strengthening with regard to the isothermal behaviour. This strengthening depends on minimum and maximum temperatures and on the timing of the cycle : it may be very important (900-1150°C) or low (725-1125°C). At a fixed temperature, the precipitates growth so that this effect is evanescent and the isothermal behaviour is finally recovered, but in thermal fatigue, the temperature cycling reaccelerates the microstructural instabilities, giving rise to an abnormal behaviour over the whole life.

Let us note that the increase of the shear stress at microscopic level (Orowan mechanism) can be represented at the macroscopic level by an increase of the yield stress. The introduction of such a variable in a viscoplasticity model using internal variables (Chaboche, 1977), leads to constitutive equations able to describe the non isothermal behaviour (Cailletaud and coll., 1980).

LIFE PREDICTION

It is now established that in our case, the temperature cycling caused the stress to be higher for the same strain range (microstructural strengthening). Jaske (1976) has made some similar observations on AISI 1010 steel ; as for the fatigue life, he concluded that the fatigue resistance was not increased, thus, he obtained shorter fatigue life in thermal fatigue, the test results being plotted in a $\Delta\varepsilon_p$ - N_F diagram. The same remark comes from a study by Taira and coll. (1979). Our conclusion will be similar for IN 100 alloy.

Application of the S.R.P. method

On the Fig. 3, the experimental results of thermo-mechanical fatigue are compared to the basic lines of S.R.P. for IN 100 alloy (Chaboche and coll., 1978) : the partitioning of the strain range for our tests leads to a - cc and a - pc part, (Manson and coll., 1971), but the location of the points indicates that the prediction made from these lines will be very unconservative for any type of partitioning (Fig. 4a). It is also worth noting that the deviation of the prediction is larger for the 900-1150°C cycling, in which the microstructural instability (and the increase of yield stress) is larger. The same holds true for any method using plastic strain range as critical variable. A modification of the method is needed in this case : changing the basic curves through the ductility, as proposed by Manson and Zab (1977) seems to be a good solution. But they have investigated the life loss due to the drop of ductility in isothermal conditions, the aim being long term predictions and environmental effect : they assume that ductility is time dependent, but that it can be considered as a constant during one cycle. On the contrary, we should have to consider that the ductility drop (which is correlated to the increase of yield stress) is variable inside of each cycle. Following such a way leads to modify the basic ideas of the S.R.P. method because an integration on one cycle is now needed. In addition, it is very difficult to characterize the variations of ductility by some special tests[1].

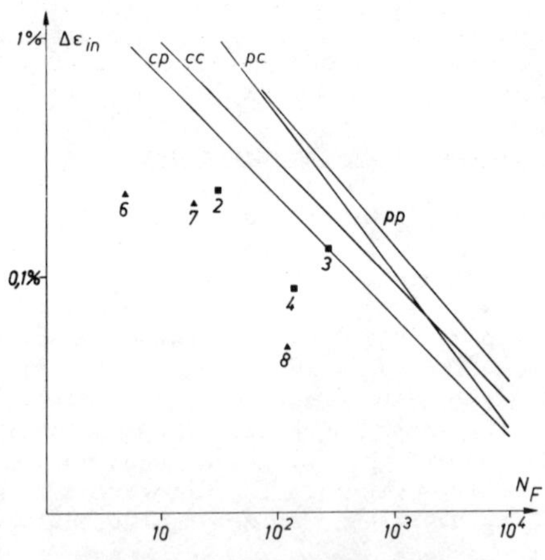

Fig. 3 — Comparison of the basic lines of S.R.P. method with the thermo-mechanical data; 725-1125°C: ■; 900-1150°C: ▲.

[1]Creep tests with a prior overheating period can be used for that purpose (Policella, 1983). But they only give information on one temperature difference. Note also that during the "hot" period, the ductility is normal.

Application of linear creep-fatigue cumulation

The use of the plastic strain as the critical variable is not very satisfactory, thus an hybrid (very classical) method is now considered : the creep contribution to failure per cycle is given (Taira, 1962) by Taira's rule (integration vs. time t of the inverse of the time to failure in creep at the instantaneous stress σ (t) and temperature T (t)) ; the fatigue contribution is expressed as the inverse of the number of cycles to failure deduced from the total strain range by the Manson-Coffin's equation. All the coefficients (2 for creep part, 4 for fatigue part) are temperature dependent (Chaboche, Stolz, 1974), the failure being reached when the sum of creep and fatigue contribution is equal to 1. As classically observed, this method gives unconservative results (Fig. 4b), but the prediction is better than the prediction by S.R.P. method. In fact, only two points (900-1150°C) are far outside of the factor of 3, they present a very abnormal tensile stress at 900°C (see Fig. 2). Finally, the introduction of stress as criticable variable for creep leads to more accurate predictions.

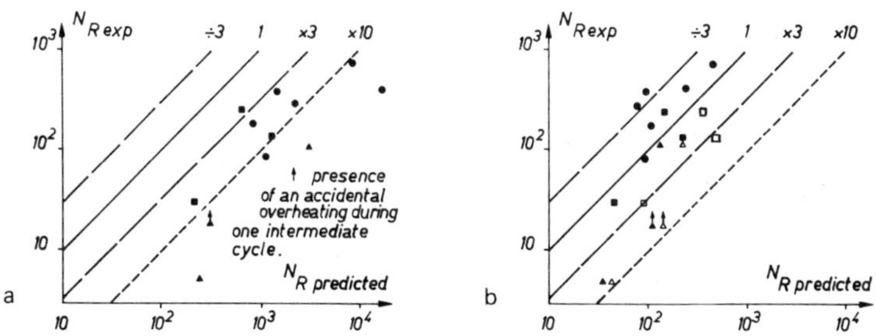

Fig. 4 – Comparison of experimental and predicted life in thermal fatigue (725-1125°C : □■ ; 900-1150°C : △▲ ; ● isothermal cycling with overheating periods). a) S.R.P. method, b) linear, non-linear creep-fatigue cumulation. □△ linear creep-fatigue cumulation ; ● ■▲ non-linear creep-fatigue cumulation.

Application of a non-linear damage cumulation model

As a method using stress as critical variable, we try Chaboche-Lemaitre's model previously developed at ONERA (Lemaitre, Chaboche, 1975 ; Chaboche, 1981). The main specificity of this model is its ability to predict very non linear interaction between creep and fatigue. A Taira's type summation is also used for the determination of creep reference ; the fatigue part depends on maximum and minimum stress, but in fact, at variable temperature, we had to introduce a reduced stress by dividing σ by the ultimate stress σ_u, the extrem points of the fatigue cycle being taken as the extrem values of the ratio γ (t) = σ (t)/σ_u (T(t)) : finally, the determination of the fatigue reference is made after calculation of N_{F_1} for γ_{max} at the corresponding temperature T (γ_{max}), and for γ_{min} at T (γ_{min}) by a linear cumulation between the two. Both creep and fatigue references are combined using damage cumulation rules using a cycle-by-cycle integration or a simplified method, as explained in the Annex I. The non-linear cumulation is an improvement with regard to the previous method ; a second favourable point is the use of stress instead plastic strain that decreases the fatigue reference, on account of strengthening effect. For these reasons, the predictions coming from the method are fairly good, only the very short test (5 cycles) being out of range.

ISOTHERMAL CYCLING WITH TEMPORARY OVERHEATING UNDER REMOTED LOAD

The observations made in thermal fatigue were confirmed in a companion study (Cailletaud and coll., 1980) conducted on a servohydraulic machine under stress or strain control : the isothermal mechanical cycling was interrupted by temporary overheating periods under remoted load, leading to an abnormal behaviour during the whole test (Fig. 5). The life prediction in such conditions shows the same trends as in thermal fatigue (Fig. 4).

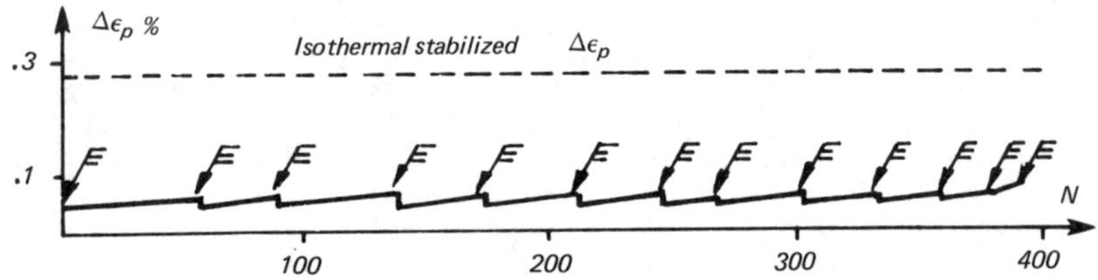

Fig. 5 — Isothermal cycling with temporary overheating periods under remoted loading. 900°C, Δσ = 684 Mpa, T + C 120 s. Overheating temperature 1150°C.

CONCLUSION

This study points out the role of a second phase of precipitates in thermal fatigue : the successive dissolution and precipitation continuously regenerate the little precipitates so that a more or less important strengthening is present during the cycling. In addition, it has been observed in the case of a Nickel Base alloy that this strengthening does not affect the number of cycles to failure : thus, a creep fatigue failure model using stress as critical variable give satisfactory results, the best predictions coming if non-linear creep-fatigue interaction is described. On the contrary, the rules using strain as critical variable (for instance S.R.P. method) lead to very unconservative results in these conditions (up to a factor 40) : the way for obtaining good predictions in this last case would be to change the ductility, but this correction needs at lot of additional tests and is not easy to perform.

ANNEX 1

The evolution of creep and fatigue damage is given by

creep : $dD_c = (\sigma/A)^r . (1 - D)^{-k} dt$ (1)

fatigue : $dD_F = [1 - (1-D)^{\beta+1}]^{\alpha} . [(\sigma_M - \bar{\sigma})/(1 - D)/M]^{\beta} dN$ (2)

with $\alpha = 1 - a((\sigma_M - \sigma_\ell)/(\sigma_u - \sigma_M))$

$\sigma_\ell = \bar{\sigma} + \sigma_{\ell 0} (1 - b\bar{\sigma})$

$M = M_0 (1 - b\bar{\sigma})$

The creep (fatigue) reference t_c (N_F) is obtained by integrating the eq. (1) (or (2)) alone. In presence of creep and fatigue simultaneously :

$$dD = dD_c + dD_F$$

$$dD = \frac{(1 - D)^{-k}}{k + 1} \frac{dt}{t_c} + \frac{(1 - (1-D)^{\beta+1})^{\alpha}}{(1 - \alpha)(1 + \beta)\ (1 - D)^{\beta}} \frac{dN}{N_F}$$

- Set $dt/t_c = dN/N_c$.

- Failure is obtained if D is equal to 1 and N to N_R :

$$\int_0^1 \left\{ \frac{(1-D)^{-k}}{k+1} + \frac{(1-(1-D)^{\beta+1})^{\alpha}}{(1-\alpha)(1+\beta)(1-D)^{\beta}} \frac{N_C}{N_F} \right\}^{-1} dD = \int_0^{N_R} dN/N_C$$

The non linearity of the cumulation is governed by N_C/N_F and the coefficients k, α and β.

REFERENCES

Cailletaud, G., and Chaboche, J.L. (1979). Macroscopic description of the microstructural changes induced by varying temperature. Proc. ICM 3. Cambridge (U.K.). ONERA-TP. 1979-112.

Cailletaud, G., Culié, J.P., and Kaczmarek, H. (1980). In A.R.S. Ponter and D.R. Hayhurst (Eds), Creep in Structures, Springer Verlag, pp. 48-71.

Cailletaud, G., Culié, J. and Kaczmarek, H. (1981). La Recherche Aérospatiale 1981-2, pp. 85-97. (English translation ESA-TT-713).

Chaboche, J.L., and Stolz, C. (1974). Revue Française de Mécanique, n° 52, pp. 37-47.

Chaboche, J.L. (1977). Bull. Acad. Pol. Sci. Sér. Sci. Tech., Vol. 25, n° 1, pp. 33, 42.

Chaboche, J.L., Policella, H., and Kaczmarek, H., (1978). Applicability of the S.R.P. method and creep-fatigue damage approach to the LCHTF life prediction of IN 100 alloy. Proc. of AGARD-SMP meeting, Aalborg, Danemark.

Chaboche, J.L. (1981). Nucl. Eng. and Design 64, pp. 233-247.

Culié, J.P., Cailletaud, G., and Lasalmonie, A. (1982). La Recherche Aérospatiale 1982-2, pp. 51-61. French and English editions.

Decker, R.F. (1969). Proc. of Symp. on Steel Strengthening Mechanisms, Zürich, pp. 147-170.

Jaske, C.E. (1976). In D.A. Spera and D.F. Mowbray (Eds.), Thermal Fatigue of Material and Components, ASTM-STP 612, pp. 170-198.

Lemaitre, J., and Chaboche, J.L. (1975). In J. Hult (ed), Mechanics of visco-plastic media and bodies, Springer, pp. 291-301.

Manson, S.S., Halford, G.R. and Hirschberg, M.H. (1971). Proc. of 1st Symp. on Design for Elevated Temperature Environment, San Francisco, California, pp. 12-28.

Manson, S.S. (1973). In fatigue at elevated temperature, ASTM-STP 520, pp. 744-782.

Manson, S.S., and Zab, R. (1977). Proc. of Conf. on Environmental Degradation of Engineering Materials, Blacksburg, Virginia, pp. 757-770.

Policella, H. (1983). Unpublished ONERA data.

Sims, C.T., and Hagel, W.C. (1972). The Superalloys. John Wiley.

Taira, S. (1962). Lifetime of structures subjected to varying load and temperature, in N.J. Hoff (ed), Creep in Structures, Academic Press.

Taira, S., Fujino, M., and Ohtani, R. (1979). Fat. of Eng. Mat. and Struct., Vol. 1, pp. 495-508.

OXIDE WEDGING EFFECTS ON NEAR THRESHOLD FATIGUE CRACK GROWTH AT ELEVATED TEMPERATURE

R. Yuuki, H. Kitagawa and Y. Miki

Institute of Industrial Science, University of Tokyo, 22-1, Roppongi 7 chome, Minato-ku, Tokyo 106, Japan

ABSTRACT

RECENTLY linear fracture mechanics has been successfully applied to the evaluation of the fatigue crack growth in the elastic creep range at elevated temperature. Many studies have been reported on the fatigue crack growth at elevated temperature mainly at intermediate stress intensity range. However, very few data have been obtained on the fatigue crack growth at near threshold stress intensity because of the difficulties of the fatigue crack growth tests at near threshold at elevated temperature. To evaluate the fatigue life of structural components based on fracture mechanics, the threshold for fatigue crack growth at elevated temperature is of practical importance.

At elevated temperature, the oxidation is known to influence fatigue crack growth. Especially at near threshold stress intensity, the influence of the oxidation seems to be very remarkable because the specimen is exposed to elevated temperature for very long time. To investigate oxide effects on near threshold fatigue crack growth, a set of ΔK decreasing fatigue tests were carried out in air for various stress ratios R at elevated temperature up to 650°C, by using a center cracked plate specimen of solution annealed type 304 stainless steel. Fatigue crack growth rate and the crack closure behaviour at near threshold stress intensity have been investigated at elevated temperature in comparison with those at room temperature.

The fatigue crack growth rate at near threshold stress intensity at elevated temperature is found to be lower and the threshold value ΔK_{th} higher than those at room temperature, although at intermediate stress intensity range the fatigue crack growth rate is known to be higher than that at room temperature. This tendency is more remarkable at low stress ratio and at high temperature. The reason for the increase of ΔK_{th} value at elevated temperature can be considered that the oxide scale formed on the crack surface has blocked the fatigue crack growth and that the oxide scale has reduced the stress intensity range effective to fatigue crack growth. This was confirmed by the measurement of crack closure behaviour and the observation of the oxide scale on the fractured surface.

This oxide wedging effects on the fatigue crack growth was analyzed quantitatively by using a new crack model with continuous springs inside. Numerical results of this model can explain the effects of the oxide thickness on the crack closure. Thus at elevated temperature the oxidation is found to be important. It is interesting that the oxidation accelerates the fatigue crack growth at intermediate stress intensity range and decelerates it at near threshold stress intensity.

CREEP IN WELDED JOINTS IN FERRITIC AND MARTENSITIC STEELS

B. Ivarsson

Swedish Institute for Metals Research, Drottning Kristinas väg 48, S-114 28 Stockholm, Sweden

ABSTRACT

The uniaxial deformation and rupture behaviour of parent metal, weld metal, and simulated HAZ structures have been determined for a 1Cr0.5Mo steel and a 12Cr steel. Constitutive equations have been developed, with which entire creep curves can be represented. These equations were then used in computer simulations of the multiaxial creep behaviour of butt-welded tubes.

For the 12Cr steel computer simulations gave results which agreed well with strains measured during internal pressure tests.

For the 1Cr0.5Mo steel no comparative tests were performed, but simulations were performed for actual load cases found in the literature. The predicted rupture behaviour was in accordance with service experience.

KEYWORDS

Chromium molybdenum steels, martensitic stainless steels, welded joints,creep strength, creep rupture strength, computer simulation, mathematical models.

INTRODUCTION

Failures occuring in constructions operating in the creep range often occur in or close to welded joints. Nevertheless weldments are treated only schematically in design codes. Within the Swedish Creep Committee the influence of welded joints on the creep behaviour of welded tubes has been studied extensively over the last few years.

Computer simulations of the creep deformation and rupture behaviour at 600°C of butt-welded cold worked tubes of type 316 stainless steel were successful; the creep strain and the time and location of the rupture could be predicted accurately (Ivarsson and Sandström, 1980; Ivarsson, 1982). These results were achieved with the use of uniaxially determined primary and secondary creep and rupture data. Furthermore, a simple graphical method for estimations of rupture time and location was developed.

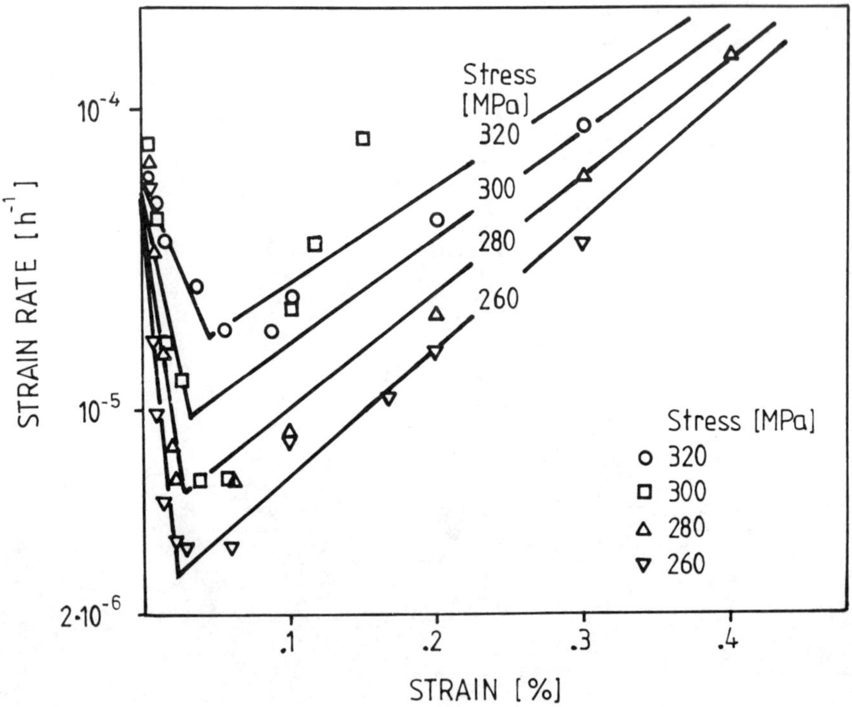

Fig. 1. Strain rate vs. strain for simulated HAZ II structure in the 1Cr0.5Mo steel.

ALLOYS STUDIED

Investigations have been carried out for a 1Cr0.5Mo steel and a 12Cr steel - two alloys commonly used in Swedish high temperature plants. The test temperatures were 525 and 550°C respectively. For both steels the structures at two locations in the HAZ were simulated by furnace heat treatment. The creep deformation and rupture properties of these structures as well as those of the parent and weld metals were determined by uniaxial testing and expressed mathematically as outlined in the next section.

CONSTITUTIVE EQUATIONS FOR CREEP DEFORMATION

In the cold worked 316 tubes tertiary creep was practically non-existent and could hence be disregarded in the computer calculations. For these two new steels, however, tertiary creep plays an important role and must thus be included in the constitutive equations used in the calculations. The creep data is fed into the computer program as the coefficients B and n in the Norton equation

$$\dot{\varepsilon} = B \cdot \sigma^n \qquad (1)$$

which originally was derived for secondary creep rates only.

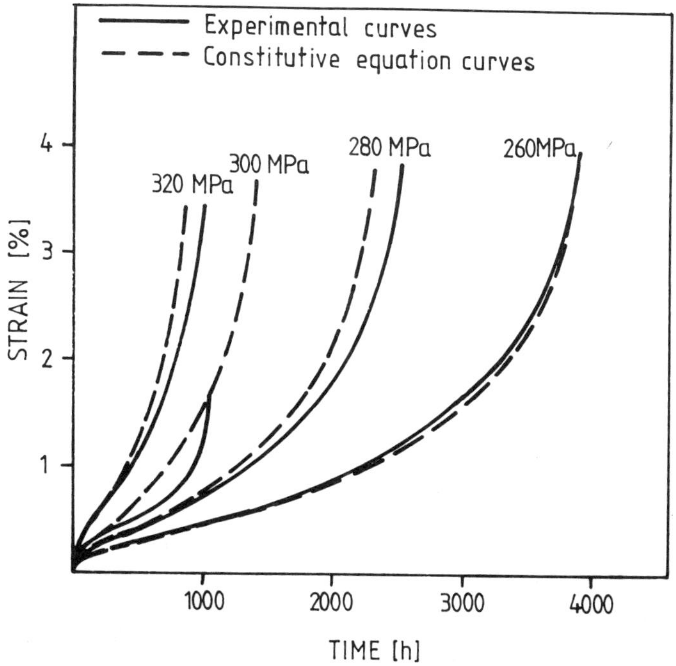

Fig. 2. Experimental and theoretical creep curves for HAZ II structure in the 1Cr0.5Mo steel.

However, in Fig. 1 it is seen that the secondary creep range has a negligible extension, and that the uniaxial data obey the following expression

$$\log\dot{\varepsilon} = C + D\cdot\varepsilon, \quad (2)$$

where the C and D values are different for primary and tertiary creep and depend linearly on log σ. This stress dependence makes it possible to transform Equation (2) into Equation (1) resulting in B and n values which depend linearly on creep strain, ε. Thus, the complete creep deformation behaviour can be accounted for in the computer calcualtions.

The straight lines in Fig. 1 are fitted to the experimental data by regression analysis. The derived sets of C and D values have been used to calculate strain-time curves which are compared with the actual creep curves in Fig. 2. This Fig. shows that, although the fit to data in Fig. 1 is far from perfect, the equations used give a satisfactory description of the creep behaviour with irregularities in single tests smoothed out.

RESULTS OF COMPUTER CALCULATIONS

In the calculations a previously used computer program has been used, i.e. a finite differences program originally developed for calculation of creep buckling in axi-symmetric shells (Samuelson, 1975).

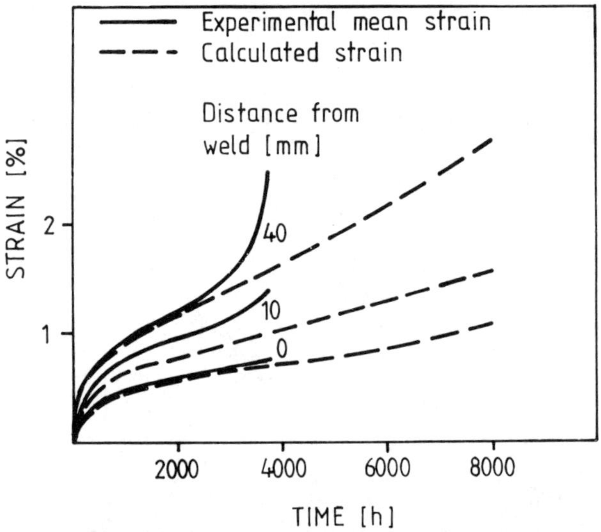

Fig. 3. Experimental and computer calculated strains. Butt-welded 12Cr steel tube, ø 37.6x1.8 mm. Internal pressure 16.2 MPa.

12Cr Steel

For this steel two internal pressure creep tests of butt-welded tubes were performed at Sandvik. For one of these tests the strains at three positions along the tube are shown as functions of time in Fig. 3. Included in this figure are also the corresponding calculated strains. The agreement between experimental and calculated strains is good up to the tertiary creep region where an accelerated swelling and bursting occurs in the parent tube - probably due to a local reduction in the wall thickness. In the tested tubes as well as in the calculations the thickness of the weld bead was about 60% greater than that of the parent tube. In Fig. 4 the variation of the calculated maximum tensile stress, i.e. the hoop stress, with time is plotted for four positions in the weldment - WM(weld metal), HAZ I (close to the fusion boundary), HAZ II (close to the parent metal), and PM (parent metal). The effective stress shows a practically identical behaviour - but at a 13 percent lower stress level. The Robinson Life Fraction Rule has been used to estimate the life fraction consumed which is marked on the curves in Fig. 4 - below the curve for the maximum tensile stress rupture criterion and above for the effective stress rupture criterion. This figure thus shows that, though the weld metal experiences the highest stress, its rupture strength is so high that the final rupture is predicted to occur in the parent metal far away from the weld - provided the same rupture criterion is valid for all structures in the welded joint. However, literature data and microscopic examinations of uniaxial test experimens indicate that the weld metal obeys the maximum stress criterion while the other parts follow a mixed criterion (Browne, Lonsdale, Flewitt, 1981; Gooch ,1982). The predicted rupture times in weld and parent metals almost coincide - the weld metal rupture time being somewhat shorter. Thus the presence of the weld has in this case little effect on the service life of the tube.

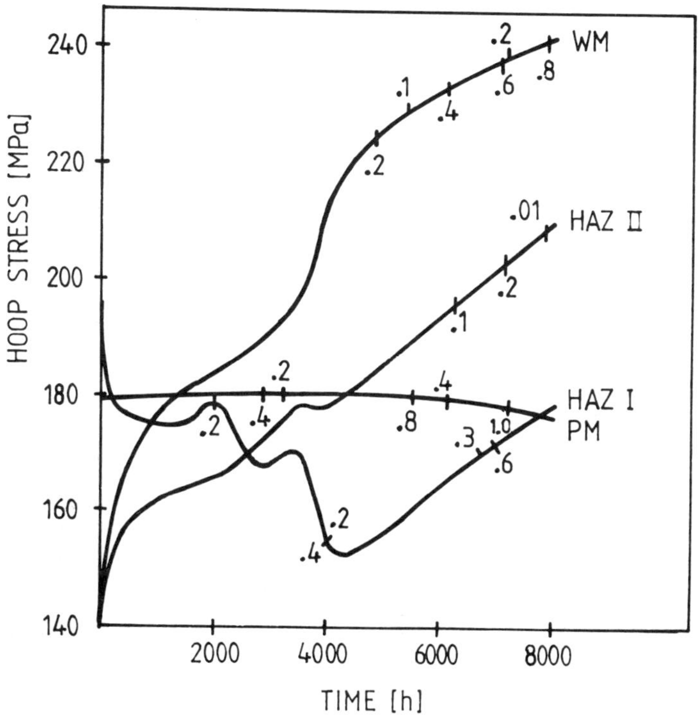

Fig. 4. Hoop stress vs. time for four positions in a butt-welded 12Cr steel tube. Cumulative life fraction consumed - see text.

<u>1Cr0.5Mo Steel</u>

No internal pressure tests were carried out for this steel. Calculations were performed for hypothetical load cases with short rupture times and for real configurations found in the literature, e.g. one reported by Weber (1977) in which the diametrical expansion of a tube was measured for more than 15 years. In this tube the nominal hoop stress (mean diameter formula) was 61 MPa, indicating a rupture time exceeding 300000 h. Since the materials data were determined for stresses between 200 and 300 MPa with rupture times up to 6000h, the calculations for long service lives represent a considerable extrapolation.

Contrary to the results for the 12Cr steel the calculated strains are virtually equal irrespective of location in the welded joint. For the higher stress levels the strain-time curves show primary and tertiary creep behaviour. The tertiary creep region decreases with decreasing stress level and for the lowest stress the creep rate diminishes throughout the entire creep life. For the case reported by Weber (1977) the measured strains after 15 years' service ranged between 0.4 and 2.0 percent. The calculated strain after 15 years (∿131500h) was somewhat lower, 0.36 percent, but must be regarded as acceptable if the excessive extrapolation is considered.

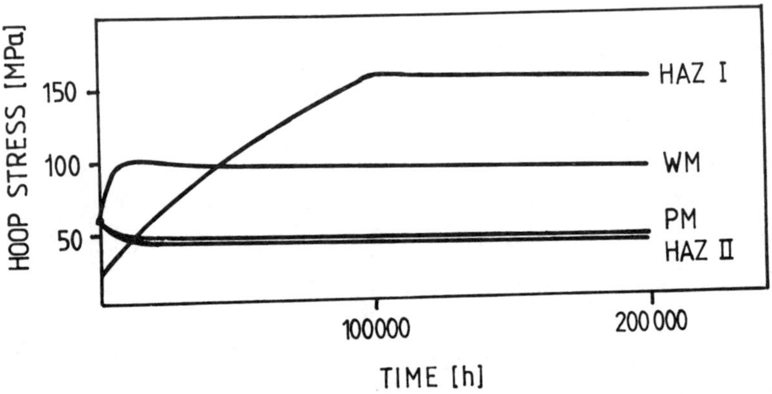

Fig. 5. Hoop stress vs. time for four positions in a butt-welded 1Cr0.5Mo steel tube, ∅ 255x22 mm. Internal pressure 11.5 MPa.

In Fig.5 the calculated hoop stresses are shown. During creep the parent metal and HAZ II (close to the parent metal) structures are off-loaded while the weld metal stress increases. Rather stable conditions are reached after a comperatively short time. The stress in the HAZ I structure (close to the fusion boundary), on the other hand, increases continuously for 100000 hours. The final rupture is also predicted to occur in this location after 225000h in spite of the high rupture strength of the structure. As a rule a mixed maximum stress/effective stress rupture criterion is valid for this steel type, but the fracture appearance indicate that for this structure rupture is controlled by the maximum stress only (Browne, Lonsdale, Flewitt, 1981). The predicted rupture position agrees with service experience, i.e. creep cracks often start in the coarse grained bainitic HAZ structure close to the fusion boundary.

SUMMARY AND CONCLUSIONS

The creep deformation and rupture properties have been simulated for butt-welded tubes of two types of steel. The following observations were made

i. Entire creep curves could be represented by rather simple constitutive equations suitable for inclusion in the computer program.

ii. The predicted creep strains in 12Cr steel tubes were in good agreement with experimental results even to some extent in the tertiary creep region. Due to premature failure in the tubes, the agreement between calculated and experimental rupture times was less satisfactory. The calculations indicate, however, that the presence of welded joints has only a minor influence on the creep life of the tubes in the investigated case.

iii. For 1Cr0.5Mo steels the calculated strains agree fairly well with service data. The welded joint has virtually no effect on the strain but leads to a considerable reduction in creep life. The weakest point of the weldment is the coarse-grained part of the HAZ.

ACKNOWLEDGEMENTS

This work has been financed by the Nordic Fund for Technological and Industrial Development, SIS (The Swedish standardization organization), and the Swedish Creep Committee which is gratefully acknowledged. Permission to publish from the Swedish Creep Committee and the Swedish Institute for Metals Research is appreciated.

REFERENCES

Browne, R.J., D.Lonsdale, and P.E.J. Flewitt (1981). The role of stress state on the creep rupture of 1%Cr1/2%Mo and 12%Cr1%MoVW tube steels. In B.Wilshire and D.R.J.Owen (Ed.), Creep and Fracture of Engineering Materials and Structures. Pineridge Press, Swansea, pp. 545-558.

Gooch,D.J. (1982). Creep fracture of 12Cr-Mo-V steels. Metal Science, 16, 79-89.

Ivarsson, B. and R.Sandström (1980). Creep deformation and rupture of butt-welded tubes of cold-worked AISI 316 steel. Met. Technol., 7, 440-448.

Ivarsson, B.(1982). Creep of butt-welded AISI 316 (18Cr-13Ni-3Mo steel) tubes: role of plastic strain and primary creep. Met. Technol, 9, 41-45.

Ivarsson, B.(1983a). Creep deformation and fracture in butt-welded 12% chromium steel tubes. Swedish Institute for Metals Research Internal Report, IM-1748, January 1983.

Ivarsson, B.(1983b). Computer simulations of creep deformation in butt-welded 1%Cr0.5%Mo steels. To be published.

Samuelson, L.Å.(1975). Creep buckling of shells of revolution. User's manual for program DSOR 06. The Aeronautical Research Institute of Sweden, Internal Report, HU-1548, part 4.

Weber, H.(1977). Service life analysis of high-temperature installations under creep conditions. In Proceedings of the international symposium, Prediction of residual lifetime of constructions operating at high temperature, the Hague. Metaalinstituut TNO, Apeldoorn. pp. 35-67.

THERMAL STRESSES, PLATE MOTION AND HOT CRACKING IN BUTT-WELDING

M. Jonsson, L. Karlsson and L-E. Lindgren

Department of Mechanical Engineering, University of Lulea, S-951 87 Lulea, Sweden

ABSTRACT

Submerged-arc welding is studied with special attention to hot cracking and to change in gap width in butt-welding of tack-welded plates. The most important welding parameters are the welding speed, the arc energy per unit length of the weld and the number of tack-welds for a given length of the weld. Eight cases are analyzed using the finite element method. The change in gap width and the risk for hot cracking are studied for two different values of each of these welding parameters. It is concluded that it is possible to find a practical combination of welding parameters that minimizes the change in gap width. The welding speed in that combination should be fairly high. The location of the maximum value of calculated change in mechanical strain across the weld, $\Delta\varepsilon^{m}_{14/10}$, agrees well with available experimental observations of the location of that crack front which is closest to the edge in the end portion of butt-welded plates. The transverse stresses become compressive at the edge and tensile at a short distance (about 0.1 m) from the edge shortly (about 30 s) after removal of the heat source. Therefore it is believed that the hot cracks nucleate at temperatures around 1200 ^{o}C due to high transverse strains ($\Delta\varepsilon^{m}_{14/10}$) and then propagate towards the interior of the plate due to the transverse tensile and compressive stresses. It is then concluded that the highest risk for hot cracking is to be found at the end portion of the butt-welded plate for low welding speeds and large values of the arc energy per unit length of the weld.

KEYWORDS

Butt-welding, gap width, hot cracking, thermal stresses, tack-weld, finite element method.

INTRODUCTION

In this paper an automatic submerged-arc welding procedure is investigated. This method of production is often used in shipyard panel production lines where several rectangular steel plates are butt-welded side by side to form a large panel. Submerged-arc welding is also used in several other butt-welding applications where a high heat input (~ 5 MJ/m) is required. In automatic butt-welding situations it is important to know how the gap width between the two tack-welded plates changes during the welding process. Given the change in gap width it may be possible to control the welding parameters such that too large variations in the gap width are avoided. Such variations are detrimental for the automatic welding procedure. The most important welding parameters are the arc energy per unit length of the weld and the welding speed. Also the distance between the tack-welds has strong influence on the change in gap width. Here this change will be studied for different values of the welding parameters. Another important issue in submerged-arc welding is the well-known tendency of hot cracking near that end of the joint where the welding is finished. The cracks usually start at some distance from the edge of the panel and extend towards the interior of the panel. In this paper transient temperature fields, change in gap width, change in mechanical strain across the weld and thermal stresses are calculated using the finite element method (FEM). Two plates of the size 1.0 x 0.5 m^2 are butt-welded. The thickness of the plates is 25 mm. The gross heat inputs studied are Q = 3.9 MJ/m and Q = 7.0 MJ/m and the welding speeds are v = 0.006 m/s and v = 0.012 m/s. The number of tack-welds is three or six. The tack-welds are placed according to Fig. 1 with equal spacing. The arc efficiency is assumed to be η = 0.9. The choice of welding parameters gives eight different cases to analyze.

TEMPERATURE FIELD

The welding configuration is shown in Fig. 1. Andersson and Karlsson (1981) showed that the selected size of the plates is large enough to model the state of stress in large plates which means that the stresses in the middle of the welded plate is not much influenced by the edges.

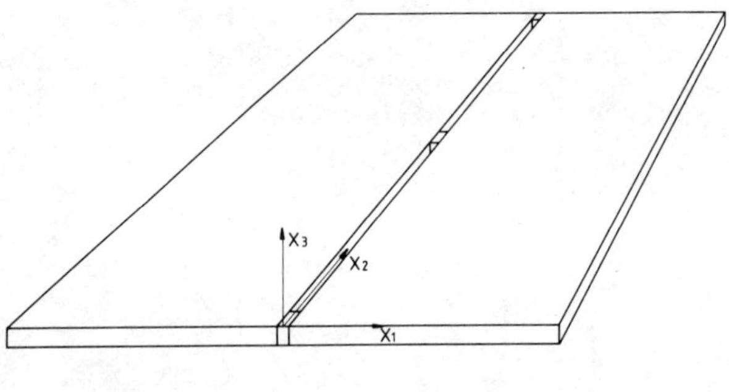

Fig. 1. Butt-welded plate to be analyzed. The heat source starts at x_2 = 0 at time t = 0 and moves along the x_2-axis with constant velocity v.

The material investigated is a fine-grain steel with a yield stress of 360 MPa. The temperature dependent thermal properties are taken from (Andersson, 1978). Uniform properties are assumed through the plate thickness.

As in (Andersson, 1978), (Andersson and Karlsson, 1981) and (Jonsson, Karlsson and Lindgren, 1983) the heat content (enthalpy) is used as the dependent variable in the heat conduction equation. The FE-method is used in the analysis. Due to symmetry only one half of the plate need be analyzed. The welding speeds v used are high enough to allow the heat flow along the weld to be neglected as compared to the heat flow transverse to the weld. Two-dimensional temperature fields are used in the analysis. These temperature fields are calculated from one-dimensional temperature fields as in (Andersson and Karlsson, 1981) and (Jonsson, Karlsson and Lindgren, 1983). An overall view of two of the calculated temperature fields is shown in Fig. 2.

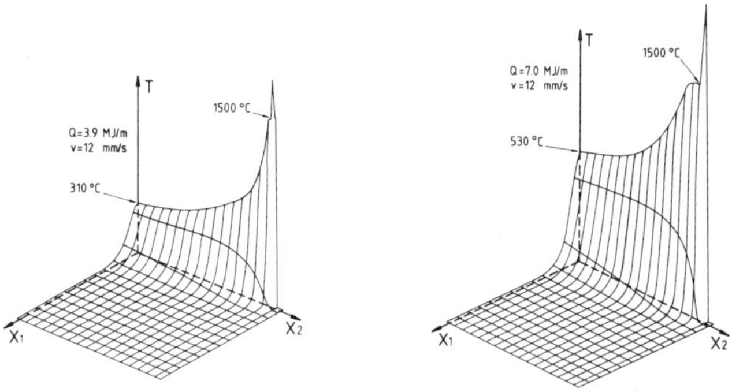

Fig. 2. Calculated temperature field at time t = 80 s.

The large gradients in the x_1-direction are noted. The gradients in the x_2-direction are much smaller except in a very small region at the heat source. The method for calculation of the temperature fields as described here was investigated by Jonsson, Karlsson and Lindgren (1983). Very good agreement was found between calculated and measured temperatures.

DISPLACEMENTS

The thermal-elastic-plastic model used by Andersson (1978), Andersson and Karlsson (1981) and Jonsson, Karlsson and Lindgren (1983) is adopted in this investigation. The material properties used here are taken from (Andersson, 1978) and modified according to (Andersson and Karlsson, 1981). Volume changes due to phase transformations are included. Mechanical properties are assumed to be equal in base metal and weld metal. The hardening modulus is set to zero.

One half of the plate is divided into 700 finite elements with cubic base functions in the x_1x_2-plane. Plane stress conditions are assumed. The model has 4000 degrees of freedom. The stress fields are calculated for times t = 0 s to 40 000 s using about 70 load (time) increments for the welding process and about 20 load (time) increments for the cooling process. The stiffness matrix is updated at every load increment. Nonlinearities are compensated for by equilibrium iterations. The stress increments during each load (time) increment are calculated using the subincrement technique.

Before the final welding the plates are tack-welded according to Fig. 1 using three or six tack-welds with one tack-weld at each of the edges and equal spacing between the tack-welds. In the case of six tack-welds the length of each tack-weld is 15 mm and in the case of three tack-welds the length of each tack-weld is 30 mm. In the mechanical model the plate is locked in the x_1-direction along the x_2-axis. In order to account for the free edges between the tack-welds the elements in front of the heat source at the line of symmetry are given a very low value of the elastic modulus except at the tack-welds. The elements at the weld line extend 2 mm in the x_1-direction. The calculated change in gap width Δg is given by two times the x_1-displacement at the nodes located 2 mm from the weld line. In Figs. 3a-d the calculated change in gap width Δg 0.1 m in front of the heat source is shown for eight different cases. Δg closer to the heat source has the same shape but has slightly larger values.

This method for calculation of the change in gap width was investigated by Jonsson, Karlsson and Lindgren (1983). Quantitatively good agreement was found between measured and calculated change in gap width during the second part in a gas-metal-arc welding process. It was also found that the residual stresses after tack-welding may influence the change in gap width causing some disagreement between calculated and measured values during the first part of the welding. The influence of residual stresses after tack-welding is not considered in this study. In Fig. 3 it is noted that Δg decrease when the number of tack-welds (ntw) is increased except for the case shown in Fig. 3b. The tack-weld (with the length 15 mm) at the edge where the welding is finished will probably break in this case. It is also noted that Δg is larger for low welding speed (v = 0.006 m/s) than it is for high welding speed (v = 0.012 m/s) during the last part of the welding. A high value of the arc energy per unit length (Q = 7.0 MJ/m) gives a smaller increase in gap width during the last part of welding than a low value (Q = 3.9 MJ/m) does. The most favourable combination of welding parameters seems to be high welding

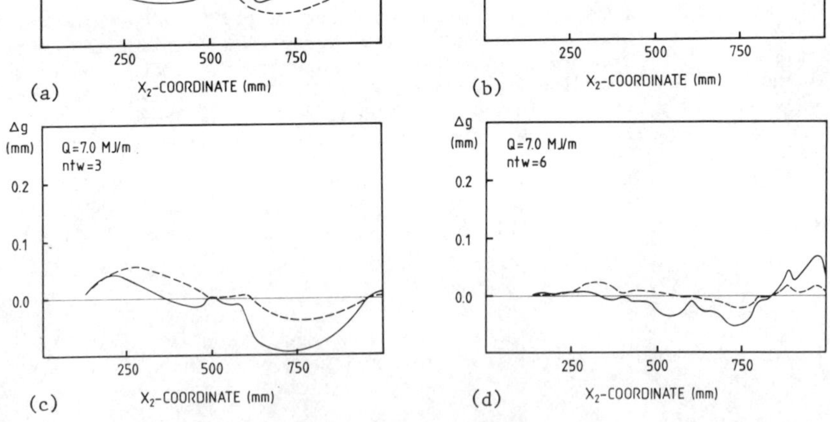

Fig. 3. Calculated change in gap width Δg at 0.1 m in front of the arc. The solid lines (——) show Δg for v=0.006 m/s and the dashed lines (– – –) show Δg for v=0.012 m/s.

speed and high arc energy per unit length of the weld. A large number of tack-welds gives of course smaller Δg than few tack-welds.

STRESSES

The most important stress in order to be able to estimate the risk for hot cracks to occur is the stress σ_{11} transverse to the weld. Therefore this stress is calculated for different times after removal of the heat source. The calculated transient transverse stress σ_{11} at the start portion and the end portion of the butt-welded plate is shown (for x_1 = 2 mm) in Figs. 4a-d.

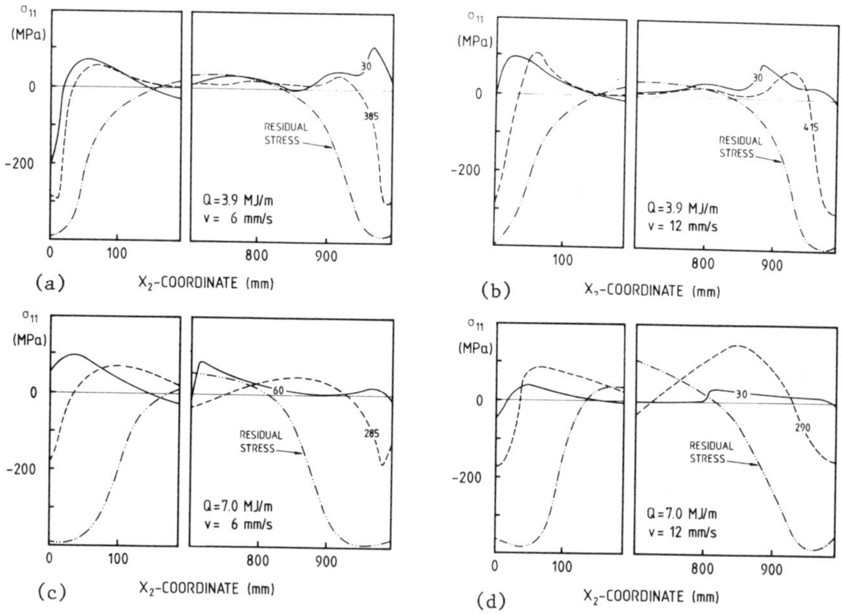

Fig. 4. Calculated transient transverse stress σ_{11} (x_1=2 mm) at the start and end portion for different times (in seconds) after removal of the heat source.

The calculated values of σ_{11} across the weld for the two different numbers of tack-welds coincide. Therefore only four cases are shown in Fig. 4. The peaks in Fig. 4a around x_2 = 0.95 m for t = 30 s, in Fig. 4b around x_2 = 0.90 m for t = 30 s, in Fig. 4c around x_2 = 0.70 m for t = 60 s and in Fig. 4d around x_2 = 0.80 m for t = 30 s are due to phase transformations. It should be noted that the temperature in the start portion (to the left in Figs. 4a-d) is much lower (about 200 °C) than in the end portion at the earliest instant shown. The peaks due to phase transformations have evened out in the start portion at the instants shown in Figs. 4a-d. The residual stresses at the start portion and end portion of the plates are very similar.

HOT CRACKING

Andersson and Karlsson (1981) introduced the change in mechanical (nonthermal) strain, $\Delta\varepsilon^{m}_{14/10}$, across the weld in the temperature range ΔT = 1400 to 1000 ^{o}C as a criterion for nucleation of hot cracks. A large positive value of $\Delta\varepsilon^{m}_{14/10}$ should mean a greater risk for hot cracks to occur. $\Delta\varepsilon^{m}_{14/10}$ is calculated as the average change in mechanical strain for the elements at the weld line and is given in Figs. 5a-d.

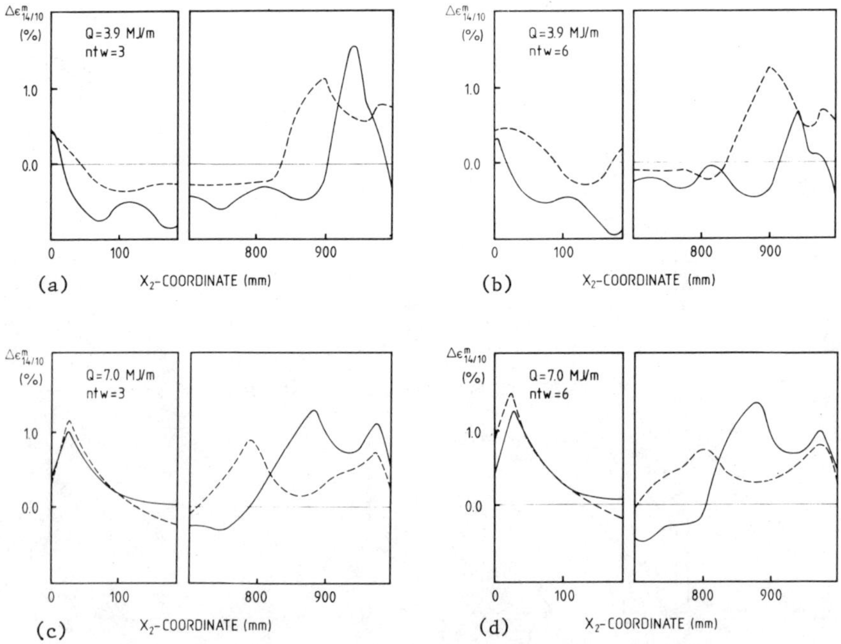

Fig. 5. Calculated change in mechanical strain $\Delta\varepsilon^{m}_{14/10}$ at the start and end portion of the butt-welded plates. The solid lines (——) show $\Delta\varepsilon^{m}_{14/10}$ for v=0.006 m/s and the dashed lines (– – –) show $\Delta\varepsilon^{m}_{14/10}$ for v=0.012 m/s.

The proposed mechanism for hot cracking in (Andersson and Karlsson, 1981) was that hot cracks nucleate at temperatures around 1200 ^{o}C due to high transverse tensile strains and they may then propagate at a temperature of about 400 ^{o}C when the maximum transverse tensile stresses arise. Due to compressive transverse stresses at the edge of the plate and tensile transverse stresses at some distance from the edge the nucleated cracks should propagate towards the interior of the plate. It is clear from Fig. 5 that $\Delta\varepsilon^{m}_{14/10}$ is larger at the end portion than at the start portion of the plate. $\Delta\varepsilon^{m}_{14/10}$ is also positive (tensile) in a larger region at the end portion than at the start portion. This means that the risk for hot cracking is larger at the end portion than it is at the start portion. A large value of Q give positive values of $\Delta\varepsilon^{m}_{14/10}$ in a larger region than a small value of Q does. A high welding speed gives smaller values for $\Delta\varepsilon^{m}_{14/10}$ than a low welding speed does except in the case shown in Fig. 5b. The influence of the number of tack-welds on $\Delta\varepsilon^{m}_{14/10}$ is negligible except in the case shown in Fig. 5b. The reason for the contrary result in Fig. 5b as compared to the results of Figs. 5a, c, d may be the very large value of the change in gap width Δg in the end portion of the plate shown in Fig. 3b.

CONCLUSIONS

It is concluded from Fig. 3 that it should be possible to find a practical combination of welding parameters that minimizes the change in gap width Δg. The welding speed in that combination of welding parameters should be fairly high.

The calculated change in mechanical strain $\Delta\varepsilon^{m}_{14/10}$ has its largest positive value around $x_2 = 0.900$ m in all cases studied here (Fig. 5) which means that most of the hot cracks should nucleate at that location. It is seen in Fig. 4 that transverse tensile stresses are formed around $x_2 = 0.900$ m shortly (about 30 s) after the removal of the heat source. Therefore it is very likely that the hot cracks start around $x_2 = 0.900$ m and extend towards the interior of the plate. The locations of the starting points of observed hot cracks in a shipyard panel production line, as reported by Andersson and Karlsson (1981), agree very well with the location of the maximum value of $\Delta\varepsilon^{m}_{14/10}$ as calculated in this paper. It is concluded that the highest risk for hot cracking is to be found at the end portion of the butt-welded plate for low welding speeds and for large values of the arc energy per unit length of the weld.

ACKNOWLEDGEMENT

Dr B. A. B. Andersson, The Aeronautical Institute of Sweden, and ASEA AB put their computer program for thermal and elasto-plastic analysis at our disposal. The project is financially supported by the Swedish Board for Technical Development (STU).

REFERENCES

Andersson, B. A. B. (1978). Thermal stresses in a submerged-arc welded joint considering phase transformations. *J. Eng. Materials and Technology*, 100, 356-362.

Andersson, B. A. B. and L. Karlsson (1981). Thermal stresses in large butt-welded plates. *J. Thermal Stresses*, 4, Nos. 3-4, 491-500.

Jonsson, M., L. Karlsson and L-E. Lindgren (1983). Thermal stresses and plate motion in butt-welding. In R. W. Lewis, J. A. Johnson, R. Smith (Eds.), *Numerical Methods in Thermal Problems*, Pineridge Press, Swansea. (To be published).

RELATIVE IMPORTANCE OF INITIATION AND PROPAGATION PHASES ON THE HIGH TEMPERATURE LOW CYCLE FATIGUE BEHAVIOUR OF A NICKEL SUPERALLOY

V. T. A. Antunes

Departamento de Engenharia Mecânica,
Faculdade de Engenharia da Universidade do Porto, Porto, Portugal

ABSTRACT

The use of isothermal tests to predict fatigue lives in situations where temperature is variable was investigated. Data on a nickel superalloy (MARM002) obtained either by temperature or cycle type variation during a test, were used to access the importance of initiation and propagation fatigue phases at two temperatures, 850°C and 1040°C and two basic types of hysteresis loops. A situation of early formation of cracks at the higher temperature with accelerated growth at lower temperatures, showed that great care is necessary when applying MARM002 isothermal tests to any situations where temperature changes either during a cycle or during a specimen or component's life.

KEYWORDS

Creep; thermal-fatigue; fatigue initiation and propagation phases; strain--range partitioning method; fatigue damage; mixed tests.

INTRODUCTION

Thermal fatigue is one of the primary failure modes which must be taken into account when designing highly stressed hot components in gas turbine engines. Thermal fatigue of these components occurs during start and shutdown periods when high strains are generated due to different rates of heating and cooling of parts of different section. The usefulness of any life prediction method in these kind of situations where fatigue and creep are superimposed, depends primarely on how good is the simulation of the service conditions. The use of isothermal strain controlled tests to simulate and study components operating in such service conditions, has been of widespread use because tests are easier to conduct, with better control of the variables of interest - plastic and creep strain, stress and temperature. This procedure becomes even more attractive with the introduction of methods like Strainrange Partitioning (Manson, 1971) offering some degree of temperature insensitivity for its basic strainrange vs. life relationships. However, an approach to a thermal fatigue situation based on isothermal information covering the temperature range of interest can be dangerous. Recent work by Taira and Fujino (1976) showed that great care is necessary in applying to thermal fatigue, data based on isothermal tests. They observed that crack growth in air above and below the range of temperature where creep becomes a significant factor for nickel superalloys, leads to the follwing pattern:

(i) Many cracks are formed early in life at high temperature but these propagate slowly.

(ii) At the lower temperatures a single dominant crack is formed late in life, but propagates rapidly.

They concluded that the detrimental effects of a thermal fatigue cycle involving tensile stress dwells a the highest temperature, were due to the early formation of cracks at the higher temperatures where growth was accelerated by lower temperature deformation processes.

The double damage of early iniciation and quicker growth of cracks under cycles with variable temperature suggest that any isothermal information is insufficient and dangerous to apply to a variable temperature case like thermal fatigue, unless the relative importance of iniation and propagation is determined throughout the temperature interval of interested. The accessment of this relative importance will be the main objective of this report dealing with an as cast nickel superalloy tested over a temperature range of 750°C - 1040°C. Earlier on, high temperature low cycle fatigue characterization of that superalloy and over that interval was reported (Antunes and Hancock, 1978). This claracterization was made using the Strainrange Partitioning (Sr P) life prediction method whose basic loops, shown in Fig. 1, were used for the series of mixed test described below.

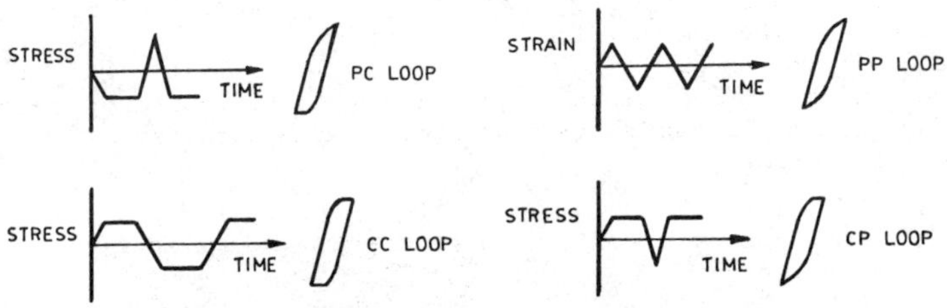

Fig. 1 Axial strain and stress is time wave forms for the basic Sr P low cycle fatigue tests

EXPERIMENTAL PROCEDURES

The experimental approach was similar to the one used to determine the strainrange vs life relationships of MARM002 and already reported (Antunes, 1980). All creep-fatigue tests were performed at a constant axial strain rate under completely reversed strain cycling. Again, either the direction of straining was reversed immediately upon reaching the predetermined strain limit, or a constant load was maintained for a period of time until the predetermined strain limit was reached. Each test was divided into two parts separated by a period of homogenization at 870°C for ½ hour without applied stress. At the end of part 1 either the type of test or temperature was changed. The mixed tests were divided into four sections:

(1) PP test at 850°C on the first part plus a CP test at 850°C on the second part.

(2) CP test plus a PP test at constant temperature of 850°C.

(3) PP test at 850°C plus PP test at 1040°C.

(4) PP test at 1040°C plus PP test at 850°C.

Inelastic strainranges were within to interval $5 \times 14^{-4} - 10^{-3}$. In each section an average of four tests were performed varying the percentage of exhausted life N_1/N_{P_1} where N_1 is the number od cycles imposed on the first part of the test and N_{p_1} the predicted total number of cycles. The damage D, done on the first part of the test was assessed during its second part using the number of cycles till rupture N_2 and the predicted number of cycles N_{p_2} for a virgin specimen. They were used to compute the damage D given as

$$D = 1 - (N_2/N_{p_2}) \qquad (1)$$

The extrapolated value of N_1/N_{p_1} for D = 0 was considered to be the iniciation point for the tests as they were defined on the first part.That value is valid for the sort of conditions (inelastic strain level and temperature) which were referred above. It is expected that for bigger inelastic strains (shorter lives) the propagation period will start earlier on life, decreasing the N_1/N_{p_1} value. The reverse is expected if smaller inelastic strains were used because this tends to extend the initiation period.

The justification for the choice of tests used previously and the clarification of the concept of initiation used in these tests will be dealt with more detail whithin the subsection dedicated to the discussion of these mixed tests.

Materials and Specimens

Details of the nickel superalloy (MARM002) and specimens used in this programme are covered elsewhere. Specimens were obtained in the as cast condition, subsquently machine to shape, heat treated for 16 hours at 870°C, furnace cooled and finally surface ground before testing. The resulting microstructure as shown in Fig. 2, consist of austenitic matrix with fine γ', $\gamma - \gamma'$ eutectic and MC carbides.

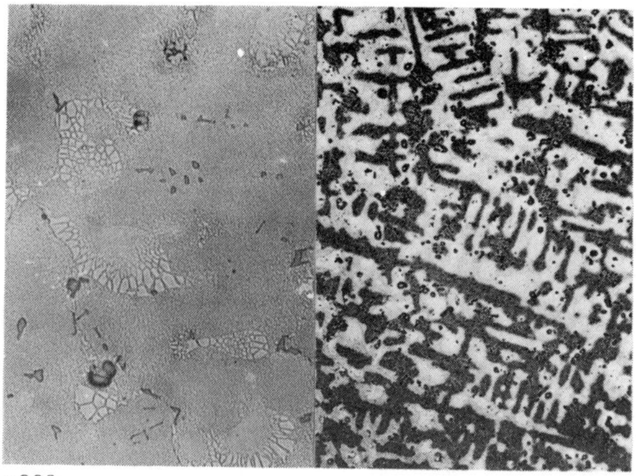

x200 x100

Fig. 2 MARM002 microestructure. Etched in Kalling's reagent (left); electrolitic etching in 15% glacial acetic acid (right).

Equipment

All specimens were tested using a servo controlled electrohydraulic system operated in closed-loop axial strain control. Axial load was applied to a specimen heated by direct resistance. Specimen longitudinal strain was measure by a variable reluctance transducer mounted on a water cooled end of a scissor type extensometer,which was located into specimen by ceramic inserts sprung into v ridges machined into the specimens. Details of this equipament, specimens preparation, tensile and creep properties are given in reports by Antunes (1980) and Clifton (1974).

RELATIVE IMPORTANCE OF INITIATION AND PROPAGATION PHASES

Studies concerned with secondary craking counting (Antunes, 1980), 5% drop detection in tensile load and interpretation of the mixed tests were the basis for the evaluation of the relative importance of initiation and propagation phases in some of the basic Sr P Cycles at two temperatures, 850°C and 1040°C.

Detection of 5% drop in Tensile Load

In all the cycles used to draw the strainrange vs life relationship the following criterium of fracture was followed: the number of cycles N_f adopted as the specimen's life is the number of cycles leading to 50% drop in tensile load or the number of cycles for separation of the specimens into two parts, whatever happens first. Within the four basic Sr P type of cycles there are two, PP and PC, where the monitoring of the tensile load can serve as a rough indication of the existence of a crack(s) within the specimen. Since the creep-plastic(CP) and creep-creep (CC) cyles employ stabilization between fixed load values the existence of a crack will only modify the partitioned strains within the cycle but not the load plateau. Hence, the initiation condition can not be determined by a load drop technique. The life period, $N_f - N_5$, which includes the last stages of propagation in each specimen, was considered to be a measure of the propagation period. This criterium was applied at both temperatures and the results for some of the PP and PC results on two intervals of temperature 750 - 850°C and 950 - 1040°C are shown in Fig. 3.

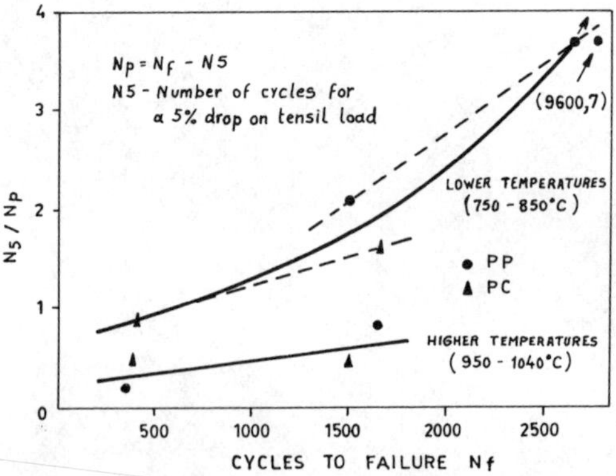

Fig. 3 Ratios N_5/N_p at two temperature ranges.

There, the ratio of the number of cycles to 5% drop in tensile load, N_5, over the number of cycles for faster propagation $N_p = N_f - N_5$, is plotted against total life N_f. The ratio N_5/N_p was found to be very small at high temperatures which is possibly an indication of a very small period of initiation at the temperature range 950 - 1040°C. The same fact is shown in Fig. 4 where the strain vs life based on 5% drop in tensile load are plotted. Here for a life of two thousand cycles there is a bigger period of propagation $N_f - N_5$ at higher temperatures (points C and D). However, the relative position of the strainrange vs life relationships at different temperatures is not altered, i.e. at the same inelastic strain, lives based on the 5% drop in tensile load confirm to be higher over the temperature range 950o - 1040°C (points A and B).

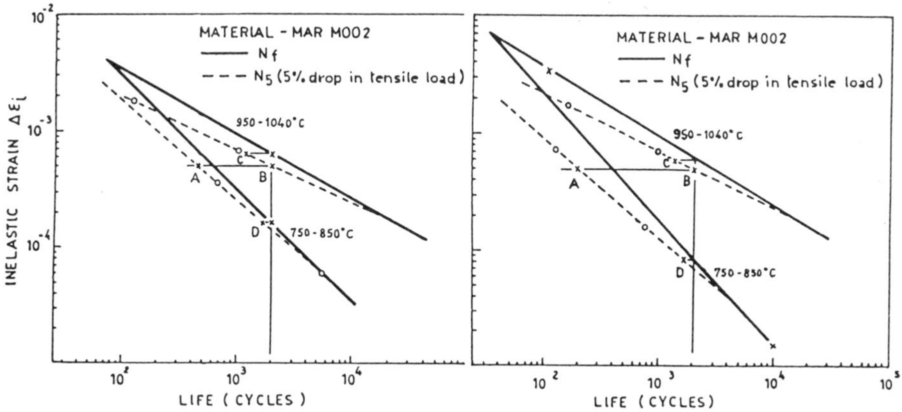

Fig. 4 Lines for fracture and 5% drop in tensile load. PP relationships (left). PC relationships (right).

Mixed Tests

In order to check the importance of initiation and propagation phases on the low cycle fatigue tests of MARM002 within the interval 750 - 1040°C, a series of mixed tests were conducted at 850°C and 1040°C. The aim was basically to determine the maximum percentage (N_1/N_{P_1}) of life that could be exhausted during the first part without damaging the specimen e.g. to determine the initiation point. This damage was accessed on the second part of the test and was considered to be $D = 1 - (N_2/N_{P_2})$ where N_2 is the experimental number of cycles till rupture and NP_2 the predicted life. Using the inelastic strains $\Delta\varepsilon_{i_1}$ and $\Delta\varepsilon_{i_2}$ of the first and second parts of the test, NP_1 and NP_2 could be calculated using the strainrange vs life relationships for the PP and CP cycles at 850°C and PP cycles at 1040°C which were determined previously. They were:

$$\Delta\varepsilon_{PP} = 0.26\ N_{PP}^{-0.94} \quad \text{(valid within the interval 750-850°C)}$$
$$\Delta\varepsilon_{CP} = 0.015\ N_{CP}^{-0.38} \quad \text{(valid within the interval 750-850°C)} \qquad (2)$$
$$\Delta\varepsilon_{PP} = 0.042\ N_{PP}^{-0.55} \quad \text{(valid within the interval 900-1040°C)}$$

The choice of these three cycles was due to:

a) Since at 1040°C there are no major differences in life for the different Sr P tests were choosed because they are easiest and least time consuming of all Sr P tests.

b) The PC and CP cycles constituted the lower and upper bounds of the strainrange vs life relationships at 850°C. Since there was no major difference between the PC and PP lives and taking into account that the PP tests are the easiest to do, the investigation on the initiation points was done on the PP line and on the upper bound of life at 850°C, i.e. the CP line. The mixed tests data one reported in Table 1. Plots of exhausted life over predicted life ratios N_1/N_{P_1} against the damage D are shown in fig. 5.

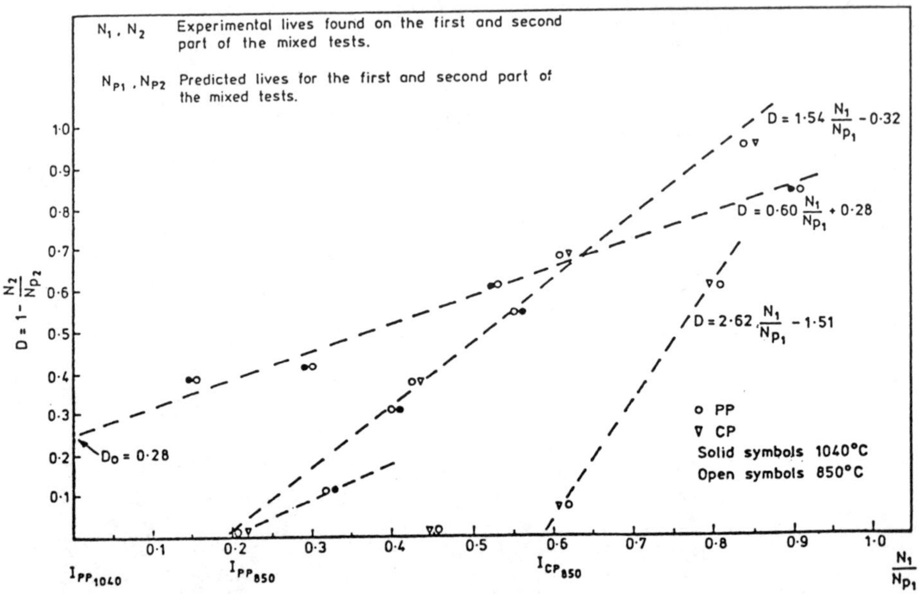

Fig. 5 Initiation points for the fatigue tests at 850 and 1040°C

Values of 21% and 58% on life were obtained in those plots as extrapolated values dor D = 0. They were considered to be the initiation points $I_{PP_{850}}$ and $I_{CP_{850}}$ for the PP-850°C tests and CP-850°C tests. The extrapolated point $I_{PP_{1040}}$ was found to be negative. That shows that even before the 1st part at 1040°C starts, there is damage imposed to the material,estimated to be here around 25% on life (point D_o). The time at 1040°C (aproximately 1/4 hr) which is used to stabilize temperature and make last checks, is sufficient to damage the material in so far as 25% of the PP 850°C lives that follow are bypassed.

DISCUSSION OF RESULTS

The results of the mixed tests involved the prediction of two lives NP_1 and NP_2. They are calculated using the strainrange vs life relationships (2). Obviously there are errors associated with these predictions. It can be shown that $I_{CP850} = 0.58 \pm 0.06$ and $I_{PP850} = 0.21 \pm 0.02$ in which the values 0.58 and 0.21 result from linear regressions performed on data from Table 1 leading to equations $D = D\left(\frac{N_1}{N_{P_1}}\right)$ reported in Fig. 5.

The change of strain levels will alter these initiation points. It is well known that the bigger the ratio of plastic to elastic strains the more important will become the propagation period.

In principle any type of cycle can be used on the second part of the mixed

TABLE 1 Mixed Tests Data

Cycle Part 1	Temp. (°C)	$\Delta\varepsilon_1$ (%)	$\Delta\varepsilon_{i_1}$ (%)	$\Delta\sigma$ (MN/m^2)	N_1	N_{P_1}	Cycle Part 2	Temp. (°C)	$\Delta\varepsilon_2$ (%)	$\Delta\varepsilon_{i_2}$ (%)	$\Delta\sigma$ (MN/m^2)	N_2	N_{P_2}
PP	850	0.777	0.048	1158	350	809	CP	850	0.785	0.114	1005	341	552
CP	850	0.691	0.111	1078	250	550	PP	850	0.788	0.124	1295	397	295
PP	850	0.596	0.033	1126	500	1206	PP	1040	0.567	0.127	626	403	579
PP	1040	0.487	0.129	587	515	563	PP	850	0.418	0.006	714	1132	6792
PP	850	0.763	0.041	1140	590	952	CP	850	0.756	0.102	1034	208	651
PP	850	0.769	0.042	1149	770	904	CP	850	0.731	0.108	984	32	583
PP	850	0.773	0.046	1155	180	850	CP	850	0.762	0.103	1042	598	604
CP	850	0.769	0.109	1043	340	562	PP	850	0.992	0.110	1394	314	334
CP	850	0.792	0.114	1072	415	514	PP	850	0.969	0.097	1378	149	382
PP	850	0.665	0.032	1000	390	1222	PP	1040	0.565	0.116	620	604	679
PP	850	0.691	0.035	1038	630	1145	PP	1040	0.578	0.126	625	268	582
PP	1040	0.562	0.115	617	360	683	PP	850	0.509	0.013	785	1263	3322
PP	1040	0.559	0.114	614	210	704	PP	850	0.534	0.020	813	1166	2011
PP	1040	0.608	0.132	658	80	541	PP	850	0.527	0.018	802	1332	2221

$\Delta\varepsilon_1$ Total strain on first part of the mixed tests.

$\Delta\varepsilon_2$ Total strain on second part of the mixed tests.

$\Delta\varepsilon_i$ Inelastic strain.

$\Delta\sigma$ Stress range.

tests but cycles with anticipated large initiation periods should be preferable, a high cycle fatigue test being ideal. In fact, any small amount of damage inflicted on the specimen during the firts part of the test will be detected more effectively and with higher sensitivity in a second part, because a big period of the specimen life - initiation period - will be bypassed. Unfortunately, cycles with big initiation periods do have a bigger scatter on life. The induced serror in N_2/N_{P_2} will be then greater and therefore the previous sensivity advantage can be nullified by the error in predicting those lives with higher initiation periods. Further investigation is suggested in this field.

Finally, it is important to reemphasize the danger of using isothermal data in situations like thermal fatigue where temperature varies during each cycle. The isothermal results of MARM002, together with initiation vs propagation studies on the same alloy, have clearly demonstrated the problems involved. This risk is implicit on the very early development of crack initiation at 1040°C, due to surface-environment interactions coupled with rapid growth at lower temperatures. Further improvement on the knowledge of the high temperature low cycle fatigue of MARM002 and other alloys will need the inclusion of thermal fatigue tests over the temperature range of interest.

CONCLUSIONS

A situation of early formation of cracks at the more elevated temperatures (950,1040°C) with accelerated growth on the lower temperature brackect (750 -850°C) showed that great care is necessary when applying MARM002 isothermal results to thermal fatigue cases or more generally to any cases where temperature changes either during a cycle or during a specimen or component's lives.

Values for the initiation points of the PP cycles at 850°C and 1040°C and CP cycles at 850°C were calculated using mixed tests. These points were estimated to be aproximately 0, 20 and 60% of the respective lives for cycles of the type PP at 1040°C, PP at 850°C and CP at 850°C respectively. The relative position of these points were confirmed by secondary crack statistical studies on longitudinal sections of the test specimens and through studies on the monitoring of the 5% drop in tensile load of some PP and PC specimens tested at 750, 850, 950 and 1040°C.

AKNOWLEDGEMENTS

The author is greatly indebted to Prof. P. Hancock at Cranfield for his helpful discussions and to the National Gas Turbine Elablishment U.K. wich conducted a three year test programm at the Cranfield Institute of Technology under Contrack AT/2028/091. Recognition is due to Mr. T. Clifton his advice on experimental procedures and Miss Aida Costa for typing the text.

REFERENCES

Antunes, V.T.A., and P. Hancock (1978). Strainrange partitioning of MARM002 over the temperature range 750-1040°C. Proc. AGARD Conf. nº 243,Aalborg, Denmark.

Antunes, V.T.A., (1980). PhD Thesis,Cranfield Institute of Technology, Beds, U.K.

Clifton, T.E. (1974). Creep and fatigue studies of Nimonic 108, Cranfield Institute of Technology Memo 94, Part 1, U.K.

Manson, S.S., G.R. Halford, and M. H. Hirscheberg (1971). Creep fatigue analysis by strainrange partitioning.NASA TMX - 67838.

Taira, S., and M. Fujino (1976). J. of the Soc. of Materials Science of Japan. 25,218-219.

INFLUENCE OF STATE OF STRESS ON CREEP FAILURE OF 2 1/4% Cr 1% Mo STEEL

K. D. Al-Faddagh*, G. A. Webster* and B. F. Dyson**

*Department of Mechanical Engineering, Imperial College, London SW7, UK
**Division of Materials Applications, National Physical Laboratory, Teddington, Middlesex, UK

ABSTRACT

Creep experiments on circumferentially notched bars of a $2\frac{1}{4}$%Cr1%Mo steel have been reported. Notched bar life first increased with increase in notch acuity but then tended to decrease for very sharp notches. This behaviour is explained in terms of the effective and hydrostatic stresses developed at a skeletal point across the notch throat. Although notch strengthening is attributed to a decrease in the effective stress at the skeletal point with increase in notch sharpness, it is argued that a progressive increase in the hydrostatic tensile stress component limits the strengthening that can be developed in very sharp notches.

KEYWORDS

Skeletal stress; creep rupture; notch strengthening.

INTRODUCTION

Equipment which operates over long periods of time at elevated temperature may suffer excessive deformation or fracture due to creep. Since many components experience triaxial stresses as a result of the applied loading conditions, or local sharp changes in section, laws describing creep behaviour under multi-axial stressing are needed in order to determine useful operating lives. Experience has shown that failure often initiates from sites of stress concentrations where there is a significant hydrostatic tensile stress component. The main purpose of this paper is to establish how state of stress influences the creep failure of engineering components.

The most frequent method of introducing a three-dimensional state of stress into a test piece in the laboratory is to subject circumferentially notched bars to an axial tensile load. Previous work by some researchers (Davis and Manjoine, 1953; Newman, Jones & Brown, 1953; Ng, Webster & Dyson, 1980) has indicated that certain materials can exhibit either notch strengthening or weakening depending upon test conditions. For example, Ng, Webster & Dyson (1980) showed that a change from notch strengthening to notch weakening can take place in a $\frac{1}{2}$%Cr$\frac{1}{2}$%Mo$\frac{1}{4}$%V steel with increase in notch sharpness. The material had a very low ductility and failed through initiation and growth of intergranular cavities. Since a triaxial tensile stress is now known to nucleate more cavities than in the corresponding uniaxial stress state (Loveday and Dyson 1979; Cane, 1980), it is not surprising that notch weakening occurred in this case. In ductile materials failing by a necking instability phenomenon, less cavitation is expected and the present investigation was initiated to determine whether lifetimes for these materials can be predicted accurately from a recently developed skeletal point stress analysis (Al-Faddagh, Fenner & Webster, 1982; Al-Faddagh, 1983).

EXPERIMENTAL TECHNIQUES AND RESULTS

The alloy tested was a 2¼%Cr1%Mo steel used in the electric power generation industries. Its chemical composition is shown in Table 1.

TABLE 1. Chemical composition of 2¼%Cr1%Mo steel

weight %

C	Si	Mn	S	P	Ni	Cr	Mo	Al	Zr	Ti	Cu	Sn
0.12	0.34	0.59	0.005	0.015	0.19	2.4	1.0	0.02	0.01	0.034	0.07	0.01

All the experiments were performed at a temperature of 538°C. Basic creep data were obtained on plain round bar specimens at both constant load and constant stress. The corresponding secondary creep rates and rupture lives are shown respectively in Figs. 1 and 2. Tests at constant stress were required because of the large ductility exhibited by the material. The constant stress was maintained by systematically reducing the load during a test on the assumption that deformation occurred at constant volume. It was found, after a small initial primary region, that the constant load tests gave a progressively increasing creep rate, whereas the constant stress tests showed an extended secondary region. Nevertheless, the secondary creep rate of both could still be characterised by the same Bailey-Norton creep expression with a stress exponent $n = 9.8$ (Fig. 1). A slight increase in creep ductility ε_R, from about 40% to 50% was obtained as failure time increased from 20 to 4000 hours. There was little evidence of cavitation which is consistent with this high ductility. The lives of the constant stress tests, t_R (see Fig. 2) were approximately five times those of the corresponding constant load tests, t_R^L, in agreement with a Hoff (1953) type of analysis which predicts $t_R = t_R^L \, n\varepsilon_R$.

Notched specimens having the same throat diameter as the plain tensile bars were machined to the dimensions shown in Fig. 3. The ratio b/a was maintained constant at 1.46 for all tests and the notch profile altered by changing the radius R only. Results of creep tests carried out at two constant nominal stress levels across the notch throat, σ_{nom}, of 200 and 175 MPa, are shown in Fig. 4. In this figure the life of the notched bars t_n has been normalised with respect to the unnotched life t_{un}. For blunt notches, notched bar life increases progressively for both stress levels as notch sharpness (a/R) increases. For sharp notches there is a slight tendency for the lives to decrease with further increase in a/R. This effect is more pronounced at 175 MPa than at 200 MPa.

Further experiments have also been conducted on two blunt and two sharply

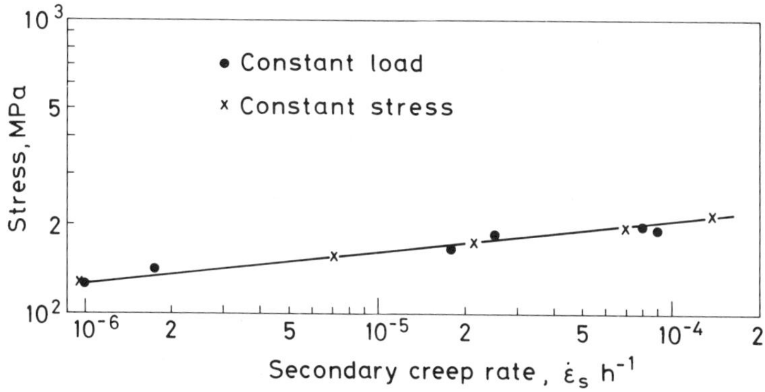

Fig. 1. Secondary creep rate data for the $2\frac{1}{4}$%Cr1%Mo steel tested at 538°C.

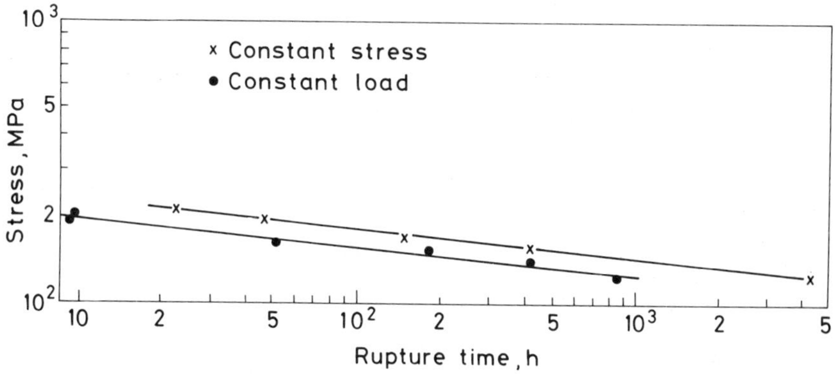

Fig. 2. Stress-rupture data for the $2\frac{1}{4}$%Cr1%Mo steel tested at 538°C.

Fig. 3. A typical circumferentially notched tensile test bar. (dimensions in mm)

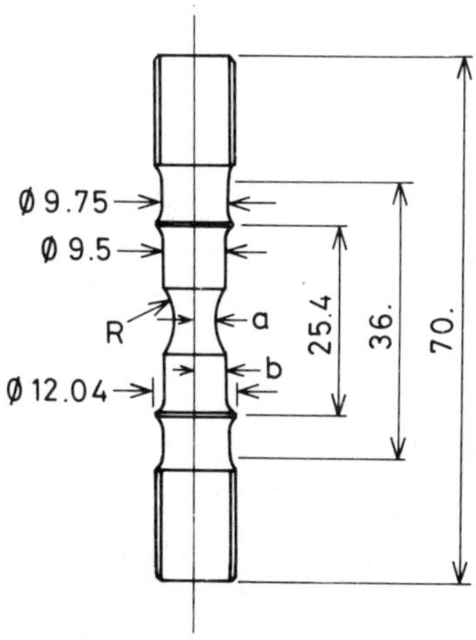

notched bars in which the stress was changed from 200 to 175 MPa and vice versa to examine cumulative damage concepts. The results of these tests

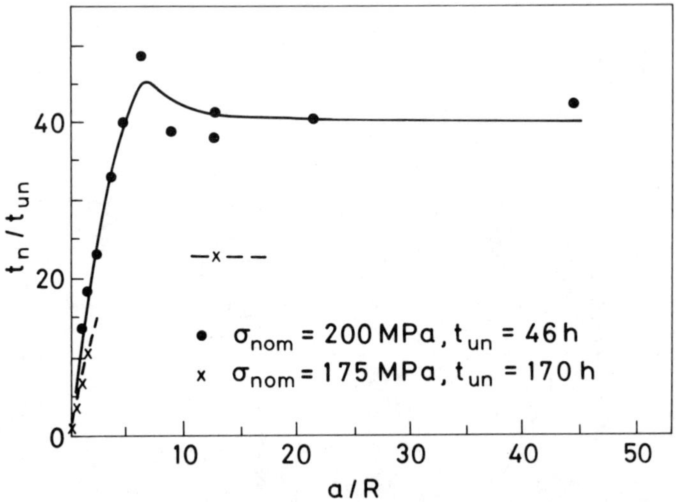

Fig. 4. Notched bar rupture life ratios as a function of notch shape.

are listed in Table 2, together with the predictions of the Robinson (1952) life fraction rule:

$$\sum \frac{t}{t_R} = D \quad (1)$$

where D is the fraction of life used up. At fracture Robinson predicts D = 1. It is clear from the series of tests that, irrespective of notch sharpness or loading history, the Robinson cumulative damage law describes failure accurately.

TABLE 2: Stepped Load Notched Bar Creep tests at 538°C

a/R	$\sigma_{nom\ 1}$ (MPa)	$\sigma_{nom\ 2}$ (MPa)	t_{R1} (h)	t_{R2} (h)	t_1 (h)	t_2 (h)	D
1.67	200	175	840	1800	420	939	1.02
1.67	175	200	1800	840	900	434	1.02
13	200	175	1800	3800	900	1892	1.0
13	175	200	3800	1800	1900	940.5	1.02

DISCUSSION

An axi-symmetric finite element stress analysis (Al-Faddagh, Fenner, Webster, 1982; Al-Faddagh, 1983) has been performed to calculate the elastic and steady-state creep stress distributions across the notch throats for various values of n. It has been found that an approximate skeletal point exists for each of the notch profiles examined. The change in effective stress (calculated according to Von-Mises criterion) at this point, $\bar{\sigma}^*$, for an applied nominal stress of 200 MPa is shown plotted as a function of a/R in Fig. 5, together with an approximate estimate due to Bridgman (1952). The approximate estimate shows a continuing decrease in effective stress with increase in a/R, whereas the finite element prediction levels off at a constant value of 124 MPa.

A further stress determined from the finite element calculations has been included in Fig. 5. This is the hydrostatic component at the skeletal point. It can be seen that this stress gradually increases with increase in notch sharpness. For $a/R > 4$ it exceeds $\bar{\sigma}^*$.

A comparison of the notched bar lives determined from the different estimates of effective stress in Fig. 5 is presented in Fig. 6. Good agreement between the experimental lives and the skeletal stress predictions is obtained for a/R up to about 6. Above this value there is some divergence as the predicted life approached a plateau but the difference is always less than a factor of two.

An upper and lower limit to notched bar failure life is also included in Fig. 6. The upper limit corresponds to failure of the gross section and the lower limit to maintenance of full constraint (Sczczepinski et al, 1966; Neimark, 1968). For the b/a ratio of 1.46 tested, full constraint could not be retained above a value of a/R of about 1.5. Above this value of a/R the Bridgman (1952) approximate analysis substantially overestimates notch lives and predicts failure of the gross section for $a/R > 10$. It is apparent therefore, that Bridgman's approach cannot be used to provide realistic estimates of notched bar lifetimes for relatively sharp notches. For these circumstances, incremental time step numerical solutions (Hayhurst and Henderson, 1977; Hayhurst, Leckie and Morrison 1978) or a skeletal stress analysis of the type proposed by Al-Faddagh, Fenner and Webster, (1982) and Al-Faddagh (1983) are needed.

A qualitative explanation of the experimental results in microstructural terms is possible by making reference to Fig. 5. This figure shows that $\bar{\sigma}^*$ decreases to a constant value with increase in a/R, whereas the hydrostatic stress continues to increase. Decreasing $\bar{\sigma}^*$ will cause an increase in rupture life, whilst an increase in hydrostatic stress will lead to void formation (Loveday and Dyson, 1979) and/or enhanced growth rates of existing voids (Cocks and Ashby, 1982) thus promoting failure at reduced ductilities and times. These trends are competing. At low a/R the fall in $\bar{\sigma}^*$ dominates and an increase in rupture life with increase in a/R is expected. For sharper notches, the increase in the hydrostatic stress predominates and notch bar life begins to decrease with increase in notch acuity. These features are consistent with the experimental results but further work is needed to quantify the effects in terms of void nucleation and growth models.

CONCLUSIONS

A skeletal point has been identified for circumferentially notched tensile bars and used to explain notched-bar rupture lifetimes in a very ductile steel. The initial increase and subsequent fall in life with increase in notch sharpness has been explained in terms of the effective and hydrostatic stress developed across the notch throat.

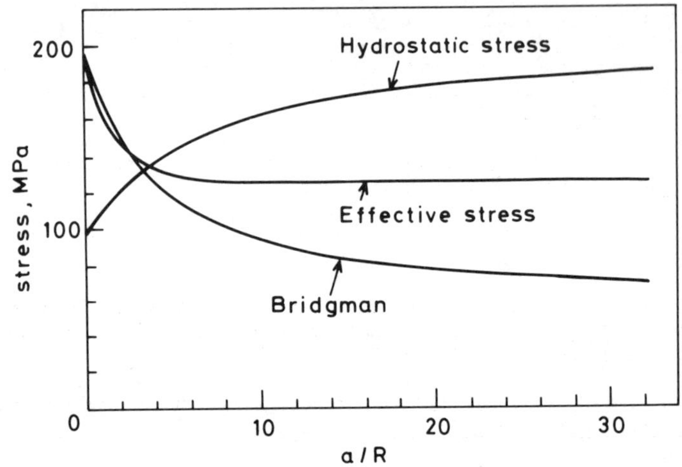

Fig. 5. Effective and hydrostatic stresses at skeletal point for notched bars.

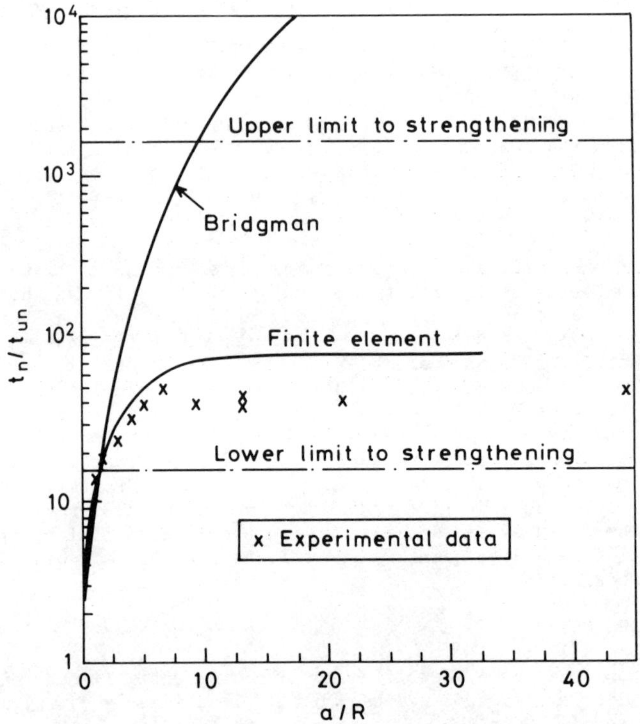

Fig. 6. Comparison of effective stress predictions with experimental data.

REFERENCES

Al-Faddagh, K.D., R.T. Fenner & G.A. Webster (1982). Steady-state stress distributions in circumferentially notched bars subjected to creep, J. Strain Analysis, Vol.17, No.3, p.123.

Al-Faddagh, K.D. (1983). Influence of state of stress on creep failure. Ph.D. thesis, Imperial College, London.

Bridgman, P.W. (1952), Studies in large plastic flow and fracture, McGraw-Hill, New York.

Cane, B.J. (1980). Creep damage accumulation and fracture under multiaxial stresses, ICF-5 on Advance in Fracture Research (edited by Francois, D, et al), Pergamon Press, Oxford/New York.

Cocks, A.C.F. & M.F. Ashby (1982). On creep fracture by void growth. Prog. Material Science, Vol.27, p.189.

Davies, E.A. & J.J. Manjoine (1953). Effect of notch geometry on rupture strength at elevated temperatures, ASTM STP 128, p.67.

Hayhurst, D.R. and J.T. Henderson (1977). Creep stress redistribution in notched bars, Int. J. Mech. Sci., Vol.19, p.133.

Hayhurst, D.R., F.A. Leckie and C.J. Morrison (1978). Continuum damage study of notch strengthening and weakening in creep rupture, Proc. Roy. Soc. A360, p.243.

Hoff, N.J. (1953). Necking and rupture of rods subjected to constant tensile load, J. App. Mech. Vol.20, p.105.

Loveday, M.S. and B.F. Dyson (1979). Creep deformation and cavitation damage in Nimonic 80A under a triaxial tensile stress, Mechanical Behaviour of Materials, 2, p.213-222, Eds. K.J. Miller and R.F. Smith, Pergamon Press.

Ng. S.E., G.A. Webster & B.F. Dyson (1980). Notch weakening and strengthening in creep of $\frac{1}{2}$ Cr $\frac{1}{2}$Mo$\frac{1}{4}$ V steel, ICF-5 on Advance in Fracture Research (edited by Francois, D, et al). Pergamon Press, Oxford/New York.

Neimark, J.E. (1968). The fully plastic plane-strain tension of a notched bar. J. Appl. Mech., Vol.35, p.111.

Newman, D.P., M.H. Jones & W.F. Brown (1953). Time-temperature dependence of the notch effect and influence of notch depth in stress rupture on a Cr-Mo-V steel, Proc. ASTM, Vol.53, p.677.

Robinson, E.L. (1952). Effect of temperature variation on the long time rupture strength of steel, Trans. ASTM, Vol.74, p.777.

Sczczepinski, W., L. Dietrich, E. Drescher & J. Miastrowiski (1966). Plastic flow of axially-symmetric notched bars pulled in tension. Int. J. Solids Structure, Vol.2, p.543.

THE USE OF FRACTURE MECHANICS APPROACH TO EVALUATE THE RELIABILITY OF A NOZZLE WELD ASSEMBLY IN POWER PLANT VESSELS

I. V. Gorynin, V. A. Ignatov and B. T. Timofeyev

Strength Problems Institute of Academy of Sciences of the Ukrainian SSR, Timirasevskaya 2, 252014 Kiev, USSR

ABSTRACT

THE EVALUATION of the reliability of power plant units cannot be based on the determined material fracture values which are governed by possible material fracture mechanism depending on in-service loading conditions, since they may not take into account the property and composition dispersion. Therefore, it is necessary to consider fracture characteristics from a statistical point of view. The fracture process of welded nozzle assembly can be devided into three stages, namely, fatigue crack initiation, stable crack growth to a critical size and brittle fracture. The low cycle fatigue strength resistance, the rate of crack propagation and the critical stress intensity factor K_{Ic} which correspond to aforementioned stages are widely used to evaluate strength and safety factor of structures in practice. The present paper deals with the results of the investigation into the reliability of the most stressed assembly in a power plant, that is the nozzle welded butt joint made by assembly welding without any post heat-treatment. Relatively small specimens and full size units were tested.

The investigation refers to a welded joint of the pipe DY-850 and the nozzle zone of a vessel made of low alloy heat-resistant 15X2HMΦA steel. Since effective heat-treatment is practically impossible in assembly welding, mild steel cladding was applied in-plant conditions to the pipe and nozzle zone edge. As a result, a butt weld was made by manual welding using electrode YOHU-13/45A (σ_{UTS} = 500 MPa, $\sigma_{0.2}$ = 450 MPa, δ = 15%, ψ = 70%, t_{ko} = +20°C) without subsequent temper. This welded joint is the weakest portion of the assembly considered. It is for this joint that the low-cycle strength resistance, the fatigue crack propagation resistance at +20°C and the critical stress intensity factor at different temperatures were determined with due regard to property dispersion.

A full size assembly of the welded nozzle DY-850 was tested under internal pressure of 4-25 MPa for 5000 cycles (maximum pressure in a cycle being higher than hydropressure of the power plant vessel). Then the full size assembly was tested in more severe conditions, namely the welded joint was circumferentially and axially notched (10-12 mm in depth) and the assembly was loaded by internal pressure of 4-28 MPa for 10^4 cycles. The notch edge opening increased by 0.005 mm after 7300 cycles, and for the following 2300 cycles the increment of the opening reached 0.012 mm, which indicated the fatigue crack initiation from the notch in the joint. The kinetics of a crack under cyclic loading was evaluated from test results for the full size assembly. At the final stage the assembly was fractured under hydrostatic pressure at -20°C and fracture toughness characteristics were determined. The fracture initiated from an axial notch with a fatigue crack 22 mm in depth. In the course of the experiment, a diagram of the load versus edge displacement prior to desealing and oscillogram of the notch edge movement were recorded. The fracture was of a quasi-brittle type.

The comparison of the number of cycles to fatigue crack initiation from the notch in a full size assembly test and the integral curve for a given characteristic of weld metal shows that the probability of the event is 50%, which proves the validity of the design technique using deformation criteria. As the number of cycles corresponding to fatigue crack growth at given loading parameters with a known value of the stress intensity factor range was recorded in a full size assembly test, it appeared possible to calculate the Paris equation power m using the relation of the parameters in this equation obtained earlier. The value of the design parameter m in the integral curve corresponds to 65% probability. Thus, the rate of fatigue crack propagation in full size assembly and in small specimens was practically equal.

The comparison of weld metal brittle fracture strength resistance for specimens and a full size assembly showed that the latter fractured at 100% fracture probability K_{Ic} according to integral distribution curve for weld metal at -20°C. It is supposed that at the brittle fracture stage the residual stresses can influence the characteristic considered. To evaluate K_{Ic} correction for residual stresses some relationships are presented in the paper.

The test data illustrate the fact that it is the combination of unique tests of full size assemblies and specimens from commercial materials followed by statistical treatment, that gives complete quantitative information on the assembly reliability.

THE EFFECT OF ORDERING ON THE HIGH TEMPERATURE DEFORMATION OF NICKEL BASED ALLOYS

A. Marucco* and B. Nath**

*Consiglio Nazionale delle Recerche,
Istituto per la Tecnologia dei Materiali Metallici non Tradizionali,
Via Induno 10, 20092 Cinisello Balsamo, Milano, Italy
**Technology Planning & Research Division,
Central Electricity Generating Board,
Central Electricity Research Laboratories, Kelvin Avenue, Leatherhead,
Surrey KT22 7SE, UK

ABSTRACT

High temperature tensile tests have been carried out on Ni_3Cr and Ni_2Cr alloys. The proof stress, the tensile strength and the ductility decrease with increasing test temperature and/or decreasing strain rate. As a consequence of ordering the work hardening coefficient shows a minimum at 748 to 873°K depending upon the test conditions. A new type of serrated flow arises from the formation and subsequent dissociation of arrays dislocation dipoles in partially ordered materials.

KEYWORDS

Ordering in Ni-Cr; Mechanical Properties; Serrated Flow; Dislocation Arrays.

INTRODUCTION

A number of nickel alloys have been found to undergo an order-disorder transformation based on the superlattice Ni_2Cr at ~823°K (Metcalfe and Nath, 1979; Tawancy, 1980; Marucco, Metcalfe and Nath, 1981; Marucco and Nath, 1983). Short range ordering (SRO) forms in these materials during cooling from the final commercial ageing temperature, usually of the order of ~973°K. This increases upon exposure at and below ~823°K, and a dispersion of very small highly ordered particles in a disordered matrix is established. On prolonged ageing the structure may, depending upon the composition, transform to a long range ordered (LRO) structure. The kinetics of the LRO transformation however, are very sluggish and in commercial materials may take tens of thousands of hours to form. The kinetics of LRO in the stoichiometric Ni_2Cr is much more rapid, and the LRO begins to form after $\sim 8.6 \times 10^4$ s ageing (Marucco, Metcalfe and Nath, 1981). An external stress during ageing enhances the kinetics of transformation significantly (Metcalfe and Nath, 1979).

Marked changes in mechanical properties have been reported to accompany an ordering transformation (Cohen and Fine, 1962; Stoloff and Davies, 1964;

Cahn, 1965). These changes include (i) the appearance of a yield point (ii) an increase in tensile strength in materials possessing SRO or a degree of LRO in the region of 0.4 to 0.5, and (iii) embrittlement. In view of these possible consequences, tensile tests have been carried out on Ni_3Cr and Ni_2Cr alloys and the results are presented here.

EXPERIMENTAL

Tensile specimens with 6.5 mm diameter gauge were machined from as-extruded bars of the two materials (Ni-30.2 wt%Cr-0.014 wt%C and Ni-22.7 wt%Cr-0.003 wt%C). The specimens were heated to the test temperature over a period of 1×10^4 to 1.5×10^4 s and then tested at constant crosshead speeds.

RESULTS

Mechanical Properties

Four types of serrated flow were observed in both alloys (Fig. 1). Three of these (A, B and C) have been seen during dynamic strain ageing of substitutional solid solutions (Baird, 1971). In Type A' yield, like Type A, there was an initial gradual increase in the flow stress. This was followed by a plateau during the first few serrations of a test and a gradual decrease at higher strains. On continued deformation, Type B serrations appeared superimposed on the Type A' flow.

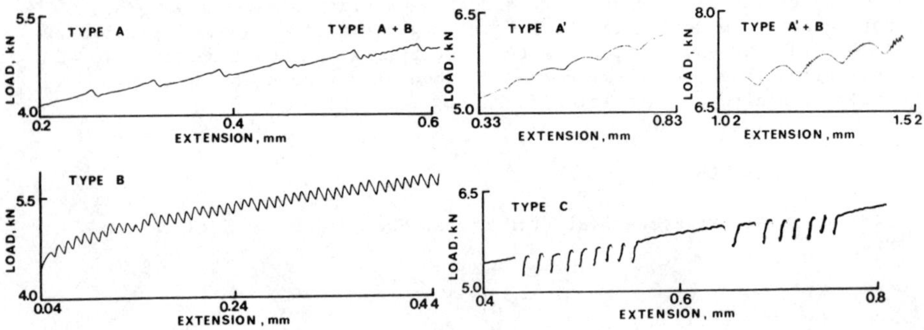

Fig. 1. Load-extension traces illustrating different types of serrations.

The 0.1% proof stresses of both materials were similar over the temperature and strain rate examined (Fig. 2). They decreased monotonically with increasing temperature, and appeared to be independent of the strain rate. For a given temperature-strain rate combination, the tensile strengths of Ni_3Cr were higher than those of Ni_2Cr. For both alloys the strengths fell as the test temperature increased and/or the strain rate decreased. In Ni_3Cr the reduction in strength was observed at all temperatures, whereas in Ni_2Cr a plateau of minimum strength appeared at 798 to 873°K depending upon the strain rate.

The ductilities of the two materials were also very similar. There was a marked reduction in this parameter as the temperature increased and/or the the strain rate decreased. Failures of most specimens tested at 748 to 873°K occurred at maximum load with virtually no sign of necking. The strain between the maximum load and the fracture decreased with increasing temperature and/or decreasing strain rate.

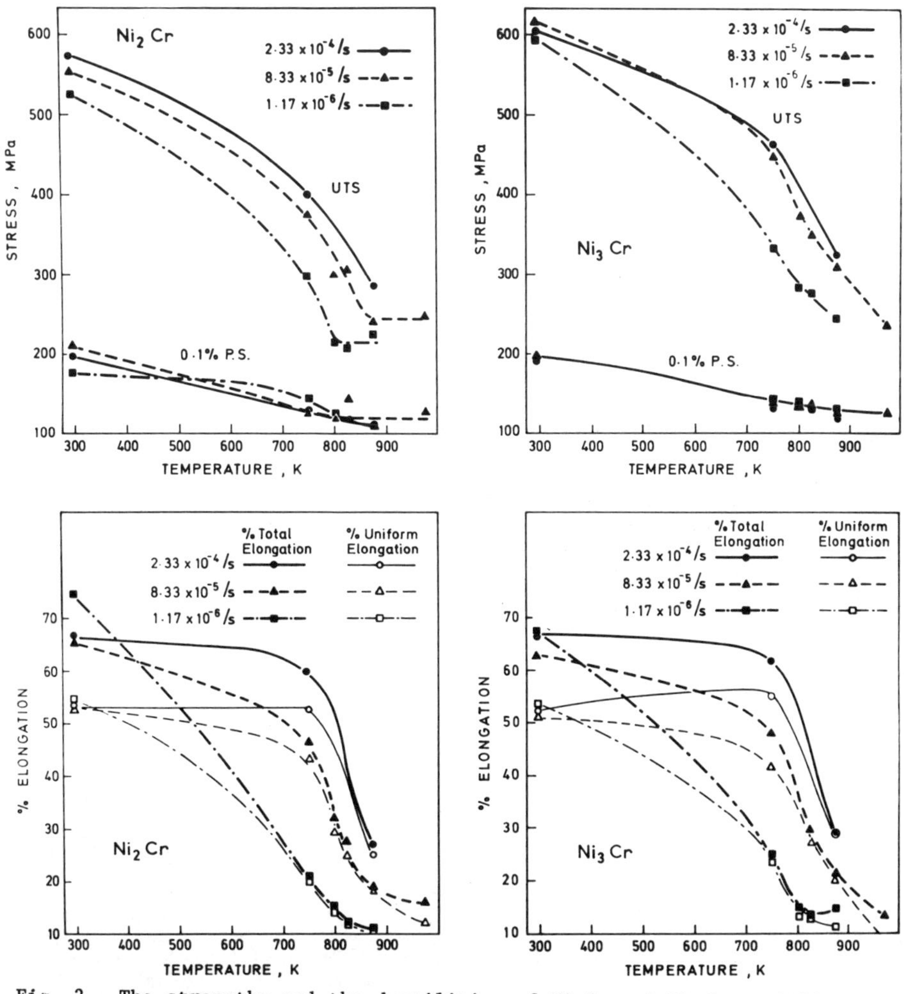

Fig. 2. The strengths and the ductilities of Ni_2Cr and Ni_3Cr at different temperatures and strain rates.

The work hardening coefficient, defined as the slope of the true stress-true strain curves at 5% strain, was measured. For both materials this parameter exhibited sharp minima at 748 to 873°K, depending upon the strain rate (Fig. 3).

Structural Examination

Ductile cup-and-cone fracture ocurred in specimens strained at 293°K. Failures of specimens tested at higher temperatures became increasingly more intercrystalline as the temperature increased and/or the strain rate decreased. Extensive grain boundary cracking was observed on the surface and through the gauge length of the test pieces (Fig. 4). Smooth facets exhibiting slip steps were observed up to 823°K (Fig. 5). Cavitation observed in all specimens increased with increasing temperature and/or decreasing strain rate.

In electron diffraction patterns weak superlattice reflections were seen only in one Ni_2Cr specimen tested at 748°K at a strain rate of 1.17×10^{-6}/s. All other test pieces possessed only SRO. A high density of dislocations was observed in specimens deformed at 293°K. Serrated grain boundaries were also seen (Fig. 6). Between 748 and 873°K planer slip had occurred and unusual dislocation pile-ups were observed. Usually dislocations in a pile-up are closely packed at the obstacle, and the dislocation spacing increases with distance from the obstacle. In the present situation rafts of evenly spaced dislocations were seen after testing between 748 and 873°K. Weak beam images (Fig. 7) revealed that the dislocations were not dissociated and that the burgers vector, determined by near invisibility criterion, was of a/2 <110> type. Imaging with opposite $\bar{g}$ established that the pile-ups were made up of dislocation dipoles. The spacings between dislocations varied from one pile-up to another (Fig. 8). It should be emphasised that such dislocation arrays were observed from the early stages of deformation; e.g. after 6% strain at 798°K and 2.33×10^{-4}/s a large number of dislocation dipole rafts were seen. A high density of dislocation tangles was observed after testing at 973°K and the deformation appeared to be homogeneous.

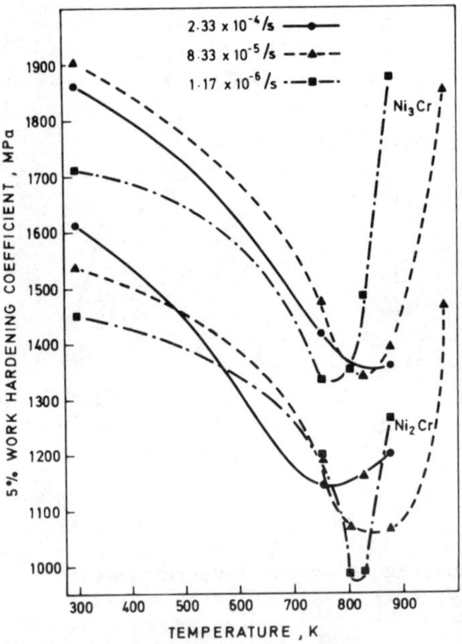

Fig. 3. The work hardening hardening coefficient as a function of temperature.

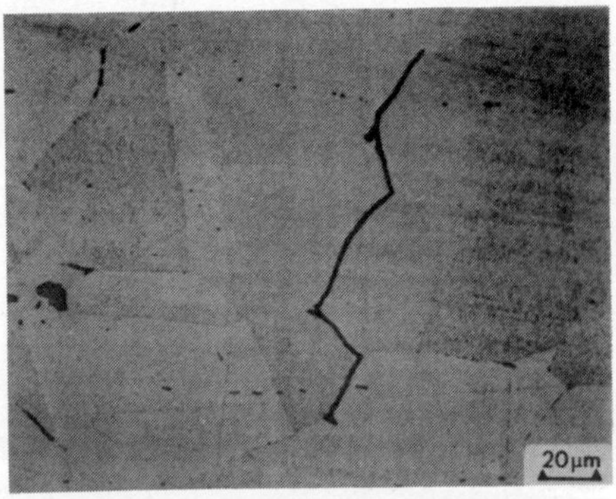

Fig. 4. W- and R- cracks in Ni_2Cr strained at 748°K, 1.17×10^{-6}/s.

DISCUSSION

Similar proof stresses of the two alloys demonstrate that the limit of solid solution strengthening in Ni is achieved by additions of <22% Cr. The tensile strengths of these alloys are determined primarily by their

Fig. 5. Slip traces and cavities on grain boundaries. Ni_2Cr tested in vacuum at 748°K, 1.83×10^{-8}/s.

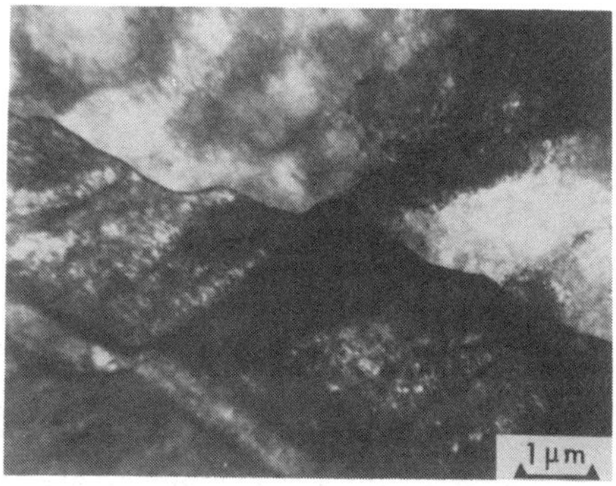

Fig. 6. Grain boundary ledges in Ni_2Cr deformed at 293°K, 1.17×10^{-6}/s

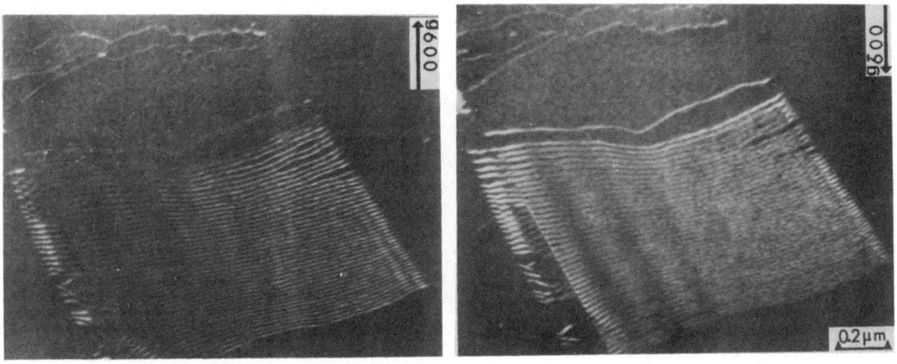

Fig. 7. Rafts of dislocation dipoles in Ni_2Cr, 748°K, 1.17×10^{-6}/s.

ductility and by their strain hardening behaviour. When the work hardening coefficients are comparable e.g. in Ni_2Cr at 748°K, then the strength is a function of the ductility, particularly of the uniform elongation and hence the strength falls with decreasing strain rate. Where the ductilities are comparable e.g. specimens tested at 293°K and 748°K and a strain rate of 2.33×10^{-4}/s, the strength is a function of the strain hardening coefficient, accordingly the strength is reduced as the temperature increases. Also, marginally superior tensile strengths of Ni_3Cr are attributable to its higher strain hardening coefficients as compared to Ni_2Cr.

Fisher (1954) was the first to suggest that the passage of one dislocation through a SRO material will destroy SRO across the slip plane and thus facilitate glide of other dislocations on the same plane. If the dislocations were absorbed at the rate at which they arrive at grain boundaries then slip will remain planar and no work hardening will occur. If, however, the annihilation was the slower process then a pile-up will form and further glide on the plane will be restricted, and other slip planes will be activated. Glide on different sets of planes will create

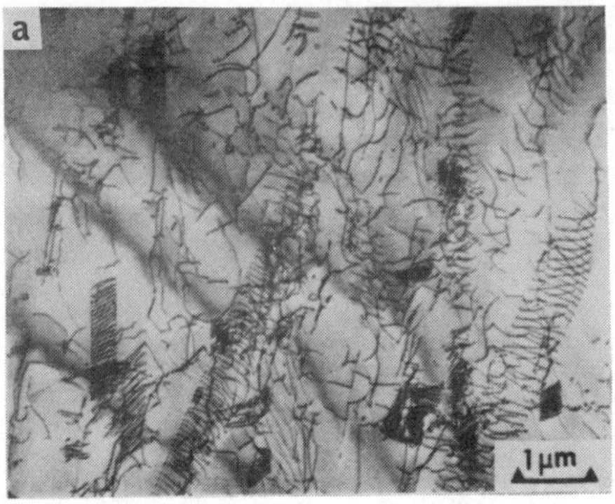

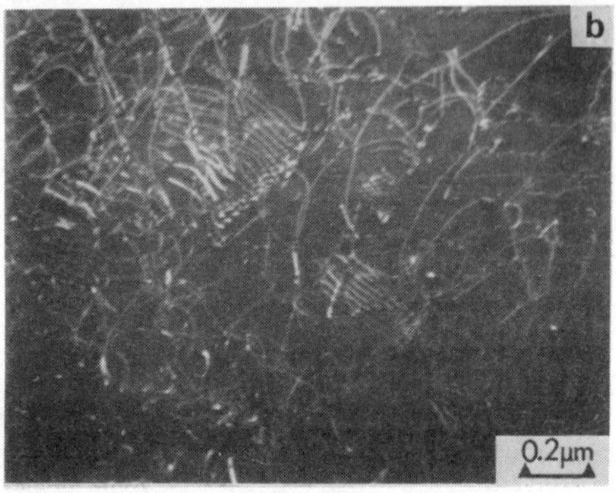

Fig. 8. Dislocation forest and dipoles in (a) Ni_2 Cr, 873°K, 1.17 × 10^{-6}/s and (b) Ni_3Cr, 748°K, 2.33 × 10^{-4}/s.

dislocation tangles and hence, strain hardening. This happens at 293°K where high rates of work hardening are observed, although initially some planar slip must have occurred to produce grain boundary ledges.

At higher temperatures higher atomic mobility will have three consequences (i) dislocation annihilation will be faster (ii) the kinetics of ordering in regions undisturbed by glide will increase (but the degree of order below T_c will be of similar magnitude) and (iii) the process of 'reordering' behind a moving dislocation will be more rapid. The first two will decrease the work hardening coefficient by encouraging planar slip while the third will have the opposite effect. For a given strain rate increasing temperature to T_c (~840°K for Ni_2Cr and ~798°K for Ni_3Cr) results in decreasing strain hardening. At temperatures above T_c the expected degrees of SRO is smaller but a metastable state of SRO may persist because of the SRO formed during heating of the specimen through T_c. This will result in a low strain hardening coefficient at high strain rates and a high strain hardening coefficient at slow strain rates, as is observed at 823°K and above in Ni_3Cr. At 973°K, neither alloy exhibits planar slip and the strain hardening is comparable to that observed at 293°K.

The other parameter which limits tensile strengths of the two alloys is the ductility. At a temperature of 748°K, and a strain rate of 2.33 × 10^{-4}/s the fracture surface exhibits a few smooth intergranular facets. The intergranular facets correspond to wedge cracks. With decreasing strain rate and/or increasing temperature the fracture became totally intergranular and consisted of varying proportion of wedge cracks and cavitated facets. The proportion of the cavitated boundaries increased as the temperature increased and/or the strain rate decreased. Similar variations in the amounts of W- and R- type cracks have been observed before (Gandhi and Raj, 1981). A specimen tested in vacuum (10^{-5} torr) and at 748°K failed in a similar manner to that tested in air, and exhibited similar ductility. Oxygen penetration is not therefore considered to be responsible for the loss of ductility (Woodford and Bricknell, 1983). The loss of ductility in Ni-Cr alloys is related to cavitational damage at grain boundaries. Cavities could nucleate on coarse grain boundary ledges which are a consequence of coplanar slip and have been observed in thin foils.

Serrated flow during dynamic strain ageing of substitutional solid solutions have been investigated extensively, and has been reviewed by Baird (1971). The Type A' serrations reported in this paper is different to the usual Type A flow. The strengthening is thought to arise from the formation of arrays dislocation dipoles. When the applied stress becomes high enough then the two components of dipoles glide over each other and a gradual weakening occurs.

CONCLUSIONS

1. The proof stress of Ni_3Cr and Ni_2Cr is reduced by ~100 MPa by increasing the temperature from 293 to 973°K and is insensitive to strain rate.

2. The tensile strength falls with increasing temperature and/or decreasing strain rate. This is determined by the ductility and the work hardening coefficient.

3. The ductility also decreases with increasing temperature and/or decreasing strain rate. It is limited primarily by cavitation of grain boundaries.

4. The work hardening coefficient exhibits a minimum at 748 to 873°K depending upon the alloy, and the strain rate. This is due to ordering and consequently planar slip.

5. Serrated flow of Type A' arises by the formation and subsequent dissociation of arrays of dislocation dipoles. This is related to ordering.

ACKNOWLEDGEMENT

The work was carried out at the Central Electricity Research Laboratories and the Istituto per la Tecnologia dei Materiali Metallici non Tradizionali. The paper is published by permission of the Central Electricity Generating Board and the Consiglio Nazionale delle Recerche.

REFERENCES

Baird, J.D. (1971). The Inhomogeneity of Plastic Deformation. ASM, 191-222.
Cahn, R.W. (1965). Local Atomic Arrangements Studied by X-ray Diffraction. AIME Met. Sci. Conf., 36, 179-412.
Cohen, J.B. and M.E. Fine (1962). J. Phys. Rad., 23, 749-762.
Fisher, J.C. (1954). Acta Met., 2, 9-10.
Gandhi, C. and R. Raj (1981). Met. Trans. A. 12A, 515-520.
Marucco, A., E. Metcalfe and B. Nath (1981). Int. Conf. on Solid-Solid Phase Transformations, Pittsburgh, August, ASM.
Marucco, A. and B. Nath (1983). Int. Conf. on Dimensional Stability and Mechanical Behaviour of Irradiated Metals and Alloys, Brighton, April, British Nuclear Energy Society.
Metcalfe, E. and B. Nath (1979). Spring Residential Conf. on Phase Transformations, York, May, Inst. of Metallurgists, 11.50-11.53.
Stoloff, N.S. and R.G. Davies (1964). Acta Met. 12, 473-485.
Tawancy, K.M. (1980). Met. Trans., 11A, 1764-1765.
Woodford, D.A. and R.H. Bricknell (1983). Embrittlement of Engineering Alloys, To be Published by Academic Press Inc.

INFLUENCE OF GRAIN ORIENTATIONS ON THE MECHANICAL PROPERTIES AND FRACTURE CHARACTERISTICS IN A DIRECTIONALLY SOLIDIFIED SUPERALLOY

Minggao Yan, Guohuan Zhang and Hengfei Ni

Institute of Aeronautical Materials, Beijing, China

ABSTRACT

In this paper, the influence of grain orientations on tensile strength, stress rupture properties and low-cycle fatigue behavior at 700° and 950°C in a directionally solidified nickel base alloy, DK5, is described. The variation of these properties could be rationalized on the basis of the Schmid's Law and the operative slip systems during deformation. The mode and mechanism of fatigue fractures in a DS alloy are discussed.

KEYWORDS

Grain orientations; directionally solidifed alloy DK5; tensile strength; stress rupture; low-cycle fatigue; fatigue fractures.

INTRODUCTION

It is well known that the elevated temperature properties of nickel base alloys could be substantially improved by directional solidification techniques and the nature and physical causes of this improvement were described as the elimination of transverse grain boundaries so that orientations not favorable for stress transference can occur (Cole, 1972). However, during solidification, the orientation of columnar grains may gradually diverge from the growth direction, which, of course, would deteriorate the mechanical properties of a DS alloy.

In the recent years, a number of reviews and technical papers in regard to the orientation dependence of mechanical properties in monocrystals and columnar grains of nickel base alloys were published (e.g. Sink, 1979; Mackay, 1980; Person, 1980; Wright, 1980). However, the fractographic aspects and their relationship to the intermediate and elevated temperature properties in DS alloys have not been investigated extensively. The purpose of present investigation is to study the effect of grain orientations on tensile, stress rupture properties and low-cycle fatigue behavior in a DS alloy. From this work, it may yield some contributions to the basic understanding of this subject , which in turn, would be beneficial to the material characterization and processing control in the course of manufacturing DS turbine blades and other related components.

MATERIAL AND EXPERIMENTAL PROCEDURE

The material used in this investigation was a directionally solidified nickel base alloy DK5 (0.12C, 10.4Cr, 9.8Co, 4.7W, 3.9Mo, 5.4Al, 2.6Ti, 0.1Zr, 0.03B, 0.004Ce and Bal. Ni). The tensile, stress rupture and fatigue specimens were machined from two sizes of test slabs (150 × 70 × 15 and 150 × 35 × 15 mm) made by D-P process along various angles from the growth direction. All specimens were solution treated at 1210°C for 4 hrs., air-quenched and aged at 970°C for 32 hrs. prior to machining.

The tensile, stress rupture and low-cycle fatigue tests were performed at temperatures 700° and 900°C. The low-cycle fatigue tests were conducted in a servo-hydroulic fatigue machine under load control at 20 CPM (triangle wave) with a stress ratio, R = 0.1. The fatigue crack propagation studies were conducted using WOL type CT specimens and crack growth rates were continously monitored with D.C. pontential drop method.

The fracture surfaces of these specimens were examined by optical, TEM and SEM techniques. The orientations of these specimens were determined by Laue back-reflected X-ray method.

EXPERIMENTAL RESULTS

Fig. 1 shows the variation of tensile and stress rupture properties at 700°C as a function of grain orientations. At 700°C, tensile strength, σ_b, decreases with increasing α, reaching to a minimum at about 20°-30° and then increasing gradually afterward. No appreciable variation of tensile elongation, δ %, and reduction of area, ψ %, was observed. A drastic reduction of the rupture life, L, at α = 10°-15° was noted.

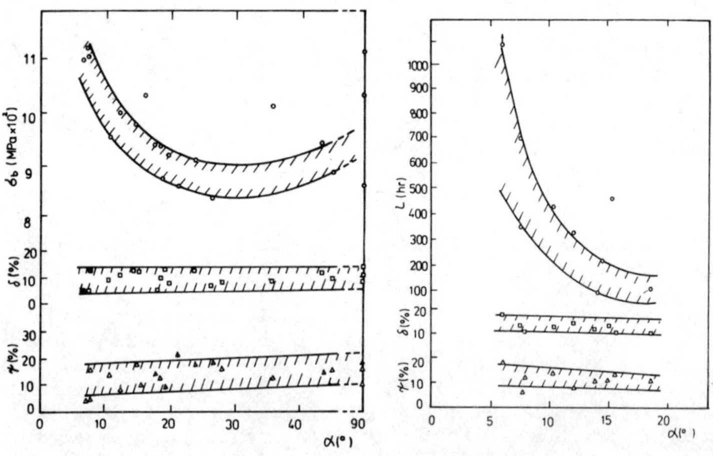

(a) (b)

Fig. 1. Effect of grain orientations on (a) tensile and (b) stress rupture properties at 700°C.

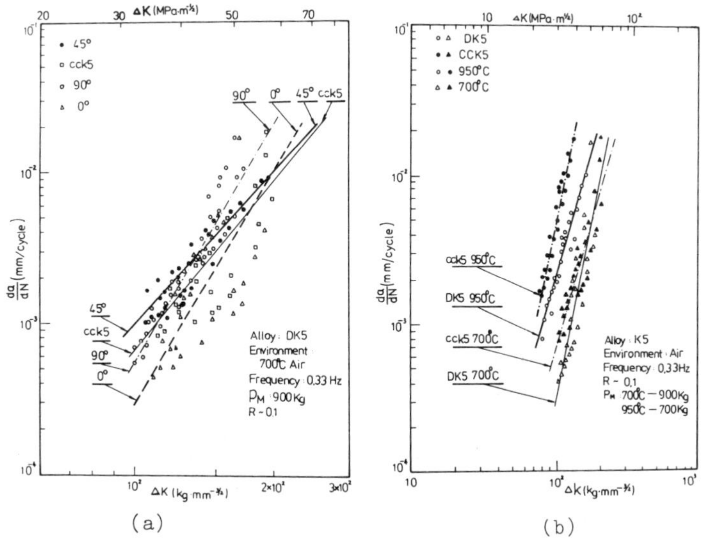

Fig. 2. Effect of grain orientations (a) and temperatures (b) on da/dN vs ΔK curves.

At 950°C, the change of σ_b and δ with α was not so obvious as that in testing at 700°C, however, a decrease of ductility with increasing α was evident. For instance, when α increases from 5° to 20°, the reduction of area, ψ%, reduces about 50%.

Fig. 2,a illustrates the dependence of da/dN on ΔK at various orientations 0°, 45° and 90° as well as conventionally cast (CCK5) specimens at 700°C. It can be seen that da/dN at lower ΔK region was considerably affected by the orientations. The crack growth rate of specimens at 0° was the lowest, then increases in a sequence of 90°, CCK5 and 45°.

The influence of temperatures on crack growth rate in DK5 specimens at various orientations is shown in Fig. 2,b. It is apparent that the temperature dependence of crack growth rate, da/dN, is rather obvious. At 950°C, the crack growth rate in DS specimens was significantly lower than that of conventionally cast ones.

The fractographic feature of tensile specimens at various orientations, as shown in Fig. 3, indicated that the amount of quasi-cleavage type of fracture increases with increasing α. At 700°C, the tensile specimens at small angle, $\alpha \simeq 7.5^{\circ}$, generally exhibits step-like cleavage fracture at the edge (Fig. 3,a) and with considerable amount of shear-dimple and dendritic fracture at the center (Fig. 3,b). Whereas for the specimen at an angle close to 90°, the interdendritic fracture is prodominate(Fig.3,d). The stress rupture specimens exhibited a similar trend in variation of of fractures with α as that of tensile specimens. For specimens with small angle α, more surface cracks and less amount of voids along interdendritic regions were observed.

The fracture surface appearance of low-cycle fatigue specimens was also considerably affected by the grain orientations. As shown in Fig. 4, the the fracture surface for 0° specimens exhibits a rather 'zig-zag' trans-

(a) (b) (c) (d)

Fig. 3. SEM micrographs of tensile specimens at 700°C. (a) $\alpha = 7.5°$(edge); (b) $\alpha = 7.5°$(center); (c) $\alpha = 90°$(edge); (d) $\alpha = 90°$(center).

(a) (b)

Fig. 4. Macro-photographs of low-cycle fatigue specimens. (a) at 0° and (b) at 90°.

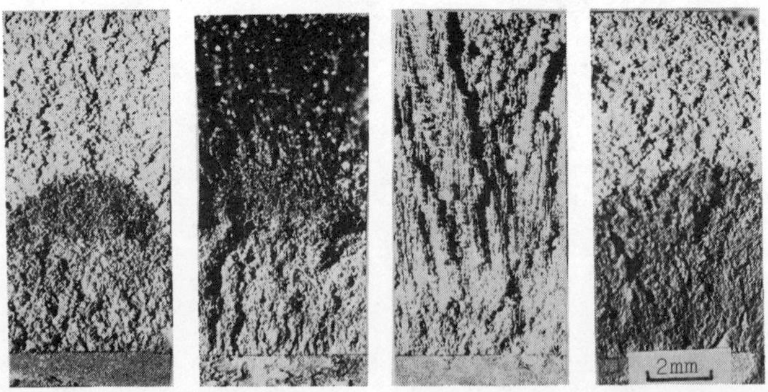

(a) (b) (c) (d)

Fig. 5. Low-cycle fatigue fracture photographs of DK5 specimens. (a) 700°C, 0°; (b) 700°C, 45°; (c) 700°C, 90°; (d) 950°C, 0°.

granular crack path (Fig. 4,a) as compared with that for 90° specimens (Fig. 4,b), where the latter, interdendritic rupture occurred predominately accompanying by a relatively flat fracture surface.

Some macro-fractographs of DK5 specimens of various orientations are illustrated in Fig. 5. It can be seen that the fracture surface of 0° orientation specimens exhibited a conventional appearance of fracture, in which a distinct contour of crack front prior to the final fracture was noted. For specimens of 90° orientation, however, a typical interdendritic rapture occurred (Fig. 5,c).

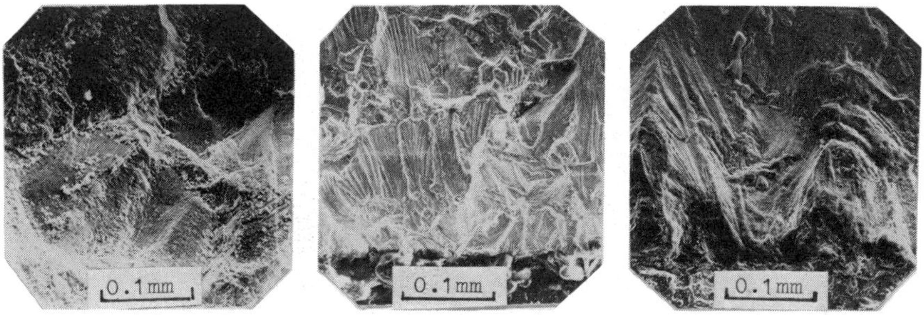

(a) (b) (c)

Fig. 6. SEM micorgraphs of low-cycle fatigue specimens in initiation stage. (a) 700°C, 0°; (b) 700°C, 45°; (c) 950°, 0°.

The fracture surfaces of DK5 specimens in the initiation stage exhibited a type of crystallographic features, as shown in Fig. 6, in which, some characteristic facets, river patterns and intersecting slip traces were revealed. The morphology of fracture surfaces of DK5 specimens in the propagation stage was examined by TEM replica techniques. Some continuous striations occurred in the specimens of 0° orientation, while in that of 45°, a quasi-cleavage fracture having discontinuous striations with some tearing ridges was observed. A typical interdendritic rupture appeared to be predominate in the specimens of 90° orientation.

DISCUSSION OF RESULTS

The orientation dependence of tensile strength of single crystals has been rationalized by the Schmid's Law and the critical resolved shear stress for the initiation of slip was expressed by (Mackay, 1980):

$$\tau = \sigma \cos\phi \cos\lambda \tag{1}$$

where σ is the normal yield strength, ϕ is the angle between normal to the slip plane and the tensile axis, λ is the angle between the slip direction and the tensile axis and $\cos\phi\cos\lambda$ is the Schmid factor. Since the relative strength, M, of a specimen at a given orientation may be described as:

$$M = \frac{\sigma}{\tau} = \frac{1}{\cos\phi\cos\lambda} \tag{2}$$

then the relative strength of a crystal is inverse proportional to the value of the Schmid factor.

Fracture Mechanism under Static Loading

The operative slip system for tensile loading in nickel alloys at intermediate temperature is {111}<101> slip system and the Schmid factor contours for this system are shown in Fig. 7,a (Mackay, 1980). Refer to Fig. 1, a good correlation of tensile strength, σ_b, and the Schmid factor contours which has a maximum at about 25°, was attained. The operative slip system for stress rupture testing during first-stage creep at intermediate temperature is {111}<112> rather {111}<101> (Leverant, 1973) and the Schmid factor contours for {111}<112> are presented in Fig. 7,b

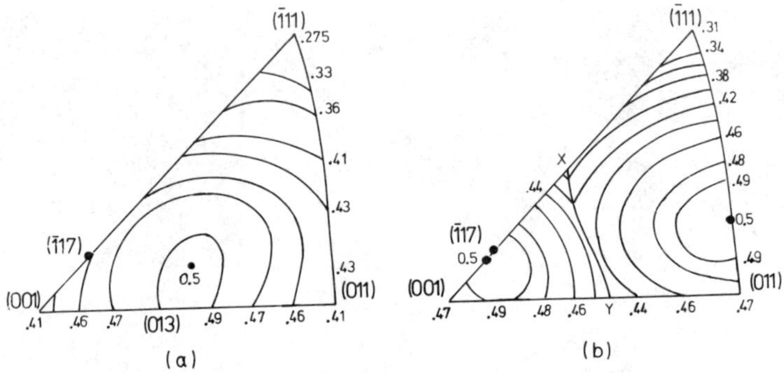

Fig. 7. Schmid factor contours for {111}<101>(a) and {111}<112> (b) slip systems.

(Mackay, 1980). The rapid decrease of stress rupture life at about 10° in DK5 alloy (Fig. 1,b) correlates well with the Schmid factor contours (Fig. 7,b) which has a maximum at about 10°.

At higher temperature, additional slip systems, such as {001}<110>, became operative (Kear, 1967), the tensile and stress rupture behaviors were relatively in sensitive to the grain orientation, however, owing to an increasing amount of interdendirtic rupture and grain boundary gliding with increasing temperature and angle α , a gradually decreasing in tensile strength and stress rupture life occurred. A drastic decrease in reduction of area with α at 950°C in DK5 alloy may be regarded as the existence of interdendritic and intergranular ruptures which promoted the amount of cleavage type of fractures during stress rupture tests.

Effect of Grain Orientations on Low-cycle Fatigue Lives

In general, the total fatigue life, Nf, may be considered as the sum of number of cycle in the initial stage, Nc, (up to a crack length 0.5 mm) and number of cycles in the propagation stage to final failure, Np, i.e.

$$Nf = Nc + Np \qquad (3)$$

Fig. 8 shows the influence of grain orientations on the total life, Nf, and the life in initial stage, Nc, in DK5 alloy at 700°C. The average value of Nf for conventionally cast specimens, CCK5, is indicated for comparison. The low-cycle fatigue lives, Nc and Nf, decrease more rapidly as the orientation deviated from 0°, reaching to a minimum at about 45° and then increasing again up to 90°. The dependence of Nf and Nc on grain orientations, is comparable to the results of tensile and stress rupture tests described in the previous section. The lower fatigue life in a DS alloy deviated from 0° orientation may be ascribed to the occurrence of interdendritic rupture along the primary dendritic columns formed during solidification.

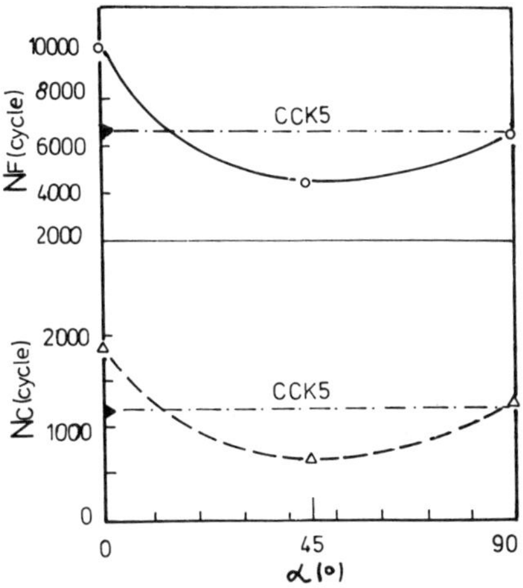

Fig. 8. Effect of orientations on Nf and Nc at 700°C in a DK5 alloy.

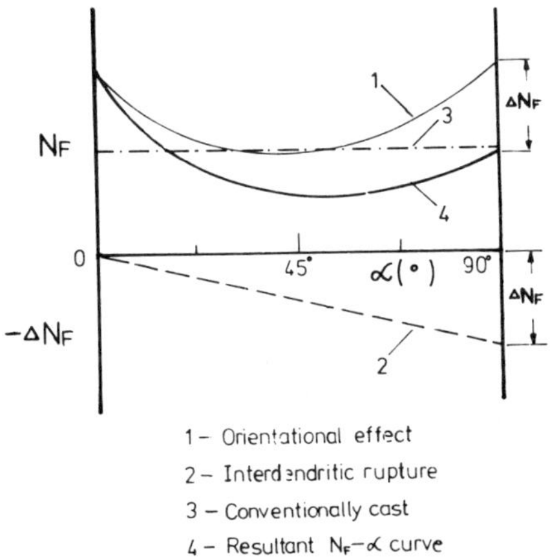

Fig. 9 Schematic of interaction of orientation effect and interdendritic rupture effect in DS alloy.

Fig. 9 illustrates a schematic diagram of the interaction of the statistical orientation effect (curve 1) and the interdendritic rupture effect (curve 2) on Nf-α curve in a DS alloy. If the variation of interdendritic rupture effect, ΔNf, is assumed approximately as a linear function with α , then a resultant curve close to the Nf-α curve (curve 3) is obtained. Hence, it may be concluded that as far as the stress rupture and low-cycle fatigue lives are concerned, no appreciable beneficial effect can be acquired in a DS alloy, when the grain orientation deviated of about 10°- 15° away from the solidification direction.

Modes of Fatigue Fractures in A DS Alloy

Based upon the above mentioned fractographic analysis, four modes of fractures in a DS alloy may be suggested:

a) Slip plane decoherence,
b) Plastic blunting process,
c) Interdendritic rupture, and
d) Intergranular fracture.

In the initial stage or under very low da/dN range, a crystallographic fracture by slip plane decoherence (Yan, 1981) usually occurs and displays amply evidence of faceting and river pattern. However, in the propagation stage, owing to the existence of dendritic structures formed during solidification, a great varieties of fracture surface appearance can be revealed as the grain orientations deviated from the growth direction. Consequently a higher crack growth rate, da/dN, and lower value of fracture toughness, Kc, in the specimens at 90° orientation would be expected. In the intermediate range, such as at 45° orientation, the lower relative strength level governed by the Schmid's Law accompanying by the existence of interdendritic rupture, yeilds a shorter fatigue life than the other orientations. Other affecting factors, such as crystal rotation (Mackay, 1980), crack-branching (Scarlin, 1976), γ' morphology (Person, 1980) and some other related variation of microstructures, may also have some additional influences on the mechanical behaviors in DS alloys.

SUMMARY AND CONCLUSIONS

As the orientation deviated from the solidification direction, the tensile strength and stress rupture life for a DS alloy, DK5, at 700°C reduced drastically and reached a minimum at about 20° - 30°. At 950°C, no appreciable effect of grain orientations on the tensile and stress rupture properties was revealed. The variation of these properties can be retionlized on the basis of the Schmid's Law and the operative slip systems during deformation. A profound effect of grain orientations on low-cycle fatigue lives was also noted, i.e. with a maximum at 0°, a minimum at about 45° and increasing to the level of the conventionally cast alloy at 90°. Based on some fractographic analysis, it was recognized that the mode of fatigue fractures for a DS alloy may be considered as one or a combination of the following mechanisms: the slip-plane decoherence, the plastic blunting process, the interdendritic rupture and the intergranular fracture.

ACKNOWLEDGEMENT

The authors wish to thank for San Chuanqi, Wang Luobao and Zhang Zuqian of the DS group who supplied the test slabs and to the assistance of our colleagues Xei Jizhou, Ouyang Jie, Wang Qingshei and Liu Caimu in fatigue testing, fractographic and X-ray differaction analysis and helpful discussions during the course of this investigation.

REFERENCES

Cole, C.S. and Cremisio, R.S. (1972). The Superalloys, Ed. by C.T. Sims, John Wiley, p.479.

Kear, B.H. and Piearcey, B.J. (1967). Trans. TMS-AIME, 239, 1209-1245.

Leverant, G.R. and Kear, B.H. (1970). Metall. Trans. 1, 491-498.

Leverant, G.R., Kear, B.H. and Oblik, J.M. (1973). Metall. Trans. 4, 355-362.

Mackay, R.A., Dreshfield, R.L. and Maier, R.D. (1980). Proceedings 4th Sym. on Superalloys, ASM, 2, 385-394.

Person, F.D., Lemfey, F.D. and Kear, B.H. (1980). Proceedings 4th Sym. on Superalloys, ASM, 3, 513-520.

Sink, L.W., Hoppin III, G.R. and Fujii, M. (1979). NASA Cr-159464.

Scarlin, R.B. (1979). Metall. Trans. 7A, 1535-1541.

Wright, P.K. and Anderson, A.F. (1980). Proceedings 4th Sym. on Superalloys, ASM, 3, 689-698.

Yan, M. and Wang, Z. (1981). Proceedings 1st China-USA Bilateral Metall. Conf. Beijing, 354-384.

INFLUENCE OF CYCLIC LOADING ON CRACK GROWTH OF A ½%Cr ½%Mo ¼%V STEEL

D. J. Smith and G. A. Webster

Department of Mechanical Engineering, Imperial College, London SW7, UK

ABSTRACT

Crack growth data under cyclic loading conditions have been presented for a ½CrMoV steel. Variables that have been examined include load amplitude, mean load, frequency and wave form. Two distinct regions of behaviour have been identified, one fatigue and the other creep dominated. In this latter region crack growth rate can be characterised by the parameter C^*. Little evidence of creep-fatigue interaction was observed.

KEYWORDS

Creep crack growth, cyclic loading, fracture mechanics.

INTRODUCTION

Much attention has been devoted recently to methods of assessing the integrity at elevated temperatures of components which may contain crack-like defects. Depending upon the type of loading experienced by the component any cracks present may extend by a creep or fatigue process, or some combination of the two. It is the purpose of this investigation to determine the parameters controlling crack growth under these conditions for a ½%Cr ½%Mo ¼%V steel of interest to the electrical power generation industries.

A considerable amount of work (Nikbin, Webster & Turner, 1976, 1977; Christian and Co-workers, 1981; Nikbin, Smith & Webster, 1983) has already been carried out at Imperial College on the growth of cracks at elevated temperatures under steady loading conditions for a range of materials, including an aluminium alloy and several low alloy steels, of varying creep ductilities. The results of over 50 such tests, on a selection of test-piece geometries, have shown that creep crack growth rate $\dot{a}$ can be characterised most satisfactorily in terms of a path independent line integral called C^* (which is a creep equivalent of the J contour integral used in plasticity) by an expression of the form

$$\dot{a} \; \alpha \; C^{*\,\phi} \qquad (1)$$

where ϕ is a number slightly less than one and the proportionality factor is chiefly governed by the material creep ductility with an increase in ductility causing a decrease in cracking rate (Nikbin & Webster, 1982). Some history of loading effects have also been examined in tests in which periodic load changes have been made (Webster & Nikbin, 1981). Although transient effects were noted after large load changes, it was again found that C^* could be used to characterise the crack growth behaviour adequately. The main aim of this investigation is to determine the most appropriate parameter for describing crack propagation when creep and fatigue effects may both be expected to be present.

EXPERIMENTS

The ½ CMV steel was heat-treated to represent stress relieved heat-affected zone weld metal. In this condition it had a mixed martensitic and bainitic structure with a creep ductility in the range 0.1 to 2% when tested at a temperature of 565°C.

All the crack growth experiments were carried out on double cantilever beam (DCB) or compact tension (CT) test-pieces at 565°C. Each specimen was provided with side grooves to control the direction of crack growth and promote flat straight fronted cracks to facilitate analysis of the data. Crack extension was measured by a combination of optical and electrical potential methods (Smith, 1983). A white heat-resistant paint was used to reduce surface oxidation and enhance contrast. It is estimated that with these procedures crack length could be measured to within an accuracy of ± 0.1mm. In most instances stable crack propagation over distances of approximately 100mm and 10mm were obtained with, respectively, the DCB and CT test-pieces.

The experimental variables which were investigated included load amplitude, mean load, frequency and waveform. In most instances a relatively small amplitude to mean load ratio was adopted. Two wave shapes, square and triangular, were used with frequencies in the range 10^{-4} to 20 Hz. The high frequency tests were carried out in a servo-hydraulic testing machine and the slow tests in dead load creep machines suitably modified to allow slow load cycling. For all the tests a continuous record of load point displacement Δ was obtained with the aid of an extensometer and electrical transducers.

RESULTS AND DISCUSSION

An example of some typical crack growth data obtained at the same mean load is shown in Fig.1. The influence of waveform and frequency at a minimum to maximum load ratio R of 0.8 can be compared with the behaviour under static loading. It is clear that an increase in frequency causes a progressive increase in cracking rate, although the effect is not in proportion to the change in frequency. For example, examination of the results at 10 Hz and 0.0001 Hz. indicate that cracking rate only increases by a factor of about two when the frequency is increased by five decades. Since number of cycles N can be related to frequency ω by

$$N = \omega t$$

where t is time, it is apparent that the number of cycles required to achieve a given crack extension increases substantially with increase in frequency. It was also observed that the square wave shape gave a slightly faster cracking rate at the same frequency than the triangular waveform consistent with proportionately more time having been spent at

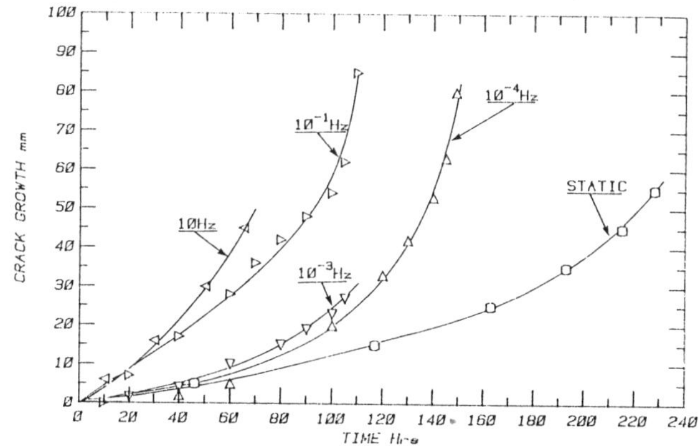

Fig. 1 Influence of frequency on crack growth in DCB test pieces at a constant mean load

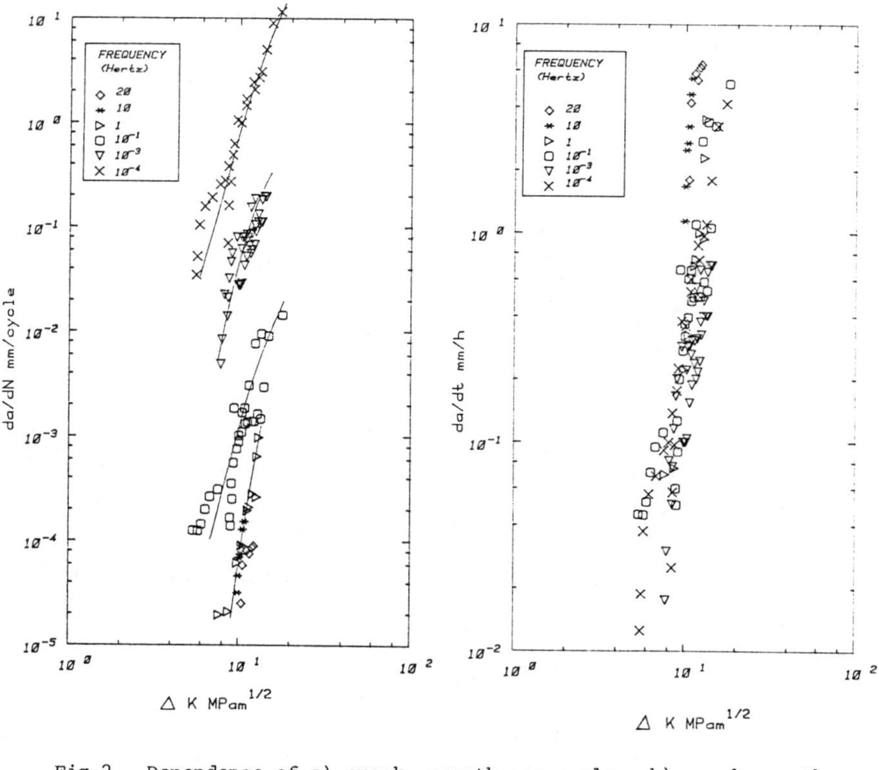

Fig.2 Dependence of a) crack growth per cycle; b) crack growth rate on stress intensity factor for DCB specimens at R = 0.8 and various mean loads.

maximum load in the former case.

The results of all the cyclic tests carried out on the DCB test-pieces at a constant R ratio of 0.8 and various mean loads, are presented in Figs. 2a and 2b. Similar features were observed for the CT specimens. Fig.2a) shows crack growth/cycle (da/dN) plotted as a function of stress intensity factor range ΔK. Approximate frequency independence is obtained between 1 and 20 Hz but as frequency decreases below this range, the crack growth per cycle increases appreciably at a constant ΔK. Fig.2b) shows the same data converted to crack growth rate $\dot{a}$ using the expression

$$\dot{a} = \omega \frac{da}{dN} \qquad (2)$$

In this case little frequency dependence is observed below 1 Hz and accelerated cracking is only apparent above about 10 Hz.

The trends of Figs. 2a) and 2b) suggest that, for these testing conditions, crack growth is controlled by cycle dependent (fatigue) processes above a frequency of about 1 Hz. and time dependent (creep) processes below this frequency. The change from time dependent to cycle dependent crack extension can be seen more clearly in Fig.3 which shows data taken from Fig.2a) at a constant ΔK of 10 MPa $m^{1/2}$. A sharp transition is apparent at 1 Hz. Above this frequency da/dN is constant, and crack growth by fatigue would be expected. Below it a slope of minus one is obtained, indicating, from Equation (2), that $\dot{a}$ is constant irrespective of frequency and crack growth is controlled by creep mechanisms.

Detailed microstructural confirmation of the mechanisms controlling failure was not possible because of oxidation of the fracture surfaces. However, predominantly intergranular fracture could be detected at the lower frequencies with some evidence of transgranular failure at the highest frequency. The sharp transition on Fig.3 indicates little creep-fatigue interaction in this material. Failure would appear to be by either fatigue or creep with little evidence of acceleration due to cumulative damage.

Fig.4 illustrates the effect of variations in the load ratio R on da/dN for loading frequencies in the regions where crack growth is creep and fatigue dominated. Similar features are apparent in both regions. An increase in crack growth per cycle with R at a given ΔK is obtained consistent with many observations (Vosikorsky, 1979; Sadananda & Shahinian, 1981). Insufficient data were obtained to establish whether the transition from creep to fatigue crack growth is sensitive to R. However, low R ratios would tend to favour fatigue dominated crack propagation and extend this process to lower frequencies. Similarly, an increase in R would be expected to give preference to creep crack growth.

At the frequencies where crack growth was controlled by time dependent processes, an attempt was made to correlate the data with the creep parameter C^*.

Estimates of C^* were obtained from the experimental data using a method proposed by Smith & Webster (1982). With this approach C^* is given by

$$C^* = \frac{P\dot{\Delta}}{B_n W} F \qquad (3)$$

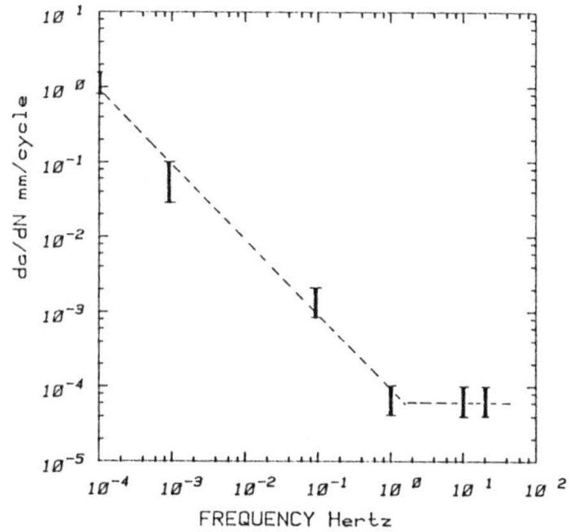

Fig. 3 Relationship between crack growth/cycle and frequency at $\Delta K = 10 MPa\ m^{1/2}$ and R = 0.8

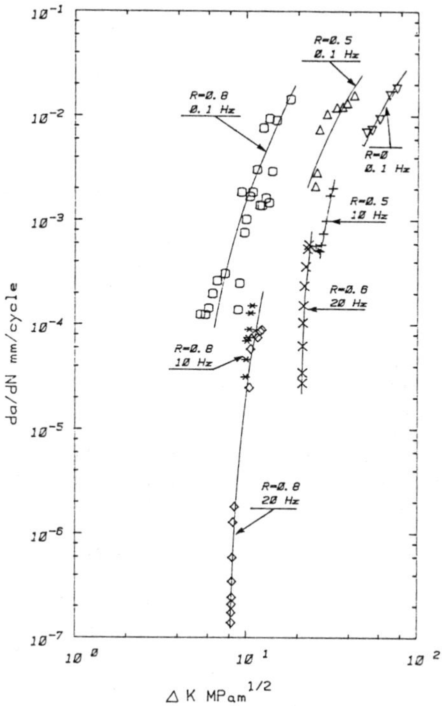

Fig. 4 Influence of load ratio R on crack growth rate as a function of ΔK for DCB specimens.

where P is load, $\dot{\Delta}$ load point displacement rate, B_n specimen thickness across the side-grooves, W is width and F is a non-dimensional factor which depends upon test-piece geometry, a/W and the creep stress index n. Values of F appropriate to this investigation have been tabulated by Smith & Webster (1981, 1982) and Smith (1983). In Equation (3) P was taken to be the maximum load of the cycle as it was anticipated that most creep damage would occur close to maximum load.

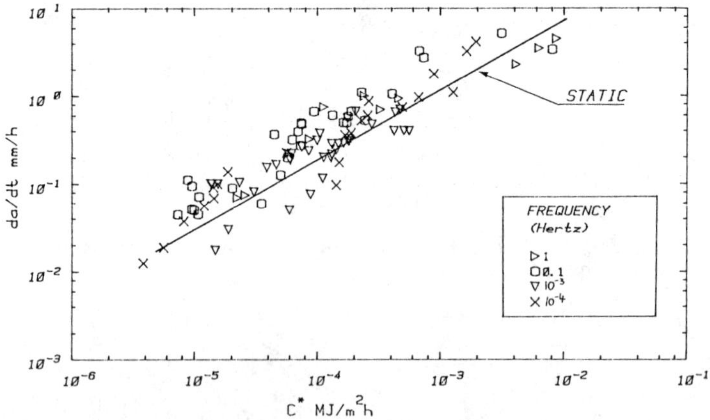

Fig. 5 Correlation of cyclic crack growth rate at R = 0.8, with static load data for DCB specimens.

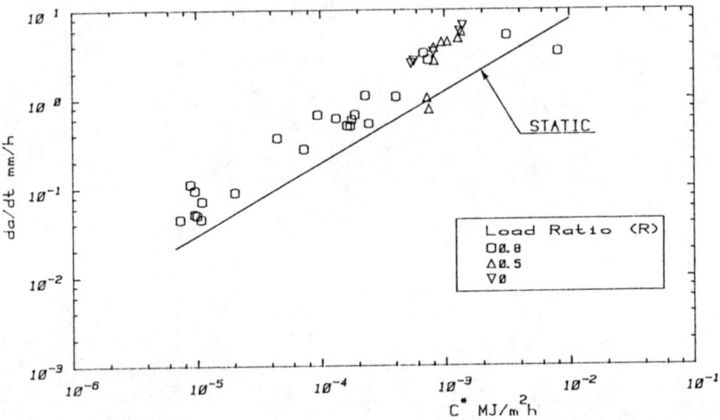

Fig.6. Correlation of cyclic crack growth at different R ratios with static load data for DCB specimens.

Illustrations of the correlations in crack growth rate obtained are shown in Figs. 5 and 6. No trend with frequency or R ratio is observed, but it is apparent that the data can be characterised satisfactorily in terms of C^*. Also included in these figures is a line representing the creep crack growth rates measured previously (Nikbin, Webster & Turner, 1977) under static loading conditions. It is evident that both the static and cyclic data show about the same slope ϕ but that slightly faster cracking rates are observed under cyclic loading than under static loading conditons for the variables covered. The observed independence on R ratio (Fig.6) is consistent with most damage occurring at the maximum load in the cycle.

Transient effects caused by load changes have been discussed previously by Webster & Nikbin (1981) in terms of the stress field local to a crack tip. On first loading an elastic stress distribution (in the absence of plasticity) will be developed. If the load is maintained, constant redistribution will take place until a steady state creep stress distribution is obtained. With cycling elastic regeneration of the stresses may prevent this happening. The amount of relaxation possible during cycling will decrease with increase in frequency as less time will be available in each cycle for creep to take place. Consequently, some acceleration in cracking rate with increase in frequency may be anticipated as was observed in practice.

CONCLUSIONS

Separate fatigue and creep dominated regions of crack growth under cyclic loading conditions have been identified for the ½CMV steel with little creep-fatigue interaction. Where crack growth was time dependent, it has been shown that the data can be correlated by C^* with the growth rates being close to those predicted from static data.

REFERENCES

Christian, E.M., Smith, D.J., Webster, G.A. & Ellison, E.G. (1980), "Critical Examination of Parameters for predicting Creep Crack Growth", in Advances in Fracture Research, Ed. Francois, D, Pergamon, Oxford,p.1283.

Nikbin, K.M., Webster, G.A. & Turner, C.E. (1976). "Relevance of non-linear Fracture Mechanics to Creep Cracking," Cracks and Fracture, ASTM STP 601, p.47.

Nikbin, K.M., Webster, G.A. & Turner, C.E. (1977), "A Comparison of Methods of correlating Crack Growth," in Fracture, 2, ICF4, Waterloo, Canada, June.

Nikbin, K.M., & Webster, G.A. (1982). "Temperature Dependence of Creep Crack Growth in Aluminium Alloy RR58", in Micro and Macro Mechanics of Crack Growth, Ed. Sadananda, K, Roth, B.B., & Michel, D.J., Met.Soc. AIME, p.137.

Nikbin, K.M., Smith, D.J. & Webster, G.A. (1983), "Influence of Creep Ductility and State of Stress on Creep Crack Growth", Int. Conf.Advances in Life Predicted Methods at Elevated Temperatures, Albany, April 1983, ASME, to be published.

Sadananda, K, & Shahinian, P. (1981) "Creep-fatigue Crack Growth," in Cavities and Cracks in Creep and Fatigue, Ed. J. Gittus, Applied Science, London.

Smith, D.J. (1983). "Creep Crack Growth in Low Alloy Steels under Steady and Cyclic Loading," PhD Thesis, University of London.

Smith, D.J., & Webster, G.A. (1981), 'Characterisations of Creep Crack growth in 1%CrMoV Steel,' J.Strain Anal., 16, p.137.

Smith, D.J. & Webster, G.A. (1982). 'Estimates of the C^* parameter for crack growth in creeping materials', Presented at Int.Conf. Elastic-Plastic Fracture Mechanics, Philadelphia, 1981, to be published ASTM STP.

Vosikovsky, O. (1979). 'The effects of stress ratio on fatigue crack growth rates in steel'. Eng.Fract.Mech, 11, 595-602.

Webster, G.A. & Nikbin, K.M. (1981). 'History of loading effects on creep crack growth in ½%CrMoV steel', in Creep in Structures, Ed.Ponter, A.R.S. & Hayhurst, D.R., Springer-Verlag, Berlin.

DAMAGE MECHANICS APPLIED TO HIGH TEMPERATURE FATIGUE

A. Plumtree* and J. Lemaitre**

*Department of Mechanical Engineering, University of Waterloo, Waterloo, Ontario N2L 3G1, Canada
**Université P. et M. Curie, Paris VI, Enset, Cachan, France

ABSTRACT

Cyclic tests on Alloy 800 at 600°C have been employed to demonstrate that fatigue damage may be related to the growth of cracks. Damage accumulation data are reported for small cylindrical smooth specimens at various strain levels. The damage accumulation rate was found to be a function of the fatigue crack growth rate.

KEYWORDS

Damage accumulation, high temperature, fatigue, fatigue crack growth rate, Alloy 800, crack propagation.

INTRODUCTION

The application of damage mechanics has proved to be most useful in life prediction at high temperatures (Lemaitre and Chaboche, 1974; Lemaitre and Plumtree, 1979). Using damage mechanics, it is the purpose of this study to establish a relationship between cyclic damage at high temperature and crack size, and to show that the damage accumulation rate is proportional to the fatigue crack growth rate.

Damage equations may be written in terms of stress for one-dimensional rupture problems. For a nominal stress of σ, the damage parameter, D, is zero for the material containing no cracks and unity when rupture takes place, and the term $\sigma/(1-D)$ is an "effective" stress ($\tilde{\sigma}$) which takes into account the weakness of the material due to the presence of voids or micro-cracks. From a thermodynamic point of view, D is an internal variable of an irreversible micro-rupture process. It appears linearly in the elastic thermodynamic potential (free energy), ψ, which is a quadratic function of elastic strain (ε) for a given temperature and may be expressed as

$$\psi = 1/2\ E\varepsilon^2\ [1-D] \tag{1}$$

where E is the elastic modulus.

For the one-dimensional case, the effective stress is written

$$\tilde{\sigma} = \sigma/(1-D) = E\varepsilon \tag{2}$$

Kachanov (1958) developed an equation for the evolution of damage in brittle materials based on thermodynamic principles. This equation has been modified for ductile materials (Rabotnov, 1969) and may be written as:

$$\frac{dD_c}{dt} = \left\{\frac{|\sigma|}{A_{\pm}[1-D_c]}\right\}^r [1-D_c]^{-q} \tag{3}$$

where t is the time and A, r, q, are temperature dependent material characteristics (A_+ for tension, A_- for compression, if damage is anisotropic). Using the assumption that damage is negligible at the commencement of testing and that damage is unity when $t = t_R$. Integration of equation (3) for the evolution of creep damage at constant stress yields

$$D_c = 1 - \{1 - t/t_R\}^{1/(r+q+1)} \tag{4}$$

The derivations of this equation and for the evolution of cyclic damage for constant stress fatigue are given by Lemaitre and Plumtree (1979).

In many practical applications, tests are carried out under strain control. Damage equations for strain control may be deduced from equations with stress control using constitutive stress-strain equations (Lemaitre and Plumtree, 1979). For pure fatigue which occurs at high frequencies resulting in time (and frequency) independent plasticity the evolution of damage (D) under completely reversed conditions may be expressed as

$$D_F = 1 - \{1 - N/N_F\}^{1/(p+1)} \tag{5}$$

where N is the number of cycles, N_F is the number of cycles to failure at the given strain range and p is an exponent which is material and temperature dependent.

In order to identify the damage, the basic function N_F may be obtained by carrying out high frequency tests (ν = 5 Hz) on round fatigue specimens subjected to a constant strain range (values of 1.54% to 2.64% were used in this present work). By recording the change in stress amplitude during each test the amount of damage after a given number of cycles N in a specimen exhibiting cyclic stability may be determined according to $D = 1 - \Delta\sigma/\Delta\sigma^*$ where $\Delta\sigma$ is the stress amplitude for the given number of cycles and $\Delta\sigma^*$ is the stress amplitude on saturation. Using equation (5) $D_F=\{1-N/N_F\}^{1/(p+1)}$, $1/(p+1)$ is the slope of a log-log plot of$(1-D_F)$ $[=(\Delta\sigma/\Delta\sigma^*)]$ versus $(1 - N/N_F)$.

Since damage is concerned with the formation of a new free surface (either internal or external) another method of assessing damage is to measure the area of this new surface (A) after a given number of cycles. The relation between the damage variable (D) and fractured area (A) in a specimen may be deduced through the strain energy release rate (G).

After a given number of cycles a load, F, acting over an undamaged area S(1-D) where S is the original area, will result in an elastic extension of u over the gauge length L_o. Equation (2) may be rewritten:

$$\frac{F}{S(1-D)} = E\,\frac{u}{L_o} \tag{6}$$

Differentiation of equation (6) gives

$$\frac{F}{S}(1-D)^{-2}\ \frac{dD}{dA} = \frac{E}{L_o}\ \frac{du}{dA}$$

or

$$\frac{du}{dA} = \frac{FL_o}{SE(1-D)^2}\ \frac{dD}{dA} \tag{7}$$

The strain energy release rate defined by a variation in compliance may be expressed

$$G = \frac{1}{2}\ F\ \frac{du}{dA} \tag{8}$$

Substituting equation (7)

$$G = \frac{F^2L_o}{2SE(1-D)^2}\ \frac{dD}{dA} \tag{9}$$

The strain energy release rate may also be calculated by applying an expression for the stress intensity factor K. Solutions are known for circular cracks of radius R and a typical formula (e.g., Harris) is given by

$$K = \frac{F}{S\ \frac{R^2}{R_o^2}}\ \frac{\pi R}{4}\left(\frac{1-R/R_o}{1-0.8\ R/R_o}\right)^{1/2} \tag{10}$$

where R_o is the original radius of the specimen ($R_o = \sqrt{S/\pi}$). The factor in parenthesis in equation (10) may be neglected which allows K to be expressed as a function of S and A

$$\frac{R^2}{R_o^2} = \frac{S-A}{S} \qquad\qquad R = \sqrt{\frac{S-A}{\pi}} \tag{11}$$

Hence

$$K = \frac{F}{S-A}\ \frac{\pi}{4}\ \sqrt{(S-A)/\pi} \tag{12}$$

Since $G = K^2/E$, substituting eqn. (12) gives:

$$G = \frac{F^2}{E}\ \frac{\sqrt{\pi}}{4}\ (S-A)^{-3/2} \tag{13}$$

Equations (9) and (13) may be equated with the result:

$$\frac{2L_o}{S\sqrt{\pi}}\ (1-D)^2\ dD = (S-A)^{-3/2}dA \tag{14}$$

Integrating equation (14):

$$\frac{2L_o}{S\sqrt{\pi}} \int_0^D (1-D)^{-2}dD = \int_0^A (S-A)^{-3/2}dA \tag{15}$$

which gives

$$\frac{2L_o}{S\sqrt{\pi}} (1-D)^{-1} \Bigg|_0^D = 2(S-A)^{-1/2} \Bigg|_0^A \qquad (16)$$

hence

$$\frac{L_o}{S\sqrt{\pi}} \left[\frac{1}{(1-D)} - 1\right] = \frac{1}{(S-A)} - \frac{1}{\sqrt{S}} \qquad (17)$$

Solving for D:

$$D = 1 - \left[\frac{1}{1 + \frac{\sqrt{\pi S}}{L_o}\left(\frac{1}{\sqrt{1-A/S}} - 1\right)}\right] \qquad (18)$$

Using equation (18) it is then possible to assess the amount of damage in a specimen by measuring the area fractured (A) knowing the original undamaged cross-sectional area (S) and gauge length (L_o).

APPLICATION TO ALLOY 800 AT 600°C

The material used in this investigation was an austenitic iron nickel chromium alloy Sanicro 31 (trade name of Sandvik AB). This is a high temperature alloy, similar to Alloy 800, which is used where corrosion resistance and high strength are required at elevated temperatures. The chemical composition of the alloy used was 0.043%C, 21.7%Cr, 33.2%Ni, 0.33%Ti and 0.20%Al.

Specimens were taken from 16 mm bar stock and solution heat treated at 1150°C for 10 minutes in an argon atmosphere followed by a water quench, resulting in a grain size of 50 $\pm$5 μm. The specimens were then machined with threaded ends for gripping in the fatigue machine, and a reduced diameter of 4.85 mm over a

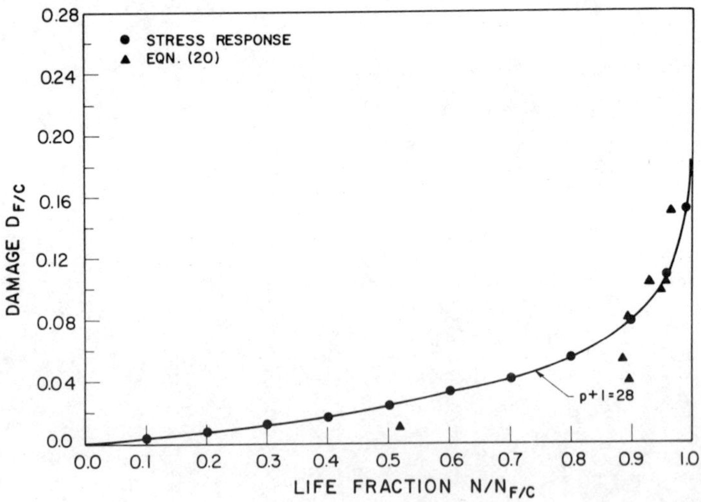

Fig. 1 Damage evolution during high frequency cycling of Alloy 800 at 600°C

gauge length of 7.5 mm.

Cyclic testing was performed on a closed loop electro-hydraulic servo-controlled testing machine. The tests were carried out under strain control. The testing procedure for controlling strain was similar to that of Abdel-Raouf, Plumtree and Topper (1973). Strain range ($\Delta\varepsilon$) and frequency (ν) were the testing variables with temperature being held constant at 600°C.

In order to show the time and frequency independency of the fatigue damage, two tests (T1 and T10) were conducted at a total strain range of 2.4%. Specimen T1 was cycled at 10 Hz and specimen T10 at 5 Hz. Both tests resulted in approximately the same lives (700 and 739 cycles respectively). For a total strain range of 2.64% cycled at a frequency of 5 Hz ($\dot{\varepsilon}$ = 2.64 x 10^{-1} sec^{-1}), the average cyclic life (N_F) was 550 cycles. Stress response was recorded during these cyclic tests which allowed the damage evolution to be assessed for any given number of cycles according to equation (5). A plot of log ($\Delta\sigma/\Delta\sigma^*$) vs. log ($1 - N/N_F$) was constructed and from the slope, the value of p(=27) was determined. The evolution of damage assessed by this method is shown in Fig. 1.

A second series of high frequency tests (5 Hz) was carried out to determine the crack size at various life fractions for two different strain ranges, $\Delta\varepsilon$ = 2.2% and 1.54% corresponding to average cyclic lives of 2700 and 900 cycles respectively ($\dot{\varepsilon} = 2\Delta\varepsilon\nu = 2.64 \times 10^{-1} sec^{-1}$). The procedure was to stop the test at a predetermined life fraction and examine the specimen for surface cracks and evaluate their size. The technique for crack detection and observation has been described by Douglas and Plumtree (1979). It was hoped that these tests would indicate if any possible relationship existed between the damage evolution derived from the stress response measurements and that derived from the crack areas according to equation (18). If the dimensions of the present specimens are considered, it will be apparent that the term $\sqrt{\pi S}/L_o$ is approximately equal to unity in which case equation (18) simplifies to

$$D \simeq 1 - \sqrt{1-A/S} \qquad (19)$$

and if A/S is relatively small compared to one, then

$$D \simeq A/2S \qquad (20)$$

This simple emprical relationship was verified for the specimens tested. By determining the cracked area and dividing by twice the original cross-sectional area, it was possible to establish a relationship between this parameter and the damage. The data points are included in Fig. 1 and fall very close to the damage evolution curve derived from the stress response measurements.

In order to clarify further the definition of damage it was necessary to determine whether a relationship existed between the damage accumulation rate (dD/dN) and the fatigue (high frequency) crack growth rate (da/dN). Crack growth data was obtained from the crack size data at various cycles by the difference method. This data was then correlated with the J integral. The log-log plot of da/dN and dD/dN versus ΔJ is shown in Fig. 2 and indicates that a Paris-type relationship is obeyed such that, $da/dN = C(\Delta J)^n$ where C and n are material constants. For the present study the values of C and n are given in Fig. 2. It follows then that D = f(a) and using the difference method to obtain the slope of the experimental damage curve corresponding to a given number of cycles (Fig. 1), the damage accumulation rate is directly proportional to the fatigue crack growth rate, which is apparent in Fig. 2.

Combining the equations for dD/dN and da/dN gives:

$$dD/dN = M\ da/dN\ (m/cycle) \quad (21)$$

where M is a constant (= 2.7 x 10^{-5}).

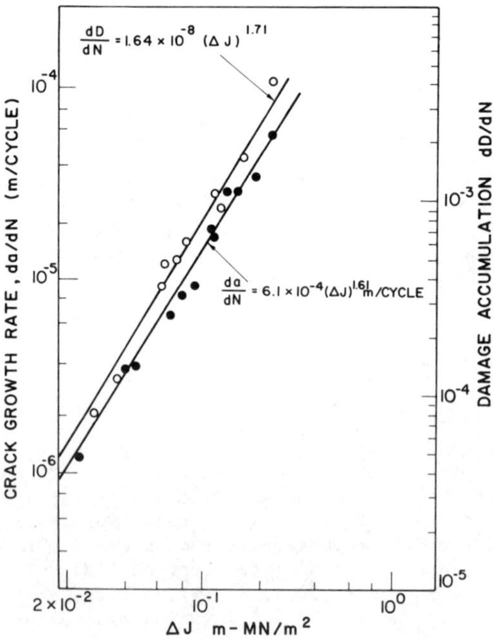

Fig. 2. Relationship between crack growth rate (da/dN), damage accumulation (dD/dN) and ΔJ for Alloy 800 at 600°C.

CONCLUSIONS

1. On the basis that damage is represented by the presence of new surfaces such as cracks, it has been shown that this damage may be related directly to the cracked area in the Sanicro 31 specimens subjected to cyclic straining at high temperature.

2. Measurements of the size of these cracks and the incorporation of an elastic-plastic fracture mechanics analysis have been used to demonstrate that the fatigue crack growth rate may be directly related to the high frequency damage accumulation rate.

ACKNOWLEDGMENTS

The authors would like to express their sincere thanks to M.J. Douglas, formerly Department of Mechanical Engineering, University of Waterloo, now Research and Development, Union Carbide of Canada Ltd., Mississauga, Ontario, for his assistance with the cyclic tests. The authors are most grateful to Sandvik, A.B. Sweden, for providing the Sanicro 31. This work was financed through a grant (A-2770) from the Natural Sciences and Engineering Research Council of Canada. The authors would like to thank Mrs. Beryl Hultin for typing this manuscript.

REFERENCES

Abdel-Raouf, H.A., A. Plumtree, and T.H. Topper (1973). Effects of temperature and deformation rate on cyclic strength and fracture of low-carbon steel. In Cyclic Stress-Strain Behaviour Analysis, Experimentation, and Failure Prediction, STP 519, ASTM, Philadelphia, 28-57.

Douglas, M.J. and A. Plumtree (1979). Application of fracture mechanics to damage accumulation in high temperature fatigue. In Fracture Mechanics, Ed. Smith, C.W., STP 677, ASTM, Philadelphia, 68-84.

Kachanov, L.M. (1958). On the time to failure under creep conditions. Izvest. Akad. Nauk. SSSR, OTN, 8, pp. 26-31.

Lemaitre, J., and J.L. Chaboche (1974). A non-linear model of creep-fatigue damage accumulation and interaction. Proc. IUTAM Symposium Mech. of Viscoelastic Media and Bodies, Gothenburg. Office National D'Etudes et de Recherches Aerospatiales, Chatillon TT No. 1394.

Lemaitre, J. and A. Plumtree (1979). Application of damage concepts to predict creep-fatigue failures. Trans. ASME, J. Engineering Mat. and Technology, 101, 284-292.

Rabotnov, Yu, N. (1969). Creep Problems in Structural Members. North-Holland Publ. Co., London.

GENERALIZATION OF MINDLIN'S KINEMATICAL PLATE THEORY FOR ANISOTROPIC MATERIALS

E. E. El-Soaly

Department of Mechanics and Elasticity, Military Technical College, Kobri El-Kobba, Cairo, Egypt

ABSTRACT

The classical plate theory, investiged by Mindlin (1951), is based on a kinematical approach which takes into account the effect of transverse shear deformations. The basic parameters in this theory depend on the plate geometry and hence it is of great importance for numerical solutions, such as finite element solutions. This theory treats the special case of isotropic plates and neglects the effect of membrane forces. Recently, as a result of technological advances, new materials, such as composite materials, are of wide applications due to their high specific stiffness and strength, damage tolerance and fatigue response characteristics, corrosion resistance and ease of fibrication. In genral, these materials have some kind of anisotropy. In the present work, a generalization of Mindlin's theory to deal with homogeneous anisotropic plates is given. Anisotropic materials, up to a monoclinic one which has one plane of symmetry and 13 independent elastic constants, can be utilised. A numerical example of orthotorpic plate under cylindrical bending, plane strain problem, is considered.

KEYWORDS

Anisotropy, bending, constitutive relations, displacements, monoclinic materials, potential energy, stresses, tensors, variational principle.

INTRODUCTION

Mindlin (1951) deduced a two dimensional theory of flexural motions of isotropic elastic plates, from the three dimensional equations of elasticity. In three-dimensional elasticity theory, there are six components of stress which are expressed in terms of six components of strain through Hooke's law. In Mindlin's theory, only five plate-stress components are considered which can be expressed in terms of the same number of the plate-strain components. The strain, and hence the stress, in the direction normal to the faces of the plate are ignored ($\varepsilon_{33}=\sigma_{33}=0$). The longitudinal displacements, u_1 and u_2 are assumed to be linearly distributed along the plate thickness, while the transversal displacement, u_3, is considered to be uniform. These displacement components lead to constant values of transverse shear stresses, σ_{13} and σ_{23}, along the plate thickness. The obtained shear stresses are violated with the boundary conditions on the upper and lower plate faces, which are free from tangential stresses. A corrective factor was introduced, by Minlin, to compensate this violation.

The main purpose of the present work is to generalize Mindlin's theory to deal with anisotropic materials. A variational approach depending on the potential energy functional is used to obtain the equilibrium equations, in terms of displacements. Using a modified constitutive relations, the stresses and the stress resultants can be determined in terms of the assumed displacement components.

At various stages of the development, a monoclinic material, which has 13 elastic moduli, is considered. Materials of anisotropy less than the monoclinic one, can be treated as special cases.

NOMENCLATURE

$i,j,k,l = 1,2,3$	Tensorial indices
$\alpha,\beta,\gamma,\delta = 1,2$	Tensorial indices
h,s,v	plate thickness, total surface and volume.
x_i	cartesian coordinates
$,i$	derivative with respect to x_i
u_i	components of displacement vector
σ_{ij}	stress tensor
ε_{ij}	strain tensor
C_{ijkl}	generalized elastic constants
$Q_{\alpha\beta\gamma\delta}$	plane stress elastic constants
$A_{\alpha\beta\gamma\delta}$	membrane rigidities
$D_{\alpha\beta\gamma\delta}$	flexural rigidities
$d_{\alpha\beta}$	shear elastic constants
$K_{\alpha\beta}$	corrective factors

GOVERNING EQUATION

For a general elastic anisotropic material, the potential energy function $F(u_i)$ is defined by :

$$F(u_i) = \frac{1}{2} C_{ijkl} u_{i,j} u_{k,l} \tag{1}$$

where C_{ijkl} are the elastic moduli defined by the general constitutive relations, of linear elasticity, between the stress tensor σ_{ij} and the strain tensor ε_{ij} in the form;

$$\sigma_{ij} = C_{ijkl} \varepsilon_{kl} \tag{2}$$

Neglecting the body forces, the total potential energy functional $\mathcal{F}(u_i)$ can be written as:

$$\mathcal{F}(u_i) = \int_v F(u_i)\,dv - \int_{s_T} \bar{T}_i\, u_i\, ds \tag{3}$$

where s_T is the part of the plate surface, s, where the stress vector $\bar{T}_i$ is defined and v is the total plate volume.

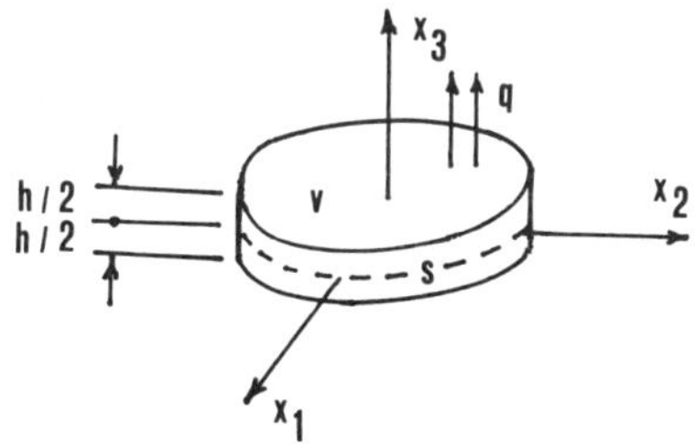

Fig. 1. Plate geometry

Among all kinematically admissible displacement fields, the functional $\mathcal{F}$ has a stationary value for the field realising the equilibrium.

The displacement functions u_i are chosen such that the longitudinal displacements u_α vary linearly and the normal transversal displacement u_3 is uniform along the plate thickness, in the following manner:

$$u_\alpha(x_1,x_2,x_3) = u^\circ_\alpha(x_1,x_2) + x_3\,\psi_\alpha(x_1,x_2) \tag{4a}$$

$$u_3(x_1,x_2) = w(x_1,x_2) \tag{4b}$$

The approximate displacement field (4) leads to a zero normal transversal strain, $\varepsilon_{33} = 0$, which is not compatible with the plane stress hypothesis. Hence; a modified constitutive relations can be used. These relations may be written, for a monoclinic material, in the form :

$$\sigma_{\alpha\beta} = \frac{1}{2}\, Q_{\alpha\beta\gamma\delta}\,(u_{\gamma,\delta} + u_{\delta,\gamma}) \tag{5a}$$

$$\sigma_{\alpha 3} = \frac{1}{2}\, K_{\alpha\delta}\, C_{\beta 3\delta 3}\,(u_{\beta,3} + u_{3,\beta}) \tag{5b}$$

$$\sigma_{33} = 0 \tag{5c}$$

where $Q_{\alpha\beta\gamma\delta} = C_{\alpha\beta\gamma\delta} - C_{\alpha\beta 33}\, C_{\gamma\delta 33} / C_{3333}$ and $K_{\alpha\beta}$ are corrective factors to take into account that the transverse shear stresses are not constant along the thickness ($0 < K_{\alpha\beta} < 1$). It may be noted that the value of the corrective factor, for isotropic material, is $\pi^2/12$. For anisotropic materials, analogous values would be obtained using the comparison of the velocity of straight-crested waves with the corresponding velocity , obtained from the three-dimensional theory.

Using equations (5), and after integration over the plate thickness; the first term of the functional $\mathcal{F}$, in equation (3), leads to:

$$\int_v F(u_i)\, dv = \frac{1}{2} \iint \left[A_{\alpha\beta\gamma\delta}\, u^\circ_{\alpha,\beta}\, u^\circ_{\gamma,\delta} + D_{\alpha\beta\gamma\delta}\, \psi_{\alpha,\beta}\, \psi_{\gamma,\delta} + \right.$$

$$\left. + d_{\alpha\beta}\, (\psi_\alpha + w_{,\alpha})\, (\psi_\beta + w_{,\beta})\right] dx_1\, dx_2 \tag{6}$$

where $A_{\alpha\beta\gamma\delta} = h\, Q_{\alpha\beta\gamma\delta}$

$$D_{\alpha\beta\gamma\delta} = \frac{h^3}{12} Q_{\alpha\beta\gamma\delta}$$

$$d_{\alpha\beta} = K_{\alpha\delta}\, C_{\beta 3\delta 3}$$

Introducing the approximate displacement field (4) into the expression (6) and then into the functional F, equation (3), an approximate functional $F^*(u^\circ_\alpha, \psi_\alpha, w)$ is reached. The variational equation

$$\delta\{ F^* (u^\circ_\alpha, \psi_\alpha, w)\} = 0 \tag{7}$$

leads to the following equilibrium equations:

$$A_{\alpha\beta\gamma\delta}\, u^\circ_{\gamma,\delta\beta} = 0 \tag{8a}$$

$$D_{\alpha\beta\gamma\delta}\, \psi_{\gamma,\delta\beta} + d_{\alpha\beta}\, (\psi_\beta + w_{,\beta}) = 0 \tag{8b}$$

$$d_{\alpha\beta}\, (\psi_{\alpha,\beta} + w_{,\alpha\beta}) = -q\, (x_1, x_2) \tag{8c}$$

where q is the lateral surface load. Equations (8) form a system of 5 partial differential equations permitting the determination of the 5 unknown displacement functions u°_α, ψ_α and w.

Once the displacement functions are determined, the stresses and hence the stress resultants can be obtained using constitutive relations (5). The membrane forces $N_{\alpha\beta}$, the bending moments $M_{\alpha\beta}$ and the shearing forces Q_α are obtained in the following form:

$$N_{\alpha\beta} = A_{\alpha\beta\gamma\delta}\, u^\circ_{\gamma,\delta} \tag{9a}$$

$$M_{\alpha\beta} = D_{\alpha\beta\gamma\delta}\, \psi_{\gamma,\delta} \tag{9b}$$

$$Q_\alpha = d_{\alpha\beta}\, (\psi_\beta + w_{,\beta}). \tag{9c}$$

The boundary conditions, obtained from the variational equation (7), require that one member of the following 5 products must be prescribed:

$$N_n u^\circ_n,\quad N_t u^\circ_t,\quad M_t \psi_n,\quad M_n \psi_t \text{ and } Q_n w,$$

where n and t are directions normal and tangential to the plate edge.

NUMERICAL EXAMPLE

As a numerical example, the cylindrical bending of a homogeneous orthotropic plate under uniform lateral load is considered. This problem was treated by

Srinivas (1970) and El-Soaly (1982). Srinivas used a three-dimensional linear elasticity solution while El-Soaly used a variational analysis of the complementary energy functional.

Consider an orthotropic plate, of thickness h and width L, which is infinitely extended along the axis x_2. The plate is hinged at both ends, $x_1 = \pm L/2$, and loaded by a uniform lateral load q_o, as shown in Fig. 2. Assuming that the 9 elastic moduli of the plate material are:

$$C_{1111}, C_{2222}, C_{3333}, C_{1122}, C_{2233}, C_{3311}, C_{1212}, C_{2323}, \text{ and } C_{3131},$$

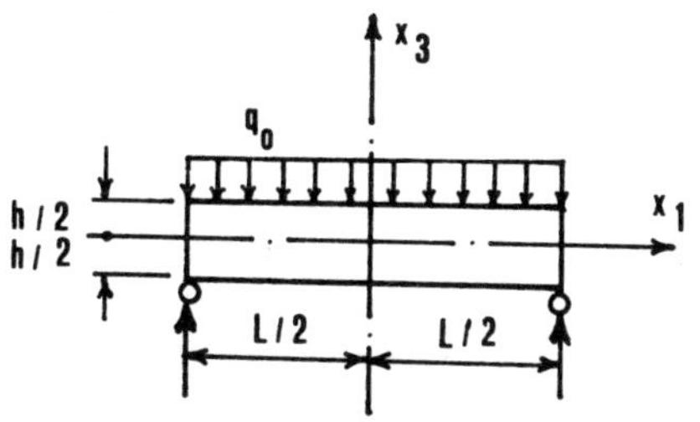

Fig. 2. Orthotropic plate under cylindrical bending.

the inclination angle ψ_1 and the deflection w are obtained, as functions of the variable x_1, in the form:

$$\psi_1 = \frac{q_o L^3}{48 D_{1111}} \left[3 \left(\frac{2x_1}{L}\right) - \left(\frac{2x_1}{L}\right)^3\right] ,$$

$$w = \frac{q_o L^4}{64 D_{1111}} \left[\frac{1}{6} \left(\frac{2x_1}{L}\right)^4 + \left(\frac{8D_{1111}}{d_{11} L^2} - 1\right) \left(\frac{2x_1}{L}\right)^2 - \frac{8D_{1111}}{d_{11} L^2} + \frac{5}{6}\right]$$

where $D_{1111} = \frac{h^3}{12} \left[C_{1111} - \frac{(C_{1133})^2}{C_{3333}}\right]$ and $d_{11} = \frac{\pi^2}{12} C_{1313}$.

The longitudinal normal stress σ_{11} is obtained, using the constitutive relation (5a), as:

$$\sigma_{11} = \frac{3\, q_o L^2}{2\, h^2} \left[1 - \left(\frac{2x_1}{L}\right)^2\right] \frac{x_3}{h} .$$

Table 1 represents the maximum values of ψ_1, w and σ_{11} in comparison with those obtained using the generalized Reissner's theory, developed by El-Soaly (1982).

The results given in Table 1 indicate that the difference, in application, of the two theories increases as the thickness ratio h/L increases and may be neglected for very small values of h/L.

TABLE 1 Comparison of Maximum Values of Inclination Angle ψ_1, Deflection w and Longitudinal Normal Stress σ_{11}

Quantity*	Present Theory	Generalized Reissner's Th.**
$\frac{24D_{1111}}{q_o L^3}\psi_1$	1	$1 - \frac{2}{5} B_{11} (\frac{h}{L})^2$
$\frac{64D_{1111}}{q_o L^4}$ w	$\frac{5}{6} - \frac{8}{\pi^2} E_{33} (\frac{h}{L})^2$	$\frac{5}{6} - \frac{4}{5} (B_{11}+E_{33}) (\frac{h}{L})^2$
$\frac{4h^2}{3q_o L^2}\sigma_{11}$	1	$1 - B_{11}(\frac{h}{L})^2$

* The maximum values are obtained for : ψ_1 at $x_1 = L/2$, w at $x_1= 0$ and for σ_{11} at $x_1=0$, $x_3=-h/2$.

** $B_{11} = C_{1133}$, $E_{33} = (C_{1111} - C_{1133} B_{11})/C_{1313}$.

CONCLUSION

A generalization of Mindlin's plate theory for homogeneous anisotropic plates has been established. The development is made for a monoclinic material. As a numerical example, the cylindrical bending of an orthotropic plate has been presented. The obtained results are compared with the generalized Reissner's theory solution. The difference between the two theories is, basicaly, due to the neglection of the normal transversal stress σ_{33} in the present theory. However, the difference is proportional to the square value of the thickness ratio h/L, which is small for thin plates.

REFERENCES

El-Soaly, E. E. (1982). Generalization of Reissner's theory for thick anisotropic plates. Second Cairo University MDP Conference. Cairo, 25-30.

Lo, K. H., R. M. Christensen, and E. M. Wu. (1977). A high-order theory of plate deformation. J. Appl. Mech., 44, 663-676.

Mindlin, R. D. (1951). Influence of rotary inertia and shear on flexural motions of isotropic, elastic plates. J. Appl. Mech., 18, 31-38.

Pagano, N. J. (1969). Exact solution for composite laminates in cylindrical bending. J. Composite Materials., 3, 398-411.

Reissner, E. (1945). The effect of transverse shear deformation on the bending of elastic plates. J. Appl. Mech., 12, 69-77.

Srinivas, S., and A. K. Rao. (1970). Bending, vibration and buckling of simply supported thick orthotropic rectangular plates and laminates. Int. J. Solid Structures., 6, 1463-1481.

Whitney, J. M. (1969). The effect of tranverse shear deformation on bending of laminated plates. J. Composite Materials., 3, 536-547.

EXPERIMENTAL CREEP MECHANISM MAPS FOR METALS

G. Malakondaiah and P. Rama Rao

Defence Metallurgical Research Laboratory, PO Kanchanbagh, Hyderabad - 500 258, India

ABSTRACT

Experimental creep mechanism maps have been constructed for five metals, namely copper, nickel, beta-cobalt, magnesium and alpha-titanium, for which fairly comprehensive creep data are available. Though data pertaining to Harper-Dorn creep are limited to alpha-titanium and beta-cobalt, Harper-Dorn creep field has been included in all the maps presented here.

KEYWORDS

Experimental creep mechanism maps; Coble creep; Nabarro-Herring creep; Harper-Dorn creep; power-law creep.

INTRODUCTION

Since the publication of the first detailed report on deformation mechanism maps by Ashby in 1972, maps for several metals have been made available by Ashby and his coworkers (Ashby, 1972; Frost and Ashby, 1973) as well as Langdon and his coworkers (Mohamed and Langdon, 1974; 1976). However, the earlier workers have made use of theoretical constitutive equations of Coble (1963) as well as Nabarro (1948) - Herring (1950) (N-H) creep processes to compute field boundaries. Recent studies on creep at low stresses (Towle and Jones, 1976; Burton, 1977) have revealed discrepancies of up to more than two orders of magnitude between the measured and predicted Coble or N-H creep rates, which the earlier maps do not account for. Further, Harper-Dorn (1957) (H-D) creep, yet another viscous creep process, has been included only in the maps for aluminium (Mohamed and Langdon, 1974) and lead (Frost and Ashby, 1973) since the data available until recently were limited. The authors' (1981, 1982) recent work on creep of alpha-titanium and beta-cobalt has demonstrated that H-D creep operates in these metals, too, and is not confined to temperatures close to the melting point as was stressed by other investigators (Yavari, Miller and Langdon, 1982; Mohamed and Ginter, 1982). With this background, an attempt has been made to construct experimental creep mechanism maps (maps based on experimental observations), that include H-D creep field, for five metals, namely copper, nickel, beta-cobalt, magnesium and alpha-titanium, for which fairly comprehensive creep data are available.

CONSTRUCTION OF CREEP MECHANISM MAPS

Creep mechanism maps of normalised grain size vs normalised stress, suggested by Langdon (Mohamed and Langdon, 1974), have been considered. Boundary between the two fields is obtained by equating the strain rates due to the creep mechanisms dominant on either side of the boundary.

The constitutive equations for N-H, Coble, H-D and climb creep processes are as follows:-

$$\dot{\varepsilon}_{NH} = A_{NH}\ (\Omega\sigma/d^2kT)\ D_{O_V}\ \exp\ (-Q_V/RT) \quad .. \quad (1)$$

$$\dot{\varepsilon}_{Co} = A_{Co}\ (W\Omega\sigma/d^3kT)\ D_{O_B}\ \exp\ (-Q_B/RT) \quad .. \quad (2)$$

$$\dot{\varepsilon}_{HD} = A_{HD}\ (b\sigma/kT)\ D_{O_V}\ \exp\ (-Q_V/RT) \quad .. \quad (3)$$

$$\dot{\varepsilon}_{c} = A_{c}\ (Gb/kT)\ (\sigma/G)^n\ D_{O_V}\ \exp\ (-Q_V/RT) \quad .. \quad (4)$$

where the subscripts NH, Co, HD and c represent N-H, Coble, H-D and climb (Mukherjee, Bird and Dorn, 1969) creep processes respectively. $\dot{\varepsilon}$ is the creep rate, σ the applied stress, G the shear modulus, n the stress exponent and A the dimensionless constant. D_{O_V} and D_{O_B} are the frequency factors and Q_V and Q_B are the activation energies for volume and grain boundary self-diffusion respectively. b, Ω, w and d respectively stand for Burgers vector, atomic volume, grain boundary width and grain diameter. R is the universal gas constant and kT has the usual meaning. A_{NH} and A_{Co} have respectively been shown to be 13.3 (Herring, 1950) and $150/\pi$ (Coble, 1963).

Data used in the construction of creep mechanism maps are given in Table 1. It is evident from Table 1 that the measured creep rates ($\dot{\varepsilon}_{Meas}$) in the N-H and Coble creep regions are several times faster than those predicted by equations 1 and 2 ($\dot{\varepsilon}_{Calc.}$) using the data given in Table 1 (only in case of beta-cobalt measured creep rates in the Coble creep region are less than those predicted by theory). These discrepancy factors, reported in Table 1, have been taken into account while computing the field boundaries.

Yet another modification that is brought out in the present maps is related to A_{HD}. As mentioned earlier, data pertaining to H-D creep (Yavari, Miller and Langdon, 1982) excepting those on alpha-titanium and beta-cobalt (Malakondaiah and Rama Rao, 1981; 1982), were confined to a narrow temperature range (0.92 to 0.99 T_m, where T_m is the absolute melting temperature) and so not much variation in A_{HD} was observed. Hence the earlier workers have assumed A_{HD} to be constant ($\sim 10^{-11}$) over the entire temperature range while constructing the maps. This appears to be invalid. Data currently available on H-D creep, presented in Fig. 1, indicate that A_{HD} is strongly dependent on test temperature. Analysis of the published data (Fig. 1) leads to the relationship

$$\sqrt{A_{HD}} = 8.35 \times 10^{-5}\ (1 - 0.97\ \frac{T}{T_m}) \quad .. \quad (5)$$

For all the metals included in Table 1 the constant A_{HD} at temperature T has been estimated using this relationship.

TABLE 1 Data used in the Construction of Creep Mechanism Maps

	Cu	Ni	β-Co	Mg	α-Ti
T_m, K	1356	1726	1768	923	1933
b, 10^{-10}m	2.56	2.49	2.51	3.2	2.89
Ω, $10^{-29}m^3$	1.18	1.09	1.11	2.32	1.77
*G_o, 10^4 MN/m^2	4.21	7.89	9.10	1.7	4.2
*$\frac{1}{G_o}\frac{dG}{dT}$, 10^{-4} K^{-1}	3.97	3.7	6.35	4.4	5.95
	(Overton and Gaffney, 1955; Chang and Himmel, 1966)	(Alers, Neighbours and Sato, 1960)	(Maringer and Marsh, 1969)	(Metals Handbook Vol.1,1961)	(Armstrong and Brown, 1964)
D_{ov}, 10^{-4} m^2/s	0.31	1.9	0.5	1.0	1.0†
Q_v, kJ/mol	200.3	279.2	273	134	241
	(Bowden and Balluffi, 1969)	(Wazzan, 1965)	(Hirano and Co-workers, 1962)	(Shewmon and Rhines 1954)	(Malakondaiah and Rama Rao, 1981)
D_{oB}, 10^{-4} m^2/s	2.0	-	-	1.0†	0.75†
w(=2b), 10^{-10}m	5.12	-	-	6.4	5.78
wD_{oB}, $10^{-14}m^3$/s	-	0.35	2.0	-	-
Q_B, kJ/mol	120	114.5	117	83.6	121
	(Burton and Greenwood, 1970)	(Wazzan, 1965)	(Brik, Larikov and Chenko, 1975)	(Jones, 1965)	(Malakondaiah and Rama Rao, 1981)
Q_B/Q_v	0.6	0.41	0.43	0.62	0.5
$(\dot{\varepsilon}_{Meas.}/\dot{\varepsilon}_{Calc.})_{NH}$	12	200	40	6	50
$(\dot{\varepsilon}_{Meas.}/\dot{\varepsilon}_{Calc.})_{Co}$	15	10	0.33	50	1.0
	(Burton and Greenwood, 1970)	(Towle, 1975; Burton, 1982) Crossland and Green-wood, 1970)	(Malakondaiah and Rama Rao,	(Jones, 1965; Crossland and Jones, 1977)	(Malakondaiah and Rama Rao, 1981)
A_c	7.4×10^5	3.8×10^5	2.7×10^{12}	2.5×10^6	2.4×10^5
n	4.8	4.6	6.5	5.0	4.3
	(Mukherjee, Bird and Dorn, 1969)	(Mukherjee, Bird and Dorn, 1969)	(Feltham and Myers, 1963)	(Crossland and Jones, 1972)	(Malakondaiah, 1980)

*$G=G_o \{1-(T-300)\frac{1}{G_o}\frac{dG}{dT}\}$ where G is the shear modulus at a given temperature T and G_o is the shear modulus at 300 K.

†estimated

Creep mechanism maps constructed for the five metals at temperatures ranging from 0.5 to 0.9 T_m are shown in Figs. 2 to 6. The chosen range of d/b from 10^3 to 10^7 corresponds to d of nearly 0.3 to 3×10^3 μm. On the normalised stress axis, the upper limit of 2×10^{-3} corresponds to the stress beyond which power-law breaks down in case of pure metals (Bird, Mukherjee and Dorn, 1969).

COMPARISON WITH EXPERIMENTAL OBSERVATIONS

A comparison of the experimentally observed creep behaviour with that predicted by the maps has been made in respect of grain size - stress - temperature regimes of dominance of Coble, N-H, H-D and climb creep processes as well as transitions from one particular creep mechanism to the other. This exercise reveals that the maps, within experimental scatter, rightly predict the creep behaviour under conditions of stress, temperature and grain size in which Coble, N-H or climb creep dominates the flow. Disparity exists only at coarse grain sizes where the maps predict dominance of H-D creep although the very limited data reported on copper (Burton and Greenwood, 1970) and nickel (Burton, Crossland and Greenwood, 1980) at coarse grain sizes indicate operation of Coble or N-H creep processes. However, in case of beta-cobalt and alpha-titanium, where data pertaining to H-D creep are available, good agreement prevails in H-D creep region, too.

GENERAL FEATURES OF THE MAPS

A general comparison of the maps shown in Figs. 2 to 6 brings out certain noteworthy features as listed below.

1. N-H creep field, in general, appears at temperatures $\gtrsim$ 0.6 T_m. With further increase in temperature N-H creep field expands at the expense of Coble creep field (e.g. see maps for nickel shown in Fig. 3). However, alpha-titanium, in which case N-H creep field appears even at 0.5 T_m (Fig. 6) is an exception.

2. Higher the ratio of Q_B/Q_V, higher is the temperature up to which Coble creep dominates over N-H creep (e.g. compare the maps for copper and nickel shown in Figs. 2 and 3 and the corresponding values of Q_B/Q_V given in Table 1. Alpha-titanium once again is an exception.

3. H-D creep can dominate the flow even at grain sizes (d) of the order of ~150 μm under specific conditions of temperature and stress (e.g. see map for alpha-titanium shown in Fig. 6).

4. At lower temperatures (less than about 0.6 T_m) transition occurs directly from Coble to H-D creep process except in the case of alpha-titanium where the transition is via N-H creep (evident from a comparison of Figs. 2 to 5 with 6).

5. The stress beyond which a transition occurs from H-D to climb creep is dependent on temperature. The transition stress decreases with increasing temperature (e.g. see maps for β-cobalt shown in Fig. 4).

REFERENCES

Alers, G.A., J.R. Neighbours, and H. Sato (1960). J.Phys.Chem.Solids, 13, 40- .

Armstrong, P.E., and H.L. Brown (1964). TMS AIME, 230, 962- .

Ashby, M.F. (1972). Acta Metall., 20, 887-897.

Bird, J.E., A.K. Mukherjee, and J.E. Dorn (1969). In D.G. Brandon and A. Rosen (Eds.), Quantitative Relation between Properties and Microstructure, Israel Universities Press, Jerusalem, pp. 255- .

Bowden, H.G., and R.W. Balluffi (1969). Phil. Mag., 19, 1001-1014.

Brik, V.E., L.N. Larikov, and V.M.F. Chenko (1975). Ukr.Fiz.Zh., 20, 397- .

Burton, B. (1977), Diffusional Creep of Polycrystalline Materials, Trans Tech Publications, Switzerland.

Burton, B., and G.W. Greenwood (1970). Metal Sci.J., 4, 215-218.

Burton, B., I.G. Crossland, and G.W. Greenwood (1980). Metal Sci., 14, 134-136.

Chang, Y.A., and L. Himmel (1966). J.Appl.Phys., 37, 3567-3572.

Coble, R.L. (1963). J.Appl.Phys., 34, 1679-1682.

Crossland, I.G., and R.B. Jones (1972). Metal Sci.J., 6, 162-166.

Crossland, I.G., and R.B. Jones (1977). Metal Sci., 11, 504-508.

Feltham, F., and T. Myers (1963). Phil.Mag., 8, 203- .

Frost, H.J., and M.F. Ashby (1973). A Second Report on Deformation Mechanism Maps, Harward University, Report NR-031-732.

Harper, J.G., and J.E. Dorn (1957). Acta Metall., 5, 654-665.

Herring, C. (1950). J.Appl.Phys., 21, 437- .

Hirano, K.I., R.P. Agarwala, B.L. Averbach, and M. Cohen (1962). J.Appl.Phys., 33, 3049- .

Jones, R.B. (1965). Nature, 207, p. 70.

Malakondaiah, G. (1980). Creep of Alpha Titanium and Beta Cobalt at Low Stress, Ph.D. Thesis, Banaras Hindu University, Varanasi, India.

Malakondaiah, G., and P. Rama Rao (1981). Acta Metall., 29, 1263-1275.

Malakondaiah, G., and P. Rama Rao (1982). Mater.Sci.Eng., 52, 207-221.

Maringer, R.E., and L.L. Marsh (1958). AIME Annual Meeting, New York (referred to by Selle, J.E. (1969). Cobalt, 45, 186-189.

Metals Handbook (1961). Vol. 1, ASM, Metals Park, Ohio.

Mohamed, F.A., and T.G. Langdon (1974). Metall.Trans., 5, 2339-2345.

Mohamed, F.A., and T.G. Langdon (1976). Scripta Met., 10, 759-762.

Mohamed, F.A., and T.J. Ginter (1982). Acta Metall., 30, 1869-1881.

Mukherjee, A.K., J.E. Bird, and J.E. Dorn (1969). Trans. ASM, 62, 155-179.

Nabarro, F.R.N. (1948). Rep.Conf. on the Strength of Solids, Physical Society, London, p. 75.

Overton, W.C. and J. Gaffney (1955). Phys.Rev., 98, 969- .

Shewmon, P.G., and F.N. Rhines (1954). TMS AIME, 200, 1021- .

Sritharan, T., and H. Jones (1979). Acta Metall., 27, 1293-1300.

Towle, D.J. (1975). Ph.D. Thesis, Sheffield University, England.

Towle, D.J. and H. Jones (1976). Acta Metall., 24, 399-407.

Wazzan, A.R. (1965). J.Appl.Phys., 36, 3596-3599.

Yavari, P., D.A. Miller, and T.G. Langdon (1982). Acta Metall., 30, 871-887.

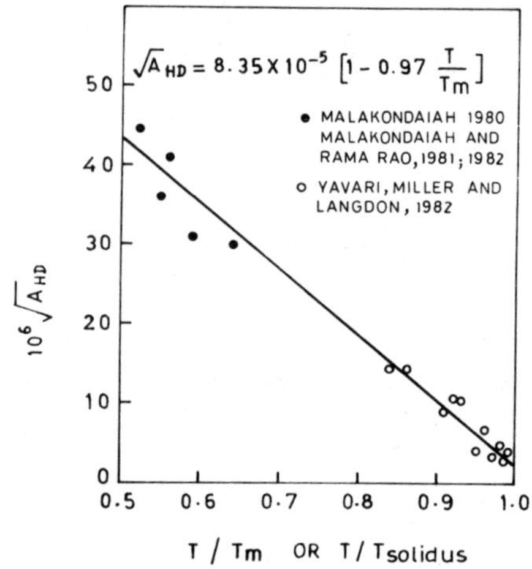

Fig. 1 Variation of A_{HD} with temperature

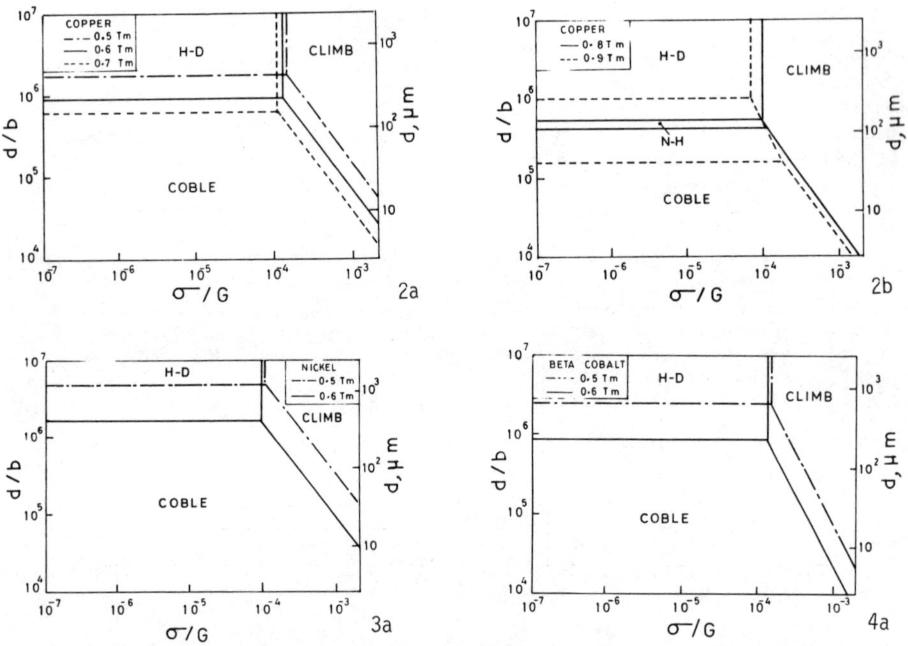

Figs. 2, 3a and 4a. Experimental creep mechanism maps for (2) Copper (3a) Nickel and (4a) beta-cobalt.

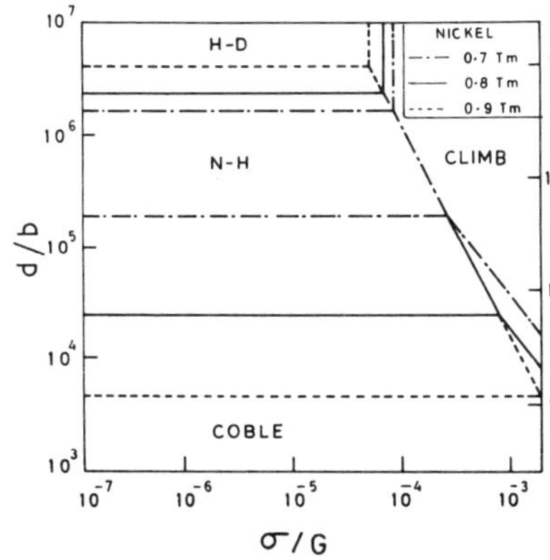

3b

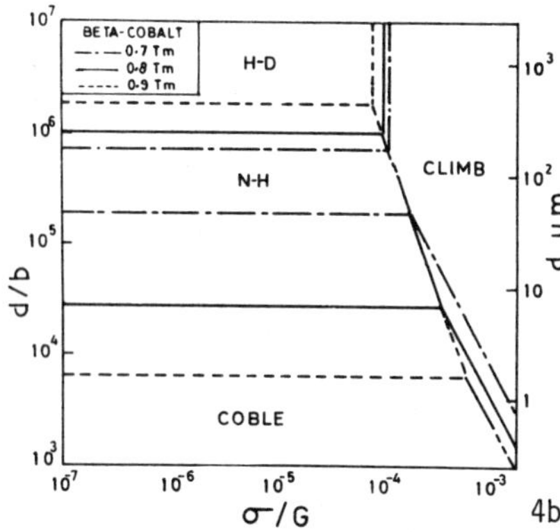

4b

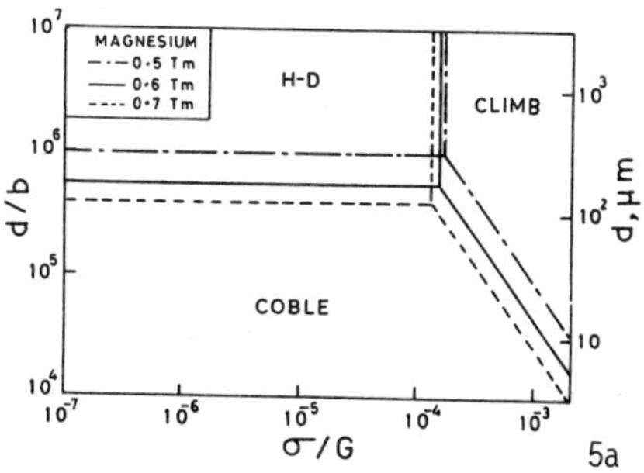

5a

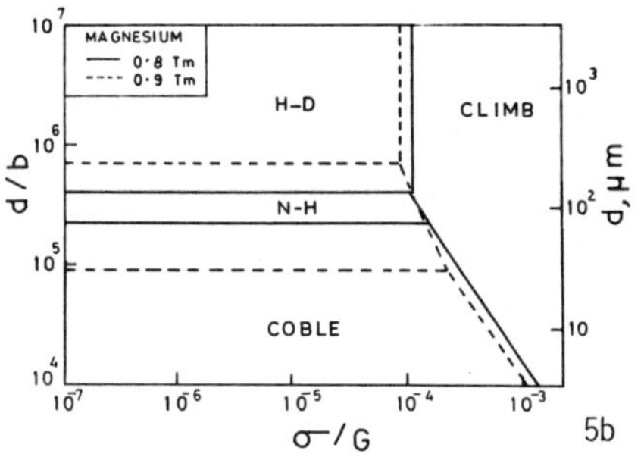

5b

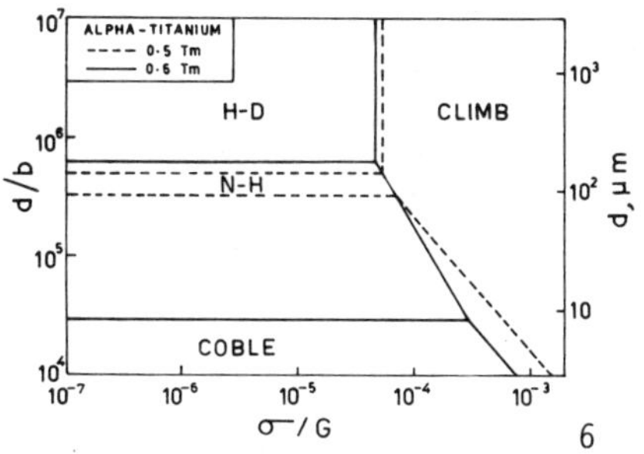

6

Figs. 3b, 4b, 5a, 5b and 6. Experimental creep mechanism maps for (3b) nickel, (4b) beta-cobalt (5a and b) magnesium and (6) alpha-titanium.

THE INFLUENCE OF MULTIAXIALITY OF STRESS AND ENVIRONMENTAL INDUCED DEGRADATION ON THE CREEP BEHAVIOUR OF ALLOY 800H TUBULAR COMPONENTS

R. C. Hurst

Commission of the European Communities, Joint Research Centre, Petten, The Netherlands

ABSTRACT

The optimum design of industrial plant operating at high temperatures relies on the extent of information concerning materials behaviour available to the designer and the accuracy of his design methods. There exists such a large division between simple materials data obtained in the laboratory and the complex conditions of operating plant that the gap must be bridged through a progressive understanding of materials behaviour with increasing complexity of stress state. This progressive understanding may be achieved through tests on simple components such as tubes, a shape of component clearly of common interest in industrial plant and ideally suited to both multiaxial stress testing and for the assessment of any potential influence of environmental degradation.

The results from the presently available facilities concentrate on the creep behaviour of Alloy 800H tubes subjected to uniaxial tensile stress and to internal pressure at 900 and 1000^{0}C in separate experiments. The stress rupture and fracture strains can be compared under both conditions and with results from conventional specimen tests. In the analysis, account is taken of the multiaxial stress state in the tube walls during the internal pressure tests and the resultant redistribution of creep stresses. Some preliminary results covering the influence of a corrosive (carburising) environment on tube creep properties are also described. Although the lack of suitable continuous deformation measurement methods limits the present potential output of the experiments, some developments are currently meriting particular attention.

KEYWORDS

Creep; tube; high temperature, Alloy 800H; multiaxial stress; environment; carburisation.

INTRODUCTION

At ambient temperatures, sufficient knowledge and expertise are available on stress analysis and materials deformation behaviour to allow designs of components to proceed from a combination of available materials data with standard analysis methods and design codes. However, at high temperatures, the time-dependent damage accumulation mechanisms through which materials degrade are not easily dealt with by analysis and design codes. In addition, the suitability of materials data, obtained from simple testing, in the design calculations for complex components is not easy to confirm. At high temperatures, not only are the components, exposed to multi-directional stress fields, subject to consequent stress redistributions but also the chemical affinity for the material with its environment is likely to be enhanced.

In order to improve the fundamental understanding of high temperature degradation processes while, at the same time, making laboratory realised data more applicable to design requirements for high temperature process plant, a progressively complex research programme has been evolved. As there is a wide, but common, interest in the performance of tubular components in high temperature processes a simple tubular test piece is chosen as the basic element for the experiments and analysis. Such a tubular specimen lends itself ideally to multiaxial stressing and provides the potential containment for environments, the aggressiveness of which can thus be assessed. The tubular material chosen is wrought Alloy 800H for which there is considerable interest and for which significant uniaxial data already exists, albeit obtained mostly in air environments. Although present testing is limited to separate axial and internal pressure creep testing of such tubes and some preliminary work using carburising atmospheres, the sequential extension to combined axial and pressure loading, to incorporation of thermal stresses using through wall temperature gradients, and to more direct simulation of process plant environments is planned.

EXPERIMENTAL FACILITIES

A special tubular testing facility has been recently designed and constructed as an extension to the Environmental Test Laboratory (Hurst and Kröckel 1980) which itself commenced operation in 1979. The main requirements of the tubular testing installation which influenced its design were the size of the tube to be tested, the pressure and chemical composition of the test gases, the test temperature range, the provision of additional axial load and the need for interruption free running of the creep tests. Experimental points are housed in four concrete cells, reinforced to withstand the worst possible type of failure of a 250 mm long, 35 mm diameter, 5 mm wall tube under a pressure of 300 bar of hydrogen at 1000°C. Hydrogen was considered in the calculation because of its presence in large proportions in the gas mixtures, simulating process gases, which are to be utilised. In the calculation of concrete strength, the chemically explosive reaction was assumed to take place but in the first operational cell the pressurised tube is in any case surrounded by inert gas thus ensuring an even larger safety factor.

The desire for continuous operation of experiments and close safety monitoring is provided locally through interlocking control systems and remotely through the link to the main laboratory and its 24 hours a day surveillance system. The Alloy 800H tubular specimens of composition Cr 19.98, Ni 31.5, C .08, Si. 29, Mn. 76, Ti. ,57, Al. 39, Co. 75, Cu. 46, S .006 and remainder iron are, in fact, 250 mm long, 25.4 mm internal diameter and 4.5 mm wall thickness with welded end caps. For internal pressure tests, these end caps are argon arc welded alloy 800H with gas inlet and outlet tubes; for axial tests on tubes these caps are electron beam welded Nimonic 105 with an internal thread for connection to the pull rods; and, finally for the axial tests with flowing gas, electron beam welded small bore stainless steel tubes are welded into the threaded Nimonic end caps and these pass down the inside of the pull rods to both the gas supply and exhaust systems.

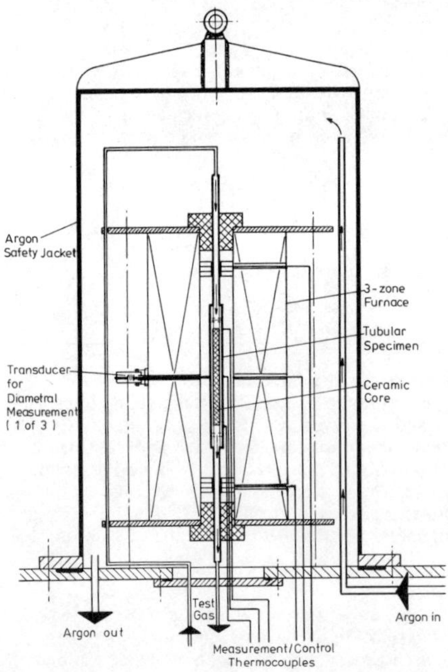

Fig. 1 Internal pressure tubular creep test rig.

The location of a tubular specimen within the furnace and argon safety jacket of the internal pressure rig is depicted in **Fig. 1.** The pressurising circuit design is complicated by the requirement for holding the test pressure constant over long term creep tests combined with maintaining a constant chemical activity of the test gas. These requirements are met by a pressure feedback design which allows a low gas flow (about 100ml/minute) to be exhausted to the atmosphere whereas the inert safety gas circuit is covered by a closed loop system.

The uniaxial testing of tubular specimens is carried out in conventional 20kN creep machines. A modified version allows axial load to be applied to a specimen through which an ambient pressure test gas simultaneously flows. In the preliminary experiments a test gas mixture of hydrogen/methane was used whilst argon safety gas blanketed the outside of the tube via a specially constructed chamber.

The results from both the internal pressure and uniaxial tests on tubes have been limited to the determination of creep rupture failure times and creep ductility measurements determined from pre- and post-test dimensions. However benchmark testing of tubular specimens under creep conditions can only satisfactorily be achieved using continuous deformation measurement and the efforts presently expended in this important aspect merit a brief description here. A thinner tube wall along the gauge length is necessitated as a result of the requirement to machine knife edges near the tube ends to support axial extensometry. This thinner tube wall will also be used in the internal pressure tests as eventually the two loads and hence the extensometry need to be combined in a single test. The diametral extensometry, already operating in a prototype form, consists of three transducers connected to the tube centre at 120^0 intervals around the circumference, one of which is shown in **Fig. 1.** As extensometry will only give an overall measurement of axial strain and perhaps a fortuitous measurement of diametral strain for such large specimens it is important to develop in parallel complementary methods aimed at ascertaining local strain measurements. Two routes are currently being followed vis the use of high temperature resistant inks in applied grids and X- and γ-radiography. The grids are only foreseen for intermittent measurements but the radiography has the potential for continuous in situ measurement, as shown in **Fig. 2.**

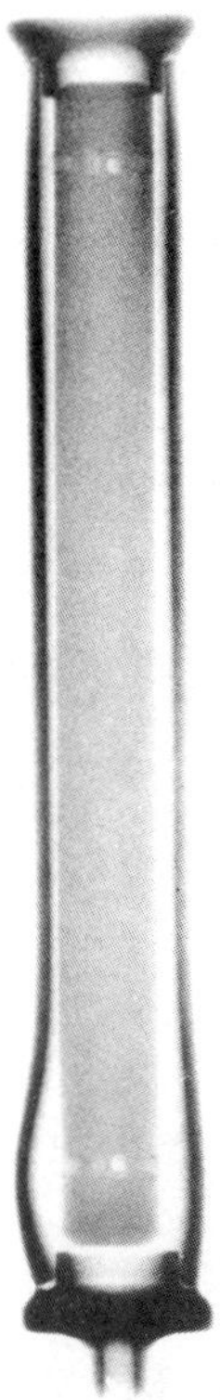

Fig. 2 Post-test X-radiograph of pressure tested Alloy 800H tube taken through the containment vessel but without the furnace.

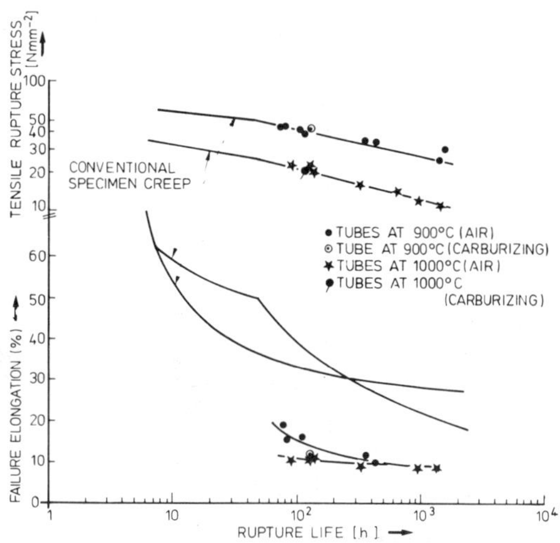

Fig. 3 Rupture strength and ductility of Alloy 800H tubes and specimens after tensile creep.

RESULTS AND DISCUSSION

A. Uniaxial tests

The stress rupture behaviour and elongations to failure under axial load have been determined up to around 1500h at both 900^0 and 1000^0C in air environment for both tubes and, previously (Bürgel 1981), conventional Alloy 800H creep specimens. The results are portrayed in **Fig. 3** where the good correlation of stress rupture behaviour is shown. Also in **Fig.3** is the apparent inferior ductility of the tubular specimens compared to that of the conventional specimens. The high temperatures and relatively short times employed in the present work unfortunately result in a paucity of comparable data from the literature. Reduction in area values reported by Persson (1978) mostly fall in the range 17-30% between 1000 and 2000h at 900^0C implying that the conventional specimen ductilities are of the right order. The cause of the lower ductility performance by the tubes is twofold. Firstly localised high elongations i.e. near the fracture area, are not fairly represented by measuring the total elongation of such a long specimen. For example a tube tested for only 90 hours at 1000^0C exhibited a total strain of 11% but platinum grid markings on the surface revealed ductilities near the fracture surface up to 25%. The second contributing factor lies in the metallurgical difference between the two types of specimen. The large grain size of the tubular Alloy 800H (around ASTM 1-2 compared to 2-3 for the conventional specimens) could be responsible for a drop in ductility according to the results of Lagneborg (1969). Some preliminary results on creep specimens cut from the tube wall confirm the metallurgical contribution to ductility reduction as their elongation to failure values fall between the tube and conventional specimen results.

The influence of a carburising environment on the uniaxial creep properties of the conventional Alloy 800H specimens has been shown (Guttmann and Bürgel 1982) to be beneficial, the stress carrying capacity of the carburised alloy being increased by around 50% compared to the air tested alloy for the same life with exposures beyond 100h at 1000^0C. If one considers the known disastrous effect of carburisation on room temperature ductility (Ennis and Lupton 1980) Guttmann and Bürgel surprisingly find about 50% enhanced creep ductility as a result of carburisation. Only two tests on tubes have been carried out to date using a carburising gas (H_2/CH_4 of $a_c = 0.8$) passing down the inside of the tube undergoing axial creep. Although no improvement in stress rupture performance or ductility is observed **(Fig. 3)** this is undoubtedly due to the much smaller proprtion of the cross section of the tube carburised compared to the conventional specimen, the depth of carburisation being approximately the same in both cases after 100 hours, **Fig. 4.**

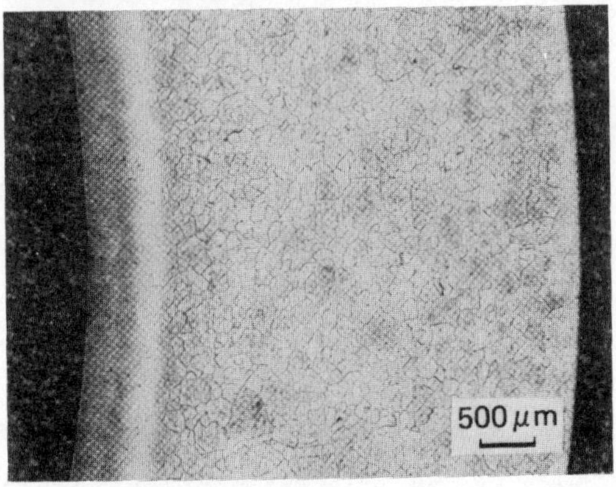

Fig.4 Part carburised Alloy 800H tube wall after $42 Nmm^{-2}$ axial creep stress test at 900^0C with H_2/CH_4 , $a_c = 0.8$ inside.

B. Multiaxial Tests

Prior to commissioning of the first tube testing rig at Petten, internal pressure tests were carried out under contract at the Electrical Research Association, Leatherhead U.K. The stress rupture results for all tubes tested, using argon as the pressurising medium, are given in **Fig. 5** along with their failure ductilities. The failure ductility values plotted are based on the maximum change in external diameter of the tube in the area of failure and these are clearly much lower than the conventional specimen values. Although the microstructural difference between the tube and conventional specimens plays again an important role, the lower ductilities exhibited by the pressure tested compared to the axially tested tubes require further explanation. If it is assumed that most of this alloy's ductility arises during secondary and tertiary creep (Bürgel 1981), then the choice of measurement position will influence the secondary creep ductilities reported and the mechanism of creep failure will affect the extent of

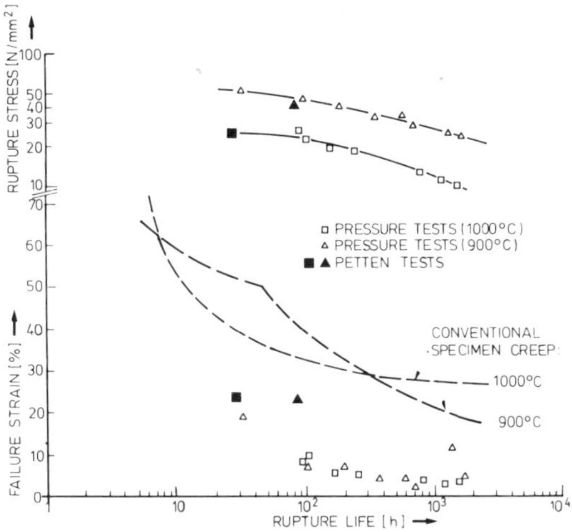

Fig. 5 Rupture strength and ductility of Alloy 800H tubes creep tested under internal pressure.

tertiary creep ductility measured. For a constant pressure the secondary creep rates in both the circumferential and radial directions are equal while the axial rate is zero due to the redistribution of stresses at the outside radius. However calculation shows that the creep rate at the inside is between 80 and 90% higher than at the outside radius and, indeed, measurements of the internal diameter taken from two sectioned tubes do show at least a 50% greater diameter change at the inside than at the outside. That total creep deformation is not only determined by the secondary creep rate is, of course, clear and near the fracture area the measurements were not at all close to a predicted ratio of 1.8-1.9, presumably due to the dominance of tertiary creep strain. Measurement of reduction in area at the fracture surface is expected to give even higher ductility values than internal diameter measurements as the non-uniform strain distribution due to a "gauge length" equal to the circumference of around 100 mm is also likely to be important. Notwithstanding the choice of measurement method the difference in fracture mode between a tubular pressure and a conventional tensile test, will also provide for a difference in total ductility. In a tensile creep test, the local stress at the fracture increases rapidly until the specimen fails, whereas in the pressure test the first crack or pinhole releases immediately the pressure and thus the creep load.

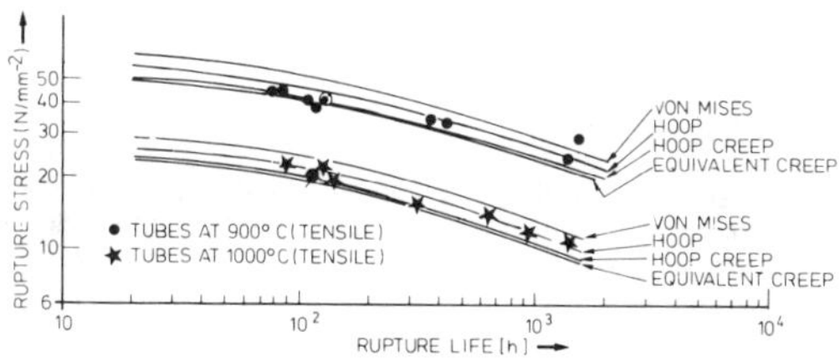

Fig. 6 Comparison of equivalent and redistributed creep stresses for pressurised tubes with tubular tensile rupture values.

In comparing multiaxial creep rupture stress properties with those obtained from uniaxial tests, care must be taken in the choice of calculation method for a representative creep rupture stress. In **Fig. 5** the maximum principal elastic stress is used to portray the pressure testing data but, whereas this method is commonly found in the literature, it ignores both the multiaxial situation and the redistribution of stresses due to creep which should both be taken into account when comparing failure criterion under tension and internal pressure. **Fig. 6** shows an alternative von Mises multiaxial equivalent stress which, for the dimensions of tube under test (considered as thick-walled), is also a maximum at the tube bore. The redistributed creep stresses are also depicted in Fig. 6 and it is important to note that although the equivalent redistributed creep stress is also a maximum at the bore of the tube, the maximum redistributed principal (hoop creep) stress is located at the outer surface of the tube. Comparison of the axial results with the different pressure test criterion point to a reasonable correlation surprisingly with maximum principle elastic (hoop) stress. Although this would fit in well with preliminary metallographic examinations of the first tubes sectioned which show that cracking is more prevalent at the tube bore **Fig. 7,** insufficient sections have been made to confirm this. Further, the calculated redistributed stress curves assume a Norton 'n' value of 5.1 (Bürgel, 1981) and as a change in n of only one unit can produce a calculated stress change of 2-3%, the n-values of the tube material must be accurately determined before any proposal for a stress rupture criterion can be proposed, if indeed a single criterion satisfies both temperatures and the range of creep lives considered. In view of the relative structural anisotropy of the tubular Alloy 800H material it would be desirable to determine the 'n' values in all three principle stress directions and, because of associated practical difficulties, establishment of the appropriate rupture criterion might best be achieved through a torsion test as advised by Henderson (1979).

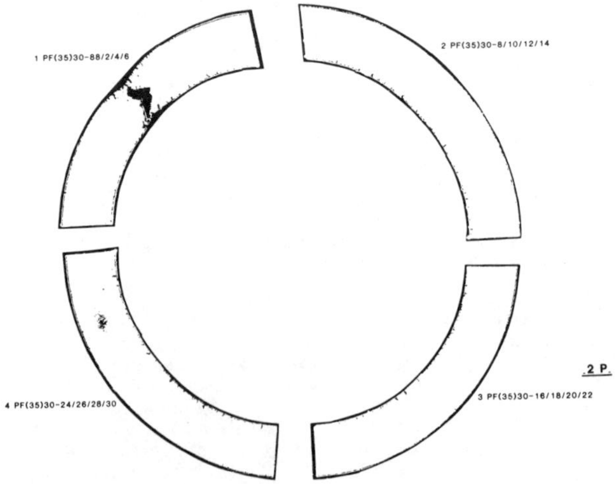

Fig. 7 Metallographic cross section of an Alloy 800H tube tested to failure with an internal argon pressure of 53 bar at 1000^{0}C.

SUMMARY AND CONCLUSIONS

1. A small but desirable contribution towards the improvement of creep design and life time prediction methods for components subjected to multiaxial stresses and aggressive environments at high temperatures is hoped to be provided via a research programme using a recently constructed facility for testing tubular specimens.

2. Results on Alloy 800H tubular specimens when tested in uniaxial creep fit well with conventional stress rupture data at 900^0 and 1000^{0}C however the ductility of the crept tubes falls significantly short of that for conventional specimens for both reasons of metallurgical structure and specimen size differences.

3. Two tests, to establish whether a high temperature carburising environment indeed enhances creep properties as reported for conventional specimens, have failed to show any improvement for tubular specimens presumably due to the limited carburisation achieved in tests of only 100h duration.

4. Comparison of calculated stress rupture values for internally pressure tested tubes compare well with uniaxial values for tubes although the determination of the influence of anisotropy and elaboration of an appropriate rupture criterion require further tests and analysis. The even lower failure ductilities reported for the pressure tested tubes can be attributed to the choice of deformation measurement position (external diameter at failure position) and to the fundamental difference between the final stages of failure in a pressure test compared to a tensile test.

ACKNOWLEDGEMENTS

The author greatly appreciates the impetus to the analysis of multiaxial creep deformation provided for by the presence of Dr. D.W.A. Rees, Trinity College Dublin during his month long secondment to J.R.C. Petten. Thanks are also due to Mr.G.Cassanelli and Mr.J.C.Hedderly for carrying out the creep tests and constructing the test facilities and to Mr. P.Frampton for metallographic services.

REFERENCES

Bürgel, R. (1981). Zeitstandverhalten der Stähle X10 NiCrAlTi 32 20 und X50 NiCrAlTi 33 20 in Luft und aufkohlender Atmosphäre. Dr. Ing. Dissertation. University of Hannover.

Ennis, P.J. and Lupton, D.F. (1980). The relationship between carburisation and ductility loss. **Behaviour of High Temperature Alloys in Aggressive Environments. EUR 6814.** The Metals Society, London. pp.979-99.

Guttmann, V., and Bürgel, R. (1982). The effect of a carburising environment on the creep behaviour of some austenitic steels. Presented at conference Corrosion in Coal Conversion Systems, C.E.G.B. London, in press.

Henderson, J. (1979) An investigation of multi-axial creep characteristics of metals. **TRANS ASME.** J. of Eng. Mat. and Techn., **101,** pp. 356-364.

Hurst, R.C., and Kröckel, H. (1980). The environmental test laboratory (ETL) of the JRC Petten Establishment. **Behaviour of High Temperature Alloys in Aggressive Environments. EUR 6814.** The Metals Society, London, pp. 335-350.

Lagneborg, R. (1969). Effect of grain size and precipitation of carbides on creep properties in Fe-20%Cr-35%Ni alloys. **J. Iron Steel Inst., 207,** pp.1503-1506.

Persson, N.G. (1978). Mechanical properties of alloy 800 above 600^{0}C. In W. Betteridge et al (eds) **Alloy 800,** North-Holland, Amsterdam. pp.135-146.

THEME II

MECHANICAL PROPERTIES OF SURFACE COATED MATERIALS

ON THE CONTACT FATIGUE RESISTANCE OF THE COBALT ALLOY COATING AND ITS MICROMECHANISMS

Fuhsing Wang, Chikung Tsi and Yungsin Chang

Department of Mechanical Engineering, Qing Hua University, Beijing, China

ABSTRACT

In order to evaluate the posibility of applying cobalt alloy coating to the sliding-rolling contact friction pair under fluid lubrication condition, some experiments on the contact fatigue resistance were conducted for a few cobalt alloy coatings. The test results show that among the cobalt alloy coatings studied, Co-1 alloy coating with lowest carbon is the best in the contact fatigue resistance. The mechanisms for the contact fatigue resistance of cobalt alloy coating were analysis and discussed based on stress analysis and microstructure observation using TEM. An experimental formula for estimating fatigue life and the depth of cobalt alloy coating is derived from the experimental results.

KEYWORDS

Cobalt alloy coating; Contact fatigue resistance; Spalling; Microstructure; Electron diffraction; Inter-martensite spacing.

INTRODUCTION

Since wear and corrosion resisting alloy coating has been vastly applied to repair and make machine parts. A large number of investigations on abrasion or corrosion resistance of alloy coating have been performed, but investigations on contact fatigue resistance are very few yet. The contact fatigue resistance of some cobalt alloy coatings was studied in this paper.

SPECIMENS AND EXPERIMENTAL CONDITIONS

The chemical composition of the experimental cobalt alloy coatings are shown in Table 1.

The rolling contact fatigue testing machine (JPM-1) of lever-loading type with two rollers was used. One of the rollers was work roller as specimen, and the other was support roller. The tests were conducted under sliding-rolling condition with lubrication oil.

TABLE 1 The Composition of Cobalt Alloy Coatings

Mark	C	B	Si	Cr	Ni	W	Fe	Co
Co-1	0.015	2.64	3.9	23.41	10.73	8.95	1.64	48.6
Co-2	0.75	2.42	3.06	25.61	12.25	9.63	2.12	41.2
Co-3			Co-1 80 % + WC 20 %					
Co-4			Co-2 80 % + WC 20 %					

Both work roller and support roller made from low carbon alloy steel 18Cr2Ni4WA are 60 mm in outer diameter. The support roller had been carburized and hardened before finishing by grinding. The work roller had been covered with one of the cobalt alloy coatings in its canal and was fluxed by a vaccum furnace.
The experimental conditions of coating specimens are shown in Table 2.

EXPERIMENTAL RESULTS

The experimental results of contact fatigue resistance and the failure form for the cobalt alloy coatings are shown in Table 2 and Fig. 1.

TABLE 2 The Experimental Conditions and Results

Specimen number	Mark	Hertzian stress P_{max} (x 10 MPa)	rev/min r.p.m.	Specific sliding (%)	Cycle number N (x10000)	Failure form
1	Co-4	147.5	3000	15	6.4	Surface radial cracking
2	"	"	"	"	14.0	"
3	"	"	2000	25	5.8	"
4	Co-1	166 + 188	"	"	400 + 116	Spalling
5	Co-2	150	"	"	94.0	"
6	"	168.5	"	"	87.0	"
7	Co-3	150	"	"	280.0	Serious pitting
8	Co-1	150	"	"	1330.0	Spalling
9	"	180	"	"	590.0	"
10	"	200	"	"	145.0	"
11	"	220	"	"	155.0	"

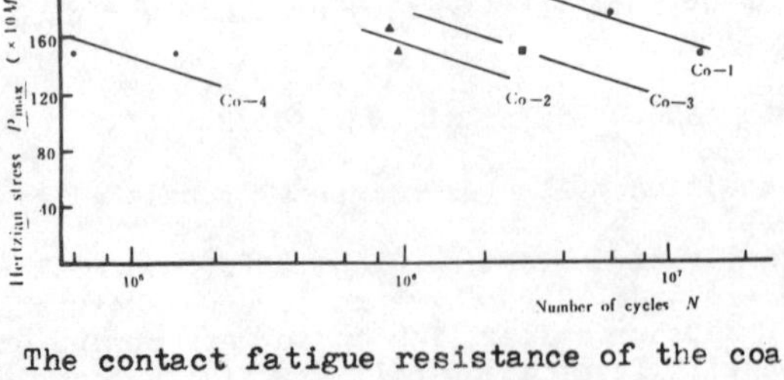

Fig. 1. The contact fatigue resistance of the coatings.

The experimental results show that Co-1 alloy coating has the best contact fatigue resistance, and the next is Co-3 and Co-2, Co-4 is the worst. The failure forms for different cobalt alloy coatings are different. The Co-4 alloy coating fails by surface radial cracking. Co-3 fails by serious pitting. But failure of Co-1 and Co-2 are caused by spalling. The spalling cracks of Co-1 alloy coating are shown in Fig. 2.

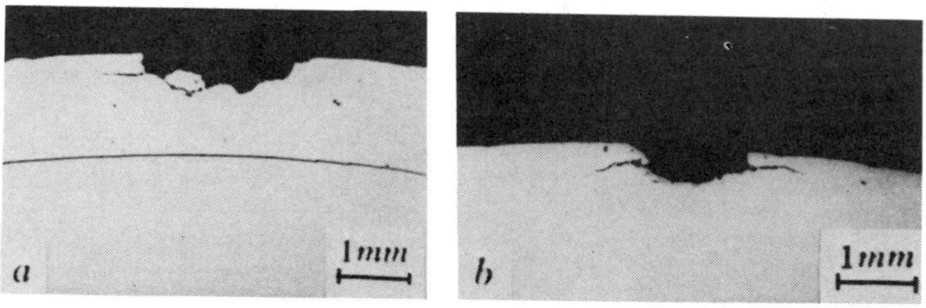

Fig. 2. Contact fatigue cracks of Co-1 alloy coating (a. specimen No.8 b. specimen No.11)

MICROSTRUCTURE AND STRESS

The thin foil samples cut at different depth (0.02 mm and 0.3 mm) from coating were prepared from specimen No.11 after contact fatigue test. Transmission electron micrographs and diffraction patterns are shown in Fig. 3 and Fig. 4. The result of analysis shows that within the F.C.C. solid-solution matrix of the cobalt alloy coating, there are H.C.P. martensite in the form of parallel laths.

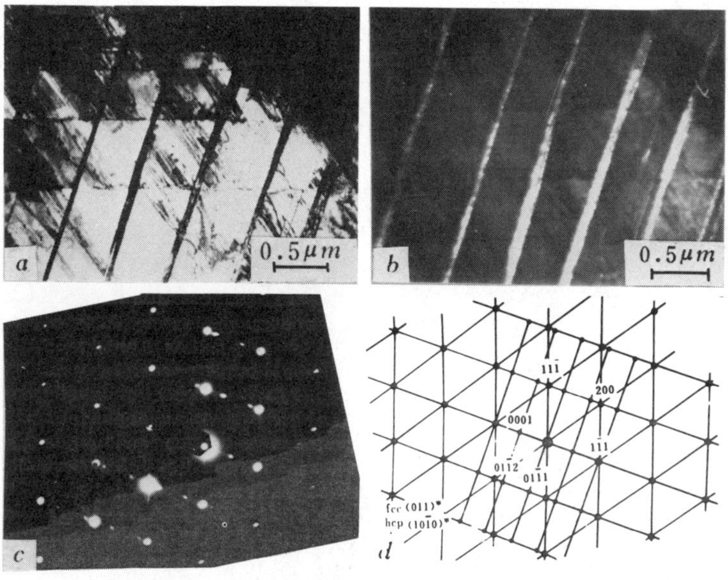

Fig. 3. Transmission electron micrographs at 0.02 mm below surface. (a. bright field; b. dark field; c. electron diffraction pattern; d. indexing.)

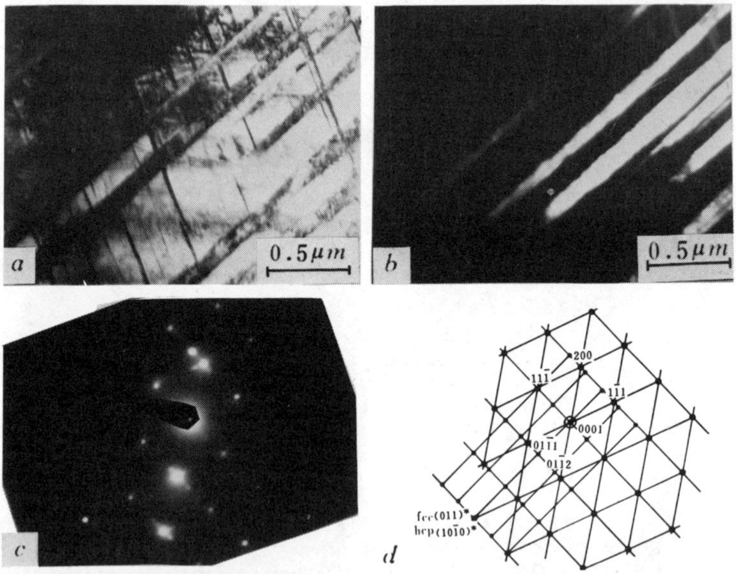

Fig. 4. Transmission electron micrographs at 0.3 mm below surface. (a. bright field; b. dark field; c. electron diffraction pattern; d. indexing.)

The failure mode of Co-1 alloy coating is spalling fatigue which is caused by the repeated orthogonal shear stress. The distribution of the orthogonal shear stress below the contact surface for the specimens No.8 to 11 was calculated according to Smith (1953). The result is given in Fig. 5. It shows that the maximum orthogonal shear stress is at 0.2 to 0.4 mm below the contact surface.

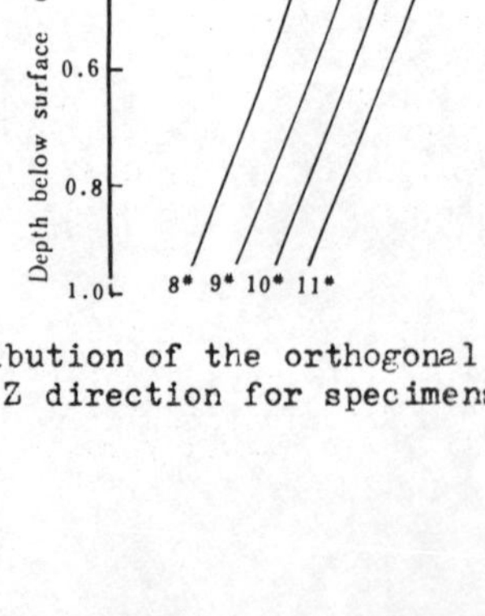

Fig. 5. Distribution of the orthogonal shear stress along Z direction for specimens No.8 to 11.

The microhardness distribution of the solid solution along the depth below contact surface was measured using 50 g load for specimen No.11. The measured result (Fig. 6.) shows that the maximum microhardness is at 0.2 to 0.4 mm region below the surface of the specimen that have been tested. In addition, the microhardness at surface layer of 0.02 mm depth is also slightly higher.

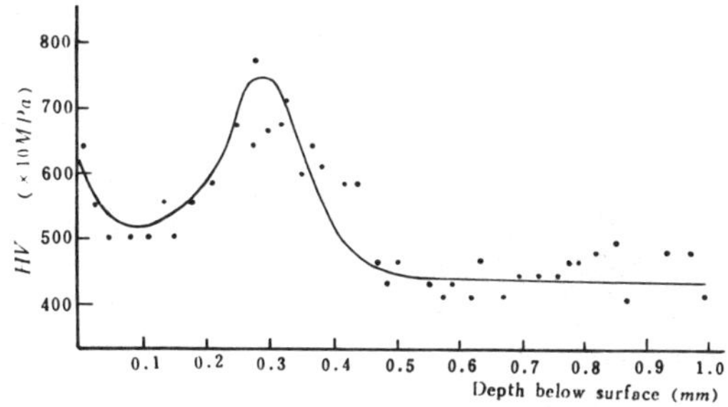

Fig. 6. Microhardness distribution of specimen No.11.

DISCUSSION

Co-1 alloy coating possesses the best contact fatigue resistance. This may relates to the appropriate composition (Table 1) and its fine and homogeneous constituent. Furthermore, the matrix of Co-1 alloy coating is the F.C.C. structure with good toughness and the deformation induced transformation (F.C.C. to H.C.P. martensite) strenthens the alloy during the contact fatigue.

The failure mode of contact fatigue in Co-1 alloy coating is spalling. The cracks are nucleated and tend to propagate at the layer 0.3-0.4 mm below the contact surface in the direction parallel to the surface (Fig. 2), where the orthogonal shear stress has the maximum values (Fig. 5). According to Merwin's (1963) plastic deformation analysis in rolling contact, the addition of residual compressive stress $(\sigma_y)_r$ will reduce the maximum shear stress to some yield strength value and eleminate any further yielding along Z-direction. But the magnitude of the orthogonal shear stress remains unchanged, this induces the reversed orthogonal shear strain γ_{yz}.

If the orthogonal shear stress is higher, the repeated orthogonal shear strain will be greater. For this reason, the cracks are nucleated and propagated at the region where the repeated orthogonal shear stress and strain have maximum value.

The layer where the orthogonal shear stress has maximum value and the repeated plastic deformation exists and where the cracks are nucleated and propagated coincides with the layer of maximum microhardness (Fig. 6). As for the higher microhardness at the surface layer of 0.02 mm depth, it is sugested that surface roughness may induce asperity contact, the stresses distribution under the contact surface are changed. A modified stress distribution proposed by Michau (1974) shows that there is high orthogonal shear stress zone very near the surface in addition to the general high orthogonal shear stress zone at 0.2-0.4 mm depth from the surface.

Analysis of the electron diffraction patterns shows the transformed martensite induced by repeated plastic deformation has the H.C.P. structure. The inter-martensite spacing relates to the repeated shear strain. Its value at 0.02 mm depth is smaller than at 0.3 mm depth. Therefore, in the surface layer of 0.02 mm depth, the inter-martensite spacing is wider, and the orientation of martensite laths is almost parallel (Fig. 3). But at high orthogonal shear stress region, the inter-martensite spacing is very small and martensite laths are distributed at different orientations (Fig. 4). So far many investigations have been performed on the relationship between inter-twin spacing and yield strength or hardness. It is considered that their relationship is similar to Hall-Petch law. Formulas described by Murr (1978) are modified into the following form:

$$\Delta\sigma_s = K_1 \left(d_1^{-\frac{1}{2}} - d_0^{-\frac{1}{2}}\right)$$

$$\Delta H = K_2 \left(d_1^{-\frac{1}{2}} - d_0^{-\frac{1}{2}}\right)$$

Where $\Delta\sigma_s$ and ΔH are the increments of yield strength and hardness produced by decreasing inter-twin spacing respectively, d_0 and d_1 are inter-twin spacings before and after deformation, K_1 and K_2 are material's constants which relate to the shear modulus and stacking fault energy, if stacking fault energy is high, K value will be small. The formulas show that a straight line relationship can be obtained between yield strength or hardness and the reciprocal square root of the inter-twin spacing. It is considered that the relationship between yield strength or hardness and inter-martensite spacing is similar to inter-twin spacing, and it was observed that the stacking fault in the matrix of Co-1 alloy coating is wide (Fig. 7), and so the stacking fault energy is low. By reference to the data given by Murr (1978), the K_2 value of 316 stainless steel was taken as the K_2 value, i.e. $K_2 = 1080\ \text{MPa}\cdot\mu\text{m}^{\frac{1}{2}}$. The inter-martensite spacing was measured from Fig. 3 and Fig. 4, i.e. $d_0 = 0.5\ \mu$m; $d_1 = 0.1\ \mu$m. The computing result is $\Delta H = 1890$ MPa, which is in good agreement with the hardness elevation data shown in Fig. 6. If d_0 is infinitely great before deformation, the maximum ΔH at hardness peak value after testing will be 3400 MPa. Of course, a lot of factors may cause material strengthening, such as grain refinement; dislocation multiplication and so on. But it is considered that the H.C.P. martensite plays an important role in the strain hardening of cobalt alloy coating.

Fig. 7. The stacking fault in the matrix of cobalt alloy coating

Fujita (1979) had proposed a relationship between the maximum amplitude of the ratio of orthogonal shear stress to the Vickers hardness and the life for spalling fatigue. Based on the above experimental data of specimens No.8 to 11, a similar relationship can be obtained:

$$(A(\tau_{yz}/HV))_{max} = -0.03\ \lg N + 0.276$$

Using this formula, the contact fatigue life can be calculated and the required depth of alloy coating can also be estimated.

CONCLUSIONS

Among the cobalt alloy coatings studied, Co-1 alloy coating has the best contact fatigue resistance. It is expected to be used for repairing certain machine parts.
In the process of the contact fatigue, H.C.P. martensite induced by the repeated orthogonal shear strain strengthens the matrix of cobalt alloy coating. Consequently the contact fatigue resistance is improved.
The relationship between the inter-martensite spacing in cobalt alloy coating and yield strength or hardness is similar to Hall-Petch law. at the layer of maximum orthogonal shear stress, the inter-martensite spacing is finest and the strength and hardness are highest.
Based on the test data, an experimental formula has been derived for the Co-1 alloy coating:

$$(A(\tau_{yz}/HV))_{max} = -0.03 \lg N + 0.276$$

Based on this formula, the contact fatigue life and permissible load can be calculated and the required depth of coating can also be estimated.

ACKNOWLEDGEMENTS

The authors are glad to acknowledge the assistance given by Mrs. Wang Baochu and Mr. Xu Binshi for the specimens preparation.

REFERENCES

Smith, J. O., and C. K. Liu (1953). Stresses due to tangential and normal loads on an elastic solid with application to some contact stress problems. J. Appl. Mech., 20 157-166.
Merwin, J. E., and K. L. Johnson (1963). An analysis of plastic deformation in rolling contact. Proc. Instn. Mech. Engrs., 177, 675-685.
Michau, B., D.Berthe, and M. Godet (1974). Influence of pressure moduration in a linear Hertzian contact on the internal stress-field. Wear, 28, 187-195.
Murr, L. E., E. Moin, and F. Greulich (1978). The contribution of deformation twins to yield stress: the Hall-Petch law for inter-twin spacing. Scripta Metallurgica, 12, 1031- 1035.
Fujita, K. and Yoshida (1979). Surface durability of case-hardened nickel-chromium steel rollers under pure rolling and sliding-rolling contacts. Wear, 52, 37-48.

TRIBOLOGICAL CHARACTERISTICS OF STEEL 20Cr2Ni4 AFTER COMPLEX THERMOCHEMICAL TREATMENT

T. C. Gao and T. C. Lei

Department of Metals and Technology, Harbin Institute of Technology, Harbin, China

ABSTRACT

The tribological characteristics of steel 20Cr2Ni4 after carburizing and carburize-sulfurizing was studied. It was shown that an excellent resistance against adhesive wear may be obtained by a complex thermochemical treatment. The coefficient of friction and the wear rate values on the surface of wearing pairs of carburize-sulfurized steel were reduced in pin-ring tests. Together with other experimental data, the carburize-sulfurizing treatment is shown to be a prospective method for improving the tribological properties of various machine parts.

KEYWORDS

Tribological characteristics; carburizing; carburize-sulfurizing; pin-ring test; coefficient of friction; wear loss; wear rate; ferrography.

INTRODUCTION

The case-hardening treatments, such as carburizing, carbonitriding, boronizing, etc, are well-known as methods for effective improvement of wear resistance of steel parts due to enhanced surface hardness. On the other hand, the sulfurizing technique can be used for the reduction of adhesive wear by decreasing the coefficient of friction between wearing surfaces (Gao, 1978; Gregory, 1978). Therefore, a complex thermochemical treatment combining both surface hardening and surface lubricating processes might be thought to be more effective for improving the anti-wear properties of steels. The aim of the present study is to clarify the tribological properties of a commercial case-hardening steel after the complex carburize-sulfurizing treatment.

EXPERIMENTAL

The material used was steel 20Cr2Ni4 having 0.17%C, 1.62%Cr, 3.70%Ni, 0.30%Si and 0.38%Mn with S, P 0.030%. A pin-ring type wear test machine was used to determine the tribological parameters on pins of 4x10x20 mm and rings of Ø(32-16)x10 mm. All pins and rings underwent a carburizing process at 900 C for 17 h for a case-depth of 1.3-1.6 mm, then quenched from 830 C and tempered at 200 C for 1 h. Half of the carburized specimens underwent an electrolytic sulfurizing at 190 C for 15 min with a sulfurized layer of 10-15 um in ticckness. The loads selected for pin-ring tests were 50,100,150,200 and 250 N and the sliding speeds were 1,1.5 and 2 ms^{-1}. The wear products were examined by ferrographic technique and the worn surfaces were observed optically.

RESULTS AND DISCUSSION

As shown in FIG.1, the coefficient of friction of carburized specimens is nearly 0.3 in average for 50 N load and increases to about 0.4 for higher loads. Sulfurizing after carburizing effectively decreases the coefficient of friction to a value of 0.1 for 50 N and 0.1-0.2 for 150 and 250 N. In the meantime, the scattering of these data for carburize-sulfurized specimens is much less than that of carburized ones.

The weight loss of pins for various sliding speeds is shown in FIG.2 from which one can clearly find that carburized pins have quickly increased weight loss with applied load, but the carburize-sulfurized ones have very slight weight loss, nearly independent on the load applied. The wear rate to load relationship, as shown in FIG.3, also indicates the superiority of the complex treatment over single carburizing.

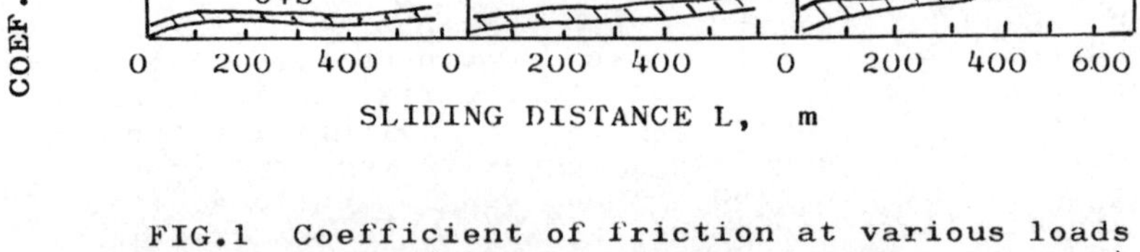

FIG.1 Coefficient of friction at various loads for carburized (C) and carburize-sulfurized (C+S) specimens. Shaded area shows scattering of data due to the variation of sliding speeds from 1 to 2 ms^{-1}.

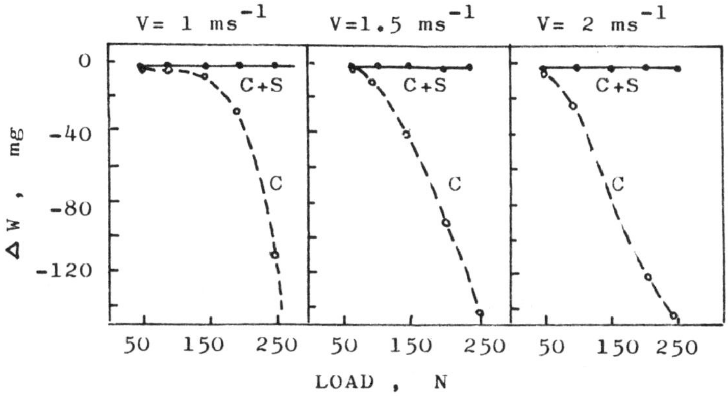

FIG.2 Weight loss of carburized (C) and carburize-sulfurized (C+S) pins after 500 m sliding at various speeds.

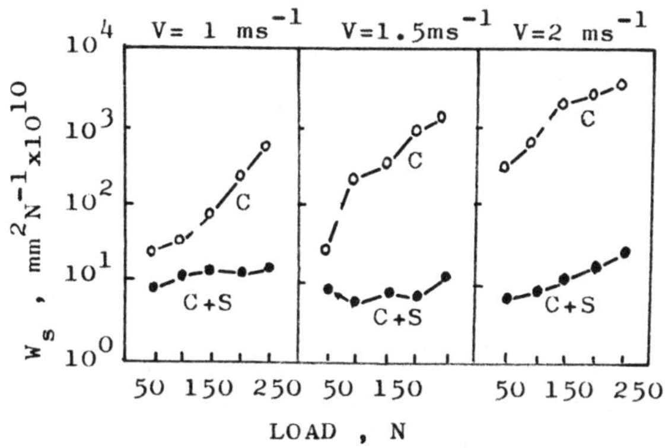

FIG.3 Wear rate of carburized (C) and carburize-sulfurized (C+S) pins at various sliding speeds. The wear rate $W_s = \Delta W/LP$, where ΔW is the weight loss, L is the sliding distance in m, and P is the applied load in N.

FIG.4 shows the morphology and the width of the worn tracks of pins after testing. It is evident that sulfurizing appreciably weakens the severity of wearing. The width of the worn tracks has been reduced nearly by a factor of 3 for the carburize-sulfurized pins in comparason with the simply carburized ones. The ferrographic picture taken with lubricating fluid, as shown in FIG.5, indicates that the debris of carburized specimens consists of particles > 50 um in size under 250 lbs (as well as 600 lbs), whilst particles of much less sizes were found for carburize-sulfurized specimens.

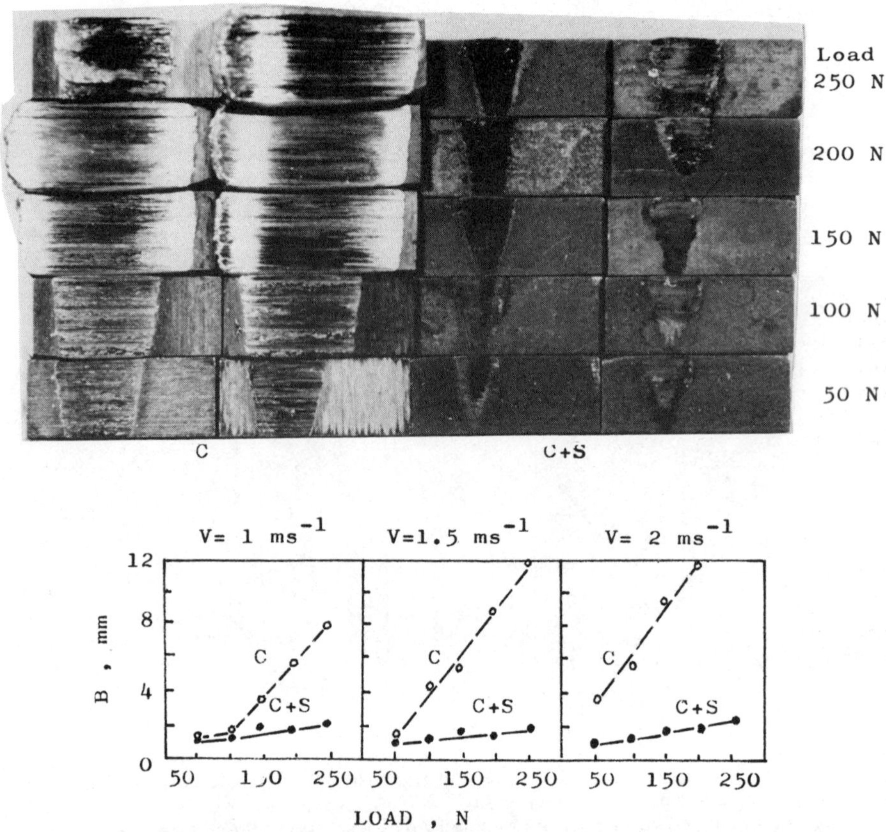

FIG.4 Morphology of worn tracks (x3.5) after 500 m sliding with a speed of 1 ms^{-1} for carburized (C) and carburize-sulfurized (C+S) pins and th the relationship between the width of worn tracks and the load applied.

C C+S

FIG.5 Ferrography of carburized (C) and carburize-sulfurized (C+S) specimens $\phi\frac{3}{4}$6.35 mm after 2 min testing in a Falex machine under 250 lbs load and a rotating velocity 300 rpm. The basic lubricating oil was used during testing. (x300)

The mechanism by which the sulfurizing improves the wear resistance of carburized specimens may be described as follows. A 10-15 um thick layer of FeS is obtained by sulfurizing which, having a hexagonal lattice structure and much less hardness,is lamellarly deposited on the carburized sublayer and very easy to slide due to reduced coefficient of friction. As a result, the surface instantaneous temperature of the wearing pair will be significantly decreased and the severe wear of carburized specimens will be replaced by a mild wear of carburize-sulfurized ones. Under high loads and high speeds of sliding,the increase of surface contacting temperature could lead to a slight smelting of the wearing material. At this time,since the FeS phase can form a low melting eutectic with the Fe matrix, this smelting would occur at relatively low temperature of surface,resulting in a partially lubricated wear condition instead of dry wear for simply carburized specimens.

The benificial effect of sulfurizing on the enhancement of wear resistance may be explained by two factors. The 1 st one is that a solid lubricating layer is soundly cohered to a hard, load carrying sublayer and thus forming an excellent from tribological point of view surface structure. The 2 nd one is that the low temperature (190 C) electrolytic sulfurizing process has no effect on the microstructure and properties of the carburized and quenched surface because tempering after quenching has been performed at 200 C,so that a high hardness value of the sublayer can be guarranteed. Indeed,there are in practice several methods for simultaneous cementation of C,N and S atoms into steels,but none of them can give sufficient surface hardness due to poor concentration of these elements compared with carburizing.

CONCLUSIONS

The complex thermochemical treatment including a low temperatureelectrolytic sulfurizing after carburizing of a commercial case-hardening steel shows benificial effect on the anti-wear properties under various loads and speeds of sliding. The carburizing process provides a hard, load carrying sublayer while sulfurizing gives a lamellarly deposited upper layer of FeS as a lubricating medium soundly cohered. The wear conditions are then changed from dry to partially lubricated and the results after wearing transformed from severe to a mild one. Therefore, the complex carburizing-sulfurizing process may be considered as one of the prospective methods for improving the wear resistance of various machine parts.

REFERENCES

Gao, T.C. (1978). Heat Treatment of Metals, No.3, 40

Gregory, J.C. (1978). Tribology International, 11, No, 2, 78

SYNTHESIS AND TESTING OF FOAM FILLED FRP SHELL FOR STATIC CRUSHING AND IMPACT

K. A. Jacob and J. Jayaraman

Structures and Materials Division, Aeronautical Development Establishment, Bangalore-560075, India

ABSTRACT

The paper describes the design details of an ogival nose cone intended to absorb energy during land impact. The energy is dissipated by progressive crushing of poly-urethane rigid foam. The crushing deformation characteristics of rigid foam under static crushing and suddenly applied loads are studied. The progressive buckling behaviour of the foam filled ogival shell is analysed by using a plastic wavefront approach and verified by static and dynamic testing. The paper provides guide lines for designing for impact energy absorption.

KEY WORDS

Composite material, rigid foam core, energy absorption, crush-deformation, impact dissipation, crushable nose cone, ogival shell.

INTRODUCTION

Certain remotely piloted vehicles (RPV) have a special requirement to have built-in energy absorption capability for the peripheral outerops such as nose, wing tips, empennage and nacelle which are likely to impact ground during recovery phase (Fig. 1.)

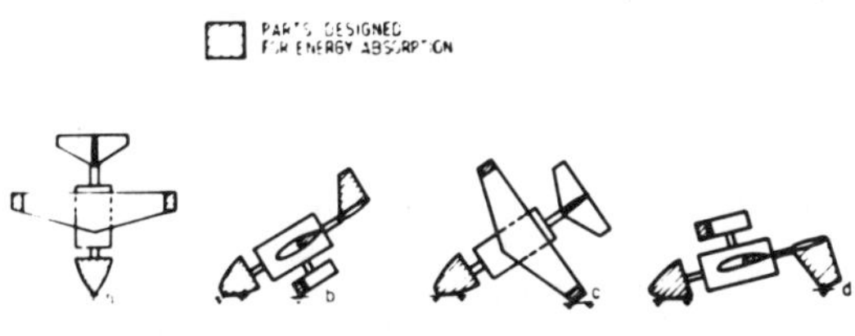

a. primary impact b. nacelle impact c. wing tip impact d. fin impact

Fig. 1. Recovery models for primary and secondary impacts

The first and primary point of impact is the nose of the vehicle in the vertical attitude recovery. The crushable nose cone is to be designed to absorb the shock of impact in such a way that the instrumentation subsystems are not subjected to shock decelerations more than that what is permitted (fragility index).

The crushable ogival nose cone consists of (i) a thin frp skin and (ii) a crushable core. One advantageous feature of crushing material energy absorber is that it is highly omni-directional and hence accommodative for oblique impacts and ground slopes (SCTM 284-57, Hexcel TSB 122, 1960).

The honeycomb material (aluminium or paper) is ideal from weight considerations. However, fabrication of multi-directional energy absorption system using this material is difficult and costly for a throw away (one time use) component. Figure 2 illustrates the different core materials considered for nose cone. The rigid foam is chosen as the crushable core material in the present study.

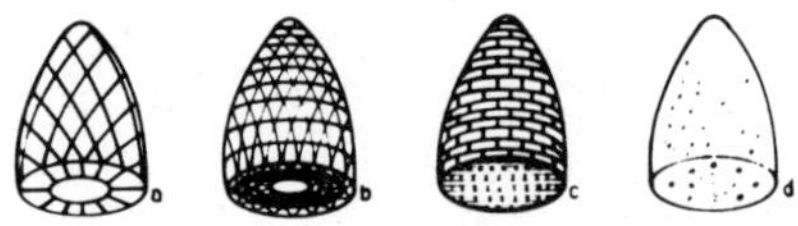

a. radial core b. tube core c. cross core d. rigid foam core.

Fig. 2. Different crushable core materials.

DESIGN CONSTRAINTS FOR ENERGY ABSORBING NOSE CONE

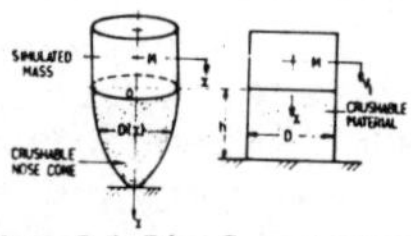

Fig. 3. Simplified parametric model.

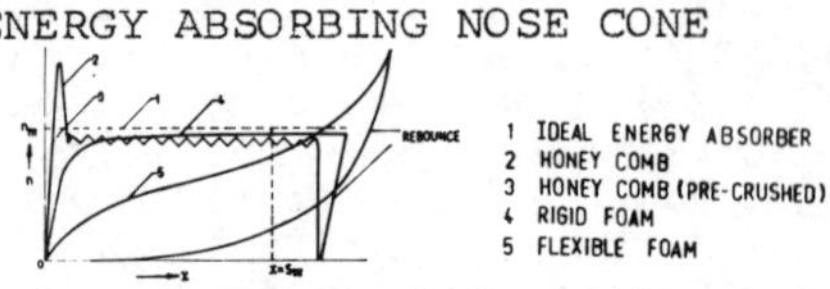

Fig. 4. Crush-deformation behavour of energy absorbing materials.

The main parameters to be considered for design are

(i) The entrance velocity (v_i), represented by free height drop $H = v_i^2/2g$

(ii) The retradation force (P), represented by deceleration coefficient n such that $P = Wn = Mng$, with maximum value n_m. W, the weight and M, the mass of vehicle respectively.

(iii) The displacement (x) during the period of decleration, with x_m the total distance to rest. By equating the kinetic energy at entrance to the energy absorbed we get

$$U = 0.5\, M v_i^2 = \int_0^{x_m} P(x,\dot{x})\, dx \tag{1}$$

(iv) The time interval (t) during the period of deceleration with t_m the time to rest.

The equation of motion of the system can be expressed as

$$M\ddot{x} + P(x,\dot{x}) = 0 \tag{2}$$

The additional design constraints are

(i) The maximum deceleration (n_m) shall be less than the fragility index (a_m) of the instrumentation, i.e., $n_m < a_m$

For the particular case considered here $n_m = 10$ is proposed.

(ii) The minimum stroke required (S_m) to stop the vehicle without exceeding the maximum deceleration limit, given by

$$S_m = v_i^2/2n_m \tag{3}$$

(iii) The jerk factor (the on set rate of deceleration dn/dt) shall be limited to a specified maximum value.

For gyro packages dn/dt = 200 is recommended.

(iv) The maximum volume of crushable material (V_m) required to absorb the kinetic energy is given by

$$V_m = 0.5\ M\, v_i^2/U_v \tag{4}$$

where $$U_v = \int_0^{\epsilon_m} \sigma(x,\dot{x})\, d\epsilon \tag{5}$$

which is the energy absorbed per unit volume of material. The gross volume (V_G) to be provided is

$$V_G = (1 + \bar{K})\ V \tag{6}$$

$\bar{K}$ is the crush factor, i.e the ratio of crushed volume (v_c) to original volume (v_o) for cellular materials. $\bar{K} = (v_c/v_o)$ varies from 20 to 30 percent for low density foams depending on density and cell size.

(v) The coefficient of retardation efficiency (η) which indicates the practical limit of stroke requirements compared to the ideal minimum associated with uniform deceleration. (Walker, 1945, Fisher, 1961, Fornhauser, 1964)

$$\eta = \int_0^1 y\,dx = S_m / x_m \text{(Refer Fig. 5.)} \tag{7}$$

where $x = x/S_m$, $y = n/n_m$, and $S_m = H/(n\eta - 1)$

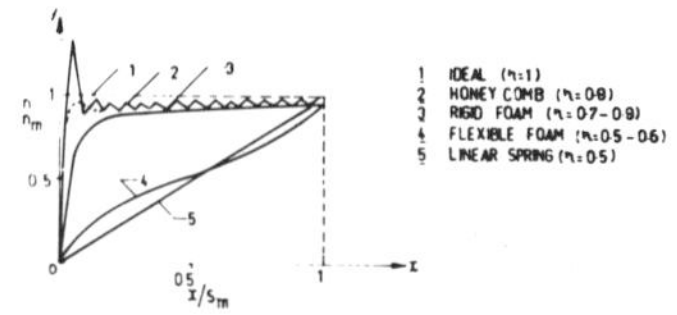

Fig. 5. Comparison of different material deformation systems

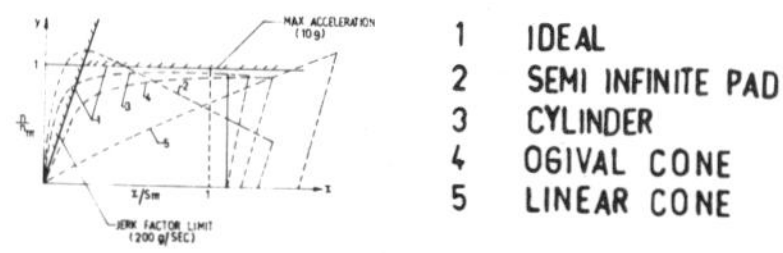

Fig. 6. System constraints for energy absorbing nose cone

CRUSHING DEFORMATION OF PUR FOAM

The load deformation curve for poly-urethane rigid foam (PUR) under static crushing can be represented by a linear elastic portion followed by a plastic crushing zone and a bottoming zone as illustrated in Fig. 7.

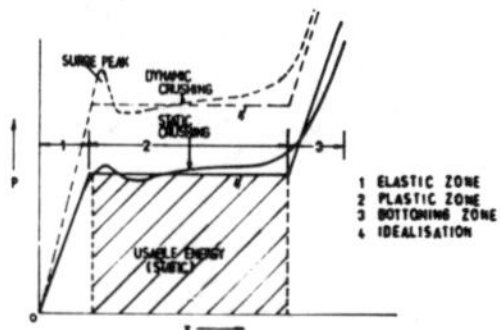

Fig. 7. The characteristic crush deformation curve for PUR foam

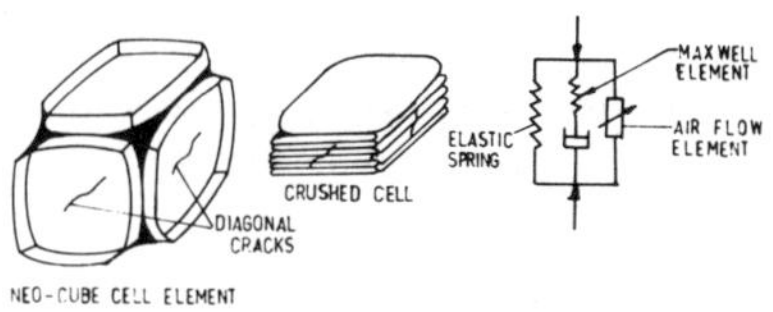

Fig. 8. Physical and idealised models

The neo-cube element model (Fig. 8.) can incorporate the variability of foam properties in rise, transverse and thickness directions. At a strain of around 5 percent the cell walls buckle and crack diagonally followed by crushing of the walls which are parallel to the loading direction. The airflow element represents the rate dependant flow of air after the cell walls are cracked.

The elastic stiffness of rigid foam (E_f) can be represented as a function of the elastic modulus of plastic matrix (E_p) and the density ratio ϕ ($\nu = \rho_f/\rho_p$)

$$E_f = E_p \phi^n \tag{8}$$

where n = 2 for rigid foam and 1.4 to 1.7 for flexible foams. The crush strength can be expressed as

$$\sigma_c = K_1 E_p \phi^2 = K_2 E_f \tag{9}$$

The constant K can be empirically determined or computed from cell geometry (Hilyard, 1982)
The strain rate effect on E_f and σ_c can be represented for the dynamic case as follows

$$(E_f)_D = (1 + \nu) E_f$$

$$(\sigma_c)_D = (1 + \nu) \sigma_c$$

where $$\nu = (E_D / E_S)_P \phi^{2/3} \tag{10}$$

E_D the dynamic (adiabatic) modulus and E_S the static modulus for the plastic material.
The sensitivity of E_f and σ_c on strain rate increases with density. (Lindholm, 1968, Andrews and Eirich, 1969) which conforms from eq.(10). The modulus of foam can be measured by ultrasonic velocity (C_f) measurement.

$$C_f = (E_f / \rho_f)^{1/2} \tag{11}$$

Most foams exhibit initial peaking and certain strain hardening after yield. The initial peaking (surge effect) is predominent at high strain rates, which may be attributed to the cyclic damping of elastic stress wave and machine reaction. The strain hardening effect can be explained by the curvature of plastic wave front and the airflow characteristics.

The temperature dependance of stiffness and crushing strength is negigible for rigid foam at normal operating temperatures (-20°C) to + 70°C) as compared to flexible foam. The temperature (τ) and humidity (θ) dependance can be incorporated in the relations for E_f and σ_c.

$$E_f(\tau,\theta) = E_p(\tau,\theta)\phi^n \tag{12}$$

$$\sigma_c(\tau,\theta) = K_1 E_p(\tau,\theta)\phi^n$$

CRUSH-DEFORMATION OF FOAM FILLED OGIVAL SHELL BY THE PLASTIC WAVE FRONT APPROACH

The total crush load is a sum of (i) the foam crushing load (ii) crushload of ogival shell and (iii) the synergistic load born by the interaction of shell and foam. The foam provides internal support for the shell and at the same time the shell provide confinement of foam; which results in an increased crush-load.

A representation of the crush deformation behaviour of the shell with foam core is given by considering the plastic wave front generated by the crushed foam. During the crush deformation the elastic energy is stored and released in a cyclic manner and dissipated by material damping. As the crush deformation increases, the already crushed foam gets distributed ahead of the advancing plane (cross head). The distribution of the crushed foam is important since the further crushing of virgin foam takes place at the interface of the crushed foam and the virgin foam. For foams with Poission's ratio nearly zero, the crushed foam distribution can be obtained by a transformation of the crushed volume into $r\,\theta\,x_1$ co-ordinates where x_1/x is the crush factor $\bar{K}$. For materials for which Poisson's ratio is greater than zero, the plastic wave front get distorted and the crushed foam distribution can be obtained by transformation into $r\,\theta\,x_2$ co-ordinates where factors r_1/r_2 and $x_2/\,x$ may be derived as a function of crush factor $(\bar{K})$ and Poisson's ratio (μ).

Here an approximate distribution of crushed foam is assumed to obtain the depth of debond, h (x) and the frontal area of the plastic wave, A (x).

The crushed volume $V\,(x) = V\,(\bar{K}\,/\,(1-\bar{K}))$ (13)

The values of h(x) and A(x) are computed by using relation (13) and the geometric considerations (Fig. 9.)

$$h\,(x) = (V_c - V_2)/\pi\left[\{r(x) + r\,(x+h)\}\,/\,2\right]^2$$

where $V_2 = \pi\,\delta^2(R - \delta/3)$

$$A\,(x) = 2\pi\,R\,(x)\,\delta\,(x) \qquad (14)$$

The crush load is then

$$P_c\,(x) = C_1\sigma_c\,A(x) + P_{pb}\,(x)\,/2 \qquad (15)$$

$$P_{pb}\,(x) = \sigma_{pb}\,(x)\,\pi\,D\,(x)\,t \qquad (16)$$

$$\sigma_{pb}(x) = \gamma\left\{E_c/[3(1-\mu_c^2)]^{\frac{1}{2}}\right\}(t/r)\,\cos\alpha\left\{(1/2)/\ln\,(S_2/S_1)\,[(1+S_1/S_2)/(1-S_1/S_2)]\right\}^{\frac{1}{2}} \qquad (17)$$

γ varies from 0.15 to 1 depending on end conditions and imperfections.

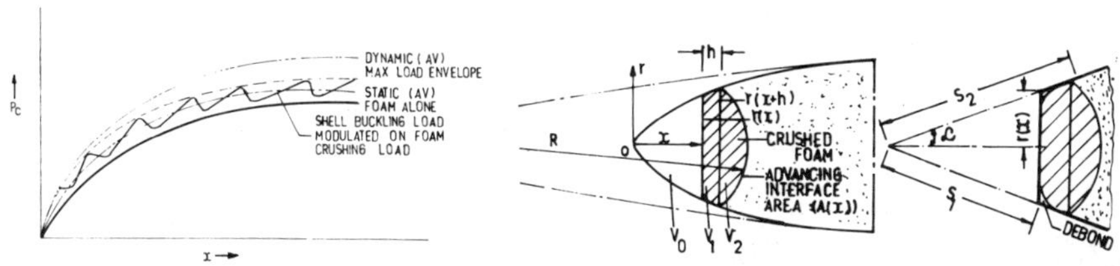

Fig. 9. Plastic wave front and depth of debond for foam filled ogival shell

It is assumed that the shell gets debonded from the foam for a height h(x) where the crushed foam moves relative to the shell. Then the shell in the debonded area can be treated as an approximate frustrum of a cone for which the buckling load is given by eq. (17). The buckling load builds up to $P_{pb}\,(x)$ and then the shell buckles and the load carried by the shell reduces. This produce a saw toothed oscillatory load which is modulated on the foam crush load. This approach gives fairly good prediction of the crush load for the foam filled ogival shell and compare well with the static crush load determined experimentally. For static crushing C_1 is unity in eq. (15) and for dynamic case $C_1 = (\sigma_c)_D/\sigma_c$

THE GEOMETRY EFFECT

In the case of crushable nose cone, the area of cross section varies with travel. The material crushing behaviour for non-constant cross sections can be better represented by a plot of crush strength versus volumetric strain as illustrated in Fig. 10.

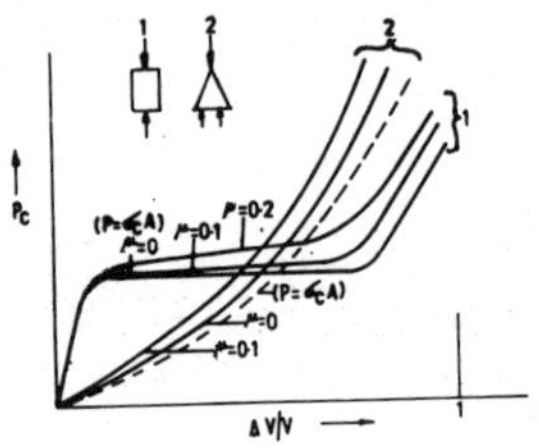

Fig. 10. Crush strength as a function of volumetric strain

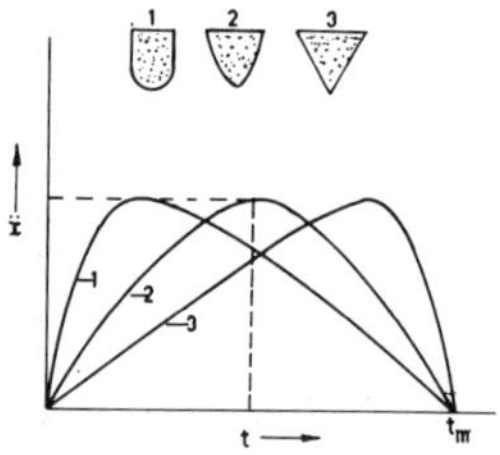

Fig. 11. Dependence of load pulse shape on geometry of cone

The shape of impact load pulse depends upon the geometry of the cone as illustrated in Fig. 11.

When the nose cone is crushed axially, the resisting force builds up proportional to the area of cross section. This assures a progressive crushing starting from the cone apex. For most efficient performance the force build up shall be limited to the value $P = W\, n_m$. This is possible by properly tailoring the core material by curtailing. This is implemented by removing extra core material by drilling holes from the base of the core to the required depths (cut off points) as illustrated in Fig. 12.

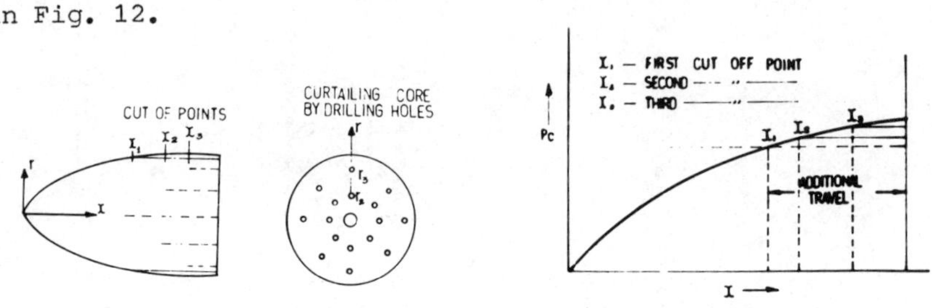

Fig. 12. Tailoring of core to avoid bottoming force

The drilling of holes also helps to avoid bulging of foam due to Poisson's effect and the resulting hoop splitting of the shell.

FABRICATION

The foam filled ogival shell is fabricated using a female mould. The glass cloth is laid up in the mould and impregnated with epoxy resin (LY 556 + HY 951). The pressure is applied using a silicone rubber pressure mandrel as illustrated in Fig. 13. Once the shell is cured the liquid components (polyol-isocynate) of PUR foam are mixed and poured to foam the core in-situ. The density of core is controlled by the weight of liquid foam ($\rho_f = W_f/V$). Casting of core under foaming pressure assured fairly uniform cell size and isotropy. The pot life of foaming compound can be increased either by cooling it before mixing or by adding small quantity of phosphoric acid (Ferrigno, 1963). After curing, the core is trimmed and a base plate is attached by epoxy bonding.

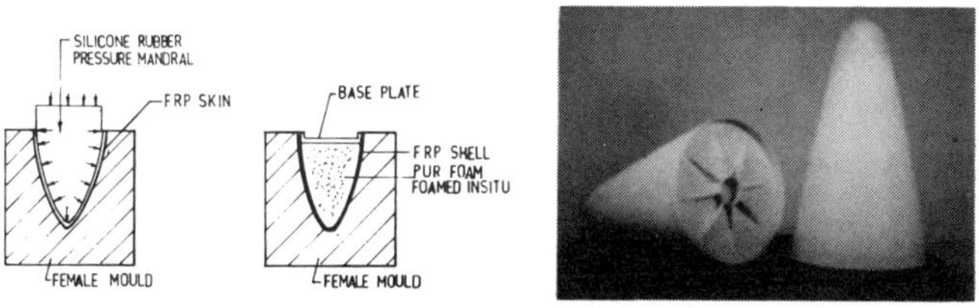

Fig. 13. Fabrication of foam filled ogival shell

TESTING

The foam filled frp shells are tested for static crushing and the load deformation curves are obtained. A representative crush deformation curve is presented in Fig. 14. The shell alone and core alone tests are also conducted to illustrate the synergistic effect due to interaction of foam core and shell.

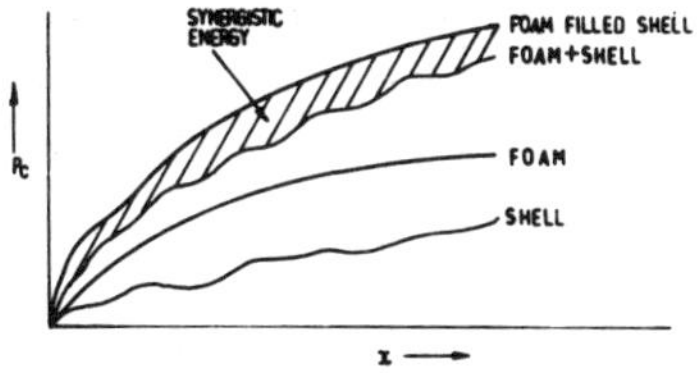

Fig. 14. Static crushing behaviour of foam filled ogival shell

For dynamic testing a weight of 400 Kg (simulating the weight of the vehicle) is attached to the nose cone and the assembly is dropped from a height of 2ms to simulate the entrance velocity of 6.2 m/sec. A shock accelerometer is mounted on the dummy weight to measure the deceleration against time. The weight is released using an electromagnetic release mechanism (EMRU) which was remotely controlled. The acceleration-time data is recorded on a storage oscilloscope. A high speed movie-400 frames/sec is also synchronised with the drop for obtaining time displacement data. The movie results were analyzed using a Vanguard motion analyser.

The dynamic deceleration history is given for a typical drop test in Figs. 15 and 16.

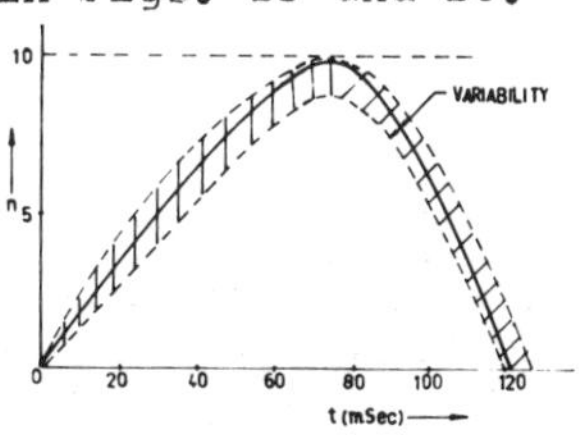

Fig. 15. Acceleration-time curve for nose cone

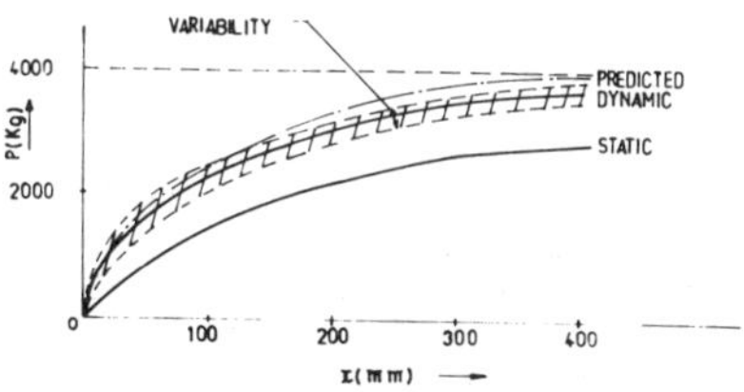

Fig. 16. The acceleration versus distance curve for nose cone

Figure 17 illustrate the crushing of nose cone under dynamic conditions.

Fig. 17. Stills from a drop test

DISCUSSION AND CONCLUSION

The static and dynamic tests show that the crushing deformation and energy dissipated by the foam filled frp shell can be predicted with reasonable accuracy. The plastic wavefront approach predicts the crushload realistically for a variable cross section cellular material compared to the equivalent area method. The frp shell provide confinement of foam which in turn increase the stiffness and strength of foam and the foam provide internal support for the shell which assures a progressive crushing. In effect, an interactive synergistic energy is predicted and which is verified experimentally. The geometry effect on shape of impact load pulse need further investigation to predict the behaviour in advance. The tailoring of foam to avoid bottoming provide a factor of safety in the system by providing additional travel.

The effect of oblique impacts, ground yielding and irregularity of surface etc need further study to incorporate these features in the design.

REFERENCES

Andrews, R.D. Jr. and Eirich, F.R., Eds., High Speed Testing, John Wiley and Sons, 1969.

Bailey, S.C., Experimental Studies on the Instability of Circular Cylindrical shells, Ph. D., Thesis, Standford University, 1967.

Ferrigno, T.H., Rigid Plastic Foams, Reinhold Publishing Corpn 1963.

Fisher, L.J.Jr., Landing Impact Dissipation Systems NASA TND-975, 1961.

Hilyard, N.C., Ed., Mechanics of Cellular Plastics, Appl. Science Publishers, 1982.

Kornhauser, M., Structural Effects of Impact, Spartan Books Inc., 1964.

Lindholm, U.S., Ed. Mechanical Behaviour of Materials Under dynamic Loads Springer Verlag, 1968.

Walker, P.B. Retardation efficiency of shock absorber and arrester gears, J. Roy. Aero. Soc. 1945, 652-657

Report SCTM-284-57, Energy Absorbing Characteristics of several Materials, (51), Sandia, Livemore, California, 1960

Hexcel TSB 122, Design Data for Preliminary Selection of Honey-Comb Energy Absorption Systems.

FRETTING IN ELECTRICAL CONTACTS

M. Braunović

IREQ, Institut de recherche d'Hydro-Québec, Varennes, Québec, JOL 2PO, Canada

INTRODUCTION

Fretting is a phenomenon that, in many instances, constitutes a problem of considerable practical importance. It has only recently been recognized as a serious factor in the degradation of electrical contacts. (Antler, 1982; Bock, 1975; Braunovic, 1977; Freitag, 1975). The process is defined as accelerated surface damage occuring at the interface of contacting materials subjected to small oscillatory displacements. Since this relative motion is of limited amplitude, it is ineffective in cleaning away the wear debris and accumulating oxides, and a highly localized, thick, insulating layer is formed in the contact interface leading to a rapid and dramatic increase in contact resistance and subsequently to virtual open circuits.

The nature of fretting depends on a large number of experimental and evironmental variables. The degree of surface damage depends on the contact material and the metallurgical state. An unified model for mechanical joints has not yet emerged in spite of considerable work over 50 years. The passage of current through the contact makes the system even more complicated. Detailed accounts of the present state of the art have been given by Waterhouse (1972) Golego (1974) and Hurricks (1970) where the detrimental effects of fretting on electrical contacts are described in some detail.

In spite of relatively scarce data on the effect of fretting on the performance of electrical contacts, the phenomenon seems to be very important in practical applications. This is because the required oscillatory movement of the contacting members can be produced by differential thermal expansion of the metals, by mechanical vibrations, load relaxation, and by junction heating as the power is turned on and off. Fretting would seem to be inherent in all contacts and with certain materials could lead to progressive deterioration of the contacting interface. This, in turn, would create the necessary conditions (higher contact resistance and heat) for the acceleration of other processes such as formation of intermetallic compounds in disimilar metal joints, creep etc., leading to failure. Hence, fretting may not lead to failure in itself, but it is definitly one of the prime factors causing the electrical instability leading to failure of a joint.

The work presented here is concerned with the effect of fretting on the contact resistance of aluminium wire in contact with different contact plating materials. An attempt has been made to explain the damage produced by fretting action under different conditions and the effect of different plating materials on aluminium/metal contacts.

EXPERIMENTAL

The materials used in this work were aluminum No. 12 AWG wire (NUAL grade) and a selection of contact materials including cadmium, tin, zinc platings and copper (ETP). The contact materials were plated on flat 1x1.5 cm coupons of 70/30 brass. The thickness of the plated materials was in the range 3 (Zn) to 30 (Sn) microns. NUAL grade aluminum wire is basically dispersion hardened Al - 0.77% Fe - 0.15% Mg alloy.

A special apparatus was constructed to produce fretting action. It has ten testing fixtures, each consisting of a specimen holder, counter-balance, and a system for producing fretting motion. The test apparatus uses a fixed rider on a movable flat specimen configuration. The rider was a U - shaped aluminum wire with a bend radius of 0.5 cm, nating with a movable flat contact plate held by teflon base mounted on the aluminum platform. Two systems were used to produce the platform motion. In one, the fretting motion was produced by a vibration system comprised of a signal generator, power amplifier, and vibrator. A displacement of the desired amplitude and frequency was produced by the signal generator, then amplified and transmitted via the vibrator to the platform. In the second, the fretting motion was produced by the thermal expansion of a copper rod heated electrically by a manganin coil would around it. Both systems were capable of producing amplitudes of a fraction of a micron. A constant load of 5 N was maintained by placing a dead-weight on the specimen holder.

The actual displacement of the platform was measured utilizing a DC/DC LVDT that was sensitive enough to detect a displacement of a fraction of a micron. The contact resistance was derived from the contact voltage measured between the aluminum wire and the flat coupons. The contact voltages of the ten fretting fixtures were measured using the cross-rod four-terminal technique. The contact voltages were recorded via datalogger every minute. The output from the datalogger was transferred to magnetic tape for computer processing and the results were plotted. A DC current of 100 mA was maintained through each contact via a constant current power supply.

Upon completion of the fretting tests, the contact areas of both wire and contact plates were examined with the Scanning Electron Microscope (SEM). Elemental compositions of the contact areas were determined by the energy dispersive analysis of X-rays (EDAX). The main advantage of this type of X-ray fluorescence analysis is the ability to identify a wide range of elements simultaneously and to obtain a quantitative measure of their concentrations in the contact regions under examination.

RESULTS

Contact resistance of the wire - contact plate material combinations as a function of fretting time are shown in Fig. 1. The contact resistance plots are the result of one run made with each material combination.

The contact resistance behavior, with the exception of Sn, is characterized by two stages. In the initial stage, the contact resistance remains virtually unaffected by the fretting action, indicating that a stable metallic contact is maintained between the contacting interfaces. In the second stage, after a certain fretting time, the contact resistance shows rather sud-

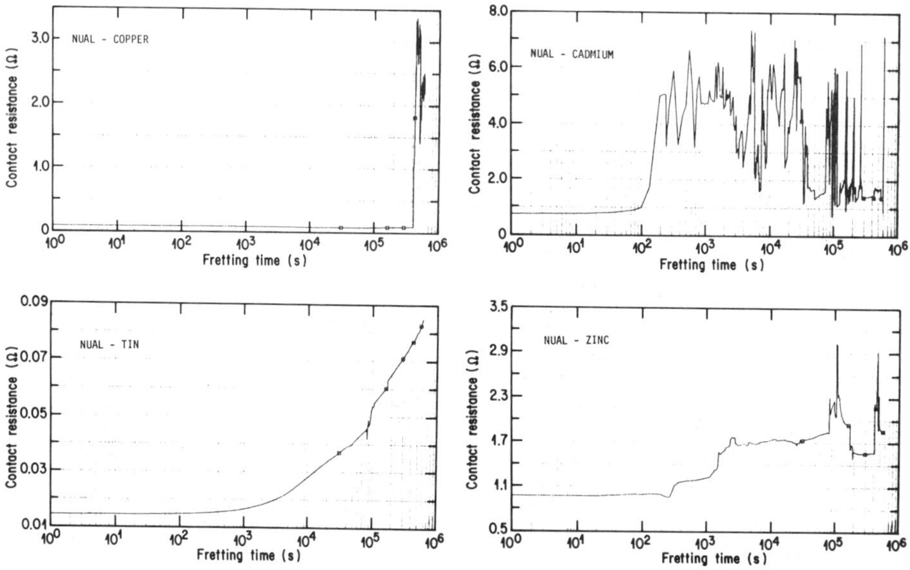

Fig. 1. Contact resistance as a function of fretting time for various combinations of aluminum wire and contact materials. Amplitude - 1 micron; Frequency - 1 Hz; Load - 5 N; Current - 100 mA (DC)

den fluctuations with time, indicating that the stable metallic contact between the contacting surfaces is lost. In the case of Sn plating, the contact resistance gradually increases with fretting time, indicating that progressive loss of a stable metallic contact is taking place, with build-up of fretting debris between the contacting members. If the contact materials are to be rated according to their performance under the fretting conditions, their ranking will be as follows: Sn - recommended, Cu - acceptable, Cd, Zn - not recommended and should be avoided.

Because of its rather stable and characteristic variation of contact resistance with fretting time, the combination of aluminum wire - tin plating was selected as the principal system for studying the effect of slip amplitude and frequency of oscillations. Furthermore, tin is widely used as a contact plating material and thus there is additional interest and importance attached to the study of the behavior of this system under different fretting conditions.

The effects of fretting slip amplitude at 1 Hz on the contact resistance is shown in Fig. 2. It is clear that the smaller the slip amplitude (1 micron) the shorter is the time to attain a given increase in the contact resistance. The observed effect can be rationalized in terms of kinetics of debris removal and also the formation and closure of the conducting spots in the contact area. In other words, the shorter slip amplitude limits the rate of debris removal and reduces the probability of formation of new metallic contacts in the contact zone thus reducing the number of conducting spots which in turn shortens the time to onset of the increase of contact resistance. On the other hand, longer slip amplitudes facilitate debris removal and increase the probability of formation of new metallic contacts thus maintaining and/or providing more conducting spots which, in turn, delays the increase of contact resistance.

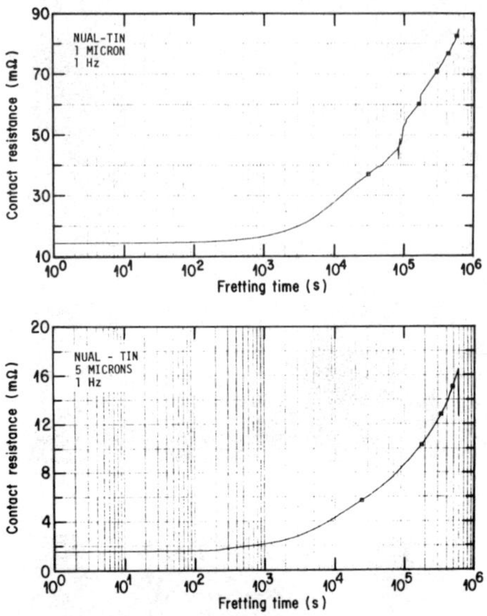

Fig. 2. Effect of fretting slip amplitude on the contact resistance of Al - Sn couple

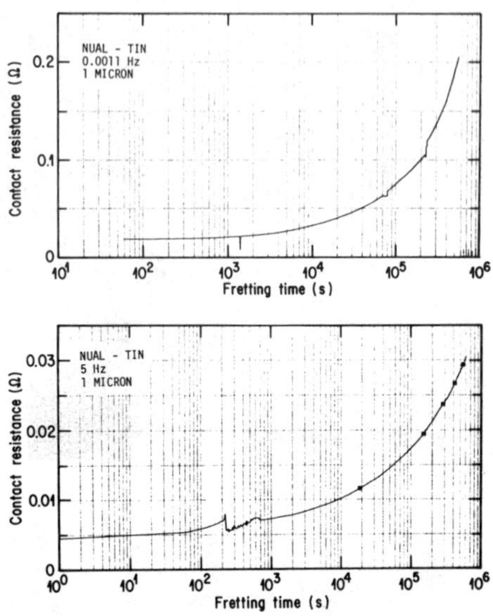

Fig. 3. Effect of frequency of oscillation on the contact resistance of Al Sn couple

The observed effect of slip amplitude on the contact resistance behavior is in accordance with the results of Golego (1974) who have found that during fretting the intensity of surface damage for different materials is higher at low slipamplitudes and that it increases and stabilizes at some higher slip lengths.

It is interesting to note that Antler and Sproles (1982) have recently found that in Pd contacts during fretting the time (or cycles) required to attain a given increase in contact resistance diminishes with increasing slip amplitude in the range 10 - 160 microns and that this is independent of frequency from 1 - 8 Hz. One possible explanation for this apparent contradiction is that the operating mechanisms responsible for the observed effects may be different. In the case of Pd contacts, frictional polymerization was cited as the operating mechanism during fretting, while in the case of aluminum contacts, it is the accumulation of wear debris and oxides that causes contact resistance to increase with fretting time. Since polymers are strongly adherent to the surface, it is possible that as the fretting distance increases more polymer is being smeared and accumulated over the contact zone thus reducing the number of conducting spots. On the other hand, oxides and wear debris are not adhering to the surface as much as polymers and thus can more effectively be removed by the wiping action of the contacting members.

The effect of oscillation frequency on the contact resistance was studied by subjecting the wire-plate couples to fretting at two different vibration frequencies, of 0.0011 Hz and 5 Hz, while keeping the slip amplitude in both cases at 1 micron. The results are shown in Fig. 3. It is evident that the lower the frequency the shorter is the time to reach a given level of eleva-

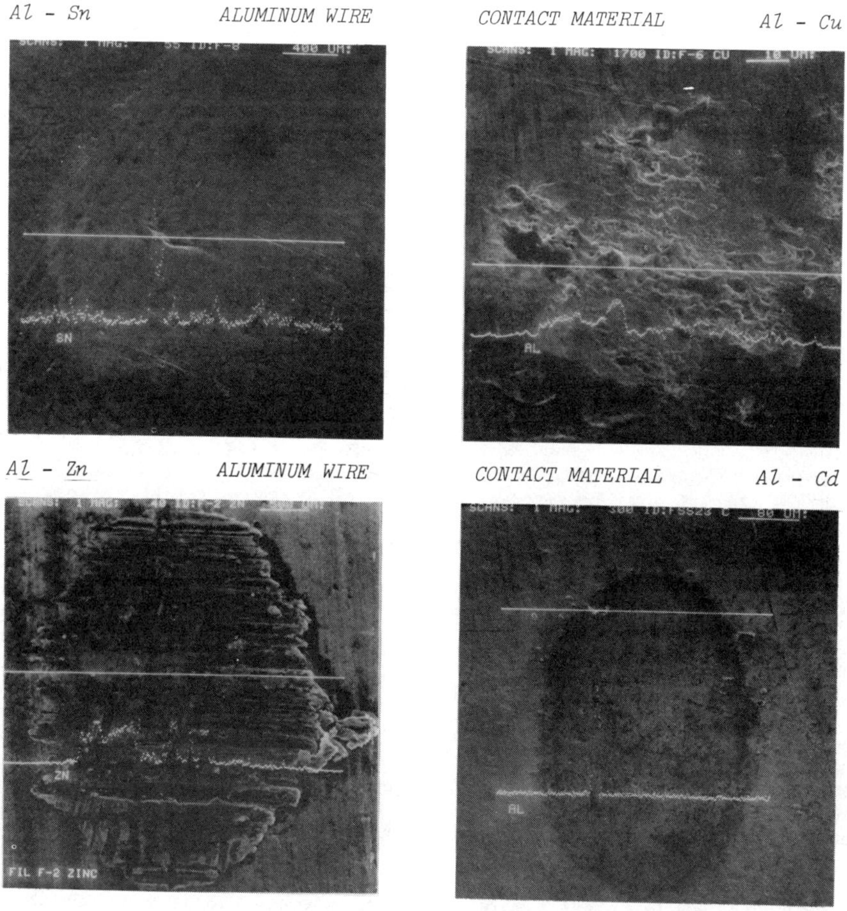

Fig. 4. Some representative SEM micrographs of the fretting damage on aluminum wire and contact materials with elemental x-ray line scans

ted contact resistance. The observed effect of frequency can be explained in terms of oxidation factor of fretting. Since oxidation is time dependent, at lower frequencies for a longer time the contact zone will be exposed by fretting to oxidation as a result of which an increasing number of conducting spots in the contact zone will be closed, thus increasing the contact resistance. The observed effect of frequency of oscillation on contact resistance is similar to weight-loss dependence as found in ferrous systems (Waterhouse, 1972).

Upon completion of the fretting tests, the contact areas of both contact materials and aluminum wire were examined and analysed by Scanning Electron Microscope (SEM). Some representative examples of the SEM examination of the fretting damage on both contact materials and wire are shown in Fig. 4 along with the elemental X-ray line scans of the damaged areas. It is clear that fretting action usually results in considerable damage to the contact areas of both contact materials and wire. Furthermore, it is also evident from the elemental X-ray line scans that a significant material transfer occurs in both directions. A detailed SEM examination of the damaged areas in the contact zones showed that fretting produced considerable wear on both wire and contact materials. Identification of a particular form of wear and associating it with a particular wire-plate combination is rather difficult since the features and effects of different types of wear were present in the contact zones of almost all wire-plate combinations. Typically, the characteristic features and effects of abrasion, adhesion and delamination wear are readily found in the contact zones of both wire and platings. However, a detailed analysis of the observed wear effects in the contact zones is beyond the scope of this work.

DISCUSSION

The results show that fretting exerts a pronounced effect on the contact resistance behavior of aluminum conductors and common contact materials. From the results, it is clear that complex physicochemical processes occur during fretting in the contact zones of metallic surfaces. The effect of fretting on the contact resistance of aluminum wires in combination with different contact materials seems to be characterized by two stages. In the initial stage, it remains virtually unaffected by the fretting action. This indicates that a stable metallic contact is made between the contacting surfaces and can be explained in the following way. When two metals are pressed against each other, surface asperities will penetrate the natural oxide films providing a good electrical contact. Also, due to highly localized contact stresses, the asperities will be severely deformed and there will be mechanical seizure of the contacting metallic surfaces. Displacement of the contact members during fretting action will shear these microscopic bridges resulting in the formation of the first wear products, a fraction of which will oxidize although the greatest proportion will remain as metallic particles. Consequently, a good metallic contact between the contacting metals will be maintained through this conducting debris. However, with further fretting action the degree of seizure between the surfaces will diminish and escape of wear products from the friction zone will be limited resulting in build-up of an insulating layer of oxides and reduction in the true contact between the metals.

In the second stage, after prolonged exposure to the fatigue-oxidation process, the metallic layers become softened and progressive separation of these layers occurs. The contact zone consists of a thick oxide layer and the remaining metallic contact is lost resulting in a rather sharp increase in the contact resistance. The cyclic nature of the contact resistance manifested by short-term fluctuations is most likely associated with temporary rupture of the oxide layer and appearance of localized metallic contacts providing good conducting paths. Subsequent wiping action and accelerated oxidation due to the high current density and thus heating in the contact zone will rapidly eliminate these conducting spots and the resistance will rise sharply. The persistance of the cyclic fluctuations in the contact resistance indicates that the generation of the conducting spots by mechanical rupture of the intermediate layer of oxides and fretting debris is counter balanced by the closure of these newly formed spots by further fretting action.

The proposed model can adequately well describe the behavior of contact resistance during fretting of aluminum wire in combination with most of the contact materials. However, in the case of Sn plating, neither sharp rise nor fluctuations but, rather, a smooth increase of the contact resistance was observed in the second stage of fretting. This could be associated with the extrusion of softer Sn into the fissures of the intermediate layer of oxides and fretting debris, thus maintaining good current conducting paths. As the thickness of the intermediate layer increases with further fretting, the number and efficiency of these conducting paths decreases resulting in a gradual augmentation of the contact resistance.

An important finding is that both fretting slip amplitude and oscillations frequency affect considerably the contact resistance behavior. It is clear that at low slip amplitudes and frequencies the intensity of fretting damage and thus the increase in contact resistance are higher than that at higher frequencies and longer slip lengths. Both effects can be rationalized in terms of the kinetics of oxidation and debris removal from the friction zone. At higher frequencies the contact zone will be exposed to oxidation for shorter times than at lower frequencies. Similarly, at lower slip amplitudes the rate of removal of fretting debris from the friction zone will be limited and thus the probability of generating new spots of metallic contact will be reduced. The effect of slip amplitude and frequency are of significant practical importance since aluminum connections during normal service lifetime are exposed to a range of intermittent and continuous movements of various lengths and durations. These movements can be caused by large thermal fluctuations or by mechanical disturbances such as repetitive vibrations produced, for instance, by operation of various electrical power apparatus, or by natural vibration of building etc.

The results of SEM surface studies showed that fretting causes considerable damage of the contact zones and that intensive exchange of materials takes place for most contact materials. It is unlikely that a single mechanism is responsible for the observed metal transfer. This is almost certain the result of the combined action of several mechanisms such as adhesion, abrasion, delamination wear, seizure of contacting metals, arcing, electroerosion etc.

In view of the ever-increasing use of aluminum conductors and the variety of interfaces with other metals, it is apparent that fretting has serious implications in the search for reliable electrical contacts.

ACKNOWLEDGEMENT

The author wishes to express his gratitude to the Canadian Electrical Association which contracted IREQ to carry out this research work. Sincere thanks go also to Dr. E.A. Dancy of IREQ who read the manuscript and made valuable comments. The technical assistance of Mr. K.J. Smith of Aluminum Company of Canada in providing the NUAL grade aluminum wire is also gratefully acknowledged. Finally, the successfull execution and completion of this project would not have been possible without the technical assistance of K. Doghramadjian, A. Joly, A. Watier, and P. Dufresne of IREQ.

REFERENCES

Antler, M. and E.S. Sproles (1982). *IEEE Trans. Components, Hybrids, and Manufacturing Technology*, CHMT-5, 158-166

Bock, E.M. and J.H. Whitley (1975). *Electrical Contacts*, IIT, Chicago, 128-138

Braunović, M. (1977). *Electrical Contacts*, IIT, Chicago, 223-232

Freitag, W.O. (1975). *Electrical Contacts*, IIT, Chicago, 17-23

Golego, N.L., A.Ya. Alyabev, and V.V. Shevelya (1974). In *"Fretting Corrosion of Metals"*, Tekhnika, Kiev. (Russian)

Hurricks, P.L. (1970). *Wear*, 15, 389-409

Waterhouse, R.B. (1972). In *"Fretting Corrosion"*, Pergamon Press, London.

COMPARING THE ABRASIVE AND EROSIVE RESISTANCE OF SOME FLAME SPRAYED COATINGS AND NITRIDED STEEL SURFACES

U. Bryggman and S. Hogmark

Institute of Technology, Uppsala University, Sweden

ABSTRACT

The influence of nitriding and coating by flame spraying on the abrasive and erosive wear resistance of some steels is evaluated by laboratory testing. A considerable increase in resistance against cyclic abrasion was achieved by nitriding and the application of flame sprayed and sintered coatings. The different surface modifications also had a positive effect on the resistance to single point grooving, whereas the responce to particle erosion was worsened for angles of impingement greater than 45 degrees, as compared to the pure substrate steels.

INTRODUCTION

On a macroscopic scale wear is typically a surface phenomenon, and wear resistant materials can successfully be produced by surface modification, for example by nitriding or coating. The present investigation aims at predicting abrasion and erosion responce of surface treated and coated materials by laboratory testing. A simple and reliable test method is desired as a complement to expensive field trials. The influence of nitriding and flame spraying on abrasive wear resistance is discussed for each test situation considered.

Abrasive wear of a surface is characterized by the cutting or plowing action of hard particles or asperities during relative sliding. Usually, machine designers use bulk material hardness as an indicator of abrasion resistance (Khrushchov, 1957; Moore 1974). In general, more sophisticated parameters are required to satisfactorily predict material behaviour in specified abrasive situations (Bryggman, 1983 a ; Hogmark and co-workers, 1979).

Material removal during erosion by impinging hard particles occurs in a way similar to abrasion, but erosion resistance is more difficult to predict. It is obvious from several investigations that a combination of high hardness and high ductility is beneficial (Sheldon and Finnie, 1966; Söderberg and co-workers, 1982; Tilly, 1973).

EXPERIMENTAL

Test Equipment

Single point grooving was performed in a pendulum tester equipped with a radially mounted cemented carbide tip (Bryggman and co-workers, 1981), see Fig. 1 a. A high cycle, low stress abrader, Fig. 1 b, was utilized to perform repetitive scratching with the aid of a cemented carbide tip on spring loaded specimens (Bryggman and co-workers, 1979). The erosion resistance was studied with the aid of a centrifugal erosion tester, Fig. 1 c, at different angles of particle impingement. Airborne alumina particles of 125-250 μm diameter and 47 m/s velocity were used as erodant. Testing was performed at room temperature.

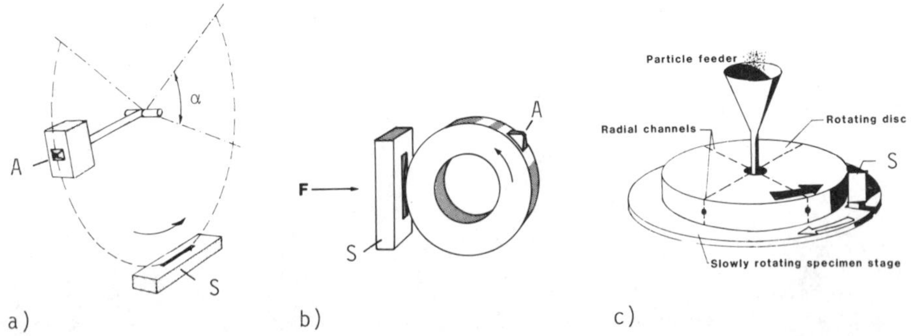

Fig. 1. Testing equipment. S = specimen, A = abrasive tip.
a) Pendulum tester. Angle α corresponds to energy consumed during grooving.
b) High cycle abrader.
c) Centrifugal erosion tester.

Materials

Five different grades of structural steel were used as substrate materials. Chemical compositions and Vickers hardness values are given in Table 1. The materials are deliberately given low hardness values corresponding to those of the core material after nitriding.

TABLE 1 Substrate materials

Swedish standard	AISI (SAE)	HV_{300N}	Chemical composition [wt%] C	Si	Mn	Cr	Mo
SIS 2258	(52100)	300	1.0	0.2	0.35	1.5	
SIS 2244	4140	300	0.40	0.3	0.8	1.1	0.2
SIS 2172	1118	300	0.18	0.3	1.3		
SIS 1312	1112	135	0.12	0.25	0.6		
OXAR 450		540	0.25	0.5	1.3	0.5	0.2

Two different nitriding processes - ion and Tenifer nitriding - were applied on four of the substrate materials. Characteristics of processes and resulting layers are given in Table 2.

TABLE 2 Nitriding parameters and layer characteristics

Process	temp $[C^o]$	time [h]	Nitride layer type	Nitride layer thickness [μm]	Diffusion layer [μm]
Ion nitriding	500	24	γ'	8	250
Tenifer nitriding	570	1.5	ε	8	100

The different combinations of nitriding process and substrate steel are given together with the resulting surface hardness in Table 3. Steel 1312 was flame sprayed with three different coating materials and conventionally flame sintered according to Table 4.

TABLE 3 Nitrided steels and surface microhardness

Substrate	Hardness HV_{150N} Ion nitrided	Hardness HV_{150N} Tenifer
2258	570	450
2244	720	530
2172	610	
OXAR	850	

TABLE 4 Flame sprayed coatings

Coating	Sintered	HV_{300N}
Molybdenum	No	515
Stellite	Yes	645
WC/Co	Yes	840

RESULTS AND DISCUSSION

Single Point Grooving

According to Bryggman (1983 a) the results from single point grooving are recorded as specific energy consumption e (Nm/mg) as a function of groove size W (mg) and numerically fit to linear curve segments in logarithmical e(W) diagrams, Fig. 2. The correlation of the experimental data to the linear segments is very good, typically with a coefficient of correlation $r > 0.98$. The investigated groove size range covers most practical applications. It is clear from the figure that e decreases monotonously with increasing W. It is also obvious that e can be used for ranking of different materials. This ranking is, however, size dependent and will differ for different groove size intervals. It has been shown previously that ranking with respect to e is closely correlated to ranking with respect to abrasion resistance (Bryggman, 1983 b).

Substrate materials. Ranking of the untreated substrate materials with respect to energy consumption is not of prime interest, but a few remarks will be given. The best abrasion resistance is displayed by different steels within different E intervals, as summarized in the e(W) diagrams of Fig. 2. 2172 proved superior for $W < 1$ mg, while the OXAR steel has a better abrasion resistance above this W value.

There is virtually no correlation with bulk hardness, a fact which is due to the extreme conditions of deformation in the material ahead of the grooving tip. Material is removed in the form of chips and the energy consumption is strongly related to the mode of lamellar chip formation. The major part of the grooving energy is expended in localized shear zones between adjacent lamellae, and material properties, such as heat conductivity and resistance to dislocation glide at elevated temperatures,are of great importance. The mechanisms of chip formation in single pass pendulum grooving are further outlined by Bryggman (1983 a).

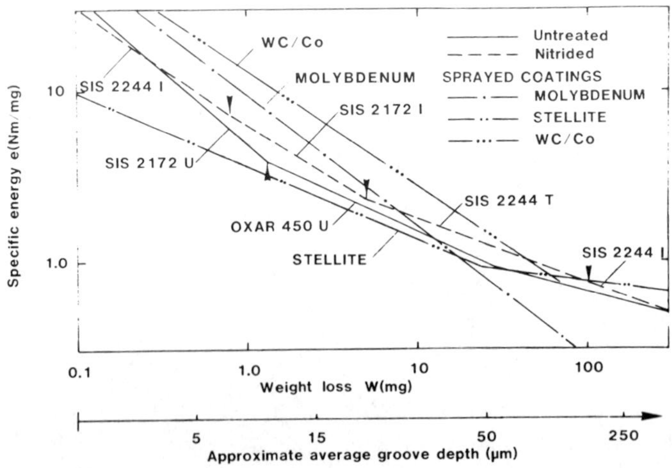

Fig. 2. Specific energy consumption versus weight loss. For the nitriding test series. Only the best untreated (U) and nitrided (Tenifer (T)), ion nitrided (I)) substrates are shown.

Nitrided substrates. Fig. 2 shows that both nitriding treatments significantly raise the energy consumption for almost all of the investigated groove size interval. Comparison with the layer thicknesses given in Table 2 indicates that the ion nitriding gives a higher abrasion resistance for groove depths in the range of the nitride layer thickness. For deeper grooves, predominantly within the diffusion layer, however, the Tenifer treatment gives the higher e values Eventually, the groove depth will exceed the 100 μm range of the Tenifer diffusion layer, and the ion nitrided substrates will dominate again. As to nitrided substrate materials 2244 generally seems superiour.

Flame sprayed coatings. Both the Mo and WC/Co coatings proved to posess a very high grooving resistance, particularly for small grooves. For the largest grooves, however, both materials lose their superiority due to the appearance of brittle fracture, see Fig. 3, and perform worse than even the untreated materials. The Stellite, on the contrary, which suffers only ductile deformation has a much lower abrasion resistance for smaller grooves but improves and ends up as the best material for the largest grooves.

High Cycle Abrasion

Resistance to combined abrasion and fatigue was tested by the high cycle abrader. Wear rate defined as the accumulated weight loss divided by the number of cycles was recorded, see Fig. 4. A decreasing initial wear rate

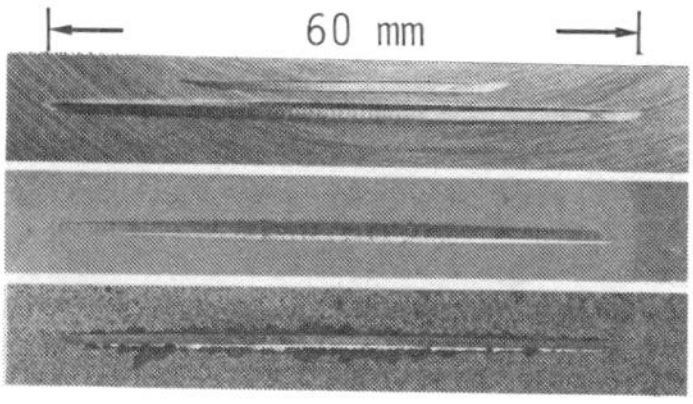

2244 Untreated

Flame sprayed Stellite

Flame sprayed Molybdenum

Fig. 3 Morphology of grooves from pendulum testing. Grooving direction: left to right.

versus number of cycles is due to self-strengthening of the tested materials according to Bryggman and co-workers (1979) and Eklund and Hogmark (1982), but also to increasing contact surface area between specimen and tip.

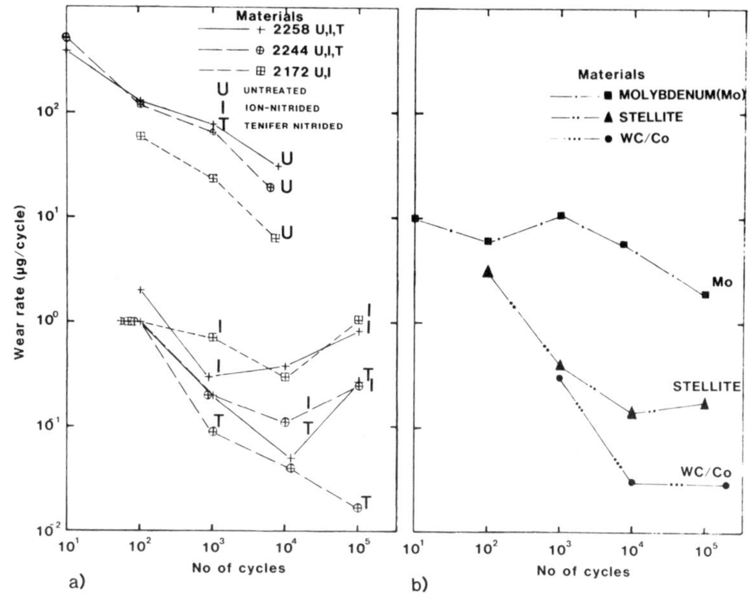

Fig. 4 Wear rate versus number of cycles as measured by high cycle, low stress abrasion.

Nitriding. Both nitriding processes result in a considerable increase in wear resistance, 1-3 orders of magnitude in this cyclic test, see Fig. 4 a. Steel 2244, though being rather poor in the untreated condition, displays the best performance after Tenifer nitriding, and no penetration of the superficial nitride layer had occured after 10^5 cycles in this material. Penetration of the nitride layer explains the positive slope of the wear curves after 10^3-10^4 cycles displayed by the other nitrided steels. Generally, Tenifer nitriding seems to be more efficient then ion nitriding, probably because it gives a more ductile nitride layer, less sensitive to surface fatigue.

Typical examples of wear scars from the high cyclic abrasion is shown in Fig. 5. Due to the elastic specimen loading, the soft and ductile substrate materials recieved a deep groove, while the hard nitrided substrates only recieved shallow scars.

Flame spraying. The cyclic abrasion resistance of flame sprayed Stellite and WC/Co is of the same order as that of the nitrided steels, see Fig. 4 b, with WC/Co giving the best performance at high cycles. As expected, the unsintered Mo layer is comparatively sensitive to this type of abrasion.

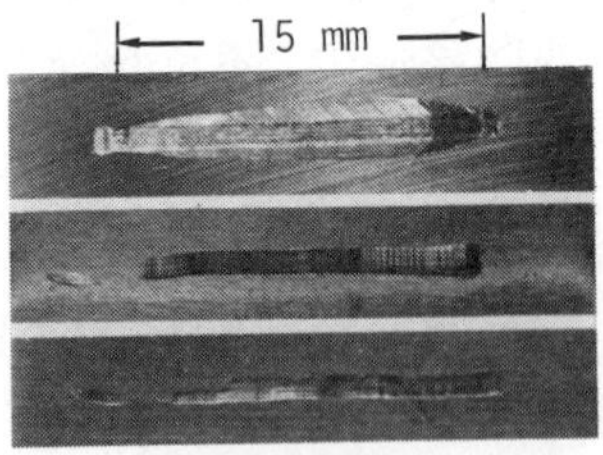

2244 Untreated

2244 Ion nitrided

2244 Tenifer nitrided

Fig. 5. Wear scars from cyclic abrasion.
Sliding direction: left to right.

Erosion

The diagrams of Fig. 6 summarize the erosion rate versus angle of impingement for all target materials tested. Both nitriding processes lead to a

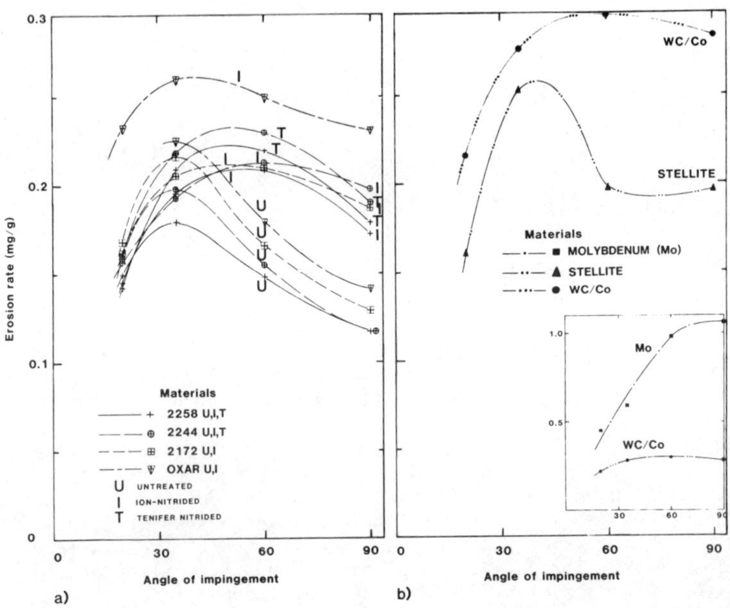

Fig. 6. Erosion rate vs angle of impingement.

considerable shift of the maximum erosion rate towards normal incidence, implying a strongly reduced erosion resistance for higher angles, see Fig. 6 a. This indicates that the increased hardness of the nitrided layers at the same time results in a brittle behaviour. This is well known from the litterature (Finnie and co-workers, 1967; Sheldon and Finnie, 1966).

The hard and somewhat porous flame sprayed and sintered materials displayed an even more reduced erosion resistance than the nitrided steels, see Fig. 4 b. The unsintered Mo layer turned out to perform extremely poorly, depending on its high porosity, and to display the most pronounced brittleness, probably because of its morphology of layered oxides.

CONCLUSIONS

For selection of materials of high abrasion resistance in service cases, where cutting mechanisms dominate over fatigue, a diagram of the type shown in Fig. 2 can be of great predictive value over a large range of damage sizes. When fatigue mechanisms dominate, however, cyclic abrasion rate and erosion rate diagrams, as exemplified in Figs 3 and 4, yield complementary information over the whole interval of angles of incidence.

REFERENCES

Bryggman, U., S. Hogmark and O. Vingsbo (1979). Evaluation of abrasion resistance by means of controlled grooving. <u>4th Int. Abrasion Coloquium</u>, Grenoble, 5-14.

Bryggman, U., S. Hogmark and O. Vingsbo (1981). Force and energy measurements during controlled grooving - A basic study of abrasive wear. <u>ASLE Trans. 24</u>, 449-458.

Bryggman, U. (1983 a). Abrasion resistance of some steels related to mechanisms of gouging wear. <u>UPTEC 8322 R</u>, Uppsala University, Sweden.

Bryggman, U. (1983 b). Comparison of test methods for prediction of gouging abrasion resistance of steel. <u>UPTEC 8321 R</u>, Uppsala University, Sweden.

Eklund, L-H. and S. Hogmark (1982). Wear resistance of friction layers on steel. <u>Scand. J. Metallurgy 11</u>, 226-232.

Finnie, I., J. Wolak and Y. Kabil (1967). Erosion of metals by solid particles. <u>J. of Mater. 2</u>, 682-700.

Hogmark, S., S. Söderberg and O. Vingsbo (1979). Surface deformation and frictional heating of steel. <u>ICM3, 2,</u> Cambridge, England, 621-631

Khrushchov, M.M. (1957). Resistance of materials to wear by abrasion as related to hardness. <u>Proc. Conf. on Lubrication and Wear</u>, London, 655-659.

Moore, M.A. (1974). The relationship between the abrasive wear resistance, hardness and microstructure of ferritic materials. <u>Wear 28</u>, 59-68.

Sheldon, G.L. and I. Finnie (1966). On the ductile behaviour of nominally brittle materials during erosive cutting. <u>J. Eng. Ind., Trans.ASME B, 88</u>, 387-392.

Söderberg, S., S. Hogmark, U. Engman and H. Swahn (1981). Erosion classification of materials using a centrifugal erosion tester. <u>Tribology Int. 14</u>, 333-343.

Söderberg, S., S. Hogmark and H. Swahn (1982). Mechanisms of material removal during erosion of a stainless steel. <u>ASLE Annual Meeting</u>, Cincinnati, Ohio.

Tilly, G.P. (1973). A two stage mechanism of ductile erosion. <u>Wear, 23</u>, 87-96.

THE SURFACE LAYER FATIGUE FAILURE OF HEAT-RESISTING STEELS AT HIGH TEMPERATURE AND THE PROTECTIVE EFFECT OF ALLOY COATINGS

Zhongxing Wu, Baochu Wang, Dashen Jin, Meiying Xie, Xin Liu and Shiqin Lu

The Central Iron and Steel Research Institute, Ministry of Metallurgical Industry, Beijing, China

ABSTRACT

The hot corrosion-fatigue resistance of heat-resisting steels and alloy coatings is studied in this paper. An examination of some components and samples has been carried out by means of metallograph, electron microprobe etc. The surface layer fatigue failing mechanism of heat-resisting steels under the hot corrosion condition and the effective protecting action of alloy coatings are discussed. These alloy coatings have been applied successfully to protect various heat-resisting steel components. It was awarded an invention prize by the Committee of Science and Technology of the People's Republic of China for the new process of "FNA coating valves" in 1980.

KEYWORDS

Hot corrosion-fatigue; surface layer fatigue failure; base metal; alloy coatings; surface protecting film.

INTRODUCTION

Heat-resisting steels are possessed of enough strength and surface chemical stability, so that they can avoid failure under the action of corrosion or cyclic stresses. But under combined action of corrosion, especially hot corrosion, and the cyclic stresses at hot burning gas, they often fail very early due to the interaction between the fatigue and corrosion. For example, the pre-burning chamber spraying nozzles usually fail in the fatigue accelerated by corrosion, the exhaust valves fail in the corrosion accelerated by fatigue. A corrosion protecting film is certainly formed at the surface of metal components operating on hot burning gas, and the protective action of this film on base metal depends mainly on its strength and physicochemical stability. In order to observe the protecting film of metal surface which fails in the combined action of hot corrosion and cyclic stresses, a "HCF" test method is used.

It is clearly shown that various heat-resisting steels suffer from serious failure while the Ni-base and Co-base alloy coatings keep in a good condition. This result provides the basis for protecting various heat-resisting steel components with alloy coatings. The coating method proposed here is a new surface-metallurgy technology. The original characteristics of coating and base metal can be retained and, indeed, the metallurgical bond between them can be formed firmly. In addition, it does no harm to the health of people and environment.

Analyzing for damage mechanism of valve and nozzle

The exhaust valves are generally made from heat-resisting steels such as $4Cr_{10}Si_2Mo$ and 21-4N etc. According to calculation, their high temperature yield strength is much greater than the cyclic stresses from which the exhaust valves suffer. Fatigue failure seldom occurs on exhaust valves, but they always fail in so called "Pockmarks" formed on their working surface. A typical "Pockmark" as shown in Fig. 1a is observed by means of scanning electron microscope. It is found that the metal itself is not immediately damaged by the cyclic stresses, but the surface protecting film is. This is a kind of corrosion damage mechanism accelerated by cyclic stresses.

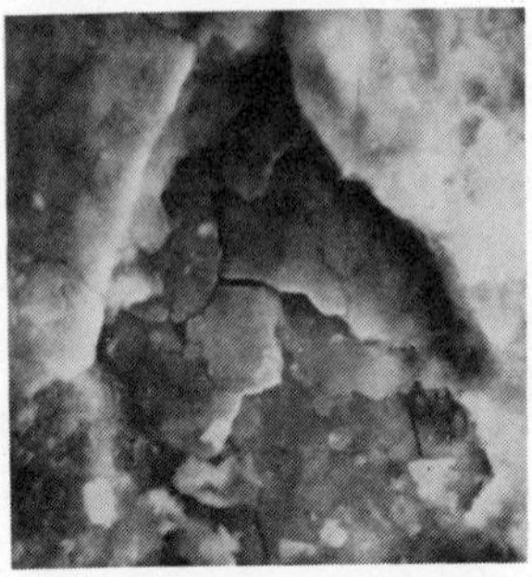

a. "Pockmark", x 2000.

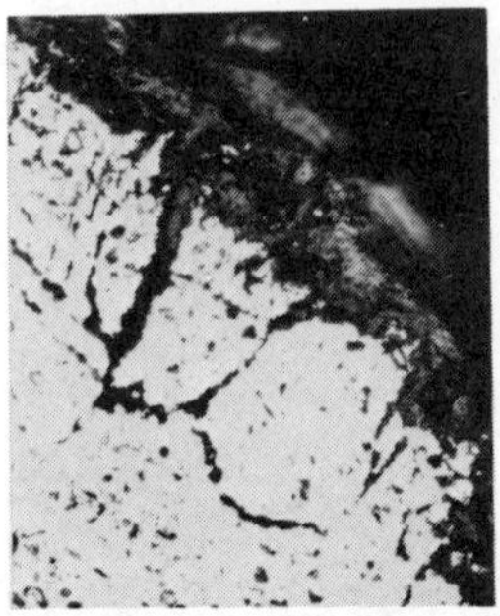

b. Intergranular corrosion cracks, x 400.

Fig. 1 Damage forms of heat-resisting steel components.

The spraying nozzles are generally made from heat-resisting steels such as $Cr_{20}Ni_{14}Si_2$ and Amco18SR etc., and suffer from a combined action of great cyclic stresses and hot burning gas corrosion. Intergranular corrosion crack as shown in Fig. 1b occurs very easily on its surface and accelerates considerably the forming of some fatigue cracking sources. The low cyclic fatigue failure occurs very early. This is the fatigue failure phenomenon accelerated by corrosion factors.

"HCF" test

Since the burning gas corrosion resistance, especially hot corrosion resistance, of Fe-base metal is very poor, it is not economical nor possible to develop a new type of steel for overcoming the above-mentioned corrosion fatigue problems. A reasonable way is to use certain alloy coatings for it, thus the "HCF" hot corrosion fatigue test to simulate the working condition of internal combustion engines must be in progress for selecting the suitable coating alloys. The base metals for the test are heat-resisting steels. The coating alloys selected for coating materials are listed in Table 1.

Table 1 Chemical constitution of coating alloys, wt.%

Coating alloys	C	Cr	B	Si	Fe	W	Ni	Co
C_{50}	0.75	22.76	1.21	3.72	2.49	8.75	11	bal.
N_{81}	0.70	25.00	2.50	3.50	<5.00	-	bal.	-

The automatic "HCF" hot corrosion-fatigue test assembly with a salt-solution-spraying device is shown in Fig. 2. In order to examine the combined failure action of hot corrosion with cyclic stresses, the red hot sample is sprayed directly with a saturated salt solution as a cooling way. The cold-hot cyclic stresses and the fierce evaporation of water in salt solution act jointly on the surface of the sample, and damage the corrosion protecting film on the surface. Both the cyclic stresses and test temperature are simulated according to the service condition.

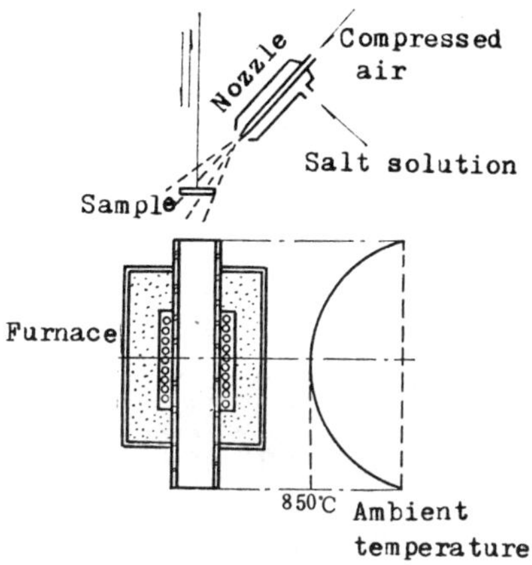

Fig. 2 "HCF" test assembly

Up to now, there is no unified criterion to evaluate the hot corrosion-fatigue test, so a sample the two halves of which can be directly compared is used as shown in Fig. 3a. The sample ($\phi 20 \times 3$ mm) is made from base metal. Its left half is coated with 170 to 250μ alloy and the right half is the original base metal. The sample is moved up and down from core to outside of the furnace during the test.

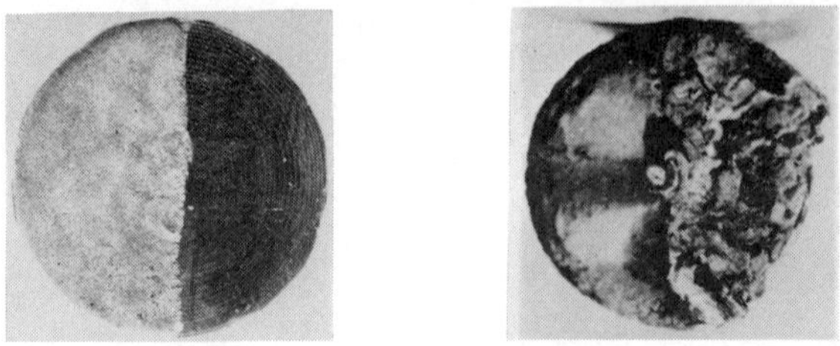

a. Before testing x 2 b. After testing x 2

Fig. 3 "HCF" sample

Every sample is taken down to be dissected and examined after 1000 cycles of hot corrosion-fatigue. The appearance of a typical sample is shown in Fig. 3b. The coating part at left half keeps in a good condition, but the corrosion protecting film on base metal part at right half appears loose, cracked and falls away layer by layer. The base metal decreases in thickness as listed in Table 2.

Table 2 The loss in metal thickness, %

Coating samples	Coating part	Base metal part
C_{50}-coated Amcol8SR	0	25
N81-coated $4Cr_{10}Si_2Mo$	0	22
N81-coated $Cr_{20}Ni_{14}Si_2$	0	23

The metallographic examination is performed on each part of the sample. The coating part on various samples all keeps in good condition as shown in Fig. 4a, but the base metal parts all suffer from serious hot corrosion-fatigue, and a lot of corrosion products fall away very early during the test. The ulcer-like corrosion appears on the surface layer of some base metal under the corrosion products as shown in Fig. 4b, and of some other as intergranular corrosion as shown in Fig. 4c.

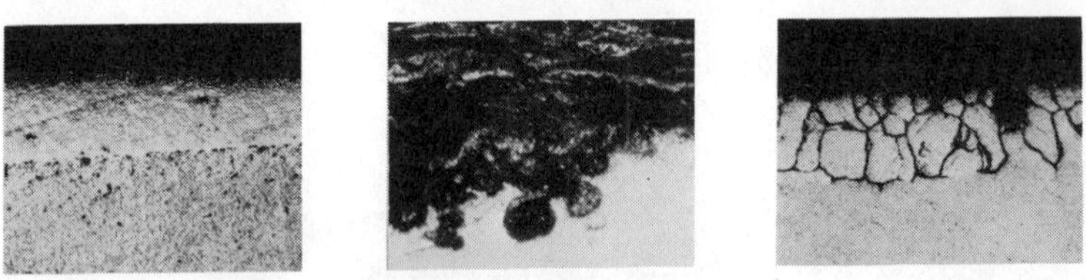

a. N81 coating part x 50 b. $4Cr_{10}Si_2Mo$ base metal part x 100 c. $Cr_{20}Ni_{14}Si_2$ base metal part x 100

Fig. 4 The metallographs of various parts of the samples

The elements distribution curves at section of various parts on the sample which are measured by electron microprobe prove that there is an about 60μ protecting film on the coating surface. This film is very thin, fine and stable because the corrosive elements, such as oxygen or sulphur, are not able to attack into coating alloy through it as shown in Fig. 5a;

but the thick corrosion products on base metal part are rich in oxygen, sulphur and chromium, and the corrosive elements all diffuse obviously into the depth of base metal as shown in Fig. 5b.

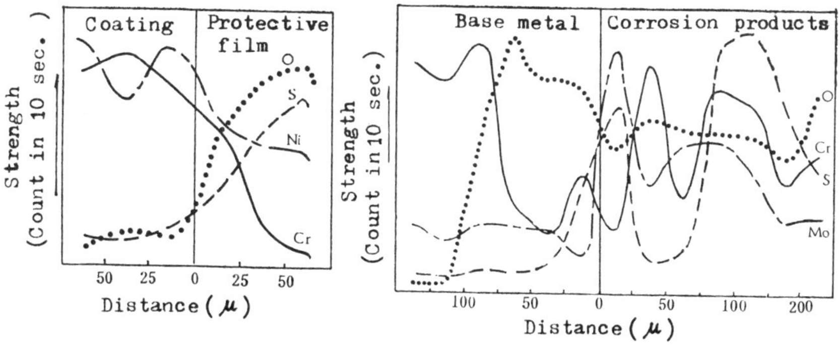

a. N81 coating part   b. $4Cr_{10}Si_2Mo$ base metal part

Fig. 5 Elements distribution curves of various parts

The results analyzed by X-ray diffraction are listed in Table 3. The fine and stable Cr_2O_3 provides the coating with excellent hot corrosion-fatigue resistance. On the other hand, the corrosion products, i.e. α-Fe_2O_3 and sodium salts, on the surface of heat-resisting steels are possessed of very loose structure and low strength. It is easy to crack and fall away under the action of cyclic stresses without any protective action.

Table 3 The results analyzed by X-ray diffraction

Analyzed samples	Distinguished main phases
Oxide film or corrosion protecting film of N81 coating	Cr_2O_3
Oxide film or corrosion protecting film of C50 coating	Cr_2O_3
"HCF" corrosion products of Amco18SR base metal	α-Fe_2O_3, Na_2SO_4-Na_2CrO_4, Na_2SO_4
"HCF" corrosion products of $4Cr_{10}Si_2Mo$ base metal	α-Fe_2O_3, Na_2SO_4

Phases are listed according to the decrease in the line intensity.

Conclusions

There are two forms of damage to heat-resisting steel components under the combined action of hot burning gas corrosion and cyclic stresses. One is the surface layer fatigue failure on metal, which is practically a corrosion failing phenomenon accelerated by cyclic stresses. In this case, the cyclic stresses are unable to damage the base metal directly but can make the protecting films on metal surface crack and fall away layer by layer, because the high temperature yield strength of heat-resisting steels is much greater than the cyclic stresses. The other is a fatigue failure accelerated by corrosion factors. In this case, the cyclic stresses can damage the protecting films on metal surface and directly damage the base metal, because the cyclic stresses are greater than the yield strength and the intergranular corrosion crackings on metal surface provide some fatigue cracking sources for fatigue failure which occurs very early on the heat-resisting steel components. Coating alloys, the Ni-base or Co-base, possess excellent hot corrosion-fatigue resistance. It is effective to use these protective coatings for heat-resisting steel components at hot burning gas.

CREEP PROPERTIES OF ELECTRON BEAM EVAPORATED $Ni\text{-}Al_2O_3$ CONDENSATES

B. E. Jacobson

Department of Physics and Measurement Technology, Linköping University of Technology, S-581 83 Linköping, Sweden

ABSTRACT

MATERIALS processing by various deposition methods offers the possibility of producing material with unique properties. The microstructural features of these materials can be varied over wide ranges, e.g., to exhibit grain sizes in the sub-micron range, voids and cavities in various distributions, and second phase material in unusual combinations. Little is known about the fundamental relationships between these microstructural features and the material properties in general and the mechanical properties in particular.

Two main features were of special interest in this study: the effect of fine grained size and the influence of second phase particle dispersions on the creep deformation behavior of evaporated nickel. Thick free-standing condensates (t = 0.25 mm) of co-evaporated Ni and alumina, Al_2O_3, were prepared by high rate (r = 4 μm/min) electron beam evaporation in the temperature range of 0.3 - 0.5 T_m. Substrates were alumina coated sheets of steel allowing subsequent stripping of the prepared condensates. The amount of alumina varied from f = 0.21vol.% to f = 0.75vol.% and it was randomly distributed as fine particles of 20-70 Å in size in the Ni matrix. The Ni grain structure was equiaxed, not columnar as would be expected with no particles present, and having a grain size of about 0.3 μm.

Creep testing was performed in a constant-stress (Andrade arm) rig in the temperature range 335-900°C and under vacuum (10^{-5} Torr or better). Isothermal straining was continued through primary creep until a steady state strain rate was observed. Then the test conditions were varied by changing the stress (increasing the load); once a new study was reached, the strain rate was again measured. The creep data were analyzed using an empirical expression for the steady state strain rate:

$$\dot{\varepsilon} = KL^{\alpha}\left(\frac{\sigma}{E}\right)^{n}\exp\left(-\frac{Q}{RT}\right)$$

where K is a structure-related constant, L the grain size, σ the applied stress, E the elastic modulus, R the gas constant and T the absolute temperature. The quantities α, n and Q are known as the grain size sensitivity, the stress sensitivity (or stress exponent) and the activation energy for creep, respectively. Microstructural investigations were performed before and after creep deformation using transmission electron microscopy (TEM).

Interesting creep properties were obtained due to both the fine grain size as such and to the presence of the fine particle dispersion. The presence of Al_2O_3 particles stabilizes the grain size, i.e. prevents grain growth to occur at elevated temperatures, produces strengthening effects, and reduces the creep ductility. However, compared with conventionally produced Ni, the evaporated Ni+Al_2O_3 material creeps at a faster rate owing to the fine grain size.

Low temperature creep deformation involves dislocation controlled processes whereas high temperature ($T > 550^{\circ}C$) deformation exhibits a change in deformation mode from dislocation glide processes at low stresses to grain boundary sliding at high stresses. In this case, to separate the contributions from each of the two processes comparisons were made between the values of macroscopic (total) creep elongation and microscopic mean grain elongation - the latter estimated statistically from measurements over about 150 grains along a line respectively parallel to and perpendicular to the creep strain direction. The total elongation was measured by an externally applied LVDT-sensor (Linear Variable Differential Transformer), the mean grain elongation by TEM analysis.

The microstructure after low temperature creep deformation is characterized by an elongated grain structure and high dislocation densities as expected from dislocation controlled deformation processes. The microstructure after high temperature deformation, however, exhibits features of more unusual types; one being a frequent distribution of extremely thin twin grains preferentially oriented in directions close to the strain direction. Furthermore, low temperature deformation gives rise to necking before fracture and has a high stress exponent whereas high temperature deformation is homogeneous, exhibits a low stress exponent and high strain rates, all indicating a superplastic like behavior.

DIFFUSION BARRIERS USED TO IMPROVE STRUCTURAL STABILITY AND MECHANICAL PROPERTIES OF W FIBRE-Ni MATRIX COMPOSITES AT ELEVATED TEMPERATURES

R. Ahlroth and P. Kettunen

Institute of Materials Science, Tampere University of Technology, Tampere, Finland

ABSTRACT

The paper reports experiments using thin coatings of TiC and TiN as diffusion barrier layers to avoid the unwanted interaction between W fibres and Ni matrix. The coatings were produced by the methods of chemical vapour deposition (CVD) and reactive sputtering.

Diffusion annealings and mechanical tensile and fatigue tests at different temperatures were done on the composites in order to check the effect of the used diffusion barriers on the stability and strength properties of the composites. The results prove that the used barrier layers increase the stability and hinder the deleterious effects of interaction on the strength properties of Ni base composites reinforced with W fibres.

KEYWORDS

Diffusion barrier, coating, metal composite, tungsten fibre, nickel matrix, tensile properties, fatigue properties, elevated temperatures.

INTRODUCTION

The tendency to improve different high temperature components cause a need to develop new materials for applications, e.g., at temperatures in the region of 800°-1100°C. One group of possible materials for this temperature range is composites in which the Ni base high temperature alloys are reinforced with W fibres. Many investigators (Ochiai, and co-workers, 1975; Cornie, Schreurs and Palmquist, 1978; Kozma, 1982) have found that the use of W fibres as reinforcing elements has met, however, serious problems as the W fibres seem not to tolerate the typical elements of high temperature superalloys, such as Ni, Co or Cr, in their surrounding at high temperatures. E.g., an interaction between W and Ni causes the formation of a brittle reaction zone at the interface. Furthermore, as Montelbano and co-workers (1968) showed, the diffusion of Ni into the cold drawn W fibre induces also recrystallization and grain growth in the fibre. As a result of this, the fibres loose their strength. Thus, the structural stability of the fibre at high temperatures is very important from the point of view of the composite.

EXPERIMENTAL DETAILS

The Ni used in the matrix was of the commercial carbonyl grade having a purity of 99.9 %. Correspondingly, the W fibres were of the general lamp grade.

Coating of the Fibres

Successful diffusion barriers for refractory-metal fibres are not available as yet. The main obstacle in developing a successful diffusion barrier is the method of coating. The coating must be well bonded, thin and uniform in thickness and free of porosity. CVD-method is suitable for producing coatings of different grades. Also sputtering must be taken as one of the most promising methods due to the flexibility to coat with a great number of different possible barrier materials.

In the present work, CVD-method was used at first to coat W fibres of the diameter of 1 mm with TiC and TiN layers. The layers were formed at 990°C by the reduction of $TiCl_4$ in the presence of H_2 and CH_4 or H_2 and N_2. The thickness of the coating was controlled by means of the deposition time. Deposition times of 20-60 minutes appeared to be sufficient to produce a coating thickness of 1-2 μm

On the thinner fibres of the diameter 100 μm and 200 μm, the TiC and TiN coatings were done by reactive r.f. sputtering in the atmosphere of Ar + CH_4 and Ar + N_2, respectively. In order to receive an uniform thickness in the coating, fibre bundles were rotated in a special jig during the sputtering process. Compositions of the coatings were controlled by the partial pressure of the reactive gas flowing through the sputtering chamber and by sputtering voltage. The temperature of the fibres during coating was 800°C.

Structures and compositions of the coatings were studied with SEM microanalysis, ESCA and SAM. Examples are shown in Fig. 1.

Composites and Their Testing

The composite specimens were produced from Ni powder and unidirectional fibres of the volume fraction of 6 %. The specimens were pressed with a pressure of 1 kN/mm^2, then sintered at 830°C in a flowing gas mixture of Ar + 3 % H_2 for 24 hours. Specimens were produced from coated as well as uncoated fibres.

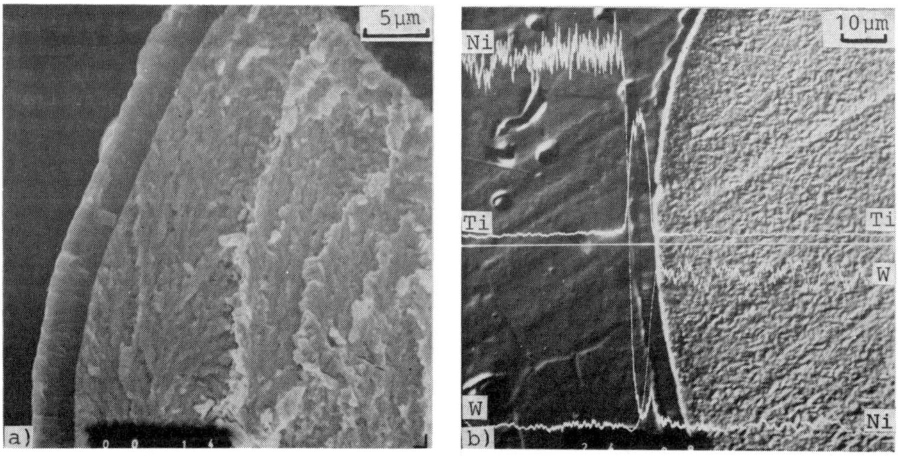

Fig. 1. Examples of the coatings used as diffusion barrier layers: a) TiC coating on W fibre formed by reactive r.f. sputtering. b) Line analyses of Ti, W and Ni along the white line across the TiC layer between Ni and W.

Special ends were brazed to the specimens meant for mechanical tests in order to ensure a good gripping in the testing machine. These specimens were ground with different grades of grinding papers, and the final polishing was done with a fine slurry of alumina.

The effect of the chosen barrier layers were tested by annealing composites with coated as well as uncoated fibres for different periods of time at different elevated temperatures in an atmosphere of flowing Ar + 3 % H_2. After annealing, the specimens were examined with SEM and electron probe microanalysis.

Tensile tests were done with an universal testing machine of the type Instron. The applied strain rate was 0.01 min^{-1}. All tensile specimens were diffusion annealed for 2.5 hours at the temperature of 1150°C. This treatment produced a diffusion zone 15 μm in thickness in the uncoated specimens. Also some fibres without and with coating were tensile tested. Fatigue tests were done with a Sonntag-type pulsator using the method of constant applied stress amplitude with a zero mean stress at the frequency of 30 cps. The tensile and fatigue tests at 20°C were done in air and those at elevated temperatures in the atmosphere of Ar.

RESULTS

Results of the diffusion tests are presented in Table 1. The effect of the used TiC and TiN coatings is clear.

In the case of CVD-coatings, the thickness of the diffusion zones between Ni and W remains at temperatures $T \leq 927°C$ smaller than in the case of uncoated fibres. On the other hand, the sputtered coatings of TiC, TiN and a douple coating of TiC + TiN appear to be able to avoid the diffusion of Ni and W through

TABLE 1 Thicknesses of the Diffusion Penetrated Zones between Ni and W. (detectibility about 1 μm)

Annealing temperature, [°C]	Annealing time, [hr]	Thickness of the diffusion zones, [μm]					
		Un-coated	CVD coatings		Sputtered coatings		
			TiC	TiN	TiC	TiN	TiC + TiN
800	202	23	4	6	<1	<1	<1
927	114	30	10	5-20	<1	<1	<1
1027	45	30	35	10-20	<1	5	<1

these layers at 927°C at least for a period of 114 hours and at 1027°C at least for 45 hours. Only in the latter case, a diffusion zone of 5 μm can be detected in TiN layer.

The results of microanalyzing show also the importance of the diffusion of Ni through the diffusion barrier. When the concentration of Ni between W and the diffusion barrier becomes high enough, an intermetallic phase begins to form. The growing of this new phase tends to destruct the barrier layer pushing it outwards from the surface of the W fibre. SAM microanalyzing profiles of Ni, Ti and Ti+N across the floating TiN barrier layer are presented in Fig. 2. We can find that no dissolving of TiN barrier layer can be detected. Furthermore, the content of Ni in the barrier layers was found to be very small although that between the barrier layer and W was considerable.

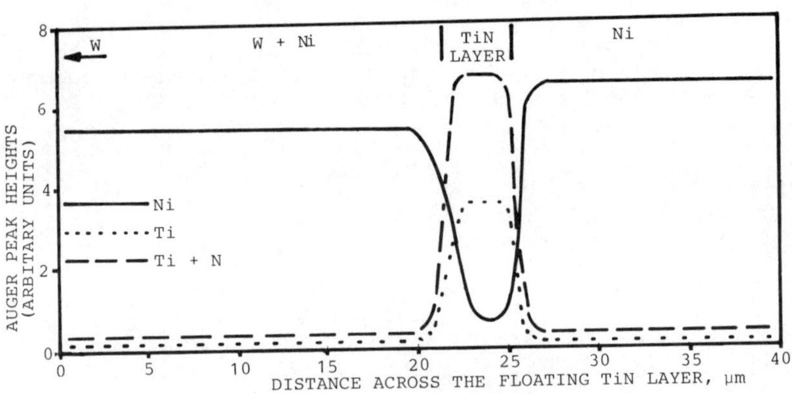

Fig. 2. Line analyses across the floating TiN layer.

The results of tensile tests on separate uncoated and coated W fibres are summarized in Fig. 3 and Table 2. Fig. 3 shows that the tensile strength of the W fibre decreases slightly with an increasing temperature and the coating as well as the coating process have no mentionable effect on the tensile strength. The reductions in area to fracture in the tensile tests, shown in Table 2, prove that these fibres have already at 650°C a high degree of ductility as indicated by their tendency to deform locally. Again here, the fibres seem not to be degraded due to the coating.

TABLE 2 The reduction in area to fracture of uncoated and coated W fibres (ϕ 200 μm) in the tensile tests at 650° and 900°C.

T, °C	Reduction of area to fracture, %		
	Uncoated W	TiN on W	TiC+TiN on W
650	68	67	-
900	79	80	80

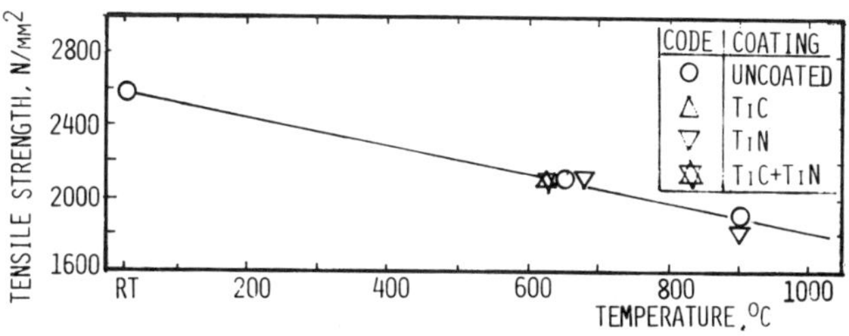

Fig. 3. Tensile strength of uncoated and coated W fibres (ϕ 200 μm).

The tensile strength of the composites at 650° and 930°C are presented in Table 3. It shows that the strength values of the composites with diffusion barriers are lower than those of the composites with uncoated fibres.

TABLE 3 Tensile strength of the composites tested at 650° and 930°C, V_f = 6 %.

STRENGTH OF COMPOSITES / COATING OF FIBRES	TEMPERATURE 650°C		TEMPERATURE 930°C	
	YIELD STRENGTH [N/mm^2]	TENSILE STRENGTH [N/mm^2]	YIELD STRENGTH [N/mm^2]	TENSILE STRENGTH [N/mm^2]
Uncoated fibres	193	270	70	127
TiC coated fibres		232		117
TiN coated fibres	121	201	60	99

The S-N curves of the composites at 20°C are presented in Fig. 4. At the endurance limit, the difference between composites with different fibres is neglible. However, at higher stress amplitudes the composites with coated fibres reveal shorter fatigue lives than those with uncoated fibres. The composites with TiN coated fibres show the shortest fatigue lives in this respect.

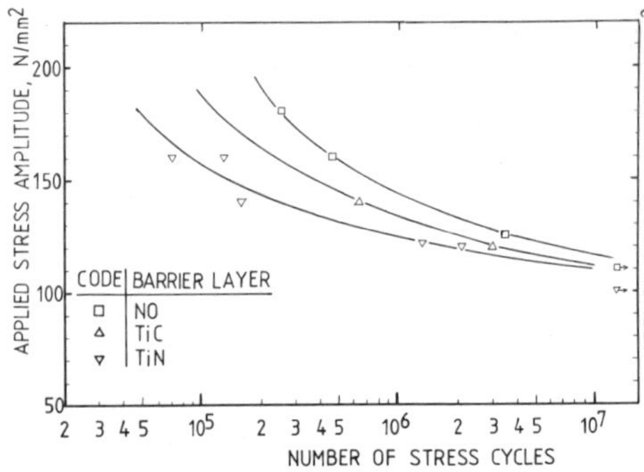

Fig. 4. S-N curves of W fibre reinforced composites at 20°C and V_f = 6 %.

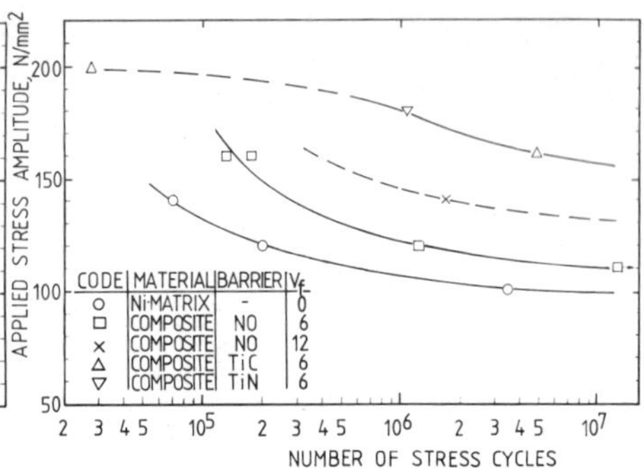

Fig. 5. S-N curves of W fibre reinforced composites at 650°C.

The S-N curves at 650°C shown in Fig. 5 indicate, however, at the volume fraction of 6 % of fibres a clear increasing effect on the fatigue strength of the composite with coated fibres when compared with the composite of uncoated fibre or with that of the matrix. One tested specimen at the volume fraction of 12 % of fibres reveals again an increased fatigue strength in function of the volume fraction. When comparing these results to those at 20°C we can find out that fatigue properties of the composites with coated fibres are considerably better at 650°C than at 20°C.

DISCUSSION

The results of the diffusion tests suggest that thin TiC and TiN barrier layers on the W fibres reduce the deleterious diffusion of Ni from matrix into the W fibres, which normally plagues the use of W fibres in reinforcing Ni base alloys.

Microanalyzing of the barrier layers and interfaces after diffusion tests show that the barrier layers retard interaction between the fibre and matrix rather well, until a temperature at which nickel penetrates the layer and an interaction zone between nickel and tungsten begins to be formed between the fibre and barrier layer. Mazzei, Van Drunen and Chung (1979) have reported similar results concerning HfN coatings on W fibres in a Ni base superalloy. The process seems thus to be controlled by the diffusion of nickel through the thin barrier layer. The grain boundaries and other defects in the layer are suggested to be diffusion paths mainly. Finally, the growing interaction zone between the barrier layer and the fibre leads to the destruction of the uniform barrier layer.

The tensile strength properties at 650°C and 930°C of the composites with the diffusion barrier layers are lower than those of the composites without diffusion barriers. The decrease is caused mainly by the poorer adherence between the coated fibre and the matrix. That is rather obvious,because in tensile test the interface layers transfer large strains emphasizing the adherence between the fibre and the matrix.

At 20°C, the used TiC and TiN barrier layers seem not to improve the fatigue strength properties. On the endurance limit, the coatings do not have any influence, but at higher stress amplitudes they clearly shorten the fatigue life. At this temperature, the fibres as well the barrier layers behave more or less brittle. At elevated temperatures, where the ductility of the fibres and barrier layers increases, the endurance limit of composites with coated fibres is higher than the one with uncoated fibres surrounded by the interaction layer.

CONCLUSIONS

The results clearly suggest that thin TiC and TiN coatings on the W fibres reduce the deleterious diffusion between W and Ni, which normally plagues the use of W fibres in reinforcement of Ni base high temperature alloys. The diffusion of Ni through these barrier layers is found to initiate the destruction of the barrier layers. Thus we must also pay attention to the structure of the barrier layer in addition to the chemical stability and mechanical requirements.

On the endurance limit, the used TiC and TiN coatings at room temperature do not have any influence, but at higher stress amplitudes they clearly shorten the fatigue life. At room temperature, the fibres as well as the coatings behave more or less brittle. At higher temperatures, where the ductility of the fibres and coatings increases, the endurance limit of the composites with coated fibres is higher than the one with uncoated fibres surrounded by the brittle interaction layer.

The results suggest that the use of ceramic diffusion barrier layers is an useful way to improve the stability of high temperature composites.

REFERENCES

Cornie, J.A., J.J. Schreurs and R.W. Palmquist (1978). A kinetic and microstructural study of oxide, carbide, and nitride diffusion barriers in HSTW (tungsten) reinforced MAR-M-200 composites. In B.R. Noton (Ed.), Proceedings of the 1978 International Conference on Composite Materials. The Metallurgical Society of AIME, New York. pp. 858-882.

Kozma, L. (1982). Study of the microstructural changes in W-wire of W-Ni and W-Fe composites during heat treatment. In H. Lilholt and R. Talreja (Ed.), Fatigue and Creep of Composite Materials. Risø National Laboratory, Roskilde. pp. 231-237.

Mazzei, P.J., G. Van Drunen and N. Chung (1979). High temperature artificial composite-alloys for industrial gas turbines. In J.A. Cornie and F.W. Crossman (Ed.), Failure Modes in Composites. The Metallurgical Society of AIME, New York. pp. 219-235.

Montelbano, T., J. Brett, L. Castleman and L. Seigle (1968). Nickel induced recrystallization of doped tungsten. Trans. TMS-AIME, 242, 1968-1979.

Ochiai, S., K. Shimomura, M. Mizuhara and Y. Murakami (1975). Effects of interfacial reaction on deformation and fracture behaviour of tungsten fibre-nickel matrix composites. Trans JIM., 16, 463-471.

MECHANICAL PROPERTIES OF PLASMA SPRAYED AND POST-TREATED FeCrAlY-COATINGS

C. Burman and T. Ericsson

Linköping Institute of Technology, S-581 83 Linköping, Sweden

ABSTRACT

Conventional plasma sprayed FeCrAlY coatings with composition Fe-20Cr-12Al-1.8Y and Fe-20Cr-9Al-1.3Y has been deposited on Alloy 800H. In order to improve the coatings hot isostatic pressing (HIP) and electron beam remelting (EB) has been used. The coatings became dense and in the case of EB the big oxide inclusions are eliminated.

The mechanical behaviour has been studied by tensile testing, LCF and creep testing at 900°C. Tensile testing showed that EB treated specimens could withstand 6% and HIP treated 3% strain prior to cracking. LCF showed a detrimental effect of grit blasting residues on the fatigue life of HIP treated and as-plasma sprayed coatings compared to untreated specimens. The EB treated coatings show a difference in LCF cracking behaviour depending on Al content, i e high Al has brittle cracks and low Al is ductile with striations. Creep ductility seems to be satisfactory. Ni diffusion in remelted coatings is very fast. Oxide inclusions in HIP treated coatings restrict the Ni diffusion into the coating.

KEYWORDS

FeCrAlY; coatings; plasma sprayed; electron beam (EB) remelting; HIP treated; tensile test; creep test; low cycle fatigue test (LCF).

INTRODUCTION

MCrAlY coatings, where M can be Ni, Co or Fe, is often used today as protection against high temperature corrosion. The Cr is forming a protective oxide scale at intermediate temperatures, while the Al forms the oxide scale at high temperatures. Yttrium is added to promote oxide scale adhesion, but the mechanism for this is still a matter for discussion. The coatings that are best known are NiCrAlY, CoCrAlY and NiCoCrAlY (1-2). FeCrAlY coatings have not been studied so much, although they show promise for combined sulphidizing and oxidizing atmospheres (3).

At the University of Linköping we have a project for studying FeCrAlY coatings deposited by conventional plasma spraying (PS) and electron beam evaporation (PVD). The emphasis in this paper is on sprayed coatings. Conventional plasma spraying has often been disregarded for high performance high temperature coatings due to porosities and slag inclusions. In an attempt to avoid these drawbacks we have tried different methods to improve the coatings. The methods used are hot isostatic pressing (HIP) and electron beam remelting (EB) of the coatings. A few trials with laser remelting was also attempted.

EXPERIMENTAL DETAILS

Coating Preparation

The substrate material used is Alloy 800 H with nominal composition Fe-31Ni-21Cr-0.6Mn-0.6Si-0.3Al-0.3Ti-0.07C.

Plasma spraying of FeCrAlY powders was carried out using a Metco 3M and a Metco 7M equipment. Spraying parameters were chosen that gave the least amount of cracking for a certain bending of a sprayed plate. Before spraying the specimens were grit blasted with Al_2O_3 particles. Al_2O_3 particles rather than steel shots were used to avoid contamination effects. Two types of FeCrAlY powders with grain size 40-60μm were used, see Table 1, and the coating thickness prior to grinding and polishing was approximately 300μm.

TABLE 1 Composition [w/o] of Powders, PVD Coatings and Observed Phases

	Fe	Cr	Al	Ni	Y
Powder type 1	bal	20	12	-	1.8
Powder type 2	bal	20	9	-	1.3
PVD 1	bal	25	15	-	3
PVD 2	bal	23	9.5	-	3
Light phase	bal	18	2.5	15	-
Dark phase	bal	24	4	8	-

Hot isostatic pressing (HIP) of coated specimens was carried out in an ASEA Quintus press. The specimen is put in a glass vessel which is evacuated and sealed. The assembly is then placed in the press, the temperature is raised to 1000°C and a pressure of about 250 MPa is applied and maintained for about 1 h. The specimen is then removed from the press and the glass layer is peeled off.

Electron beam (EB) remelting of the sprayed coatings was carried out in a Steigerwald EB welding equipment. The goal is to get complete fusion of the coating while minimizing the melting of the base material. Treatment with a 2.5 kW CO_2 laser was tried in a few cases as well. For both EB and laser treatments the parameters are given by the equation

$$P = A\ b\ t\ v \qquad (1)$$

where P is the power (W), A the energy per volume (Ws/mm^3), b the width of each melting pass (mm), t the thickness (mm) and v the travelling speed (mm/s). In the laser case the width is controlled by the distance between the focus and surface while, in the EB case, it is controlled by the amount of transverse oscillation of the beam. The oscillation had also to be matched to the coating thickness because of the tendency to get more heat input at the turning points of the oscillations. The maximum width per pass was in both cases approximately 2.5 mm since the coating material for larger widths tended to be blown away and form droplets due to degassing. The two treatments are very sensitive to the coating thickness which must not vary more than 0.05 mm. To overcome this problem the coatings were ground to 200μm thickness prior to EB or laser treatment.

The work on the PVD coatings is reported elsewhere (4).

Testing of the Coating

To evaluate the mechanical properties of these coatings tensile, low cycle fatigue (LCF) and creep testing in air at 900°C have been performed. Evaluation of the tests was made using optical microscopy, SEM and EDS. The tensile testing equipment consists of a computer controlled MTS 160 kN machine with a furnace with three separately controlled heating zones, that is attached to the machine frame. Strain measurement was made by mounting an extensometer on the ram and calibrating the strain versus ram displacement under cold conditions, when stretching of the pullrods would be more severe, to see if the relationship was linear. The tensile testing was made under strain control and a constant strain rate of 0.1%/min was used. To see at which strain cracking of the coating started, the testing was stopped at different amounts of strain and the specimens were allowed to cool down and were then examined. If no cracks were detected, the specimen was put back into the rig and the testing continued.

Creep testing of some PVD and PS coatings has been performed in air at 900°C. The applied initial stress was in most cases 23 MPa. For the PS and post-treated coatings the load was increased to 28 MPa after ~400 h, see Table 2, in order to reach the preset strain in a reasonable time. At the preset strain the testing was stopped, the specimen was cooled down and removed from the rig.

Some low cycle fatigue testing at 900°C has been performed to study the crack propagation and the effect of the coatings on the fatigue life of the specimens. Strain measurement during the fatigue testing was made using side entrance quartz rods. The testing was made under strain control, the plastic strain range was 1% and the frequency was 0.01 Hz. When the peak load was reduced to 50% of the initial peak load, the test stopped automatically. After cooling down the specimens were pulled to fracture and the fracture surface was studied in SEM.

Earlier tests (5) have shown evidence that the big oxide inclusions in the HIP treated coatings act as barriers against Ni diffusion into the coating. In order to investigate this some coatings have been applied on IN 738 and then HIP treated. These coatings are then oxidized at 950°C.

RESULTS AND DISCUSSION

The microstructure of the as-plasma sprayed coating is shown in Fig 1. The hardness is approximately 250 mHV. The HIP treatment produced a dense coating without voids, but the oxide inclusions remain intact and there is a diffusion zone to a depth of about 0.1 mm with nickel aluminide precipitates in the base material, see Fig 2. There was some fear that there would be silicon diffusion from the glass vessel during the HIP treatment but investigations with EDS have shown no such tendencies.

TABLE 2 Creep testing in air at 900°C

Specimen	Coating	Thickness of coating [µm]	Stress MPa	Testing time[h]	Elongation [%]	Comments
1	HIP type 1	300	23/28	1198	4.5	No penetrating cracks
2	EB type 2	190	23/28	679	7.5	No penetrating cracks
3	PVD 1	21	23	498	4	Oxide at the interface
4	PVD 1	23	23	419	48	Oxide at the interface
5	PVD 2	45	30	79	3	Oxide at the interface
6	PVD 2	45	30	92	10	Oxide at the interface

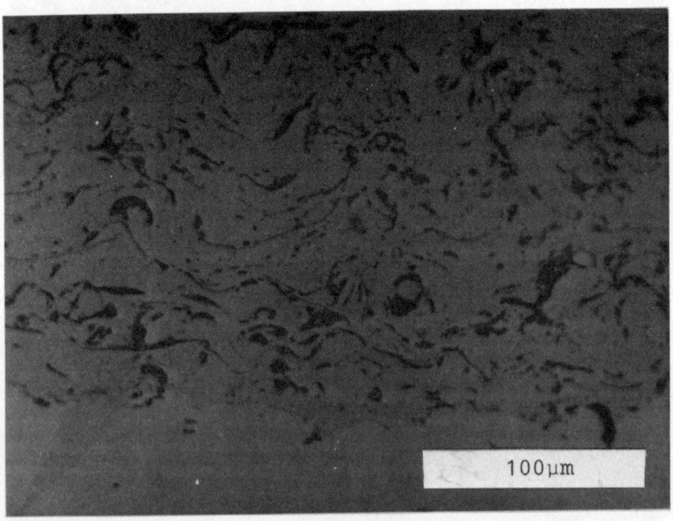

Fig 1 Sprayed coating with oxide inclusions. Unetched.

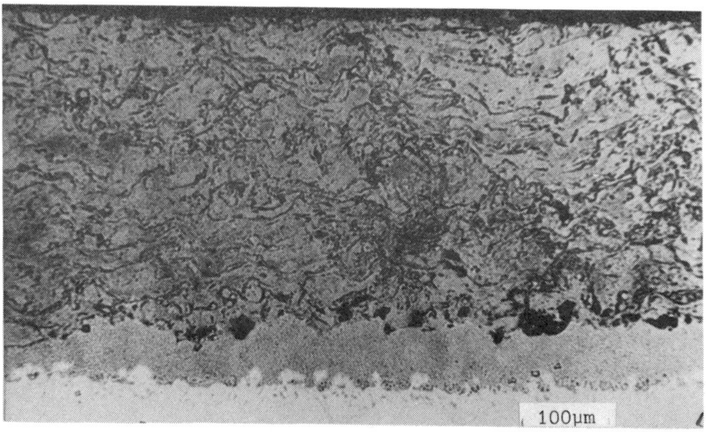

Fig 2 Sprayed coating after HIP treatment. Oxide inclusions are intact and there is a diffusion zone in the base material. Etched.

The parameter A in eq 1 was specific for the material and was found to be 50 Ws/mm for complete remelting of the coating with minimal melting of the base material. The main difference between laser and EB treatment seems to be the cooling rate. During laser treatment an inert gas shield is needed around the beam to avoid excessive oxidation of the molten material, while the EB treatment is done in vacuum. The gas flow in the shield gives a very high cooling rate which causes cracks in the coating. Therefore the work has been concentrated on EB treatments.

The EB treatment produced a dense coating with very little melting of the base material. The large oxide inclusions that were present in the as-sprayed condition were broken up and reformed as small spheroidal particles evenly distributed in the coating. As can be seen from Fig 3 the coating has a typical weld structure. The hardness is about 600 mHV. The surface of the remelted coating consisted of weld strings that had to be removed by grinding in order to get a smooth surface. Thermal distortion of the specimens sometimes caused problems to grind and polish in a controlled manner. The distortion can be avoided to some extent by minimizing the heat input into the specimens, i e by having thin coatings and fast translation. A good combination was t=200 m and V=7 mm/s.

The tensile testing showed that the HIP treated specimens could withstand about 3% strain prior to cracking of the coating. Only a big crack in the middle of the gage section was formed. Prior to the cracking of the coating itself microcracking of the oxide scale was observed at about 1% strain. The starting point and the crack path was difficult to identify, but there are some indications that the crack started at or near the outer surface and that the crack preferentially follows oxides present in the grain boundaries of the coating.

The EB treated coatings could stand 6-7% strain prior to cracking. Also in this case microcracks were observed in the oxide scale after about 1% strain and several small cracks instead of a large one were observed. The crackpath is transgranular and the cracks are found where big oxide inclusions survived the remelting treatment due to a slight miscalculation of the coating thickness. The tips of the cracks show evidence of plastic strain, see Fig 4.

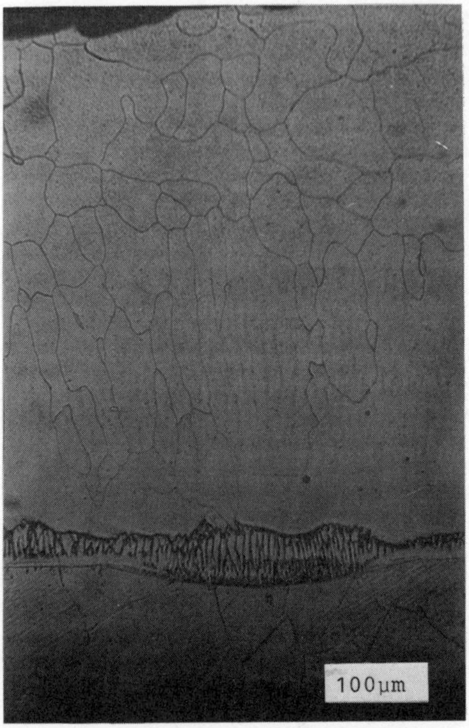

Fig 3 Coating after EB treatment. Weld structure. Oxide inclusions are broken up. Etched.

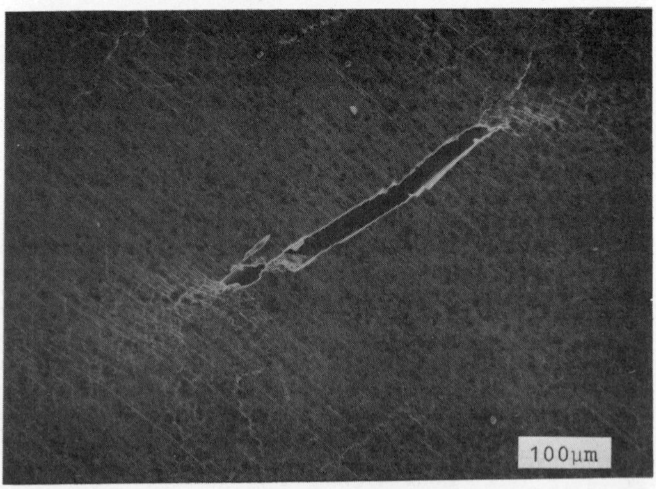

Fig 4 EB treated coating after 7% strain. Crack with signs of plastic strain at the tip.

The smaller strain to crack formation for the HIP treated coatings is probably due to the big oxide inclusions present which act as stress raisers.

The creep test showed that there was only surface cracks in the HIP treated coating after 4.5% and in the EB treated coating after 7.5% strain, see Table 2. There was no tendency for cracks to extend through the coatings. In the PVD coatings there seemed to be a tendency for oxide formation at the interface between coating and base material. At higher strains there is seen some indications of separation at the grain boundaries and of oxygen diffusing along the boundaries, see Fig 5. The surface cracks seem to be shorter and wider on EB and HIP treated coatings compared to PVD coatings, see Figs 6-7. This is probably a result of the greater thickness of the sprayed coatings which allows easier stress relieving.

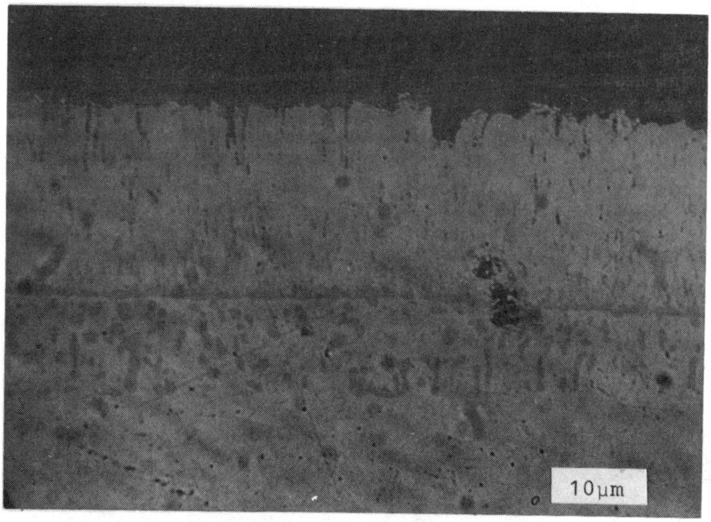

Fig 5 Coating PVD 2 after 10% strain. Oxides in grain boundaries and at the coating-base-material interface. Unetched.

The LCF testing is the only case displaying any difference between the two different Al contents, see Table 3. There is a definite benefical effect on fatigue life from the EB treated coating with lower Al content, while the EB treated with higher Al content, HIP treated, and untreated plasma sprayed coatings have a shorter fatigue life compared to the uncoated specimens. In the case of HIP treated and untreated coatings this detrimental effect seems most likely to be due to big oxide inclusions in the interface between coating and base material, see Fig 2. These oxides are residues from the grit blasting prior to spraying and they are very good starting points for crack development. The fracture surface of the coatings appears to be fairly ductile, but in the untreated case we got separation between coating and base material, indicating that the bonding was poor. In the case of EB treated coatings with higher Al content, there is a clevage-like appearance, see Fig 8, indicating a fast fracture. The EB coatings with low Al content, on the other hand, show striations and a more ductile appearance, see Fig 9.

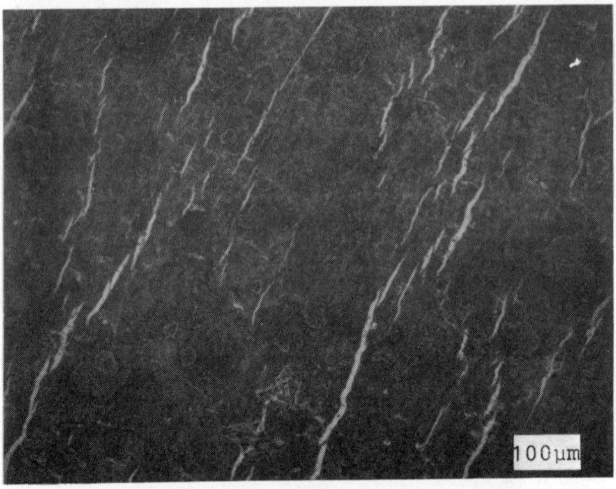

Fig 6 HIP treated coating after 4.5% strain. Short wide cracks.

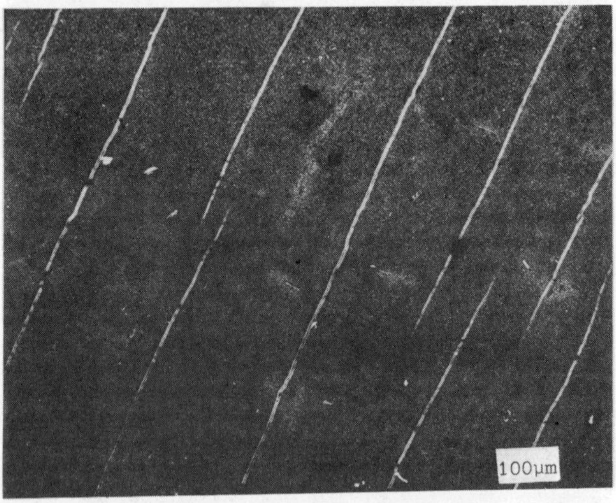

Fig 7 Coating PVD 1 after 4% strain. Long sharp cracks.

TABLE 3 Low cycle fatigue testing in air at 900°C

Speci-men	Coating	Strain range	Cycles to failure	Comment
1	Uncoated	1.04%	720	Dummies
2	Uncoated	1.04%	760	
3	EB type 2	1.02%	1030	Ductile coating
4	EB type 2	1.01%	1160	Ductile coating
5	EB type 1	1.01%	410	Brittle coating
6	EB type 1	1.03%	482	Brittle coating
7	EB type 1	1.01%	690	Brittle coating
8	HIP type 2	1.02%	450	
9	Untreated type	1.01%	255	Separation between coating and base mamerial
10	HIP type 1	1.05%	300-500	(Computer trouble)

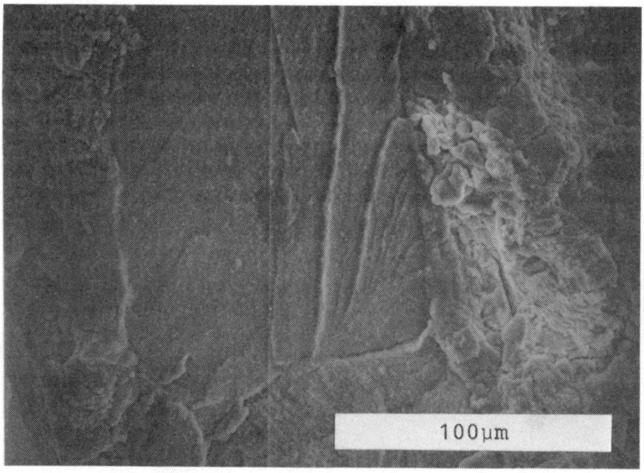

Fig 8 EB treated coating with high Al content after LCF at 900°C. Cleavage-like appearance.

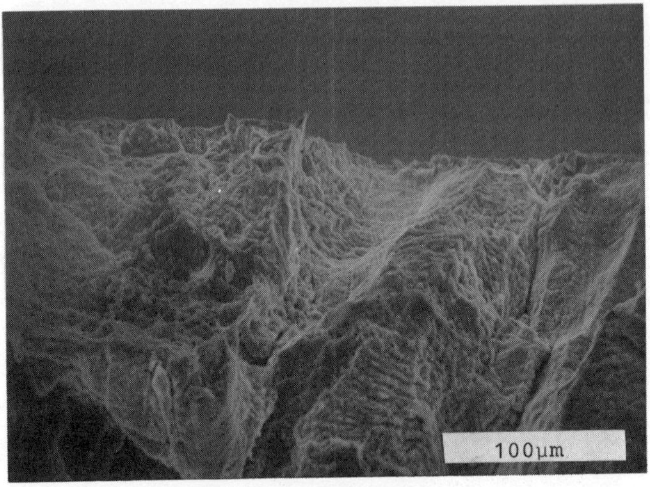

Fig 9 EB treated coating with low Al content after LCF at 900°C. Ductile appearance with striations.

Optical microscopy shows that the EB treated coatings after testing consist of two phases, see Fig 10, and that the difference between the two Al contents is in the relative amount of the phases. The coatings with higher Al content have more of the dark phase. Analysis with EDS shows that the dark phase has relatively low Ni and high Al and Cr content compared to the light phase, see Table 1. The appearance seems to indicate that the light phase consists of areas around grain boundaries that have been enriched in Ni and depleted in Al and Cr due to interdiffusion. Micro-hardness-testing showed that the hardness of the light phase is approximately 270 mHV and the hardness of the dark phase is approximately 360 mHV.

Since there was more dark phase in the coating with higher Al content, this seems to give an explanation of the more brittle behaviour of this coating. The diffusion test of HIP treated coatings on IN 738 has so far shown that the oxide inclusions do not completely prohibit but very much restrict the Ni diffusion to the surface.

CONCLUSIONS

Creep testing of the plasma sprayed and post treated coatings show that they are as ductile as the PVD coatings and that the ductility should be enough for use as protective coatings. The appearance of the surface cracks indicate that the greater thickness of these coatings compared to PVD coatings, enables easier stress relieving by plastic flow, i e we get more blunting of the cracks. LCF testing indicates that there is a detrimental effect from grit blasting residues on the fatigue life for HIP treated and untreated plasma sprayed coatings compared to uncoated specimens. This problem would be possible to avoid with other blasting materials and more controlled blasting parameters. The EB treated coatings show longer or shorter fatigue life depending on Al content. The crack surfaces indicate that there is a difference in fracture mode between the two coatings. This difference seems to be due to the difference in the relative amounts of Ni rich and Ni poor phases.

Diffusion studies show that the long oxide inclusions present in the HIP treated coating inhibit Ni diffusion into the coating. The interdiffusion for EB treated coatings is very fast, probably because it is ferritic.

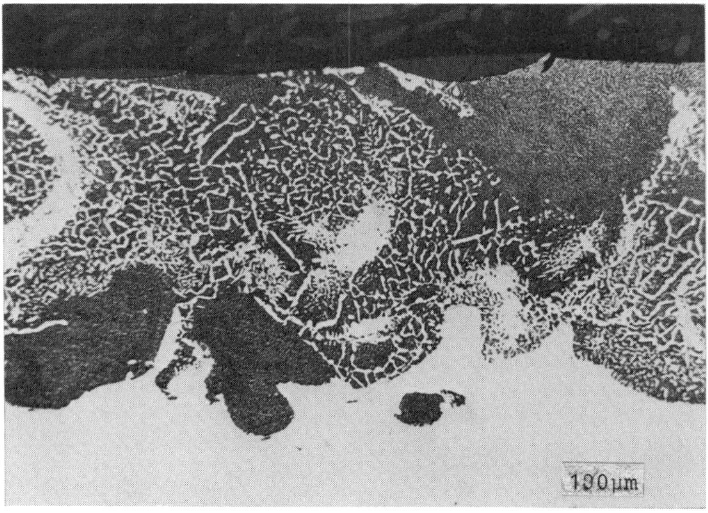

Fig 10 EB treated coating with high Al content after 24 h at 900°C. Etched.

Since there is no corrosion test data available as yet, it is difficult to say which coating is the best, but the post treatment that is easiest to use is HIP since it is insensitive to coating thickness and does not require so much grinding and polishing.. Furthermore it inhibits Ni diffusion to the surface, which should be a good thing in sulphidizing atmospheres.

ACKNOWLEDGMENTS

The project has been supported by the National Swedish Board for Technical Development (STU) and several Swedish industries: Bulten-Kanthal, Sandvik, FFV-maintenance division, Stal-Laval and Volvo Flygmotor.

REFERENCES

1 A Kumar, M Nasrallah, D L Douglass, The Effect of Yttrium and Thorium on the Oxidation Behaviour of NiCrAl Alloys, Oxidation of Metals, Vol 8, No 4, 1974, pp 227-263.

2 Giggins, Pettit, Oxide Scale Adherence Mechanisms and the Effects of Yttrium, Oxide Particles and Externally Applied Loads on the Oxidation of NiCrAl and CoCrAl Alloys, Report ARL 75-0234, June 1975.

3 F Pettit, High Temperature Corrosion and Use of Coating for Protection, Course, Aug 16-20, 1982, Linköping, Sweden.

4 B Andersson, C Burman, T Ericsson, H Henzell, B Jacobson and N G Persson, A Study of Electron Beam Evaporated and Plasma Sprayed FeCrAlY coatings for High Temperature Corrosion Protection, Report LiTH-IFM-IS-75, Nov 81.

5 C Burman and T Ericsson, A Study of Plasma Sprayed FeCrAlY Coatings after Various Post Treatments. In S C Singhal (Ed), High-Temperature Protective Coatings, AIME, 1983, pp 51-59, Atlanta, Georgia, March 7-8, 1983.

THE CAVITATION EROSION OF COATED AND SURFACE TREATED STEELS

A. Barletta and A. Ball

Department of Metallurgy and Materials Science, University of Cape Town, South Africa

ABSTRACT

A laboratory investigation has been made of the resistance to cavitation erosion of a large variety of surface treatments and coatings which can be applied to steels. The performance of these specimens has been ranked according to the cumulative volume loss over a five hour period of testing and the rate of volume loss. The mode of erosion has been characterised by optical sectioning and scanning electron microscopy. The work has emphasised the importance of the nature and strength of the interface, together with mechanical property compatability between the surface region and the base metal. Thus treatments such as carburised steels show excellent potential for erosion resistant materials, whereas chemically vapour deposited and electroplated coatings with distinct interfaces frequently show catastrophic spalling behaviour. Post heat treatments of samples which had received an electroless nickel coating can optimise their erosion resistance by both a hardening mechanism and an improvement in the nature of the interface. Furthermore, it has been found that thickness of the electroless nickel coating must exceed a minimum for a given intensity of cavitation.

KEY WORDS

Cavitation erosion; coated steel; surface treated steel; electroless nickel coated steel; mechanical compatability.

INTRODUCTION

The work described herein is concerned with the alleviation of the problem of the cavitation erosion of components of hydraulic machinery which is being developed for the South African Gold Mining Industry (Protheroe, 1979; Heathcock, Protheroe and Ball, 1979a). This machinery is operated on water based emulsions containing approximately 5 percent additive in water. These fluids are used to reduce pollution and the high costs associated with pipe bursts. However, a considerable disadvantage of these fluids is their susceptibility to cavitation and as a consequence severe erosion damage occurs in the hydraulic components. Although small items such as valves can be manufactured from special high cost alloys, this solution is not economically viable for the larger pump components such as casings, swash plates and impellors. For these it was considered that surface treatments and coatings could be used to combat the erosion and research has been undertaken to evaluate the performance and suitability of a wide variety of different treatment and coatings (Heathcock, Ball and Protheroe, 1981-82). The present paper is a comparative study of the performance of the more promising treatments; these include carburising, nitriding and electroless nickel coating of steel.

EXPERIMENTAL

The method of cavitation testing was that used by Vyas and Preece (1977), Singer and Harvey (1979), and Heathcock (1980). The apparatus consists of a modified industrial ultrasonic drill operating at a frequency of 20kHz. An exponential velocity transformer or 'horn' amplifies the oscillations and creates cavities in the distilled water test bath which is maintained at a temperature of 25°C. Disc specimens (14mm diameter x 3mm), polished to 1 μm surface finish, were weighed and positioned at a distance of 0,35mm below the drill tip. Resonant oscillations at an amplitude of 75μm creates severe cavitation at the surface of the specimen. The extent of the damage is determined at regular intervals by removal and measurement of volume loss. The cumulative volume loss is recorded as a function of time and the rate of volume loss can be calculated from this data. Volume losses of the samples were compared with those of an uncoated sample of the base metal.

Scanning electron microscopy was used extensively to investigate the nature of the erosion and the perforations of the surface layer. Sectioning of the eroded samples and examination by optical microscopy also provided information about the mode of erosion and the extent of any fracture or spalling of the interface below the damaged surface.

The surface treated and coated test samples were produced under industrial conditions and the details of the base metal, thickness of coating or diffusion layer and their respective hardnesses are given in Table 1.

RESULTS

On the basis of the five hour cumulative volume loss of the various surface treatments and coatings, a specimen of 655M13 steel which had been carburised for 3 hours showed the smallest erosion loss. This amounted to 0,03mm^3. Furthermore, as can be seen from Fig. 1, the steady state erosion rate remains low for very long periods of time and the scanning and optical microscopy confirms that a semi-ductile mode of erosion occurs and no massive spalling or brittle fracture takes place. Although chromium coated steels show excellent erosion resistance after periods of less than six hours, catastrophic brittle removal ensues and as a consequence it has little potential for components in hydraulic systems.

Samples of 817M40 steel which had been quenched, hardened and tempered at 600°C and then given an electroless nickel coating show excellent performance particularly those that had received a post heat treatment at a temperature of 400°C. A cumulative volume loss of 0,2mm^3 was measured for these samples.

Samples of the steel 905M39 which had received nitriding treatments by

TABLE 1 Characteristics of the Surface Treated and Coated Steels

Treatment	Base Metal	Hardness of Base VHN (micro-hardness)	Max Hardness of Coating VHN (micro-hardness)	Approx. Thickness of Coating (µm)	5 hour CVL (mm^3)	Steady State Erosion Rate of Coating (mm^3/hr) **	Comments about Erosion
Carburised	655M13	402 916	916	250	0,03	0,06	Slow uniform erosion
Gas nitrided	905M39 (Q&T 600°C)	400 950	950	300	0,29	0,03	Rapid initial loss of 'white' layer
Ion nitrided	905M39 (Q&T 600°C)	400 1100	1100	300	0,95	0,06	Rapid initial loss of 'white' layer
Boronized	817M40 (Q&T 600°C) + 220707	422	2022	100	3,05	0,80	Rapid loss of hard outer layer
Hard chrome plated	835M30 (Q&T 500°C)	279	1038	25	0,06	0,32	Low erosion initially
Electroless nickel plated	817M40 (Q&T 600°C)	402	680	25 - 100	0,5-1,7	0,15 - 0,25	Thickness above a critical value required
Post heat treated electroless nickel plated	817M40 (Q&T 200-600°)	402	680 - 1100*	25	0,21-2,2	0,1 - 0,25	Large improvement for heat treatments above 200°C
Carbide coated	070M20	238	1769	150	2,91	0,65	Disintegration of porous and brittle coating
Chemically vapour deposited tungsten carbide	817M40 (Q&T 500°C)	334	1819	20	1,36	0,4	Disintegration of brittle coating

* varying hardness with varying PHT temperature (200 - 600°C)
** value not given where coating removed before a period of 5 hours.

gaseous and ion bombardment techniques, showed equivalent performance to electroless nickel coated samples. A sample gas nitrided for 60 hours gave a cumulative volume loss after five hours of 0,29mm^3. A disadvantage of nitrided specimens is the formation of a very brittle 'white' outer layer on the sample. This is shown in Fig. 2. The initial rate of erosion can be seen in Fig. 3 to be very high during the brittle removal of this layer and then the samples show a very low steady state erosion rate. This steady state erosion rate is comparable to that for the carburised steels.

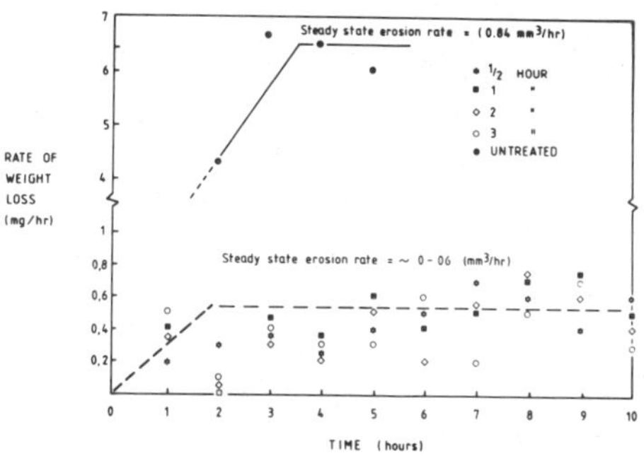

Fig. 1 The effect of different carburising times on the rate of weight loss of a 655M13 steel over a ten hour period of erosion.

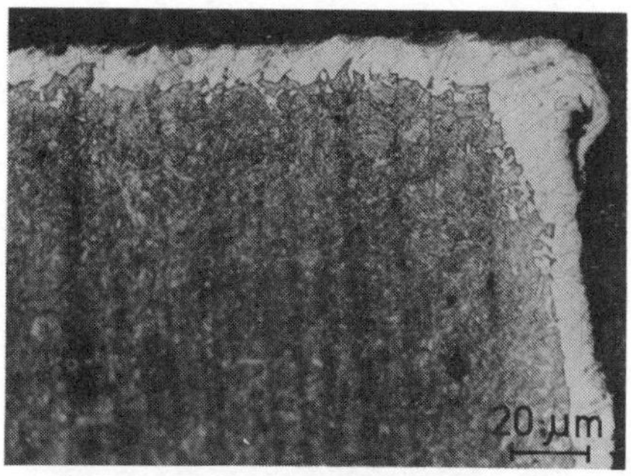

Fig. 2 The microstructure of a 905M39 steel which had been quenched and tempered and then ion nitrided for 72 hrs at 470°C. Note the outer 'white' layer of complex nitride.

Boronised samples, although created by a diffusion process, show very disappointing performance and this can be explained in terms of the sharp interface between the alloy layer and the base meterial. Furthermore, the surface layer is extremely hard and brittle and erosion rates are higher than that of the base metal. The nature of the surface layer and the incipient cracks are shown in Fig. 4.

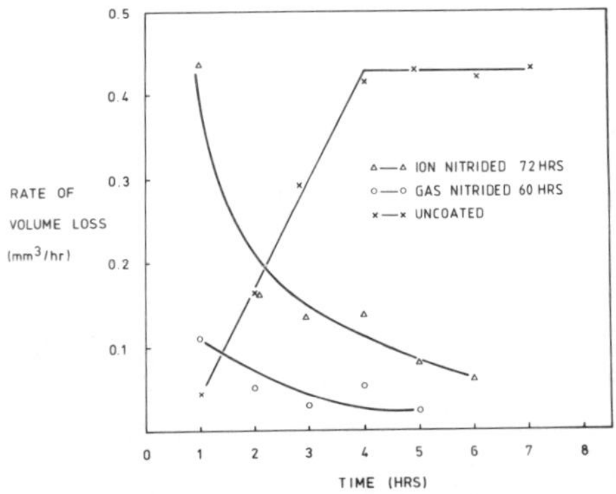

Fig. 3 The effect of nitriding on the rate of volume loss of a quenched and tempered 905M39 steel which had received gas and ion nitriding treatments over a period of 7 hours.

Fig. 4 The microstructure of a 817M40 steel which had been quenched and tempered and then boronised for 5½ hrs at 930°C. Note the sharp interface between the surface layer and base metal and the incipient cracking of the surface layer.

Samples of the steel 070M20, which had been subjected to a high velocity explosive deposition of mixed carbides by a detonation method or D-Gun technique, did improve the performance of the base metal as indicated in Fig. 5. However, the steady state rate of erosion of the coating is quite rapid (0,6mm^3/hr) and the micrograph Fig. 6 shows extensive brittle fracturing around the particles within the coating. The remains of the polymeric sealant can be seen on the eroded surface.

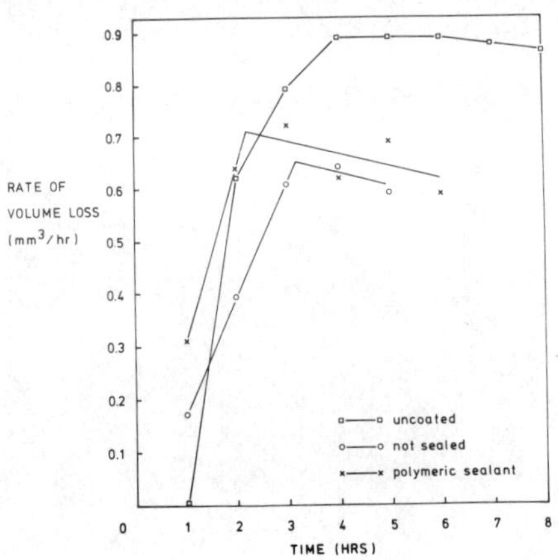

Fig. 5 The rate of volume loss of D-Gun coated 070M20 steel.

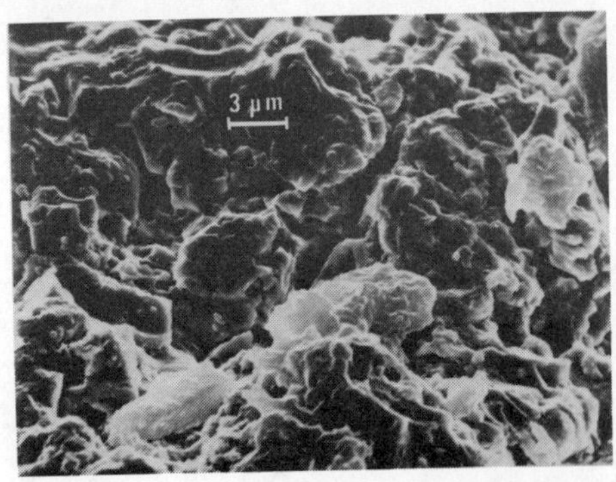

Fig. 6 A scanning electron micrograph of the eroded D-Gun carbide coating showing the disintegration of the porous material.

The samples of the steel 817M40, which had received a coating of tungsten carbide by a chemical vapour deposition technique (CVD), showed very poor performance. The specimens tested displayed very inhomogeneous properties and severe attack was localised (Fig. 7). This localised attack led to an undercutting of the coating and large scale material removal. The rate of erosion loss is shown in Fig. 8. It appears that the stresses created by cavitation are concentrated by defects within the coating and rapid deterioration ensues.

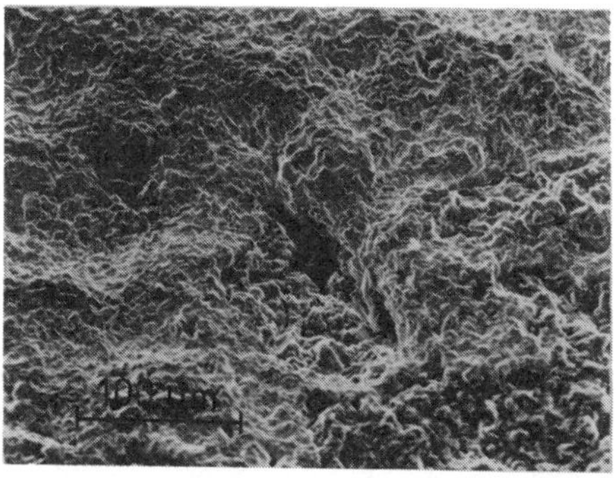

Fig. 7 A scanning electron micrograph of the eroded chemical vapour deposited tungsten carbide coating. Localised severe erosion is evident.

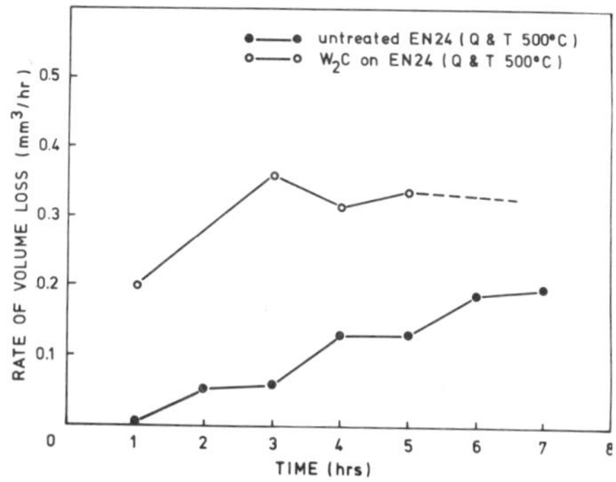

Fig. 8 The rate of volume loss of achemical vapour deposited tungsten carbide coating on a 817M40 quenched and tempered steel.

The results obtained for electroless nickel on 817M40 which had been quenched and tempered at 600°C were particularly interesting in that the erosion resistance was found to be a function of coating thickness. The rate of volume loss as displayed in Fig. 9 shows that coatings of thickness less than 25μm spall in a catastrophic manner after very short periods of time. Thicker coatings (>25μm) show much improved resistance. A further interesting feature of these coatings was the marked influence of a heat treatment subsequent to the coating treatment. The five hour cumulative volume loss due to erosion of a 25μm thick coating of electroless nickel is shown as a function of the post heat treatment temperature in Fig. 10. The corresponding microhardness of the electroless nickel coating is also shown in this figure. The optimum erosion resistance corresponds to the maximum microhardness. It will be noted that the hardness curve shows a maximum and it is suggested that this behaviour is due to precipitation of Ni_3P and the coarsening or overaging of the precipitate at temperatures above 400°C. This explanation has been supported by the monitoring of the X-ray diffractometer signal. The intensity of an Ni_3P diffraction peak increased linearly as a function of heat treatment. This indicates continuous growth of the Ni_3P precipitates in the electroless nickel matrix.

If the erosion resistance was controlled by hardness alone, then the minima in the cumulative loss curve would be more pronounced. The relatively good performance of samples which had been overaged at temperatures of 500° and 600°C must be a consequence of the increased degree of interdiffusion between the base metal and electroless nickel coating. The results of the energy dispersive X-ray analysis shown in Fig. 11 indicate that an appreciable degree in diffusion of nickel has occurred after a treatment at

600°C. The scanning electron micrography of Fig. 12 shows the catastrophic spalling phenomenon which occurred for both thin coatings and coatings which have not received a post heat treatment at a sufficiently high temperature.

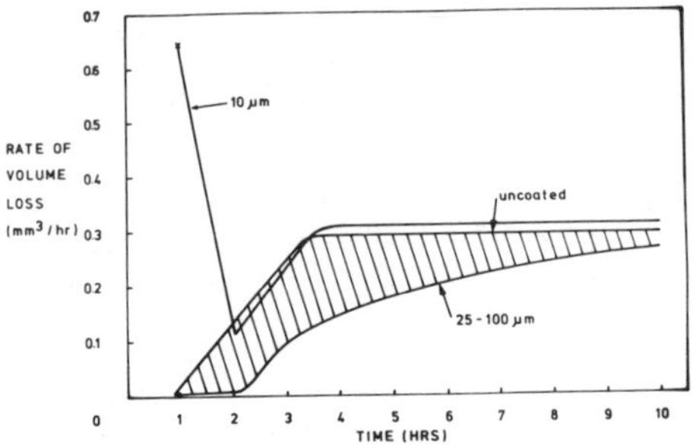

Fig. 9 The rate of volume loss of electroless nickel coated 817M40 steel over a period of 10 hrs. The thickness of the coatings are indicated. All specimens were given post heat treatment of 10 hrs at 300°C.

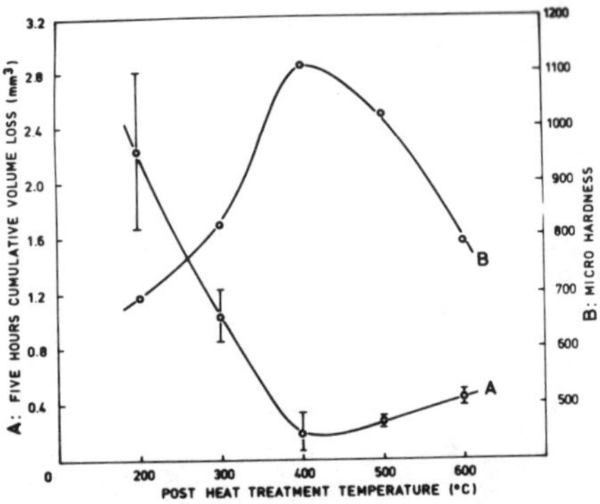

Fig. 10 The five hour cumulative volume loss and microhardness ofelectroless nickel coated 817M40 samples as a function of the temperature of the 3 hr post heat treatment.

Fig. 11 The structure of a sectioned sample of a 817M40 steel which had been coated with electroless nickel and then given a post heat treatment at 600°C for 3 hrs. The elemental compositions at the positions indicated are given by the energy dispersive x-ray analysis spectra.

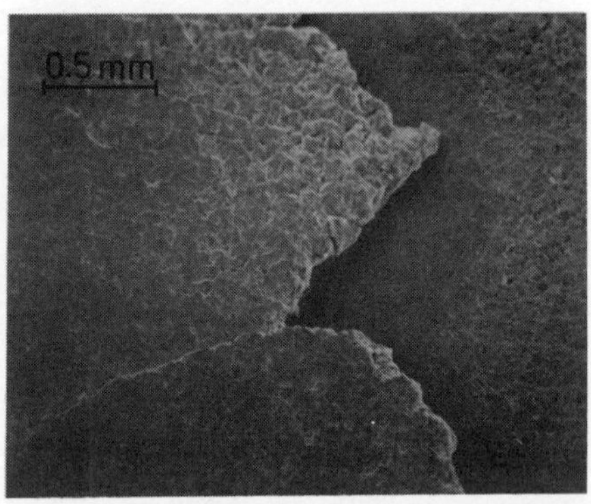

Fig. 12 The erosion by delamination of a thin electroless nickel coating which had received an insufficient heat treatment.

DISCUSSION AND CONCLUSIONS

Although erosion is caused by the collapse of small individual cavities at or close to the surface of a material, the regular pulsing causes the simultaneous collapse of a large number of cavities over the whole surface of the specimen. The concerted pulse produces high stresses and strains in the near surface regions during each cycle. These impact conditions will seek out any defects which may exist either on the surface or in the sub-surface regions. These defects will concentrate the stresses and erosion damage will be nucleated. During subsequent pulses these regions of damage will grow and erosion will proceed. It is, therefore, important that any surface treatment or coating is homogeneous and free from microscopic defects. In particular any interface between the coating and base metal must be coherent and defect free.

It must be assumed that any real material will contain defects of some size and form and the erosion resistance will depend upon the material's ability to absorb energy and resist microfracture. The compromise between hardness and microtoughness has been discussed in the reviews edited by Preece (1979) and by Heathcock, Protheroe and Ball (1979b). The poor performance of steel specimens which have been boronised, hard chrome plated, D-Gun carbide coated and chemically deposited with tungsten carbide, can be ascribed to the presence of defects both within the surface region and at the interface together with the low microtoughness. The low microtoughness is associated with the extreme surface hardness of these specimens (Table 1).

The study of the electroless nickel coating as a function of both thickness of coating and post heat treatment has highlighted two important features of a coated system. The interface between the base metal and the electroless nickel coating represents a region of weakness. Defects may be present and differences in elastic and plastic response to an applied stress will generate decohesive shear stresses. The magnitude of stress pulse generated at the interface by cavitation will be attenuated with distance below the specimen surface as depicted in Fig. 13.

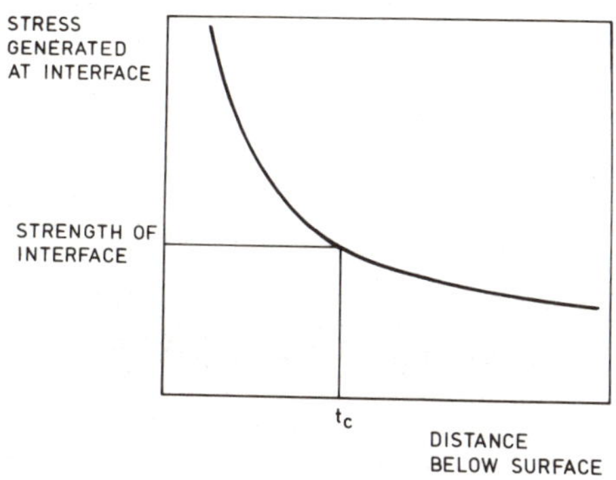

Fig. 13 A schematic representation of the stress generated at the interface between a coating as a function of distance below the surface. The critical thickness of a coating, t_c, below which spalling occurs, is a function of the strength of the interface.

For a given strength of the interface there will be a critical thickness above which the imposed stress will create no damage. Coatings of thickness less than this critical value will fail by delamination. For our specimens which had received a standard post heat treatment for 10 hours at 300°C, this critical thickness was approximately 25μm (Fig. 9). Thus this work has demonstrated that the provision of thicker coatings may have an advantage with respect to erosion resistance.

An alternative approach to improving the performance of a coated specimen is to increase the strength of the interface between the base metal and surface layer. If this can be achieved by diffusion, then a smaller critical thickness will be appropriate. The results of post heat treatments at various temperatures up to 600°C are shown in Fig. 10 and the provision of a diffused interface is demonstrated in Fig. 11. Associated with this interdiffusion is an increase in hardness for annealing temperatures up to 400°C and a decrease in hardness for annealing temperatures in excess of 400°C. The optimum erosion resistance at 400°C can be ascribed to both an increased hardness of the coating and an improved interfacial strength. The overriding importance of the interfacial strength is emphasised by the small depreciation in erosion resistance, as compared with the sharp fall in hardness for post heat treatments above 400°C. Failure to provide a sufficiently strong interface or a sufficiently thick coating will result in serious erosion by delamination.

Surface treatments of steels which involve the diffusion of the interstitial atoms carbon, nitrogen and boron should provide excellent erosion resistance on account of the diffuse nature of the interface. There will be no distinct change in the mechanical response of the surface and subsurface regions and in addition the fineness and homogeneity of the resultant microstructures will provide good microtoughness. With the exception of the boronised steel these predictions have proved to be correct and carburised steel outperformed all of the samples. The inferior performances of the boronised sample is associated with a sharp diffusion boundary and the initial poor performance of the nitrided specimens was a consequence of the removal of an outer brittle layer of complex nitrides. Carefully controlled carburising and nitriding treatments show potential for the provision of erosion resistant steel components.

ACKNOWLEDGEMENTS AND REFERENCES

This work forms part of the technical collaboration between the University of Cape Town and the Chamber of Mines of South Africa. Mr B. Greeves and Ms P. Ellis are thanked for their assistance in the preparation of the manuscript.

Heathcock, C.J. (1980). Cavitation erosion of materials. Ph.D. Thesis, University of Cape Town.

Heathcock, C.J., B.E. Protheroe, and A. Ball (1979a). The influence of external variables and microstructure on the cavitation erosion of materials. Proceedings 5th Int. Conf. on Erosion by Liquid and Solid Impact, Cambridge, England, 63-1 - 63-8.

Heathcock, C.J., B.E. Protheroe and A. Ball (1979b). The influence of microstructure on the cavitation erosion of materials. Proceedings 5th Int. Conf. on Strength of Metals and Alloys, Aachen, W. Germany, 219-224.

Heathcock, C.J., A. Ball, and B.E. Protheroe (1981/1982). Cavitation erosion of cobalt-based stellite alloys, cemented carbides and surface-treated low alloy steels. Wear, 74, 11-26.

Preece, C.M. (1979). Treatise on Material Science and Technology, Erosion, 16. Academic Press, New York.

Protheroe, B.E. (1979). The prevention of erosion damage in water based hydraulic machines. Proceedings 5th Int. Conf. on Erosion by Liquid and Solid Impact, Cambridge, England, 64-1 - 64-8.

Singer, B.G. and S.J. Harvey (1979). Gas content and temperature effects in vibratory cavitation tests. Wear, 52, 147-160.

Vyas, B. and C.M. Preece (1976). Stress produced in a solid by cavitation. J. Appl. Phys., 47, 5133-5138.

THEME III

DESIGN WITH FIBRE REINFORCED (COMPOSITE) MATERIALS

A CORRELATION BETWEEN FRACTURE ENERGY AND FRACTURE MORPHOLOGY IN MIXED-MODE FRACTURE OF COMPOSITES

H. T. Hahn* and T. Johannesson**

*Washington University, St. Louis, Missouri, USA
**Linköping Intitute of Technology, S-581 83 Linköping, Sweden

ABSTRACT

The fracture toughness in terms of the critical strain energy release rate G_c and the critical stress intensity factor K_c were measured for mixed mode crack propagation along the fibers in two unidirectional graphite/epoxy composites, T300/C-69 and T300/1034. When the state of stress was varied, both G_c and K_c increased considerably with increasing proportion of mode II loading. The calculated values of G_c were fitted to a recently suggested linear relation between G_c and K_I/K_{Ic}. The fracture surfaces showed corresponding increases of hackles and resin debris. An approximate analysis is provided for the relation between the fracture toughness measured and the fracture surface morphology.

INTRODUCTION

The ultimate failure of a graphite/epoxy laminate is usually fiber dominated. However, it takes place after a considerable amount of subcritical damage growth has occurred. The subcritical damage growth mostly consists of ply cracking and delamination. These two components thus contribute to the apparent toughness of a laminate.

Ply cracking in general occurs under mixed mode (K_I and K_{II}) loading. The critical strain energy release rate, G_c, is found to vary with the K_I/K_{II} ratio. Vanderkley (1981) measured a five-fold increase of G_c as the proportion of mode II loading was increased to $G_{II}/G = 0.5$. However the opposite trend was observed when a tough resin was used (Schapery (1982)). Vankerkley (1981) found the increase of G_c to be associated with an increase of fiber damage roughening of the fracture surface. Sinclair and Chamis, (1978) and later, Awerbuch and Hahn (1981) observed an increase of roughness of the fracture surface in off-axis unidirectional specimens as the off axis angle was varied away from 90°.

Mixed-mode fracture toughness was measured by several investigators: Wu and Reuter (1965) for Scotchply 1002 and balsa wood, and McKinney (1972) and Jurf and Pipes (1982) for graphite/epoxy composites. It was found that the critical energy release rate criterion

$$\left(\frac{K_I}{K_{Ic}}\right)^2 + \left(\frac{K_{II}}{K_{IIc}}\right)^2 = 1$$
$$\frac{K_{IIc}}{K_{Ic}} = \left(\frac{S_{22}}{S_{11}}\right)^{1/2} \tag{1}$$

underestimated K_{IIc} from K_{Ic}.

Wu (1963) suggested an empirical mixed-mode fracture criterion

$$\frac{K_I}{K_{Ic}} + \left(\frac{K_{II}}{K_{IIc}}\right)^2 = 1 \tag{2}$$

where K_{Ic} and K_{IIc} are independent of each other.

Hahn (1983) suggested that the energy release rate approach could be used provided the critical energy release rate was allowed to depend on K_I by

$$G_c = G_{IIc} - (G_{IIc} - G_{Ic}) \frac{K_I}{K_{Ic}} \tag{3}$$

This criterion reduces to (1) if $G_{Ic} = G_{IIc}$ and to (2) if $G_{Ic} << G_{IIc}$.

The objective of this paper is to determine G_c for unidirectional graphite/epoxy composites under mixed-mode loading. The results are then used to test the validity of the mixed-mode fracture criterion. Also, the fracture surface morphology is examined in light of the corresponding G_c-value, and an approximate analysis is performed to relate G_c to the fracture surface morphology.

EXPERIMENTAL PROCEDURES

Unidirectional laminates $[0]_8$ were prepared from two types of prepregs, T300/C-69 and T300/1034 according to manufacturers' recommended procedures. Rectangular tensile coupen specimens were cut at various angles to the fiber direction. Small cracks were then produced parallel to the fibers at the center of each specimen using a water jet saw. The specimens were then loaded to failure in tension in a servo-hydraulic testing machine at a strain rate of $3 \cdot 10^{-4}\ s^{-1}$. The load-displacement relation and the failure load were recorded. Pieces of the fracture surface close to the original crack tips were cut out, coated with a thin Pt layer and then studied in a scanning electron microscope.

RESULTS AND DISCUSSION

Measured values of crack size, a, far-field stress, σ_∞, and crack angle θ were used to calculate K_I and K_{II} at failure. The critical strain energy release rate G was then calculated from the following relations:

$$G = \frac{n}{2}\left[(S_{11}\cdot S_{22})^{1/2}\cdot K_I^{\,2} + S_{11}\cdot K_{II}^{\,2}\right],$$

$$n = \left[2\left(k + \frac{S_{12}}{S_{11}}\right) + \left(\frac{S_{66}}{S_{11}}\right)\right]^{1/2}, \qquad (4)$$

$$k = \left(\frac{S_{22}}{S_{11}}\right)^{1/2}$$

where S_{11}, S_{22}, and S_{12}, and S_{66} are the on-axis compliance coefficients for the unidirectional composite. The material constants used for the calculation are shown in Table 1 and the results of the calculations in Table 2.

TABLE 1 Materials Constants

Material	S_{11}	S_{22}	S_{12}	S_{66}	
T300/C-69	7.25	112	-2.17	141	$(TPa)^{-1}$
T300/1034	7.24	100	-2.54	280	$(TPa)^{-1}$

The specimens used have a finite width and therefore finite width correction factors Y_I, Y_{II} should be used so that

$$K_I = Y_I \sigma_y \sqrt{\pi a}$$

$$K_{II} = Y_{II} \sigma_{xy} \sqrt{\pi a} \qquad (5)$$

However, no attempt was made to calculate Y, and, therefore, the apparent critical stress intensity factors with $Y_I = Y_{II} = 1$ are given in Table 2. By the same token, the values of G_c given in Table 2 are the apparent critical strain energy release rates.

The G_c values are plotted vs. the normalized mode I stress intensity factor, K_I/K_{Ic}, in Fig. 1. Regression analysis of the data with respect to the mixed mode failure criterion in eq. (3) are shown as straight lines in Fig. 1. The regression analysis results for G_{Ic} and G_{IIc} are given in Table 3. When comparing the results, it should be bore in mind that specimens of T300/C-69 had almost constant crack size for various angles while for specimens of T300/1034 the ratio between the specimen width and the projection of the crack on a plane perpendicular to the loading was kept constant. The T300/1034 specimens all had a larger crack size-to-width ratio and thus will require different finite width correction factors.

Examinations of fracture surfaces in the scanning electron microscope reveal a consistent variation of fracture surface morphology with K_I/K_{II} ratio. At pure mode I loading ($\theta = 90°$, Fig. 2a), the resin parts of the fracture surfaces are smooth and rather typical of a cleavage type of fracture. As the

TABLE 2 Calculated Values for K_I, K_{II} and G_c (apparent) at failure

θ	2a	$\frac{2a\sin\theta}{w}$	σ_∞	K_I	K_{II}	G_c
o	mm		MPa	$MPam^{1/2}$	$MPam^{1/2}$	J/m^2
			T300/C-69			
10.0	5.9	0.03	210.0	0.608	3.45	250
45.0	6.2	0.09	28.8	1.27	1.27	149
90.0	5.6	0.12	18.2	1.33	0.0	133
			T300/1034			
7.5	76.6	0.33	93.3	0.551	4.18	455
15.0	26.6	0.33	55.3	0.912	3.40	357
30.0	20.0	0.33	26.1	1.16	2.01	219
45.0	14.2	0.33	14.6	1.09	1.09	137
60.0	11.6	0.33	10.8	1.09	0.63	117
75.0	10.4	0.33	9.6	1.15	0.31	121
90.0	10.0	0.33	10.8	1.21	0.0	132

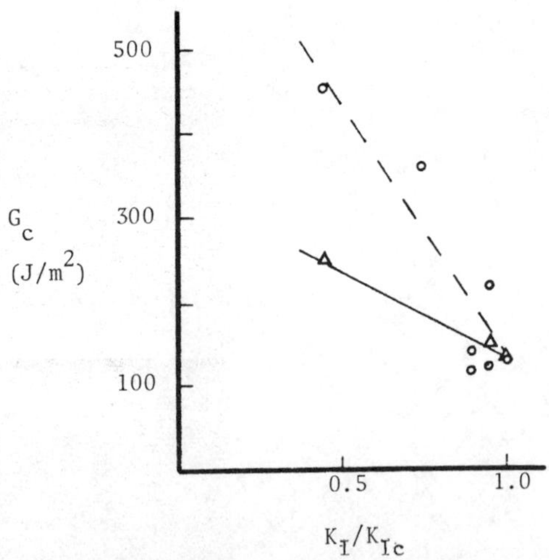

Fig. 1. The apparent critical strain energy release rate, G_c, vs. the normalized mode I stress intensity factor, K_I/K_{Ic} for unidirectional graphite/epoxy laminates.

Δ - T300/C-69

o - T300/1034

proportion of mode II loading is increased (Fig. 2b-e), the amount of hackle formation steadily increases and the hackles end to a more upright position. At very high proportions of mode II loading, however, the hackles are gradually replaced by resin debris. One possible explanation of this is that in a more upright position they are more easily torn away in the fracture process, thus forming resin debris.

The fracture surfaces also show a certain amount of fiber failures as was reported by Vankerkley (1981) but no definite relation to the off-axis angle could be detected.

The hackle formation in the matrix phase under a higher proportion of mode II loading results in an increased surface area, and hence leads to a higher critical strain energy release rate.

A simple model, based on the fracture surface morphology, that represents the variation of G_c with increasing mode II loading is the following. Let us assume that the matrix phase of the composite can be approximated by a thin slab of the resin material between two fibers. Assume further that the hackle orientation angle α is the complementary angle to the off-axis angle θ (Fig. 3). This assumption is supported by the results of Sih and Chen (1973). Finally, assume that hackles are formed with the geometry defined in Fig. 4.

For a unit area of macroscopic crack extension, the actual surface area A due to the hackles is given by

$$A = 2(1-v_f)H \qquad (6)$$

TABLE 3 Results of the Regression Analysis

Material	G_{Ic} J/m^2	G_{IIc} J/m^2
T300/C-69	136.5	347
T300/1034	120.0	767

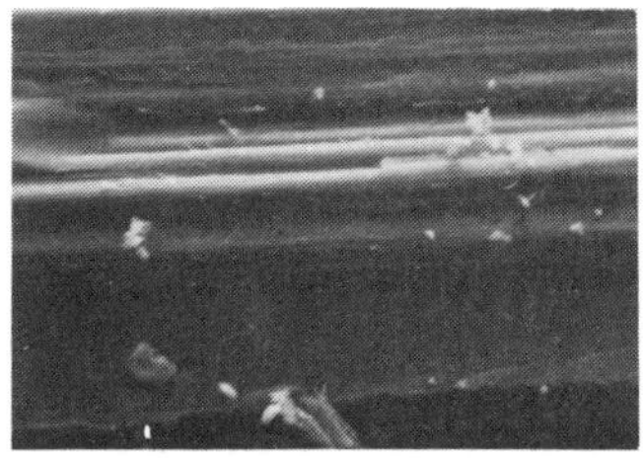

a) Off-axis angle θ = 90°

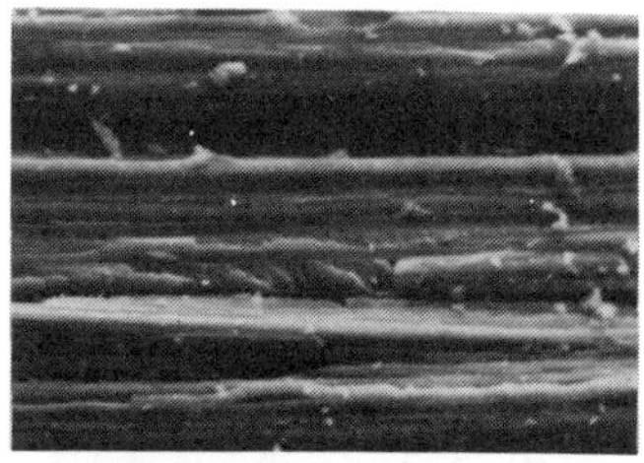

b) Off-axis angle θ = 45°

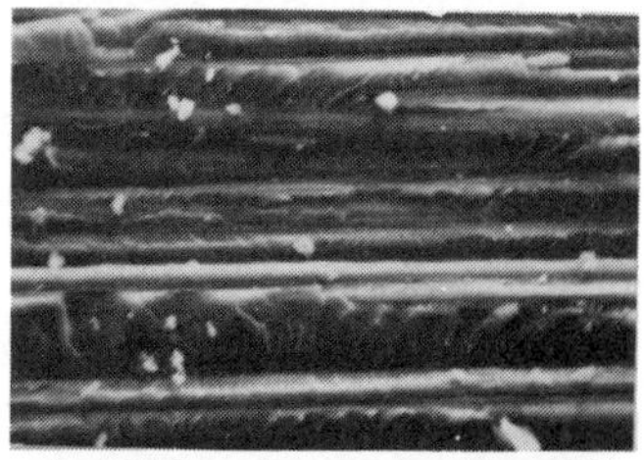

c) Off-axis angle θ = 30°

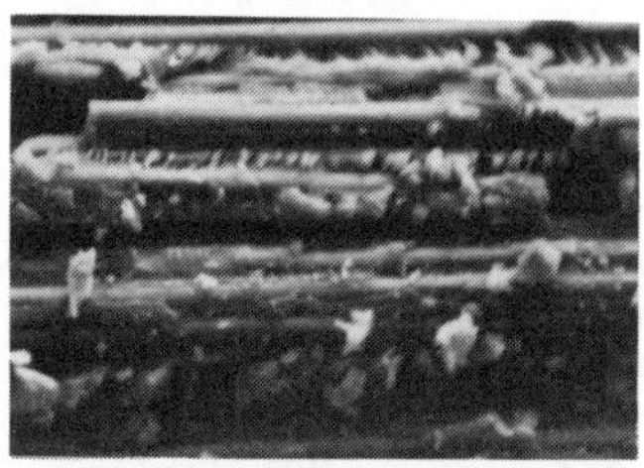

d) Off-axis angle θ = 15°

e) Off-axis angle θ = 7.5°

Fig. 2. Fracture surfaces. SEM 500X

where

$$H = \frac{\tan\alpha}{d_o} \cdot \frac{d_o}{\sin\alpha} = \frac{1}{\cos\alpha} = \frac{1}{\sin\theta} = \sqrt{1+\tan\theta^{-2}} \quad (7)$$

Since $\tan\theta = K_I/K_{II}$, we have

$$A = 2(1-v_f)\ \sqrt{1+(K_{II}/K_I)^2} \quad (8)$$

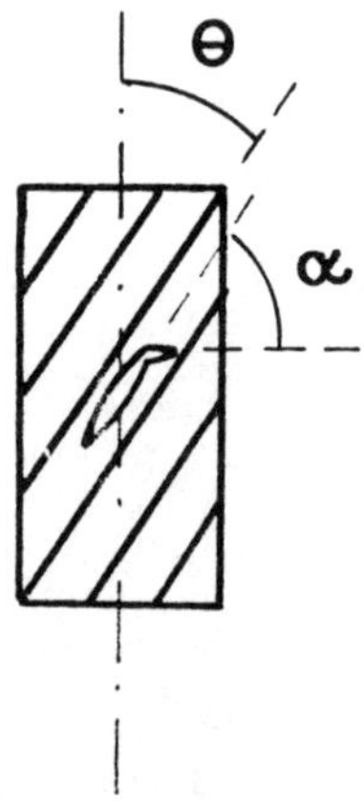

Fig. 3. Crack propagation forming hackles.

Assuming further that the surface area required to connect the hackles is small compared with A, we finally derive

$$G_c = G' + G''_m\ \sqrt{1+(K_{II}/K_I)^2} \quad (9)$$

Here G' may be taken as the critical energy release rate due to fiber-matrix debonding, and G" is related to the surface energy of the resin, ∂_m, by

$$G'' = 2\partial_m(1-v_f) \quad (10)$$

Note that $G_{Ic} = G' + G''$.

A plot of the experimental result in terms of the above equation is shown in Fig. 5. Linear regression analysis yields G' = 82.8 J/m^2 and G" = 52.9 J/m^2 for T300/1034 and G' = 112.2 J/m and G" = 24.0 J/m^2 for T300/C-69.

When $K_{II} = 0$, G_{Ic}=(G' + G") is 135.7 and 136.2 J/m^2 for T300/1034 and T300/C-69, respectively. The G_c vs. $\sqrt{1+(K_{II}/K_I)^2}$ relation shows a larger slope for T300/1034 than for T300/C-69. This is due to the larger crack size to width ratio of the T300/1034 specimens. Additional data not reported here indicate that G_c increases with the crack size when the off-axis angle is small. Whether or not such behavior can be remedied by using appropriate finite width correction factors is under investigation.

Ply cracking in a laminate takes place under mixed-mode (I/II) loading. The K_I/K_{II}-ratio for a unidirectional ply in a loaded laminate varies with the properties of the surrounding plies and with previous delamination at the interfaces of the ply. As the critical strain energy release rate G_c for ply cracking in the ply varies strongly with K_I/K_{II} ratio, the contribution from that ply to the apparent laminate toughness will depend on neighboring ply properties, possible delamination as well as the overall loading condition. Thus, the energy absorbing ability of a laminate depends strongly on the stacking sequence and the loading conditions as well as on the type of material.

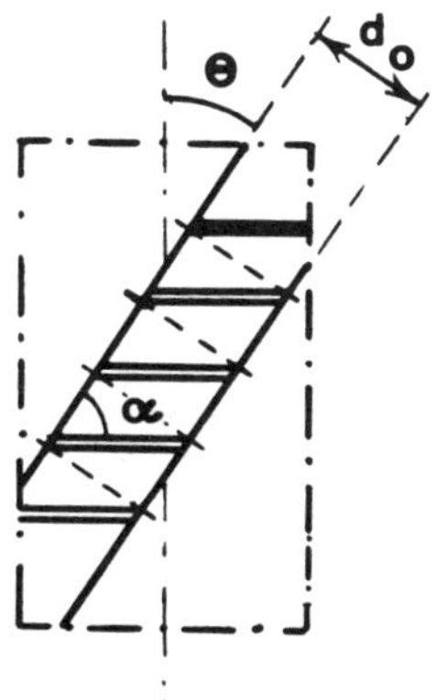

Fig. 4. Hackle formation in a thin slab of the resin material (thickness d_o) of a unidirectional laminate loaded at an angle θ^o to the fiber direction.

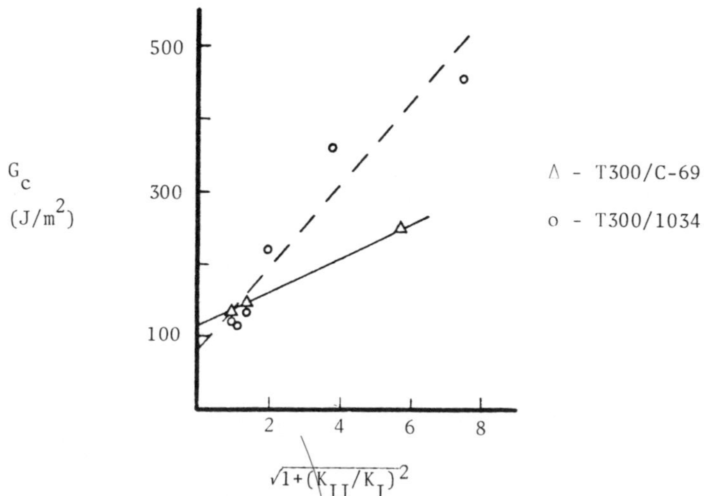

Fig. 5. The apparent critical strain energy release rate, G_c, vs. fracture surface area enlargement factor due to hackle formation $H = \sqrt{1+(K_{II}/K_I)^2}$.

CONCLUSION

The fracture toughness measured as G_c was found to vary with the mode of loading according to

$$G_c = G_{IIc} - (G_{IIc} - G_{Ic})\, K_I/K_{Ic}.$$

The value of G_{IIc} calculated from a regression analysis was found to be almost five times larger than G_{Ic}.

The fracture process as evident from the fracture surface appearance changes considerably, showing more formation of hackles and resin debris, as the proportion of mode II loading increases.

Ply cracking parallel to the fibers provide an essential contribution to subcritical failure of a laminate. The ply cracking in identical laminates will obviously proceed differently with different specific energy absorption rate for different modes of loading of the ply.

ACKNOWLEDGEMENTS

This paper is based on work supported in part by Swedish National Board for Technical Development (STU) under Grant No. 82-5074 (TJ) and also by National Science Foundation (USNSF) under Grant No. MEA-8110777 (HTH). The authors wish to thank Dr. G. Grüninger of Institut für Bauweisen-und Konstruktionsforschung, DFVLR (W. Germany) where the data for T300/C-69 specimens were obtained, and P. Sjöblom and M. Blikstad for fabrication and testing of T300/1034 specimens.

REFERENCES

Awerbuch, J., and H. T. Hahn (1981). "Off-Axis Fatigue of Graphite/Epoxy Composite in Fatigue of Fibrous Composite Materials," STP 723, American Society for Testing and Materials, Philadelphia, 248-273.

Hahn, H. T. (1983). "A Mixed-Mode Fracture Criterion for Composite Materials," to appear in Composites Technology Review.

Jurf, A. R., and R. B. Pipes (1982). "Interlaminar Fracture of Composite Materials," J. Composite Materials 16, 386.

Schapery, R. A. (1982). "Research on Composite Materials for Structural Design," in Mechanics of Composites Review, AFWAL/AFOSR/NASA/NADC, 109.

Sih, G. C., and E. P. Chen (1973). "Fracture Analysis of Unidirectional Composites," J. Composite Materials 7, 230.

Sinclair, J. H., and C. C. Chamis (1978). "Fracture Surface Characterization of Off-Axis Composites," NASA TM-73-200 (NASA).

Vanderkley, D. S. (1981). "Mode I-Mode II Delamination Fracture Toughness of a Unidirectional Graphite/Epoxy Composite," Texas A & M University, MM 3724-81-15.

Wu, E. M. (1963). "Application of Fracture Mechanics to Orthotropic Plates," University of Illinois, T & AM Report No. 248.

Wu, E. M., and R. C. Reuter, Jr. (1965). "Crack Extension in Fiberglass Reinforced Plastics," University of Illinois, T & AM Report No. 275.

OPTIMUM DESIGN OF LAMINATED PLATES WITH FIBRE REINFORCED (COMPOSITE) MATERIALS

Yongqiu Jiang and Yeong-Chiou Chiang

Department of Engineering Mechanics, Xian Jiaotong University, China

ABSTRACT

MANY design criteria must be considered in the design of composite materials for structures; among them the displacement, stability and vibration are considered in this paper. The deflection, buckling stress (under axial compression and shear) and natural frequency of laminated plates with fibre reinforced (composite) materials are discussed in detail.

The plates are assumed to be simply supported at four edges, each layer of the plate is having the same thickness h and an equal number of fibres of the same material in the $+\alpha_i$ and $-\alpha_i$ directions with the X-axis in the same type of matrix, as shown in Fig. 1.

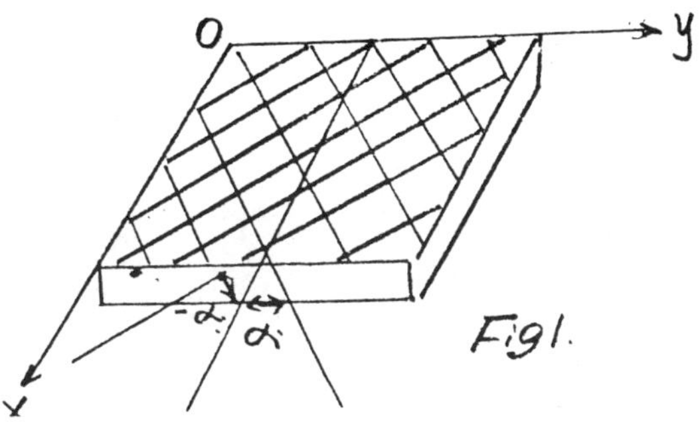

Fig1.

The present problem is to find the fibre directions of all the layers that give the lowest deflection, highest buckling stress (under axial compression and shear) and highest natural frequency (lowest mode). Therefore one has to solve the unconstrained minimization (or maximization) problems. The objective functions are the deflection, the critical buckling stress and natural frequency, and the design variables are the fibre directions. Preassigned parameters are the material properties, the thickness of each layer, the number of layers and the aspect ratio of the plates. Since the objective functions are rather complicated functions of the design variables, optimization methods that do not use the derivatives are suitable for solving the problem. As the objective functions for deflection, buckling stress under axial compression and natural frequency are explicit functions of the design variables, Variable Metrical Method (DFP Method) is selected and the golden section search method for solving one-dimensional minimization problem is used; while the objective function for buckling stress under shear is implicit function (a system of linear algebraical homogeneous equations), Powell's method (conjugate direction technique) is chosen and the golden section search method for solving the one-dimensional minimization problem is used too.

Four computer programs using FORTRAM IV were written and were checked for the case of isotropic plates. The computer used was Siemens 7760.

Numerical examples

Numerical calculations were made for fibre reinforced laminated plates with three, four and six layers. The plates considered have various aspect ratios. The material considered is boron/epoxy.

$$E_1 = 2.11 \times 10^4 \text{ kg/mm}^2$$

$$E_2 = 2.11 \times 10^3 \text{ kg/mm}^2$$

$$\gamma_{12} = 0.3, \quad G_{12} = 7.03 \times 10^2 \text{ kg/mm}^2$$

The thickness of each layer is assumed to be 0.254 mm.

A summary of all the numerical results is given in Tables 1-4.

Table 1 Optimum fibre directions for deflections of the plates under uniformly distributed load P_o

Number of layers	R=a/b	Fibre directions degrees α_1	α_2	α_3	α_4	Objective function $\phi = \frac{\pi W}{16 P_o a^4} \times 10^6$	m	n
4	0.5	0	0	0	0	5.233	1	1
4	0.8	38	38	38	38	3.723	1	1
4	1.0	45	45	45	45	2.508	1	1

Table 2 Optimum fibre directions for natural frequency

Number of layers	R=a/b	Fibre directions degrees α_1	α_2	α_3	α_4	α_5	α_6	Objective function $\phi = \omega b^2 \sqrt{\frac{\rho}{Q_{22} t^3}}$	m	n
3	1	45	45	45				13.36	1	1
4	1	45	45	45	45			13.36	1	1
6	0.5	0	0	0	0	0	0	37.00	1	1
6	0.8	38	38	38	38	38	38	17.14	1	1
6	1	45	45	45	45	45	45	13.36	1	1
6	2	90	90	90	90	90	90	9.25		

Table 3 Optimum fibre directions for buckling stress under axial compression

Number of layers	R=a/b	Fibre directions degrees α_1	α_2	α_3	α_4	α_5	α_6	Objective function $\phi = \frac{12 N_x b^2}{\pi^2 t^3 Q_{22}}$	m	n
3	1	45	45	45				22.000	1	1
4	1	45	45	45	45			22.000	1	1
6	0.5	0	0	0	0	0	0	42.127	1	1
6	0.8	38	38	38	38	38	38	23.154	1	1
6	1	45	45	45	45	45	45	22.000	1	1

Table 4 Optimum fibre directions for buckling stress under shear

Number of layers	R=a/b	Fibre directions degrees α_1	α_2	α_3	α_4	α_5	α_6	Objective function $\phi^* = \frac{\pi^4}{32a^2 R N_{xy}} \times 10^5$	m	n
3	1.0	45	45	45				7.834	3	3
6	0.3	22.7	23.7	23.7	24.2	23.5	22.6	4.816	4	4
6	0.6	33.7	33.5	33.4	33.5	33.6	33.6	2.285	4	4
6	0.8	40.1	40.1	40.1	40.1	40.1	40.1	1.495	4	4
6	1.0	45	45	45	45	45	45	0.992	4	4
6	1.5	55.2	54.9	54.9	55.0	55.1	55.0	0.307	5	5

* Objective function $\phi = \frac{\pi^4}{32a^2 R N_{xy}}$ has to satisfy the following system of homogeneous linear algebraical equations:

$$\phi \, \overline{W}_{pq} - \frac{pq}{V_{pq}} \sum_{m=1}^{\infty} \sum_{n=1}^{\infty} H_{mn} \, \overline{W}_{mn} = 0$$

From these numerical results we conclude that the optimum fibre directions for deflection, critical stress (under axial compression and shear) and natural frequency are those shown in Table 5.

Table 5 Optimum fibre directions

Number of layers	R=a/b	Fibre directions degrees α_1	α_2	α_3	α_4	α_5	α_6
3	1.0	45	45	45			
4	0.5	0	0	0	0		
4	0.8	38	38	38	38		
4	1.0	45	45	45	45		
6	0.5	0	0	0	0	0	0
6	0.8	38	38	38	38	38	38
6	1.0	45	45	45	45	45	45
6	1.5	55	55	55	55	55	55
6	2	90	90	90	90	90	90

MECHANICAL PROPERTIES OF BAMBOO-MESH REINFORCED CEMENT COMPOSITES

M. A. Mansur and M. A. Aziz

Department of Civil Engineering, National University of Singapore, Kent Ridge, Singapore

ABSTRACT

The results of an experimental investigation exploring the feasibility of using bamboo in the form of woven mesh as reinforcement for cement mortar, similar to ferrocement, is reported. Test results have shown that inclusion of bamboo mesh imparts considerable ductility to the mortar and increases significantly its tensile and flexural strengths. However, such improvements, particularly in tension, are associated with wide cracking. Poor bond resulting from volume changes of bamboo under wet and dry conditions of the matrix, and low elastic modulus of bamboo are the two main reasons held responsible. But use of some cheap water repellent coating has been found to be effective to overcome high water absorption of bamboo and to improve the bond strength significantly.

KEYWORDS

Bamboo mesh; composite material; cracking; ductility; failure; flexural strength; tensile strength; sealer treatment; tests.

INTRODUCTION

Bamboo is a natural vegetation available in abundance in the tropical and subtropical regions. It has been in use for centuries in the construction of low-cost rural housing in developing countries. The use of bamboo as a reinforcement for cement mortar or concrete is comparatively a new idea that has evolved as a result of ever increasing demand for low-cost building materials. Some research work has already been carried out on the properties of bamboo and its potential use in various structural concrete elements.

A comprehensive and detailed investigation of the properties of bamboo in tension, compression and flexure was reported by Glen (1950). Mehra et al. (1951), Narayan and Rahman (1962), and Youssef (1976) investigated the water absorption and volume changes of bamboo. Fang and Mehta (1978), and Fang and Fay (1978) found that suitable sulphur treatment of bamboo can significantly reduce the water absorption potential during concrete curing process. Tests carried out by Mentzinger and Plourde (1966) on untreated, varnish treated and sealer treated bamboo indicated substantial improvement in bond strength between bamboo and concrete. Investigations were also reported on bamboo-reinforced concrete structural elements, such as beams (Glen, 1950), slabs (Ali and Pama, 1978), columns (Cook et al., 1978), and walls (Kalita et al., 1978).

Although bamboo is remarkably strong in tension, previous investigations showed little promise of the possible replacement of steel by bamboo mainly due to its low elastic modulus, poor bond with concrete and high water absorption potential. Untreated bamboo absorbs a significant amount of water from wet concrete resulting in swelling, and it subsequently shrinks as the concrete dries out. The contact between bamboo and concrete is thus lost resulting in little or no composite action. The interface bond may however be improved by suitable treatment of bamboo to minimize swelling. In addition, an increase in specific surface area of bamboo may help in improving the bond.

The specific surface area of bamboo may be increased if a bamboo culm is cut into thin strips. These strips can be woven into reasonably flexible meshes which have been traditionally used in handicraft industries for making fishing traps, cages and other decorative and utility products. The feasibility of using such bamboo meshes as reinforcement for cement mortar, similar to ferrocement, through a study of the mechanical properties of the composite is the primary objective of the present investigation. Specimens were tested in direct tension and flexure. The volume fraction of bamboo was varied by varying the size and number of layers of meshes. Both treated and untreated bamboo strips were employed. The results of these tests are presented and discussed in this paper.

EXPERIMENTAL PROGRAMME

In the present study, tensile and flexural strengths of mortar and the composites were determined. The specimens for the composite were divided into two series, A and B, according to the treatment of bamboo meshes. In series A, they were in virgin condition, but those in series B were coated with a thin layer of water sealer under the trade name 'throseal'. Each series comprised of a group of specimens (three for each type of test), each with a different quantity of bamboo. They were designated by a letter symbol indicating the series name, followed by two numbers. The first and second numbers denote respectively the size of opening and number of layers of meshes. As the dimensions of test specimens and opening size of meshes were fixed, area fractions of bamboo strips in the direction of applied stress were different for tensile and flexural specimens in the same group as indicated in Figs. 2 and 4.

Materials

Local bamboo having a light yellow colour was used. The bamboo culm was approximately 10 cm in diameter and 9 mm in thickness. Strips, 5.5 mm wide and 0.88 mm average thickness, were produced manually from split bamboo culms which were seasoned for at least two months after splitting.

Tensile tests were carried out in an Instron testing machine on three representative samples of bamboo strips. The stress-strain relationship was found to be linear up to fracture. Water absorption and bond pull-out tests were also conducted on treated and untreated bamboo strips. The results of these tests are summarized in Table 1.

Ordinary Portland cement conforming to BS12 was chosen and saturated surface dry sand passing through sieve #14 and retained on sieve #100 was used for the mortar. The mix ratio was 1:1.5, and the water-cement ratio was 0.4. An admixure, 'rapidard' was added to the matrix at 60 cc/kg of cement to obtain 28-day strength in seven days. The properties of the mortar are also furnished in Table 1.

Preparation of Test Specimens

Two types of specimens were cast to determine tensile and flexural strengths of the matrix and the composites. The tensile specimens were 304 x 50 x 19 mm. The ends of these specimens were enlarged to a width of 76 mm to facilitate proper gripping. The dimensions of flexural specimens were 375 x 100 x 19 mm.

Bamboo strips prepared earlier were woven manually into meshes keeping the desired opening size constant as far as practicable. A completed bamboo mesh is shown in Fig. 1. These meshes were then cut into the shapes and sizes of the test specimens. For B-series, a thin layer of water sealer was applied using a painting brush and the meshes were allowed to dry in the air for at least 24 hours.

The specimens were cast in plywood moulds. The bamboo meshes were tied to the moulds by means of wires which were stretched and positioned on top and bottom of the meshes to prevent them either from sinking or from floating, and to maintain uniform spacing across the thickness. The clear cover was 4 mm. The mortar was first pressed manually and final compaction was achieved by vibrating the moulds for about 30 seconds on a vibrating table. The specimens were demoulded after 24 hrs of casting and cured under water for 5 days. At the end of the curing period, they were allowed to dry in the air for 24 hrs. before testing.

Test Procedure

Tests were conducted in an Instron testing machine. Tensile specimens were subjected to direct tension. Electrical resistance strain gauges were used to measure the strains before cracking. In the post-cracking range, the deformations of the specimens over a gauge length of 100 mm were measured by using a Demec gauge. The flexural specimens were tested under third-point loading on a span of 300 mm. Midspan diflections of the specimens were measured by a dial gauge (1 div. = 0.01 mm).

RESULTS AND DISCUSSION

General Behaviour

The typical stress-strain curves for tensile specimens are shown in Fig. 2. Each curve represents the average of three specimens. It can be seen that the stress-strain relationship was approximately linear up to the cracking load. At cracking, there was a rapid unloading of the specimens. A redistribution of internal forces from the matrix to the bamboo reinforcement took place until a new mechanism had been fully mobilized. Then the load (i.e., stress) continued to increase again. Figure 2 shows that the amount of drop after cracking decreased with an increase in the area fraction of bamboo strips. The attainment of ultimate strength was accompanied by wide cracking (Fig. 3). None of the bamboo strips did fracture at ultimate load indicating that the failure was due to destruction of bond between bamboo and matrix.

Figure 4 shows the load-deflection curves for flexural specimens in B-series. Unlike tensile specimens, the load after cracking remained nearly at a constant value with increasing deflection until the new mechanism had been established. Thereafter, the load continued to increase. In contrast to tensile specimens in which there was only one major crack, several cracks were observed in flexural specimens at failure (Fig. 5). This indicates that the use of bamboo in the form of mesh helps distributing the cracks in flexural specimens. It also imparts significant ductility to the matrix.

Tensile Strength

Test results under direct tension are presented in Fig. 6. Cracking of the specimens with untreated bamboo (series A) occurred at a stress lower than the ultimate strength of mortar (2 N/mm^2). Though insignificant, cracking strength decreases with increasing quantity of bamboo as can be seen in Fig. 6. The presence of any reinforcement with inadequate bond has the effect of reducing the effective cross section, and hence the reduced strength.

A comparison of the specimens in A and B-series in Fig. 6 shows that the use of water sealer improved both the cracking and ultimate strengths. This may attributed to the improvement in bond as indicated by the results of the bond pull-out tests shown in Table 1. For obvious reasons, the ultimate strength of the composite increased with increasing quantity of bamboo strips.

If, after cracking, the applied load is assumed to be resisted only by the bamboo strips in the direction of applied load, then the ultimate tensile strength, σ_{tu}, of the composite is given by

$$\sigma_{tu} = \sigma_{bu} A_b \tag{1}$$

in which σ_{bu} and A_b denote the ultimate tensile strength and area fraction of bamboo, respectively.

The predictions of Eq. 1 are compared with the present test results in Fig. 6. In general, the predicted values are higher than the test results because of inadequate bond. However, with the use of water sealer, the experimental results approached the predicted values.

Flexural Strength

The results of the flexural tests are shown in Fig. 8. It can be seen that the amount of bamboo has apparantly no influence on cracking strength. But an increase in the area fraction of bamboo improves the ultimate strength significantly. It can also be observed in Fig. 8 that the use of water sealer improves both the cracking and ultimate strengths.

The present test results are compared also in Fig. 8 with the ultimate strength predictions of the equation proposed by Ali and Pama (1978). It was based on full composite action between bamboo and mortar and different moduli of elasticity for tension and compression in the cracking range. The final expression was obtained from equilibrium consideration and can be written as

$$M_u = 0.85\ f'_c\ A_m \frac{h_c^2}{2} + \frac{1}{3}\ \sigma_{bu}\ A_b\ (h - h_c^2)^2 \tag{2}$$

in which M_u = ultimate moment; f'_c = cylinder compressive strength of mortar; A_m = area fraction of mortar; h = overall depth; and h_c = depth of compression zone as given by

$$h_c = \frac{0.5\ \sigma_{bu}\ A_b\ h}{(0.85\ f'_c\ A_m + 0.5\ \sigma_{bu}\ A_b)} \tag{3}$$

Figure 8 shows that the theoretical predictions are unsafe both for series A and B, with untreated and treated bamboo, respectively. The deviation is more for specimens having higher area fraction of bamboo. However, for smaller volume fraction of treated bamboo, the theory gives good predictions of ultimate bending strength.

CONCLUDING REMARKS

The feasibility of using bamboo in the form of woven mesh as reinforcement for cement mortar has been experimentally investigated. Test results have shown that inclusion of bamboo mesh imparted considerable ductility to the mortar and increased its tensile and flexural strengths. However, such improvements, particularly in tension, are associated with wide cracking. Poor bond strength between bamboo and mortar and low elastic modulus of bamboo are the two main factors held responsible. Poor bond is due to volume changes of bamboo. It absorbs a significant amount of water from wet matrix resulting in swelling and shrinks as the matrix dries out. But, use of some cheap water sealing agent has been found to be effective in reducing the water absorption potential of bamboo, thereby improving the bond strength. Application of casting pressure may improve the bond further. This aspect is currently being investigated and will be reported in a further paper.

The results of the present tests are compared with the theoretical predictions of Ali and Pama (1978). In general, the theory gives unsafe predictions. However, test results of specimens containing sealer treated bamboo mesh, irrespective of its quantity, are very close to the predicted values in case of tension. For bending, closer agreement has been obtained only for specimens containing small area fraction (about 2%) of bamboo.

REFERENCES

Ali, Z., and Pama, R.P. (1978). Mechanical properties of bamboo reinforced slabs. Proc. Int. Conf. on Material of Construction for Developing Countries, AIT, Bangkok, pp. 49-66.

Cook, D.J., Pama, R.P., and Singh, R.V. (1978): The behaviour of bamboo-reinforced concrete columns subjected to eccentric loads. Magazine of concrete Research, Vol. 30, No. 104, pp 145-151.

Fang, H.Y., and Mehta, H.C. (1978). Sulfur-sand treated bamboo rod for reinforcing structural concrete. Advances in Concrete, Vol. 165, Am. Chem. Soc., pp.241-254.

Fang, H.Y., and Fay, S.M. (1978). Mechanism of bamboo-water-concrete interaction. Proc. Int. Conf. on Materials of Construction for Developing Countries, AIT, Bangkok, pp.37-48.

Glen, H.E. (1950). Bamboo reinforcement in Portland cement concrete. Bulletin No. 4, Clemson Agricultural College Clemson, S.C. pp 123-127.

Kalita U.C., Khazanchi, A.C., and Thyagarajan, G. (1978). Bamboo-crele wall panels and roofing elements for Row cost housing. Proc. Int. Conf. of Materials of Construction for Developing Countries, AIT, Bangkok, pp.21-36.

Mehra, S.R., Uppal, H.L., and Chadda, L.R. (1951). Some preliminary investigation in the use of bamboo for reinforcing concrete. Indian Concrete Journal, Vol. 25, No. 1, pp.20-21.

Mentzinger, R.J., and Plourde, R.P. (1966). Investigation of treated and untreated bamboo as reinforcement in concrete. Research Report, Villanova University, Pennsylvania.

Narayana, S.K., and Rahman, P.M.A. (1962). Bamboo concrete composite construction, Journal of the Institution of Engineers, India, Vol. XLII, No. 9, pp.25-36.

Youssef, M.A.R. (1976). Bamboo as a substitute for steel reinforcement in structural concrete, New Horozons in Construction Materials, Vol. 1, Envo Publishing Co. Inc., pp. 525-554.

TABLE 1 Properties of Constituent Materials

Material	Properties	Mean Value
Bamboo strips	Cross sectional dimensions	
	(a) Width	5.5 mm
	(b) Thickness	0.88 mm
	Ultimate strength	190 N/mm^2
	Modulus of elasticity	29.5 kN/mm^2
	Water absorption (24 hrs. soaking)	
	(a) Without treatment	44.9 %
	(b) With water sealer	6.0 %
	Ultimate bond stress	
	(a) Without treatment	1.95 N/mm^2
	(b) With water sealer	2.60 N/mm^2
Mortar (1:1.5)	Cylinder compressive strength	42.5 N/mm^2
	Modulus of rupture	3.80 N/mm^2
	Direct tensile strength	2.00 N/mm^2
	Modulus of elasticity	
	(a) In compression	19.7 kN/mm^2
	(b) In tension	19.2 kN/mm^2

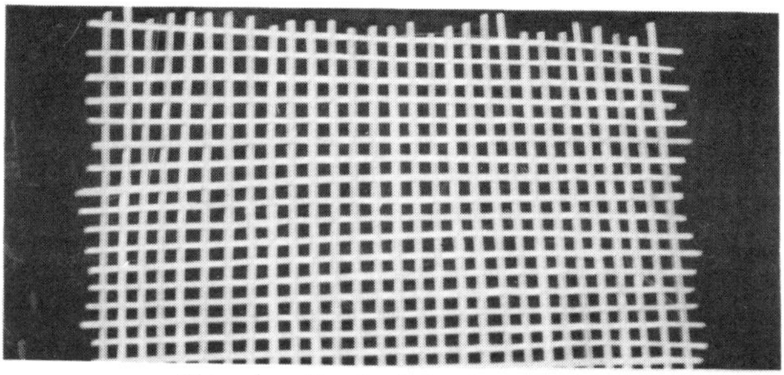

Fig. 1. Typical bamboo-mesh

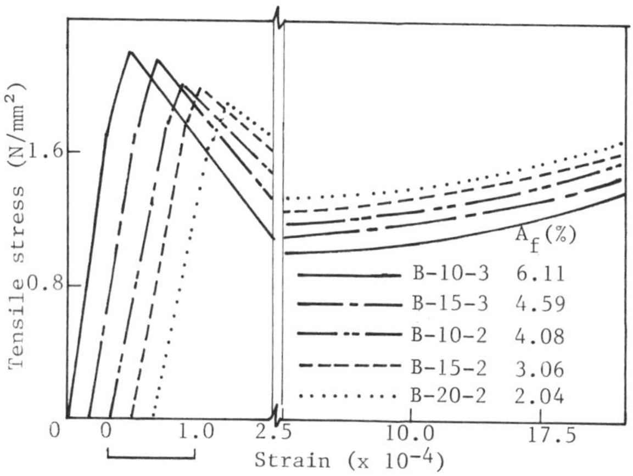

Fig. 2. Stress-strain curves for tensile specimens in B-series

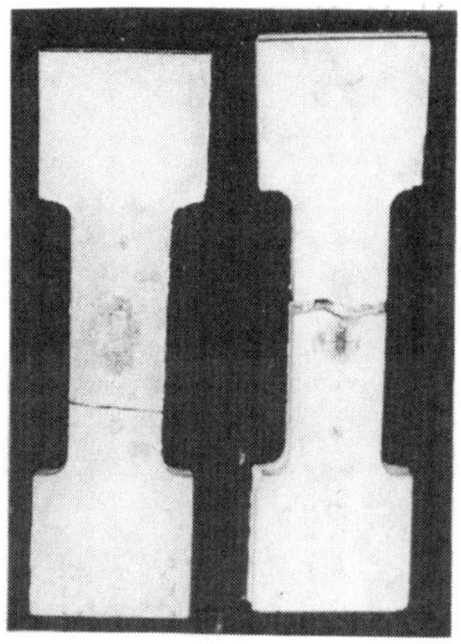

Fig. 3. Tensile specimens after failure

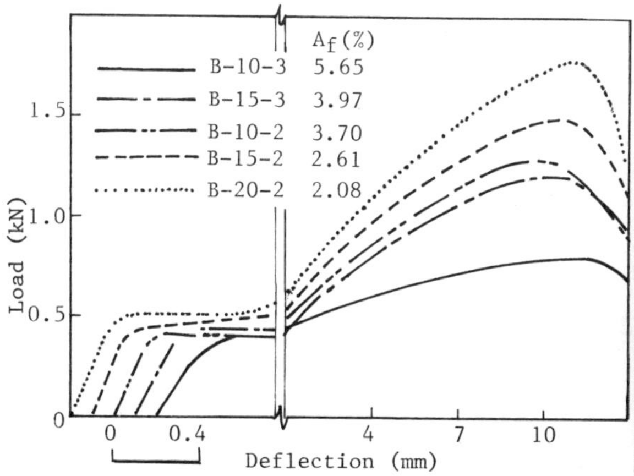

Fig. 4. Load-deflection curves for flexural specimens in B-series

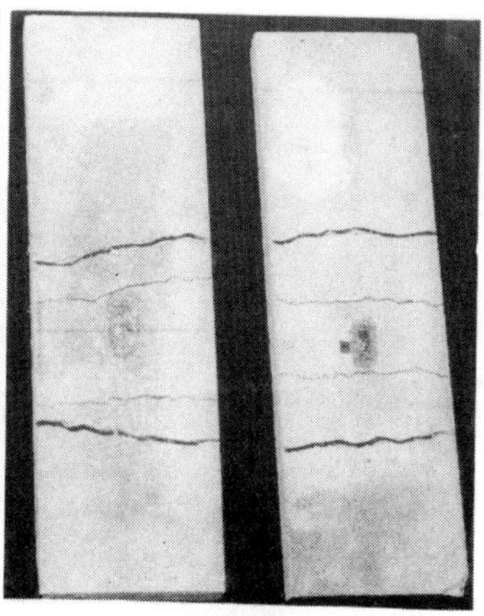

Fig. 5. Flexural Specimens after failure

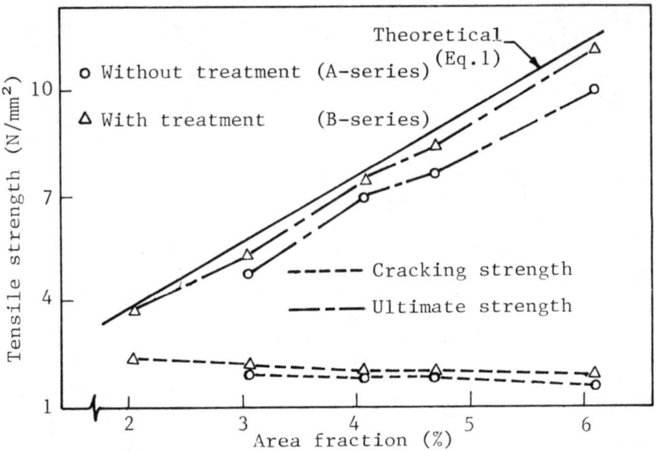

Fig. 6. Tensile test results

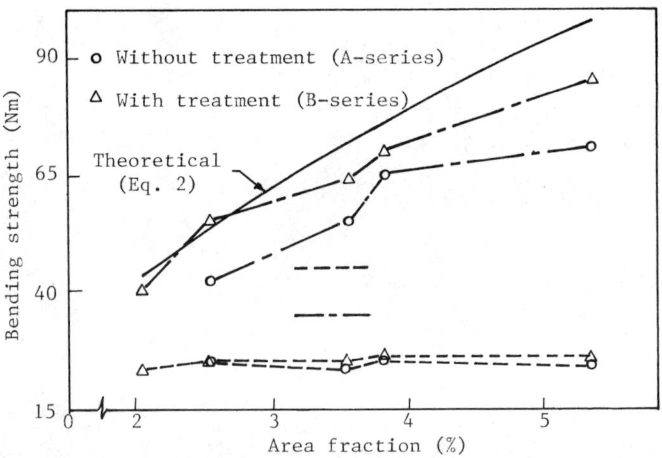

Fig. 7. Flexural test results

THE INFLUENCE OF MICROSTRUCTURE ON COMPOSITE STRENGTHENING*

R. J. Arsenault and R. M. Fisher*****

****Engineering Materials Group, University of Maryland College Park, Maryland, USA**

*****U.S.S. Research Lab., Monroeville, Pa., USA**

ABSTRACT

A transmission electron microscopy investigation of SiC of two different types in an Al alloy matrix was undertaken. The results obtained indicate that in the as received and short time annealed condition these composites have a high dislocation density ($10^{10}cm^{-2}$) and a small subgrain size (0.5 to 1.5 μm). The observed strength of the composites can be explained in terms of the high dislocation density and small sub grain size.

KEYWORDS

SiC; Al alloy; composites; electron transmission microscopy; high dislocation density; small sub grains.

INTRODUCTION

Numerous investigators have applied the concepts of continuum mechanics to predict the strengthening of discontinuous metal matrix composites (Chou and Kelly, 1980; Kelly, 1973; Piggot, 1980). However, when these concepts are applied to discontinuous fiber and or particulate composites there is no agreement between the predicted and experimentally observed strengthening. An example of this disagreement are composites of SiC in Al alloy matrix.

If fiber strengthening is considered first, it is possible to calculate the theoretical strength of this material, by the formulation of Piggot (1980) and Chou and Kelly (1980);

*This research was supported by the Office of Naval Research under Grant No. N00014-82-K-0493.

$$\sigma_{uc} = \sum^{\ell_i < \ell_c} \frac{\sigma_{m\varepsilon} \ell_i V_i}{d} + \sum^{\ell_j > \ell_c} \sigma_{uf} \left(1 - \frac{\sigma_{uf} d}{4 \ell_j \sigma_{m\varepsilon}}\right) V_j + \sigma_{m\varepsilon} (1 - V_f) \tag{1}$$

$$\sigma_{m\varepsilon} = 104 \text{ MPa}, \ \sigma_{uf} = 6900 \text{ MPa}, \ \frac{\ell_c}{d} = 66 \text{ and } V_f = 0.2$$

where $\sigma_{m\varepsilon}$ is the strength of the matrix at the failure strain of composite, σ_{uf} is the ultimate strength of the composite, ℓ_c is the critical length of the fiber and d is the diameter. The assumptions involved in the derivation of the above equation are as follows: There is perfect alignment of the fibers, there are no voids in the composite material and there is perfect bonding between matrix and the fiber. In order to calculate the σ_{uc} of the composite it is necessary to determine the distribution of fiber lengths within the processed composite. We cannot use the lengths of the fiber in their as manufactured form, due to the fact that there is a large amount of fiber breakage during the processing of the composite. Using data described elsewhere (Arsenault and Newborn, 1983) and taking the handbook value for Al alloy 6061 in the annealed condition, the σ_{uc} obtained is 186 MPa. This is a surprising result, for the measured strength of 20 % V_f SiC in 6061 in the annealed condition is 448 MPa, and even more surprising is the fact that there is a high density of voids in the composite material (Arsenault and Newborn, 1983).

If we now consider the particulate strengthening case the theoretical strength can be calculated by a formulation of Piggot (1980), which is based on the assumption that the particles are plates,

$$\sigma_{uc} = \frac{\sigma_{m\varepsilon} S V_p}{4} + V_m \sigma_{m\varepsilon} \tag{2}$$

$$S = \frac{2\ell}{t} = 8, \ \sigma_{m\varepsilon} = 124 \text{ MPa}$$

where ℓ is the length of the long axis of platelet, t is the thickness, v_p is the volume fraction of platelets and v_m is the volume fraction of the matrix. The assumptions involved in the derivation of this equation are as follows: The plates are aligned, i.e. the long dimension of the plate is in the direction of tensile loading, there is perfect bonding and there are no voids in the composite. The σ_{uc} calculated for 20 V_p% SiC in 6061 in the annealed condition is 160 MPa, whereas, the experimental value is 276 MPa (Arsenault and Newborn, 1983). From a microscopic examination it is evident that the particles are not all plates, they are not all aligned and there is a significant number of voids (Arsenault and Newborn, 1983).

The fact that the experimental values of strength of these composite materials is above that predicted by the continuum mechanics theories is more than likely due to a synergistic effect. There are two possible candidates for this synergistic effect; a dislocation generation mechanism, i.e. a higher dislocation density than in the annealed wrought alloy, and secondly, the possibility of having a very fine sub grain or grain size matrix material.

The high dislocation density is based on the difference in thermal coefficient of expansion of SiC and Aℓ. There is a difference in the thermal coefficient of expansion between SiC and Aℓ of about a factor of 10 and as a result of cooling down from the annealing or processing temperature, dislocations are generated at the fiber-matrix interface or particulate-matrix interface.

An investigation was undertaken to determine the dislocation density in SiC-Aℓ alloy matrix (6061).

EXPERIMENTAL PROCEDURE

Fiber composite material containing 20 V % beta SiC fiber in the shape of 1.27 cm dia. rod, and a plate of 1.3 cm thick containing 20 V % alpha SiC particulate were obtained. From both of these composite materials 3mm diameter disks by 0.40mm thick were prepared by electrical discharge machining. Transparent foils were produced by an ion thinning technique, and examined in the U.S. Steel 1 MV microscope.

RESULTS

The electron transmission micrographs obtained for the fiber composite material exhibit a high dislocation density i.e. approximately 10^{10} cm^{-2}, as can be seen in Fig. 1. The lines or banding of the fibers is due to microtwins (Feng and Newborn, 1981). The dislocation density in the particulate composite material as shown in Fig. 2, and is about the same as the fiber composite materials. The high dislocation density is evident in Figs. 3 and 4. In Fig. 4 fine precipitates are evident on the dislocations. From selected area diffraction patterns it was concluded that these were Mg_2Si precipitates.

The subgrain structure is evident in Fig. 5. This micrograph shows the range of subgrain size, i.e. from 1.0 μm to 2.5 μm.

DISCUSSIONS AND CONCLUSIONS

In the Introduction it was pointed out that the continuum mechanics models of composite strengthening were not capable of predicting the ultimate strength of the composite.

The increase in strength of the composite, as compared to the base Al alloy produced in the same manner, can be described as follows:

$$\Delta\sigma_{uc} = \sigma_{DIS} + \sigma_{WH} + \sigma_{SG} + \sigma_{COMP} \qquad (3)$$

where σ_{DIS} is the increase in strength due a higher dislocation density, σ_{WH} is the difference in the rate of work hardening between the 20 V % composite and the 0 V % Al alloy, σ_{SG} is the strengthening due to the small sub-grain size, and σ_{COMP} is the composite strengthening. However, the last two terms' probably do not have to be considered.

If we consider the fiber case, reasonable values for the terms on the right hand side of eqn. 3, are $\sigma_{DIS} \simeq 138$ MPa, $\sigma_{WH} \simeq 138$ MPa, which results in $\Delta\sigma_{uc} \simeq 376$ MPa. This predicted value is slightly greater than that observed, however, what has not been considered, is the presence of voids which exist in the samples prior to testing. A similar result is obtained for the particulate, i.e. the platelet, composite. In conclusion the strengthening observed is not "composite" strengthening, but strengthening of the matrix. In terms of failure modes, the evidence strongly indicates that failure mode is not the breaking of fibers or platelets, but failure starts at pre-existing voids or "clumps" of wiskers (i.e. fibers) in the sample.

REFERENCES

Arsenault, R.J., and H. Newborn (1983). To be published in Met. Trans.
Chou, T.W., and A. Kelly (1980). Ann. Rev. Mat. Sci., 10, 229.
Feng, C.R., and H. Newborn (1981). Private communication.
Kelly, A. (1973). Strong Solids, 2nd ed. Clarendon Press, Oxford.
Piggot, M.R. (1980). Load Bearing Fiber Composites, Pergammon Press.

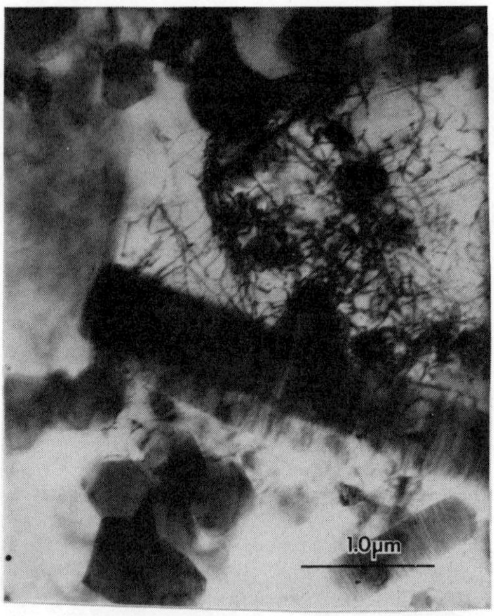

Fig. 1 - High-voltage-electron micrograph of a comprising 6061 Al-20 wt.% β-SiC particles.

Fig. 2

- High-voltage-electron micrograph of a composite comprising 6061 Al-20 wt.% α-SiC fibers.

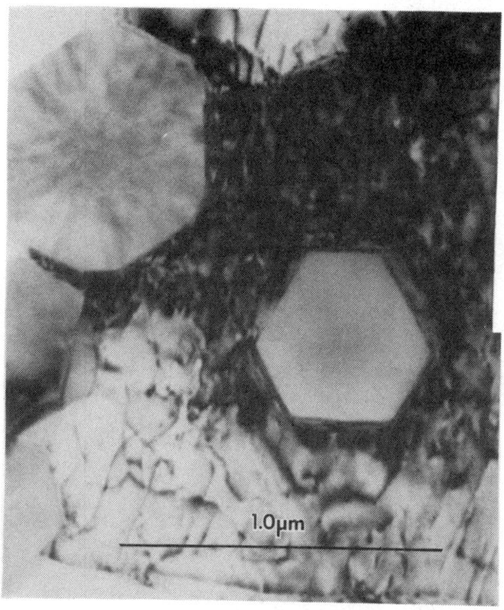

Fig. 3

- Higher magnification of sample in Fig. 1 showing high dislocation density and very fine precipitates around SiC particles.

Fig. 4

- Higher magnification of sample in Fig. 2 showing high dislocation density and very fine precipitates around SiC fibers.

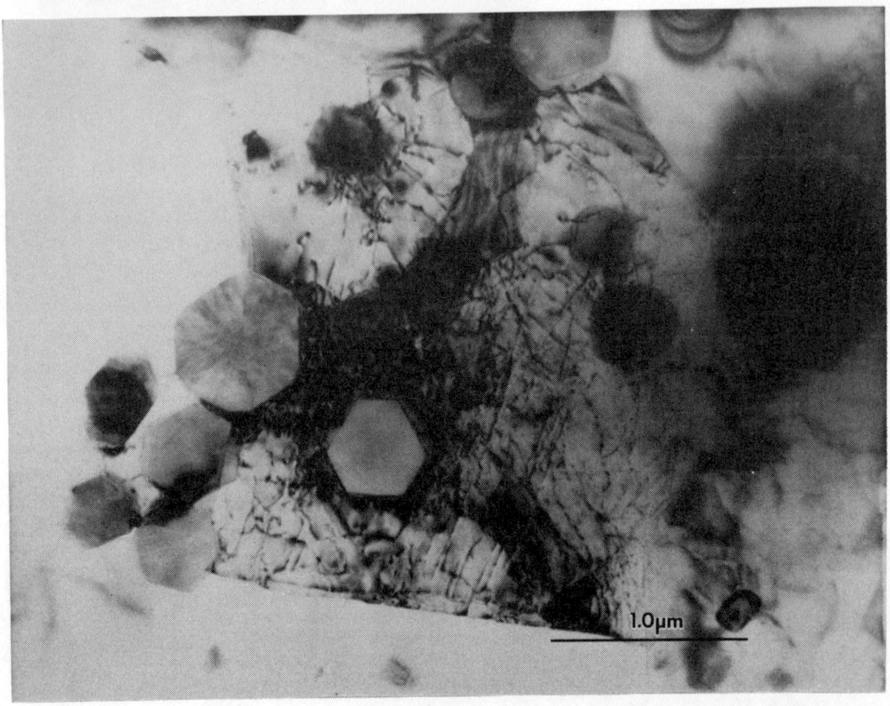

Fig. 5 A fiber composite sample showing the subgrain size.

MICROSTRESS REDISTRIBUTION IN FIBRE-REINFORCED COMPOSITES DUE TO TIME-DEPENDENT PROCESSES

V. Kafka

Institute of Theoretical and Applied Mechanics,
Czechoslovak Academy of Sciences, Prague, Czechoslovakia

ABSTRACT

A new theoretical approach to the problem of time-dependent deformation and the consequent micro-stress redistribution in fibre-reinforced composites is presented. The identification problem is solved with macro-behaviour as the main source of information.

KEYWORDS

Fibre-reinforced composites; time-dependent processes, micro-stresses, identification problems, structural parameters, internal variables.

INTRODUCTION

The classical approach to the analysis of the stress- and strain-state in composites starts from the notion that we know the material properties of the constituents and the geometry of their composition and the macroscopic properties are what is sought. This is a paradox in some sense as both the properties of the material constituents and the geometry of composition are usually very difficult to find out and to describe exactly whereas the macroscopic properties are easily determinable. Thus e.g. Varschavsky (1972) found a substantial disagreement (up to 100%) between the "in situ" and the "homogeneous sample" properties of the aluminium matrix. Another complication are the special interface properties.

It seems therefore reasonable to consider the macroscopic properties (moduli, stress-strain curves and flow-curves in simple loading conditions) as the main source of information for the quantitative description of composites. The mathematical analysis of these curves together with some supplementary qualitative and quantitative information on the microstructure enables the formulation of a quantitative structural mathematical model of the composite and the prediction of the macroscopic behaviour in other loading conditions than in the testing ones.

The supplementary qualitative and quantitative information is usually: the type of reinforcement, the type of the rheological equations of the constituent materials, the volume fractions and the stable material properties - e.g. the moduli of the reinforcing material.

The so-called semi-empirical solutions are generally disliked - especially by pure theoreticians - and the approach outlined above could be perhaps classified so, but in fact it is as natural as the determination of the Young´s modulus - the difference is only that in non-elastic deformations we need more information and use therefore not only the elastic response, but also the non-linear curves.

The idea outlined above was elaborated earlier for elastic--plastic deformation of fibre-reinforced composite materials (Kafka, 1973, 1974). In the presented paper it is elaborated for the time-dependent processes without the limitation on the reinforcement to be unidirectional and continuous.

The general ideas of the concept were published elsewhere (Kafka, 1979). A composite material is described by the constitutive equations of its constituents, by their volume fractions and by structural parameters descriptive of the geometry of composition. Some problems of composites with anisotropy resulting from the structure of composition have

been solved too (Kafka, 1976).
The following analysis is based on the results of the quoted papers.

ANALYSIS

As a simple example let us consider the case of a continuous but not unidirectional elastic reinforcement (subscript e) in a matrix which has immediate elastic response with subsequent linear creep (subscript v).
Then - according to the earlier author's papers (Kafka, 1976, 1979) - the deformation process is described by the following set of equations:

$$V_e \sigma_{ije} + V_v \sigma_{ijv} = \bar{\sigma}_{ij} \tag{1}$$

$$V_e \varepsilon_{ije} + V_v \varepsilon_{ijv} = \bar{\varepsilon}_{ij} \tag{2}$$

$$e_{ije} = \mu_e s_{ije} \quad , \qquad \varepsilon_e = \varrho_e \sigma_e \tag{3}$$

$$\dot{e}_{ijv} = \mu_v \dot{s}_{ijv} + \frac{1}{2H} s_{ijv} \quad , \qquad \dot{\varepsilon}_v = \varrho_v \dot{\sigma}_v \tag{4}$$

$$\varepsilon'_{ije} = \varepsilon_{ije} - \bar{\varepsilon}_{ij} \tag{5}$$

$$\varepsilon'_{ijv} = \varepsilon_{ijv} - \bar{\varepsilon}_{ij} \tag{6}$$

$$e'_{ije} = \mu_e s'_{ije} \quad , \qquad \varepsilon'_e = \varrho_e \sigma'_e \tag{7}$$

$$\dot{e}'_{ijv} = \mu_v \dot{s}'_{ijv} + \frac{1}{2H} s'_{ijv} \quad , \qquad \dot{\varepsilon}'_v = \varrho_v \dot{\sigma}'_v \tag{8}$$

$$\sigma_{ije} - \sigma_{ijv} + \frac{\sigma'_{ije}}{[\eta]_e} - \frac{\sigma'_{ijv}}{[\eta]_v} = 0 \tag{9}$$

where $V_e = 1 - V_v$ are the volume fractions,

$\sigma_{ije} / \varepsilon_{ije}$ are the average values of the stress/strain tensors in the e-material constituent,

$\bar{\sigma}_{ij} / \bar{\varepsilon}_{ij}$ are the macroscopic stress/strain tensors,

$s_{ij} / \sigma (e_{ij} / \varepsilon)$ are the deviatoric/isotropic parts of the stress (strain) tensors

$$\mu_e = (1+\nu_e)/E_e \quad , \qquad \varrho_e = (1-2\nu_e)/E_e \tag{10}$$

are the elastic constants of the e-material.

H is the coefficient of viscosity of the v-material

ε'_{ije} , ε'_{ijv} are defined by (5) and (6)

σ'_{ije} , σ'_{ijv} are defined by (7) and (8)

$[\eta]_e$, $[\eta]_v$ are non-negative structural parameters which

- for the case of anisotropic composition - have different values for different indices ij in (9).
- For the case of transversal isotropy it holds (Kafka,1976):

$$[\eta]_e = \eta_e \ , \quad [\eta]_v = \eta_v \quad \text{for} \quad ij = 11$$

$$[\eta]_e = \eta_e^{\perp} \ , \quad [\eta]_v = \eta_v^{\perp} \quad \text{for} \quad ij = 22, 33, 23$$

$$[\eta]_e = \eta_e^{*} \ , \quad [\eta]_v = \eta_v^{*} \quad \text{for} \quad ij = 12, 13$$

The set of 9 tensorial equations (1) - (9) corresponds to 9 tensiorial unknows:

$$\dot{\sigma}_{ije} \ , \ \dot{\sigma}_{ijv} \ , \ \dot{\varepsilon}_{ije} \ , \ \dot{\varepsilon}_{ijv} \ , \ \dot{\sigma}'_{ije} \ , \ \dot{\sigma}'_{ijv} \ , \ \dot{\varepsilon}'_{ije} \ , \ \dot{\varepsilon}'_{ijv} \ , \ \dot{\bar{\varepsilon}}_{ij}$$

(with $\dot{\bar{\sigma}}_{ij}$ being given).

From this set the following macroscopic constitutive equation can be deduced:

$$\dot{\bar{\varepsilon}}_{ij} = \mu_e \dot{\bar{s}}_{ij} + \delta_{ij} \varrho_e \dot{\bar{\sigma}} + V_v (\mu_v - \mu_e) \dot{s}_{ijv} + \\ + \delta_{ij} V_v (\varrho_v - \varrho_e) \dot{\sigma}_v + \frac{1}{2H} V_v s_{ijv} \tag{11}$$

with the evolution equations for its internal variables (for the sake of simplicity only the case $\nu_e = \nu_v$ is presented here):

$$\dot{\sigma}_{ijv} = [M] \dot{\bar{\sigma}}_{ij} - \frac{1}{2H} \left([N] s_{ijv} - [P] s'_{ijv} \right) \tag{12}$$

where

$$[M] = \frac{\mu_e \mu_v [\eta]_e [\eta]_v + \mu_e (V_e \mu_e [\eta]_e + V_v \mu_v [\eta]_v)}{\mu_e \mu_v [\eta]_e [\eta]_v + (V_e \mu_v + V_v \mu_e)(V_e \mu_e [\eta]_e + V_v \mu_v [\eta]_v)}$$

$$[N] = (1 - [M]) / (\mu_v - \mu_e)$$

$$[P] = \frac{V_e \mu_e [\eta]_e}{\mu_e \mu_v [\eta]_e [\eta]_v + (V_e \mu_v + V_v \mu_e)(V_e \mu_e [\eta]_e + V_v \mu_v [\eta]_v)}$$

$$\dot{\sigma}'_{ijv} = \frac{1}{\mu_v} \Big[-\mu_e \dot{\bar{\sigma}}_{ij} + (V_e \mu_v + V_v \mu_e) \dot{\sigma}_{ijv} +$$

$$+ \frac{1}{2H} (V_e s_{ijv} - s'_{ijv}) \Big] \tag{13}$$

From these equations we can easily deduce the expressions for the macroscopic Young-modulus in the x_1 -direction (axis of transversal isotropy) $\bar{E}$ and in the transverse direction - $\bar{E}^{\perp}$:

$$\bar{E} = \frac{\eta_e \eta_v \varkappa + (V_e + V_v \varkappa) (V_e \eta_e \varkappa + V_v \eta_v)}{\eta_e \eta_v (V_e \varkappa + V_v) + V_e \eta_e \varkappa + V_v \eta_v} E_e =$$

$$= \frac{E_v}{\varkappa + V_v M (1 - \varkappa)} \tag{14}$$

$$\bar{E}^{\perp} = \frac{\eta_e^{\perp} \eta_v^{\perp} \varkappa + (V_e + V_v \varkappa)(V_e \eta_e^{\perp} \varkappa + V_v \eta_v^{\perp})}{\eta_e^{\perp} \eta_v^{\perp} (V_e \varkappa + V_v) + V_e \eta_e^{\perp} \varkappa + V_v \eta_v^{\perp}} E_e =$$

$$= \frac{E_v}{\varkappa + V_v M^{\perp} (1 - \varkappa)} \tag{15}$$

where $\varkappa = E_v / E_e$

Now let us demonstrate the simplest procedure of identification of the structural parameters.

Let us suppose that we know $V_e, V_v, \nu_e = \nu_v, E_e, E_v$. Under constant loading $\bar{\sigma}_{11}$ or $\bar{\sigma}_{22}$ we can measure the immediate response and determine $\bar{E}$ and $\bar{E}^{\perp}$ and the asymptotic value of deformation after an infinite time. Let us suppose that the load is high enough and the matrix viscous enough so that the stress in the matrix can be considered as completely relaxed in the asymptotic state. From the asymptotic deformation we can determine the asymptotic moduli $\bar{E}_{(as)}$, $\bar{E}^{\perp}_{(as)}$ and the respective expressions follow from (14) and (15) with $\varkappa = 0$:

$$\bar{E}_{(as)} = \frac{V_e}{1 + \eta_e} E_e \tag{16}$$

$$\bar{E}^{\perp}_{(as)} = \frac{V_e}{1 + \eta_e^{\perp}} E_e \tag{17}$$

From (16) and (17) we can determine η_e and $\eta_e^{\perp}$ and from (14) and (15) η_v and $\eta_v^{\perp}$.

This was the case with continuous reinforcement. A discontinuous reinforcement is described (Kafka, 1981) by a model with $\eta_e = \eta_e^{\perp} = \infty$ which transforms (14) and (15) into:

$$\bar{E} = \frac{\eta_v + V_e (V_e + V_v \varkappa)}{\eta_v (V_e \varkappa + V_v) + V_e \varkappa} E_v \tag{18}$$

$$\bar{E}^{\perp} = \frac{\eta_v^{\perp} + V_e (V_e + V_v \varkappa)}{\eta_v^{\perp} (V_e \varkappa + V_v) + V_e \varkappa} E_v \tag{19}$$

and the asymptotic values are zero. Values of η_v, $\eta_v^{\perp}$ can be determined from (18) and (19) in this case.

More information can be drawn from the flow-curve. Its differential equation for $\bar{\sigma}_{11}$ = const. can be deduced from (11), (12) and (13). In the case $\nu_e = \nu_v = 0{,}5$ its form is:

$$\ddot{\bar{\varepsilon}}_{11} + A\dot{\bar{\varepsilon}}_{11} + B\bar{\varepsilon}_{11} = C\ddot{\bar{s}}_{11} + D\dot{\bar{s}}_{11} + F\bar{s}_{11} \tag{20}$$

with

$$A = \left[1 + \mu_v \hat{N} - (V_e \mu_v + V_v \mu_e)\hat{P}\right] / 2H\mu_v$$

$$B = (\hat{N} - V_e \hat{P}) / 4H^2 \mu_v$$

$$C = \mu_e + V_v(\mu_v - \mu_e)M$$

$$D = \left[\mu_e + V_v(2\mu_v - \mu_e)M + \mu_e \mu_v(\hat{N} - \hat{P})\right] / 2H\mu_v$$

$$F = \left[V_v M + \mu_e(\hat{N} - \hat{P})\right] / 4H^2 \mu_v$$

$$\hat{N} = \frac{2N + N^{\perp}}{3}$$

$$\hat{P} = \frac{2P + P^{\perp}}{3}$$

For constant loading ($\ddot{S}_{11} = \dot{S}_{11} = 0$) only A,B and F appear and they can be determined when identifying the differential equation with the measured flow-curve. The calculation of further material parameters follows.

With all the parameters known we can predict the behaviour of the composite in other loading conditions and calculate the respective microstress redistributions and its meaning for strength of the material.

REFERENCES

Kafka,V.(1973,1974).Structural theory of elastic-plastic deformation of unidirectional fibre-reinforced materials. Acta Technica ČSAV,18,418-436,19,87-95.

Kafka,V.(1976).On elastic-plastic deformation of composite materials with anisotropic structure.Acta Technica,21,685-706.

Kafka,V.(1979).On constitutive equations of quasihomogeneous materials.ZAMM,59,423-430.

Kafka,V.(1981). On mechanics of two- and multi-phase solids. Proc.-Mechanics-IV.Congr.,423-430,Varna,Bulgaria.

Varschavsky,A.(1972).The matrix-fatigue behaviour of fibre composites.J.of Materials Science,7,159-167.

MATERIAL DESIGN OF FIBROUS LAMINATED COMPOSITES WITH REQUIRED FLEXURAL STIFFNESS

M. Miki

Materials System Research Laboratory, Kanazawa Institute of Technology, Nonoichi Ishikawa, Japan

ABSTRACT

A new material design method of fibrous laminated composites with required flexural stiffness is proposed. The selection of kinds of fibers and matrices and the determination of fiber orientation angles can be done directly from the imposed conditions on the flexural stiffness of laminated plates.

KEYWORDS

Composite materials; laminated plates; material design; flexural stiffness; flexural lamination parameter; stacking sequence.

INTRODUCTION

In recent years, composite materials are referred to as "Tailored materials". This means that the physical properties of composite materials such as stiffness, strength and the coefficient of thermal expansion can be varied by changing the kinds and the orientation of reinforcements. However, the material design method for designing the composites with required physical properties are few and no approaches to the design of fibrous laminated composites with required stiffness have been proposed.

Tsai and Hahn (1980) developed a new evaluation method of mechanical properties of laminated composites. Since stiffness and strength of laminated plates can be evaluated easily by using their expressions, we can choose the optimal constructions from many proposed ones.

Chamis (1980) proposed a simple evaluation method of stiffness and strength by using some charts. This method can also be used to choose the optimal constructions.

When these evaluation procedures are used repeatedly, the best construction can be found from many trials. However, the material design methods using these evaluation procedures are essentially methods of trial and error. And it is very difficult to propose the trial construction of laminates for a certain application. To avoid this difficulty, the author (1982) has proposed a new approach for designing laminated composites with required in-plane stiffness. In this paper, a new material design method of fibrous laminated composites with required flexural stiffness is proposed.

FLEXURAL STIFFNESS OF LAMINATED COMPOSITES

The composite plates intended for this study are multiple balanced angle ply symmetric laminates with $2N_h$ plies of unidirectional laminae of a kind. The stacking sequence of the laminates is represented by the following code.

$$[(+\theta_1/-\theta_1)_{N_1}/(+\theta_2/-\theta_2)_{N_2}/\cdots/(+\theta_n/-\theta_n)_{N_n}]_s \,, \quad N_h=\sum_{k=1}^{n} N_k \tag{1}$$

This type of laminates are free from couplings between extension and bending, and between shear and extension, and have very little coupling between bending and twisting. That is, the coupling stiffness $B_{ij}=0$ and the element D_{16} and D_{26} of bending stiffness are nearly equal to zero when the number of plies is large.

The flexural effective engineering constants, that is, E_1^f:effective Young's modulus along the 1-axis, E_2^f:effective Young's modulus along the 2-axis, E_6^f:effective shear modulus for the twisting along the 1- or 2-axis, and ν_{21}^f: coupling coefficient between bendings along 1- and 2-axes, are considered as flexural stiffness of the laminates. The subscript 1 and 2 represent the major and minor principal material axes of the laminates.

Tsai and Hahn (1980) gave the expression for the normalized flexural stiffness D_{ij}^* $(=12D_{ij}/h^3)$ as follows.

$$\begin{bmatrix} D_{11}^* \\ D_{22}^* \\ D_{12}^* \\ D_{66}^* \end{bmatrix} = \begin{bmatrix} U_1 & W_1^* & W_2^* \\ U_1 & -W_1^* & W_2^* \\ U_4 & 0 & -W_2^* \\ U_5 & 0 & -W_2^* \end{bmatrix} \begin{bmatrix} 1 \\ U_2 \\ U_3 \end{bmatrix} \tag{2}$$

where U_m $(m=1,\cdots,5)$ is the linear combination of on-axis modulus of a lamina and can be considered as a material constant, and W^*'s are the factors represented as follows.

$$W_1^*=\frac{24}{h^3}\int_0^{h/2}\cos 2\theta\, z^2 dz \,, \quad W_2^*=\frac{24}{h^3}\int_0^{h/2}\cos 4\theta\, z^2 dz \tag{3}$$

where h is the thickness of the laminate, θ is the fiber orientation angle, and z is the direction of the normal to the middle surface of the laminates.

The flexural effective engineering constants are obtained from D_{ij}^* as follows.

$$E_1^f=\Delta/D_{22}^*,\ E_2^f=\Delta/D_{11}^*,\ E_6^f=D_{66}^*,\ \nu_{21}^f=D_{11}^*/D_{22}^*,\ \Delta=D_{11}^*D_{22}^*-D_{12}^{*2} \tag{4}$$

Table 1 shows the values of U's of typical composites.

TABLE 1 The Values of U's of Typical Composites (U in GPa)

Tsai & Hahn (1980)

Material	U_1	U_2	U_3	U_4	U_5
T300/5208	76.37	85.73	19.71	22.61	26.88
Kevlar49/epoxy	32.44	35.55	8.65	10.54	10.95

FLEXURAL LAMINATION PARAMETER DIAGRAM

Flexural Lamination Parameters

In this paper, W_1^* and W_2^* in Eq. (3) are referred to as flexural lamination parameters. Figure 1 shows the section of the laminates represented by Eq. (1). In Fig. 1, z_k (k=1,•••,n) represents the distance from the middle surface to the upper surface of k-th ply group in which the fiber orientation is $+\theta_k/-\theta_k$. Then W_1^* can be rewritten as follows.

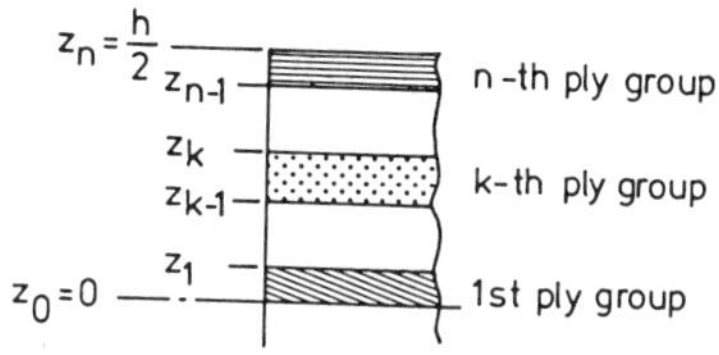

Fig. 1 The sectional view of laminates.

$$W_1^* = \frac{24}{h^3}\int_0^{h/2} \cos 2\theta \; z^2 dz = \sum_{k=1}^{n} \xi_k \cos 2\theta_k \tag{5}$$

where $z_0=0$, $z_n=h/2$ and

$$\xi_k = (2z_k/h)^3 - (2z_{k-1}/h)^3 \tag{6}$$

Similarly,

$$W_2^* = \sum_{k=1}^{n} \xi_k \cos 4\theta_k \tag{7}$$

The summation of ξ_k equals to unity.

Allowable Region of Flexural Lamination Parameters

The design of composite laminates with required flexural stiffness begins with the determination of the design specifications for the flexural effective engineering constants. The author propose a lamination parameter diagram to determine the possible design specifications.

The diagram of the flexural lamination parameters W_1^* (as abscissa) and W_2^* (as ordinate) is considered. The lamination parameters are evaluated by Eqs. (5) and (7) when the stacking sequence of a laminate is given, and then the point on the flexural lamination parameter diagram is determined. This point is referred to as lamination point. In order to investigate the possibility of flexural stiffness of composite laminates, the allowable region of flexural lamination parameters should be clarified.

From Eqs. (5) and (7), the following condition holds.

$$-1 \leqq W_r^* \leqq 1 \quad (r=1,2) \tag{8}$$

When the number of kind of fiber orientation angle is one, Eqs. (5) and (7) are reduced to the following.

$$W_1^* = \cos 2\theta \;, \quad W_2^* = \cos 4\theta \tag{9}$$

In this case, the following relation holds.

$$W_2^* = 2W_1^{*2} - 1 \tag{10}$$

Consequently, the allowable region of lamination parameter is represented by the curve in Fig. 2.

When the number of kinds of fiber orientation angles is two, the flexural lamination parameters are given as follows.

$$W_1^*=\xi_1\cos2\theta_1+\xi_2\cos2\theta_2$$
$$W_2^*=\xi_1\cos4\theta_1+\xi_2\cos4\theta_2 \qquad (11)$$

The necessary and sufficient condition to obtain the orientation angles θ_1 and θ_2 are

$$-1 \leqq \cos2\theta_r \leqq 1 \quad (r=1,2) \qquad (12)$$

From Eqs. (11) and (12), the following inequality can be obtained.

$$W_2^* \geqq 2W_1^{*2} - 1 \qquad (13)$$

Then the allowable region of the flexural lamination parameters is shown as the hatched region in Fig. 3.

When the number of kinds of orientation is n, the evaluation of $W_2^*-(2W_1^{*2}-1)$ yields the allowable region of lamination parameters.

$$W_2^*-(2W_1^{*2}-1)$$
$$=\sum_{k=1}^{n}\xi_k\cos4\theta_k-[2(\sum_{k=1}^{n}\xi_k\cos2\theta_k)^2-1]$$
$$=2\sum_{k=1}^{n}\xi_k(\cos2\theta_k-\sum_{k=1}^{n}\xi_k\cos2\theta_k)^2 \geqq 0 \qquad (14)$$

From Eq. (14), the term $W_2^*-(2W_1^{*2}-1)$ is non-negative, and is zero only in the case that the orientation angle θ_k is the same. Therefore, the allowable region of the flexural lamination parameters is the same region for the laminates with two kinds of orientation angles. Consequently, two kinds of orientation angles are necessary and sufficient for designing the laminates with required stiffness.

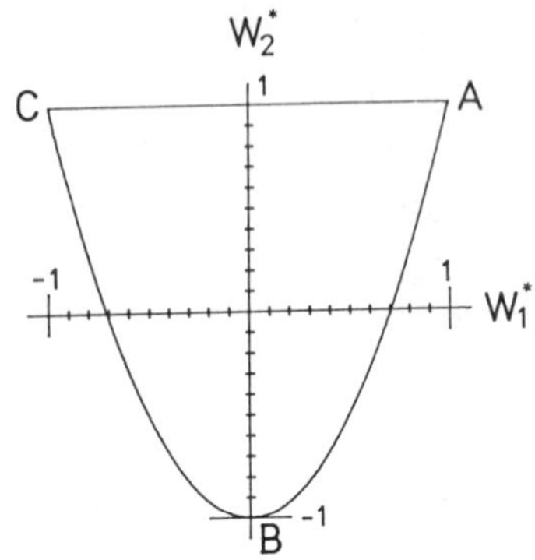

Fig. 2 Allowable region (curve ABC) of the flexural lamination parameters. The stacking sequence is $[(+\theta/-\theta)_{N_h}]_s$.

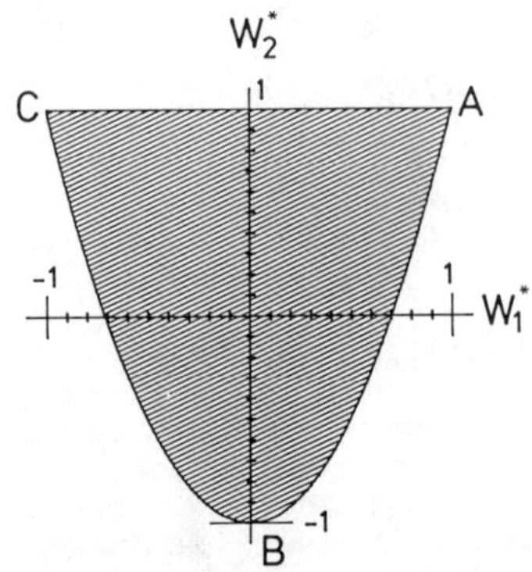

Fig. 3 Allowable region (hatched region) of flexural lamination parameters. The number of the kinds of fiber orientation is two.

Contours of Flexural Effective Engineering Constants on the Flexural Lamination Parameter Diagram

When the flexural lamination parameters W_1^* and W_2^* are determined, the flexural effective engineering constants are evaluated by using Eqs. (2) and (4). The contours of effective Young's modulus E_1^f can be drawn from those relations. The expression for E_1^f-contour is obtained as follows.

$$E_1^f=\frac{D_{11}^*D_{22}^*-D_{12}^{*2}}{D_{22}^*}=\frac{(U_1+U_2W_1^*+U_3W_2^*)(U_1-U_2W_1^*+U_3W_2^*)-(U_4-U_3W_2^*)^2}{U_1-U_2W_1^*+U_3W_2^*}$$

Then

$$W_2^* = \frac{U_2^2 W_1^{*2} - U_2 E_1^f W_1^* + U_1 E_1^f + U_4^2 - U_1^2}{U_3(2U_1 + 2U_4 - E_1^f)} \qquad (E_1^f\text{-contour}) \qquad (15.1)$$

Similarly, E_2^f-contour, E_6^f-contour, and ν_{21}^f-contour are obtained as follows.

$$W_2^* = \frac{U_2^2 W_1^{*2} - U_2 E_2^f W_1^* + U_1 E_2^f + U_4^2 - U_1^2}{U_3(2U_1 + 2U_4 - E_2^f)} \qquad (E_2^f\text{-contour}) \qquad (15.2)$$

$$W_2^* = \frac{U_5 - E_6^f}{U_3} \; (E_6^f\text{-contour}), \quad W_2^* = \frac{U_2 \nu_{21}^f W_1^* + U_4 - U_1 \nu_{21}^f}{U_3(1+\nu_{21}^f)} \; (\nu_{21}^f\text{-contour}) \qquad (15.3, 15.4)$$

Fig. 4 shows the contours of flexural effective engineering constants on the flexural lamination parameter diagram of Kevlar/epoxy. The design specifications for flexural stiffness can be easily determined by using the diagrams. It should be noted that there is E_6^f-scale instead of E_6^f-contour because the E_6^f-contour becomes parallel to the abscissa.

DETERMINATION OF THE STACKING SEQUENCE OF COMPOSITE LAMINATES

Suppose that the following conditions are given for flexural effective engineering constants.

$$E_1^f \geqq 120, \quad E_2^f \geqq 40, \quad E_6^f \geqq 15 \text{ (GPa)}, \quad \nu_{21}^f \leqq 0.3 \qquad (16)$$

Since a region which satisfied the above conditions is obtained in the lamination parameter diagram of T300/5208, this material can be used.

Figure 5 shows the region which is referred to as the feasible design region. One design point which gives the values of W_1^* and W_2^* is determined when another evaluation condition for effective engineering constants is imposed. For example, the maximization of E_2^f yields the design point as the point P in Fig. 5. The values of the flexural lamination parameters W_1^* and W_2^* are obtained easily by using the W^*-scales of the diagram, or from the calculation of the intersecting point of contours exactly. In this case, the values

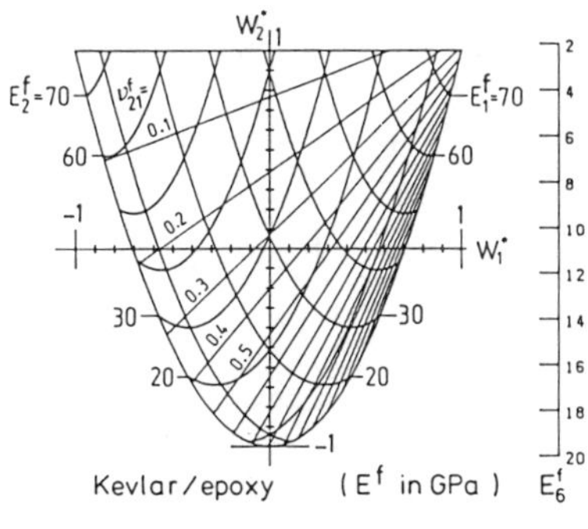

Fig. 4 Flexural lamination parameter diagram of Kevlar/epoxy.

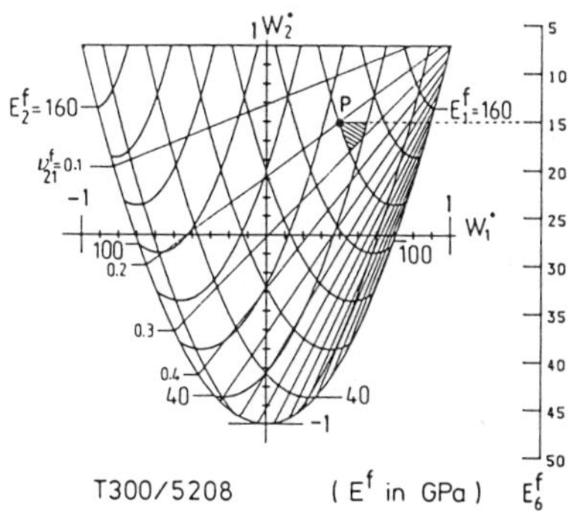

Fig. 5 Feasible design region on the flexural lamination parameter diagram.

of W_1^* and W_2^* for the point P are obtained as follows.

$$W_1^*=0.395 \ , \quad W_2^*=0.603 \tag{17}$$

The stacking sequence of the laminate can be determined from these values.

Since the allowable region of the flexural lamination parameters of the laminates with two kinds of fiber orientation angles is the same as that of the laminates with n kinds of orientation angles, two kinds of orientation angle is necessary and sufficient for designing laminated composites with required flexural stiffness.

When the number of kinds of orientation angles is two, the flexural lamination parameters are evaluated by Eq. (11). The design variables are ξ_1, ξ_2, θ_1 and θ_2 in the equation, but the number of independent design variables is three because ξ_1 and ξ_2 are not independent with each other. On the other hand, the number of equations is two. Therefore, the value of ξ_1 or ξ_2 can be given arbitrarily when Eq. (11) is solved for θ_1 and θ_2 under the conditions in Eq. (17). This is very important because the value of ξ is discrete when the number of plies is finite.

The orientation angles θ_1 and θ_2 are obtained from Eq. (11), as follows.

$$\theta_1=\frac{1}{2}\cos^{-1}T_1 \ , \quad \theta_2=\frac{1}{2}\cos^{-1}T_2$$

$$T_1=\{2\xi_1W_1^*\pm\sqrt{[4\xi_1^2W_1^{*2}-2\xi_1(2W_1^{*2}-\xi_2W_2^*-\xi_2)]}\}/2\xi_1 \tag{18}$$

$$T_2=\{2\xi_2W_1^*\pm\sqrt{[4\xi_2^2W_1^{*2}-2\xi_2(2W_1^{*2}-\xi_1W_2^*-\xi_1)]}\}/2\xi_2$$

The factor ξ_1 and ξ_2 are given from Eq. (6), as follows.

$$\xi_1=(2z_1/h)^3 \ , \quad \xi_2=1-\xi_1 \tag{19}$$

Suppose that the total number of plies $2N_h$ equals to twenty. In this case, the possible value of $2z_1/h$ is a multiple of 0.1. Table 2 shows the values of design variables which satisfy the condition of Eq. (17). Then, the stacking sequence is represented as follows.

$$[(+6.35/-6.35)_3/(+67.90/-67.90)_7]_s \tag{20.1}$$

$$[(+78.66/-78.66)/(+13.88/-13.88)_9]_s \tag{20.2}$$

TABLE 2 Design Variables

$2z_1/h$	ξ_1	ξ_2	θ_1(deg.)	θ_2(deg.)
0.7	0.343	0.657	6.35	67.90
0.9	0.729	0.271	78.66	13.88

Table 3 shows the detail of flexural stiffness of the laminate of which stacking sequence is represented by Eq. (20). It can be seen from the table that the design specifications given in Eq. (16) are satisfied.

TABLE 3 Flexural Properties of the Laminate

Flexural stiffness (GPa)		Flexural compliance (GPa^{-1})		Flexural effective engineering constants (GPa)	
D^*_{11}	122	d^*_{11}	0.00834	E^f_1	120
D^*_{22}	54.4	d^*_{22}	0.0186	E^f_2	54.4
D^*_{12}	10.7	d^*_{12}	-0.00163	E^f_6	15.0
D^*_{66}	15.0	d^*_{66}	0.0670	ν^f_{21}	0.196

CONCLUDING REMARKS

A new material design method of fibrous laminated composites with required flexural stiffness is proposed in this paper. The selection of kinds of unidirectional lamina and the determination of fiber orientation angles can be done directly from the imposed conditions on the flexural stiffness of laminated plates.

A newly constructed diagram for flexural lamination parameters made possible to determine the design specifications and to specify the fiber orientation angle of each ply and the stacking sequence.

The designing procedure is summarized as follows.

Step 1 : Draw the flexural lamination parameter diagram of available materials.
Step 2 : Make a design specifications in relation to flexural stiffness.
Step 3 : Find the feasible design region on flexural lamination parameter diagram. When no design region is found on the diagram of a material, change the material. In case no design region can be found for all available materials, the design specification should be relaxed. In this step the ply material is determined.
Step 4 : Determine one point in the feasible design region (referred to as a design point) with another condition such as the maximization of E^f_2.
Step 5 : Give the number of plies and the number of the kinds of orientation angles. Two kinds of orientation angles are enough for the plate with large number of plies.
Step 6 : Obtain the fiber orientation angle of each ply from the flexural lamination parameter of the design point.

REFERENCES

Chamis, C. C. (1980). Design Engineering, May, 39-44.
Miki, M. (1982). Proceedings 4th ICCM, Tokyo Japan, 1725-1731.
Tsai, S. W. and Hahn, H. T. (1980). Introduction to Composite Materials, Technomic, Westport.

INFLUENCE OF MATRIX RESIN ON ULTIMATE STRENGTH OF UNIDIRECTIONAL CFRP

Y. Miyano*, M. Kanemitsu*, T. Kunio and M. Miki***

*Materials System Research Laboratory, Kanazawa Institute of Technology, 7-1 Ohgigaoka, Nonoich, Kanazawa 921, Japan

**Department of Mechanical Engineering, Keio University, 3-1-14 Hiyoshi, Kohoku, Yokohama 223, Japan

ABSTRACT

The time and temperature dependent strengths, that is, the longitudinal tensile and compressive strengths, the transverse tensile and compressive strengths and the interlaminar shear strength, of unidirectional CFRP are investigated in connection with epoxy resin of its matrix. As results, these strengths show thermo-rheologically simple behavior in the same manner as the tensile strength of epoxy resin used as the matrix of CFRP.
A method for evaluation of the strengths of CFRP at any strain rate and temperature is proposed, in which the shift factor of matrix resin and the strengths of CFRP at several temperatures under a constant strain rate are used.

KEYWORDS

Composite materials; carbon fiber reinforced plastics; strengths; viscoelastic behavior; time and temperature dependence.

INTRODUCTION

It has been well known that the carbon fiber reinforced plastics (CFRP) shows a remarkable anisotropy in their mechanical properties depending on the combination of carbon fiber and matrix resin, since the carbon fiber has high strength and high modulus. In general, the strength and stiffness of CFRP are very high in the longitudinal direction of fibers, while they are extremely low in the transversal direction of fibers.

On the other hand, the matrix resin of CFRP shows remarkably time and temperature dependent mechanical behaviors, that is, viscoelastic behaviors. Thus, the CFRP also displays viscoelastic behaviors.

This paper is concerned with the time and temperature dependent strengths, that is, the longitudinal tensile and compressive strengths, the transverse tensile and compressive strengths and the interlaminar shear strength, of unidirectional CFRP with respect to various loading directions.

SPECIMEN AND TEST PROCEDURES

Preparation of specimen

TABLE 1 Composition and cure condition of resin and CFRP

Composition	
Resin	CFRP
Epoxy EPIKOTE 828/100*	Matrix: Epoxy resin
Hardener: Methyl himic anhydride / 104*	Carbon fiber: TORAYCA T300 3000fil.
Cure accelerator: 2-ethyl-4-methylimidazol / 1*	Volume fraction of fibers: 53.2 - 58.9%
Cure condition	
12h at 70°C (5MPa), 4h at 150°C and 2h at 190°C	

* : 100, 104 and 1 show the weight ratio.

TABLE 2 Mechanical properties of resin and unidirectional CFRP

Item			Resin	CFRP L*	CFRP T*
Density		kg/m^3	1.22×10^3	1.53×10^3	
Young's modulus		GPa	3.3	140	8.8
Tensile strength		MPa	76	1970	65
Flexural strength		MPa	153	1470	67
Compressive strength		MPa	153	941	226
Coefficient of thermal expansion	$°C^{-1}$	< 125°C	69×10^{-6}	0×10^{-6}	44×10^{-6}
		> 125°C	140×10^{-6}	0×10^{-6}	111×10^{-6}
Glass transition temperature		°C	125	——	125

* : L and T represent the longitudinal and the transversal direction, respectively.

TABLE 3 Testing conditions of epoxy resin and CFRP

Test \ Condition	Rate of displacement mm/min (Strain rate min^{-1})
Tension test for epoxy resin	0.05, 0.5, 5, 50 ($4.25\times10^{-4} \sim 4.25\times10^{-1}$)
Longitudinal tension test for CFRP	0.05, 0.5, 5, 50 ($3.95\times10^{-4} \sim 3.95\times10^{-1}$)
Longitudinal compression test for CFRP	0.05, 0.5, 5, 50 ($6.35\times10^{-4} \sim 6.35\times10^{-1}$)
Transverse tension test for CFRP	0.05, 0.5, 5, 50 ($5.1\times10^{-4} \sim 5.1\times10^{-1}$)
Transverse compression test for CFRP	0.05, 0.5, 5, 50 ($2.5\times10^{-3} \sim 2.5$)
Interlaminar shear test for CFRP	0.01, 0.5, 5, 50 ($1.33\times10^{-3} \sim 1.33$)

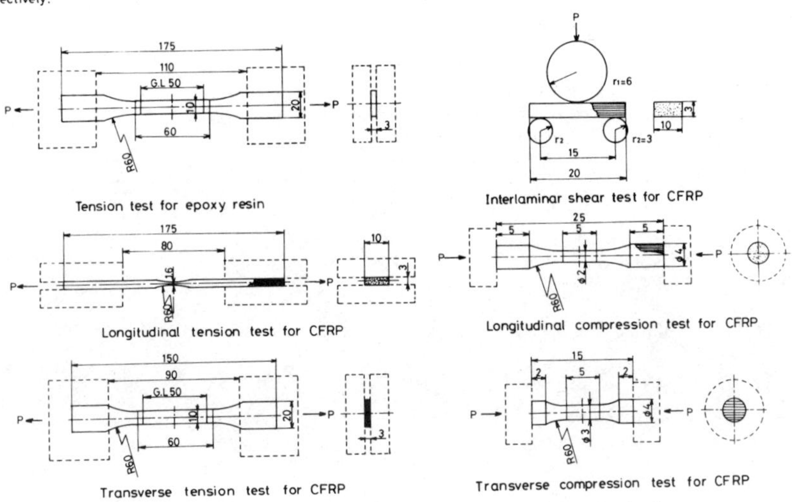

Fig. 1. Specimen geometries and testing methods. (all dimensions in mm)

The compositions and the cure conditions for the epoxy resin used as the matrix and the CFRP unidirectionally reinforced are given in TABLE 1 (Miyano, 1983). The unidirectional CFRP plate was made by use of so called a filament winding technique. A sheet of CFRP laminate plate (150mm x 150mm x 3mm or 5mm) with no void which was confirmed under a microscope was removed from the mold, then was machined into five kinds of specimen as shown in Fig. 1 for the tests described later. A sheet of epoxy plate (180mm x 250mm x 3mm) employed as the matrix of the CFRP was also machined into the geometry as shown in Fig. 1.

TABLE 2 shows the mechanical properties of the epoxy resin and the unidirectional CFRP.

Test and evaluation procedures

The tension test for the epoxy resin and five kinds of test for the CFRP, that is, the tension and compression tests for the longitudinal and transversal directions of fibers and the interlaminar shear test for the unidirectional CFRP, were conducted by using the Instron type universal testing machine with a constant temperature chamber.

This testing method are illustrated in Fig. 1. All of these tests were performed at 7 steps of the constant temperature T = 60, 80, 100, 120, 140, 160 and 180°C under 4 kinds of the constant rate of the displacement V which were tabulated in TABLE 3.

The displacement rates V were converted into the strain rate R according to the following relation :

$$R = k \cdot V \tag{1}$$

The factor k for each test was calibrated by measuring the strain at the center of specimen and the relative displacement of two loading points at room temperature.

The tensile strengths of epoxy resin $F_M(R,T)$, the longitudinal tensile and compressive strengths of CFRP $F_{LT}(R,T)$ and $F_{LC}(R,T)$, and the transverse tensile and compressive strengths of CFRP $F_{TT}(R,T)$ and $F_{TC}(R,T)$, which are represented by the functions of strain rate R and temperature T, were evaluated from the breaking loads $W_b(R,T)$ by using the following equation :

$$F(R,T) = \frac{W_b(R,T)}{b \cdot h} \tag{2}$$

where b and h are the width and thickness of specimen, respectively.

The interlaminar shear strength of CFRP $F_{LS}(R,T)$ was also obtained from the breaking loads $W_b(R,T)$ by the following equation :

$$F_{LS}(R,T) = \frac{3}{4}\,\frac{W_b(R,T)}{b \cdot h} \tag{3}$$

RESULTS AND DISCUSSIONS

Time and temperature dependence of the strengths in CFRP

The left side of Fig. 2 gives the $F_M(R,T)$. It is recognized from the figure that the $F_M(R,T)$ remains almost unchanged irrespective of different strain rate in the region of the temperature below T = 100°C, while it decreases markedly as the temperature rises and/or the strain rate decreases in the region of temperature above T = 100°C. Therefore, it is clear that the epoxy resin changes from a glassy state to a rubbery state with the rise of temperature and the decrease of strain rate.

The above mentioned five kinds of strength, $F_{LT}(R,T)$, $F_{LC}(R,T)$, $F_{TT}(R,T)$, $F_{TC}(R,T)$ and $F_{LS}(R,T)$, of the unidirectional CFRP are given on the left side of Fig. 3 through Fig. 7 as a function of strain rate R at various constant temperature T. From these figures, they also change with the strain rate R and temperature T in the same manner as the above stated tensile strength of epoxy resin used as the matrix of CFRP.

Reciprocation law between time and temperature

The master curve of strength of epoxy resin $F_M(R',T_0)$ against the reduced strain rate R' at the reference temperature T_0 = 100°C was constructed by shifting the $F_M(R,T)$ at various constant temperatures T along the log scale of strain rate R so that they overlapped each other as shown in Fig. 2. Since the $F_M(R,T)$ at various temperatures can be superposed smoothly, the reciprocation law between time and temperature (Schwarzl, 1952; Williams, 1955) is applicable to $F_M(R,T)$. And, this strength $F_M(R',T_0)$ increases as the reduced strain rate R' increases. Therefore, it was found that the tensile strength of epoxy resin shows thermo-rheologically simple behaviors.

As for the five kinds of strength for the CFRP, their master curves given in Fig. 3 through Fig. 7 at the reference temperature T_o = 100°C were constructed using the same procedure as mentioned above for the epoxy resin of matrix. The completion of smooth master curves shows that the reciprocation law between time and temperature also holds for these strengths of CFRP. Thus, it is clear that the strengths of CFRP also show thermo-rheologically simple behaviors. Looking at the curves, it can be found that the respective strengths of CFRP increase with the reduced strain rate R', but they show a little different shape one another.

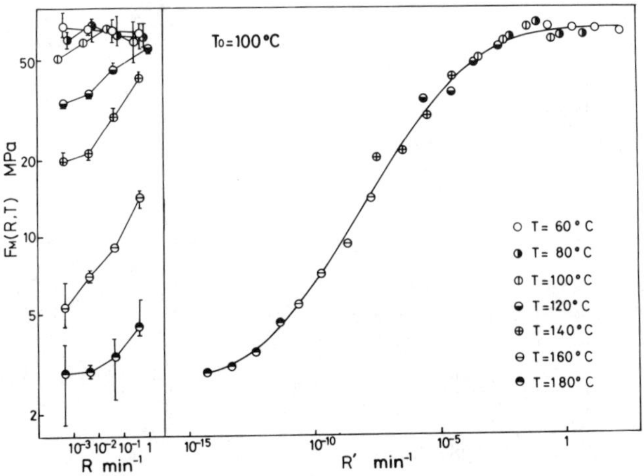

Fig. 2. Master curve for tensile strength of epoxy resin.

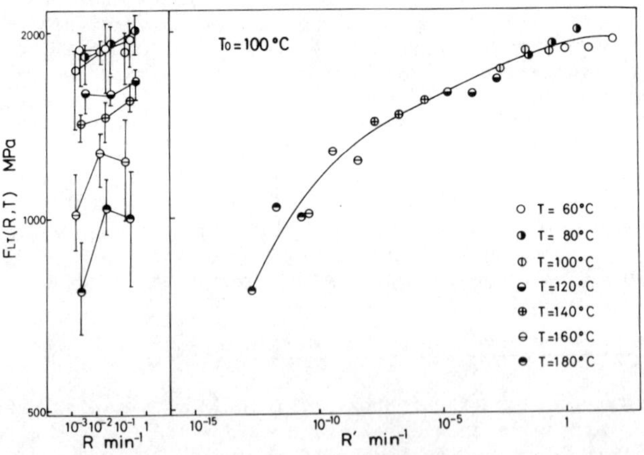

Fig. 3. Master curve for longitudinal tensile strength of CFRP.

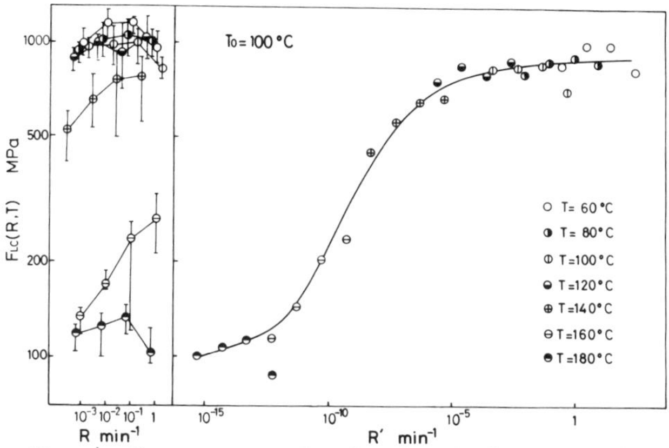

Fig. 4. Master curve for longitudinal compressive strength of CFRP.

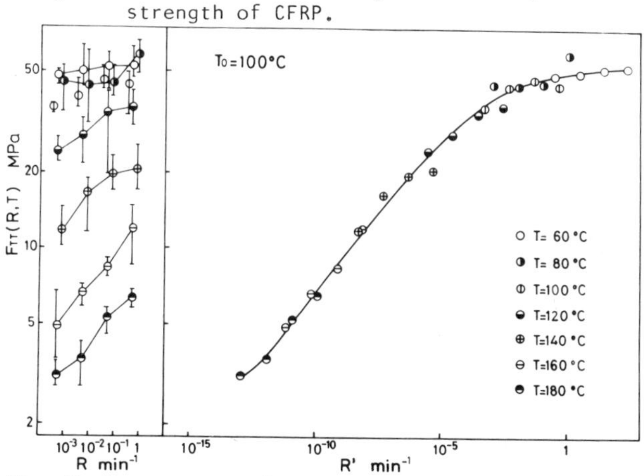

Fig. 5. Master curve for transverse tensile strength of CFRP.

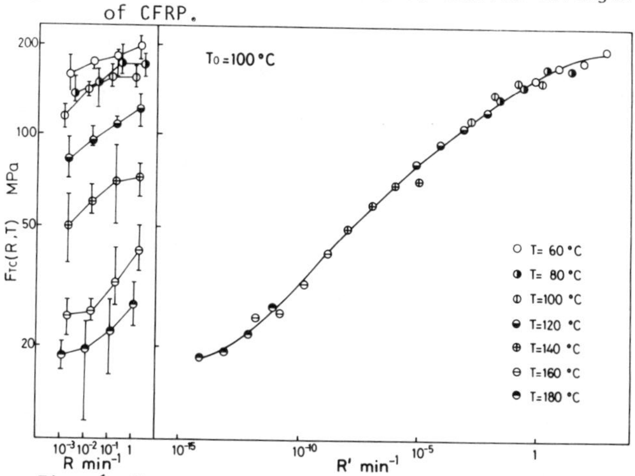

Fig. 6. Master curve for transverse compressive strength of CFRP.

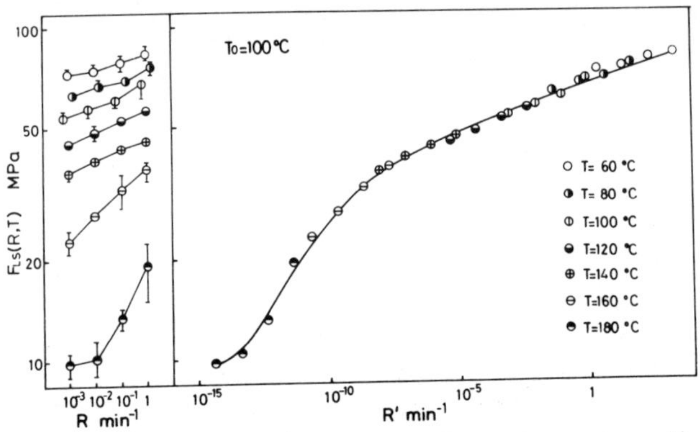

Fig. 7. Master curve for interlaminar shear strength of CFRP.

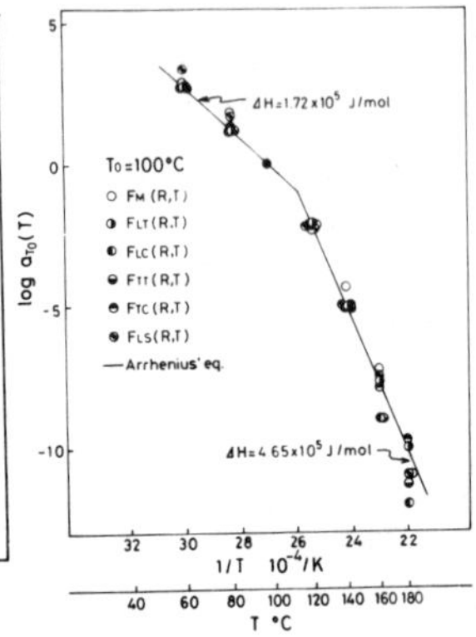

Fig. 8. Time - temperature shift factors $a_{T_0}(T)$.

Fig. 8 shows the time - temperature shift factors $a_{T_0}(T)$ obtained experimentally for the master curves concerning the strengths of both CFRP and matrix resin. All of these shift factors are found to be quantitatively in good agreement and can be approximated using the Arrhenius' equation having two different activation energies :

$$\log \ a_{T_0}(T) = \frac{\Delta H}{2.303 \cdot G}\left(\frac{1}{T} - \frac{1}{T_o}\right) \tag{4}$$

where ΔH is the activation energy J/mol, G is the gas constant 8.31 J/K·mol, and T and T_o are the testing and reference temperature in K, respectively.

The temperature at the joining point of two Arrhenius' equations corresponds closely to the glass transition temperature of the matrix resin.

Estimation method for the strengths of CFRP

Based on the results obtained in previous sections a estimation method is proposed, through which the five kinds of strength of CFRP at any strain rate R and temperature T could be evaluated from the strengths of CFRP obtained at several temperatures under a single constant strain rate, using the time - temperature shift factor of matrix resin.

The procedure shown in Fig. 9 will be explained as follows :

Step 1 : To measure the $F_M(R,T)$ against strain rate R at various temperatures T.

Step 2 : To obtain the master curve $F_M(R',T_o)$ and the shift factors $a_{T_0}(T)$ of matrix resin by use of the reciprocation law between time and temperature.

Step 3 : To measure the strength of CFRP $F(R_o,T)$ against T under a certain value of R_o.

Step 4 : To construct the master curve of strength of CFRP $F(R',T_o)$ against the reduced strain rate R' at the reference temperature T_o using the $F(R_o,T)$ (in Step 3) and the $a_{T_0}(T)$ (in Step 2).

Step 5 : To estimate the strengths of CFRP $F(R,T)$ at any temperature T under any strain rate R using the $F(R',T_o)$ (in Step 4) and the $a_{T_0}(T)$ (in Step 2).

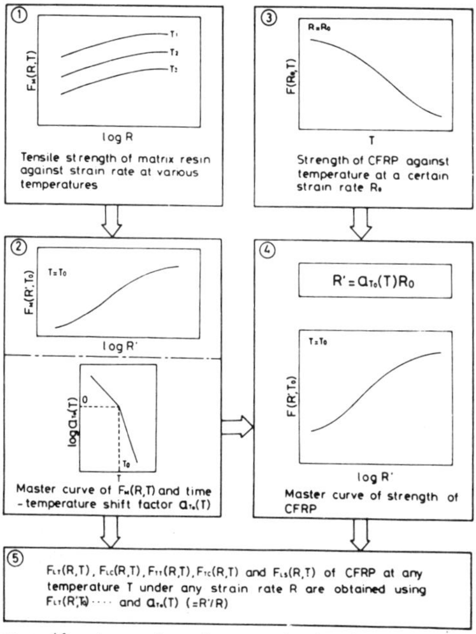

Fig. 9. Expendient estimation method of various strengths of CFRP. (In this figure, $F_{LT}(R,T)$ ··· $F_{LS}(R,T)$ are designated simply by F(R,T).)

CONCLUSIONS

The time and temperature dependent strengths, that is, the longitudinal tensile and compressive strengths, the transverse tensile and compressive strengths and the interlaminar shear strength, of unidirectional CFRP were investigated in connection with epoxy resin of its matrix.

The results obtained will be summarized as follows :

1) The master curves of strength can be constructed by using the reciprocation law between time and temperature.
2) The time - temperature shift factors of respective strengths of CFRP are quantitatively in good agreement with that of the matrix resin.
3) A method for evaluation of the strengths of CFRP at any strain rate and temperature from the shift factor of matrix resin and the strengths of CFRP at several temperatures under a constant strain rate is proposed.

ACKNOWLEDGEMENTS

The authors wish to thank Professor M. Takashi, Dr. H. Nakayasu and Dr. T. Isogai for many helpful discussions. The assistances of Mr. M. Shimbo and Mr. S. Sugimori in performing the experiments is gratefully acknowledged.

REFERENCES

Miyano, Y., Kanemitsu, M. and Kunio, T. (1983). Fib. Sci. Tech. 18, 65-79.
Schwarzl, F. and Staverman, A. (1952). J. Appl. Phys. 23, 838-843.
Williams, M. L., Landel, R.F. and Ferry, J. D. (1955). J. Am. Chem. Soc. 77, 3701-3707.

EFFECT OF FIBER ORIENTATION ON STABLE CRACK GROWTH IN FRP

T. Kunio, M. Shimizu and S. Sohmiya

Department of Mechanical Engineering, Keio University, Yokohama, Japan

ABSTRACT

3-point bending fracture toughness tests have been carried out on the FRP specimens with low fiber volume fraction in 4 different fiber orientations, together with the direct observation of the local fracture event at the growing crack front. It is shown that a macro crack growth becomes stable when the micro unstable crack growth due to micro pop-in in the matrix resin is restrained by the pinning effect of fibers and the anti-crack opening effect of unbroken fiber behind the crack tip. Dependence of crack growth resistance on fiber orientation is evaluated by the R - curve method with special attention to the intermittent crack growth behavior.

KEYWORDS

FRP, Fiber orientation, Fracture mechanisms, Fracture toughness, R - curve and Stable crack growth.

INTRODUCTION

Fiber reinforced plastics (FRP) have been extensively used for the structural components because of the superiority of their toughness. Although many theoretical and experimental studies (Beaumont, 1974; Bhatia, 1975; Bowling, 1979; Griffith, 1979; Witt, 1979) have so far been carried out on the clarification of fracture mechanisms in FRP involving the estimation of critical fracture energy for the onset of final fracture, there remain many subjects unsolved concerning the basic nature of the strengthening effect of fibers in the fracture toughness of FRP. The aim of the present study is to make clear the role of fiber orientation in the crack growth process, especially in the local fracture behavior at the stationary and growing crack fronts and also to evaluate the pinning effect of fibers against the crack propagation in the FRP using the R - curve method in fracture mechanics. Discussion has also been made on the appearance and its cause of non-linear behaviors in load-deflection (P - δ) curve obtained from the fracture toughness tests.

MATERIAL AND TEST METHOD

The laminated plate of FRP was employed as a specimen material, in which roving glass fibers have been embeded uniderectionally into the matrix of the epoxy resin (EPICOTE 828, HHPA and K 61 B). The fiber volume fraction Vf of the present material is equal to 2.6%, which is greater than the critical fiber volume fraction (Vfc) in the Gc - Vf relation (Kunio, Shimizu and Sohmiya, 1981). The plate were machined into the plate specimen having the fiber orientation angles θ of 0°, 30°, 45°and 90°. Figure 1 shows the angle θ and the geometry of specimen for 3-point bending fracture toughness test. The precrack was introduced to the specimen in the following manner. Firstly, a deep slit was made by the use of a thin grinder edge (thickness = 0.3mm), and then the specimen was heated up to about 400 K. With application of sharp razor edge into the slit, then a sharp precrack was formed. The specimen prepared in this manner was subjected to the 3-point in plane bending with the crosshead speed of 1 mm/min. The crack extension resistance in the fracture toughness test was evaluated by the following equation:

$$G = (P^2 / 2t) \times (dc/ da)$$

where, G : Energy release rate, t : Thickness of specimen, c :Compliance,
a : Crack length and P : Applied load.

As it was difficult to record the amount of crack extension accurately at each stage of load increment up to the final fracture of a single test specimen, the current length of growing crack was estimated from the amount of variation in compliance, as an equivalent crack length obtained from the calibration curve given the relation between Compliance "c" and a crack legth "a" for the present specimen geometry.

TEST RESULTS AND CONSIDERATION

Non-linear behavior of load-deflection (P -δ) curves

Figure 2. shows the load-deflection curve (P -δ) obtained from the test with respect to the specimens having 4 different angles of fiber orientation. The non-linear behavior can be observed in the P-δ curves depending on the fiber

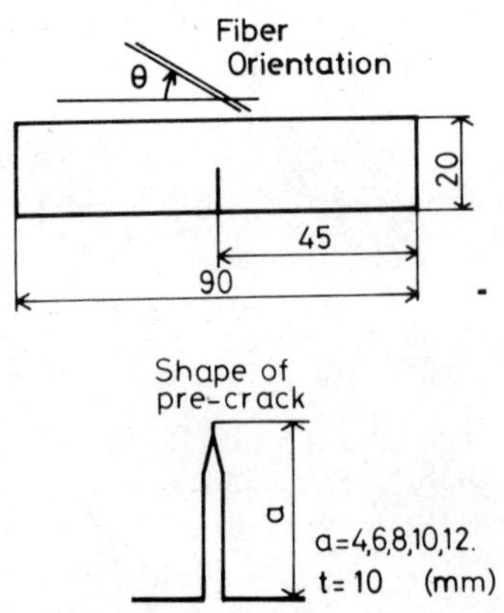

Fig. 1. Geometry of specimen.

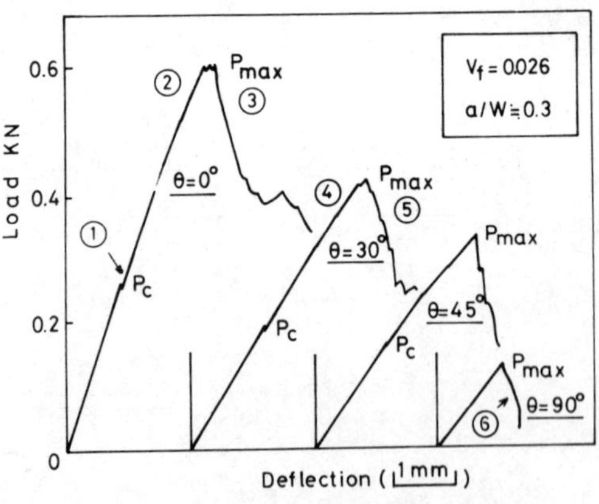

Fig. 2. Typical load-delection curves.

orientation. An attention will be given to such a non-linear behavior of the P - δ curves, especially to that appearing at the loads between Pc and Pmax. The characteristics of this behavior are summerized as follows.

(i) The linearity of the P - δ curves breaks by a sudden drop of the load at the Pc point, irrespective of fiber orientations.

(ii) This phenomenon occurs remarkably in the specimen having the fiber orientation of $\theta = 90°$, where the Pc point agrees with the Pmax point at which the load reaches the maximum value on the P - δ curve.

(iii) A lot of small load drops similar to that at the Pc point are produced everywhere on the P - δ curve beyond Pc point and as a result, an obvious change occurs in the slope of the curve beyond the Pc point, thus the non-linearity of the P - δ curve is resulted.

(iv) The non-linear portion of P - δ curve from Pc to Pmax tends to disappear with the increase of fiber orientation angle θ. Non-linear behavior is not recognized at all at $\theta = 90°$.

Observation of Stable Crack Growth Phenomenon

Crack initiation behavior. The non-linear behavior which has been discussed so far is attributable to crack growth which occurs in the material along with the increment of applied loads. Fig. 3 gives the photos showing a series of local fracture events at the precrack front. The numbers (1)/(6) attached to the photos refer to the location on the P - δ curve in Fig. 2 where the photographic records were made. The number (1) photo shows the crack initiation behavior at the precrack front just after the point Pc where a micro fracture has been initiated from the resin part of the precrack tip. More detailed observations were made on such a local fracture at the precrack front by the new technique described as follows: - After unloading of the specimen which has been loaded up to the point Pc, a cut was introduced along cross section A' B' C' D' parallel to the precrack surface as shown in Fig. 4. Then,utilizing the transparent property of the test material a fracture condition at the precrack tip was directly observed by the microscope through the A' B' C' D' plane. Fig. 5(a) shows an example of this observation, where the half penny shaped crack with both sides pinned by the fibers can be clearly recognized. This fact indicates that the fibers has a strong blocking effect against the crack propagation in the matrix.

Crack propagation behavior in the non-linear domain (Pc - Pmax). The direct observations of the stable crack growth were made at the load level between Pc and Pmax using the same technique. A large number of micro pop-ins similar to that occuring at the Pc point are generated successively at other location along the precrack front with increase in load. The onset of macro crack growth occurs when large portion of precrack edge has been occupied by such micro pop-ins. The further growth of the macro crack occurs in the same manner as in the case of crack initiation. This can be confirmed by the fact that the macro crack tip is composed of the aggregate of small half penny shaped cracks as shown in Fig. 5(b). From previous study (Kunio, 1981, 1983), such small half penny shaped crack has been known to be resulted from the generation of micro pop-in at crack tip. Further observation of the generation of micro pop-in revealed that each small load drop on the non-linear portion of P - δ curve is in correspondence to a single micro pop-in. Then, thorough investigations were made into the cause of arrest of the micro crack growth induced by the micro pop-in. As a result, it was confirmed that there exist fibers at both ends of the half penny shaped micro crack as shown in the magnified view in Fig. 5(c), showing that the fibers have a strong pinning effect against the enlargement of a minute unstable fracture caused by the micro pop-in. Further observation of the micro crack behavior showed that the unbroken fiber is left behind the

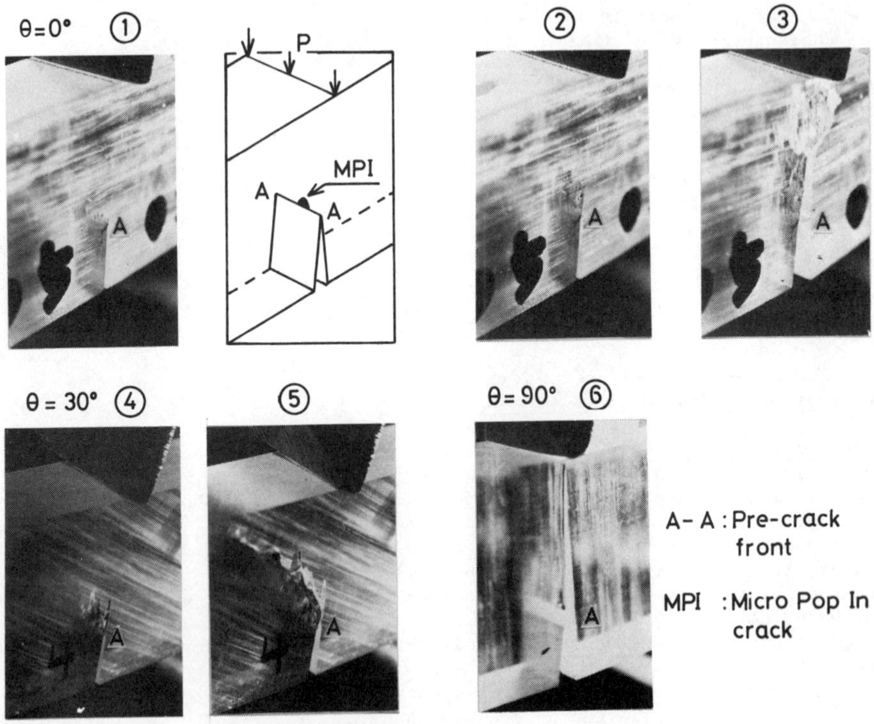

Fig. 3 (1)/(6). Local fracture events at precrack front.

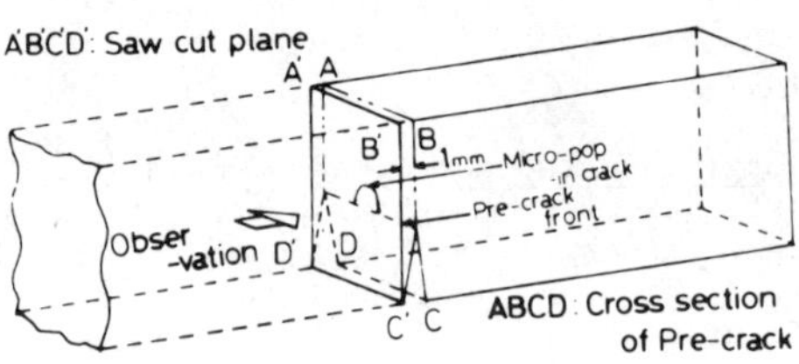

Fig. 4. Illustration of the direct observation technique for the local fracture at crack front.

front giving the effect of restraining the opening of macro crack front which results in a stable growth of cracks.

Further study was made of fiber orientation dependency of crack propagation behavior. As a result, the following points were made clear; i) The increment of a crack length per single micro pop-in increases as the θ increases. ii) Even if the fiber is on the tilt against the precrack, the initial crack growth occurs along the precrack direction (see (4) in Fig. 3). iii) At the maximum load on $P-\delta$ curve where the fiber at the precrak tip is broken in avalanche, the macro crack changes its propagation direction to the

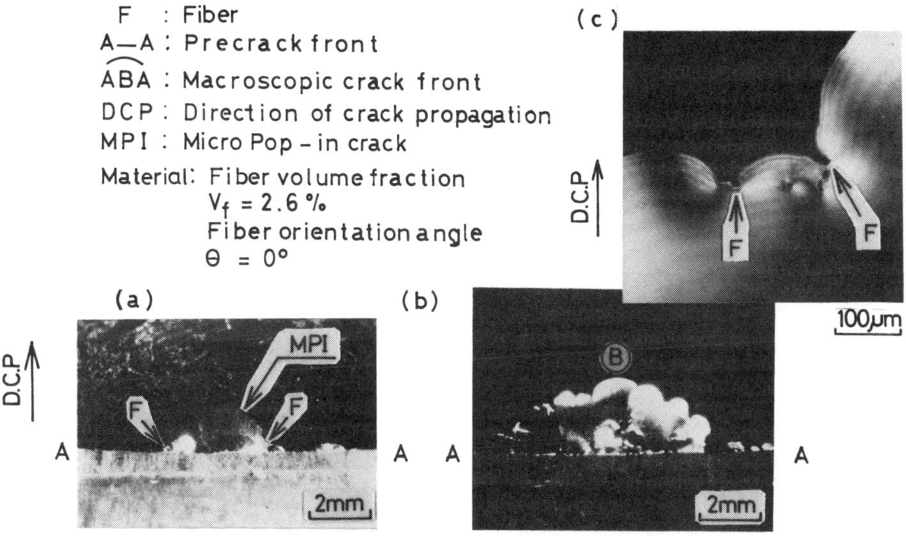

Fig. 5 (a), (b), (c). Results of direct observation of local fracture at crack front; (a) a half penny shaped crack (at Pc point), (b) macro scopic crack (in non-linear domain) (c) magnifited view of (b).

fiber orientation (see (3) and (5) in Fig. 3). As a result, delaminating fracture occures. In case of θ = 90°, the delaminating crack appears immediately after the generation of micro pop-in at the point Pc (= Pmax) on P - δ curve (see (6) in Fig. 3).

Evaluation of Crack Extension Resistance by the R-curve

Crack extension resistance in the non-linear domain. Figure 6 shows the R-curve for the specimens having different fiber orientations, where the equivalent crack length has been taken as the representation of crack length. From this, it is evident that when θ = 0° a macroscopic unstable fracture takes place at the point Pc and Pmax and also that in the non-linear domain from Pc to Pmax a stable fracture takes place. In this stable fracture process, a large number of very small steps are formed on the R-curve in correspondence to the generation of micro pop-in at the precrack tip or at the front of growing macro crack (Fig. 6). Then, as observed in the magnified view "A", it is confirmed that these steps had been formed by fine unstable cracks which stopped their growth owing to the pinning effect of fiber against the crack propagation. Such a phenomenon was generally observed in both samples of θ = 30° and 45°.

Dependence of fracture toughness on fiber orientation. Figure 7 was rearranged to Fig. 8 to examine the dependence of fracture toughness of present FRP on the fiber orientation. From this result, it is evident that the value of Gc (Pmax) decreases with increase in the fiber orientation angle θ, while Gc (Pc) remains constant irrespective of θ. This shows that there exists a critical angle of fiber orientation θc, at which both fracture toughness values coincide with each other. Gc(Pc) agrees with Gc(Pmax). This means that in the FRP having the angle θ larger than θc, the micro fracture initiation at the precrack tip immediately leads to a final fracture of the specimen. In order to obtain the θc value of the present FRP, additional fracture toughness tests were conducted on the specimen having different fiber orientation angle θ = 50°, 55°, 60°, 65° and 70°. As a result, the critical fiber orientation was found to be θc = 60° for the present FRP.

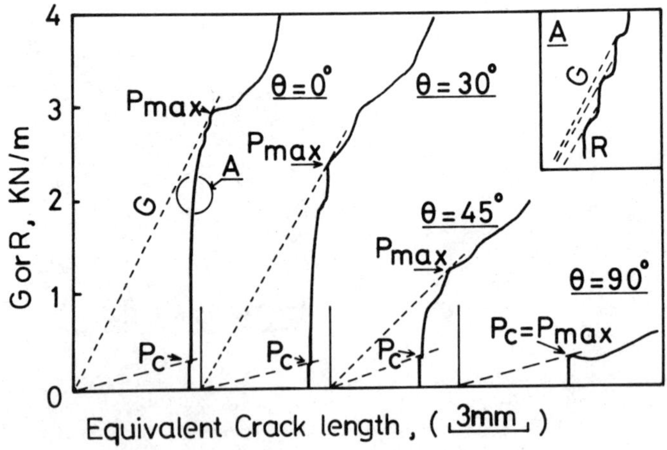

Fig. 6. R-curves of FRP with different fiber orientations.

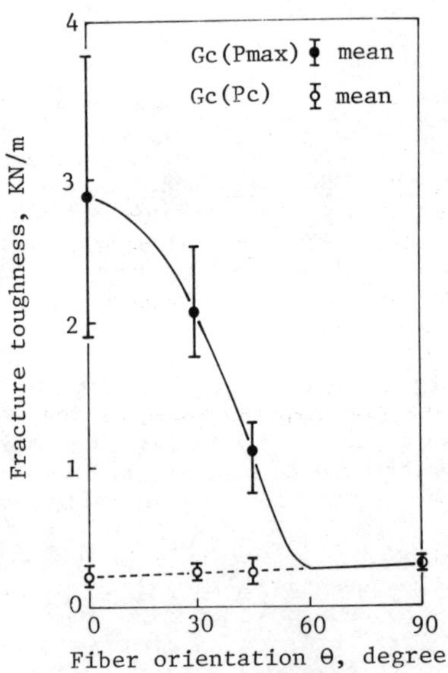

Fig. 7. Dependence of fracture toughness on fiber orientation.

CONCLUSION

Based on the results of direct observation of a local fracture behavior at the crack tip with special interests in the relation between fiber orientation and crack growth behavior and also results of the analysis of R-curve obtained from the fracture toughness test, discussions has been made on the strengthening effect of fibers in fracture resistance of FRP. Results obtained are surmmarized as in the followings.

(1) Although macro crack growth occurs easily as the result of generation of a micro pop-in at the crack front, characteristics of which is essentially an unstable fracture, such crack growth becomes stable when the following two effects are produced in the crack growth process;

(i) the pinning effect of fibers against the micro crack growth in the resin part and (ii) the restraining effect of unbroken fibers which was left behind the crack front, on opening of the macro crack tip.

(2) There exists the critical fiber orientation angle θc : in the FRP having fiber orientation angle $\theta > \theta c$, the generation of micro pop-in at the precrack front immediately leads to the final fracture of specimen, while having fiber orientation angle $\theta < \theta c$, fracture occurs after the maximum load (Pmax) is attained accompanying the stable crack growth.

ACKNOWLEDGMENT

The authors are indebted to Dr. K. Yamada for his valuable discussion, and finacial supports from the grant in aid for scientific research by the gratefully ministry of education, Japan, and Fukuzawa memorial fund in Keio University are gratefully acknowledged. Thanks are also due to messrs. K. Hotta and H. Higashino for their assistance in conducting the experimental work.

REFERENCES

Beaumont, P. W. R. (1974). A fracture mechanics approach to failure in fibrows composites. J. Adhesion, 6, 107-115.

Bhatia, N. M.,and R. M. Verrette (1975). Crack arrestment of laminated composite. ASTM STP, 593, 200-214.

Bowing, J., and G. W. Groves (1979). The debonding and pull-out of ductile wires from a brittle matrix. J. Material Scie., 14, 431-442.

Griffith, W. I., M. F. Kanninen and E. F. Rybicki (1979). A fracture mechanics approach to the analysis of Graphith / Epoxv laminated precrack tension panels. Nondestructive Evaluation and flaw critical for composite materials. ASTM STP,696, 185.

Kunio, T., M. Shimizu and S. Sohmiya (1981). Role of local fiber distribution at noch tip in the fracture toughness of FRP. In D. Francois (Ed.), Advances in fracture research, Vol. I. Pergamon Press, Oxford. pp. 495-502.

Kunio, T., M. Shimizu and S. Sohmiya (1983), Effect of local fiber distribution ahead of crack on the fracture toughness of FRP, Int. J. Fiber, Scie. and Tech., 18, 151-161.

Witt, W. P., A. N. Palazotto and H. T. Hahn (1979). Numerical and experimental comparision of the notch tip stress in a laminated plate. AIAA journal, 17-4, 500.

NONDESTRUCTIVE EVALUATION OF COMPOSITE MATERIALS — A DESIGN PHILOSOPHY

J. C. Duke, Jr., E. G. Henneke, II, W. W. Stinchcomb and K. L. Reifsnider

Materials Response Group, Engineering Science and Mechanics Department, Virginia Polytechnic Institute and State University, Blacksburg, VA 24061-4899, USA

ABSTRACT

Efficient and reliable structural design utilizing fiber reinforced composite materials may only be accomplished if the materials used may be nondestructively evaluated. There are two major reasons for this requirement: i) composite materials are formed at the time the structure is fabricated and ii) at practical strain levels damage, changes in the condition of the material, that influence the structure's mechanical performance is present.

The fundamental basis of such a nondestructive evaluation capability is presented. A discussion of means of assessing nondestructively the material condition as well as a damage mechanics theory that interprets the material condition in terms of its influence on the mechanical response, stiffness, strength and life is provided.

KEY WORDS

Composite materials; nondestructive evaluation; damage mechanics; stiffness; strength; life; ultrasonics; radiography; stress wave factor; moire interferometry; replication; thermography; mechanical testing; modeling; cumulative damage.

INTRODUCTION

Structural fiber reinforced composite materials are inherently imperfect. Efficient and reliable design involving such materials must therefore recognize this sobering fact. As part of the preliminary design process it is important to anticipate the existence of material imperfections and allow for their presence. In the final stage, however, it is imperative that such imperfections be located, described, and their significance as regards the subsequent fulfillment of the design requirements determined. To nondestructively evaluate the response of fiber reinforced composite materials to expected service conditions is an important design requirement.

There are three distinct, but related, aspects of this task. First, the condition of the material must be determined nondestructively. Second, the implications of the service requirements must be rationalized in terms of the mechanical loading of the material. Finally, an analysis of the material's ability to satisfy the service requirements must be made. The Materials Response Group at Virginia Tech has been studying this topic for some time. The approach followed has been to develop techniques and methodologies for nondestructively describing the condition of advanced fiber reinforced composite laminates in both the undamaged and damaged condition. In addition, based on observation of variations in material condition as a result of mechanical loading, the formulation of a damage mechanics theory has been undertaken that is capable of describing the material's stiffness, strength and life.

DETERMINATION OF MATERIAL CONDITION

In general, the condition of a material is a function of the fabrication and mechanical loading history and is different from point to point in the material component. However at any instant in time this condition determines how the material will respond to imposed loads. In other words, the material's present collective condition determines its response. Furthermore, the condition of the material alone is insufficient to determine the material's strength and life. This is because both the strength and life are determined in part by the future mechanical loading spectrum and as such may not be uniquely defined. Consider, for example, two different mechanical loading conditions--sustained load and monotonically increasing load with time. As regards the strength, the load at failure will undoubtedly be different even if the initial material conditions were the same. As regards the life, the time to failure would be expected in general to be shorter for the case of increasing load. Clearly the approach must be taken that given the initial material condition and the mechanical loading condition, how does the material condition change so as to reduce the material's ability to satisfy the design requirements? Consequently the approach to nondestructively characterizing the material's condition must be predicated on the nature of the damage mechanic theory used to describe the condition change.

Nondestructive Inspection Methods

Extensive use has been made in our laboratory of ultrasonic and X-ray radiographic techniques and replication for visualizing the internal material condition. In addition, efforts to quantify, directly and nondestructively, the effects of the material condition have included acoustic emission, ultrasonic stress wave factor monitoring, and thermography.

Ultrasonic inspection. Using ultrasound and examining the variation in the way the material transmits this mechanical disturbance makes it possible to obtain a geometric description of regions of material imperfection. Figure 1 displays the results of ultrasonically C-scanning a glass epoxy laminate a) before loading, b) after loading to 75% of the ultimate load, and c) after additional loading that caused extensive delamination. The gage region that has become damaged can be easily distinguished in b) and the delamination in c) as well; however, it is

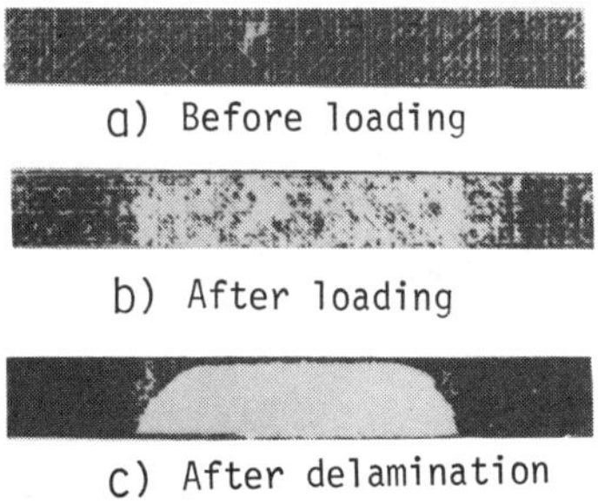

Fig. 1. Ultrasonic C-scan.

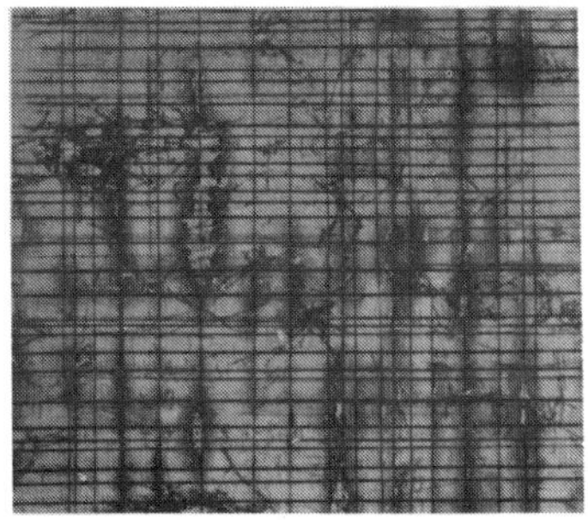

Fig. 2. Penetrant enhanced radiograph.

not possible to distinguish by this form of ultrasonic inspection the nature of the imperfection.

X-ray radiography. In general, the difference in X-ray absorption caused by regions of imperfection resulting from manufacture or damage are insufficient to allow for detection without the use of X-ray opaque penetrants. Figure 2 shows a portion of an X-ray radiography of a graphite epoxy laminate that has been cyclically loaded. Easily discernable are matrix cracks and delaminations; utilization of stereoscopic radiography allows for accurate location of this damage through the thickness (Reifsnider and Jamison, 1982).

Edge replication. The application of the surface replication to the study of damage development was first performed in the Virginia Tech Materials Response Laboratory. Figure 3 is a photograph of such a replica that was formed by pressing acetone softened acetate tape against the surface of the composite specimen, allowing it to harden in place, and then peeling it away for examination. Replication provides a simple means of following the development of damage during loading.

Acoustic emission. Monitoring of the sounds emitted during the deformation of composite materials possesses the potential for detecting the occurrence and possibly quantifying damage. Figure 4 displays the

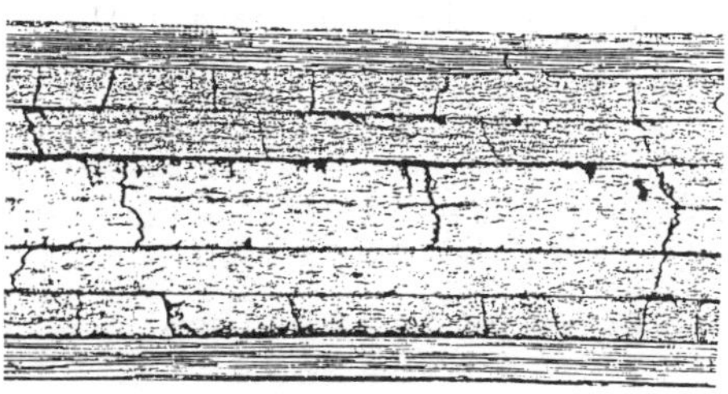

Fig. 3. Edge replica

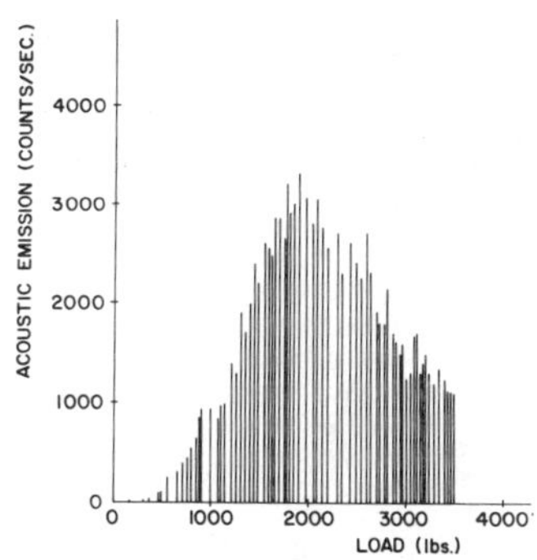

Fig. 4. Acoustic emission (1 lb=4.45 N)

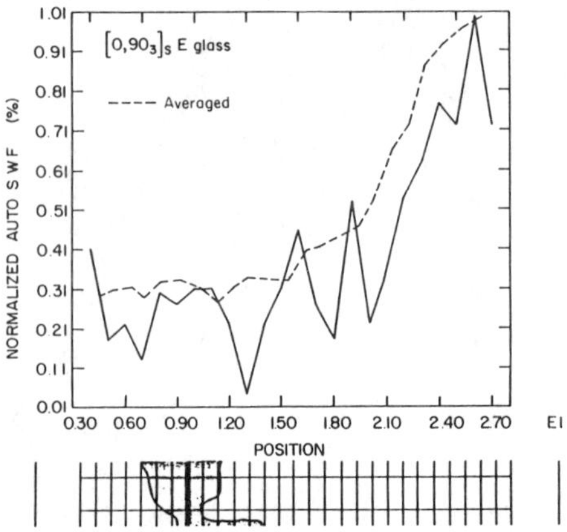

Fig. 5. SWF versus position

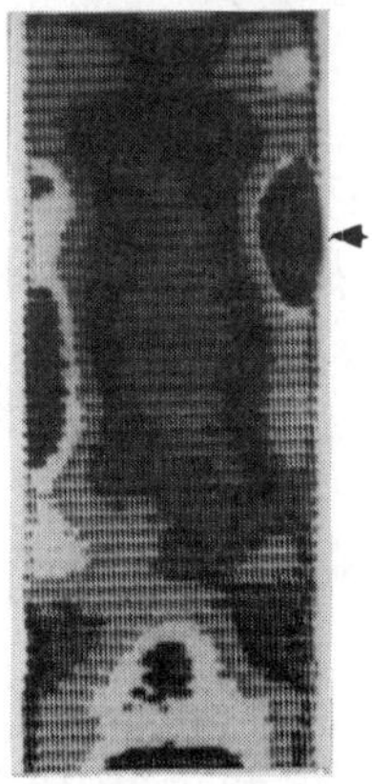

Fig. 6. Vibrothermographic pattern

nature of the acoustic emission activity monitored during the quasi-static loading of a boron fiber reinforced titanium specimen transverse to the fibers (Duke and co-workers, 1982). The ability to continuously monitor large structures makes this technique especially attractive for detecting the initiation of damage in a structure.

Ultrasonic stress wave factor (SWF). By introducing an ultrasonic pulse into the material and detecting it from the same surface as introduced at a different point, it is possible to interrogate the mechanical properties in a direction parallel to this surface. The major advantage of this technique is that it examines the properties in the directions in which loads are normally applied. It is possible to quantify this examination by means of a stress wave factor obtained by counting the oscillations of the detected signal above a prescribed threshold level for a specific time interval. The larger this factor the better the transmission of the ultrasonic energy. By scanning an object it is possible to determine the region of lowest SWF, which would be suggested by the technique as the weakest region. The validity of this suggestion is demonstrated in Fig. 5 which shows the SWF versus position along the length of the specimen; below the graph is a schematic of the specimen where the shaded region indicates the area of failure resulting from quasi-static loading (Henneke and co-workers, 1983).

Thermography. In many circumstances the thermal properties of fiber reinforced laminated composites are inhomogeneous. Passive heating will lead to variations in temperature from place to place; often the variation is not directly associated with mechanical performance. However, by high frequency low amplitude vibration of composite material, heating occurs that is directly related to the material's mechanical performance. It is possible to observe such heating in real time using a thermographic camera. This unique combination of mechanical vibration and thermal monitoring, vibrothermography, possesses great potential for directly assessing the criticality of the condition of composite materials. Figure 6 is an example of the heating observed in a damaged composite specimen; the dark area is the highest temperature and is located at a delaminated region. The technique was originated at Virginia Tech and is under development by Reifsnider, Henneke, and Stinchcomb (1980).

DETERMINATION OF MECHANICAL RESPONSE

Damage Mechanics

As defined here, damage is any change in the material condition which decreases the stiffness, strength, or life. Further it is important to note that the "amount" of damage which occurs before failure is dependent on the mechanical loading condition. Consequently any attempt to quantify damage must reflect this peculiarity. The philosophy of the Materials Response Group regarding the formulation of a damage mechanics theory for fiber reinforced composite materials recognizes that damage is a natural consequence of the mechanical loading of a composite material and as such must be incorporated into any analysis procedure. In addition, it is important to emphasize that this, in the context of the definition of damage, seeks to differentiate between changes in material condition which affect stiffness, strength and life, and those which do not.

Mechanical testing. At present it is not possible to reliably predict the performance of a structure made of composite material based only on mechanical properties of quasi-static strength, fatigue life and so on. Any predictive methodology for these materials must reflect the nature of the damage mechanisms that control the performance. The approach of the Materials Response Group has been to examine before, during, and after mechanical loading, the condition of the material. The previously discussed nondestructive techniques have been essential to determine the nature of the damage. But of equal importance is the determination of the response of the damaged material in the presence of mechanical loading. Monitoring of the laminate stiffness has been deliberately used to quantify the influence of the damage. As an example, Fig. 7 displays the variation of secant modulus (which mimics the stiffness) with increasing cyclic load. The figure also displays the

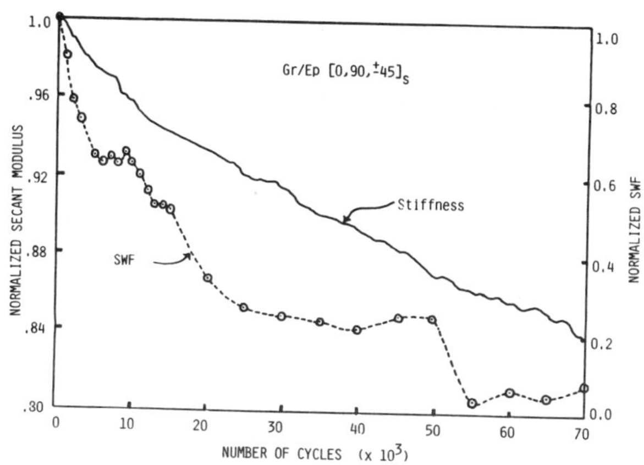

Fig. 7. Variation of secant modulus (and SWF) with cyclic loading.

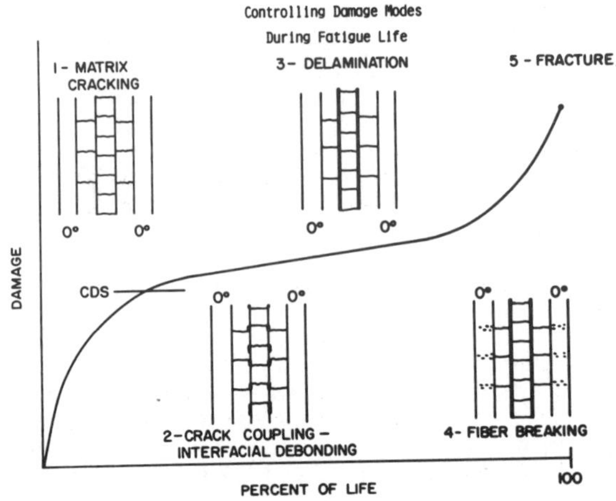

Fig. 8. Summary of damage accumulation.

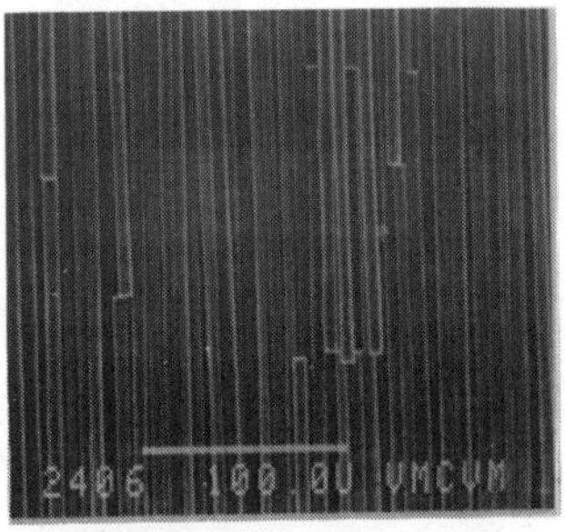

Fig. 9. SEM of fiber breaks.

sensitivity of the SWF technique to the damage development as well. In addition to determine the local influence of damage, full field displacement by means of moiré interferometry has been utilized (Jamison and co-workers, 1981). Such an approach is necessitated by the extensive damage that interacts under load. A major emphasis of the mechanical testing activity has been and is being directed at establishing experimentally the variation of laminate strength and stiffness with applied load history.

Damage documentation. Through a combination of nondestructive and destructive analysis techniques, it has been possible to carefully document the damage at various stages during the life of the composite materials being studied. Figure 8 summarizes schematically these findings. Aside from the nondestructive techniques described above, extensive use has been made of sectioning and deplying (a procedure involving pyrolyzing the matrix material) of the composite specimen and examination by scanning electron microscopy. Figure 9 shows a scanning electron micrograph of a section of a deplied 0° ply of a graphite epoxy laminate that was neighboring a transverse matrix crack; the fiber breaks are clearly visible (Jamison and co-workers, 1982).

Cumulative damage model. Ultimately, to design, manufacture, and maintain composite material structures, it is necessary to account for the effects of complex loading spectrum. As regards composite materials, any model must clearly incorporate the interaction of the various types of damage mechanisms that occur. Recently, such a modeling procedure has been conceived that rationalizes the phenomenological aspects of the damage development of fiber reinforced laminated composites (Reifsnider and Stinchcomb, 1983). The generalized form of the damage accumulation is:

$$(1-F^r(n)) = \int_0^x (1-F^L(n))\, f(n/N)\, d(n/N) \qquad 0<x\leq 1$$

$F^r(n)$ -- residual strength $f(n/N)$ -- degradation function
$N(n)$ -- life locus $F^L(n)$ -- localized stress intensity

Both $F^r(n)$ and $F^L(n)$ are normalized and life is obtained by integrating this expression over the range 0 to 1; evaluation at intermediate values provides the residual strength. It is important to note that the expression describes the behavior of the critical elements, those elements that control the final fracture of the laminate. Damage occurring in noncritical elements influences this expression because it gives rise to local stress redistribution that influences the value of N(n) and thereby f(n/N). Phenomenologically the stiffness as a function of damage must be determined and used as an input into various models for calculating $F^L(n)$. Additionally the life locus N(n) of the critical element under arbitrary load must be a phenomenological representation of experimentally determined behavior.

For evaluating the strength or life for imperfect conditions of the material when the mechanical history is unknown, the condition of the critical element is used to establish the new starting point, that is, the present strength. In such circumstances the ability to nondestructively evaluate the material condition is essential.

SUMMARY

In order to nondestructively evaluate the mechanical performance of a composite material it is necessary to nondestructively determine the material condition particularly of the critical element and to assess the influence of the expected service conditions as regards the degradation of the strength of the material; the end of life occurs when the time at which the applied load and the condition of the noncritical elements causes the stress acting on the critical element to equal its degraded strength.

ACKNOWLEDGEMENTS

Appreciation is expressed for their sponsorship: U.S. Army, Navy, Air Force, and NASA.

REFERENCES

Duke, J. C., A. Govada, and A. Lemascon (1982). NDE of boron carbide coated boron fiber reinforced titanium. Proc. TMS-AIME Symp. on Failure Modes in Composites, Dallas, Texas.

Henneke, E. G., J. C. Duke, W. W. Stinchcomb, A. Govada, and A. Lemascon (1983). NASA Contractor Report 3670, Grant NSG-3-172.

Jamison, R. D., A. L. Highsmith, and K. L. Reifsnider (1981). Comp. Tech. Rev., 3, 158-159.

Jamison, R. D., K. Schulte, K. L. Reifsnider, and W. W. Stinchcomb (1982). Characterization and analysis of damage mechanisms in fatigue of graphite/epoxy laminates. Proc. of ASTM Symp. on the Effects of Defects in Composites, San Francisco, California.

Reifsnider, K. L., E. G. Henneke, and W. W. Stinchcomb (1980). In W. W. Stinchcomb (Ed.), Chap. III-1, Mechanics of Nondestructive Testing, Plenum Press, New York, pp. 249-276.

Reifsnider, K. L. and R. D. Jamison (1982). Int'l. J. Fatigue, 4, 187-198.

Reifsnider, K. L. and W. W. Stinchcomb (1983). Semi-annual report of work performed under subcontract to General Dynamics for the Air Force Materials Laboratory, No. F 33615-81-C-5049.

CREEP BEHAVIOUR OF REINFORCED AND NON-REINFORCED RESINS

M. W. Darlington* and A. T. Marques**

***School of Industrial Science - C.I.T., Cranfield, MK43 0AL, UK**
****Department of Mechanical Engineering, F.E.U.P., Rua dos Bragas, 4099 Porto Codex, Portugal**

ABSTRACT

The creep behaviour of two types of glass fibre reinforced polyester (GRP) laminates was studied over a range of temperatures. A series of four-point bending creep tests were made to compare the behaviour of laminates in water and in air at 23°C, 40°C and 60°C. Creep test durations of up to two months were used. It was found that the presence of water slightly increases the time-dependence of GRP samples within this time scale.

Tensile creep data, at 23°C, of the above-mentioned GRP laminates were obtained. The presence of water had a smaller effect on the tensile creep behaviour of GRP specimens than that observed in flexure.

Tensile creep tests were carried out on specimens made from resin only, in air and immersed in water, at a temperature of 23°C. A simple analysis, based on the 'rule of mixtures' prediction of composite stiffness, was made.

KEYWORDS

glass fibre reinforced polyester (GRP); creep; composite stiffness; prediction

INTRODUCTION

Glass fibre reinforced polyester laminates (GRP) are often used in load-bearing applications under conditions which approximate long-term static loading. Many of these applications also involve contact with a liquid environment whilst the material is being stressed. Despite this, relatively little published creep data on GRP in air can be found. Furthermore, virtually no experiments appear to have been carried out to determine whether or not the creep behaviour of GRP is affected by immersion in liquid environments whilst the sample is under load.

The purpose of this paper is to examine the effect of water on creep behaviour of two types of glass fibre polyester laminates over a range of temperatures, the water being in contact with one side of the specimen only.

EXPERIMENTAL PROCEDURE

Materials

The materials used in the present study are as follows:

TABLE 1 Characteristics of Laminates

Board*	Layers	Resin Crystic 272E/Glass	Cure	Post-cure	Thickness (mm)
I	$7\times300g/m^2$ cfm**	2.1 to 1	24 h at 23°C	16h at 40°C	4.874 +0.54 −0.93
II	$3\times300g/m^2$ cfm $2\times600g/m^2$ wr***	1.3 to 1	"	"	3.286 +0.29 −0.32
III	$7\times300g/m^2$ cfm	2.1 to 1	"	3h at 80°C	4.958 +0.47 −0.94
IV	$3\times300g/m^2$ cfm $2\times600g/m^2$ wr	1.3 to 1	"	"	3.611 +0.25 −0.82

* Made by vacuum-assisted resin injection
** cfm - continuous filament mat
*** wr - woven-roving

Additional information such as fibre, resin and void contents were determined as reported by Marques (1981, 1982). The resin specimens were cut from 3 mm-thick cast resin sheets.

Test Programme

Long-term flexure tests were carried out with six especially-designed, accurately-built and fully-proven four-point bending creep rigs as described by Marques (1981).

Strips, approximately 25 mm wide and 300 mm long, were cut from the boards mentioned in Table 1. A very small tank had to be built within the pure bending region of the specimen in order to place the water in contact with the upper surface of said specimen.

A serie of trials were made to check that there was no leakage at the edges of the specimen and that the use of the tank did not affect the creep behaviour in air of the specimens tested. The characteristics of the tests were as follows:

TABLE 2 Characteristics of GRP Tests

Board	(°C) Temperature	Applied Stress (MN/m^2)	100-sec Strain (%)	Environment	Duration
I	23	28.7	0.4	Tank with + w/o water	Up to 2 months
	40	*23.2	0.4	"	"
II**	23	25.6 +0.4 −0.1	0.3	"	"

	40	*21.5±2.5	0.3	"	"
III	23	33.52	0.4	"	"
	60	*20.0	0.4	"	"
IV**	23	26.0 +1.9 −1.0	0.3	"	"
	60	*19.26±0.06	0.3	"	"

* Long-term zero-load monitor tests were also carried out for correction purposes
** Different stresses were used to match a 100-second strain.

Long-term tensile creep tests were made with two specially-designed, carefully-built and fully-proven heavy-duty tensile creep rigs as described by Marques (1981). Only specimens with a post-cure of 16h at 40°C were tested as this post-cure results in a poorer flexure creep response and is more often used in the GRP industry.

The specimens were 400 mm long and 25 mm wide (38 mm in the gripping area). Tests were carried out at 23°C, with and without water being in contact with one face only. Applied stresses were 23.2 MN/m^2 and 31.6±1 MN/m^2 for Board I and II specimens respectively. As was the case for the flexure tests, a small tank had to be built in order for the water to be in contact with the upper surface of the specimen. It must be noted that the tank was attached to each pair of specimens tested (with and without water), in order that any small increase in stiffness due to its presence (previously checked by 100-second creep tests at low strains, at less than 1%) would occur in both.

Tensile creep tests were also carried out with resin crystic 272 E specimens. The post-cures previously mentioned were applied to different sets of specimens. The tests in air were performed in rigs described by Darlington and Saunders (1970). The tensile creep tests in water were made using the environmental tensile creep rig described by Smith (1976).

TABLE 3 Characteristics of Resin Specimen Tests

Post-cure	Temperature (°C)	Applied Stress (MN/m2)**	Environment	Duration
16h at 40°C	23	12	Air	Up to 2 months
	23	12	*Water	"
3h at 80°C	23	12	Air	"
	23	12	*Water	"

* Long-term zero-load monitor tests were also carried out for correction purposes.
** To give 100-second strains between 0.3% and 0.4%.

RESULTS AND DISCUSSION

The comparison between the flexure creep behaviour in air and water at 23°C, of Board I and III specimens and of Board II and IV, is made in Figs. 1 and 2 respectively.

The effect of water on the creep behaviour at 23°C of Board III specimens seems to be greater than on Board I specimens. Thus, whereas Board III specimens show a significant improvement in creep behaviour in air at 23°C as compared to those from Board I, the same comparison of specimens in contact with water shows a smaller improvement.

It is apparent that the relative behaviour of Boards II and IV in air and water follows the pattern observed for Boards I and III. The more pronounced long-term effect of water on Board II as compared to Board I, and on Board IV as compared to Board III, could be detected and is attributed to a greater percentage of voids in Board II and IV laminates.

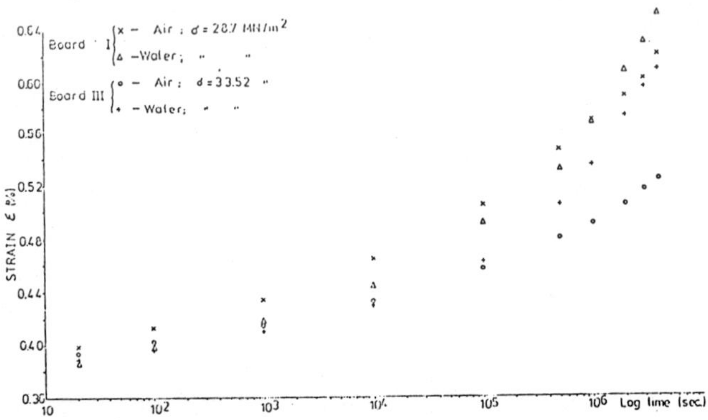

Fig. 1. Comparison of flexural creep behaviour - Board I and III. (T=23°C)

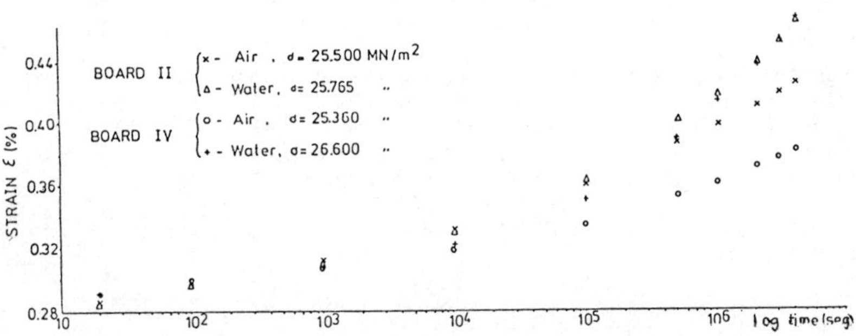

Fig. 2. Comparison of flexural creep behaviour - Board II and IV. (T=23°C)

The flexure creep behaviour in air and water, at 40°C of Board I and II specimens and at 60°C of Board III and IV specimens, is presented in Figs. 3-6 after zero-load correction (swelling strains of 0.02% to 0.04% at 2×10^6 seconds, were evident).

A comparison of the results of the tests made at 40°C with those made at room temperature, showed that at 40°C there was a slight increase in creep strains but a reduction in the effect of water on the creep behaviour of Board II specimens. A comparison of the results of the tests made at 60°C showed an appreciable increase in creep strain but no significant increase in the effect of water on the creep behaviour of Board IV specimens.

On the whole, and regarding the creep behaviour of specimens with and without water at 40°C and 60°C, the increase in temperature can be said to create a slight increase in the effect of water on Boards I and II, but none on Boards II and IV.

It is apparent that a softening of the resin takes place in specimens with water, under stress, and tested at 40°C and 60°C.

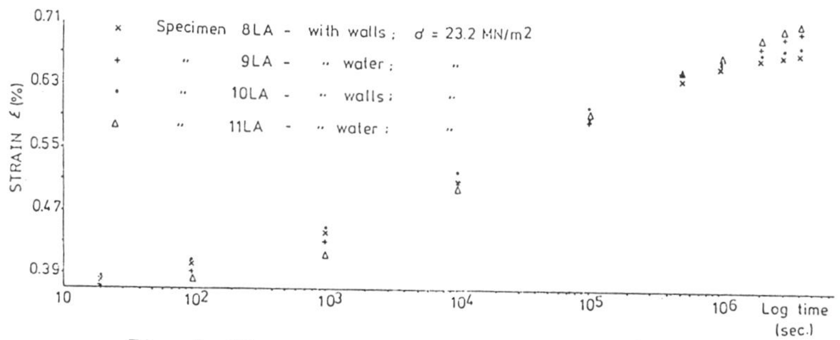

Fig. 3. Flexure creep curves - Board I (T=40°C)

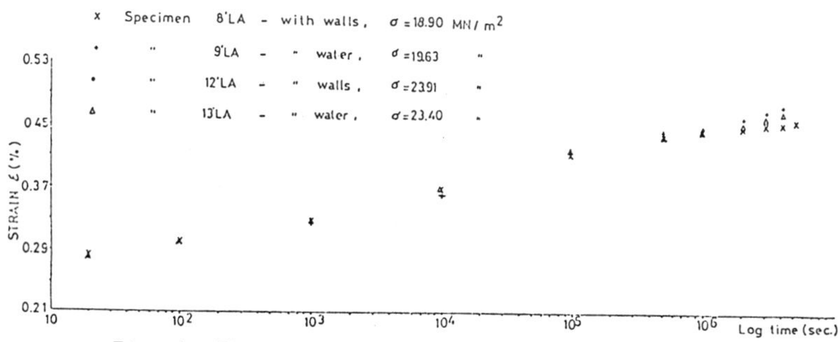

Fig. 4. Flexure creep curves - Board II (T=40°C)

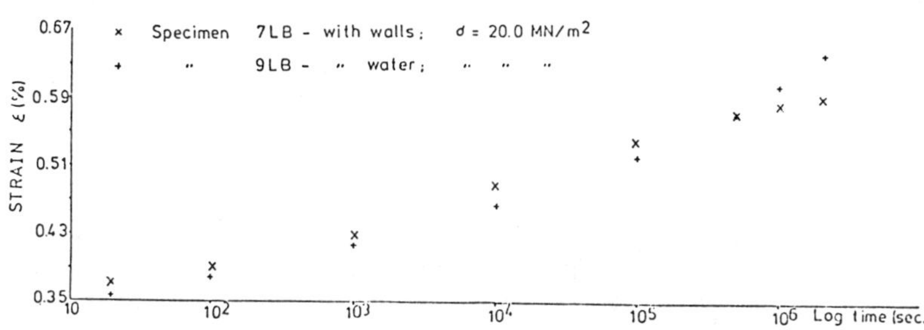

Fig. 5. Flexure creep curves - Board III (T=60°C)

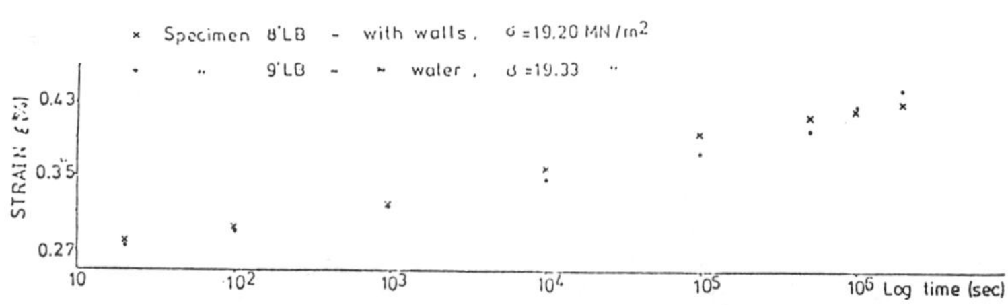

Fig. 6. Flexure creep curves - Board IV (T=60°C)

Although the results of the creep tests at 40°C and 60°C appear to have no dramatic effect on the creep behaviour of the specimens from either of the boards, visual observation of the surface of the specimens with water showed a whitening effect around the fibres. At 60°C, this whitening effect was also observed in the specimens in contact with water used for long-term zero load monitoring.

The tensile creep curves for Board I and II specimens are presented in Fig. 7. An analysis of these curves leads to the conclusion that the water had a marked effect on the creep behaviour of the specimens tested. The general creep behaviour in air, of Board II specimens was better than that of those from Board I (even allowing for the fact that different 100-second strains were used, i.e., 0.3% and 0.42% for Board II and I specimens respectively).

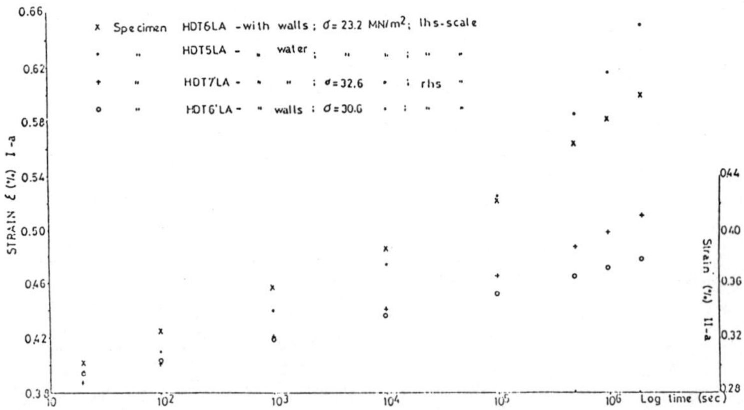

Fig. 7. Tensile creep tests - Boards I and II (T=23°C)

The analysis of the results of tensile creep tests, at 23°C, with resin samples only showed an enormous difference between the creep in water and that in air. For 16h post-cure specimens at 40°C, the specimens in water showed an increase in strain of 0.99% from 20 to 3×10^6 seconds, whereas those in air showed a just 0.29% increase in strain over the same period. The zero-load creep tests for specimens immersed in water showed swelling strains of 0.13% and 0.10% (at 2×10^6 seconds) for 16h-40°C and 3h-80°C post-cure specimens respectively. A softening of the resin is seen to take place when the water is in contact with the resin specimens while under stress.

Tensile creep tests of previously saturated samples were also carried out, the results being presented in Figs. 8 and 9.

An analysis of Fig. 8 shows that at short-creep times the saturated sample is softer than the initially dry specimens. The time-dependence of the saturated specimen is similar to that of the initially dry specimens tested in water (corrected for zero load) over much of the time scale.

Figure 9 shows that the time dependence of the saturated specimens is greater than that of the initially dry samples tested in air, even when one allows for the fact that the saturated sample starts at a higher strain. It is suggested that the higher creep strain values at longer creep times of saturated specimens creep-tested in water, as compared to the air creep curve, are due to an additional softening of the resin caused by some interaction of the applied stress and the already absorbed water. Further work is necessary to substantiate this somewhat puzzling point.

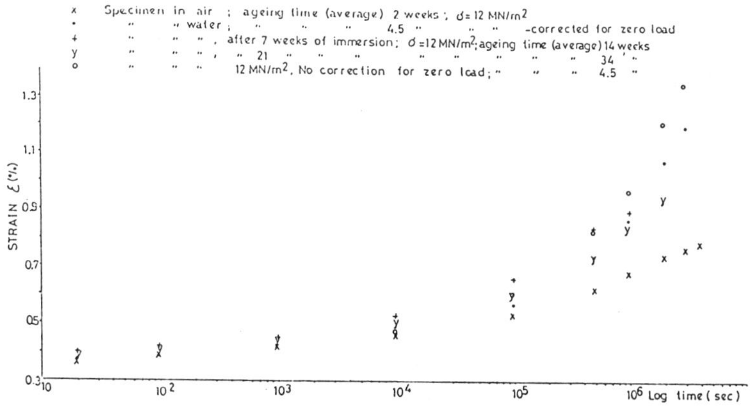

Fig. 8. Tensile creep test: -Average axial creep strain of resin specimens (T=23°C)

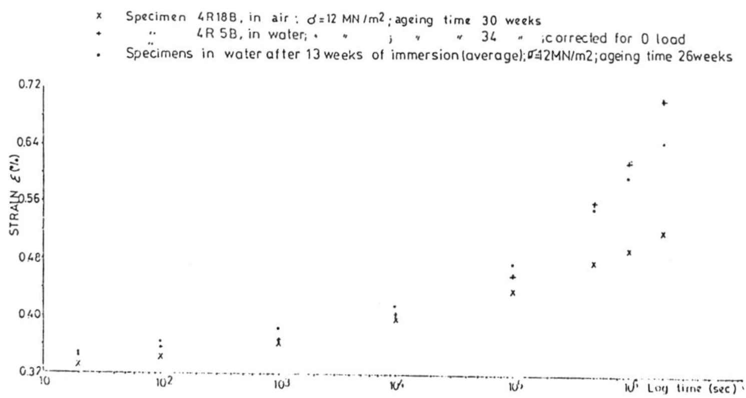

Fig. 9. Tensile creep tests: -Axial creep strains of resin specimens in air and water (T=23°C)

PREDICTION OF GRP CREEP BEHAVIOUR

In order to establish whether or not the effects of water on the creep behaviour of GRP can be entirely explained by the 'softening effect' of the water on the resin, or whether effects of glass-resin interface are dominant, as well as for prediction purposes, simple calculations were made using the 'rule of mixtures'.

For a linear viscoelastic material the 'rule of mixtures' can be written as:

$$E_c(t) = \beta(t).E_g.V_g + E_m(t).(1-V_g)$$

Where E_c, E_g and E_m are the Young's moduli of the composite, fibre and matrix respectively; V_g is the fibre volume fraction; is the fibre efficiency factor and (t) states for time-dependence. Note that the time-dependence of arises through its dependence on parameters such as the matrix shear modules.

The values of β (t) were calculated for GRP specimens in air and in contact with water (one face only). The values of β (t) for each particular time and for each of the GRP boards were estimated by inserting the average experimental values of $E_c(t)$ and $E_m(t)$ into the afore-mentioned equation; E_g was considered as constant. A regression line was used to fit the β values and to predict values of β for time over 10^6 seconds (i.e., up to 4×10^6 seconds).

The prediction of the GRP creep behaviour in air and water is presented in Figs. 10 and 11. It can be noted that a reasonably accurate prediction was obtained.

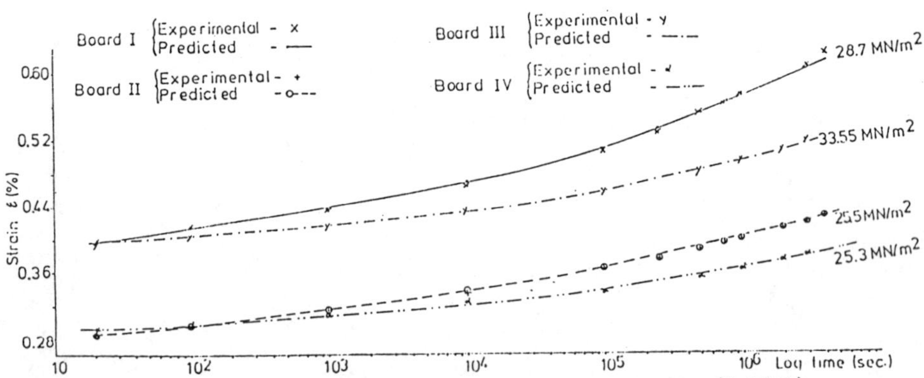

Fig. 10. Prediction of GRP creep behaviour in air (T=23°C)

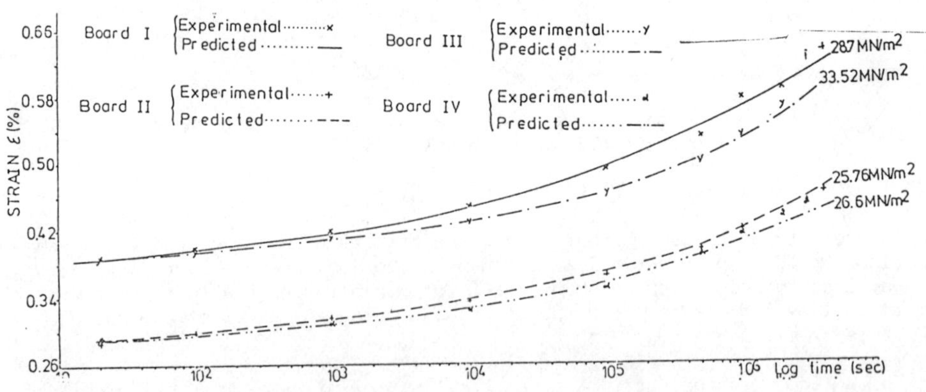

Fig. 11. Prediction of GRP creep behaviour with water (T=23°C)

Marques (1981) analysed the values of β (t) for specimens tested (Boards I, II, III, IV) in air and in contact with water. It seems that the presence of water may cause a reduction of the fibre efficiency factor. In fact, the presence of water causes swelling stresses, thus giving a possible reduction of β. However, the softening effect of water on resin may lead to a better transfer of load to the fibres as the resin becomes more ductile. The balance of these two effects will determine whether or not there is an improvement of the fibre efficiency factor of specimens tested in water as compared to those tested in air.

CONCLUSIONS

1) As a whole, the presence of water did not have a marked effect on the creep behaviour, at 23°C, of GRP specimens for creep times of up to two months.

2) At 23°C, water had a more pronounced effect on the creep behaviour of laminates with a combination of woven-roving and continuous filament mat and with a greater percentage of volume void content.

3) The temperature increase produced higher creep strains in air and water for both types of laminates and post-cures. The effect of water on the creep behaviour at temperatures higher than 23°C was only enhanced over that at 23°C in Board I and III specimens, whereas there was no increase in the effect of water for Board II and IV specimens. The reduction of inter-laminar residual stresses as temperatures increase lends a greater importance to the fibre dominant layers (woven-roving) of Boards II and IV, thus reducing the effect of water on the laminate as a whole.

4) The change from a post-cure of 16h at 40°C to that of 3h at 80°C produced a considerably better creep behaviour of GRP specimens tested in air at 23°C, although the improvement in behaviour of specimens tested in water was very slight.

5) The creep tests of resin specimens showed that water had a greater softening effect on samples post-cured for 16h at 40°C.

6) A double-role effect was attributed to the presence of water, thus: (a) a negative effect introduced by the swelling strains, the possible formation of voids or microcracks, and the lowering of the glass-transition temperature T_g of the composite as a whole, and (b) a positive effect at the fibre/matrix interface which permitted stress relaxation and contributed to ductility at the interfaces, thence a better transfer of load to the fibres.

REFERENCES

Marques, A. T. (1981). PhD. Thesis, Cranfield

Darlington, M. W., and Marques, A. T. (1982). Paper presented at the 1982 FRP Symposium, N.A.C.E., Niagara Falls

Darlington, M. W., and Saunders, D. W. (1970). J. Phys. E. Sci. Inst., 3, 6.

Smith, G. R. (1976). PhD Thesis, Cranfield.

INTERFACIAL FAILURE IN STEEL FIBER REINFORCED POLYSTYRENE IMPREGNATED MORTAR

T. Ertürk* and M. Tokyay**

*Department of Metallurgical Engineering, Middle East Technical University, Ankara, Turkey

**Department of Civil Engineering, Middle East Technical University, Ankara, Turkey

ABSTRACT

Interfacial bond strength, failure modes, and pull-out behavior in steel fiber reinforced plain and polystyrene impregnated mortar have been studied using single-fiber pull-out tests. In plain mortar cohesive failure occurred in the calcium hydroxide (portlandite) layer adjacent to the fiber/matrix interface; matrix cracking was also observed. In contrast, polystyrene impregnated mortar exhibited adhesional separation at the interface, with the matrix being intact. An adhering layer of matrix material covered the relatively rough steel wire surfaces in unimpregnated mortar specimens whereas the fiber surfaces were clean in polystyrene impregnated mortar specimens. A fourfold increase in the interfacial bond strength has been achieved by polymer impregnation. Impregnated specimens also showed a slower decay in fiber frictional shear traction during pull-out; pull-out work improved by 640 percent.

KEYWORDS

Steel fiber reinforced mortar, polymer impregnated mortar, concrete, composites, polystyrene impregnation, interfacial bond strength, interface failure, fiber pull-out, fiber debonding, frictional shear traction.

INTRODUCTION

Unreinforced cement, mortar and concrete are inherently brittle materials, exhibiting low tensile strength and low fracture toughness. They are always reinforced and often prestressed as well by large diameter steel bars in large constructions. In thin sections, panels, surface coatings, and in applications where protection against small scale fragmentation is needed, however, it is preferable to reinforce them with short, strong and stiff elastic fibers such as asbestos, glass, or steel. The goal in incorporating these fibers into cementitious materials is to improve the tensile strength, but more importantly to provide a degree of postcracking ductility for increasing the toughness. Due to the necessarily low volume fractions used (∿2-5%, for workability considerations in the fresh state), the gain in tensile strength is not dramatic, being on the order of about 50%. Neither is the increase in elastic modulus significant, being approximately 5%. On the other hand, the work of fracture, as measured for example by the area bounded by the tensile stress-strain curve, is enhanced by several orders of magnitude. Thus, the real improvement is in toughness, not in strength or modulus.

The reinforcement of cement and concrete with asbestos fibers has been practiced for many years. Due to the depletion of high quality asbestos, though, attention is now being directed toward the replacement of asbestos with man-made fibers. Currently, steel fiber is a widely used substitute material for the reinforcement of cementitious materials.

The toughness of fiber reinforced concrete could be enhanced even further if the strength of the matrix were increased considerably above that achievable with fiber reinforcement alone. A favorable postcracking ductility provided by the pull-out of properly designed fibers should impart extraordinary toughness to a cementitious composite having a strong matrix. One means to accomplish this, though expensive, is by impregnating the fiber reinforced cementitious material with monomers, followed by polymerization. A combination of polymer impregnation and steel fiber reinforcement has been the subject of the present investigation.

The behavior of fiber reinforced cementitious composite materials is dependent on properties of the fibers, the concrete matrix, and the fiber/matrix interface. Of these, the interface between the fiber and cement is of prime importance: the shape of the tensile stress-strain curve, the strength of the composite, and the ability to dissipate energy during pull-out, depend critically on the mechanism of stress transfer between the reinforcement and the matrix. Thus, an understanding of the interfacial strength and failure phenomena in steel fiber reinforced polymer impregnated concrete is essential to the understanding of the overall composite behavior.

Recent studies, as reviewed by Page (1982), on the composition, microstructure, debonding, and pull-out characteristics of interfaces in steel fiber/Portland cement matrix composite systems have led to a consensus that a weak layer of hexagonal lamellar crystals of calcium hydroxide (portlandite), extending to a depth of several tens of micrometers, is precipitated in intimate contact with the embedded surfaces. Electron microscopy, x-ray analytical examination, and microhardness traverses of the near-interface material (Al Khalaf and Page, 1979; Morris and Garrett, 1979; Pinchin and Tabor, 1978a) indicated that the plate-like portlandite crystals, exhibiting a dendritic growth morphology with the c-axis roughly perpendicular to the steel surfaces, decreased in proportion and lost their preferred orientation with increasing distance from the interface. The transitional zone away from the dense interfacial layers exhibited a higher degree of porosity than the bulk matrix.

Debonding of steel/mortar interfaces showed a dual character; both adhesional failures at the interface and cohesive failures of the portlandite ($Ca(OH)_2$) layer away from the interface were observed. By coating the surfaces of the fibers, Al Khalaf, Page and Ritchie (1980) reduced the modes of failure to a purely cohesive one, thereby achieving a certain degree of improvement in the interfacial bond strength. It is clear, however, that significant gains in bond strength can be achieved by improving both the adhesion at and cohesion near the interface. Polymer impregnation could be effective in accomplishing this; hence, a characterization of failure and pull-out behavior of steel/mortar interfaces reinforced with polymer impregnation has been the objective of the present study.

EXPERIMENTAL PROCEDURE

Single-fiber pull-out experiments have been carried out in order to determine the interfacial bond strength, failure modes, and pull-out behavior in plain and polystyrene impregnated cylindrical mortar specimens containing an axially embedded steel fiber of 1.5 mm diameter. The samples were cast vertically into molds 5 cm in diameter and 10 cm long, using ordinary Portland cement and graded natural sand with a maximum size of 2.38 mm. The mortar, consisting of 61% aggregate, 25% cement and 14% water by weight, was compacted by rodding. Hard drawn, high carbon steel wires with a 0.2% offset yield strength of 1500 MPa, were polished using 400 grade emery paper and cleansed with alcohol prior to embedding. Through preliminary experiments, the optimum pull-out distance for the impregnated condition (where the total pull-out traction is just short of fiber yield strength) was determined to be approximately 50 mm (Argon, Hawkins, and Kuo, 1979). The fibers were embedded at this distance in all specimens.

Cast specimens were cured for 28 days at 20° C and under 100% relative humidity, followed by a 3-day air cure prior to testing. A portion of the 28-day-old samples were dried at 110° C for 24 h. They were then impregnated for 6 h with polystyrene monomers, and polymerized for 40 h in an oven at 95° C, followed by air cooling. No catalyst was used. Monomer loading reached a nearly constant value of 6% after 6 h, Fig. 1, and a monomer conversion greater than 96% was achieved in all samples.

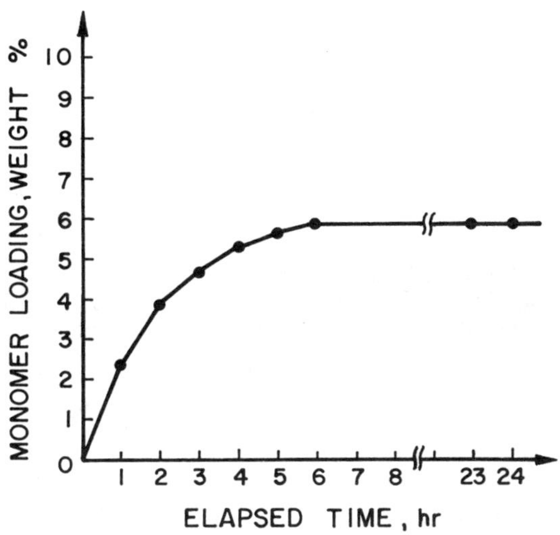

Fig. 1.

Influence of soaking time on monomer (styrene) impregnation of 50 mm in diameter, 100 mm long mortar cylinders dried at 110° C.

For pull-out tests, specimens were held freely in a cylindrical casing with the top end slotted, allowing fibers to protrude. The bottom end, having a welded connecting rod, was screwed to the casing, and attached to the testing apparatus by means of a universal joint. The protruding steel fibers were gripped as close as possible to the casing and pulled out in an Instron universal tester at a crosshead speed of 10 mm/min. Some of the tests were carried out at a velocity of 0.05 mm/min, stopped at load maximum, and unloaded. These specimens were then split diametrically into two pieces using the secondary tension technique (split tension test) in order to expose the failed interfacial regions for SEM fractographic examination. Morris and Garrett (1979) have reported that the near-surface material is not disturbed by this technique.

RESULTS

Typical fiber pull-out force—displacement curves for the plain and impregnated specimens are depicted in Fig. 2. From each group, 30 specimens were tested. The maximum pull-out force (debonding load) has been measured as 0.72 kN with a coefficient of variation of 13.2%, and 2.93 kN with a coefficient of variation of 6.7% for the plain and polystyrene impregnated mortars, respectively. From simple equilibrium considerations, the interfacial shear traction is given by

$$\tau_b = \frac{P}{\pi d \ell} \tag{1}$$

where P is the pull-out force, and d and ℓ are the diameter and embedded length of the fiber. Using Eq. 1, the maximum shear tractions (at initial debonding) for the plain and polystyrene impregnated mortars have been

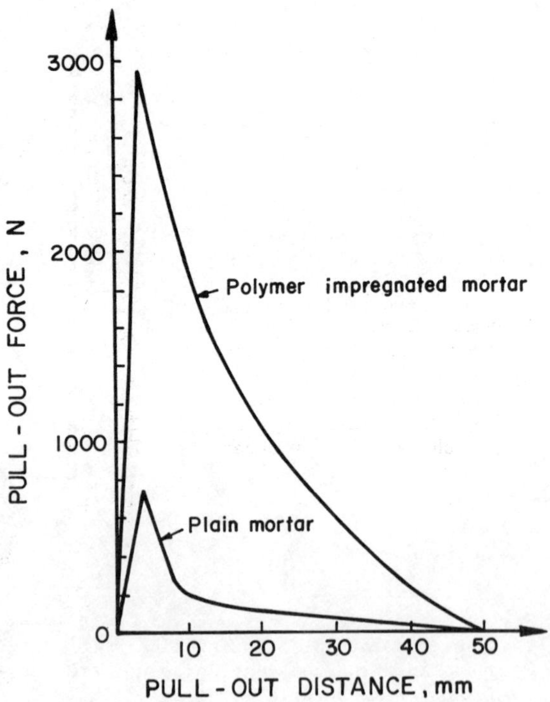

Fig. 2.

Force—displacement curve of pull-out of 1.5 mm diameter steel wire from plain and polystyrene impregnated mortar. Complete debonding occurs at load maximum.

determined as 3.1 MPa and 12.7 MPa, respectively, a fourfold increase in interfacial bond strength. Frictional work of fiber pull-out obtained from the area bounded by the force—displacement curves of Fig. 2 are 8.0 MNm and 51.5 MNm for the unimpregnated and impregnated specimens, respectively, signifying an increase of 6.4 times.

Loads necessary to split the 60 specimens were also recorded. Even though the tensile strengths obtained using these values are invalid due to the presence of a relatively large fiber in the centers of specimens, they are reported here for comparison purposes. Splitting strengths were determined to be 1.7 MPa and 7.2 MPa for the plain and impregnated specimens, respectively. This represents a fourfold increase.

From the measured pull-out force and known fiber length, Fig. 2, and using Eq. 1, the frictional shear tractions during pull-out were calculated and then plotted in Fig. 3. A rapid decay in fiber shear traction, τ_b, is seen for the plain mortar specimen; at a pull-out displacement of 12 mm frictional traction decreases by about 65% and attains a nearly constant value of 1 MPa in the plain mortar specimens, whereas the frictional traction falls off by approximately 25% at this distance in impregnated specimens. In the latter, the drop in τ_b is about 45% at the midpoint of fiber pull-out.

Fractographic examination of interfacial failure regions at initial debonding in ordinary mortar indicated that cohesive failure occurred adjacent to the

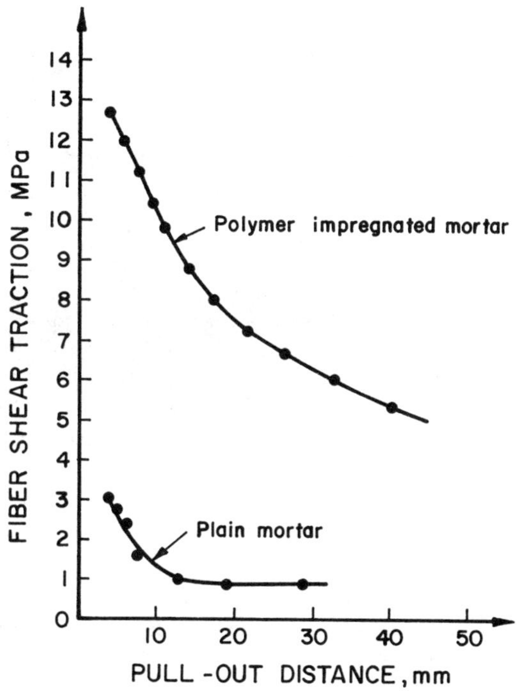

Fig. 3.

Degradation of frictional shear traction during pull-out of 1.5 mm diameter steel fiber from plain and polystyrene impregnated mortar. Plain mortar shows a faster decay.

Fig. 4. Cohesive failure near the wire/matrix interface and adherence of matrix material to the fiber (at left) in plain mortar. Nearly all of the fiber surface is covered with a thin layer of mortar.

Fig. 5. Microstructure of the near-interface material in plain mortar, showing small, porous crystals of calcium silicate and large crystals of calcium hydroxide (portlandite).

interface, Fig. 4. As can be seen from this figure, a thin layer of matrix material continued to adhere to the wire surface. Nearly complete adherence of a thin matrix layer to the exposed wire surfaces was observed in all samples. The morphology of the near-interface material consisted of small porous calcium silicate and large plate-like portlandite crystals, Fig. 5. Cracks adjacent to the interface, which were observed frequently in ordinary

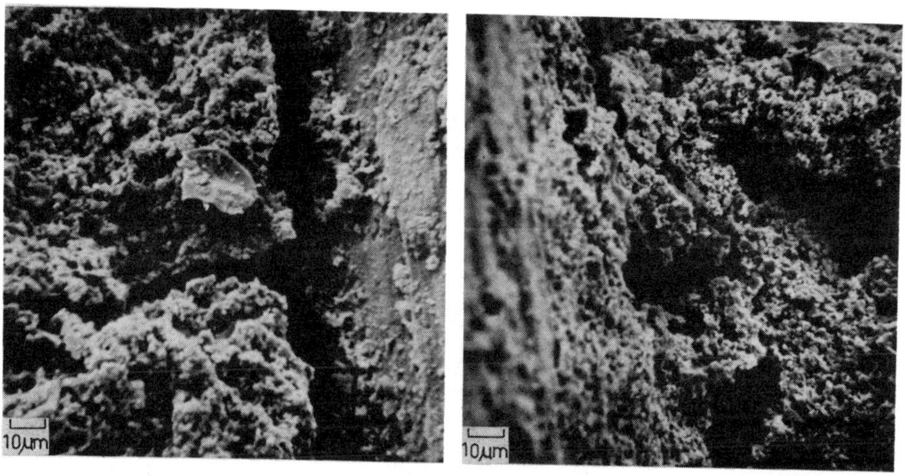

(a) (b)

Fig. 6. Matrix failure adjacent to the steel fiber in plain mortar. (a) Debonding occurred away from the interface and a thin layer remained adhering to the wire surface, (b) Matrix failure with interfacial bond still intact. Note the porous structure of the matrix.

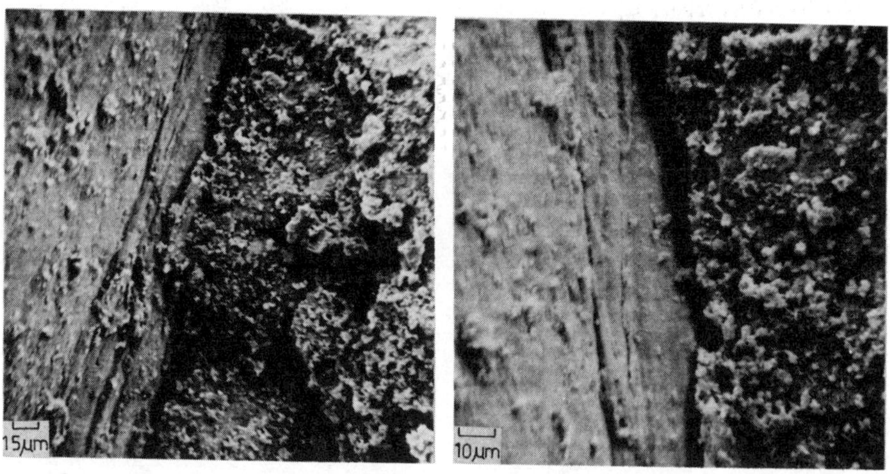

(a) (b)

Fig. 7. (a) and (b) Adhesive failure at the fiber/matrix interface in polystyrene impregnated mortar. Relatively rough fiber surfaces are clean except for some fragmented matrix particles entrapped in asperities. Matrix is denser.

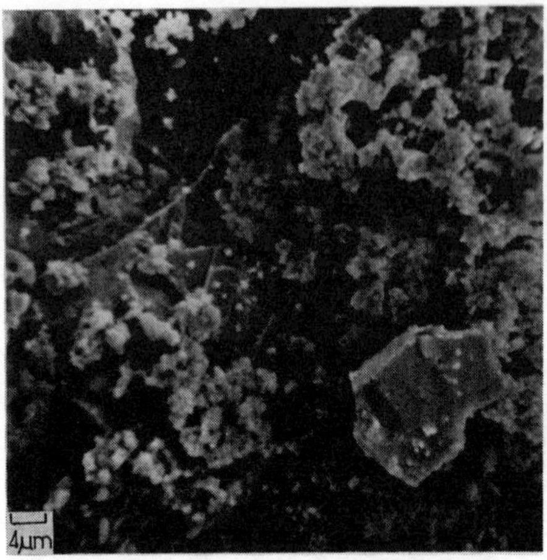

Fig. 8.

Microstructure of the matrix near the interface in polystyrene impregnated mortar, showing increased density and $Ca(OH)_2$ (portlandite) crystals.

mortar specimens, are depicted in Fig. 6. Figures 6(a) and 6(b) show matrix cracking with interface debonding and intact interface bond, respectively. A layer of matrix material adhering to the wire surface can be seen in Fig. 6(a). In addition, Fig. 6 shows that a relatively rough wire surface is obtained using the emery paper.

In the polystyrene impregnated specimens, on the other hand, adhesive failure of the interfaces was observed exclusively, shown in Fig. 7. Steel fiber surfaces were clean except for secluded particles which had fragmented away from the matrix and were entrapped in the asperities of the fiber. The near-interface material was intact; occasionally, small cracks were observed which could not be related to the initial debonding process. A dense microstructure was obtained adjacent to the interface, as seen in Fig. 7. Large holes, however, were not filled with polymer, shown in Fig. 7(a). Morphology was similar to that of unimpregnated samples (Fig. 5), except for reduced porosity, depicted in Fig. 8.

Examination of the split fracture surfaces revealed that fracture passed through the sand grains, implying a much improved interfacial bond, whereas in unimpregnated samples sand grains were circumvented by fracture.

DISCUSSION AND CONCLUSIONS

Strengthening the matrix and the bond between the fiber and matrix by polymer impregnation changed the locus of failure from decohesion adjacent to the interface to bond separation at the interface. This increased the bond strength by a factor of four.

Polystyrene impregnated mortar exhibited a slower drop in the frictional shear traction during pull-out in comparison to plain mortar, even though the surface roughness of the wires was the same in both cases. As noted by Pinchin and Tabor (1978b) and Argon, Hawkins and Kuo (1979), a fragmented layer of matrix material with roller-like particles which form upon debonding applies pressure on the matrix during the initial stages of pull-out. The contact pressure is soon lost, however, because the cement near the wire surface is compacted, leading to degradation of shear traction. This effect is more pronounced with rough surfaces because a larger volume of material adheres to the asperities. In the present case, substantial adherence of matrix material to the wire surface in plain mortar causes a rapid drop in frictional traction. On the other hand, the wire in impregnated mortar, having less adhering material to build up contact pressure, shows a slower drop in interfacial shear traction.

A combination of a fourfold increase in interfacial bond strength and slow drop in fiber shear traction led to a 640 percent improvement in fiber pull-out work. These improvements, coupled with a fourfold increase in matrix tensile strength, indicate that exceptional toughness and strength should be obtained with steel fiber reinforced polystyrene impregnated mortar.

REFERENCES

Al Khalaf, M.N., and C.L. Page (1979). Steel/mortar interfaces: microstructural features and mode of failure. *Cement and Concrete Research, 9,* 197-208.

Al Khalaf, M.N., C.L. Page, and A.G.B. Ritchie (1980). Effects of fiber surface composition on mechanical properties of steel fibre reinforced mortars. *Cement and Concrete Research, 10,* 71-77.

Argon, A.S., G.W. Hawkins, and H.Y. Kuo (1979). Reinforcement of mortar with Metglas fibers. *J. Mater. Sci., 14,* 1707-1716.

Morris, A.D., and G.G. Garrett (1979). Structural variations affecting interfacial failure in steel fibre reinforced cement. In K.J. Miller and R.F. Smith (Eds.), *Mechanical Behaviour of Materials,* Vol. 3. Pergamon Press, London. pp. 285-292.

Page, C.L. (1982). Microstructural features of interfaces in fibre cement composites. *Composites, 13,* 140-144.

Pinchin, D.J., and D. Tabor (1978a). Interfacial phenomena in steel fibre reinforced cement. I. Structure and strength of interfacial region. *Cement and Concrete Research, 8,* 15-24.

Pinchin, D.J., and D. Tabor (1978b). Interfacial phenomena in steel fibre reinforced cement. II. Pull-out behaviour of steel wires. *Cement and Concrete Research, 8,* 139-150.

THE EFFECT OF EXTREME TEMPERATURES ON THE ELASTIC PROPERTIES AND FRACTURE BEHAVIOR OF GRAPHITE/POLYMIDE COMPOSITES

D. H. Morris and R. A. Simonds

Department of Engineering Science and Mechanics,
Virginia Polytechnic Institute and State University,
Blacksburg, Virginia 24061, USA

ABSTRACT

The influence of elevated, room and cryogenic temperatures on the elastic moduli and fracture strengths of two graphite/polyimide composites was studied. The fracture behavior of notched specimens was modeled using an average stress failure criterion.

KEYWORDS

Composites; graphite/polyimide; temperature effects; notched strength

INTRODUCTION

Graphite/polyimide is a composite material being developed to extend the useful temperature range of composite materials for such applications as jet engines, supersonic cruise aircraft, and space shuttle structures. All structures, whether they are in an air or space environment, must be fastened to other structures. Since this fastening often results in holes being introduced into the structure, the notch behavior of the basic material must be understood. The primary objective of this work was to develop an understanding of the notch behavior of two graphite/polyimide composites over the temperature range that would be experienced by part of the space shuttle structure.

In order to characterize the notch behavior of the graphite/polyimide materials studied in this program, both notched and unnotched tension tests were conducted. The data reported in this paper are the results of the second phase of a two-phase program. The first phase covered tests on notched specimens that were 102 mm (4 in) wide or smaller, and has been previously reported (Garber, Morris and Everett, 1982). The second phase of the program investigated the fracture behavior of test specimens that were up to 305 mm (12 in) wide. In addition, the second phase considered graphite/polyimide composites with two matrix materials.

MATERIAL AND SPECIMENS

The material used for this study consisted of Celion 6000 graphite fibers in a matrix of PMR-15 or NR-150B2 polyimide. Specimens were cut from panels of $[\pm45]_{2s}$ and $[0/45/90/-45]_s$ laminates. Ply thicknesses in the cured laminates ranged between 0.127 mm (0.005 in) and 0.152 mm (0.006 in). Fiber volume fractions varied from 54 to 69 percent.

Unnotched tension specimens were 25.4 mm (1 in) wide by 406 mm (16 in) long. Fracture specimens were 406 mm (16 in) long with widths ranging from 51 mm (2 in) to 305 mm (12 in). Fracture specimens had center circular holes that ranged between 1.6 mm (1/16 in) and 76.2 mm (3 in) in diameter.

TEST PROCEDURES

Tension and fracture tests were performed at 316, 24, and -157°C (600, 75, -250°F). An environmental chamber, which was small enough to permit the specimens to be gripped outside the chamber, was used to control the test temperature. The maximum variation in temperature was ±5°C (±9°F) at the extreme temperatures. Heat was provided by electrical resistance elements, and cooling was provided by liquid nitrogen.

All tests were performed at a constant cross-head speed. For tension tests the cross-head speed was 0.008 mm/s (0.002 in/min), and for fracture tests it was 0.002 mm/s (0.005 in/min). For all experiments the test section within the environmental chamber was 203 mm (8 in). It took approximately 1 h to reach 316°C (600°F), and 15 min to reach -157°C (-250°F). Specimens were allowed to soak at temperature for 10 to 15 min before being tested. It was assumed that mechanical properties were not affected by the total time in the environmental chamber (time to reach temperature plus soak time plus testing time). This assumption was based on previous results (Shyprykevich and Wolter, 1982) where it was found that the failure stress of a $[0/\pm45°/90]_s$ laminate of AS/3501-5A graphite/epoxy was independent of soak time for the times encountered in the work reported herein. Similar results were reported (Rummler and Clark, 1979) for $[0]_6$ and $[0/\pm45°]_s$ laminates of HTS/710 graphite/polyimide.

Tension modulus values were determined from strain gage measurements. The accuracy and repeatability of the gages for determining elastic modulus has been established (Chapman, 1979). It was generally not possible to measure ultimate strains due to gage failures. Strains were not measured for the fracture tests.

RESULTS AND DISCUSSION

Unnotched Tension Tests

The results of the unnotched tension tests are shown in Figs. 1 and 2. Three replicate tests were performed at each temperature for each laminate. The results shown in the figures are average values of the replicate tests. It is seen from Figs. 1 and 2 that both axial tensile modulus and tensile strength are nearly independent of temperature for the fiber-dominated laminate, except for the strength of the laminate with the NR-150B2 matrix.

The situation is quite different for the matrix-dominated laminate. Figures 1 and 2 reveal that both modulus and strength decrease with increasing

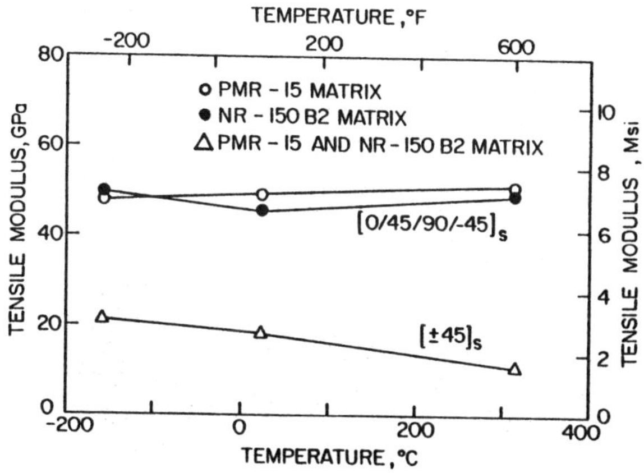

Fig. 1. Axial tensile modulus versus temperature for fiber and matrix-dominated laminates.

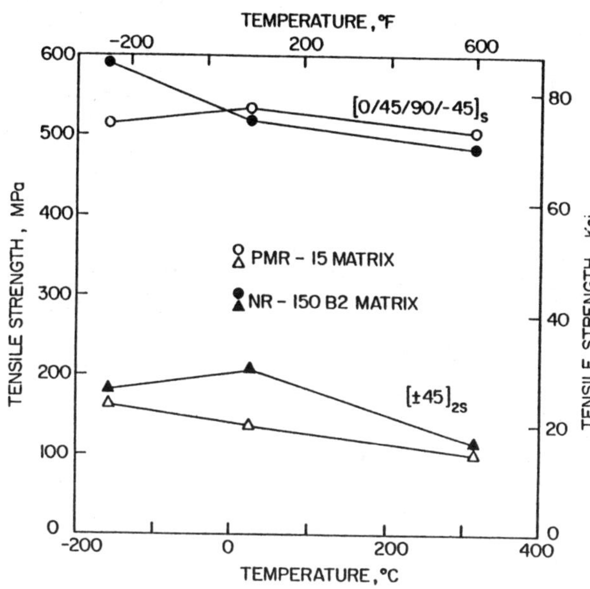

Fig. 2. Tensile strength versus temperature for fiber and matrix-dominated laminates.

temperature, except for the strength of the laminate with the NR-150B2 matrix. These figures also show that the modulus (initial slope of a stress-strain curve) is practically independent of the matrix material, but the strength appears to be more dependent upon the type of matrix.

Notched Tension Tests

The effect of temperature on the notched strength of the $[0/45/90/-45]_s$ quasi-isotropic laminate is shown in Fig. 3. For the laminate with the PMR-15 matrix, and for a given hole size, the effect of temperature is practically insignificant. A similar trend is noted for the unnotched laminate, Fig. 2.

On the other hand, the notched strength of the quasi-isotropic laminate with the NR-150B2 matrix is practically constant between room and elevated temperature, but the strength increases at cryogenic temperature. A similar trend is noted from Fig. 2 for the strength of the unnotched laminate. Regardless of the matrix material, Fig. 3 reveals that the effect of hole size is significant.

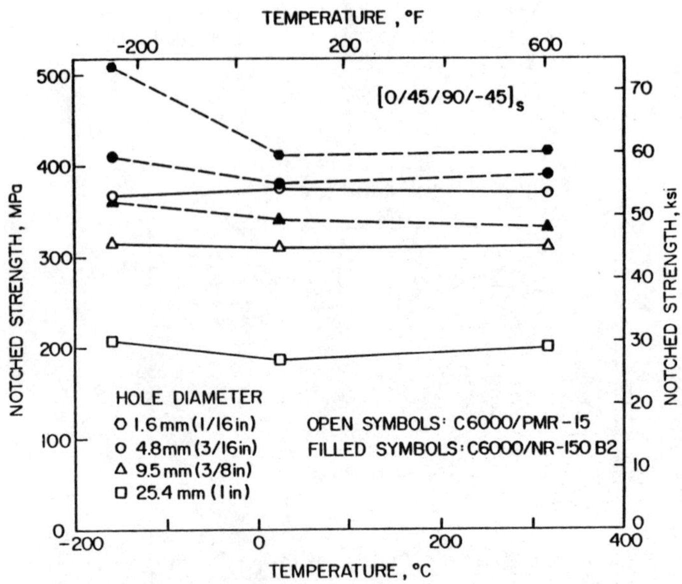

Fig. 3. Notched strength versus temperature for the $[0/45/90/-45]_s$ laminates, width = 63.5 mm (2.5 in).

Failure Model

A failure model was selected that would explain the effect of hole size and also predict the notched strength. The model selected was the average stress failure criterion (Whitney and Nuismer, 1974; Nuismer and Whitney, 1975). This model states that failure of a notched specimen will occur when the average stress over some distance ahead of the notch, a_0, equals the unnotched tensile strength. For a quasi-isotropic laminate this failure criterion may be written

$$\frac{\sigma_N^\infty}{\sigma_0} = \frac{2(1 - \xi)}{2 - \xi^2 - \xi^4} \tag{1}$$

where $\xi = R/(R + a_0)$, σ_N^∞ is the failure stress for a notched infinite width plate, σ_0 is the unnotched tensile strength, and R is hole radius. The infinite width plate notched strength is found by multiplying the experimentally determined notched strength for a finite width plate by a finite width correction factor (Nuismer and Whitney, 1975). The distance a_0 is called a characteristic length.

Using Eqn. (1), the unnotched strength, the notched strength, and the finite width correction factor, a value of a_0 = 5.1 mm (0.20 in) was found to model the fracture behavior of the C6000/PMR-15 quasi-isotropic laminate. As can be seen from Fig. 4, one value of a_0 models the data at the three temperatures.

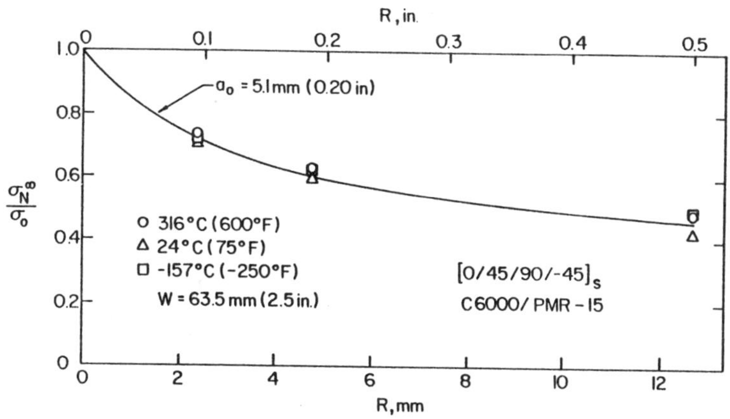

Fig. 4. Comparison of experimental and predicted strengths for the C6000/PMR-15 laminate.

Using Eqn. (1) with a_0 = 5.1 mm (0.20 in) failure stresses were predicted for plates that had widths (W) of 102 mm (4 in), 203 mm (8 in) and 305 mm (12 in). The plates were tested at temperatures (T) of 316°C (600°F) and 24°C (75°F). The results are shown in Table 1, where the predicted values are compared with experimental results. Good agreement was found.

A comparison between notched strengths for the C6000/PMR-15 and C6000/NR-150B2 quasi-isotropic laminates is shown in Fig. 5. Some of the specimens with the NR-150B2 matrix failed at the grips of the testing machine, while others failed between the hole and grips. Thus, the data for this laminate should be viewed with caution. More data are needed to establish a definite trend, and to determine if the average stress failure criterion will model the notched behavior.

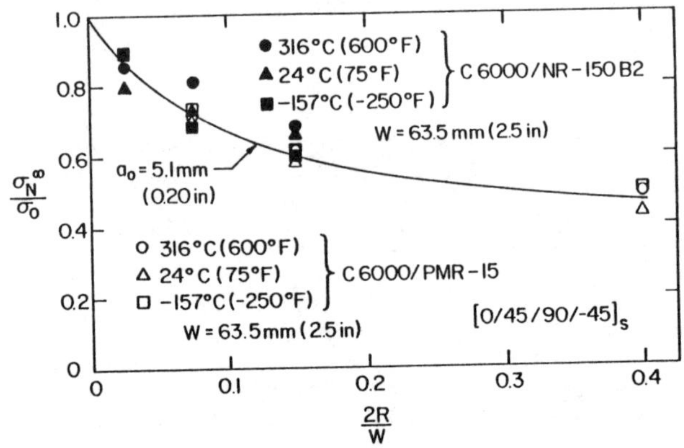

Fig. 5. Comparison of experimental and predicted strengths for the C6000/PMR-15 and C6000/NR-150B2 laminates.

TABLE 1 Comparison of Failure Stresses

T °C (°F)	W mm(in)	R mm(in)	σ_N(Eqn.1) MPa(Ksi)	σ_N(Exp.) MPa(Ksi)	Difference %
316(600)	102 (4)	12.7(0.50)	180(26.2)	199(28.8)	9.0
		4.8(0.19)	250(36.3)	281(40.7)	10.8
		2.4(0.09)	301(43.7)	297(43.1)	-1.4
24(75)	102 (4)	12.7(0.50)	159(23.1)	175(25.4)	9.1
		4.8(0.19)	220(32.0)	234(34.0)	5.9
		2.4(0.09)	265(38.5)	274(39.7)	3.0
	203 (8)	12.7(0.05)	244(35.4)	254(36.8)	3.8
		25.4(1.00)	201(29.2)	186(27.0)	-8.1
	305(12)	38.1(1.50)	190(27.5)	186(27.0)	-1.9

SUMMARY AND CONCLUSIONS

It has been shown that elastic modulus and unnotched tensile strength for the fiber-dominated laminate are not highly temperature dependent. On the other hand, the elastic moduli and tensile strengths of the matrix-dominated laminates are quite sensitive to changes in temperature.

Notched tensile strength is dependent on hole size, but is practically independent of temperature for the C6000/PMR-15 laminate. The notched strength of the C6000/NR-150B2 laminate is also dependent on hole size, and appears to be temperature dependent. However, some of the specimens failed away from the hole, and the data should be viewed with caution.
The average stress failure criterion gives reasonable predictions of notched strength for the C6000/PMR-15 quasi-isotropic laminate. Due to un-

certainty in some of the data for the C6000/NR-150B2 laminate, no attempt was made to model the data. However, the data tend to indicate that the characteristic length, a_0, may be temperature dependent.

ACKNOWLEDGEMENT

The authors would like to acknowledge the support of the NASA Langley Research Center through NASA Grant NSG 1571. Sincere appreciation is extended to R. A. Everett, Jr. of NASA Langley for his encouragement and helpful discussions.

REFERENCES

Chapman, A. J. (1979). Graphite/polyimide tension tests at elevated and cryogenic temperatures. In H. B. Dexter and J. G. Davis, Jr. (Eds.), Graphite/Polyimide Composites, NASA Conference Publication 2079. National Aeronautics and Space Administration, Hampton, Virginia. pp. 175-187.

Garber, D. P., D. H. Morris, and R. A. Everett, Jr. (1982). Elastic properties and fracture behavior of graphite/polyimide composites at extreme temperatures. In N. R. Adsit (Ed.), Composites for Extreme Environments, ASTM STP 768. pp. 73-91.

Nuismer, R. J., and J. M. Whitney (1975). Uniaxial failure of composite laminates containing stress concentrations. In G. P. Sendeckyj (Ed.), Fracture Mechanics of Composites, ASTM STP 593. pp. 117-142.

Rummler, D. R., and R. K. Clark (1979). Mechanical and thermophysical properties of graphite/polyimide composite materials. In H. B. Dexter and J. G. Davis, Jr. (Eds.), Graphite/Polyimide Composites, NASA Conference Publication 2079. National Aeronautics and Space Administration, Hampton, Virginia. pp. 289-301.

Shyprykevich, P., and W. Wolter (1982). Effects of extreme aircraft storage and flight environments on graphite/epoxy. In N. R. Adsit (Ed.), Composites for Extreme Environments, ASTM STP 768. pp. 118-134.

Whitney, J. M., and R. J. Nuismer (1974). Stress fracture criteria for laminated composites containing stress concentrations. J. Comp. Matls., 8, 253-265.

DEFORMATION AND FAILURE OF THIN WALLED FILAMENT WOUND GRP TUBES SUBJECTED TO TRANSVERSE IDENTIFICATION

M. G. Bader*, R. P. Harrison and M. J. Hinton*****

*Department of Metallurgy & Materials Technology,
University of Surrey, Guildford, UK
**Fulmer Research Institute, Stoke Poges, Slough, UK
***Royal Armament Research and Development Establishment, Fort Halstead,
Sevenoaks, Kent, UK

ABSTRACT

Thin walled tubes 100mm diameter x 1mm thickness manufactured from E-glass fibres and an epoxy resin by filament winding, at helix angles between ±45 and ±85°, have been subjected to transverse indentation by a 25mm radius indenter. The initial tube deformation is elastic but non-linear. Failure occurs by a complex interaction of localised matrix cracking, fibre/matrix debonding and an instability which results in the formation of a crease in the wall some 40° away from the centre of indentation. The deformation has been monitored by strain gauges applied to both the outside and inside of the tube. Deformation maps showing the strain distribution over the tube surfaces are presented.

INTRODUCTION

Filament wound fibre reinforced plastics tubes have a number of potential applications for which the prime design requirements are pressure containment together with adequate longitudinal stiffness. These requirements may be met by helical winding at angles in the range of 45° to 75° to the tube axis. Thin walled tubes of about 100mm diameter and 1mm wall thickness have been demonstrated to offer significant weight savings over steel tubes designed for similar performance. A secondary requirement for such tubes is that they should be sufficiently robust to resist accidental damage caused by rough handling and minor impacts. The work described has been concerned with establishing the behaviour of thin wall, filament wound, glass-fibre/epoxy resin tubes subjected to radial indentation by a blunt indenter.

Initially the work was concerned with the effects of low velocity (up to 10m/s) impact damage but it has been established that the deformation and damage caused by slow indentation is identical to that produced by low velocity, drop-weight impact testing. Consequently in our current studies we have used a fully instrumented slow indentation technique to monitor the deformation and damage produced by impact.

There are a number of features of the mode of deformation and failure of a radially indented thin walled FRP tube which make it difficult to model by conventional analytical techniques. The wall of the tube is essentially a thin $\pm\theta$ laminate. The tubes tested were of "two cover" construction, so that the laminate sequence at any point was either $[+\theta, -\theta, +\theta, -\theta]$ or $[-\theta, +\theta, -\theta, +\theta]$ according to the winding programme. In fact the two sequences alternate on a diamond grid at intervals of approximately 25mm, which is the filament tape width. Thus the laminate is locally unbalanced but is effectively balanced in terms of the overall skin. The tube wall is, of course, highly anisotropic according to the winding angle used (see Table 1).

Fig. 1 ±65° tube indented 10mm

Fig. 2 ±65° tube indented 30mm showing failure mode

The tube wall is very flexible and large indentations are needed to cause local failure. A ±65° tube, for instance, must be indented 25-30mm by a 25mm radius indenter. Most of this deformation is "elastic" in the sense that it is reversible but the large deformations cause the tube to change shape significantly and this causes the force vs. indentation depth relationship to become highly non-linear. Figure 1 shows a typical tube indented to 10mm. At this stage only localised damage has occurred and the deformation would be fully recovered and the elastic properties of the tube unaffected. Figure 2 shows the same tube at failure. The indentation depth was 30mm and the failure occurred as a result of a sudden instability resulting in the formation of a pair of distinct creases some 40° around the circumference from the point of indentation. Severe damage occurs in the laminate at the crease and this is the effective failure event.

The details of this failure process were described in a previous paper (Manders et al, 1979). In the present work, we have attempted to establish the distribution of strain in the deformed region of the tube during the whole process of indentation up to the occurrence of this failure event. In particular we wished to establish whether the failure was initiated essentially by an elastic instability or whether it was due to tensile failure of the fibres in that region.

EXPERIMENTAL

The tubes were manufactured by a conventional wet winding technique. The fibre was a 1200tex E-glass roving (Silenka 051 L) and the epoxy resin was the Ciba-Geigy MY750/HY917/DY063 system. Full details of the materials and cure system are given in the reference quoted above.

The tubes as manufactured were approximately 1.2m long and from these were cut test pieces each 500mm long. This length was a compromise as the indentation resistance of the tube is a function of its length. This effect is very marked at short lengths but the 500mm length chosen was only about 5% less stiff than the full 1.2m tube. The principal experimental technique was to measure the strain distribution in the tube wall by means of electrical resistance strain gauges. This also involved a number of compromises. It was necessary to measure the strain over a large area of the tube; and to affix strain gauges at all the required locations would have significantly increased the stiffness of the tube. We therefore chose to strain gauge a few points and to vary the point of indentation relative to the gauges. In this way it was possible to build up a complete strain map. It was decided to use stacked rosette gauges. It was necesary to use rosettes as the strain field was very complex and the 0°, 45°, 90° rosette allowed the full state of strain to be determined. The stacked gauge has the disadvantage that errors can arise if the flexural strains are high due to the different distances of each gauge element from the neutral axis. On the other hand the planar rosette may be inaccurate where the in-plane strain gradients are high. On balance, the stacked gauges were considered to be the most appropriate. Five pairs of rosettes were fixed to each tube at 20mm intervals along the tube axis with a gauge on the outside and inside surfaces of the tube at each of the 5 locations as shown in Fig 3. The gauges were of 120 Ω resistance and 3mm gauge length.

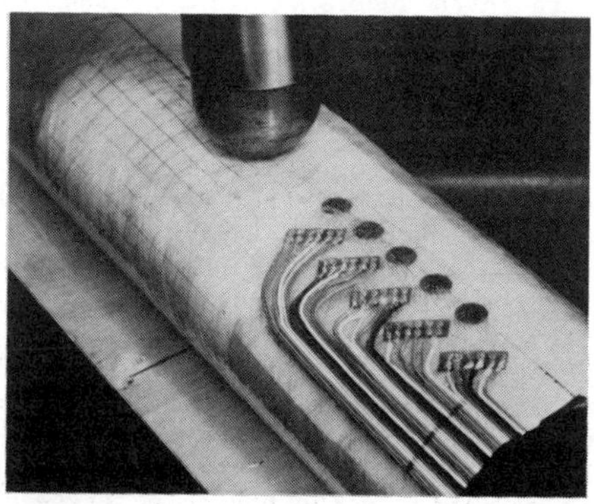

Fig. 3 Arrangement of 5 pairs of rosette strain gauges on outer and inner surfaces of tube. Note also the indentation grid.

An indentation grid was set up relative to the gauge positions. This grid was at 10mm intervals from 0 to 200mm in the axial direction and 10° intervals in the hoop direction. This grid is shown in Fig 4 together with the system of coordinates used. The strain map was built up by sequentially indenting the tube at each grid point and then transforming the strain data into a format using the indentation point as a reference. In this way the positions of the gauges were systematically varied relative to the actual indentation point. Since there were 5 pairs of rosettes it was necessary to indent at only a fifth of the grid positions to obtain the full strain distribution. In practice a deliberate overlap was built into the indentation sequence so that any damage to the gauges or changes in the laminate properties due to local damage could be detected. It was unfortunately impossible to indent at a number of grid positions very close to the gauges as the indenter actually touched the gauges as the tube deformed. Likewise the positions where the gauges would lie along the tube support were also unavailable. These points are indicated by the dots at grid intersections in Fig 4.

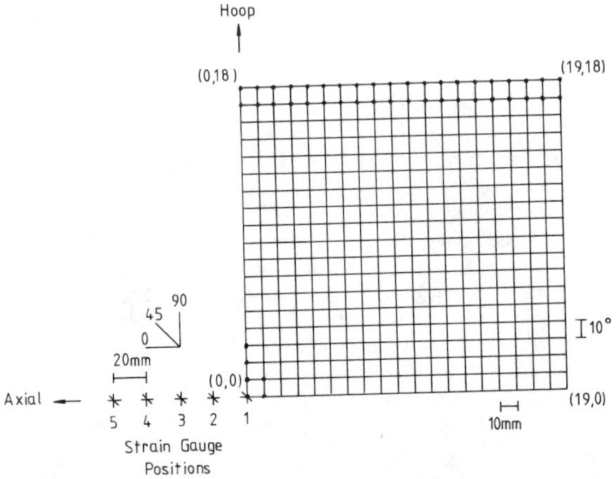

Fig. 4
Indentation grid relative to strain gauge positions. Note that the intersections marked by bold dots are those at which indentation was not possible.

This technique was very satisfactory for small indentations (up to 15mm) when local damage to the tube was minimal and did not affect the tube stiffness. However at higher indentations significant damage occurred and it was necessary to use several tubes before the complete set of strain data could be collected.

The indentation experiment was carried out on a bench model Instron testing machine. The indentor was hemispherical with a radius of 25mm (see Figs 1-3). In most cases indentations in the sequence 5, 10, 15 and 20mm were applied at each point. In some cases indentations at intervals of 1mm were recorded in order to obtain more data at critical locations.

The strain data was collected by means of a high speed data acquisition system (mDas; manufactured by Base 10 Instruments Ltd). This was capable of accepting 32 input voltages which it would scan at up to 1000 channels per second. The strain gauges were connected to a bridge completion network, which allowed each gauge element to be individually balanced. The individual gauge factor and excitation voltage were also input to the data unit so that calibration could be effected during subsequent computation operations.The function of the data logger was to record the required information onto a magnetic tape cartridge for subsequent transfer to a main frame computer.

DATA PROCESSING

Each package of raw data contained tube and grid reference, indentation depth, strain gauge outputs and gauge factors. This was then processed to obtain the six values of strain for each grid point (i.e. 0°, 45°, 90° inside and outside surfaces). From this it was possible to calculate the magnitude and direction of the principal strains on the tube surfaces and also the membrane and bending components of strain. We have not yet attempted to calculate the stress components due to an inadequate knowledge of the stiffness anisotropy of the tube wall. At this point it is useful to emphasize the volum of data collected. At each of the 380 grid points, representing one quarter of the tube surface, there are six components ofstrain for each increment of indentation. In the subsequent results section it is possible to present only a small selection of the total information.

RESULTS AND DISCUSSION

The elastic properties of the laminate are given in Table 1. The values for the unidirectional material were obtained from uniaxial coupon specimens fabricated from similar glass-fibre and resin to that used for the tubes. The 75° tube values were determined experimentally on specimens cut from a tube, whilst the 65° values were calculated using standard laminate theory. E_1 and E_2 are the moduli parallel and normal to the fibres respectively whilst Ex and Ey are those for the axial and hoop directions in the tubes.

TABLE 1 Mechanical Properties of Laminate

Unidirectional Material		75° Tube		65° Tube	
E_1	45.0 GPa	Ex	16.5	16.0	GPa
E_2	14.0 GPa	Ey	40.0	31.0	GPa
G_{12}	5.0 GPa	Gxy	7.4	9.7	GPa
ν_{12}	0.3	ν_{xy}	0.15	0.25	
ν_{21}	0.1	ν_{yx}	0.37	0.48	

$V_f \approx 0.54$, wall thickness $\approx$ 1mm

75 Degrees.Outer Surface,Principal Strains.

25mm Deflection.

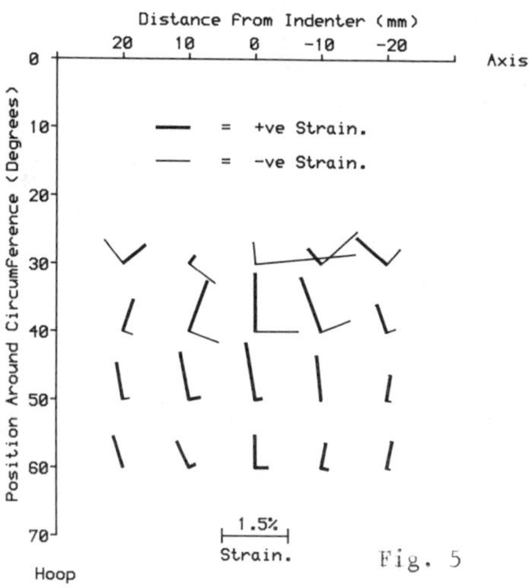

Fig. 5

75 Degrees.Inner Surface,Principal Strains.

25mm Deflection.

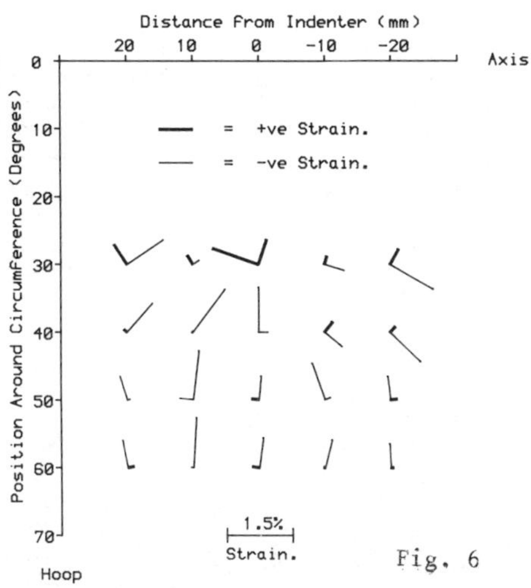

Fig. 6

Figures 5 and 6 show the magnitude and orientation of the two principal strains at selected grid locations around the region of failure in a 75° tube at 25mm indentation. The bolder lines indicate tensile strains and the fainter lines compression. The strain pattern is approximately symmetric about a meridian passing through the centre of indentation (0,0). On the outer surface large tensile hoop strains are indicated in the region 40°-60°

around the hoop direction, whilst on the inner surface the hoop strains are large and compressive at these locations. This is due to the development of the very sharp curvature which preceeds failure. This may be seen on Figs 1 and 2. Strains in the axial direction are generally much smaller in this region but increase towards the indenter. Failure occurred at 30-35mm indentation approximately 50° around the hoop direction from the centre of indentation. The maximum tensile strains recorded prior to failure in the fibre directions were of the order of 1.5%. A similar general pattern was observed for the 65° tubes but in this case the maximum strain in the fibre direction was almost 2%. This suggests that the failure event was probably initiated by localised fibre failure in this highly deformed region.

Localised delamination is associated with this failure event and it was observed to be more extensive in the 65° tubes. This may be reconciled with the observation that the strain distribution on the inner and outer surfaces indicated a higher interlaminar shear stress in these tubes.

Figures 7-10 show the changes in strain at selected locations as the indentation was increased. In Fig 7, 8 and 9 the locations are (0,3), (0,4) and (0,5) respectively. It will be observed that strain components change sign as the indentation proceeds. The 90° outside value is first tensile reaching a maximum at 13mm and then decreases to become compressive at indentations greater than 20mm. An opposite trend is observed for the 90° inside gauge.

A smaller effect is seen for the 0° gauges but the outer strain becomes highly compressive and the inner tensile. Thus at high indentations the tube wall is in a state of biaxial bending. Although the maximum strains indicated on this figure exceed 2% the strain resolved in the fibre directions would be less than this (Fig 10). Note that this tube failed at 27mm indentation, which is lower than the average for this configuration. Figures 8 and 9 show similar information for the other two locations. Note that the point of strain inflexion moves to a higher indentation at location (0,4) and there is no inflexion at (0,5) before tube failure intervenes. This location is close to the actual point of failure initiation. In Fig 10 the strains have been resolved to the fibre directions and their normals for a 65° tube. The maximum strain in the outer surface exceeds 2% at failure. All other strains are well below this level. Tensile strains normal to the fibre directions on the outer suface are very small and are compressive on the inner surface. This indicates that extensive transverse cracking would not be expected at this point. This is indeed our observation.

CONCLUSIONS

A methodology has been developed for the acquisition of full strain data from a tube subjected to a complex deformation. Our initial analysis indicates that failure is due to tensile failure of fibres at critical locations. It is our intention to develop a technique to described the shape of the deformed surfaces and to obtain the stress resultants and bending components within the wall.

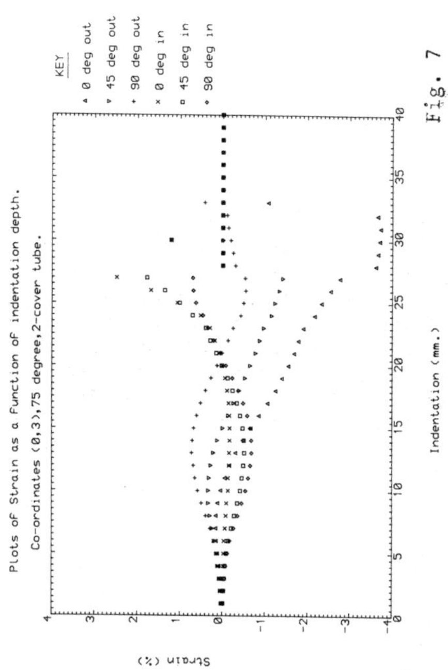

Fig. 7

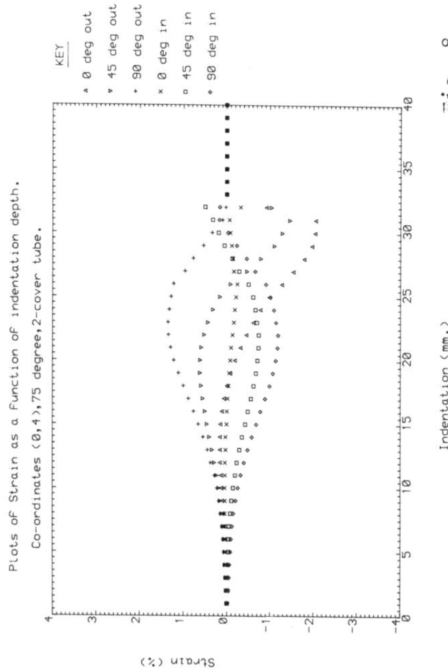

Fig. 8

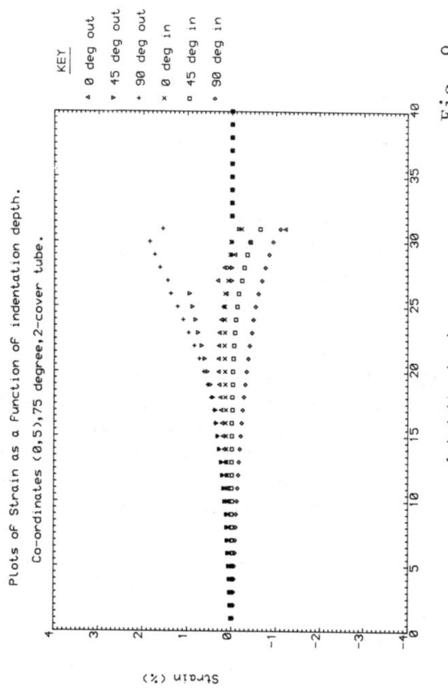

Fig. 9

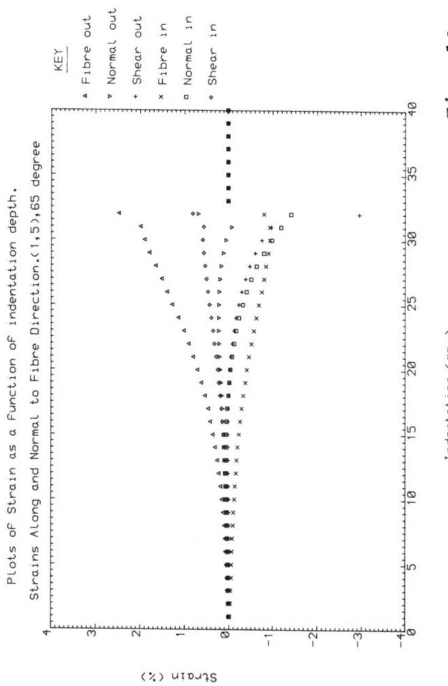

Fig. 10

ACKNOWLEDGEMENTS

This work was carried out in the Department of Metallurgy and Materials Technology, University of Surrey with support from Royal Armament Research and Development Establishment.

REFERENCE

Manders, P.W., M.G. Bader, M.J. Hinton, P.J. Flower, (1979) in K.J. Miller and R.F. Smith (Eds.), Mechanical Behaviour of Materials, Pergamon Press, Oxford. Vol. 3, pp. 275-284.

RELIABILITY-BASED FATIGUE-PROOF DESIGN OF COMPOSITE MATERIALS

H. Ishikawa*, A. Tsurui, T. Tanimoto*** and H. Kimura***

*Department of Information Science, Kagawa University, Takamatsu 760, Japan

**Department of Mathematical Engineering, Kyoto University, Kyoto 606, Japan

***Department of Mechanical Engineering, Doshisha University, Kyoto 606, Japan

EXTENDED ABSTRACT

I. Introduction

Composite materials, especially glass fiber reinforced plastics (referred to as FRP), have been recently in more extensive use as machine parts and/or structural components. Accordingly, the appropriate estimation of fatigue life of FRP has come to be of great importance for those concerned with design of composite materials. Undoubtedly, fatigue life of the material would have an inherent scatter and, in additon, FRP possess less material uniformity in comparison to conventional metallic materials. Hence, a good understanding of the statistical aspect of fatigue properties of FRP is deemed essential to their successful application in the practical design.

In view of the above, the first phase of the present paper deals with the theoretical modelling of the statistical failure probability in the sense that the failure probability can be evaluated only through the limited number of experimental data, which would, of course, vary from sample to sample. The distribution function for the statistical failure probability under service loading has been derived by making use of the linear damage accumulation hypothesis. Also, a simple but important policy for fatigue-proof design is proposed. This design policy requires that a value smaller than a pre-specified allowable value for the failure probability be correct with a pre-specified reliability level.

In the second phase of the paper, the statistical fatigue life data of satin woven glass cloth FRP were obtained experimentally under rotating bending of constant stress amplitude with not less than twenty replication tests at each stress level. Then they were analyzed with the aid of a log-normal probability model to clarify the statistical fatigue properties of composite materials. Finally, the reliability-based design safe lives of composite materials have been investigated so as to meet the requirement in the practical design from the viewpoint of the proposed reliability-based fatigue-proof design principle.

** Department of Mathematical Engineering, Kyoto University, Kyoto 606 Japan
*** Department of Mechanical Engineering, Doshisha University, Kyoto 606 Japan

II. Theoretical modelling of failure probability under service loading

Failure probability has to be treated as a statistic since it can be evaluated only through experimental data.

Let Z_n be a normalized damage accumulated during n cycles of stressing of not necessarily the same level. Z_n is assumed of the following form by Miner's rule:

$$Z_n = \sum_{i=1}^{n} \left(\frac{S_i}{s_0}\right)^b \tag{1}$$

where S_i = applied i-th stress level, s_0 = normalization stress level, b = a constant and S_i's are statistically independent, each having a distribution function R(s).

Let Z_c be the critical value of Z_n. Introducing stress distribution R(s) of the form:

$$R(s) = \begin{cases} 0 & \text{for } s < s_0 \\ 1 & \text{otherwise} \end{cases} \tag{2}$$

we have, from Eq.(1),

$$Z_n = \sum_{i=1}^{n} \left(\frac{S_0}{s_0}\right)^b = \sum_{i=1}^{n} 1 = n$$

Therefore, Z_c is deemed to be the critical value of Z_n under constant amplitude of s_0. In other words, Z_c gives a measure associated with fatigue life under constant stress amplitude s_0.

Then, the distribution $F_N(n)$ of fatigue life N under service load with arbitrary distribution R(s) can be derived in the following way. For $n \gg 1$, the distribution of Z_n can be well approximated by a normal distribution with the aid of central limit theorem.

$$P[Z_n \le z_n] \simeq \frac{1}{\sqrt{2\pi n}\,\nu} \int_{-\infty}^{z_n} \exp\left[-\frac{(\xi - n\lambda)^2}{2n\nu^2}\right] d\xi$$

$$= \Phi\left[\frac{z_n - n\lambda}{\sqrt{n}\,\nu}\right] \tag{3}$$

where

$$\Phi(u) = \frac{1}{\sqrt{2\pi}} \int_{-\infty}^{u} \exp\left[-\frac{u^2}{2}\right] du \tag{4}$$

$$\lambda = E[Z_n]/n = \frac{1}{s_0^b} \int_0^{\infty} s^b dR(s) \tag{5}$$

$$\nu^2 = E\left[\left\{Z_n - E[Z_n]\right\}^2\right] = \frac{1}{s_0^{2b}} \left[\int_0^{\infty} s^{2b} dR(s) - \left\{ \int_0^{\infty} s^b dR(s) \right\}^2 \right] \tag{6}$$

Consequently, $F_N(n)$ is given in the following form:

$$F_N(n) = P[N \le n] = P[Z_n > Z_c]$$

$$= \int_{-\infty}^{\infty} F_{Z_c}(z_c) \cdot \frac{1}{\sqrt{2\pi n}\,\nu} \exp\left[-\frac{(z_c - n\lambda)^2}{2n\nu^2}\right] dz_c \tag{7}$$

where $F_{Z_c}(z_c)$ is the distribution function of Z_c. Making use of Taylor's series expansion of $F_{Z_c}(z_c)$ around $z_c = n\lambda$, we have

$$F_N(n) = F_{Z_c}(n\lambda) + \frac{n\nu^2}{2} F''_{Z_c}(n\lambda) + O(1/n^2) \tag{8}$$

Generally, since the second and higher terms of the right-hand side of Eq.(8) is sufficiently small enough in comparison with the first term, $F_N(n)$ can be approximated by

$$F_N(n) \simeq F_{Z_c}(n\lambda) \tag{9}$$

Fatigue life distribution has often been reported to be well approximated by a log-normal probability model. Therefore, it is reasonable to assume that Z_c follows a log-normal distribution. With Z_c thus assumed, $\ln Z_c$ becomes a normal random variable with mean μ and variance σ^2, and consequently $(\ln Z_c - \mu)/\sigma$ follows a standard normal distribution. Hence,

$$F_N(n) = F_{Z_c}(n\lambda) = P[Z_c \le n\lambda] = P\left[\frac{\ln Z_c - \mu}{\sigma} \le \frac{\ln(n\lambda) - \mu}{\sigma}\right]$$

$$= \begin{cases} \Phi\left[\dfrac{\ln(n\lambda) - \mu}{\sigma}\right] & \text{for } n\lambda > 0 \\ 0 & \text{otherwise} \end{cases} \tag{10}$$

Fatigue life distribution can be given by the above equation. However, since the true value of parameters μ and σ are usually unknown, they need to be replaced by their maximum likelihood estimates $\hat{\mu}$ and $\hat{\sigma}$ from a sample of size k, thus $P = \Phi[\{\ln(n\lambda) - \mu\}/\sigma]$ by its estimate $\hat{P}$. $\hat{\mu}$ and $\hat{\sigma}$ are considered as random variables in the sense that they vary from sample to sample, and so is $\hat{P}$.

Putting

$$\ln(n\lambda) = q \tag{11}$$

we have

$$\hat{P} = \Phi\left[\frac{q - \hat{\mu}}{\hat{\sigma}}\right] \tag{12}$$

Joint distribution $Q(\xi,\eta)$ of maximum likelihood estimates $\hat{\mu}$ and $\hat{\sigma}$ from a sample of size k (assuming normal population) can be expressed in the following form with the fact that $(\hat{\mu} - \mu)/\sigma$ and $k\hat{\sigma}^2/\sigma^2$ are statistically independent and that they have, respectively, a normal distribution with mean zero and variance $1/k$ and χ^2 distribution of k-1 degrees of freedom:

$$Q(\xi,\eta) = P[\hat{\mu} \le \xi \ , \ \hat{\sigma} \le \eta] = P\left[\frac{\hat{\mu}-\mu}{\sigma} \le \frac{\xi-\mu}{\sigma} \ , \ \frac{k\hat{\sigma}^2}{\sigma^2} \le \frac{k\eta^2}{\sigma^2}\right]$$

$$= \left(\frac{k}{2\pi}\right)^{1/2} \int_{-\infty}^{(\xi-\mu)/\sigma} \exp\left[-\frac{ku^2}{2}\right] du \cdot \left[2^{(k-1)/2}\Gamma\left(\frac{k-1}{2}\right)\right]^{-1} \int_0^{(k/\sigma^2)\eta^2} v^{(k-3)/2} e^{-v/2} dv \qquad (13)$$

From the above, the distribution function H(p) of the statistical failure probability $\hat{P}$ can be derived:

$$H(p) = P[\hat{P} \le p] = P[\Phi\left[\frac{q-\hat{\mu}}{\hat{\sigma}}\right] \le p] = P[q-\hat{\mu} \le \Phi^{-1}(p)\hat{\sigma}]$$

$$= \iint_{q-\xi \le \Phi^{-1}(p)\eta} dQ(\xi,\eta)$$

$$= 1 - \int_0^{\infty} \frac{1}{\Gamma\left(\frac{k-1}{2}\right)} \Phi\left[\frac{\sqrt{k}}{\sigma}(q-\mu) - \Phi^{-1}(p)\sqrt{2w}\right] w^{(k-3)/2} e^{-w} dw \qquad (14)$$

where $\Phi^{-1}(\cdot)$ is inverse function of $\Phi(\cdot)$. (If $\Phi(u) = p$, then $u = \Phi^{-1}(p)$.)

III. Reliability-based fatigue-proof design principle

The integrand in Eq.(14) is exemplified in Fig. 1 as a function of w, with

$$x = (q-\mu)/\sigma \qquad (15)$$

One example of simulated results of H(p) is shown in Fig. 2 in the case of sample size k = 20. H(p) thus defined is now utilized for the reliability-based fatigue-proof design of composite materials, whose principle requires that the decision of the failure probability being less than p during design life be correct with a pre-specified level of reliability. In other words, fatigue-proof design has to be made so as to obtain as large value of $n\lambda$ as possible under the condition that the sample size k, reliability level H and allowable value of failure probability p_H are given. Schematic presentation of this design principle is described in Fig. 3, where the variation of p_H is given as a function of x for, for instance, H = 0.99, 0.90 and 0.50. The value $n\lambda$ can be determined as follows, for one example of design requirement of $p_H = 5\times10^{-4}$ with H = 0.90, getting the value of x = -3.8 from the figure:

$$n\lambda = \exp[\hat{\mu} - 3.8\hat{\sigma}]$$

where $\hat{\mu}$ and $\hat{\sigma}$ are maximum likelihood estimates in the sample of size k = 20 obtained experimentally.

IV. Reliability-based design safe lives of composite materials

Rotating bending fatigue tests were carried out on the satin woven glass cloth FRP. Figure 4 illustrates fatigue life distributions at several stress levels on a log-normal probability paper. These results were analyzed to determine reliability-based design safe lives on the basis of the proposed design principle. Valuable and interesting details will be presented at the conference.

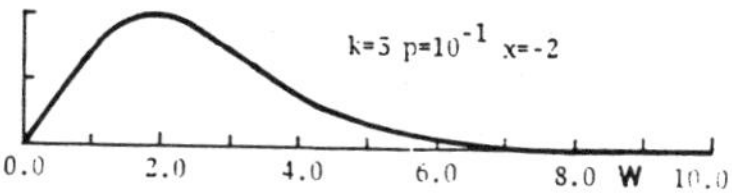

Fig. 1. Integrand in Eq. (13) as a function of w.

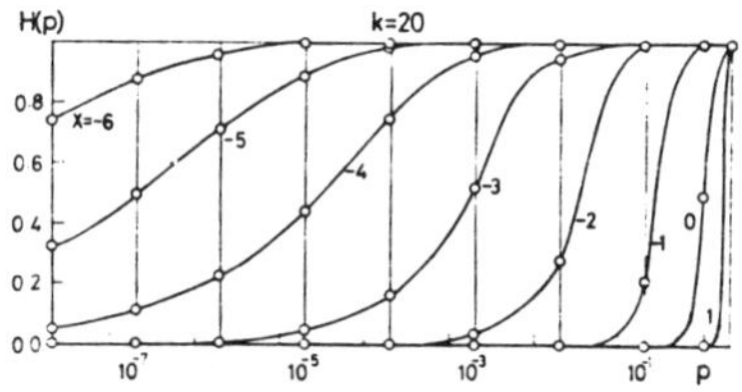

(d) In case of sample size $k=20$.

Fig. 2. Profiles of the distribution function $H(p)$ of the statistic $\hat{P}$.

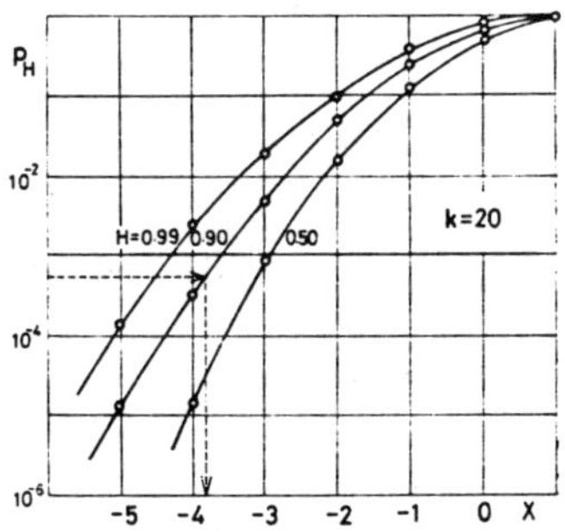

Fig. 3. Plots of the allowable failure probability p_H versus x, when the values of the reliability H are pre-specified.

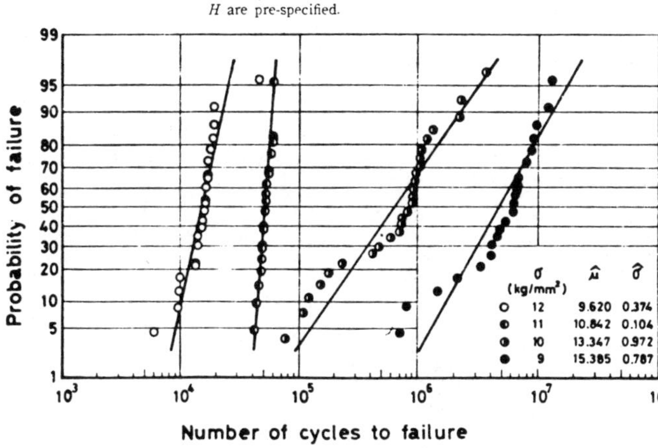

Fig. 4. Fatigue life distribution of satin woven FRP (log-normal probability model)

RESIDUAL STRENGTH DEGRADATION MODEL FOR GLASS/POLYESTER LAMINATES UNDER REPEATED TENSION AND COMPRESSION LOADINGS

T. Tanimoto*, H. Ishikawa, S. Amijima* and H. Kimura****

*Department of Mechanical Engineering, Doshisha University, Kyoto, Japan
**Department of Information Science, Kagawa University, Kagawa, Japan

ABSTRACT

Experimental studies were first performed on the residual strength degradation process of Glass/Polyester laminates under repeated tension and compression. Then, a new residual strength degradation model was devised for this material and the validity of the proposed model was examined through the comparisons with the sufficient amount of test data. It is of great interest to note that the statistical distribution of fatigue life has been derived, from the proposed model, to be a well-known double exponential distribution. It is also shown that the model can be easily extended to reflect the engineering reality for other kinds of composite and different loading conditions.

KEYWORDS

Residual strength; composite materials; fatigue; statistical distribution; reliability; Glass/Polyester laminates; residual strength degradation model.

INTRODUCTION

The fatigue damage process in composite materials is a very complicated phenomenon due to the complex microstructure of the material, being a combination of fibers, matrix, voids and a rather poorly understood interface between the fibers and matrix. Much effort has been made to evaluate the residual strength under service loading, because the degradation process of residual strength is of vital importance in the fatigue design of, for instance, aircraft structures. Halpin et al (1973) proposed a degradation equation which is based upon the crack propagation of homogeneous materials. Hahn and Kim (1975), Yang and Liu (1977), Chou and Croman (1978), and other investigators derived residual strength distributions and compared them with experimental data. However, according to our experimental work (Tanimoto, 1975) with Glass/Polyester laminates, the tendency in the decreasing rate of residual strength with increasing number of cycles appeared quite different from those results for Graphite/Epoxy as reported by the aforementioned investigators.

Therefore, in the present paper, a new residual strength degradation model for composite materials has been proposed on the basis of the assumption that the residual strength (treated as a random variable) decreases monotonically in exponential form with increasing number of stress cycles. An experimental test program on the plain woven glass cloth laminated FRP has been carried out to generate statistically meaningful data for the verification of the validity of the proposed model.

SPECIMEN PREPARATION AND TESTING PROCEDURES

The material consists of polyester resin as matrix and satin woven glass cloth as reinforcement. It was made by means of a hand-lay-up method at room temperature and then cured for four hours at 353K (80°C). After being removed from the oven, it was cut in the direction parallel to the warp of the glass fabric into dimensions shown in Fig. 1. The pieces were then kept for two days under constant conditions of 296K (23°C) and 65% humidity. Table 1 shows the composition of the material tested and the glass fiber content of the specimen. Fatigue tests were carried out under zero-mean-stress at a speed of 2000 r.p.m.. Specimens were tested for different numbers of cycles at a given stress level; these cycles represent different fractions of fatigue life. After cycling to the proper number of cycles, the machine was stopped and the specimen was taken out carefully to avoid any additional damage. Then the tensile strength was measured to examine the change of properties in the material at a given fraction of the fatigue life.

TABLE 1 Material of the Specimen

Matrix	Polyester resin (Polylite FG-284)
Reinforcement	Plain woven glass cloth (MG-252)
Glass Content by Volume	34.24 (%)

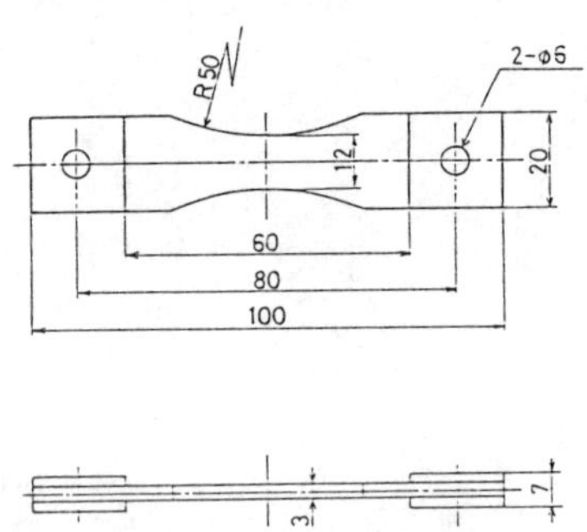

Fig. 1. Test specimen.

Residual Strength Degradation Model

According to our previous experimental research on the residual strength (Tanimoto, 1975) it will be reasonable to consider that the residual strength for Glass/Polyester laminates decreases monotonically with an increase of the number of cycles. As shown in Fig. 2, the ultimate strength, $\mathfrak{R}(0)$, residual strength, $\mathfrak{R}(n)$, and fatigue life, $\mathfrak{N}_f$, as well, should be treated as a random variable. And therefore in this paper, the random variable is represented by a Gothic letter.

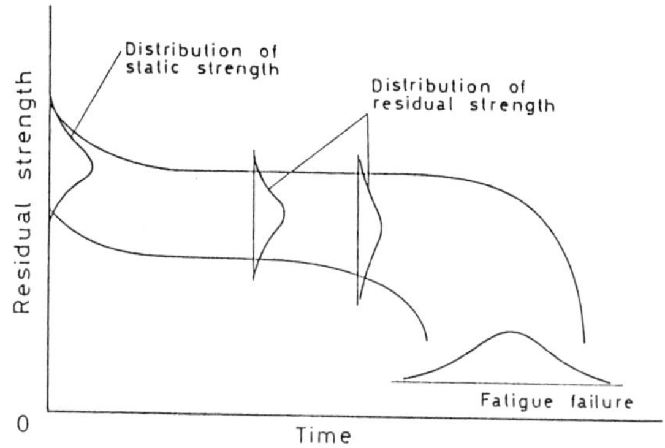

Fig. 2. Schematic representation of the degradation of residual strength with time as well as distributions of ultimate strength and fatigue life.

Now, it is assumed that the degradation process of residual strength is represented by the following rate equation:

$$\frac{dR(n)}{dn} = -f(s,\ w,\ r)\cdot R(n) \tag{1}$$

in which f(s, w, r) is a function of the stress level s, the loading frequency w, and the stress ratio r.

For simplicity, if the loading frequency w and the stress ratio r is fixed constant, then we can write

$$f(s,\ w,\ r) = f(s) \tag{2}$$

From Eqs. (1) and (2) we get

$$\frac{dR(n)}{dn} = -f(s)\cdot R(n) \tag{3}$$

Integration of both sides of Eq. (3) from n_0 to n_1 cycles leads to

$$\ln R(n_1) - \ln R(n_0) = -f(s)(n_1 - n_0) \tag{4}$$

In other form,

$$R(n_1) = R(n_0)\exp[-f(s)(n_1 - n_0)] \tag{5}$$

With $n_0 = 0$ and $n_1 = n$, Eq. (5) reduces to

$$R(n) = R(0)\exp[-f(s)\cdot n] \tag{6}$$

This equation implies that the residual strength R(n) after n cycles of cyclic stress s decreases monotonically in the exponential form from the ultimate strength R(0) with an increase of n.

Statistical Distribution of Ultimate Strength and Residual Strength

The statistical distribution of the ultimate strength, R(0), of virgin material is assumed to follow a two-parameter Weibull distribution in the following form:

$$F_{R(0)}(x) = P[R(0) \le x] = 1 - \exp[-(\frac{x}{\beta})^{\alpha}] \tag{7}$$

where α is the shape parameter and β is the scale parameter.

Similarly, the distribution function of residual strength R(n) is given in the following form with the aid of Eq. (6).

$$F_{R(n)}(x) = P[R(n) \le x] = P[R(0)\exp[-f(s)n] \le x]$$
$$= P[R(0) \le x\cdot\exp[f(s)n]]$$
$$= F_{R(0)}(x\cdot\exp[f(s)n]) = 1 - \exp[-(\frac{x\cdot\exp[f(s)n]}{\beta})^{\alpha}]$$
$$= 1 - \exp[-(\frac{x}{\beta^*(s)})^{\alpha}] \quad (8)$$

where

$$\beta^*(s) = \beta\cdot\exp[-f(s)n] \quad (9)$$

Eq. (8) indicates that the statistical distribution of residual strength is approximately a two-parameter Weibull distribution with the shape parameter α and the scale parameter $\beta^*(s)$. The equation reduces to Eq. (7) when we put $n = 0$, which is the distribution of the ultimate strength independent of applied stress level s.

Statistical Distribution of Fatigue Life

It is obvious that the event $\{N_f \le n\}$ is probabilistically identical to the event $\{R(n) \le s\}$ according to the so-called S-R Model (Stress-Residual Strength Model). Hence, the distribution function of N_f is obtained as

$$F_{N_f}(n) = P[N_f \le n]$$
$$= P[R(n) \le s] = F_{R(n)}(s)$$
$$= 1 - \exp[-(\frac{s}{\beta^*(s)})^{\alpha}]$$
$$= 1 - \exp[-\exp[\alpha f(s)n + \alpha\ln(s/\beta)]]$$
$$= 1 - \exp[-\exp[\xi(s)\{n - \eta(s)\}]] \quad (10)$$

where

$$\xi(s) = \alpha f(s) \quad (11)$$
$$\eta(s) = -\frac{\ln(s/\beta)}{f(s)}$$

It is quite interesting to note that the statistical distribution of fatigue life has been derived, based upon the proposed model, to be a well-known double exponential distribution. In Eq. (10), $1/\xi(s)$ is the scale parameter and $\eta(s)$ is the location parameter in a double exponential distribution.

EXPERIMENTAL RESULTS AND DISCUSSIONS

Basic S-N Diagram

Basic S-N diagram of the specimen is shown in Fig. 3. Ten specimens were used for each stress level. The expected value of fatigue life was calculated with the aid of a two-parameter Weibull distribution, and was represented as N_f^*. The value of cycle ratio indicating the different fraction of fatigue life was determined in the present paper based upon such an expected value, N_f^*, for convenience. It is reasonable to consider from Fig. 3 that the relation between the applied stress and the mean value of the life is practically linear in log-log representation, thus providing the following equation:

$$S^{14.11}\cdot N_f^* = 6.70 \times 10^{16} \quad (12)$$

Determination of Parameters

Figure 4 shows the distribution of ultimate tensile strength of the specimen which was plotted on a Weibull probability paper. Estimation of parameters,

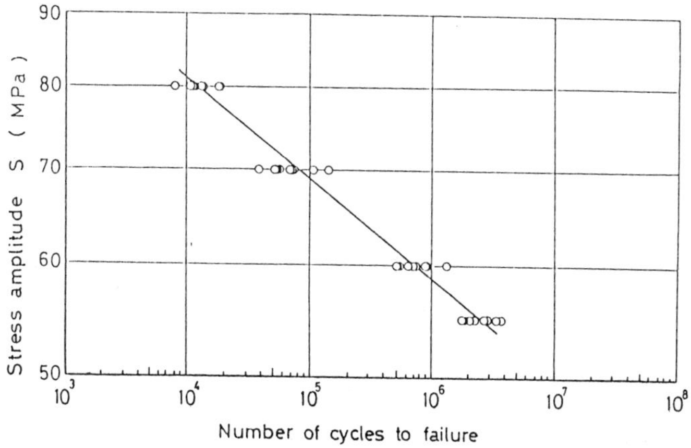

Fig. 3. Basic S-N diagram.

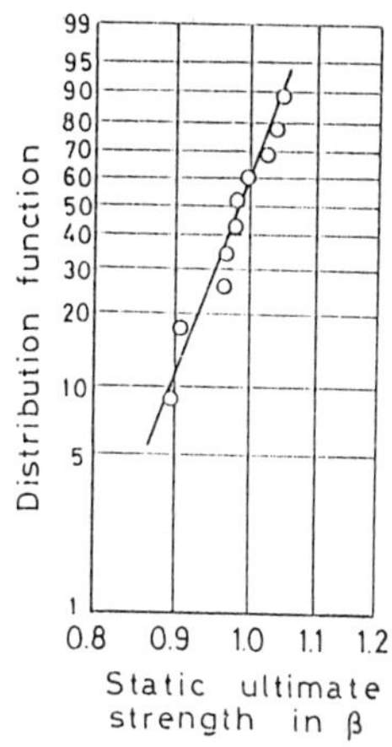

Fig. 4. Static tensile strength distribution of the virgin material.

shape parameter α and scale parameter β, was performed by the least square error method. Estimated parameters, $\hat{\alpha}$ and $\hat{\beta}$, were found to be 30.34 and 225.4 MPa, respectively.

Figure 5 shows the residual strength variation under cyclic loading with zero-mean-stress and stress amplitude of 80 MPa. In the figure, the ordinate is normalized residual strength as the ratio of ultimate strength after cyclic loading to original ultimate strength, σ/σ_{ult}, and the abscissa is the cycle ratio representing various fractions of fatigue life, n/N_f^*. Residual strength was measured using ten specimens respectively at each cycle ratio of 0.1, 0.2, 0.4, 0.6 and 0.8.

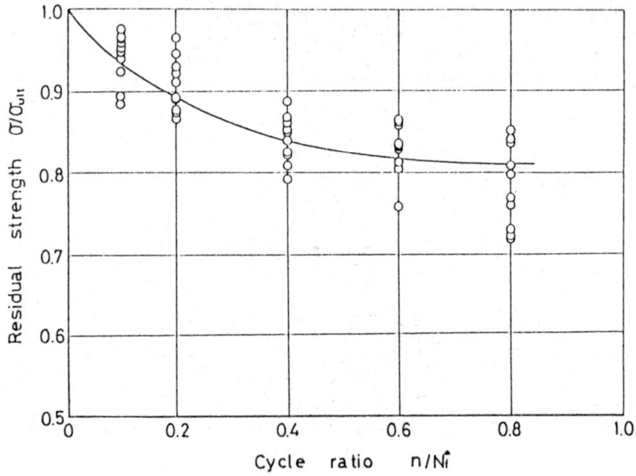

Fig. 5. Variation of residual strength as a function of cycle ratio. (for stress amplitude s = 80 MPa)

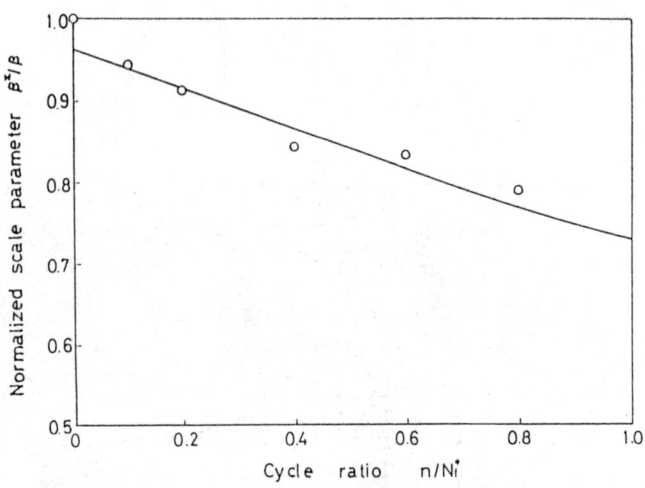

Fig. 6. The best fit exponential curve for the relationship between expected residual strength and cycle ratio.

It is mentioned previously that according to the present model, the residual strength distribution has the same shape parameter as that for the ultimate strength of virgin material, while the scale parameter has to be determined for each cycle ratio. For simplicity, therefore, it was assumed here that α is independent of stress level and cycle ratio, and f(s) could be determined with the aid of the best fit exponential curve for the relationship between normalized scale parameter β^*/β and cycle ratio n/N_f^*.

An example of the best fit curve is shown in Fig. 6 for s = 80 MPa. The residual strength distribution at the cycle ratio of 0.6 is shown in Fig. 7 as an example. The theoretical prediction using Eq. (8) is drawn by a solid line and the experimental data by circles. From the figure we can see a reasonable agreement between test data and theoretical prediction.

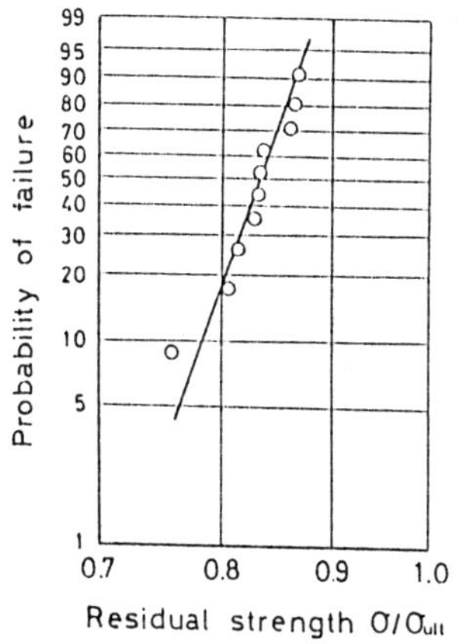

Fig. 7. Comparison between experimental and theoretical distributions of residual strength ($n/N_f^* = 0.6$; s = 80 MPa)

Residual Strength Degradation Curve

Since R(0) follows a two-parameter Weibull distribution, the expected value can be written as

$$\sigma ult = E[R(0)] = \hat{\beta}\Gamma(1 + \frac{1}{\hat{\alpha}}) \tag{13}$$

Suppose that the material has the ultimate strength of σult, it follows from Eq. (6) that the residual strength, after being subjected to n cycles of the cyclic stress s_1, is given by

$$R(n) = \sigma ult \cdot \exp[-f(s_1)\cdot n] \tag{14}$$

Hence, the residual strength at the moment of fatigue failure is given by

$$R(N_f^*) = \sigma ult \cdot \exp[-f(s_1)\cdot N_f^*] \tag{15}$$

The values of $R(N_f^*)$ calculated on the basis of f(s) which was obtained from the best fit exponential curve were 161.5, 121.7 and 123.9 MPa, respectively, for s = 80, 60 and 55 MPa. These results might suggest that $R(N_f^*)$ could be approximately equal to stress range, 2s, under zero-mean-stress condition. Theoretical calculation of residual strength was actually made in such a way that (i) $R(N_f^*)$ is put equally to 2s for convenience, (ii) f(s) is determined from Eq. (15), and (iii) residual strength R(n) for different stress cycles could be estimated from Eq. (14). Comparison between experimental data and predicted results of residual strength were performed for the various cyclic stress values. The result for s = 80 MPa is shown in Fig. 8. Theoretical prediction in accordance with the proposed model is presented in the figure as a solid curve, along with the experimental data indicated by circles. Again the agreement between test data and theoretical prediction is deemed reasonable. Our experimental results for Glass/Polyester showed that the tendency in decreasing rate of residual strength with increasing number of cycles appears quite different from the results derived for Graphite/Epoxy as reported

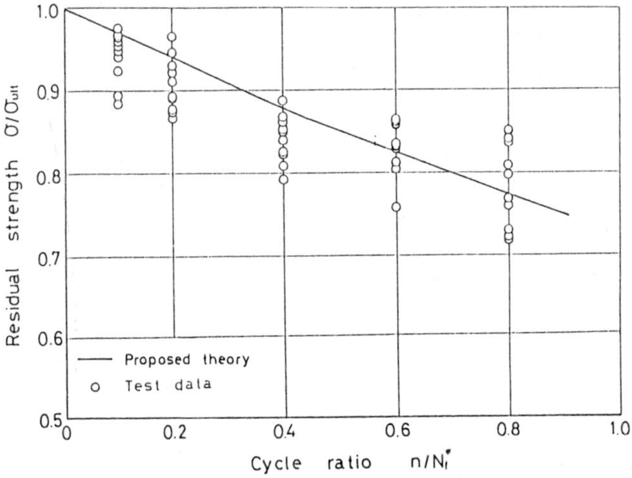

Fig. 8. Comparison between experimental data and theoretical curve for the degradation of residual strength (s = 80 MPa).

by Yang and other investigators. In their approaches on Graphite/Epoxy, the time rate decrease of residual strength is inversely proportional to the residual strength to a certain power. It is clear from our test results that such an approach is not suitable for Glass/Polyester.

Extended Model

The residual strength degradation model proposed for Glass/Polyester laminates in this paper could be extended easily to reflect the engineering reality for other kinds of composites and different loading conditions as follows.

The rate of change in the residual strength, with an increase of the number of cycles, is assumed to be represented by

$$dR(n) \,/\, dn = -f(s)\cdot c\cdot n^{c-1}R(n) \tag{16}$$

in which c is a constant which can be adjusted to fit the various residual strength data for a particular kind of compostes involved. Hence, the degradation process is indicated by

$$R(n) = R(0)\exp[-f(s)\cdot n^{c}] \tag{17}$$

The distribution function of residual strength is written as

$$F_{R(n)}(x) = 1 - \exp[-(\frac{x}{\beta_1^*(s)})^{\alpha}] \tag{18}$$

where

$$\beta_1^*(s) = \beta\cdot\exp[-f(s)\cdot n^{c}] \tag{19}$$

These equations based upon the extended model reduce to the original model when c = 1.

In Fig. 9 which shows a result for s = 60 MPa, the theoretical prediction by the extended model is shown as a chain curve, where the value of c was assumed to be 0.6. It is apparent that better fit can be obtained by employing the

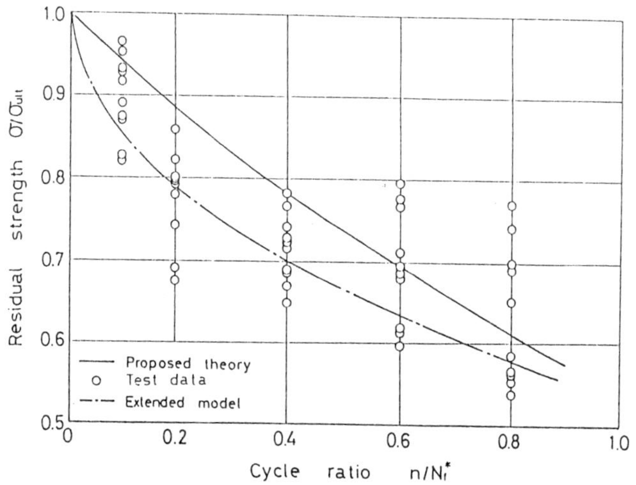

Fig. 9. Comparison between experimental data and theoretical curves for the degradation of residual strength (s = 60 MPa).

extended model. Furthermore, it will be possible to make a conservative estimation, as is often desired in the practical design, by the adequate adjustment of constant value c.

CONCLUDING REMARKS

A new residual strength degradation model for composite materials has been proposed in the present paper. The validity of the model has been verified through the comparisons with the sufficient amount of experimental data for Glass/Polyester laminates. It was also shown that the model could be extended easily for other kinds of composites and different loading conditions.

REFERENCES

Halpin, J. C., K. L. Jerina, and T. A. Johnson (1973). ASTM STP 521, ASTM, p. 5.

Hahn, H. T., and R. Y. Kim (1975). J. of Composite Materials, 9, 297-311.

Yang, J. N., and M. D. Liu (1977). J. of Composite Materials, 12, 176-203.

Chou, P., and R. Croman (1978). J. of Composite Materials, 12, 177-194.

Tanimoto, T, and S. Amijima (1975). J. of Composite Materials, 9, 380-390.

ON THE USE OF THE THREE POINT BEND AND COMPACT TENSION SPECIMENS TO MEASURE FRACTURE TOUGHNESS OF COMPOSITE LAMINATES

C. E. Harris and D. H. Morris

Department of Engineering Science and Mechanics,
Virginia Polytechnic Institute and State University, Blacksburg,
Virginia 24061, USA

ABSTRACT

The influence of specimen thickness on the fracture toughness of two laminates and three specimen geometries was investigated. As thickness increased the toughness decreased and approached an asymptotic value that was dependent upon the type of laminate but was practically independent of specimen geometry. Enchanced X-ray photographs and removal of an outside ply revealed that most of the delaminations were surface effects.

KEYWORDS

Fracture toughness; three-point bend specimen; compact tension specimen; center-cracked tension specimen; thick laminates.

INTRODUCTION

The first specimen configurations utilized for fracture toughness testing of metals were the center-cracked panel and double-edge notched tension specimen. Problems associated with testing these configurations began to surface when test results identified fracture toughness as a decreasing function of increasing specimen thickness. The thick specimens required higher loads that exceeded the load capacity of many testing machines; also, problems arose in gripping the specimens. In addition, these specimens utilized a large quantity of material. The edge-notched bend specimen was the first alternative configuration to be utilized because it was analogous to half the double-edge notched tension specimen and had the same crack-tip stress state. The bend specimen was smaller than the double-edge notched specimen and required substantially lower test loads. The acceptability of the bend specimen was verified by a series of round-robin tests conducted by the ASTM (Brown, 1970). A further savings in material could be achieved by the use of the compact tension specimen. Approximate expressions for s ress intensity factor generated by the boundary collocation method and fini.e element method have been experimentally verified by photoelastic measurements (Brown and Srawley, 1966). Consequently, the three-point bend and compact tension specimens were adopted by the ASTM as the standard fracture toughness test specimens for metals as specified in test specification E-399 (1982 Annual Book of ASTM Standards).

Most fracture studies of laminated composite materials have utilized the center-cracked and edge notched tension specimens to measure fracture toughness. A substantial body of test data has been generated for thin laminates (4 to 12 plies thick) utilizing these specimen configurations. To date, these studies have not conclusively demonstrated the existence of a material property for general design applications. Furthermore, it has not been demonstrated that the thin laminate data can be extrapolated to thick laminate applications. Since the same experimental problems exist in testing thick center-cracked tension specimens of laminated composites as existed with thick isotropic materials, there is a need to conduct fracture tests of composites utilizing the bend and compact tension specimens. Furthermore, an experimental study of the fracture behavior of the three specimen geometries may be useful in identifying a material property and laminate design approach. The subject of this paper is limited to a comparison of fracture toughness test results obtained from the three specimen configurations, for several laminate thicknesses.

Several fracture mechanics investigations (Cruse and Osias, 1974; Owen and Cann, 1979; Prewo, 1978; Reedy, 1980; Shih and Logsdon, 1981) have utilized the center-cracked tension and either the three-point bend and/or the compact tension specimens to compare measured fracture toughness of laminated composites. At first glance these programs appear to yield contradictory test results. The test data from some programs indicate that the different specimen types yield similar results while the data from other programs indicate just the opposite. However, there are differences in the details of these programs that may contribute to the different results. Also, it should be noted that none of the previous fracture studies had specimen thickness as a test parameter.

EXPERIMENTAL PROCEDURES

Fracture Tests

The center-cracked tension specimens were 51 mm (2 in.) wide and 203 mm (8 in.) long. The three-point bend specimen was 102 mm (4 in.) long and 25.4 mm (1 in.) wide, while the compact tension specimen was 63.5 mm (2.5 in.) by 61.0 mm (2.4 in.) with 12.1 mm (0.5 in.) diameter pin loading holes. The crack length-to-width ratio was 0.50 for all specimens. For each laminate type the specimens were oriented such that the specimen and laminate panel thickness direction coincided and the machined slot (crack) was perpendicular to the 0° fiber direction. The width of the machined slot was 0.40 mm (0.016 in.) in the center-cracked tension specimen and 0.48 mm (0.019 in.) in the other two specimen types.

The fracture tests were conducted at a constant crosshead displacement rate of 0.02 mm/s (0.05 in./min). The center-cracked tension specimen ends were held in 51 mm (2 in.) wide wedge-action friction grips such that the specimen length between grip ends was about 127 mm (5 in.). The three-point bend fixture utilized roller supports and a loading ram with a 19.1 mm (0.75 in.) radius. The recorded test data included a plot of crack-opening displacement (COD) versus load. The COD clip gage was held directly in the crack in the specimen by machined knife-edge tabs. Four replicate tests were conducted.

Determination of Fracture Toughness

A finite element analysis of each specimen geometry was performed. The computer code (Chu and colleagues, 1976) utilizes a special crack element that contains the crack tip and performs the singular zone stress analysis. Plane strain and plane stress analyses were performed with the laminated composite treated as homogeneous and anisotropic. The computer output included mode I and II stress intensity factors for an applied unit load. For a given test condition the fracture toughness at the maximum test load was then determined by multiplying the maximum load by the stress intensity factor for a unit load. Further details may be found in a previous article (Harris and Morris, 1983).

RESULTS AND DISCUSSION

Typical load versus COD for each specimen type are shown in Figs. 1 and 2 for the $[0/\pm45/90]_{8s}$ and $[0/90]_{16s}$ laminates, respectively. The load records of the three-point bend (TPB) and compact tension (CT) specimens are similar while the record of the center-cracked tension (CCT) specimen is different from the other two. The CCT specimen failed catastrophically once the maximum load was reached. The discontinuity in the load-COD record of the CCT specimen corresponded to the formation of crack tip damage. The initial load reductions in the records for the TPB and CT specimens corresponded to damage formation at the crack tip. Enhanced x-ray photographs of the damage in each specimen type is shown in Figs. 3 and 4 for the $[0/\pm45/90]_{8s}$ and $[0/90]_{16s}$ laminates, respectively. In each case the

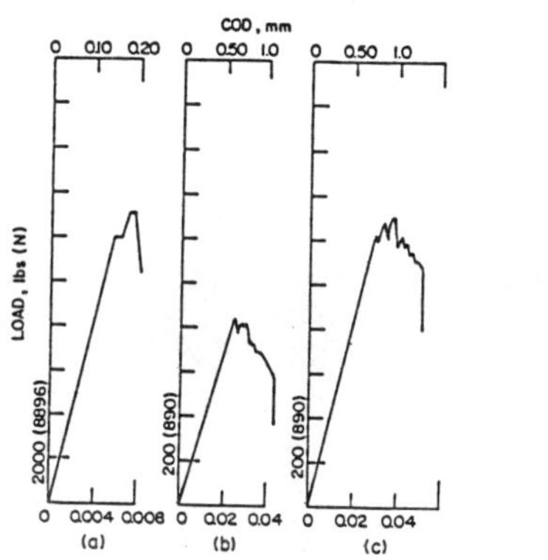

Fig. 1 Comparison of load versus COD for (a) Center-Cracked, (b) Three-Point Bend and (c) Compact Tension Specimens for the $[0/\pm45/90]_{8s}$ laminate.

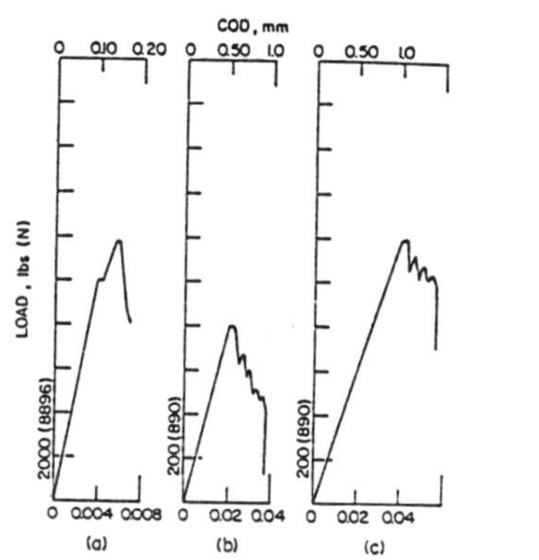

Fig. 2 Comparison of load versus COD for (a) Center-Cracked, (b) Three-Point Bend and (c) Compact Tension Specimens for the $[0/90]_{16s}$ laminate.

a. Center-cracked Tension

b. Compact Tension

c. Three Point Bend

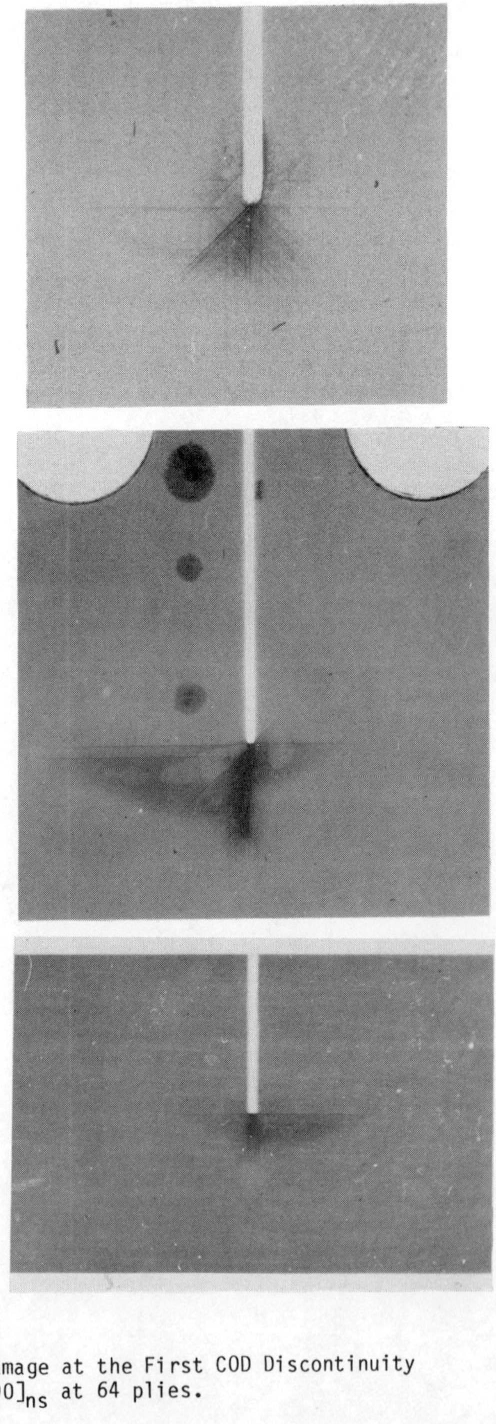

Fig. 3 Comparison of Damage at the First COD Discontinuity for the $[0/\pm 45/90]_{ns}$ at 64 plies.

a. Center-Cracked Tensio

b. Compact Tension

c. Three Point Bend

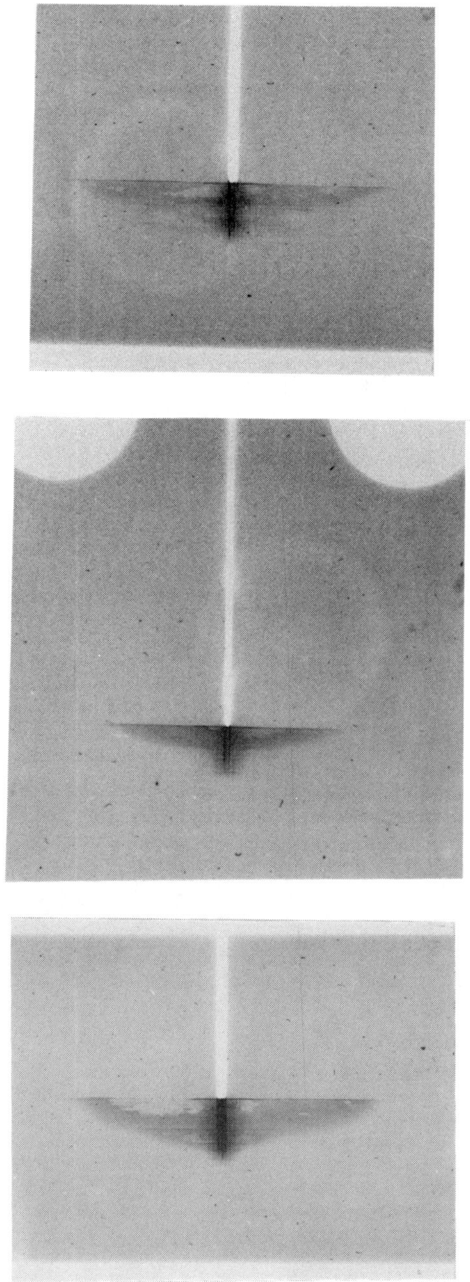

Fig. 4 Comparison of Damage at the first COD discontinuity for the $[0/90]_{ns}$ at 64 plies.

damage is characterized by a macrocrack colinear with the machined crack with an associated region of delamination and matrix cracks in the fiber directions. The TPB and CT specimens did not catastrophically fracture into two specimen halves as did the CCT specimen. When substantial damage was reached in the TPB and CT specimens the load dropped off to a very low value at which time the test was terminated. Only a small compression region on the back edge held the specimens intact. The macrocracks in the TPB and CT specimens were colinear with the machined crack. Broad regions of delamination were also present, as seen in Figs. 3 and 4. Preliminary destructive examinations indicate that these broad delamination regions are generally associated with delamination of the 0° surface plies.

The fracture toughness at the maximum load is shown in Figs. 5 and 6 as a function of specimen thickness for the $[0/\pm45/90]_{ns}$ and $[0/90]_{ns}$ laminates, respectively. Each toughness value is the average of 4 replicate tests. Data scatter was generally less than 10%. The maximum difference in the toughness values between the specimen types is 8.1% for the $[0/\pm45/90]_{ns}$ laminate and 6.5% for the $[0/90]_{ns}$ laminate.

A comparison of the toughnesses for the three specimen types is for the thicker specimens. For these specimens, damage modes such as delaminations and axial splits in the matrix appear to be primarily surface effects. In thick specimens this effect on specimen fracture would be much less significant than would be the case for the 8 ply laminates. This may in part be responsible for the change in toughness for increasing specimen thickness. Work is currently being conducted to study damage development through the specimen thickness, and will be reported in a later paper.

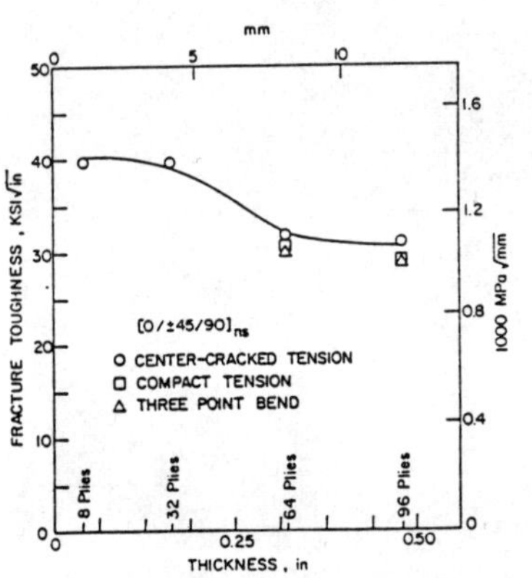

Fig. 5 Fracture toughness as a function of specimen thickness for the $[0/\pm45/90]_{ns}$ laminate.

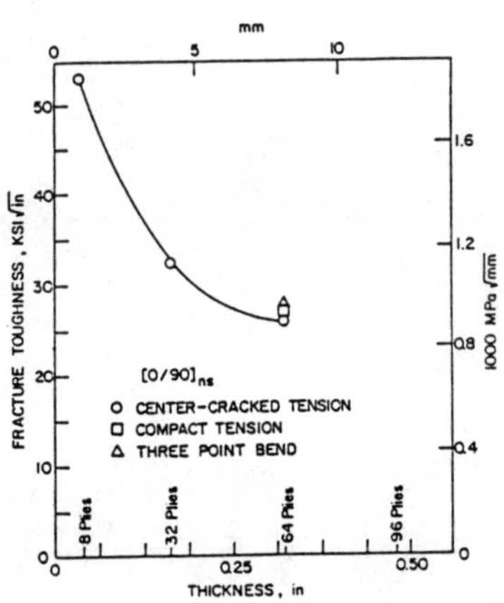

Fig. 6 Fracture toughness as a function of specimen thickness for the $[0/90]_{ns}$ laminate.

SUMMARY

Fracture toughness as a function of thickness was investigated for two laminates and three specimen geometries. Toughness decreased and approached an asymptotic value as laminate thickness increased. For thick laminates the toughness was practically independent of specimen geometry.

Enhanced x-ray photographs and removal of the outside plies revealed that a significant portion of the ply delamination was a surface effect. Work is underway to study damage development in the two laminates.

ACKNOWLEDGEMENT

The financial support provided for this work by NASA Grant NAG-1-264 from the Fatigue and Fracture Branch of NASA-Langley is gratefully acknowledged. Further, sincere appreciation is extended to C. C. Poe, Jr. of NASA-Langley for his encouragement and helpful discussions.

REFERENCES

Brown, W. F., Jr. (1970). Review of Developments in Plane Strain Fracture Toughness Testing, ASTM STP 463. American Society for Testing and Materials, Philadelphia.

Brown, W. F., Jr. and J. E. Srawley (1966). Plane Strain crack Toughness Testing of High Strength Metallic Materials, ASTM STP 410. American Society for Testing and Materials, Philadelphia.

1982 Annual Book of ASTM Standards, Part 10, American Society for Testing and Materials, Philadelphia.

Chu, C. S., J. M. Anderson, W. J. Batdorf and J. A. Aberson (1976) Finite Element Computer Program to Analyze Cracked Orthotropic Sheets, NASA Contractor Report CR-2698, National Aeronautics and Space Administration.

Cruse, T. A. and J. R. Osias (1974). Exploratory Development on Fracture Mechanics of Composite Materials, AFML-TR-74-111, Air Force Materials Laboratory, Wright-Patterson air Force Base, Ohio.

Harris, C. E., and D. H. Morris (1983). A comparison of several methods of calculating fracture toughness from composite laminate test data. Proc. of the 1983 SESA Spring Meeting (to appear).

Owen, M. J. and R. J. Cann (1979). J. of Material Science, 14, 1982-1996.

Prewo, K. M. (1978). J. of Composite Materials, 12, 40-52.

Reedy, E. D., Jr. (1980). J. of Composite Materials Supplement, 14, 118-131.

Shih, T. T., and W. A. Logsdon (1981). In R. Roberts (Ed.), Fracture Mechanics: Thirteenth Conference, ASTM STP 743, American Society for Testing and Materials, Philadelphia, pp. 316-337.

ACOUSTIC EMISSION FROM FILAMENT WOUND PIPES UNDER LONG TERM LOADING CONDITIONS

L. Gołaski*, D. Hull and M. Kumosa***

*Institute of Materials Science and Applied Mechanics, Technical University of Wrocław, Wrockław, Poland
**Department of Metallurgy and Materials Science, The University of Liverpool, Liverpool, UK

ABSTRACT

The failure processes in filament wound pipes subjected to uniaxial and biaxial loading has been studied with the help of acoustic emission. Under constant applied stress condition the total counts increased with logarithm of time. Effect of periodic unloading on acoustic emission activity has been described.

KEYWORDS

Acoustic emission; glass reinforced plastics; creep.

INTRODUCTION

Fibre composite materials are highly anisotropic and so the prediction of design limits and the analysis of failure processes is complicated. The critical loading criteria are difficult to determine because they are strongly dependent on composite structure. Because of the limitations inherent in the mechanical characterisation of fibre composite materials, new methods of testing are necessary. Acoustic emission /AE/ has become an important tool for studying the integrity and failure of components made from fibre composites (Wiliams and others, 1978; Gołaski and others, 1982; Hamstad, 1982). The summation of AE counts gives information about the accumulation of damage in the material and the rms voltage gives an indication of the severity of the damage.

Although it is known that the strength of composite materials is time dependent, the effects of time under load have not been studied extensively and many unresolved questions remain.

This paper is concerned with the application of AE techniques to fibre composites tested under plane stress long term loading conditions. The tests relate to the conditions found in pipe-lines and pressure vessels. To simulate these service conditions more fully some tests were made which included periodic unloading. A critical condition in relation to service applications is the stress at which weepage occurs. All the tests were extended to weepage.

MATERIALS AND LOADING CONDITIONS

The tests were made on filament wound pipes with a winding angle $\phi = \pm 54^{o}44'$. Each pipe was made up of four complete layers. The pipe diameter was 50 mm and the wall thickness 1.5 mm. The pipes were fabricated from Fibreglass Ltd. Suprewind 20/70 E glass fibres, 2400 Tex roving, and a terephthalic polyester resin, Impol T400. The glass had been treated with a resin compatible size containing a silane coupling agent. Full details of the method of fabrication and materials used are given elsewhere (Hull and others, 1978).

Two internal pressure tests were used which are designated Mode II and Mode III respectively. Mode II is a closed-end condition so that the hoop stress σ_H is twice the axial stress σ_A. In Mode III the ends of the tube are not constrained, being free to slide on rubber seals, so that the axial stress is zero. The pipes were pressurised smoothly up to the required pressure and then held at constant pressure. Some of the pipes were unloaded to zero pressure and reloaded to the test pressure.

The AE activity during smooth pressurisation was strong and the signal levels are best evaluated by the rms voltage. However, at constant pressure the AE activity was relatively weak and a summation of counts was used. A signal processor, Acoustic Emission Technology Corporation Model 201, was used together with a sensor with a nominal frequence of 175 kHz. In the tests the total gain was 90 dB for Mode II and 98 dB for Mode III.

RESULTS

The total counts versus time for a pipe tested in Mode II at σ_H = 136 MPa is shown in fig. 1. After 8h no leak was observed. The count rate decreased with time and during the last hour the total counts increased by only 2%. When the pressure was increased to σ_H = 170 MPa the count rate increased rapidly and after 37 min. weepage occurred. The total counts in the first 8h was 7.65×10^5 and during the next 37 min. was 6.9×10^5. Fig. 2 shows that at both pressures the total counts can be described by the relation

$$N = A + B \log t$$

A similar set of observations was obtained from a second pipe tested under similar conditions. Weepage occurred after 26 min. at σ_H = 170 MPa.

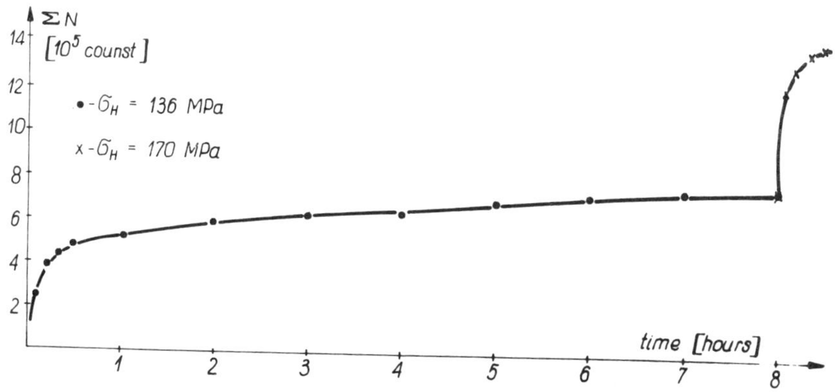

Fig. 1. AE summation from pipe loaded in M II.

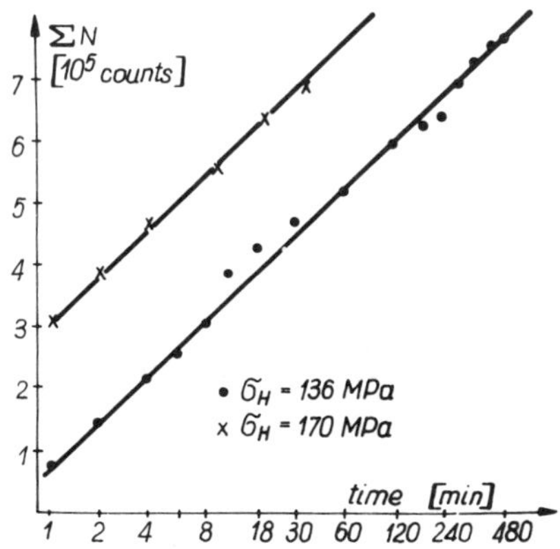

Fig. 2. AE summation from pipe loaded in M II.

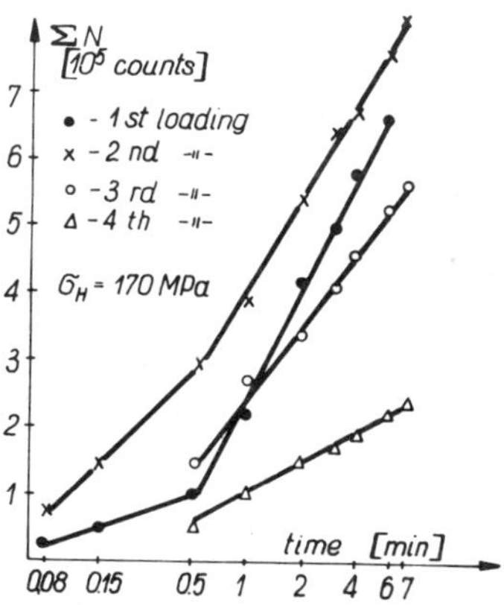

Fig. 3. AE summation from pipe loaded repeatedly in M II.

The results for a pipe loaded in Mode II periodically to σ_{II} = 170 MPa are shown in fig. 3. The pressure was applied for seven minutes before unloading and then re-applied. The cycle was repeated four times. The slope of the N - logt curve was approximately linear for each loading time but decreased at each successive application of pressure. Evidence for a delay time of about 10 to 30 s can be seen at the beginning of each cycle. This test was repeated at σ_{II} = 170 MPa on another pipe without any unloading intervals. Weepage occurred after 20 min. The results are shown in fig. 4 and the AE data for all the pipe tests is summarised in Table 1.

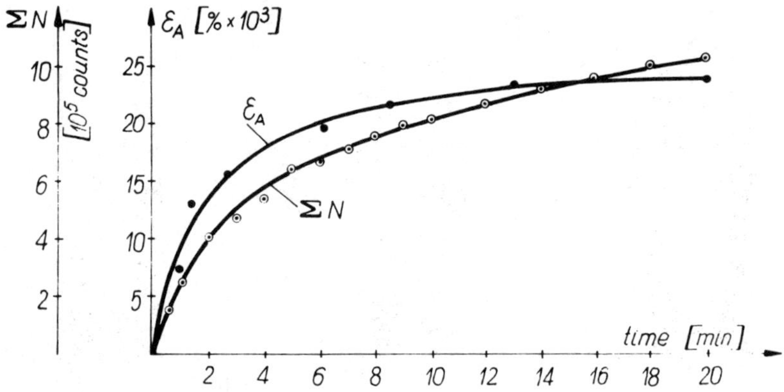

Fig. 4. Axial strain and AE summation from pipe loaded in M II σ_H = 170 MPa.

Table I. AE data for pipes tested in Mode II.

Pipe No.	ΣN after 1h at σ_H=136MPa	ΣN after 2 min. at σ_H=170MPa	Time of load at σ_H=136MPa hours	Time to weepage at σ_H=170MPa min.
1	5.3×10^5	3.9×10^5	8	37
2	6.4×10^5	4.1×10^5	24	26
3	-	4.2×10^5	-	25
4	-	4.2×10^5	-	20

The number of counts during the first 20 min. of test at σ_H = 170 MPa is 10.25×10^5 /pipe no.4, fig. 4/ which compares with 20.4×10^5 counts in a pipe loaded intermittently for the same total time. The slope of the N - logt curve for pipe no.4 is the same as for pipe no.3 during the first pressure cycle. Fig.4 shows that the total counts for a pipe loaded continuously follows the measured axial strain. Weepage appeared in the form of tiny droplets all over the surface of the pipe (fig.5.) No discontinuity in the N-t curve was observed at the onset of weepage.

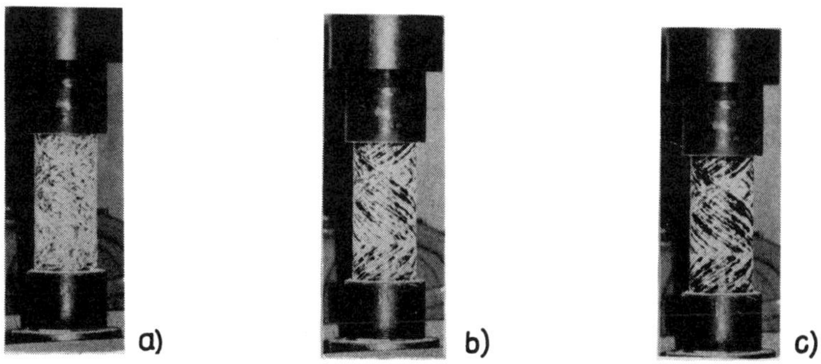

Fig. 5. Weepage in pipe pressurized at σ_H = 170 MPa
a/ at the onset of weepage
b/ 2 min. after the onset of weepage
c/ 6 min. after the onset of weepage.

The Kaiser effect was observed (Gołaski and others, 1982) at low pressures in Mode II tests during continuously increasing pressurisation. In the present tests it was found that the Kaiser effect occurred after periods up to 24h at constant pressure of σ_H = 136 MPa. However, for pipes loaded to σ_H = 170 MPa the Kaiser effect was not observed although the amount of AE activity during each loading cycle decreased progressively.

The AE activity from pipes tested in Mode III was not as strong as in Mode II tests. To obtain the same level of total counts an amplification of 98 dB was used. In contrast to the Mode II tests AE tended to occur in bursts as illustrated in fig.6. The time interval between bursts varied with pressure level. The relationship between N and logt was approximately linear (fig.7) confirming the N = A + B logt relationship.

No weepage was observed in pipes pressurised at σ_H = 210 MPa for 18h or at σ_H = 250 MPa for 6h. Pipes continuously pressurised started to weep at σ_H = 260 MPa. Weepage occurred in the form of fine jets (Hull and others, 1978).

The effect of pressure on the composite materials in the wall of the pipes is to produce transverse cracking in the Mode II tests and intralaminar shear cracking in Mode III tests. Laminar cracking is followed by interlaminar fracture. Hull and others (1978) found that all these processes lead to progressive decoupling of the fibre and the matrix so that the contribution of the resin to the composite properties decreases. This has a substantial effect on the stiffness moduli which approach the values predicted by netting analysis. The change of stiffness modulus $S_A = \frac{\sigma_H}{\varepsilon_A}$ with loading time for Mode II and Mode III tests is shown in fig.8. The stiffnesses vary linearly with time and the changes are strongly dependent on

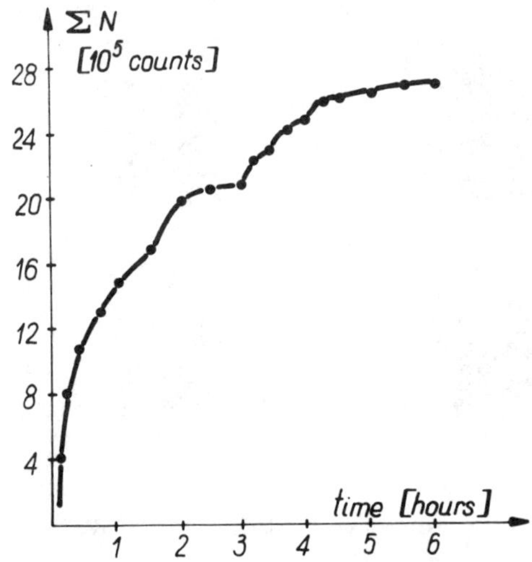

Fig. 6. AE summation, pipe loaded in M III, σ_H = 250 MPa.

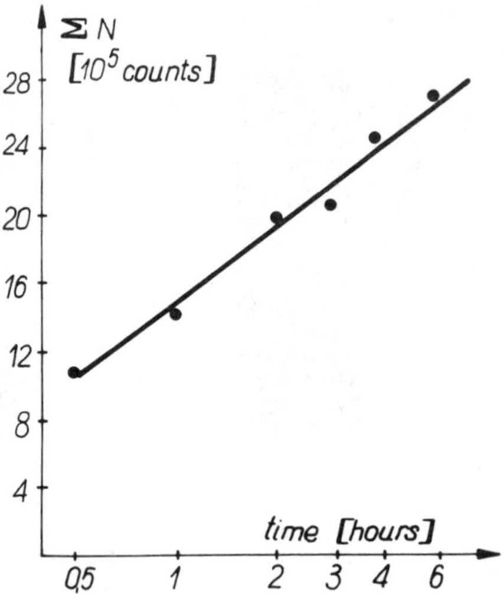

Fig. 7. AE summation, pipe loaded in M III, σ_H = 250 MPa.

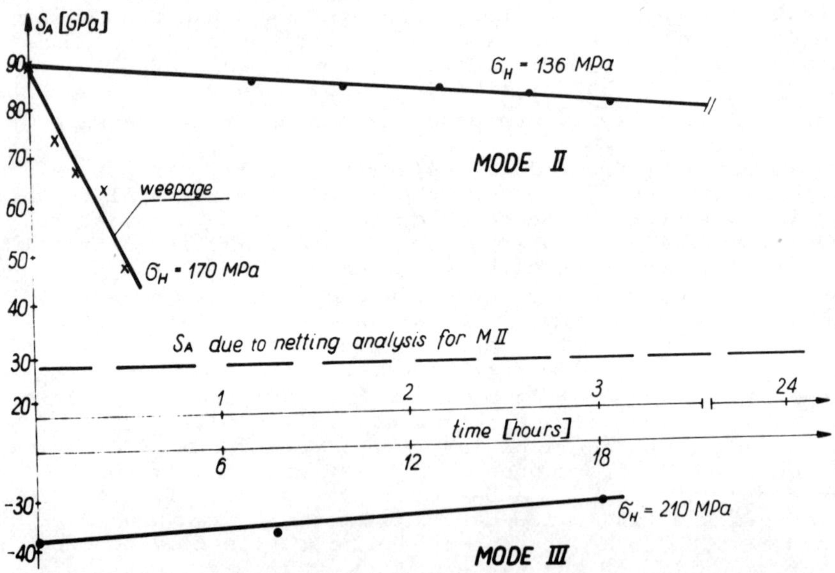

Fig. 8. Changes of axial stiffness for pipes tested in M II and M III.

the applied pressure. There is no discontinuity in the stiffness-time curves at the onset of weepage.

CONCLUSIONS

a/ AE occurs during long term loading.

b/ The total counts increase linearly with log t.

c/ The slope of the N - logt curve depends on applied pressure.

d/ Unloading during a constant pressure test results in a change in the slope of the N - logt curve which decreases progressively with each unload.

e/ The total counts, in a given time, for a pipe tested in repeated loading is greater than in a pipe tested without interruption.

f/ There is no discontinuity in the N - t curve when weepage occurs.

g/ Changes in the Kaiser effect may lead to a prediction of weepage time and pressure.

h/ The shape of the N - t curve is simular to the axial strain - t curve at constant pressure.

i/ Periodic bursts of AE activity occur during Mode III testing.

j/ Long term loading results in a decrease in stiffness which is dependent on the pressure mode und value.

REFERENCES

Gołaski, L., M. Kumosa, and D. Hull (1982). J. Acoustic Emission, 1, 95-101.

Hamstad, M.A. (1982). J. Acoustic Emission, 1, 151-156.

Hull, D., M.L. Legg and B. Spencer (1978). Composites, 9, 17-24.

Wiliams, J.H. and S.S. Lee (1978). J. Composite Mat., 12, 348-370.

A STUDY ON INTERLAMINAR SHEAR STRENGTH OF COMPOSITES

Shuangyin Zhang

Institute of Mechanics, Chinese Academy of Sciences, Beijing, China

ABSTRACT

An evaluation about the validity of the double-grooved specimen tensile-shear testing method to measure the interlaminar shear strength of the laminated composites is presented. The paper is divided into three parts: experimental investigation, FEM calculation and shear strength analysis. The experimental results exhibit that the shear strength q_0 varies with the length of the shear surface. The FEM calculations show that there is a severe stress concentration at the grooves. Based on the results of the experiments and calculations, a shear strength analysis was made. The shear fracture mechanism was analyzed and the concept of critical shear damage dimension was proposed.

KEYWORDS

Interlaminar shear strength; laminated composites; critical damage dimension; grooved specimen; tensile-shear test; FEM.

INTRODUCTION

Interlaminar shear failure is often found in laminated composites. Interlaminar shear strength is an important material parameter, which must be considered in the design of composite structures. Several testing and analyzing methods have been developed so far to characterize and evaluate this strength.

Chiao and co-workers(1977) published a set of experimental results obtained by using double-grooved tensile-shear specimen. The data given by them lie in a wide scattering band from 7.7 MPa through 23.5 MPa. Markham and Dawson (1975) proposed a simple analytic model for calculating the shear stress distribution and obtained the following formula:

$$\frac{q}{q_o} = \frac{\theta \ Ch\ (\ \theta - \beta x\ )}{Sh\ \theta} \tag{1}$$

where $\theta = (\ L/t\ )\sqrt{(\ G/2E\ )}$, $\beta = (\ 1/t\ )\sqrt{2G/E}$; G and E are equivalent shear and tensile moduli respectively; q_0 is the mean shear stress and can be calculated from formula (2).

$$q_o = 2p_o t\ /\ L \tag{2}$$

where x, t and L are shown in Fig. 1. p_o is the applied tensile stress.

The experimental shear strength given by Chiao and her colleagues (1977) is the mean shear stress, q_o. The shear stress calculated from formula (1) does not distribute uniformly along the shear surface. If the maximum shear stress given by formula (1) is regarded as the interlaminar shear strength, it will be a better approximation to the accurate value than q_o. But formula (1) takes no account of the stress concentration in the immediate vicinity of the groove. The shear strength value calculated from it is by no means the exact one.

In the present paper, in order to evaluate the validity of the double-grooved specimen tensile-shear method, experiments and FEM calculations were carried out. Based on the results of the experiments and FEM calculations, an analysis on the shear fracture mechanism was made and a modified Markham formula and the concept of critical damage dimension were proposed.

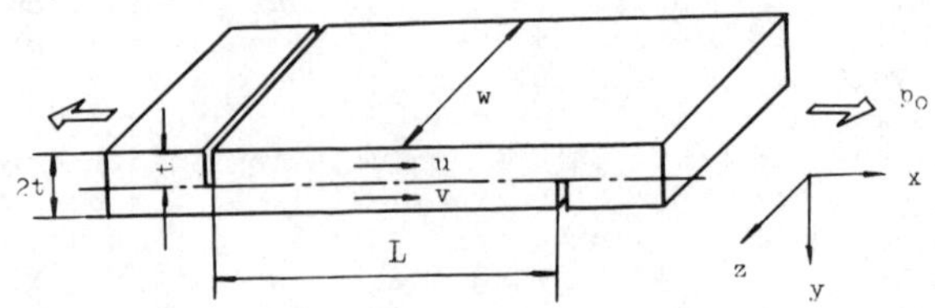

Fig. 1. The configuration of double-grooved specimen.

EXPERIMENTAL PROCEDURE AND RESULTS

Test Material

The test material is one kind of GFRP-roving glass cloth reinforced phenolics resin. The ratio of volume fraction of warp fibres to that of weft fibres is 1:1. Before doing tensile-shear test, two additional tests were conducted. The first test is to measure the volume fractions of the constituent materials and the second to determine the elastic constants, which were used in the FEM calculations and the shear strength analysis. The volume fraction of the glass fibre V_g is equal to 55.8% and that of the phenolics resin V_{ph} is equal to 44.2%.

The Results of the Tensile-shear Tests

The experimental data are listed in Table 1. The specimens are divided into two categories according to the length of the shear surface. The long specimens were tested by tensile-shear load, whereas the short ones were loaded by compressive-shear load. The values in the column of shear strength are the mean shear stress q_0 as given by formula (2).

TABLE 1 Specimen Configuration and Experimental Results

Specimen Dimension (mm)					Experimental Results			
No.	Length of shear surface	Depth of groove	Width	Thick-ness	Test method	Failure load KN	Shear strength q_0 (MPa)	Average value of q_0 (MPa)
1	100.4	11.6	40.1	20.9	Tensile shear	13.9	3.47	
2	100.0	11.7	40.5	20.7		15.8	3.90	3.62
3	101.0	11.7	40.5	20.8		14.3	3.50	
4	79.6	11.8	40.5	20.8		14.9	4.60	
5	80.1	11.5	39.8	20.8		13.9	4.36	4.43
6	79.5	10.7	40.4	20.8		13.9	4.33	
7	49.0	11.9	40.4	20.8		12.6	6.40	
8	48.8	11.7	40.3	20.9		11.6	5.90	6.10
9	49.0	11.8	40.5	20.8		11.7	5.90	
10	99.1	11.7	20.1	21.0		5.70	2.85	
11	99.1	11.8	20.6	20.8		7.40	3.82	3.34
12	79.1	11.8	20.2	20.2		7.60	4.77	
13	79.1	11.6	19.8	20.9		5.60	3.56	4.20
14	29.9	11.7	40.0	20.8	Compres-sive shear	23.8	19.9	
15	29.1	11.3	39.9	21.0		21.5	18.5	19.2
16	19.3	11.6	40.0	20.8		16.6	21.5	
17	19.4	11.7	40.0	20.8		13.8	17.8	20.7
18	19.4	11.7	40.1	20.9		17.6	22.7	
19	29.2	11.8	20.0	21.1		6.50	11.1	
20	29.2	11.7	20.0	21.0		7.92	13.6	12.3
21	18.6	11.65	20.0	21.7		9.40	25.3	25.3

The test results of the shear strength q_0 are considerably influenced by the length of the shear surface, L. The larger the length of the shear surface is, the smaller the value of q_0 will be.

The values obtained from the short specimens by using compressive-shear load is much larger than those from the long specimens by using tensile-shear load. The reason for this phenomenon is twofold. One is due to the effect of the length of shear surface, and the other is due to the existence of normal stress component on the shear surface. In the case of tensile-shear, the normal stress is tensile stress, whereas in the case of compressive-shear, it is compressive. The values of these normal stresses can be of the same order of magnitude as the shear stress component. Their influence is not ignorable. The tensile stress component promotes the shear failure of the specimens and renders a lower shear strength. On the contrary, the effect of the compressive stress component is just the opposite.

FEM CALCULATIONS

Plane Strain Problem

When the width of the specimen shown in Fig. 1 is much larger than its thickness, 2t, the stress distribution is independent of z coordinate. This problem can be reduced to the plane strain problem. The general Hook's law can be expressed in the matrix form.

$$\{ \varepsilon_{ij} \} = [R_{ij}] \{ \sigma_{ij} \} \tag{3}$$

$$R_{ij} = S_{ij} - S_{i3} \cdot S_{j3} / S_{33} \tag{4}$$

where S_{ij} are the stiffness coefficients in three-dimension case.

By using formula (4), R_{ij} can be derived from S_{ij} measured through aforementioned tests. The engineering elastic constants which correspond to R_{ij} are: E_L = 19.5 GPa; E_T = 4.1 GPa; G_{LT} = 1.02 GPa; ν_{LT} = 0.059.

Finite Element Divisions

Two different finite element divisions are shown in Fig. 2 and Fig. 3. The grids of one mesh pattern, Fig. 2, are more uniform than the other one, Fig. 3. The calculated results of the latter mesh pattern exhibit larger stress concentration than the former one.

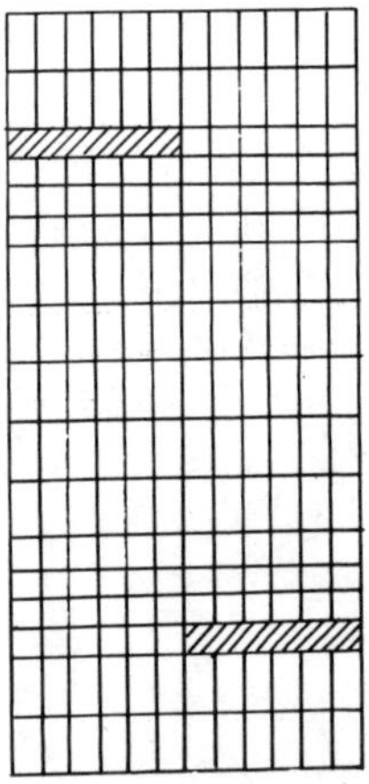

Fig. 2. Finite element mesh (I).

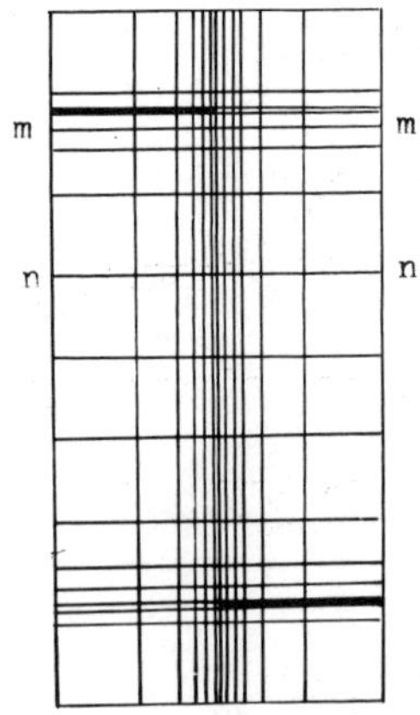

Fig. 3. Finite element mesh (II).

Results and Discussion

The curve of the shear stress obtained by using the mesh pattern II was drawn in Fig. 4. The curve of formula (1) was also given in this figure. It can be seen that there is a marked difference between them. The curve of FEM shows a large stress concentration in the zone near the groove. The curve of formula (1) does not show stress concentration.

The distribution of the longitudinal displacement along the cross section of the specimen is shown in Fig. 5. It indicates that under the tensile-shear load, the cross section of the specimen is no longer plane, but curves tremendously. The nearer to the root of the groove the cross section lies, the larger it curves. This fact rejects the following assumption made by Markham (1975), in deriving the formula (1).

$$(u - v) / t = q / G \tag{5}$$

where u and v are the average displacements over the two halves of the thickness of the cross section, which positioned on the two sides of the central line of the cross section respectively, as shown in Fig, 1. This is the reason why the curve of FEM differs from that of formula (1).

Figure 6 illustrates the comparison between the normal stress component and the shear one. In the vicinity of the groove, the normal stress component is quite large.

In Fig. 6, we can also see the influence of the equivalent shear modulus, G_{LT}, on the stress distributions. The larger G_{LT} renders a higher concentration of shear stress. The effect of G_{LT} on the normal stress σ_y has the similar trend.

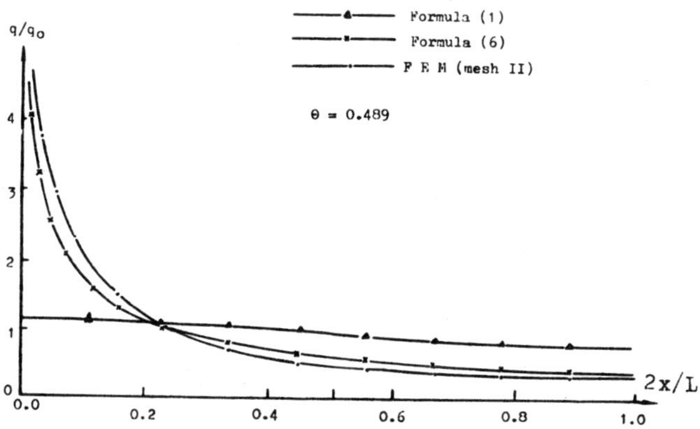

Fig. 4. Comparison of FEM results with formula (1) and formula (6).

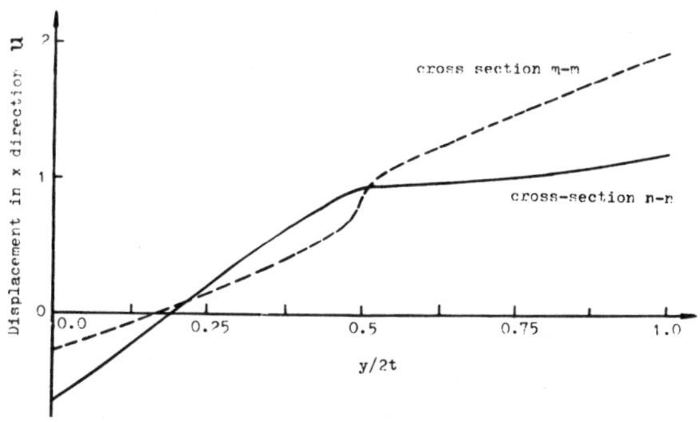

Fig. 5. Distribution of longitudinal displacement on two cross sections (mesh pattern II).

SHEAR STRENGTH ANALYSIS

Modified Markham Formula

The following modified Markham formula was proposed to take the stress concentration into consideration.

$$\frac{q}{q_0} = A\cdot\left(\frac{2\theta x}{L}\right)^{-1/2}\cdot\frac{\theta}{\mathrm{Ch}\,\theta}\cdot\mathrm{Ch}\left(\theta-\frac{2\theta x}{L}\right) \tag{6}$$

where $(2\theta x/L)^{-1/2}$ expresses the singularity of the shear stress at the root of the groove. The coefficient A can be determined by using formula (7).

$$(2/l)\cdot\int_0^{L/2} q/q_0\cdot dx = 1 \tag{7}$$

The coefficient A depends on the parameter θ. Table 2 lists the values of A for three different shear surface lengths of the specimens which all have the same thickness 2t = 12 mm.

TABLE 2 The Coefficient A for Three Different L

L	100 mm	50 mm	30 mm
θ	1.63	0.815	0.489
A	0.54	0.426	0.343

In the FEM calculations described in the section 2 of this paper, the value of θ is equal to 0.489. Substituting θ = 0.489 into formula (6), the shear stress distribution was calculated, the curve of which is given in Fig. 4. It can be seen that this curve coincides with that of FEM fairly well.

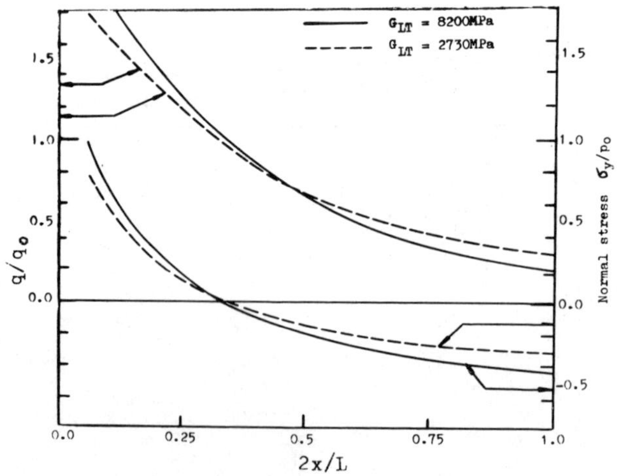

Fig. 6. Comparison of normal stress component with shear stress component (FEM mesh I).

Shear Strength and Critical Damage Dimension

The stress concentration near the groove incurs the material damage prematurely. The shear stress in the damage zone can only reach its critical value q_c, just like mild steel yields at the yield stress. As the load increases, the damage zone enlarges gradually. When the size of this damage zone reaches its critical value, named as critical damage dimension, r_c, the material fractures catastrophically.

Based on this idea, the following equilibrium equation can be written:

$$t \cdot p_o = P_b / 2w = q_c \cdot r_c + \int_{r_c}^{L/2} q \cdot dx \tag{8}$$

or

$$t \cdot p_o = q_c r_c + \int_{r_c}^{L/2} A \cdot \left(\frac{2\theta x}{L}\right)^{-1/2} \cdot \frac{\theta}{\mathrm{Sh}\theta} \cdot \mathrm{Ch}\left(\theta - \frac{2\theta x}{L}\right) \cdot dx \tag{9}$$

where P_b is the failure load.

Equation (8) has two unknown parameters, q_c and r_c. If one of them is known, the other one can easily be obtained. For example, we assumed that the shear strength q_c of the laminated composite equals to the shear strength of the bulk resin, say q_c = 15 MPa, we obtained r_c = 2.8 mm.

According to the failure mechanism described above, we can sketch the stress distribution along the shear surface just before fracture of the material, see Fig. 7.

It is clear that the shorter the shear surface is, the closer the mean shear stress q_0 approaches to the shear strength q_c. When the length of the shear surface, L, equals to $2r_c$, q_0 will be the true shear strength, q_c.

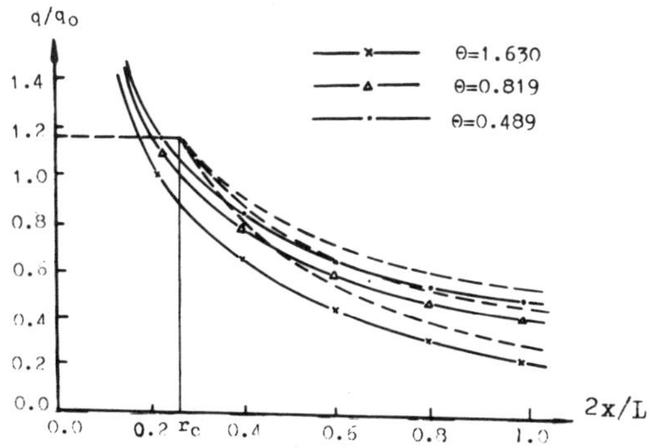

Fig. 7. Shear stress distribution on the shear surface.

CONCLUDING REMARKS

The shear strength measured by double-grooved specimen tensile-shear testing method depends on the shear surface length. Markham's formula (1) takes no account of the stress concentration at the root of groove. The modified formula (6) proposed here coincides the results of FEM fairly well. The concept of critical damage dimension extends some insight into the mechanism of the shear failure of laminated composite in this test.

ACKNOWLEDGEMENT

The author would like to express his gratitude to Mr. Jiang Canxing, who extended his kind help to the author in the experiments.

REFERENCES

Chiao, C. C., R. L. Moore, and T. T. Chiao (1977). Measurement of shear properties of fibre composites, Part 1. Evaluation of test methods. Composites, 8, 161-169.

Markham, M. F., and D. Dawson (1975). Interlaminar shear strength of fibre reinforced composites. Composites, 6, 173-176.

Zhang, Shuangyin, (1982). A hybrid FEM approach to stress distribution in non-homogeneous materials. Proceedings of the International Conference on Finite Element Methods. Science Press, Beijing, China, pp. 318-323.

TOUGHNESS OF THIN SLABS UNDER FLEXURAL OR IMPACT LOADING

N. I. Fattuhi

Civil Engineering Department, Kuwait University,
College of Engineering and Petroleum, Kuwait

ABSTRACT

Different size mortar slabs were reinforced with monofilament polypropylene and hooked steel fibres, fibrillated polypropylene network and deformed steel bars.

The 1300 x 300 x 20 mm slabs were tested in flexure under central point loading, and continuous load-central deflection curves were obtained. The tests indicated that the highest flexural strength and toughness resulted when slabs were reinforced with polypropylene network. These slabs also exhibited multiple cracking unlike slabs reinforced with short fibres which failed mainly by a single crack.

Impact tests were performed on 750 x 750 x 60 mm slabs supported on four sides on a steel frame. A steel ball positioned at a predetermined height was dropped at the centre of the slabs, and deflections at different positions were measured after different numbers of blows. The results indicated that with the different types of fibre reinforcement used, the slab reinforced with hooked steel fibres exhibited the least deflection under the condition of varying numbers of blows.

KEYWORDS

Fibre reinforced concrete; polypropylene fibres; steel fibres; polypropylene network; cracking; deflection; toughness; flexural strength; impact resistance.

INTRODUCTION

Recent concern about health risks associated with the use of asbestos fibres has directed attention towards the utilization of other types of fibres such as glass, polypropylene and steel. One of the principal uses of asbestos fibres is in the manufacture of thin asbestos cement sheets, in which the fibres considerably enhance the strength, toughness and fire resistance of the composite.

The use of the glass fibres as a reinforcement for cement has increased in recent years (Kempster, 1981), after the introduction of alkali-resistant glass fibres. However, the durability of the composites made with these fibres after prolonged exposure to moisture and high temperature is under question. Despite their low modulus of elasticity, fibrillated polypropylene fibres have been used for many years in the manufacture of concrete piles to increase their impact resistance (Fairweather, 1971). Also, the increase in impact resistance of concrete stair treads has been demonstrated (Bailey and co-workers, 1975) by the inclusion of polypropylene and steel fibres. The geometry of the steel fibres has a significant effect on the flexural strength (Hughes and Fattuhi, 1977) and impact resistance (Bailey and co-workers, 1975) of the composite. However, steel fibres are relatively expensive and susceptible to corrosion.

The recent introduction of fibrillated polypropylene network (Hannant, Zonsveld and Hughes, 1978) has made possible many practical applications. One of these is their use in the production of fibre reinforced cement permanent formwork (Raithby, Galloway and Williams, 1981). However, earlier investigations (Hannant, Zonsveld and Hughes, 1978; Dave and Ellis, 1979) were mainly concerned with the reinforcement of small thin slabs with fibrillated polypropylene network. Recent investigation (Fattuhi and Hughes, 1982) has shown that the efficiency of fibre reinforcement would significantly be reduced when the beam's span to depth ratio is increased.

The present investigation is concerned with the reinforcement of relatively large cement mortar slabs with monofilament polypropylene and hooked steel fibres and fibrillated polypropylene network. Their benefits in increasing the toughness of the composite under flexural and impact loading is experimentally evaluated.

MATERIALS AND PREPARATION OF SPECIMENS

A mortar mix with the proportions of 1:2:0.56 ordinary portland cement:sand zone 2:water by weight was used for making the different slabs. Polypropylene and steel fibres with dimensions shown in Table 1 were used to reinforce the impact and flexural slabs. Continuous fibrillated polypropylene network opened and stabilized was used to reinforce two flexural slabs, while straight deformed bars 8 mm in diameter (F_y = 460 N/mm^2) were used to reinforce one impact slab in two directions.

TABLE 1 Fibre Properties

Fibre Type	Nominal Length* (mm)	Measured Length** (mm)	Measured Cross-Section** (mm)
Monofilament	20	19.9	0.22x0.30
Polypropylene	35	35.8	0.22x0.30
410 Denier	50	49.4	0.22x0.30
Hooked Steel Dramix ZD 40/40	40	41.0	0.40

* Specified by the manufacturer

** Minimum of 30 samples

A drum mixer was used to prepare the mortar matrix and fibres with a volume fraction (V_f) of 1% were added continuously to the mix until uniform fibre distribution was achieved. For the fibrillated polypropylene network reinforced slabs, a thin layer of mortar matrix was first placed at the bottom of the mould. Then a set of four network layers was placed on top and was embedded into the mortar. Mortar and sets of network layers were then successively added to form a slab of the required thickness and network content. Steel moulds 1300 x 300 x 20 mm and 750 x 750 x 60 mm were used for making the flexural and impact slabs respectively.

All specimens were compacted on a table vibrator, then covered with polythene sheets until demoulding 24 hours later. The flexural slabs were placed in the laboratory covered with wet burlap at 24 $\pm$ 2°C until the age of 14 days when they were tested. The impact slabs were placed in a water tank outside the laboratory where the temperature varied between 20-37°C until they were tested at the age of 14 to 17 days.

FLEXURAL TESTING

The flexural slabs were loaded centrally over a 1000 mm span in a purpose made steel frame. Steel rollers were used for supporting and loading the slabs, and the load was applied on the cast face upwards at an average rate of 2.2 mm/min, using a manually operated loading jack. The load and central deflections were measured by a 10 kN capacity tension-compression load cell and an LVDT transducer, respectively. The load cell and the transducer were connected to an amplifier, and the output was fed into an x-y plotter which produced complete load-central deflection curves for the tested slabs.

IMPACT TESTING

The impact slabs were placed in a purpose made steel frame and were supported on their four sides. Softwood strips 12.6 mm in thickness were placed between the slabs and the steel frame to ensure full contact between the slab and the frame before testing. A circular aluminium plate 120 mm in diameter and 29 mm in thickness with shamphered edges was placed centrally on the slab. The plate prevented the crushing of the surface of the slab as a result of the fall of the 7.140 kg steel ball. A height of drop of 500 mm was adopted in order to allow the non-reinforced slab to fail after more than one drop. The height of drop was changed from 500 mm to 1000 mm after the number of blows reached 20. When the number of blows reached 30, the aluminium plate was removed, and the steel ball was allowed to fall directly on the surface of the slab. However, the height of drop for the slab reinforced with deformed bars was maintained at 1000 mm throughout the test. The deflections of the top surface of the slabs at the different points shown in Fig. 1 were measured (relative to the supporting steel frame) with a set of dial gauges fixed on a mounting frame. Deflections were measured after different numbers of blows.

FLEXURAL TEST RESULTS

Details of the flexural slabs are presented in Table 2, while the results of the flexural tests are shown in Table 3. The calculation of the flexural strengths was based on the applied loads and the selfweight of the slabs, since the latter had significant contribution. The tests indicated that slabs (F61 and F62) reinforced with fibrillated polypropylene network resulted in the

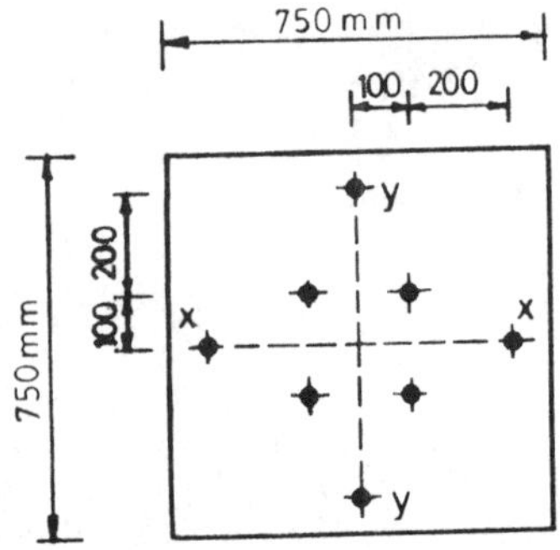

Fig. 1. Location of deflection measurement points for impact slabs.

highest flexural strengths, followed by slabs (F21 and F22) reinforced with 20 mm monofilament polypropylene. The moderate increases in strength over the non-reinforced slabs can be attributed to high slab span to depth ratio

TABLE 2 Details of Flexural and Impact Slab

Flexural Slab No.	Impact Slab No.	Reinforcement Content - V_f (%)	Reinforcement Type
F11, F12	11	0	None
F21, F22	12	1	20 mm Polypropylene fibre
F31, F32	13	1	35 mm Polypropylene fibre
F41, F42	14	1	50 mm Polypropylene fibre
F51, F52	15	1	40 mm Hooked steel fibre
F61, F62	--	1	Polypropylene network
-	16	1.2	Deformed bars

(Fattuhi and Hughes, 1982) and relatively low volume fraction of fibres. The relatively low strengths of slabs (F51 and F52) reinforced with hooked steel fibres could be attributed to poor distribution and low volume fraction of fibres.

TABLE 3 Flexural Strength Results

Slab No.	MOR* (N/mm^2)
F1	6.32
F2	7.03
F3	6.04
F4	5.30
F5	5.71
F6	7.26

*Modulus of rupture is average of two slabs.

Most slabs failed by a single crack. However, both slabs reinforced with fibrillated polypropylene network and one slab reinforced with hooked steel fibres exhibited extensive cracking after reaching their ultimate loads indicating considerable ductility. Examination of the failure surfaces indicated that all types of fibres failed by pulling out of the mortar matrix, while the fibrillated network fractured after the slabs underwent considerable deformations.

Load-central deflection curves for the highest strength from selected pairs of slabs are shown in Fig. 2. If the areas under these curves are assumed to

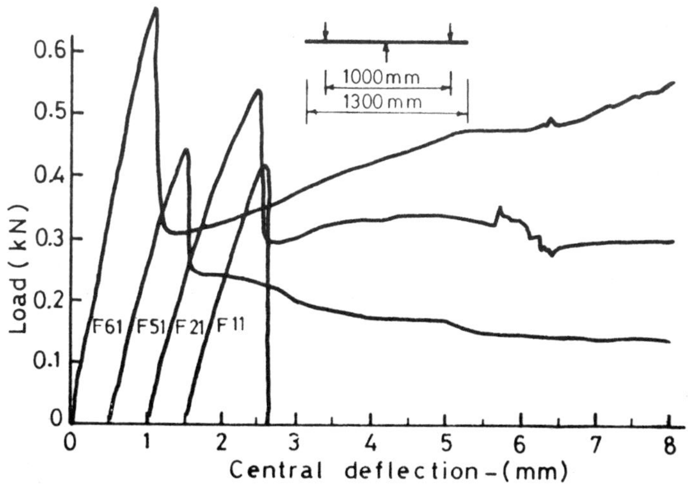

Fig. 2. Load central deflection curves for flexural slabs.

be a measure of toughness, then the benefits of fibre and network reinforcement are clearly demonstrated by the portions of the curves beyond the ultimate applied loads. However, the toughness of each slab in the pair varied as shown in Table 4.

TABLE 4 Load at 6 and 8 mm Deflections as a % of Maximum Applied Load for Selected Flexural Slabs

Slab No.	Load at 6 mm Deflection (%)	Load at 8 mm Deflection (%)
F21	53	54
F22	36	35
F51	34	29
F52	96	86
F61	71	83
F62	69	80

The results indicate the superior toughness exhibited by the slabs reinforced with the fibrillated polypropylene network. These slabs maintained nearly 85% of the ultimate load after undergoing 20 mm deflections. The tail of the curve shown in Fig. 2 demonstrates the gradual transfer of stress from the mortar matrix to the network reinforcement. Differences in the initial slopes of the curves were mainly due to different slab thicknesses. The average slab thickness near the failure surface varied between 19.8 and 23.1 mm, being thickest for slabs F61 and F62.

IMPACT TEST RESULTS

Details of the impact slabs are presented in Table 2. The deflections of the top surface of different slabs at the different points (Fig. 1) in the x-y directions are shown in Fig. 3.

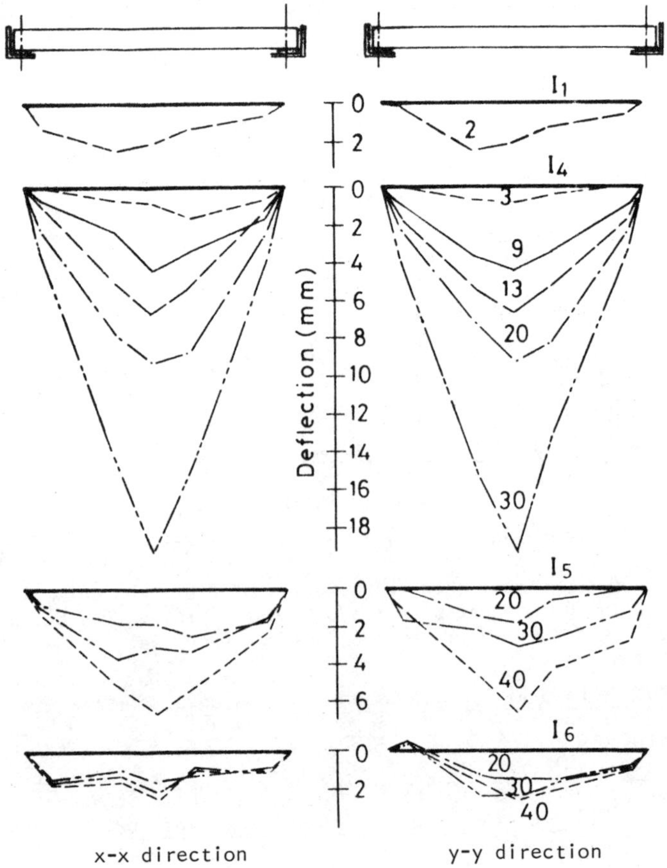

Fig. 3. Deflections of different slabs in x-y directions.

The results indicated that the non-reinforced concrete slab (I1) failed after two blows only. Among the specimens with different lengths of monofilament polypropylene reinforcement, the slab reinforced with 50 mm long fibres (I4) demonstrated the least deflections under the conditions of varying numbers of blows. All slabs reinforced with monofilament polypropylene fibres developed wide cracks and hairline cracking was not visible.

The slab reinforced with hooked steel fibres (I5) showed considerable resistance to impact loading by producing smaller deflections than slabs reinforced with monofilament polypropylene fibres (Table 5). This is thought to have been achieved by the formation of extensive hairline cracking in the vicinity of the impact area.

The results also indicated that the slab reinforced with deformed bars demon-

TABLE 5 Deflection of Impact Slabs

Slab No.	No. of Blows to 1st Visual Crack	Deflection of Central Point at (mm)			
		9 Blows	20 Blows	30 Blows	40 Blows
11	2	--	--	--	--
12	3	6.83	17.76	--	--
13	3	6.98	17.69	--	--
14	3	4.52	9.31	19.68	--
15	6	0.18	1.88	3.16	6.60
16	11	0.67	1.58	2.25	2.59

strated the least deflections indicating their superior efficiency. This was partly due to their higher volume fraction (V_f = 1.2%). However, the combination of deformed bars and fibre reinforcement should produce composites of considerable toughness and ductility.

The inclusion of all types of reinforcement delayed the appearance of the visual cracks. The hooked steel fibres were superior to the polypropylene fibres in inhibiting the appearance of the first crack.

CONCLUSIONS

The tests indicated that the addition of 1% volume fraction of monofilament polypropylene and hooked steel fibres and fibrillated polypropylene network resulted in little increase (if any) in the flexural strength of thin mortar slabs. The largest increase of 15% over the nonreinforced slabs was that obtained by using fibrillated polypropylene network. These slabs also exhibited extensive cracking after reaching their ultimate loads indicating considerable ductility. The slabs also maintained nearly 85% of their ultimate loads after undergoing 20 mm deflections, indicating considerable toughness.

The impact resistance of the slab reinforced with hooked steel fibres was superior to those reinforced with different lengths of monofilament polypropylene fibres. Also, the addition of the hooked steel fibres resulted in considerably smaller deflections under all testing conditions, but the extent of hairline cracking was greater than in the case of polypropylene fibre reinforcement. Generally, the inclusion of fibres maintained the integrity of the slabs after failure under flexural or impact loading.

ACKNOWLEDGEMENT

Thanks are due to the fibre suppliers, namely Plasticisers Ltd. (U.K.), and Bekaert (Belgium). The assistance of Mr. M. Wafa in carrying out the experimental work is gratefully acknowledged.

REFERENCES

Bailey, J.H., S. Bentley, B. Mayfield, and P.S. Pell (1975). Mag. Concr. Res., 27, 167-170.

Dave, N.J., and D.G. Ellis (1979). Int. J. Cement Compos., 1, 19-28.

Fairweather, A.D. (1971). Prospects for Fibre Reinforced Construction Materials, Exhibition Conf., London, 41-44.

Fattuhi, N.I., and B.P. Hughes (1982). Int. J. Cem. Compos. and Lightweight Concr., 4, 117-122.

Hannant, D.J., J.J. Zonsveld, and D.C. Hughes (1978). Composites, 9, 83-88.

Hughes, B.P., and N.I. Fattuhi (1977). Mag. Concr. Res., 29, 199-206.

Kempster, E. (1981). Civ. Engng, 57-63.

Raithby, K.D., J.W. Galloway, and R.I.T. Williams (1981). Int. J. Compos. and Lightweight Concr., 3, 237-246.

MEASUREMENT OF INTERFACIAL SHEAR STRENGTH IN GRAPHITE FIBER/ALUMINUM MATRIX COMPOSITES

V. Provenzano, R. J. Weimer and S. C. Sanday

U.S. Naval Research Laboratory, Washington, DC 20375, USA

ABSTRACT

The first quantitative measures of the interfacial shear strength of the bond between an aluminum matrix and graphite fibers have been completed for composites fabricated from ion-plated composite precursor tapes.

KEYWORDS

Metal matrix composites; interfacial bond strength; fiber critical length; aluminum; graphite; ion-plating.

INTRODUCTION

Ceramic fiber/metal matrix composites are of great interest for structural applications because of their high density-normalized stiffness and strength. However, successful performance of such composites depends largely on attaining satisfactory interfacial bond strength between a typically brittle fiber and a ductile metal matrix, because it is the interfacial bond that transfers loads between matrix and fibers to produce composite behavior. Quantitative determination of interfacial bond strength and its effect on macro-mechanical composite properties is essential to the rational development of composite materials and, ultimately, to the design of composite structural components.

Standardized test methods that provide a quantitative measure of fiber/matrix interfacial strength are not available for brittle-fiber-reinforced metal matrix composites. Conventional adhesive tests, such as the lap shear test and the T-peel test, have limited applicability because they are model tests, and interfacial failure in composites is a complex process not easily correlated to results from a model test. A direct measure of the strength of the bond between fiber and matrix would be provided by a fiber pullout test, which is both simple and straightforward. However, major difficulties arise when attempting to apply the test to graphite-fiber/aluminum-matrix composites because of intrinsic properties of the constituent materials. First, the graphite fibers are so small in diameter and so brittle that specimen preparation and test procedures appear impractical. Second, the aluminum matrix work-hardens considerably in the vicinity of the fiber; this process has been shown to artificially inflate the magnitude of interfacial shear strength measured in pullout tests on other composites (Kelly and Davies, 1965).

Two other techniques presently being developed appear promising to provide direct quantitative measures of the bond strength between a ductile matrix and a small-diameter brittle fiber. The first technique consists of diffusion or adhesion bonding a thin sheet of unidirectionally reinforced Al/Gr between two cantilever beams so shaped that their compliance remains constant when a crack is propagated along the bond line. This specimen configuration yields both the transverse fracture toughness and, if the crack propagates in interface failure, an estimate of the tensile strength of the fiber/matrix interface bond. Progress in this development will be reported elsewhere in the near future. The subject of this paper is a second technique, which is a modification of a procedure first suggested by Kelly and Tyson, (1964), whereby the interfacial shear strength is calculated from measured values of fiber critical aspect ratio in specimens loaded in tension. A brief description of the pertinent theory is presented in next section.

APPLICABLE THEORY

The experimental technique used by Kelly and co-workers, and other researchers (Kelly and Tyson, 1964; Drzal and co-workers, 1980) to measure the interfacial shear strength of ductile matrix composites takes advantage of the mechanics of deformation and failure to reduce the test to a few simple measurements. When a composite material made of stiff, continuous, brittle fibers reinforcing a less stiff, ductile matrix is stretched in the direction parallel to the fibers, tensile stresses arise in both the fibers and the matrix. As the applied load is increased, tensile stresses continue to increase both in the matrix and in the fibers until the tensile stress in a fiber reaches its fracture strength. Load transfer between a broken fiber and the matrix occurs by interface shear at the broken end of the fiber. The fiber tensile stress increases linearly with distance from the broken end until the fiber segment is carrying full load again, provided it is still long enough to develop the full shear stress. As the applied load is increased, this failure process repeats itself in all fibers, and then fiber segments, until no segment is long enough to develop a tensile stress large enough to fracture the segment. This failure process leads to the notion of a critical fiber length, which is the minimum fiber length required for developing a tensile stress in the fiber equal to its fracture strength, when load is transferred from matrix to fiber by interfacial shear. A straightforward strength-of-materials analysis yields the basic relationship between critical fiber length and average interfacial shear strength, namely:

$$l_c/d = \sigma_f/2\tau_i \tag{1}$$

where l_c = fiber critical length, d = fiber diameter, σ_f = fiber fracture strength, and τ_i = interfacial shear strength between the fiber and the matrix. Therefore, if l_c/d and σ_f can be independently determined, the average interfacial shear strength, τ_i, can be calculated. Details of the procedure used to measure the fiber critical length are given next.

CONSIDERATIONS FOR EXPERIMENTAL PROCEDURE

The diameter and fracture strength of a fiber may be readily measured, but measurement of the fiber critical length requires special techniques due to the nature of the composite material being investigated. This composite is a graphite-fiber-reinforced aluminum manufactured by the ion plating process. This process consists of ion-plating aluminum onto the graphite fibers of a tow that has been spread to a ribbon-like configuration approximately 25 μm thick to produce a precursor tape, which is then plied and consolidated by hot-press/diffusion-bonding to form Al/Gr composite sheets. While the ion-plating technique is being developed to enable the eventual deposition of alloys, to date, only commercially pure aluminum has been ion-plated onto graphite fibers. The fiber coating is approximately one μm thick, and it was reasonable to expect that hot-press/diffusion-bonding a precursor tape between plates of 2024-aluminum alloy would produce a laminated composite with a core lamina approximating a graphite-fiber-reinforced 2024-aluminum composite. Furthermore, the plate thickness required to ensure complete tensile fragmentation of the fibers without exhausting the ductility of the plate could be determined a priori, and specimens with at least two different plate thicknesses would have to be tested.

EXPERIMENTAL PROCEDURE

Aluminum ion-plated graphite precursor tapes, nominally 25 μm thick, 1 cm wide, and 12 cm long, were aligned and sandwiched between pairs of either 1.5 or 3.0 mm thick plates of 2024-aluminum alloy and hot pressed at 520°C at a pressure of 69 MPa for 30 minutes to produce good consolidation and diffusion bonding between the plates and the precursor tape. Two conventional flat tensile specimens were machined from each of the resulting 3 mm and 6 mm thick laminates to have reduced width gage sections 36 mm long and 9 mm wide. Two specimens, one of each thickness, were strained to failure to determine their ultimate tensile strengths. The other two specimens were loaded to approximately 80 percent of their corresponding ultimate tensile strengths and then unloaded. It should be noted that this stress level was chosen so that the material would be stressed well above its proportional limit. Sections at and near the fracture surface of the failed specimens and at several locations along the gage length of the homogeneously deformed specimens were cut with a diamond saw and swabbed with a 30 percent hydrochloric acid solution to dissolve the aluminum without attacking the graphite fibers. The exposed fiber segments were examined and photographed using a Coates and Welter 106A field emission scanning electron microscope (SEM). These micrographs were used to measure the lengths of the fiber segments and obtain length distribution histograms using a Zeiss Videoplan image anlaysis system.

RESULTS AND DISCUSSION

Representative SEM micrographs of several sections of the 3 mm thick tensile specimens are shown in Fig. 1. Fig. 1(a) is a micrograph of the fracture surface of the specimen pulled to failure which shows extensive non-homogeneous plastic deformation of the aluminum both in the adherent plates and in the composite tape prior to failure. Necking produced large transverse stresses on the graphite fibers, which caused the fragmentation of the fiber into small and irregular pieces. Fig. 1(b), shows that at some distance (approximately 8 mm) away from the fracture region, the plastic deformation of the aluminum and resulting transverse stresses were significantly smaller than in the fracture region. The average length of the fiber segments in this region is longer, and the fiber pieces appear less distorted and less fragmented. Fig. 1(c) presents a typical view of material 13 mm from the center of the gage section of the specimen loaded to approximately 80 percent of ultimate strength. The stresses in this region near the fillet of the gage section never reached fracture strength of the fibers. The micrograph presented in Fig. 1(d) was typical of the material chosen for this analysis, primarily because it yielded the shortest fiber segments. Moreover, the regularity of fiber breaks, as well as the absence of the fiber distortion and fragmentation evident in Figs. 1(a) and (b), indicate that this region of material was nearly uniformly strained. Results from the two 6 mm thick specimens, not included here for the sake of brevity, paralleled the results shown for the 3 mm thick specimens.

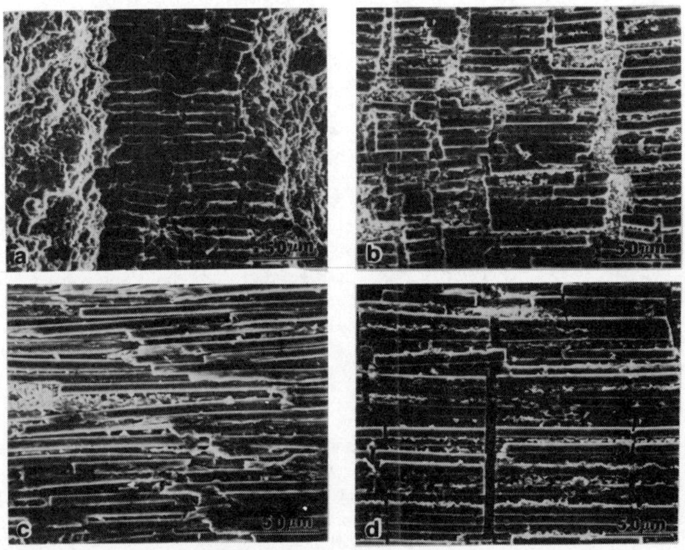

Fig. 1. SEM micrographs of ion-plated Al/Gr: (a) and (b) are from tensile specimen strained to failure; (c) and (d) are from tensile specimen loaded to 80 percent of ultimate strength.

The histogram shown in Fig. 2 was constructed from SEM micrographs of the type shown in Fig. 1(d) and is approximately a Gaussian distribution with a mean fiber segment length of 88 μm and a standard deviation of 25 μm. The as-received fibers were 7 μm in diameter and had averaged tensile strength of 2758 MPa (400 ksi). Because rule-of-mixture (ROM) tensile properties were obtained in bulk composites fabricated from the same constituents by the same consolidation process, it may be assumed that the fibers suffered no significant degradation. Using the preliminary statistics described above to define the critical fiber length, taken to be the mean value of the distribution, Eq. 1 may now be solved for τ_i, the interfacial shear stress. This computation results in a value of 120 (± 34) MPa. Confidence in the accuracy of this test method may be gained by an indirect consideration. Specifically, the calculated interfacial shear strength should not exceed the shear strength of the matrix, which was approximately 124 MPa.

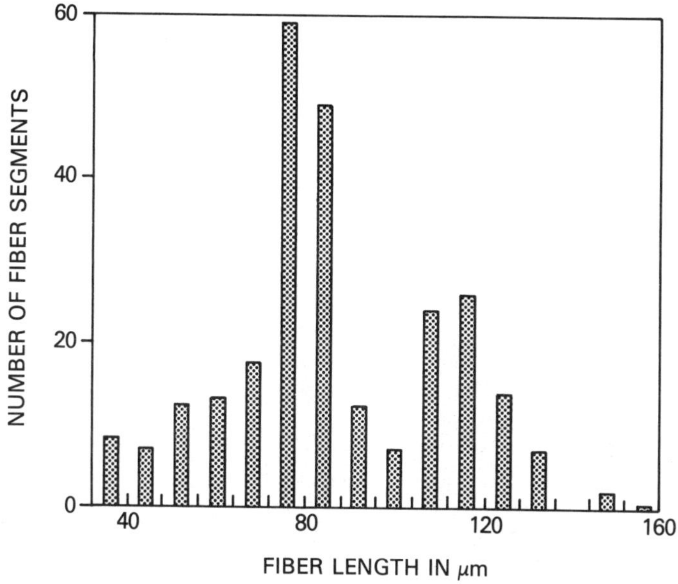

Fig. 2. Histogram of number of fibers broken to indicated lengths for tensile specimen loaded to 80 percent of ultimate strength.

CONCLUSIONS

Using the critical fiber length method, the interfacial shear strength, τ_i, between graphite fibers and an aluminum matrix has been determined in composites fabricated from ion-plated composite precursor tapes. Some uncertainty regarding the in-situ tensile strength of the fibers remain, and in future work, the strength of fibers extracted from the consolidated composite will be used in the analysis to provide more reliable statistics.

REFERENCES

Drzal, L.T., M.J. Rich, J.D. Camping, and W.J. Park (1980). Interfacial Shear Strength and Failure Mechanisms in Graphite Fiber Composites, 35th Annual Technical Conference, 1980 Reinforced Plastics/Composites Institute, The Society of the Plastic Industry, Inc.

Kelly, A. and G.J. Davies (1965). Met. Review 10, 37, 1, 1965.

Kelly, A. and W.R. Tyson (1964). Proc. 2nd Intn. Mat. Symposium, Berkeley, CA, John Wiley and Sons.

MECHANICAL BEHAVIOR OF A SHORT FIBER COMPOSITE

Göran Tolf

Department of Mechanics, The Royal Institute of Technology, S-100 44 Stockholm, Sweden

ABSTRACT

A theoretical investigation of the stressfield in a short fiber composite is performed. The stresses are calculated by integrating Navier´s equations. Knowing this stressfield,macroscopic properties are calculated and conclusions are drawn about the mechanical behaviour of the composite, like critical fiber length and fracture toughness, when the geometry and mechanical properties of the constituents are varied.

KEYWORDS

Composites, shortfibers, micro stresses.

INTRODUCTION

The short fiber composite has become a very common construction material in the engineering of today. Thus, one could think, great efforts are made to understand the behaviour of these materials, not only on the macrolevel, but also on the microlevel. Not only the ability to optimize the material usage, but also the technical possibility to create materials with certain mechanical properties, is dependent on this understanding.

This paper is based on a model presented in (1) and consists of calculations performed on composites with different mechanical properties and geometrical configuration of the constituents. From the calculations we have tried to make conclusions on critical fiber length and how fracture will occur in the composite. Also the effective elastic moduli are calculated.

It is found that the critical fiber length increases with increasing overlap between the fibers at constant volume fraction and decreases with increasing volume fraction at constant overlap, and that long fibers will give a low ductility of the composite. Due to the limitations of the model, the results are only qualitative but still show some interesting features.

2 MODEL

The material is modelled by a regular two-dimensional pattern of fibers embedded in a matrix, as is shown in figure 1.

A unit cell consists of one fiber in the middle and of surrounding matrix. It has the length 2a and width 2b. The length of the fiber is 2l and the width 2h. With this cell the entire composite can be built. The bond between fiber and matrix is assumed to be perfect and the constituents isotropic and linearly elastic.

In the plane strain case the non-vanishing displacements are u(x,y) in the x-direction and v(x,y) in the y-direction. In each constituent we assume that these displacements obey Navier´s equations in their static form.

If we assume simple macroscopic boundary conditions of the form

$$\bar{u}_x = \bar{\varepsilon} x, \ \bar{\sigma}_{xy} = 0, \ \bar{\varepsilon} \text{ is a constant}$$

, on lines parallel to the y-axis, and

$$\bar{\sigma}_{xy} = \bar{\sigma}_{yy} = 0$$

, on lines parallel to the x-axis, where bared symbols mean macroscopic quantities, we can reduce the microscopic problem by symmetry, to solving Navier´s equations over the region in figure 2. The boundary conditions then take the simple form

$$x = 0: \ u = 0, \ \frac{\partial v}{\partial x} = 0 \ , \ x = l: \quad u = a\bar{\varepsilon} \ , \ \frac{\partial v}{\partial x} = 0$$

$$y = 0: \ v = 0, \ \frac{\partial u}{\partial y} = 0 \ , \ y = 2b: \quad v = -2b\bar{\nu}\bar{\varepsilon}, \ \frac{\partial u}{\partial y} = 0$$

, where $\bar{\nu}$ is the macroscopic Poisson´s ratio

$$\bar{\nu} = - \bar{\varepsilon}_{xx}/\bar{\varepsilon}_{yy}$$

This is an unknown quantity, determined by the condition $\bar{\sigma}_{yy} =$ $= 0$.

For a thorough derivation of these boundary conditions, the reader is referred to (1). Furthermore it is required that the displacements and stresses are continuous across the fiber-matrix interface.

A finite difference approximation is made of the problem and the results are presented and discussed in the following sections.

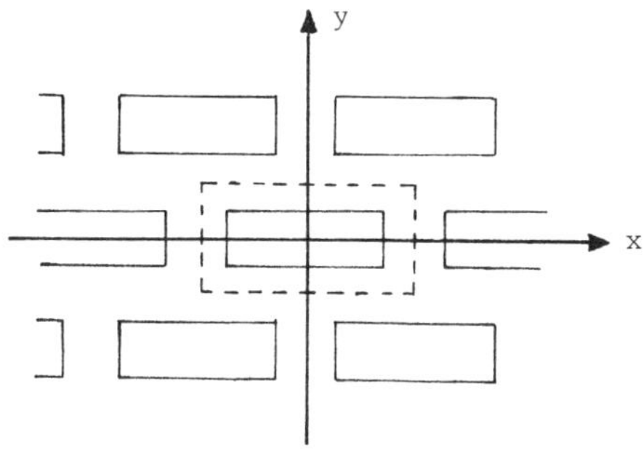

Figure 1. Model of the fiber composite. The unit cell is shown by the dashed line.

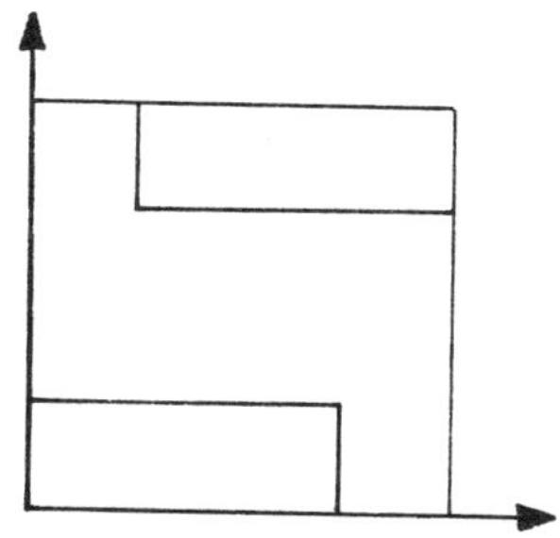

Figure 2. Representative domain for calculation of the field.

3 RESULTS

Two kind of calculations were performed. The first to find the dependence of the stressfield on the elastic properties of the constituents and the second to find the dependence on the geometrical configuration.

For the first part, see (1)!

In the second kind of calculation we have chosen three kinds of unit cell, characterized by different overlap between the fibers, i.e. with different values on the ratio a/l. Cell 1 with $a/l = 2$, cell 2 with $a/l = 8/5$, and cell 3 with $a/l = 4/3$. The volumefraction is taken to be 40 % for each cell, and for comparison also 60 % for cell 3. The length to widthratio of the fibers is then altered, and four calculations are performed for each cell:

Cell 1:	$l/h = 3.1, 6.3, 31, 63$
Cell 2:	$l/h = 5.2, 10.4, 52, 104$
Cell 3, 40 %:	$l/h = 7.5, 15, 75, 150$
Cell 3, 60 %:	$l/h = 4.7, 9.4, 47, 94$

In all these calculations the elastic constants were given the values

$$E_f/E_m = 10 \ , \quad \nu_f = 0.3 \ , \quad \nu_m = 0.4.$$

These values should apply to a glass-epoxy composite.

Diagram 1 shows the principal stress concentration factor, defined as the maximum principal stress, i.e. the highest tension in the fiber, normated by the macroscopic stress, and diagram 2 shows the same factor in the matrix.

We can see that for high 1/h-values these properties become constant, which can be expected, in the cells with overlap of the fibers. In cell 1 though the matrix will take a higher load with increasing 1/h.

Diagram 3 shows the shear stress concentration factor, defined as the maximum shear stress along the horizontal interface between fiber and matrix, normated by the macroscopic stress. It can be noted, that for low 1/h-values the maximum shear stress appears at the fiberending, while for higher values on 1/h it appears right under the neighbouring fiberending. Also in this case we can see that when overlap exists the maximum shear stress becomes constant with increasing 1/h, while for cell 1 we have a growing dependence on 1/h.

In diagram 4 the effective Young´s modulus normated by E_m is drawn. Even here we see that when 1/h grows we will have a constant value on $\bar{E}$, also for cell 1.

That $\bar{E}$ is decreasing with the overlap is probably due to the fact, that the high stresses that are obtained in the matrix of the cells with low overlap, are not to be expected in a real composite, since plastic deformation is likely to take place at a lower stress level.

In table 1 the ratios between the maximum tensile stress in the fiber, $\sigma_{max,f}$, and respectively maximum tensile stress in matrix, $\sigma_{max,m}$, and maximum shearstress along the horizontal interface, $\sigma_{max,i}$, are listed, along with figures over where these maxima occur. Since the tensile strength of glassfibers is about 1000 - 4000 MPa and of epoxy 60 - 80 Mpa, we will have a fiber dominated fracture of the composite if $\sigma_{max,f}/\sigma_{max,m} \gtrsim 10 - 70$.

If we assume the bondstrength between fiber and matrix to be around 50 % of the tensile strength of epoxy, debonding will take place if

$$\sigma_{max,f}/\sigma_{max,i} \lesssim 20 - 100 \text{ and } \sigma_{max,m}/\sigma_{max,i} \lesssim 2 .$$

We can by table 1 conclude that in cell 1 we will always have matrix dominated fracture and when 1/h grows debonding may occur. In cell 2 we will also have fracture in the matrix, but when 1/h grows we may here have fracture in the fiber instead. This is also valid for cell 3, 40 %. In cell 3, 60 %, we see that for small 1/h we will have both fracture in the matrix and debonding, but for high 1/h fracture of the fibers will take place. Though, this discussion is only valid when the Coulomb-Mohr fracture criteria can be applied, i.e. when the constituents are brittle. If the matrix is ductile, we can at most compare $\sigma_{max,f}$ and $\sigma_{max,i}$ and say that when $\sigma_{max,f}/\sigma_{max,i} \lesssim 20$ debonding will take place, else the fibers will fracture.

In the case of overlap between the fibers, i.e. cell 2 and 3, the greatest tension in the matrix is situated along the vertical interface between fiber and matrix, according to the pictures in table 1. Therefore debonding is likely to take place at this interface.

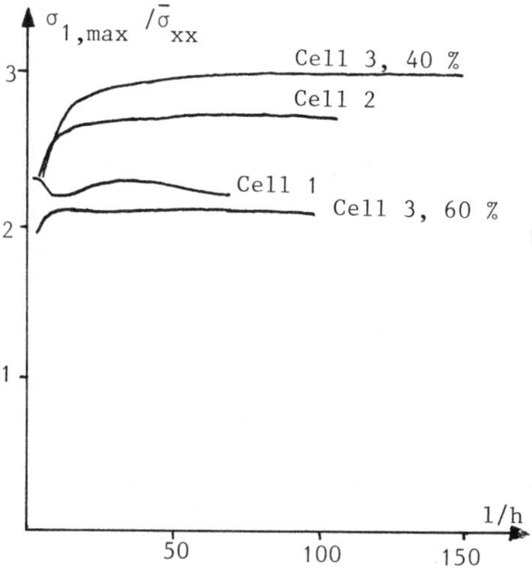

Diagram 1
Maximum principle stress in fiber normated by $\bar{\sigma}_{xx}$. E_f/E_m =10, ν_f = 0.3, ν_m = 0.4. The cells are defined in table 1.

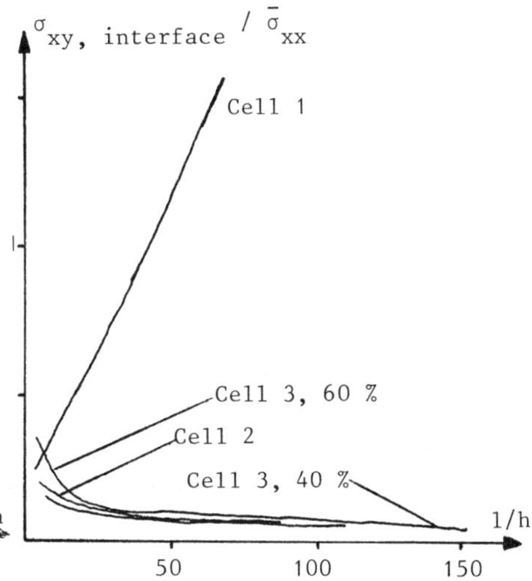

Diagram 3
Maximum interfacial shearstress, between fiber and matrix, normated by $\bar{\sigma}_{xx}$. E_f/E_m = 10 , ν_f = 0.3 , ν_m = 0.4. The cells are defined in table 1.

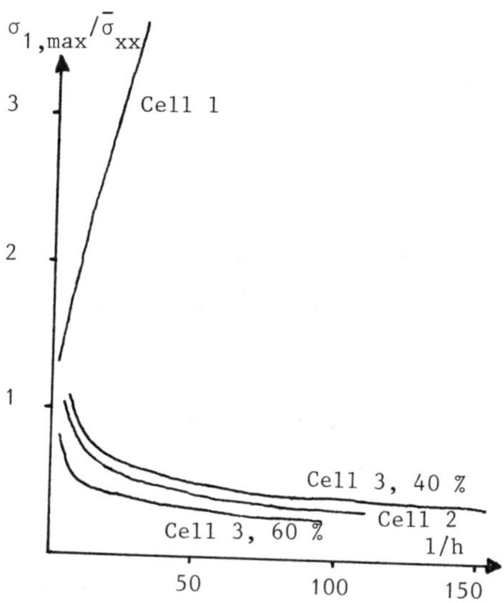

Diagram 2
Maximum principle stress in matrix normated by $\bar{\sigma}_{xx}$. E_f/E_m = 10, ν_f = = 0.3 , ν_m = 0.4. The cells are defined in table 1.

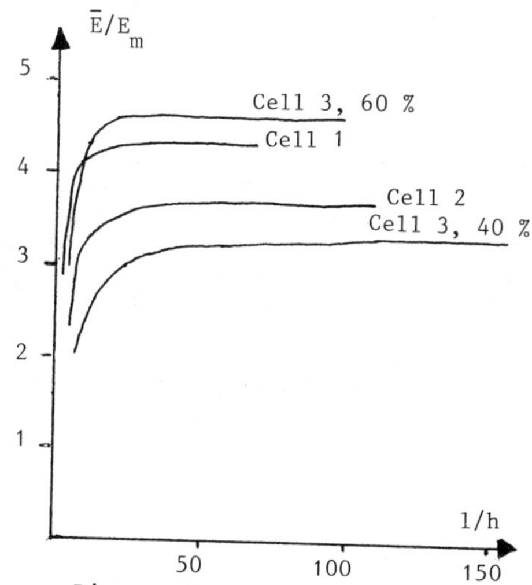

Diagram 4
Effective elastic modulus in the fiberdirection normated by E_m . E_f/E_m = 10 , ν_f = 0.3 , ν_m = 0.4. The cells are defined in table 1.

4 DISCUSSION

By looking at diagram 1 we can study how the critical fiber length, i.e. the least length that will give optimal usage of the fiber, depends on the l/h-ratio for the different types of cell. We see that the maximum stress concentration in cell 1 is nearly constant while for cell 2 it becomes constant when $l/h \approx 10 - 20$, cell 3, 40 % when $l/h \approx 20 - 30$, and cell 3, 60 %, when $l/h \approx 5 - 10$. Thus we can conclude that the critical fiber length increases with the overlap of fibers at constant volume fraction, and decreases with the volume fraction at constant overlap. This seems quite plausible since by increasing the overlap at constant volume fraction, the vertical distance between the fibers gets wider, which makes it harder to transport the stresses between the fibers. In the same way the stress transportation is helped by increasing the volume fraction.

These results can also be used to predict the mechanical properties of the composite qualitatively. We can first conclude by looking at the results for cell 1 that small overlap between the fibers is highly undesired, since this leads to bad usage of the fibers and high stressconcentrations in the matrix.

If we want the composite to be brittle, we want a fiber dominated fracture, since this leads to that the matrix must take most of the load in the fractured region, which it is not able to do, so that a fast crack propagation takes place. If instead debonding between fiber and matrix, and fracture of the matrix take place, a more ductile behaviour can be expected. Thus, from table 1, we can conclude that by using fibers with high l/h we get a brittle composite and with low l/h, a ductile composite. But by using fibers with low l/h we lose stiffness, although we see from diagram 9 that we lose only 20 % of the stiffness when changing l/h from 150 to 15 (cell 3, 40 %). We can also increase the volume fraction and then maintain the stiffness. Another possibility would be to mix short and long fibers and so combine the different properties.

These calculations show that a better understanding of the behaviour of composite materials can be achieved by investigations of the microstresses. The limits on the memory of the computer used as mentioned in (1), makes the calculation a bit too rough, especially around the corner of the fiber, where great displacement- and stressgradients can be expected. It can also be mentioned that the photoelastic pattern, corresponding to the calculated stressfield, has been drawn (2) and compared to experimental results and showed very good agreement. Thus the results presented here should be qualitatively within acceptance.

TABLE 1
Maximal Stress Ratios

Cell no.	*Volume fraction %*	*a/l*	*l/h*	$\sigma_{max.f}/\sigma_{max.m}$	$\sigma_{max.f}/\sigma_{max.i}$	○ = $\sigma_{max.f}$ and $\sigma_{max.m}$; × = $\sigma_{max.i}$
1	40	2	3·1	1·8	4·6	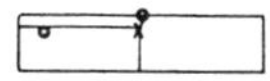
1	40	2	6·3	1·4	3·6	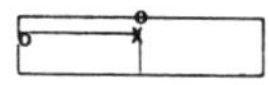
1	40	2	31	0·60	1·5	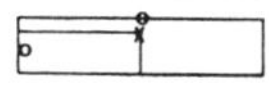
1	40	2	63	0·36	0·78	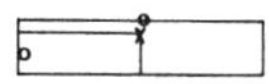
2	40	8/5	5·2	2·2	6·2	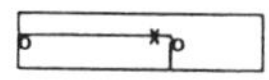
2	40	8/5	10·4	3·5	9·0	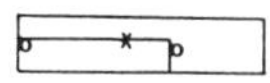
2	40	8/5	52	6·9	17·9	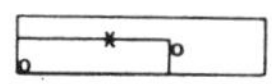
2	40	8/5	104	7·8	26·6	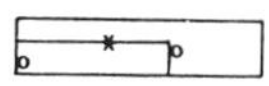
3	40	4/3	7·5	2·3	8·5	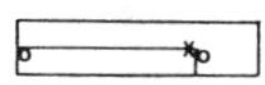
3	40	4/3	15	3·8	15·2	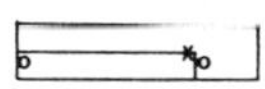
3	40	4/3	75	6·8	21·7	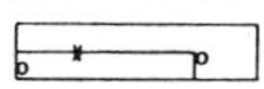
3	40	4/3	150	8·0	31·3	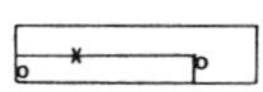
3	60	4/3	4·7	2·6	5·6	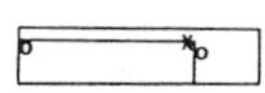
3	60	4/3	9·4	4·3	9·0	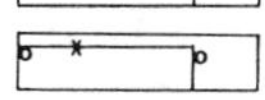
3	60	4/3	47	6·6	10·4	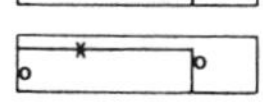
3	60	4/3	94	7·9	13·8	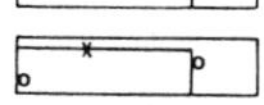

ACKNOWLEDGEMENT

The author is indebted to Professor Stig Hjalmars for proposing this investigation, to Tekn. Dr. Kurt Berglund for rewarding discussions during the performance and to Tekn. Dr. Bengt Lindberg for help with the programming.

REFERENCES

(1) Tolf, G., Mechanical Behaviour of a Short Fibre Composite, accepted for publication in Fibre Science and Technology

(2) Tolf, G., Stressconcentrations on the microlevel in a Short Fibre Composite, TRITA-MEK-80-14.

DELAMINATION AND TRANSVERSE FRACTURE IN GRAPHITE/EPOXY MATERIALS

W. L. Bradley* and R. N. Cohen**

*Mechanical Engineering, Texas A&M University, College Station, Texas, USA
**General Dynamics Corporation, San Diego, USA

ABSTRACT

Four graphite/epoxy composite materials systems have been studied to determine the transverse and delamination fracture behavior. Measured critical energy release rates have been correlated with micromechanisms of fracture determined using in-situ fracture in a SEM and post-mortem fractographic examinations.

KEYWORDS

Delamination fracture; transverse fracture; graphite/epoxy; fractography; in-situ fracture.

INTRODUCTION

Graphite/epoxy materials systems have exciting potential in weight critical applications because of their exceptional strength to weight ratio when loaded in a direction parallel to the fibers. However, in many practical applications of fiber reinforced plastics, significant loads may arise in the component in directions not reinforced by fibers, leading to component failure at relatively low loads. Out-of-plane stresses resulting from local buckling or geometric irregularities can lead to delamination. Loading perpendicular to the fiber direction in a lamina or a unidirectional laminate can also produce matrix cracking parallel to fibers, and therefore as with delamination, result in failure at relatively low stress levels by transverse cracking.

The initial development of graphite/epoxy systems emphasized composite stiffness and high glass transition temperature, T_g, to allow higher service temperatures. The resins which optimize these properties, however are generally quite brittle, making the composite very vulnerable to transverse and/or delamination cracking. More recent development work has sought to replace these brittle resins with somewhat tougher ones that still have a reasonably high stiffness and glass transition temperature.

In this research program, we have studied the delamination and transverse fracture behavior of four graphite/epoxy systems. Our approach has been to characterize the macroscopic fracture behavior using a fracture mechanics approach, and then, relate this behavior to micromechanisms of fracture as observed using in-situ fracture in a scanning electron microscope as well as post-mortem fractography of the fracture mechanics test coupons.

TRANSVERSE AND DELAMINATION FRACTURE

This study of the fracture behavior of graphite/epoxy composite materials emphasized first the macroscopic measurement of energy release rate followed by the microscopic observation of the various energy dissipative process.

Macroscopic Fracture Toughness Measurements Using Fracture Mechanics Approach

Four different graphite/epoxy systems were chosen for this study: Hercules AS1/3502 and AS4/3502 and Hexcel T6T145/F185 and T6C190/F155. The comparison of the behavior of the two Hercules systems allowed us to determine how much effect the fiber toughness and interfacial strength have on transverse and delamination fracture toughness since the AS4 fiber is significantly tougher than the AS1 fiber (higher strength and strain to failure) but less adherent to resin because of its smooth finish. The Hexcel F185 and F155 are rubber toughened epoxies with neat specimen critical energy release rates of 6000 J/m^2 and 730 J/m^2 (Moulton, 1982), respectively, compared to the Hercules 3502 whose G_{1c} is 70 J/m^2 (Williams, 1981).

Transverse cracking was studied using compact tension specimens with $W = 5$cm. and $B = 0.5$cm. These were tested in a 20Kip MTS machine equipped with 100 lb. load cell . LVDT's were attached to the ends of specimens and were used to measure crack opening displacement. They were also used to control the closed loop system, allowing testing to be done under crack tip opening displacement control rather than simply actuator displacement control. This arrangement permitted very stable crack growth completely across the specimen. Crack size was measured directly using resistance gages (Krag gages) and unloading compliances were measured at intervals of crack growth of about 0.1cm. Critical energy release rates were determined in the usual way using measured changes in compliance with crack size (dC/da) and the loads required at each crack length to continue crack extension. Corrections were made in the analysis to account for the fact that displacement was measured at the edge of the specimen rather than along the load line.

Delamination cracking was studied using a split laminate (double cantilevered beam) arrangement. Crack growth was stable throughout the test. Mode I fracture was produced using symmetric loading while a mixed mode I/mode II fracture was produced using asymmetric loading of the split ends of the laminate while supporting the uncracked end of the specimen. The ratio of $G_{II}/(G_I + G_{II})$ varied from 0.0 to 0.40 in this study. Cracking in these tests was continuous (no interuption for unloading compliance measurements) with the critical energy release rate calculated using an analysis previously developed by Devitt, Schapery, and Bradley (1980). This analysis involved the measurement of load, displacement and beam stiffness EI, allowing the calculation of the crack length and energy release rate using linear beam theory.

The results of the mode I delamination and transverse fracture testing are summarized in Table 1. In the very brittle 3502 resin, the introduction of fibers is seen to increase the energy absorbed in transverse and delamination fracture while in the more ductile F155 and F185, the introduction of fibers decreased the transverse and delamination fracture toughness. These results may be interpreted to imply that introduction of strong brittle fibers into a weaker, brittle resin enhances the energy absorbed in fracture. This enhancement is a result of the fact that the resin rich region between plies is very thin and minor fiber misalignment or waviness will result in some interaction of the crack tip with the fibers. The details of this interaction will be discussed further in the fractography section. The introduction of fibers into a more ductile resin decreases the energy absorbed in fracture by decreasing the volume of material ahead of the crack tip capable of significant deformation and constraining the deformation of the available resin. The result is a net decrease in energy absorbed in fracture, the increase in energy dissipation via crack tip/fiber interaction being less than the decrease in energy dissipation resulting from the reduced resin deformation.

It is further worth noting that the transverse fracture toughness is less than the delamination fracture toughness in the tougher resin systems. If these systems derive their toughness principally from resin deformation, then one would expect delamination fracture to be more difficult than transverse fracture. This is because the crack tip extension in delamination occurs through a thicker resin rich region in delamination than in transverse cracking. The AS1/3502 system shows increasing resistance to crack propagation comparing delamination to transverse cracking, as one might expect in a system where crack tip interaction with the fibers enhances resistance to crack propagation. That this trend is not noted in the AS4/3502 may be a result of the weaker interfacial bonding in AS4 than in AS1 due to the much smoother fiber surface finish.

TABLE 1 Transverse and Delamination Fracture Toughness

Material	Type of Fracture	Ave. $\dot{a}$ $(\frac{m}{s})$	$G_{IC}(\frac{J}{m^2})$
Neat 3502 Resin			69 [10]
AS1/3502	Delamination	———	155
AS1/3502	Transverse	———	225
AS4/3502	Delamination	2.8×10^{-5}	225
AS4/3502	Transverse	1.74×10^{-5}	120
Neat F155 Resin			730
T6C190/F155	Delamination	3.5×10^{-5}	600
T6C190/F155	Transverse	3.85×10^{-5}	410
Neat F185 Resin			6040
T6T145/F185	Delamination	4.5×10^{-5}	2600
T6T145/F185	Transverse	1.39×10^{-5}	1525

The critical energy release rate versus degree of mode II loading is summarized in Fig. 1. The toughest resin (F185) shows a monotonically decreasing G_{Ic} with increasing degree of mode II loading, while the 3502 systems showed an increased energy absorbed with increasing mode II shear. The change for T6C190/F155 at various ratios of mode II shear was not very significant.

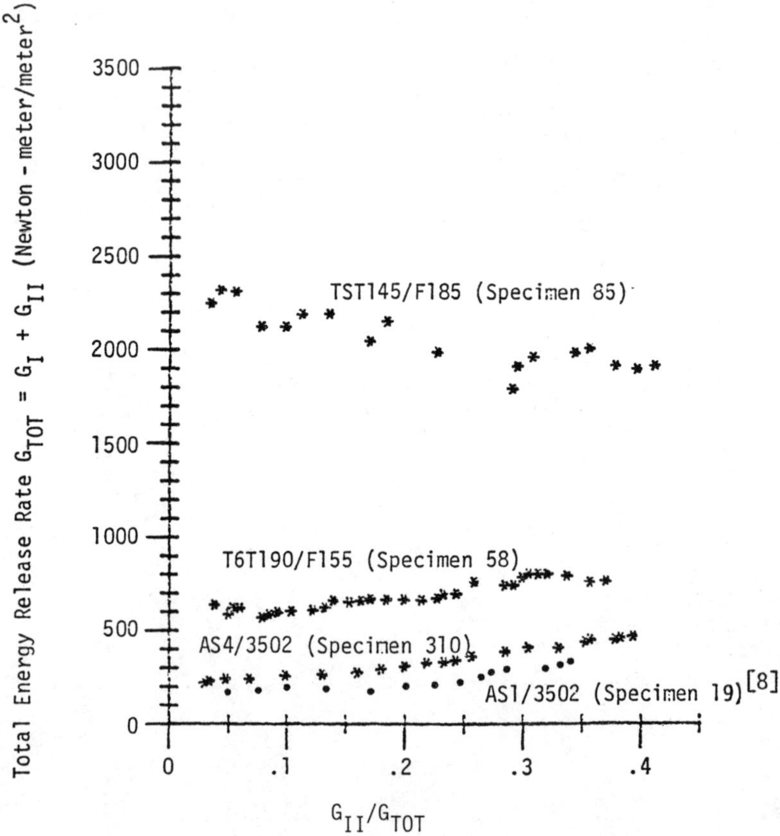

Fig. 1. Total critical energy release rate for mixed mode delamination for increasing fraction of mode II shear as indicated by the ratio of mode II energy release rate to total energy release rate.

Fractographic results to be presented presently seem to suggest that mode II shear simply rotates the microscopic plane of fracture which continually redirects the crack tip into the fibers. Thus, the mixed mode loading results in a series of mini mode I cracks joined by tear ridges, giving a scalloped appearance to the fracture surface. As previously explained, interaction of the crack tip with the fibers enhances the energy absorbed in fracture in a system in which the resin is brittle. Thus, the increasing mode II loading would be expected to give an increasing total critical energy release rate for the 3502, and it does. The very tough F185 which derives its toughness from matrix deformation decreases in toughness with increasing crack tip interaction with the fibers since this probably reduces the matrix deformation locally.

A history effect has also been observed, particularly in the systems with the brittle 3502 resin. When a high fracture of mode II shear is followed by pure mode I loading, the observed energy release is greater than it would be for a history of only mode I, as seen in Fig. 2. This appears to be associated with fiber and bundle pullout, with occasional local ply jumping which gives a significant roughening of the slip plane which is not reversible when pure mode I loading is used subsequently. This rougher fracture plane increases the actual surface area being created in fracture (as distinct from the nominal projected area). The bundle pullout can also lead to tie zone formation which does dissipate additional energy.

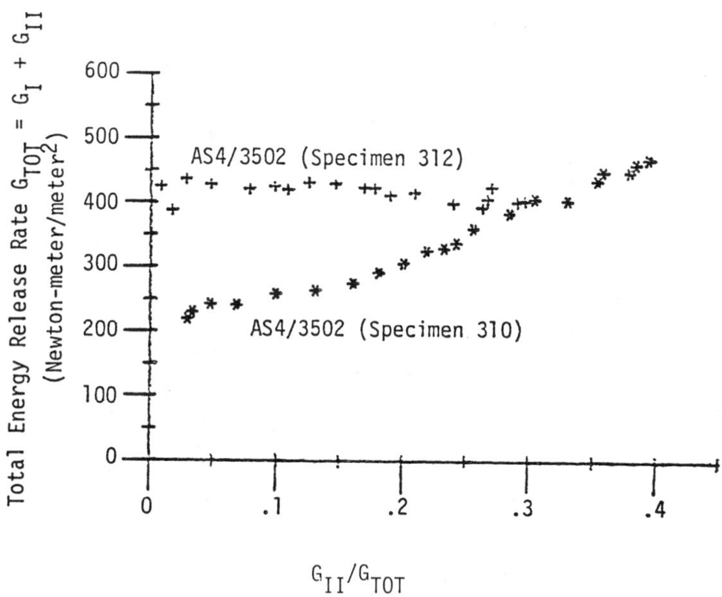

Fig. 2. Total critical energy release rate for delamination fracture for increasing fraction of mode II shear (specimen 310) and decreasing fraction of mode II shear (specimen 312) for AS4/3502.

Micromechanisms of Fracture Using In-Situ Fracture in SEM

In-situ delamination fracture results are seen in Fig. 3 for the tough F185 and the brittle 3502. Deformation and microcracking in the F185 resin is seen to occur over 4-5 fiber diameters on either side of the primary crack, while in the 3502 the deformation and resultant energy dissipation is limited to the resin between fibers bounding the primary crack.

Fractographic results on the F185 and 3502 are seen in Fig. 4 for mode I delamination fracture. Gross matrix deformation around the rubber particles in the local resin rich regions is seen in the F185 system. The fractured surface of the 3502 is seen to be brittle and fairly featureless with the only interesting artifact being the scalloping in the resin rich regions which results from crack tip interaction with the fibers. This interaction occasionally results in fiber breakage beginning with decohesion at the fiber/matrix interface as shown in Fig. 5A. More typically it results in initiation of a series of microcracks in the matrix just ahead of the crack tip as seen in Fig. 3. Small tear ridges then join the new microcracks to the primary crack producing a series of scallops. This repeated scalloping pattern is clearly seen in Fig. 4.

The distribution of rubber particles as well as the cross-link density in the F155 resin is apparently somewhat heterogeneous as suggested by the in-situ fractography shown in Fig. 5B. The resin ahead of the primary crack tip is brittle in some areas but quite soft and ductile in others.

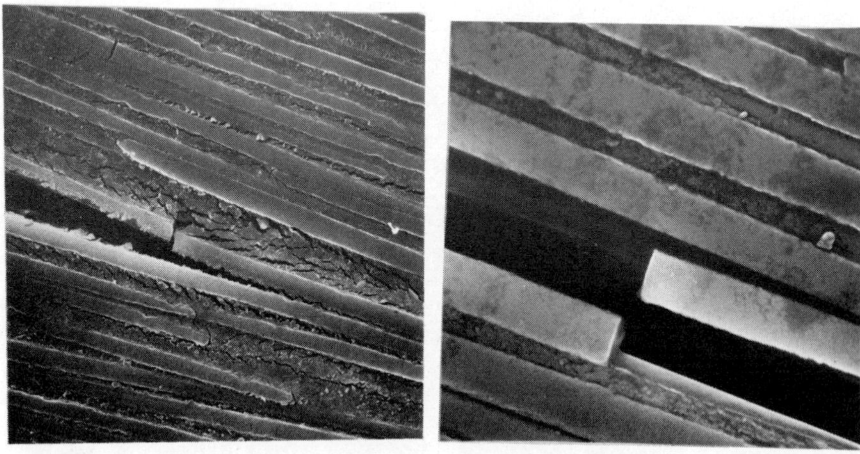

Fig. 3. In-situ fracture of T6T145/F185 (left, 500X) and AS4/3502 (right, 1300X).

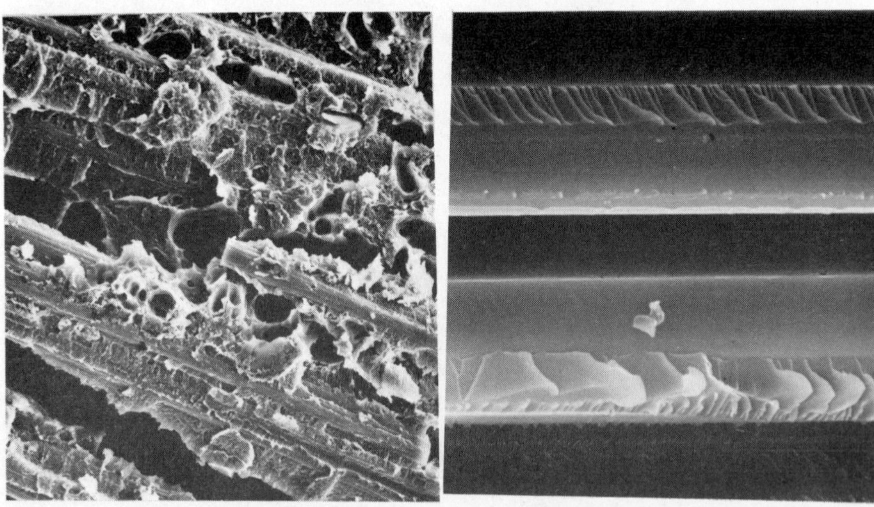

Fig. 4. Post-mortem fractography of T6T145/F185 (left, 700X) and AS4/3502 (right, 2000X).

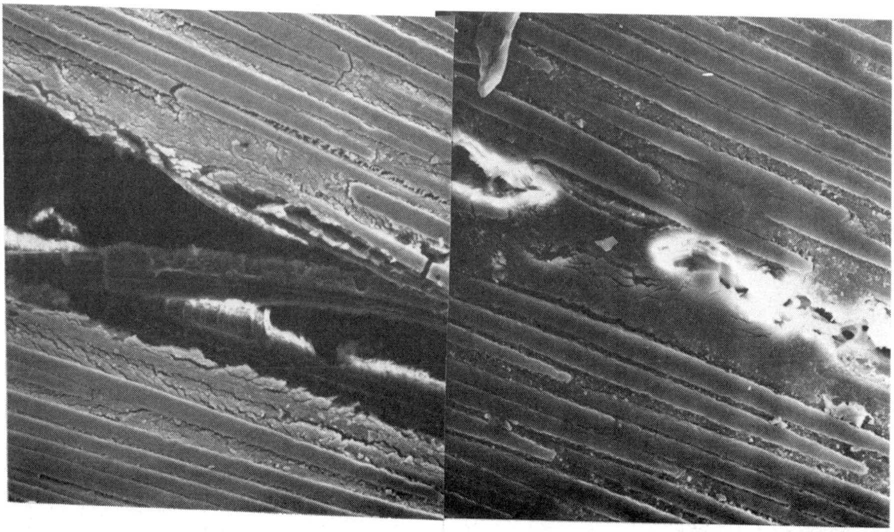

Fig. 5A and 5B. In-situ fracture showing decohesion at interface leading to fiber breakage in T6T145/F185 (left, 500X); T6C190/F155 showing heterogeneity of resin ahead of crack tip (right, 600X).

SUMMARY

The transverse and delamination fracture have been observed to depend on the matrix ductility, and in more brittle systems, to the incidence of crack tip interaction with the fiber. In such systems, mixed mode loading increases the total energy dissipated in fracture. History effects have been noted. In-situ fracture clearly shows the large volume of matrix which may deform during delamination fracture, including material 3-4 fiber diameters removed from the primary fracture planes.

ACKNOWLEDGEMENTS

The generous support of the Air Force Office of Scientific Research (contract F49620-82-C-0057) is gratefully acknowledged.

REFERENCES

Devitt, D.F., R.A. Schapery and W.L. Bradley (1980). A method for determining the mode I delamination fracture toughness of elastic and viscoelastic composite materials. Report MM 3724-80-6, Texas A&M University.

Moulton, R.J. (1982). Personal communication and customer bulletins from Hexcel Corporation.

Williams, D. (1981). Mode I cracking in an epoxy and a graphite fiber reinforced epoxy. Master's thesis, Texas A&M University, August 1981.

MECHANICAL BEHAVIOUR OF MATERIALS — IV

(in two volumes)

Volume 2

INTERNATIONAL SERIES ON THE
STRENGTH AND FRACTURE OF MATERIALS AND STRUCTURES

Other Titles in the Series

BUNSELL *et al.*	Advances in Composite Materials, 2 Volumes
EASTERLING	Mechanisms of Deformation & Fracture
FRANCOIS	Advances in Fracture Research, 6 Volumes
GARRETT & MARRIOTT	Engineering Applications of Fracture Analysis
GIFKINS	Strength of Metals & Alloys (ICSMA 6), 3 Volumes
HAASEN *et al.*	Strength of Metals & Alloys (ICSMA 5), 3 Volumes
MILLER & SMITH	Mechanical Behaviour of Materials (ICM 3), 3 Volumes
PIGGOTT	Load Bearing Fibre Composites
OSGOOD	Fatigue Design, 2nd Edition
RADON	Fracture & Fatigue: Elasto-Plasticity, Thin Sheet & Micromechanisms
SIH	Progress in Fracture Mechanics
SMITH	Fracture Mechanics: Current Status, Future Prospects
TAPLIN	Advances in Research on the Strength & Fracture of Materials, 6 Volumes

Related Pergamon Journals
(free specimen copy gladly sent on request)

Acta Metallurgica
Canadian Metallurgical Quarterly
Computers & Structures
Corrosion Science
Engineering Fracture Mechanics
Fatigue of Engineering Materials & Structures
International Journal of Solids & Structures
Materials Research Bulletin
Metals Forum
Physics of Metals & Metallography
Scripta Metallurgica

MECHANICAL BEHAVIOUR OF MATERIALS — IV

Proceedings of the Fourth International Conference, Stockholm, Sweden, 15–19 August 1983

Edited by

J. CARLSSON

and

N. G. OHLSON

Royal Institute of Technology, Stockholm, Sweden

Volume 2

PERGAMON PRESS

OXFORD · NEW YORK · TORONTO · SYDNEY · PARIS · FRANKFURT

U.K.	Pergamon Press Ltd., Headington Hill Hall. Oxford OX3 0BW, England
U.S.A.	Pergamon Press Inc., Maxwell House, Fairview Park, Elmsford, New York 10523, U.S.A.
CANADA	Pergamon Press Canada Ltd., Suite 104, 150 Consumers Road, Willowdale, Ontario M2J 1P9, Canada
AUSTRALIA	Pergamon Press (Aust.) Pty. Ltd., P.O. Box 544, Potts Point, N.S.W. 2011, Australia
FRANCE	Pergamon Press SARL, 24 rue des Ecoles, 75240 Paris, Cedex 05, France
FEDERAL REPUBLIC OF GERMANY	Pergamon Press GmbH, Hammerweg 6, D-6242 Kronberg-Taunus, Federal Republic of Germany

---

First edition 1984

Library of Congress Cataloging in Publication Data

Main entry under title:
Mechanical behaviour of materials-IV.
(International series on the strength and fracture of materials and structures)
Papers submitted for the Fourth International Conference on Mechanical Behaviour of Materials. Cf. Pref.
Includes indexes.
1. Materials—Mechanical properties—Congresses.
I. Carlsson, Janne, 1932- . II. Ohlson, N. G.
III. International Conference on Mechanical Behavior of Materials (4th : 1983 : Stockholm, Sweden)
IV. Series. V. Title: Mechanical behaviour of materials-4.
TA417.6.M42 1983 620.1'1 83-12166

British Library Cataloguing in Publication Data

Mechanical behaviour of materials-IV.—(International series on the strength and fracture of materials and structures)
1. Materials—Congresses
I. Carlsson. J. II. Ohlson, N. G. III. Series
620.1'1292 TA401.3

ISBN 0-08-029340-9

In order to make this volume available as economically and as rapidly as possible the authors' typescripts have been reproduced in their original forms. This method unfortunately has its typographical limitations but it is hoped that they in no way distract the reader.

Printed in Great Britain by A. Wheaton & Co. Ltd., Exeter

CONTENTS OF VOLUME 2

CONTENTS OF VOLUME 1

THEME I: High Temperature Deformation and Failure

THEME IV

SHEET FORMING OF HIGH-STRENGTH MATERIALS

COMPUTER-AIDED PREDICTION OF SHEET METAL MANUFACTURING PROCESSES AND ITS EXPERIMENTAL VERIFICATION

D. Lee* and P. A. Stine**

***General Electric Corporate Research and Development, PO Box 8, Schenectady, NY 12301, USA**
****General Electric Major Appliance Business Group, Appliance Park, Louisville, KY 40225, USA**

ABSTRACT

An interactive computer software package that predicts the degree of success or failure of formed sheet metal parts at the design stage is applied to several production parts. Given the material, the rate of deformation, the geometry and loading boundary conditions, the program computes the distribution of major and minor strains in the formed part. Concurrently, the limiting strain forming limit diagrams corresponding to the particular material are computed based on the defect growth model. The computed strain distribution is then directly displayed graphically on the forming limit diagram so that any design and processing changes could be readily made at the computer terminal. Several examples of computed results are presented which, in turn, are directly compared with the measured strain distributions obtained from the circular grid technique. Included in the experimental work are an aluminum-killed steel, enameling iron, and high strength steel sheets. Although the present analysis is limited to axisymmetric cases, the results demonstrate the utility of such an integrated computer analysis method to wider ranges of geometries.

KEYWORDS

Metal stamping analysis, computer-aided-engineering, sheet metal forming, software package, forming limits, steels, material modeling.

INTRODUCTION

Although sheet metal forming is a widely used manufacturing method, an interdisciplinary engineering approach had not been adopted by the industry until in the 1960s when the concept of the forming limit diagram (FLD) was introduced by Keeler (1965) and Goodwin (1968). The forming limits, which represent the relationship between limiting major and minor strains in the plane of the deformed sheet, however, were used as a diagnostic tool rather than a predictive method.

Predicting such a failure in the formed sheet metal part at the design stage is, however, not an easy task. It was not until the 1970s that important advances were made in the development of constitutive relations, localized plasticity theory and computational methods that could be utilized in the modeling of sheet forming processes (Hutchinson and Neale, 1977; Needleman and Rice, 1977; Levy et al., 1978). For example, the idea that necks develop from local regions of initial nonuniformity (Marciniak and Kuczyński, 1967) was applied by a number of authors to predict entire

FLDs, as reported by Hutchinson and Neale (1977) and Lee and Zaverl (1982). Recently, however, an interactive computer system that is capable of predicting success or failure of sheet metal parts at the design phase has been developed (Lee, 1982) by combining the FLD computation method with a finite element analysis procedure.

The purpose of the present work is to apply the integrated predictive method to a number of sheet metal parts using a variety of materials to further verify the analytical method. At the same time, the sensitivity of the analysis program was examined.

OUTLINE OF THE GENERAL APPROACH

The main elements of the computer code consist of an analysis program, CAD (computer-aided design) program, and several file conversion programs. The analysis program consists of constitutive relations, a FLD analysis program, and a finite element analysis program. The CAD program contains all the information necessary for a full geometric description of the parts. In addition, the material data base as well as details of loading boundary conditions are specified.

The rate dependent flow theory of plasticity is used in the FLD analysis where the inelastic strain rate, $\dot{\epsilon}_i$, may be expressed in terms of the gradient of the yield function, f, and the scalar multiplier, $\dot{\lambda}$. The yield function is written in terms of the distortion matrix, M_{ij}, which describes the variation of the flow stress with orientation (Shih and Lee, 1978).

$$f = M_{ij}\sigma_i\sigma_j \tag{1}$$

where σ_i and σ_j denote the stress vectors corresponding to the appropriate tensor counterparts, $\hat{\sigma}_{ij}$. The well-known anisotropy parameter, R, is related to M_{33} by $R = 2/M_{33}-l$ for a planar isotropic material under the plane stress loading condition.

The strain hardening and strain rate hardening behavior of materials was described by expressing the equivalent strain rate, $\dot{\bar{\epsilon}}$, in terms of the equivalent stress, $\bar{\sigma}$, or

$$\dot{\bar{\epsilon}} = \epsilon_o \cdot \left[\frac{\bar{\sigma}}{k} \right]^{1/m} \tag{2}$$

where m in Eq. 2 is known as the strain rate sensitivity of flow stress. The effective flow stress, k, and its dependence on equivalent plastic strain, $\bar{\epsilon}$, can be determined from a uniaxial tension test at a reference strain rate, $\dot{\epsilon}_o$. It has been shown that a number of experimental data could be represented by the well-known Swift-type equation (Swift, 1952), or $k = k_o\,(\epsilon_o + \bar{\epsilon})^n$, where k_o and ϵ_o are material constants and n is the well-known strain hardening exponent. All the material parameters were obtained from uniaxial tension tests (Lee, 1982).

The detailed method of computing the limiting and elongation FLDs has been recently published (Lee and Zaverl, 1982). The approach is based on the idea that necking develops from local regions of initial nonuniformity. In addition to the inhomogeneity index, η, the input material parameters that are required for the calculation of FLD are m, n, ϵ_o, and M_{33}. An experience based "shop" FLD which describes the boundary between the "safe" and "marginal" forming limit is also included. In general, the predicted FLDs are reasonably consistent with experimental observations (Lee and Zaverl, 1982; Hecker, 1975).

Finally, the software package utilizes a modified version of the finite element analysis code (Levy et al., 1978). The finite element analysis code uses a shell-type element, where the nodal point coordinates are specified along a line representing the inner surface contour of the particular axisymmetric part. Therefore, no thickness elements are present in the construction of the nodal points describing the particular geometry. Aside from using the deformation theory of time-independent plasticity, the computational method involves the determination of the nodal positions by the strain energy minimization process.

CASE STUDIES

A total of three cases will be examined to illustrate various features of the analysis program. All the cases represent actual forming, for which laboratory experiments or forming of production parts have been carried out to verify the predictions. Some of the special features of the formed parts are summarized below:

Part Name	Material	Features
(a) Range Vent	EIDQ Steel	Non-axisymmetric part
(b) Dryer Inner Door	AKDQ Steel	Anisotropic and non-axisymmetric
(c) Test Cup	High Strength Steel	Axisymmetric

A brief summary of the analytical and experimental results are summarized in the following section.

Range Vent Forming

A photograph of the formed range vent and the corresponding finite element representation of the circular section of the vent geometry are shown in Figs. 1a and 1b, respectively. The imposed boundary conditions include the description of frictional characteristics ($\mu = 0.14$) and the edge clamping conditions. The average forming strain rate was assumed to be 0.00067 sec^{-1}. Enameling iron drawing quality steel was used.

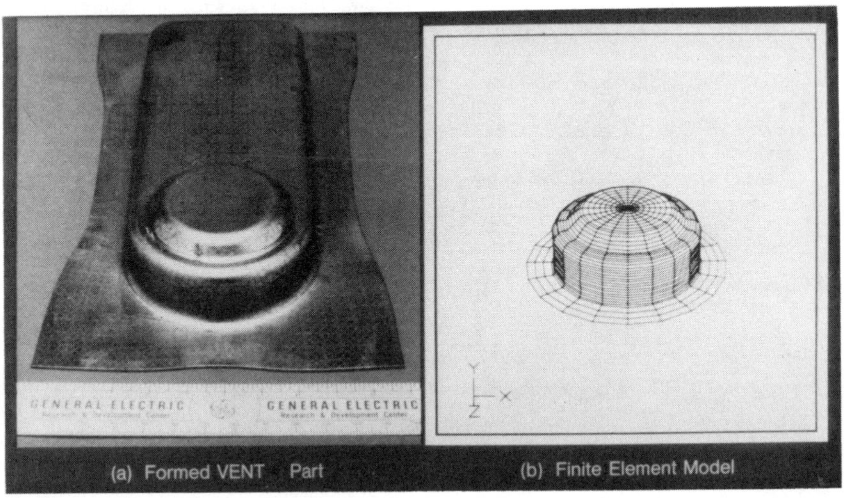

Fig. 1. (a) A photograph of the formed part and (b) a finite element model for the circular section in (a) for the range vent.

Given the material properties, the geometry file, and loading boundary conditions, the next step is to compute the strain distributions in the formed vent using the finite element analysis program. A parallel analysis starting from the material data file yields a computed limiting strain FLD from the given set of material parameters. Plotting the calculated strain distribution on the composite computed limiting FLD and shop FLD diagrams gives all the information that is necessary to summarize the final result, as shown in Fig. 2a. The nodal positions are identified by the nodal point numbers in the insert, and the computed radial (major) and hoop (minor) strains at each nodal point are specified in the composite FLD by the corresponding nodal point numbers. The maximum strain levels at the specified nodal points in the range vent are in the "safe" region. The computed major and minor strains are directly compared with experimental data obtained from the

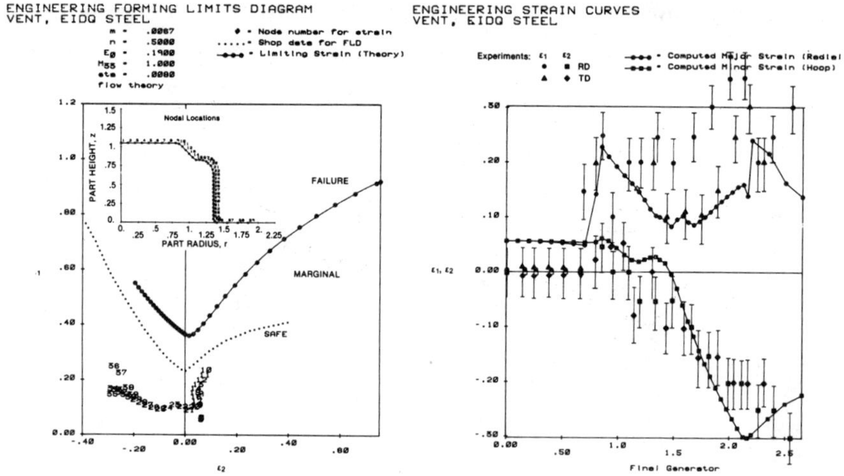

(a) (b)

Fig. 2. (a) A description of the nodal points (insert) and the result of the finite element analysis superimposed on the computed FLD for the range vent, and (b) a comparison of computed major and minor strains along the new generator with experimental data.

circular grid method, as shown in Fig. 2b. Production forming of the actual range vent did not produce any localized necking, which is consistent with the summary plot of Fig. 2a. The computed results show some discrepancies from the experimental data in Fig. 2b; computed results are in better agreement with the experimental data measured along the transverse direction than those obtained along the rolling direction. Another cause for the discrepancy is believed to be due to the assumption that the initial sheet geometry is circular, instead of being rectangular, as shown in Fig. 1a.

Dryer Inner Door Forming

The production dryer door was formed with the aluminum-killed drawing quality steel which is plastically anisotropic. The actual part was semi-circular, and only the corner was analyzed using the axisymmetric model. The friction coefficient was 0.06 and the average forming strain rate was 2.87 sec^{-1}.

Results of the strain distribution analysis and forming limit computation for the dryer inner door part are summarized in Fig. 3a; the specific nodal location for the corner position is again given in the insert. The computed strain levels at some of the nodal locations are close to the "safe"/"marginal" boundary, identifying potential problem areas in the actual forming process. Computed and experimental strain distributions along the length of final generator, as shown in Fig. 3b, confirm the result of analysis. The experimental data showed that some strains exceeded the computed strain levels. Aside from the necking that occurred in the selected locations, the computed strains are in good agreement with the experimental results.

Cup Forming

An axisymmetric cup was formed in the laboratory using a high strength steel sheet at an average strain rate of 0.006 sec^{-1}. The material is isotropic and the friction coefficient of 0.018 obtained from a separate test is used in the analysis. The geometry of the part is shown in the insert of Fig. 4a.

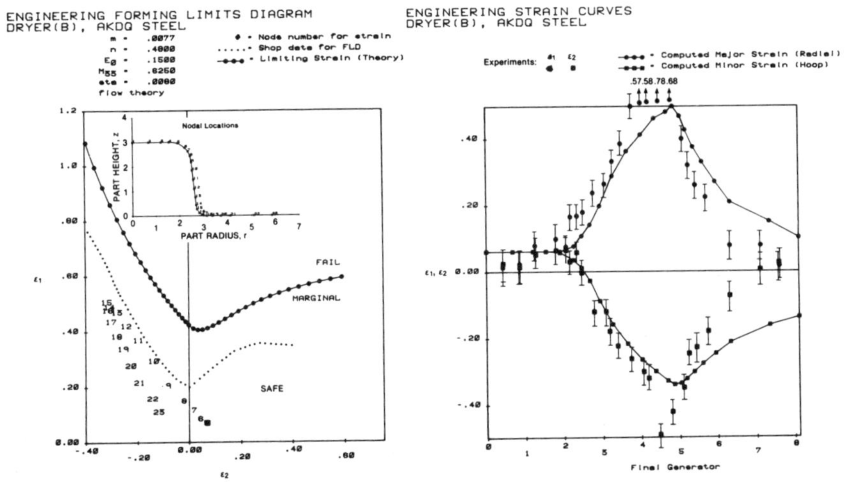

(a) (b)

Fig. 3. (a) A description of the nodal points (insert) and the result of the finite element analysis superimposed on the computed FLD for the dryer inner door and (b) a comparison of computed major and minor strains along the new generator with experimental data.

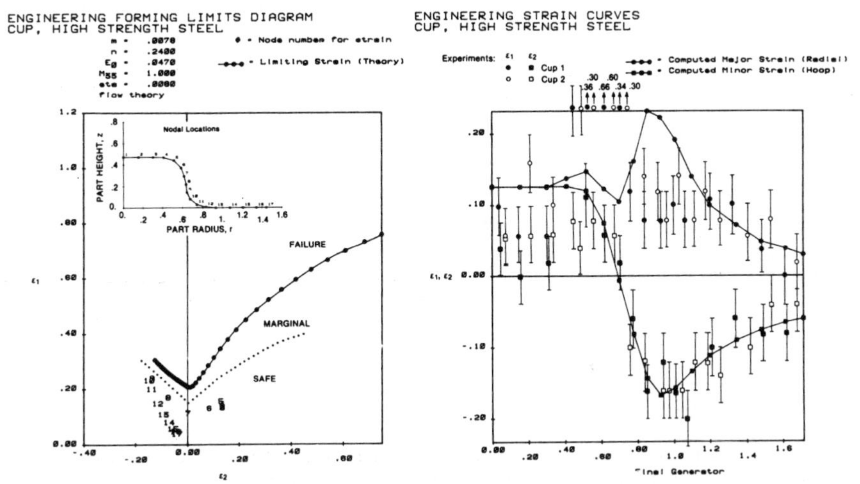

(a) (b)

Fig. 4. (a) A description of the nodal points (insert) and the result of the finite element analysis superimposed on the computed FLD for the test cup, and (b) a comparison of computed major and minor strains along the new generator with experimental data.

Results of the finite element analysis and forming limit computation are summarized in Fig. 4a, indicating that the particular cup may experience some difficulties during the forming process, especially near nodal points 8 and 9. Major discrepancies were noted when the computed strain distribution was compared directly with the experimental observations, as shown in Fig. 4b. For example, the measured radial strains in the vicinity of punch radius exceeded 30% engineering strain, causing the material to develop localized neck. Except for the major strain in the punch radius region, computed strain values are in excellent agreement with the experimental data.

In order to examine the cause of the discrepancy between the analytical and experimental results, several sensitivity analysis were made to isolate effects of input variables. Some of these variables included friction coefficient, punch force, and the nodal position of the applied boundary conditions. Increasing the friction coefficient at the punch radius raised the radial strain in this region; the punch force and the location of the imposed boundary condition had no major effect on the level of computed radial strain.

SUMMARY AND CONCLUSIONS

A computer-aided predictive method has been applied to a number of production sheet metal parts. The integrated code is capable of providing an instantaneous graphic representation of possible failure sites during the forming of sheet metal parts. Key elements of the analysis program consists of constitutive equation, computed and "shop" data FLDs, and finite element analysis programs.

In order to emphasize some of the key points in the analysis program, several production and laboratory parts were examined. Axisymmetric analysis is indeed simpler to use as compared to a full non-axisymmetric, three-dimensional analysis. From a practical point of view, a large fraction of complex parts could be approximated as axisymmetric problems. Such cases were illustrated with the examples of range vent and dryer inner door. Results of the analysis and experimental verification of the formed parts demonstrate the utility of such a computer-aided sheet metal formability analysis package.

ACKNOWLEDGMENTS

The authors would like to express their appreciation to C. M. Forth who has made important contributions to the development of systems and graphics software. High strength steel tension tests were made by C. P. Palmer.

REFERENCES

Goodwin, G.M. (1968). *La Metallurgica Italian, 60,* 767-774.

Hecker, S.S. (1975). *Sheet Metal Industries, Nov.,* 671-676.

Hutchinson, J.W. and K.W. Neale (1977). *Mechanics of Sheet Metal Forming,* Edited by D.P. Koistinen and N. Wang, GM Res. Lab., Plenum Press, 269-285.

Keeler, S.P. (1965). *Sheet Metal Industries, 42,* 683-691.

Lee, D. (1982). *J. Metals, 34,* 20-29.

Lee, D. and F. Zaverl, Jr. (1982). *Int. J. Mech. Sci., 24,* 157-173.

Levy, S., C.F. Shih, J.P.D. Wilkinson, P.A. Stine and R. McWilson (1978). *Formability Topics-Metallic Materials,* Edited by B.A. Niemeier, A.K. Schmieder and J.R. Newby, STP647, ASTM, Phil., PA., 238-260.

Marciniak, Z. and K. Kuczyński (1967). *Int. J. Mech. Sci., 9,* 609-620.

Needleman, A. and J.R. Rice, (1977). *Mechanics of Sheet Metal Forming,* Edited by D.P. Koistinen and N. Wang, GM Res. Lab., Plenum Press, 237-267.

Shih, C.F. and D. Lee (1978). *J. Eng. Mat. Tech., Trans. ASME, 100,* 294-302.

Swift, H.W. Swift (1952). *J. Mech. Phy. Solids, 1,* 1-18.

ON THE LAW OF STRENGTH OF DUAL-PHASE PLAIN CARBON STEELS AS NATURAL COMPOSITE MATERIALS

T. C. Lei and G. K. Liu

Department of Metals and Technology, Harbin Institute of Technology, Harbin, China

ABSTRACT

Tensile strength of martensite-ferrite dual-phase steels with various carbon content were measured after intercritical quenching from $\alpha + \gamma$ temperature range and 200 C tempering. The phase-hardening of ferrite and the phase-softening of martensite are found to be interrelated phenomena and quite important in forming the strength of dual-phase steels. The law of mixtures commonly used for evaluating the strength of fiber-reinforced composite materials was proved logically not suitable and quantitativelt not applicable to dual-phase steels.

KEYWORDS

Dual-phase; phase-hardening; phase-softening; law of mixtures; phase-hardening coefficient; phase-softening coefficient; rate of phase-hardening; rate of phase-softening; structural nonconformity; ptoperty nonconformity.

INTRODUCTION

Martensite plus ferrite (M+F) dual-phase steels have been widely studied since last decade of years. As a group of high strength and highly formable sheet forming material these steels are quite useful in automobile and other industries. However, the law of strength of dual-phase steels is not well understood and most investigators(e.g.Davies,1979;Ramos and others,1979) consider that the strength of dual-phase steels obeys the law of mixtures as a composite material :

$$\sigma_C = \sigma_M V_M + \sigma_F V_F \tag{1}$$

in which σ_C, σ_M and σ_F are the strengths of the steel, the martensite and the ferrite phases respectively, V_M and V_F are the volume fractions of the two phases. However, the nonconformity of the law of mixtures has been found repeatedly (e.g.Koo and others,1980;Lei and Shen,1982). The aim of the present study is to examine the properties of the M and F constituents and their contribution to the strength of dual-phase steels and on basis of experimental data to clarify the suitability of the law of mixtures.

EXPERIMENTAL

Plain carbon steels with 0.06, 0.12, 0.20, 0.29, 0.38 and 0.49 % C were used. Tensile specimens 1 mm thick and 25 mm gage length were 920 C normalized for 15 min and then intercritically quenched from 740, 770, 800 and 830 C (10 min) in 5%NaOH water solution for obtaining M+F dual-phase structure with various V_M and carious carbon content in martensite phase. The specimens were finally tempered at 200 C for 2 h. Tensile tests were carried out at a strain rate of $10^{-3}s^{-1}$. Microstructures were examined optically and by means of electron microscopy. Microhardness of the two phases were measured with 5 g load in 10 s.

RESULTS AND DISCUSSION

Typical dual-phase structure with ferrite matrix and martensite islands has been obtained for all steels. Volume fraction of martensite, V_M, depends on the carbon content in steel and temperature of intercritical quenching, as shown in TABLE 1.

TABLE 1 Volume fraction of martensite (V_M)

Quenching temperature, C	Carbon content in steel, %C						%C in martensite
	0.06	0.12	0.20	0.29	0.38	0.49	
740	17	23	28	64	77	83	0.65
770	18	28	40	74	93		0.48
800	32	55	72				0.34
830	53	65					0.25

It can be seen from FIG.1 that the microhardness of ferrite, $(HM)_F$, in dual-phase steels is much higher than that in normalized state and increases with increasing V_M. The microhardness of martensite, $(HM)_M$, on the contrary, is much lower than that of pure martensite with the same %C and decreases with increasing V_F (or increases with increasing V_M since $V_M + V_F = 1$). As has been shown earlier (Lei and Shen, 1982), the increased microhardness of ferrite in dual-phase steels is caused by the phase-hardening effect of the surrounding martensite having volumetric expansion during transformation from austenite. The decreased microhardness of martensite is caused by the surrounding ferrite which introduces a decreased dislocation density in martensite resulting in a phase-softening effect on the later. Thus, the phase-hardening of ferrite and the phase-softening of martensite are interrelated phenomena and are the most specific features in microstructure of dual-phase steels.

Linear functions of the microhardness values can be summarized from data in FIG.1 in the following form:

$$(HM)_F = (HM)_{100\%F} + K_F V_M \qquad (2)$$

$$(HM)_M = (HM)_{100\%M} - K_M(1-V_M) \qquad (3)$$

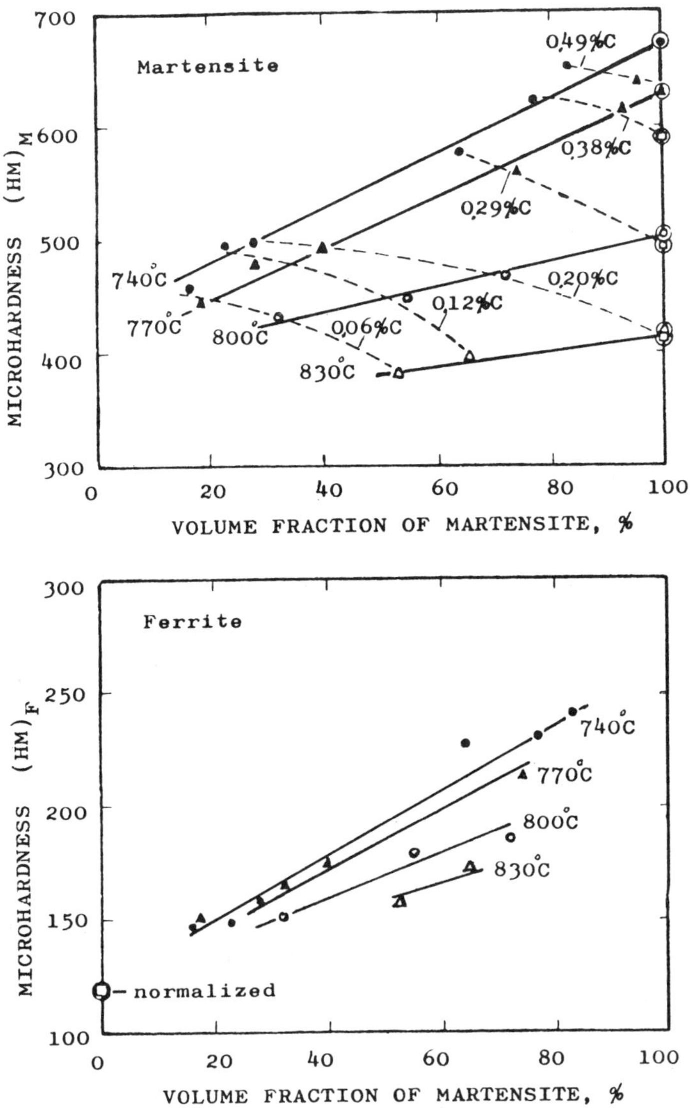

FIG.1 Microhardness values of the ferrite and martensite phases, $(HM)_F$ and $(HM)_M$, vs volume fraction of martensite for steels with various carbon content and quenched from different intercritical temperatures (after 200 C tempering).

in which $(HM)_{100\%F}$ and $(HM)_M$ are microhardness values of pure ferrite and pure martensite; K_F is the phase-hardening coefficient of ferrite and the K_M is the phase-softening coefficient of martensite; $(1-V_M)$ represents the volume fraction of ferrite. It is clear that the values of K_F and K_M are the slopes of the straight lines in FIG.1 and both of them decreases with decreasing carbon content in martensite.

The carbon content in martensite depends on the temperature of quenching(see TABLE 1) and the relationship between K_F and K_M values with carbon content in martensite, $(\%C)_M$, can be deduced from FIG.1 and shown in FIG.2. Here one can find that the K_F-$(\%C)_M$ and K_M-$(\%C)_M$ relationships are practically straight lines passing through the origin when prolonged. Thus, they may be written in the form :

$$K_F = C_F \ (\%C)_M \tag{4}$$

$$K_M = C_M \ (\%C)_M \tag{5}$$

in which C_F is the phase-hardening rate of ferrite and C_M is the phase-softening rate of martensite which indicate the effect of unit carbon content in martensite. It is rational to think that the phase-hardening or phase-softening effect will diminish to zero when the carbon content in martensite gradually decreases.

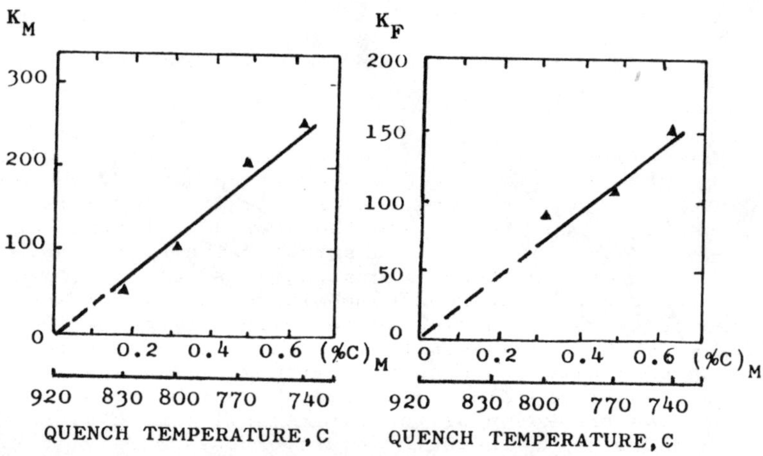

FIG.2 Phas-hardening coefficient of ferrite (K_F) and phase-softening coefficient of martensite (K_M) vs carbon content in martensite$(\%C)_M$ or temperature of intercritical quenching.

The phase-hardening and phase-softening rates, C_F and C_M, can be easily found by taking the slopes of the straight lines in FIG. 2. Thus, for plain carbon dual-phase steels tempered at 200 C, $C_F=250$ and $C_M=390$. Therefore, the eq.(4) and (5) may be used to estimate the coefficients of phase-hardening of ferrite and of phase-spftening of martensite, K_F and K_M, with which the microhardness values of the two phases in dual-phase steels can be calculated by using eq.(2) and (3).

The strength of both M and F phases can be calculated by converting the microhardness data shown in FIG.1 (or estimated as mentioned above) with the aid of the ASTM HV to UTS conversion chart. FIG.3 shows the UTS of M and F phases, the true (measured) UTS of dual-phase steels and the predicted UTS values by using the law of mixtures. It is clear that the true UTS values of dual-phase steels are far apart from predicted ones, so that the law of mixtures is quantitatively not applicable for evaluating the strength of dual-phase steels. Further, it can be seen from FIG.3 that at lower values of V_M ($<$ 50%), the UTS of dual-phase steels is very close to that of the ferrite phase. Since for practice the most interesting values of V_M are in the range 10-40%, so the UTS depends primarily on the strength of ferrite phase. This on no account means that the martensite phase in dual-phase steels has no contribution to the strength of the material. In fact, the martensite phase plays very important, but indirect role in strengthening the steel by its phase-hardening effect on ferrite.

The difference between the UTS values predicted by law of mixtures using the strength values of pure M and pure F (curve 3 in FIG.3) and the measured ones (curve 1 in FIG.3) may be divided into two parts by curve 2 which indicates the UTS values of dual-phase steels predicted by law of mixtures using true values of the M and F phases converted from microhardness data. The first part Δ_1 (see bottom drawings in FIG.3) is caused by structural nonconformity and the second part Δ_2 is caused by the property nonconformity . The structural nonconformity means that the microstructure of dual-phase steels is quite different from the fiber-reinforced composite material for which the law of mixtures is often used succesfully. It can be seen that the difference Δ_1 decreases with increasing quenching temperature and changes with V_M through a maximum at $V_M \approx 75\%$. This tells us that in fiber-reinforced composites the frature is controlled by the breakdown of the fibers, whilst in dual-phase steels the M phase joins in the fracture processes only at $V_M > 75\%$.

The property nonconformity Δ_2 indicates that the strength of M and F phases are quite different from that of pure M or pure F because of the phase-hardening or phase-softening effect and vary with V_M. Furthermore, it can be found from FIG.3 that the structural nonconformity Δ_1 is greater than the property nonconformity Δ_2 for all temperatures of quenching. So, the structural nonconformity should be more important in the argument of the nonsuitability of the law of mixtures.

CONCLUSIONS

It has been proved that in dual-phase steels the ferrite and martensite phases are quite different from those in common

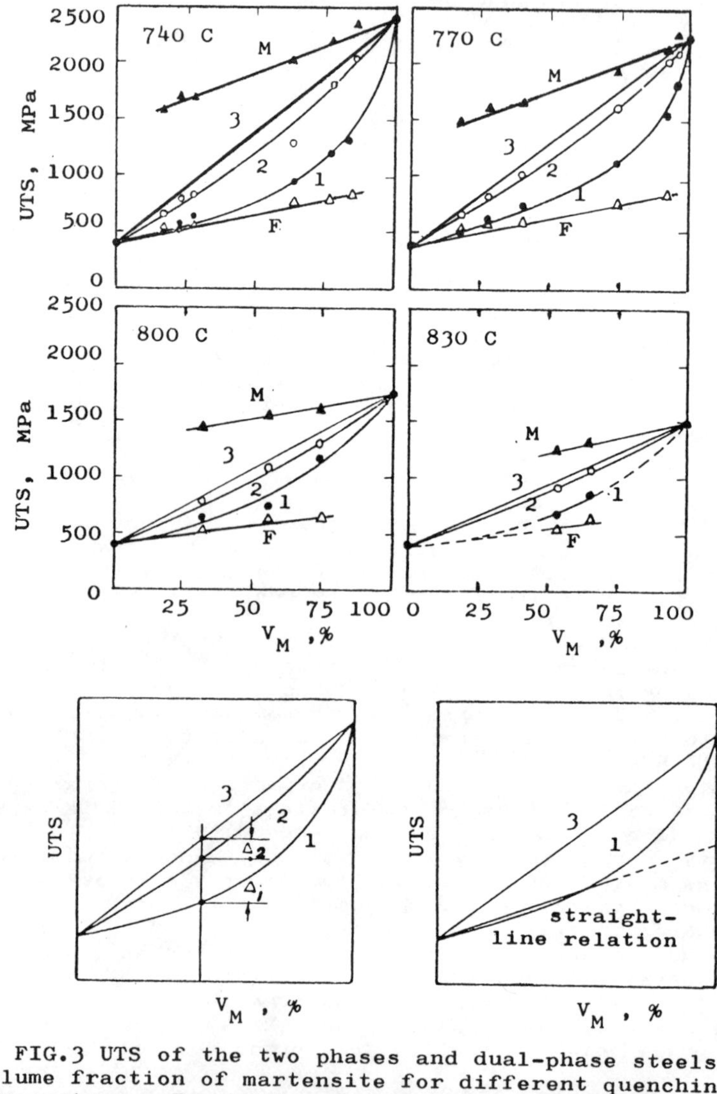

FIG.3 UTS of the two phases and dual-phase steels vs volume fraction of martensite for different quenching temperatures. Two bottom drawings are for analysis.

F,M - strength of ferrite and martensite phases;
1 - true(measured) strength of dual-phase steels;
2 - strength of steels calculated by law of mixtures using true strength values of two phases;
3 - strength of steels calculated by law of mixtures using strength values of pure M and pure F.

steels both structurally and in properties. Ferrite in dual-phase steels is phase-hardened by martensite and martensite is phase-softened by ferrite. The coefficients and rates of phase-hardening of ferrite and of phase-softening of martensite are shown by experiments. The law of mixtures commonly used for evaluating the strength of fiber-reinforced composite materials and often utilized by many investigators of dual-phase steels, has been shown logically not suitable and quantitatively not applicable. There exist the structural nonconformity and the property nonconformity of the law of mixtures when applied to dual-phase steels. The straight line relationship between the strength of dual-phase steels with low V_M values,repeatedly used as a basis for supporting the law of mixtures,is in fact a misunderstanding of the experimental data,principally differs from the straight line relationship of the law of mixtures.

Since the initial microstructure and mechanical properties of a sheet forming material is important in designing the technological procedure and estimating the properties of a stamped part, it seems necessary to study in more detail the structure and properties of the M and F constituents before and after processing and in this sense the prediction of the microhardness of these phases (and also their strength) in plain carbon as well as in low-alloyed dual-phase steels should be taken into consideration.

REFERENCES

Davies,R.G.(1978). Met.Trans.,9A,671
Davies,R.G.(1979). Met.Trans.,10A,113
Koo,J.,Young,M.J.,and Thomas,G.(1980). Met.Trans.,11A,852
Lei,T.C.,and Shen H.P.(1982). Acta Met.Sinica,18,75
Ramos,L.F.,Matlock,D.K.,and Krauss,G.(1979). Met.Trans.,10A,259

MATERIAL INHOMOGENEITY AND SURFACE ROUGHENING DURING PLASTIC DEFORMATION

T. Abe and S. Nagaki

Department of Applied Mechanics, Okayama University, Okayama 700, Japan

ABSTRACT

Analytical study is made on the relation between the roughening of free surface and inhomogeneity of material during plastic deformation. In order to simulate surface roughening, a plane model of inhomogeneous material is employed and elastic-plastic deformation of the model is calculated by the finite element method. Based on the result of the calculation, an equation is proposed which describes the effects of strain, the size and shape of inhomogeneous region and the mode of deformation on the surface roughening.

KEYWORDS

Plastic deformation; surface roughness; inhomogeneous material; polycrystalline metal; constraint ratio.

INTRODUCTION

The roughening of surface during plastic forming of metals, such as sheet metal forming, is an important problem in relation to the surface finishing of the products as well as to the fracture of sheet metal. Yamaguchi and Mellor (1976) pointed out, for instance, that the formability in deep drawing is affected by the surface roughening of sheet metal. The surface roughening during plastic deformation is closely related to the inhomogeneity of materials.

Experiments on several commercial metals (Osakada and Oyane, 1971) show that surface roughness R_{max} which appears during plastic deformation is expressed by

$$R_{max} = c_{max} \ \varepsilon \ d \tag{1}$$

where ε and d are plastic strain and grain size of polycrystalline metals, respectively, and c_{max} is a non-dimensional constant.[1] The theoretical background for Eq.(1) has not been given so far.

It is obvious that surface roughness would not be produced when (A) the material is homogeneous or (B) the deformation proceeds under constant strain condition[2], even if the material is inhomogeneous. In other words, the roughening of surface is caused by (i) inhomogeneity of materials and (ii) its non-uniform deformation.

In the present paper, analytical study is made on the relation between surface roughening and inhomogeneity of plastic deformation. In order to simulate surface roughening, a model of inhomogeneous material is employed and elastic-plastic deformation of the model is calculated by the finite element method. The value of d in Eq.(1) is considered to be the size of each inhomogeneous region in the following analysis.

MODEL OF INHOMOGENEOUS MATERIAL

The geometrical configuration of inhomogenity in materials is usually quite complex. For instance, we may consider grains in polycrystalline metals. In order to analyse the effect of inhomogeneity, however, it is necessary to adopt a model of inhomogeneous material. A simple plane model is used in the following, which is composed of two kinds of rectangular regions with different stress-strain relations, that is, different work-hardening rates. The two kinds of regions are shown with the signs + and − in Fig. 1.

The boundary BC is assumed to be free surface. The applied stress is supposed to be uniaxial in x-direction. Owing to the symmetric nature of geometrical configuration of the model, we may then assume that the broken lines BOAC remain straight after deformation. It is characteristic of the model that strict boundary conditions can be given in the numerical calculation. Meahwhile, it is often difficult for models with another geometrical configurations to give strict boundary conditions.

Two kinds of materials are assumed to have linear work-hardening rates with inclinations H_{p1} and H_{p2}, which are shown in Fig. 2 and expressed as follows.

$$H_{p1} = H_{po} (1 + \psi) \quad , \quad H_{p2} = H_{po} (1 - \psi) \qquad (2)$$

It is also supposed that the stress-strain relations for respective regions are not affected by the deformation of the neighbouring regions.

---

1 The maximum peak-to-valley height R_{max} is discussed below, though similar discussion is possible for the average roughness R_a.

2 Constant strain (or constant stress) implies that strain (or stress) is constant throughout the inhomogeneous material. It is also called local uniform strain (or stress).

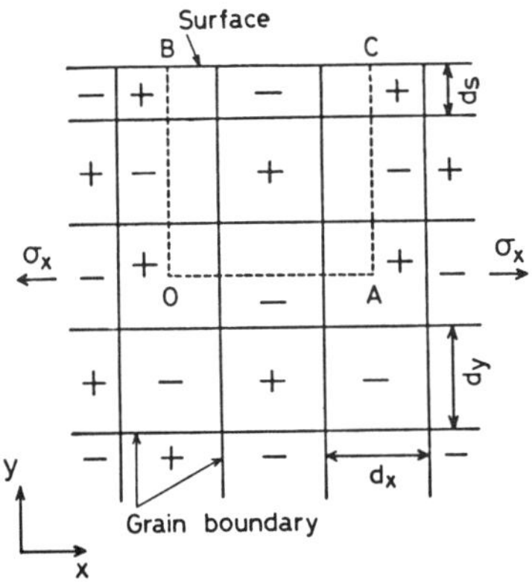

Fig. 1 Plane model of inhomogeneous material and its surface

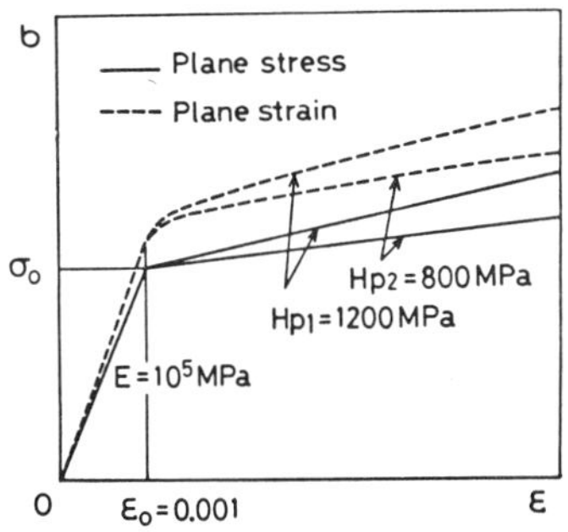

Fig. 2 Assumed stress-strain relations

As a representative value, the mean work-hardening rate H_{po}=1000 MPa and ψ=0.2 are assumed, while elastic constants are assumed to be the same for both regions (E=10^5 MPa, ν =0.3).

For the numerical calculation, the region OACB in Fig. 1 is divided into 256 triangular elements as shown in Fig. 3 and finite element elastic-plastic analysis is performed as usual. Prandtl-Reuss equation of plastic flow and Mises yield criterion are employed. Plane stress is assumed in the calculation unless otherwise stated.

In order to express local mode of deformation quantitatively, constraint ratio η is defined as follows (Abe, 1969, 1980).

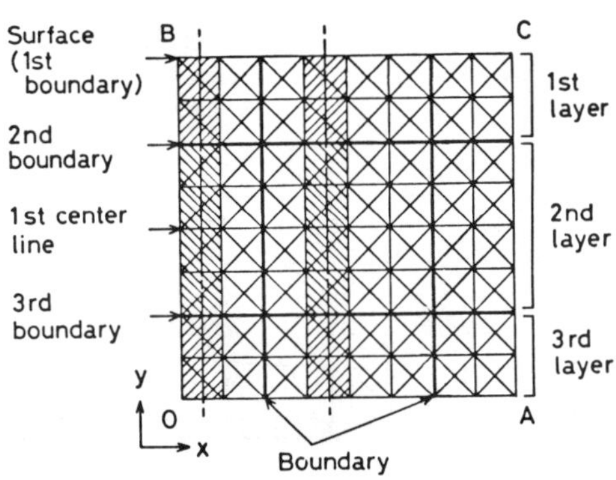

Fig. 3 Finite element mesh and numbers of layers and boundaries for the region OACB

$$\eta = \Delta\varepsilon / (\Delta\varepsilon)_\sigma \quad , \quad \Delta\varepsilon = \varepsilon - \tilde{\varepsilon} \tag{3}$$

where ε and $\tilde{\varepsilon}$ are local and the nean strain and $(\Delta\varepsilon)_\sigma$ is the value of $\Delta\varepsilon$ under constant stress condition. In the plastic range, $(\Delta\varepsilon)_\sigma$ is given by

$$(\Delta\varepsilon)_\sigma = \frac{(\Delta H_p)_\varepsilon}{H_{po}} (\tilde{\varepsilon} - \varepsilon_o) \ , \quad (\Delta H_p)_\varepsilon = H_p - H_{po} \tag{4}$$

where Hp is the work-hardening rate in respective regions. It is obvious from Eq.(3) that $\eta=0$ and $\eta=1$ correspond to the deformation under constant strain and constant stress, respectively.

The aspect ratio R of the inner region is defined by

$$R = d_y / d_x \tag{5}$$

where d_x and d_y are the sizes of the region in x- and y-directions respectively (Fig. 1). Meanwhile, the thickness of surface grain is denoted by d_s as shown in Fig. 1, and the ratio R_{surf} is defined as follows.

$$R_{surf} = 2\ d_s / d_y \tag{6}$$

CALCULATED RESULTS AND DISCUSSION

(i) Influence of thickness Rsurf of surface layer

Fig. 4 shows an example of the enlarged deformation patterns of boundaries, center lines and surfaces of the model (the magnification of displacement is 100). It is seen that the roughness R_{max} of the surface, as well as the roughness of inner boundaries (i.e. peak-to-valley height), increases with strain $\tilde{\varepsilon}_x$.

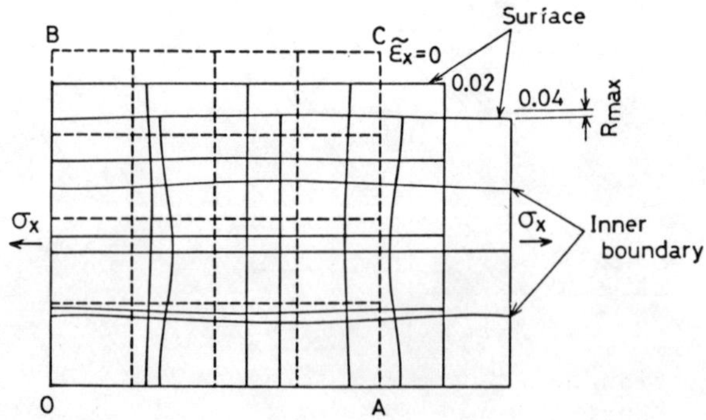

Fig. 4 Displacements of free surface and inner boundaries during plastic deformation ($R=R_{surf}=1$)

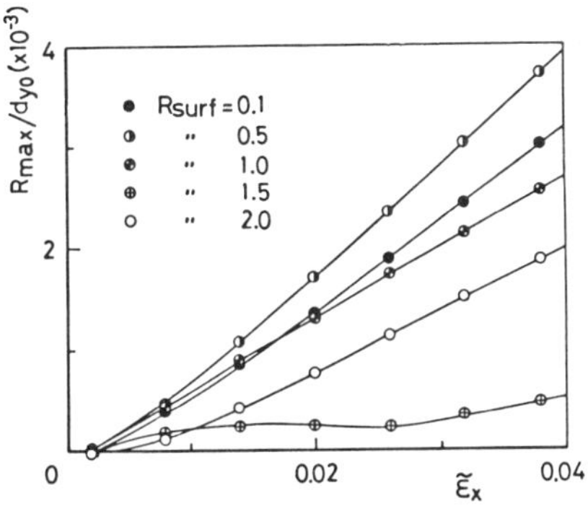

Fig. 5 Ratio R_{max}/d_{yo} versus mean strain $\tilde{\varepsilon}_x$ (R=1)

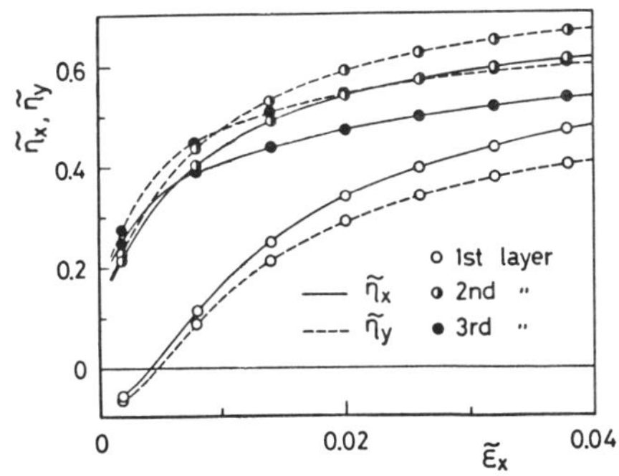

Fig. 6 Averaged constraint ratio $\tilde{\eta}_x$ and $\tilde{\eta}_y$ versus strain $\tilde{\varepsilon}_x$ (R= R_{surf}=1.0)

Fig. 5 shows the relation between the ratio R_{max}/d_{yo} and the mean strain $\tilde{\varepsilon}_x$ for various values of R_{surf}, where d_{yo} is the initial value of the size dy. The surface roughness increases almost linearly with strain, which supports the relation obtained experimentally (Eq.(1)). Fig. 6 shows the change in the mean constraint ratios $\tilde{\eta}_x$ and $\tilde{\eta}_y$ of each layer with the strain $\tilde{\varepsilon}_x$. It is seen that the changes of constraint ratio $\tilde{\eta}_x$ in x-direction and $\tilde{\eta}_y$ in y-direction are quite similar. The calculated result between the constant c_{max} in Eq.(1) and R_{surf} is plotted in Fig. 7 ($\tilde{\varepsilon}_x$=0.04).

Analytical consideration is made on the surface roughening during plastic deformation based on the numerical calculations.

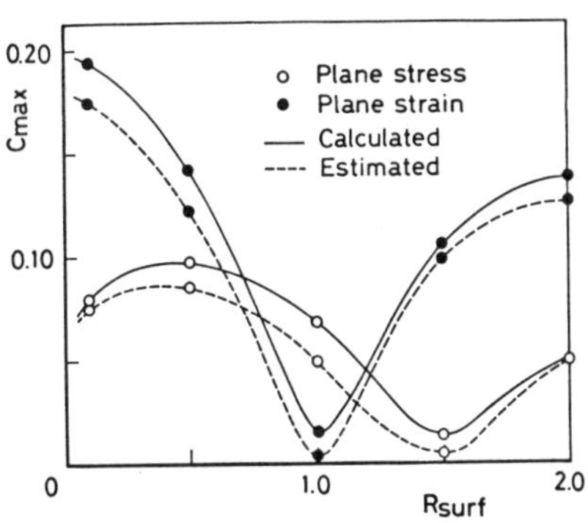

Fig. 7 Relation between c_{max} and R_{surf} (R=1), where broken lines show estimated values from Eq.(7)

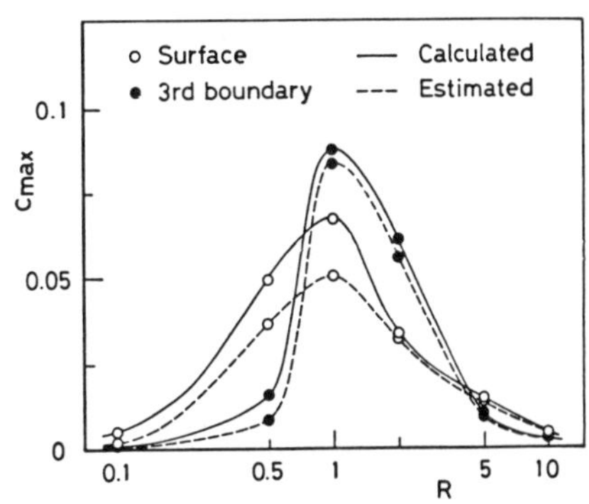

Fig. 8 Relation between c_{max} and R (R_{surf}=1), where broken lines show estimated values from Eq.(7)

Assuming that the boundary OA in Fig. 1 is straight after deformation and considering the displacement in y-direction of the surface, the roughness R_{max} is found to be given by [3]

$$R_{max} = (\bar{\eta}_{y1} R_{surf} - 2 \bar{\eta}_{y2} + \bar{\eta}_{y3})\psi \tilde{\varepsilon}_x d_y / 2$$

$$\equiv c_{max} \tilde{\varepsilon}_x d_y \qquad (7)$$

where $\bar{\eta}_{yi}$ is the constraint ratio of i-th layer in y-direction at the positions of the peak and the valley ($\bar{\eta}_{yi}$ is averaged only in y-direction, e.g. the shaded area in Fig. 3). Eq.(7) indicates that no roughening of surface occurs ($R_{max}=0$) if deformation proceeds under constant strain ($\bar{\eta}_{yi}=0$) or if material is homogeneous ($\psi=0$). The values of the coefficient c_{max} in Eq.(7) are estimated from the calculated constraint ratio $\bar{\eta}_{yi}$ and are also plotted against R_{surf} in Fig. 7 (the broken lines). The estimated curve agrees well with the calculated one.

(ii) Influence of aspect ratio R of inner region

The roughening of surface is also affected by the aspect ratio R of the inhomogeneous region. The solid lines in Fig. 8 show the changes in the coefficient c_{max} of the surface and the inner 3rd boundary for various values of R, where R_{surf} is kept constant ($R_{surf}=1$, $\tilde{\varepsilon}_x=0.04$). It is seen that the surface roughening is highest for $R\cong 1$. The broken line shows the curve estimated from Eq.(7). It is seen that the both changes agree well. It is also found that the variation of R_{max} with R is resulted mainly from the changes in constraint ratio $\bar{\eta}_{yi}$ with R.

(iii) Surface roughening in tension and in compression

Experimental results (Osakada and Oyane, 1971) show that surface roughening is greater in compressive plastic deformation than in tensile deformation. The roughening R_{max}/d_{yo} of surface of the present model in tension and in compression is shown in Fig.9(a), where d_{yo} is the initial size of inhomogeneous region. The calculated surface roughening is larger in compression than in tension, as is seen from the figure.

Fig. 9(b) shows the roughness R_{max} divided by d_y, where d_y is the size of inhomogeneous region at respective stages of deformation. The difference between the surface roughness in tension and that in compression is not seen here. Futhermore, the value of constraint ratio $\bar{\eta}_{yi}$ is nearly equal in tension and in compression. Therefore, in contrast to the former case (ii), the difference is mainly attributed to the change of the aspect ratio (i.e. the size d_y in Eq.(7)) in tensile and in compressive deformations.

The calculated surface roughening mentioned above, however, is far smaller than the experimental value, so far as the size d of inhomogeneous region is taken as the mean grain diameter of polycrystalline metals (c.f. $c_{max}\cong 1$ for compression test of copper specimen). Therefore, it may be speculated that the size

---

[3] The derivation of Eq.(7) will be reported in a separate paper.

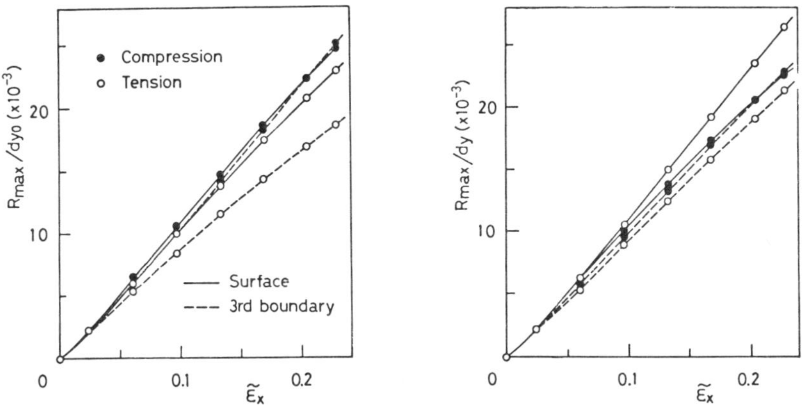

(a) R_{max}/dyo (b) R_{max}/dy

Fig. 9 Surface roughening in tension and in compression (R_{surf}=0.5, R=1)

d or d_y in the present analysis corresponds to some group of grains, not to a single grain in polycrystalline metals.

CONCLUSION

An analytical expression (eq.(7)) is proposed to express the roughening of free surface during plastic deformation of inhomogeneous materials, based on the numerical calculation of the plastic deformation of the plane model of inhomogeneous material. The influence of various factors affecting the roughening of free surface is well described by the equation.

ACKNOWLEDGEMENT

The authors would like to express their gratitude to Messrs. T. Akase, T. Yoshikawa and K. Maruyama, Miss. K. Yamamoto and Okayama University Computer Center for their help in the numerical calculations. The present work was supported by a Grant-in-Aid for Scientific Research from the Japan Ministry of Education.

REFERENCES

Abe, T. (1969). Bulletin of JSME, 12-50, 165-171.
Abe, T. (1980). Bulletin of JSME, 23-176, 177-184.
Osakada, K. and M. Oyane (1971). Bulletin of JSME, 14-68, 171-177.
Yamaguchi, K. and P.B. Mellor (1976). Int. J. Mech. Sci., 18, 85-90.

THE INFLUENCE OF MICROSTRUCTURE AND TEXTURE ON THE FORMING PROPERTIES OF A DUAL PHASE Al-KILLED STEEL

C. S. da Costa Viana and L. P. Mendonça Brandao

Instituto Militar de Engenharia, Seçao de Engenharia e Ciência dos Materiais, Pça Gen Tiburcio no 80, Urca, CEP 22290 Rio de Janeiro, RJ, Brazil

ABSTRACT

AT PRESENT, high strength sheet steel is more and more being used as automotive material for improving safety and for reducing fuel consumption by weight reduction. However, as far as thin cold rolled sheets for outer body panels are concerned, there has been few detailed investigations about the influences of processing conditions and microstructures on the forming properties required for those materials [1]. Among these properties, low initial yield strength, high work hardening rate, absence of discontinuous yielding and high R-value can be mentioned as important ones, together with ease of production and low cost. It is now well known that many of these properties can be found in dual-phase steels produced by intercritical annealing [2] except the high R-value which is texture dependent and therefore requires a proper processing for its development [3]. Al-killed steel has traditionally been the high-R-valued material used in autobodies, its texture being obtained by a near-70% cold rolling followed by a low heating rate (~40°C/h) long (~48 h) subcritical (~700°C) box annealing. Recent works [1,4] have explored the possibility of developing a high R-value dual-phase Al-killed steel with claimed success.

In the present work a commercial Al-killed hot rolled steel sheet is used in a similar attempt. Cold rolling reductions between 70 and 90% were used followed by several long (~6 h) and short (~5 min) annealing cycles reaching temperatures between 760 and 800°C. Work hardening, R and stretching properties are analysed and related to microstructure, second phase volume fraction and distribution and crystallographic texture, the latter being measured quantitatively by Crystallite Orientation Distribution Function (CODF) methods (see figure).

In general, the results seem to indicate a marked influence of the second phase on the R-value though the final textures do not differ appreciably. This seems to be mainly due to differences in the heat treatments which influence the recovery and recrystallization behaviour responsible for the texture development in Al-killed steels [5]. The general behaviour is compared to the box annealed material.

REFERENCES

1. Kurihara, K., Hosoya, Y. and Nakaoka, K. (1981) Proceedings of the 6th ICOTOM, Tokyo, p. 1066.
2. Davies, R.G. (1978) Met. Trans., 9A, May, p. 671.
3. Grumbach, M., Parnière, P., Roesch, L. and Sauzay, C. (1975) Mem. Sci. Rev. Met., Mars, p. 241.
4. Nishimoto, A., Hosoya, Y. and Nakaoka, K. (1981) Trans. ISIJ, Vol. 21, p. 778.
5. Parnière, P. (1973) Mem. Sci. Rev. Met., LXX no 6, p. 445.

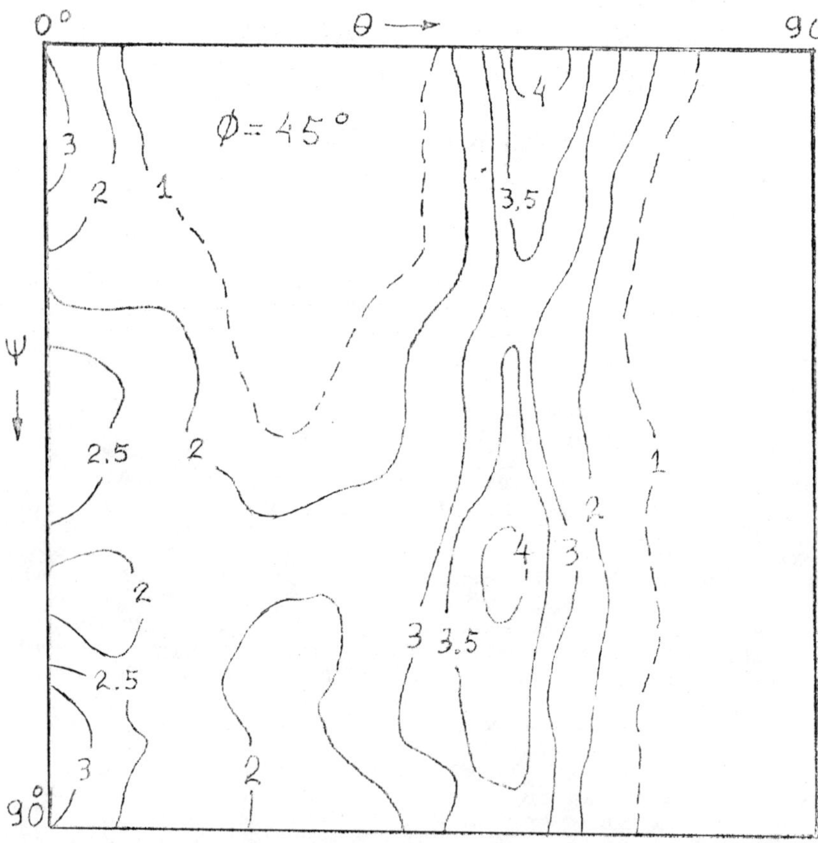

Fig. 1. CODF of a steel intercritically annealed at 760°C and water quenched.

LOCALIZED FLOW (SHEAR BANDS) IN Al AND Al-3%Mg ALLOY

Y. Nakayama and K. Morii

College of Engineering, University of Osaka Prefecture, Sakai 591, Japan

ABSTRACT

IT IS WELL known that if a sheet material such as used in sheet forming processes is deformed by tension, following behavior will be observed: the load increases with increasing elongation, diffuse necking (geometrical instability) begins at the maximum load, and then localized necking (flow localization) followed by the fracture of the material occurs. The diffuse necking is not always connected directly with the fracture, since the area of the deformation extends to the width of the sheet material. On the other hand, the localized necking is accompanied by the through-thickness strain-concentration, so that the deformation in the necked region governs the ultimate strain of the material as a whole. This is true in the sheet forming processes, because the strain at which the flow localization occurs will give the forming limit of the material. Therefore it can be pointed out that the plastic flow localization in the through-thickness direction of the sheet material has a greater significance as for the practical problems of metal working processes.

Though many theoretical studies concerning with the condition for the onset of the unstable flow have been reported, the microscopic mechanisms of the flow localization were hardly investigated from the metallurgical viewpoint. Various materials used in the sheet forming applications would show various deformation characteristics depending on their crystal structure, alloy composition and microstructure. Then, it may be quite valuable for a deeper understanding of the flow localization to reveal the relation between these factors and the unstable flow.

From this point of view, the present study deals with the localized flow in sheet materials, and aims to investigate experimentally the microscopic mechanisms of the flow localization in relation to the materials properties.

It may be hard in a tensile deformation to discriminate the plastic unstable behavior inherent in the material properties from that caused by the geometrical instability (diffuse necking). Then, in the present study, the localized flow in pure shear, i.e. plane strain deformation, is mainly investigated, in which the diffuse necking would not occur, but the flow localization along the direction of zero extension may be observed. The materials used were sheets of pure aluminum and aluminum alloy containing

3wt% magnesium, both which have higher stacking fault energy*. It is noted that during deformation at room temperature, the Al-Mg alloy will exhibit some particular deformation mode, i.e. dynamic strain aging or Portevin-Le Chatelier effect, and that the sign of strain rate sensitivity ($m=\partial \ln\sigma/\partial \ln\dot{\varepsilon}$) of the flow stress is negative.

In order to correlate the mechanisms of the localized flow to the crystallographic slip deformation, single crystals were mainly used as the specimens. The specimens have [111] or [001] initial orientation for the tensile deformation, and $(211)[\bar{1}11]$ or (101)[010] ((sheet plane) [direction of elongation]) orientation for the plane strain deformation. These specimens were deformed by uniaxial tension, plane strain compression and rolling at the temperatures between 77 - 673K, and stress-strain relations were measured. The development of slip bands and localized flow in the deformed specimens were observed by optical microscopy, and detailed dislocation structures were examined by transmission electron microscopy.

The major results are summarized as follows:

1. The occurrence of the localized flow (shear instability) depends both on the initial orientation of the specimens and on the temperature of deformation. The (101)[010] specimens, irrespective of Al or Al-Mg alloy, did not show flow localization in the present experimental condition. The localized flow was observed in the $(211)[\bar{1}11]$ Al specimen deformed at 77K and in the $(211)[\bar{1}11]$ Al-Mg specimen deformed at 77K - RT. However, the flow localization tended to disappear with increasing deformation temperature: no localized flow was found in the Al specimen at RT and in the Al-Mg specimen at 473K.
2. No systematic correlation can be obtained between the onset of the flow localization and the sign of the strain rate sensitivity m. During deformation at about room temperature, both the $(211)[\bar{1}11]$ and (101)[010] Al-Mg specimens showed serration on their load-displacement curves, and negative strain rate sensitivity was measured by strain rate change tests. When the deformation temperature was lowered to 77K or raised to 473K, the serration disappeared, and the sign of m changed to positive. Accordingly, both the orientation and temperature dependence of flow localization given in "1." is hard to explain on the basis of the strain rate sensitivity or dynamic strain aging.
3. The flow localization is closely connected with the dislocation structures developed during deformation of the specimens. In the $(211)[\bar{1}11]$ Al-Mg specimens deformed by accompanying with the localized flow, layer-like dislocation-cell structures were formed along one of active slip planes prior to the onset of flow localization. On the other hand, the microstructure of the $(211)[\bar{1}11]$ specimens in which no flow localization occurred, e.g. Al-Mg deformed at 473K and Al deformed at RT, was composed of well developed subgrains, which may suggest that extensive dynamic recovery was taking place during deformation. The (101)[010] Al-Mg specimen showed equiaxed cell structure at the room temperature deformation. The dislocation structure formed in the course of deformation depended greatly on both the crystallographic orientation and the deformation temperature.

---
* The localized flow in the materials with lower stacking fault energy has already been studied by the present authors.

INSTABILITY IN PLASTIC BENDING OF *THIN SHEET METAL*

J. L. Duncan*, M. Sue-Chu and X. J. Wang*****

*Department of Mechanical Engineering, McMaster University, Hamilton, Ontario L8S 4L7, Canada

**Steel Company of Canada, Hamilton, Ontario, Canada

***Beijing University of Iron and Steel Technology, Beijing, China

ABSTRACT

Instability in pure bending of aged steel is examined experimentally and theoretically. Unlike previous theories which employ the lower yield stress, the proposed bending model is based on a quasistatic strain hardening curve. A mechanical model of Lüders straining in the tensile test is presented which indicates that the lower yield stress is associated with a much higher strain rate than that achieved in typical bending.

The study of instability in pure bending is important for technological reasons and also as a means of understanding the basic mechanics of deformation of material exhibiting discontinuous yield phenomena.

KEYWORDS

Plastic bending of sheet, ageing of sheet steel, discontinuous yielding, Lüders strain, upper and lower yield stress, instability in bending, coil breaks.

INTRODUCTION

Plastic bending occurs either by shear deformation following the model of Green (1954), Fig. 1(a), or by more uniform straining, in which the Bernoulli assumption of converging plane sections, Fig. 1(b), is well satisfied. For pure bending, the latter leads to a linear strain distribution, shown in Fig. 2, in which

$$\varepsilon = y/R \qquad (1)$$

for both elastic and plastic deformation. In aged mild steel having the stress strain diagram shown in Fig. 3, the stress distribution associated with the bending strain in Fig. 2 is unresolved. The subject was reviewed at length by Roderick and Phillipps in 1949, by Campus in 1966 and by LeBlois

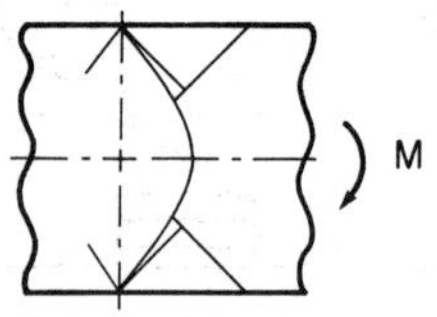

Fig. 1(a). Slip-line field solution for bending, Green (1954).

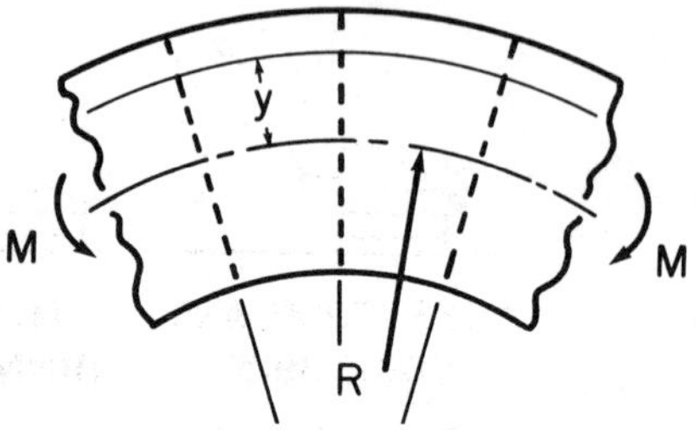

Fig. 1(b). Pure bending in which plane sections converge.

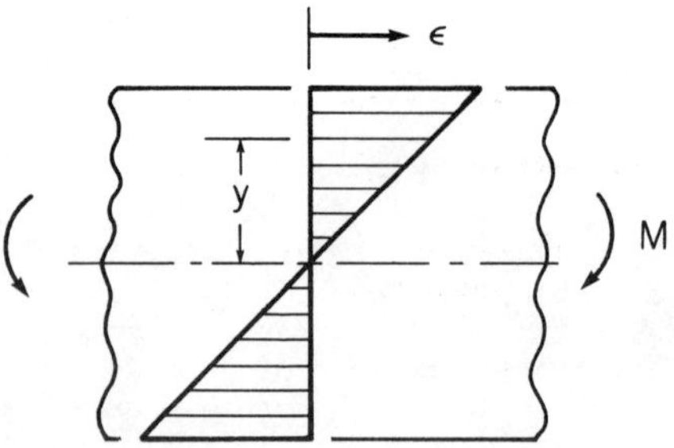

Fig. 2. Linear strain distribution in both elastic and plastic bending.

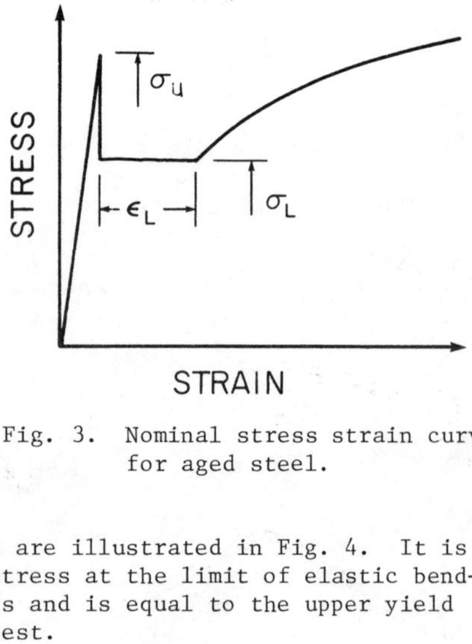

Fig. 3. Nominal stress strain curve for aged steel.

and Massonnet in 1972; the major theories are illustrated in Fig. 4. It is generally accepted that the outer fibre stress at the limit of elastic bending greatly exceeds the lower yield stress and is equal to the upper yield stress in a carefully conducted tensile test.

The authors investigated the kinking phenomena in sheet in pure bending in which, after reaching a maximum elastic curvature, $1/R_e$, Fig. 5(a), bending is localized in one region of greatly increased curvature, $1/R_p$, while other regions recover elastically, Fig. 5(b). It is postulated that this is the consequence of a moment-curvature characteristic that attains a maximum at the elastic limit, as shown in Fig. 6. The phenomenon depends critically on the post-elastic bending stress distribution but it was found that none of the classical theories accounted for the results observed. By considering the dynamic nature of Lüders deformation, as was also done by Kocks in a lengthy article published in 1981, it is suggested that the use of the lower yield stress is inappropriate in a bending theory. The description of Lüders straining in the tensile test which is given below is not new, but it is not widely known and is included here to help to explain the proposed bending hypothesis.

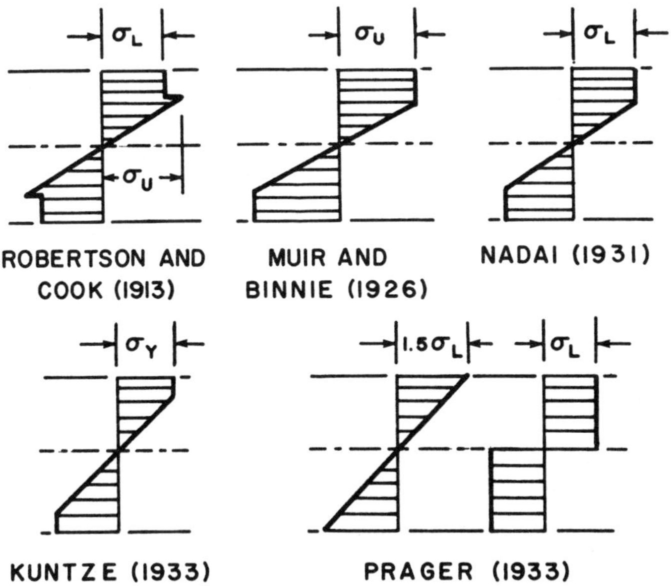

Fig. 4. Various theories of bending stresses associated with the linear strain distribution in Fig. 2.

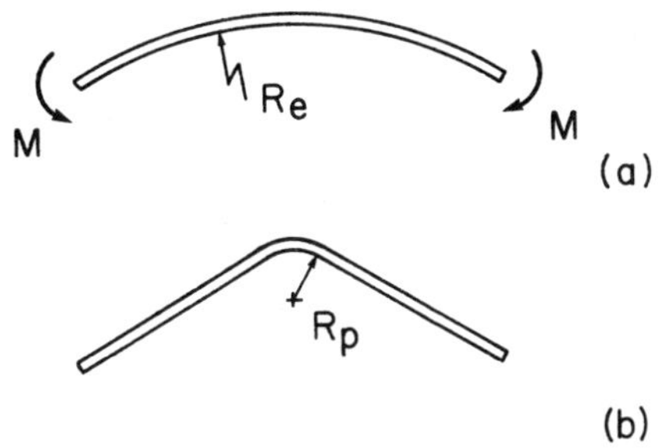

Fig. 5. Unstable transition from the limiting elastic curvature (a) to localized plastic bending (b).

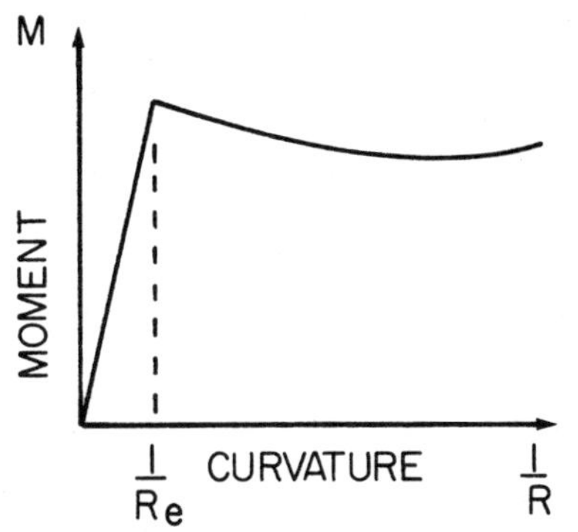

Fig. 6. Proposed moment curvature characteristic in unstable bending.

THE TENSILE TEST

The propagation of a Lüders band is shown in Fig. 7(a). Under the action of a constant lower yield load, P_L, and crosshead speed V, the band moves with a speed v into the unstrained region A. Magnifying the region shown, the strain increases from zero to the Lüders strain, ε_L, as in (b), over a short

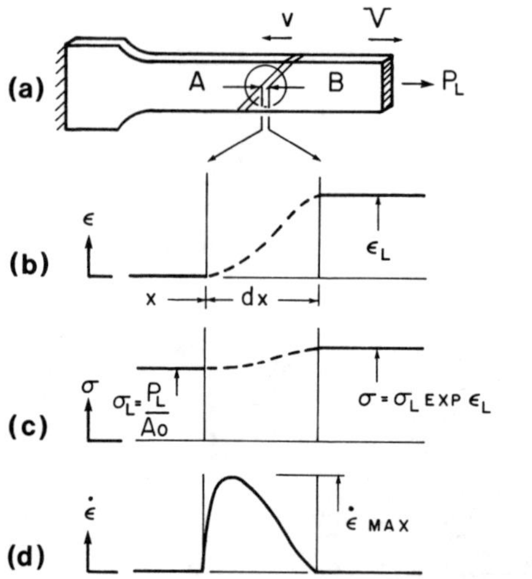

Fig. 7. Lüders strain in the tension test (a) showing the plastic strain (b), true stress (c) and strain rate (d) in a front of width dx.

distance, dx, which may be only several hundredths of a millimeter. The equilibrium condition throughout the testpiece is,

$$P_L = \sigma A = \sigma A_0 \exp(-\varepsilon) \qquad (2)$$

where A_0 is the initial cross-sectional area; this indicates that the stress rises above the nominal lower yield stress as the cross-section diminishes and at any point may be expressed as

$$\sigma = \sigma_L \exp \varepsilon \qquad (3)$$

as in Fig. 7(c). In a sharp-fronted band, the strain rate, $\dot{\varepsilon}$, is zero on either side of the band, but is high within the band and reaches a peak, $\dot{\varepsilon}_{max}$, as shown in Fig. 7(d). If the nominal stress strain curve of Fig. 3 is replaced by a true stress strain curve as in Fig. 8, the constant load region follows equation (3) as shown. Fig. 8, however, is only a two-dimensional view of the three-dimensional diagram, Fig. 9, in which strain rate, $\dot{\varepsilon}$, is the third axis.

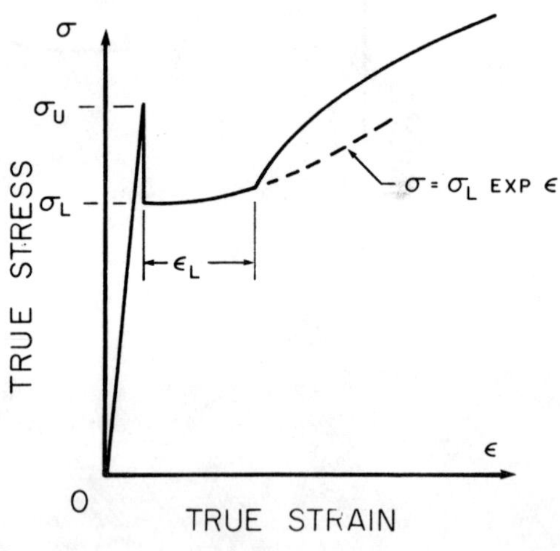

Fig. 8. True stress strain curve for a material element.

The material behaviour is represented in Fig. 9 by a surface $\sigma = \sigma(\varepsilon,\dot{\varepsilon})$. The deformation process for any element can now be illustrated by a path which starts at the origin, rises to the upper yield, σ_u, at some point and then drops to the constant load surface, $\sigma_L \exp \varepsilon$; deformation starts at a strain rate $\dot{\varepsilon}_{max}$ and the path follows the intersection of the constant load surface

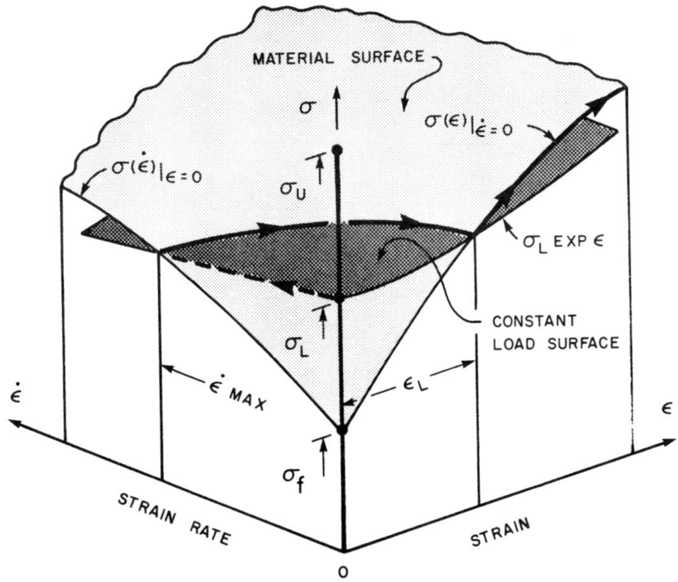

Fig. 9. Diagram illustrating the deformation path for a material element in the stress, strain, strain rate space.

and the material property surface until the strain rate has fallen to zero and the total strain is equal to the Lüders strain, ε_L. Strain hardening flow then occurs under quasistatic conditions, $\dot{\varepsilon} \approx 0$, with the load rising above P_L.

The above description is highly simplistic in that it assumes that stress is a function of state variables, ε and $\dot{\varepsilon}$. It is well known however, that stress depends also on the deformation history, nevertheless, the simple model illustrates that the equilibrium conditions that exist in a tensile test lead to large changes of strain rate within the Lüders band. (The diagram also helps to visualize other yield point phenomena. If, for example, the extension speed V is increased, it is likely that the maximum strain rate, $\dot{\varepsilon}_{max}$, will increase; the diagram then indicates both the lower yield stress, σ_L, and the Lüders strain, ε_L will also increase, as is observed.)

BENDING THEORY

In bending a strip, differentiation of equation (1) gives the strain rate at some distance y from the neutral axis, i.e.

$$\dot{\varepsilon} = y\,\frac{d(1/R)}{dt} \tag{4}$$

There is no discontinuity in strain rate at the elastic/plastic interface in bending the strip. The rate of change of curvature in plastic bending, even when kinking occurs, is not likely to be very high, and this suggests that the bending stress distribution should follow the quasistatic stress strain curve, as shown in Fig. 10. The moment-curvature diagram may be calculated

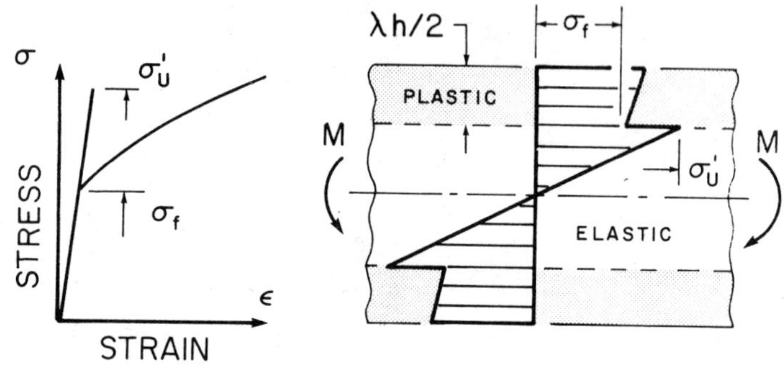

Fig. 10. Proposed stress distribution for an aged material in plastic bending.

numerically for the elastic-plastic stress distribution illustrated in Fig. 10, however, the critical factor governing stability is the slope of the curve immediately beyond the elastic limit. It is shown in the appendix that if $\sigma_f > 2\sigma_u'/3$ the moment curve is positive and stable bending should occur. If $\sigma_f < 2\sigma_u'/3$, the curve would be as shown in Fig. 6 and kinking may take place.

EXPERIMENTS AND RESULTS

Strip tensile tests were performed on a number of samples of rimming, killed and micro-alloyed steel. True stress-strain curves were computed and the plastic strain hardening data fitted by the relation

$$\sigma = K(\varepsilon_0 + \varepsilon_p)^n \tag{5}$$

where ε_p is the plastic strain. The initial flow stress was taken as

$$\sigma_f = K\,\varepsilon_o^n \tag{6}$$

Experiments were performed in a testing machine using the bending equipment and test-piece shape shown in Fig. 11. The moment-curvature diagram could be obtained from the autographic diagram as described by Sue-Chu (1980). Kinking, when it occurred, was evident both in the testpiece and in the autographic diagram.

For many materials, the maximum elastic curvature, $1/R_e$, was not well predicted by the tensile test upper yield stress and a new value of upper yield stress was calculated from the elastic curvature, i.e.,

$$\sigma_u' = Eh/2R_e \tag{7}$$

It is considered that this is a better indication of the upper yield stress of the material than the tensile test result, and this computed value is used in subsequent diagrams.

The theory predicts that bending is unstable only if $\sigma_f < 2\sigma_u'/3$. In Fig. 12, the results of all tests are plotted in a diagram of σ_u' versus σ_f. Those tests in which bending is stable are indicated by open circles; those in whic kinking occurred by solid points. The line, $\sigma_u' = 3\sigma_f/2$ divides the two

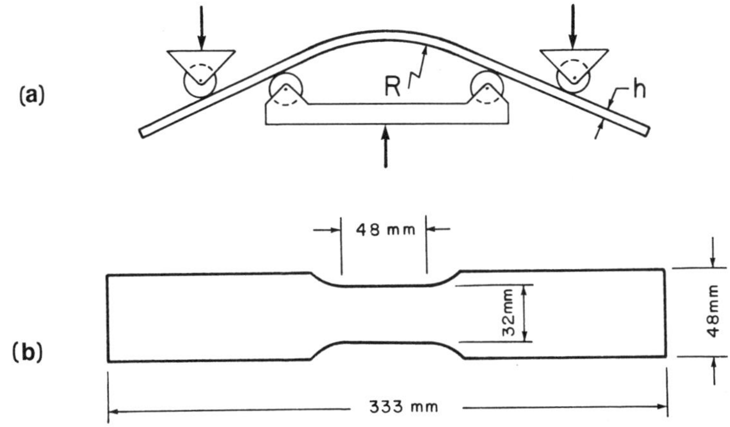

Fig. 11. Diagram of the four-point bending arrangement (a) and test-piece, (b).

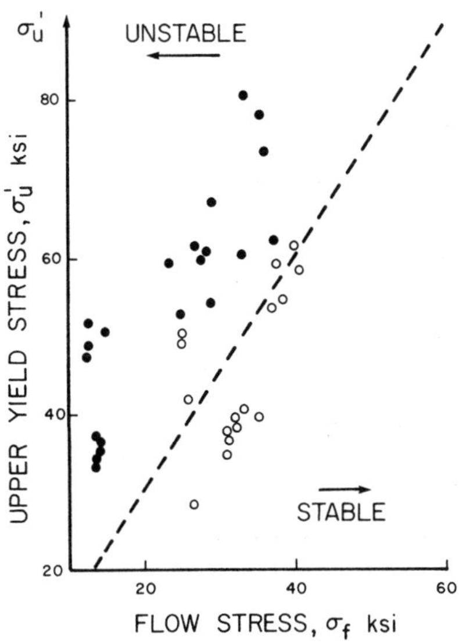

Fig. 12. Bending stability diagram following proposed theory. Solid circles indicate experiments in which kinking was observed, open circles indicate stable bending.

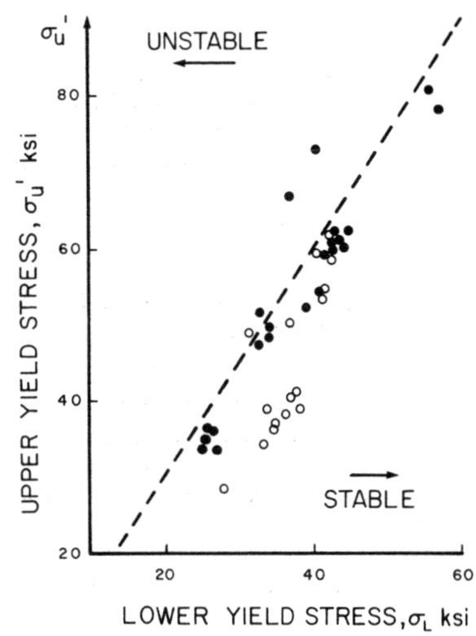

Fig. 13. Bending stability diagram following the stress distribution of Robertson and Cook. Solid circles indicate experiments in which kinking occurred, open circles indicate stable bending.

regions and it is apparent that all but five tests agree with the predictions. Following Robertson and Cook, the instability condition would depend on the lower yield point, i.e., $\sigma_L < 2\sigma'_u/3$. A diagram of σ_L versus σ'_u is shown in Fig. 13.

DISCUSSION

It is suggested here that the lower yield stress observed in the tension test of aged mild steel is associated with deformation at a much higher strain rate than that in bending; consequently, bending theories ought to employ a quasi-static flow stress, σ_f. In this work, σ_f is obtained by the admittedly unsatisfactory process of back-extrapolation of the tensile strain hardening curve and some values appear unrealistically low. The bending theory proposed however is based on σ_f and an upper yield stress derived from the measured curvature at the elastic limit. The experiments presented are insufficiently discriminating to prove the theory, but agreement appears better than with previous bending theories. The results support the suggestion made by Southwell in discussion of the paper of Roderick and Phillipps, that "the time had now come when the pure bending test should supersede the tension test as a standard means of determining the properties of a material."

REFERENCES

Robertson, A. and G. Cook (1913). Proc. R. Soc. (Lond.), A 88, 462.
Muir, J. and D. Binnie (1926). Engineering, 122, 743.
Nadai, A. (1931). Plasticity, McGraw-Hill, N. Y.
Kuntze, W. (1933). Stahlbau, 6, 49.
Prager, W. (1933). Forsch. Ing. Wes., 4, 95.
Roderick, J.W. and I.H. Phillipps (1949). Engineering Structures, Butterworths, London, 9.
Green, A.P. (1954). J. Mech. Phys. Solids, 3, 1.
Campus, F. (1966). I.A.B.S.E. Pubn., Zurich, 26, 97.
LeBlois, Cl. and Ch. Massonnet. (1972). Int. J. Mech. Sci., 14, 94.
Sue-Chu, M. (1980). M.Eng. Thesis, McMaster University.
Kocks, U.F. (1981). Chalmers Anniversary Vol., Progress in Material Sci., 185.

APPENDIX

If the zone of plasticity has penetrated a strip of width, b, to a fractional depth, λ, as in Fig. 10, then, neglecting strain hardening the applied moment is

$$M = \sigma'_u b[(1 - \lambda) h]^2/6 + \sigma_f b \lambda(2 - \lambda) h^2/4$$

Writing $\sigma_f = c\,\sigma'_u$, and differentiating, we obtain

$$(6/\sigma'_u bh^2) \frac{dM}{d\lambda} = (3c - 2)(1 - \lambda)$$

Beyond the elastic limit, the slope of the moment curve will be positive if $c > 2/3$. If $c < 2/3$, the slope is negative and the bending will be unstable; the moment curve may eventually rise due to strain hardening, however kinking will have already occurred.

ACKNOWLEDGEMENTS

The authors thank J. Madronich and P. Gatt for their assistance with experiments and the preparation of the manuscript and J. D. Embury, U. F. Kocks and R. Sowerby for their contributions to the ideas expressed. The authors also thank the Natural Sciences and Engineering Research Council of Canada for their support of this work.

AGE hr.	TENSILE TEST σ_u ksi	σ_L ksi	σ_f ksi	ε_L	BEND σ'_u ksi	$\frac{\sigma_L}{\sigma_u}$	$\frac{\sigma_f}{\sigma_u}$	STABILITY
1	2	3	4	5	6	7	8	9
Rimmed Steel A, h = 0.034 in, Aged at 100°C								
0.00	42.3	37.9	25.7	.031	41.9	.90	.61	S
0.50	42.4	39.2	24.8	.036	52.8	.74	.50	U
1.00	42.7	40.6	29.4	.038	54.4	.74	.54	U
5.00	43.4	41.6	23.4	.041	59.3	.70	.39	U
13.25	48.5	42.5	28.0	.045	60.0	.71	.47	U
13.75	45.0	42.6	33.5	.049	60.1	.71	.56	U
14.25	45.2	42.3	27.5	.052	60.2	.70	.46	U
14.75	49.0	42.8	28.4	.055	60.4	.71	.47	U
15.75	49.7	42.1	27.0	.056	61.7	.68	.44	U
18.00	46.6	44.4	37.2	.066	62.2	.71	.60	U
Rimmed Steel B, h = 0.04 in, Aged at 200 to 270°C								
0.00	43.4	32.1	12.6	.060	47.8	.67	.26	U
0.75	42.2	33.6	13.0	.078	48.3	.70	.27	U
0.50	44.3	34.0	14.9	.068	50.5	.67	.30	U
1.00	48.1	32.2	12.9	.070	51.6	.62	.25	U
Rimmed Steel B', h = 0.039 in, Aged at 100°C								
0.00	28.7	28.0	26.2	.009	28.6	.98	.92	S
0.25	35.0	33.2	31.3	.024	34.6	.96	.90	S
0.50	36.7	34.9	31.3	.034	36.9	.95	.85	S
1.00	41.7	36.7	32.7	.041	38.9	.94	.84	S
1.50	42.9	38.4	35.2	.062	39.7	.97	.89	S
2.00	41.8	37.3	33.2	.065	40.5	.92	.82	S
Rimmed Steel C, h = 0.058 in, Aged at 100°C								
0.00	35.4	34.1	32.0	.008	37.0	.92	.86	S
2.00	44.3	41.5	37.0	.038	53.2	.78	.70	S
4.00	45.5	41.7	38.7	.039	55.0	.76	.70	S
13.00	46.6	42.8	40.9	.040	58.3	.73	.70	S
19.50	44.9	40.6	37.5	.051	59.6	.68	.63	S
23.00	45.5	42.5	40.1	.068	61.5	.69	.65	S
Mild-Steel, h = 0.056 in, Aged at 100°C								
0.00	28.1	27.0	13.5	.018	33.5	.81	.40	U
2.00	28.4	25.6	13.6	.018	34.3	.75	.40	U
4.00	27.7	25.7	14.2	.018	35.2	.73	.40	U
23.00	27.7	27.0	14.1	.019	36.5	.74	.39	U
36.00	27.5	26.3	13.4	.019	37.0	.71	.36	U
HSLA, h= 0.076 in, Aged at 225°C								
0.00	71.3	57.1	35.8	.043	78.3	.73	.46	U
8.00	60.4	55.9	33.2	.050	81.0	.69	.41	U
DQ Steel, h = 0.031 in, Aged at 100°C								
0.00*	35.0	31.3	25.9	.028	49.3	.63	.53	S
0.00**	41.5	37.4	25.3	.040	50.6	.74	.50	S
22.00*	39.8	37.1	29.3	.046	67.2	.55	.44	U
22.00**	42.7	40.7	36.0	.063	73.4	.55	.49	U
DQ Galvanized Steel, h = 0.08 in								
0.00	34.1	34.1	32.3	.000	39.0	.87	.83	S

U indicates Kinking S indicates Stable Bending

* longitudinal direction ** transverse direction

TABLE I
TENSILE AND BENDING TEST DATA

TEST METHODS FOR FORMABILITY OF HOT ROLLED STEEL SHEETS

M. Lindberg and L. Melin

University of Luleå, Sweden

ABSTRACT

Several methods for testing formability of hot rolled sheets and plates have been investigated. The best distinction between different materials was obtained by a combination of a stretch forming test and a hole expansion test. The test results may be affected by cut edges, and a cutting practice is recommended here to minimize this influence.

KEYWORDS

Hot rolled steel sheets and plates; formability test; cutting clearance; hole expansion test; stretch forming test; tensile test; deep drawing test; bend test.

INTRODUCTION

Hot rolled sheets and plates are manufactured with increasing mechanical strength and are used with smaller thicknesses. This affords new possibilities to the designer to save weight and costs but at the same time production might be more difficult because of the lower formability of the high strength steel. The demand for good methods for testing formability have thus been more pronounced.

There are today no standardised test methods to predict the formability of hot rolled plates or compare the formability of different materials. This fact makes it difficult to introduce new materials in the workshop and to utilize the advantages of the new materials.

It is well known that edge condition has a great influence on formability, both in laboratory tests and in production. It was therefore obvious to the authors that test methods for hot rolled plates must be divided into two categories, those which only measure the pure material and those which measure the influence of defects in edges. In this work tests from both categories have been studied. Methods from the first group include: the tensile test, the stretch forming test and the limit drawing ratio (LDR) test. From the second group, the bend test, notched tensile test and hole expansion test (Lindberg 1982) were studied.

MATERIAL

When establishing test methods it is essential to carry them out on materials with widely varying properties. In this case normal standardized steels as well as new company grades were used. Mechanical properties of materials used in the tests are given in Table 1 and chemical analyses in Table 2.

TABLE 1 Mechanical Properties

Material	Thickness (mm)	Rel (MPa)	Rm (MPa)	A5 (%)
SIS 1312	2	337	435	36
SIS 1312	4	286	381	39
SIS 1312	8	286	423	40
SIS 2132	4	382	545	21
SIS 2132	8	359	519	28
Steel A	2	304	386	32
Steel B	4	280	339	44
Steel C	4	397	488	36
Steel D	8	374	465	31
Steel E	4	538	607	27
Steel F	8	530	614	22
Steel G	3	447	680	28

TABLE 2 Chemical Analyses

Material	Thickness mm	C %	Si %	Mn %	P %	S %	N %	Al %	Nb %
SIS 1312	2	.08	.02	.46	.020	.014	.009	.028	<.005
	4	.07	.05	.39	.006	.028	.008	.036	<.005
	8	.09	.06	.64	.015	.023	.008	.035	<.005
SIS 2132	4	.14	.30	1.44	.019	.015	.003	.067	<.005
	8	.15	.24	1.34	.018	.013	.005	.027	<.005
Steel A	2	.06	-	.23	.007	.012	.008	.042	<.005
B	4	.06	.02	.21	.007	.014	.007	.058	<.005
C	4	.08	.05	.42	.013	.004	.006	.038	.016
D	8	.07	.07	.39	.009	.005	.006	.045	.017
E	4	.06	.21	1.52	.024	.004	.014	.047	.030
F	8	.05	.21	1.45	.018	.005	.012	.058	.037
G	3	.13	.50	1.56	.014	.005	.007	.062	-

PROCEDURE AND RESULTS

In order to obtain a measure of the effectiveness in distinguishing the materials, a statistical model has been applied to the tests. The test results are assumed to be normally distributed with a mean value, μ, and a standard deviation, σ. The estimated means, $\bar{x}$, and standard deviations, s, were calculated. Two series of measured values were then compared by means of the calculated t-value. The standard t-distribution then gives the appropriate probability.

Formulae:

$$s_0 = \sqrt{\frac{(n_1 - 1)s_1^2 + (n_2 - 1)s_2^2}{n_1 + n_2 - 2}} \qquad (1)$$

$$t = \pm \frac{\mu_1 - \mu_2^2 - (\bar{x}_1 - \bar{x}_2)}{s_0 \sqrt{\frac{n_1 + n_2}{n_1 \cdot n_2}}} \qquad (2)$$

Hypothes: $\mu_1 - \mu_2 = 0$

n_1 and n_2 are the numbers of specimens in each test.

Stretch forming test

In stretch forming the sheet or plate is locked between a die and a blank holder and a punch is forced against the material, Fig. 1. In a stretch forming test standardized tools with a spherical punch are used and the maximum dome height when fracture occurs is measured and related to the stretch formability of the material. The method suffers from the fact that friction between punch and test material affects the result. It is also almost impossible to lock the plate completely between die and holder and insufficient locking may also affect the test result.

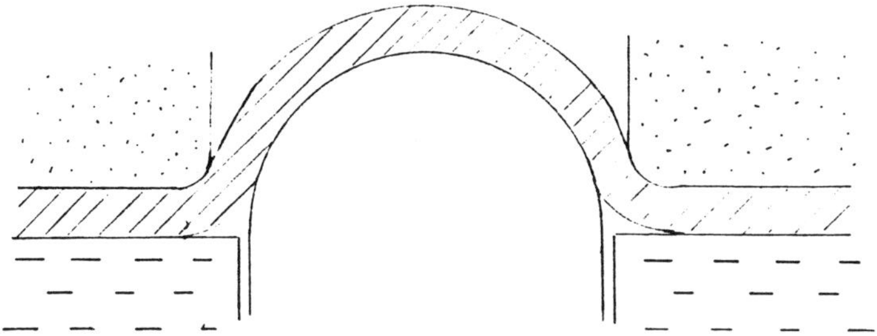

Fig. 1. Stretch Forming

Standardized stretch forming tests known as Erichsen and Olsen tests are widely used on cold rolled materials but the small geometrical dimensions of these tools make them unsuitable for testing hot rolled material. Instead a larger punch, diameter 100 mm, was used in this case. To minimize friction between punch and test sample, polythene film was used as a lubricant.

The method has a very good distinction capability as shown in Table 3.

TABLE 3 Probability that Observed Differences in Dome Height Distinguish Different Materials. Thickness = 4 mm.

	SIS 2132	Steel B	Steel C	Steel E
SIS 1312	99,9 %	90 %	99.9 %	99.9 %
SIS 2132		99.9 %	95 %	98 %
Steel B			99 %	99.9 %
Steel C				99.9 %

Hole Expansion Test

The hole expansion test measures how much material at an edge can be elongated bofore cracking occurs. A hole in the test piece is expanded over a flat punch until the edge cracks, Fig. 2.

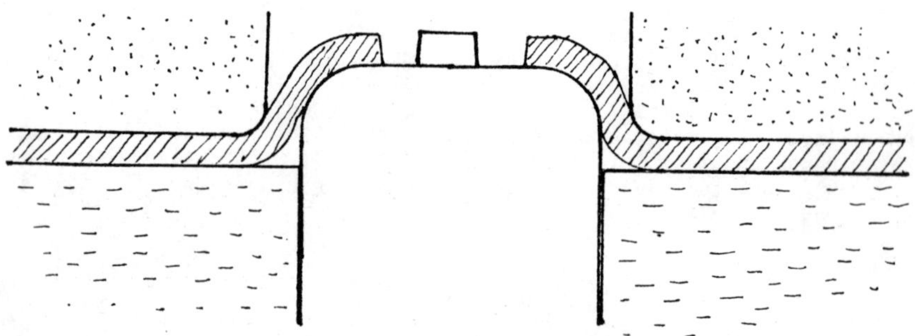

Fig. 2. Hole expansion test.

Our experiments showed clearly that the result of the hole expansion test is affected both by the material tested and quality of the edge. If holes were drilled better results were obtained than with punched holes. Also the method of punching had a great influence on the results. Only with properly selected clearances was it possible to obtain results which could be reproduced.

With correctly punched holes, the hole expansion test has a good distinction capability which is shown in Table 4.

The sensitivty given by the hole expansion test is smaller than that which stretch forming gives. However, hole expansion tests give information concerning how sensitive a material is to deformation hardening and other influences from the punching process, and thus can be complement to the strech forming test.

TABLE 4 Probability that Observed Differences in the Result from Hole Expansion test Distinguishes Different Materials. Thickness = 4 mm.

	SIS 2132	Steel B	Steel C	Steel E
SIS 1312	99.9 %	99 %	70 %	80 %
SIS 2132		99.9 %	98 %	99.9 %
Steel B			99 %	99.9 %
Steel C				30 %

Tensile test and notched tensile test

It is known from steel producers that the normal tensile test does not always give adequate information about the formability of sheets and plates. One reason for this might be that the tensile test does not measure the notch sensitivity of the edge. In order to determine whether notch sensitivity could be measured, tensile test pieces were provided with notches. These notches were 2 mm in depth, had an opening angle of 45 deg and a base radius of 0.25 mm (Charpy V notch). Two notches opposite each other were machined on each test piece. The width at gauge length was 30 mm.

Only ultimate strength could be measured on the notched test pieces, their elongations were too small to be registered. In this respect our results differ from those reported by Matsudo and co-workers (1974). Test pieces taken parallel and perpendicular to rolling direction gave different results, Table 5.

TABLE 5 Ultimate Strength of Notched Test Pieces

Steel	Thick-ness mm	Strength parallel MPa	Strength transverse MPa
SIS 1312	2	327	392
	4	350	363
	8	399	400
SIS 2132	4	505	500
	8	468	494
Steel A	2	354	351
B	4	323	317
C	4	460	456
D	8	456	447
E	4	557	562
F	8	579	585
G	3	573	614

It was not possible to correlate the result in Table 4 with the formability of the materials and so the distinction power is low.

Bend test

Bend tests were carried out in V-bend tools up to a 120^0 bending angle and with different punch radii and die widths. The edges on the test pieces were cut with different clearances and it was observed that clearance had a significant influence on the results. The test gives only positive or negative information which cannot be treated statistically. It is our opinion that bend tests can be used on material which is to be used in bending operations with the proviso that the edges are cut properly.

Limit drawing ration (LDR)

A limit drawing ratio is normally used on cold rolled products but can be defined also for hot rolled material. However, the normal standardized tools can not be used in this case because of unfavourable metal thickness/punch diameter ratio. In this case when a punch diameter of 100 mm was used, a good quantitative estimate of the drawability was obtained. Results from 4 mm material are given in Table 5. However, only one value per quality is not sufficient for a statistical analysis.

TABLE 6 Limit Drawing Ratio. Thickness = 4 mm

Steel	LDR
SIS 1312	1.83
SIS 2132	1.82
Steel B	1.76
Steel C	1.71
Steel E	1.72

Although LDR gives good information about the formability, this test method is not recommended because of the large number of discs required to get a significant result and because normal presses in laboratories do not have sufficient power for suitable tools.

CUTTING AND PUNCHING CLEARANCES

The size of the clearance is very important for the properties in the cut edge and for tool wear. An earlier investigation has shown that the clearance has to be chosen with regard to sheet thickness and tensile strength (Lindberg 1982). Large burr indicates a bad cut which gives poor forming capacity. The relation between burr height and clearance is shown in Fig. 3. Our experience is that clearances indicated by the second minimum on the curve in Fig. 3 give best formability. It is also known that tool wear decreases with increasing clearance which gives us another reason for choosing the clearance at the second minimum.

The optimum clearance indicated in Fig. 3 depends on the tensile strength as shown in Fig. 4.

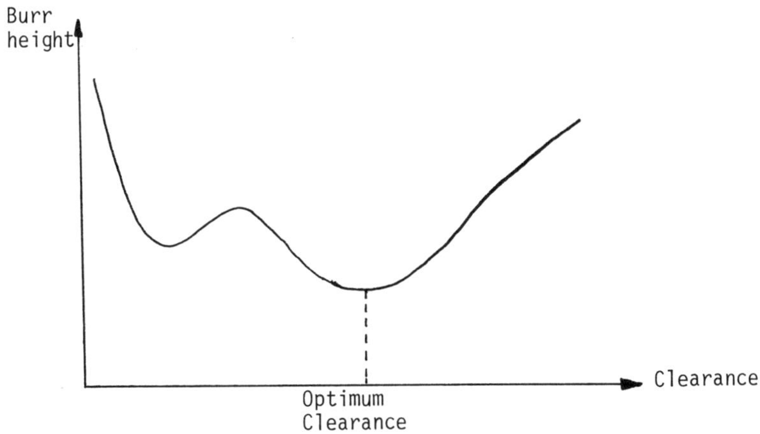

Fig. 3. Relation between burr height and cutting clearance.

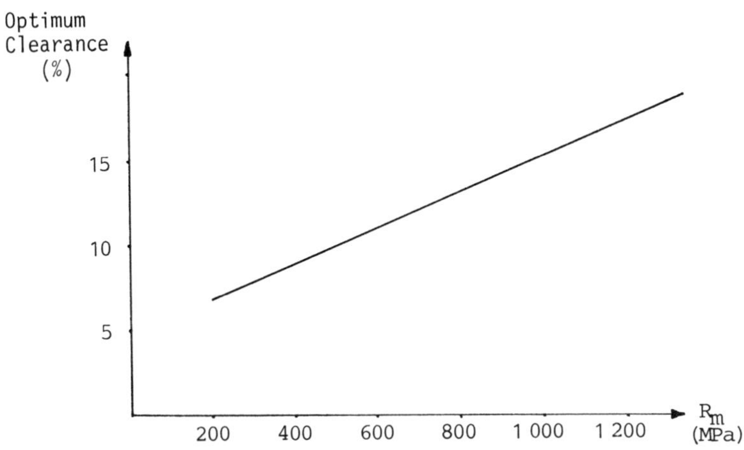

Fig. 4. Relation between optimum cutting clearance and ultimate strength.

It must be pointed out that the clearance interval which gives acceptable edge quality is different for different materials, i.e. the common structural steels are less sensitive to changes in clearances than the new micro alloyed high strength steels.

CONCLUSIONS

The combination of test methods chosen; stretch forming test and hole expansion test, give good information about the formability of a material. The stretch forming test shows how much deformation the material can undergo on parts which are not affected by cutting or other mechanical treatments, while the whole expansion test indicates the notch sensitivity at edges. It is also shown that cutting clearance has great influence on the forming operations following cutting.

ACKNOWLEDGEMENT

The authors would like to express their gratitude to the National Swedish Board for Technical Development (STU) which has made this work possible.

REFERENCES

Lindberg, M. (1982). Provmetoder för varmvalsad plåts formbarhet. University of Luleå, 1982:23T.

Matsudo, K., T. Shimomura, K. Osawa, M. Yoshida and Y. Uchida (1974). Nippon Kokan Tech. Rep. Overseas, June, 1-14.

ANALYSIS OF PLASTIC INSTABILITY IN SHEET-METAL FORMING

G. Ferron and M. Mliha-Touati

Laboratoire de Mécanique et de Physique des Matériaux, E.R.A. au C.N.R.S. n° 123, E.N.S.M.A., Rue Guillaume VII, 86034 Poitiers Cedex, France

ABSTRACT

A first-order approach was developed to analyse the growth of a pre-existing defect in the form of a band across a sheet under biaxial tension. Owing to the flexibility of the method, it was possible to pay attention to the details of strain-hardening and strain-rate hardening behaviours of"brass-type" and "steel-type" materials. In particular the low limit strains observed for soft aluminium in the stretching range can be explained in terms of a rapid decrease of the work-hardening rate at high strains, together with a distortion of the yield surface. It is also possible to account for the low limit strains obtained for A-K steel in the sequence : balanced biaxial tension → uniaxial tension.

KEYWORDS

Sheet-metal formability ; localized necking ; limits to ductility ; strain-hardening ; strain-rate hardening ; yield surface ; plastic equation of state ; plastic anisotropy.

INTRODUCTION

The capability of predicting the success of a sheet-metal forming operation is of considerable technological importance. When the strain ratio is maintained nearly constant throughout the process, the Forming Limit Diagram (FLD), i.e. the plot of the limit value ε_1^* of the larger surface strain against the limit value ε_2^* of the smaller surface strain, furnishes a valuable aid for engineering applications. However, a number of issues still remain unexplained. In particular, the FLD in the stretching domain ($\varepsilon_2 > 0$) can present positive, negative or zero slopes, depending on the material (Azrin and Backofen, 1970 ; Backofen, 1973). Also, the theoretical determination ot the limit strains under complex strain paths is a difficult and debated question (Bergström and Olund, 1982 ; Laukonis and Ghosh, 1978 ; Müschenborn and Sonne 1975 ; Ranta-Eskola, 1980 ; Ronde-Oustau and Baudelet, 1977).

This work was motivated by the recent advances in the formulation of the constitutive relations and in the analysis of plastic instability in simple tension. It should be recalled, first, that localized necking in sheets under biaxial tension occurs at larger (effective) strain levels than diffuse necking in simple tension. Therefore, it is highly desirable to use work-hardening laws that provide a realistic account of the behaviour of the material beyond the usual range of tensile testing. Secondly, it has been observed in several materials that the classical flow - theory of plasticity developed by Hill (1950) sometimes fails to accommodate the stress-strain curves obtained under different loading conditions (uniaxial tension, plane strain deformation, balanced biaxial tension). As an attempt to overcome such discrepancies, the new yield criterion proposed by Hill and presented by Mellor and Parmar (1978) will be used. Thirdly, it has been shown (Kocks and co-workers, 1979 ; Ferron and Mliha-Touati, 1982) that the necking behaviour in simple tension is markedly changed when a state-parameter formulation of constitutive laws is adopted, in place of the usual descriptions where the accumulated plastic strain is treated as a "state variable". These improvements should also be added to the analysis of plastic instability in sheet-metal forming.

In the present paper, inhomogeneous plastic flow is assumed to develop from a pre-existing defect in the form of a band across the sheet. In comparison with a Marciniak-type model (Hutchinson and Neale, 1978 ; Marciniak and co-workers, 1973 ; Parmar and Mellor, 1978) the most obvious difference is that a linearized (first-order) treatment is adopted here. This position is motivated by the versatility of the method. In fact, the development of an exact load equilibrium relation is tied down *ab initio* to the use of a particular form of the constitutive law. In contrast, it will be shown that it is possible to set up linearized equations for non-uniform plastic flow, *which apply for any form of one-parameter equation of state*. In these equations, the constitutive law is specified by the current values of the work-hardening coefficient and of two (independent) rate sensitivity coefficients.

For completeness, it should finally be noted that the analysis is limited to plane stress conditions, i.e. the stress σ_3 in the thickness direction is assumed to be negligible. Likewise, bending moments, transverse shear forces, inertial effects and thermal effects are assumed to be absent in the sheet.

ANALYTICAL FRAMEWORK

Yield criterion and associated plastic flow rule

In order to reconcile the stress-strain curves obtained under different loading conditions, Hill's new criterion is used. Under plane stress conditions ($\sigma_3 = 0$), and assuming that there is only normal anisotropy (i.e., no planar anisotropy), the effective stress $\bar{\sigma}$ is given by (Mellor and Parmar, 1978)

$$\bar{\sigma} = \left\{ \frac{1}{2(1+r)} \left[(1+2r)(1-\alpha)^m + (1+\alpha)^m \right] \right\}^{1/m} \sigma_1 \qquad (1)$$

where $\alpha = \sigma_2/\sigma_1$ is the stress ratio in the plane of the sheet, r is the ratio of the width strain to the thickness strain in uniaxial tension tests and m is a parameter which allows distorsion of the yield surface.

Besides, the assumption of normality of the plastic strain increments leads to

$$\rho_i = \frac{(1+\alpha)^{m-1} - (1+2r)(1-\alpha)^{m-1}}{(1+\alpha)^{m-1} + (1+2r)(1-\alpha)^{m-1}} \qquad (2)$$

where $\rho_i = \dot{\varepsilon}_2/\dot{\varepsilon}_1$ is the strain rate ratio in the plane of the sheet.

Lastly it follows from the assumption of equivalence of plastic work that the effective strain rate $\dot{\bar{\varepsilon}}$ is given by

$$\dot{\bar{\varepsilon}} = A\dot{\varepsilon}_1 \tag{3}$$

with

$$A = \frac{[2(1+r)]^{1/m}}{2}\left[\frac{1}{(1+2r)^{1/(m-1)}}(1-\rho_i)^{m/(m-1)}+(1+\rho_i)^{m/(m-1)}\right]^{(m-1)/m} \tag{4}$$

The above equations reduce to the equations of Hill's "old" theory with $m = 2$, and to the Von-Mises yield criterion and its associated plastic flow rule with $r = 1$.

Mechanical equation of state

It is now generally recognized that the plastic behaviour of metallic materials is history-dependent, and that the accumulated plastic strain is not a pertinent state-variable. However, a single parameter is sufficient in most cases (excluding the transient following a radical alteration in the external conditions applied to the specimen) to characterize the current state of the material. Accordingly the concept of mechanical equation of state proposed by Hart (1970) is adopted. Although Hart's theory has so far been developed for uniaxial behaviour, the results can be tentatively expected to hold for triaxial configurations as well. We postulate, then, that *any isothermal deformation history* can be described in effective terms by the differential form

$$d\mathrm{Ln}\,\bar{\sigma} = \gamma d\bar{\varepsilon} + \nu d\mathrm{Ln}\dot{\bar{\varepsilon}} \tag{5}$$

where the work-hardening coefficient γ and the strain rate sensitivity coefficient ν are only functions of the current values of the two variables $\bar{\sigma}$ and $\dot{\bar{\varepsilon}}$.

It has been shown (Ferron and Mliha-touati, 1982), that the evolutionary equ.(5) cannot be used to describe spatial differences when $\bar{\varepsilon}$ is not a state-variable. The reason for this lies in the fact that two material elements that exhibit developing differences follow different deformation paths. Using the operator δ to express these spatial differences, the appropriate first-order equation between the gradients of $\bar{\sigma}$, $\bar{\varepsilon}$ and $\dot{\bar{\varepsilon}}$ should then be written

$$\delta\,\mathrm{Ln}\,\bar{\sigma} = \gamma\,(\,\delta\bar{\varepsilon} - \Delta\bar{\varepsilon}_y) + \nu\,\delta\mathrm{Ln}\dot{\bar{\varepsilon}} \tag{6}$$

where $\Delta\bar{\varepsilon}_y$ is the difference in the $\bar{\varepsilon}$-values of the two material elements of interest when they attain the same hardness y. The reader is referred to Hart (1970) for the definition of y, and to Ferron and co-workers (1983) for the calculation of $\Delta\bar{\varepsilon}_y$ in a constant (average) strain rate test.

Sheet imperfections

The analysis is now based on the growth of a pre-existing defect in the form of a band which is inclined at an angle ψ with respect to the principal axes. Using the operator δ to express any spatial difference between the defective and homogeneous regions, we can write : $\varepsilon_3 = \mathrm{Ln}\,(h/h_o)$; $\dot{\varepsilon}_3 = \dot{h}/h$;

$\delta\varepsilon_3 = \delta\mathrm{Ln}\,h - \delta\,\mathrm{Ln}\,h_o$; $\delta\dot{\varepsilon}_3/\dot{\varepsilon}_3 = \delta\,\dot{h}/\dot{h} - \delta\,\mathrm{Ln}\,h.$

The above relations can be handled to treat the development of the rate gradients for an initial defect of "geometric" or "mechanical" character (or both together : a "mixed" defect). To this end, it is appropriate to define the outset of straining ($\bar{\varepsilon} = 0$) *after the application of the local prestrain that produces a deformation defect* (Ferron and co-workers, 1983). Thus, the thickness gradient resulting from the local prestrain is included in the $\delta \mathrm{Ln}\, h_o$-value. Besides, it is needed to assign a value to the initial mechanical heterogeneity. This heterogeneity is most suitably characterized by the value of the hardness gradient, $\delta \mathrm{Ln}\, y_o$. Finally, a mixed defect is fully specified by the three values $\delta \mathrm{Ln}\, h_o$, $\delta \mathrm{Ln}\, y_o$ and ψ_o.

For lack of room, we simply give the broad lines of calculation in the case of a band perpendicular to the direction of the larger principal strain ε_1 in the plane of the sheet. Then, the basic relations are derived from the equilibrium condition across the band ($\delta\ \mathrm{Ln}\, \sigma_1 + \delta \mathrm{Ln}\, h = 0$) and from the compatibility condition ($\delta \mathrm{Ln}\, \dot{\varepsilon}_2 = 0$). With $\rho_i = \dot{\varepsilon}_2/\dot{\varepsilon}_1$, the compatibility condition becomes

$$\delta\, \mathrm{Ln}\ \rho_i = -\,\delta\, \mathrm{Ln}\ \dot{\varepsilon}_1 \tag{7}$$

or, with the assumption of constancy of volume, $\delta\, \mathrm{Ln}\, \rho_i = -\,(1+\rho_i)(\delta\dot{\varepsilon}_3/\dot{\varepsilon}_3)$. Lastly, the absence of strain gradients in the direction of the band gives : $\delta\varepsilon_1 = -\,\delta\varepsilon_3$.

EQUATIONS FOR NON-UNIFORM PLASTIC FLOW

Equation for $\delta \mathrm{Ln}\bar{\sigma}$

On combining equs.(1) and (2) the effective stress $\bar{\sigma}$ can be expressed as a function of ρ_i and σ_1. Furthermore, using equ.(7) and omitting intermediate calculations, the difference $\delta\, \mathrm{Ln}\bar{\sigma}$ can be expressed as follows

$$\delta \mathrm{Ln}\bar{\sigma} = -\,B\ \delta\, \mathrm{Ln}\ \dot{\varepsilon}_1 + \delta\, \mathrm{Ln}\ \sigma_1 \tag{8}$$

with

$$B = \frac{4\rho_i^2\ (1-\rho_i^2)^{\frac{2-m}{m-1}}\ (1+2r)^{\frac{1}{m-1}}}{(m-1)\left[(1+2r)^{\frac{1}{m-1}}(1+\rho_i)^{\frac{m}{m-1}}+(1-\rho_i)^{\frac{m}{m-1}}\right]\left[(1+2r)^{\frac{1}{m-1}}(1+\rho_i)^{\frac{1}{m-1}}+(1-\rho_i)^{\frac{1}{m-1}}\right]} \tag{9}$$

Equation for $\delta \mathrm{Ln}\dot{\bar{\varepsilon}}$

According to equs.(3) and (4) the difference $\delta\, \mathrm{Ln}\dot{\bar{\varepsilon}}$ results from gradients in ρ_i and $\dot{\varepsilon}_1$. Furthermore, using equ.(7), we obtain

$$\delta\, \mathrm{Ln}\ \dot{\bar{\varepsilon}} = C\ \delta\, \mathrm{Ln}\ \dot{\varepsilon}_1 \tag{10}$$

$$\text{with } C = \frac{\dfrac{1}{(1+2r)^{1/(m-1)}}(1-\rho_i)^{1/(m-1)} + (1+\rho_i)^{1/(m-1)}}{\dfrac{1}{(1+2r)^{1/(m-1)}}(1-\rho_i)^{m/(m-1)} + (1+\rho_i)^{m/(m-1)}} \tag{11}$$

Equation for $\delta\bar{\varepsilon}$

The difference $\delta\bar{\varepsilon}$ is obtained on integrating $\delta\dot{\bar{\varepsilon}}$ over the time of straining. Under a linear strain-path the nominal ρ_i-value is constant. Thus, recognizing that $\delta\varepsilon_1 = \int_0^{\tau} \delta\dot{\varepsilon}_1\ d\xi$, the use of equs.(3) and (10) leads to

$$\delta\bar{\varepsilon} = AC\ \delta\varepsilon_1 \tag{12}$$

Equations for flow localization under linear strain-paths

It is now possible to derive the equations describing the development of the rate gradients throughout a proportional straining process. Appropriate use of the above equations leads to four equivalent forms, according as $\bar{\varepsilon}, \varepsilon_1$, ε_3 or h is chosen as the relevant variable. The most useful equation should be expressed in terms of the thickness

$$[1-AC\gamma-(\nu C+B)(\rho+1)]\delta \text{Ln } h+(\nu C+B)(\rho+1)(\delta\dot{h}/\dot{h}) - \gamma\Delta\bar{\varepsilon}_y+AC\gamma\ \delta\text{Ln } h_o = 0 \qquad (13)$$

The expressions for A, B, and C given in equs. (4), (9) and (11), respectively, are utilized here with $\rho_i = \rho$ = constant.

The general form of $\Delta\bar{\varepsilon}_y$ remains to be specified. As shown by Ferron and co-workers (1983), $\Delta\bar{\varepsilon}_y$ in a constant strain rate process can be expressed as

$$\Delta\bar{\varepsilon}_y = - f\ (a_m\ , y_o, \dot{\bar{\varepsilon}}, \bar{\varepsilon})\ \delta\bar{\varepsilon} - g\ (a_m, y_o, \dot{\bar{\varepsilon}})\ \delta \text{Ln } y_o \qquad (14)$$

where f and g are two functions of positive values, which depend on the particular form of the work-hardening law adopted ; a_m are the material parameters.

Extension to complex strain-paths

Equ.(13) can be handled to treat the development of the rate gradients for a piecewise-linear strain-path as well. To this end, it is appropriate to regard the sheet at the end of the preceding (linear) stage of straining as a "new" sheet that exhibits a defect characterized by the two values δLn h_n and δ Ln y_n. Equ.(13) should thus be re-expressed with δLn h_n instead of δLn h_o. Besides, one should bear in mind that the initialization of the problem at the onset of the "new" stage of straining must be made by assigning zero values to the strain, both inside and outside the band, so that $\Delta\bar{\varepsilon}_y$ must be taken as

$$\Delta\bar{\varepsilon}_y = -f(a_m;y_n;\dot{\bar{\varepsilon}};\bar{\varepsilon}-\bar{\varepsilon}_n)(\delta\bar{\varepsilon}-\delta\bar{\varepsilon}_n)-g(a_m;y_n;\dot{\bar{\varepsilon}})\delta \text{Ln } y_n \qquad (15)$$

where y_n is the hardness, and $\bar{\varepsilon}_n$ and $\delta\bar{\varepsilon}_n$ are the effective strain and strain gradient, respectively, at the end of the preceding stage.

The above formulation enables us to examine the effect of a change in strain rate instead of - or together with - a change in the strain ratio. This is a direct consequence of the one-parameter equation of state adopted. It is reflected in the above equations by the strain rate dependence of the two functions f and g ; also, the γ-value is affected by a change in strain rate.

QUANTITATIVE PREDICTIONS AND DATA COMPARISONS

The above analysis is primarily intended for the prediction of the limit strains for "no obvious tear". Since the industrial definitions of these limit strains (ε*) are closely related to the assessment of an acceptable value of the strain gradients, the limit strains were determined by assigning a critical value to the thickness gradient, δ ln h*. The calculations were made by assuming the pre-existence of a defect of geometric character (δ Ln h_o = - 0.002 ; δ Ln y_o = 0). The δ Ln h* - value was taken equal to δln h* = 10 δln h_o = - 0.02.

Then, it was verified that the quantitative predictions do not suffer from a loss of accuracy resulting from the linearization : in particular, the present approach gives substantially the same results as a Marciniak-type approach with a Swift-type constitutive relation of the form :

$\bar{\sigma} = k\ (\bar{\varepsilon}_o + \bar{\varepsilon})^N\ \dot{\bar{\varepsilon}}^M$, where k and $\bar{\varepsilon}_o$ are constants, N is the work-hardening exponent and M is the strain rate sensitivity coefficient. On the other hand, significant improvements in the prediction of the limit strains under linear or complex strain-paths can be obtained as a result of the flexibility of the method. Two illustrations are given below.

<u>FLD of aluminium</u>

It is known that the FLD of aluminium predicted by the theory of Marciniak and co-workers (1973) in the stretching range is much higher than the experimental ones. Part of this discrepancy may be due to a m-value less than 2 for this material (Parmar and Mellor, 1978). However,there is also increasing evidence that a Voce-type hardening law is more appropriate to f.c.c. metals than a Swift-law. Accordingly, the FLD of aluminium was calculated using the Voce-parameters determined experimentally by the present authors. It can be seen (Fig. 1) that a satisfactory account of the FLD determined by Ghosh and Backofen (1973) is obtained when the effects of i) the m-value and ii) the work-hardening law adopted, are taken into consideration. These two modifications to the constitutive relations are of comparable importance with regard to the improvement in the prediction of the FLD.

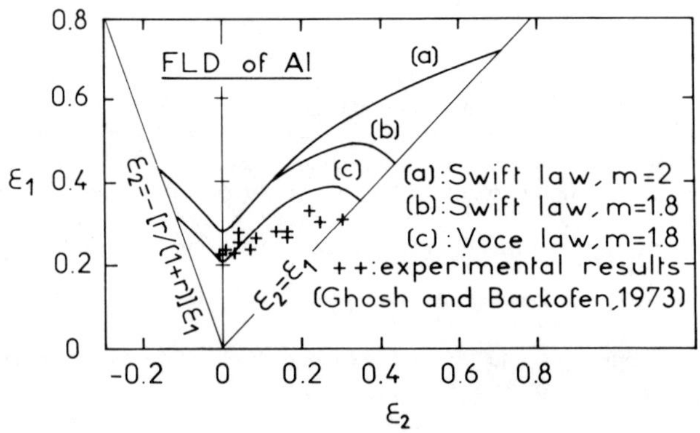

Fig.1. Comparison between experimental and calculated FLDs of aluminium.

<u>Limit strains of A-K steel under complex strain-paths</u>

To our knowledge, a theoretical prediction of the limit strains under complex strain-paths is still wanting. In effect, it is not possible to avoid considering a defect of both "geometric" and "mechanical" character in the re-initialization of these problems. This may partly be the reason for the lack of such predictions.

The limit strains of A-K steel were determined under linear strain-paths, and in the sequences [$\rho = -r/(1+r) \rightarrow \rho = 1$] ; [$\rho = 1 \rightarrow \rho = -r/(1+r)$] .The material parameters utilized were average values taken from the literature. In particular and owing to the conflicting experimental results comparing the stress-strain curves under different loading conditions (Wagoner, 1981), the m-value was chosen equal to 2. Quite a good agreement with experimental data was obtained (Fig.2). *Particularly, the low limit strains observed in the sequence* [$\rho = 1 \rightarrow \rho = -r/(1+r)$] *are readily predicted, evenwhen the possible* ρ *- dependence of the work-hardening behaviour is left out of consideration (Ronde-Oustau and Baudelet, 1977).*

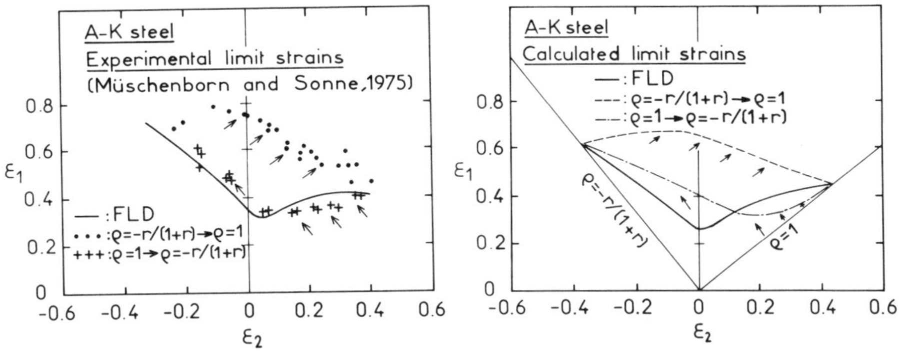

Fig.2. Comparison between experimental and calculated forming limits of A.K. steel under complex strain-paths.

CONCLUSION

A linearized analysis for imperfection growth in biaxially loaded sheets has been developed. The advantages of the proposed approach come from i) its applicability to any form of one-parameter equation of state which describes at best the plastic behaviour of the material at large strains and ii) its ability to predict the limit strains under complex strain-paths. Two effects, however, were neglected. One is volume damage associated with the growth of cavities (Jalinier and Schmitt, 1982). The other is the development of thermal gradients during non-uniform plastic flow (Ferron, 1982).These two effects have a detrimental influence on the stability of plastic flow.

REFERENCES

Azrin, M., and W.A. Backofen (1970). Metall. Trans.,1A, 2857-2865.
Backofen, W.A. (1973). Metall.Trans., 4A, 2679-2699.
Bergström, Y. and S. Olund (1982).Mater. Sci. Eng., 56, 47-61.
Ferron, G. (1982). Mat. Sci. Eng., 52, 133-138.
Ferron, G., and M. Mliha-Touati (1982). Scripta Metall., 16, 911-916.
Ferron, G., M. Mliha-Touati and P. Tanti (1983). Mat. Sci. Eng., in press.
Ghosh, A.K., and W. A. Backofen (1973). Metall. Trans., 4A, 1113-1123.
Hart, E.W. (1970). Acta Metall. ; 18, 599-610.
Hill, R. (1950). The Mathematical Theory of Plasticity, Clarendon Press, Oxford.
Hutchinson, J.W., and K.W. Neale (1978). in D.P. Koistinen and N.M. Wang (Ed), Mechanics of Sheet-Metal Forming, Plenum, New-York. pp.269-285.
Jalinier, J.M., and J.H. Schmitt (1982). Acta Metall. , 30, 1799-1809.
Kocks, U.F., J.J. Jonas and H. Mecking (1979). Acta Metall. , 27, 419-432.
Laukonis, J.V., and A.K. Ghosh (1978). Metall. Trans., 9A, 1849-1856.
Marciniak, Z., K. Kuczynski and T. Pokora (1973). Int.J.Mech.Sci.,15, 789-805.
Mellor, P.B., and A. Parmar (1978). In D.P. Koistinen and N.M. Wang (Ed.), Mechanics of Sheet-Metal Forming, Plenum, New-York. pp.53-77.
Müschenborn, W., and H.M. Sonne (1975). Arc. Eisenhüttenwesens, 46, 597-602.
Parmar, A., and P.B. Mellor (1978). Int. J. Mech. Sci., 20, 385-391.
Ranta-Eskola, A. (1980). Mem. Sci. Revue Metall., 4, 543-552.
Ronde-Oustau, F., and B. Baudelet (1977). Acta Metall., 25, 1523-1529.
Wagoner, R.H. (1981). Metall. Trans., 12A, 877-882.

THE EFFECT OF NON-LINEAR STRAIN PATHS ON THE PLASTIC PROPERTIES OF STEELS

T. Davis and W. B. Hutchinson

Department of Metallurgy and Materials Engineering, The University of Aston, Gosta Green, Birmingham B4 7ET, UK

ABSTRACT

THE CONDITION of proportional straining (linear strain paths) is seldom relevant to sheet forming processes even when these consist of a single operation. For multi-stage forming processes there are abrupt and substantial changes in strain path. Such changes can substantially modify the plastic flow properties of the material relative to the behaviour in proportional straining. The conditions of metal flow and strain distribution are important to the successful production of pressed components but local microscopic behaviour may also be influenced as in the development of plane strain necks in biaxial stretching.

Previous investigations of this area have tended to concentrate on three different aspects. These are

(i) the Bauschinger effect in reverse loading,

(ii) the distorsion of the yield locus due to pre-straining (isotropic or kinematic descriptions of hardening), and, more recently,

(iii) the effect of strain path changes on limit strains at failure.

The present investigation is more general in nature and aims to analyse the flow stress and work hardening behaviour of sheet material pre-strained according to a range of different procedures.

Three different steels are being investigated. These are a plain low carbon deep drawing quality material and two higher strength steels employing both solid solution hardening (rephosphorised steel) and dispersion hardening (dual phase steel). The latter two steels have comparable strength levels. Pre-straining has been carried out in simple tension, in plane strain tension using a modified tensile testing procedure with box-section specimens, and in equi-biaxial stretching using Marciniak tooling in a large press. A wide range of pre-strain levels is being used in each case. Final testing is done in simple tension with continuous monitoring of flow stress, work-hardening behaviour and plastic anisotropy throughout the test. These tensile tests are being carried out in several directions in the pre-strained sheet.

The experimental programme is still in progress but results achieved so far show that the work-hardening characteristics and instability conditions can be greatly modified by the changes in the strain path. Results obtained with these limited combinations of tests will be analysed to try to determine general functions of the initial and final strain states. The significance of different microstructural strengthening mechanisms will also be discussed.

FORMING LIMITS OF TITANIUM SHEETS

S. Kohara*, M. Katsuta** and K. Aoki**

*Institute of Interdisciplinary Research, University of Tokyo, Tokyo, Japan
**College of Industrial Technology, Nihon University, Narashino, Japan

ABSTRACT

The forming limit curves for titanium sheets were determined and the effect of strain path on the curves was investigated. Large limit strains in the region near plane strain deformation were achieved by using a two stage strain path.

KEYWORDS

Titanium sheet; formability; forming limit curve; limit strain; strain path.

INTRODUCTION

The forming limit curve (FLC) (Keeler, 1965) has been used to asses the formability of sheet materials. The FLC is usually determined experimentally from the specimens deformed by a monotonic strain path. In practice, however, most of forming processes of metal sheets are not single, but multi stage deformation. It was shown that a proper two stage strain path yielded the higher limit strains in the FLC for aluminium sheet than those by a single stage strain path (Kohara, 1982a, 1982b, 1982c).

The purpose of this study is to determine the FLCs for titanium sheets and to investigate the effect of strain path on the curves.

EXPERIMENTAL

Two grades of commercially pure titanium sheets were used in the present study. The one contains 0.057% oxygen and the other, 0.103% oxygen (hereafter denoted as Ti(A) and Ti(B)), while the contents of the other impurity elements are almost same in both grades of sheets. The as-received sheets were in cold-rolled state and the nominal thickness of the sheets was 0.7 mm.

Softening curves of the used sheets were obtained by isothermal annealing and the conditions for temper annealing of specimens were determined from those curves. Tensile specimens and blanks for other tests were cut from the sheets and annealed at 673, 773, 873 and 973 K for 1.8 ks in vacuum. From the results of tensile test, ultimate tensile strength, total elongation, anisotropic coefficient (r-value) and strain hardening exponent (n-value) were determined.

The FLCs for titanium sheets were determined by punch stretching, hydraulic bulging and tensile testing. To determine the limit strains in various strain states between unidirectional tension and equibiaxial tension, rectangular blanks (Nakazima, 1968) (Hecker, 1975) with various widths and three different lubrication conditions were used in punch stretching with a die of 100 mm dia and a punch of 60 mm dia. Furthermore, three types of elliptical dies with aspect ratios of 100:65, 100:75 and 100:85 (mm), in addition to a circular die of 100 mm dia, were used in hydraulic bulging.

The effect of strain path on the FLC was investigated using a two stage strain path; equibiaxial tensile deformation followed by uniaxial tensile deformation. Circular blanks were bulged into cups with different heights as the primary stage, and the specimens for uniaxial tensile deformation were cut from the bulged cups and then, elongated uniaxially as the second stage. As titanium sheet has high degree of anisotropic properties, the direction of deformation must be clarified. In the second stage, the specimens were cut and elongated parallel to the rolling direction of the sheet. Throughout this study, limit strains were determined with a circulr grid of 2.5 mm dia printed on the surface of a blank in a way that major strain, ε_1 is parallel to, and minor strain, ε_2 is perpendicular to the rolling direction of the sheet.

RESULTS

The softening curves for Ti(A) and Ti(B) are shown in Figs. 1 and 2. The results show that Ti(B) sheets are much harder than Ti(A) sheets even after the same temper annealing. The FLCs for Ti(A) determined by a single stage strain path are shown in Fig. 3. The curves shift to the upper position with increasing annealing temperature. The limit strains are very high in the regions near uniaxial tensile deformation and also, near equibiaxial tensile deformation. However, limit strains are fairly low in the region near plane strain deformation. The FLCs for Ti(B) determined by a single stage strain path are shown in Fig. 4. The shape of the FLC for Ti(B) differ from that for Ti(A) in that the limit strains are comparably high at the uniaxial tensile deformation side, but those are not so high as the case of Ti(A) at the equibiaxial tensile deformation side. The results show that Ti(B) sheet is inferior to Ti(A) sheet for biaxial stretch forming. The FLCs determined by a two stage strain path are shown in Figs. 5 and 6. In both materials, the limit strains shift to larger values compared with those by a single stage strain path. Especially the increase in limit strains in the region near the plane strain deformation is remarkable.

DISCUSSION

It has been reported that the work hardening due to the primary straining was reduced by the secondary straining which was in different mode from the primary one. (Hutchinson, 1976) (Sang, 1979) (Renta-Eskola, 1980) In this experiment on the effect of strain path, the blanks were equibiaxially expanded in the primary stage and then, uniaxially stretched in the secondary stage. The specimens were elongated in the direction of tension axis, but simultaneously contracted in width in the secondary stage. So, the secondary deformation in the reverse direction may have the effect of relieving a part of work hardening due to primary deformation. It results in the increased limit strains in the two stage strain path forming.

Fig. 7 shows changes in local strains obtained from four circles around the center of a specimen during two stage strain path deformation. In the primary stage, strains increase equibiaxially (0→1). In the early secondary stage, only the strains in the direction of tension axis increase (1→2) and then, elongation in the tension direction and contraction in the lateral direction occure simultaneously (2→3). In the last secondary stage, the specimen deforms just as in unidirectional tensile test (3→4). It seems likely that this result support the above cosideration. Thus, a higher forming limit can be achieved by selecting an adequate strain path in sheet forming of titanium.

SUMMARY

(1) The forming limit curve for low grade titanium sheet differs from that for high grade titanium sheet in that limit strains are low especially in the equibiaxial tensile deformation side.
(2) For both grades of titanium sheets, the limit strains shift to higher values by using the two stage strain path deformation.
(3) A higher forming limit can be achieved by selecting a proper strain path in the forming process of titanium sheet.

REFERENCES

Hecker, S. S. (1975). Sheet Metal Industries, 52, 11, 671-676.
Hutchinson, W. B., B. R. Arthey and P. Marmström (1976). Scripta Met., 10, 8, 673-675.
Keeler, S. P. (1965). SAE Paper No. 650 535.
Kohara, S., M. Katsuta and K. Aoki (1982). Aluminium, 58, 2, 122-124; 8, 487-488; 12, 736-738.
Nakazima, K., T. Kikuma and K. Hasuka (1968). Yawata Technical Report, No. 264, 141-154.
Renta-Eskola, A. J. (1980). Met. Tech., 7, 2, 45-49.
Sang, H. and D.J. Lloyd (1979). Met. Trans., 11, 1773-1776.

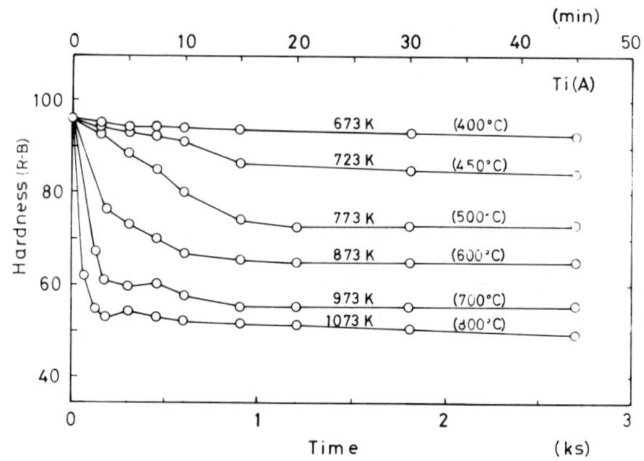

Fig. 1 Softening curves for Ti(A) sheet.

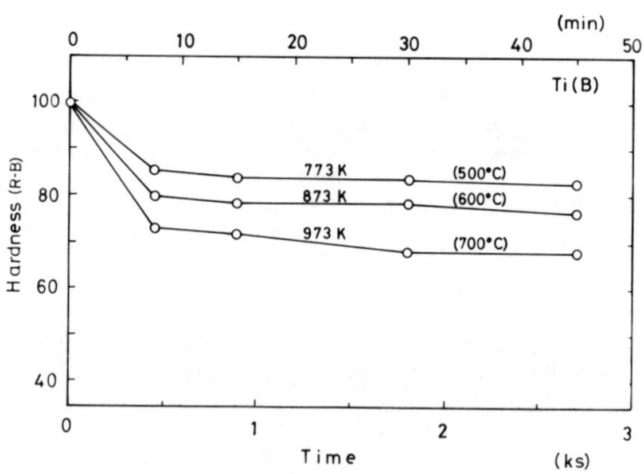

Fig. 2 Softening curves for Ti(B) sheet.

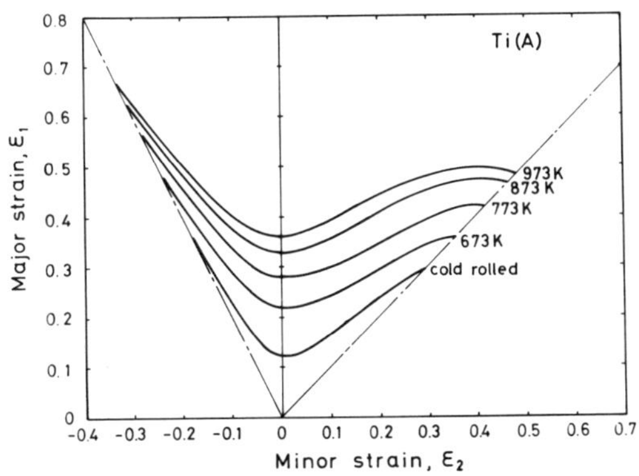

Fig. 3 Forming limit curves for Ti(A) sheet.

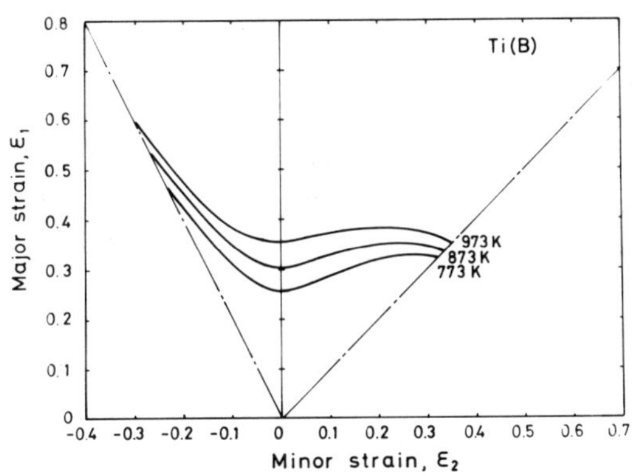

Fig. 4 Forming limit curves for Ti(B) sheet.

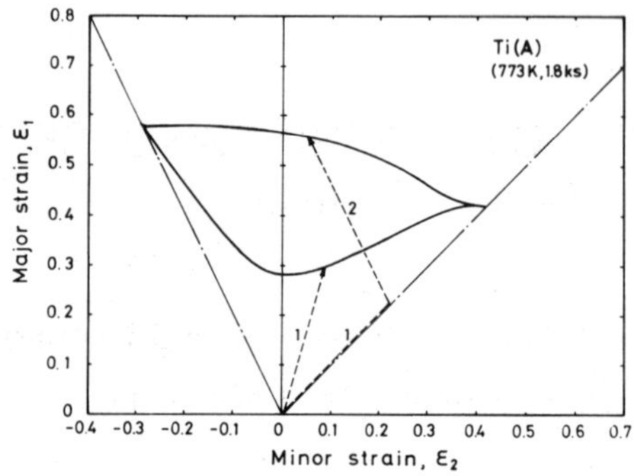

Fig. 5 Forming limit curves for Ti(A) sheet determined by two stage strain path.

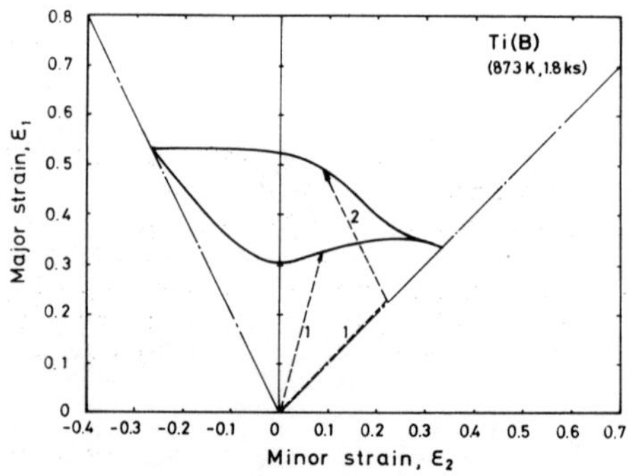

Fig. 6 Forming limit curves for Ti(B) sheet determined by two stage strain path.

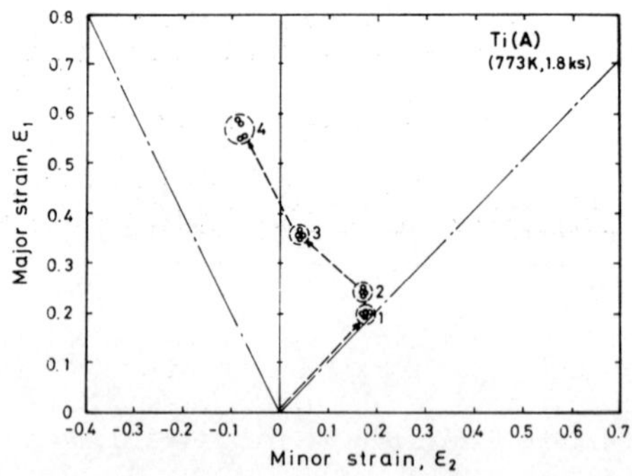

Fig. 7 Change in local strains around the center of a specimen during two stage strain path.

EFFECT OF MATRIX AND SECOND PHASE ON THE MAXIMUM UNIFORM STRAIN OF METALS

Mingzhi Huang, Jingxi Luo and Baoping He

Research Institute for Strength of Metals, Xi'an Jiaotong University, China

ABSTRACT

The effects of the matrix and the second phase of metals on $d\sigma/d\varepsilon-\varepsilon$ curve and $\sigma-\varepsilon$ curve are studied in six kinds of mormalized steels, Armco iron, copper and brass. It is shown that $d\sigma/d\varepsilon-\varepsilon$ curve mainly depends on the matrix of the metals and is not very much influenced by the second phase. While the $\sigma-\varepsilon$ curve is very sensitive to the second phase. So that the strain hardening rate $d\sigma/d\varepsilon$ can be changed by changing the structure of the matrix. On the other hand, if the strength level and plastic strain ability need to be changed one has to study the effect of the second phase. It has been found that $d\sigma/d\varepsilon$ may be conveniently expressed in terms of the $\sigma_u \cdot \varepsilon_u/\varepsilon$.

KEYWORDS

Formability, strain hardening rate, matrix, second phase, maximum uniform strain.

INTRODUCTION

The maximum uniform strain as well as the strain hardening rate are important parameters for cold working formability of metals. However, the relation of this parameters with the matrix and the second phase is not quite clear yet. We have studied the effects of the matrix and the second phase on strain hardening exponents in $\lg\sigma-\lg\varepsilon$ curve. The result showed the strain hardening exponent could be divided into three stages, except yeilding stage (Huang Mingzhi, Luo Juogxi, He Baoping; 1983). This paper presents some results about the effects of the matrix and the second phase on the strain hardening rate $d\sigma/d\varepsilon$ and the maximum uniform strain.

It is well known that, in the load-strain plot, the necking of a specimen starts at the maximum load, where:

$$d\sigma/d\varepsilon = \sigma$$

This is so called necking criterion. The maximum uniform strain corresponds to the intersection of the flow stress curve ($\sigma - \varepsilon$) and the strain hardening rate curve ($d\sigma/d\varepsilon - \varepsilon$). Therefore ε_u should be influenced by the factors which control the shapes and the relative positions of this two curves. Pickering F. B. (1971) suggested a schematic diagram about the effects of grain size, volume fraction of peralite on this two curves, but failed to give enough experimental evidence. In this study, we try to examine the effects of the matrix and the second phase on $d\sigma/d\varepsilon - \varepsilon$ and $\sigma - \varepsilon$ curves.

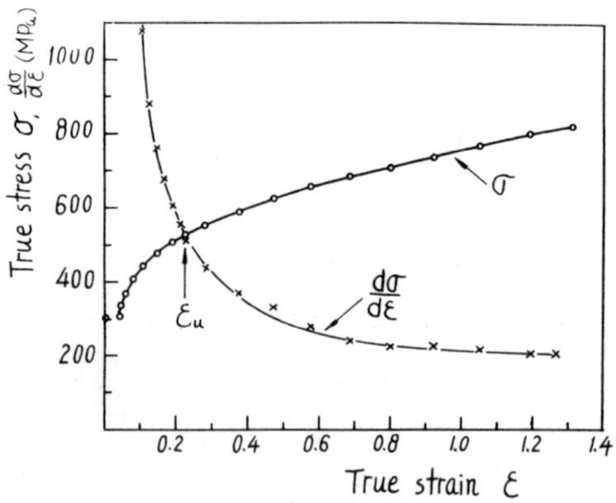

Fig. 1 The σ-ε and $d\sigma/d\varepsilon$-ε curves of 10# steel. The intersection of this two curves is maximum uniform strain.

During plastic deformation, the microstructure and the submicrostructure of metal change drastically, for example, multiplication and tangling of dislocations, the formation of the cell structure etc.. All of the process are tightly connected with the matrix and the second phase. In order to investigate the deformation behavior, three groups of specimens in various matrix and second phase have been employed for this study.

1. The matrix of the specimens is the same but their second phase is different;
2. The second phase of specimens is the same but their matrix is different;
3. Both the matrix and the second phase are different.

EXPERIMENTAL MATIERALS AND PROCESS

See (Huang Mingzhi, Luo Jingxi, He Baoping; 1983)

RESULTS AND ANALYSIS

There are six kinds of steels in the first group: 10#, 15Mn, 25Mn, 30Mn,

45Mn, and 60Mn. All of them are normalized. They have the same matrix with various volume fraction of cementite (Huang Mingzhi, Luo Jingxi, He Baoping; 1983).

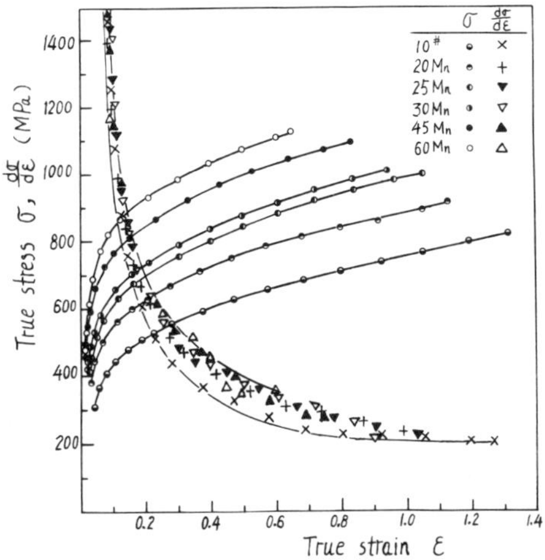

Fig. 2 The ture stress σ vs ture strain ε and the strain hardening rate $d\sigma/d\varepsilon$ vs true strain ε plot of six normalized carbon steels.

Fig. 2 shows that the σ-ε curves of six specimens are seperated according with the carbon content, while the $d\sigma/d\varepsilon$-ε curves are almost within a narrow band, especially in the uniform strain stage. It shows that the strain hardening rate, in small strain range is mainly related to the conditions of the matrix. The effect of the second phase on $d\sigma/d\varepsilon$ is only brought about in large strain range.

Our study in the stages of hardening exponent of the steels showed that n_1 — the hardening exponent in the first stage— is sensitive to the matrix and n_2 is sensitive to the second phase (Huang Mingzhi, Luo Jingxi, He Bao Baopin, 1983). A knee strain ε_1 from the first stage to the second one has been found. ε_1 is a critical strain from which the effect of the second phase on $\lg\sigma$-$\lg\varepsilon$ plot takes place. Thus, in small strain range $\varepsilon < \varepsilon_1$, the experimental data from the same matrix specimens overlap in $d\sigma/d\varepsilon$ vs ε plot of Fig. 2. In large strain range the data in scattering band of $d\sigma/d\varepsilon$ is connected with the carbon content at certain ε. For the high carbon content steels the experimental data lie in the upper part of the scatter band. In other words, $d\sigma/d\varepsilon$ increases with the increasing of the second phase. The flow stree curve σ-ε is sensitive to the structure including matrix and second phase. Since in this group the matrix of six steels is the same, it is comprehensible, that σ-εcurve is largely affected by the second phase.

The matrix of six kinds of steels are the same, their $d\sigma/d\varepsilon$-ε curves are nearly overlaped. The volume fractions of the second phase are different.

With the increase of carbon content, the yielding point increases while the yielding point elongation decreases. It makes the σ-ε curves move to the left and upwards. The $d\sigma/d\varepsilon$ curves are not influenced by carbon content. So the intersection of this two curves moves towards the left, i.e., the maximum uniform strain decreases. F. B. Pickering claimed the volume fraction of peralite did not influenced the yielding strength, it only influenced the yielding point elongation. Thus, the σ-ε curves move to the left only. But this experiment shows that the lower yielding point is changed with the increase of carbon content while the upper yielding point maintains constant in certain range of carbon content. With the increase of the carbon content the lower yielding point increases (Fig. 2) and the ε_u decreases. As a result that the σ-ε curve moves to the left and upwards simultaneouly.

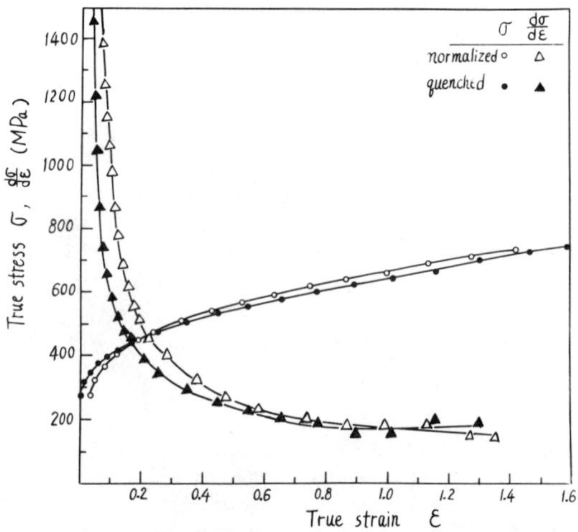

Fig. 3 The σ-ε and $d\sigma/d\varepsilon$-ε curve of Armco iron 940°C, 1 hr. normalized, and Armco iron quenched at 750°C then 180°C tempered after normalization.

Two kinds of Armco iron (0.035%C) were served as a second group. One was 940°C, 1 hr. normalized, the other was 750°C quenched then 180°C tempered after normalization. The quenched and tempered Armco iron corresponds to solid solution treatment, the ferrite grain has been oversaturated by carbon but the grain size is the same as the normalized one. The second phase is the same for both of normalized and quenched specimens.

Fig. 3 shows that the yield point elongation of quenched and tempered Armco is disappeared. The two σ-ε curves are nearly overlaped while the $d\sigma/d\varepsilon$-ε curves are seperated. It is associated with that the $d\sigma/d\varepsilon$-ε curve mainly depends on the matrix and the σ-ε curve is mainly affected by the second phase. As a result that the ε_u of quenched and tempered Armco is lower than that of normalized one.

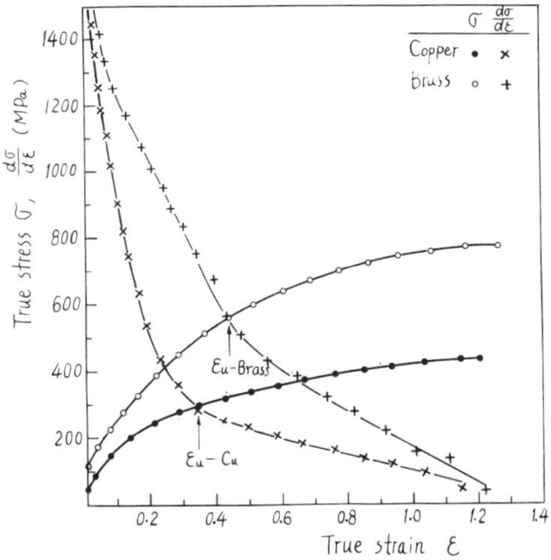

Fig. 4 The dσ/dε-ε and σ-ε curves of brass and copper

For the third group,both the matrix and the second phase are changed, the two curves shift simultaneously, Fig. 4 shows the dσ/dε-ε and σ-ε curves of copper (99.94%Cu) and brass (36.36%Zn). The matrix of brass is α-solid solution which is different from the pure copper. A second phass-β-phase is involved in brass. Thus, the intersection ε_u of the two metals are different.

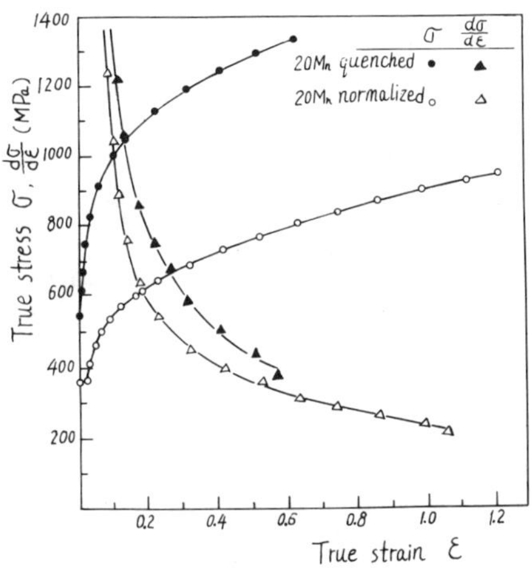

Fig. 5 The dσ/dε-ε curve and σ-ε curve of normalized 20Mn steel and quenched-tempered 20Mn after normalization.

Fig. 5 shows the $d\sigma/d\varepsilon-\varepsilon$ curve and $\sigma-\varepsilon$ curve of normalized 20Mn steel and quenched and tempered (750°C quenched, 180°C tempered) 20Mn after normalization. The matrix of normalized 20Mn is ferrite and the second phase is pearlite. The matrix of quenched and tempered 20Mn is carbon oversaturated ferrite and the second phase is martensite. Both the matrix and the second phase are different for the two specimens. Their $d\sigma/d\varepsilon-\varepsilon$ and the $\sigma-\varepsilon$ curves are also seperated. So that the ε_u are different.

It can be summerized that the $d\sigma/d\varepsilon-\varepsilon$ curve mainly depends on the matrix and is less affected by the second phase, especially in small strain range. The $\sigma-\varepsilon$ curve is very sensitive to the second phase. Therefore, one has to concern the matrix if the strain hardening rate needs to be changed, on the other hand, if the strength level and the plastic strain ability is going to be changed, attention has to be paid on the effects of the second phase. It will give some clue in changing the formability of the metals.

THE RELATIONSHIP OF $d\sigma/d\varepsilon$ WITH σ_u, ε_u

Generally, the strain hardening rate $d\sigma/d\varepsilon$ is calculated by means of numerical differentiation in $\sigma-\varepsilon$ plot. It is tedious and time consuming. If the true stress-true strain curve is fitted with $\sigma=\sigma_o+H\ln\varepsilon$ then:

$$d\sigma/d\varepsilon = H/\varepsilon$$

where H may be refered to as strain hardening rate factor.

when $\varepsilon = \varepsilon_u$

$$d\sigma/d\varepsilon = \sigma_u$$

Then: $H = \sigma_u \cdot \varepsilon_u$

We have calculated the correlation coefficients of $\sigma-\varepsilon$ curves with the equation $\sigma=\sigma_o+H\ln\varepsilon$ for six normalized and four spheroidized carbon steels (He Baoping, 1981). The correlation coefficients are all better than 0.99. The equation $\sigma=\sigma_o+H\ln\varepsilon$ is not good in the small strain range, it fails at $\varepsilon=0$. However, it is very accurate in the strain range near ε_u. The $d\sigma/d\varepsilon$ $d\sigma/d\varepsilon-\varepsilon$ curve depends on H. $H=\sigma_u\cdot\varepsilon_u$ would be very convenient to determine the strain hardening rate, if we compare the experimental date $d\sigma/d\varepsilon$ with the H value ($H=\sigma_u\cdot\varepsilon_u$). Fig. 6 shows the calculated H ($=\sigma_u\cdot\varepsilon_u$) is well coincident with experimental results in the range of $\varepsilon<\varepsilon_u$. Therefore the $\sigma_u\cdot\varepsilon_u$ can be employed to determine the strain hardening rate $d\sigma/d\varepsilon$ by the equation:

$$d\sigma/d\varepsilon = H/\varepsilon$$

or

$$d\sigma/d\varepsilon = \sigma_u \cdot \varepsilon_u/\varepsilon$$

in the strain range $\varepsilon < \varepsilon_u$.

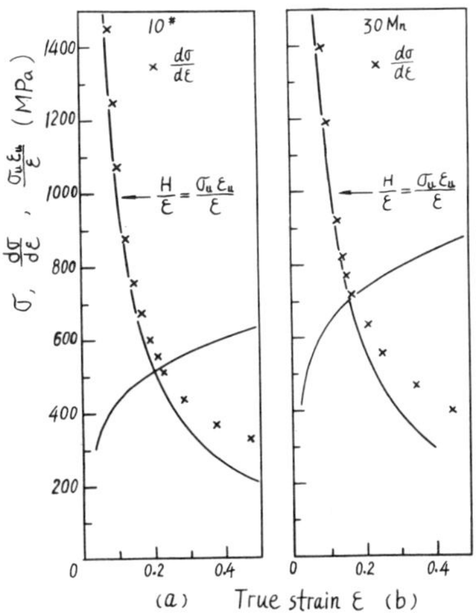

Fig. 6 **The constrast of calculated H(=$\sigma_u \cdot \varepsilon_u$) to experimental $d\sigma/d\varepsilon$ in 10# and 30Mn steels**

CONCLUSION

1. **The $d\sigma/d\varepsilon-\varepsilon$ curve mainly depends on the matrix of metals and is less affected by the second phase. It is significant, especially, in small strain range (before ε_u). Therefore, in order to change the strain hardening rate $d\sigma/d\varepsilon$, the matrix of the metals should be changed;**

2. **$\sigma-\varepsilon$ curve is very sensitive to the second phase. If the strength level and the plastic strain ability of the metal need to be changed, investigation should be carried out mainly on the effect of the second phase or the interaction of the second phase and the matrix;**

3. **The formula $d\sigma/d\varepsilon= \sigma_u \cdot \varepsilon_u/\varepsilon$ can be applied approximately to determine the strain hardening rate $d\sigma/d\varepsilon$ in the range of $\varepsilon < \varepsilon_u$.**

REFERENCES

He Baoping (1981), Master Degree Dissertation, Xian Jiaotong University, China.

Huang Mingzhi, Luo Jingxi, He Baoping (1983). Some Problems About Strain Hardening Exponent of Metals, ICM 4, Stockholm.

Pickering F.B., Toward Improved Ductility and Toughness, Kyoto International Conference 25, 26, 1971.

SOME PROBLEMS ABOUT STRAIN HARDENING EXPONENT OF METALS

Mingzhi Huang, Jingxi Luo and Baoping He

Research Institute for Strength of Metals, Xi'an Jiaotong University, China

ABSTRACT

Armco iron, copper, aluminium, brass and steels were employed in this study. Steels were treated to various microstructures and strength levels. The true stress-true strain curves are determined by rolling edge diametrical extensometer. The result shows that the strain hardening curve can be divided into three straight line sections in $\lg\sigma-\lg\varepsilon$ plot and each section agrees with Hollomon's equation. The hardening exponent n_1 of the first stage is sensitive to the matrix, the hardening exponent n_2 of the second stage and the knee strain ε_1 from n_1 to n_2 are both influenced by the second phase. Actually, the relation $n=\varepsilon_u$ is not true, even though it can be deduced from $\sigma=k\varepsilon^n$ and $d\sigma/d\varepsilon=\sigma$. It is better to use the approximate relation $n_2=\varepsilon_u$ when $\varepsilon_1<\varepsilon_u<\varepsilon_2$. The existence of the third stage of strain hardening is probably an important phenomenom worth studying.

KEYWORDS

Formability; strain hardening; hardening exponent; matrix; second phase; **maximum uniform strain; stages of strain hardening.**

INTRODUCTION

The strain hardening exponent n and maximum uniform strain ε_u are important parameters for formability of metallic plate, usually, their relation is expressed as $n=\varepsilon_u$ which can be deduced from $\sigma=k\varepsilon^n$ and $d\sigma/d\varepsilon=\sigma$. However, a lot of experiments showed that several stages existed in the strain hardening curve, more than one n could be found, so that the relation $n=\varepsilon_u$ would not be true. The factors which influence n_1 and n_2 have been investigated, such as temperature and strain rate by Bergström Y. and Morrison (1971), grain size by Morrison (1966). But the effects of matrix and second phase on n_1 and n_2 have not yet been reported. In metal forming, matrix and second phase usually play important roles, for instance, dual phase material is now broadly applied in plate manufacturing. Thus, the effect of matrix and second phase on n_1 and n_2 would be important in pratice.

EXPERIMENTAL MATERIALS AND PROCEDURE

Six different types of commercial normalized carbon steels (Table 1), 45Cr steel treated into several strength levels (Table 2), spheroidized carbon steels (Table 3) and commercial copper, aluminium, brass were studied.

TABLE 1 The Composition of Normalized Carbon Steels

Steel	10#	20Mn	25Mn	30Mn	45Mn	60Mn
C (wt%)	0.10	0.18	0.27	0.33	0.44	0.62
Si (wt%)	0.33	0.30	0.28	0.25	0.31	0.33
Mn (wt%)	0.47	0.85	0.86	0.81	0.82	0.86

TABLE 2 The Tempering Temperature of 45Cr Steel (0.46%C, 0.26%Si, 0.66%Mn, 1.11%Cr) quenched at 840°C

200	240	280	320	360	400	450	500	600	680

TABLE 3 The Composition of Spheroidized Carbon Steels

Steel	C(wt.%)	Mn(wt.%)	Si(wt.%)	Heat treatment
40	0.40	0.67	0.34	880°C quenched, 715°C held 24 hrs.
T8	0.81	0.28	0.23	780°C quenched, 715°C held 24 hrs.
T12	1.22	0.23	0.23	880°C quenched, 750°C held 1 hr. then 715°C held 16 hrs.

TABLE 4 The Composition of Other Experimental Materials

Material	C	Si	Mn	Cu	Zn	Fe
Armco iron	0.035	0.018	0.13			
Copper				99.94		0.008
Brass				other	36.35	0.008
Aluminium		0.16		0.01		0.21

Round tensile specimens (10mm diameter and 50mm length) were used in this experiment. The load-diameter curves were determined on INSTRON 1195 and 1196 machine by a rolling edge diametrical extensometer which was invented by one of the authors (Luo Jingxi, 1983). The rolling edge diametrical extensometer could automatically search the necking area and measure the minimum diameter of the specimen during tensile test. Its maximum error was less than 0.015mm. The true stress-true strain curves were calculated according to the load-diameter curves. The curvature radius of necking contour were determined with continuous photographs. The results showed that the contour of necking was closer to a hyperbola rather than a circle. Since Chen-Chi's equation (1978) and Bridgeman's equation (1952) dealt with hyperbola and spheric contour respectively, Chen-Chi's equation was employed in this study. Even though the difference between the two equations was very small (the difference was about 0.37%, for a/R=0.3; and about 1.4%, for a/R=0.8) (Fig. 1).

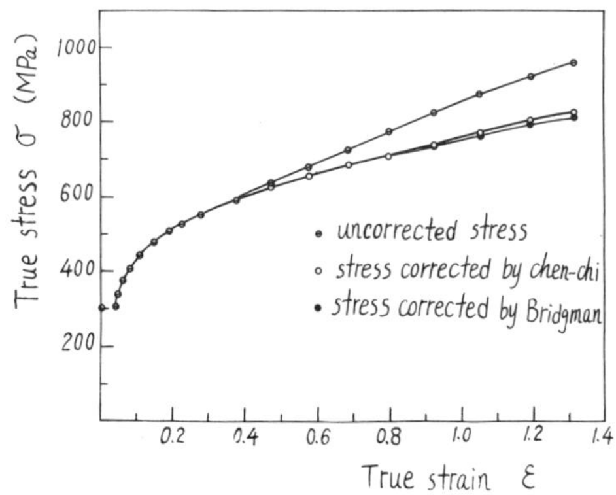

Fig. 1 Corrected true stress-true strain curve by Chen-Chi's and Bridgeman's equations

RESULTS AND ANALYSIS

We studied six empirical equations which described true stress-true strain curves (He Baoping, 1981). It showed that for both the whole curve during uniform deformation and the local deformation stage the power law fitted experimental data better than others. If the power law equations were ranked by correlation coefficient, the order would be $\sigma = k(\varepsilon_0 + \varepsilon)^n$ (Swift), $\sigma = \sigma_0 + k\,\varepsilon^n$ (Ludwik) and $\sigma = k\,\varepsilon^n$ (Hollomon). Since all of the correlation coefficients were about 0.99, and the Hollomon's equation was simpler and being used extensively, we applied Hollomon's equation in this study.

It is known that the strain hardening curve can be divided stage-by-stage in $\lg\sigma - \lg\varepsilon$ plot. The double-n behavior of iron, steels, aluminium alloys and titanium alloys has been reported frequently (Bowen and Partridge, 1974; Monteiro and Reed-Hill, 1971, 1973). However, in our study, three sections could be found in strain hardening curves on normalized plain carbon steels, even though the true stress higher than that at the necking point had been corrected with Chen-Chi's equation. From Fig. 2, all of the curves start from the end of yielding stage to the beginning of fracture.

The slope of each stage represents the hardening exponent. They are n_1, n_2, and n_3. The intersection points --knee strain-- are marked as ε_1, ε_2 respectively. The maximum uniform strain ε_u is situated somewhere between ε_1 and ε_2. The first stages of all steels are parallel with each other, i.e. the n_1 is approximatly a constant, while the n_2 and ε_1 decrease with increasing carbon content. The knee atrains at the intersection of two stages on $\lg\sigma$-$\lg\varepsilon$ plot, become significant with increasing of carbon content in steels.

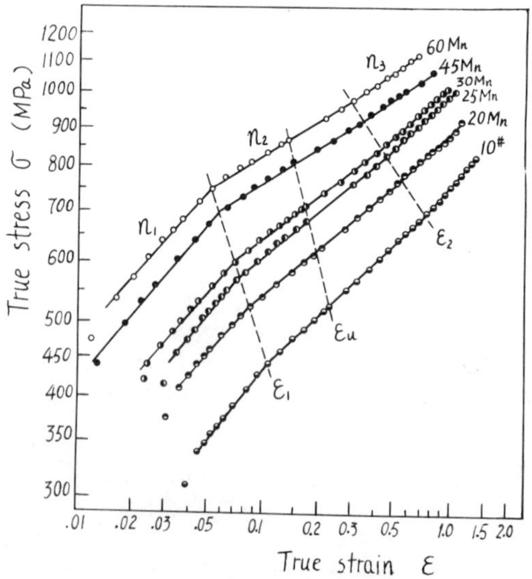

Fig. 2 The $\lg\sigma$-$\lg\varepsilon$ plot displayes stages of six different kinds of commercial normalized carbon steels.

In our specimens, the structure of normalized plain carbon steel consists of pre-eutectoid ferrite and pearlite. Its matrix can be considered as ferrite which includes pre-eutectoidic and eutectoidic ferrite. The second phase would be cementite. It is thought that the matrix of six steels are the same while the volume fractions of the second phase are different. Because n_1 values of six steels are nearly the same according to Fig. 1. It seems to be that n_1 is only connected with the matrix of a material. It is unlikely influenced by the second phase. n_2 and ε_1 are changed with carbon content or the volume fraction of pearlite, so n_2 and ε_1 are both sensitive to the second phase. This point of view is supported by the data of sphroidized carbon steels (40#, T8, T12) (Table 5). These specimens have the same matrix but various volume fractions of second phase (spherical cementite), their n_1 are nearly equal while n_2 are different.

The stages of strain hardening curve is related to the changes of micro and submicro structure during deformation. Generally, the knee strain ε_1 is considered to be correlated with the initiation of dynamic recovery of cross slip (Vöhringer, Macherauch; 1967) and the formation of cell structure (Christ, Smith; 1967). In some paper ε_1 is also treated as a deformation threshold of second phase (Soyama, 1966). It may also be related to the nonuniform deformation among grains. Aramd and Gurland (1976) suggested that the transition from n_1 to n_2 at ε_1 (double-n behavior) "corresponds to the attainment of critical particle or particle-matrix interface stresses which are sufficient of nucleate secondry processes to attainment of this critical stress level". n_3 and ε_2 have not yet been reported in the literature, perhaps, they are the parameters related to the fracture behavior.

TABLE 5 The Values of Strain Hardening Exponent of Several Quasi-Stable Materials

Materials	Armco	Al	Cu	Brass	40#	T8	T12
n_{total}	0.237	0.250	0.443	0.423	0.229	0.209	0.170
n_{unif}	0.245	0.234	0.484	0.409	0.227	0.210	0.192
n_1	0.243	0.263	0.523	0.563	0.236	0.233	0.215
n_2	0.201	0.219	0.329	0.350	0.194	0.191	0.148
u	0.220	0.215	0.348	0.433	0.197	0.180	0.139

n_{total} the hardening exponent of whole curve

n_{unif} the hardening exponent before maximum uniform strain

n_1 the hardening exponent of 1st stage

n_2 the hardening exponent of 2nd stage

ε_u the maximum uniform strain

It would be interesting to see if n_3 and ε_2 will present some useful informations about the strain hardening effect on fracture process. It is worth noting that not all of metals have three stages of strain hardening (without including the yielding stage). Among our samples, commercial aluminium, Armco iron, steels have three stages. Nevertheless, copper and brass only have two stages, their ε_1 — knee strain from n_1 to n_2 — are close to the maximum uniform strain ε_u i.e. $\varepsilon_1 \doteq \varepsilon_u$ or even $\varepsilon_1 > \varepsilon_u$. On the contrary, ε_1 are far less than ε_u in Armco iron and aluminium for example ε_1=0.1, ε_u=0.22 for Armco iron, ε_1=0.036, ε_u=0.215 for aluminium. The difference between this two kinds of metals could be explained in terms of dynamic recovery of cross slip (Vöhringer, Macherauch, 1967). It would be correlated with the difference of stacking fault energy. Since the dynamic recovery is related not only to the temperature and strain rate but also to the stacking fault energy. The lower the stacking fault energy, the larger the width of extended dislocation. Thus, it is more difficult to activate screw dislocation. As a result, in the metals with low stacking fault energy, the n_1 can be maintained to such a high strain that ε_1 close to ε_u.

THE RELATIONSHIP BETWEEN HARDENING EXPONENTS AND THE MAXIMUM UNIFORM STRAIN ε_u

$n = \varepsilon_u$ can be deduced from Hollomon's equation $\sigma = k\varepsilon^n$ and the necking criterion $d\sigma/d\varepsilon = \sigma$. This relations are usually employed for investigations on mechanical behavior of meterials, such as K_{1C}, da/dN etc. As a matter of fact, the relation $n = \varepsilon_u$ would not be true, especially, for unstable microstructure materials like low temperature tempered steels. The strain hardening curve may be divided into several stages in $\lg\sigma - \lg\varepsilon$ plot, so that more than one hardening exponent value n can be found.

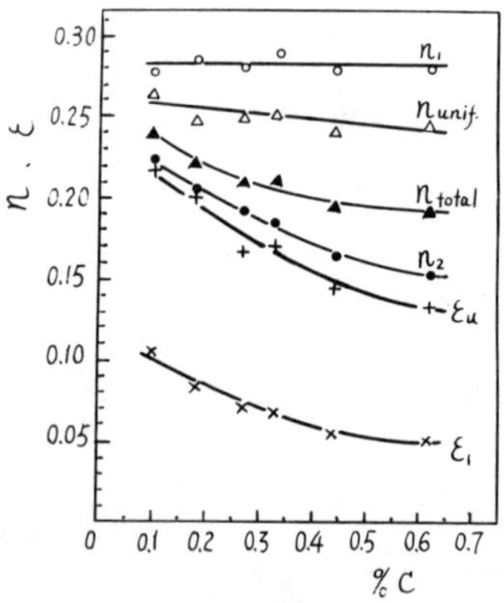

Fig. 3 The plot of n_1, n_2, n_{unif}, n_{total}, ε_1, ε_u vs the carbon content in investigated six normalized steels.

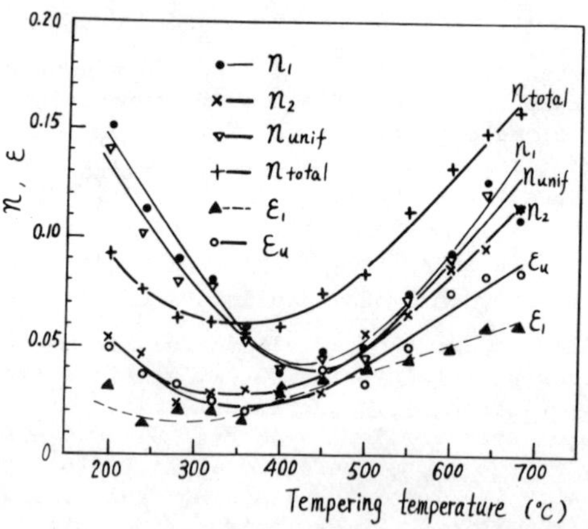

Fig. 4 The plot of n_1, n_2, n_{unif}, n_{total}, ε_1, ε_u vs tempering temperture in 45 Cr steel.

Fig. 3 shows the parameters of n_1, n_2, $n_{unif.}$, n_{total}, ε_1, ε_u vs the carbon content in six normalized steels. It is shown that all hardening exponents except n_2 are much larger than ε_u, while n_2 is the only one close to ε_u. It is significant, especially, in low carbon steels. The difference between n_2 and ε_u becomes large with the increasing of the carbon content. Thus, the transition from n_1 to n_2 is getting apparent in high carbon content steels.

Fig. 4 shows the n_1, n_2, n_{unif}, n_{total}, ε_1, ε_u vs the tempering temperture in 45Cr steel. Again, n_2 is close to the ε_u. The relation $n_2 = \varepsilon_u$ is valid in specimens at various tempering temperture.

Table 5 shows some data in several quasi-stable metals: Armco iron, aluminimum, copper, brass, T8, T12, 40#. It is shown that the values of n are not equal to ε_u. The relation $n_2 \doteq \varepsilon_u$ is more remarkable than that of normalized steels. However, there are some exceptions which show $n = \varepsilon_u$, such as brass for example. This is because in brass ε_1 is larger than ε_u. It can be concluded that the relation $n = \varepsilon_u$ is not appropriate. $n_2 = \varepsilon_u$ is better than $n = \varepsilon_u$, but it is valid only under the presumption of $\varepsilon_1 < \varepsilon_u$. The experimental strain hardening curve can be fitted with Hollomon's equation in seperate sections. For those materials having high stacking fault energy the relation $n_2 \doteq \varepsilon_u$ is valid at $\varepsilon_1 < \varepsilon_u < \varepsilon_2$.

CONCLUSIONS

1. The strain hardening curves of metals can be fitted with power law of Hollomon's equation $\sigma = k\varepsilon^n$ in different sections. There are three straight line sections in $\lg\sigma - \lg\varepsilon$ plot. So the hardening exponents in Hollomon's equation should be seperated into n_1, n_2, n_3 and two knee strain ε_1 and ε_2 in $\lg\sigma - \lg\varepsilon$ plot. Our investigation shows that n_1 seems to be affected mainly by the matrix of the material rather than the second phase. However, n_2 is sensitive to the second phase. The knee strain ε_1 from the first stage to the second stage is also sensitive to the second phase.

2. There are n_3 and ε_2 in the strain hardening curves. Now, the physical meaning is not clear, and the factors which influence n_3 and ε_2 are worth to be clarified. Probably, it is an important subject to be studied further.

3. The relation $n_2 = \varepsilon_u$ is suggested to replace $n = \varepsilon_u$ in most metals when $\varepsilon_1 < \varepsilon_u < \varepsilon_2$.

REFERENCES

Aramd, L., and J. Gurland (1976). Acta Met. 24, 901
Bergströn, Y., B. Aronsson (1970). Met. Trans., 1, 1029

Bowen, A. W., and P. G. Partridge (1974). Jounal of Physics D. Applied Physics, 7, 7, 969-978

Bridgeman, P. W., (1952). Studies in Large Plastic Flow and Fracture

Chen Chi (1978). Proceeding of Metallic Fracture, Met. Ind. Press, 169

陈篪 金属断裂研究文集 冶金出版社 (1978)

Christ, B. W. and G. V. Smith (1967). Acta Met. 15, 809

He Baoping, (1981). Master Degree Dissertation, Xi'an Jiaotong Uni., China

贺保平 (1981) 硕士论文 西安交通大学 中国

Soyama, Y. (1966). Journal of The Society of Materials Science, Japan 15.17

Luo Jingxi, and Li Guangxin (1983) Physical Testing and Chemical Analysis, Physical Testing, 19. 1. 42-45

駱競晞 李光新 (1983) 理化检验，物理分册 19.1. 42–45

Monteiro, S. N. and R. E. Reed-Hill (1971). Met. Trans. 2. 2947

Monteiro, S. N. and R. E. Reed-Hill (1973). Met. Trans. 4. 1011

Morrison, W. B. (1966). Trans. ASM, 59. 824

Morrison, W. B. (1971). Met. Trans. 2. 331

Vöhringer, O., E. Macherauch, (1967). Z. Metallkde Bd. 58. 21

ON THE PREDICTION OF FORMING LIMIT DIAGRAMS FOR ANISOTROPIC SHEETS

K. W. Neale* and E. Chater**

*Faculty of Applied Science, Université de Sherbrooke, Sherbrooke, Québec, Canada

**Division of Applied Science, Harvard University, Cambridge, Mass., USA

ABSTRACT

A transversely isotropic, plane stress yield function recently proposed by Budiansky (1982) is used to characterize yielding for typical sheet metals. Its implications on the prediction of forming limit diagrams are examined.

KEYWORDS

Forming limit diagrams; localized necking; plastic anisotropy; sheet metals.

INTRODUCTION

The prediction of localized necking failures in thin sheets is a topic which has attracted a great deal of interest in recent years. This stems mainly from the need to obtain formability properties for the new sheet metal alloys currently entering the automobile and home appliance industries. Although much progress has been made with regard to the analysis of sheet necking phenomena, certain issues remain unsettled. This paper focuses on one of these open questions, related to necking in anisotropic sheets. It concerns the fact that predicted necking strains have been found to be unusually sensitive to small variations in material parameters used to characterize anisotropy (Bassani, Hutchinson and Neale, 1979; Neale and Chater, 1980, 1981).

The results of sheet necking calculations are usually presented as "forming limit diagrams," which depict the combinations of major principal strain ε_1^* and minor principal strain ε_2^* at necking. In a typical analysis, we consider a sheet element subjected to proportional straining where $\rho = \varepsilon_2/\varepsilon_1$ is the prescribed strain ratio. For isotropic sheets the range of interest is $-1/2 \le \rho \le 1$; where $\rho = -1/2$ corresponds to uniaxial tension, $\rho = 0$ signifies in-plane plane strain and $\rho = 1$ denotes equi-biaxial stretching.

A troublesome observation at present in the prediction of forming limit curves is the unrealistic sensitivity of the results obtained with the existing characterizations of anisotropy. This is especially evident in the biaxial stretching range ($0 \leq \rho \leq 1$). Consider, for example, the anisotropic yield function representation suggested by Bassani (1977):

$$|\sigma_1 + \sigma_2|^n + \frac{n}{m}(1+2R)\,\sigma_u^{\,n-m}\,|\sigma_1 - \sigma_2|^m = [1 + \frac{n}{m}(1+2R)]\,\sigma_u^{\,n} \quad (1)$$

Here σ_1 and σ_2 are the principal (Cauchy) stresses in the plane of the sheet (σ_3=0); σ_u is the equivalent uniaxial stress; and m, n and R are anisotropic parameters. This equation describes *transversely isotropic* behaviour ("normal anisotropy") where the normal to the sheet corresponds to the axis of transverse isotropy. For m=n and m=n=2, (1) reduces to the yield functions previously given by Hill (1950, 1979).

Typical sheet necking calculations based on (1) give rise to the forming limit curves shown in Fig. 1 (Neale and Chater, 1980). These results were

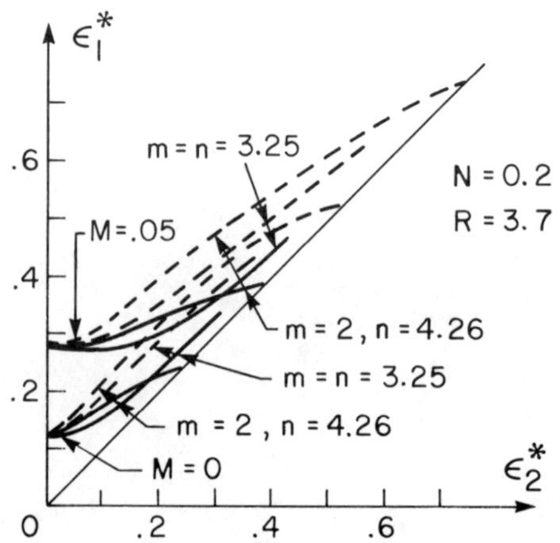

Fig. 1. Forming limit diagrams

obtained from a Marciniak-type analysis with an initial imperfection amplitude ξ = 0.02. A uniaxial stress-strain-strain rate curve of the form

$$\sigma_u = K\varepsilon_u^{\,N}\,\dot{\varepsilon}_u^{\,M}$$

was assumed, where N and M denote the strain-hardening and strain-rate sensitivity exponents, respectively. Solid curves represent results obtained with deformation theory of plasticity, whereas dashed curves refer to flow theory.

The combinations of m and n given in Fig. 1 are such that the ratio of equibiaxial to uniaxial yield stress σ_b/σ_u remains fixed. The variations of m and n here thus represent changes in the detailed shape of the yield locus. These results show that the predicted trends for the forming limit curves are significantly dependent on the anisotropic parameters m and n. This observation holds whether we use flow theory or deformation theory of plasticity, and whether we assume rate-independent material behaviour (M=0) or account for strain-rate sensitivity (M=0.05). Since the extreme sensitivity evident in Fig. 1 is not likely to be representative of the real behaviour of sheet metals, alternate phenomenological descriptions of anisotropy would perhaps be more suitable for forming limit diagram computations. One yield function of the general formulation recently proposed by Budiansky (1982) deserves consideration. In this paper we investigate its implications on the predicted shapes of forming limit diagrams.

FORMING LIMIT DIAGRAM SHAPES

We shall restrict attention to the prediction of forming limit diagrams in the biaxial stretching range ($0 \leq \rho \leq 1$) as it is in this region where the predicted trends are particularly sensitive to the anisotropic yield function parameters. The usual assumptions of plane stress, transverse isotropy, incompressibility and proportional straining (ρ=const.) prior to localization are adopted.

The type of anisotropy considered here has no influence on the plane strain result ($\rho=0$). It can, however, greatly affect the predicted necking strains in the equi-biaxial stretching limit ($\rho \to 1$). To provide an indication of the overall shape of the forming limit curve in the biaxial stretching range, we introduce a "shape parameter" S^* defined as follows: $S^* = \varepsilon_b^*/\varepsilon_{ps}^*$. This is simply the ratio of the equi-biaxial limit strain $\varepsilon_1^* = \varepsilon_2^* \equiv \varepsilon_b^*$ to the plane-strain value ε_{ps}^*. Some insight into the effects of anisotropy on this parameter can be gained from the following simple relation (Neale and Chater, 1981).

$$S^* \equiv \frac{\varepsilon_b^*}{\varepsilon_{ps}^*} = \frac{1}{2}\left[1 - \frac{1}{NC_b}\right] \tag{3}$$

This expression is a result of a necking-band bifurcation analysis assuming rate-independent deformation theory of plasticity and isotropic hardening. In (3), N is the strain-hardening exponent and

$$C_b = \sigma_b \left.\frac{d^2\sigma_2}{d\sigma_1^2}\right|_{\sigma_1=\sigma_2=\sigma_b} \tag{4}$$

is a measure of the yield locus curvature at the equi-biaxial state ($C_b<0$ for convexity). With the yield function (1), C_b can vary from $C_b=-\infty$ (for $m<2$) to $C_b=0$ (when $m>2$) giving rise to values of $S^*=1/2$ and $S^*=\infty$, respectively. This explains the source of the strong sensitivity of the forming limit diagrams (Fig. 1) to the anisotropic parameter m.

Recently, Budiansky (1982) has developed a general formulation for the plastic behaviour of transversely isotropic sheets which includes the previous theories (e.g. Bassani, 1977; Hill, 1950, 1979) as special cases. Budiansky's yield function is expressed parametrically as follows:

$$\frac{\sigma_1 - \sigma_2}{2\sigma_s} = g(\alpha)\sin\alpha \qquad \frac{\sigma_1 + \sigma_2}{2\sigma_b} = g(\alpha)\cos\alpha \tag{5}$$

where, with $g(0) = g(\pm\pi/2) = 1$, σ_s and σ_b become the current yield stresses in pure shear and equi-biaxial tension, respectively. The yield stress in uniaxial tension is given by

$$\sigma_u = 2\sigma_b\, g(\alpha_u)\cos\alpha_u \quad , \quad \phi_u = \tan^{-1}(\sigma_b/\sigma_s) \tag{6}$$

Furthermore, we have

$$\frac{g'(\alpha_u)}{g(\alpha_u)} = \frac{Y^2 - (1+2R)}{2Y\ (1+R)} \tag{7}$$

where $Y = \sigma_b/\sigma_s$.

A possible yield function put forward by Budiansky (1982), which we consider in this study, is the explicit form

$$g(\alpha) = \sum_{n=0}^{p} a_n \cos(2n\alpha) \tag{8}$$

The constants a_n can be chosen, for example, to fit experimental data. Convexity of the yield locus requires that

$$g^2 + 2(g')^2 - gg'' \geq 0 \tag{9}$$

for $0 \leq \alpha \leq \pi/2$. From (4), (5) and (8) it follows that

$$C_b = \left(\frac{\sigma_b}{\sigma_s}\right)^2 [g''(0) - 1] = -\left(\frac{\sigma_b}{\sigma_s}\right)^2 [1 + 4\ \Sigma\ n^2\ a_n] \tag{10}$$

Substituting this into (3) gives the value of the shape parameter S^* of the forming limit diagram.

RESULTS AND DISCUSSION

We first considered using (8) to characterize some yield functions for typical sheet metals. As "experimental data" we took the yield loci that Bassani (1977) computed using the Bishop-Hill procedure. We shall discuss the particular case corresponding to Bassani's Fig. 7, for which $R=2.7$, $\sigma_b/\sigma_u=1.02$ and $\sigma_b/\sigma_s=1.94$.

Two approximate fits of the yield function (1) reported by Bassani (1977) for this case are m=2, n=3.52 and m=n=2.81. These give rise to $S^*=1.66$ and $S^*=\infty$, respectively (N=0.2). The results obtained with Budiansky's function (8) are summarized in the following table.

TABLE 1 FLD Shape Parameter S^*

P	C_b	S^* (N=0.2)
7	- 5.92	0.92
8	- 7.45	0.84
9	- 8.73	0.79
10	- 11.01	0.73
11	- 12.41	0.70
12	- 13.73	0.68

(For p<7 and p>12 it was not possible to satisfy convexity of the yield locus.) These results indicate that forming limit diagram shapes (e.g. S^*) obtained with the anisotropic yield function characterization (8) do not exhibit the same sensitivity observed with the representation (1).

ACKNOWLEDGEMENT

This work was supported in part by the Natural Sciences and Engineering Research Council of Canada (Grant A-8584) and by le Ministère de l'éducation du Québec (Programme FCAC).

REFERENCES

Bassani, J.L.(1977). Int. J. Mech. Sci., 19, 651-660.
Bassani, J.L., J.W. Hutchinson, and K.W. Neale (1979). Metal Forming Plasticity, edit. H. Lippmann, Springer-Verlag.
Budiansky, B.(1982). Harvard Univ. Rept. MECH-26.
Hill, R.(1950). The Mathematical Theory of Plasticity. Oxford University Press.
Hill, R.(1979). Math. Proc. Camb. Phil. Soc., 85, 179-191.
Neale, K.W., and E. Chater (1980). Int. J. Mech. Sci., 22, 563-574.
Neale, K.W., and E. Chater (1981). Proc. 8th CANCAM, Moncton, Canada, 255-256.

A MATERIALS MODEL FOR STRAIN SOFTENING WITH APPLICATIONS IN MODELING STRAIN LOCALIZATION

T. C. Lowe[1] and A. K. Miller

Department of Materials Science and Engineering, Stanford University, Stanford, California, USA

ABSTRACT

An advanced model of nonelastic deformation is proposed which is capable of predicting strain softening in metals during complex non-proportional loading histories. The model, an extension of the MATMOD constitutive equations, now includes representations of the effects of microstructural heterogeneity, strain-activated recovery, and colinearity of internal stress and strain rate; these additions underly the model's ability to predict strain softening in aluminum. This ability is utilized by coupling the new model, MATMOD-4V, with a model of strain localization. Model predictions of strain softening and strain localization caused by strain softening demonstrate some of the capabilities of MATMOD-4V.

KEYWORDS

Strain softening; strain localization; mathematical modeling; constitutive equations; nonelastic deformation.

INTRODUCTION

Strain softening (also known as work softening) is simply defined as any process by which nonelastic straining causes or enhances softening of metals and alloys that have been hardened by cold work, solutes, second phases, or other strengthening mechanisms. This behavior is contrary to the usual expectation of strain hardening during nonelastic deformation. However, strain softening occurs in many engineering materials and is of considerable technological importance.

The technological significance of strain softening lies primarily in its relationship to strain localization, an important concern in metal forming processes and other areas. Plastic instabilities associated with strain softening and subsequent strain localization may affect material selection

---

[1]Presently at Sandia National Laboratories, Livermore, California, USA 94550

and processing, particularly in sheet forming. For example, efforts to use lighter weight, high-strength aluminum alloys or thinner gauge iron-based alloys in automobile bodies may be limited by difficulties with strain localization and formability due to strain softening. However, in some forming operations, strain softening may have beneficial effects in restoring ductility, eliminating the need for intermediate annealing treatments (Coffin, 1967).

The objectives of this work revolve around the need for a quantitative model of strain softening and its effects upon strain localization. A unified model of nonelastic deformation is developed which is capable of predicting strain softening behavior during complex non-proportional loading histories over a wide temperature range. Because of the complexity of strain softening processes, the model is initially restricted to representing softening resulting from instabilities of dislocation substructures. Aluminum, which has a strong tendency for such instabilities, is chosen as a test material for evaluating the capabilities of the model. Computer simulations are run to show predictions of strain softening and applications in predicting strain localization.

NEW CONCEPTS IN MODELING STRAIN SOFTENING

Physically based and phenomenological representations of the various sub-processes controlling nonelastic deformation are used to develop a model which may be applied to a wide range of materials and deformation conditions. The model, an extension of the MATMOD constitutive equations developed by Miller (1976), provides a quantitative representation of material properties so that it is immediately useful in engineering applications. As in MATMOD, the new model, MATMOD-4V, describes the relationship between stress, strain rate, and structure in terms of state variables which represent the isotropic and directional strengthening effects of dislocation substructures. However, new modeling concepts are introduced which lead to a physically meaningful representation of strain softening. A detailed description of the model and its development is omitted (due to space limitations) but can be found elsewhere (Lowe, 1983). Only modeling concepts which are important in representing strain softening are summarized here.

Microstructural Heterogeneity

Though the inhomogeneity of plastic deformation is well known, it is rarely explicity represented in deformation models. The effects of dislocation substructure heterogeneity upon the stability of deformation are important in strain softening processes and are therefore directly represented in MATMOD-4V. State variables $F_{def,\rho}$ and $F_{def,\lambda}$ are defined to represent the separate isotropic strengthening contributions due to relatively homogeneous dislocation distributions ("forest" dislocations) and heterogeneous substructural features such as cell and subgrain boundaries. The heterogeneous substructural hardening variable ($F_{def,\lambda}$) is particularly important since changes in cell and subgrain boundary features are commonly associated with strain softening (Raymond and Dorn, 1964; Ferreira and Stang, 1979). Additional effects of microstructural heterogeneity are included in MATMOD-4V by recognizing the distributions that exist in both the spatial extent and the kinetics of change of internal stresses. State variables are defined to represent the rapid rearrangement kinetics of shorter range internal stresses (R_A) and the slower kinetics of longer range internal stresses (R_B). The four state variables R_A, R_B, $F_{def,\rho}$, and $F_{def,\lambda}$ thus provide a framework for quantitatively representing the important physical processes that occur during strain softening. Structure evolution equations describe the changes in each variable during deformation.

Strain-Activated Recovery

The concept of strain-activated recovery is critical in explaining unidirectional strain softening. This behavior often occurs during unidirectional deformation following strain rate or temperature changes and is commonly attributed to thermally activated recovery of unstable dislocation substructures by processes involving cross slip and climb of dislocations (Longo, 1970; Miura and Hamashima, 1980). Experimental evidence exists which shows that recovery during unidirectional strain softening is at least partially strain-activated, particularly at low temperatures where thermal recovery is negligible (Hammad, 1965). The magnitude of strain-activated recovery during unidirectional strain softening in a given material increases with temperature and amount of prestrain (Hammad, 1965; Longo, 1970). The temperature, strain rate, and microstructure dependence of strain-activated recovery is represented in the MATMOD-4V structure evolution equations.

The effects of unidirectional strain softening are commonly overlooked in analyses of strain rate effects upon metal forming. Usually, only the strain rate sensitivity is accounted for in plasticity analyses of instabilities during nonelastic deformation (Campbell, 1967; Hutchinson and Neale, 1977). However, microstructural instabilities leading to strain softening may occur due to strain rate changes, independent of the strain rate sensitivity of the flow stress. Material instabilities due to unidirectional strain softening may occur in multi-step forming processes where substantial differences in strain rate or temperature exist between steps (Howard, 1973).

State Variable Interactions and Colinearity of Internal Stress and Strain Rate

Most physically based constitutive laws for describing deformation behavior are derived for one-dimensional solids subject to simple stress states (Miller, 1976; Hart, 1976; Gittus, 1978). In MATMOD-4V, important effects of multiaxial stress states and non-proportional loading upon the stability of heterogeneous dislocation substructures and rearrangements of internal stress are represented in terms of two concepts: 1) state variable interactions, and 2) the colinearity of internal stress and strain rate. Changes in straining direction during deformation may result in microstructural instability and a decrease in the flow stress (Yoshida and Miyauchi, 1978; Laukonis and Ghosh, 1978); this behavior is termed directional strain softening. Experimental studies have shown that this softening is primarily isotropic (Lowe and Miller, 1983) and is due to changes in heterogeneous dislocation substructures (Hasegawa, Yakou, and Karashima, 1975). These changes are represented in MATMOD-4V in terms of an interaction between the long range internal stress variable R_B and the heterogenous substructural hardening variable $F_{def,\lambda}$. Additional interactions are defined which represent other couplings between internal stresses and the heterogeneous microstructural features.

The rate of change of internal stress variables R_A and R_B following a strain direction change depends upon the colinearity of the internal stress and strain rate vectors. Rapid rearrangements in internal stress are predicted when the internal stress and strain rate are not colinear. This feature of the model influences the kinetics of directional strain softening through the state variable interactions.

MODEL PREDICTIONS

Model predictions are generated using a two-dimensional computer code with material constants for pure aluminum. Though the model is capable of predicting a wide range of deformation behaviors, only predictions of strain softening or other behaviors relevant to strain localization are presented. Unidirectional strain softening of a cold-rolled sheet is shown in Fig. 1 in a simulation of slow uniaxial stretching of the sheet in the rolling direction. The predicted softening is a result of the difference in the strain rates during the rolling and stretching operations. Directional strain softening is shown in Fig. 2 in simulations of stretching of cold-rolled sheets transverse to the prior rolling direction. Softening is not observed in similar simulations of stretching parallel to the rolling direction since there is no change in straining direction to trigger microstructural instability. It is important to recognize that these predictions represent idealized material responses to deformation and are not directly comparable to similar experimental data. Sheet metal forming experiments in particular are strongly influenced by factors other than the intrinsic properties of the deforming material. For example, the resumption of hardening following directional strain softening (Fig. 2) would not necessarily be observed experimentally since geometric instability and failure could occur during the softening transient.

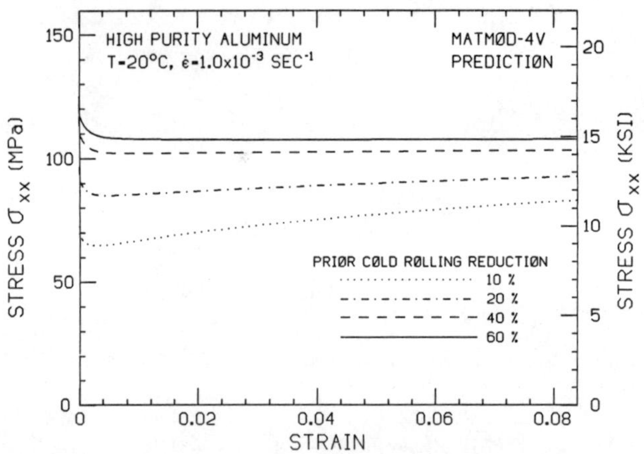

Fig. 1 Unidirectional strain softening of a 60% cold-rolled sheet during slow uniaxial stretching in the rolling direction (x-direction). Persistent softening is not predicted for rolling reductions less than 60%.

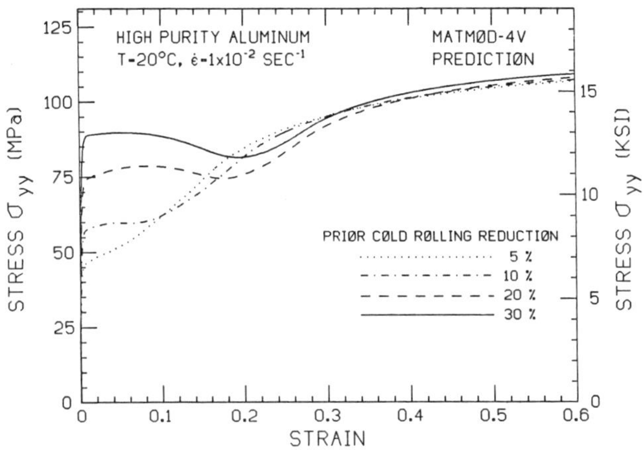

Fig. 2. Directional strain softening is predicted during transverse stretching (y-direction) of sheets cold-rolled to large strains in the x-direction. No softening is seen after small rolling prestrains.

APPLICATION OF MATMOD-4V TO PREDICTING STRAIN LOCALIZATION

Existing analyses of strain localization are limited by their dependence upon simple empirical descriptions of material behavior (Campbell, 1967; Hutchinson and Neale, 1977). Multiple benefits can be gained by using MATMOD-4V to more realistically describe material behavior. MATMOD-4V can provide instantaneous values of important materials parameters such as the strain rate sensitivity m and the strain hardening coefficient γ. This is shown in Figs. 3 and 4 for simple monotonic loading. Both m and γ can also be predicted for complex histories. This capability is particularly important in predicting strain localization during complex thermo-mechanical processing of materials. The model also defines material behavior at a microscopic level, so that it may be applied independently in separate regions of a deforming body. Thus, the development of heterogeneities in materials properties during strain localization may be represented. An additional benefit offered by MATMOD-4V is its unique ability to predict material instabilities that result in strain softening.

A simple simulation, based upon Argon's (1973) analysis of plastic instabilities in shear, is performed to demonstrate how the ability of MATMOD-4V to predict strain softening may be utilized in predicting strain localization. A geometry-independent criterion for plastic instability is adopted which predicts that strain localization will begin when $d\varepsilon/d\varepsilon\ \sigma$ is greater than zero. This criterion has been verified in experimental studies by Conrad, Demeri, and Bhatt (1973), and Bowden and Raha (1970). Strain localization in shear bands during directional strain softening is simulated in Fig. 5 using this stability criterion and Argon's localization model. The local increase in the strain rate in shear bands and corresponding decrease in strain rate in the surrounding matrix are indicated. Additional simulations of strain localization in other deformation modes and during complex deformation histories are in progress.

ACKNOWLEDGEMENT

This work was sponsored by the U.S. Department of Energy, Office of Basic Energy Sciences, Contract PA# DE-AT03-76ER 70057.

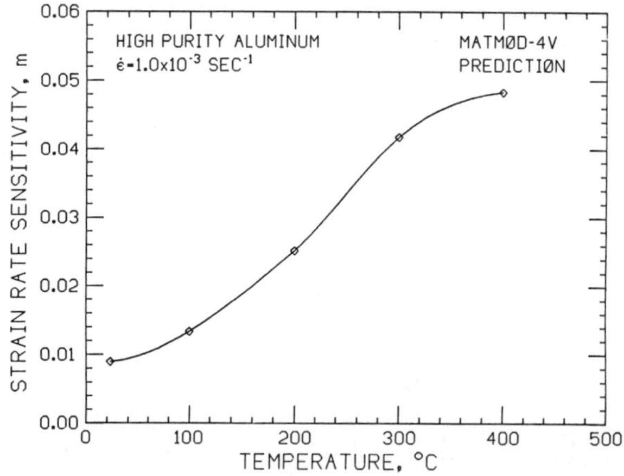

Fig. 3. MATMOD-4V prediction of the temperature dependence of the strain rate sensitivity at a constant strain of 0.6 during uniaxial deformation.

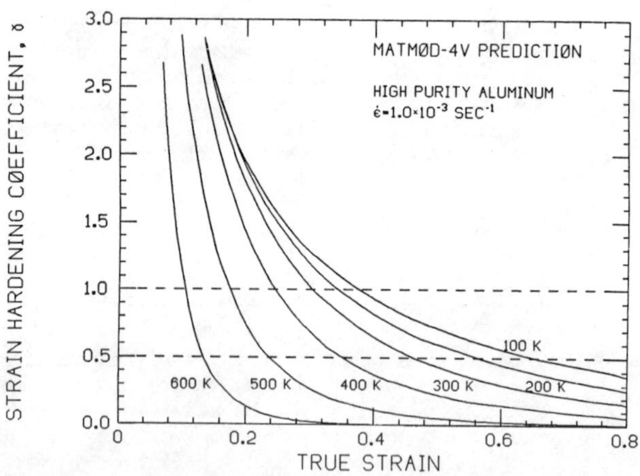

Fig. 4. MATMOD-4V prediction of the temperature and strain dependence of the strain hardening coefficient during uniaxial straining.

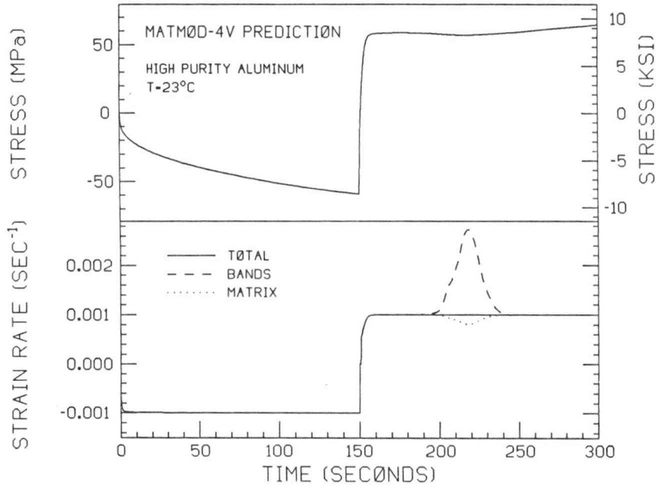

Fig. 5. Localization of strain is predicted during directional strain softening at room temperature. The local strain rate in the shear bands increases while the strain rate in the surrounding matrix decreases.

REFERENCES

Argon, A. S. (1973). In, The Inhomogeneity of Plastic Deformation. American Society for Metals, Metals Park, Ohio. pp. 161-190.
Bowden, P. B. and Raha, S. (1970). Phil. Mag., 22, 463-482.
Campbell, J. D. (1967). J. Mech. Phys. Solids, 15, 359-370.
Coffin, L. F. (1967). ASM Trans. Quart., 60, 160-175.
Conrad, H., Demeri, M. Y., and Bhatt, D. (1973). In S. S. Hecker, A. K. Ghosh, and H. L. Gegel (Eds.), Formability: Analysis, Modeling, and Experimentation. Met. Soc. AIME, New York. pp. 208-231.
Ferreira, I. and Stang, R. G. (1979). Mat. Sci. & Engng., 38, 169-174.
Gittus, J. H. (1978). Acta Met., 26, 305-317.
Hammad, F. H. (1965). Ph.D. dissertation, Stanford University.
Hart, E. W. (1976). J. Engng. Mat. & Tech., 96H, 193-202.
Hasegawa, T., Yakou, T., and Karashima, S. (1975). Mat. Sci & Engng., 20, 267-276.
Howard, J. (1973). Scripta Met., 7, 295-300.
Hutchinson, J. W. and Neale, K. W. (1977). Acta Met., 25, 839-846.
Laukonis, J. V. and Ghosh, A. K. (1978). Met. Trans., 9A, 1849-1856.
Longo, W. P. (1970). Ph.D. dissertation, University of Florida.
Lowe, T. C. (1983). Ph.D. dissertation, Stanford University.
Lowe, T. C. and Miller, A. K. (1983), to be published.
Miller, A. (1976) J. Engng. Mat. & Tech., 96H, 95-105.
Miura, S. and Hamashima, K. (1980). J. Mat. Sci., 15, 2550-2558.
Raymond, L. and Dorn, J. E. (1964). Trans. AIME, 20, 560-567.

RELATIONSHIP BETWEEN STRUCTURE AND MECHANICAL BEHAVIOUR OF Al-10wt%Cu ALLOYS PRODUCED BY RHEOCASTING

N. A. El-Mahallawy*, N. Fat-Halla and M. A. Taha***

*Department of Mechanical Design and Production Engineering, Faculty of Engineering, Ain-Shams University, Abbaseia, Cairo, Egypt
**Department of Mechanical Enginering, Faculty of Engineering, Al Azhar University, Abbaseia, Cairo, Egypt

ABSTRACT

The mechanical properties of rheocast Al-10 wt% Cu alloy have been investigated. The influence of rheocasting processing variables on strength, ductility, malleability, hardness and impact toughness is presented. Due to the formation of a novel non-dendritic structure with nearly spherical primary particles, the rheocasting material exhibits better and homogeneous mechanical properties in comparison to the conventionally cast material having dendritic structure. The rheocasting material shows strength and ductility differential phenomena where the compressive strength was four times the tensile strength, additionally, the compressive malleability was seven times the tensile ductility. A model is suggested in the present paper for clarification of these anomalous phenomena.

KEYWORDS

Rheocasting; stir casting; Al-Cu alloys; microstructure and mechanical properties; strength differential phenomenon; ductility differential phenomenon.

INTRODUCTION

The rheocasting process has been recently developed, in which a novel structure consisting of primary non-dendritic particles embedded in a eutectic matrix (Flemings and Mehrabian, 1973; Spencer, Mehrabian and Flemings,1972; Ramati, Abbaschian and Mehrabian, 1978; Flemings and Mehrabian, 1976). This type of structure has proved to result in striking thixotropic behaviour,and different possibilities of application of the process have been discussed . Materials prepared by rheocasting were applied in die-casting of ferrous and non-ferrous alloys where good casting properties are combined with longer die life (Flemings and Mehrabian, 1973; Flemings and Mehrabian, 1976; Flemings and Mehrabian, 1972; Fascetta and others, 1973; Young, Riek and Flemings , 1977). Shaped products obtained by suction casting of the slurry have been successfully prepared in our laboratories with homogeneous structure and properties. Ingots prepared by rheocasting (RC) also showed promising results

with improved properties, homogeneous structure, and almost negligible shrinkage porosity (Taha and El Mahallawy, 1979). These ingots could be used for further metal forming operations having improved formability and better product properties (Taha and Suery, 1983).
Up till now, a lack of information about the mechanical behaviour of RC materials exists. A previous work (Vogel, Doherty and Cantor, 1977) showed that in hypoeutectic Al-Cu alloys, the presence of small volume fractions of primary aluminium appeared to reduce the fracture toughness due to the formation of $CuAl_2$, which is a brittle phase, around the primary ductile particles allowing the crack path to be around the primary ductile phase . Tensile tests conducted on some carbon- and stainless-steel ingots prepared by rheocasting exhibited interesting results (Taha and El Mahallawy,1980). Improvement in the tensile strength and ductility of RC ingots was obtained and was mainly attributed to the almost negligible amount of interdendritic shrinkage which is usually present in conventionally cast (CC) ingots . Additionally, homogeneous properties was found in RC ingots with values falling between the maximum (at mould wall) and minimum (at centre line) values measured in the CC ingots. A strength differential phenomenon was noticed, i.e., the compressive strength was 7 times the tensile strength,and the malleability in compression was 20 times that in tension. Such a remarkable differential phenomenon was not present in the CC alloy.
The purpose of the present investigation is to study the mechanical properties of RC Al-10 wt% Cu alloy, and to correlate them with the processing variables and with the microstructural parameters. The strength and ductility differential phenomena are also discussed.

EXPERIMENTAL PROCEDURE

Al-10 wt% Cu was prepared from commercial purity materials. Rheocasting experiments were carried out under protective gas atmosphere using a batch type stir-caster. The details of the apparatus are described elsewhere (Assar , El Mahallawy and Taha, 1982). The alloy was thoroughly stirred during cooling from above the liquidus to a predetermined temperature within the liquid/solid range. A series of experiments was made under continuous cooling conditions, with constant primary cooling rate of $3.33 \times 10^{-2} \pm 0.04 \times 10^{-2}$ K/s (El Sawy , El Mahallawy and Taha, 1981). Another series of experiments was made with continuous cooling followed by isothermal stirring in the liquid/solid temperature range. In both cases, a slurry with relatively good fluidity was formed (Assar, El Mahallawy and Taha, 1981), and was immediately sucked after stopping stirring action into cylindrical copper tube (11.2 mm and 10 mm, external and internal diameters respectively) and then rapidly quenched into cold water achieving high secondary cooling rate of the remaining liquid (El Sawy, El Mahallawy and Taha, 1981). The influence of the following processing variables was investigated; stirring speed'V' (ranging from 340 to 992 rpm), stirring temperature T (from 863K to 893K) and isothermal stirring time t (from zero to 35 minutes).
Under all experimental conditions, both the partially solidified rheocast slurry inside the crucible and the completely solidified sucked rheocast rods showed a high degree of chemical homogeneity where Cu concentration at different positions was 9.894 (+ 0.006 - 0.064) wt%. For mechanical testing , different tensile and compressive test specimens were machined along the RC and CC rods. The CC rods were prepared by suction of superheated liquid followed by water quenching. Microstructural and chemical analysis was made on the quenched rods and the results show complete chemical and microstructural homogeneity along the rod axis. The tensile specimens were 4 mm in diameter and 15 mm gauge length, while, the compressive specimens were 6 mm in diameter and 12 mm length. A Hounsfield tensometer was used where a constant strain rate

of 1.94×10^{-3} was applied. The compression tests were carried out under frictionless conditions.

MICROSTRUCTURE

In general, the conventional dendritic morphology was observed in the CC rods which were cast from a completely liquid state and solidified in the copper mould. Such structure was not observed in the RC rods, which were cast from a partially solidified stirred slurry. In the RC alloy, the structure was generally composed of nearly rounded primary particles embedded in a matrix (solidified remaining liquid) consisting of fine primary dendrites in a much finer eutectic (Cf. Fig.1). The structure was always homogeneous throughout the RC specimens. It was found that the stirring speed V has the main influence on the morphology and size of the primary alpha-particles. At low speed V, primary phase was in the form of coarse (some what rounded) particles agglomerated together as shown in Fig.1(a). At high speed, more rounded (nearly spherical) separated primary particles were obtained (Fig.1(b)). Other processing variables (stirring temperature and stirring time) did not change the morphology and size of the alpha-particles. As can be detected from the phase-diagram, the fraction solid (F_s), defined as the volume fraction of

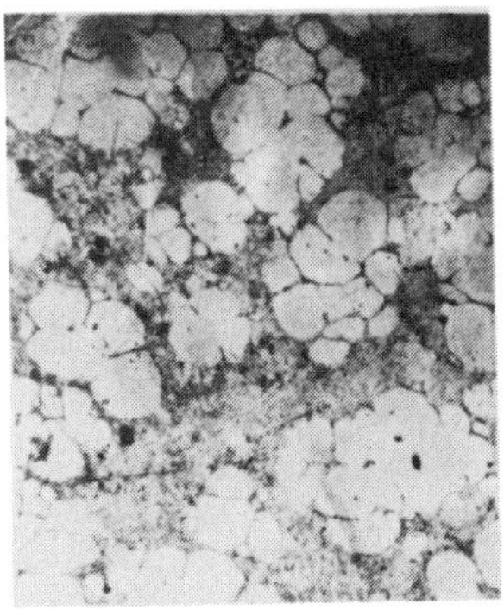

V=482r.p.m.

500 μm

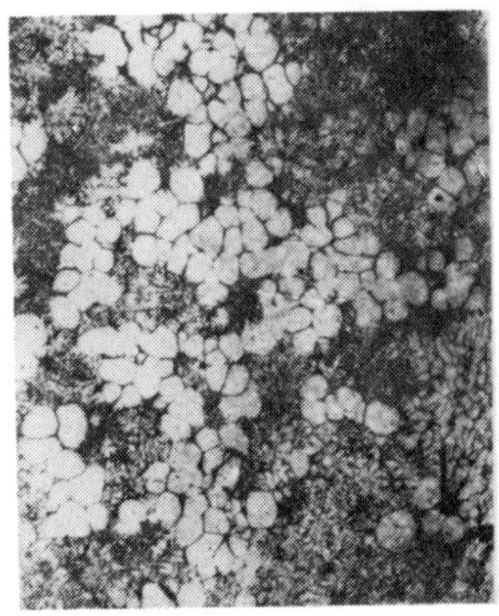

V=992r.p.m.

Fig.1. Typical microstructures of RC Al-10 wt% Cu alloy at low and high V.

the alpha-particles solidified during stirring in the primary cooling stage, increased as the stirring temperature T decreased. More quantitative details about the structure is found elsewhere (Assar, El Mahallawy and Taha,1982). Small amount of microporosity could exist in RC Al-10 wt% Cu alloy with more or less spherical form and existing either in the matrix, inside the agglomeration or adjacent to agglomeration boundary (Assar, El Mahallawy and Taha, 1982). Generally, the porosity occurred due to atmosphere entrapment which did not exceed 1.5% under processing conditions t = 0, V not more than 700 rpm and different T. Higher stirring speeds drastically increased porosity entrapped in the matrix which could reach a value of 7.5%.

RESULTS ON MECHANICAL PROPERTIES

The processing variables which showed influence on the mechanical properties of RC Al-10 wt% Cu were the stir temperature T, and stir speed V. Isothermal stirring time, t, did not cause any sensible variation in the mechanical properties since the microstructure did not change. The increase in microporosity percent as t increased (7.5% at t = 30 min) did not reduce the strength since the porosity formed was of the entrapped matrix type with a fine (average size 50 μm) spherical shape and uniformly distributed in the matrix. The mechanical properties investigated are; ultimate tensile strength (σ_t), ultimate compressive strength (σ_c) (in the present research σ_c was taken as that stress at which cracks appeared on the surface of the specimen), elongation percent (ϵ_t), malleability in compression test (ϵ_c), hardness (VHN), and impact strength (I). The influence of T and V on these properties is shown in Figs. 2, 3, and 4.

The increase in strength and decrease in ductility with T, shown in Fig.2, could be attributed to the volume fraction of primary ductile particles, F_s, which decreases as T increases (Assar, El Mahallawy and Taha, 1982). The variation of strength and ductility with V is also similar. Figs. 2(c) and (d) show that σ_t and σ_c increase with V, while, $_t$ and $_c$ decrease with V. It is observed that the influence of V is more remarkable at high T. Both VHN and I increase with V. The improvement in strength with V could be attributed to microstructural parameters. With increasing V, the primary particles decrease in size and the number of their agglomeration also decrease (Assar, El Mahallawy and Taha, 1982).

The compression tests showed anomalous results, that is to say, the strength and malleability are much higher than those obtained in tension test (four times in case of strength and seven times in case of ductility). The compressive properties are more sensitive to variations in T and V than the tensile. In all cases, both strength and ductility have been improved in RC alloy (Fig. 5 and 6) compared to the CC alloy. It is to be noticed that the CC alloy was highly degasified using physical treatment (presolidification method), then sucked by vacuum, from relatively low temperature, into the copper mould. The rest of the melt alloy was then used for rheocastin . Therefore, the improvement in mechanical properties of RC alloy is related to the novel microstructure.

DISCUSSION

Variation of mechanical properties of RC alloy with processing variables can be best understood if correlated to the microstructural parameters. The parameters considered in the present discussion are; fraction solid of the primary alpha-particles, F_s, and their size λ.

Tables 1 and 2 summarize the results obtained, concerning the effect of F_s and λ on the mechanical properties of the RC alloy. Fig. 5 shows a gradual decrease in tensile and compressive strength with increasing F_s. Simultaneously, the tensile ductility and compressive malleability increase gradually. These results stem, obviously, from the gradual increase in F_s, the softer alpha-particles. Similar behaviour can be seen in Fig. 6 which illustrates a decrease in strength and an increase in ductility as λ increases.

In order to calculate the tensile strength of the RC alloy, the rule of mixture is assumed to be applicable. This assumption is based on the fact that the fracture strain of the RC alloy (3-7%) is not much different from that of the matrix (4%) (Ochiai, Shimomura and Murakami, 1975; Taha, 1981), considering the matrix (remained liquid) composition ranging between 15 and 25 wt% Cu. therefore, the ultimate tensile stress (σ_t) can be represented by;

$$\sigma_t = F_s . \sigma_s + (1 - F_s) . \sigma_m \quad (1)$$

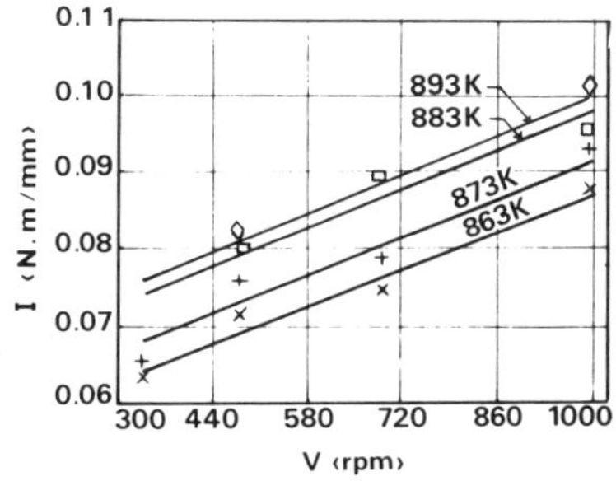

Fig.4. Influence of V on I.

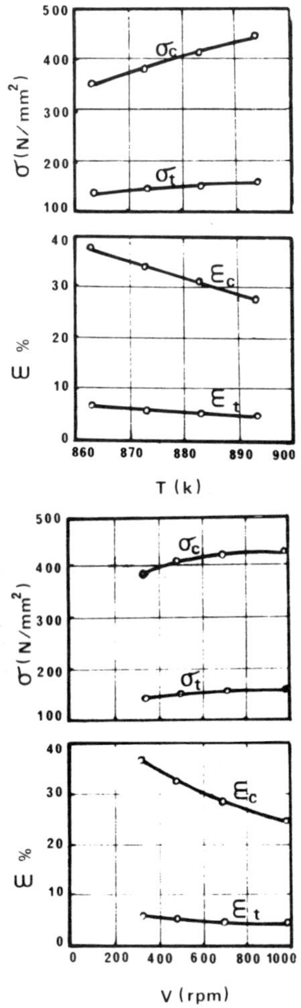

Fig.2. Influence of T and V on σ and ε.

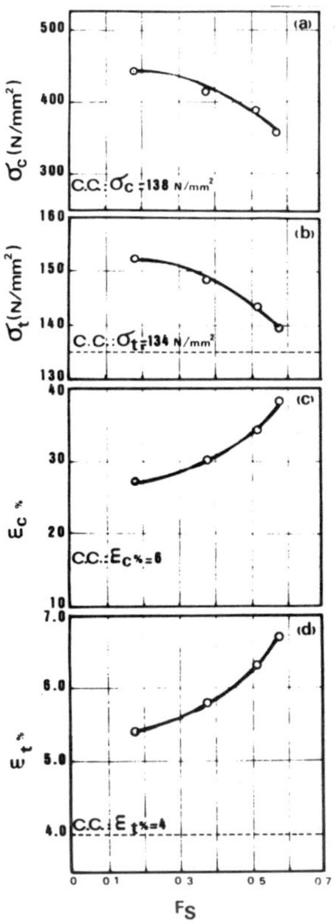

Fig.5. Influence of F_s on σ and ε.

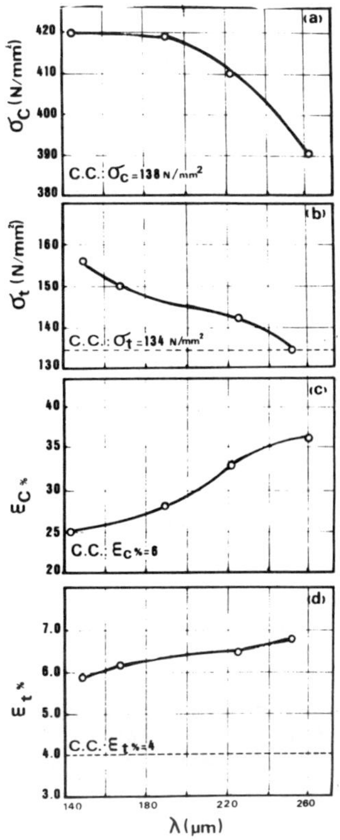

Fig.6. Influence of λ on σ and ε.

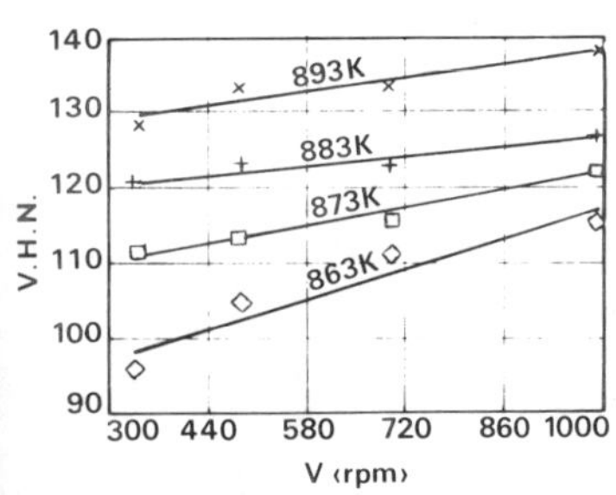

Fig.3. Influence of V on VHN.

TABLE 1 Effect of F_s on the Mechanical Properties of RC Al-10 wt% Cu

F_s	σ_t (N/mm^2)	ϵ_t%	σ_c (N/mm^2)	ϵ_c%
0.18	152.3	2.6	424.0	26.7
0.38	148.4	3.7	397.7	29.7
0.52	143.1	5.1	371.3	33.0
0.58	139.2	6.2	346.5	37.2
CC	134.0	4.0	138.0	6.0

TABLE 2 Effect of λ on mechanical Properties of RC Al-10 wt% Cu

λ μm	σ_t (N/mm^2)	ϵ_t%	σ_c (N/mm^2)	ϵ_c%
158.7	155.7	5.88	420.0	24.6
175.7	149.4	6.20	418.3	27.7
228.3	141.4	6.50	409.1	32.3
252.0	134.0	6.70	389.4	36.0

where σ_s and σ_m refer to the strength of the primary solid and that of the matr respectively. Since the strength of particles is a function of their size , equation (1) can be rewritten in the form;

$$\sigma_t = F_s K \lambda^{-n} (1-F_s).\sigma_m \quad (2)$$

or in the form;

$$\sigma_t = F_s \sigma_{s0} (\frac{\lambda}{\lambda})^{-n} + (1 - F_s). \sigma_m \quad (3)$$

where σ_{s0} is the theoretical strength of primary solid alpha-particles when they reach a size λ_0. However, the effect of changes in the particles and matri strengthes due to variation in their composition when different T are employed, are asided for simplicity. In order to obtain the exponent n, the results in Table 2 are used in making log-log plot using a value of σ_m = 120 MN/mm sq., where n was found to be 0.45 .

It is of interest at this stage to compare the mechanical properties obtained for RC and CC alloys. Figs. 5 and 6 in addition to Tables 1 and 2 illustrate that the tensile strength of the RC alloy is 4-13% greater than that of the CC alloy, while the ductility is fluctuating around that of the CC alloy. Considering the tensile strength and ductility of the CC alloy, it is noticed that two opposing factors are present; the first is the amount of soft primary phase (about 0.86 under equilibrium conditions) (Hansen, 1958), which lowers the strength and increases the ductility; the second, is the fine dendritic structure obtained by quenching from the liquid, where small dendrite arm spacing ar found. This latter factor should raise the strength of the CC alloy.However , considering equation (3) , it can be seen that the strength is more sensitive to variation of F_s than to a similar variation in λ. This explains the lower strength of the CC alloy (F_s = 0.86) compared to that of the RC alloy. On the other hand, the ductility of the CC alloy did not exceed 4% while that of the RC alloy reached more than 6% for an F_s of only 0.58. This is due to the following reasons;i) in RC alloy, primary ductile particles are almost spherical in shape and more uniformly distributed in the matrix, while in CC alloy the primary phase is dendritic thus resulting in a higher ductility of the forr (Dekazinczy and Backofen, 1960); ii) in RC alloy a better dispersion of the non-metallic inclusions is obtained,iii) in RC specimens,the microvoids formed at the particle/matrix interface follow more or less the smooth roundness of the particles with no sharp corners while in CC specimens the voids are larger and may fall in corners between dendrite arms,iv) in RC specimens,the concentration of Cu inside the alpha-particles is uniform (Assar,El Mahallawy and Taha, 1982), while it is not the case in CC alloy, and v) in RC specimens the matrix contains an amount of fine soft dendrites while in the CC specimens the matrix is purely a brittle eutectic (Assar,El Mahallawy and Taha,1982).

Now,the anomalous results of the compression test includes remarkable increase in strength and malleability not only relative to CC alloy but also to the σ_t and ϵ_t of the same RC alloy. It is now well established (Pittinato, Kerlins and Russo,1975)that crack origin on a microscopic scale whether occurring at non-metallic inclusions,at grain boundaries or at zones of heavy deformation will coalesce with other voids and then grow under the action of hydrostatic tension until complete fracture occurs.It is also known that application of a hydrostatic compressive stress on a tensile specimen results in delayed fracture ,i.e., increase of ductility (Pittinato and others,1975; French and others , 1975). Previous investigation on the fracture of Al-Si alloys showed that the matrix (which was the softer phase) deformed formerly and then when heavy deformation interferred with the hard Si particles,cracking of the latter occurred followed by catastrophic failure(Guarland and others,1963).Moreover,through an investigation on the mechanical properties of unidirectionally solidified Al-Si eutectic alloys a strength differential phenomenon was observed,however, this increase in strength was on the expense of malleability(Sahoo,and Smith, 1975) In the present case the matrix is the hard phase and the particles are the softer one.The model shown in Fig. 7 was proposed in the present study in order to explain the anomalous results. Since the softer phase is distributed more uniformly in a relatively hard matrix and possessing a spherical morphology, it can be argued that the compression stress transmitted to the softer particles through the matrix surrounding it from every where,and leading to a state of hydrostatic compression depending on the perfection of roundness of the particles.This state of hydrostatic stress delays the crack growth thus leading to increase in ductility and consequently to an increase in strength.The same model fits reasonably to the tensile results.The presence of hydrostatic tensile stress on the ductile particles will help in rapid crack growth.In general,the ductility during tensile test is controlled by two opposing factors being;i)increase due to uniform distribution of non-metallic inclusions,less voids,and more uniform concentration profile inside the grains, ii) decrease in ductility due to hydrostatic tensile state of stress applied.

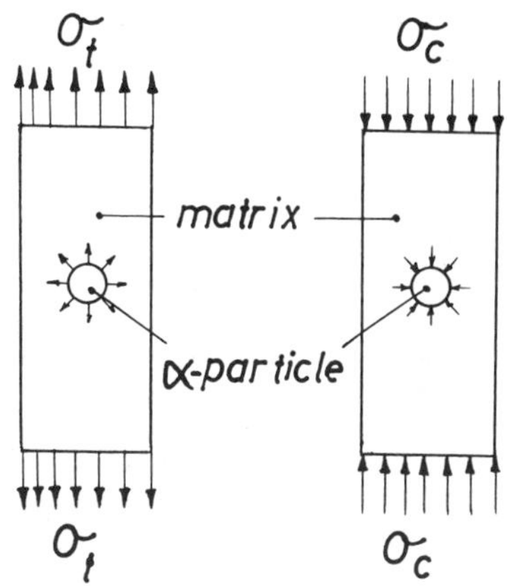

Fig.7 Model proposed for tension and compression

CONCLUSION

RC Al-10 wt% Cu alloy possesses compressive strength about 4 times that in tension, and a strain to fracture in compression about 10 times that in tension. A model is proposed to explain these results depending on the existence of hydrostatic stress on the primary particles which are almost spherical.

ACKNOWLEDGEMENT

A portion of the experimental measurements was obtained by A.Assar in his M.Sc. thesis, Ain-Shams University, 1980, under the supervision of two of the present authors; N.A.El Mahallawy and M.A.Taha.

REFERENCES

Assar, A., N. A. El-Mahallawy and M. A. Taha(1981). Aluminium, 57, 807-810.

Assar, A., N.El Mahallawy and M.A.Taha (1982). Met.Technology, 9, 165-170.

Dekazinczy, F., and W.A. Backoften (1960). Trans. ASM, 52, 55.

El-Sawy, A., N. A. El-Mahallawy and M. A. Taha (1981). Proc. 7th Int. Light Metals Congress (Leoben-Wien), 114-115.

Fascetta, E. F., R. G. Riek, R. Mehrabian and M. C. Flemings (1973). AFS Cast Metals Research J., 9, 167-171.

Flemings, M. C. and R. Mehrabian (1972). Trans. AFS, 80, 173-182.

Flemings, M. C. and R.Mehrabian (1973). Trans. AFS, 81, 81-88

Flemings, M. C. and R. Mehrabian (1976). New Trends in Mat. Processing, ASM, Metals Park, Ohio, 98.

French, I. A., and P. E. Weinrich (1975). Met. Trans., 6A, 785-790

Guarland, J., and J. Plateau (1963). Trans. ASM, 55, 442-455.

Hansen, M. (1958). Constitution of Binary Alloys, 2nd ed., McGraw Hill Book Co., New York-Toronto-London, 85.

Ochiai, S., K. Shimomura and Y. Murakami (1975). Met. Sc. , 9, 535.

Pittinato, G., V. Kerlins and A.Russo (1975). SEM/TEM Fractography Handbook, McDonnell Douglas Astronautics Co., Huntigon Beach, California.

Ramati, S. D. E., G. J. Abbaschian and R. Mehrabian (1978). Met. Trans.,9B, 241-245.

Sahoo, M. and R.W. Smith (1975). Met. Sc. , 9, 217-222.

Spencer, D. B., R. Mehrabian and M.C. Flemings (1972). Met. Trans., 3, 1925-1932.

Taha, M.A. (1981). Proc. 7th Int. Light Metals Congress (Leoben Wien), 110-111.

Taha, M. A. , and N. A. EL-Mahallawy (1979). Proc. 46th Int. Foundry Congress, Madrid, "Egypt-official exchange paper", paper no. 15, 15/1-15/14.

Taha, M. A., and N. A. El-Mahallawy (1980). Mechanical Behaviour of Materials, v. 2, ed. by K. J. Miller and R. F. Smith, Pergamon Press, 537-546

Taha, M. A., and M. Suery , to be published.

Vogel, A., R. D. Doherty and B. Cantor (1977). Proc. Sheffield Int. Conf. on Solidification and Casting, U. K.

Young, K. P., R. G. Riek, and M. C. Flemings (1977). Proc. Sheffield Int. Conf. on Solidification and Casting, U. K.

THEME V

MATERIAL ASPECTS OF FRACTURE MECHANICS IN ENGINEERING PRACTICE

DIRECT OBSERVATIONS OF THE MICROMECHANISMS OF FRACTURE IN POLYMERIC SOLIDS USING THE SCANNING ELECTRON MICROSCOPE

D. G. Gilbert, P. W. R. Beaumont and W. C. Nixon

Cambridge University Engineering Department, Trumpington Street, Cambridge CB2 1PZ, UK

ABSTRACT

Dynamic fracture mechanics tests on polymeric solids have been carried out inside the scanning electron microscope. Direct observations of the micromechanisms of fracture at slowly advancing crack tips have been recorded on video tape. A selection of photomicrographs showing fracture processes, are presented for each polymer tested. Where possible, connections are made between the direct observations of fracture events and proposed modes of failure based on conventional fractographic examinations of broken test pieces.

KEYWORDS

Polymers; crazing; dynamic fracture; fracture mechanisms; scanning electron microscopy; crack propagation; fractography.

INTRODUCTION

For brittle thermoplastics, a cavitation process is recognised as the initial stage of the nucleation and growth of a crack. Under a dilational stress, small cavities form at inhomogeneities and grow. These voids are stabilised by fibrils of polymer chains drawn across the void. This microscopic defect is known as a "craze". Under increasing stress the craze extends and finally fails either by rupture of the fibrils or by failure of the craze/bulk polymer interface. Such a sequence of failure events enables crack initiation and propagation at a stress much below that required for shear yielding. The result is brittle fracture. In some polymers, a dispersion of elastomeric particles can induce wide-spread crazing. Each particle acts as a stress concentrator and nucleates a craze away from the crack tip. A larger deformation zone is formed and the fracture toughness of the polymer is therefore increased.

This paper describes a number of fracture mechanics experiments based on the compact tension (CT) test (Brown and Srawley, 1966) and double torsion (DT) test (Evans, 1973) which were performed inside a scanning electron microscope. Direct observations of the micromechanisms of fracture of polystyrene (PS), polymethylmethacrylate (PMMA), high impact polystyrene (HIPS) and toughened polymethylmethacrylate were recorded onto video tape. A specially edited video tape will be shown at the Conference and a selection of photomicrographs showing various micromechanisms of crack extension are presented in the paper. These direct observations of the micromechanisms of fracture are described and, where possible, related to previously proposed modes of failure based on conventional fractographic examination of broken test-pieces.

EXPERIMENTAL TECHNIQUE

A specially designed, instrumented loading rig was fitted to the stage of a Cambridge Instruments Stereoscan 100 electron microscope. Samples of commercially extruded PMMA, toughened PMMA and PS, and injection moulded HIPS were machined into CT and DT specimens. The specimens were loaded via a flexible drive shaft driven by a stepper motor outside the SEM. A schematic diagram (Fig. 1) shows the two specimens under load and the orientation of the test-piece with respect to the observer. Each specimen was coated with a Au/Pd alloy to reduce charging effects. An acceleration voltage of 5 or 10 kV was used to prevent beam damage of the polymer. Direct observations of the micromechanisms of fracture in the vicinity of a slowly advancing crack front were recorded onto video tape. The photomicrographs presented in the next section are taken from these video recordings.

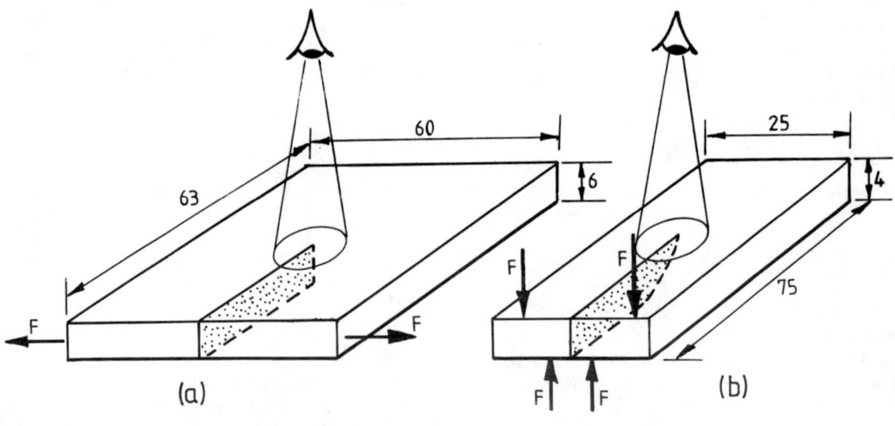

Fig. 1. Schematic diagram of the (a) compact tension and (b) double torsion test pieces and orientation with respect to observer. Dimensions in mm.

DIRECT OBSERVATIONS OF FRACTURE

Polystyrene

Under increasing load, crazes nucleate and grow at the crack/craze tip and begin to overlap. Microcracks form in front of the main crack by fibril rupture leaving ligaments of undeformed matrix material between the microcracks, Fig. 2 (a). The main crack extends by the ligaments failing by further craze formation, but now the crazes are observed approximately

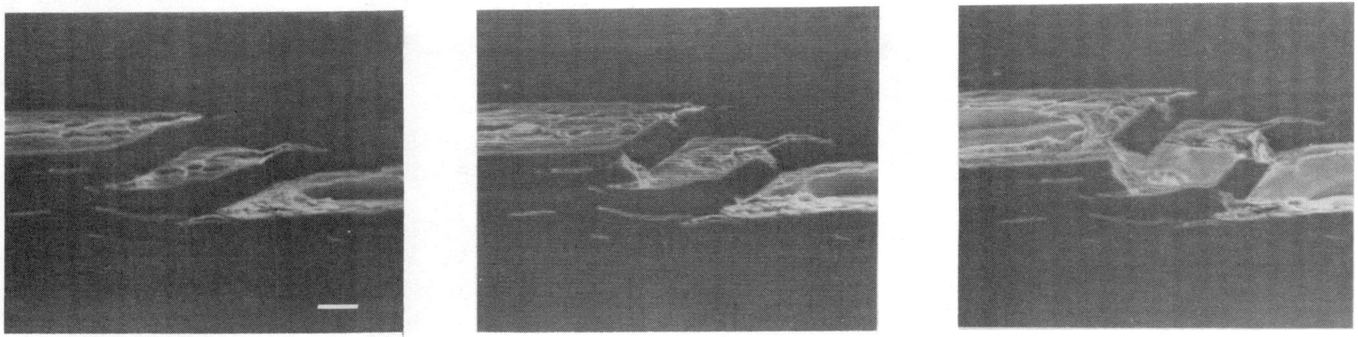

Fig.2. Microcrack coalescence in PS , broken in DT . Bar size 25μm

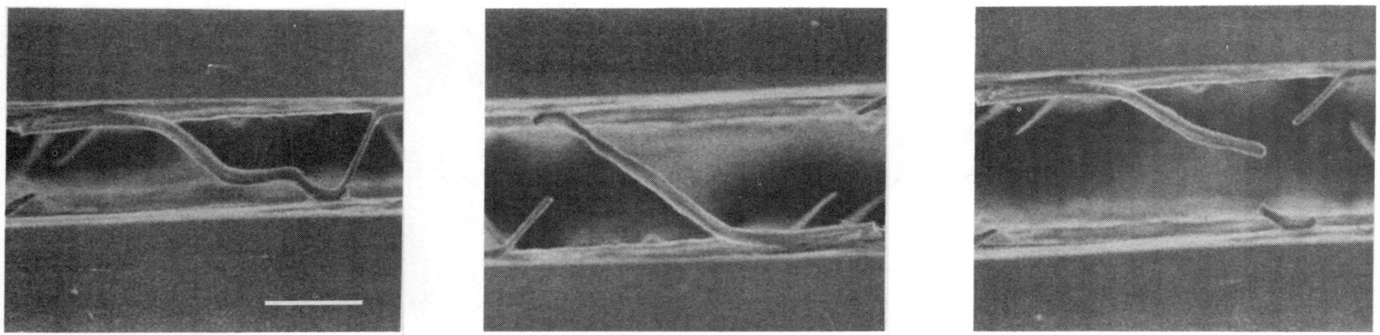

Fig.3. Ligament formation in PMMA , broken in CT . Bar size 20μm

perpendicular to the original crazes and concentrated close to the original craze tip. Two crazes fail and produce V-shaped wedges, Fig. 2 (b). Further crazing and extension takes place until finally the ligament is detached from one side and "springs" apart, Fig. 2 (c). This mechanism produces the characteristic step-like feature on the fracture surface observed by Hull (1970), and causes subcritical crack growth to take place in a series of crack jumps (Mostowitz and Turner, 1974).

Polymethylmethacrylate

PMMA undergoes stable crack advance with the formation and rupture of a single narrow craze ahead of the crack tip. As the crack velocity increases, new crazes form parallel to the direction of propagation at a few microns from the main craze. These crazes close as soon as the crack tip passes. The length of craze is sensitive to the crack growth rate; they shorten as the velocity increases. A shallow surface scratch may cause the crack/craze tip to deviate and slow down, but almost at once, the polymer directly ahead of the primary crack fractures and the crack continues to propagate in its original plane. Such crack detours leave a characteristic protruding and indented V-shape on the surface of the specimen. This disturbance to the advance of the main crack may produce ligaments between the crack faces. Ligaments were also observed to span the surfaces of an accelerating crack, Fig. 3 (a)-(c). Similar ligaments have been observed by Atsuta and Turner (1972) in the fracture of epoxy. These ligaments are formed when a crack that is propagating on two adjacent planes transfers to a single plane. In our study, ligaments were only observed as the crack began to accelerate.

High Impact Polystyrene

An intense region of multiple crazing is observed in the vicinity of a slowly advancing crack front, Fig. 4. This crazed zone extends for a few millimetres in front and around the crack tip, absorbing energy in formation and producing a reduced modulus zone at the crack tip. A large strain is required to initiate crack propagation and the corresponding crack tip opening displacement is large, Fig. 5. This can be directly measured. Under increasing displacement, crack propagation takes place by craze breakdown producing microcracks which the main crack follows. Microcracks join by the ductile tearing of the remaining polymeric material. This process produces parabolas on the fracture surface, like those observed in fatigue by Rimnac, Hertzberg and Manson (1982).

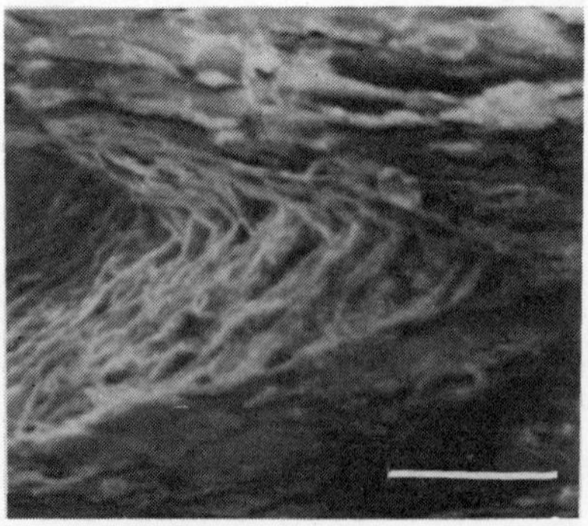

Fig. 4. Rupture of HIPS at crack tip, broken in CT. Bar size 20 μm.

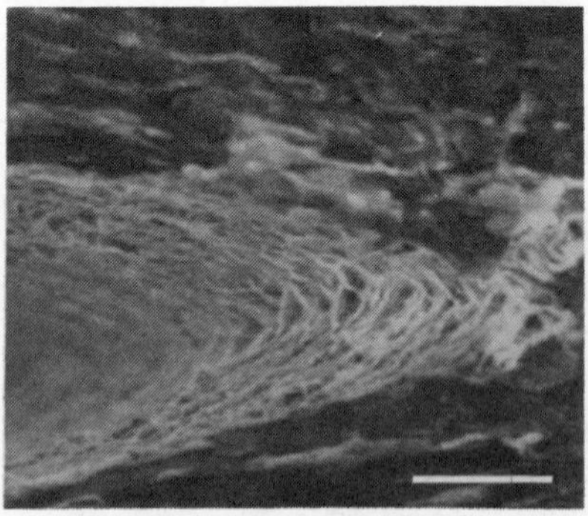

Fig. 5. View into crack tip of HIPS, broken in CT. Bar size 25 μm.

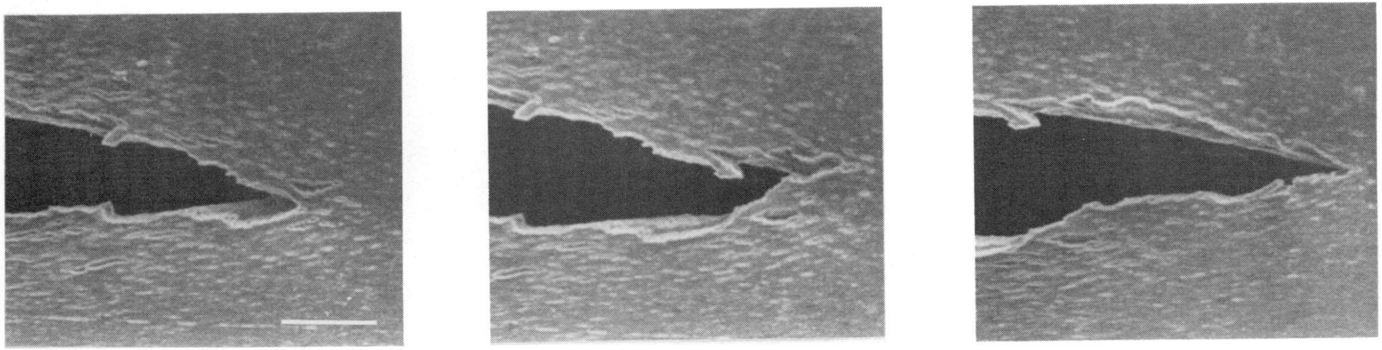

(a) (b) (c)

Fig.6. Crack propagation in toughened PMMA , broken in CT . Bar size 20μm

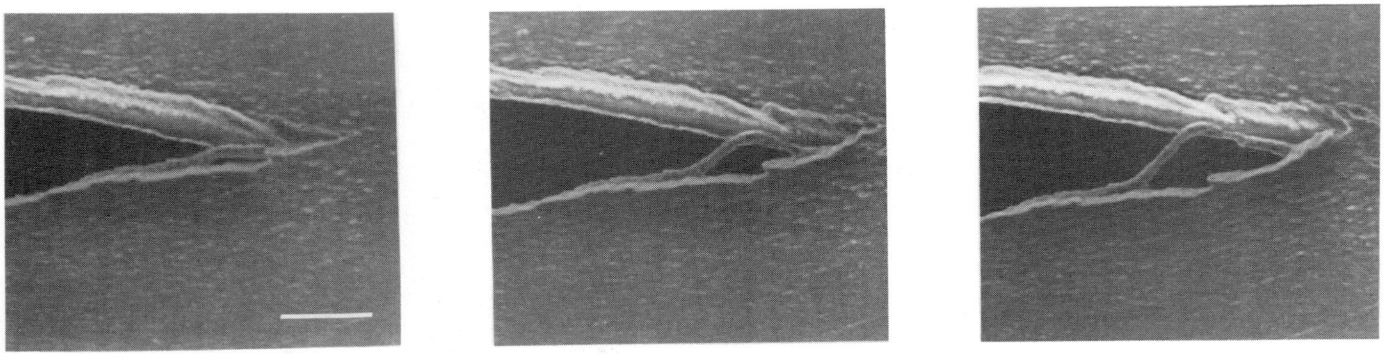

(a) (b) (c)

Fig.7. Ligament formation in toughened PMMA , broken in CT . Bar size 20μm

Toughened Polymethylmethacrylate

Crack advance in toughened PMMA is preceded by the formation of a less crazed deformation zone than observed in HIPS. Crack extension takes place by the opening of microcracks in front of the main crack and the subsequent failure of the material between them, Fig. 6 (a)-(c). Microcracks on either side of the propagating crack tend to remain open even after the crack tip has passed. As observed with untoughened PMMA, a surface scratch or similar imperfection may cause the crack tip to change direction. Fracture then occurs in the bulk of the polymer on the original failure plane and the deviated surface crack becomes arrested. The start of such a crack deviation or "branching" is the origin of the unusual ligament seen forming in Fig. 7 (a)-(c).

CONCLUSIONS

Slow crack growth in polymeric solids has been studied by direct observation using the SEM. Observations of crack initiation and growth provide insight into the dynamic micromechanisms of fracture which are consistent with proposed modes of failure based on conventional fractographic methods. The technique, together with a knowledge of applied load, may be used to obtain directly the crack velocity, surface crack tip opening displacement and the localised strain field.

ACKNOWLEDGEMENTS

One of us (D.G.G.) would like to acknowledge the financial support of the S.E.R.C. of Great Britain for the provision of a research studentship. We would like to thank Professor M.F. Ashby and Dr. A.M. Donald for many helpful discussions.

REFERENCES

Atsuta, M. and D.T. Turner (1982). Fractography of highly cross-linked polymers. *J. Mat. Sci. Letters*, 1, 167-169.

Brown, W.F. and J.E. Srawley (1966). Plane strain crack toughness testing. *A.S.T.M. S.T.P. 410.*

Evans, A.G. (1973). A simple method for evaluating slow crack growth in brittle materials. *Int. J. Fract.*, 9, 267-275

Hull, D. (1970). Effect of crazes on the propagation of cracks in polystyrene. *J. Mat. Sci.*, 5, 357-363.

Moskowitz, M.D. and D.T. Turner (1974). Fractography of a styrene-acrylonitrile co-polymer-microcracks and band structure. *J. Mat. Sci.*, 9, 861-864.

Rimnac, C.M., R.W. Hertzberg and J.A. Manson (1982). The fracture behaviour of a PXE/HIPS polyblend. *Polymer*, 23, 1977-1982.

FATIGUE CRACK PROPAGATION IN VISCOUS ENVIRONMENTS

J. L. Tzou, S. Suresh and R. O. Ritchie

**Materials and Molecular Research Division,
Lawrence Berkeley Laboratory and Department of Materials Science and
Mineral Engineering, University of California, Berkeley, CA 94720, USA**

ABSTRACT

The effect of dehumidified silicone and paraffin oils with viscosities from 5 to 60,000 cS, on fatigue crack propagation in a low alloy steel is examined at both near-threshold ($\lesssim 10^{-6}$ mm/cycle) and higher ($\sim 10^{-6}$ to 10^{-3} mm/cycle) growth rates. Results show that at low load ratios crack growth rates in oil *exceed* those in moist air at near-threshold levels, whereas at higher growth rates crack propagation in moist air is considerably faster than in oil. Such observations are discussed in terms of three mechanisms specific to dry oil environments: suppression of moisture-induced hydrogen embrittlement and/or metal dissolution, minimization of oxide-induced crack closure and hydrodynamic wedging effects of the viscous fluid within the crack.

INTRODUCTION

It is well recognized that the fatigue resistance of cyclically-stressed components can be markedly affected by the presence of oil environments. For example, early studies (Cordiano, Cochran, Wolfe, 1956) on rolling contact fatigue indicated that the use of water-glycol based hydraulic fluids in place of mineral oils could reduce bearing lives by up to 10%. In numerous subsequent investigations, fatigue lifetimes have been reported to be increased, decreased or unaffected by the presence of such apparently non-corrosive fluids, for environments ranging from medicinal oils to synthetic lubricants (Galvin, Naylor, 1965). Despite the lack of understanding of the role of such environments, oils (particularly silicone oils) are regularly used in fatigue studies to simulate reference inert environments or in multi-axial fatigue testing to provide a medium for transmitting superimposed hydrostatic pressure.

The objective of the present work is to examine mechanistically the role of silicone and paraffin oils in influencing the fatigue behavior of a low alloy steel. Specifically, the effect of dehumidified oils of viscosities ranging from 5 to 60,000 cS is studied on rates of fatigue crack propagation, and behavior compared with that for moist and dry gaseous atmospheres. It is found that oil environments can lead to both faster or slower propagation rates, compared to behavior in room air, depending upon oil chemistry, viscosity and principally the growth rate regime.

EXPERIMENTAL PROCEDURES

Tests were performed on a fully bainitic 2¼Cr-1Mo steel (ASTM A542 Class 3) of 500 MPa monotonic yield strength, 400 MPa cyclic yield strength, 610 MPa tensile strength and 77% reduction in area at room temperature. Fatigue crack propagation testing was conducted with 12.7 mm compact specimens (T-L orientation) cycled at 50 Hz frequency (sine wave) at load ratios ($R = K_{min}/K_{max}$) of 0.05 and 0.75, using d.c. electrical potential techniques to monitor crack length. Room temperature oil environments were provided by immersing the specimen in an enclosed oil bath. The oil was dehumidified by continuously bubbling dry helium gas and by maintaining a small positive pressure of helium within the system. Two types of oils were investigated, namely a series of chemically similar silicone oils (dimethyl polysiloxane) with kinematic viscosities of 5, 1000, 12,500 and 60,000 cS, and two paraffin oils (hydrocarbon mixtures) with viscosities of 25 and 75 cS.

RESULTS

The variation of fatigue crack propagation rates (da/dN) with stress intensity range ($\Delta K = K_{max} - K_{min}$) for A542-3 steel tested in the various oil environments is shown in Figs. 1-3. Results are compared with previously obtained data (Suresh, Zamiski, Ritchie, 1981) for this steel in room temperature moist air (30% relative humidity), dehumidified gaseous hydrogen (138 kPa pressure) and dehumidified gaseous helium (138 kPa pressure). In Fig. 1 where the low load ratio ($R = 0.05$) results for the silicone oils are plotted, it is apparent that crack growth rates in oil are *slower* than in air *above* $\sim 10^{-6}$ mm/cycle yet *faster* than in air at near-threshold levels *below* $\sim 10^{-6}$ mm/cycle. The near-threshold effect of oil is absent, however, at higher load ratios ($R = 0.75$) where threshold ΔK_0 values for both high and low viscosity oil environments as well as moist air and dry hydrogen atmospheres are virtually identical (Fig. 2). Similar low load ratio results were found for paraffin oils (Fig. 3), although the effect of these oils in enhancing growth rates at near-threshold levels and reducing growth rates above $\sim 10^{-6}$ mm/cycle was somewhat greater than in silicone oils. Also in Fig. 3 the highest and lowest viscosity silicone oils (5 and 60,000 cS) are compared with the two paraffin oils (25 and 75 cS) at $R = 0.05$. Above $\sim 10^{-6}$ mm/cycle the data for oils show the following trends: a) growth rates are faster at low load ratios with higher viscosities, b) growth rates in silicone oil are faster than those in paraffin oil and c) crack growth is slower in both oils than in air or hydrogen. The effect of viscosity on the value of the fatigue threshold ΔK_0 is negligible at $R = 0.75$ whereas at $R = 0.05$ there is a small but finite trend of decreasing ΔK_0 with increasing viscosity in both oils (Fig. 4).

Scanning electron microscopy of the samples tested in oil revealed fracture morphologies similar to that observed in room air, e.g., at near-threshold levels a fine-scale transgranular mode was seen with isolated evidence of intergranular facets (Fig. 5). The extent of fracture surface oxidation determined from Auger measurements, however, was quite different to that observed in gaseous environments (Fig. 6). Whereas the thickness of the near-threshold crack surface oxide layer (Fe_2O_3) reached 0.2 μm at $R = 0.05$ in moist air, measurements in both types of oil indicated oxide films which were only $\sim$ 3 nm thick.

DISCUSSION

Similar to previous work (Suresh, Ritchie, 1982) on the role of gaseous and aqueous environments in influencing fatigue crack propagation in lower strength steels, it is clear from the present study that the effect of viscous environments is also markedly different at near-threshold as opposed to higher growth rates (i.e., below and above $\sim 10^{-6}$ mm/cycle, respectively). Specifically both dehumidified silicone and paraffin oils are shown to give rise to *faster* growth rates at near-threshold levels and *slower* growth rates above 10^{-6} mm/cycle, compared to crack growth in moist air at R = 0.05.

In general terms, such contrasting effects of environment have been rationalized in terms of a competition between two concurrent yet mutually competitive processes, namely corrosion fatigue mechanisms, e.g., hydrogen embrittlement and/or active path corrosion, which *increase* crack growth rates, and crack closure mechanisms, e.g., resulting from oxidation products at the crack tip (oxide-induced) or irregular fracture morphologies (roughness-induced), which *decrease* crack growth rates. Whereas corrosion fatigue mechanisms tend to dominate environmentally-influenced crack growth behavior at higher growth rates and at high load ratios, the role of closure mechanisms become increasingly more important at low load ratios as growth rates approach threshold levels (Suresh, Ritchie, 1982).

In the present study on viscous environments, the observed effects of dehumidified oils can be interpreted in terms of similar considerations, specifically involving three distinct mechanisms, i.e., i) exclusion of moisture from the crack tip thereby minimizing hydrogen embrittlement/metal dissolution processes, ii) exclusion of moisture and oxygen from crack tip thereby minimizing oxide-induced crack closure, and iii) partial penetration of the oil films within the crack to create additional closure from the hydrodynamic wedge effect of the viscous fluid. Since each of these mechanisms has a differing effect on crack growth rates depending upon the regime of crack propagation rate behavior, each mechanism is now discussed in turn.

Since both dry silicone and paraffin oils do not chemically react with steels, the principal function of such media is to act as inert environments for crack growth, as evidenced by the fact that their behavior is similar to that of dry gaseous helium (Figs. 1-3). Thus, the exclusion of moisture from the crack tip region, resulting from immersion in dehumidified oil, will minimize any enhancement in crack growth rates due to metal dissolution (oxidation) at the crack tip (Fig. 6) and/or hydrogen embrittlement from the consequent release of hydrogen. This reduced contribution to crack advance from corrosion fatigue processes will be effective over the entire range of growth rates and load ratios, but will dominate behavior above $\sim 10^{-6}$ mm/cycle since closure mechanisms here are ineffective. Thus it is to be expected that crack growth rates in oil in this regime will be slower than in moist air or hydrogen gas, and be comparable with rates in helium (Figs. 1, 3). Similar explanations in terms of shielding the crack tip from detrimental environmental species have been used in the past to account for the reduction of crack growth rates in several alloys tested in inert liquid environments such as dodecyl alcohol, silicone oils and grease (Ryder, Martin, Abdullah, 1977).

For the current steel at the high cyclic frequencies tested, the exclusion of moisture and oxygen afforded by the oil environment has the reverse effect on crack growth rates at lower, near-threshold levels. As shown in Fig. 6, analysis of crack growth in moist air at low load ratios has revealed oxidation deposits on near-threshold fracture surfaces to be $\sim$ 20-40 times larger than the naturally-occurring oxide thickness. Such oxide films result in significant oxide-induced crack closure when they build up to thicknesses comparable with crack tip displacements. Accordingly, growth rates below $\sim 10^{-6}$ mm/cycle are progressively slower in moist environments at low load ratios due to an increasing influence of oxide-induced closure, since in this steel the near-threshold contribution to crack advance from corrosion fatigue processes is generally small at 50 Hz. The exclusion of moisture and oxygen to the crack tip in oil tests, however, results in minimal crack flank oxidation (Fig. 6) such that near-threshold growth rates in oil exceed those in moist air primarily due to a smaller influence of oxide-induced closure (Figs. 1-4). At high load ratios, however, where closure effects are minimal, growth rates in oil are almost identical to those in air (Fig. 2). The relative effect of oxide-induced crack closure on near-threshold growth rates in dry oil compared to moist air (at R = 0.05) can be appreciated by considering the simple model for such closure of a rigid wedge inside a linear elastic crack (Suresh, Parks, Ritchie, 1982). Here closure is modelled to reduce the stress intensity range from its nominal value ($\Delta K = K_{max} - K_{min}$) to some near tip effective value ($\Delta K_{eff} = K_{max} - K_{cl}$), where K_{cl} the closure stress intensity is given in terms of the maximum oxide wedge thickness d located at distance 2ℓ behind the crack tip and the elastic modulus in plane strain E' (Tada, Paris, Irwin, 1973):

$$K_{cl} \approx \frac{d\ E'}{4\sqrt{\pi \ell}} \quad . \tag{1}$$

Taking $\ell \sim 2$ μm (Suresh, Parks, Ritchie, 1982), the maximum excess oxide (d) in moist air close to ΔK_o is ~ 0.2 μm compared to ~ 3 nm in oil (Fig. 6). From Eq. (1) the corresponding values of K_{cl} at R = 0.05 are ~ 4.5 MPa$\sqrt{m}$ in air and less than 0.1 MPa$\sqrt{m}$ in oil, clearly showing that the driving force for crack extension in oil is far less restricted by oxide-induced closure.

Both protective effects provided by the oil environments described above are essentially independent of oil chemistry and viscosity, provided of course the oil is chemically inert. However, the current results in Figs. 1-4 indicate growth rates to be marginally higher in the higher viscosity oils at both near-threshold (R = 0.05) and higher growth rates. Moreover, at comparable viscosities, paraffin oils showed the faster growth rates near ΔK_o and the slower growth rates above $\sim 10^{-6}$ mm/cycle (at R = 0.05). Such observations can be rationalized by considering the extent of penetration of the oil films into the crack and the resultant closure which arises due to the hydrodynamic wedge effect of the viscous fluid (Endo, Okada, Hariya, 1972). The penetration of the oil is governed by the capillary action of the crack walls and related to the physical properties of the oil (notably the viscosity, surface tension, wetting angle, etc.). The subsequent total hydrodynamic pressure generated by such an oil is then given by (Endo and co-workers, 1972):

$$p = - \frac{6\eta\, x^4}{h^3} \frac{d\theta}{dt} \quad , \tag{2}$$

where h is the crack mouth displacement, $d\theta/dt$ the angular closing velocity of the crack walls, η the absolute viscosity and x the penetration depth ($\propto \eta^{-}$ Using superposition methods, the linear elastic stress intensity resulting from such pressure (K_p) is directly proportional to p, such that at constant frequency, crack length and stress amplitude, K_p will be proportional to the fourth power of the penetration distance. However, as x is dependent upon η^{-1}, K_p will be governed by the inverse third power of viscosity. Since paraffin oils wet steel, the penetration of these fluids into the crack is far more effective than for silicone oils. In fact, based on experimental observations and consideration of the above analyses (Tzou, Suresh, Ritchie, 1983), it appears that oil film penetration is minimal at 50 Hz at near-threshold growth rates, particularly for the 1000 to 60,000 cS viscosity oils, such that crack closure induced by oil pressure at these levels is relatively insignificant. At higher growth rates, on the other hand, although penetration into the crack is still restricted for the higher viscosity oils (at R = 0.05) due to the small crack mouth displacements, partial oil penetration into the crack can result in additional crack closure for oils of lower viscosity. Accordingly, crack growth rates between $\sim 10^{-6}$ to 10^{-5} mm/cycle in the paraffin oils (Figs. 3 and 4) and in the 5 cS silicone oil (Fig. 4) were actually slower than in dehumidified helium gas indicating that both environmental shielding and viscous fluid-induced closure mechanisms were active. Thus, effects of oil chemistry and viscosity on fatigue crack propagation behavior can be rationalized in terms of corrosion fatigue processes, crack closure and hydrodynamic wedging. However, whereas closure induced by corrosion debris is significant at near-threshold levels, for the frequencies and oils tested, closure induced by viscous oil pressure is only relevant at higher growth rates.

CONCLUSIONS

Silicone and paraffin oils are shown to have significant but opposite effects on fatigue crack growth behavior of A542-3 steel above and below 10^{-6} mm/cycle. In the near-threshold region ($< 10^{-6}$ mm/cycle), crack growth rates at low load ratios are *higher* and threshold ΔK_0 values lower in oil than in moist air although little effect is seen at high load ratios. There is a small but finite decrease in ΔK_0 with increasing viscosities. At higher growth rates (10^{-6} to 10^{-3} mm/cycle), growth rates in oil are *lower* than those in moist air or dry hydrogen, with the lowest viscosity and paraffin oils showing the slowest rates. Such complex crack growth characteristics in dry viscous environments can be rationalized in terms of three mutually competitive mechanisms: suppression of corrosion fatigue processes, minimization of oxide-induced closure and development of a wedging action of the oil due to partial penetration within the crack.

ACKNOWLEDGMENTS

The work was supported by the Director, Office of Energy Research, Office of Basic Energy Sciences, Materials Science Division of the U.S. Department of Energy, under Contract No. DE-AC03-76SF00098.

REFERENCES

Cordiano, H. V., E. P. Cochran, and R. J. Wolfe (1956). A study of combustion resistant hydraulic fluids as ball bearing lubricants. Lubric. Engng., 12, 261-274.

Endo, K., T. Okada, and T. Hariya (1972). Fatigue crack propagation in bearing metals lining on steel plates in lubricating oil. Bull. JSME, 15, 439-445.

Endo, K., T. Okada, K. Komai, and M. Kiyota (1972). Fatigue crack propagation of steel in oil. Bull. JSME, 15, 1316-1323.

Galvin, G. D., and H. Naylor (1965). Effect of lubricants on the fatigue of steel and other metals. Proc. Instn. Mech. Engrs., 179, 857-875.

Ryder, D. A., M. Martin, and M. Abdullah (1977). Some factors influencing stage I fatigue-crack growth. Metal Science, 11, 340-344.
Suresh, S., and R. O. Ritchie (1982). Mechanistic dissimilarities between environmentally-influenced fatigue crack propagation at near-threshold and higher growth rates. Metal Science, 16, 529-538.
Suresh, S., D. M. Parks, and R. O. Ritchie (1982). Crack tip oxide formation and its influence on fatigue thresholds. In J. Bäcklund (Ed.), Fatigue Thresholds, Vol. 1, EMAS Ltd., Warley. pp. 391-408.
Suresh, S., G. F. Zamiski, and R. O. Ritchie (1981). Oxide-induced crack closure: An explanation for near-threshold corrosion fatigue behavior. Metall. Trans. A, 12A, 1435-1443.
Tada, H., P. C. Paris, and G. R. Irwin (1973). In Stress Analysis of Cracks Handbook, Del Research Corp., Hellertown, PA.
Tzou, J. L., S. Suresh, and R. O. Ritchie (1983). Role of oil chemistry and viscosity in influencing fatigue crack propagation in viscous fluids. Metall. Trans. A, submitted to.

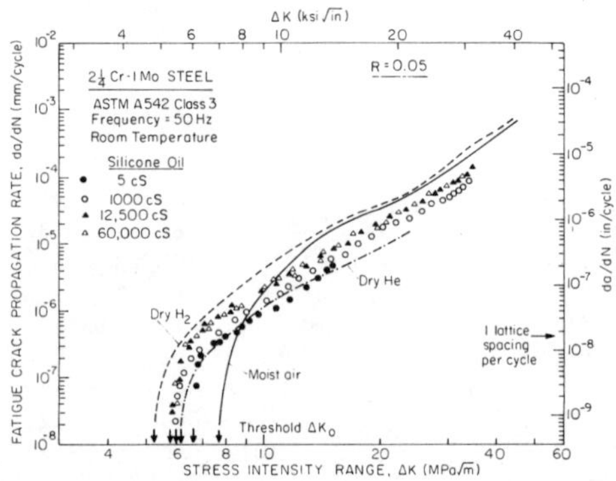

Fig. 1. Fatigue crack growth behavior in silicone oils (5-60,000 cS viscosities) at R = 0.05 compared to data for moist air and dry gaseous hydrogen and helium.

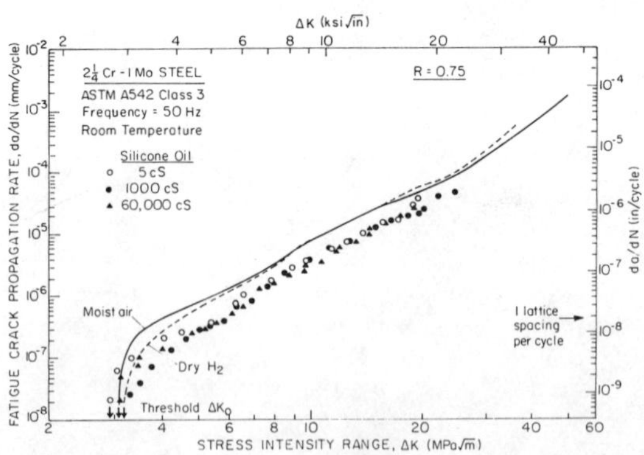

Fig. 2. Fatigue crack growth behavior in silicone oils (5-60,000 cS viscosities) at R = 0.75 compared to data for moist air and dry gaseous hydrogen.

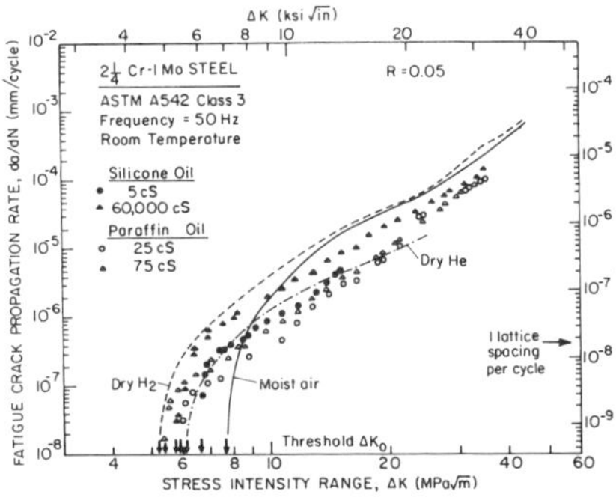

Fig. 3. Comparison of fatigue crack propagation behavior at R = 0.05 in high and low viscosity silicone and paraffin oils.

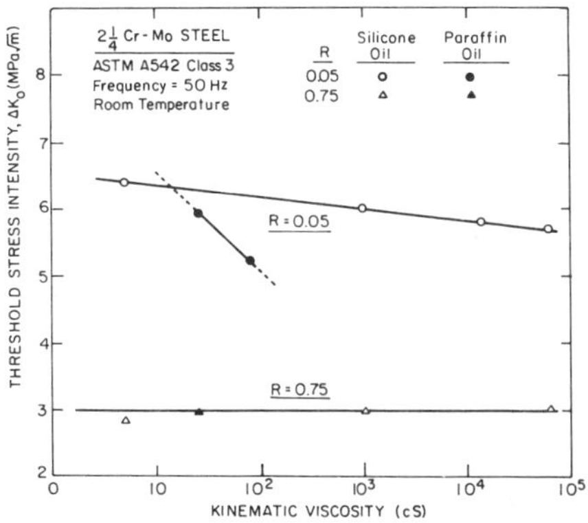

Fig. 4. Variation of threshold stress intensity range (ΔK_o) at R = 0.05 and 0.75 with kinematic viscosity for silicone and paraffin oils.

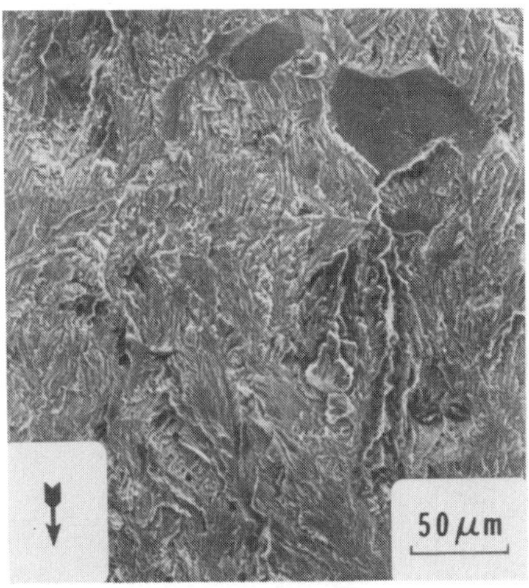

Fig. 5. Fractography of near-threshold fatigue in oil, showing fracture surface in 25 cS paraffin oil at $\Delta K \approx 6.0$ MPa$\sqrt{m}$ (R = 0.05). Arrow indicates crack growth direction.

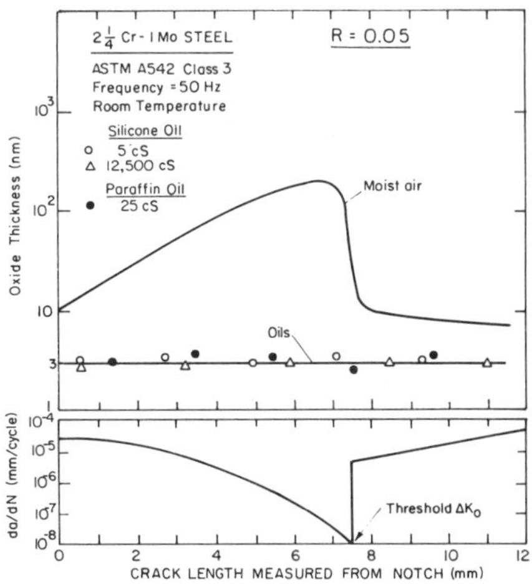

Fig. 6. Variation in thickness of crack surface oxide debris with crack length and growth rate for oil environments compared to previous results in moist air.

AN INVESTIGATION OF VOID DISTRIBUTION IN NECKING REGION AND TENSILE FRACTURE CHARACTERISTICS OF DUAL-PHASE STEEL 1010

B. D. Hong and T. C. Lei

Department of Metals and Technology, Harbin Institute of Technology, Harbin, China

ABSTRACT

The void distribution in necking region and tensile fracture characteristics of a dual-phase steel containing 0.11% C were studied by measuring the void density, average void size and their changes with the distances from fractured surface. Experimental results show that quenching and tempering at various temperatures has obvious effect on the void distribution and fracture processes. The data obtained indicate that the fracture characteristics of the steel can be appreciably changed by varying the conditions of heat treatment.

KEYWORDS

Dual-phase steel; necking region; void distribution; void density; average void diameter; fracture.

INTRODUCTION

Recently, the plain carbon and low alloy steels with martensite plus ferrite dual-phases have been widely studied for their good formability and high work-hardening properties (Davies, 1978; Koo and Thomas, 1979; Lei and Shen, 1982). Some investigators (Tamura and others, 1973; Hong and Lei, 1982) have found that the volume fractions and the strength ratios of the two phases in dual-phase steels can be changed by varying the conditions of heat treatment, so the matching between strength and ductility of the steels may be widely altered. The strain behavior of the martensite and ferrite grains in the region of uniform elongation of tensile specimens have been studied earlier by photoetching technique (Hong and Lei, 1982), but little information has been reported concerning the after-necking phenomenon. The aim of the present study is to clarify the effect of microstructure formed after various heat treatmeats on void distribution in necking region and tensile fracture characteristics of dual-phase steel 1010.

EXPERIMENTAL PROCEDURES

The material used was steel 1010 with 0.11%C. Specimens 4x6 mm in cross-section and 50 mm gage length underwent a step-quenching procedure, with heated specimens to 950°C for 15 min, intermediately cooled to 770, 800, or 830°C with 15 min holding and then quenched into 10% NaOH water solution. The volume fraction of martensite phase was determined by a structure analyzer EPIQUANT and is equal to 16, 25 and 40%, respectively. Part of the quenched specimens were tempered at 150, 300 and 450°C for 1.5 h.

A scanning electron microscope model S-550 was used for observing the tensile fractography and the void distribution in the necking region of fractured specimens. The planar void density and average void diameter were measured on transverse sections by a layer-by-layer grinding and polishing technique in order to obtain the changes of these two parameters as a function of the distance from the fractured surface.

RESULTS AND DISCUSSION

Void Distribution

Some of the micrographs taken from transverse sections in the necking region are shown in Fig.1, which clearly indicates the predominate distribution of voids in ferrite phase for quenched and low temperature tempered specimens. The void distribution parameters, including void density (ρ) and average void diameter (d) vs distance from the fractured surface (Z) are shown in Fig.2. It can be seen that the void density of both quenched and tempered specimens decreases with distance Z. This is obviously consistent with the decreased degree of plastic deformation in sections apart from the fractured surface. The void density of tempered specimens is much higher in the vicinity of fractured

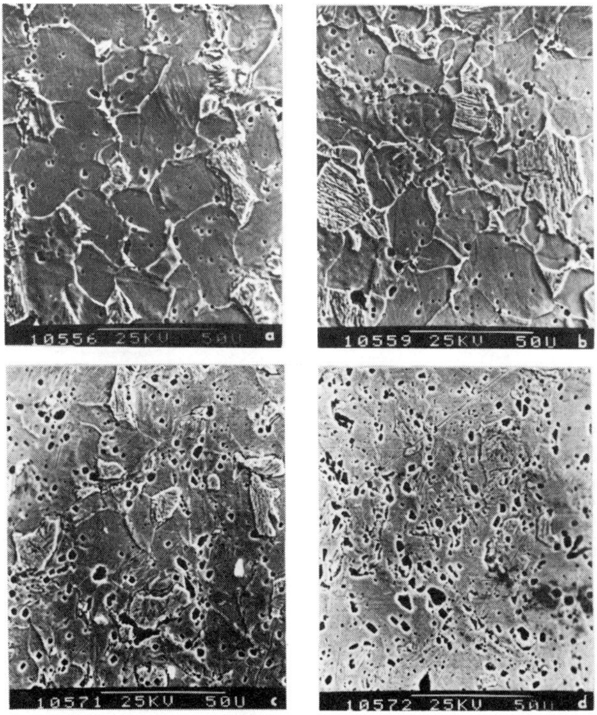

Fig.1. Void distribution on transverse sections of dual-phase steel 1010 after various heat treatment:
a, 930-770°C quenched, Z=1 mm;
b, 930-830°C quenched, Z=1 mm;
c, 930-800°C quenched and 150°C tempered, Z=0.1 mm;
d, 930-800°C quenched and 450°C tempered, Z=0.1 mm.

surface (Z=0.1mm) by comparison with the quenched ones, but much lower in farther sections (Z=2 and 4mm), resulting in a more steeper decrease than that of the later. Such a distribution of void density is caused, probably, by the increased and more concentrated plastic deformation in the necking region, resulting in a decreased length of necking region and increased reduction of area of fractured surface for the tempered specimens.

It must be noted that the 930-830°C step-quenched specimens (curve 2 in Fig. 2) have higher void densities than those of 930-770°C quenched ones (curve 1). The carbon content and volume fraction of ferrite (which can be estimated by using the Fe-C phase diagram) of 930-830°C quenched specimens are both less than the 930-770°C quenched. This makes larger plastic deforma-

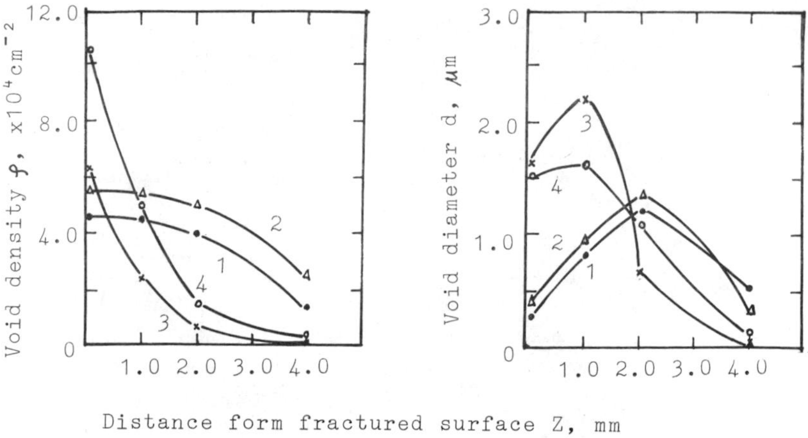

Fig.2. Planar void density (ρ) and average void diameter (d) on transverse sections of dual-phase steel 1010 vs distance from fractured surface (Z):
1, 930-770 °C quenched; 2, 930-830 °C quenched;
3, 930-800 °C quenched, and 150 °C tempered;
4, 930-800 °C quenched and 450 °C tempered.

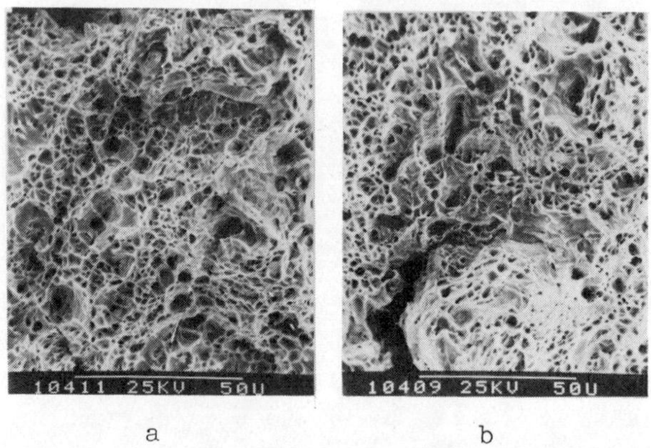

Fig.3. SEM fractographs of quenched dual-phase steel 1010:
a, 930-770 °C quenched (16% martensite);
b, 930-830 °C quenched (40% martensite).

tion of ferrite in the former and, probably, leads to more nucleation sites of voids due to nonmetallic inclusions with smaller sizes. The much higher void densities of tempered specimens (curves 3, 4) than quenched (curves 1, 2) in the vicility of fractured surface is, certainly, by the additional nucleation sites caused by precipitation of carbide particles during tempering.

The average void diameter, as shown in the right part of Fig. 2, changes with the distance from fractured surface through a maximum value for both quenched and tempered specimens. This results from the rapid reduction of area in fractured surface causing serious extention of voids along specimen axis and thus compression in transverse direction. The maximum void diameter appears in a section of shorter distance for tempered (Z=1mm) than quenched (Z=2mm) specimens. Higher values of void diameter for tempered specimens in the vicility of fractured surface may be explained by the existance of larger particles of carbide precipitates coalesced during tempering.

Fracture Characteristics

The fractographic pictures of 930-770 and 930-830 °C step-quenched specimens are shown in Fig.3. It is clear that the fracture is mainly of the dimpled type of ferrite phase with a small part of quasicleavage areas which agrees with the volum fraction of martensite. The tempered specimens gave entirely dimpled fracture (not shown). From the observations of void distribution

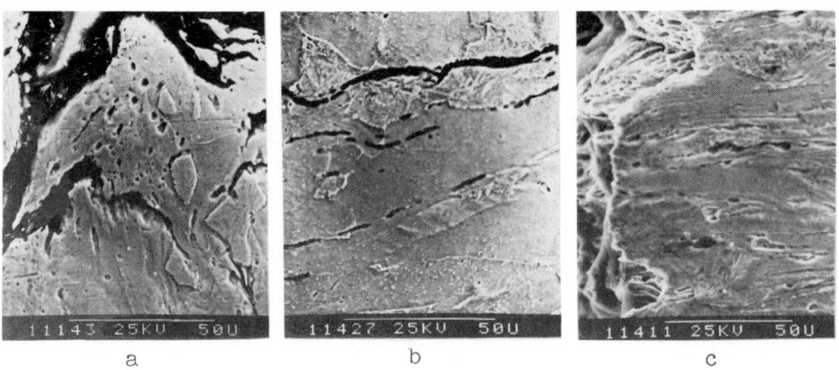

Fig.4. SEM microstructures on longitudinal sections of specimens tempered at:(a)150 °C, (b) 300 °C and (c) 450 °C. Photograph (b) was taken at Z=0.2 mm.

and crack propagation on longitudinal sections as shown in Fig. 4, it may be seen that for quenched and low temperature (150°C) tempered specimens (Fig.4a) the microcracks are formed by the connection of voids within or along the martensite-ferrite interface but also on the ferrite side. Such observations have been reported earliar (Lei and Shen, 1982) for dual-phase steel 1020. For specimens tempered at higher temperatures (300 and 450°C) voids can nucleate in both ferrite and martensite grains (Fig.4b and c) and the fracture is completely of the dimpled type, and the crack path frequently goes through martensite (in fact, it is now troostite) regions.

CONCLUSIONS

1. In the necking region of tensile spesimens of dual-phase steel 1010 the maximum void density is found in the vicility of fractured surface, whereas the maximum void diameter appears at a certain distance from it.

2. The distribution of voids, including their density and average diameter, are strongly affected by the conditions of heat treatment. Larger void densities and diameters for tempered specimens are related to the decreased strength and increased plasticity of the two phases caused by tempering.

3. The quenched dual-phase steel 1010 are fractured during tensile tests by a mixed mode with predominate part of dimples and quasi-clevage areas according to the volume fraction of martensite. The tempered specimens fractured completely by a dimple mode.

REFERENCES

Davis, R. G. (1978). Metall. Trans., 9A, 41.
Hong, B. D. and T.C.Lei(1982). J.Harbin Inst. Technol., No.4,82.
Koo, J. Y. and G. Thomas(1979). Metal Progress, 116, No.4,66.
Lei, T.C. and S.P. Shen (1982). Acta Met. Sinica, 18, 75.
Tamura, E.,Y.Tomota, Y. Yamaoka, S. Kanatani, M. Ozawa and A. Akao (1973). Iron and Steel (Japan), No.3,454.

FRACTURE TOUGHNESS AND X-RAY DIFFRACTION OBSERVATION OF FRACTURE SURFACE OF DUCTILE CAST IRON

Z. Yajima*, Y. Hirose and K. Tanaka*****

*Faculty of Enginering, Kanazawa Institute of Technology, Kanazawa, Japan
**Faculty of Education, Kanazawa University, Kanazawa, Japan
***Faculty of Engineering, Kyoto University, Kyoto, Japan

ABSTRACT

Tha fracture toughness tests of ductile cast iron (JIS FCD 60) were conducted at room temperature by using blunt-notched compact tension specimens with different notch-tip radii. The dc potential method was successfully used to detect crack extension. The effect of the notch-tip radius ρ on the fracture toughness K_i was divided into three regions: Region I where ρ is smaller than the radius of spheroidal graphite ρ_G(= 30 μm), Region II where ρ is above ρ_G and below 2 mm, and Region III where ρ is larger than 2 mm. K_i is constant in Region I and III, while it increases linearly with $\sqrt{\rho}$ in Region II. Based on electron fractography and X-ray fractography, micromechanisms of crack extension as a function of notch-tip radius in cast iron are proposed.

KEYWORDS

Ductile cast iron; fracture toughness; notch effect; fractography; X-ray diffraction; plastic strain; fracture strain

INTRODUCTION

Cast iron is widely used as machine parts in fields. Fracture mechanics analysis of fracture initiation from existing flaws or notches in cast iron has not yet been well developed. In the present paper, the fracture toughness tests of ductile cast iron (JIS FCD 60) were conducted by using compact tension (CT) specimens with blunt notches. Crack extension is detected by the electrical potential method. The effect of notch-tip radius on the fracture toughness of cast iron was determined experimentally. Based on fractographic observations by scanning electron microscopy and X-ray diffraction technique, the mechanisms of fracture were discussed as a function of notch-tip radius.

EXPERIMENTAL PROCEDURE

Experimental Conditions

The specimen was machined from as-cast plates of spheroidal graphite cast iron (JIS FCD 60), which were 50 mm thick, 300 mm wide and 250 mm long. The chemical compositions of the material are as follows: (wt%) 3.70C, 2.40Si, 0.40Mn, 0.04P, 0.018S. The material has a yield strength of 392 MPa and a tensile strength of 579 MPa. The matrix structure of the material is mostly ferrite with a small amount of pearlite. The grain size of ferrite is 90 μm. The diameter of spheroidal graphite is 60 μm.

The blunt notch was introduced to CT specimens with an electro-discharge machine. The notch-tip radius was varied from 0.12 to 3 mm. Some specimens were pre-cracked by fatigue according to ASTM standard E399-78 (1978). The details of specimen preparation and dimensions are given elsewhere (Yajima, Hirose and Tanaka, 1983a).

Fracture Toughness Test

The dc electrical potential method (EPM) was used to detect the point of crack extension. A direct current of 30A was flown in and the potential difference was measured at notch mouth (Yajima, Hirose and Tanaka, 1983a,b). The data obtained by this method were compared with those by *R*-curve method given in ASTM standard E813-81 (1981).

Inspection of Fracture Surface

The topography of the fracture surface was examined with scanning electron microscopy (SEM). The distribution of the half-value breadth of X-ray diffraction profiles was measured with an X-ray stress analyser. X-rays of Cr-K_α radiation was irradiated on an area of 1 mm width and 10 mm length at the central region of the fracture surface close to a pre-crack or a notch, and the diffraction of Fe (211) plane was recorded (Yajima, Hirose and Tanaka, 1983b).

EXPERIMENTAL RESULTS

Detection of Crack Initiation Point

In Fig. 1 are shown the curves of displacement u and electrical potential change ΔV plotted against load P for a CT specimen with a fatigue pre-crack. Several specimens were unloaded from loads such as indicated by a, b, c and d, and then fractured by fatigue. Detail observation of fracture surfaces indicated that crack extension occurred at the inflection point B shown in P-ΔV curve in the figure (Yajima, Hirose and Tanaka, 1983a). Point A corresponds to the opening of a pre-crack. The J value at B was calculated from P-u curve (ASTM E813, 1981) to be 0.45×10^4 N/m. Figure 2 shows the *R*-curve constructed based on ASTM E813 (1981). As a blunting line, the vertical line of the critical stretched zone width S_c(= 6 μm) is used. The J value at the intersection, J_{IC}, is 0.44×10^4 N/m, which agree well with the value determined by the EPM method. In Fig. 2, the *R*-curve of bluntly notched CT specimen is also shown. No stretched zone was detected on the fracture surface. In the latter tests, the EPM method was used to detect the crack extension point.

The critical stress intensity factor K_i is used as the fracture toughness because P-u curves were linear before crack extension.

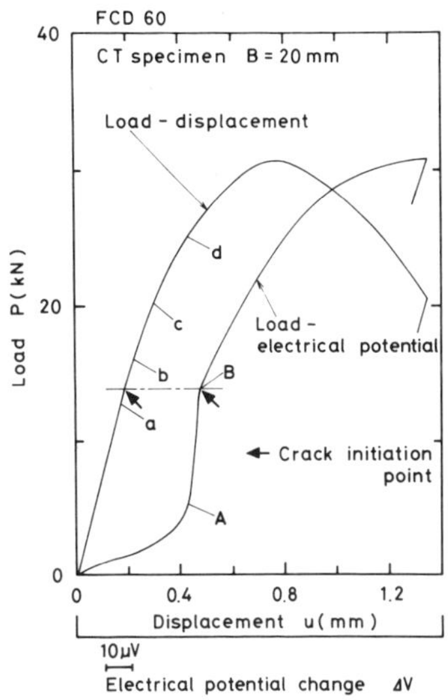

Fig. 1. Load versus load point displacement and electrical potential difference.

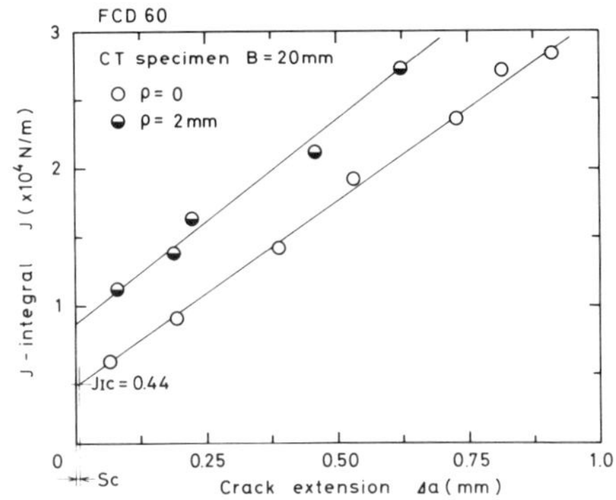

Fig. 2. Variation of J-integral with crack extension.

Relation between Fracture Toughness and Notch-Tip Radius

Figure 3 shows the variation of K_i with the notch-tip radius ρ. The value of K_i for blunt notches was calculated by assuming $\rho = 0$. The relation between K_i and ρ can be divided into three regions of I, II and III. The notch-tip

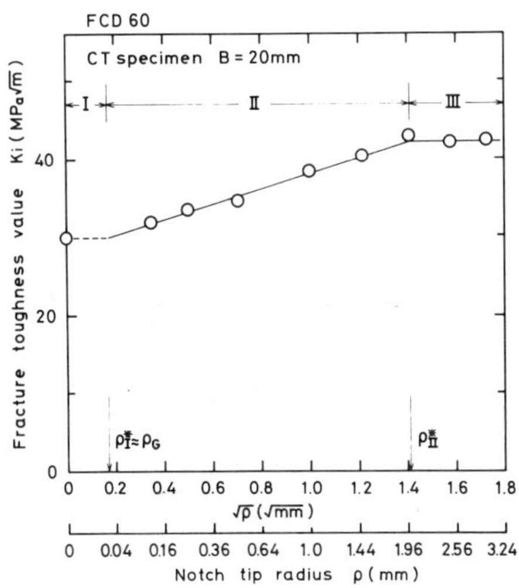

Fig. 3. Relation between fracture toughness and square root of notch-tip radius.

radius ρ_I* at the transition from Region I to II is almost identical to the graphite (ρ_G = 30 μm). The notch radius ρ_{II}* at the boundary between Regions II and III is 2 mm.

Electron Fractography

Scanning electron micrographs taken from fracture surfaces in the vicinity of the tip of a pre-crack or a notch are shown in Fig. 4, where fracture occurs from left to right. For the case of a pre-cracked specimen, the stretched zone of about 6 μm width is seen between fatigue fracture surface and monotonic fracture surface. When ρ is above 0.12 mm, no stretched zone is detected. In the case of ρ = 2.5 mm, a microcrack can be seen on the notch surface as indicated by the arrow in Fig. 4(d). The typical feature of slow fracture surfaces is dimples in the matrix, fracture of graphite and separation of graphite from the matrix. Referring to Fig. 3, the crack extension is accompanied by stretched zone formation in Region I, and several microcracks are formed along the periphery of the notch in Region III.

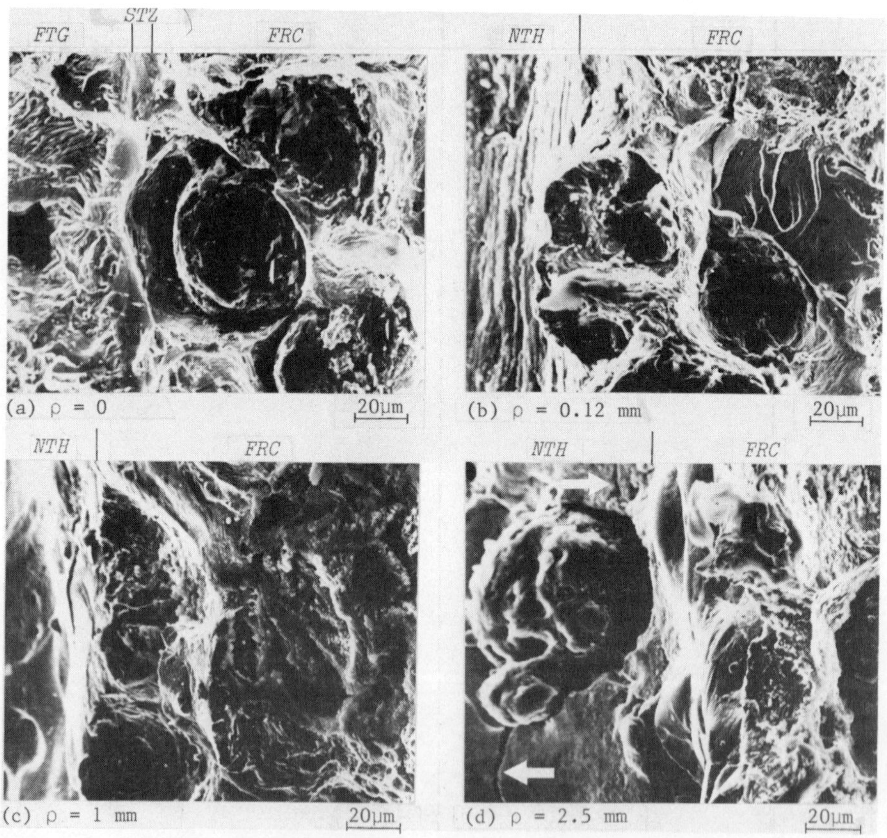

Fig. 4. Scanning electron micrographs of fracture surfaces. (Crack growth direction is from left to right).
FTG : Fatigue surface. *STZ* : Stretched zone.
NTH : Notch face. *FRC* : Fracture toughness surface.

X-Ray Observation of Fracture Surfaces

The distribution of the half-value breadth of X-ray diffraction profiles beneath the slow fracture surface is shown in Fig. 5 where the half-value breadth B is divided by its initial value B_o. B/B_o tends to decreases approaching one as the depth increases. In the previous study (Yajima, Hirose and Tanaka, 1983b), the depth at $B/B_o \approx 1$ indicates the size of the plastic zone and its relation to K_i agrees with the fracture mechanics relation. The value of B/B_o is a unique function of plastic strain ε_p as

$$B/B_o = 0.77 \log \varepsilon_p + 2.10 \quad (1)$$

This relation was found to be independent of the loading mode when von Mises equivalent strain is taken as ε_p in the above equation (Tanaka and co-workers, 1980; Yajima, Hirose and Tanaka, 1983b). The plastic strain on the fracture surface can be evaluated from the value of B/B_o by using equation (1). The fracture surface strain ε_f thus evaluated is plotted as a function of ρ in Fig. 6. It is noted that the variation can also be divided into three region like K_i-$\sqrt{\rho}$ relation. ε_f decreases with increasing $\sqrt{\rho}$ in Region II, and it is constant in Regions I and III.

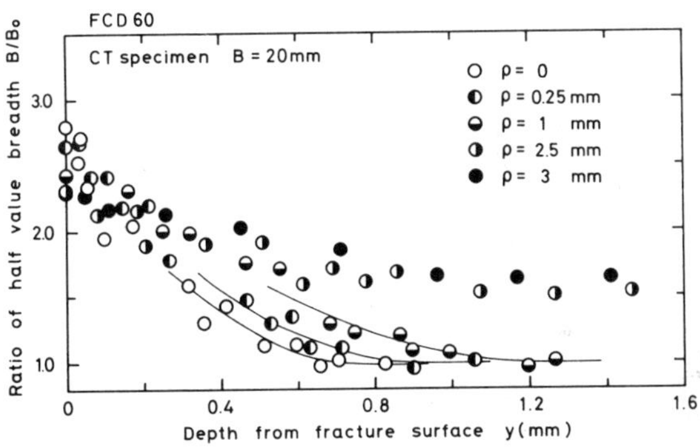

Fig. 5. Half-value breadth distribution near fracture surface.

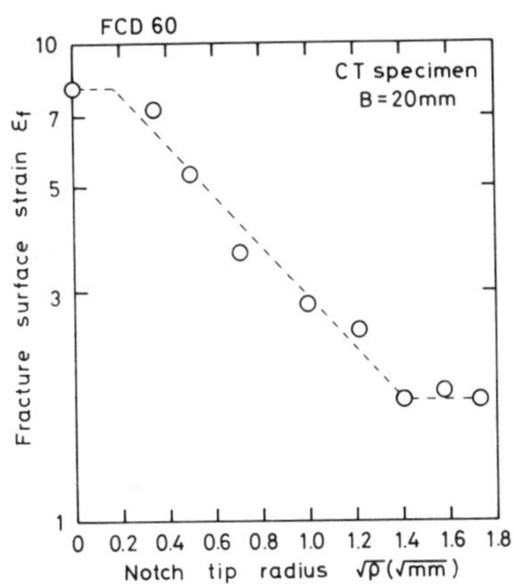

Fig. 6. Relation between fracture surface strain and square root of notch-tip radius.

DISCUSSION

The relation between K_i and $\sqrt{\rho}$ obtained for the present material is very different from that obtained for a high strength steel (AISI 4340) by the authors (Yajima, Hirose and Tanaka, 1983c). In the cases of high strength steels tempered at several temperatures ranging from 200 to 600°C, the value of K_i increases proportional to $\sqrt{\rho}$ when ρ is above the effective notch-tip radius ρ_{eff}.

The elastic maximum stress σ_{max} at the tip of deep notches with the tip radius ρ is expressed by (Creager and Paris, 1967)

$$\sigma_{max} = 2K / \sqrt{\pi\rho} \tag{2}$$

where K is the stress intensity factor calculated by assuming $\rho = 0$. When plastic deformation occurs at the notch-tip in elastic perfectly plastic material, the maximum plastic strain ε_{max} is given by the following equation proposed by Rice (1968):

$$\varepsilon_{max} = (\ 3/4\ )(\ 1/E\ )(\ K^2/\rho\sigma_Y\ ) \tag{3}$$

where E and σ_Y are Young's modulus and the yield strength, respectively. It should be noted that both σ_{max} and ε_{max} are determined by a parameter of $K/\sqrt{\rho}$. Therefore, if the fracture occurs at a constant value of σ_{max} or ε_{max}, K_i becomes proportional to $\sqrt{\rho}$. The results of high strength steels agree with this prediction. On the other hand, both decrease as ρ increases in cast iron. In Fig. 7, σ_{max} and ε_{max} calculated from the data given in Fig. 3 by using equations (2) and (3) are plotted as a function of ρ.

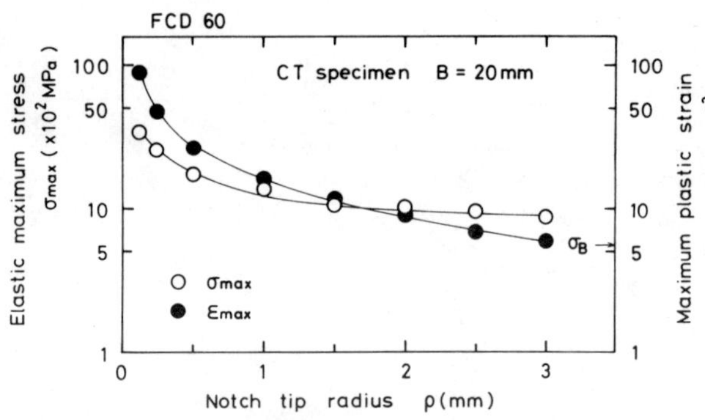

Fig. 7. Relation between elastic maximum stress, maximum plastic strain and notch-tip radius.

Figure 8 illustrates the mechanisms of crack extension from various notches. The crack extension in Region I, where ρ is less than the radius of graphite, takes place after the stretching of the pre-existing flaw tip like in high strength steels. The effective notch-tip radius in cast iron is equal to the radius of graphite. In Region III, the K_i value is invariable with ρ; σ_{max} is nearly equal to the tensile strength σ_B (see Fig. 7). This corresponds to the SEM finding that cracks are formed at several locations on the notch periphery surface. Fracture may start from the weakest graphite in the highly stressed zone near notch bottom.

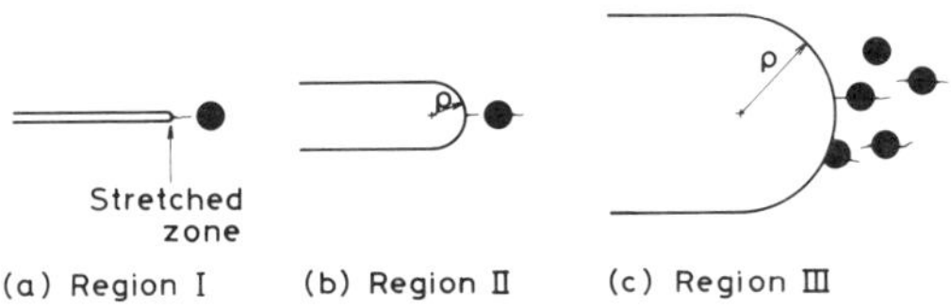

Fig. 8. Model for crack extension from notch.

Region II is regarded as a transitional region. Fracture strain is low in Region III and high in Region I. In Region II, ε_{max} calculated by equation (3) decreases with increasing ρ as shown in Fig. 7. This trend is also clearly seen in the variation of the fracture surface strain measured by the X-ray diffraction technique (see Fig. 6). The fracture surface strain further substantiates the changing fracture strain with the notch-tip radius. To analyse the transitional behavior, the statistical treatment of graphite in the highly stressed region in the vicinity of the tip of a pre-crack or notch will be necessary.

CONCLUSIONS

(1) The dc potential method can be applied to detect the point of crack extension in bluntly notched compact tension specimens of cast iron. The critical stress intensity factor K_i for crack extension determined by the potential method is almost identical to the value determined by the R-curve method.
(2) The relation between K_i and notch-tip radius ρ can be divided into three regions: Region I where ρ is smaller than the radius of spheroidal graphite ρ_G(= 30 μm), Region II where ρ is above ρ_G and below 2 mm, and Region III where ρ is larger than 2 mm. K_i is constant in Region I and III, while it increases linearly with $\sqrt{\rho}$ in Region II.
(3) The stretched zone was formed at the pre-crack tip in Region I and several microcracks were formed along the periphery of the notch in Region III. The fracture surface strain measured by the X-ray diffraction method was large in Region I and low in Region III. In Region II, it decreased with increasing notch-tip radius. Based on the observation results, micromechanisms of crack extension as a function of notch-tip radius in cast iron are proposed.

REFERENCES

ASTM Standard, Part 10, E399-78 (1978).
ASTM Standard, Part 10, E813-81 (1981).
Yajima, Z., Y. Hirose, and K. Tanaka (1983a), "Fracture toughness of blunt-notched CT specimens of ductile cast iron," J. Soci. Mat. Sci. Japan, 32. Accepted for publication.
Yajima, Z., Y. Hirose, and K. Tanaka (1983b), "X-ray diffraction observation of fracture surfaces of ductile cast iron," Advances in X-Ray Analysis, 26. Accepted for publication.
Yajima, Z., Y. Hirose, and K. Tanaka (1983c), "X-ray diffraction study on fracture surface in fracture toughness tests of blunt notched CT specimen of high strength steel," J. Soci. Mat. Sci. Japan, 32. Accepted for publication.
Tanaka, K., K. Fujiyama, Y. Nakagawa, K. Nakamura, and Y. Kawabe (1980). J. Soci. Mat. Sci. Japan, 29, 710-716.
Creager, M., and P. C. Paris (1967). Int. J. Frac., 3, 247-252.
Rice, J. R. (1968). "Fracture," H. Liebowitz, ed., II, 191-311, Academic Press, New York.

STRESS AND STRAIN FIELDS AROUND INCLUSIONS IN A PLASTICALLY DEFORMING MATRIX

R. D. Thomson and J. W. Hancock

Department of Mechanical Engineering, University of Glasgow, Scotland, UK

ABSTRACT

Fracture often starts in the vicinity of inclusions in a deforming matrix where the local stress and strain conditions may lead to either failure of the inclusion/matrix interface or of the particle itself. The solutions to this problem depend on the way in which the mechanical properties of the material are idealised and on the chosen mathematical techniques. Analytic solutions are available for an elastic inclusion in an elastically deforming matrix but for plastic deformation it is necessary to resort to numerical analyses. Numerical solutions are presented for spherical rigid and elastic inclusions in elastic/plastic matrices.

KEYWORDS

Inclusion, interface, radial stress

ELASTIC SOLUTIONS

Although solutions for spherical and cylindrical inclusions may be derived as special cases of Eshelby's (1957) results, it is more convenient to refer to earlier work by Goodier (1933) whose results show that the maximum radial stress on the interface of a cylindrical inclusion under a remote uniaxial tensile stress σ occurs in the loading direction and has a magnitude

$$\sigma_{rrmax} = 1.5\sigma.$$

The solution for a general multiaxial loading can be obtained from the uniaxial solution by superposition. The maximum interfacial radial stress for any plane strain triaxial loading may be obtained by summing the pure shear field and an appropriate hydrostatic component σ_m to give

$$\sigma_{rrmax} = 1.15\bar{\sigma} + \sigma_m$$

where $\bar{\sigma}$ is the flow stress. For a rigid spherical inclusion in uniaxial tension, Goodier (1933) finds that

$$\sigma_{rrmax} = 2\sigma$$

due to a remote uniaxial stress σ. This can also be generalised to axisymmetric loading by superimposing an appropriate hydrostatic stress.

ELASTOPLASTIC SOLUTIONS

The problem of a cylindrical elastic inclusion in hardening and nonhardening elastic/plastic matrices was addressed by Orr and Brown (1974) whose results imply that the stress distribution around the interface for triaxial loading is the sum of the stresses due to the pure shear component and the hydrostatic or mean component. This numerical result is of considerable practical use. A useful feature of the Orr and Brown (1974) solution is their plot of the directions of maximum shear strain around the inclusion from which the form of a possible slip-line field may be inferred. This slip-line field predicts a maximum interfacial radial stress of

$$\sigma_{rrmax} = 0.5\pi k = 0.9\sigma_0$$

in the direction of the maximum principal stress. Superposition of a hydrostatic stress does not change the nature of the slip-line field for an incompressible material and leads finally to an expression of the form

$$\sigma_{rrmax} = 0.9\sigma_0 + \sigma_m$$

The finite element analysis of Argon, Im and Safoglu (1975) gives the maximum interfacial radial stress in pure shear k as

$$\sigma_{rr} = 1.5k$$

in accord with the $(\pi/2)k$ from the slip-line field. In contrast with the work of Orr and Brown (1974), this result is interpreted as a steady state solution. Following McClintock and Rhee (1962), strain hardening materials are considered to be intermediate between non-hardening and purely elastic materials. Argon, Im and Safoglu (1975) then replace the shear stress k by the equivalent flow stress to predict a maximum value

$$\sigma_{rrmax} = \bar{\sigma} + \sigma_m$$

NUMERICAL ANALYSIS

In the present work, the problem of a spherical inclusion in a matrix subject to large remote plastic strains has been analysed using the MARC finite element program described elsewhere (Thomson and Hancock (1983)). The finite element model was subjected to increments of displacement controlled uniaxial tensile loading along the remote boundary parallel to the x-axis, reaching large final remote plastic strains of the order of several hundred times the initial yield strain. Solutions were obtained for a spherical inclusion under remote uniaxial tensile loading for different combinations of inclusion and matrix response.

RESULTS

The numerical results typically showed the greatest stresses and strains in elements near the interface. Significant stress and strain gradients only developed near the inclusion-matrix interface which justified the representation of an isolated inclusion in an infinite matrix by a remote boundary at 6 times the inclusion radius. Yielding first occurred at the pole of the inclusion at a small distance from the interface. The resulting plastic zone spread with each increment of remotely applied load but the polar element did not yield until some time after its neighbours. This plastic zone extends to leave an elastic lobe at about 60^{0} to the loading direction before full plasticity is finally reached (fig 1).

No large plastic strain concentrations were found on the interface and at large remote plastic strains the maximum plastic strain over the whole field was found to occur a small distance away from the interface in the loading direction. The maximum plastic strain on the interface developed at approximately $\pi/4$ to the loading direction. Figure 2 compares the effective plastic strain ($\bar{e}^{p}$) in a number of significant elements for a power-hardening matrix. From fig 2 it is apparent that the plastic strain became progressively smaller as the pole of the inclusion was approached along the loading direction. These moderate strain concentrations became almost constant at large remote plastic strains, tending, for example to approximately 1.1 times $\bar{e}^{p}_{\infty}$ at $\pi/4$ around the interface and to 0.8 in the polar element.

The maximum interfacial normal or radial stress was found to develop initially at the pole of the inclusion. As the deformation continued however, the greatest interfacial radial stress developed in the adjacent element. This feature is also shown in the work of Orr and Brown (1974) and Argon, Im and Safoglu (1975). Figure 3 shows the maximum radial interfacial stress for the power-hardening matrix normalised with respect to the remote flow stress. The mean stress continued to rise as the uniaxial load was applied but following an observation by Argon on the results of Orr and Brown (1974) the maximum interfacial radial stress was assumed to be of the form

$$\sigma_{rrmax} = C\bar{\sigma}_{\infty} + \sigma_{m\infty}$$

in which the appropriate remote mean stress $\sigma_{m\infty}$ is simply superimposed on the local interfacial stress field resulting from a purely deviatoric remote stress state of magnitude $\bar{\sigma}_{\infty}$. The maximum interfacial radial stress was then calculated for a purely deviatoric remote stress field. These calculations showed that for increasing remote strain the interfacial radial stress concentration, C, is not a constant. This is in contrast to the approximation of Argon, Im and Safoglu (1975) that the interfacial radial stress concentration is constant for non-linearly hardening materials.

The nature of the inclusion/matrix interface in Swedish Iron is of relevance to experimental work by Thomson and Hancock (1983) and has been investigated by a finite element analysis of an elastic inclusion in an elastic/strain-hardening matrix. Again the radial stress concentration on the interface does not reach a steady state. The simplest expression which gave a reasonable description of the interfacial radial stress concentration in this case was found to comprise a quadratic segment and two linear segments such that

$$C = -3.89.10^{-4}(\bar{e}^{p}/e_{0})^{2}+4.83.10^{-2}(\bar{e}^{p}/e_{0})+0.8 \qquad 0<(\bar{e}^{p}/e_{0})< 60$$

$$C = 9.79.10^{-3}((\bar{e}^{p}/e_{0}) - 60) +2.3 \qquad 60<(\bar{e}^{p}/e_{0})<530$$

$$C = 1.40.10^{-2}((\bar{e}^{p}/e_{0}) -530) +6.9 \qquad 530<(\bar{e}^{p}/e_{0})$$

DISCUSSION

In the large strain solutions presented, the maximum interfacial radial stress concentration continued to rise as the remote strain increased. A similar result was reached by Orr and Brown (1974) in contrast to the conclusions of Argon, Im and Safoglu (1975). Their analysis was however restricted to effective plastic strains of the order of the matrix yield strain and close examination of fig 3 shows that at such strains the interfacial stress concentration passes through a minimum turning point and thereafter rises monotonically. This minimum turning point and the slow approach to steady state are also features of the analysis of non-hardening materials and of Swedish Iron. For this low-yield, strongly strain hardening material, the development of the interfacial radial stress has been approximated by the piecewise analytic expressions presented. These expressions have been used by Thomson and Hancock (1983) to determine the interfacial conditions leading to void nucleation at second-phase particles.

REFERENCES

Argon A S, Im J and Safoglu R (1975). Cavity formation from inclusions in ductile fracture. Metal Trans, 6A, 825-837.
Eshelby J D (1957). The determination of the elastic field of an ellipsoidal inclusion and related problems. Proc R Soc, A241, 376-396.
Goodier J N (1933). Concentration of stress around spherical and cylindrical inclusions and flaws. ASME Appl Mech Mag, 55, 39-44.
McClintock F A and Rhee S S (1962). On effects of strain hardening on strain concentrations. Proc 4th US Nat Congress on Appl Mech, 2, 1002.
Orr J and Brown D K (1974). Elastoplastic solution for a cylindrical inclusion in plane strain. Engg Fract Mech, 6, 261-274.
Thomson R D and Hancock J W (1983). To be published.

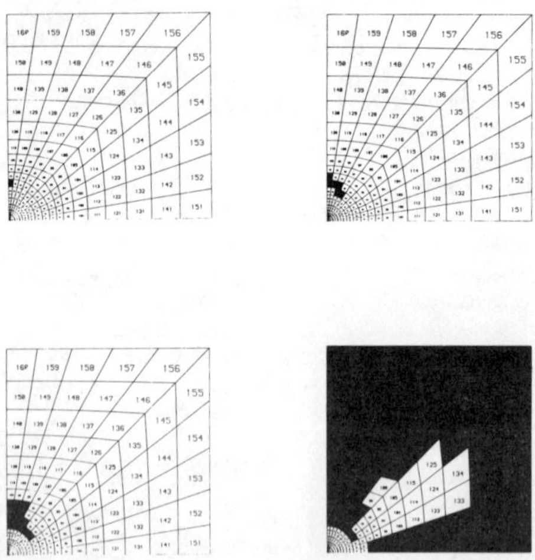

Fig 1. Plastic zone around a rigid inclusion: power hardening matrix with $(\bar{\sigma}/\sigma_0)^{(1/n)} - (\bar{\sigma}/\sigma_0) = (\bar{e}^p/e_0)$: σ_0=0.14 GPa, n=0.28, e_0=0.00067

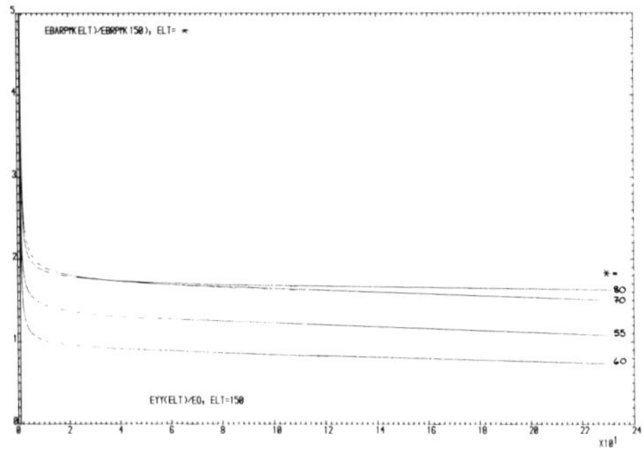

Fig 2. Plastic strain concentrations near inclusions in a power-hardening matrix (elements 80, 70, 60 in loading direction, element 55 at $\pi/4$).

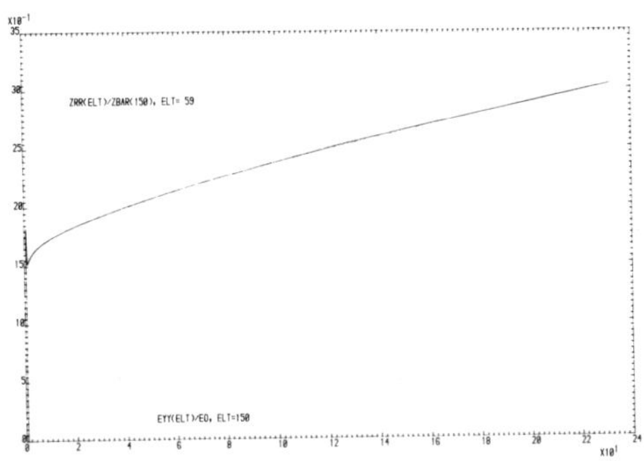

Fig 3. Maximum interfacial radial stress around a rigid inclusion in a power-hardening matrix.

THE USE OF J-RESISTANCE CURVES TO CHARACTERIZE THE EFFECTS OF MICROSTRUCTURE, STRENGTH AND IRRADIATION ON FRACTURE MICROMECHANISMS IN NUCLEAR REACTOR PRESSURE TUBES

L. A. Simpson

Materials Science Branch, Whiteshell Nuclear Research Establishment, Pinawa, Manitoba R0E 1L0, Canada

INTRODUCTION

In the CANDU[1] nuclear reactor, the pressurized heavy-water coolant and fuel bundles are contained in a lattice of zirconium alloy pressure tubes, usually cold-worked (c.w.) Zr-2.5% Nb. An important parameter in such a demonstration is the critical crack length (CCL) or the length at which the growing crack will become unstable. In the past, critical crack length has been determined by burst tests on actual tube sections containing through-thickness axial slits (Langford and Mooder, 1971). While this test method has provided a good basis for leak-before-break assurance under normal operating and material conditions, experience with the reactors has identified some new material conditions, which may be attained near "end-of-life" of the pressure tubes, for which critical crack length data are incomplete. These include long-term irradiation effects and end-of-life hydrogen levels. A better understanding of the effects of hydrogen and irradiation on the fracture micromechanisms would improve our ability to predict critical crack length for end-of-life conditions.

We have been developing a framework by which CCL in a pressure tube may be estimated from tests on small fracture-mechanics specimens. Previous work concentrated on the development of a test procedure to determine J-resistance (J_R) curves (Simpson and Clarke, 1979; Simpson and Wilkins, 1980; Simpson, 1981). In this paper we address the application of the resistance-curve approach to material assessment, that is, "what can the resistance curve reveal about the fracture process?" In particular, we investigate the effects of strength, irradiation and hydrogen content on the resistance curve. Simpson and Wilkins (1980) noted that small variations in flow stress in cold-worked Zr-2.5% Nb resulted in significant changes in the slope of the J_R curve at 20°C, with the strongest material showing the lowest slope. Because irradiation strengthens this alloy further, we would expect a further loss of toughness in such material. Burst testing has not been able to clearly define the effect of irradiation on CCL, especially at temperatures in the range 170 to 300°C, which are important in the operation of a reactor. Similarly, the influence of zirconium hydrides on toughness, while significant and well documented at 20°C, has not been clearly defined for the temperature range 170 to 300°C.

---

1 CANada Deuterium Uranium

EXPERIMENTAL

The role of microstructure and strength on tearing resistance has been assessed by

(a) determining J_R curves for several zirconium alloys at 20°C, to compare with previous results on c.w. Zr-2.5% Nb,

(b) determining J_R curves for irradiated c.w. Zr-2.5% Nb and irradiated and hydrided c.w. Zr-2.5% Nb.

For task (a), the three alloys chosen included two cold-worked ones, Zircaloy-2 and Excel, and a heat-treated Zr-2.5% Nb. The latter has a somewhat different microstructure from the others, consisting of α-zirconium and a martensitic α' phase. The flow stresses, $\bar{\sigma}$, for these alloys, defined as 1/2 (yield + ultimate), range from 700 MPa to 900 MPa at 20°C and are included in Table 1 in the Results section. Compact specimens of width W = 34 mm, cut from the pressure tubing, were tested using the procedures described by Simpson and Clarke (1979) for obtaining J_R curves. Displacement was measured by photographing the specimen during the test, and crack extension by the D.C. potential-drop method.

For task (b), forty compact specimens were fabricated from c.w. Zr-2.5% Nb and completely prepared for testing, prior to irradiation. They were then irradiated in the WR-1 reactor for one year at 350°C to a fluence of approximately 6.5×10^{20} neutrons/cm^2. Displacement and crack growth were determined by the same procedures used for unirradiated specimens.

Scanning electron fractography and optical metallography were used to relate fracture micromechanisms to the resistance-curve behaviour.

RESULTS

J_R curves were calculated for each specimen tested using Ernst and colleagues' (1981) equation. Figure 1(a) shows the results for irradiated c.w. Zr-2.5% Nb at four temperatures. The curves for the three temperatures from 170 to 300°C seem indistinguishable when plotted this way, although it will be shown later that the initial slopes are distinguishable. Every point for the irradiated data falls below the solid line, which is the mean value of a very narrow distribution of data for unirradiated material at 240°C (Simpson and Wilkins, 1980). Similarly, the dashed lines show the limits for room-temperature data obtained from unirradiated material of several strength levels from Simpson and Wilkins (1980). The pre-irradiation flow stress of the irradiated alloy was 800 MPa, close to that for the unirradiated specimens giving the highest toughness (upper dashed line). Thus, we have demonstrated irradiation reduces toughness over the complete temperature range.

Figure 2(b) contains the results for specimens that were both irradiated and hydrided. The very drastic reduction in toughness by hydrides at 30°C is well known, but the higher temperature results also show a toughness reduced from that for specimens irradiated only.

All of the J_R curves showed an approximately linear region following crack initiation, followed by a more variable behaviour. The steeper curves tended to rise continuously while others reached a steady-state value of J_R. To analyze these curves in more detail, the initial portions were plotted on an expanded scale in accordance with ASTM 813, Standard Test Method for J_{IC}, A Measure of Toughness. Values of J_{IC}, the initiation toughness, are tabulated in Table 1 along with the slope of the R-curve regression line (a linear fit to the points on the R-curve between the 0.15 and 1.5 mm offset lines). The latter is expressed as the initial tearing modulus T_o, defined as

$$T_o = \frac{E}{\bar{\sigma}^2} \frac{dJ_R}{da} \tag{1}$$

where E = Young's modulus. The tearing modulus is a concept developed by Paris and colleagues (1979), and can be viewed as a toughening rate.

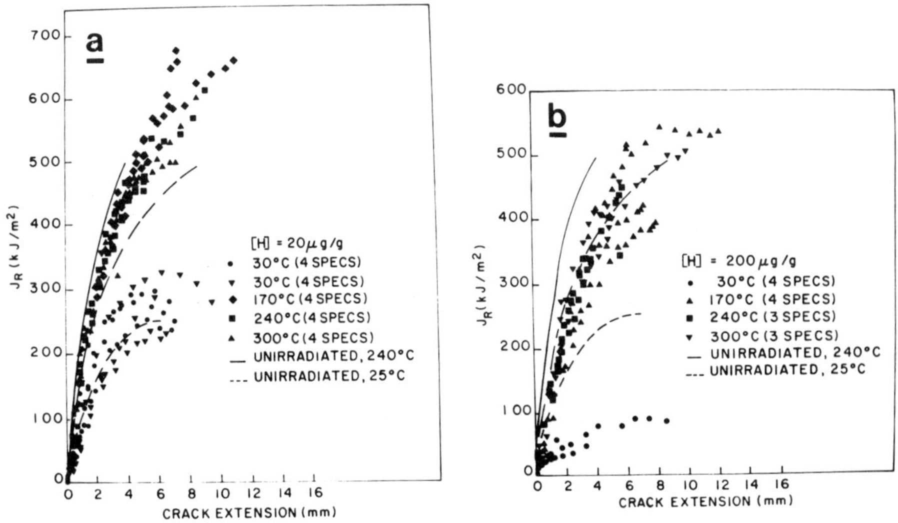

Fig. 1. J-resistance curves: a: irradiated c.w. Zr-2.5% nb,
b: irradiated and hydrided Zr-2.5% Nb.

The J_{IC} results were quite variable in some specimens, due to difficulties in detecting initiation, and are used here only for ranking the materials. Conversely, the measurements of T_o were highly reproduceable for a given material.

In Fig. 2, the tearing modulus is plotted against the flow stress. There appears to be a linear correlation between T_o and $\bar{\sigma}$ for all the cold-worked alloys, irrespective of temperature. The hydrided and irradiated material also correlates linearly on a separate line. The heat-treated alloy is an exception to this behaviour and has been omitted from the diagram.

Fractography and Micromechanics

Macroscopically, in these specimens the crack initiates under predominantly plane-strain conditions, but rapidly develops shear lips, so that after

Table 1 Results of J_R Testing

Alloy	Temperature (^{o}C)	[H] (μg/g)	J_{IC2} (kJ/m^2)	$\bar{\sigma}$ (MPa)	Irradiated	T_o
Zr-2	20	20	45	700	No	27*
h.t. Zr-2.5% Nb	20	20	29	700	No	8.9*
EXCEL	20	20	20	900	No	12.0*
c.w. Zr-2.5% Nb	20	20	20	791	No	22.2
c.w. Zr-2.5% Nb	20	20	20	834	No	18.0
c.w. Zr-2.5% Nb	20	20	20	864	No	12.7
c.w. Zr-2.5% Nb	30	20	14	875†	Yes	10.9*
c.w. Zr-2.5% Nb	170	20	23	714†	Yes	24.4
c.w. Zr-2.5% Nb	240	20	22	639†	Yes	33.3
c.w. Zr-2.5% Nb	300	20	21	575†	Yes	31.0
c.w. Zr-2.5% Nb	30	200	9	875†	Yes	3.18*
c.w. Zr-2.5% Nb	175	200	7	714†	Yes	16.4
c.w. Zr-2.5% Nb	240	200	13	639†	Yes	24.1
c.w. Zr-2.5% Nb	300	200	15	575†	Yes	31.0

* Attained a steady state value of J_R
† Estimated from increase in hardness

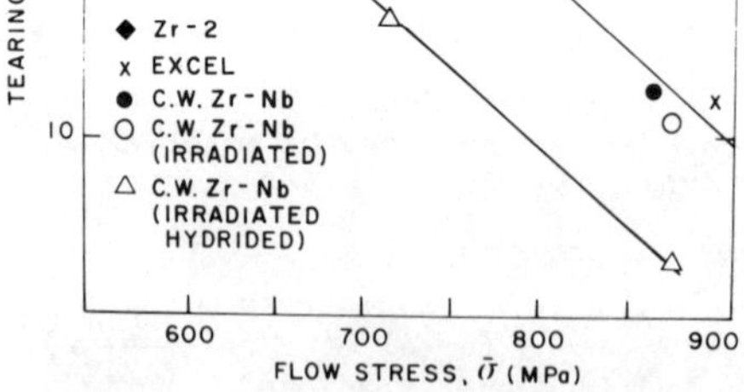

Fig. 2. Effect of flow stress on initial tearing modulus.

about 2 mm of crack extension the stress state is primarily one of plane stress.

All fracture surfaces showed that crack propagation was by ductile void growth and coalescence; however, special features of each alloy determined the details of the fracture process. The microstructure of the Excel and c.w. Zr-2.5% Nb alloys consisted of elongated grains of α-zirconium (several micrometres long by several tenths of a micrometre in the radial and circumferential directions) enclosed in a grain boundary network of β-zirconium. Voids tended to nucleate in columns directed ahead of the crack front, colinear with the grain structure (see Fig. 3(a)). As deformation proceeded, coalescence of these columns into tube-shaped voids occurred, with a final shearing-off of the ligaments separating the tubes. This is evident in Fig. 3(b) which shows the fracture surface of irradiated Zr-2.5% Nb tested at 240°C, and also in the surface of the Excel alloy in Fig. 3(c), where the voids are aligned in rows.

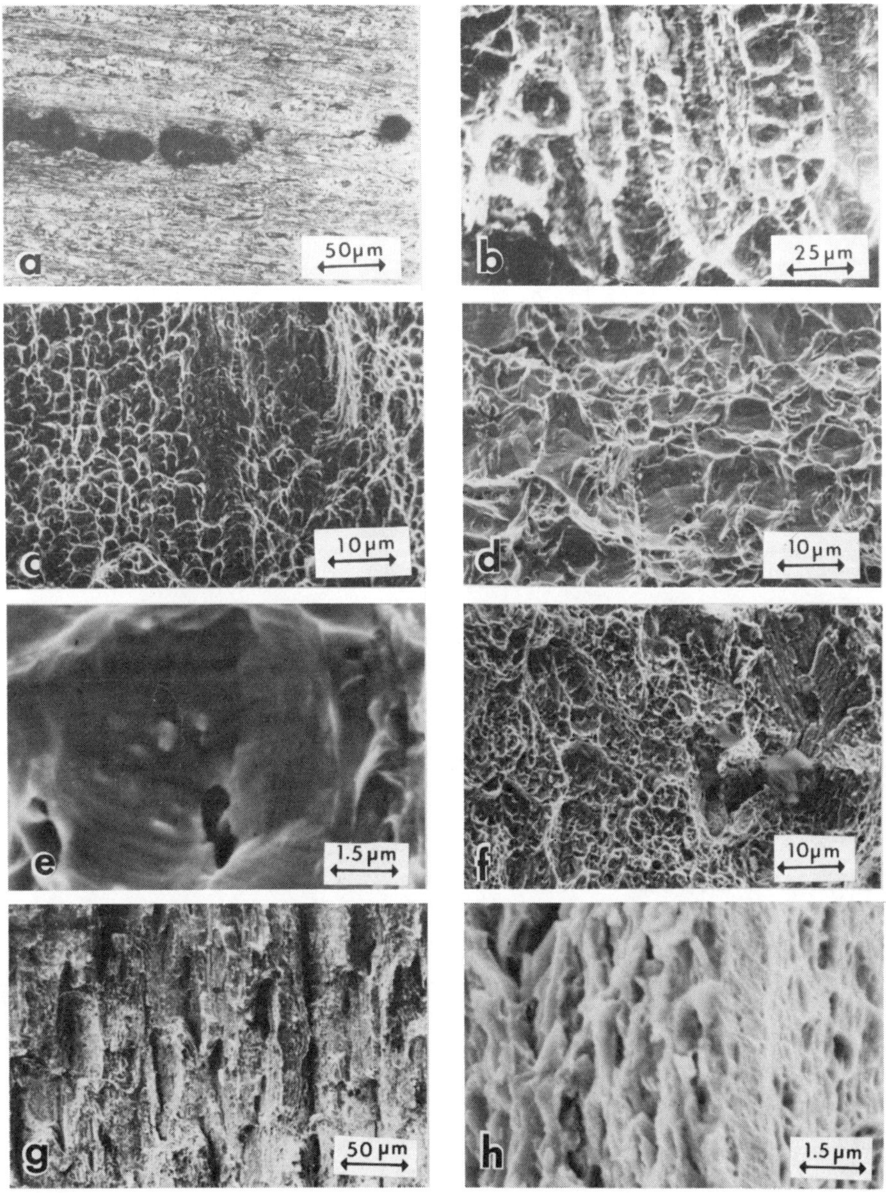

Fig. 3. Fracture microstructures: a: void column ahead of crack tip in c.w. Zr-2.5% Nb, b: fracture surface, irradiated c.w. Zr-2.5% Nb tested at 240°C, c: fracture surface, Excel, d: fracture surface, Zircaloy-2, e: fracture surface, Zircaloy-2, f: heat-treated Zr-2.5% Nb, g: hydrided irradiated c.w. Zr-2.5% Nb tested at 25°C, h: hydrided c.w. Zr-2.5% Nb tested at 25°C.

The Zircaloy-2 surface, Fig. 3(d), shows the more equiaxed void distribution reflecting the larger equiaxed grain size. The voids are also much larger than for the other two alloys, and there appear to be small impurity precipitates associated with many of them (see Fig. 3(e)).

The heat-treated Zr-2.5% Nb fracture surface has a very fine void size with some areas of cleavage-type fracture (see Fig. 3(f)).

Hydrides exist as platelets in the parent zirconium phase, in a preferred orientation with respect to the tube; that is, the platelet normals tend to line up in the radial direction such that the platelets themselves are parallel to the circumferential surface of the tube. Their effect on through-thickness crack propagation is minimal in this orientation, however, the transverse component of the triaxial, crack-tip stress field is sufficient to fracture the hydride in the plane of the platelet. These fractures produce cracks that penetrate down into the primary fracture surface, as shown in Fig. 3(g). This behaviour was seen only at room temperature and has the effect of dividing the crack front into a number of thin, tensile ligaments, which ultimately fail by a ductile shearing-off process. This behaviour was associated with very flat J_R-curves (see Fig. 3(e)).

DISCUSSION

The results of these experiments indicate that the J_R curve can be a very sensitive probe into the effect of metallurgical variables on toughness. In particular, this work has identified effects of irradiation and hydride content on toughness in the temperature range 170 to 300°C in c.w. Zr-2.5% Nb. Earlier work (Langford and Mooder, 1971), which used burst testing with its associated scatter and limited numbers of specimens, did not resolve these effects.

Crack Initiation

Because the single specimen method using potential drop for crack extension measurement did not yield reliable J_{IC} determinations, it is only possible to state that Zircaloy-2 had the highest J_{IC}, followed by the other pressure-tube alloys, which are in the range 20 to 30 MPa $m^{1/2}$. Also, both irradiation and the presence of hydrides appear to reduce J_{IC}.

J-Controlled Growth

It was pointed out by Hutchinson and Paris (1979) that for J-controlled growth, i.e. for the J-integral to describe the deformation fields near the advancing crack tip, two conditions must be satisfied. First, the amount of crack growth must be a small fraction of the remaining ligament in the specimen, and second,

$$\omega = \frac{W-a}{J}\frac{dJ}{da} >> 1 \tag{2}$$

In our previous work (Simpson 1981), it was shown that these criteria are violated soon after 1 mm of crack growth has occurred. Since T_o was determined on the portion of the J_R curve between 0.15 and 1.5 mm of crack growth, most of this curve would relate to J-controlled growth.

Rice, Druggan and Sham (1980) developed a theory of stable crack growth based on the concept that the crack tip moves through the material with a constant crack opening angle. Although their calculation assumed plane strain, and our material went through a transition from plane strain to plane stress, it is interesting to look at trends. For contained yielding, they derived an equation for the steady-state value of J_R:

$$J_{SS} = J_{IC} \exp (0.128 \cdot T_o) \tag{3}$$

They showed that steady-state values of J_R would only be attained in practical experiments for low values of T_o, that is, for $T_o < 20$. Our results, given in Table 1, when assessed with this equation, are qualitatively consistent in that only specimens with low T_o values show a steady-state J_R, with the possible exception of Zircaloy-2. Unlike the other alloys, crack initiation in Zircaloy-2 did not occur until the specimens were very close to their plastic-limit loads for plane stress. In this case, the assumption of contained yielding does not apply.

Quantitatively, the steady-state values of J_R calculated by applying Eqn. (2) to our data are much lower than our experimental ones. This reflects the high shear-lip contribution to J_R. (Eqn. (2) was developed for full plane-strain conditions.)

Relationship Between T_o and Microstructure

The heat-treated alloy showed considerably less toughness than the cold-worked alloys. This is consistent with the results of early burst tests (Langford and Mooder, 1971), and a comparison of burst test ductility between Zircaloy-2 and heat-treated tubes by Cheadle and Ells (1965).

More generally, the highest values of T_o were associated with the larger void sizes on the fracture surfaces. This, in part, may reflect the large grain sizes in the high-T_o alloys (except for the heat-treated alloy) and the distribution of nucleation sites. While impurity particles are seen in many voids in Zircaloy-2 (Fig. 3(e)), the mechanics of void nucleation in c.w. Zr-2.5% Nb and Excel are not known (Simpson 1981). However, the way in which the voids line up with the microstructure suggests that they nucleate at the grain boundaries, possibly in the β-phase.

The very low toughness observed in hydrided material at room temperature is directly related to the splitting of circumferential hydrides. The remaining tensile ligaments then shear off, with the production of very fine elongated voids. Knott (1973) describes a mechanism by which plane-stress fracture toughness decreases with thickness for very thin specimens. Such a mechanism is consistent with our observation of very low toughness when the hydrides split.

CONCLUSIONS

(1). J_R curves have identified very subtle effects of irradiation and hydride content on fracture toughness in zirconium alloys.

(2). The initial tearing modulus, T_o, decreases linearly with flow stress $\bar{\sigma}$, whether it is altered by temperature, irradiation or alloy fabrication.

(3). High T_o values are associated with large voids on the fracture surface.

(4). J_R curves have been successfully obtained from radioactive specimens using conventional methods and remote handling.

ACKNOWLEDGEMENTS

The author is grateful for experimental assistance from B. Ellis, A. Shillinglaw, D. Stark and W. Stanley.

REFERENCES

Cheadle, B. A., and C. E. Ells (1965). Effect of crystallographic orientation of the fracture ductility of Zr-2.5 wt. pct. Nb and Zircaloy-2 tubular products. Met Soc. Trans AIME., 233, 1044-1052.

Ernst, H. A., P. C. Paris and J. D. Landes (1981). Estimations on J-integral and tearing modulus T from a single specimen record. In R. Roberts (Ed.), Fracture Mechanics, ASTM STP 743, Philadelphia. pp. 476-502.

Hutchinson, J. W. and P. C. Paris (1979). Stability analysis of J-controlled crack growth. In J. D. Landes, J. A. Begley and G. A. Clarke, (Eds.), Elastic-Plastic Fracture, ASTM STP 668, American Society for Testing and Materials, Philadelphia. pp. 37-64.

Knott, J. F. (1973). Fundamentals of Fracture Mechanics. Butterworths, London. pp. 116-119.

Langford, W. J. and L. E. J. Mooder (1971). Metallurgical properties of irradiated cold-worked Zr-2.5 wt% Nb pressure tubes. J. Nuclear Mater., 39, 292-302.

Paris, P. C., H. Tada, A. Zahoor, and H. Ernst (1979). The theory of instability of the tearing mode of elastic-plastic crack growth. In J. D. Landes, J. A. Begley, and G. A. Clarke, (Eds.), Elastic-Plastic Fracture, ASTM STP 668, Philadelphia. pp. 5-36.

Rice, J. R., W. J. Druggan and T. L. Sham (1980). Elastic-plastic analysis of growing cracks. In P. C. Paris (Ed.), Fracture Mechanics, ASTM STP 700. Philadelphia. pp. 189-221.

Simpson, L. A. (1981). Effects of specimen geometry on elastic-plastic R-curves for Zr-2.5% Nb. In D. Francois (Ed.), Advances in Fracture Research, Vol. 2, Pergamon Press, Oxford. pp. 833-841.

Simpson, L. A. and C. F. Clarke (1979). An elastic-plastic R-curve description of fracture in Zr-2.15% Nb pressure tube alloy. In J. D. Landes, J. A. Begley, and G. A. Clarke, (Eds.), Elastic-Plastic Fracture, ASTM STP 688, Philadelphia. pp. 643-662.

Simpson, L. A. and B. J. S. Wilkins (1980). Prediction of fast fracture in Zr-2.5% Nb pressure tubes using elastic-plastic fracture mechanics. In K. J. Miller and R. F. Smith (Eds.), Mechanical Behavior of Materials, Vol. III, Pergamon, Oxford. pp. 563-572.

STANDARD METHOD OF TEST FOR ELASTIC-PLASTIC FRACTURE TOUGHNESS J_{IC} RECOMMENDED IN JAPAN

H. Miyamoto*, H. Kobayashi and N. Ohtsuka*****

*Science University of Tokyo, Japan
**Tokyo Institute of Technology, Japan
***Chiyoda Chemical Engineering and Construction Co., Ltd., Japan

ABSTRACT

The J_{IC} test method recommended by the Japan Society of Mechanical Engineers, JSME S001-1981, is outlined. Its applicability and availability compared with the ASTM Standard, ASTM E813-81, are discussed in this paper.

KEYWORDS

Fracture toughness; standard; J_{IC} test; stretched zone width; R-curve; blunting line; ductile tearing; plane strain; metallic material.

INTRODUCTION

In Japan, the Japan Society of Mechanical Engineers (JSME) Committee S781 on "Standard Method of Test for Elastic-Plastic Fracture Toughness J_{IC}" (Chairman: H. Miyamoto, Vice-chairman: H. Kobayashi) standardized the J_{IC} test method and the JSME Standard was published in October 1981 under the designation JSME S001-1981.

The objective of the J_{IC} test method recommended by JSME is to determine J_{IC}, the value of J at the onset of actual ductile (dimple) tearing for metallic materials under the mode I and the plane-strain conditions. The recommended test specimens are compact tension or three-point bend type that contain deep initial cracks. The JSME Standard includes two multiple specimen techniques and three single specimen techniques. In the former, the J_{IC} value is determined by the stretched zone width (SZW) technique or the R-curve technique. In the latter, the electrical potential, ultrasonic or acoustic emission technique can be applied.

On the other hand, the American Society for Testing and Materials (ASTM) Standard for the J_{IC} test method was published in August 1981 under the designation ASTM E813-1981. There are several differenes between the JSME and ASTM methods.

The purpose of this paper is to give a brief description of the J_{IC} test method recommended by JSME and to discuss its applicability and usefulness with special attention given to a comparison of the two methods recommended by JSME and ASTM.

STRECTCHED ZONE WIDTH (SZW) TECHNIQUE

Two or more specimens are loaded to selected different displacement levels which are lower than that at the initiation of slow stable crack growth. The J-integral of each specimen is calculated by a modified Merkle-Corten equation (Clarke and Landes, 1979) in terms of an area under load versus load-line displacement record. The specimen is unloaded and crack extension, Δa, that occurred during loading is marked by such an appropriate way as subsequent fatigue cycling. After breaking the specimen open, microscopic observation of the fatigue precrack tip is made on the fracture surface. The sub-critical stretched zone width, SZW, is determined as the average of three or more measurements of apparent crack growth at evenly spaced locations from 3/8 to 5/8 of the thickness of the specimen as shown in Fig.1. To establish a crack tip plastic blunting line, J values are plotted against the corresponding measured SZW values and a best fit line between the relations is determined as illustrated in Fig.2.

In order to determine the critical stretched zone width, SZW_C, three or more specimens are pulled apart by overloading. SZW_C is defined as a critical value of SZW at the initiation of ductile tearing and is not affected by any following stable crack growth. So, it is possible to measure SZW_C on the overload fracture surface by the same procedure as the measurement of SZW. Using a method of least squares on the relation of J vs. SZW_C, a best fit line parallel to the J-axis is determined as shown in Fig.2.

J_{in} is defined as the intersection of the line $SZW=SZW_C$ and the blunting line. The J_{in} value equals J_{IC}, provided the requirements on fatigue precracking and the following validity requirements are satisfied.

$$b_o = W - a_o \geq 25J_{in}/\sigma_{fs} \tag{1}$$

$$B \geq 25J_{in}/\sigma_{fs} \tag{2}$$

where b_o = initial uncracked ligament, W = specimen width, a_o = original crack size, B = specimen thickness. σ_{fs} = effective yield strength which is the average of tensile-yield strength (σ_{ys}) and the ultimate- tensile strength (σ_B). Eq.(2) is not necessarily required, provided J_{in} is confirmed to be constant irrespective of the thickness by an additional test for specimens which have a different thickness from the original one. It is desired to change the thickness more than twice as large or less than half as small as the original one.

R-CURVE TECHNIQUE

A plastic blunting line is determined experimentally in the same way as the SZW technique. Four or more specimens are loaded up to displacement levels so as to cause ductile tearing. By following the procedure described in the SZW technique, the physical crack growth, Δa, is determined as the average of the measurements which are made at three or more locations spaced evenly between 3/8 and 5/8 of the thickness of the specimen as shown in Fig.3. A best fit straight line of J upon Δa that is less than 1 mm is determined as

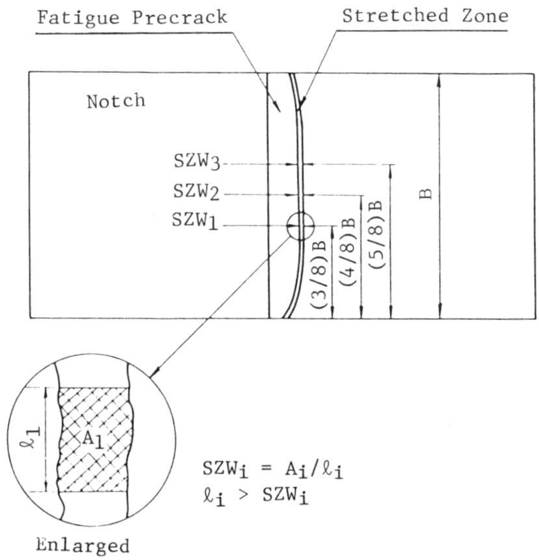

Fig.1 Schematic Fracture Appearance of J_{IC} Specimen and Measurement of SZW in the JSME Standard

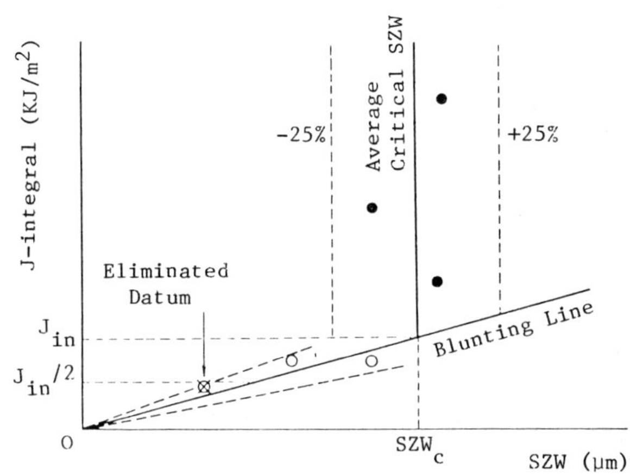

Fig.2 Schematic Illustration of the SZW Technique in the JSME J_{IC} Test Method

the material resistance (R-curve) to crack growth. The intersection point between the regression line (R-curve) and the blunting line defines J_{in} as illustrated in Fig.4. J_{in} equals J_{IC} if the following equation is satisfied in addition to Eq.(1) and Eq.(2).

$$(dJ/da)_R \leq (1/2)(dJ/da)_B \qquad (3)$$

where $(dJ/da)_R$ = the slope of the R-curve, and $(dJ/da)_B$ = the slope of the blunting line.

SINGLE SPECIMEN TECHNIQUES

The electrical potential, ultrasonic or acoustic emission technique can be used to make the following measurement non-destructively during loading of specimens; (1) the difference of electrical potential, (2) the difference of ultrasonic echo, or (3) the variation of acoustic-emission event count, accumulated energy count or amplitude distribution of the event.

The single specimen techniques require actually three specimens, namely A, B and C, to complement the uncertainty of the techniques. A load-line displacement, δ_{in}(A), of the first specimen A at the initiation of ductile tearing is determined by one of the non-destructive techniques. The second specimen B is loaded up to a displacement level which is larger than δ_{in}(A) but smaller than $1.1\delta_{in}$(A). The third specimen C is loaded up to a displacement level which is larger than $0.9\delta_{in}$(A) but is smaller than δ_{in}(A). During the loading, specimens B and C are monitored by one of the non-destructive techniques so as to confirm the initiation of ductile tearing on specimen B but no initiation on specimen C. δ_{in}(B) is determined as a value of displacement of specimen B at the initiation of ductile tearing. After unloading, fractographic examination of the three specimens is made on the fracture surface. When the initiation of ductile tearing is observed specimens A and B but not on specimen C, the J_{in} value is determined as an average of the two J values corresponding to δ_{in}(A) and δ_{in}(B). J_{in} equals J_{IC}, provided Eq.(1) and Eq.(2) are satisfied.

COMPARISON OF TWO STANDARDS ON J_{IC} TEST METHOD

Comparison of the J_{IC} test methods recommended by JSME and ASTM was made according to actual test results. As summarized in Table 1, the results obtained are as follows:

(1) If we assume a J versus SZW (=Δa) blunting line to be of the form

$$J = \lambda\, \sigma_{fs}\, SZW \qquad (4)$$

where σ_{fs} is the effective yield strength, the experimentally determined values of constant λ for alloy steels and aluminum alloys show a tendency to become smaller as σ_{fs} becomes larger. For intermediate-strength materials, however, Eq.(4) can stand and the value of λ is 4. This experimentally determined blunting line (Kobayashi and co-workers, 1977, 1980) does not coincide with the specified equation in the ASTM Standard (i.e., λ=2). In the case of some low- and intermediate-strength materials, the intersection of the assumed blunting line in the ASTM Standard and the R-curve can not be obtained. Therefore, the JSME Standard recommends that the blunting line should be determined experimentally for each material.

(2) In some materials, the J_{IC} value determined by the single specimen techniques of the JSME method have wide variation due to material inhomogeneity. For this kind of material, difficulty arises in determining

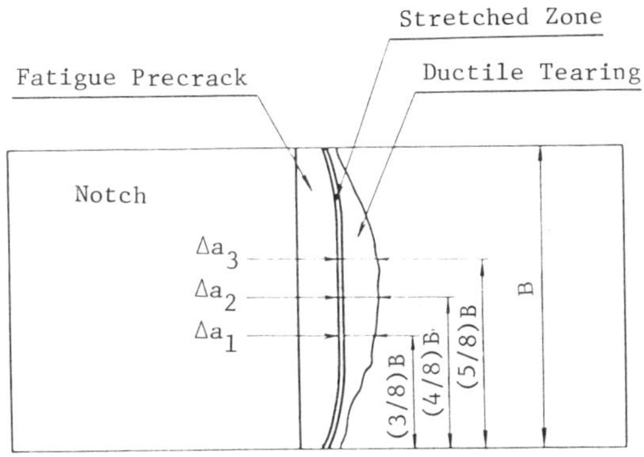

Fig.3 Schematic Fracture Appearance of J_{IC} Specimen and Measurement of Δa in the JSME Standard

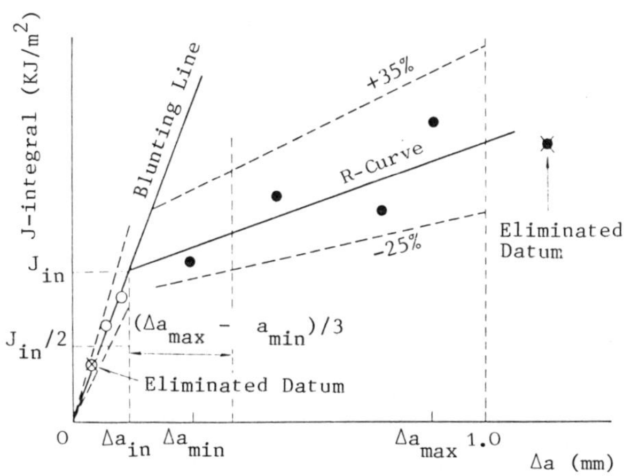

Fig.4 Schematic Illustration of the R-curve Technique in the JSME J_{IC} Test Method

one R-curve by the R-curve technique (both JSME and ASTM methods). It is possible, however, to determine accurately J_{IC} for each specimen as the intersection of the J-SZW blunting line and the line of SZW=SZWc according to the JSME method.

(3) The JSME Standard defines crack extension as the mid-thickness average, Δa_{max}, of inner three or more measurements instead of the through-thickness average, Δa_p, at nine or more locations as defined in the ASTM Standard. If we define the crack extension as Δa_p (ASTM method), the slope of the R-curve shows a tendency to become smaller as the specimen size becomes larger. On the other hand, use of Δa_{max} (JSME method) at the mid-thickness of the specimen where the plane-strain flat type fracture precedes (Miyamoto and Kikuchi, 1983; Kikuchi and co-workers, 1983) can give a unique and lowest slope of the R-curve regardless of specimen thickness. This is the reason why Eq.(2) on thickness requirement is not always necessary in the JSME Standard. According to the ASTM method, the valid plane-strain R-curve can not be evaluated as a material property.

(4) Qualification of Δa data is different between the two standards. In the JSME Standard, the 0.15 mm lower limit offset line is eliminated and the maximum crack extension is limited to 1.0 mm instead of the 1.5 mm offset line in the ASTM Standard. The R-curve being nonlinear, the slope decreases with increasing Δa and tends to show a plateau. A linear fit of R-curve for large values of Δa (such as the data between the two offset lines in the ASTM method) can overestimate J_{IC}. The JSME method recommends to approximate the R-curve as a straight line using the Δa data just after the initiation of ductile tearing. So, according to the JSME method, the J_{IC} value can be evaluated more accurately by using a smaller specimen size compared with the ASTM method.

(5) The JSME Standard accepts the three single specimen techniques. The electrical potential and ultrasonic techniques yield fairly good results in some intermediate- and high-strength materials. The accoustic emission technique is affected strongly by the micro-structure of materials, so that the applicability of the technique to J_{IC} test method depends on the type of materials (Ohtsuka, 1979). The unloading compliance technique in the ASTM Standard is not particularly recommended in the JSME Standard due to the inaccuracy of the technique for small values of Δa.

CONCLUSION

The J_{IC} test method recommended by the Japan Society of Mechanical Engineers, JSME S001-1981, was outlined. Two multiple specimen techniques and three single specimen techniques are recommended in the JSME standard. The former are the stretched zone width (SZW) technique and the R-curve technique. The electrical potential, ultrasonic and acoustic emission technique are the latter ones. The applicability and availability of the standard test method compared with the ASTM Standard, ASTM E813-81, were described in this paper.

REFERENCES

ASTM (1981). Standard Test for J_{IC}, A Measure of Fracture Toughness, ASTM Standard, E813-81.

Clarke, G.A. and Landes, J.D. (1979). Evaluation of J Integral for the Compact Specimen. Journal of Testing and Evaluation, Vol.7, No.5, 264-269.

JSME (1981). Standard Method of Test for Elastic-Plastic Fracture Toughness J_{IC}, JSME Standard, S001-1981.

Kikuchi, M., Sugawara S. and Miyamoto, H. (1983). The Thickness Effect of Side Grooved CT Specimens. To be presented at 7th Int. Conf. on SMiRT, Chicago, Ill.

Kobayashi, H., Hirano, K. Nakamura, H. and Nakazawa, H. (1977). A Fractographic Study on Evaluation of Fracture Toughness. Proc. ICF 4, Vol. 3, Pergamon Press, 583-592.

Kobayashi, H., Nakamura, H. and Nakazawa, H. (1980). The Elastic Plastic Fracture Toughness and the Fatigue Crack Growth Resistance in Base Metals and Weldments of A533B-1 and 304 Steels. Proc. 4th Int. Conf. on Pressure Vessel Technology, Vol.I, 251-256.

Miyamoto, H. and Kikuchi, M. (1983). Studies on Size Effect and Crack Growth of the Side Grooved CT Specimens. To be presented at 16th ASTM Symposium on Fracture Mechanics, Columbus, Ohio.

Ohtsuka, N., Nakano, M. and Ueyama, H. (1979). An Evaluation of Procedures for J_{IC} Determination. Preprints for a Joint USA/Japan Seminar, Hayama, Japan, 71-80.

Table 1 Comparison of Two Standards on J_{IC} Test Method

Item	JSME S001-1981	ASTM E813-81
Thickness of specimen, B	$(B \geq 25\ J_{in} / \sigma_{fs})^{*1}$	$B > 25\ J_Q / \sigma_Y$
Location for measurement of Δa	Mid-thickness average at 3 or more locations in (3/8-5/8)B	Through-thickness average at 9 or more locations
Limit of Δa	$\leq$1.0 mm	Between the 0.15mm and 1.5mm offset lines
Blunting line	To be determined experimentally*2	$\Delta a_B = J / 2\sigma_Y$
Applicable multiple-specimen techniques	SZW technique or R-curve technique	R-curve technique
Applicable single-specimen techniques	Electrical potential, ultrasonic, or acoustic emission technique	Unloading compliance technique

(Note) *1 : Not necessarily required, provided J_{in} is confirmed to be constant regardless of thickness B.

*2 : Recommended equations on blunting line can be used for some specified materials.

EFFECT OF CRACK CLOSURE ON NEAR-THRESHOLD CRACK GROWTH BEHAVIOUR IN A HIGH STRENGTH Al-ALLOY UP TO ULTRASONIC FREQUENCIES

A. F. Blom*, A. Hadrboletz and B. Weiss****

***Department of Aeronautical Structures and Materials, Royal Institute of Technology, Stockholm and Structures Department, Aeronautical Research Institute, Bromma, Sweden**
****Institute of Solid State Physics, University of Vienna, Austria**

ABSTRACT

Experiments were performed to study the effect of crack closure on near-threshold fatigue crack growth rates under R-values ranging from −1 to 0.7. It was found that crack closure levels increase rapidly as ΔK_{th} is approached. This is explained by a shear-mode type of crack growth and the formation of oxide debris. An effective stress intensity factor range $\Delta K_{eff} = K_{max} - K_{op}$ correlates the data for various R-values. At R = −1 both low frequency (50 Hz) tests and high-frequency (20 kHz) tests were carried out. No significant difference in crack growth behaviour due to the frequency effects could be found.

KEYWORDS

Fatigue, thresholds, crack-growth, crack-closure, aluminium alloy, frequency, oxide, shear mode, room-temperature, micro-structure, effective stress intensity, stress ratio.

INTRODUCTION

During the past decade considerable efforts have been devoted to the study of near-threshold fatigue crack growth behaviour (1). Especially the influence of various experimental parameters and test procedures have been investigated in detail. One objective of the present study was an evaluation of the effects of test frequency on the fatigue crack growth behaviour. In view of the considerable time required for threshold-measurements using conventional test equipments a considerable time-saving could be achieved by raising the frequency (2). A second objective consisted in an investigation of the stress ratio effect on the growth rate of fatigue cracks. A high-strength Al-alloy (Al 2024-T3) was selected for the experimental work. Using conventional test machines fatigue crack growth rates were determined at test frequencies between 40 and 80 Hz for R-ratios between −1 and +0.7. High frequency tests were performed at 20 kHz at an R-ratio of −1. For the evaluation of the test results various mechanisms proposed in the literature based on crack closure concepts were taken into consideration.

MATERIAL AND EXPERIMENTAL TECHNIQUES

In the present investigation specimens of the wrought high-strength Al-alloy 2024-T3 were tested. The specimen material was supplied as a plate of 6 mm thickness of the following composition: 4.5% Cu, 1.5% Mg, 0.6% Mn, traces of Zn, Si, Cr, Fe, Ti and Zr. This plate had been given a thermomechanical pre-treatment consisting of solution heat treatment, cold working and naturally aging to a substantially stable condition. This treatment resulted in room-temperature mechanical properties as follows: yield strength $R_{p0.2} = 345$ MPa; ultimate tensile strength $R_m = 450$ MPa; dynamic Young's modulus $E = 72\,500$ MPa.

The microstructure of the wrought and heat treated Al 2024-T3 consists of a typically pancake-like grain structure with a grain-size of $120 \times 70 \times 20$ μm. Two types of specimens used for low-frequency tests were machined from the plate material: CCT-specimens, $W = 50$ mm and CT-specimens, $W = 60$ mm. For the tests at high frequencies, 20 kHz, CCT specimens with $W = 20$ mm and $L = 125$ mm were used. All specimens were machined to achieve a crack-plane in LT orientation (ASTM E 399) (3).

In CCT specimens the starting notch was introduced by drilling a center hole of 1.5 mm diam. and then introducing two opposing side cuts of 0.5 mm length be electro-discharge machining, while CT-specimens were machined to achieve a chevron-notch. The specimen surface in the vicinity of the notch was metallographically polished and/or finally electropolished, (electrolyte: $HClO_4 - C_2H_5OH$).

Low frequency (LF) test data were obtained by using servohydraulic testing machines operated at a frequency between 40 and 80 Hz in laboratory air at ambient temperature. Experiments were performed with stress ratios R varying from −1 to +0.7.

For the high frequency (HF) tests a resonance system operating at a frequency of approximately 20 kHz was employed. A detailed description of the system is given in Ref. (4). In this system a properly tuned bar-shaped specimen is excited to longitudinal push-pull vibrations at zero mean-stress ($R = -1$). The cyclic stress varies sinusoidally along the length axis with maximum in the mid-section of the specimen. The differences in the nature of loading between conventional and resonance specimens were described in detail in Ref. (5).

For the LF-experiments fatigue crack growth data above and below 10^{-8} m/cycle were measured according to the proposed ASTM standard technique (6). Crack lengths during the LF-tests were determined by means of a travelling light microscope with a resolution of better than 50 μm. For the 20 kHz test method a strain-shedding technique (4) was used to determine ΔK_{th} values. As soon as growth of a fatigue crack could be detected the cyclic loading was reduced after consecutive load intervals of $N = 10^6$ in steps of approx. 5% of the previous total strain amplitude. The strain amplitude at which after $N = 10^7$ no further advance of the crack tip could be observed was recorded as the threshold-strain amplitude. The crack was observed by means of a travelling light microscope with a resolution better than 2 μm, thus the threshold corresponds to a crack growth rate of less than 10^{-12} m/cycle. It is possible to repeat the threshold measurements in a single specimen with cracks of increasing length at least 4 times, with intervals of slow fatigue crack growth used for da/dN measurements.

Determination of crack closure effects, here defined as the load P_{op} at which the crack opens completely, was performed in the LF-tests with compliance techniques involving either back-face strain (BFS) or crack opening displacement (COD) measurements. No difference in the P_{op} values could be found for the two techniques. The estimated relative error is in the order of ±10%. The closure-measurements were performed statically during an interruption of the dynamic crack growth experiments.

The determination of the P_{op} value in specimens under resonance loading was carried out with the aid of a strain gauge technique similar to that described by Schmidt and Paris (7). A detailed description of our procedure will be given (8). For each specimen the P_{op}-value was determined after completion of the final threshold measurements. A small circular depression centered on the fatigue crack near the tip of the crack was introduced by electro-discharge machining with a diameter chosen to be slightly larger than the active area of the strain gauge (e.g. for a strain gauge type HB LY 11 a square of 0.6 mm). Considerable care was devoted to avoid the penetration of any adhesive into the crack. This strain gauge technique measures P_{op} during cyclic loading whereby it is assumed that the measured strain value is directly proportional to the crack opening (5). A typical test result is shown in Fig. 1.

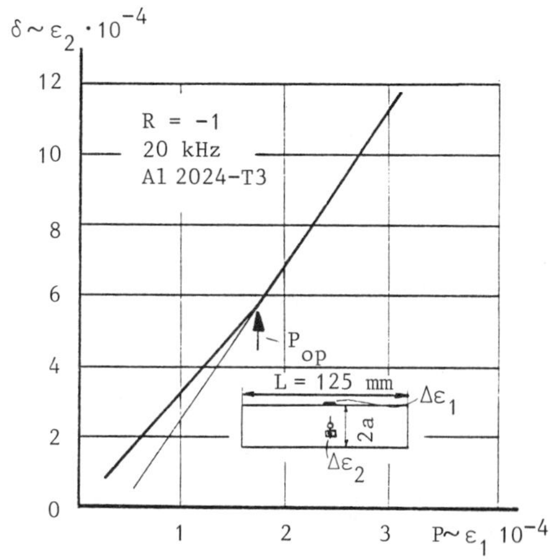

Fig. 1 Crack opening-load measurements

For the LF-tests the stress intensity factors K were calculated according to the recommendations in ASTM standard E-647 (6). For R = -1 only the tensile part of the load cycle was used to calculate ΔK both at LF- and HF-testing.

For the HF-tests ΔK was calculated using the following formula (5):

$$\Delta K = \frac{\Delta\varepsilon}{2}\, E\, \sqrt{\pi a}\; Y\left(\frac{2a}{W}\right)\, \phi\left(\frac{2a}{W}, \frac{2\rho}{W}\right)$$

where $\Delta\varepsilon$ = strain range, E = dynamical Young's modulus, 2a = total crack length, W = specimen width, 2ρ = initial notch diameter.

$$Y\left(\frac{2a}{W}\right)\quad \phi\left(\frac{2a}{W}, \frac{2\rho}{W}\right)$$

= calibration function for the resonance test method. As discussed in a recent publication (5), at a stress-ratio of R = -1 a different dynamical behaviour occurs under tension and compression. The calibration function was determined by means of an experimental technique and a dynamical finite element analysis (5). For the present investigation a refined finite element model incorporating the effect of initial notch sizes was developed (9). The correction function is listed in Table 1.

2ρ/W	Exp	FEM	Exp	FEM	Exp	FEM	Exp	FEM	Exp	FEM
0.2	-	-	1.0	-	0.96	-	0.92	-	0.88	-
0.15	-	-	0.98	1.0	0.94	0.94	0.90	0.88	0.85	0.81
0.1	1.0	1.0	0.96	0.97	0.91	0.92	0.87	0.87	0.82	0.80
2a/W	0.1		0.2		0.3		0.4		0.5	

Table 1. Correction function Y ϕ as function of initial notch size and crack length

RESULTS AND DISCUSSION

The fatigue crack growth rates near the threshold ΔK_{th}, obtained for the Al-alloy with both test frequencies at $R = -1$, laboratory air and RT and various stress ratios, are plotted in the da/dN vs ΔK diagram in Fig. 2. Data for various R values obtained by the LF-tests reveal marked effect of the stress ratio on da/dN and ΔK_{th}.

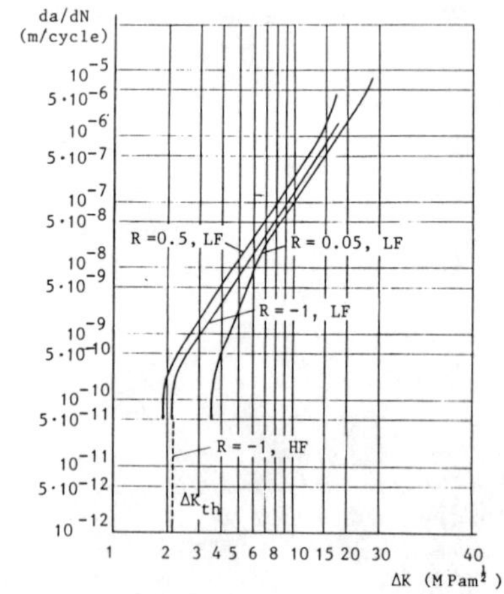

Fig. 2 FCG-data for Al 2024-T3 at RT and laboratory air

The influence of the stress ratio on ΔK_{th} is shown in Fig. 3 for the low test frequencies. It can be seen that the ΔK_{th} values decrease with increasing positive R-values, in agreement with data reported (1) for various materials. For the stress ratio of $R = -1$ the values of ΔK_{th} are $2.1 \pm 5\%$ and $2.15 \pm 5\%$ $MPam^{\frac{1}{2}}$, determined with the HF and LF methods, respectively. The fact that the ΔK_{th} values for $R = -1$ are less than at $R = 0$ is in agreement with published data for steel (10) and with data published by Pook for various metals and alloys (11).

In recent publications (12, 13) the R-dependence of ΔK_{th} values is explained by crack closure effects. In the present investigation the stress intensity range for crack closure was determined for both test frequencies. From these data the effective stress intensity range $\Delta K_{eff} = K_{max} - K_{op}$ was determined and plotted in Fig. 3. It can be deduced that ΔK_{eff} markedly reduces the effect of stress ratio. These findings are in agreement with data reported for Nimonic 105 (14) and Cu (15) which indicate that $\Delta K_{th,eff}$ is not affected by the R-value. Our data may be considered as a further support of the hypothesis that the R-dependence of ΔK_{th} can be

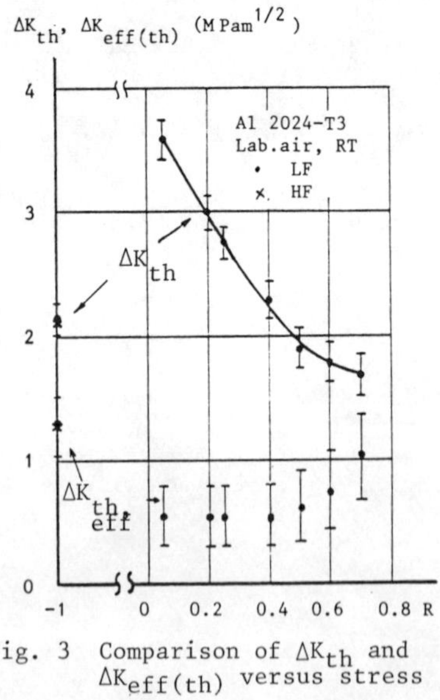

Fig. 3 Comparison of ΔK_{th} and $\Delta K_{eff(th)}$ versus stress ratio R

explained on the basis of the crack closure effect. This closure effect cannot, however, be explained solely in terms of Elber's plasticity induced closure (16) which is dominant under plane stress conditions only. Near threshold, however, the crack grows predominantly in a state of plane strain. For the tested specimens with a thickness of 6 mm it is expected that plane strain will prevail also at higher ΔK-values until the plastic zone size aquires a magnitude in the order of the specimen thickness. In Fig. 4 the change of P_{op}/P_{max} as function of ΔK for a stress ratio $R = 0.05$ is shown. Similar trends are found for the other R-values. This implies that crack closure is dependent not only on R as stated by Elber for plasticity induced closure (16), but also on microstructural features at least at near-threshold ΔK values. The decreasing trend of P_{op}/P_{max} versus ΔK in Fig. 4 may be explained by comparing the plastic zone size for various ΔK-values to the grain size. The grain size in the crack growth direction is approximately 70 µm. When the plastic zone is smaller than this value continuum considerations are not sufficient and crack growth is governed by microstructural features. Equalizing the monotonic plane strain plastic zone size (17) to the grain size yields a value of $K_{max} = 12.5\ \text{MPam}^{\frac{1}{2}}$. A similar calculation for the reversed plastic zone size yields $\Delta K = 25\ \text{MPam}^{\frac{1}{2}}$.

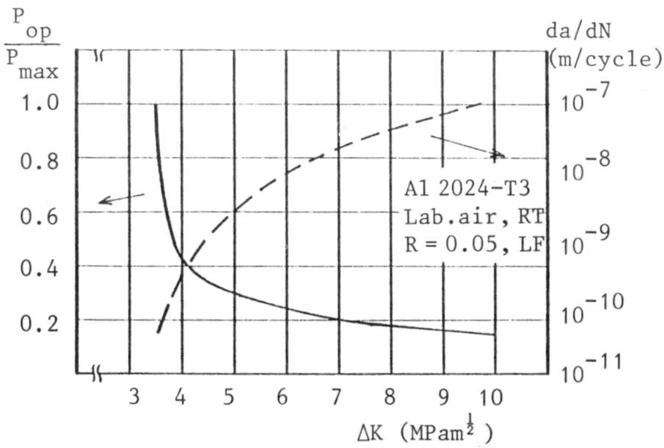

Fig. 4 Variation of crack closure level and crack growth rates versus ΔK

Thus the trend in Fig. 4 is to be expected and this conclusion is further merited by the results given in Refs. (12, 18). For the tests with $R = -1$ the values of P_{op}/P_{max} at threshold were measured to 0.37 and 0.40 with HF and LF techniques, respectively, i.e. substantially lower than for positive R-values. It is suggested that this behaviour is due to the fact that negative load cycles tend to flatten the crack surfaces and thus reducing the effect of roughness-induced closure as discussed in the following.

The high closure levels observed at near-threshold fcg-rates have been explained recently to be due to oxide-induced crack closure (13) and/or roughness-induced crack closure (12,19). In Fig. 5 fracture surfaces are shown of specimens tested with $R = 0.05$, 0.25, 0.5 and -1.0 at LF and also with $R = -1.0$ at HF. As was suggested by Ritchie et al (13), corrosion debris formed on freshly exposed surfaces at the crack tip may wedge-open the crack at stress intensities above K_{min} when the oxide deposits reach a thickness comparable to the crack tip opening displacement. This effect should be strongest at low stress intensities were the difference between K_{op} and K_{min} is largest and oxidation may be enhanced by fretting. A similar effect can be deduced from the dark deposits shown in the fracture surfaces in Fig. 5. Most debris is observed at $R = 0.05$, while at $R = 0.5$ no significant amount of debris can be seen. Thus, oxide induced crack closure may partly explain the influence of the stress ratio on ΔK_{th}. This effect is more pronounced in the present

R: 0.05 0.25 0.5 -1 -1

f: 40 - 80 Hz 20 kHz

Fig. 5 Fracture surfaces of Al 2024-T3 indicating oxide debris at threshold

R = 0.25; LF R = 0.5; LF

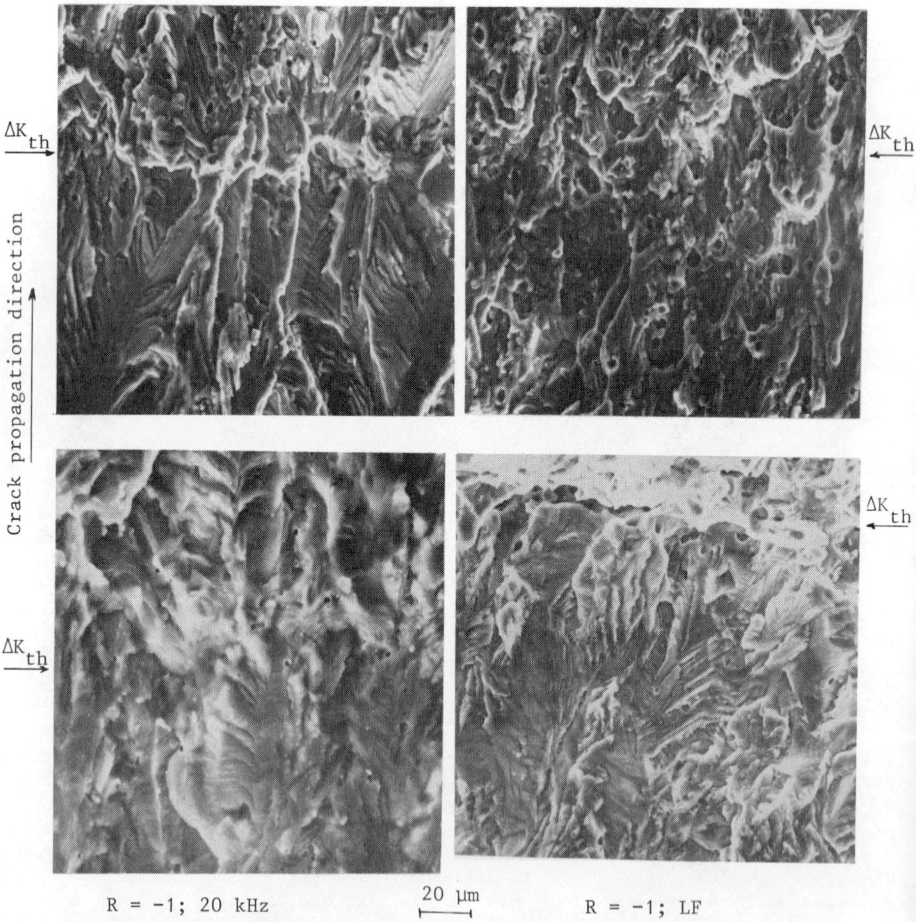

R = -1; 20 kHz R = -1; LF

Fig. 6 SEM micrographs of fracture surfaces of Al 2024-T3 at threshold under various conditions investigation than observed by Suresh et al (20) for high strength Al-alloys. The other effect that can be observed is roughness induced closure in terms of a shear mode type of crack growth indicated by shear facets.

In Fig. 6 such facets can be seen in specimens tested at R = 0.25 and R = -1, whereas for R = 0.5 the effect of mode II displacements seems negligible. A similar evaluation of fracture surfaces of specimens tested at other R-values, indicate an increase in shear facets with a reduction of the stress ratio. These results agree with observations made by McEvily et al (12) and Davidson (21). The fracture surface topography is predominantly transcrystalline as can be seen from the SEM-fractographs shown in Fig. 6. Mode II displacements can also be deduced from the serrated crack path shown in Fig. 7. Similar observations have been reported by Beevers et al for Ti-alloys (22). Contrary to observations in ductile materials no indications of large amounts of plasticity, for instance fatigue slip bands, can be seen in this Al-alloy, Fig. 7. Furthermore, microhardness measurements did not reveal any hardening in the region close to the crack path. Only with a special etching technique a small number of slip bands can be revealed as shown in Fig. 7. The extension of these slip bands corresponds to a calculated plastic zone size of a few µm at ΔK_{th}. A large amount of small secondary cracks initiated at iron and silicon containing inclusions could be observed.

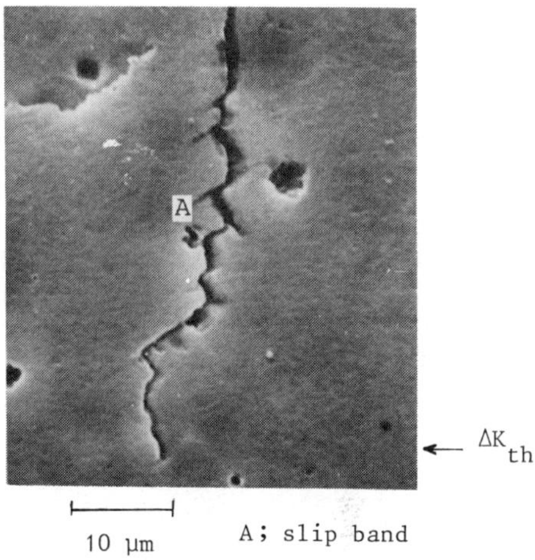

Fig. 7 SEM micrograph showing a serrated crack path in Al 2024-T3 at threshold conditions

No influence of the test frequency can be deduced, neither from the FCG data, Fig. 2, nor from the fracture surface topography as shown in the SEM-fractographs, Fig. 6. The lack of a frequency effect may be explained by the fact that no significant plastic zones can be observed in the specimens, as already discussed. A further indication of the absence of significant cyclic plastic deformation is the fact that at both test frequencies the temperature of the specimen in the region around the crack tip did not reveal any measureable increase.

CONCLUSIONS

The determination of near-threshold fcg-rates in a wrought high strength Al-alloy for stress ratios varying from -1 to 0.7 revealed that increasing R-values generally increase the fcg-rates and decrease ΔK_{th} except for R = -1, at which slightly higher fcg-rates than at R = 0.05 were observed.

The influence of R on ΔK_{th} can be discussed in terms of crack closure. Oxide-induced closure and roughness-induced closure appear to play an important role, whereas plasticity induced closure may have little effect. Oxide induced closure is attributed to debris on the crack surfaces. Roughness induced closure is indicated by shear facets and a serrated crack path.

Crack closure effects are more pronounced at low positive stress ratios. At R = -1 the crack closure effects are less than at R = 0.05. This explains the higher fcg-rates for R = -1 compared to R = 0.05. A suggested explanation for this effect is the flattening of the crack surfaces due to repeated impact during the compressive load cycles at R = -1.

The value of P_{op}/P_{max} decreases continuously when ΔK is increased. This behaviour is explained by the plastic zone size being less than the grain size at the observed ΔK-values. For higher ΔK-values the fatigue crack growth is insensitive to microstructure and plasticity induced closure is prevalent.

No significant frequency effects on near-threshold crack growth behaviour could be found. This is attributed to the very small amount of plastic deformation in the particular Al-alloy.

ACKNOWLEDGEMENT

The authors would like to thank Professor R. Stickler for helpful discussions.

REFERENCES

(1) Bäcklund, J., Blom, A.F. and Beevers, C.J., Proc. Int. Conf. on Fatigue Thresholds, EMAS, U.K., 1982
(2) Wells, J., Buck, O., Roth, L.D. and Tim, J.K., Proc. Int. Conf. on Ultrasonic Fatigue, AIME, U.S.A., 1982
(3) ASTM E 399, in 1982 Annual book of ASTM Standards, Vol. 10.
(4) Stickler, R. and Weiss, B. in Ref. 2, p. 135
(5) Blom, A.F., Hadrboletz, A. and Weiss, B. in Proc. ECF 4, Eds. Maurer, K.L. and Matzer, F.E., EMAS, U.K., 1982, p. 462
(6) Proposed ASTM Test Method For Measurement of Fatigue Crack Growth Rates, Appendix II in ASTM STP 738, 1981
(7) Schmidt, R.A. and Paris, P.C. in ASTM STP 536, 1973, p. 79
(8) Hadrboletz, A.; To be published
(9) Blom, A.F. and Glemberg, R.; To be published
(10) Lindley, T.C. and Richards, C.E., in Ref. 1, p. 1087
(11) Pook, L.P. in ASTM STP 513, 1972, p. 106
(12) Minakawa, K. and McEvily, A.J. in Ref. 1, p. 373
(13) Suresh, S., Zaminski, C.F. and Ritchie, R.O., Met. Trans. A, Vol. 12 A, 1981, p. 1435
(14) Byrne, J. and Duggan, T.V., in Ref. 1, p. 759
(15) Liaw, P.K., Leax, T.R., William, R.S. and Peck, M.C., Met. Trans. A, Vol. 13 A, 1982, p. 1607
(16) Elber, W. in ASTM STP 486, 1971, p. 280
(17) Rice, J.R. in ASTM STP 415, 1967, p. 247
(18) Asaro, R.J., Hermann, L. and Barik, J.M. Met. Trans. A, Vol. 12 A, 1981, p. 1133
(19) Suresh, S. and Ritchie, R.O., Met. Trans. A, Vol. 13 A, 1982. p. 1627
(20) Suresh, S., Palmer, I.J.C. and Lewis, R.E., Fat. Eng. Mat. Struct., Vol. 5, 1982, p. 133
(21) Davidson, D.L., Fat. Eng. Mat. Struct., Vol. 3, 1981, p. 229
(22) Walker, N. and Beevers, C.J., Fat. Eng. Mat. Struct., Vol. 1, 1979, p. 135

CREEP CRACK INITIATION AND GROWTH IN AN AUSTENITIC STAINLESS STEEL

E. Maas and A. Pineau

Centre des Matériaux, Ecole des Mines de Paris, ERA CNRS N° 767, B.P. 87, 91003 Every, Cedex, France

ABSTRACT

The initiation and growth behavior of creep cracks are investigated in a Type 316L stainless steel. Two specimen geometries, CT type and DEN type specimens, containing different initial crack lengths were tested at 600°C with low applied loads leading to low crack growth rates $\dot{a}$ ($\dot{a} < 10^{-3}$ mm/hr). The results show that there is no unique relationship between $\dot{a}$ and a given loading parameter whatever it is, either K, σ_{net}, σ_{Ref} or C*. It is found that it is possible to relate uniquely the time for crack initiation (t_i) and the initial value of C*. This relationship is used in conjunction with the Riedel and Rice stress-strain field to model creep crack growth. The model which is based upon the accumulation of creep strain ahead of the growing crack is shown to be able to account for the effect of load history. Calculated and experimental curves a(t) are shown to be in reasonable agreement, especially in the regime corresponding to low crack growth rates.

INTRODUCTION

The initiation and growth behavior of creep cracks is a technical and theoretical important problem which has received the attention of an increasing number of researchers over the last past decade, see eg.[1,2,3]. This research effort can be explained by the demand for the development of methods for assessing acceptable defect sizes for plants in high temperature service. A large number of problems have not yet been completely solved. In particular the relevant characterizing parameters are not fully established. Furthermore the physical damage micromechanisms taking place ahead of a growing creep crack are not completely understood. Our study deals with these two aspects of the problem in an austenitic stainless steel. A large experimental program was instigated to test systematically the applicability of several loading parameters which are candidate for characterizing creep crack growth behavior. The emphasis was laid on very low crack growth rates which are more relevant to practical situations. The present paper gives a short account of the results.

EXPERIMENTAL PROCEDURES

The experiments were conducted on one heat of Type 316L stainless steel. The material composition is given in Table 1.

TABLE 1 Chemical composition (Wt %)

C	N	Mn	Si	S	P	Cu	Co	B	Ni	Cr	Mo
.03	.075	1.84	.44	.002	.021	.175	.11	.0011	12.3	17.5	2.47

Two specimen geometries were employed : compact tension (CT) type and double edge notched (DEN) type specimens. The specimen dimensions were for the DEN geometry B = 10 mm, W = 30 mm; for the CT specimens, B = 10 mm, W = 40 mm. In the latter case a number of tests were carried out on 20 pct side-grooved specimens. After fatigue precracking at room temperature and at low $\Delta K (\simeq 11 \text{ MPa} \sqrt{m})$ the specimens were subjected to a constant load at 600°C. In order to test the applicability of different characterizing parameters the effect of precrack length was investigated over the range between a/w = 0.30 and 0.60. The specimens were instrumented to measure both the crack length a(t) using a DC potential drop technique and the load line displacement $\Delta(t)$ using an extensometer. This enabled us to determine the time for crack initiation (t_i) and to correlate a with various parameters such as K, the net stress (σ_{net}), the reference stress (σ_{Ref}) and the creep integral C*. The reference stress was calculated using the results published by Haigh and Richards [4]. Two procedures were adopted for the determination of C*. The first one, which is essentially "experimental", relies upon the relationship proposed by Harper and Ellison [5] and utilizes directly the experimental measurement of Δ . It is noticed C^*_{HE}. The second one derives from the numerical calculations by Kumar and Shih [6] and from the knowledge of the parameters describing the stationary creep law $\dot{\varepsilon}_s = B\,\sigma^n$. In our material, at 600°C, $B = 1.63\ 10^{-23}$, n = 7.8 when σ is given in MPa and ε is expressed in hr^{-1}. Hereafter this second definition for C* is noticed C^*_{KS}.

RESULTS AND DISCUSSION

A large number of tests showed that there exists a close relationship between t_i and the initial value of C^*_{HE}, noticed C^*_{iHE} (Fig. 1). Such a correlation could not be obtained with the other three tested parameters. Fig. 1 indicates that the relationship can be written as :

$$t_i\ C^{*\alpha}_{iHE} = \text{Cste, where } \alpha = 0.6 \qquad (1)$$

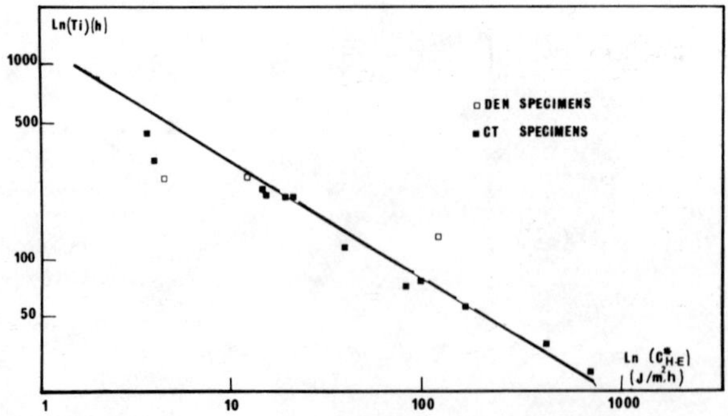

Fig. 1. Time for crack initiation (ti) as a function of the initial value of the creep integral C^*_{HE}.

In this expression t_i was defined as the time to propagate the initial creep crack over a characteristic distance X_c = 50 µm. Sections of interrupted specimens showed that this distance which is close to the grain size in our material corresponded to a real physical crack advance of about 50 µm. It is worth noting that the relationship (1) applies to all the experimental results over the range $10h < t_i < 700h$ and is similar to the theoretical expression derived from the Riedel and Rice analytical calculations [7] which leads to :

$$\varepsilon_c = B \left(\frac{C^*}{B I_n X_c}\right)^{\frac{n}{n+1}} t_i \qquad (2)$$

if it is assumed that crack initiation occurs when the creep ductility (ε_c) is exhausted over a distance (X_c). Assuming that ε_c is constant the relationship (2) gives $\alpha = n/n + 1 \simeq 0{,}9$, which is different from the experimental value. It felt that this difference is associated with the fact that ε_c could be a decreasing function of time.

It is well to remember that in the relationship (1) the determination of C^*_{HE} is largely "experimental". For predictive purposes, it is crucial to calculate completely C* and, therefore, to compare the experimental curves $\Delta(a)$ with the "theoretical" ones derived from the numerical calculations by Kumar and Shih [6]. As a general rule this comparison (Fig. 2) shows that, at crack initiation, the values $\dot{\Delta}_{exp}$ and $\dot{\Delta}_{KS}$ are very close, if a factor of 4 is accepted, except in the case of lightly loaded tests (specimen 23). On the other hand, it was observed that at the onset of crack propagation, $\dot{\Delta}_{exp} \approx$ cste. This is at variance with $\Delta_{KS}(a)$ which is found to be a steadily increasing function of a, as shown in Fig. 3. This figure shows also that, at the end of the test, $\dot{\Delta}_{KS}$ is much smaller than $\dot{\Delta}_{exp}$. These differences can be attributed to the effect of (i) primary creep at the onset of crack propagation and (ii) tertiary creep at the end of the test. These effects are not taken into account in the numerical calculations which are based upon stationary creep. More detailed calculations have shown that primary creep is partly responsible for the fact that at small crack extension, $\dot{\Delta}_{exp}$ is almost constant [8].

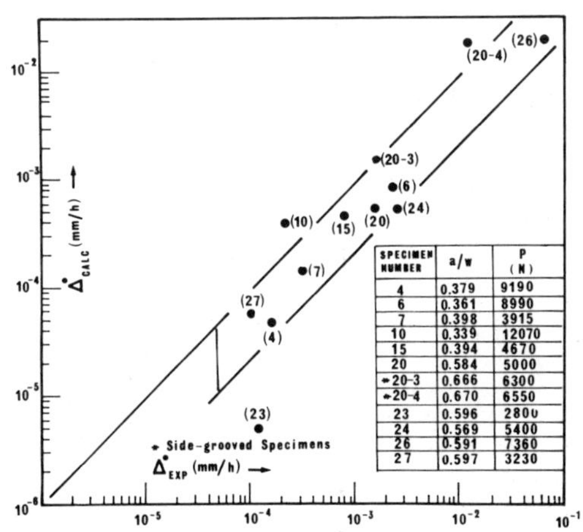

SPECIMEN NUMBER	a/w	P (N)
4	0.379	9190
6	0.361	8990
7	0.398	3915
10	0.339	12070
15	0.394	4670
20	0.584	5000
•20-3	0.666	6300
•20-4	0.670	6550
23	0.596	2800
24	0.569	5400
26	0.591	7360
27	0.597	3230

Fig. 2. Comparison between the calculated displacement rate and the experimental results. The loading conditions corresponding to each specimen are indicated.

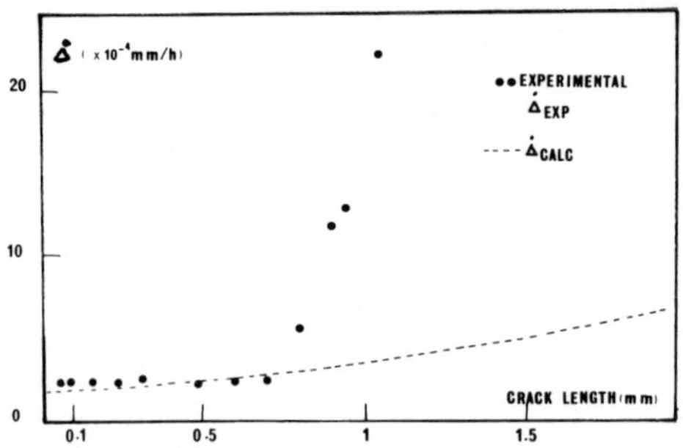

Fig. 3. Comparison between the curves $\dot{\Delta}_{calc.}$ and $\dot{\Delta}_{exp.}$ for the CT specimen n°10 a/w = 0,34, P = 12070 N

Crack propagation results are given in Fig. 4 where $\dot{a}$ is plotted versus C^*_{HE}. This figure shows that there is not a unique relationship. The same behavior was observed when $\dot{a}$ was plotted as a function of the other three investigated parameters. In Fig. 4, two stages can be distinguished. The first one, stage I, which corresponds to very low crack growth rates, ie $\dot{a} < 2.10^{-3}$mm/hr has not been largely investigated in the litterature in spite of its practical importance. During this stage it was noticed that $\dot{\Delta}_{exp}$ was almost constant. This corresponds to a constant value for C^*_{HE} for a given test. On the other hand, during the second stage, $\dot{a}$ is an increasing function of the loading parameter. In this regime, if a large "scatter band" in crack growth rate is accepted, $\dot{a}$ is approximately proportional to C^*_{HE}, as already observed by other authors [eg.1]. Here it is worth noting that the apparent linear correlation between $\dot{a}$ and C^*_{HE} in stage II may be artificial. This arises from the fact that C^*_{HE} is calculated from $\dot{\Delta}_{exp}$. In stage II and in our material detailed calculations [8] showed that, at high crack growth rates, $\dot{\Delta}_{exp}$ is essentially related to the variations in crack length and not to the overall behavior of the specimens. In other words, in this regime the $\dot{a}$ — C^*_{HE} correlation is doubtful since it is based largely upon a correlation between $\dot{a}$ and $\dot{a}$.

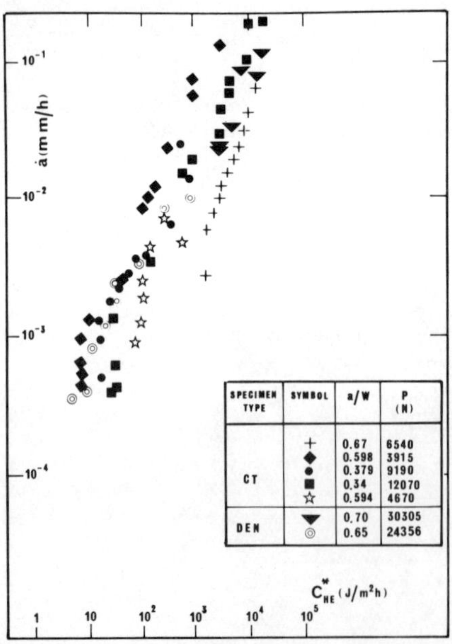

Fig. 4. Variation of the creep crack growth rate $\dot{a}$ with the creep integral C^*_{HE}.

Scanning electron microscopy observations showed that in all cases the fracture mode was intergranular. Detailed optical microscopy observations of polished sections taken in the vicinity of the crack surfaces showed that abundant grain boundary damage was predominantly associated to stage II crack growth (Fig. 5). This indicates that the accumulation of creep damage ahead of a growing crack must be taken into account to model creep crack growth.

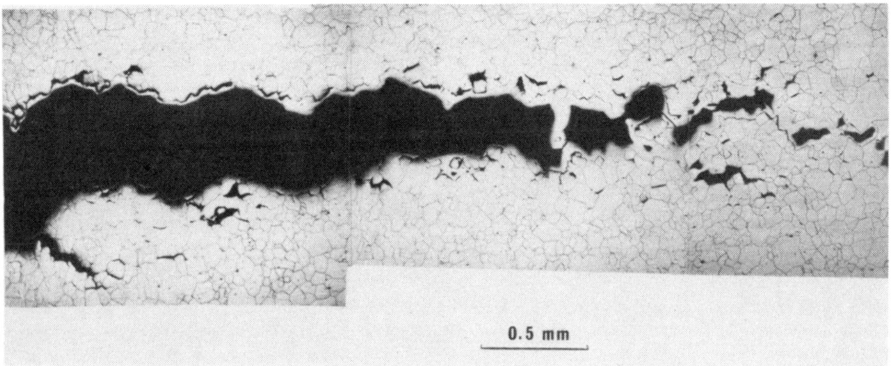

Fig. 5. Optical microscopy of a sectioned specimen (specimen n°13.a/w=0.6,P=3933N) illustrating the accumulation of grain boundary damage along the crack growth.

An attempt was made to model the experimental results in terms of a creep ductility exhaustion concept. The basis of the model relies upon the following analysis.

(i) In the crack plane, the strain rate ahead of the crack tip is assumed to be given by Riedel and Rice expression [7] for a stationary crack, ie :

$$\dot{\varepsilon} = k \ (C^*/r)^{n/n+1} \qquad (3)$$

where k can be calculated from tables.

(ii) Crack initiation is assumed to take place when the creep strain reaches a critical value ε_c, function of t_i, over a characteristic distance X_c = 50μm. The value of ε_c is determined from the relationships (1) and (2).

(iii) Crack growth is modelled as successive initiation steps governed by the stationary strain rate field (equation 3). The observed acceleration in crack growth rate is attributed to the accumulation of creep strain ahead of the crack tip. With this respect our model is very similar to a recent approach proposed by Kubo et al [9]. The creep strain accumulation is calculated using equation (3). This model shows that at a distance pX_c from the initial crack position the crack growth rate can be written as expression (4) where t_i is given by the experimental relationship (1). This model accounts for the effect of loading history since it incorporates the initial value of C*, ie C_i^*.

$$\dot{a}(pX_c) = \frac{X_c}{t_i} \cdot \frac{C^*(pX_c)}{C_i^*} \cdot \frac{1}{1 - \frac{X_c}{t_i} \sum_{j=0}^{p-1} \left(\frac{C^*(jX_c)}{C_i^*}\right)^{\frac{n}{n+1}} \cdot \frac{1}{\dot{a}(jX_c)} \cdot \left(\frac{1}{p+1-j}\right)^{\frac{n}{n+1}}} \qquad (4)$$

The main difficulty in the application of this model lies in the proper calculation of C* and therefore the proper calculation of the displacement rate $\dot{\Delta}$. This is illustrated in Fig. 6 and 7 where curve (1) is calculated with the values of $C^*_{KS}(a)$ while curve (2) is obtained with $C^*_{HE}(a)$. Curve (2) is not completely "theoretical" since it incorporates $\dot{\Delta}_{exp}$. In spite of this difficulty Fig. 6 shows that our model gives consistent results expecially for the regime corresponding to low crack growth rates. In stage II, our model which does not take into account the effect of tertiary creep underestimates the crack growth rates. The agreement is poorer in Fig. 7. It is believed that this situation arises partly from the fact that curve (1) was calculated using stationary creep while in this lightly loaded test, the importance of primary creep cannot be neglected.

Fig. 6 - 7. Creep crack length as a function of time. Comparison between the experimental results and the two calculated curves.

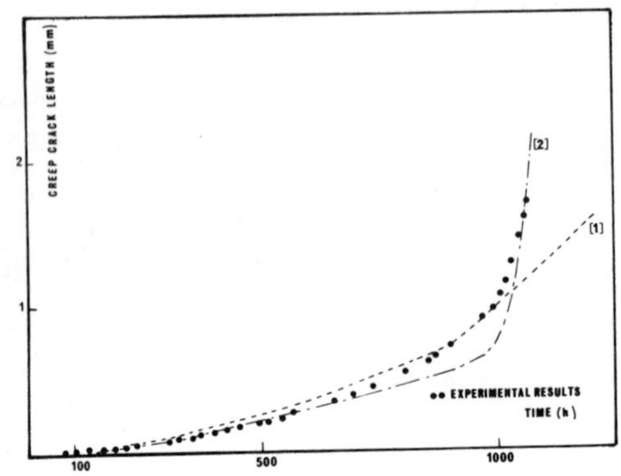

Fig. 6. Specimen n° 10 (a/w = 0.34 , P = 12070 N).

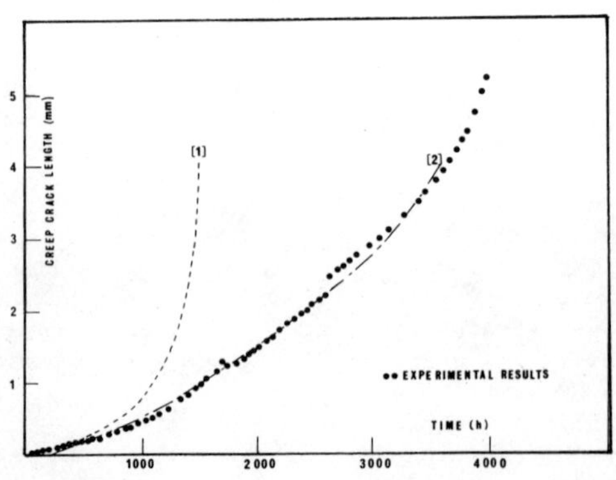

Fig. 7. Specimen n° 4 (a/w = 0.379 , P = 9190 N).

CONCLUSIONS

(1) There exists a close relationship between the time for crack initiation and the initial value of the creep integral C_i^*.

(2) It is not possible to relate uniquely the crack growth rate $\dot{a}$ with a given loading parameter because of the loading history.

(3) A simple model based upon the creep ductility exhaustion concept and the Riedel and Rice strain rate field gives consistent results, especially for low crack growth rates corresponding to $\dot{a} < 2.10^{-3}$ mm/hr.

REFERENCES

1. Ohji, K., K. Ogura, S. Kubo and Y. Katada 15-19 Sept. 1980). Int. Conf. on Engineering Aspects of Creep, Sheffield (England), C240/80, 9.
2. Saxena, A. (1980). Fracture Mechanics, 12th Conference ASTM STP 700, 131.
3. Landes, J.D. and J.A. Begley (1976). Mechanics of Crack Growth ASTM STP 590, 128.
4. Haigh, J.R. and C.E. Richards (1974). Central Electricity Research Laboratories Lab. Memorandum N° RD/L/M461.
5. Ellison, E.G. and M.P. Harper (1978). J. of Strain Analysis, 13, 35.
6. Kumar, V. and C.F. Shih (15-19 sept. 1980). Int. Conf. on Engineering Aspects of Creep, Sheffield, C228/80, 211.
7. Riedel, H. and J. Rice (1980). Fracture Mechanics ASTM STP 700, 112.
8. Maas, E, and A. Pineau. Unpublished research.
9. Kubo, S. K. Ohji and K. Ogura (1979). Eng. Fract. Mechanics, 11, 315.

FLAWS, FRACTURE MECHANISMS AND HERTZIAN CRACKS IN BRITTLE SOLIDS

I. Finnie and D. Dolev

Department of Mechanical Engineering, University of California, Berkeley, CA 94720, USA

ABSTRACT

This paper treats the problem of predicting fracture in a brittle solid loaded by a spherical indenter. Two quite different approaches have been take in the literature. One is based on the Weibull probabalistic treatment of brittle strength while the other makes use of concepts from linear elastic fracture mechanics. It is shown that both approaches provide useful information for this fracture problem and their advantages and limitations are discussed. Finally, it is shown that the classical Auerbach's law which relates mean fracture load $\overline{P}$ to the indenter radius R in the form $\overline{P} \sim R$ is a simple consequence of linear elastic fracture mechanics if it is assumed that the brittle solid contains a large range of flaw sizes at all locations.

KEYWORDS

Auerbach's Law, flaw distribtuion, fracture under spheres, glass strength, Hertzian cracks, ring cracks, Weibull distribution.

INTRODUCTION

Over one hundred years ago, Hertz (1881) presented a general solution for contact stresses between elastic solids. The result is well known and shows that the stresses on the surface of the semi-finite solid are compressive within the contact area, while outside on the free surface the radial stress is tensile, and the compressive stress is on equal magnitude but compressive. The radial tensile stress σ_r on the free surface has its maximum value σ_a at the rim of the contact circle of radius r = a and for larger radii is given by

$$\sigma_r = \sigma_a(a/r)^2 \tag{1}$$

The radial stress also decreases rapidly with depth below the surface and tensile stresses exist only in a shallow region shown schematically in Fig. 1. This description of the stress state applies strictly only to the case of frictionless contact which occurs when the indenter and surface have the same elastic constants. A correction which has to be applied when this is not the case will be discussed later.

The fracture pattern which develops when a brittle solid is loaded by a spherical indenter is similar in many material. The crack appears as a ring concentric, but outside the contact area, on the contact surface. It also extends downward and outward as shown schematically in Fig. 1. With increasing load the crack extends to form the frustum of a cone and additional concentric cracks may form.

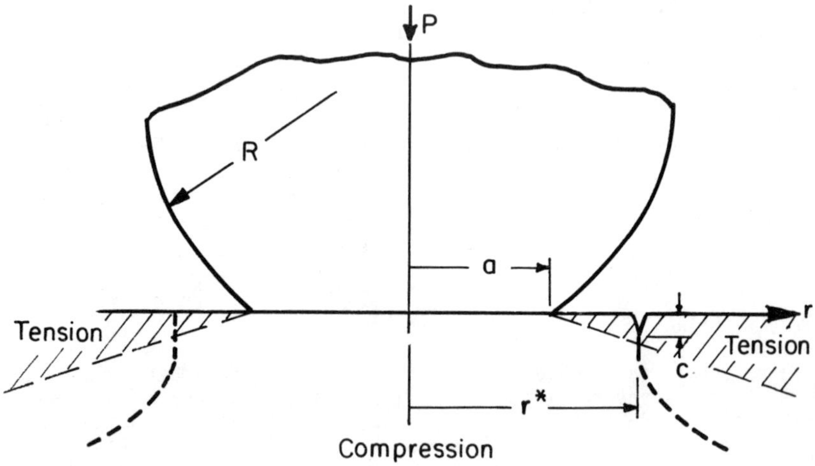

Fig. 1. Schematic view of a semi-infinite solid loaded by a spherical indenter

If brittle solids failed at a critical value of the maximum tensile stress, the Hertz solution would predict a relation between fracture load and indenter radius of the form $P \sim R^2$. Also fracture would be expected to occur at the rim of the contact area with little variability in fracture location or fracture load. In fact this simple picture is far from realized and as a result a vast amount of literature has appeared on this topic of Hertzian cracking. Starting with Auerbach (1891) it has been observed that the relation between fracture load and indenter radius is more nearly $P \sim R$, at least for small indenters. This leads to a strong "size-effect" with the value σ_a of the maximum tensile stress at fracture ranging as $R^{-1/3}$. Also the fracture load and fracture location exhibit a great deal of variability with the mean location of the crack on the surface (r* in Fig. 1) lying well outside the radius a of the contact circle.

It is interesting that it was the variability in the load to produce Hertzian cracking which motivated the pioneering studies of Weibull (1939 a,b,c) on the probablistic treatment of brittle strength. He pointed out that departures from the idealized relation $P \sim R^2$ in Hertzian fracture could be explained using the procedures he developed but did not pursue this problem in any detail. Rather he extended his work in many other directions such as fatigue and reliability. The variety of fracture and other problems to which Weibull's concepts have been applied are truly remarkable (Finnie, 1977). His 1939 papers not only represent one of the greatest contributions to the study of the fracture of brittle solids but also are of fundamental value in many other fields.

With this background of success it is curious that the very test which motivated Weibull's work has been used to question the validity of his approach. The origin of this criticism appears to start with the work of Tillet (1956) who tested glass using indenters with a radius ranging from 0.375 cm to 12.7 cm. Tillet drew lines $\bar{P} \sim R$, $R < 4$ cm; $\bar{P} \sim R^2$, $R > 4$ cm through her data points and showed the standard deviation by bars above and below the mean value $\bar{P}$ of the load. Several aspects of Tillet's work appear to have been misunderstood by subsequent investigators. First, it has been though that relations such as $P \sim R$ and $P \sim R^2$ would not be expected from something as random as the flaw size distribution invoked by Weibull. In fact, Tillet's data are fitted better by expressions of the form $P \sim R^{1.1} < 2.54$ cm, $P \sim R^{1.7}$ $R > 2.54$ cm so this criticism is not entirely valid. A more serious misinterpretation made originally by Roseler (1956), and repeated by later writers, is that the scatter in Tillet's results is almost the same for small and large indenters, a result which would not be expected from the Weibull approach. However, if it is noted that Tillet's data are in a log-log plot it is seen that her results show a constant ratio of the standard deviation to the mean. This is exactly what would be predicted by a two parameter Weibull distribution. In the years following Tillet and Roesler's work an extensive and one might say biomodal distribution of literature has appeared on Hertzian cracking. One has followed the probabalistic approach of Weibull while the other has been based on linear elastic fracture mechanics (LEFM) in an attempt to show that Hertzian cracking results can be explained as a deterministic basis. Generally, the approaches have been presented as mutually exclusive with no attempt having been made to reconcile the two viewpoints.

In the present paper we will point out the predictions which can be made using Weibull's approach. We will then consider the approach based on LEFM to illustrate the predictions which can be made. Depending on the choice of inherent flaw geometry and fracture toughness that is made, these deterministic predictions agree quite well with those based on flaw statistics. Thus the Hertzian fracture problem is seen to be a rather unique case in which both deterministic and probabalistic approaches can be used to provide useful predictions. Finally Auerbach's Law is shown to be a simple consequence of LEFM if one assumed that all sizes of flaws are available at all locations. This approach emphasizes the physical basis of Auerbach's Law and does not appear to have been presented in the present explicit form by previous investigators.

THE WEIBULL APPROACH

The procedures developed by Weibull are well known and have been applied to predict ring cracking by H.L. Oh and Finnie (1967, 1970) and other authors. For this reason we give only an outline at this time.

In essence the Weibull approach involves integrating a function of the stress over the region stressed in tension to determine the probability of failure for a given stress distribution and for a stressed region of a given size. For glass which invariably fails from surface flaws the integration is taken over the surface area stress in tension while for polycrystalline ceramics a volume integral may be more appropriate. Following Wiebull the probability of failure of elemental area dA stressed in uniform tension is taken for glass as

$$F(\sigma) = 1 - \exp[-\{(\sigma - \sigma_u)/\sigma_o\}^m \, dA] \quad \sigma > \sigma_u; \quad F(\sigma) = 0 \quad \sigma \leqslant \sigma_u \qquad (2)$$

where the parameters m, σ_o and σ_u are assumed to be properties of the material and are related to the distribution of microscopic flaws in the material as discussed by Pankow and Finnie (1979). Equation 2 is Weibull's three parameter distribution for uniaxial tensile stresses. Taking $\sigma_u = 0$ we obtain the more tractable two parameter distribution which has been used to study a wide variety of fracture problems. Weibull extended his uniaxial treatment to multiaxial stresses. As shown by Vardar and Finnie (1975) the three parameter Weibull distribution presents some difficulties for multiaxial tensile stresses and most computations have been made with the simpler two parameter distributions. For the present case it may be shown that the error produced by neglecting the compressive hoop stress and using Weibull's uniaxial treatment is negligible.

The application of Eq. 2 to predict the probability distribution for the fracture of glass under a spherical indenter involves an integration over the region stressed in tension as discussed by H.L. Oh and Finnie (1967, 1970). The second paper corrects an omission in the first paper which was pointed out be Dr. K.P.L. Oh. That is, the probability that the material under the indenter survives the radial tensile stresses it experienced prior to submergence under the indenter must be included in the calculations. The correction decreases the predicted fracture load for the parameters considered by a factor of about 1.5. A similar factor was observed by Finnie and Vaidyanathan (1974) in comparing the mean fracture load for cylindrical punches and spheres with the same contact radius. The stress state outside the contact surface is indentical for the two cases but for the punch the surface under the indenter never experiences tensile stresses. As a result the mean fracture load is greater for a cylindrical punch than for a spherical indenter with the same contact area.

Returning to the predictions for spheres, the Weibull parameters used were obtained from 38 tests on glass speciments tested in pure bending. To emphasize the uncertainty that is always present in determining parameters from experimental data H.L. Oh and Finnie (1967) presented two independent estimations of m, σ_o and σ_u. The predictions based on these parameters for the mean fracture load for steel indenters on Pyrex 7740 glass plate are shown by the solid lines in Fig. 2 (H.L. Oh and Finnie 1970). Also shown are the mean values of fracture load obtained from over fifty observations at each radius using hardened steel spheres as indenters. The solid circles and squares represent tests on 0.6 cm and 3.2 cm thick plates. A few years after this work was carried out, Johnson, O'Connor and Woodward (1973) pointed out the importance of frictional forces when the indenter and surface have different elastic constants. They tested steel spheres on glass and glass spheres on glass. From their results it is possible to convert the experimental values for steel spheres on a glass surface to glass spheres on glass (open circles). Then, allowing only for the change in elastic constants, it is possible to compute data points that would correspond to the fictitious case of frictionless steel spheres contracting a glass surface (open squares). These values are the ones that should be compared with the analysis of H.L. Oh and Finnie (1970) which neglects friction at the contact surface. The agreement of predictions with the corrected experimental values is quite satisfactory. However, one might expect that any predictions based on surface stresses would underestimate fracture loads. That is, because the stresses decay rapidly below the surface in the Hertzian test while in pure bending they are essentially constant over the length of any small surface flow. H.L. Oh and Finnie (1970) also compared predicted and experimental values for the standard deviation of fracture load and for the mean

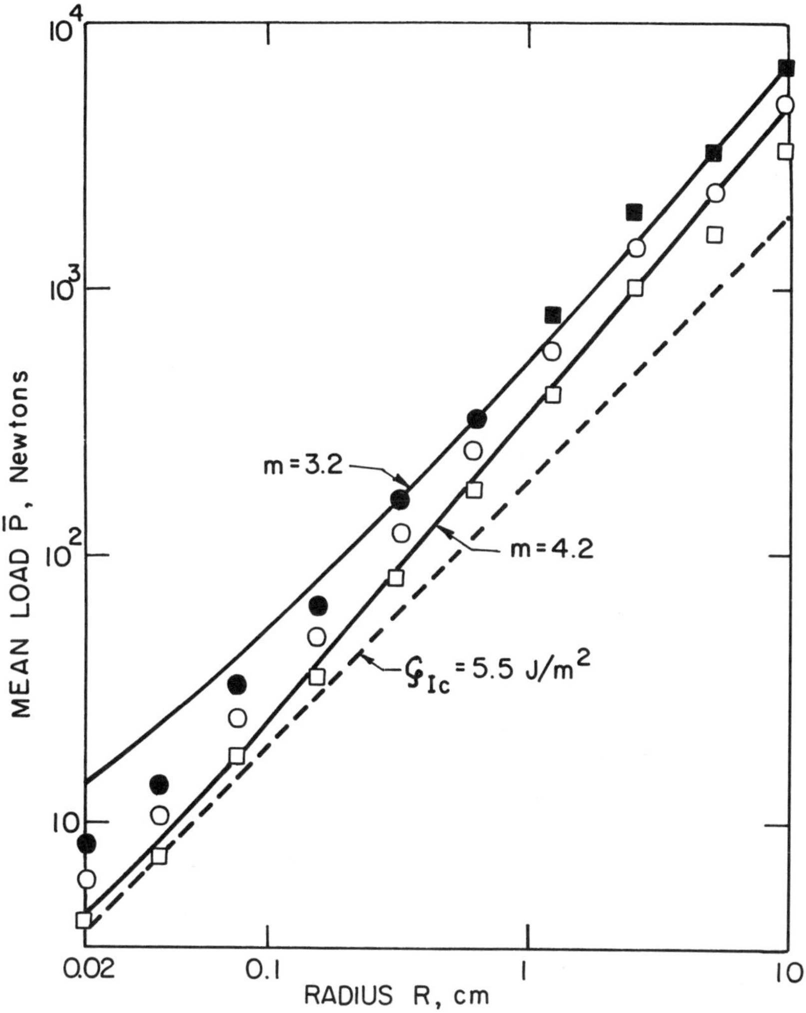

Fig. 2 Experimental data and predictions for fracture of glass under a indenter. Symbols and curves are described in text.

crack location on the surface. The mean crack location was predicted to be 15 to 25 percent greater than the contact circle radius while the experiments with steel spheres on glass showed values ranging from about 25 to 30 percent. It is interesting that Johnson, O'Connor and Woodward (1973) observed a mean crack location about 17 percent greater than the contact circle radius for glass spheres on glass plates and a comparable figure of about 25 percent for steel spheres on glass plates. Thus we would conclude that, if allowance is made for frictional effects, the Weibull treatment allows the prediction of many features of Hertzian cracking under spherical indenters to be made from the results of other types of strength tests.

APPROACHES BASED ON LEFM

Two basic routes have been taken to explain "Auerbach's Law" on a deterministic basis. One based on energy arguments is due to Roessler (1956) and the other which considers the stress intensity factor for a crack in a Hertzian stress field originated with the work of Frank and Lawn (1967). Limitations of these approaches have been pointed out by Finnie and Vaidyanathan (1974) and by Wilshaw (1971). A recent and perhaps a most complete treatment of the Hertzian cracking problem from an LEFM point of view has been given by Wilshaw (1971). Assuming that the inherent flaw can be treated using the two dimensional solution for an edge crack in a semi-infinite solid he obtained a plot of fracture force as a function of normalized crack size c/a for different r*/a (using the notation of Fig. 1). The minimum force is observed for values of $c/a \simeq 0.06$ and $r^*/a \simeq 1.2$. An essentially similar approach was followed by Dolev (1976) who used a slightly different expression for the stress intensity factor K_I and found a maximum K_I for a given load for $c/a = 0.043$ and $r^*/a = 1.18$. Dolev's results can perhaps be visualized more clearly if one plots a dimensionless stress intensity factor as shown in Fig. 3. Here K_I is the mode I stress intensity factor, $k_1 = (1 - \nu_1{}^2)/\pi E_1$ where ν_1 and E_1 are Poisson's ratio and the Elastic Modulus of the surface and k_2 is the corresponding value for the indenter. If one assumes that all sizes of cracks are available at all radii then fracture is predicted to occur at $r^*/a = 1.2$ for cracks of a dimensionless depth $c/a = 0.043$. Since K_I has the constant value K_{Ic} (the plane strain fracture toughness) for crack extension, a constant value of the coordinates in Fig. 3 leads to Auerbach's Law. A similar conclusion can be reached from the equations given by Wilshaw. For quantitative prediction we assumed a value for glass of $G_{Ic} = K_{Ic}(1 - \nu^2)/E$ of 5.5 J/m^2 based on values given in the literature for glass in air. The resulting prediction for glass indenters on a glass surface is shown by the dashed line in Fig. 3 and should be compared to the open circles which are experimental values for steel indenters connected for the case of glass on glass. It should be recalled that the prediction is based on a two dimensional solution and would be appropriate if the inherent flaw is a very shallow semi-elliptical crack. It seems more likely that the inherent flaws would more closely resemble the semi-circular flaws induced by short indenters. However, to our knowledge no solution exists for a semi-circular or semi-elliptical flaw in a Hertzian stress field.

At this stage we should point out an uncertainty that appears to exist in our understanding of Hertzian crack formations. Wilshaw (1971) and other writers postulate that initial fracture occurs at a flaw size c/a below that corresponding to the maximum K_I in Fig. 3 for a given radius. As a result the flaw is assumed to grow unstably until it reaches the same value of K_I at a larger c/a. However if a wide range of flaw sizes is present one should expect initiation to occur as often for flaws with c/a above the value for maximum K_I as for flaws below this value. It appears that a detailed study of the stress intensity solution for surface flaws in a Hertzian stress field is needed to determine whether surface length or depth is the dominant variable. Without such a solution any discussion of the growth of inherent flaws is highly speculative.

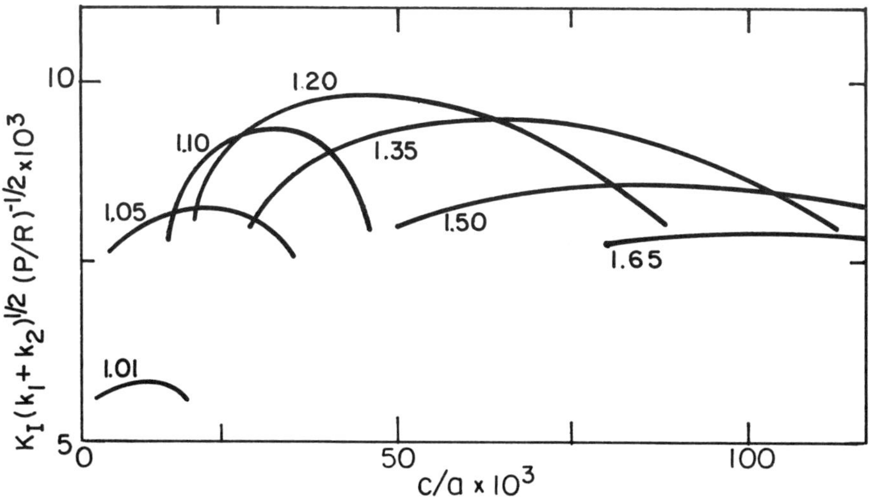

Fig. 3. Dimensionless stress intensity factor as a function of dimensionless crack size for different values of the dimensionless crack location r/a

THE PHYSICAL BASIS OF AUERBACH'S LAW

In earlier work (Finnie, Dolev and Khatibloo, 1982) we pointed out that Auerbach's Law may be deduced directly from basic LEFM considerations and the unusual nature of the Hertzian stress field detailed calculation if it assumed that all sizes of inherent flaws are available at all locations. This approach we believe emphasizes the physical basis of Auerbach's Law and is worth restating.

REFERENCES

Auerbach, F. (1891). Annln. Phys. Chem., 43, 61-100.
Dolev, D. (1976). D.Sc thesis, Technion, Israel Institue of Technology.
Finnie, I. and Vaidyanathan, S. (1974). Fracture mechanics of ceramics, 1, Plenum Press, 231-244.
Finnie, I., Dolev, D. and Khatibloo, M. (1982). Trans. ASME, J. Eng. Mat. and Tech., 103, 183-184.
Frank, F.C. and Lawn, B.R. (1967). Proc. Roy. Soc. (London) 299A, 291-306.
Hertz, H., The original 1881 and 1882 papers are published in "Gesammelte Werke von Heinrich Hertz", 1, 155-173, 174-196, Leipzig 1895.
Johnson, K.K., O'Connor, J.J. and Woodward, A.C. (1973). Proc. Roy. Soc. (London) A334, 95-117.
Oh, H.L. and Finnie, I. (1967). J. Mech. Phy. Solids 15, 401-411.
Oh, H.L. and Finnie, I. (1970). Int. J. Fract. Mech. 6, 187-300.
Pankow, D. and Finnie, I. (1979). Proc. 3rd Int. Conf. on Mat., 3, 67-76.
Roesler, F.C. (1956). Roc. Phy. Soc. B69, 981-992.
Tillet, J.P.A. (1956). Proc. Phy Soc. B69, 47.
Vardar, O. and Finnie, I. (1975). Int. J. Fract. 11, 495-508.
Weibull, W. (1939). Ingvetenskakad, Handl No. 149, 151, 153, Stockholm.
Wilshaw, T.R. (1971). J. Phys. D: Appl. Phys. 4, 1567-1581.

ON THE WEIBULL MODEL OF CLEAVAGE FRACTURE

D. Qu, X. W. Wang and C. K. Tsai

Central Iron and Steel Research Institute, Beijing, China

ABSTRACT

A Weibull type cleavage fracture probability function is used to describe the cleavage fracture of notched and cracked specimens. Method of evaluating the Weibull parameters using the experimental data of cleavage stress σ_f and cleavage K_{IC} is discussed and the results are used to analyse the scatter band of cleavage K_{IC} and cleavage stress σ_f in order to show that the Weibull type fracture criterion is a general and unified one for analysing the cleavage fracture of specimens with different geometries.

KEYWORDS

Cleavage fracture, Weibull statistical model. Fracture probability. Cleavage stress. Cleavage K_{IC} . Carbide crack size distribution function.

INTRODUCTION

Cleavage fracture analysis is an important task in characterizing the fracture resistance of structural steels in low temperature applications. It is commonly accepted that the cleavage resistance can be defined by a single critical stress parameter----the cleavage fracture stress σ_f measured by fracturing a notched specimen at the temperature range of cleavage initiation. In order to apply the single parameter σ_f to predict the cleavage K_{IC} of a cracked specimen, the well-known RKR model postulated that the stress σ_1 in front of the crack tip has to exceed σ_f over a characteristic distance X_0 before fracture could occur. This RKR criterion leads to an expression for cleavage K_{IC} which predicts the temperature dependence of cleavage K_{IC} . However, the model gives no insight into the scatter band of measured cleavage K_{IC} which is an intrinsic properties of the cleavage fracture of structural steels. Recently, it was shown by Pineau(1981) that the temperature dependence of cleavage K_{IC} and its intrinsic large scatter band can be sucessfully described by the Weibull model of cleavage K_{IC} expressed by the following formula:

$$B\,\sigma_y^m\left(\frac{K_{Ic}}{\sigma_y}\right)^4\iint\left[\frac{\sigma_1(u,0)}{\sigma_y}\right]^m\left[\frac{\tilde{\sigma}_1(\theta)}{\tilde{\sigma}_1(0)}\right]^m u\,du\,d\theta = \sigma_u^m V_u \ln\left(\frac{1}{1-p}\right) \quad (1)$$

The parameters m and $\sigma_u^m V_u$ which are related to the material microstructures and cleavage properties (σ_f) were empirically determined to give the right prediction of the temperature dependence and the scatter band of cleavage K_{Ic}. However, the above Weibull parameters have not been corelated directly with the experimental data (σ_f) of notched specimens. Thus, the appropriateness of the Weibull model for notched and cracked specimens and also for smooth specimen is remained to be verified.
In this paper, the theoretical basis of the Weibull model of cleavage fracture is briefly introduced. The method of estimating Weibull parameters m and $\sigma_u^m V_u$ by using the experimental data (σ_f and K_{Ic}) of cracked and notched specimens is described and the results are used to predict the temperature dependence and scatter band of cleavage K_{Ic} in order to show the appropriateness of the model to the cleavage fracture of specimens with different geometries.

BRIEF DESCRIPTION OF THE THEORETICAL BASIS OF THE WEIBULL MODEL

Jayatilaka, Trustrum (1977) had derived a Weibull type expression for the failure probability of the brittle materials based on a proposed defect size distribution function and the strain energy density criterion of fracture for an inclined crack. The proposed defect size distribution function is

$$f(a) = \frac{c^{n-1}}{(n-2)!} e^{-\frac{c}{a}} a^{-n} \tag{2}$$

which is consistent with the result of metallographic analysis of the carbide size distribution. However, the analysis is two-dimensional and the strain energy density criterion is in-applicable for the cleavage fracture of steels. By using the same carbide crack size distribution function (eq.(2)) and the commonly accepted Griffith criterion for cleavage propagation of an inclined carbide crack under the applied principal stress σ_1 , a three-dimensionao analysis for the failure probability had been performed and the following expression for the failure probability $F(\sigma_1)$ of a crack under the applied principal stress σ_1 had been obtained:

$$F(\sigma_1) = \frac{1}{(n-1)!(4n-3)}\left[\frac{(1-\nu^2)\pi c\sigma_1^2}{2E\gamma_p}\right]^{n-1} = \left(\frac{\sigma_1}{\sigma_u}\right)^m \tag{3}$$

with

$$m = 2(n-1) \tag{4}$$

and

$$\sigma_u = \left[(n-1)!(4n-3)\right]^{\frac{1}{m}}\left[\frac{2E\gamma_p}{(1-\nu^2)\pi c}\right]^{\frac{1}{2}} \tag{5}$$

where c and n are the parameters in equation (1), describing the carbide crack size distribution; and γ_p is the effective surface energy of ferrite. Using equation(3), the fracture probability P of a specimen with N carbide cracks under the applied stress σ_1 can be derived based on the weakest link theory:

$$P = 1 - \prod_1^N \left[1 - F(\sigma_1)\right] = 1 - \exp\left[-\frac{V}{V_u}\left(\frac{\sigma_1}{\sigma_u}\right)^m\right] \tag{6}$$

where $V_u = \frac{V}{N}$ is the average volume shared by a carbide crack and V is the volume of the specimen.
For notched and cracked specimen, where the stress is non-uniform. We may apply the equation (6) to give the failure probability P_i of each volume element ΔV_i under the local maximum principal stress σ_{1i} :

$$P_i = 1 - \exp\left[-\sigma_{1i}^m \Delta V_i / \sigma_u^m V_u\right]$$

then the fracture probability of the notched and cracked specimen based on weakest link theory is

$$P = 1 - \prod_{i=1}^{N} (1 - P_i) = 1 - \exp\left[- \frac{\int \sigma_1^m dV}{\sigma_u^m V_u} \right]$$

and the statistical criterion of cleavage fracture for notched and cracked specimen is therefore

$$\int_{V_p} \sigma_1^m \, dV = \sigma_u^m V_u \ln\left(\frac{1}{1-p}\right) \qquad (7)$$

where V_p is the volume of plastic zone at the notch or crack-tip region, since the carbide crack forms only after plastic deformation.

DETERMINATION OF THE WEIBULL PARAMETERS

The brief derivation shown above leads to a general criterion of cleavage fracture, equation (7), in which the parameters m, σ_u and V_u are defined by the carbide crack size distribution, equation (2), and the material properties by equation (5) and the average volume shared by a crack carbide. They are independent of the specimen geometry.
For crack specimens, let $u=r/(K_{Ic}/\sigma_y)^2$ and $dV=Brd\theta dr=B(K_{Ic}/\sigma_y)^4 u du d\theta$. By combining the maximum principal stress distribution along the crack extension line given by McMeeking's FEM computations (1977) $\sigma_1(r,0) = \sigma_1(u,0)$ and the angular factor

$$\tilde{\sigma}_1(\theta) = \frac{1}{2}[\tilde{\sigma}_\theta(\theta) + \tilde{\sigma}_r(\theta)] + \{\frac{1}{4}[\tilde{\sigma}_\theta(\theta) - \tilde{\sigma}_r(\theta)]^2 + \tilde{\sigma}_{r\theta}(\theta)^2\}^{1/2}$$

calculated from HRR fields to give the maximum principal stress distribution $\sigma_1(r,\theta) = \sigma_1(u,0)\,\tilde{\sigma}_1(\theta)$. the criterion (7) can be rewritten as

$$B\sigma_y^m \left(\frac{K_{Ic}}{\sigma_y}\right)^4 \iint \left[\frac{\sigma_1(u,0)}{\sigma_y}\right]^m \left[\frac{\tilde{\sigma}_1(\theta)}{\tilde{\sigma}_1(0)}\right]^m u\,du\,d\theta = \sigma_u^m V_u \ln\left(\frac{1}{1-p}\right) \qquad (8)$$

From this criterion, the scatter of cleavage K_{Ic} is described by the factor $\left[\ln\left(\frac{1}{1-p}\right)\right]^{\frac{1}{4}}$ If we use the mean value of K_{Ic}:

$$\bar{K}_{Ic} \equiv \int_0^1 K_{Ic}(P)dp = \int_0^\infty (1-p)\, dK_{Ic}(P) \qquad (9)$$

the criterion (8) can be rewritten as:

$$C_{cr} \equiv \left[\frac{\bar{K}_{Ic}}{\Gamma\left(1+\frac{1}{4}\right)\sigma_y}\right]^4 B_{cr}\sigma_y^m \iint \left[\frac{\sigma_1(u,0)}{\sigma_y}\right]^m \left[\frac{\tilde{\sigma}_1(\theta)}{\tilde{\sigma}_1(0)}\right]^m u\,du\,d\theta = \sigma_u^m V_u \qquad (10)$$

Similarly, for notched specimens, we have the criterion (7) rewritten in the form:

$$\sigma_f^m \frac{B_n\rho^2}{Q^m} \iint \left[\frac{\sigma_1(\xi,0)}{\sigma_y}\right]^m \varphi_1^m(\theta)\,\xi\,d\xi\,d\theta = \sigma_u^m V_u \ln\left(\frac{1}{1-p}\right) \qquad (11)$$

and

$$C_n \equiv \left[\frac{\bar{\sigma}_f}{\Gamma\left(1+\frac{1}{m}\right)}\right]^m \left[\frac{B_n\rho^2}{Q_m}\right] \iint \left[\frac{\sigma_1(\xi,0)}{\sigma_y}\right]^m \varphi_1^m(\theta)\,\xi\,d\xi\,d\theta = \sigma_u^m V_u \qquad (12)$$

Here, $\xi = 1+\frac{x}{\rho}$, and x is the distance from the notch-tip, $\sigma_1(\xi,0)$ is the maximum principal stress distribution along $\theta=0$ and had been calculated by Griffiths and Owen (1971) using FEM. $\varphi_1(\theta)$ is angular factor of $\sigma_1(r,\theta)$ and can be estimated by $\varphi_1(\theta) = \left[\frac{\cos(\omega/2+\theta)}{\cos\omega/2}\right]^n$ according to Bates (1979), where ω is the flank angle of the notch. $Q=\sigma_1^{max}/\sigma_y$ is the stress intensification of the notched specimen at fracture. B_n and ρ are the thickness and notch-root radius of the notched specimen.
Now, we can determine the Weibull parameter m by comparing the computing results of C_{cr} and C_n using(10) and (12) and the measured σ_f and K_{Ic}. Since

$\sigma_u^m V_u$ and m=2(n-1) are material constants according to (2),(4) and (5), thus we should have $C_{cr}=C_n$ at fracture. By computing C_{cr}/C_n for different m values and finding the intersecting point of (C_{cr}/C_n~m) curve with the horizontal line $C_{cr}/C_n=1$, the value of m at the intersecting point is then its right value, and the value of the parameter $\sigma_u^m V_u = C_n = C_{cr}$ is also determined.

EXPERIMENTAL AND COMPUTATIONAL RESULT AND DISCUSSION

Cleavage stress σ_f and cleavage K_{Ic} were measured at various temperature for a 4% Si silicon steel with grain sizi of 93μm. Since temperature may have some effect on cleavage stress, especially at very low temperature where deformation twinning prevails. The comparison of C_{cr} with C_n for determining the parameter m was conducted at temperature range in which deformation by dislocation glide prevails. The yield strength of the steel at various temperature and strain-rate $\dot{\varepsilon}$ can be described by the following formula(Du,1982)

$$\sigma_y = 343 + 591\ \exp\left[-\left(1.94\times10^{-3} + 0.121\times10^{-3}\ \ln\frac{\dot{\varepsilon}_o}{\dot{\varepsilon}}\right)T\right]\ \text{MPa}$$

with $\dot{\varepsilon}_o=10^5\,\text{sec}^{-1}$.

The angular factors of HRR field for the steel with the strain hardening exponent n=0.1 were calculated by using the following formulas (Bilek,1977; Rice, 1968)

$$\tilde{\sigma}_\theta(\theta)/\tilde{\sigma}_\theta(0) = f(\theta,n)$$

$$\tilde{\sigma}_r(\theta)/\tilde{\sigma}_\theta(0) = (1+n)f(\theta,n) + \left[(1+n)^2/(2+n)\right]f''(\theta,n)$$

$$\tilde{\sigma}_{r\theta}(\theta)/\tilde{\sigma}_\theta(0) = -\left[(1+n)/(2+n)\right]f'(\theta,n)$$

$f(\theta,n=0.1)=0.4968+0.4819\cos\theta+0.00037\cos2\theta+0.01703\cos3\theta+0.0023\cos4\theta$ The computing result is shown in Fig.1. It should be pointed out that the maximum principal stress $\sigma_1(r,\theta)$ retains its highest value until $\theta\approx\pi/3$. The limits of integration for the angular factor in equation(10) were taken as $-\pi$ to $+\pi$.

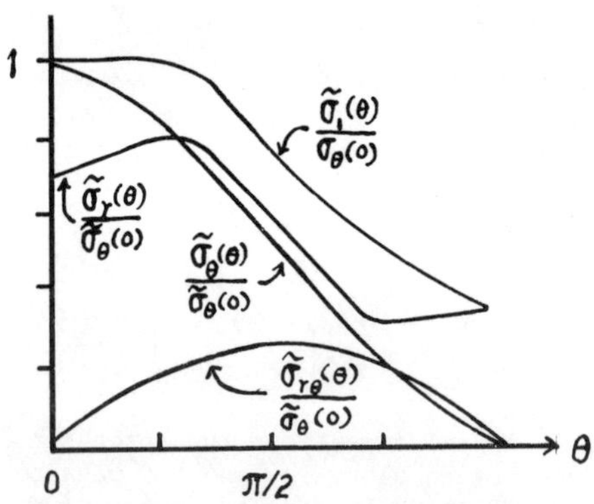

Fig.1. the angular factors for n=0.1

The crack-tip stress distribution along the crack plane (θ=0) calculated by large deformation FEM (McMeeking, 1977) for n=0.1 is shown in Fig.2. where the abscisa had been changed into $r/(K_{Ic}/\sigma_y)^2$ by the transformation formula $r/b=800r/(K_{Ic}/\sigma_y)^2$, which can be derived from McMeeling's formula by inserting n=0.1, $\sigma_y/E=1/300$ and $\nu=0.3$.

$$b \approx \delta_t = 0.55\left[\frac{2}{\sqrt{3}}(1+\nu)(1+n)\frac{\sigma_y}{nE}\right]^n \frac{K_{Ic}}{E}\,\frac{1-\nu^2}{\sigma_y}$$

For simplicity, calculations for notched specimen were conducted at the onset of general yielding by using the stress distribution curve for $L/L_{Gy}=0.953$ and $\sigma_{nom}/\sigma_y=2.051$ as shown in fig.3, and correspondingly the stress intensification factor in equation (12) should be taken as Q=2.55. The angular range of integration for notched specimen is taken from $-3\pi/8$ to $+3\pi/8$.

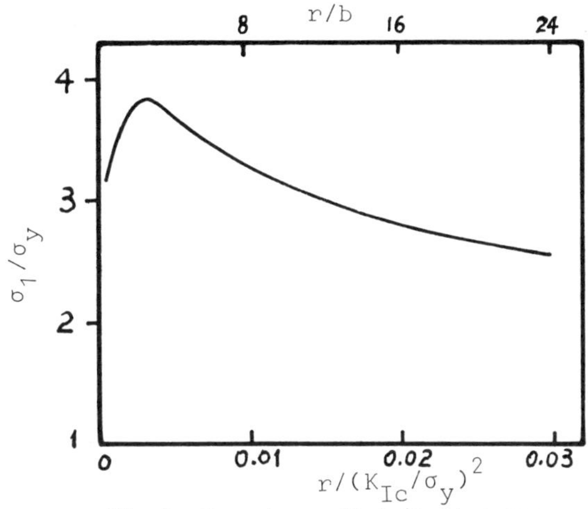

Fig.2. the stress distribution curve $\sigma_1/\sigma_y - r/(K_{Ic}/\sigma_y)^2$ of blunted crack for n=0.1, $\sigma_y/E=1/300$

Using the cleavage fracture stress σ_f and the cleavage fracture tonghness K_{IC} measured for 4% Si silicon steel and the yield strength at the crack tip region σ_y = 538MPa under the strain rate $\dot{\varepsilon}=2\varepsilon/t=2.1\times10^{-3}\,sec^{-1}$ and temperature T=263°K, the computed results of Ccrand C_n for different values of m were shown in table 1. The curve of C_{cr}/C_n versus m was plotted and the Weibull parameters m and $\sigma_u^m V_u$ corresponding to $C_{cr}/C_n=1$ is determined to be m=15.8 and $\sigma_u^m V_u=2.4\times10^{41}$ $MPa^{15.8}$ m^3

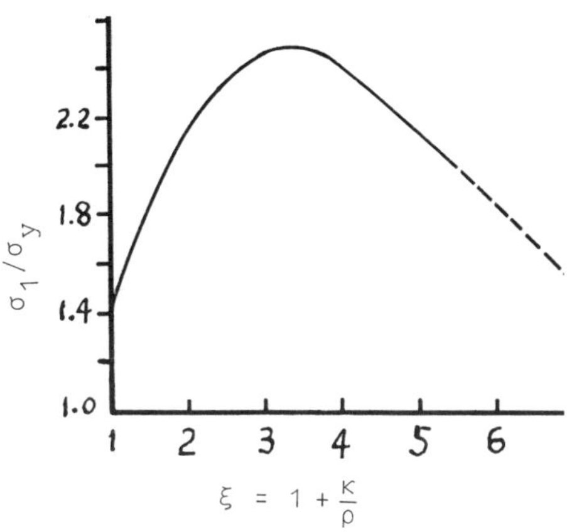

Fig.3. the maximum principal stress distribution σ_1/σ_y for notched specimen

Based on this result, the K_{IC} -T curves for different cumulative fracture probability P=0.1;0.61 and 0.9 are plotted using the equation (10), the result is shown in Fig.4. Most of the measured K_{IC} falls into the mode corresponding to the cumulative fracture probability P=0.61.

Computations for SA533B-1 steel were also conducted based on the above formulations and procedures. The results were shown in table 2. and the comparison of the prediction using computed result (m=15.2, $\sigma_u^m V_u$ $=2.94\times10^{41}MPa^{15.2}\,m^3$) with the experimental data is shown in Fig.5.

Table 1. The variations of the values of C and C with m in 4Si-Fe

m	cracked specimen			notched specimen			$\frac{C_{cr}}{C_n}$
	$A=\int_0^{0.03}[\frac{\tilde{\sigma}_1(u,0)}{\sigma_y}]^m u\,du$	$B=\int_{-\pi}^{\pi}[\frac{\tilde{\sigma}_1(\theta)}{\tilde{\sigma}_1(0)}]^m d\theta$	$C_{cr}=B_{cr}\bar{\sigma}_y^{m-4}[\frac{\bar{K}_{Ic}}{\Gamma(1+4/m)}]^4 AB$	$D=\int_1^{5.5}[\frac{\sigma_1(\xi,0)}{\sigma_y}]^m \xi\,d\xi$	$E=\int_{-3\pi/8}^{3\pi/8}\varphi^m(\theta)d\theta$	$C_n=[\frac{\bar{\sigma}_f}{\Gamma(1+\frac{1}{m})}]^m\frac{B_n\rho^2}{Q^m}DE$	
20.0	5.58×10^6	2.350	1.377×10^{64}	5.74×10^8	0.971	2.825×10^{63}	4.87
18.0	4.46×10^5	2.353	3.807×10^{57}	9.38×10^7	1.025	1.685×10^{57}	2.26
15.8	2.84×10^4	2.365	2.394×10^{50}	1.29×10^7	1.093	2.404×10^{50}	0.99
15.0	1.06×10^4	2.370	5.819×10^{47}	6.30×10^6	1.120	7.861×10^{47}	0.74
12.0	2.74×10^2	2.390	9.867×10^{37}	4.32×10^5	1.238	3.759×10^{38}	0.26

where $\bar{\sigma}_f$=1372 MPa Q=2.55 σ_y=538 MPa T=263°K ρ=.25x10^{-3} m
$\bar{K}_{Ic}$=38.75 MPa$\sqrt{m}$ B_n=11.5x10^{-3}m B_{cr}=6.6x10^{-3}m

Table 2. The variations of the values of C and C with m in SA533-B

m	cracked specimen			notched specimen			$\frac{C_{cr}}{C_n}$
	A	B	C_{cr}	D	E	C_n	
21.0	2.02×10^7	2.35	1.63×10^{61}	1.42×10^9	0.946	1.73×10^{60}	9.42
18.0	4.46×10^5	2.35	9.78×10^{50}	9.38×10^7	1.025	3.33×10^{50}	2.94
15.2	1.34×10^4	2.37	2.98×10^{41}	7.54×10^6	1.110	2.89×10^{41}	1.03
12.0	2.74×10^2	2.39	4.50×10^3	4.32×10^5	1.240	1.32×10^{31}	0.34

where $\bar{\sigma}_f$=1830 MPa Q=2.55 σ_y=718MPa T=145°K ρ=0.25x10^{-3} m
B_n=12.7x10^{-3}m $\bar{K}_{Ic}$=40MPa$\sqrt{m}$ B_{cr}=25.4x10^{-3}m

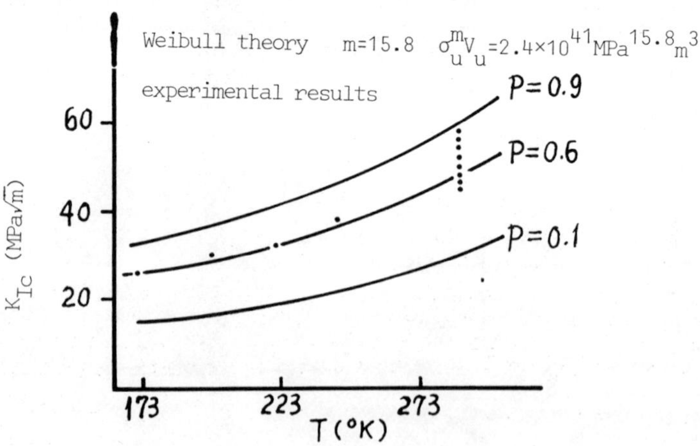

Fig. 4. Comparison of the Weibull model predictions with cleavage K_{Ic} for 4Si-Fe.

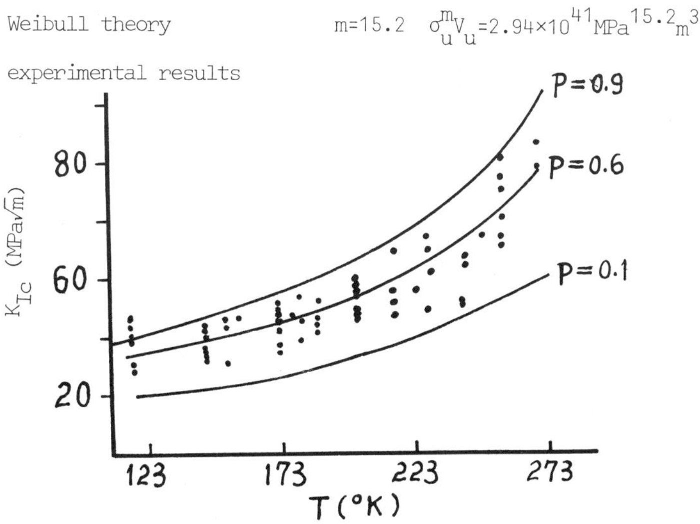

Fig. 5. Comparison of the Weibull model predictions with cleavage K_{Ic} for SA533-B.

CONCLUSSION

The theoretical basis of the Weibull statistical model of cleavage fracture for structural steel is briefly discussed. Based on the unified fracture criterion $\int_{V_p} \sigma_1^m dV = \sigma_u^m V_u \ln\left(\frac{1}{1-p}\right)$ method of estimating the Weibull parameters m and $\sigma_u^m V_u$ by using the experimental data of σ_f and K_{IC} measured by notched and cracked specimens was described. Based on the above Weibull formulations and the computed results of m and $\sigma_u^m V_u$, the temperature dependence and the scatter band of cleavage K_{IC} can be predicted satisfactorily and the appropriatness of the model can thus be concluded.

REFERENCES

Bates,R.C., A.T.Santhanam (1977). Mater. Sci. and Eng., 41, 243-250.
Bilek,Z., W.Uhlmann, Z.Kne'sl, M.Kuna (1977). Inter. J. Fract., 12, No.3, 507-509.
Du,W.Q. (1982). M.S.Thesis, Central Iron and Steel Res. Inst. Beijing, China.
Griffiths,J.R., D.R.J.Owen (1971). Mech. Phys. Solids., 19, 419-431.
Jayatilaka, A.Des., and K.Trustrum (1977). J. Mater. Sci. 12, 1426-1430.
McMeeking, R.M. (1977). Mech. Phys. Solids, 25, 357-381.
Pineau,A. (1981). Advances in Fracture Research Vol.2. 553-577.
Ritchie,R.Q, W.L.Server, and R.A.Wullaert (1979). Metal. Trans. 10A, 1557-1570.

THE ASSESSMENT OF THE FRACTURE TOUGHNESS OF Fe-Ni-C ALLOYS WITH AUSTENITIC AND AUSTENITE-MARTENSITE STRUCTURES

N. J. Hurd* and R. Brook**

***GKN Technology Research Centre, Wolverhampton, UK**
****Department of Metallurgy, University of Sheffield, UK**

ABSTRACT

The fracture toughness of Fe-Ni-C alloys having either a austenitic or a duplex austenite-martensite matrix has been assessed. The toughness was measured using the J-integral, Crack Tip Opening Displacement and lower-bound fracture toughness (K_{Icd}) criteria. Fracture was influenced by the occurrence of transformation-induced plasticity and segregation of impurities to grain boundaries. Slow, stable, crack growth occurred under monotonic loading by brittle modes (intergranular or microcleavage, depending on the precise composition); dimpled rupture was only found after tempering the duplex material. The significance of these findings on the assessment of toughness by the techniques mentioned above is discussed. It is shown that the fracture toughness, based on the initiation of crack growth, could be over-estimated in these circumstances.

KEYWORDS

High-nickel steels; fracture toughness; K_{Id}; J-integral testing; C.O.D. testing; transformation-induced plasticity (TRIP); brittle fracture; fractography.

INTRODUCTION

The aim of this work was to investigate the factors which influence the fracture toughness of austenite-martensite structures and, at the same time, compare the various methods of assessing toughness in situations where appreciable crack tip plasticity occurs. A 0.5%C-24%Ni-Fe alloy was studied because it is austenitic at room temperature, which greatly facilitated control of the martensitic transformation by sub-zero cooling. The martensitic forms as large, well-defined plates and much is known about the transformation and its kinetics (Brook Entwisle and Ibrahim 1960, Brook and Entwisle, 1965, Maki and co-workers, 1972, and Lee and co-workers, 1977). Furthermore, the study of such a simple alloy should provide a basis for the understanding of more complex systems.

It was expected that, after transformation, linear elastic fracture mechanics would be applicable whereas the austenitic material would fail only after gross plasticity had developed. For the latter condition it was decided to evaluate three elastic-plastic fracture criteria, two of which are the subject of National Standard test procedures.

EXPERIMENTAL DETAILS

The analysis of the Fe-0.5%C-24%Ni alloy used is given in Table 1, together with that of a similar alloy containing 0.25%Mn. The addition of manganese was made to eliminate intergranular fracture that was found subsequently to otherwise dominate the behaviour of the base alloy.

TABLE 1 Chemical Analyses

Alloy	Ni	C	Mn	S	Others
5C24N	24.2	0.48	<0.02	0.001	P, Sn <0.005; Sb < 0.01
5C24NM	24.0	0.49	0.25	0.002	Si 0.04, Al 0.023

Fracture toughness tests were carried out on an 5000 Kg Instron machine; single-edge notched bend specimens, 10mm thick, were used. The specimens were pre-cracked to $\frac{a}{W} \sim 0.6$, due regard being paid to the pre-cracking requirements laid down in the standards. Load (P), load-point-displacement (LPD) and clip-gauge-opening across the notch (Vg) were measured continuously during the tests.

Values of J were evaluated at the various points of interest - such as initiation and maximum load - using the expression in the appropriate standard (ASTM, 1977), viz.

$$J = \frac{2A}{B(W-a)} \qquad (1)$$

where A is the area under the load-LPD curve, a the crack length, W is the specimen width and B its thickness.

Three formulae were used to obtain values of the crack-tip opening displacement (CTOD, δ) from the notch opening:-

(a) the 'rotation' formula in its initial form

$$\delta = \frac{Vg(W-a)}{(n-1)a+W+nz} \qquad (2)$$

where n, is the rotational constant, (taken as 3) and z is the height of the knife edges.

(b) the current expression (BS.5762,1979)

$$\delta = \frac{K^2(1-v^2)}{2\sigma yE} + \frac{0.4(W-a)Vgp}{0.4W+0.6a+z} \qquad (3)$$

where v is Poisson's ratio, σy is the yield strength, E is the elastic modulus and Vgp is the plastic component of the clip gauge opening.

(c) the relationship due to Venzi (1973) was also employed

$$\delta = \alpha Vg + {}^{\alpha}/_{\beta} \exp\{-\beta Vg\} + Co \qquad (4)$$

where $\alpha = \dfrac{0.4(W-a)}{a+z+n(W-a)}$, $\beta = \dfrac{.150\alpha}{(W-a)}$; $Co = -\dfrac{\alpha}{\beta}$

The lower bound fracture toughness (K_{Icd}) was calculated using the expression given by Witt and Mager (1972)

$$K_{Icd} = \frac{P_2 \; ^{A_1}/_{A_2} \; Y(\frac{a}{W})}{BW^{\frac{1}{2}}} \tag{5}$$

Here, A_1 was taken simply to be the area up to maximum load; A_2 is the area under the load-displacement up to any load P_2 on the linear portion of the curve; Y is the appropriate compliance function.

Resistance curves were generated using the heat-tint technique. In addition, a DC potential-drop (DCPD) method, a compliance method (Chipperfield, 1976) and a simple acoustic emission technique were used to detect the onset of slow, stable, crack growth.

Standard tensile tests and compressive yield strength tests were also made. Optical metallography was carried out on normal metallographic sections and also on sections taken through the nickel-plated fracture surfaces. These sections were electropolished to eliminate any surface martensite due to mechanical polishing. Fractography was carried out on a Cambridge Mark IIA Stereoscan.

RESULTS AND DISCUSSION

The fracture toughness of the fully-austenitic material was found to be somewhat lower than expected for such a low-strength (compressive yield strength = 228MPa) material. In addition, although slow, stable, crack growth occurred through the ligament during the toughness tests, the fracture mode was found to be intergranular with a small component (less than 10%) of quasi-cleavage, instead of the dimpled rupture mode usually associated with slow crack growth. Despite the brittle nature of the fracture, a steeply-rising J-Δa resistance curve was found (Fig. 1). The interpretation of the resistance curve data using the ASTM 'J' standard method was made difficult by the fact that no stretch-zone formed at the crack tip as expected from the blunting line construction ($J = 2\sigma_{flow}\Delta a$, where σ_{flow} is the flow stress); intergranular fracture occurred directly from the tip of the fatigue pre-crack. In some areas along the pre-crack front a small region of dimpled rupture was followed by quasi-cleavage, (Fig. 2). However, this occurred in places where the fatigue crack has stopped part-way through a grain, and on subsequent monotonic loading the crack had therefore to propagate through the remainder of the grain before it could follow the grain boundaries.

Employing the blunting line construction is therefore inappropriate for this material since it does not describe the physical mechanism occurring, While the construction has previously been found to incorrectly describe the precise extent of the stretch zone (Blauel and Hollstein, 1978), in this case an entirely different fracture mode has been found to occur from that identified with the stretch zone construction. Furthermore, employing the construction would lead to non-conservative estimates of the initiation toughness of the material. Employing the method of measuring Δa suggested in the CTOD standard is able to account for this unusual physical behaviour since the measurement excludes any stretch zone.

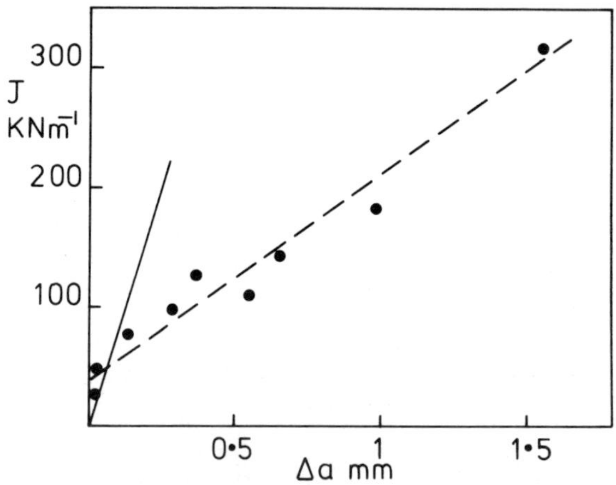

Fig. 1. J-Δa resistance curve. Austenitic SC24N alloy.

Fig. 2. Fatigue crack/ductile initiation/quasi-cleavage/ intergranular fracture. Austenitic SC24N alloy Mag.x150.

Bearing these points in mind, J was therefore evaluated at the onset of growth, a value of $J_i = 25$ kNm^{-1} being obtained. The slope of the resistance curve was found to be 167 kNm^{-1}/mm. The average value of J at maximum load was found to be 197 kNm^{-1}. The DCPD method yielded values at initiation in good agreement with that deduced from the heat-tint method, and was well able to detect the onset of growth by this micromode. The large-scale crack-tip blunting effects which complicate the interpretation of the records in other low-strength materials were not present here. The Chipperfield compliance method and the acoustic technique were somewhat less sensitive in detecting initiation, as summarised in Table 2.

TABLE 2 Detection of Initiation

Method	Average value of J_i deduced (nNm^{-1})	Standard Deviation
Heat tint	25	-
Potential drop	26	1.4
Chipperfield	32	4.5
Acoustic	36	11

A comparison of the various methods considered for assessing toughness will now be made. When compared on the basis of an "equivalent K", J_i was found to give a lower toughness for the material than K_{Icd} (69 $MNm^{-3/2}$ compared with 115 $MNm^{-3/2}$, respectively). This is not surprising since J_i is evaluated at initiation while K_{Icd} as evaluated here involves a contribution up to maximum load.

J and the CTOD have been taken to be related by an expression of the form

$$J = M\sigma_y \delta \tag{6}$$

where M is held to be material dependent.

In this case, values of CTOD of initiation, and hence M, were found to depend on the formula used to infer CTOD from the opening across the notch. For CTOD values evaluated from the original 'rotation' formula (equation 2), the relation between J_i and CTOD at initiation was found to have $M \sim 1.3$, using the tensile properties for the bulk material for σ_y in the expression. Whether or not the effect on the local σ_y of the formation of high-strength martensite local to the crack tip should be taken into account is not clear. For CTOD values at initiation evaluated using equations (3) and (4), M values of 4.51 and 2.39 were obtained respectively. Comparing the parameters at maximum load (with the associated larger notch openings) gave M values which were relatively insensitive to the expression used. In this way, values of 2, 2 and 2.2 were obtained from equations (2) to (4) respectively.

Consider now the transformed material. The maximum proportion of martensite that could be formed by refrigeration was found to be 80 per cent. This microstructure also exhibited slow crack growth by a brittle intergranular mode during the toughness test. In this case, all the requirements for a valid K_{Ic} toughness test (BSI, 1977) were fulfilled except the $P_{MAX}/P_Q < 1.1$ criterion and the fact that the crack length was slightly longer than speci- fied. To observe a brittle mode of slow crack growth under these conditions is, again, unusual. In this case, however, the resistance curve was very shallow ($J_i = 10kNm^{-1}$ and $dJ/da = 4kNm^{-1}/mm$).

The average value of J at maximum load for this duplex matrix was found to be 25 kNm^{-1}, shich is, again, appreciably above J_i. Initiation was, once more, easily detected by the DCPD method, an average value of 11 kNm^{-1} being obtained with relatively little scatter (standard deviation 1.2). Here, the acoustic method also yielded an average value (11 kNm^{-1}) in good agreement with that deduced from heat testing, but with considerably more scatter (standard deviation 4.5).

The J-integral and lower bound fracture toughness approaches yielded similar results of 41 and 43 $MNm^{-3/2}$, respectively, when compared on the basis of an "equivalent K". The value of M in equation (6) was found to be approximately 0.7 when using J and CTOD evaluated at initiation, and using equation (2) to evaluate CTOD.

Tempering the 80% martensite microstructure for 3 hours at 300°C reduced the yield strength (from 1217 MPa to 968 MPa) and promoted a dimpled-rupture fracture mode during the toughness tests. The value of J at initiation was found to be comparable to that of the austenitic material (J_i of 29 kNm^{-1}) but the post-initiation behaviour was considerably different (Fig. 3). The

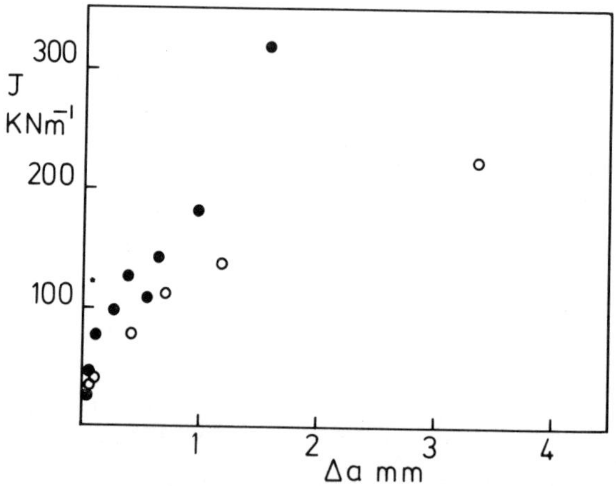

Fig. 3. J-Δa resistance curve. 5C24N alloy, tempered. (80% martensite (O), austenite (O)).

DCPD again detected initiation clearly to give a mean value for J_i of 26 kNm^{-1}. K_{Icd} once more provided a higher toughness than J_i when compared in terms of K (96 $MNm^{-3/2}$ compared with 70 $MNm^{-3/2}$). Equation (2) when used to calculate the CTOD at initiation gave a value for M equal to unity.

Overall, the J-CTOD relation is affected by the relationship used to determine crack-tip opening from clip-gauge opening. In particular, the values of M obtained on the basis of J_i differ significantly from that ($M \approx 2$) estimated from finite-element studies.

Comparing the resistance curves of the austenitic and tempered duplex microstructures in the Fe 0.5C 24Ni alloy shows that while the initiation toughness is similar, the post-initiation toughness is considerably different. This is reflected in the disparity (found generally) or the equivalence (found sometimes) between the different parameters, used in assessing toughness in this work. Since the initiation and post-initiation toughness may vary independently, the use of a single parameter will be inadequate in general when making an overall assessment of material toughness. Hence, two parameters need to be considered. While this has been considered generally to be the case for ductile materials failing by dimpled rupture, this work shows that this approach may also be necessary for materials failing by micromodes - intergranular or quasi-cleavage - which are generally considered to be brittle in metallurgical terms.

The dominant intergranular mode in the Fe-Ni-C alloy was identified as being due to a combination of sulphur segregation to the austenite grain boundaries (as revealed by Auger spectroscopy) together with a high hardness local to the crack tip. The high local hardness could be caused either by the matrix hardness (in the 80% martensite structure) or by the deformation-induced martensite (in the austenitic structure). In the latter instance, sections taken perpendicular to the fracture showed that deformation-induced martensite forms at the crack tip. Midribbed plates of $\{259\}_\gamma$ habit plane and 'obtuse' plates of $\{225\}_\gamma$ habit were observed. The martensite was extremely localised to the fracture surface ($\sim$ 80μm deep) and this deformation-induced martensite is held to be responsible for the abnormal behaviour of this material in this condition.

In order to assess the effect of eliminating the intergranular-embrittlement, an Fe 0.5C 24Ni 0.25Mn alloy was tested in the austenitic and duplex austenite-80% martensite conditions. Slow crack growth was now found to occur by a quasi-cleavage mode in both cases. This emphasises the generality of these observations regarding the occurrence of slow crack growth by brittle micromodes. Deformation-induced martensite again occurred in the austenitic material, and the toughness (as deduced by the method using DC PD) was found to be similar to that of the manganese-free material, whereas, in a similar comparison, the toughness of the duplex structure was slightly higher (J_i = 17 kNm^{-1}).

CONCLUSIONS

(1) This work has shown that brittle fracture modes - both intergranular or quasi-cleavage - can initiate directly at the tip of a fatigue precrack in relatively low-strength materials when loaded monotonically. In these cases, the blunting line construction is inappropriate, and it leads to non-conservative estimates of the toughness at initiation. The reason for this deviation from the expected behaviour of the bulk material in this case is that the fracture behaviour is influenced by deformation-induced martensite which forms local to the crack tip. This serves to emphasise the important effect that relatively small-scale metallurgical effects have on toughness.

(2) Macroscopic slow crack growth has been observed to occur in these alloys by brittle micromodes. No comparable evidence of true slow crack growth by these brittle modes has been reported so far as we are aware.

(3) In some cases where these brittle modes occur the slope of the resistance curve, dJ/da, was appreciable, so that failure to detect initiation accurately would lead to overestimates of the initiation toughness. This highlights the importance of detecting slow crack growth in toughness tests, whatever the micromode.

(4) The DC PD method is able to detect slow crack growth by these brittle modes, even in the low strength materials. The effects of large scale crack-tip blunting, which generally complicate interpretation, were not found.

ACKNOWLEDGEMENTS

The authors thank Professor G. W. Greenwood for providing research facilities in the Department of Metallurgy, University of Sheffield. NJH is also grateful to the Science Research Council for the financial support which made the work possible.

REFERENCES

Brook, R., Entwisle, A.R., and Ibrahim, E.F., (1960). Jnl. Iron Steel Inst., 195, 292-298.

Brook, R., and Entwisle, A.R., (1965). Jnl. Iron Steel Inst., 203, 905-912.

Maki, T., Shimooka, S., Unemoto, M., and Tamura, I., (1972). Trans. Jap. Int. Met., 13, 400-407.

ASTM, (1977). Recommended Practice for J_{IC} Determination, Document E24-01-09. American Society for Testing Materials, USA.

BSI, (1979). British Standard for COD Testing, BS 5762, British Standards Institution, London.

Venzi, S., (1973). Report No. 1547R, Centro Sperimentale Metallurgio, Rome, Italy.

Witt, F.J., and Mager, T.R., (1972). A procedure for determining boundary values on fracture toughness K_{IC} at any temperature, HSST Program. Tech. Manuscript 19.

Chipperfield, C.G., (1976). Int. Jnl. Fract., 12, 873-886.

Blauel, J.G., and Hollstein, T., (1978). 2nd. European Colloquium on Fracture, VDl-Verlag, Dusseldorf, pp178-185.

Methods of Test for Plane Strain Fracture Toughness (K_{IC}) Testing of Metallic Materials, BS 5447, (1977). British Standards Institution, London.

FATIGUE CRACK GROWTH RETARDATION IN AN ULTRA HIGH CARBON STEEL

H. Toyoda, M. Saito and S. Ohta

Central Research Laboratory, Kobe Steel, Ltd., Chuo-ku, Kobe, Japan

ABSTRACT

The effect of carbide morphology on the fatigue crack propagation resistance in an ultra high carbon steel was investigated. It was found in experiments and microscopic observation that fatigue crack growth retardation occurred when massive eutectic carbides are extended in the direction perpendicular to the macroscopic crack propagation direction. The retardation was caused by bending or branching of the crack when it encountered massive carbides. Reduced stress intensity factors at the tip of bent or branched crack corresponded with the experimental results.

KEYWORDS

Fatigue crack propagation; ultra high carbon steel; cabide morphology; microstructure; fracture mechanics; stress intensity factor

INTRODUCTION

An ultra high carbon Cr-Mo steel is used for rolls of hot rolling such as blooming and billetting mills. Microstructure of the steel consists of bainitic or pearlitic matrix and eutectic carbides; the latter are formed during the solidification between dendritic arms and are usually gross in size. While the steel is selected mainly for its good wear resistance, fatigue crack properties are also critical in the service life of hot rolling rolls since surface thermal fatigue cracks often called fire cracks or heat checks are formed in the very early stage of the service and act as the stress raiser for subsequent crack growth. The authors have reported that the size and distribution of eutectic carbides had a great influence on fatigue crack propagation properties in this steel (1981a) and that the finding was carroborated when it was applied to an actual rolling mill (1981b). In the present study, the effect of brittle second phase, or eutectic carbides, on the fatigue crack propagation behaviour in the ultra high carbon steel was investigated more in detail experimentally and results were analysed with the linear facture mechanics.

EXPERIMENTAL PROCEDURE

Material

Material tested is a forged ultra high carbon Cr-Mo steel. Chemical compositions are given in TABLE 1. Microstructure of the steel normally consists of the matrix of bainite or pearlite and eutectic carbides. Size and distribution of the eutectics may vary with different casting and forging practices. Carbon content naturally has a strong influence on the size.

Photo-micrographs of the four heats of material are shown in Fig. 1. In Material A, eutectics are dispersed into clusters of fine carbides due to rapid solidification. For Material B to D, low solidification rate was taken to allow to form massive carbides. Distribution of the carbides were controlled by forging. Forging ratio (cross-sectional area ratio) was chosen as 3 for Material A, B and D, and as 1.5 for Material C. All the materials were then normalised and tempered in the same condition to make the microstructure of their matrices essentially the same.

TABLE 1 Chemical Compositions (wt.%)

Material	C	Cr	Mo
A,B and C	1.70	3.0	0.30
D	1.90	3.0	0.30

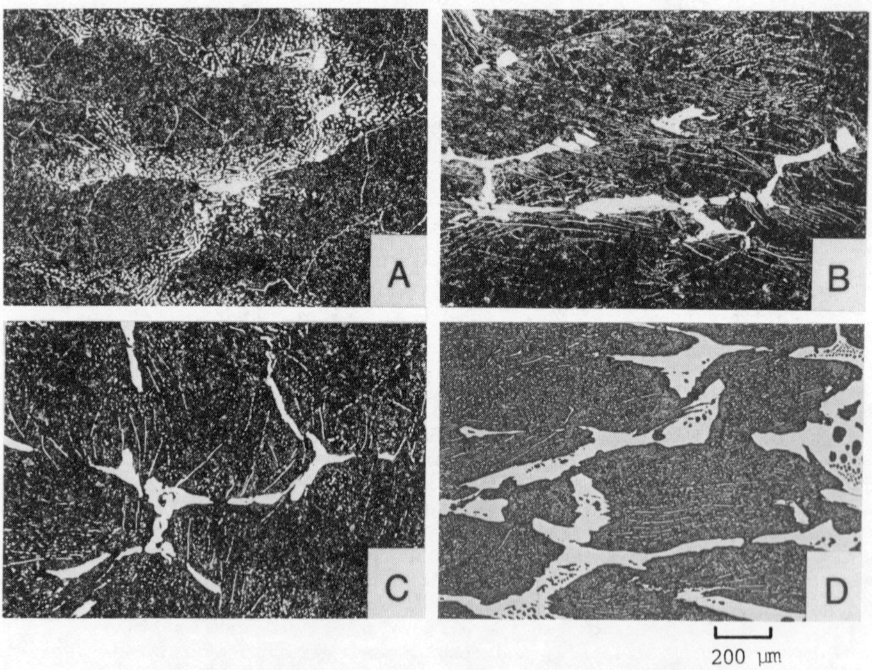

Fig. 1 Microstructure of the materials tested (Forging direcions are horizontal. Carbides are seen in light contrast.)

Fatigue Crack Propagation Tests

Fatigue crack propagation tests were carried out by using compact specimens at the stress ratio approximately zero. The experimental work was divided into two parts. In the first part, Material A to D were tested to examine the effect of carbide morphology on the crack growth. All the specimens were taken so that the expected crack plane was perpendicular to the longitudinal direction of the forgings, hence the crack propagated across arrays of extended carbides. In the second part, the effect of carbide orientation on the crack growth was studied to investigate the role of massive carbides in further extent. Only Material D was tested for the part. Specimens were cut out in various angles with respect to the forging direction. Schematic illustration of the specimen with respect to carbide orientation is shown in Fig. 5 together with test results.

EXPERIMENTAL

Effect of Carbide Morphology

Results of fatigue crack propagation tests for Material A to D are shown in Fig. 2 in a usual ΔK vs. da/dN plot. Substantial differences were observed among the materials tested. The highest resistance to crack growth was obtained with Material D, which had massive carbides well extended across the crack plane. Material B with less carbon content came second. Properties of the two materials were even better than those of a medium-carbon alloy steel with a similar matrix structure but without the eutectic carbide phase, of which the fatigue crack propagation data are included in Fig. 2. The material with clusters of fine carbides (Material A) exhibited the poorest resistance. Material C with massive carbides but not extended so well as Material B because of lower forging ratio came inbetween.

The results shown above indicate that massive carbides have a crack growth retardation effect when they are extended across the macroscopic crack propagation direction. Microscopic observation revealed that massive carbides were preferential paths of crack propagation as shown in Fig. 3 whereas the crack was hardly affected when it went through clusters of fine carbides as shown in Fig. 4.

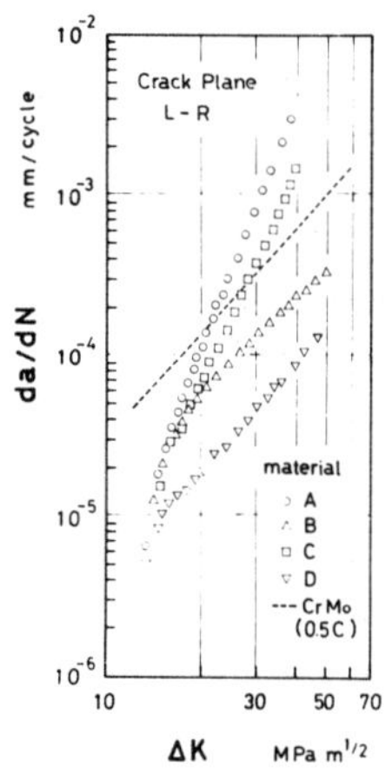

Fig. 2 Effect of carbide morphology on fatigue crack propagation

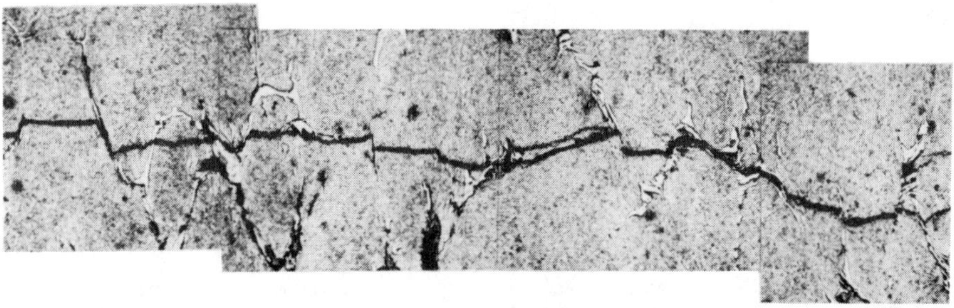

Fig. 3 Micrograph showing crack path (Material D) 400 μm

Fig. 4 Micrograph showing crack path (Material A) 200 μm

Effect of Carbide Orientation

Shown in Fig. 5 are the results of fatigue crack propagation tests for the effect of carbide orientation with respect to macroscopic crack propagation direction. Progressive retardation of crack growth was observed as the angle between the two directions, θ, was increased from 0 to 90 degree.

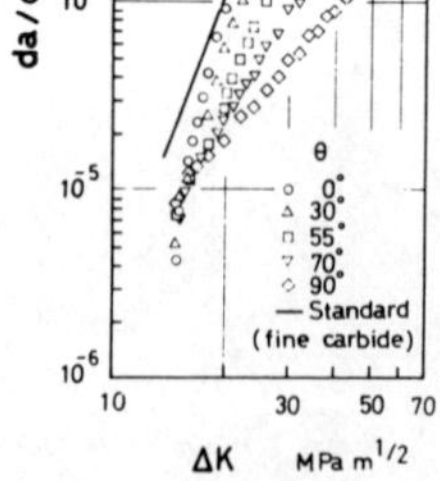

Fig. 5 Effect of carbide orientation on fatigue crack propagation

DISCUSSION

It was found that fatigue crack growth retardation occurred if massive carbides were present in proper shape and orientation. If carbides were not massive enough, crack paths were not affected with them as was seen in Fig. 4, Since massive carbides are much more brittle than the matrix, a crack would be bent toward the direction in which carbides extended every time it met one. The crack would then proceed into the matrix when the crack tip came to the end of the carbide. The rate and direction to which the crack proceeds would be determined by combined modes of the stress intensity factor at the tip.

Three possible models for the crack configuration found in the present materials are shown in Fig. 6. Solutions for the stress intensity factor at the tip of the bent crack are given by Kitagawa, Yuuki and Ohira (1975). Stress intensity factors of modes I and II exist when the crack is inclined to the tensile axis. Using their results and letting b/a' be as small as 0.01 and θ=60 degree, one can get stress intensity factors as,

$$F_I=0.55 \text{ and } F_{II}=0.43 \quad (1)$$

where $F_I=K_I/K_I'$, $F_{II}=K_{II}/K_I'$ (2)

K_I, K_{II}: stress intensity factors for the bent crack

K_I' : stress intensity factor for the straight crack of the length a'

For the forked crack, solution is also given in the same literature. Let 2θ=90 degree and b/a' <<1, then,

$$F_I=0.55 \text{ and } F_{II}=0.35 \quad (3)$$

For the parallel crack shown in Fig. 6, solution for parallel edge cracks seems to be appropreate. When two parallel cracks are of the same length and d/a' <<1, according to the analysis by Isida (1979),

$$F_I=0.77 \text{ and } F_{II}=0 \quad (4)$$

In the discussion on the fatigue crack propagation for inclined cracks, Tanaka (1974) derived the effective stress intensity factor to be applied for the crack propagation law, $da/dN=C\,K_{eff}^m$, on the basis of the model proposed by Weertman (1966), as follows,

$$K_{eff}=(K_I^4+8K_{II}^4)^{1/4} \quad (5)$$

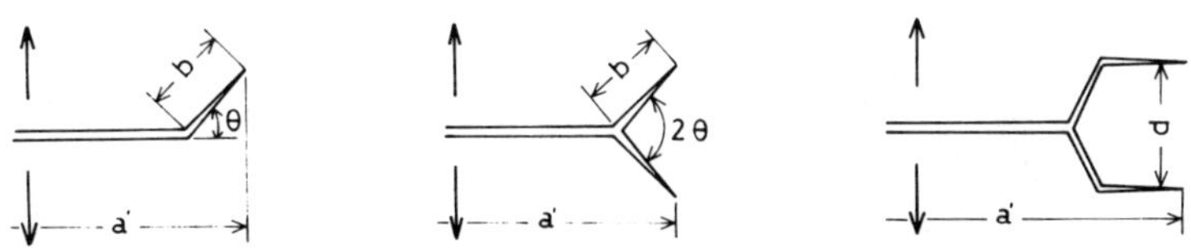

(a) Bent crack (b) Forked crack (c) Parallel cracks

Fig. 6 Crack propagation models

Using this formula, one can get the reduction ratio for the effective stress intensity factor, $F_{eff}=K_{eff}/K_I'$, for the three crack models shown in Fig. 6, namely bent, forked and parallel cracks with particular dimensions mentioned above to be 0.78, 0.68 and 0.77, respectively.

Observed retardation of fatigue crack growth may be expressed in terms of the reduced stress intensity factor range as schematically shown in Fig. 7 and,

$$F^*_{eff}=\Delta K_A/\Delta K_B \tag{6}$$

This means that the effective stress intensity factor for the apparent value, ΔK_B, in testing on Material B is ΔK_A. Since the fatigue crack propagation line for Material A represents the properties of the matrix of Material B. The value of F^*_{eff} varies with da/dN level when the two lines are not parallel. For the results of Material A and B shown in Fig. 2, values for F^*_{eff} are obtained as 0.88 and 0.66 for da/dN values of 3×10^{-5} and 1×10^{-4} mm/cycle, respectively, which are close to F_{eff} values obtained above.

Shown in Fig. 8 is a schematic illustration for fatigue crack propagation when the array of carbides is inclined to the tensile axis. When the crack front reaches a carbide at the point B, it will break to extend the crack to its end (point C), where the crack will again be bent and proceed into the matrix until it meets the next carbide. If one assumes the crack to grow instantaneously from the point B to C, only Δb is the distance the crack grows with cyclic loading while the apparent length of the crack increases by Δa. Thence the fatigue crack propagation rate, da/dN, within the matrix is obtained by multiplying $\Delta b/\Delta a$ on that experimentally observed. Value of $\Delta b/\Delta a$ may be geometrically derived as,

$$\Delta b/\Delta a=(\ell/2d\ \sin\beta\cdot\cos\theta+\cos(\beta-\theta))^{-1} \qquad (0<\theta\leqq\pi/2) \tag{7}$$

$$\Delta b/\Delta a=1-\ell/2p \qquad (\theta=0) \tag{8}$$

Notation is as defined in Fig. 8 where the crack is, for simplicity, assumed to meet carbides at their midlength all the time. The angle β is determined by solving the following equation derived for inclined cracks by Erdogan and Sih (1963).

$$\sin\beta+(3\cos\beta-1)\cdot\cot\theta=0 \qquad (0<\theta<\pi/2) \tag{9}$$

$$\beta=-70.5 \text{ degree} \qquad (\theta=\pi/2) \tag{10}$$

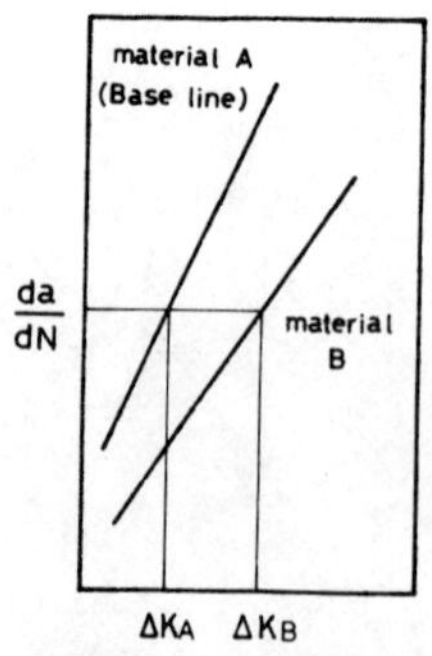

Fig. 7 Retardation due to stress intensity factor reduction

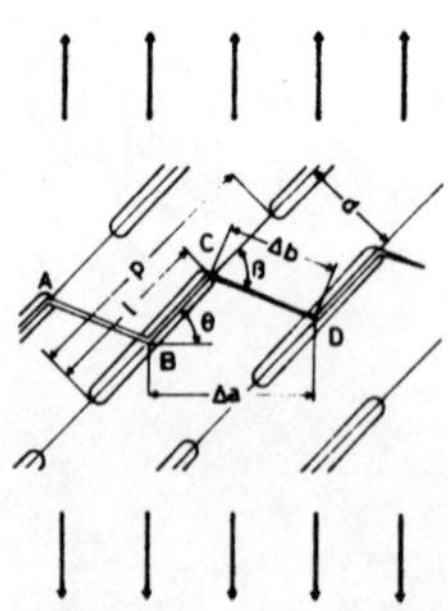

Fig. 8 Fatigue crack propagation in the array of carbides

Relation between the correction factor $\Delta b/\Delta a$ and the angle θ was calculated and shown in Fig. 9, where the values $\ell/d=0.5$ and $\ell/p=0.5$ were used according to microscopic observation of the shape and distribution of carbides. The effective stress intensity factor at the point C can be obtained for various angles θ by using the scheme already mentioned for bent cracks. Relation between F_{eff} and θ is shown if Fig. 10 together with experimentally determined values of the stress intensity factor reduction ratio, F_{eff}^*, which were obtained from the results shown in Fig. 5 to which the correction factor shown in Fig. 9 were applied. Similar trends are observed for F_{eff}^* with the latter slightly lower where the angle θ is small.

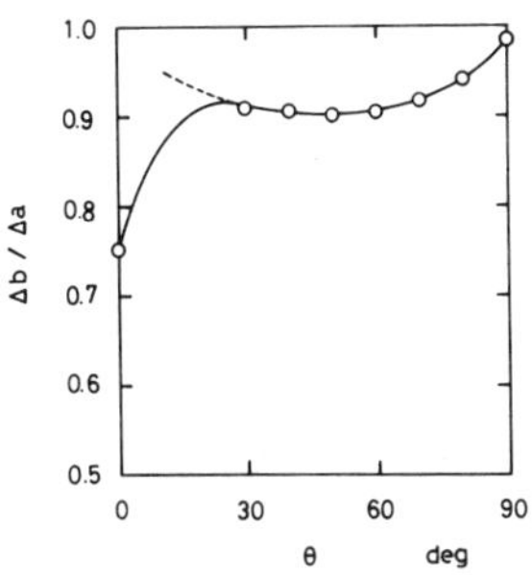

Fig. 9 Crack path correction factor

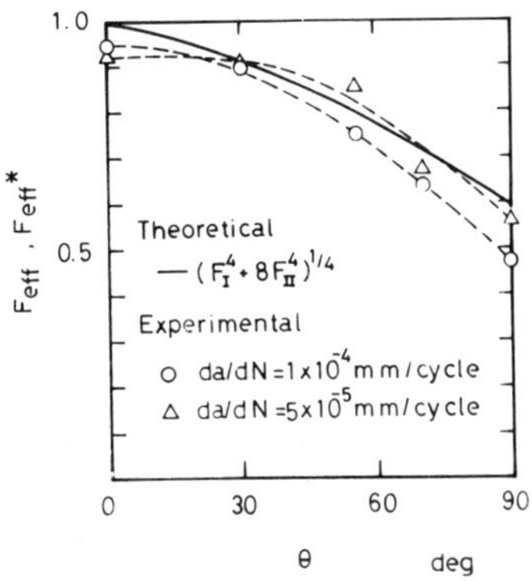

Fig. 10 Reduction ratio of effective stress intencity factor

CONCLUSION

The fatigue crack growth in an ultra high carbon steel was found to be retarded when the brittle carbide phase was sufficiently massive and well extended in the direction perpendicular to the macroscopic crack growth direction. Retardation was caused by bending or branching of the crack due to massive carbides, which resulted in a reduction of the effective stress intensity factor at the crack tip.

The retardation effect was even more pronounced than expected from the analysis made herein as the crack became longer, which could be ascribed to three dimensional distribution of carbides, more irregular crack plane and more complicated sress state.

REFERENCES

Erdogan, F and G.C. Sih (1963). J. Bas. Eng., Trans. ASME, 85D, 519-527
Isida, M. (1979). Trans. Japan Soc. Mech. Engr., 45A, 306-317
Kitagawa, H.,R.Yuuki and T.Ohira (1975). Eng. Fract. Mech., 7, 515-529
Ohta,S.,H.Toyoda and M.Saito(1981a).Trans.Iron Steel Inst.Japan,21,B-197
Ohta,S.,H.Toyoda and M.Saito(1981b).Trans.Iron Steel Inst.Japan,21,B-198
Tanaka, K. (1974). Eng. Fract. Mech., 6, 493-507
Weertman, J. (1966). Int. J. Fract. Mech., 2, 460-467

STRETCH ZONE, DUCTILITY AND TOUGHNESS STUDIES IN THE TRANSITION REGION OF 1045 STEEL

M. R. Bayoumi and M. N. Bassim

Department of Mechanical Engineering, University of Manitoba, Winnipeg, Manitoba, Canada

ABSTRACT

The variation of fracture toughness expressed by the J-integral, tensile and bulge ductility of 1045 steel which exhibits a transition from linear elastic behavior to elastic-plastic behavior, with temperature was investigated. The toughness study also included measurements of the stretch zone width from fractured surfaces to determine the process of crack blunting before the onset of crack growth. Three-point bending specimens were used for determination of the J-integral while square coupons of the material were tested for evaluation of the bulge ductility. Stretch zone width studies were carried out using scanning electron microscopy. The results show that the variation of J_{IC} with temperature can be analyzed in terms of two distinctive regions with a transition temperature occuring where the material behavior changes from linear elastic to elastic-plastic at the crack tip. There is also a linear correlation between bulge ductility and stable crack growth. The transition from linear elastic to elastic-plastic behavior at the crack tip was confirmed from the stretch zone measurements.

KEYWORDS

J-integral, bulge ductility, transition region, stretch zone, stable crack growth, AISI 1045 steel.

INTRODUCTION

Considerable effort has been undertaken in recent years for characterization, both from the macroscopic and microstructural aspects of the fracture toughness of metallic materials. Particularly, research has recently concentrated on development of criteria for these materials which accurately describe their behavior as a function of testing temperatures. Of these criteria, the use of the J-integral, which is well-suited for describing elastic-plastic behavior has been predominant (Turner, 1981). Methods of determination of the J-resistance curves are presently standardized where the onset of stable crack propagation is determined using partial unloading and compliance measurement procedures. Generally, J_{IC} increases as a function of temperature. The shape of the J_{IC}-temperature curve seems, however, to vary depending on the material tested and the experimental procedure for determination of J_{IC}. While Bassim and co-workers (1977) found that the variation of J_{IC} with temperature can be described by the intersection of two straight lines defining a transition between completely linear elastic behavior and elastic-plastic behavior, other investigators report different tendencies where J_{IC} reaches a maximum at a certain temperature followed by a plateau of J_{IC} as temperature is further increased (Tobler and co-workers, 1979). Such curves, however, still define to a lesser degree, this transition temperature mentioned earlier.

Correlation between fracture toughness and the stretch zone ahead of the crack tip has been reported. Robinson and Tetelman (1976) have described a procedure for evaluation of the J-integral from measurements of the width of the stretch zone while Nguyen-duy and Bayard (1981), have shown that the stretch zone width is an indication of the extent of plastic blunting at the crack tip and thus can be related to the change in microstructure due to the introduction and multiplication of dislocations in the plastic zone. More recently, Amouzouvi and Bassim (1982) have reported that there is a correlation between the stretch zone and the extent of prior straining in 4340 steel and calculated from the stretch zone width J_{IC} values which are in agreement with those measured from J-resistance curves in these predeformed 4340 steel specimens.

Several studies are available in the literature on the effect of temperature on ductility of engineering materials. The tensile ductility expressed as the reduction in area at fracture as well as the percentage noncrystalline fracture, as a function of temperature has been described by Tetelman and McEvily (1967) in materials with different microstructures. Generally, ductility increases from zero at the brittleness transition temperature and reaches a near constant value once a temperature T_b, defined as the propagation transition temperature in initially flaw free materials is reached.

Other parameters describing the ductility of materials are available. The most notable of such parameters is the bulge ductility which has been determined both theoretically and in experimental studies by Weiss and co-workers (1976) and by Azrin and Backofen (1970). Such ductility has been correlated with the fracture toughness parameters K_{IC} and G_{IC} since plane strain is obtained in bulge ductility measurements.

A study relating toughness, ductility and stretch zone width measurements in the transition region in an engineering material such as 1045 steel is of great importance in characterization of the fracture toughness of this steel as a function of testing and/or operating temperatures. Such a study provides an understanding of the crack blunting and propagation process and combines the two approaches of mechanics of materials as well as the microstructural aspect of fracture. The objective of this present investigation is thus to study experimentally the effect of temperature on toughness, ductility and stretch zone width in 1045 steel. Of particular concern is the transition region where the behavior of this steel changes from completely elastic to elastic-plastic and stable crack growth would hence take place.

EXPERIMENTAL PROCEDURE

Fracture toughness, expressed by the parameter J_{IC} and ductility, both under tensile and biaxial bulge loading were determined as a function of temperature for AISI 1045 steel in the annealed condition. Three types of specimens were used. The fracture specimens were three-point bending specimens with a thickness of 12.4 mm and with dimensions conforming to ASTM specifications E-399-78, while the tensile specimens were of the flat type with a thickness of 3 mm and width of the reduced section corresponding to 6.5 mm. The bulge ductility specimens were square coupons of 114.3 mm and a thickness of 1.9 mm and contained a reduced circular area with a diameter of 12.7 mm and a thickness of 1.0 mm. The bulge ductility specimens were obtained using several passes of hot rolling and annealing to obtain the same hardness as that of the fracture and tensile specimens. The initial notch in the three point bending specimens was taken to be in the longitudinal-transverse orientation.

Initially, to determine the transition region for this steel, standard charpy specimens, having the same hardness and notch orientation, were fractured and the absorbed energy was obtained as a function of temperature. The transition region was found to be between -60 to 25°C. The 15 ft - lb transition for this steel is about 0°C. J_{IC} was obtained experimentally by following the procedures of ASTM specification E 813-81. Specimens were fatigue precracked to a final crack length corresponding to a/w = 0.57. The J-resistance curve was determined by following the method of 10% unloading. Tests at low temperatures were performed in an environmental chamber where low temperatures were obtained by blowing cold nitrogen gas. All tests were performed at 0.25 mm/min displacement rate on a servo-hydraulic testing machine.

Tensile and bulge ductility tests were performed at displacement rate of 0.5 mm/min using a screw driven testing machine. This machine was equipped with a specially fabricated punch and specimen holder for performance of the bulge ductility tests (Bayoumi and Bassim, 1983).

After fracture of the three-point bending specimens, samples were taken from the fracture surface and examined using a scanning electron microscope (super Mini SEM ISI) with a tilt angle = 30°. The length of the stretch zone d was measured at different points along the crack front and the results were averaged. Also determined was the length of stable crack growth in each specimen. The value of J_{IC} was calculated using the equation

$$J_{IC} = \frac{2m\sigma_f d}{\cos\delta + \sin\delta} \times \frac{1}{G} \qquad (1)$$

where $_f$ is the flow stress of the material (equal to $(\sigma_y + \sigma_{ultimate})/2$) determined at each temperature in the tensile tests, G is the magnification and m is a constant factor which is generally taken as equal to 2.0, but may depend on the testing conditions and temperature.

RESULTS AND DISCUSSION

Scanning electron micrographs of the fractured surface of the specimen tested in the temperature range from -60°C to 25°C are shown in Fig. 1. On these micrographs are identified the stretch zone width as well as the size of stable crack growth. It is evident that both these parameters increase as a function of temperature. This is shown in Figs. 2 and 3 for the stretch zone width and the stable crack growth respectively. While there is a marked transition in the variation of the stretch zone width with temperature occurring at about -20°C, a change in behavior of the stable crack growth with temperature occurs at a lower temperature, around -40°C for this material.

The width of the stretch zone obtained was used to calculate the variation of fracture toughness J_{IC} with temperature. This is shown in Fig. 4. As expected, there is a transition from mostly linear elastic to elastic-plastic behavior taking place at -20°C. Such a tendency in the presence of a transition is in agreement with that reported from direct measurements of J_{IC} using the partial unloading method and reported earlier (Bayoumi and Bassim, 1983). However, the values of J_{IC} from stretch zone measurements are generally higher by about 25 to 40% depending on the temperature. This increase in J_{IC} is explained in terms of using a value of 2 for constraint factor in Eq. 1. However, if such a parameter is adjusted for the variation in temperature, it was found to range from 1.02 to 1.6 as shown in Fig. 5 where J_{IC} is taken from J-resistance curves obtained from partial unloading and COD is evaluated from stretch zone width measurements. Fig. 6 shows the relationship between bulge ductility and stretch zone width as well as stable crack growth. The temperatures at which the bulge ductility tests were performed are indicated on the curves in Fig. 6. From -40°C to 25°C, the variation of bulge ductility with stable crack growth is linear, while the variation of bulge ductility with stretch zone width indicates a transition occuring at -20°C. It is noted that at the low temperature of -60°C, the material exhibits an entirely elastic behavior with unstable crack growth. In this case J_{IC} is calculated from

$$J_{IC} = G_{IC} = \frac{K_{IC}^2(1 - \nu^2)}{E} \qquad (2)$$

where E and ν are Young's modulus and Poisson's ratio respectively.

The described results can be used to analyze the fracture process of a typical BCC material as a function of temperature. Considering that this process occurs in two steps, namely the blunting process followed by stable crack growth, it can be concluded that the measurement of the stretch zone width is indicative of the process of blunting ahead of the crack. For AISI 1045 steel, the blunting becomes very limited, if at all existent below the temperature of -20°C. Thus, a transition in the stretch zone width is present at this particular temperature. On the other hand, the process of stable crack growth is evident in this material down to a temperature of -40°C. Thus, the variation of stable crack growth with temperature shows changes only at -40°C or below where effectively, no stable crack growth is observed.

Combining the change in the two process of crack propagation as a function of bulge ductility, it can be concluded that bulge ductility produces a state of stress equivalent to carrying out fracture to the stable crack growth process. Hence, the variation of bulge ductility with stable crack growth is linear down to a temperature of -40°C. Modelling of the relationship between bulge ductility and toughness as well as between toughness and temperature has been described elsewhere (Bayoumi and Bassim, 1982,1983).

CONCLUSION

The correlation between fracture toughness, bulge ductility and stretch zone width of fractured specimens has been investigated as a function of temperature in the transition region of AISI 1045 steel. The results show the existence of a transition in the variation of fracture toughness and stretch zone width with temperature. Such a transition is indicative of the blunting process at the crack tip. However, the bulge ductility appears to be related to the process of stable crack growth process during fracture.

ACKNOWLEDGEMENT

The authors would like to acknowledge the support of the Natural Sciences and Engineering Research Council of Canada and the Faculty of Engineering of the University of Manitoba.

REFERENCES

Amouzouvi, K.F. and M.N. Bassim (1982). Mat. Sci. Eng., 55, 257-262.
Azrin, M. and W.A. Backofen (1970). Met. Trans., 1, 2857-2865.
Bassim, M.N., S. Mikhail and D.R. Hay (1977). Can. Met. Quart., 16, 210-213.
Bayoumi, M.R. and M.N. Bassim (1983). Eng. Fract. Mech., in press.
Bayoumi, M.R. and M.N. Bassim (1983). Int. J. Fract., in press.
Bayoumi, M.R. and M.N. Bassim (1983). Int. J. Fract., submitted.
Nguyen-Duy, P. and S. Bayard (1981). J. Eng. Mat. Tech., 103, 55-61.
Robinson, J.N. and A.S. Tetelman (1976). Eng. Fract. Mech., 8, 301-313.
Tetelman, A.S. and A.J. McEvily (1967. Fracture of Structural Materials, John Wiley & Sons.
Tobler, T.L., R.P. Mikesell and R.P. Reed (1979). ASTM STP677, American Society for Testing and Materials.
Turner, C.E. (1981). Proceedings of the 5th International Conf. on Fracture, Pergamon Press.
Weiss, V., Y. Kasai and K. Sieradzki (1976). ASTM STP605, American Society for Testing and Materials.

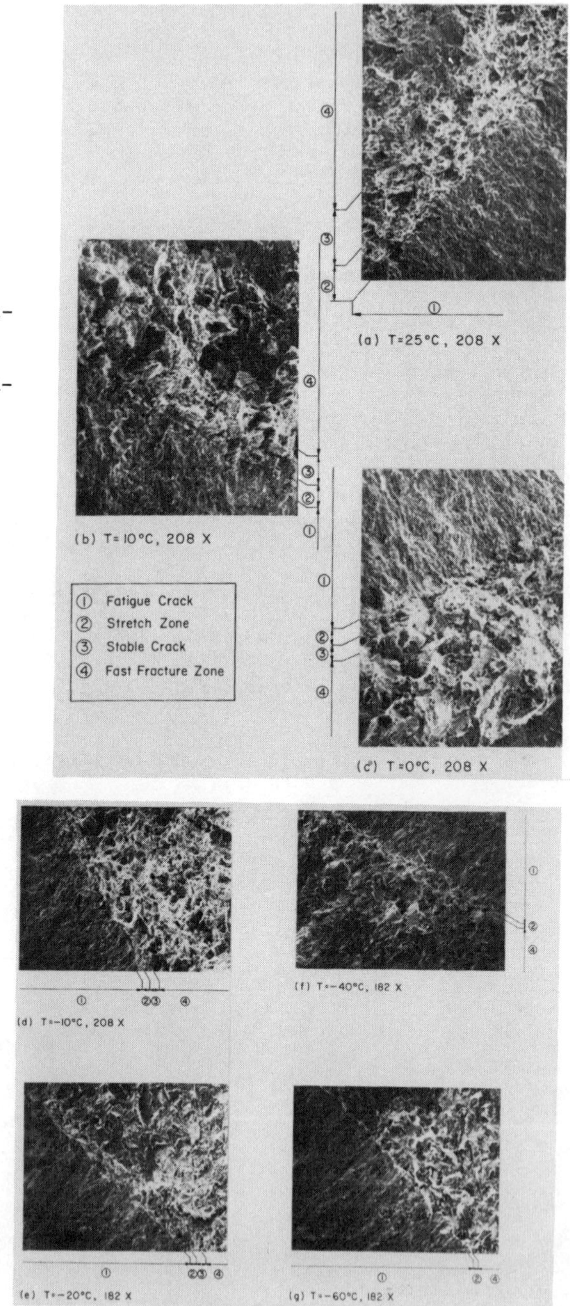

Fig. 1: Scanning electron micrographs of fractured specimens at different temperatures.

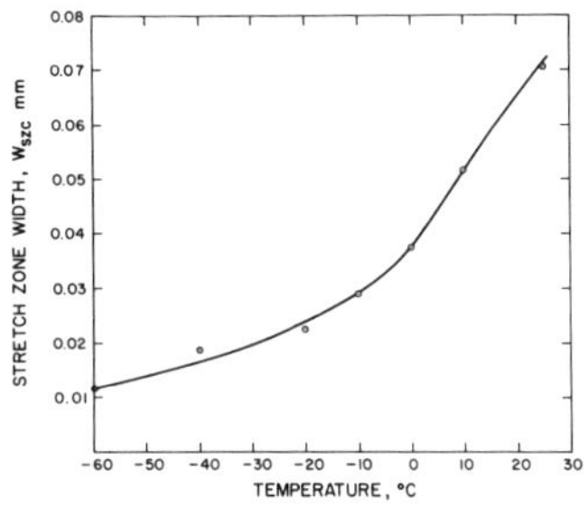

Fig. 2: Stretch zone width as a function of temperature.

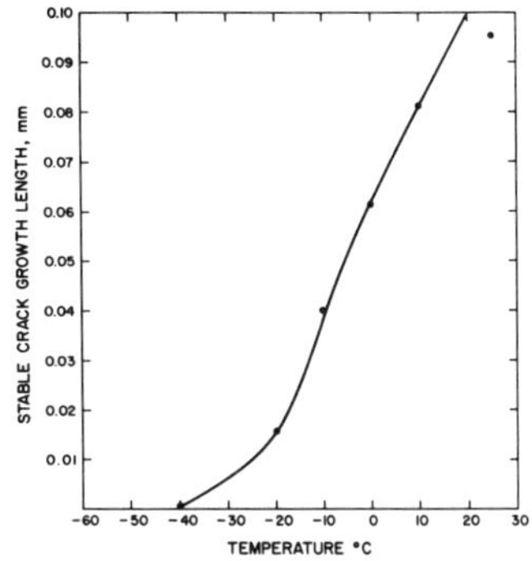

Fig. 3: Stable crack growth as a function of temperature.

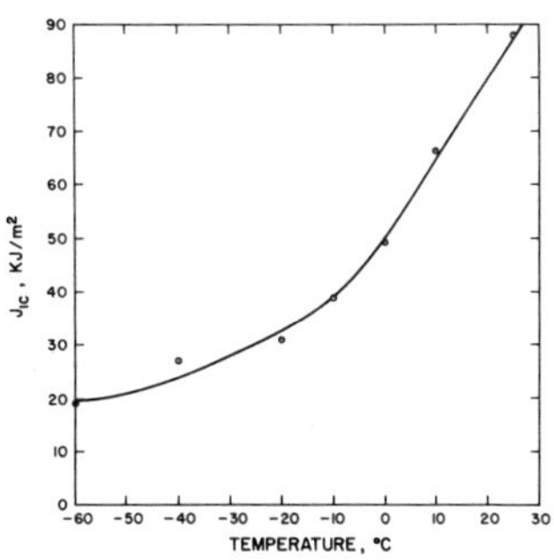

Fig. 4: The variation of J_{IC} with temperature.

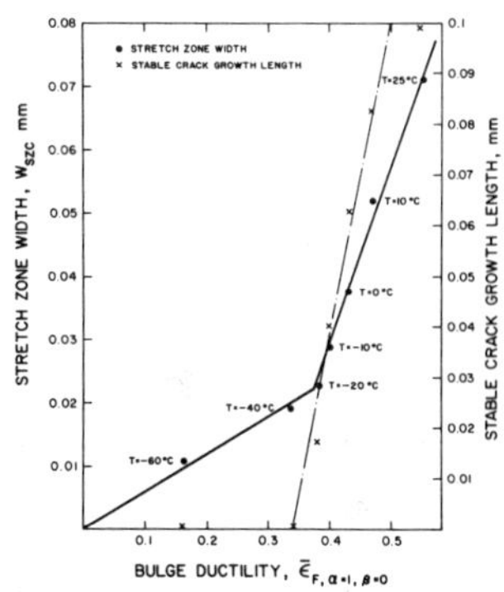

Fig. 5: Stretch zone width, stable crack growth vs bulge ductility.

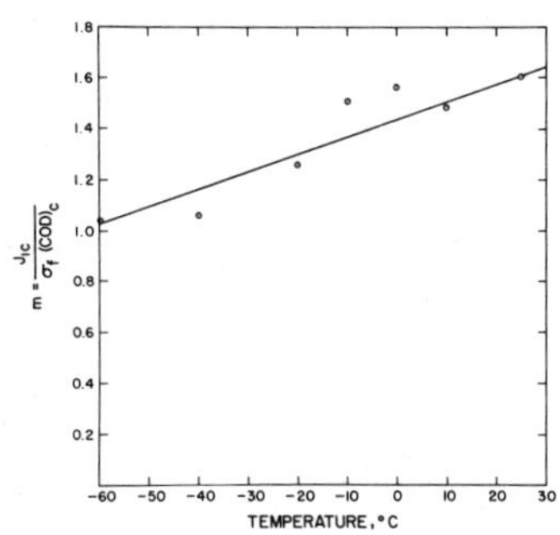

Fig. 6: Dependence of constraint factor m with temperature.

STUDY ON FRACTURE PROCESS OF NITRIDED CASES OF STEEL 30CrMo WITH ACOUSTIC EMISSION TECHNIQUE

Shouhua Zhang*, Qingfu He*, Shihui Song*, Weiming Huang*, Yang Deng* and Yaoxin Zeng**

*Beijing University of Iron and Steel Technology, Beijing, China
**China Academy of Railway Sciences, Beijing, China

ABSTRACT

In this paper, initiation and propagation of cracks in nitrided cases of two different structures on 30CrMo steel are studied on notched three-point bending specimens with the acoustic emission (AE) technique. It has been found that the formation of the first macrocrack in specimens with an ε-$Fe_{2-3}N$ case occurs at lower loads and cracks propagate much faster than in specimens with a γ'-Fe_4N case. For both structures, the number of AE counts per unit length of crack is nearly the same. Study on AE amplitude distribution shows that the fracture process proceeds first with initiation of microcracks and then linkage with the main crack or between themselves occurs. Optical microscopic and SEM observations support the findings of the AE amplitude study.

KEYWORDS

Crack initiation; crack propagation; ion-nitriding; nitrided cases; acoustic emission; cumulative count; amplitude; cleavage; deformability.

INTRODUCTION

Steel machine parts are frequently nitrided in order to enhance their wear resistance and fatigue resistance. Steel nitrided becomes in fact a composite material, composed of a ductile core and a brittle case. It is known that when a nitrided specimen is subjected to loading, no appreciable plastic deformation occurs in the ductile core, owing to the lateral restraints imposed by the rigid case, even if the yield strength of the core seems to have been appreciably exceeded. And the sustained load drops immediately when the brittle case cracks. Therefore, an assessment of the resistance to the initiation and propagation of cracks in the nitrided case is of prime importance for the evaluation of the load bearing capacity of the nitrided part. The acoustic emission technique is a useful tool in investigating the fracture process of the nitrided cases on steels, as shown in our previous papers (Zhang, 1977, 1978). Not only does the method catch the instant of cracking unmistakenly, but it also evaluates the brittleness of cases of various microstructures, so long as the AE signals are recorded simultaneously with the load-displacement measurement in the bending of the nitrided specimens. The shape of the AE rate curves varies considerably with the brittleness of the nitrided cases. It has been established that the load bearing capacity of a specimen depends largely on the deformability of its nitrided case.

MATERIALS AND METHODS

The chemical composition of the steel used in this work is as follows: C 0.29%, Si 0.22%, Mn 0.40%, Cr 0.95%, Mo 0.20%, P $<$0.02%, S $<$0.02% and Fe bal. Dimensions of the three-point bending specimens are 20x10x120 mm (depth x width x length), having a 90° notch 2 mm deep with a 0.5 mm radius at the root. The span between the bending supports is 80 mm. The specimens were quenched from 880°C and tempered at 580°C. Half of the specimens were then ion-nitrided in a $N_2 + H_2$ atmosphere at 530°C, to obtain a layer of γ'-Fe_4N nitride on the surface, and the other half in $NH_3 + C_3H_8$ at 550°C, to obtain a layer of ε-$Fe_{2-3}N$ nitride. The typical microstructure of a nitrided case is shown in Fig. 1. The depth and the hardness distribution in both cases were so controlled as to be nearly the same, Fig. 2. The nitrided case consists of a nitride layer and a transition zone which is composed of nitrogen-containing α-grains, occassionally with nitride or carbide at the grain boundary. The physical characteristics of the nitrided cases are listed in Table 1.

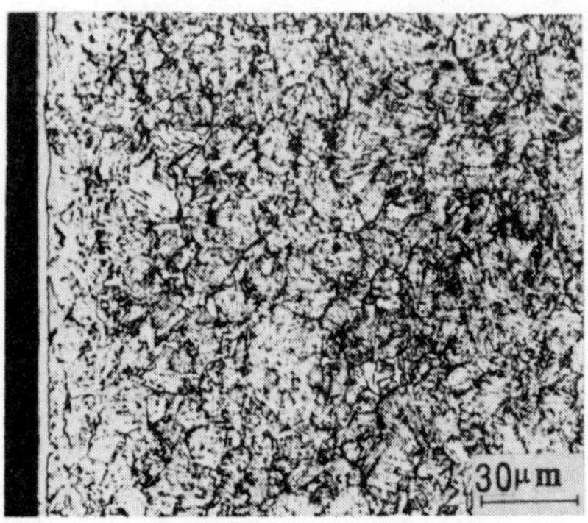

Fig. 1. Typical microstructure of a nitrided case.

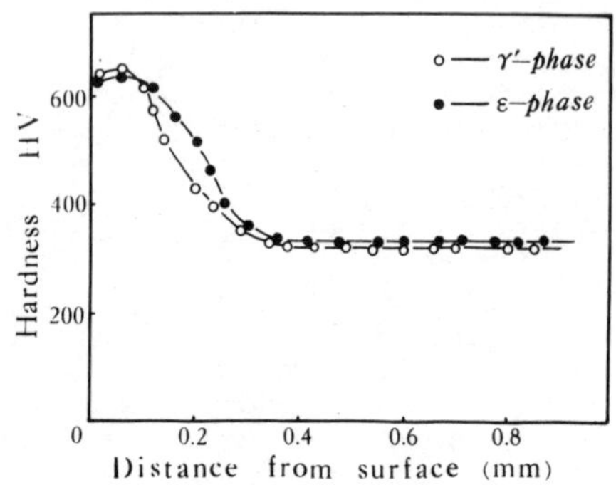

Fig. 2. Hardness distribution in nitrided cases.

TABLE 1 Physical Characteristics of Nitrided Cases

Specimen	The outmost nitride layer Structure	Thickness, μm	Transition zone mm	Surface hardness Hv
A	ε 94% γ' 6%	9	0.26	650
B	γ' 96% ε 4%	8	0.28	640

The three-point bending tests were carried out on a universal testing machine and the load-NOD (notch opening displacement) curves were recorded on an X-Y recorder. While the fracture process was being followed by an optical micrometer, the AE signals emitted during the bending fracture were received by a PZT-5 transducer with a frequency band 120±10 KHz. They were preamplified, filtered, amplified again, transformed and finally registered on the same X-Y recorder and were also recorded on an optical oscillograph directly after the second amplification. The parameters of AE recorded are: the counts per unit time, the cumulative counts and the amplitude of the signals. Microscopic studies were also made on the specimens after fracture. Before the bending tests, the structure of the nitride layers had been determined with an X-ray diffractometer.

RESULTS AND DISCUSSIONS

As the bending load was gradually increased, at a certain instant, a short crack appeared at the root of the notch and a large AE signal on the count rate curve occurred simultaneously, which was preceded by several small signals. It implies that minute cracks were initiated in the case before the occurrence of the macrocrack. It was found that specimens with an ε-nitride outmost layer fracture at much lower loads than γ'-nitride specimens, as revealed by the first large signal that appeared. The average value of loads for the former was 3826 N at a NOD value of 0.0176 mm, and for the latter, 5000 N at 0.0215 mm. It is obvious that γ'-nitride cases endured large deformations and consequently higher loads before fracture than ε-nitride ones.

As the bending load was further increased, the already existing short crack at the root of the notch grew on the lateral surface of the specimen. When the central crack had grown up to several millimeters, there appeared, all of a sudden, two arc-shaped cracks, one on each side of the central crack. The process repeated itself until several pairs of such arc-shaped cracks were formed on the surface. It is believed that the central crack was opened up by the tensile stresses at its tip, while the formation of the side arc-shaped cracks was induced by the plastic yielding of the core.

The measured lengths of the central crack prior to the appearance of the side arc-shaped cracks were correlated to the bending loads or to the NOD values. It was established that the length of the central crack varies linearly with NOD values for both types of the nitrided cases. But the slope of the line for ε-nitride specimens is much steeper than that for γ'-nitride specimens, Fig. 3. Their regression equations are:

for γ'-nitride, $y = 21x - 0.678$, $\gamma = 0.891$,
for ε-nitride, $y = 47x - 0.047$, $\gamma = 0.923$

respectively, where y is the length of the central crack in mm, x the NOD value in mm and γ the correlation coefficient. Experimental results show that cracks in ε-nitride cases propagate much faster than the γ'-nitride ones.

The cumulative counts in AE emitted during the growth of the central crack were found to vary linearly with the lengths of the crack in the first several millimeters. However, the slopes of the curves for both structural types, be it γ'-nitride or ε-nitride, are nearly the same. These curves deviate from linearity in the later stages of crack growth. The regression equations are:

for γ'-nitride, $y = 2.68x - 0.607$, $\gamma = 0.962$,
for ε-nitride, $y = 2.29x - 0.210$, $\gamma = 0.979$, and
for both structures, $y = 2.51x - 0.375$, $\gamma = 0.960$,

where y is the cumulative counts of AE in 10^3, x the length of crack in mm and γ the correlation coefficient. The relations are shown in Fig. 4.

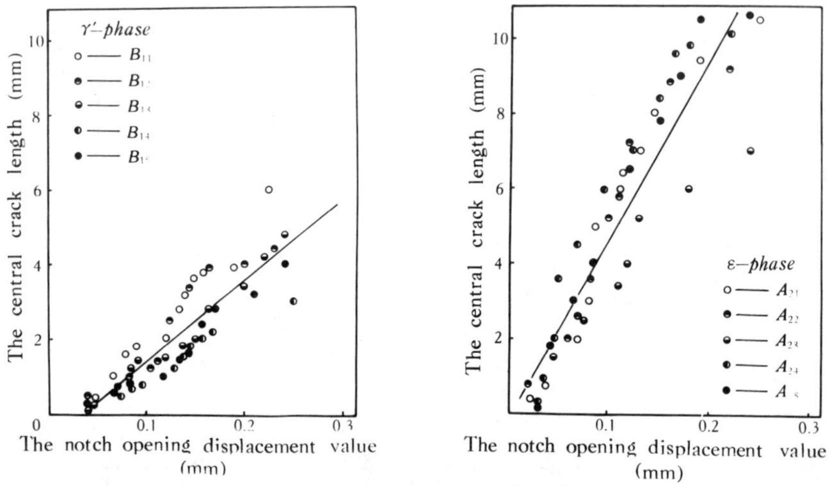

Fig. 3. Crack length vs NOD in bending of nitrided specimens.

The distribution of the amplitude of AE signals, the number of which indicates the number of events occurring during the fracturing process, was studied. The histograms of the amplitude distribution reveal the presence of two maxima, one at the lowest range of amplitudes and the other at the range of larger amplitudes, Fig. 5. If the amplitude distribution is analyzed on the basis of different crack intervals, it is seen from Table 2 that the distributions in intervals also show the two maxima feature. When the fracture appearance

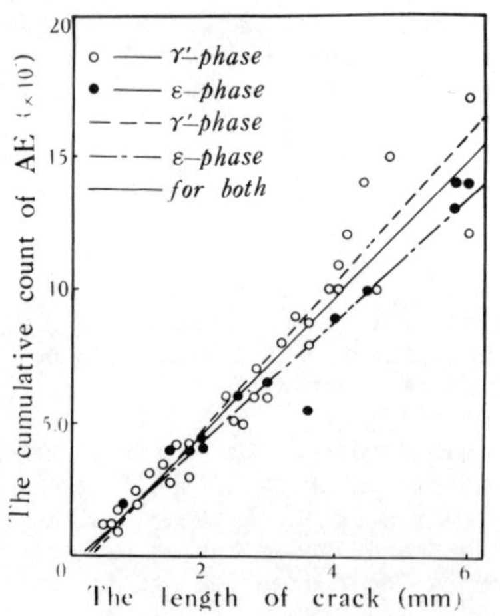

Fig. 4. Cumulative counts of AE vs lengths of central crack.

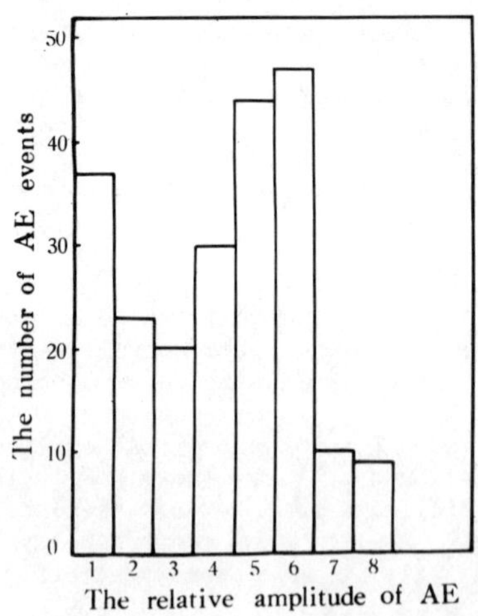

Fig. 5. Histogram of AE amplitude distribution in fracturing a γ'-nitride specimen.

is examined with SEM, the contours of the crack front are clearly defined, Fig. 6. The cracks in the nitrided case on the lateral surfaces lead much ahead of the crack in the core. The AE signals received along with the propagation of the central crack are actually caused by the events occurring both in the case and in the core. Since the depth of the nitrided case is about 0.30 mm, the AE amplitude distribution in the first half millimeter of crack characterizes the crack process in the nitrided case. The first peak in this interval consists of numerous cracking events with small AE amplitudes. And the cracking process in the case seems to proceed first with the formation of microcracks and then linkages between themselves or with the main crack. This is confirmed by SEM observations, Fig. 7. As the crack front

TABLE 2 AE Amplitude Distribution During Growth of Central Crack in a γ′-nitride Specimen

Relative amplitude	Number of events in crack intervals, mm 0-0.5	0.5-1.0	1.0-2.0	2.0-3.0	3.0-5.0
1	9	3	6	12	7
2	1	5	4	7	6
3	1	4	4	9	2
4	6	10	3	9	2
5	5	10	9	13	7
6	1	9	3	23	11
7	0	3	0	4	3
8	0	0	0	2	7
Sum	23	44	29	79	45
Events/mm	67		29	79	$22\frac{1}{2}$

propagates into the core, the contribution of the core cracking to AE is getting more and more significant and the second peak of large amplitudes is probably due to the core crackings. The cracking events at the second peak increase in number and the peak moves towards higher amplitude values in the later intervals of crack propagation, as shown in Table 2. It is interesting to note that the cracking events are not evenly distributed among the intervals and the number of events per unit length changes high and low alternately as the crack grows. It is not unusual in burst phenomena.

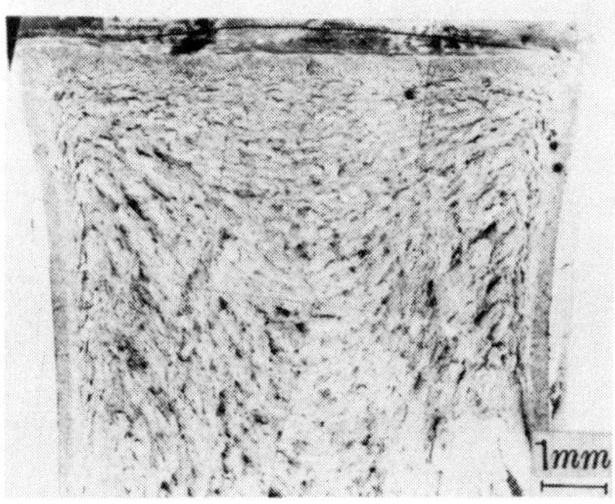

Fig. 6. Fracture surface under SEM

There are indications that when the crack fornt in the nitrided case propagates, it first starts at or close to the boundary between the outmost layer and the transition zone where there is the highest incompatibility. This is due to the large difference in elastic constants and in plastic properties of the nitride phase and the α-grains. Then the crack starts to grow in the lateral directions both inwards and outwards. Owing to the plasticity of the nitride compounds, however small they may be, the outmost layer will persist a short duration before it breaks. The crack keeps growing inwards. The speculation is amply supported by the metallographic observations that the crack opening is widest at or close to the boundary between the nitride layer and the transition zone and it is narrower in the nitride layer, Fig. 8. The mode of fracture is cleavage in the nitride layer and it is mixed transgranular and intergranular in the transition zone.

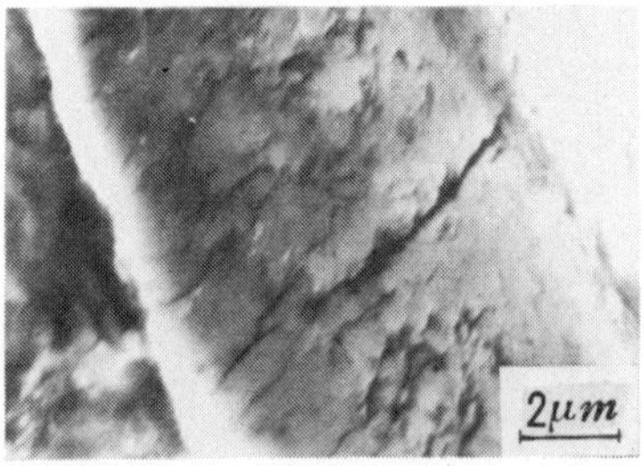

Fig. 7. Cracks in nitride layer, SEM.

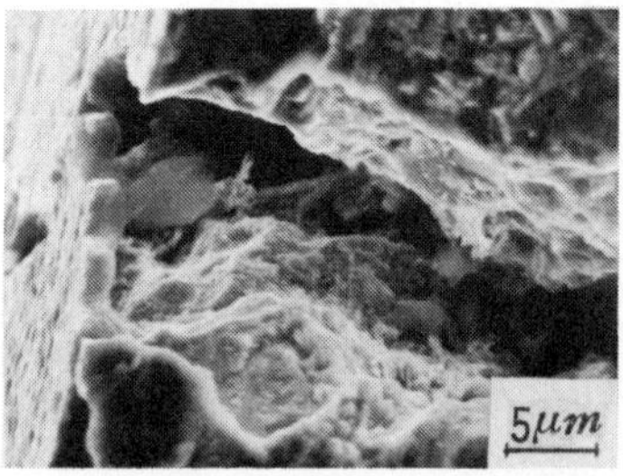

Fig. 8. Crack tip in nitrided case, SEM.

In conclusion, we are certain that, of the two types of nitrided cases studied, the γ'-nitride exhibits better deformability and toughness. Cracks will appear less frequently in the γ'-case and, once initiated, it will propagate at lower rates than in the ε-case. The phenomena observed can be explained by the fact that γ'-Fe_4N is cubic in crystal structure, while ε-$Fe_{2-3}N$ is hexagonal (Hawkins 1973), the latter being believed to have less slip systems and consequently smaller microplasticity, thus more liable to crack. As the nitride cleaves, the bonds between atoms break suddenly. Part of the elastic strain energy excites lattice vibration that appears as acoustic emission (Gillis, 1971; Lord, 1975; Tetelman, 1971). Since the types of bonding, hence the energy of bonding, between atoms in these two structures are similar, it is expected that the AE counts per unit length of crack both for γ'-nitride and ε-nitride should be close to each other.

REFERENCES

Gillis, P. P. (1971). Acoustic Emission, ASTM STP 505, 20-29.

Hawkins, D. T. (1973). In T. Lyman and others (Ed.), Metals Handbook, Vol. 8, 8th ed. ASM, Metal Park. 243-250

Lord, A. E., Jr. (1975). In W. P. Mason and R. N. Thurston (Ed.), Physical Acoustics, Vol. XI. Academic Press, New York. Chap. 6, 290-345.

Tetelman, A. S., and R. Chow (1971). Acoustic Emission, ASTM STP 505, 30-40.

Zhang, S., Y. Chen and S. Liu (1977). Acta Metallurgica Sinica, Vol. 13, 139-144.

Zhang, S., Y. Chen, S. Liu and Z. Yuan (1978). Paper presented at the 17th Internat. Conf. on Heat Treatment of Materials, Barcelona, 9-12 May, 1978.

FINITE ELEMENT ANALYSIS OF FATIGUE CRACK CLOSURE IN COMPACT SPECIMEN

H. Nakamura, H. Kobayashi, S. Yanase and H. Nakazawa

Department of Physical Engineering, Tokyo Institute of Technology, Tokyo, Japan

ABSTRACT

A finite element analysis was carried out to evaluate the fatigue crack closure behavior in the compact specimen. The quantitative discussion was made with special attention to the comparison between ideal and fatigue cracks. The transition behavior from the fatigue crack to the ideal crack was also evaluated.

KEYWORDS

Crack closure; fatigue crack; ideal crack; stable crack growth; single peak overload; finite element analysis.

INTRODUCTION

It is well known that a crack closure phenomenon is caused by residual plastic deformations remaining in the wake of the crack during fatigue crack growth (Elber, 1971). The phenomenon has been analyzed using finite element methods (Newman, 1976; Ogura, 1977; Shiratori, 1977). An attempt to develop an analytical method based on the Dugdale model has been also made (Budiansky, 1978). However, according to the recent experiments at high stress ratio R, the crack opening stress intensity factor K_{op} becomes smaller than the minimum stress intensity factor K_{min} (Kobayashi, 1983), which shows the quite different trend

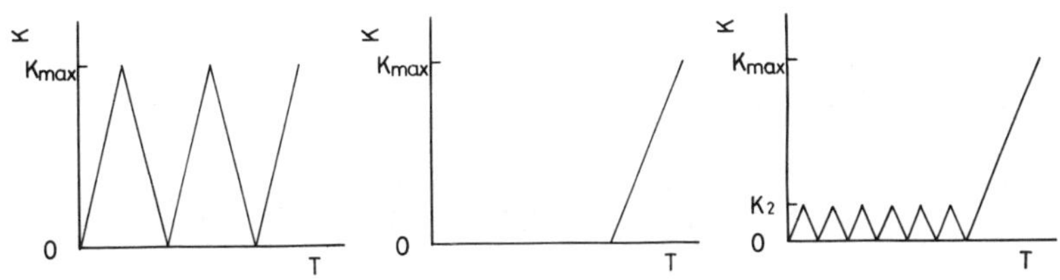

(a) fatigue crack (b) ideal crack (c) transitional crack

Fig. 1 Fatigue, ideal and transitional cracks.

from the conventional analytical results.

On the other hand, the crack closure phenomenon is also caused during stable crack growth. The phenomenon corresponds to that of fatigue crack growth at R = 1. A comparison between the characteristics of the ideal, the fatigue and the stable growing cracks (see Fig. 1 (a) and (b)) was made in this study. The justification concerning the conventional crack closure measurement techniques, the compliance method using the crack opening displacement (Elber, 1971) and the back face strain technique (Kikukawa, 1976), was also included. Additionally, the results were applied to assess the transitional behavior from the fatigue crack to the ideal crack (see Fig. 1 (c)).

NOMENCLATURE

	Crack opening displacement	Crack tip opening displacement
Fatigue crack	$\Delta\delta$	$\Delta\delta_0$
Ideal crack	δ	δ_0
Transition	δ_p	δ_{p0}

METHOD OF ANALYSIS

A two dimensional, elastic-plastic finite element analysis was used to investigate crack closure in a 1 inch compact type specimen (289 nodes, 508 elements) under cyclic loading (R = 0, 0.6) and stable crack growth (R = 1). All the results were compared with those of the ideal crack. In the analysis, the material stress-strain behavior was assumed to be bilinear with a yield stress, σ_{ys}, of 505 MPa, Young's modulus, E, of 2.06 x 10^5 MPa and a strain hardening rate, H' = 0.013E. The analysis was made under the plane stress condition. To account for the Bauschinger effect, kinematic hardening was assumed. K_{max} in Fig.1 was set at 62.0 MPa$\sqrt{m}$. The crack tip node was released and allowed to displace at K_{op} or K_{min} (Ogura, 1977). The crack growth increment was 0.15 mm.

The crack tip opening displacement was defined as the node displacement just before the crack tip. K_{cl} was determined as the K value when this displacement became zero. On the other hand, K_{op} was determined as the K value when a reaction force of the crack tip node came up to zero. It should be noted that the K_{op} value gives a correct value when crack closure occurs. However, it becomes an incorrect value when crack closure does not occur.

To assess the transitional behavior from the fatigue crack to the ideal crack, an analysis of the single peak overload tests was carried out by applying the single peak overload, overload ratio $\gamma = K_{max} / K_2 = 1.5 \sim 3.0$, to the constant stress intensity, K_2 = 24.8 MPa$\sqrt{m}$ (R = 0).

RESULTS AND DISCUSSION

Residual Plastic Deformation and Crack Closure

Figure 2 shows the relation between the residual plastic stretch, $(\delta - \Delta\delta)$, at $K = K_{max}$ and the nondimensional distance from a crack tip, x/a_0, where δ and $\Delta\delta$ are the crack opening displacements of the ideal and fatigue cracks, δ_{0max} is the crack tip opening displacement of the ideal crack at $K = K_{max}$, and a_0 (= 21 mm) is the initial crack length. $(\delta - \Delta\delta)$ shows a tendency to become

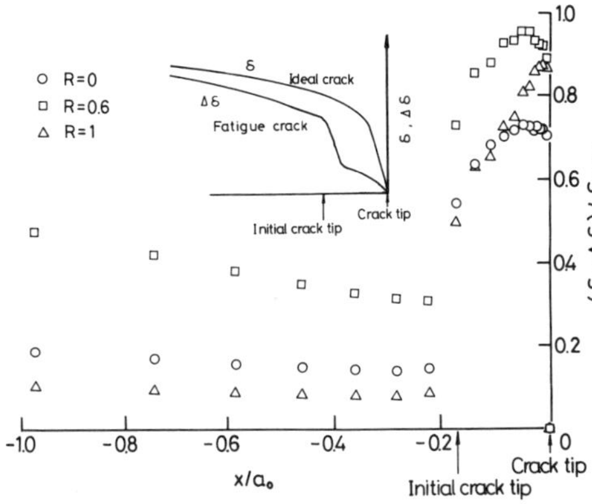

Fig. 2 Relation between (δ − Δδ) and distance from crack tip.

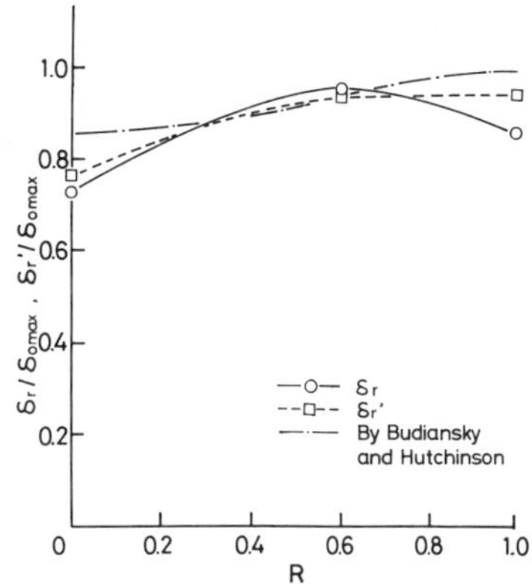

Fig. 3 Dependence of δ_r and δ'_r on R.

larger as reaching the crack tip and it takes a constant value, defined as δ_r, near the crack tip. Then, it decreases to the value just before the crack tip, defined as δ'_r.

Figure 3 shows the dependence of δ_r and δ'_r on R. δ_r and δ'_r increase with increasing R for R < 0.6. For R > 0.6, they become constant (δ_r) or decrease slightly (δ'_r) with increasing R. The analytical result of Budiansky and Hutchinson (1978) shows a different tendency from the present result for R > 0.6. In their analysis, the crack tip is postulated being closed under stable crack growth, which means $\delta_r = \delta_{0max}$ at R = 1. The assumption does not reflect the actual tip characteristics as stated below.

Figure 4 shows the dependence of K_{op} and K_{cl} on R. K_{op} increases monotonically with increasing R. On the other hand, K_{cl} could not be obtained for R > 0.6 because the crack tip was still open at $K = K_{min}$. So, unloading was

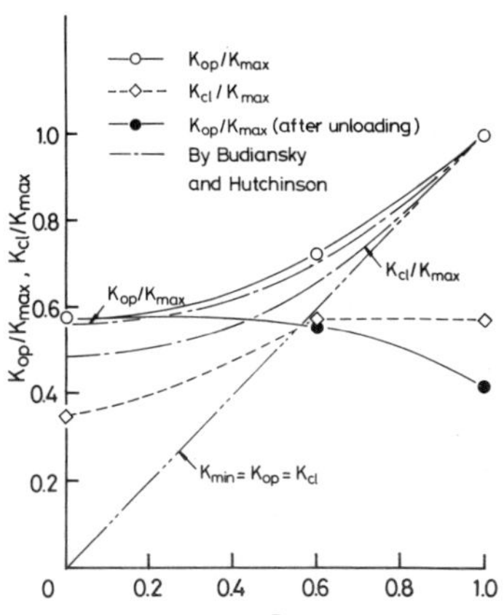

Fig. 4 Dependence of K_{op} and K_{cl} on R

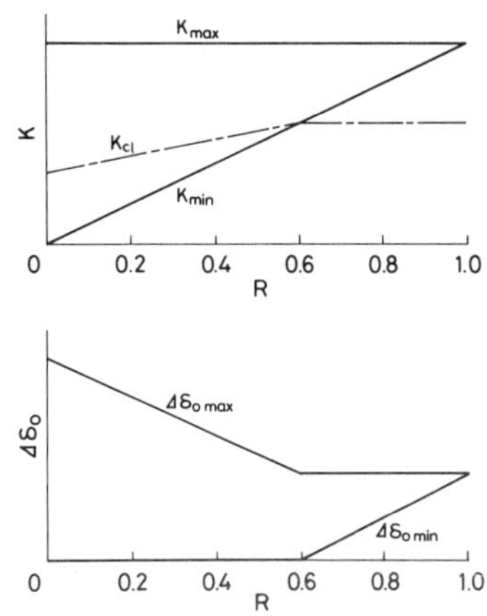

Fig. 5 Dependence of K_{cl} and $\Delta\delta_0$ at at $K = K_{max}$ and $K = K_{min}$ on R

attempted to evaluate an apparent K_{cl} for R > 0.6. The results show that K_{cl} becomes constant in these region which shows the same tendency with the experimental results (Kobayashi, 1983).

Figure 5 shows the schematic illustration of the dependence of the crack tip opening displacement, $\Delta\delta_0$, at $K = K_{max}$ and $K = K_{min}$ on R. $\Delta\delta_0$ at $K = K_{max}$ decreases with increasing R and it becomes constant for R > 0.6. On the other hand, $\Delta\delta_0$ at $K = K_{min}$ is zero for R < 0.6 and it increases with increasing R for R > 0.6. It is well known that the near-threshold characteristics in fatigue crack growth are insensitive to R at high R values. The reason should be attributed to the fact that $\Delta K = \Delta K_{eff} = K_{max} - K_{min}$ stands in these R ranges, where ΔK_{eff} is the effective stress intensity factor range. Moreover, fretting oxide induced crack closure does not occur because crack surface contact is difficult to be caused, which also promotes the above mentioned fact. In the analysis of Budiansky and Hutchinson (1978), these characteristics are not taken into consideration. So, their results give incorrect tendency at high R values in Fig. 4.

Relation between $\Delta\delta_0$ and K

$\Delta\delta_0$ obtained in the present analysis contains an elastic component which is not proportional to $(K - K_{op})^2$ but to $(K - K_{op})$ and which does not contribute the actual crack tip opening displacement caused by crack tip plastic blunting. So, this component was excluded. The results of $\Delta\delta_0/\delta_0$ for a given K can be expressed as follows;

$$\text{Loading } (K_{op} < K < K_{max});\ \frac{\Delta\delta_0}{\delta_0} = A\left(\frac{K - K_{op}}{K}\right)^2,\ A = 0.76 \tag{1}$$

$$\text{Unloading } (K_{cl} < K < K_{max});\ \frac{\delta_{0max} - \delta_r - \Delta\delta_0}{\delta_0} = B\left(\frac{K - K_{cl}}{K}\right)^2,\ B = 0.52 \tag{2}$$

The values of A and B shows a good agreement with the results, A = 0.73 and B = 0.54, obtained by Budiansky and Hutchinson (1978).

Recently, authors have provided experimental support to evaluate K_{op} accurately (Kobayashi, 1981). This method was based on the measurements of tip opening displacements of the fatigue crack, a striation spacing S, and the ideal crack, a sub-critical stretched zone width SZW. The experiments on alloy steels, aluminum alloys and a titanium alloy show that ratios of S to SZW are given by the following expression (Kobayashi, 1981);

$$S/SZW = \begin{cases} 0.12 & \text{(alloy steels)} \\ 0.18 & \text{(aluminum alloys)} \\ 0.11 & \text{(titanium alloy)} \\ 0.13 & \text{(total average).} \end{cases} \tag{3}$$

Substituting S/SZW into $\Delta\delta_0/\delta_0$ in Eq.(1), K_{op}/K_{max} can be evaluated as follows;

$$K_{op}/K_{max} = \begin{cases} 0.60 & \text{(alloy steels)} \\ 0.51 & \text{(aluminum alloys)} \\ 0.62 & \text{(titanium alloy)} \\ 0.59 & \text{(total average).} \end{cases} \tag{4}$$

It should be noted that K_{op}/K_{max} in Eq.(4) evaluated by using the experimental results agree well with other analytical results (Newman, 1976; Ogura, 1977; Shiratori, 1977; Budiansky, 1978) as well as with present study (see Fig. 4) at R = 0.

The results of Eq.(3) are in the case of small scale yielding. In large scale yielding, S/SZW for a given K_{max} becomes larger. So, K_{op}/K_{max} would decrease with increasing K_{max}. For comparison, the analysis at $K_{max} = 24.8$ MPa√m was carried out. Actually, K_{op}/K_{max} obtained show such a tendency;

$$K_{op}/K_{max} = \begin{array}{l} 0.58 \ (K_{max} = 24.8 \text{ MPa}\sqrt{m}, \ R = 0) \\ 0.45 \ (K_{max} = 62.0 \text{ MPa}\sqrt{m}, \ R = 0). \end{array} \quad (5)$$

Detection of Crack Closure

Figure 6 shows the relation between the strain, ε_b, on the back face (the face opposite that from which the notch is machined) and K at R = 0, where ε_0 is the value of ε_b at $K = K_{max}$. The relation between ε_b and K does not show clear detection of crack closure.

However, if the elastic strain component ε_e is subtracted from ε_b (Kikukawa, 1976) as shown in Fig. 7, the relation between $(\varepsilon_b - \varepsilon_e)$ and K shows clear detection of crack closure. K_{op} determined from Fig. 7 shows a good agreement with K_{op} determined analytically by using a reaction force at the crack tip. On the other hand, in the unloading process, $(\varepsilon_b - \varepsilon_e)$ decreases with decreasing K due to reverse plastic flow at the crack tip. Then, it increases with increasing K due to occurrence of crack closure. The K value at the intersection of these two curves corresponds with K_{cl} accurately. The relation between the crack mouth displacement and K gave the same results. These results may give support to the justification of measuring K_{op} or K_{cl} according to such techniques.

Transition from Fatigue Crack to Ideal Crack

The crack tip opening displacement, δ_{p0}, formed during a single peak overload as shown in Fig. 1 (c), is identified as a giant striation spacing, GS, fractographically. For a given K_{max} as shown in Fig. 1, GS takes an intermediate

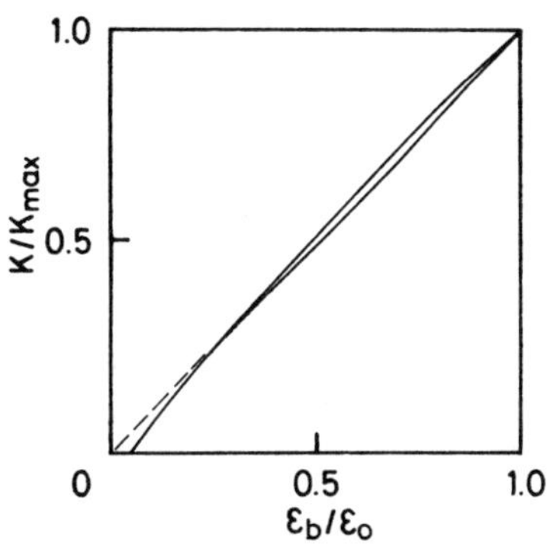

Fig. 6 Relation between K and back face strain, ε_b.

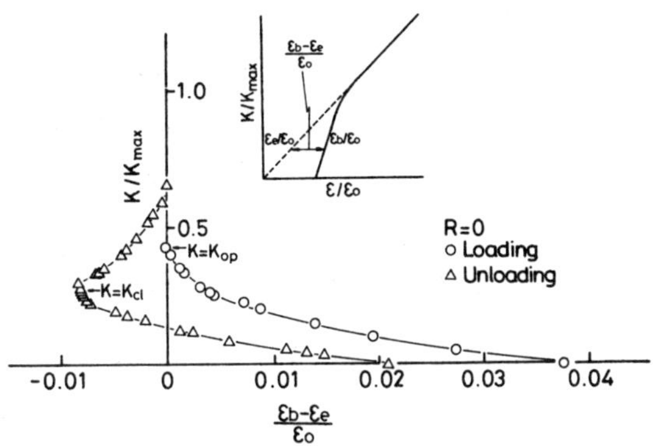

Fig. 7 Relation between K and back face strain, ε_b, obtained by using subtracting circuit.

value between S and SZW, i.e.

$$S < GS < SZW \tag{6}$$

And it reaches to SZW as the overload ratio, γ, increases. The transitional behavior from the fatigue crack to the ideal crack has been studied by some researchers, and several theories have been advanced (McMillan, 1970; Hertzberg, 1979; Kobayashi, 1981).

Hertzberg (1979) has directly related the crack closure concept to the evaluation of δ_{po}. The result is given by the following expression;

$$\frac{\delta_{p0}}{\delta_{0max}} = \frac{(K_{max} - K_{op})^2}{K_{max}^2} \tag{7}$$

where K_{op} is the value under constant amplitude loading ($K = K_2$). However, Eq.(7) does not agree with the present analysis.

If we assume that the fatigue loading history during constant amplitude loading ($K = K_2$) does not affect the tip opening formed between K_2 and K_{max} (McMillan, 1970), δ_{p0} is given by the following expression;

$$\frac{\delta_{p0} - \Delta\delta_2}{\delta_{0max}} = \frac{K_{max}^2 - K_2^2}{K_{max}^2} \tag{8}$$

where $\Delta\delta_2$ is $\Delta\delta_0$ at $K = K_2$. From Eq.(1) and Eq.(8), $\delta_{p0}/\delta_{0max}$ is given as follows (Kobayashi; 1981);

$$\frac{\delta_{p0}}{\delta_{0max}} = \frac{A\ (K_2 - K_{op})^2 + K_{max}^2 - K_2^2}{K_{max}^2}\ , \quad A = 0.76 \tag{9}$$

Figure 8 shows the results of δ_{p0}/δ_{02} as a function of $\gamma = K_{max}/K_2$ (K_2 = 24.8 MPa$\sqrt{m}$) obtained by the present analysis of single peak overload tests, where δ_{02} is δ_0 at $K = K_2$. In Fig. 8, prediction from Eq.(9) is also shown. It is

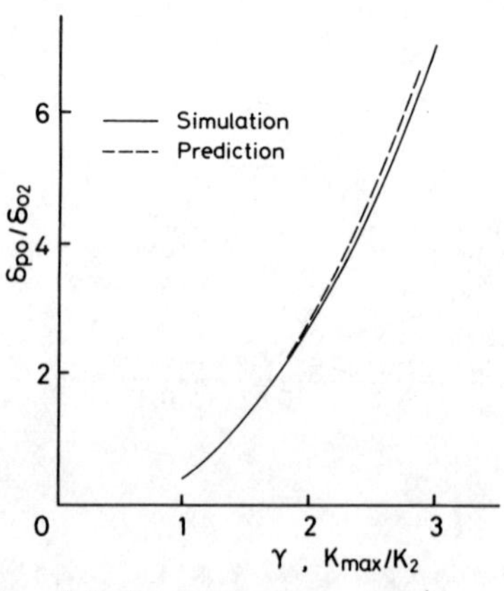

Fig. 8 Relation between δ_{p0}/δ_{02} and γ (= K_{max}/K_2).

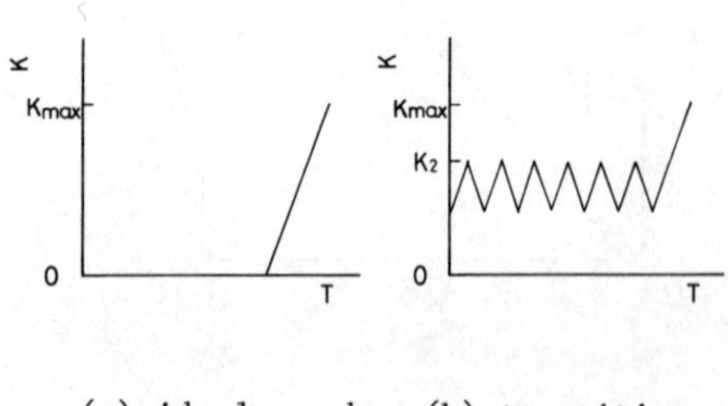

(a) ideal crack (b) transition

Fig. 9 Transition from fatigue crack (R > 0) to ideal crack.

clear that Eq.(9) gives good prediction for the result. It should be noted that Eq.(9) can hold for $R > 0$ as shown in Fig. 9.

CONCLUSION

A finite element analysis was performed to investigate the crack closure phenomenon under cyclic loading and stable crack growth. The results were compared with those of ideal crack. The results obtained are as follows;

1. The relation between $\Delta\delta_0$ (fatigue crack) and δ_0 (ideal crack) for a given stress intensity K_{max} is given by the following expression;

$$\left(\frac{K_{max} - K_{op}}{K_{max}}\right)^2 = \frac{1}{A}\frac{\Delta\delta_0}{\delta_0} , \quad A = 0.76 \tag{a}$$

   From Eq.(a), K_{op} can be determined as $K_{op}/K_{max} = 0.59$ for $R = 0$, by using experimentally obtained $\Delta\delta_0/\delta_0$.
2. As R increases, the crack surface contact becomes not be caused.
3. The analysis gives justification to the measurements of crack closure by the methods using back face strain and crack mouth displacement.
4. The transitional behavior from the fatigue crack to the ideal crack is evaluated quantitatively by using Eq.(a).

REFERENCE

Budiansky, B. and J. W. Hutchinson (1978). Analysis of Closure in Fatigue Crack Growth, Journal of Applied Mechanics, ASME, 45, 267-276.

Elber, W (1971). The Significance of Fatigue Crack Closure, Damage Tolerance in Aircraft Structures, ASTM STP 486, 230-242.

Hertzberg, R. W (1979). On the Relationship between Fatigue Striation Spacings and Stretch Zone Widths, Int. Journ. of Fracture, 15, R69-R72.

Kikukawa, M., M. Jono and K. Tanaka (1976). Fatigue Crack Closure Behavior at Low Stress Intensity Level, Proc. of Second International Conference on Mechanichal Behavior of Materials, ASM, special volume, 254-277.

Kobayashi, H., H. Nakamura and H. Nakazawa (1981). Mechanics of Fatigue Crack Growth, Mechanics of Fatigue, ASME, AMD-47, 133-150.

Kobayashi, H., T. Ogawa, H. Nakamura and H. Nakazawa (1983). Oxide Induced Crack Closure and Near-Threshold Characteristics in A508-3 Steel, Proc. Inter. Symp. on Fracture Mechanics, Beijing, China, in press.

McMillan, J. C. and R. M. Pelloux (1970). Fatigue Crack Propagation under Programmed Loads and Crack Tip Opening Displacements, Engng. Fracture Mechanics, 2, 81-84.

Newman, J. C., Jr.(1976). A Finite-Element Analysis of Fatigue Crack Closure, Mechanics of Crack Growth, ASTM STP 590, 281-301.

Ogura, K. and K. Ohji (1977). FEM Analysis of Crack Closure and Delay Effect in Fatigue Crack Growth under Variable Amplitude Loading, Engng. Fracture Mechanics, 9, 471-480.

Shiratori, M., T. Miyoshi and H. Miyamoto (1977). Analysis of Fatigue Crack Growth and Closure behavior, Trans. of the Japan Society of Mechanical Engineers, 43, 3577-3593 (in Japanese).

THE EFFECT OF INTERMITTENT REST PERIODS AT ELEVATED TEMPERATURES ON THE FATIGUE PROPERTIES OF 4140 STEEL

Song-Hee Kim

Department of Materials Engineering, Monash University, Clayton, Victoria, 3168, Australia

ABSTRACT

The effect of intermittent rest periods on fatigue properties of 4140 steel have been investigated. It was observed that the fatigue life was increased, in some cases by a factor of two. It is proposed that the increase in fatigue life is due to a delay of fatigue damage accumulation and to an increase in fatigue resistance. This hypothesis is based on generally accepted concepts of cumulative damage rules which are expressed in terms of various fatigue damage parameters. The present research shows the second of the above two factors is the major one.

KEYWORDS

Low cycle fatigue; rest periods; cumulative damage laws; fatigue damage recovery; fatigue damage delay.

INTRODUCTION

1. Relationships Between Fatigue Life and Fatigue Damage Parameters

The fatigue life of material components is usually determined in terms of either one of three parameters; i.e. applied stress, applied strain or dissipated plastic work. Logarithmic plots of the fatigue parameters against fatigue life usually result in linear relationships (Miner, 1945; Manson, 1953; Morrow, 1965). Although modified relationships have been tried (Manson, Freche and Ensign, 1966 and Miller, 1976) it is generally found that fatigue data can be represented by the following equations:

$$\Delta\sigma = \sigma_f'(N_f)^b \text{ or } (N_f)(\Delta\sigma)^{-1/b} = \sigma_f^{-1/b} \tag{1}$$

$$\Delta\varepsilon_p = \varepsilon_f'(N_f)^c \text{ or } (N_f)(\Delta\varepsilon_p)^{-1/c} = \varepsilon_f^{-1/c} \tag{2}$$

$$\Delta W_p = W_f'(N_f)^f \text{ or } (N_f)(\Delta W_p)^{-1/f} = W_f^{-1/d} \tag{3}$$

where $\Delta\sigma$, $\Delta\varepsilon_p$ and ΔW_p represent the applied cyclic stress, plastic strain or plastic work, respectively, and will be referred to as the "fatigue damage parameters", N_f is the life to failure, σ_f' is called the fatigue strength, ε_f' is called the fatigue ductility and W_f' is called the fatigue toughness, and b, c and f are exponents of the equations.

2. Linear Cumulative Fatigue Damage Rules

It has been proposed by Miner (1945) that the fatigue life of a component can be predicted on the basis of a cumulative damage rule. The rule assumes that, irrespective of the magnitude of stress, each and every stress cycle has some influence of the ultimate endurance of the specimen, and is responsible for a certain amount of fatigue damage. It is also held that the accumulated fatigue damage up to failure is a material constant. This rule can be written:

$$\sum_{i=1}^{N_f} (\Delta\sigma_i)^{-1/b} = C_1 \tag{4}$$

Sandor (1968) has proposed a similar rule in terms of plastic strain:

$$\sum_{i=1}^{N_f} (\Delta\varepsilon_{p,i})^{-1/c} = C_2 \tag{5}$$

Kim, Griffiths and Stachurski (1982) have extended these rules to include dissipated plastic work:

$$\sum_{i=1}^{N_f} (\Delta W_{p,i})^{-1/f} = C_3 \tag{6}$$

In equations (4) to (6) the constants C_1, C_2 and C_3 represent the fatigue damage constants for any of the fatigue parameters: they will be different from each other, although constant for any given material. The constants are called the "fatigue resistance".

EXPERIMENTAL METHODS

1. Material

Experiments were done on an SAE 4140 steel supplied as 12.5 mm diameter rod with a chemical composition of C,0.44%; Mn,0.80%; Si,0.4%; S,0.03%; P,0.025%; Ni,0.08%; Cr,0.83%;Mo,0.19%. The steel had been heat treated by the supplier by oil-quenching from 840°C and tempering at ∿590°C: this as-received condition is termed the "quench-and-tempered" condition. Specimens were cut from this rod, and some of them were annealed in vacuum at 900°C for one hour and furnace cooled to give a ferrite/pearlite microstructure - the so-called "annealed" condition. Metallographic examination of transverse sections for annealed specimens could reveal no evidence for surface decarburization although in any event they were, prior to testing, mechanically polished, first with emery paper and finally with γ-Al_2O_3 powder (<3 μm particle size).

The mechanical properties of the annealed 4140 steel were: $\sigma_{Y0.2\%}$, 465 MPa; σ_{UTS}, 915 MPa; reduction in area (R/A), 37%; strain hardening exponent (n), 0.24; cyclic hardening exponent (n'), 0.19. And those of the quench-tempered steel were: $\sigma_{Y0.2\%}$, 725 MPa; σ_{UTS}, 964 MPa; reduction in area (R/A, 39%; strain hardening exponent (n), 0.08; cyclic hardening exponent (n'), 0.13.

2. Specimen Design and Test Method

Specimens were cylindrical, with a gauge length of 10 mm and diameter 7 mm, and were loaded axially in an MTS servo-hydraulic machine. Strain was measured by a knife-edge extensometer (resolution 0.01% strain) attached to the gauge length. Both load- and strain-controlled tests were done, with continuous on-line computer recording of load and strain. Tests were done at zero mean load (or strain, where appropriate), using a sinusoidal waveform with a frequency of 0.05 Hz.

3. Intermittent Heat Treatments

These were done in a tube furnace in a nitrogen atmosphere at 300°C. Some oxidation was observed, with the steel developing a blue colour. Several control experiments established that the act of removing the specimen from the machine and then replacing it had no effect on its subsequent behaviour.

RESULTS

1. Continuous Tests

Basic fatigue properties were measured in continuous tests and the data were evaluated using least square regression analysis. The characteristic constants obtained from such analysis are shown in Table 1.

TABLE 1 Basic Fatigue Properties Obtained from Continuous Tests (see equations (4) to (6))

Conditions of Material	b	c	f	C_1	C_2	C_3
Quench-and-Tempered	-.086	-.64	-.71	1.44×10^{39}	0.847	2.93×10^4
Annealed	-.137	-.645	-.767	1.44×10^{25}	1.41	2.52×10^4

2. Fatigue Tests with Rest Periods

It was arbitrarily decided to concentrate attention on conditions which, for a continuous test, gave a life of about 500 and 400 cycles respectively in stress and strain control. Stress control tests were interrupted at 33% and 67% of the predictive life in continuous test - i.e. after 180 cycles and again after 360 cycles. Strain control tests were interrupted at 50% and 100% of the predictive life in continuous test - i.e. after 200 cycles and again after 400 cycles. These specimens were heat treated as described above, the duration of the heat treatment being the same each time. The results of these tests, showing fatigue life as a function of total rest time at 300°C, are shown in Fig. 1 for the annealed steel. Notice the general increase of fatigue life with ageing time. Some differences between tests under stress and strain control are evident below 10 hours for total rest time. (The discussion on these are reported in Kim, 1983.) Fatigue resistance, as defined by equations (4) to (6) has been calculated for all tests. The effect of rest periods on fatigue resistance calculated from equations (5) and (6) are shown in Figs. 2 and 4.

In the case of quench-and-tempered steel, the effect of rest period on fatigue life is shown in Fig. 3 where a less remarkable increase in fatigue life was observed. Note that the effect of rest periods on the measured fatigue resistance based on equation (4) would be very similar in constant stress control tests with the effect on fatigue life shown in Figs. 1 and 3 because cumulative damage in terms of stress is proportional to number of cycles to failure. Figures 5 and 6 are shown to illustrate the effect of cyclic hardening and cyclic softening on fatigue damage accumulation. Both specimens were tested under stress control, but specimen no. 40 was in annealed condition, whereas specimen no. 13 was quench-and-tempered. This difference manifested itself most clearly in the first 100 cycles. The rest periods brought about softening of specimens on resumption of fatigue cycling.

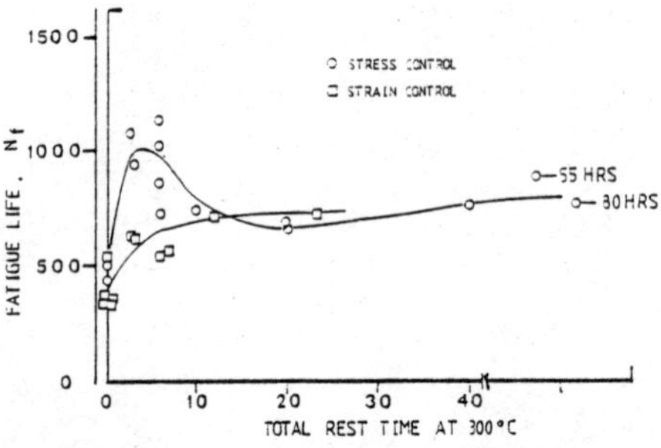

Fig. 1. The effect of rest periods at 300°C on fatigue life of annealed 4140 steel under stress and strain control (i.e. σ_a = 590 MPa and $\Delta\varepsilon_{t/2}$ = 1.55%).

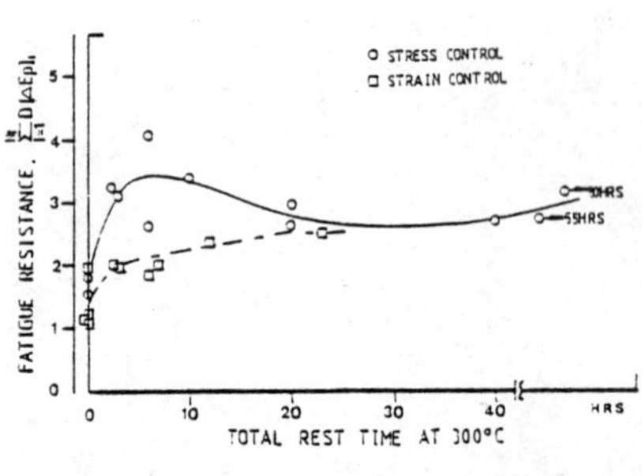

Fig. 2. The effect of rest periods at 300°C on fatigue resistance of annealed 4140 steel in terms of plastic strain under stress and strain control.

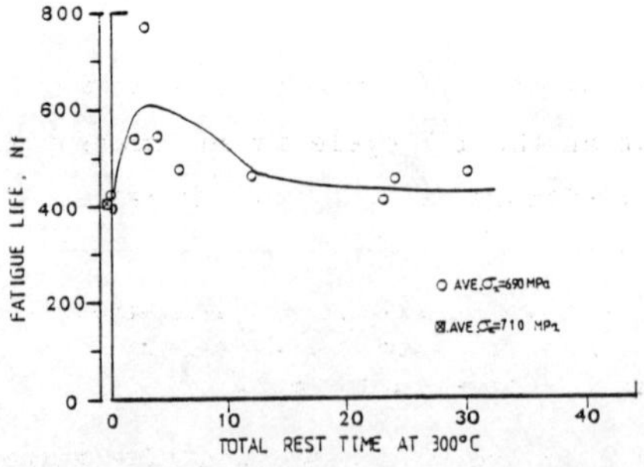

Fig. 3. The effect of rest periods at 300°C on fatigue life of quench-and-tempered 4140 steel under stress control (i.e. σ_a = 690 MPa).

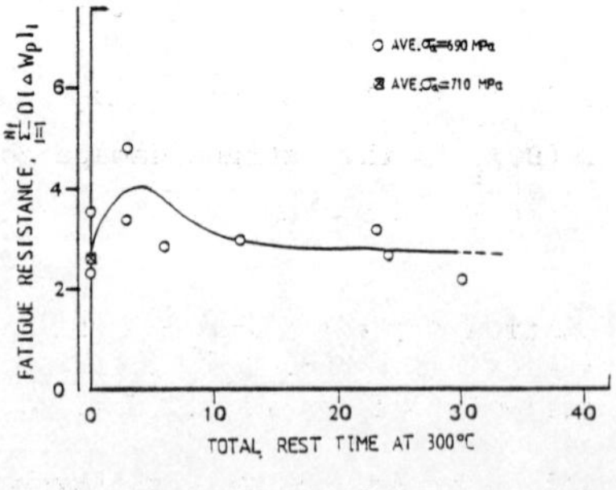

Fig. 4. The effect of rest periods at 300°C on fatigue resistance of quench-and-tempered 4140 steel in terms of dissipated plastic work under stress control (i.e. σ_a = 690 MPa).

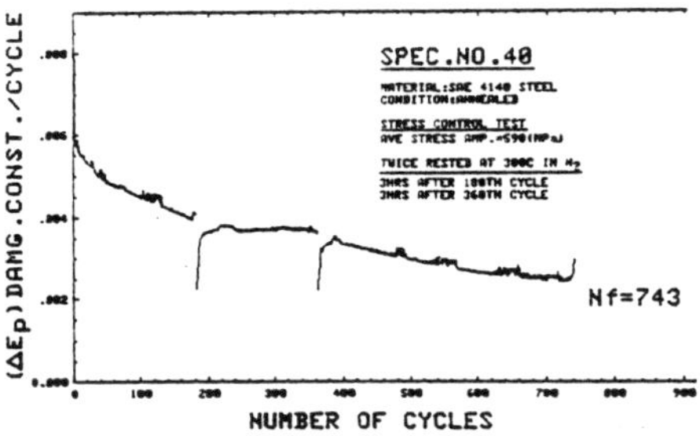

Fig. 5. The effect of rest periods at 300°C on fatigue damage accumulation in terms of plastic strain in cyclic hardening material (annealed 4140 steel) under stress control.

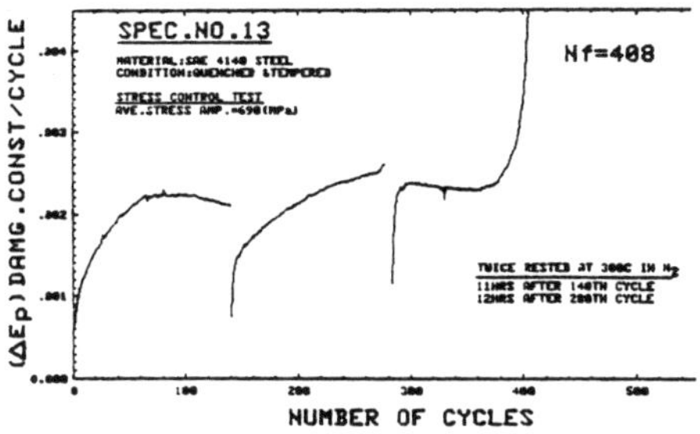

Fig. 6. The effect of rest periods at 300°C on fatigue damage accumulation in terms of plastic strain in cyclic softening material (quench-and-tempered 4140 steel) under stress control.

DISCUSSION

Knowing the value of fatigue resistance (as in Table 1), and using the cumulative damage rules, equations (4) to (6), one can predict the fatigue life of any given specimen under given conditions. Introduction of rest periods at elevated temperature alters the process of accumulation of fatigue damage. This leads to the concepts of "acceleration" and "delay" of damage accumulation. Such effects were observed and reported by Sandor (1968, 1970) for the case of mild steel. Rest periods causing strain-ageing resulted in a delay of fatigue damage accumulation as deduced from the increased fatigue life. More frequent insertion of rest periods also gave proportionally increased fatigue life. Miller and Hatter (1972) also reported that retardation in plastic strain accumulation after rest period could increase fatigue life.

Assuming that fatigue resistance is not altered by the rest periods it may be possible to say that the increase in fatigue life is partly due to delay of damage accumulation, and its effect can be expressed in a generalised form:

$$\sum_{i=1}^{N_k} (D_C)_i - \sum_{i=1}^{N_k} (D_I)_i = A \tag{7}$$

where $(D_C)_i$ is the fatigue damage constant at the ith cycle in a continuous test, $(D_I)_i$ is the fatigue damage constant at the ith cycle in an interrupted test, and N_k is the ordinal number of an arbitrary cycle after the rest period.

In equation (7) the first and the second term represent the accumulation of fatigue damage to the $N_k{}^{th}$ cycle in continuous and interrupted tests respectively and can be based on any fatigue damage parameters.

From the experimental results shown in Figs. 5 and 6, A has a positive value, defined as "fatigue damage delay". If A in equation (7) is negative this indicates "acceleration of fatigue damage accumulation". Furthermore, increase in fatigue resistance is observed here with application of various cumulative damage laws. The amount of increase is directly measured in the present work following the expression below and defined here as "fatigue damage recovery":

$$\sum_{i=1}^{N_{f_2}} (D_I)_i - \sum_{i=1}^{N_{f_1}} (D_C)_i = R \tag{8}$$

where N_{f_2} and N_{f_1} are the actual fatigue lives in interrupted and continuous tests, respectively.

The first and second terms in equation (8) represent fatigue resistance measured respectively from interrupted and continuous tests.

Therefore it is believed that the reason for the increase in fatigue life of our material consists of the effect of delay of fatigue damage accumulation and increase in fatigue resistance (recovery of fatigue damage). The latter is found to play a major role in extension of fatigue life. However it has not been investigated how fatigue damage recovery is related to physical meaning of fatigue damage.

SUMMARY

(i) Intermittent rest periods at 300°C increased the fatigue life of annealed and quench-and-tempered 4140 steel.

(ii) Two new concepts, those of delay of fatigue damage accumulation and of fatigue damage recovery have been proposed in quantitative forms.

(iii) Delay of fatigue damage accumulation due to rest periods has been experimentally observed but it is believed to contribute only marginally to the increase in fatigue life in these experiments.

(iv) Fatigue damage recovery has been measured, and was observed to be the major factor in the increased fatigue life in these experiments.

(v) Further study is required to define the microstructural mechanisms for the phenomena described in this paper.

ACKNOWLEDGEMENT

The work was carried out as part of a Ph.D. programme at Monash University under the supervision of Drs Z.H. Stachurski and J.R. Griffiths. The author is grateful to them for their comments on this paper.

REFERENCES

Kim, S.H. (1983). Ph.D. thesis, to be submitted to Monash University in May.

Kim, S.H., J.R. Griffiths and Z.H. Stachurski (1982). Proc. of 8th Australasian Conf. on Mechanics of Structures and Materials, Newcastle, Aug., Paper No. 19.

Manson, S.S. (1953). Heat Transfer Symposium, University of Michigan, Eng.Res.Inst., 9-75.

Manson, S.S., J.C. Freche and C.R. Ensign (1966). Fatigue Crack Propagation, ASTM STP415, ASTM, 384-412.
Miller, K.J. and D.J. Hatter (1972). J.Strain Analysis, 7, 69-73.
Miller K.J. (1976). Proc.2nd Int.Conf.Mech. Behavior of Materials, Boston, Special Volume, 236-253.
Miner, M.A. (1945). J.Appl.Mech., Am.Soc.Mech.Engrs, 67, 159-164.
Morrow, J. (1965). Internal Friction, Damping and Cyclic Plasticity, ASTM STP378, ASTM, 45-87.
Sandor, B.I. and J. Morrow (1968). Proc.3rd Conf. on Dimensioning, Budapest, 215-225.
Sandor, B.I. (1970). Achievement of High Fatigue Resistance in Metals and Alloys, ASTM STP467, ASTM, 254-275.

FAILURE PREDICTIONS BASED ON TEARING INSTABILITY AND LIMIT LOAD — A PARAMETER STUDY

L. Hodulak

Fraunhofer-Institut für Werkstoffmechanik, Wöhlerstr. 11, 7800 Freiburg, Federal Republic of Germany

ABSTRACT

Differences in failure predictions based on the tearing instability and on the limit load are shown for simple specimens (CCP, SECP, DECP, C) under load control. In the plane strain calculations performed by approximate methods for two steels with the upper shelf Charpy energies of 90 J and 160 J specimen sizes W and normalized crack lengths a/W have been varied in a wide range.

KEYWORDS

failure assessment; ductile fracture; tearing instability; limit load, stable crack growth

INTRODUCTION

Failures of metal structure components such as pressure vessels and pipings occur usually in a ductile manner. For an assessment of failure resistance in the high ductility range the limit load concept and concepts of elastic-plastic fracture mechanics (EPFM) are available. Within the EPFM approach especially the concept of ductile tearing (Hutchinson and Paris, 1979) based on the J-integral as a fracture parameter is well developed. It enables to describe the ductile failure process from the onset of crack extension up to crack instability. The evaluation procedures are, however, rather complicated and there are limitations for the use of the J-Integral from the theoretical point of view. It is necessary to check, whether the J-integral is meaningfull for the analysed case and whether the crack growth is J-controlled.

The limit load concept is easier to use, but it is suitable rather for obtaining lower bound solutions, since strain hardening can be incorporated in the calculation only in a very approximate way. Being unable to handle properly cases with some crack extension occuring prior to plastic collaps, the concept will not garantee conservativity in each case.

From the pragmatical point of view it is important to know, whether and how much failure predictions resulting from each of these methods differ for typical structure components. In situations, in which the difference is not significant, the simpler method (limit load) should be preferred otherwise the use of the more sophisticated method (tearing instability) may be necessary.

Appreciable progress has been achieved in the development of methods for calculation of the crack loading parameter $J_{applied}$ and in methods for testing of material response in terms of $J_{material}$ (J_R-curve), as well. Though theoretical validity limits of the tearing instability concept are exceeded in many practical applications, the concept may supply reasonable results also beyond its theoretical limits as some tests indicate.

In Fig. 1 some experimental and calculated results are shown for three large-scale and two small specimens tested in the framework of the German reactor safety program (Kussmaul and Issler, 1983). The maximum loads obtained from tearing instability analysis seem to correlate with the experimental failure loads better than the limit loads for most of these specimens.

Nevertheless, further verification experiments with large-scale specimens and structure components are needed to show the virtues and limitations of the concept and of the evaluation procedures, too. In planning such experiments, which are usually very expensive, it is important to know, what kind of specimen can reveal the investigated phenomena most clearly despite of possible uncertainties and scatter of data.

The goal of this study is to show quantitatively the differences in failure predictions based on the limit load and tearing instability concepts, to supply data for an assessment of the significance of some qualitative statements occuring in discussions in ductile failure predictions and to assist in planning of further experiments.

SPECIMENS AND CALCULATION METHODS

Tensile specimens shown in Fig. 2 ranging in size W from 50 mm to 400 mm and in normalized crack length a/W from 0.1 to 0.9 were analyzed using two sets of material data specified in Fig. 3. Particulary, loads at crack initiation (F_i), crack instability (F_{max}), and at the plastic collaps evaluated with initial ($F_L(a_0)$) or final crack length ($F_L(a_0 + \Delta a)$) and the amount of stable crack extension Δa at instability were calculated.

The values of $J_{applied}$ for the EPFM analyses were calculated by the engineering approach (EA) according to the EPRI fracture handbook (Kumar, German, and Shih, 1981). In the Ramberg-Osgood approximations of the stress-strain curves the constants to be determined were taken as $\alpha = 1.35$, $n = 10$ and $\alpha = 0.91$, $n = 10.6$ for materials I and II, respectively.

In addition the failure assessment diagram (FAD) with the assessment line modified for strain hardening by Milne (Chell and Milne, 1983) was used for failure analysis of some specimens. This method is not based on the J-integral, but it can be interpreted in terms of J and considered as an approximate method of tearing instability analysis. The stress intensity factor expressions were taken from the handbook by Tada and co-worker (1973). The collaps loads were calculated according to lower bound formulae (Kumar, German and Shih, 1981) for CCP, SECP, and C specimens and a formula by Issler and co-workers (1980) for DECP specimens.

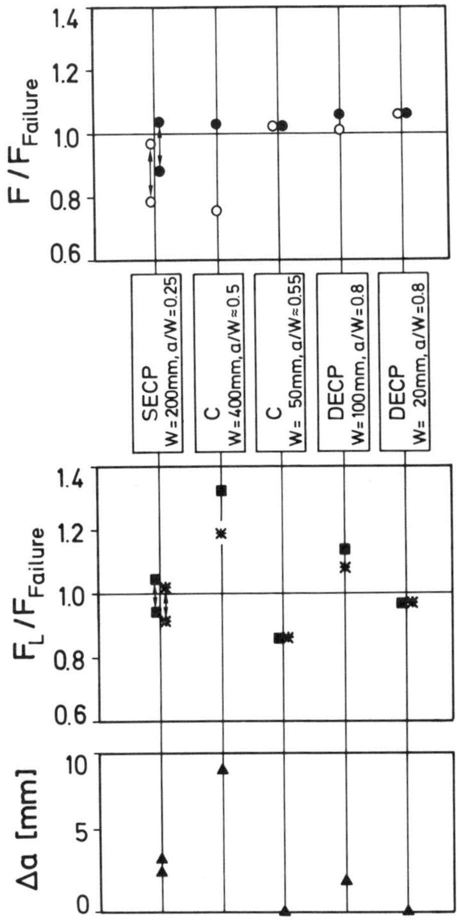

Fig. 1. Comparison of calculation results with experimental failure loads $F_{Failure}$ for several specimens of material I. EA calculation.

- o load at the onset of the crack extension
- ● maximum load
- ■ $F_L(\sigma_y, a_0)$
- ✱ $F_L(\sigma_y, a_0 + \Delta a)$
- ▲ stable crack extension Δa

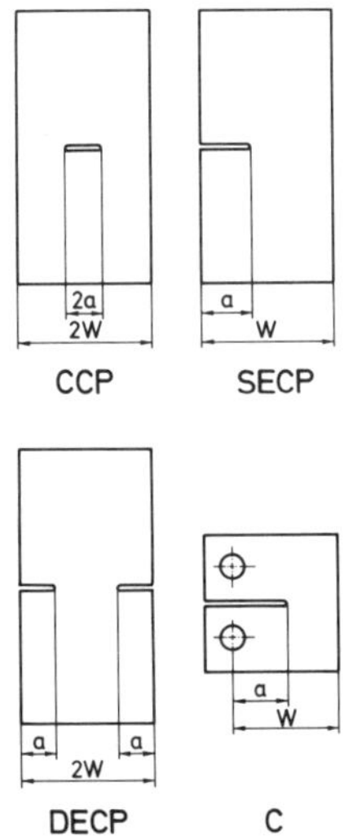

Fig. 2. Specimen geometries

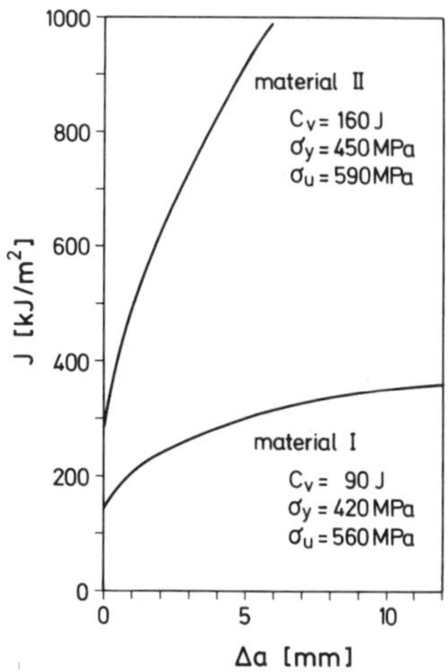

Fig. 3. Material data (Hodulak and Blauel, 1983; Krägeloh and co-workers, 1980)

The limit load results in Figs. 4,5 and 8 are obtained from the same formulae using, however, the flow stress $\sigma_F = (\sigma_y + \sigma_u)/2$ as reference stress.

In all calculations the state of plane strain, the independence of the J_R-curves of specimen geometry, and J-controlled crack growth are assumed.

RESULTS

In Fig. 4 the ratio of the crack instability load F_{max} and the limit load F_L calculated with the flow stress σ_F is plotted vs. normalized initial crack length a_o/W. The crack instability was calculated by the EA method.

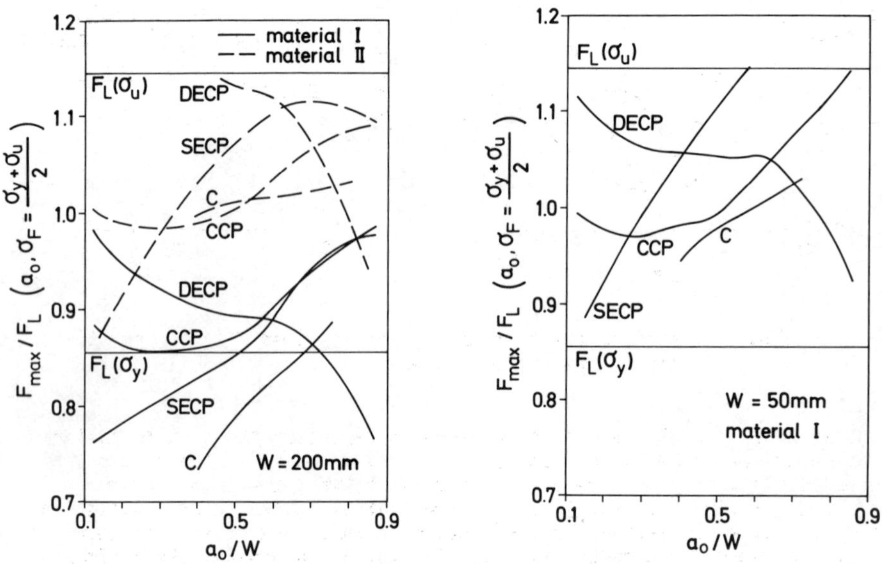

Fig. 4. Ratio of the tearing instability load F_{max} and the limit load F_L. EA calculation.

Depending on the specimen geometry and on the material toughness the limit load $F_L(\sigma_y)$ underestimates or overestimates the value of F_{max}. Provided that F_{max} corresponds reasonably to the failure load as Fig. 1 indicates, the limit loads $F_L(\sigma_y)$ will not assess the failure load conservatively for SECP and C specimens mashined from material I in sizes exceeding $W \approx 200$ mm.

The value of F_{max}/F_L increases with decreasing specimen size, increasing material toughness (material I → material II), and decreasing bending of the specimen (C → SECP → CCP/DECP). It can be concluded from Fig. 4, which flow stress $\sigma_F = f(\sigma_y, \sigma_u)$ should be used in the plastic collaps calculations to obtain a good estimation of failure load for given specimen size and material toughness.

The F_{max} in Fig. 5 was calculated by the FAD method and the results are qualitatively similar to those of Fig. 4. Compared to the EA method the FAD method seems to supply conservativer values of F_{max}.

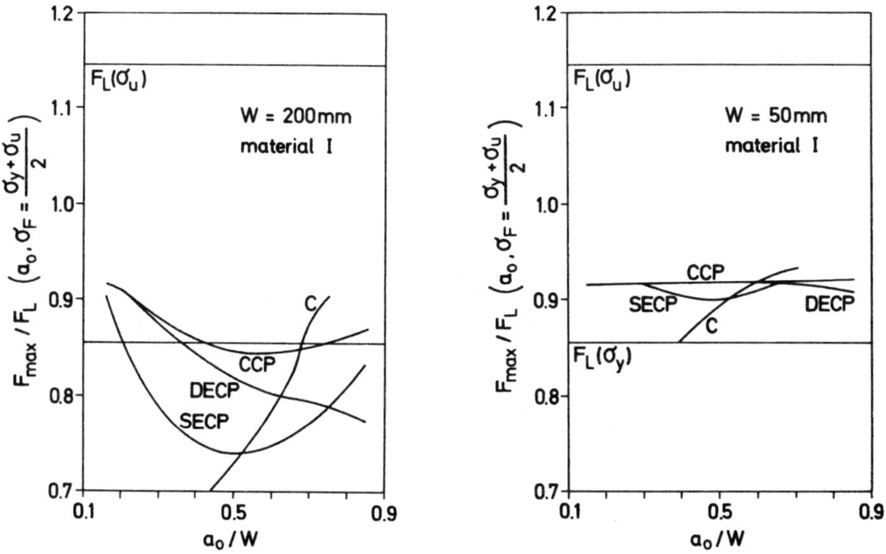

Fig. 5. Ratio of the tearing instability load F_{max} and the limit load F_L. FAD calculation.

Similar to linear elastic fracture mechanics approach the ductile failure assessment may be based just on J_{Ic} as material constant. Being sure that there is still an appreciable load increase between crack initiation (F_i) and crack instability (F_{max}) some uncerntainties in the (crack initiation) analysis may be tolerated. The dependence of this load reserve in terms of F_{max}/F_i on specimen geometry is shown in Fig. 6. In general the ratio F_{max}/F_i increases with specimen size.

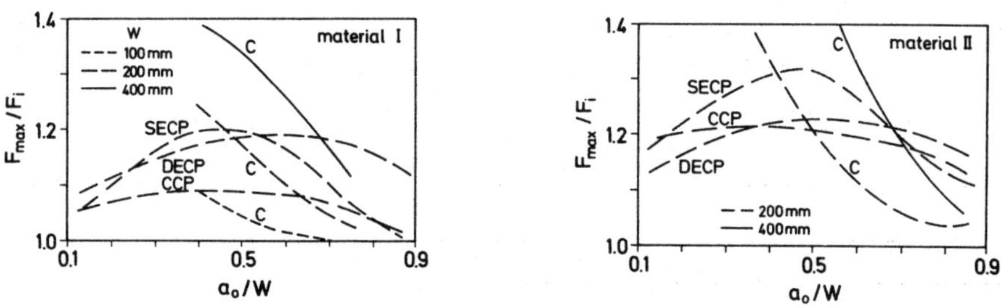

Fig. 6. Ratio of loads at the tearing instability (F_{max}) and at the onset of crack extension (F_i). EA calculation.

In Fig. 7 results of stable crack extension Δa for a wide range of specimen geometries are compiled. With the exception of the DECP Δa does not exceed 2 percent of the specimen width W for material I. In general the normalized stable crack extension $\Delta a/W$ increases somewhat with increasing specimen

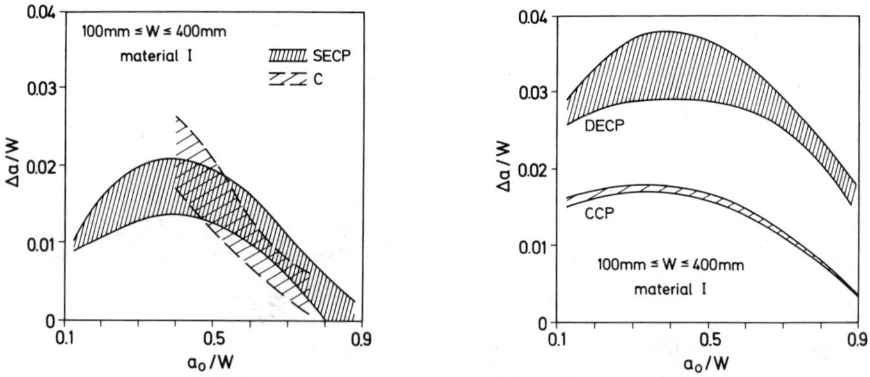

Fig. 7. Normalized stable crack extension $\Delta a/W$. EA calculation.

size. As a consequence the underestimation of the limit load due to neglected stable crack extension increases with increasing specimen size, too. Anyway, for material I this underestimation is below 10 percent up to specimen size W = 400 mm (see Fig. 8).

Fig. 8. Limit load F_L with and without correction for stable crack extension Δa. EA calculation.

ACKNOWLODGEMENT

The author acknowledges financial support of the German Ministery of Research and Technology under contract number RS 150 488

REFERENCES

Chell, G.G. and I. Milne (1983). Ductile tearing instability analysis. In Elastic-Plastic Fracture: Second Symposium. ASTM STP 803

Hodulak, L. and J.G. Blauel (1983). Application of two approximate methods for tearing instability. In Elastic-Plastic Fracture: Second Symposium. ASTM STP 803

Hutchinson, J.W. and P.C. Paris (1979). Stability analysis of J-controlled crack growth. In Elastic-Plastic Fracture. ASTM STP 669, pp. 37-64

Issler, L., R. Birk, and R. Hund (1980). Anwendung eines neuen Näherungsverfahrens aus den USA zur Zähbruchanalyse. In Proceedings 6th Seminar of the Staatliche Materialprüfungsanstalt, Stuttgart, FRG

Krägeloh, E., L. Issler, and R. Zirn (1980). Analyse von Großprobenversuchen mit Zähbruchkonzepten. Proceedings 6th Seminar of the Staatliche Materialprüfungsanstalt, Stuttgart, FRG

Kumar, V., M.D. German, and C.F. Shih (1981). An Engineering Approach for Elastic-Plastic Fracture Analysis. EPRI report NP-1931. Palo Alto

Kussmaul, K. and L. Issler (1983). Evaluation of the elastic-plastic fracture mechanics methodology on the basis of large scale specimens. In Elastic-Plastic Fracture: Second Symposium, ASTM STP 803

Tada, H., P.C. Paris, and G.R. Irwin (1973). The Stress Analysis of Cracks Handbook. Del Research Corp., Hellertown

FATIGUE CRACK PROPAGATION AND THRESHOLD OF TYPE 304L AUSTENITIC STAINLESS STEEL

N. E. Atanasiu* and B. R. Irimescu**

*Polytechnic Institute, 77206 Bucharest, Romania
**Metallurgical Research Institute, 77749 Bucharest, Romania

ABSTRACT

The fatique crack propagation (FCP) behaviour of Type 304 L stainless steel was investigated at room temperature in the range of 5 x 10^{-8} to 5 x 10^{-4} mm/cycle with differentstress ratios for frequencies between 1 and 20 Hz. The low FCP region which leads to threshold is considered and an increasing effect of load ratio was observed as the threshold level was approached, the control of crack growth from ΔK to K_{max}. The results obtained by testing compact-tension specimens were correlated with empirical equations known in the literature to obtain the best fit. At low growth rates the fracture mode is sensitive to the scale and configuration of the microstructure/crack tip plasticity interaction.
The fracture surfaces has been examined by a scanning electronic microscope (SEM) of the JSM 50A type, such as they resulted from breaking at the test specimens.
Scanning electron fractography obtained have been used to characterise the rezultind failure mode of cracked surfaces. The fractography revealed fracture mode consisting of ductile transgranular fracture to the striation-type mode at higher rates.

KEYWORDS

Fatique (materials); crack propagation; stress ratio; threshold level; stainless steel; fractography; transgranular fracture; striation; void; britlle carbide; matrix; intermetallic phase.

INTRODUCTION

Austenitic stainless steels are among the prime candidates for structural applications in the nuclear reactor industry. Since many structural components will be subjected to cyclic loading fluctuations during lifetime it is necessary to know how the fatique crack propagation affects the reactor safety or operation. Using the linear-elastic fracture mechanics approach, the fatigue crack growth (FCG) of Type 304 L stainless steel was characterized for a range of different mean stresses. The FCG rate da/dN was related to the stress intensity range ΔK presenting a different evolution, which permits the distinctions of three regions of FCGidentifiedas low -, intermediate - and high FCG stages. It was found that at a given value of stress intensity factor range, ΔK, the FCG rate increased with increasing values of the stress ratio. However, ΔK did not provide the best basis for corelating FCG rate values for various stress ratios. In adition to the ΔK, other factors as maximum stress intensity factor, K_{max}, fracture toughness, K_c, the stress intensity factor range at the threshold level, ΔK_{Th}, or the effective stress intensity K_{eff}, can be successfully incorporated into a general FCG equation, therefore more useful in the analysis and design of a wide variety of structures. The different crack propagation "lows" are discussed to provide a much better correlation for the experimental results.

At very low rates, the existence of a threshold stress condition for the growth of an existing crack has been estabilished, level below which mode I FCP becomes insignificant. This threshold, defined as ΔK_{Th}, and the slow fatique crack rates show an increased sensitivity to the stress ratio R(ratio of $F_{min}/F_{max} = K_{min}/K_{max}$). This suggests that at very low rates a K_{max} control is more possible than a ΔK dependence.

For the high FCG stage (da/dN $> 10^{-4}$ mm/cycle) it has been shown that this mode of growth can be initiated at lower ΔK values for the higher levels of R ratio, K_{max} being a strong element which controls the FCG process when K_{max} approaches K_c.

The matrix of austenitie stainless steel is ductile, but the hard brittle carbides and the intermetallic phases cause void formation because they are unable accommodate the large plastic strain amplitudes imposed by the more ductile matrix.

Void formation are caused either by breakage of the hard brittle particles an separation of the broken parts, ar by separation of the particle-matrix interface. This process of void and crack formation can be very well observed by following up the behaviour of individual brittle carbides or intermetallic phases.

EXPERIMENTAL

The material studied was a solution-annealed (1050°C/2,5 min/mm/water) AISI Type 304 L austenitic staintess steel forged to ф 70 mm. The chemical composition and the mechanical propreties are given in Tables 1 and 2.

TABLE 1 Chemical composition of AISI 304 L SS (%)

C	Mn	P	S	Si	Cr	Ni	Mo	Co	Nb
0,021	1,88	0,030	0,009	0,51	19,10	9,85	0,50	0,20	0,34

TABELE 2 Mechanical Properties of AISI 304 L (solution-annealed)

0,2% Proof stress MPa	UTS MPa	Elongation* %	R of A %
235	570	51,4	69

*gage lenght = 25,4 mm

The specimens employed were the ASTM standard 25 mm thick Compact-Tension (1978). The fatique tests were carried out on 100KN MTS closed-loop servo-hydraulic testing machine under load control to within ± 1 %. The frequency of testing was between 1 and 20 Hz, and the wave form was sinusoidal. The tests were carried in air (50 pct.relative humidity) at room temperature at R values of 0,05; 0,25 and 0,70. The specimens were carefully polished to give a finish suitable for visual examination of the crack lengths using an ajusted 40 x traveling microscope (the least division 0,001 mm). The FCG rates (da/dN) were determined by dividing each increment of crack extension by the number of cycles producing that increment. The strees intensity factor (K) was calculated using the compliance function for this testing mode given by Brown and Srawley (1966).
Additional tests were conducted to determine the threshold condition for stress ratio R = 0,05; 0,25; 0,40; 0,55; 0,70 and 0,80 by the following procedure. The cracks were initially fatique grown 2 mm from the V-starter notch, which was made by means of an electroerasimate machine (maximum radius of 0,15 mm) then the load was shed in 10 pct. decrements at the R ratio of interest. At any given load level, the crack was allowed to grow beyond the plastic zone size. This procedure was repeated until no crack extension was observed during 10^{-8} cycles of loading. The limit of detectability was therefore less than 10 mm/cycle. All fatique test were at least duplicated in an attempt to minimize errors due to scatter.
The different areas of the fracture surface resulting from test conditions was examined by scanning electron microscope (SEM) operated at 25 KV.
The fractografic analysis has shown that there is difference between the modes of fracture of the austenitic stainless steel at different values of ΔK.
Figure 1 illustrates ductile transgranular fracture of the low K_{max} values and figure 2 indicates ductile striations of the higher K_{max} values.

RESULTS

Conventional double logarithm plots of da/dN versus ΔK for R ratios of 0,05; 0,025; 0,70 are presented in Fig.3. It will be seen that ΔK does not correlate the crack growth rate results over the entire R range, and in general, at a given value of ΔK, da/dN increases with increasing R value, becoming more evident as the threshold is approached. In Fig.4 the same points were

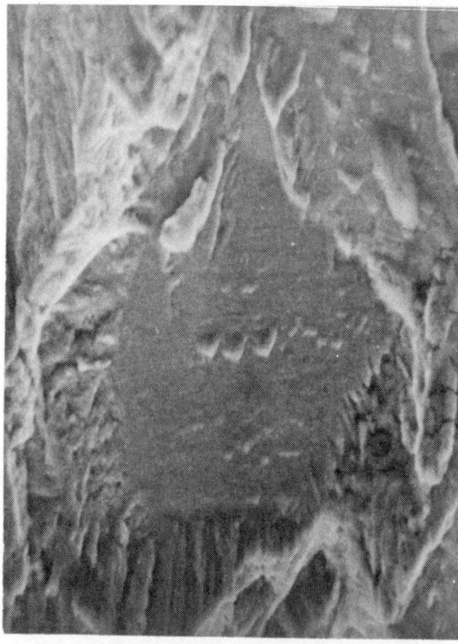

Fig.1. Ductile transgranular fracture (2000 x).

Fig.2. Ductile striations (3000 x).

interpreted in terms of K_{max} rather then ΔK. Thus appears that K_{max} becomes a more important stress parameter for low growth rates (da/dN $< 10^{-6}$ mm/cycle).

The most frequently used correlation letween FCG rate and the stress intensity factor is the "power low" proposed by Paris (1963):

$$da/dN = C \, \Delta K^{m} \qquad (1)$$

where C and m are material constants. Equation (1) predicts a linear relationship between log (da/dN) and log (ΔK), but it holds only for intermediate FCG region. To satisfy the physical reality that unstable crack growth occurs when maximum stress intensity approaches the critical value of fracture toughness K_c, Forman and co-workers (1967) proposed the relation:

$$da/dN = C \, (\Delta K)^{m} / \, (1-R) \, K_c - \Delta K \qquad (2)$$

where the stress ratio efects are taking into acount. However special care must be taken to define the fracture toughnesc K_c (e.g., the austenitic stainless steels are sufficiently "Tough"). For this case, an experimental value of K_c = 120 MPa$\sqrt{m}$ was chosen, agreement with Landes (1981).

A more general equation was proposed by Sperr (1977) and can be written as:

$$da/dN = C \, (\Delta K - K_{TH})^{m} \, / \, (1 - R) \, K_c - \Delta K \qquad (3)$$

in which the threshold stress intensity range ΔK_{Th} is introduced. The K_{max} and stress ratio R effects were employed by Walker (1970) to describe the FCG through the relation:

$$da/dN = C \left[K_{max} \, (1 - R)^{n} \right]^{m} \qquad (4)$$

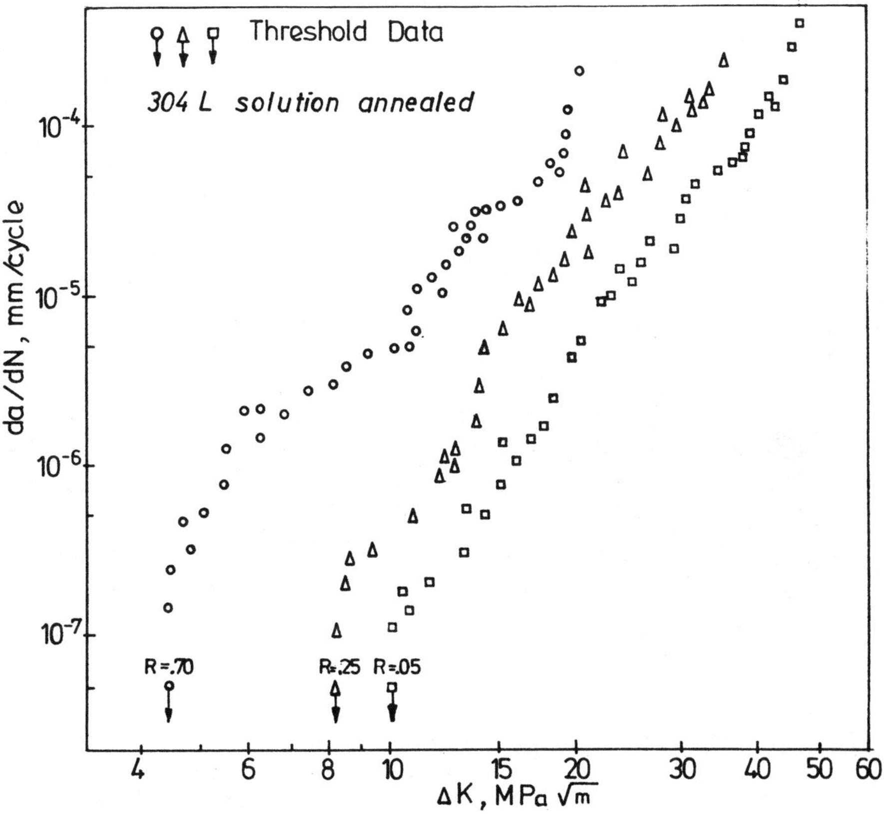

Fig.3. Crack growth rate vs.stress intensity factor range for Type 304 L SS.

where the term K_{max} $(1 - R)^n$ was nomed the "efective stress intensity. A value of n = 0,667 was sugested by Walker for Type 301 steel and a value of n = 0,50 was proposed by James (1972) for Type 304 steel at 538°C. The present results show n = 0,2336 to correlate for Type 304 L steel at room temperature.
A least-squares regression was used to obtain the best fit through the data and to compute the empirical constants from eq.1-4. The results are given in table 3, and shows the best fit for eq. 3 and 4.

TABLE 3 Constants of eq.1-4 by least-squares regression

R	Eq.	Range	C*	m*	Nr.of date	Correlation r
0,05	1	$5 \times 10^{-7} - 10$	$5,2711 \times 10^{-12}$	4,5352	63	0,9567
0,25	2	$5 \times 10^{-7} - 2\times10$	$6,3365 \times 10^{-10}$	4,7757	46	0,9350
0,70	3	$5 \times 10^{-8} - 3\times10$	$3,0929 \times 10^{-5}$	1,2919	68	o,9921
0,70	4	$5 \times 10^{-8} - 10$	$4,5246 \times 10^{-12}$	4,5108	221	0,9828

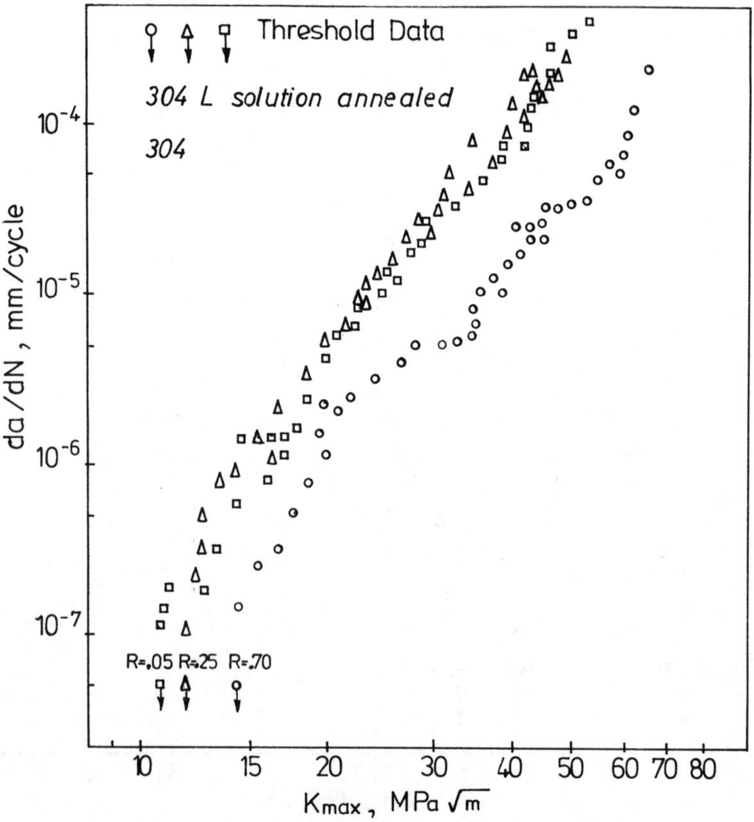

Fig. 4. Crack growth rate vs. maximum stress intensity factor for Type 304 L ss

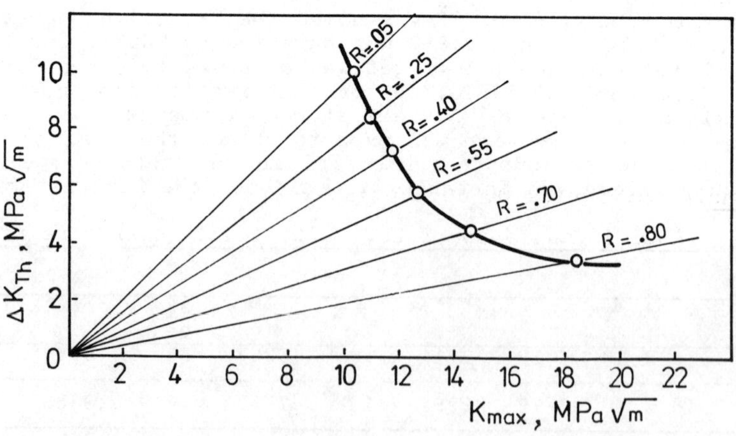

Fig.5. Effect of the maximum stress intensity on the threshold values for Type 304 L SS

*for da/dN in mm/cycle and ΔK in $MPa\sqrt{m}$.
The dependency of the threshold with the stress ratio R was suggested among others by Klesnil and Lucas (1971) by a function of the from:

$$\Delta K_{Th(o)} = \Delta K_{Th(R)} / (1-R)^{\gamma} \quad (5)$$

that is ΔK_{Th} tends to decrease with increasing R. For Type 304L a value of $\gamma = 0,665$ is suggested. The result is in agreement with other published data (Cooke and Beevers, 1974).
In Fig.5 the ΔK_{Th} is represented versus K_{max} for different R values. It seens that K_{max} is a strong controlling parameter. In addition the threshold properties appear to be sensitive to stress history, mechanical properties and microstructure (Atanasiu, 1982, Irimescu, 1983).

CONCLUSIONS

The results lead to the following conclusions:

1. At a given value of ΔK, the FCG rate increased with increasing the R value.
2. ΔK did not provide the best fit in correlating the experimental data over the range of R values. The best results were obtained using a general equation (which includes K_c, K_{Th}, R) or a parameter $K_{max}(1 - R)^n$ named "effective stress intensity".
3. The present results bring evidence of the existence of a threshold for AISI 304 L steel. Its magnitude is strongly dependent on stress ratio R, and might be influenced by K_{max}.
4. The fracture mode at low rates is sensitive to the scale and configuration of the microstructure/crack tip plasticity interaction.
5. Scanning electron fractography revealedfracture modes consisting of ductile transgranular fracture to the striation - type mode at higher rates depinding of the K_{max} values.

REFERENCES

ASTM E 399 (1978). Annual Book of ASTM,Port. 10.pp.540 - 561.

Brown,W.F., and J.E.Srawley. (1966) ASTM STP 410. Am.Soc.Testing and Materials Philadd.

Paris,P.C. and F.Erdogan (1963). J.Basic Engng., Trans.ASME, Series D, 85, p.528 -539.

Forman , R.G., V.E.Kearney, and R.M.Engle (1967). J.Basic Engng. Trans.ASME, Series D, 89., p.459-464.

Sperr, W.R.(1977). In D.M.R. Taplin (ed.), Fracture 1977, Vol.28. Pergamon Press, New York.p.1049 - 1056.

Walker, E.K.(1970). ASTM STP 462.Am.Soc.Testing and Materials, Philadelphia.p.1 - 15.

James, L.A. (1972). Nucl.Tech., 14, p.163 - 170.

Klesnil, M., and P. Lukas (1971). Mater.Sci.Engng., 9,p.231-140

Cooke,R.J., and C.J. Beevers (1974). Mat.Sci.Engng., 13,p.201-210.

Landes,J.D. (1981).int.J.Fracture, 17,p.R 47 - R 51.

Atanasiu, N.v(1982).Electron Microscopy,Vol.1,p.701-703,Vol.2, p. 167 - 169.

Irimescu, B.R. (1983). St.cerc.Mec.Apl., 42, p.161 - 177. (in Romanian).

THE EFFECTS OF THE SIZE OF THE SECOND PHASE PARTICLES ON THE MECHANICAL PROPERTIES OF A LOW ALLOY STEEL

Youshi Hong, Yuhgong Yeh and Xiaoxin Xia

Institute of Mechanics, Chinese Academy of Sciences, Beijing, China

ABSTRACT

The effects of the size of second phase particle on the tensile and fracture properties have been studied for a high carbon low alloy steel. The specimens were treated by different heat treatments and four groupings of particle size distribution of nearly equiaxed carbide were obtained. The range of the mean particle spacing, λ, is 0.525-2.17 μm. It is shown that the yield strength, σ_y, ultimate strength, σ_u, and fracture strength, σ_f, are inversely proportional to λ. It is indicated that $J_{0.05}$ or $\delta_{0.05}$ does not vary with the particle size monotonically. An expression which relates tensile properties and a microstructure parameter to fracture property is suggested to describe the present results. Moreover, there is a quite stable relation between $J_{0.05}$ and $(\sigma_y \cdot \delta_{0.05})$.

KEYWORDS

Second phase particle; spheroidite; carbide size; microstructure parameter; tensile properties; J integral; COD.

INTRODUCTION

Deformation and fracture behaviour of alloys are affected markedly by the second phase particles. These effects may vary as the volume fraction and dimension of second phase particles are changed. The investigations on the effects of the size of second phase particles on the mechanical behaviour have been done to some extent (Edelson and Baldwin, 1962; Gladman, Holmes and McIvor, 1971; Hodgson and Tetelman, 1969; Roesch, 1973). They dealt primarily with the effects of second phase particles on the brittle fracture behaviour of alloys in lower temperature. The roles of second phase particles were usually evaluated by observing the effects of the particles on the tensile properties, impact energy, transition temperature, etc. There seem to be few papers, however, which study the effects of particle size on the ductile fracture properties by the values of J integral and COD. Furthermore, from the dimensional analysis recently, it is also believed that particle size is a necessary factor in evaluating the fracture property of materials (Zheng 1982).

The present work attempts to study the effects of the particle size on the fracture and tensile properties by means of determining the critical values of J integral and COD as well as tensile properties on a high carbon low alloy steel which was heat-treated into four groupings of particle size distribution of nearly equiaxed carbide.

EXPERIMENTAL PROCEDURE AND RESULTS

The steel tested is a high carbon low alloy steel (1.02% C, 0.24% Si, 0.27% Mn, 0.011% P, 0.0086% S, 1.57% Cr). The specimens were austenitized at 844°C and quenched in oil to room temperature. Then the specimens were divided into four groups and tempered to different values of Hollomon-Jaffe tempering parameter (P) which ranges from 16300 to 21100, respectively (Table 1). The microstructure of each kind of the specimens was observed using scanning electron microscope. The microstructures of four groups of the specimens are spheroidites. The carbide is nearly equiaxed. Four groupings of carbide size distribution can be distinguished clearly (Fig. 1).

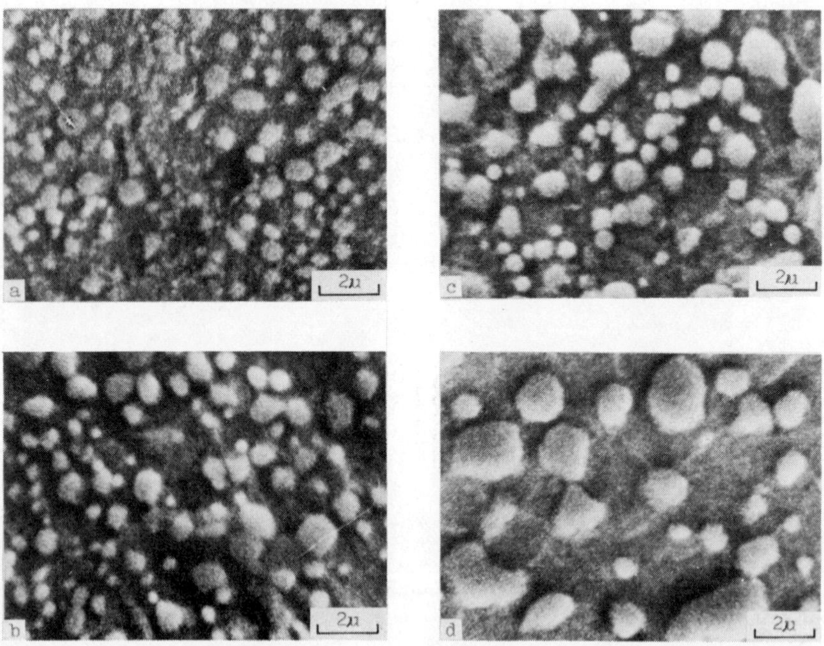

Fig. 1. Four groupings of spheroidal carbide size distribution of the steel tested. (a) Group A, (b) Group B, (c) Group C, (d) Group D.

The microstructure parameters were measured according to the following formulas (Edelson and Baldwin, 1962):

$$\lambda_f = (1-f)/N \quad (1)$$

$$D = \frac{3\cdot f\cdot\lambda_f}{2(1-f)} = \frac{3\cdot f}{2\cdot N} \quad (2)$$

$$\lambda = D(1-f)\sqrt{\frac{2}{3\cdot f}} = \frac{(1-f)}{N}\sqrt{\frac{3\cdot f}{2}} \quad (3)$$

where λ_f is mean free path between particles, D mean particle diameter, λ mean particle spacing, f the volume fraction of particle, and N the number of particles per unit length of random line. In the present case, f was calculated from the carbon content and N was measured on SEM photographs. The results of four kinds of carbide size distribution are also given in Table 1. It is seen that the carbide size becomes larger as the value of tempering parameter increases.

TABLE 1 Heat Treatments and Carbide Dimension

Group	Tempering System*	P**	λ_f, µm	D, µm	λ, µm
A	550°C for 5 hr.	16260	1.08	0.302	0.525
B	710°C for 5 hr.	19420	1.91	0.532	0.926
C	710°C for 50 hr.	20400	2.85	0.794	1.38
D	710°C for 250 hr.	21090	4.48	1.25	2.17

*Austenitized at 844°C.
**P = T(15.5+log t), where T is absolute temperature, and t time (in sec.).

Tensile properties were measured on cylindrical tensile specimens (ø8x80mm) at room temperature. The results are shown in Fig. 2, which shows that the yield strength, σ_y, the ultimate strength, σ_u, and the fracture strength, σ_f, decrease, and that the true fracture strain, ε_f, and the strain hardening exponent, n, increase and then both tend towards stable, with the increasing of λ.

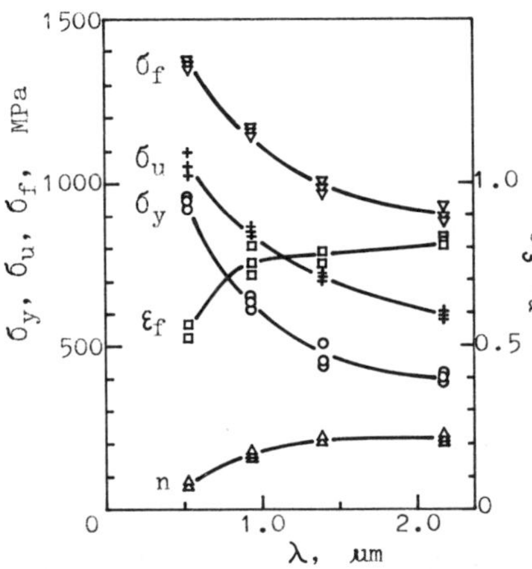

Fig. 2. Variation of tensile properties with λ.

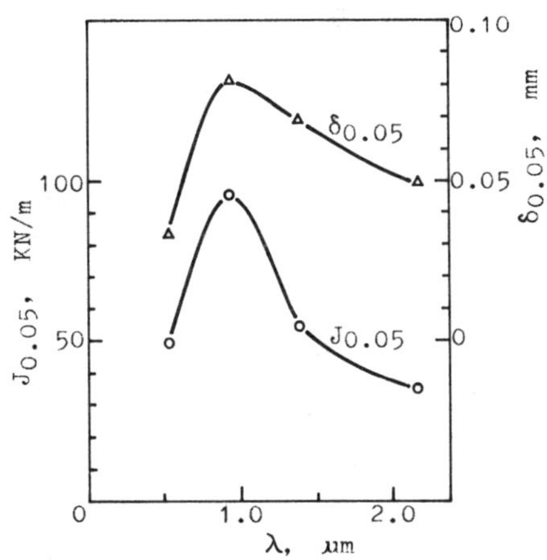

Fig. 3. Variation of $J_{0.05}$ and $\delta_{0.05}$ with λ.

Fracture properties, i.e. $J_{0.05}$ and $\delta_{0.05}$ (the conditional critical value of J integral and COD, respectively), were determined on three point bend specimens (20x24x96mm) by means of the resistant curve method according to the Chinese National Testing Standards: GB2038-80 and GB2358-80. This experiment was done at room temperature. The experimental data of $J_{0.05}$ and $\delta_{0.05}$ satisfy the criteria which are stipulated in the standards. Figure 3 presents that the variation of $J_{0.05}$ and $\delta_{0.05}$ with λ is not monotonic. When λ approaches 1 μm (P=19420) both $J_{0.05}$ and $\delta_{0.05}$ reach their maximum values.

Scanning electron microscope was used to examine the fracture surface of the bend specimens and tensile specimens. It is shown that, from the preliminary observation, there are two kinds of dimple dimensions on the fracture surface. One is smaller, and its diameter is of the order of micron, which is approximately equal to the corresponding carbide dimension. The other is bigger, and its diameter is of the order of tens of micra. The size and depth of the bigger dimple between groups of specimens are different.

DISCUSSION

It has been seen that, as λ increases, σ_y, σ_u, and σ_f all decrease distinctly in a similar tendency (Fig. 2). Considering the Hall-Petch relation:

$$\sigma_y = \sigma_o + K \cdot d^{-\frac{1}{2}} \tag{4}$$

the resistant force inside the grain, σ_o, involves the role of particle, and the parameter, K, is also concerned with the particle dimension (Feng, Wang and Qiu, 1975). The microstructure of the steel tested is composed of ferrite and spheroidized carbide. With the raising of tempering parameter, the carbides spheroidize and grow. Besides, the recrystallization of ferrite may also take place. The size of recrystallized grain is related to the particle size (Hu and Qian, 1980). Thus it may be considered that, in this case, the role of grain size, d, in Eqn. (4) can be expressed approximately in particle size. Therefore, in view of the Hall-Petch relation, it is reasonable to assume that:

$$\sigma_y = \sigma_o' + K' \cdot \lambda^{-1} \tag{5}$$

where σ_o' is a parameter which may be regarded as the yield strength of the matrix without particles, and K' is a material parameter. For a given structural state, σ_o' and K' may be considered as constants. So, there will be a linear relation between σ_y and λ^{-1}. Figure 4 shows that the present results do obey such a relation. According to the least square method, it is obtained that:

$$\sigma_y = 210 \text{ (MPa)} + 0.384 \text{ (N/mm)} \cdot \lambda^{-1} \tag{6}$$

The correlation coefficient r_1 = 0.99, and the standard deviation s_1=32MPa. In addition, there is also a quite good linear relation between σ_u or σ_f and λ^{-1} (Fig. 4). The following expressions are also obtained:

$$\sigma_u = 476 \text{ (MPa)} + 0.314 \text{ (N/mm)} \cdot \lambda^{-1} \tag{7}$$

The correlation coefficient r_2=0.98, and the standard deviation s_2=36MPa. and

$$\sigma_f = 767 \text{ (MPa)} + 0.321 \text{ (N/mm)} \cdot \lambda^{-1} \tag{8}$$

The correlation coefficient r_3=0.99, and the standard deviation s_3=32MPa. It is worth while pointing out that Hahn and Rosenfield (1966), Kelly (1973), and Törrönen and Kotilainen (1977) have reported their results and opinions of the effects of particle size on the strength. All of them had similar opinions which imply that yield strength is a function of the reciprocal of some particle parameter. But in the present work, the effect of particle

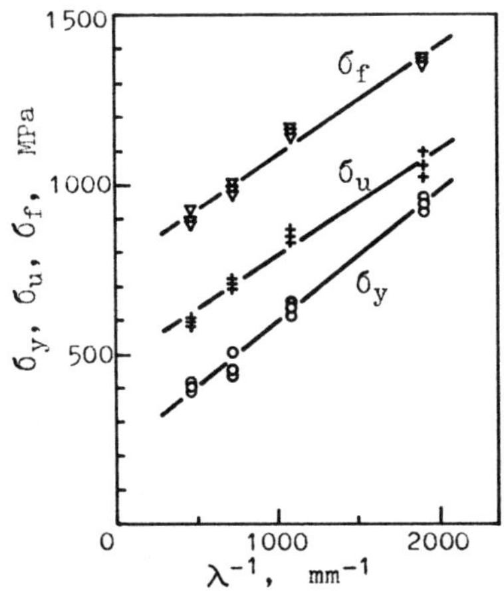

Fig. 4. σ_y, σ_u and σ_f as a function of λ^{-1}.

size on the yield strength is referred to the classical Hall-Petch relation. In other words, Eqn. (5) is a modified Hall-Petch relation which is adapted to such a condition when strength is markedly influenced by particle size and grain size may also be expressed in terms of particle size.

Figure 3 shows that $J_{0.05}$ or $\delta_{0.05}$ does not vary monotonically with the particle size. When λ is about 1 μm (P=19420), both $J_{0.05}$ and $\delta_{0.05}$ reach their maximum values. Several models have been advanced (Hahn and Rosenfield, 1975; Krafft, 1964; Schwalbe and Backfisch, 1977) which relate certain tensile properties and microstructural parameters to fracture property. It appears that these previous models, in which the fracture property is directly proportional to particle dimension, cannot describe this non-monotonicity of the present results reasonably. Recently, Zheng (1982) proposed a model in principle which relates the fracture toughness, K_{1c}, to tensile property and material parameters:

$$K_{1c} = \sqrt{\frac{E}{1-v^2}(C\cdot\sigma_y\cdot\varepsilon_u\cdot\lambda + C'\cdot R_{pc}\cdot\sigma_y\cdot\bar{\varepsilon})} \qquad (9)$$

where E is the Young's modulus, v the Poisson's ratio, ε_u the ultimate strain, R_{pc} the dimension of plastic-zone, $\bar{\varepsilon}$ the average strain within the plastic-zone, C and C' non-dimensional parameters. The first term in the brackets results from the plastic work required to form a unit length of the process zone and the second term results from the plastic work spent in forming plastic-zone around the crack tip. According to the present results and Zheng's idea, it is assumed that in our case the fracture property such as $J_{0.05}$ or $\delta_{0.05}$ can be divided into two parts. One of them is directly proportional to the particle size, and the other inversely proportional to the particle size, that is:

$$J = J_1(\lambda^{\alpha}) + J_2(\lambda^{-\beta}) \qquad (10)$$

and

$$J_1(\lambda^{\alpha}) = A'\cdot\lambda\cdot\int_{\varepsilon_s}^{\varepsilon_f}\sigma\cdot d\varepsilon \approx A\cdot\frac{k}{1+n}\cdot\varepsilon_f^{1+n}\cdot\lambda \qquad (11)$$

$$J_2(\lambda^{-\beta}) = B'\cdot\frac{w^2}{\lambda}\cdot\int_0^{\varepsilon_s}\sigma\cdot d\varepsilon \approx B\cdot\frac{\sigma_s^2}{E}\cdot\frac{\delta_e^2}{\lambda} \qquad (12)$$

where k and n are determined from $\sigma=k\varepsilon^n$, ε_s is the yield strain, ε_f is the fracture strain, w is the stretch-zone width and is represented by δ_e, the "elastic" part of COD. Thus, Eqn. (10) can be written as:

$$J = A \cdot \frac{k}{1+n} \cdot \varepsilon_f^{1+n} \cdot \lambda + B \cdot \frac{\sigma_s^2}{E} \cdot \frac{\delta_e^2}{\lambda} \tag{13}$$

From our experimental data, it is obtained that A=23.8, B=36.5, and the correlation coefficient r=0.97. Equation (13) can describe the non-monotonic effect of particle size on fracture property and relates the fracture property to macro and micro parameters of material.

It is known that there is a certain relation between J and $(\sigma_y \cdot \delta)$, that is:

$$J = M \cdot \sigma_y \cdot \delta \tag{14}$$

From the theoretical analysis and practical determination, the constant, M, may be 1 to 3 and it has been pointed out (Robinson, 1976) that this relation may be material dependent and M is possibly related to the work-hardening rate. In our results, although $J_{0.05}$ and $\delta_{0.05}$ do not vary with the particle size monotonically, they still conform to Eqn. (14) (Fig. 5). The mean value of M is 1.72.

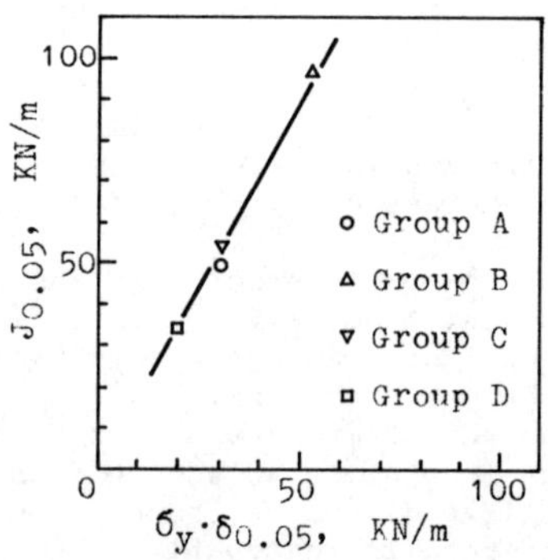

Fig. 5. The relationship between $J_{0.05}$ and $(\sigma_y \cdot \delta_{0.05})$.

CONCLUSIONS

1. The yield strength, σ_y, increases with the decreasing of the particle spacing, λ, and fits Eqn. (5):

$$\sigma_y = \sigma_0' + K' \cdot \lambda^{-1} \tag{5}$$

The variation of the ultimate strength, σ_u, and fracture strength, σ_f, with the particle size follows a similar relation as Eqn. (5).

2. The fracture properties, $J_{0.05}$ and $\delta_{0.05}$, do not vary with the particle size monotonically. The following expression:

$$J = A \cdot \frac{k}{1+n} \varepsilon_f^{1+n} \cdot \lambda + B \cdot \frac{\sigma_y}{E} \cdot \frac{\delta_e}{\lambda} \tag{13}$$

can describe the tendency of the present results comparatively well.

3. Although there is a non-monotonicity between fracture properties and particle size, the present results also conform to the relation:

$$J = M \cdot \sigma_y \cdot \delta \tag{14}$$

The mean value of M is 1.72.

ACKNOWLEDGEMENT

The authors are grateful to Professor Zheng Zhemin[1] for his continual guidance in every stage of this work.

REFERENCES

Edelson, B. I., and W. M. Baldwin, Jr. (1962). The effect of second phases on the mechanical properties of alloys. Trans. ASM, 55, 230-250.

Feng Duan, Wang Yening and Qiu Dilong (1975). Metal Physics. Chinese Science Press. Chap. 14, pp. 623-733. (In Chinese).

Gladman, T., B. Holmes and I. D. McIvor (1971). Effects of second phase particles on strength, toughness and ductility. In Effect of Second-Phase Particles on the Mechanical Properties of Steel. The Iron and Steel Institute, London. pp. 68-78.

Hahn, G. T., and A. R. Rosenfield (1966). A modified double-pile-up treatment of the influence on grain size and dispersed particles on brittle fracture. Acta Met., 14, 1815-1825.

Hahn, G. T., and A. R. Rosenfield (1975). Metallurgical factors affecting fracture toughness of aluminum alloys. Met. Trans., 6A, 653-668.

Hodgson, D. E., and A. S. Tetelman (1969). The effect of microstructure on the cleavage strength of quenched and tempered steels. In Fracture 1969, Proc. of the Second International Conference on Fracture, Brighton, April 1969. Chapman and Hall Ltd, Great Britain. pp. 266-277.

Hu Gengxiang and Qian Miaogen (1980). Metallography. Shanghai Science and Technology Press. Chap. 8, pp. 307-345. (In Chinese).

Kelly, P. M. (1973). The quantitative relationship between microstructure and properties on two-phase alloys. Int. Met. Rev., 18, 31-36.

Krafft, J. M. (1964). Correlation of plane strain crack toughness with strain hardening characteristics of a low, a medium, and a high strength steel. Appl. Mat. Res., 3, 88-101.

Robinson, J. N. (1976). An experimental investigation of the effect of specimen type on the crack tip opening displacement and J-integral fracture criteria. International Journal of Fracture, 12, 723-737.

Roesch, L. (1973). Influence of second-phase particles on the low-temperature fracture behaviour of iron. In Papers Presented at the Third International Conference on Fracture, Vol. IV. No. III-344.

Schwalbe, K-H., and W. Backfisch (1977). Correlations between fracture toughness, tensile properties, fracture morphology, and microstructure of a high-strength steel. In M. R. Taplin (Ed.), Advances in Research on the Strength and Fracture of Materials, (ICF4), Vol. 2A. Pergamon Press. pp. 73-78.

Törrönen, K., and H. Kotilainen (1977). Effect of tempering treatments on the fracture behaviour of a Cr-Mo-V pressure vessel steel. In M. R. Taplin (Ed.), Advances in Research on the Strength and Fracture of Materials, (ICF4), Vol. 2A. Pergamon Press. pp. 141-147.

Zheng Zhemin (1982). Continuum mechanics and fracture. Advances in Mech., 12, 133-140. (In Chinese).

[1] Professor Zheng Zhemin (Professor Che-Min Cheng).

FRACTURE TOUGHNESS OF CAST IRON

J. Pirs

Materials Department, Technical Faculty Rijeka, University of Rijeka, 51000 Rijeka, Yugoslavia

ABSTRACT

Fracture behaviour of flake graphite cast irons using microfractographic methods and the application of fracture toughness testing to these low strength, brittle materials.

KEYWORDS

General engineering type of Meehanite iron ; fracture behaviour ; fracture toughness testing ; fractography of fractured surfaces ; scanning electron microscopy .

INTRODUCTION

It has been accepted for many years that the fracture mechanisms of flake graphite cast irons in both the as cast and annealed condition is controlled by the brittle behaviour of graphite flakes. This phenomen was investigated by Meyersberg (1936) Clough and Shank (1957) and finally by Gilbert (1963). The amount, size, shape and the distribution of graphite flakes in grey cast iron greatly influence its properties. The graphite flakes interrupt the continuity of the matrix and also introduce notches at their apexes, as shown by Huetter and Stadelmaier (1958). Intensive studies have been performed by Angus and Pearce (1946) on the influence of chemical composition , by Gabel (1954) on the influence of the rate of cooling of the casting in the mould and finally by Namur (1961) of the type of graphite flakes formed, all these investigations being devoted to the influence of these factors on the impact fracture characteristics of flake graphite cast irons.

Recent developments have occurred in the observation of fracture surfaces by the electron microscopy of replicas and the direct observation with scanning electron microscope, as shown by Beacham and Pelloux (1964). At the same time the analysis and testing of grey cast irons performed by Glover and Pollard (1969, 1970, 1971) and Grüter (1975) has developed with the increased interest in fracture toughness.
In this paper results are reported of the study of the fracture behaviour of flake graphite cast iron in the as cast condition, using microfractographic methods of investigation and the application of fracture toughness testing to these low strength, brittle materials.

EXPERIMENTAL PROCEDURE

General Engineering Type GD 250 Meehanite cast iron with flake graphite was investigated during this investigation. The investigated flake cast iron meets the requirements of Grade 14, B.S.S. 1452 and may be used to meet Grade 17, B.S.S. 1452 .

The chemical analysis of the investigated cast iron is given in Table I.

TABLE I Chemical Composition of Investigated Iron

C	Si	Mn	P	S
3.44	1.35	0.60	0.10	0.106

In the as cast condition the minimum tensile strength of the investigated Meehanite GD 250 iron is 250 N/mm^2 with hardness values varying from 180 to 220 . Minimum impact strength is 4 Joules. The tensile strength values are valid for 30 mm diameter separately cast test bars , tested in air at 20°C.The hardness values on section thickness of casting, the testing taking place either on the test bar or at predetermined location on the casting. The impact strength is determined on unnotched D.V.M. test piece and we determined the impact values on instrumented pendulum machine.

On the investigated iron we obtained tensile strength value of 305 N/mm^2 with Brinell Hardness of 207 N/mm^2 and with impact strength values which were all below the prescribed minimum .

The metallographic structure of the investigated cast iron is shown in Fig. 1. From the results shown in Fig. 1 it may be observed that we obtained pearlitic and pearlitic - sorbitic base matrix . In Fig. 1 b the inclusions of MnS and small quantities of free ferrite are shown.

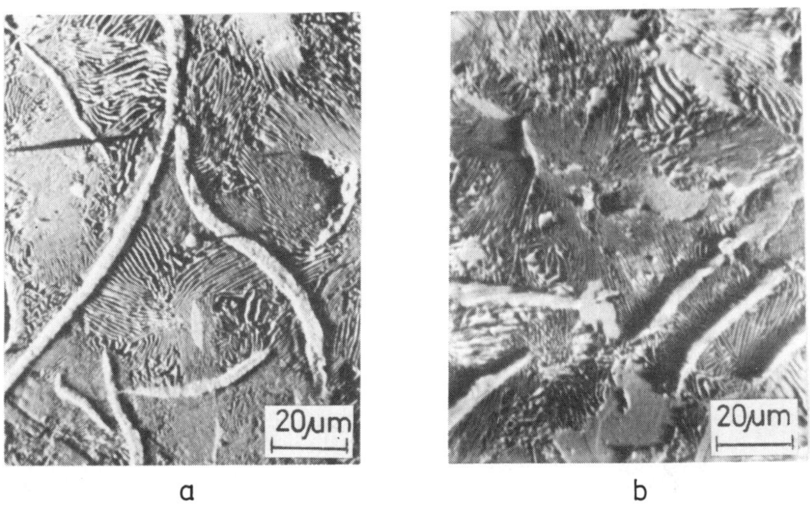

Fig. 1 Metallographic structure of cast iron
(a) pearlitic base matrix
(b) pearlitic-sorbitic base matrix

Stress - intensity factors K were determined on behalf of the Double - Cantilever - Beam Probe which is shown in Fig. 2 .

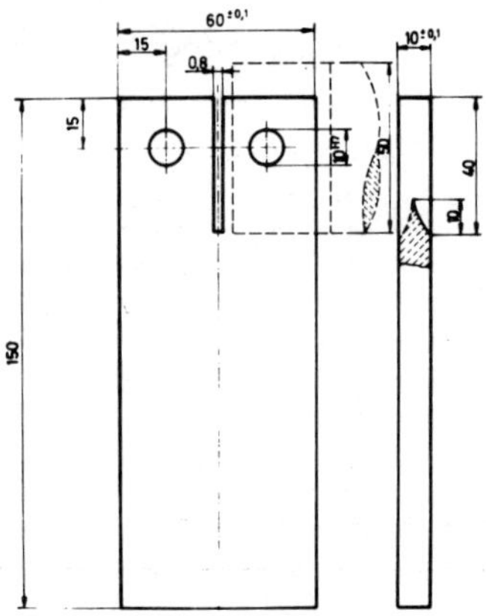

Fig. 2 Double - Cantilever - Beam Probe

The dynamic crack was produced on behalf of Amsler HFP - IO pulsating machine and the crack extension was followed by optical method, thus using a contrast liquid . After the dynamic crack was produced , the derivation of K_Q and K_{Ic} was performed on behalf of the " offset procedure ". Thestatic breaking of the sample was performed on Amsler 600 kN tensile testing machine, using electronic registration of F and crack opening v , by simultaneous drawing of force/displacement test records. Stress-intensity factor was determined on behalf of the method of 5 % secant line with tangential line on the force/displacement record.

Aftere fracture toughness determinations , the fractured surfaces of the specimens were studied by direct observation on JEOL Stereoscan microscope.

EXPERIMENTAL RESULTS

The values of K_Q were determined using the following equation derived by Stellwag and Mostova (1978)

$$K_Q = \frac{3,464\ F_Q}{b\sqrt{h}} \sqrt{\left(\frac{a}{h}\right)^2 + 1,2\left(\frac{a}{h}\right) + 0,693} \sqrt{\frac{b}{b_n\,(1-\nu^2)}}$$

where b is specimen thicknes , b_n reduced specimen thickness h specimen width and ν Poisson's Ratio.

TABLE II Results of Fracture Toughness Determinations

		DCB-1	DCB-2	DCB-3	DCB-4	DCB-5
b,	mm	10	10	9,6	9,6	9,6
bn,	mm				6,5	6,5
a,	mm	31,1	29,5	39,1	40,5	40,5
h,	mm	30	30	30	30	30
F_Q,	N	7500	7200	6250	5400	4900
K_Q,	$MN/m^{3/2}$	25,19	24,25	25,94	25,48	26,04

Results obtained are given in Table II where the samples are designated with DCB . The evaluation of the condition of validity of the relation , that K_Q equals to K_{Ic} was not performed in spite of the unknown data about $R_{p\ 0.2}$. From the results in Table II it can be concluded that we obtained practical identical results of fracture toughness on all samples , and most identical results we obtained also during fractographic investigations , shown in Fig. 3 .

The results of fractographic investigations of the fractured surfaces showed that there are small differences in the morphological characteristics , where at small magnifications more or less coarser grained fractures can be observed . Other results showed two other characteristic features of the fracture , fracture between graphite and metal base matrix and the fracture through the metal base matrix . The surface of graphite flakes is relatively clean and brighter lines may be seen on the surface , which are sometimes connected to irregular polygons . For the fracture through the metal base matrix it is characteristic that the fracture surface showed chanels , which followed lamellae of pearlite in the structure , und dimples on places , where the angle between fracture and lamellae of pearlite was different . This feature is characteristic for the fracture in pearlite , where microcrack proceeds separately in lamellae of cementite and ferrite, and stays on the same plane if geometrical conditions allow such a position of the microcrack . On all fractured surfaces there were no signs of brittle fracture through the metal base matrix .

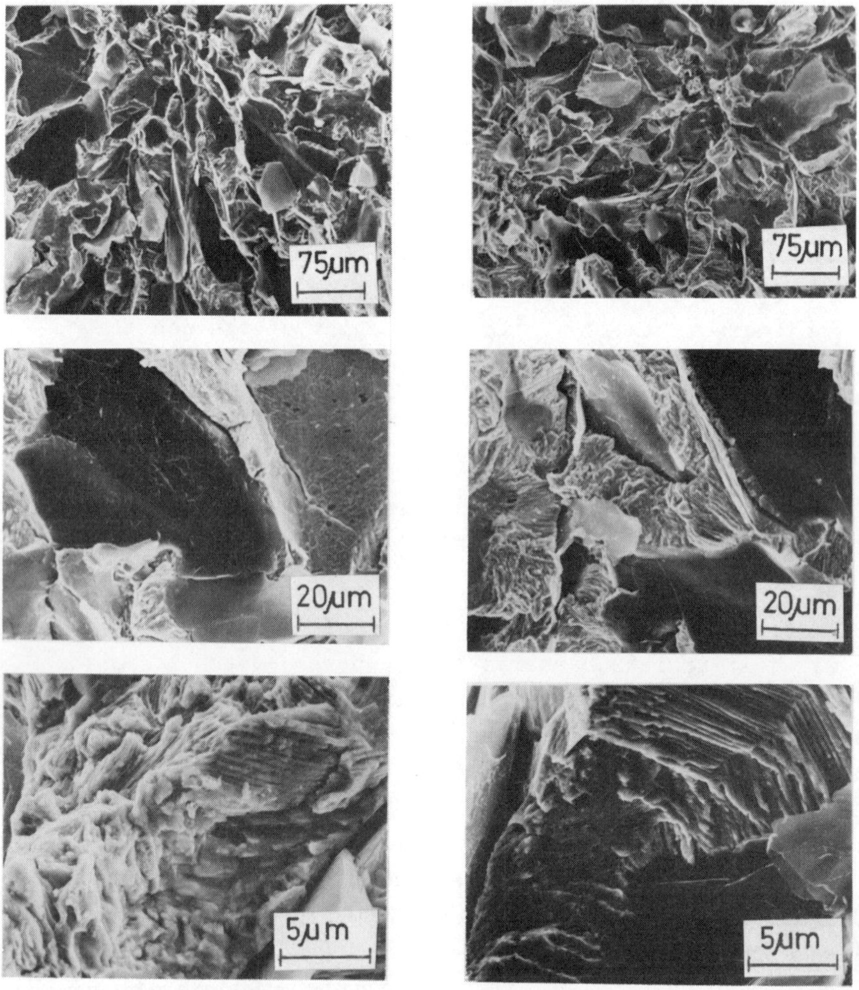

a b

Fig. 3 Results of fractographic surface investigations

CONCLUSIONS

The plane strain fracture toughness value of flake graphite cast iron has been determined on irons having pearlitic and pearlitic - sorbitic base matrix . The variations in K_Q values depend very much of the fracture through the matrix whi which showed very small variations . In all cases there was no brittle fracture character of the investigated surfaces .

ACKNOWLEDGEMENTS

The author wish to thank to the Directory of the Torpedo factory of diesel engines and foundry in Rijeka for the provision of the test material being produced in licency of the International Meehanite Metal Company . Thanks are due also to the Selfgoverning Interest Community for Scientific work of the Socialist Republic Croatia for financial assistance for the research theme No. 1.3.2.6 of the research project Materials.

REFERENCES

Angus, H. T., and J. G. Pearce (1946). B.C.I.R.A. Bulletin , 8 , 323.
Beacham, C. D., and R. M. N. Pelloux (1964). A.S.T.M. S.T.B. 381 , 210.
Clough, W. R., and M. E. Shank (1957). Trans. ASM , 49, 241.
Gabel, E. (1954). Fonderie, 107 , 4272-4279.
Gilbert, G. N. J. (1963). B.C.I.R.A. Journal, 11 , 512.
Glover, A. G., and G. Pollard (1969). Proc. 2nd Int. Conference on Fracture , Paper 29, 250-259.
Glover, A. G., and G. Pollard, (1970). I.S.I. Publ. 120 ,90-94.
Glover, A. G. and G. Pollard (1971). J.I.S.I., Feb. 138-141.
Grüter, L., Rissausbreitung in Gusseisen, Dr-Ing. Dissertation , Bochum (1975).
Huetter, L. J. and H. H. Stadelmaier, (1958). Engineering School Bulletin 66 Feb. , North Carolina State College.
Meyersberg, G. (1936). Giesserei, 23 , 285.
Namur, R. (1961). Fonderie Belge , 31 , 33-44.
Stellwag, B., (1978). Int. J. of Fracture Mech. 14, 303-305.

FRACTURE TOUGHNESS OF TUNGSTEN-FREE TiC- AND (Ti)C,N-BASE HARD ALLOYS AND ITS RELATION TO THE MICROSTRUCTURE

M. Loshak, N. Konovalenko, T. Shapoval and M. Dudkin

Institute for Superhard Materials of the Ukrainian Academy of Sciences, Kiev, USSR

ABSTRACT

Applicability of K_{Ic} parameter as a criterion for evaluation of reliability of tools made from TiC-Ni-Mo and (Ti)C,N-Ni-Mo sintered alloys was investigated in cutting metals. The effect of microstructure and test temperature on K_{Ic} value is shown. The alloys with grain size of carbide phase of about 1.5 to 2.0 μm and with approximately equal TiC/(Ti,Mo)C ratio possess the highest K_{Ic} value. At high cutting speeds and low feed rates the alloys with high K_{Ic} values are characterized with a higher wear rate. At higher feed rates and depths of cut as well as in milling, the alloy inserts with high K_{Ic} values are more reliable.

KEYWORDS

Hard alloys, microstructure, strength, fracture toughness, temperature, wear resistance, reliability, cutting.

INTRODUCTION

In recent years TiC-Ni-Mo and (Ti)C,N-Ni-Mo alloys have been widely used instead of WC-TiC-Co alloys for finish and semi-finish turning of steels. In turning which is characterized by continuous cutting these tool materials should exhibit a high wear resistance which is specified by the ability of the alloy when machining carbon and alloyed steels to resist to diffusion dissolution of hard alloy components in steels and to diffusion penetration of steel components into the tool material. In milling and in other machining operations characterized by intermittent cutting and variations in removed layer thickness it is rather essential to impart sufficient strength and durability to the hard alloy. At the same time it is known that the substantial scattering of the transverse rupture strength values (TRS) is observed within various batches of commercial alloys of TiC-Ni-Mo (TH20 grade) and (Ti)C,N-Ni-Mo (KHT16 grade) systems (800 to 1370 MPa and 1000 to 1900 MPa, respectively) (Tumanov, Ochkasov, 1981).

The TRS value being an integral characteristic influenced by a number of factors (chemical composition, microstructure, specimen surface integrity, internal defects) does not uniquely define the material toughness and hence the life of the products made of it. At the same time up-to-date state of fracture mechanics allows quantitative definition of some important aspects of macrofracture of carbide products and those of microfracture that defines largely the wear of a composite. Some regularities in dependence of stress intensity factor K_{Ic} on microstructure of tungsten carbides are found (Chermant, Osterstock, 1976 ; Ingelstrom, Nordberg, 1974 ; Nakamura, Gurland, 1980). The paper includes results of K_{Ic} measurement for W-free alloys of some batches different in their microstructures. Study of microstructure and some other parameters (hardness HV, tranverse rupture strength TRS, coercivity Hc, wear resistance in cutting) offers the possibility to compare the K_{Ic} values and these parameters. Temperature dependence of K_{Ic} and HV was also studied.

EXPERIMENTAL

Investigated were TH20 alloy of several commercial batches and KHT16 alloy of a single batch. Mechanical tests were performed on sintered 5 mm x 5 mm x 35 mm specimens. Non-ground specimens were transverse rupture tested in 3-point bending, the loading points spacing being 30 mm. The same technique was used for fracture toughness (K_{Ic}) measurement. In both cases no less than 10 specimens were tested. A device for K_{Ic} measurement was inserted in a vacuum chamber which allowed testing in a vacuum up to $130 \cdot 10^{-8}$ MPa in a temperature range from 290 to 1300 ^{o}K. Fatigue cracks which were grown on the pre-notched specimens were taken as initiators for failure (Konovalenko, Maistrenko, Mitlikin, 1980).

The K_{Ic} measurements directly on cutting inserts were performed by measuring the crack length at the indentor impression produced in a Vickers test (Lawn, Fuller, 1975), P=150 N(indirect method). There is a 95 per cent-guarantee that for W-free alloys the K_{Ic} measurement error in the case of indirect method as compared with direct one does not exceed 8 per cent. Vickers diamond pyramid hardness was measured at room temperature. The temperature dependence of hardness was studied on a testing machine (Bakul and others, 1978) using Knoop diamond pyramid with a rhombic base and an apex angle between edges of 130° and 172°30', load on the indentor being 5 N.

The structure of the materials was studied on the УЭМВ-100К transmission electron microscope and on the Stereoscan S4-10 scanning electron microscope (Pilyankevich and others, 1979). The average grain size of the carbide phase as well as specific surface of each of the phase of the TH20 alloy with various K_{Ic} values were evaluated quantitatively. The mode of crack propagation across the phase components was determined qualitatively.

Relationship between the cutting efficiency and K_{Ic} value has been studied when turning and milling 45, 30ХГСА and 30ХНМА steels under the following conditions: cutting speed of 50-200 m/min, feed rate of 0.1-0.4 mm/rev. and depth of cut of 1-2 mm. The inserts parameters were as follows: back rake angle $\gamma = 0^o$, back clearance angle $\alpha = 8^o$, working approach angle $\varphi = 45^o$. The path of cutting for each of tested inserts was 3000 m. Cutting efficiency of the inserts in turning was evaluated by a flank wear land h_3, wear intensity $J_L = \frac{h_3(\mu m)}{3000\ m}$ and descent of the cutting edge due to the plastic deformation h_α; in milling the strength of the inserts was evaluated using a "fracturing feed" technique.

EXPERIMENTAL RESULTS

It can be seen from Table 1 that the TH20 alloy batches of similar composition and density differ essentially in mechanical properties. This is attributed to the differences in their microstructures. Measurements of magnetic properties (Hc) and the average grain size of the carbide phase offer the possibility to divide the TH20 alloy batches under study into 3 groups. The first group (batch 4) is characterized by a small grain size of the carbide phase: more than 50 per cent of carbide grains are of 0.5 to 1 μm in diameter*. The coercivity of such specimens

TABLE 1 Characteristics of TH20 Alloy

Batch number	γ, g/cm^3	Hc, kA/m	HV*30/30, MPa	TRS, MPa	K_{Ic}, $MPa.m^{1/2}$	Average grain size of the carbide phase, μm
1	5.69	4.0-5.5	14500	1120	8.8±0.2	2.0
2	5.63	2.2-3.2	14300	1100	7.4±0.4	3.8
3	5.65	2.3-2.6	14420	1080	7.4±0.4	4.2
4	5.67	5.0-6.5	14700	1140	8.4±0.3	1.0
5	5.60	2.3-3.5	15160	680	7.6±0.5	3.8
6	5.63	3.7-4.4	13420	1050	10.3±0.5	1.8
7	5.62	3.8-4.4	13270	990	9.9±0.4	1.9
8	5.66	2.2-2.6	14960	1010	6.3±0.5	4.0

*The average value error in hardness and bending tests did not exceed 3 per cent and 5 per cent, respectively.

*The maximum grain size in plan.

is over 4 kA/m. The second group (batches 1,6,7) is characterized by a more uniform distribution of grain diameters around the average value of 1.8 to 2.0 µm. The third group (batches 2,3,5, 8) is defined by the coarsest crystalline structure with conglomerates. The coercivity of these specimens lies within 2-3 kA/m.

The batches of the third group (Fig.1) are characterized by a considerable increase in surface area of a (Ti,Mo)C coating ring as compared with a grain TiC core surface area, meanwhile in the case of the second group batches the surface areas of TiC and (Ti,Mo)C portions in the carbide phase are approximately equal (Table 2)*.

TABLE 2 Specific Surface Area of Phases Defined in Stereological Analysis

Batch number	Phase relative specific surface area on a metallographic specimen, %		
	TiC	(Ti, Mo)C	Binder
1	38	48	14
3	18	72	10
4	30	59	11
6	42	46	12
8	13	75	12

Figure 2 shows the temperature dependence of the K_{Ic} and HV values for TH20 and KHT16 alloys together with data on changes in TRS (Tumanov, Ochkasov, 1981). In the range of 273 to 900 ^oK the hardness decrease rate is somewhat lower than in the range of higner temperatures. At any temperature the hardness of TH20 alloy is higher than that of KHT16 alloy, while the fracture toughness of KHT16 alloy is higher than that of TH20 alloy.

With temperature increase up to 900 ^oK little changes in fracture toughness of the two alloys are observed, while higher temperatures cause its rapid increase. Tumanov and Ochkasov (1981) have shown that TRS value firstly lowers with temperature and subsequently increases at temperatures above 900 ^oK.

Figure 3 shows the relationship between the K_{Ic} values and wear rate of TH20 inserts in turning 30XHMA steel at cutting speed of 200 m/min and feed rate of 0.15 mm/rev. Under these working conditions the higher the K_{Ic} values the higher wear rate of the cutting inserts and the lower their durability. When cutting 30XГCA steel at feed rate of 0.4 mm/rev. and depth of cut of 2 mm the amount of chipping of the inserts with K_{Ic}=6-8 MPa.m$^{1/2}$ appeared to be 30 per cent as high as the amount of chipping of the inserts with K_{Ic} =8-10 MPa.m$^{1/2}$. Moreover, when cutting at increased feed rate, the TH20 inserts tend to deform plastically

*The crack propagation mostly occurs at the (Ti,Mo)C phase boundaries (metallographic specimens of the third group batches) and at the phase boundaries (those of the second group batches).

the cutting wedge. For example, when machining 30ХГСА steel at speed rate of 150 m/min and feed rate of 0.3 to 0.4 mm/rev. a considerable descent of the cutting edge (h_α =0.10 to 0.12 mm) is observed in the inserts with high K_{Ic} values (8.5 to 10.0). For inserts with K_{Ic}=7.0-8.5 the value of the cutting edge descent is smaller and values at 0.04-0.08 mm; for the cutting inserts with K_{Ic}=5.5-7.0 the cutting edge descent is not observed.

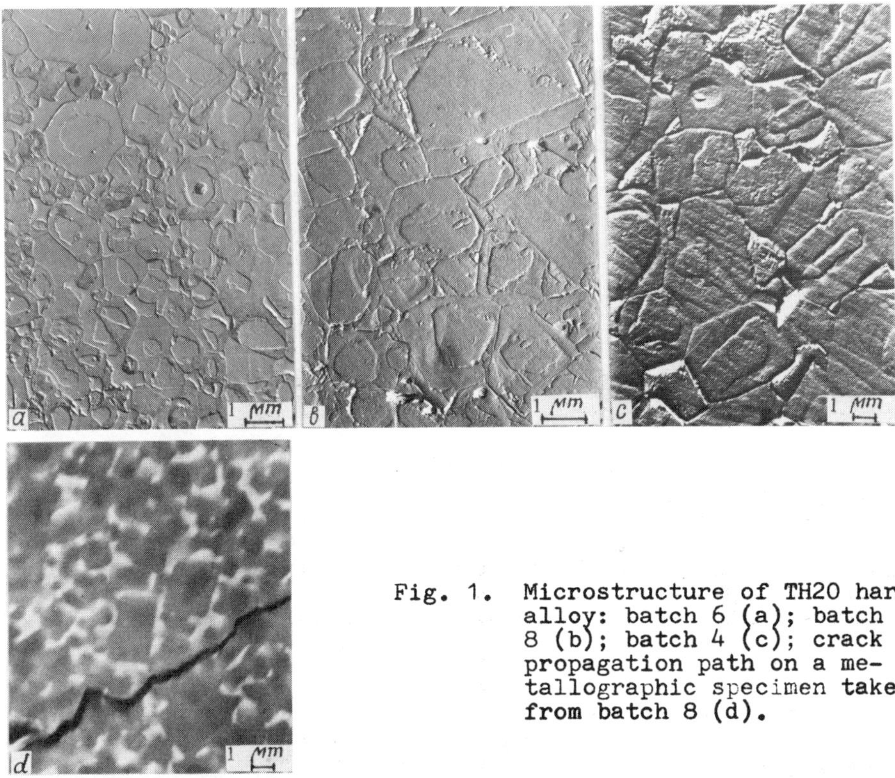

Fig. 1. Microstructure of TH20 hard alloy: batch 6 (a); batch 8 (b); batch 4 (c); crack propagation path on a metallographic specimen taken from batch 8 (d).

When milling 100 mm x 100 mm x 300 mm blanks from 45 steel using a single-point milling cutter at cutting speed of 50 m/min and depth of cut of 1.5 mm the feed rate has been increased stepwise from 0.2 mm/rev. until chipping of the cutting edge occured. The feed rate at which failure occured was within 0.25 to 0.79 mm/tooth, its value having been increased proportionally with K_{Ic}.

DISCUSSION

Tungsten-free hard alloys of the same quality when sintered may differ in their microstructure due to variations in production technology and chemical composition. Variations in microstructures of the alloys are responsible for essential difference in their mechanical properties. The critical stress intensity factor K_{Ic} (which does not depend on other factors) has been chosen

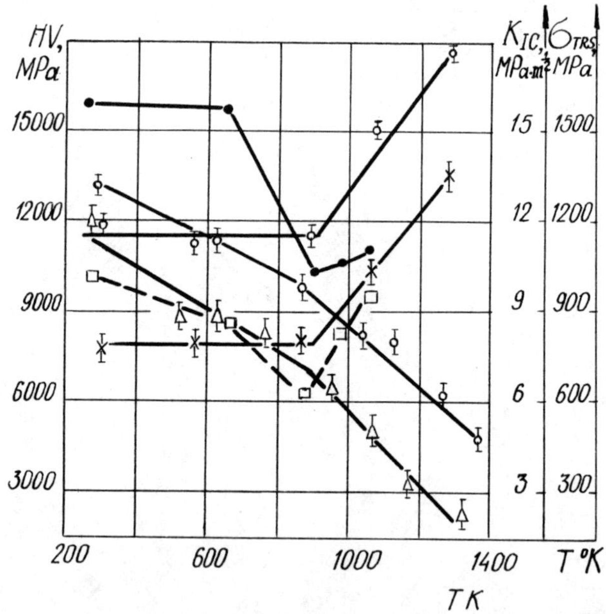

Fig. 2. Fracture toughness (2,6), hardness (3,4) and strength (1,5) as a function of temperature; 1,2,4 - KHT16 alloy; 3,5,6 - TH20 alloy.

as the main parameter for the evaluation of the hard alloy fracture toughness specified by the properties of the structure components.

The highest K_{Ic} values are characteristic for the alloys having the microstructure with average grain size of the carbide phase being of 1.5 to 2 µm and with approximately equal ratio of surface areas of the grain core from TiC and the coating from (Ti, Mo)C.

There is a definite relationship between HV and K_{Ic} values at room temperatures: the alloys with higher K_{Ic} values have lower hardness. K_{Ic}/TRS relation for the alloys of various batches does not exist because TRS depends simultaneously on a number of factors. Within a single batch, however, the specimens with higher TRS have higher fracture toughness as well.

Under high cutting speeds and moderate feed rates the wear rate increases for high K_{Ic} values. At higher feed rates and depths of cut as well as in milling the cutting inserts with higher K_{Ic} values are more reliable. At the same time it should be taken into account that increase in feed rate at higher cutting speed results in the cutting edge descent. In this case the increase in K_{Ic} value is accompanied by the lowering of hardness and decreasing in resistance to plastic deformation of the hard alloy at running temperatures.

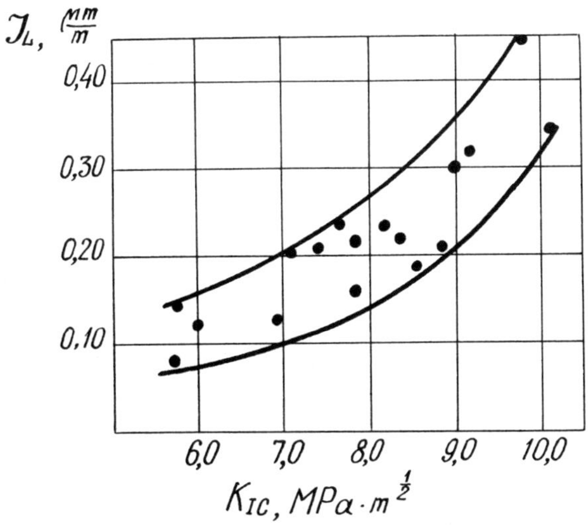

Fig. 3. Intensity of wear vs. K_{Ic} value for TH20 alloy inserts. Workpiece material: 30XHMA steel; V=200 m/min; S=0.15 mm/rev.; t=1 mm; L=3000 m.

Thus, the choice of an optimum alloys structure to obtain the highest tool life is specified by temperature and force parameters. The K_{Ic} value which uniquely defines the material toughness may be used as the durability criterion for the choice of either the alloy grade or working conditions as well as for the qualitative assessment of the microstructure in controlling the hard alloy production technology.

REFERENCES

Bakul, V., V.Yevdokimenko, L. Litvinova, M. Loshak, and V. Malnev (1978). Synteticheskiye Almazy, N1, 21-22.

Chermant, J., and F. Osterstock (1976). J. Mater. Sci., 11, 1939-1951.

Ingelstrom, N., and H. Norberg (1974). Eng. Fract. Mech., N6, 597-607.

Konovalenko, N., A. Maistrenko, and M. Mitlikin (1980). Sverkhtverdye Mater., N5, 29-34.

Lawn, B., and E. Fuller (1975). J. Mater. Sci., 10, 2016-2024.

Nakamura, M., and J. Gurland (1980). Met. Trans., 11A, N1, 141-146.

Pilyankevich, A., T. Shapoval, G. Dzodziyev, and V. Paderno (1979). Porosh. Metal., N10, 73-79.

Tumanov, V., and V. Ochkasov (1981). In Tverdye Splavy. Trudy VNIITS. Metallurgiya, Moskva, 14-18.

EXPERIMENTAL DETERMINATION OF HOLE INITIATION CRITERIA AT INCLUSIONS IN LOW ALLOY STEEL AND SPHEROIDAL GRAPHITE CAST IRON

Sung Kang Hur*, Sun Yao Qing and D. Francois*****

*CNRS ERA 910, Université de Technologie de Compiègne, 60206 Compiègne, France
**Huazhong University of Science and Technology, Wuhan, China
***Ecole Centrale des Arts et Manufactures, 92290 Chatenay Malabry, France

ABSTRACT

The stress and the strain distribution in the notched section of cylindrical tensile specimens of spheroīdal graphite ferritic cast iron and of 30 CD4 steel containing globular or elongated sulfide inclusions were calculated by the finite element method and they were compared with experimental measurements. Interrupted tests were performed and the microscopic examination of the longitudinal sections of the notched area allowed to find the crack initiation stress criteria.

KEYWORDS

Spheroīdal graphite cast iron, low alloy steel, sulfide inclusions, notched tensile specimens, ductile fracture, cavity initiation.

INTRODUCTION

Some theoritical models for the initiation of holes at inclusions in ductile fracture exist (Brown and Embury,1973; Fisher and Gurland,1981; Goods and Brown,1979; Gurland and Plateau,1963; Kurita, Roesch and Sausay,1977; Mc Clintock,1967; Melander,1981; Murdry,1982; Puttick,1959) but there are only few precise experimental verifications (Brown and Embury,1973; Mudry,1982; Rosenfiled,1968). We chose to study this problem on spheroīdal graphite cast iron and on a chromium molybdenum steel containing manganese sulfide inclusions either globular or elongated. In order to obtain criteria which take into account the effect of the hydrostatic stress, notched cylindrical specimens were used with different notch root radii (Mackensie, Hancock and Brown,1977). This method requires a knowledge of the evolution of the stresses and of the strains in the whole notched section. They were calculated by the finite element method and the results were checked against experimental measurements. Interrupted tests were performed and quantitative micrographic analyses of the longitudinal sections of the specimens were undertaken.

EXPERIMENTAL METHOD

Materials Tested

The spheroïdal graphite cast iron was prepared at Centre Technique des Industries de la Fonderie (CTIF). Table 1 gives the chemical composition.

TABLE 1 Chemical Composition (wt %)

Material \ Element	C	Mn	Si	S	P	Mg	Ni	Cr	Mo	Al
Cast iron	3.55	0.11	2.75	0.006	0.4	0.52-0.046	0.20	-	-	-
Steel	0.19	0.76	0.25	0.029	0.022	-	-	1.10	0.27	0.03

Bars were cast between 1420 and 1280°C in sodium silicate sand molds hardened with CO_2. The 200 mm long and 25 mm in diameter cast bars were heat treated to obtain a ferritic matrix. It consisted in a 23 h hold time at 870°C followed by a 10°C/h furnace cooling down to 600°C and finaly by air cooling. Table 2 gives the dimensions of the structural elements : the mean grain diameter d, the mean diameter R of the graphite spheroids, the volume fraction of the graphite f_G and the mean distance between the graphite spheroids, Λ_{bb} side to side and Λ_{cc} center to center.

TABLE 2 Structural parameters of the cast iron

α (μm)	R (μm)	f^G_V (%)	Λ_{bb} (μm)	Λ_{cc} (μm)
55	34.9	11.12	47.2	113.6

The chemical composition of the 30 CD 4 steel is given in table 1. The matrix consisted of ferrite containing globular cementite with an average diameter of about 0.5 μm. In one material (33 G) the manganese sulfides were elongated with a length between 10 and 100 μm and a width of about 0.5 μm. In the other material (34 G) the sulfides were globular with an ellipsoïdal shape 24 μm long, 5 μm wide and 2 to 5 μm thick.

TABLE 3 Mechanical properties

	$\sigma_{0.2}$ (MPa)	σ_{UTS} (MPa)	ε_f (%)	n
Cast iron	313	448	24.9	0.128
Steel	373*	538	47.1	0.204

n is the stress hardening exponent
* lower yield point

Tensile Tests on Notched Specimens

The notched specimens were 18 mm in diameter, 115 mm long, the diameter of the notched section was 10 mm and the notch radii were 10 mm (AE 10), 4 mm (AE 04) and 2 mm (AE 02). The tests were performed at a cross bar velocity of 0.5 mm/mn.

The change of the diameter Ø of the notched section was measured during the tests by photography and with a LVDT extensometer. For some steel specimens a circular grid was inscribed on the notch surface with a lathe, the depth and the width of the grooves remaining less than 0.03 mm. Photographs were taken of the deformed specimens at various stages of loading.

Micrographic Examinations

The specimens were cut along the axis. They were polished with a diamond paste and cleaned with an ultrasonic cleaner. They were observed without etching or for the steel specimens in order to reveal the decohesion between inclusions and matrix after etching for 20 seconds in 4% picral followed by a light polishing.

For the spheroïdal cast iron the volume fraction of holes was measured on a micrograph at a X 200 magnification using a 20 x 29 cm^2 grid of 300 randomly scattered points.

For the steel specimens the fraction of the damaged sulfides was measured by counting in 0.8 by 0.8 mm squares.

CALCULATIONS OF STRESS AND STRAIN DISTRIBUTIONS IN THE NOTCHED SPECIMENS

The calculations were carried out by the finite element method using 8 nodes quadrangular elements of the MEF code by J.M. Detraux and G. Touzot (Université de Technologie de Compiègne). The formulation was in large deformations incremental elastoplasticity with Von Mises criterion and isotropic hardening. The better than 3% agreement between the experimental stress (load divided by the notched area) versus mean deformation ($\varepsilon = 2\text{Ln}\,\frac{\phi_o}{\phi}$) behavior of the notched specimens and the calculated one demonstrated the quality of the calculation which was further ascertained by the comparison between the experimental and the calculated deformations of the grids.

As the load is increased the maximum of the stress shifts from the surface to the center of the notched section. When the whole section is in plastic deformation the stress distribution is not very different from the Bridgman one (1964). A higher hydrostatic stress is achieved in our materials owing to the strain hardening which is different from zero contrary to the Bridgman theory. Table 4 gives the triaxiality stress factor σ_m/σ_{eq} in the center of the specimens.

TABLE 4 Triaxiality stress factor σ_m/σ_{eq}

	Unnotched specimen	ρ = 10	ρ = 4	ρ = 2
Finite element	0.33	0.6	1.02	1.25
Bridgman's formule	-	0.56	0.82	1.14

the stress and the strain gradients decrease when the notch root radius increases.

CRACK INITIATION

In the spheroïdal graphite cast iron the initiation of cracks takes place at the interface between the graphite and the matrix. In the steel it differs according to the shape of the sulfides. The crack initiation for the nodular sulfides is by decohesion at the interface, for the elongated sulfides by fracture of the sulfides themselves.

For the cast iron the conditions of crack initiation were found by extrapolation at zero of the volume fraction of holes as a function of the elongation of the specimens (fig. 1)

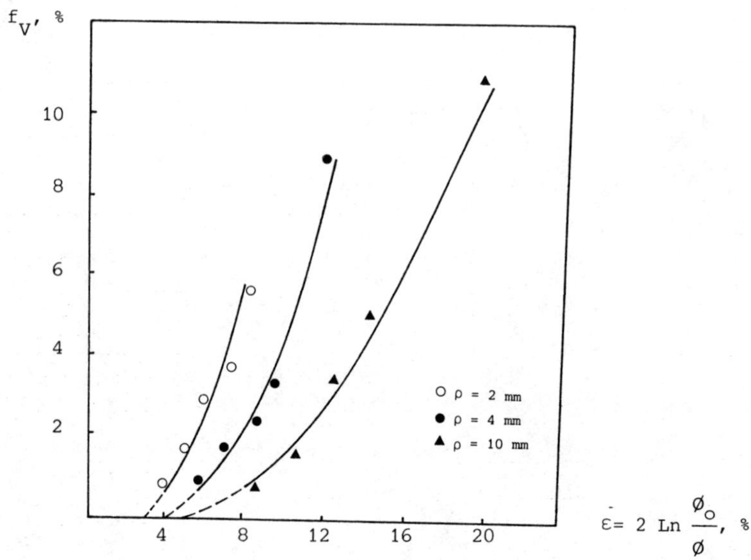

Fig. 1. Evolution of the volume fraction of holes in the cast iron as a function of the mean strain in the notched section ε.

For the steels maps of the percentage of damaged inclusions were drawn as described earlier. The conditions of crack initiation were taken along the 20% damage boundary (fig. 2).

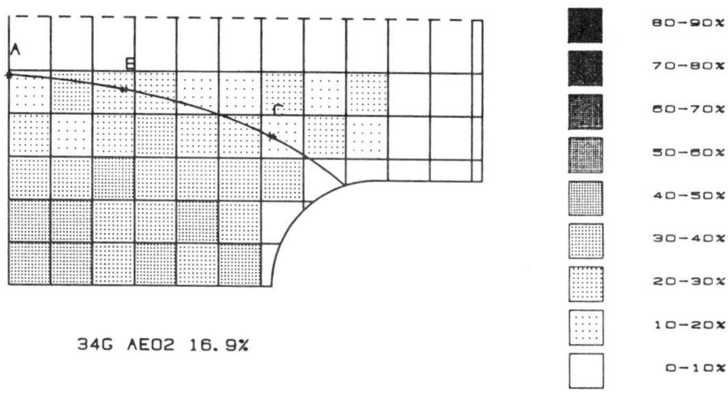

Fig. 2. Distrubution of the damaged inclusions

According to Argon and Co-workers (1975a, 1975b), using a model of spherical inclusions in a perfectly plastic or elastic matrix the decohesion condition at the interface is given by $\sigma_m + \sigma_{eq} = \sigma_c$ where σ_m is the hydrostatic stress, σ_{eq} the equivalent stress at the inclusion site and σ_c the fracture stress.

Beremin (1981) following Berveiller and Zaoui (1979), proposed as a decohesion condition at the interface $\sigma + k\ (\sigma_{eq} - \sigma_o) = \sigma_c$ where σ is the maximum principal stress, σ_o the elastic limit and k a shape parameter which is equal to 1 for a spherical inclusion.

Table 5 shows the fracture stress of the interface for the spheroïdal graphite cast iron according to the two preceding theories.

TABLE 5 Critical local stress of cavity formation at the inclusion

Specimen type	$\sigma_c = \sigma_m + \sigma_{eq}$ (MPa)	$\sigma_c = \sigma + \sigma_{eq} - \sigma_o$ (MPa)
ρ = 2 mm	658	595
ρ = 4 mm	710	623
ρ = 10 mm	672	634

Figure 3 shows that for nodular sulfides in the steel. Beremin's criterium is roughly followed with the following values : K = 0.5 and $\sigma_c = 730 \pm 50$ MPa.

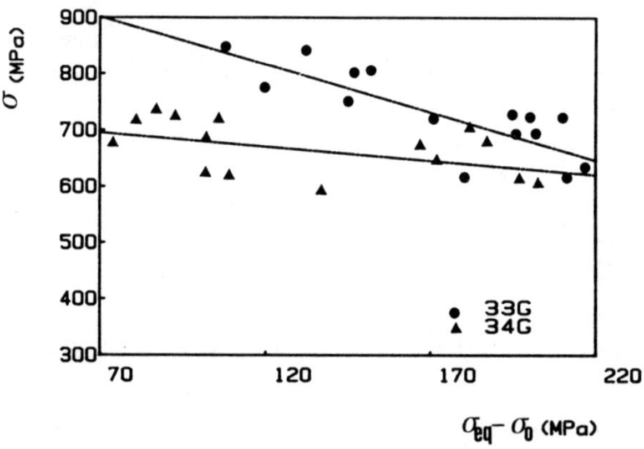

Fig. 3. Crack initiation condition for globular and elongated sulfides in steel

For the elongated sulfides K = 1.7 and σ_c = 1020 $\pm$ 50 MPa. According to the theory the k factor should indeed increase when the excentricity of the inclusion increases.

CONCLUSION

Using for the initiation of hole at inclusions fracture the criterion $\sigma + k\,(\sigma_{eq} - \sigma_O) = \sigma_c$ where σ is the maximum principal stress, σ_{eq} the Von Mises equivalent stress, σ_O the elastic limit, the following values were found for k and σ_c.

In the spheroïdal graphite cast iron k = 1 ; σ_c = 620 $\pm$ 20 MPa.

In the 30 CD 4 steel with globular sulfides k = 0.5 ; σ_c = 730 $\pm$ 50 MPa and with elongated sulfides k = 1.7 ; σ_c = 1620 $\pm$ 50 MPa.

REFERENCES

Brown, L.M., and J.D.Emburry (1973). Proc. 3rd Int. Conf. on The Strengths of metals and alloys, (Cambridge, U.K.) 164

Fisher, J.R., and J. Gurland (1981). J. Met. Sci., 185-202

Goods, S.H., and L.M. Brown (1979). Acta Met., 27, 1-15

Gurland, J., and J. Plateau (1963), J. Trans. Amer. Soc. Metals, 56, 442

Kurita, Y., L. Roesch, C. Sauzay (1977) Mem. Sci. Rev. Met., 12, 731-756

McClintock, F.A. (1967). Ductility ASM Metals, Park Ohio, 199-225

Mudry, F. (1982). Thèse de doctorat d'état

Puttick, K.E. (1959). Phil. Mag., 4, 964-969

Rosenfield, A.R. (1968). Met. Rev., 3, 121, 29-40

Mackensie, A.C., J.W. Hancok and D.K. Brown (1977). Eng. Fract. Mech., 9, 167-188

Bridgman, P.W. (1964). Studies in large plastic flow and fracture, Harvard University, Cambridge, Mass.
Argon, A.S., J. Im and R. Safoglu (1975). Met. Trans. 6A, 825-837
Argon, A.S., and J. Im (1975). Met. Trans. 6A, 839-851
Beremin, F.M. (1981). Cavity formation from inclusions in ductile fracture of A 508 steel
Berveiller, M., and A. Zaoui (1979). J. Mech. Phys. Solids, 26, 325-345

A COMPUTER SIMULATION OF DUCTILE BEHAVIOUR IN TWO STEELS OF MEDIUM AND LOW STRENGTH

I. C. Howard* and Guo Chen Li**

***Department of Mechanical Engineering, University of Sheffield, Mappin Street, Sheffield S1 3JD, UK**
****Institute of Mechanics, Academia Sinica, Beijing, China**

INTRODUCTION

The ductile behaviour of two steels (EN8 representing many medium strength steels, and Swedish iron representing low strength steels) have been simulated by comparing the point-wise behaviour of the material in a necking tensile specimen or in notched bars with the macroscopic response of a void model developed by Li and Howard (1983a). The displacement of the external boundary of the cell is regulated through the relationship

$$\frac{\dot{R}}{R} = -\alpha \frac{\dot{L}}{L} \qquad (1)$$

$$(\text{or, } \dot{\bar{\varepsilon}}_r = -\alpha \dot{\bar{\varepsilon}}_z),$$

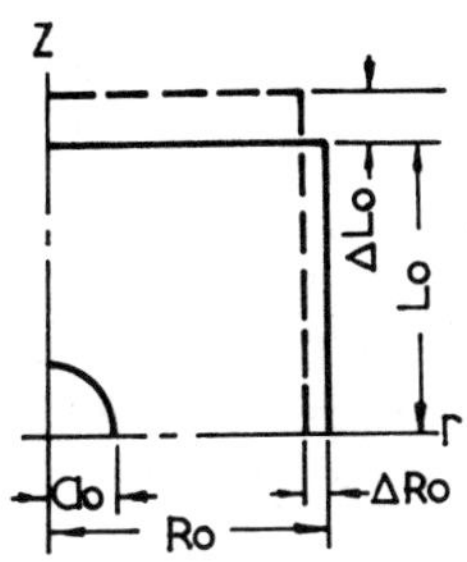

Fig. 1. A quadrant of the axisymmetric cell.

which controls the rate at which the current radius R of the cell contracts as the length L of the cell increases. Different amounts of radial contraction can be assigned to the cell as it extends through the choice of α. This, in effect determines the degree of stress triaxiality that the cell experiences. (α = 0.5 corresponds to incompressible behaviour of the material in which the cell is embedded, whilst α = 0.2 is consistent with the triaxiality computed by McMeeking (1977) in material with moderate hardening close to the tip of a blunting crack.)

The response of the matrix material in the cell is initially determined by a computer simulation of a tensile test (Li, 1983), the tangent modulus of the material being continuously adjusted so that the computed overall response of a necking bar agreed with the results of experimental tests. The test data could not be followed with the required degree of accuracy unless the material was allowed to soften on the attainment of some critical condition, and we have found that in these two ductile materials this condition is controlled by strain. Our previous work (Li and Howard, 1983a,b,c) on higher strength steels used either stress control or a mixture of stress and strain control (according to the amount of strain accumulated in the element of material) of the onset of softening.

The stress and strain developed within the cell are most conveniently expressed in terms of the following quantities:-

$$\begin{aligned} \bar{\varepsilon}_e &= \tfrac{2}{3}(\bar{\varepsilon}_z - \bar{\varepsilon}_r) && \text{- the equivalent strain} \\ \bar{\varepsilon}_m &= \tfrac{1}{3}(2\bar{\varepsilon}_r + \bar{\varepsilon}_z) && \text{- the mean strain} \\ \bar{\sigma}_e &= \bar{\sigma}_z - \bar{\sigma}_r && \text{- the equivalent stress} \\ \bar{\sigma}_m &= \tfrac{1}{3}(\bar{\sigma}_z + 2\bar{\sigma}_r) && \text{- the mean stress} \end{aligned} \qquad (2)$$

in which $\bar{\sigma}_z$ and $\bar{\sigma}_r$ are the average normal stresses along the external boundary of the cell. Similar quantities associated with the point-wise response of the matrix are expressed without the overbar.

RESULTS OF THE COMPUTATIONS

Simulation of the experimental results of Mackenzie, Hancock and Brown (1977) produced the data of Tables 1 and 2 for EN8 steel and Swedish Iron.

TABLE 1 $\bar{\varepsilon}_e$ - $(\bar{E}/\bar{E}_t)$ for EN8 Steel

$\bar{\varepsilon}_e$	0.010	0.030	0.040	0.050	0.065
$\bar{E}/\bar{E}_t$	35	53	75	105	132
$\bar{\varepsilon}_e$	0.080	0.100	0.120	0.140	0.500
$\bar{E}/\bar{E}_t$	145	230	340	580	700
$\bar{\varepsilon}_e > 0.52$	$\bar{E}/\bar{E}_t = -400$				

TABLE 2 $\bar{\varepsilon}_e$ - $(\bar{E}/\bar{E}_t)$ for Swedish Iron

$\bar{\varepsilon}_e$	0.010	0.040	0.050	0.065	0.080
$\bar{E}/\bar{E}_t$	98	140	180	230	280
$\bar{\varepsilon}_e$	0.100	0.120	0.140	0.500	0.700
$\bar{E}/\bar{E}_t$	410	540	650	710	810
$\bar{\varepsilon}_e > 0.72$	$\bar{E}/\bar{E}_t = -800$				

A comparison between the simulations and the experiments is shown in Fig. 2. The material of both simulations <u>must</u> soften if the observed behaviour in the neck is to be simulated at all well.

Figure 3 shows the stress-strain response of the continua compared with that of the void model for various values of α. The values of the other parameters that control the behaviour of the cell are detailed below.

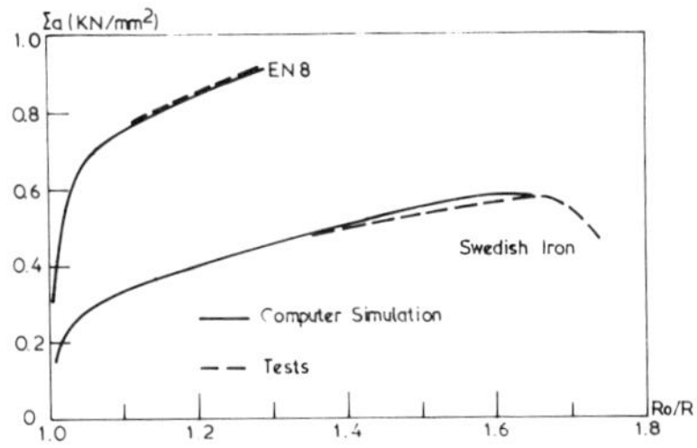

Fig. 2. Average stress-deformation curves at the neck of a bar in a tension test. Test results from Mackenzie, Hancock and Brown (1977).

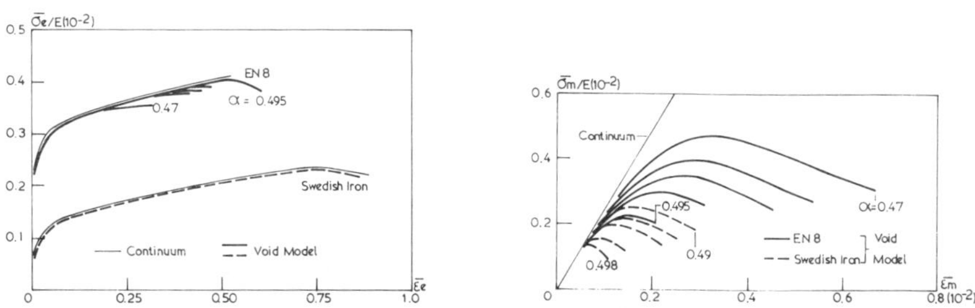

(a) Deviatoric response. (b) Dilatational response.

Fig. 3. Stress-strain curves.

(EN8:- α = 0.47, 0.48, 0.485, 0.49, 0.495
Swedish iron:- α = 0.49, 0.493, 0.495, 0.497, 0.498)

<u>EN8</u>

The tangent modulus was taken to be that of the simulated continuum until the softening point $\varepsilon_e = 0.52$ was reached. Thereafter, the rate of softening was increased according to the following scheme:-

$$e_f = -400 \quad \text{for} \quad \varepsilon_e \geqq 0.52 \,,$$

$$e_f = -400 + \frac{(\varepsilon_e - 0.80)}{0.20} \times 350 \quad \text{for} \quad \varepsilon_e > 0.80 \,, \qquad (3)$$

$$e_f = -50 \quad \text{for} \quad \varepsilon_e > 1.0$$

Here $e_f = E/(-E_f)$ where the softening modulus E_f is given in the usual way by $E_f = (\ \sigma_e/\ t)/\dot{\varepsilon}_e$.

The geometry of the cell was specified by $r_0 = a_0/R_0 = 0.19$ and $L_0/R_0 = 1$.

Swedish Iron

Again, the hardening of the matrix was that of the simulated continuum, with the softening specified by:-

$$e_f = -800 \quad \text{for} \quad \varepsilon_e \geqq 0.72,$$

$$e_f = -800 + \frac{(\varepsilon_e - 2.00)}{0.20} \times 700 \quad \text{for} \quad \varepsilon_e > 2.00 \qquad (4)$$

$$e_f = -100 \quad \text{for} \quad \varepsilon_e > 2.20 \,.$$

For this material $r_0 = a_0/R_0 = 0.13$ and $L_0/R_0 = 1$.

Figure 4 shows the variation of axial stress with elongation in the cell, each curve displaying a peak point after which the cell may be considered to have failed. As the general ductility increases (that is, as α approaches 0.5 or Swedish iron replaces EN8) the region over which $\bar{\sigma}_z$ remains sensibly constant increases, and there is no definite point of maximum stress.

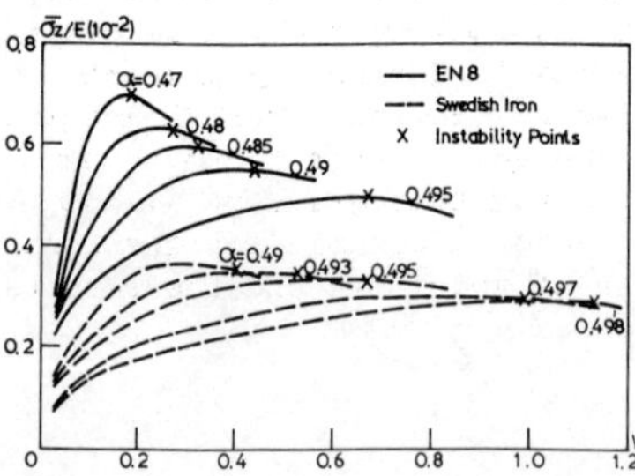

Fig. 4. Normalized axial stress versus elongation in the void model.

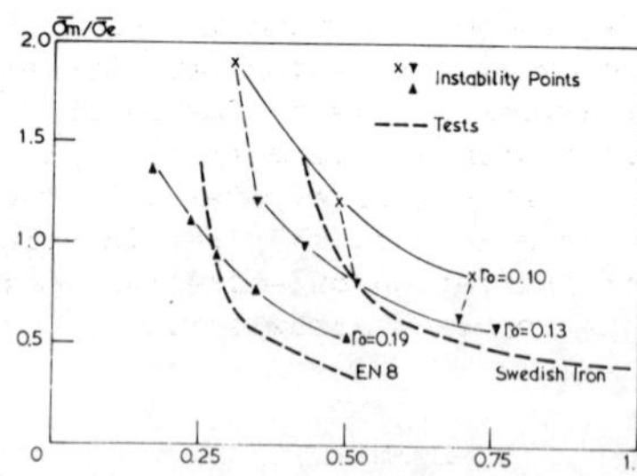

Fig. 5. The relationship between a triaxiality parameter and the equivalent strain at the initiation of fracture. Test data from Mackenzie and colleagues (1977).

We have taken the peak point to be that where the plateau or quasi-plateau begins to drop, this being the best estimate of the point where unstable deformation of the cell sets in. These results can be used to predict a ductility failure locus based on the relationship between triaxiality and strain at failure initiation. The results appear in Fig. 5 where a comparison with the experimental data of Mackenzie and colleagues (1977) is also made.

DISCUSSION

The major difference between the work reported here on low strength steels and previous work (Li and Howard, 1983a,b,c) on higher strength steels is the change to strain control from our previous use of stress-control of the softening in the continuum and in the matrix of the void model. This was anticipated (Li and Howard, 1983c) in work attempting to simulate the behaviour of high strength Ql steel, where it was found that the softening point was insensitive to stress when the material was sufficiently strained. In the very ductile materials investigated here we have found that the softening is controlled by strain at all levels of deformation. A further difference lies in the number of control parameters. In materials that are sensitive to stress control we have used a two-parameter criterion to initiate softening whilst the criteria of equations (3) and (4) for the more ductile steels involve just one parameter of strain.

The agreement between our work and the experimental results that were simulated is encouraging. There are, however, some differences that cannot be explained by making changes in the values of the microstructural parameters of the model. For example, if the value of r_0 for Swedish iron is changed slightly the ductility curve of Fig. 5 is shifted bodily upwards, but it is not rotated towards the experimental data. There may be several reasons for this. Firstly, Mackenzie and his colleagues (1977) used Bridgman's (1952) analysis to deduce from their experiments the state of stress within the neck. Bridgman's field of stress and deformation is somewhat at variance with those of modern finite element studies (for example, the work of Hancock and Brown (1983) and of Norris, Moran, Scudder and Quinones (1978)), and this may make the experimental points shown in our Fig. 5 slightly different from those that actually existed in the tests. However, and probably more importantly, it is likely that the discrepancy is also associated with the neglect of plastic dilatancy and the resultant single-parameter control of softening in the matrix during the present computations. The inclusion of plastic dilatancy in the material response of the continuum has been accomplished with the result that there is an increase of softening when all other control parameters are kept fixed. There is also a further strain dilatancy parameter available for the control of the initiation of softening. The details of this work and the results will be reported elsewhere.

REFERENCES

Bridgman, P. W. (1952). Studies in large plastic flow and fracture. McGraw-Hill, New York.

Hancock, J. W. and D. K. Brown (1983). On the role of strain and stress state in ductile failure. J. Mech. Phys. Solids, 31, 1-24.

Li, G. C. (1983). Necking in uniaxial tension. Int. J. Mech. Sci., 25, 47-57.

Li, G. C. and I. C. Howard (1983a). The effect of strain softening in the matrix material during void growth. J. Mech. Phys. Solids, 31, 85-102.

Li, G. C. and I. C. Howard (1983b). The sensitivity of the macroscopic consequences of void growth in ductile materials to various mechanical and geometrical microparameters. Int. J. Solids Structures in the press.

Li, G. C. and I. C. Howard (1983c). A computer simulation of ductile behaviour in a high strength material. International Symposium on Fracture Mechanics, Beijing, China. 22-25 November.

Mackenzie, A. C., J. W. Hancock and D. K. Brown (1977). On the influence of state of stress on ductile failure initiation in high strength steels. Engng Fracture Mech., 9, 167-188.

McMeeking, R. M. (1977). Finite deformation analysis of crack-tip softening in elastic-plastic materials and implications for fracture. J. Mech. Phys. Solids, 25, 357-381.

Norris, D. M. Jr., B. Moran, J. K. Scudder and D. F. Quinones (1978). A computer simulation of the tension test. J. Mech. Phys. Solids, 26, 1-19.

A CRITICAL ASSESSMENT OF THE THRESHOLD STRESS INTENSITY FACTOR

B. Cina

Materials and Process Engineering, Engineering Division, Israel Aircraft Industries Ltd., Ben-Gurion International Airfield, Israel

ABSTRACT

The commonly observed range of values of the threshold stress intensity factor, ΔK_{th} can be several orders of magnitude greater than the values of stress intensity factor as calculated by L.E.F.M. for very small cracks. From a critical assessment of the literature it is concluded that there may be a minimum viable size of about 10-100μm at which fatigue cracks can form and propagate. Possible crack nuclei are identified. ΔK_{th} as conventionally measured may represent only a residual retardation effect and does not seem to be a material property. For a given alloy, ΔK_{th} can have a large range of values.

KEY WORDS

Fatigue ; threshold stress intensity factor ; cracks ; crack nuclei ; retardation.

INTRODUCTION

The experimental observation of a so-called threshold stress intensity factor or range is well established and would be of considerable practical importance if ways could be found of increasing its absolute value. There may, however, be reason to question its existence as a material property. In this paper, some fundamental implications of the concept ΔK_{th} are considered.

VALUES OF Δ K IMMEDIATELY FOLLOWING CRACK INITIATION

It will be shown that for the absolute values of ΔK_{th} as measured in the laboratory, it should be impossible for fatigue cracks of atomic dimensions to propagate in a test piece of so-called sound metal. It is the accepted belief that fatigue cracks, as first formed, are of atomic dimensions. Let us assume progressive crack lengths of 100Å, 1,000Å, 10,000Å, 100,000Å and even 1,000,000Å. Let the material in question be an aluminium alloy plate, 2024-T351, 25.4 mm thick. Such material tested in pulsating tension fatigue (R=0) will have a fatigue limit of about 34Ksi (234.6 MN/m^2) for un-notched test pieces - see MIL-HDBK-5C (1978). For a sample stressed in pulsating tension and at an applied stress range of 0 to 35 Ksi (241.5 MN/m^2) failure will occur probably after about 10^7 to 2.10^7 cycles. Assuming a plate sample of finite width, W, having an edge crack of length "a" such that $\frac{a}{W}$ is small, and that crack lengths as above exist during the progress of the crack, the stress intensity factor can then be calculated from the equation given by Broek (1978),

$$K_I = 1.12\,\sigma\sqrt{\pi a}$$

where σ is the applied stress and "a" is the instantaneous crack length. The values of K_I so obtained for an applied stress range of 0 to 35 Ksi (241.5 MN/m^2) are given in Table 1.

TABLE 1 Calculated Stress Intensity Factors for an Aluminium Alloy.

Crack length,"a",mms (Å)	$10^{-5}(10^2)$	$10^{-4}(10^3)$	$10^{-3}(10^4)$	$10^{-2}(10^5)$	$10^{-1}(10^6)$
K_I, MPa $m^{\frac{1}{2}}$.	0.05	0.15	0.48	1.51	4.74
K_I, Ksi $in^{\frac{1}{2}}$.	0.04	0.14	0.44	1.37	4.31

It will be seen that only when the crack length is of the order of 10^{-2} to 10^{-1} mms are stress intensity values reached which are akin to the threshold value, $\sim$3Ksi $in^{\frac{1}{2}}$, obtained by extrapolation from data by Campbell and co-workers (1975).

Similar calculations have been carried out for a 4340 steel of ultimate tensile strength (U.T.S.) 200Ksi (1380 MN/m^2) as used by Lankford and Kusenberger (1973) as referred to later. Such a steel has a fatigue limit of 116Ksi (800.4 MN/m^2) when tested in pulsating tension at R=0 using un-notched test pieces - see MIL-HDBK-5C (1978). It will fail in 10^5 cycles at a stress range of 0 to 130 Ksi (897 MN/m^2). The values of the stress intensity factor, K_I, for the same instantaneous crack lengths as chosen before and for an applied stress range of 0 to 130 Ksi (897 MN/m^2) are given in Table 2.

TABLE 2 Calculated Stress Intensity Factors for a Low Alloy Steel.

Crack length,"a",mms (Å)	$10^{-5}(10^2)$	$10^{-4}(10^3)$	$10^{-3}(10^4)$	$10^{-2}(10^5)$	$10^{-1}(10^6)$
K_I , MPa $m^{\frac{1}{2}}$.	0.18	0.56	1.78	5.6	17.6
K_I , Ksi $in^{\frac{1}{2}}$.	0.16	0.51	1.62	5.1	16.0

From data by Campbell and co-workers (1975), 4340 steel would be expected to have a ΔK_{th} of about 11 to 16 MPa $m^{\frac{1}{2}}$ (10 to 15 Ksi $in^{\frac{1}{2}}$) at a U.T.S. of 200Ksi (1380 MN/m^2). It can be concluded that cracks or crack-like defects of the order of 10^{-2} to 10^{-1} mm in depth would have to be present to obtain a level of stress intensity akin to that of the threshold stress.

Hertzberg (1976) performed a similar calculation when considering the possible correlation between the threshold stress intensity range and the fatigue limit. For a T-1 steel he found that for an observed value of ΔK_{th} the material would have had to contain a penny shaped crack of length about 0.14 mm.

Melton, Scarlin and Hoffelner (1982) found that for fatigue tests carried out at R.T. and 500°C on a Ni-base superalloy, Nimonic 901, failure initiated invariably at carbide particles intersecting the surface. From the present author's measurements of these carbides from the published work, they would appear to have been of about 0.02 mm in maximum dimension. Melton and others (1982) performed a calculation similar to those described above based on the threshold stress intensity and fatigue limit observed in their work and found that there had to be a defect of 0.23 mm radius present to correlate the threshold and fatigue values. This was of the order of the largest grain size present. They considered the latter as the critical crack nucleus. However since in their work a fatigue crack was demonstrated to progress crystallographically from the inclusion at the surface to a grain boundary, the critical crack size of about 0.02 mm is probably the more appropriate figure to be used.

Melton and others (1982) performed similar calculations for a 12% Cr steel based on data of Friedl, Scarlin and Zelizko (1981). Here again a critical defect size of about 20μm was observed which corresponded with the size of surface intersected inclusions at which the cracks clearly initiated.

These several calculations imply that in the absence of prior cracks or crack-like defects of the order of 10-100μm in length for aluminium, low alloy steel, martensitic stainless steel, and a Ni-base superalloy, fatigue cracks of atomic dimensions will not propagate. Since it is known that fatigue cracking will in fact be obtained under the above test conditions, and it is commonly believed that the cracks thus formed will initially be of atomic dimensions, then either there is no such thing as a threshold fatigue stress intensity, or there are crack-like defects of 10-100μm in length already present in our materials. In the latter case the concept of fatigue cracks of atomic dimensions being formed from sound metal then becomes questionable with all that that implies vis à vis the several proposed mechanisms of fatigue. As a further alternative, fatigue cracks are formed independently of crack-like defects and are of minimum instantaneous size of 10-100μm.

The Minimum Observed Size of Fatigue Cracks.

It is difficult to establish from the literature the minimum sizes at which fatigue cracks have been detected. Using a replica technique, De Lange (1964) observed microcracks as small as 1μm in length on the surface of test pieces in a L.A.S. (35CD4) and in an aluminium alloy (26 St). Their depth was unknown. All these cracks subsequently propagated. Their origin is not defined.

The Initiation of Fatigue Cracks From Prior Defects.

Lankford and Kusenberger (1973) studied the initiation of fatigue cracks formed in pulsating tension in a 4340 steel. They found that cracks initiated at surface and subsurface inclusions due to a debonding occurring at the inclusion-matrix interface, specifically at the tensile pole of the inclusion. The debond seam gradually grew leading to nucleation of microcracks within the steel matrix. These authors in turn refer to three other works all of which likewise showed that fatigue cracks initiated exclusively or almost exclusively at surface inclusions. Exclusive initiation was observed in 2024-T4 (Erhardt, Pelloux and Grant, 1969) and 7075-T6

(Grosskreutz, Show and Benson, 1969) aluminium alloys. In the case where almost exclusive initiation was observed (Cummings, Stulon and Schulte,1958) of 540 specimens of 4340 and 4350 type steels having a single fracture nucleus, 99% were attributable to inclusion-related crack nucleation. In Lankford and Kusenberger's (1973) work the inclusions were spherical or slightly ellipsoidal, and their sizes ranged from 0.0002 to 0.003 in. (0.005 to 0.076 mms). The smaller of these values is close to that calculated, 0.01 mm., as being necessary to justify the existence of a fatigue threshold stress intensity factor in the 4340 steel, and both values, 0.005 and 0.076 mms., are close to those calculated for aluminium alloys, 0.01 to 0.1 mm.

The invariable nucleation of fatigue cracks at carbide inclusions in a Ni-base superalloy (Melton and co-workers, 1982) has already been referred to, as has the similar effect in a 12% Cr martensitic stainless steel (Friedl and co-workers, 1981).

Schijve (1982) has likewise drawn attention to the fact that from the literature and at least for some materials, micro cracks start at inclusions, and refers to the work of Stone and Swift (1979) who found fatigue crack initiation at sub-surface intermetallic particles on flight simulation loadings. One such particle was 4 μm across.

There would seem to be strong evidence for fatigue cracks to initiate at inclusions (oxide particles, carbides, intermetallic compounds, etc.) where such are present, and for the minimum crack size so obtained to be of the order of 10-100μm. It is not clear where De Lange's (1964) cracks of 1μm surface length initiated. Several questions arise here. Does the fatigue crack form only at or above a minimum size, that of the inclusion, or does it first grow to this size ? Secondly, where do fatigue cracks initiate in materials free of inclusions, and do they too have a minimum viable size ? These questions are possibly related. In inclusion-free material cracks are known to nucleate at persistent slip bands, intrusions, and grain or twin boundaries, dependent on the alloy and test conditions. Cina (1960) observed extrusions as small as 1μm and up to 40μm in length lying on crystallographic planes in a 20% Cr-iron containing only 0.03% C. The extrusions did not seem to be associated with the few carbide particles present. Cracks lying at 45° to the surface subsequently grew from the associated intrusions, although not necessarily from the smallest ones. Significantly, extrusions smaller than 1μm were not observed. The depth of the intrusions is believed to be less than their length.

The very fact that fatigue failures have been observed to initiate at inclusions suggests that the level of stress intensity thereby associated is essential for crack formation i.e. a minimum dimension of the inclusion is a determining factor. If cracks formed elsewhere than at inclusions they could not be of atomic dimensions since the stress intensity factor would be orders of magnitude less than that associated with cracks at inclusions and therefore inadequate to allow their growth. It follows that for many metallic materials, fatigue cracks form only above a minimum size which would seem to be of the order of 1-10μm.

A case would therefore seem to exist for a threshold stress intensity to have at least an empirical significance and for it to imply the existence in our materials of significant defects, inherent or formed above a minimum size during stressing.

The Effect of Microstructure and Static Mechanical Properties on ΔK_{th}

The effect of grain size and yield stress on ΔK_{th} in different alloy systems is inconsistent and seems to be alloy dependent. Thus an increase in grain size can be accompanied by a decrease in ΔK_{th} as in Cu-Al alloys (Higo, Pickard and Knott, 1981) and pearlitic steels (Kao and Byrne, 1982), an increase in ΔK_{th} as in α brass (Beevers, 1980) and a titanium alloy (Brown and Smith, 1982), and can have no effect on ΔK_{th} as in copper (Mullner, Weiss and Stickler, 1982). Likewise an increase in yield strength can be accompanied by a decrease in ΔK_{th} as in α brass (Beevers, 1980), low alloy steels (Ritchie, 1977, and Minakawa and McEvily, 1982) and aluminium alloys (Minakawa and McEvily, 1982) and an increase in ΔK_{th} as in Cu-Al alloys (Higo and co-workers, 1981) and in a Ni-Al bronze (Pickard, Ritchie and Knott, 1976).

If ΔK_{th} were a material property, it would not be expected to vary both directly and indirectly with grain size or yield strength as has been observed. Furthermore if it does so vary, any one determination of it would seem to be of very limited practical value for a given alloy system.

THE METHOD BY WHICH ΔK_{th} IS MEASURED

ΔK_{th} is generally measured on cracks of considerable length, often of the order of millimeters, so that the possibility of a retardation effect is a real one. Whereas in the initial stages of reducing the level of applied stress, so as to approach the so-called threshold stress intensity factor, it may be possible for a high stress to eliminate the residual stress effects of the previous even higher applied stress, it may become progressively more difficult to do so as the applied stress levels decrease. As the stress level becomes very low, residual compressive stresses may still be of significant level and sufficient to suppress further crack propagation. On such a basis, the threshold stress intensity is a real effect but could simply represent the inability of the final applied stress to overcome the residual effects of the prior stress history. It would then not represent any intrinsic property of the material i.e. there would then be no absolute threshold stress intensity factor but an infinite series of such factors dependent on the previous stress history and crack length. Cadman, Brook and Nicholson (1982) recently reached a similar conclusion after employing very careful control of the step-down sequence in their measurements of ΔK_{th} on a C-Mn low alloy steel. Where ΔK_{th} is approached by raising Kmin, retardation effects may be even more pronounced.

Smith (1982a) is extremely critical of the accuracy with which threshold intensity values are quoted - often to three significant figures. He points to a test variability of $\pm$5% on the same batch and an inter-laboratory variability of $\pm$15%. He attributes the prime variability to mean stress and corrosion effects. He concludes that the best method of threshold determination is one which simulates service conditions and that for variable amplitude loading there may not even be a threshold.

The Effect of the Initial Crack Length on ΔK_{th}

Not only is the measured value of ΔK_{th} affected by the previous stress history but it has also been found to be affected by the initial crack length. Kitagawa and Takahashi (1976) were perhaps the first systematically to study the subject. They found that for uni-axial cyclic tension on a low alloy steel, ΔK_{th} was constant for initial crack lengths in the range 0.5 to 5 mms. For cracks $<$0.5 mm, $\Delta \sigma_{th}$ increased but at a decreasing rate and asymptotically approached the fatigue limit for unnotched samples. This would imply that ΔK_{th} would decrease as the crack length decreased below 0.5 mm. Kitagawa and Takahashi (1976) concluded that for such small cracks conventional linear fracture mechanics can not be applied. Since the bulk stress field in a sample with a small crack becomes predominantly elastic for decreasing crack size, the conditions would seem to be ideal for the application of L.E.F.M. One possible reason for the averred inapplicability would be the existence of a minimum viable crack size for initiation. From Kitagawa and Takahashi's (1976) work this would possibly be somewhat less than 0.06 mm. which is in reasonable agreement with the results of Lankford (1973). In even more recent work, Smith (1982b), like Kitagawa and Takahashi (1976), concluded that L.E.F.M. is inadequate for short cracks and that some form of elastic-plastic crack growth parameter is needed.

On a double logarithmic plot of ΔK_{th} versus initial crack size, Romaniv and co-workers (1982) found that for mild steel and for an Al-Mg-Mn alloy, ΔK_{th} was constant for initial crack lengths (As) greater than 1.2 and 0.5 mm. respectively. These results compare fovourably with those of Kitagawa and Takahashi (1976). For shorter initial crack lengths, ΔK_{th} and As were linearly related. The smallest initial crack they investigated for these two materials was about 0.2 mms. long. In contrast to the above results, Romaniv and co-workers (1982) found no critical initial crack length for an 0.45% C stainless steel for cracks in the range 0.3 to 2.0 mm. Perhaps cracks smaller than 0.3 mm. should also have been investigated.

An extremely interesting observation in the work of Romaniv and co-workers (1982) was that ΔK_{th} increased as specimen thickness decreased from 6 to 1.5 mm. for an initial crack size of 2 mms.

The conclusions from all the foregoing work on the variation of the value of ΔK_{th} with initial crack length and specimen dimensions are that it would be hard to defend a claim that ΔK_{th} is a material property, and that McEvily's (1979) recommendation for more fundamental work on crack size effect should be adopted.

The Effect of R Ratio on ΔK_{th}

The consistently observed decrease of ΔK_{th} with increasing R ratio has been discussed by Cadman, Nicholson and Brook (1982). They concluded that the effect can be attributed either to closure or to oxidation. In the former case ΔK_{th} will be dependent on the previous load history and therefore will be almost impossible to assess for variable stress history. If oxidation of the crack surface is the explanation, (there is no R ratio effect in vacuum - see Beevers, 1982), the extent of this will likewise be quantitativel unpredictable.

The conclusion again to be drawn would seem to be that threshold is a highly complex phenomenon but not a material property.

CONCLUSIONS

There is evidence that, at least for many alloys of industrial importance, fatigue cracks invariably initiate from crack-like defects, (inclusions, intermetallic particles, etc.).

Doubts are raised as to whether fatigue cracks of atomic dimensions can propagate in material free of detectable defects. Their formation may also be in question.

For materials free of crack-like defects there would seem to be a possibility that fatigue cracks form instantaneously only at or above a minimum viable size, of the order of 10-100μm.

ΔK_{th} as conventionally measured may represent only a residual retardation effect and not an inherent material property.

The implications of the concept of ΔK_{th} with respect to fatigue crack initiation and initial propagation deserves further study.

REFERENCES

Beevers, C. J. (1980). Metal Science, Vol. 14, 418.

Beevers, C. J. (1982). Some Aspects of the Influence of Microstructure and Environment on Δ K Thresholds. In J. Bäcklund, A. F. Blom and C. J. Beevers (Eds.), Fatigue Thresholds, Proc. Internat. Conf., Stockholm, Sweden, 1981, Vol. 1, pp. 257-275.

Broek, D. (1978). Elementary Engineering Fracture Mechanics. Sitjhoff and Noordhoff. The Netherlands. p. 76.

Brown, C. W. and G. C. Smith (1982). The Effect of Microstructure and Texture on the Fatigue Crack Growth Threshold in Ti-6Al-4V. In J. Bäcklund, A. F. Blom and C. J. Beevers (Eds.), Fatigue Thresholds, Proc. Internat. Conf., Stockholm, Sweden, 1981, Vol. 1, pp. 329-343.

Cadman, A. J., R. Brook and C. E. Nicholson (1982). Effect of Test Technique on the Fatigue Threshold ΔK_{th}. In J. Bäcklund, A. F. Blom and C. J. Beevers (Eds.), Fatigue Thresholds, Proc. Internat. Conf., Stockholm, Sweden, 1981, pp. 59-75.

Cadman, A. J., C. E. Nicholson and R. Brook (1982). Microstructural Effects at and near the Fatigue Thresholds. In J. Bäcklund, A. F. Bloom and C. J. Beevers (Eds.), Fatigue Thresholds, Proc. Internat. Conf., Stockholm, Sweden, 1981, Vol. 1, pp. 293-311.

Campbell, J. E., W. E. Barry and C. E. Feddersen (1975). Damage Tolerant Design Handbook. MCIC-HB-01. Part 2. Battelle, Columbus, Ohio, U.S.A.

Cina, B. (1960). The Effect of Mode of Stressing in Fatigue Testing. J. Iron & Steel Inst., Vol. 194, 324-336.

Cummings, H. N., F. B. Stulon and W. C. Schulte (1958). Proc. Amer. Soc. Test. Mat., Vol. 58, 505.

Erhardt, K., R. M. N. Pelloux and N. J. Grant (1969). A.F.M.L. Technical Report No. AFML-TR 69-85. Wright Patterson Air Force Base, Dayton, Ohio U.S.A.

Friedl, K. H., R. B. Scarlin and V. Zelizko (1981). In D. Francois (Ed.), Advances in Fracture Research, 5th. Internat. Conf. on Fracture, Cannes, p. 923.

Grosskreutz, J. C., G. C. Show and D. K. Benson (1969). A.F.M.L. Technical Report No. AFML-TR 69-121, Wright Patterson Air Force Base, Dayton, Ohio, U. S. A.

Hertzberg, R. W., (1976). Deformation and Fracture Mechanics of Engineering Materials. John Wiley and Sons, New York.

Higo, Y., A. C. Pickard and J. F. Knott (1981). Effects of Grain Size and Stacking Fault Energy on Fatigue Crack Propagation Thresholds in Cu-Al Alloys. Metal Science, Vol. 15, No. 6, 233-240.

Kao, Po-We, and J. G. Byrne (1982). Microstructural Influences on Fatigue Crack Propagation in Pearlitic Steels. In J. Bäcklund, A. F. Blom and C. J. Beevers (Eds.), Fatigue Thresholds, Proc. Internat. Conf., Stockholm, Sweden, 1981, Vol. 1, pp. 313-327.

Kitagawa, H., and S. Takahashi (1976). Applicability of Fracture Mechanics to Very Small Cracks or the Cracks in the Early Stage. In Proc. Second Internat. Conf. on Mechanical Behaviour of Materials, Boston, U.S.A. Amer. Soc. Metals, pp. 627-631.

De-Lange, R. G. (1964). Plastic Replica Methods Applied to a Study of Fatigue Crack Propagation in Steel 35CD4 and 26ST Aluminium Alloy, Trans. Met. Soc., A.I.M.E., Vol. 230, 644-648.

Lankford, J., and F. N. Kusenberger (1973). Initiation of Fatigue Cracks. Met. Trans., A, Vol. 4, 553-559.

McEvily, A. J., (1979). On Notch Sensitivity and the Threshold Level. In Recent Research on Mechanical Behaviour of Solids, Uni. Of Tokyo Press, 329-340.

Melton, K. N., R. B. Scarlin and W. Hoffelner (1982). In R. Brunetaud and others (Eds.), Proc. Conf., High Temperature Alloys for Gas Turbines, Liege, Belgium. D. Reidel Pub. Co., Dordrecht, Holland, pp. 659-671.

MIL-HDBK-5C (1978). Dept. of Defence, Washington, D. C., U.S.A.

Minakawa, K., and A. J. McEvily (1982). On Near-Threshold Fatigue Crack Growth in Steels and Aluminium Alloys. In J. Bäcklund, A. F. Blom and C. J. Beevers (Eds.), Fatigue Thresholds, Proc. Internat. Conf., Stockholm. Sweden, 1981, Vol. 1, pp. 373-390.

Mullner, H., B. Weiss and R. Stickler (1982). The Effect of Microstructure on the Fatigue Threshold in Copper. In J. Bäcklund, A. F. Blom and C. J. Beevers (Eds.), Fatigue Thresholds, Proc. Internat. Conf., Stockholm, Sweden, 1981, Vol. 1, pp. 423-440.

Pickard, A. C., R. O. Ritchie and J. F. Knott (1976). In Proc. Fourth Internat. Conf. on Strength of Metals and Alloys, Nancy, France, Vol. 2, p. 473.

Ritchie, R. O. (1977). Metal Science, Vol. 11, 368.

Romaniv, O. N., V. N. Siminkovich and A. N. Tkach (1982). Near Threshold Short Fatigue Cracks,in J. Bäcklund, A. F. Blom and C. J. Beevers (Eds.), Fatigue Thresholds, Proc. Internat. Conf., Stockholm, Sweden, 1981, Vol. 2, pp. 799-807.

Schijve, J. (1982). Differences between the Growth of Small and Large Fatigue Cracks in Relation to Threshold K Values. In J. Bäcklund, A. F. Blom and C. J. Beevers (Eds.), Fatigue Thresholds, Proc. Internat. Conf., Stockholm, Sweden, 1981, Vol. 2, pp. 881-908.

Smith, R. A. (1982a). Fatigue Thresholds - A Design Engineer's Guide through the Jungle. In J. Bäcklund, A. F. Blom and C. J. Beevers (Eds.), Fatigue Thresholds, Proc. Internat. Conf., Stockholm, Sweden, 1981, Vol. 1, pp. 34-44.

Smith, R. A. (1982b). Short Fatigue Cracks. In Quantitative Measurement of Fatigue Damage, Internat. Conf., Dearborn, PA., U.S.A. Amer. Soc. Test. Mat. To be published.

Stone, M. and T. Swift (1979). Future Damage Tolerance Approach to Air-Worthiness Certification. In Proc. 11th. Internat. Conf. Aeronautical Fatigue, Brussels, Belgium, paper 2.9.

A COMPARISON OF J_{Ic}, STRETCHED ZONE SIZE AND CRACK OPENING DISPLACEMENT IN A FERRITIC STEEL

V. Provenzano, C. M. Gilmore, F. A. Smidt and J. R. Hawthorne

Thermostructural Materials Branch, Material Science & Technology Division, Naval Research Laboratory, Washington DC 20375, USA

ABSTRACT

A comparison was made between the J-integral value, the crack opening displacement (COD) computed from the stretched zone height, and the COD from the load line displacement on HT-9 martensitic stainless steel. The fracture toughness for two specimen thicknesses (2.5 and 12.7 mm) and two test temperatures (24 and 200°C) was determined by conducting J-integral tests on compact tension specimens using the single specimen compliance method. The stretched zone-COD was determined from stereographic measurements on scanning electron micrographs of the crack tip region, while the load line displacement-COD was computed from the J-integral test data. The J_{Ic} for each specimen was related to the product of the COD value, the constraint factor, m, and the flow stress, σ_f. For the 12.7-mm specimens that approximated conditions of plain strain, the stretched zone-COD and the load line-COD were quite comparable and they yielded a constraint factor of about 2. This value of m is consistent with the theory that relates COD to J_{Ic}. However, for the 2.5-mm specimens, where conditions approximating plain stress were present, the stretched zone-COD was 3 to 4 times smaller than the corresponding load line-COD. For the thin specimens, consistent with theory, the load line-COD values yielded a constraint factor of 1.6 while the much smaller stretched zone-COD yielded a constraint factor of about 5 whereas theory predicts a value close to 1. A rationale for the lack of agreement with the theory of the stretched zone-COD in the thin specimens and a detailed analysis of the crack tip microprocesses are presented.

KEYWORDS

Stretched zone, crack opening displacement, J_{Ic}, scanning electron microscopy, stereography, crack tip microprocesses, plain stress, plain strain, and HT-9 martensitic stainless steel.

INTRODUCTION

Linear Elastic Fracture Mechanics (LEFM) provides a rational framework for relating the fracture stress to the size of cracks existing in a structure. However, valid LEFM tests require that the specimen thickness (B) is such that:

$$B \geq 2.5\left(\frac{K_{Ic}}{\sigma_y}\right)^2 \quad . \; . \; . \; . \; . \; . \; . \; . \quad (1)$$

In many engineering applications the material thickness is less than that required by Eq. 1 and the stress to cause fracture of this thinner material would be greater than that required to fracture a LEFM specimen. It is desirable, therefore, to be able to predict the fracture stress in specimens with less than the thickness required by Eq. 1. Further, in many test programs it is not possible or desirable to test specimens with the thickness required by Eq. 1, as an example, in testing of materials for advanced nuclear reactor systems, where insufficient space is available in the irradiation facilities. Thus, thinner specimens must be tested and, from the tests of thinner specimens, it is desired to predict the fracture stress of thicker structures.

The J-integral method has been proposed to treat elastic-plastic fracture problems. The J-integral method utilizes specimens of thicknesses less than that required by Eq. 1. Rice (1968) proposed the J-integral relationship of the form:

$$J = -\frac{1}{B}\,\frac{dU}{da}\,, \quad . \; . \; . \; . \; . \; . \; . \; . \quad (2)$$

where U is the energy obtained from the area under the load-displacement curve and a is the crack length. The critical value of J (J_{Ic}) occurs when crack extension begins. It has also been proposed that the crack opening displacement (COD) at the crack tip is a measure of the resistance to fracture. Knott (1980) for example, has related the COD to J_{Ic} using the relationship

$$J_{Ic} = m\,\sigma_f\,\mathrm{COD}, \quad . \; . \; . \; . \; . \; . \; . \; . \quad (3)$$

where m is a constant related to the constraint in the specimen, and σ_f is the flow stress. According to Knott's model, the values of the constraint factor, m, can range from 0.79 to 3. The higher values of m correspond to higher degrees of constraint as observed in plain strain fracture; the lower values of m correspond to lower degrees of constraint as in plane stress fracture.

A number of techniques and approximations have been employed to measure the magnitude of the COD at the crack tip . For example, Burdekin and Stone (1966) utilized a paddle wheel COD meter, and Fields and Miller (1977) made use of silicone rubber to infiltrate cracks under load. Broek (1974) equated the crack opening displacement to the size of the stretched zone that was measured with large angle stereo pairs produced from transmission electron micrographs of a surface replica. Broek found that the constraint factor m averaged to 2.5 for a group of 14 aluminum alloys. Employing a model developed by Wells (1971), the COD at the crack tip can also be determined from measurements of load line displacement corrected for geometrical effects and rotation due to plastic deformation.

The purpose of this paper is to compare the values of COD calculated from the stretched zone to the COD values obtained for the load line displacement and to test the applicability of Knotts model (Eq. 3) using the load line displacement-CODs, stretched zone-CODs and J_{Ic} values determined from the single specimen compliance method. In a previous study, Gilmore and co-workers (1983) have presented additional details on the experimental procedure and results of the stretched zone measurements. Therefore, for the sake of brevity these details are not given in the present work.

EXPERIMENTAL PROCEDURE

The HT-9 martensitic stainless steel used in the present study was supplied in the form of 3.3 cm diameter rod stock. Details about the chemical composition and the heat treatment are described elsewhere (Gilmore and co-workers, 1983). In the present study, four conditions of alloy HT-9 were tested: two thicknesses of 12.7 mm and 2.5 mm, and two specimens of each thickness were tested at 24°C and 200°C. the specimens were cut with the fracture plane parallel to the axis of the rod. The 12.7-mm specimens conformed to ASTM E399-74 geometry, however, the notch region had been modified to permit the mounting of razor knife edges for measurement of the load line displacement with a clip gage. The 2.5-mm specimens had the same geometry as the 12.7-mm specimens except their thickness was reduced to one fifth of the 12.7-mm value. Additional details about the specimen design may be found in the studies of Loss and Angelino (1980) and Loss and co-workers (1981).

The fracture toughness of the rod was measured by conducting J-integral tests. The J-R curves for the fracture toughness tests were obtained by means of the single specimen compliance technique. Prior to the J-integral tests, the specimens were fatigue precracked to 1.78 mm crack length. The J_{Ic} values were corrected for rotation (Loss and co-workers, 1981) and for crack extension as proposed by Ernst (1980).

Fractographic examinations on the failed specimens were performed on a Coates and Welter 106A field emission scanning electron microscope (SEM). The stereographic measurements were performed by employing a semi-automatic image analysis system that incorporates a Hewlett-Packard Model 9830A mini-computer for data recording and reduction. More details about the system may be found in the papers by Thomas (1972) and by Sprague (1980).

The size of the stretched zone in each specimen was determined stereographically by making several measurements in the center portion of the crack front and then taking an average of these measurements. Multiple measurements of the same crack profiles yielded a standard deviation of 7 percent. The stretched zone was defined as the featureless region that is bounded by the fatigue precrack on one side and by the stable crack extension region on the other side.

As mentioned earlier, the COD at the crack tip was determined also from the load-displacement data using the model formulated by Wells (1971) and given by the following relations.

$$\mathrm{COD} = \frac{0.45\ (W-a)}{0.45W + 0.55a}\left\{V_L - \frac{\gamma\sigma_y\ (1-\nu^2)}{E}\right\} \quad . \quad (4)$$

$$\text{for} \qquad V_L \geq \frac{\gamma\sigma_y W(1-\nu)}{E},$$

where V_L = mouth opening, σ_y = yield stress, E = Young's modulus, ν = Poisson's ratio, γ = non-dimensional limiting value of elastic mouth opening, $\gamma \cong 2.5$, a = crack depth, W = specimen width.

RESULTS AND DISCUSSION

The optical micrographs shown in Fig. 1 provide an overall view of the fracture surfaces of the J-integral specimens of the two thicknesses tested at 24°C. From the figure it can be seen that conditions approximating plane strain were present in the 12.7-mm specimen because no significant lateral contraction or shear lips were observed, while conditions approximating plane stress were present in the 2.5-mm specimen as indicated by the presence of large lateral contraction and extensive shear lips. Though not shown for sake of brevity, approximate conditions of plane strain were also present in the 12.7-mm specimens tested at 200°C, while conditions close to plane stress persisted in the 2.5-mm specimen tested at 200°C. As discussed below, this basic difference in the fracture mode observed in the two specimen thicknesses at both temperatures is focal point of the present work.

Fig.1. Fracture surfaces of 2.5-mm (a) and 12.7-mm (b) specimens of alloy HT-9 tested at 24°C.

The results from the J-integral tests on HT-9 steel, the stretched zone-COD and the load line displacement-COD are presented in Table 1. Table 1 shows that for 12.7 mm specimens there was very good agreement between the stretched zone-COD and COD computed on the basis of the load line displacement. The two sets of COD values when substituted in Eq. 3 yield a constraint factor, m, close to 2. This value of m is consistent with the theory, since for conditions of plain strain the value of m should be about 3. However, Table 1 shows that for the 2.5-mm thick specimens the COD values from the load-displacement were more than a factor of 3 larger than the COD determined from the stretched zone height. The m factor calculated from Eq. 3 for the stretched zone COD was about 5. This result is not a reasonable result, since the constraint factor should be closer to 1 for conditions of plane stress. However, the m factor computed from the load line displacement COD was 1.6, which is a reasonable result for a thin specimen.

TABLE 1 - Summary of J-Integral Values, Corresponding Stretched Zone Height and Constraint Factor for HT-9 Alloy Tested at 24 and 200°C

	J_{Ic} in KJ/m^2	Flow Stress in MPa	Str. Zone COD in μm	Load Line COD in μm	Factor-m from Str. Zone COD	Factor-m from load line COD
HT-9 - 1 12.7-mm at 24°C	91	738	76	69	1.6	1.8
HT-9 - 2 12.7mm at 200°C	20	660	96	95	1.9	1.9
HT-9 - 1 2.5mm at 24°C	147	738	38	122	5.2	1.6
HT-9 - 2 2.5mm at 200°C	170	660	46	161	5.6	1.6

The reason for the significant difference in the two sets of COD values for the 2.5-mm specimens can be best understood by considering the micromechanics of the crack opening process in plane strain and in plane stress conditions.

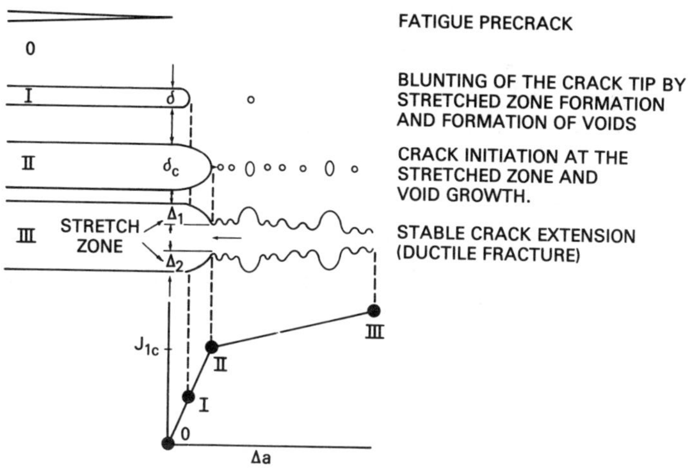

Fig. 2. A schematic model of crack propagation in fracture toughness specimens together with the corresponding regime in J-R curve. The magnitude of the crack opening displacement (COD) at fractue ($δ_c$) is equal to the sum of $Δ_1$ and $Δ_2$.

Figure 2 is a schematic representation of a model proposed by Landes and McCabe (1979) to describe the micromechanical deformation that occurs for a ductile metal at the crack tip during J-integral testing. According to this model, the micromechanical deformation involves the following sequence of steps. First, as the external load is applied to a J-integral specimen the tensile stresses open the fatigue pre-crack and the plastic deformation blunts the crack tip giving rise to a stretched zone while voids ahead of the crack begin to form. Secondly, the stretched zone continues to increase and also the voids increase in size until a critical stretched zone size, δc, is reached and the crack begins to propagate by linking the voids with the blunted crack. This model appears to successfully explain the formation of the stretched zone and the crack initiation process in the 12.7-mm specimens as shown by the typical crack profile shown in Fig. 3. The model, however, does not appear to apply to the fracture process that occurred in the 2.5-mm specimens. In these latter specimens the observed stretched zone was much smaller than the 12.7-mm specimens and large shear lips were present as illustrated by the SEM micrographs of Fig. 4. The importance of these observations on the thin specimens is discussed below.

A proposed reason for the smaller stretched zone in the thin specimen with a correspondingly large load line displacement is that shear fracture occurred rather than stretching at the crack tip. Shear fracure would not produce a stretched zone, but it would produce a crack opening displacement of the crack faces. In the thin specimens the fracture at the side of the specimen was predominately shear fracture and there was no stretched zone observed at the edge of the specimen. Progressing towards the center of the specimen the amount of slant deformation (shear deformation in the plane of the crack) decreased because the amount of constraint increased towards the center of the specimen (Fig. 4). At the center of the specimen the shear deformation was a minimum and the maximum stretched zone was observed at the center of the specimen. In the thin specimens shear deformation extended to the center of the specimen as experimentally verified by the observation that the stretched zone continuously increased to the center of the specimen. Thus, even in the center of the thin specimen, shear deformation could be expected to be competing with the stretching deformation, and since shear deformation produces no stretched zone the full stretched zone height as observed in thick specimens was not realized.

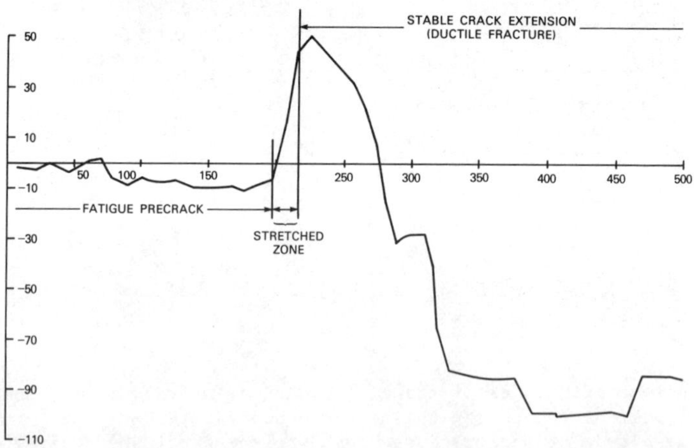

Fig. 3. Typical crack profile used to measure the size of the stretched zone and the depth of large voids in the fracture surfaces of the CT specimens

The observation that the stretched zone height and the amount of shear lip was different in the 2.5-mm than in the 12.7-mm specimens has implications concerning the use of 2.5-mm thick specimens for fracture toughness determination. The fracture toughness of the 2.5-mm specimens was significantly higher than in the 12.7-mm specimens; the higher fracture toughness was most likely the result of the extensive shear deformation observed in the thin specimens. Since the fracture microprocesses were different in the thick and in the thin specimens, it is possible that fracture in the thick and and thin specimens would have different response to changes in test parameters such as temperature. In these tests the value of J_{Ic} in the thicker specimens exhibited a greater temperature dependence than the thinner specimens. By changing the temperature from 200°C to room temperature J_{Ic} in the 12.7-mm specimen dropped 24 percent, whereas J_{Ic} in the 2.5-mm specimen dropped only 13 percent.

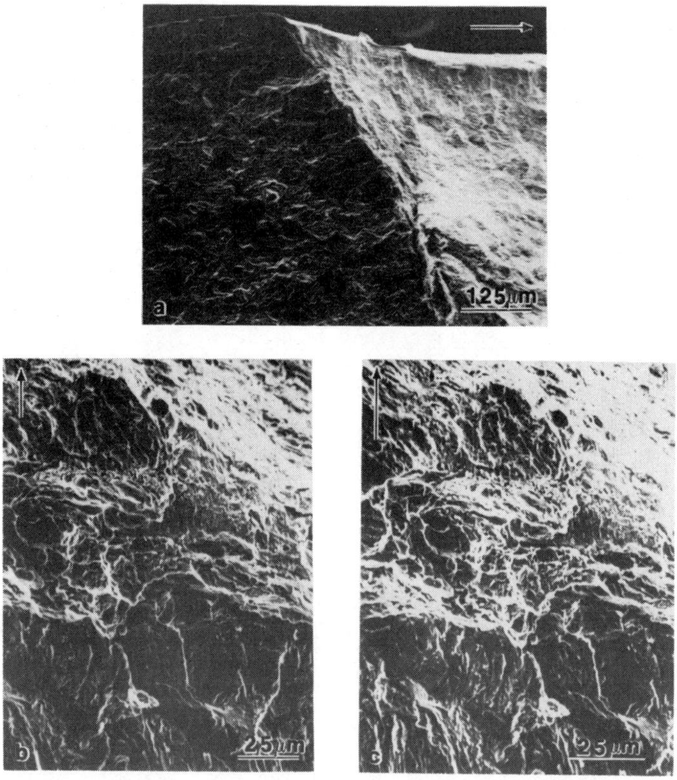

Fig. 4. Fracture Surfaces (SEM) of 2.5-mm specimen of alloy HT-9, tested at 24°C. Arrows show direction of microscopic crack propagation. The low magnification micrograph (a) shows the extensive shear lips at the edge of the specimen, and the stereo pair (panels b and c) show the irregular stretched zone.

The minimum thickness, B, for J test specimens recommended by ASTM E813 is:

$$B \geq 25 \frac{J_{Ic}}{\sigma_y}$$

Substitution of the values of J_{Ic} and σ_y from Table 1 into this equation results in a critical thickness of about 6.4 mm. The 12.7-mm specimens satisfy the critical thickness criterion, but the 2.5-mm specimens were less than half the critical thickness. These results support the necessity of satisfying the standard specification and indicate that specimens that have less than the recommended thickness fracture by a different mechanism than specimens that safisfy the thickness criterion.

CONCLUSIONS

1. The J_{Ic} values for 12.7 mm specimens was 30 percent less than the value for the 2.5-mm specimens.
2. The COD values based upon stretched zone and loading displacement were nearly equal in the 12.7-mm specimens.
3. In the 2.5-mm specimens the stretched zone COD was much less than the COD based upon load line displacement.
4. The stretched zone model of the micromechanical fracture process during a J test is not valid for thin specimens; for the fracture of thin J specimens a new model should be developed based upon a shearing process.

ACKNOWLEDGEMENTS

This work was supported by the Office of Naval Research. The authors wish to express their appreciation to Dr. James Sprague for assistance with the stereographic measurements and to Blaine Menke and Allen Hiser for their help in the J-integral tests.

REFERENCES

Broek, D. (1974). Engineering Fracture Mechanics, Vol. 6, 173-181.

Burdekin, F. M., and D. E. Stone (1966). J. Strain Analysis, Vol. 1 (No. 2), 145-153.

Ernst, H., P. C. Paris, and T. D. Landes, Estimation of J-integral and Tearing Modulus from a Single Specimen Test Record, ASTM 13th National Symposium on Fracture Mechanics, Philadelphia, PA., 16-18 June 1980.

Fields, B. A., and K. J. Miller (1977). Engineering Fracture Mechanics, Vol. 9, 137-146.

Gilmore, C. M., V. Provenzano, F. A. Smidt, Jr., and J. R. Hawthorne (1983). The influence of Thickness and Temperature on Stretched Zone Size in J_{Ic} Tests, Metal Science. Accepted for publication.

Knott, J. F. (1980). Metal Science, Vol. 14, No. 8 & 9, 327-336.

Landes, J. D., and D. E. McCabe (1979). Proceedings of the Third International Conference on Materials, Cambridge, England, 539-547.

Loss, F. J., and G. Angelino (1980). Proceedings International Conference on Analytical and Experimental Fracture Mechanics, Rome, Italy, G. C. Sih, and M. Mirabile (Eds.), 291-304.

Loss, F. J., B. H. Menke, R. A. Gray, J. R. Hawthorne, and H. E. Watson (1981). Effects of Radiation Materials: Tenth Conference, ASTM STP 725. David Kramer, H. R. Brager, and J. S. Perrin (Eds.), 76-91.

Rice, J. R. (1968). J. Appl. Mech., 379-386.

Sprague, J. A. (1980). Automated Data Reduction from Transmission Electron Micrographs, AIME Symposium on Advanced Techniques for the Characterization of Microstructures, Las Vegas, NV.

Thomas, L. E. (1972). Proceedings of the Thirteenth Annual Meeting of the Electron Microscopy Society of America, G. M. Bailey (Ed.), 460-461.

Wells, A. A. The Status of COD in Fracture Mechanics, in Third Canadian Congress of Applied Mechanics, Calgary 1971, 59-77.

ASSESSMENT OF DUCTILE FRACTURE RESISTANCE BASED ON AN ENGINEERING APPROACH

V. Kumar, M. D. German and D. F. Mowbray

General Electric Company, Corporate Research and Development, Schenectady, New York 12301, USA

ABSTRACT

This paper presents the summary of an engineering approach for analyzing fracture behavior in the ductile range and its application to flawed 304 stainless steel piping. Following a brief description of the engineering approach, crack initiation, stable growth and instability studies are presented for several cases of flawed pipe geometry under various loading conditions. The effect of flaw type and size, pipe diameter and length, and applied loading on the failure load and mechanism is investigated. Results are examined against predictions of the net-section stress failure criterion which is an alternate approach for examining the integrity of flawed 304 SS piping.

KEY WORDS

Ductile fracture, 304 stainless steel, J-integral, crack opening displacement, crack initiation, stable growth, instability, elastic-plastic estimation method, tearing modulus, net-section stress criterion.

INTRODUCTION

In ductile materials such as pressure vessel and piping steels large plastic strains are required for crack initiation and propagation. In addition, crack initiation is usually followed by stable growth prior to fast unstable fracture. A comprehensive investigation of methods for treating these problems has been carried out at the General Electric Research and Development Center over the past several years under Electric Power Research Institute sponsorship. This has led to the development of an engineering approach for the analysis of fracture behavior in the ductile range [1]. The objective of this paper will be to present an up-to-date report on the state of development of the engineering approach and show applications to the fracture analysis of 304 stainless steel piping [2].

THE ENGINEERING APPROACH

Recent analytical, experimental and computational studies have suggested that the J-integral and the crack opening displacement (COD), denoted by δ, are the most viable parameters for characterizing crack initiation and stable growth in ductile materials. The rates of change of J and δ with respect to crack length (dJ/da and dδ/da) appear to be the controlling parameters. J and δ as well as their rates have also been found to be related through a simple relationship. Furthermore, resistance curves based on J and δ can be regarded as the material property for a limited range of crack growth provided certain requirements are satisfied. These findings, discussed in References [3 - 8] constitute the concept of J- or COD-controlled growth.

Applications of the J- or COD-based methodology to practical problems of interest require very sophisticated analysis, usually accomplished with the aid of advanced finite element methods. This is normally expensive, time-consuming and requires a high degree of expertise to apply. It clearly represents a major obstacle to the potential user of ductile fracture mechanics concepts. To overcome these difficulties, an engineering approach [1] has been developed for predicting fracture of flawed structural components in the elastic-plastic and fully plastic regimes of material deformation. The approach is simple to use, provides accurate results, and does not require any finite element analysis. The key elements of the approach are: (i) a catalogue of fully plastic solutions for the J-integral, COD and other relevant fracture parameters for various crack configurations, (ii) an estimation procedure which enables easy calculation of elastic-plastic solutions by appropriately combining the elastic and fully plastic solutions, and (iii) a simple method of predicting crack initiation, stable growth and instability by combining the crack driving force estimates with a J- or COD-resistance curve determined from standard specimen tests.

The fully plastic solutions correspond to various planar and a number of common structural crack configurations. The specimens include the center cracked plate, double edge cracked plate, single edge cracked plate (tension and bending) and the compact specimen. These solutions are tabulated for both plane stress and plane strain. The two-dimensional structural configurations include cylinders with axial and circumferential cracks, and a two-dimensional model of a nozzle corner flaw. The most recent analyses have been directed towards 3-dimensional crack problems such as through-wall cracks in cylindrical shells subjected to tension or bending loads, axial surface flaws in internally pressurized cylinders and circumferential surface flaws in cylinders under remote tension [11]. A broad range of values for the crack length, relevant structural dimensions, and strain hardening exponent are considered.

The estimation procedure provides an interpolation over the entire range of elastic-plastic deformation from small-scale yielding to fully yielded conditions including contained plasticity. In this, the elastic and fully plastic solutions are basically superposed to arrive at the elastic-plastic estimates. Several verification studies have been completed in which the estimation procedure solutions are compared with experimental data and full finite element crack solutions. Figure 1, for example, illustrates that for a through-wall flawed cylinder of 304 SS subjected to a tension load the values of the J-integral obtained by the estimation method is in close correspondence with results obtained by an elastic-plastic thin shell finite element analysis. The approach clearly yields accurate predictions of quantities of interest

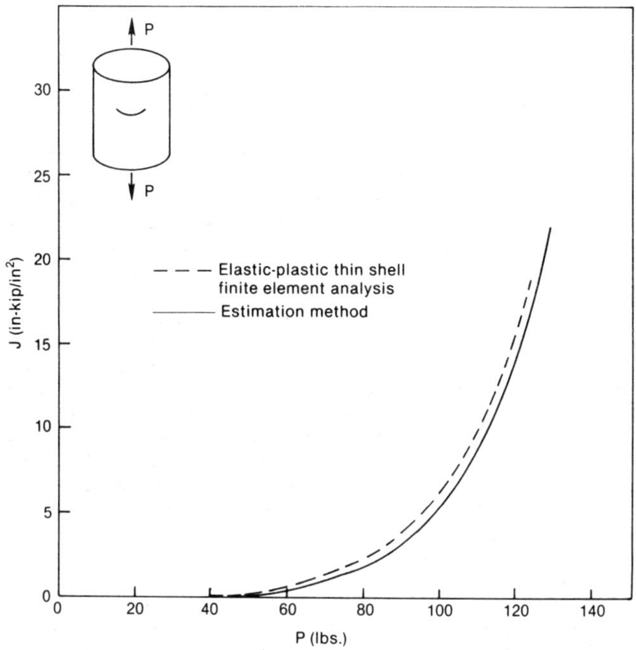

Fig. 1. Comparison of J-integral results for a through-wall flawed cylinder obtained by elastic-plastic estimation method and elastic-plastic thin shell finite element analysis.

which would otherwise have been computed by elaborate finite element calculations. With the aid of this estimation procedure, the tearing modulus can also be easily computed for crack growth instability predictions. The estimation method has recently been extended to include 3-dimensional crack geometries, effect of thermal/residual stresses and alternate stress-strain laws [11].

Analysis of a flawed structure is carried out by constructing the J-integral (or δ_t) crack driving force solutions and superposing on it the experimentally determined material J- (or δ_t-) resistance curve. Assuming that the conditions for J-controlled growth are satisfied, one can predict crack initiation, stable growth, instability and other quantities relevant to overall fracture behavior. Depending upon the application one can also draw upon the so-called J-integral versus tearing modulus stability assessment diagrams or the failure assessment diagrams. The crack driving force diagram allows one to analyze in detail all stages of the fracture process, from initiation to instability. The remaining two are specialized: the stability assessment diagram defines regions of loading where stable and unstable behavior exist and the failure assessment diagram allows one to perform a quick analysis to determine whether or not a fracture problem exists. Several illustrations on the use of such diagrams for analyzing ductile fracture resistance of flawed structures can be found in References [1, 2, 11].

Most recently, studies have been carried out to extend the engineering approach to 3-dimensional surface crack problems in plates and cylinders. This involves developing the fully plastic solutions and the associated elastic-plastic estimation procedure for surface crack problems. Two approaches have been pursued to address the 3-D fully plastic crack problem. The first one employs a 3-D incompressible finite element method, the development of which is discussed in detail in Reference [11]. The second approach is based on a newly developed continuum line-spring model [11]. This new approach is found to give more accurate results than the traditional shell line-spring model [12] especially in the cases of axially flawed cylinders subjected to internal pressure. For further details the reader is referred to Reference [11].

APPLICATIONS TO 304 STAINLESS STEEL PIPING

The approach described above has been applied to the problems of assessing the fracture stability and crack opening response of flawed 304 SS piping [2, 13]. Flawed pipe geometries investigated include the cases of a remote axial loading of a part-through, full circumferential crack and a through-wall, part circumferential crack subjected to remote bending or axial loads. The engineering approach solution method is used to study crack mouth opening profile, crack initiation and propagation and crack growth instability as a function of material flow properties, pipe diameter (2R), flaw size (a) and pipe length (L).

Several crack driving force diagrams in terms of the J-integral versus crack length with applied load (or displacement) as parameters were obtained for 4-inch (0.102 m) and 26-inch (0.660 m) diameter pipes. By superposing the material J-resistance curve on such diagrams predictions were made for crack initiation, extent of stable growth and the instability load. A typical crack driving force diagram is shown in Fig. 2; several other diagrams of this type are given in References [2, 13]. The effects of pipe length and diameter, flaw size and type of applied loading on the crack initiation, stable growth and instability behavior were examined with the aid of such diagrams.

Stability diagrams in terms of J-integral versus tearing modulus T_J were also constructed [2, 13]. An example stability assessment diagram is illustrated in Fig. 3 for a deep crack, $a/t = 1/2$, and 26-inch (0.660 m) diameter pipe. There is a strong effect of pipe length at this depth of crack. Applied J versus T_J curves are shown for various lengths of L/R between 1 and ∞. Superimposed on the diagrams are three lines which establish several regimes of behavior. The two horizontal lines indicate the onset of crack extension (J_{Ic}) and plastic collapse; the vertical line indicates the material resistance to tearing (T_{JR}). J_{Ic} and T_{JR} are indexed at 4000 in-lb/in^2 (700 kNm/m^2) and 1030, respectively. The plastic collapse line is based on a flow stress of 60 ksi (414 MPa) and is configuration dependent (i.e., it depends upon the geometry, crack size and type of loading) and will vary in magnitude from example to example. Referring to Fig. 3, the various regimes of behavior are as follows: (1) Below the J_{Ic} line there is no crack growth or failure

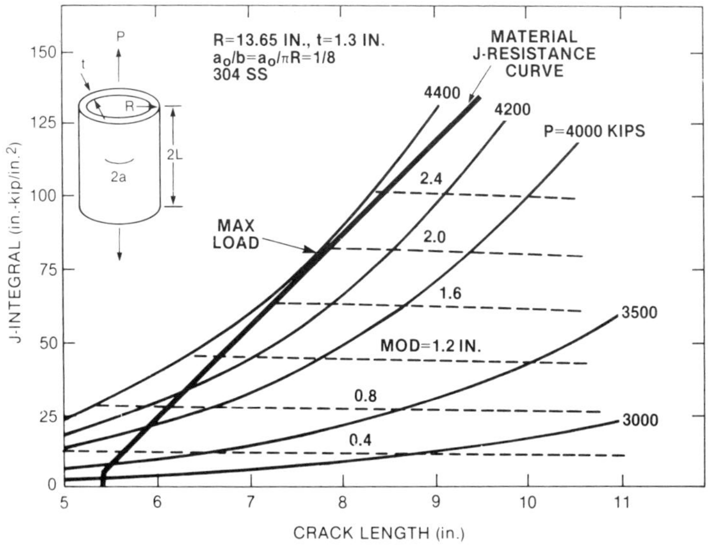

Fig. 2. Crack driving force diagram for a 26-inch (0.660 m) diameter pipe in tension containing a through-wall crack (1 in.-kip/in.^{2} = 175 kNm/m^2, 1 in. = 0.0254 m).

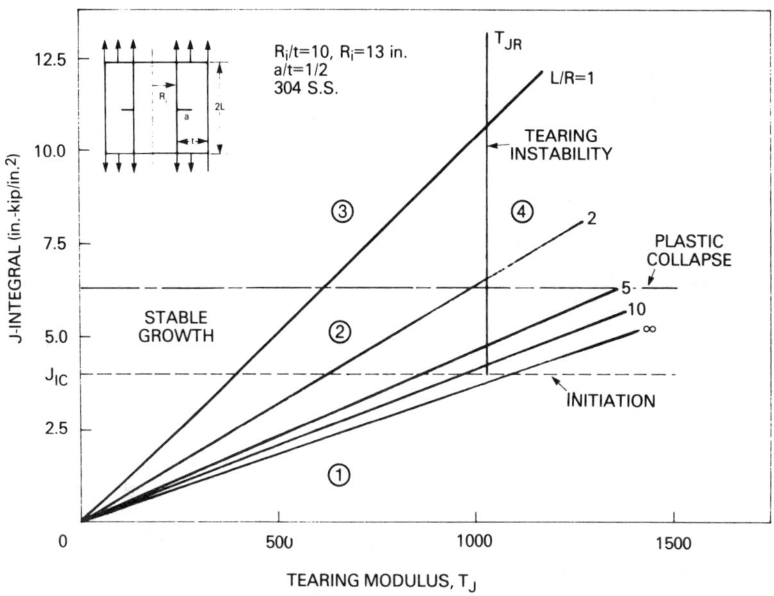

Fig. 3. Stability assessment diagram for a part-through, complete circumferentially cracked pipe in tension with a 26-inch (0.660 m) diameter and a/t = 1/2 (1 in.-kip/in.^{2} = 175 kNm/m^2).

possibility; (2) Stable crack growth is defined in the region between the J_{Ic} and plastic collapse lines and to the left of the T_{JR} line; (3) Above the plastic collapse line and to the left of the T_{JR} line failure occurs by plastic collapse; (4) To the right of the T_{JR} line and above the J_{Ic} line failure occurs by rapid crack extension. A similar diagram for the case of a through-wall flawed cylinder subjected to remote tension loading is shown in Fig. 4. The effect of pipe length on various failure mechanisms is clearly indicated.

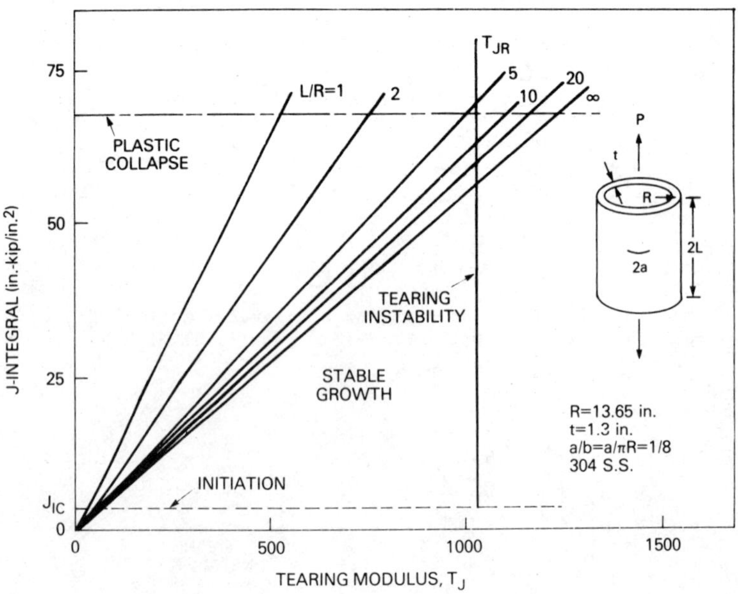

Fig. 4. Stability assessment diagram for a 26-inch (0.660 m) diameter, through-wall flawed pipe in tension with a/b = 1/8 (1 in.-kip/in.^{2} = 175 kNm/m^2).

Fracture analysis methods for type-304 stainless steel piping currently proposed for inclusion in the ASME code employ a net-section stress fracture criterion. This criterion assumes that the failure occurs when the average stress on the unflawed section reaches the collapse value as defined by limit loads. Our investigations show that most of the initiation load predictions for part-through flaws fall on or above the plastic collapse line indicating the net-section stress criterion to be conservative. For a through-wall cracked pipe in a tensile or a bending field the results show a strong dependence on pipe diameter. The failure load associated with tearing instability are in general equal to or higher than the net-section collapse load; this shows that the net-section stress criterion would be conservative for pipe size - loading combinations investigated here. The above observations are based on a 60,000 psi (414 MPa) flow stress of the 304 SS pipe material. Choosing the flow stress at a lower value would of course serve to introduce further conservatism in the net-section stress criterion. A typical result comparing the net-section stress criterion against failure predictions which include crack initiation, stable growth and instability events is shown in Fig. 5. Although the net-section stress criterion gives conservative predictions for the failure load, it cannot take into account the effect of flaw type and size, pipe diameter and length and applied loading. For further details the reader is referred to References [2, 13].

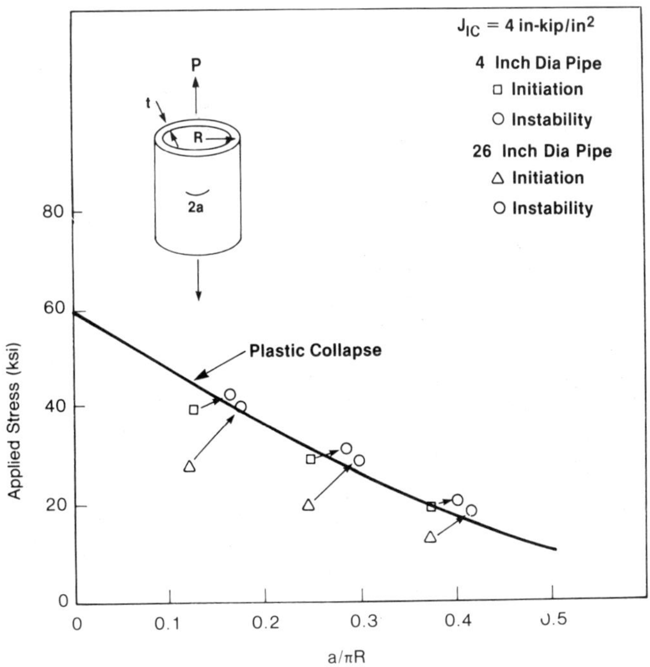

Fig. 5. Comparison of failure load (with stable growth accounted for) by the J_R curve approach against the net-section stress criterion for a through-wall flawed pipe in tension (1 ksi = 6.895 MPa, 1 in.-kip/in.^{2} = 175 kNm/m^2, 1 in. = 0.0254 m).

ACKNOWLEDGMENT

This work was supported by the Electric Power Research Institute (EPRI) under the projects RP1237-1 and T118-8 administered by Dr. D. M. Norris, Jr.

REFERENCES

[1] Kumar, V., German, M. D., and Shih, C. F., "An Engineering Approach for Elastic-Plastic Fracture Analysis," EPRI NP-1931, Topical Report,

Research Project 1237-1, General Electric Company, Schenectady, NY, July 1981.

[2] German, M. D., Andrews, W. R., Kumar, V., Shih, C. F., deLorenzi, H. G., and Mowbray, D. F., "Elastic-Plastic Fracture Analysis of Flawed Stainless Steel Pipes," Final Report EPRI NP-2608-LD, Research Project T118-8, General Electric Company, Schenectady, NY, September 1982.

[3] Rice, J. R., "A Path Independent Integral and the Approximate Analysis of Strain Concentration by Notches and Cracks," J. Appl. Mech., Vol. 35, 1968, pp. 379-386.

[4] Hutchinson, J. W., "Singular Behavior at End of Tensile Crack in Hardening Material," J. Mech. Phys. Solids, Vol. 16, No. 1, 1968, pp. 13-31.

[5] Rice, J. R. and Rosengren, G. F., "Plane Strain Deformation Near Crack Tip in Power-Law Hardening Material," J. Mech. Phys. Solids, Vol. 16, No. 1, 1968, pp. 1-12.

[6] Begley, J. A. and Landes, J. D., "The J-Integral as a Fracture Criterion in Fracture Toughness Testing," in Fracture Toughness ASTM Special Technical Publication 514, 1972, pp. 1-23, 24-29.

[7] Shih, C. F. Andrews, W. R., deLorenzi, H. G., German, M. D., Van Stone, R. H., and Mowbray, D. F., "Methodology for Plastic Fracture," Final Report EPRI NP-1735, Research Project 601-2, General Electric Company, Schenectady, NY, March 1981.

[8] Kanninen, M. F., et al., "Development of a Plastic Fracture Methodology," Final Report EPRI NP-1734, Research Project 601-1, Battelle's Columbus Laboratories, March 1981.

[9] Paris, P. C., Tada, H., Zahoor, A., and Ernst, H., "The Theory of Instability of the Tearing Mode of Elastic-Plastic Crack Growth," in Elastic-Plastic Fracture, ASTM Special Technical Publication 668, 1979, pp. 5-36 and 251-265.

[10] Hutchinson, J. W. and Paris, P. C., "Stability Analysis of J-Controlled Crack Growth," in Elastic-Plastic Fracture, ASTM Special Technical Publication 668, 1979, pp. 37-64.

[11] Kumar, V., et al., "Estimation Technique for the Prediction of Elastic-Plastic Fracture of Structural Components of Nuclear Systems," Combined Seventh and Eighth Semiannual Report to EPRI, Contract No. RP1237-1, General Electric Company, Schenectady, NY, February 1, 1982 - January 31, 1983.

[12] Rice, J. R., "The Line-Spring Model for Surface Flaws," in The Crack Surface: Physical Problems and Computational Solutions (edited by J. L. Swedlow), ASTM, NY, 1972.

[13] German, M. D., Kumar, V., and deLorenzi, H. G., "Analysis of Through Cracks in 304 SS Piping and Comparisons with Experimental Results," presented at ASME Winter Meeting, Phoenix, AZ, November 19, 1982.

FATIGUE CRACK GROWTH AT HIGH STRAIN AMPLITUDES

C. Wüthrich and W. Hoffelner

Brown Boveri Research Center, CH-5405 Baden, Switzerland

ABSTRACT

In order to correlate fatigue crack growth data measured at high strain amplitudes, the linear-elastic stress intensity factor has to be replaced by another load parameter. It is shown that a path independent integral Z, similar to the J-integral, is the appropriate parameter for materials for which the stress increments are a unique function of the strain increments. The Z parameter was used to interprete fatigue crack growth data measured on three steels over a wide range of strain amplitudes. Experimental evidence shows that the range where the present concept may be applied is limited but extends far beyond the range of the linear-elastic stress intensity factor.

KEYWORDS

Fatigue crack growth; stress intensity factor; J-integral; stress-strain hysteresis.

INTRODUCTION

The limitations of linear-elastic fracture mechanics have been investigated in detail in the field of fracture toughness testing. In order to account for the nonlinear effects, elastic-plastic parameters like the J-integral and the crack opening displacement have been developed and successfully applied to describe initiation of crack growth and stable crack growth. As far as fatigue crack growth is concerned, the limitations of linear-elastic fracture mechanics are not as clear. Even the parameters governing fatigue crack growth at high strain amplitudes where nonlinear effects become predominant are not well established. In order to overcome the deficiencies of the ΔK concept in the regime of high amplitudes, a quantity ΔJ was introduced by Dowling and Begley (1976) on an empirical basis. This quantity has been used successfully to correlate fatigue crack growth data (Dowling and Begley, 1976; Huang and Pelloux, 1980). The theoretical basis of the experimental procedure, however, has been unclear. It is the aim of the present paper to show under what condiditions the cyclic deformation at the crack tip may be related to a path independent integral similar to the well known J-integral. If these conditions are fulfilled, the method of Dowling and Begley (1976) gives a correct description of fatigue crack growth. Experimental investigations will show if the basic assumptions of the theory are valid. Fatigue crack growth measurements on different materials and specimen types will be used to find out if the path independent integral is an appropriate load parameter.

DEFINITION OF A CYCLIC J-INTEGRAL

The J-integral, introduced by Rice (1968) to describe the behaviour of cracks under monotonic loading, cannot be applied if unloading occurs. The reason is that the J-integral is based on the theory of deformation plasticity which represents plastic deformation by nonlinear elastic behaviour. In the case of cyclic loading a quantity similar to the J-integral may be introduced if the stress, σ_{ij}, the strain, ε_{ij}, the traction vector, T_i, and the displacement, u_i, are replaced by their increments Δ, related to the minimum or maximum of the load wave (Lamba, 1975; Wüthrich, 1982). This quantity has the form

$$Z = \int_{\Gamma} (V\, dy - \Delta T_i \frac{\partial \Delta u_i}{\partial x} ds) \tag{1}$$

with

$$V = \int_{o}^{\Delta\varepsilon} \Delta\sigma_{ij}\, d\Delta\varepsilon_{ij} , \tag{2}$$

where the x and y axes are parallel and normal to the crack surface, respectively. The quantity ds is an element of the integration path Γ. For the moment it is assumed that the crack stays open during the load cycle. The quantity Z has the character of a path independent integral if the stress increments are given by a unique function of the strain increments, i.e.:

$$\Delta\sigma = f\,(\Delta\varepsilon) . \tag{3}$$

Under the assumption that this relation is valid for all strain ranges, the function f is identical to the cyclic stress-strain curve. The validity of this assumption has to be tested by cyclic stress-strain measurements. Only practical experience will show if the Z parameter correlates fatigue crack growth data even in cases where significant deviations of the ideal material behaviour occur. In the rest of the paper Z designates the maximum value of the path independent integral and then the ranges of stress and strain close to the crack tip are determined by this quantity alone. In the case that Eq. (3) can be represented by a power law, the ranges of stress and strain are given by a field of the HRR type (McClintock, 1971). The experimental evaluation of Z is completly identical to the method introduced by Dowling and Begley (1976) for the determination of ΔJ. However, Z is not a difference of two J values and the designation ΔJ commonly used in literature is not correct. For linear-elastic materials the relation between the equivalent values of Z, K and J is given by

$$Z = \frac{(\Delta K)^2}{E} = (\sqrt{J_{max}} - \sqrt{J_{min}})^2 , \tag{4}$$

with Young's modulus E. The range of specimen or component size for which the cyclic path independent integral Z may be used is limited and has to be determined by experimental or theoretical investigations. This range includes the linear-elastic regime where fatigue crack growth is controlled by ΔK.

EXPERIMENTAL INVESTIGATIONS

Experimental investigations were performed on three different steels: A heat resistent low alloy steel (13CrMo44), a martensitic stainless steel (X20CrMoV121) and an austenitic stainless steel (X10CrNiTi189). The specimen types used for crack propagation studies were double cantilever beam specimens (DCB, dimensions: 60 x 90 x 10 mm), compact tension specimens (CT, dimensions: 60 x 61.25 x 25 mm) and center cracked specimens (CC, cross section: 4.5 x 11.5 mm).

The existence of an unique relation between the stress and strain increment is the basic requirement for the application of a path independent integral in fatigue crack growth studies. Therefore, the saturated stress-strain hysteresis loops were measured for the three materials using the incremental step technique. The comparison of the hysteresis cycles determined at different strain ranges showed that the condition mentioned above is not exactly fulfilled, but may be taken as a reasonable approximation. The stress-strain hysteresis loops of the martensitic stainless steel are given in Fig. 1. Figure 2 shows that the increasing parts of the loops do not perfectly follow the same line. Similar deviations of the ideal cyclic behaviour given by Eq. (3) were observed for the other two steels. Data for the austenitic steel were already reported in a preceding paper (Wüthrich and Hoffelner, 1981). The evaluation of the crack growth measurements was done under the assuption that the Z-integral is a reasonable parameter even if the assumptions of the theory are not perfectly fulfilled.

In a first set of experiments, DCB specimens were tested under load and displacement control. If only the positive part of the load wave was taken into account to calculate ΔK, the crack propagation rates measured under load and displacement control showed significant differences at high ΔK values. A better coincidence was obtained if the compressive part of the load wave was included in the calculation of ΔK. This procedure, however, is not consistent with the concept of linear elastic fracture mechanics. Therefore, a nonlinear crack parameter has to be introduced. During the rest of the paper only those crack propagation data measured on DCB specimens will be used which could be related to ΔK unambiguously.

To investigate crack propagation at high amplitudes in more detail, experiments on CT specimens under displacement control were performed. For the determination of Z the analogy to the J-integral and the well known calibration for the CT specimen was used (ASTM E-813, 1982). According the procedure introduced by Dowling and Begley (1976), only that part of the load displacement curve was taken into account which corresponds to an open crack. The data measured on DCB specimens and evaluated with the ΔK parameter agree reasonably well with the data measured on CT specimens and evaluated with Z (Figs. 3 - 5). In order to compare the results, the ΔK values were converted to Z using Eq. (4). In the case of the martensitic stainless steel the diagram of propagation rate versus Z shows an inflection point at high Z values. At the inflection point the value of Z agrees approximately with the value of the fracture toughness J_{IC}. In this range a change of fracture mechanism seems to take place.

Bending is predominant in CT specimens, whereas center cracked specimens are loaded in pure tension. The calculation of Z for center cracked specimens was done using the analogy to J and results reported in literature (Kumar, German and Shih, 1981). The calculation of J or Z, respectively, is based on the assumption of a power law hardening material. In the case of the J-integral the stress-strain curve determined in a tensile test has to be approximated by a power law. For the calculation of Z, on the other hand, the increasing parts of the stress-strain hysteresis cycles were taken as basis for the evaluation. Since the assumption of a unique relation between stress and strain increments is not fulfilled exactly for the materials investigated, averaged curves were defined. The parameterization was done in the form of a Ramberg-Osgood relation separating linear and nonlinear contributions. The experiments on the center cracked specimens were done under displacement control keeping the displacement range Δx constant (gage length: 10 mm). The crack growth data agree in almost all cases with the results determined with the two other types of specimens (Figs. 3 - 5). Only for the austenitic stainless steel at extremely high amplitudes was a significant deviation from the expected behaviour observed. Obviously the range of validity of the Z-integral was exceeded in that case.

DISCUSSION

The present investigation shows that the concept of path independent integrals is not confined to uniaxial loading. The method introduced by Dowling and Begley (1976) may, therefore, be put on a theoretical basis. In practice several restrictions have to be taken into account. The assumption of a function relating stress and strain increments in an unambiguous way will never be fulfilled exactly. Nevertheless, it seems to be a reasonable approximation for many materials. Deviations observed in the range of very high strain amplitudes do not play a significant role. Such high amplitudes occur only in an area very close to the crack tip. In this area material separation takes place and the continuum description breaks down anyway. The assumption that the crack remains open during the load cycle will almost never be valid. The Z-integral may be still be applicable if the cut-off procedure proposed in literature is used (Dowling and Begley, 1976). The range of validity of the Z-integral has not yet been investigated in detail. It might be expected that the minimum dimensions of the specimens are determined by similar rules as for the J-integral (ASTM E-813). It is well known from literature that the size requirements for specimens loaded in tension are more restrictive than for bending specimens (Shih, de Lorenzi and Andrews, 1979). This is in agreement with the results of the present investigation where significant deviations for the tensile specimen have been observed. The introduction of a nonlinear crack parameter is only meaningfull if it gives the same results as the linear-elastic parameter ΔK in the regime of small amplitudes. This requirement was fulfilled for all materials of the present investigation. The introduction of the Z-integral extends considerably the range in which crack propagation rates can be measured, even if the restrictions for the specimen dimensions are not completely known. It can be concluded that the Z parameter is an appropriate parameter to describe fatigue crack growth at high amplitudes.

ACKNOWLEDGEMENTS

Thanks are due to Dr. A. Atrens for useful comments and to Mr. H.-U. Küng for experimental assistance.

REFERENCES

ASTM E-399 (1982). Annual Book of ASTM Standards, Philadelphia.
ASTM E-813 (1982). Annual Book of ASTM Standards, Philadelphia.
Dowling, N. E., and J. A. Begley (1976). Mechanics of Crack Growth, ASTM STP 590, ASTM, Philadelphia, pp. 82-103.
Huang, J. S., and R. M. Pelloux (1980). Met. Trans., 11A, 899-904.
Kumar, V., M. D. German and C. F. Shih (1981). An Engineering Approach for Elastic-Plastic Fracture, Report NP-1931, Electric Power Research Institute, Palo Alto.
Lamba, H. S. (1975). Eng. Fract. Mech., 7, 693-703.
McClintock, F. A. (1971). In H. Liebowitz (Ed.), Fracture - An Advanced Treatise, Vol. III, Academic Press, New York and London, pp. 47-225.
Rice, J. R. (1968). J. Appl. Mech., 35, 379-386.
Shih, C. F., H. G. de Lorenzi and W. R. Andrews (1979). Mechanics of Crack Growth, ASTM STP 668, ASTM, Philadelphia, pp. 65-120.
Wüthrich, C. (1982). Int. J. Fract., 20, R35-37.
Wüthrich, C., and W. Hoffelner (1981). 13. Sitzung des Arbeitskreises Bruchvorgänge, Deutscher Verband für Materialprüfung DVM, Berlin, pp. 214-221.

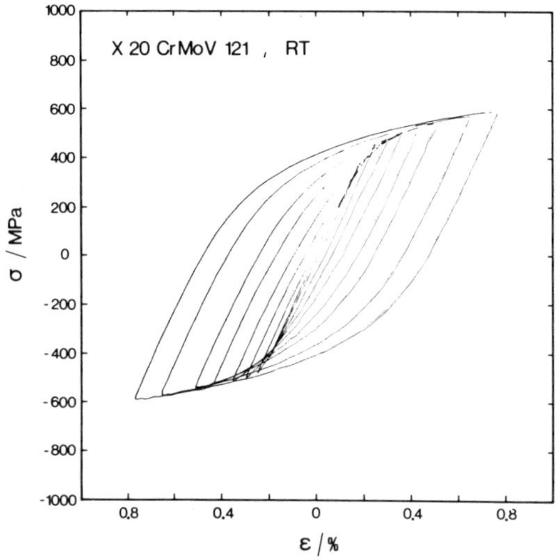

Fig. 1. Cyclic deformation of X20CrMoV121. Saturated stress strain cycles.

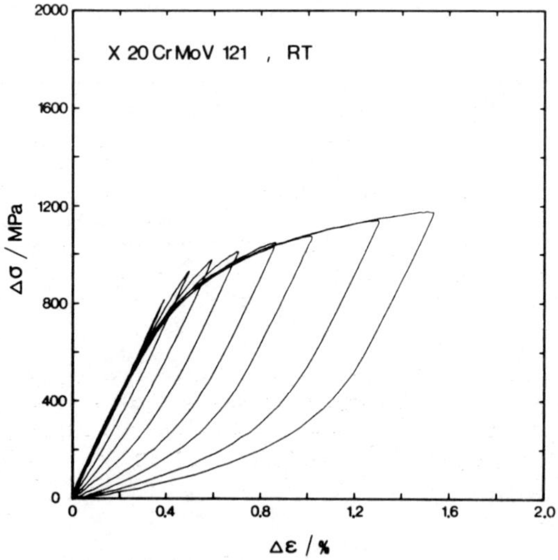

Fig. 2. Cyclic deformation of X20CrMoV121. Stress strain cycles related to the load minimum.

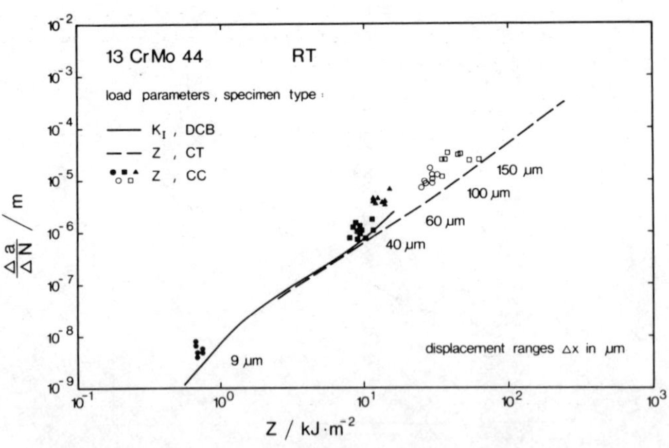

Fig. 3. Crack propagation per load cycle versus Z for the steel 13CrMo44.

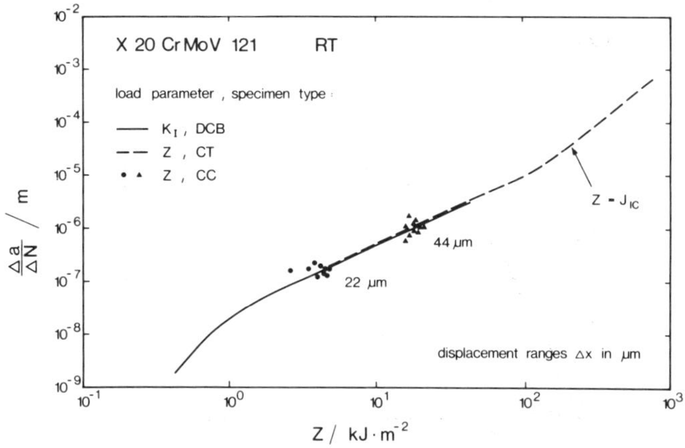

Fig. 4. Crack propagation per load cycle versus Z for the steel X20CrMoV121.

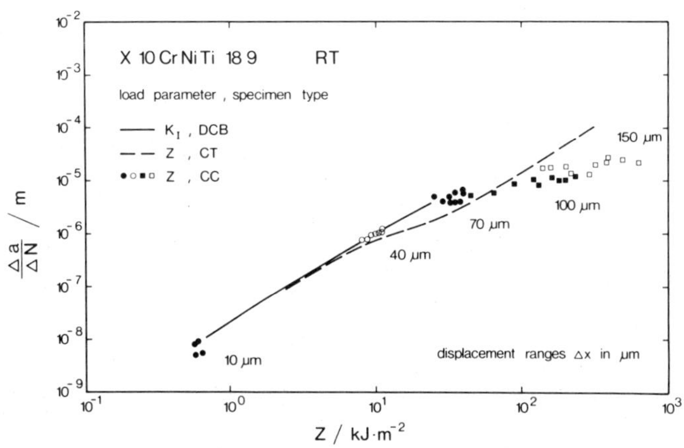

Fig. 5. Crack propagation per load cycle versus Z for the steel X10CrNiTi189.

COMPARATIVE STUDY OF CRACK RESISTANCE PROPERTIES OF THREE DIFFERENTLY ALLOYED HSLA STEELS

A. Radović, S. Sedmak and Lj. Nedeljković

Faculty of Technology and Metallurgy, University of Belgrade, Yugoslavia

ABSTRACT

Alloy design of three quenched and tempered HSLA steels (differently produced to 700 MPa yield strength) is analysed in terms of alloying, microalloying and cleanliness levels and the effect of each on the steel crack resistance properties measured by Charpy V and J integral values is assessed.

KEYWORDS

Quenched and tempered HSLA steels; alloy design; microalloying; cleanliness; crack resistance; fracture toughness; J_{Ic}; J_R curve.

INTRODUCTION

Quenched and tempered grades among HSLA steels have the best combination of high mechanical properties and good toughness. They are applied because of a relatively higher production cost beyond the strength range attainable by controlled rolling, the upper limit of which lies at about 550 MPa yield strength. In this way quenched and tempered HSLA steels close the gap between controlled rolled and high-alloyed high-strength steels. After a great impetus in early seventies, the production and application of quenched and tempered HSLA steels experienced a substantial fall because of decreasing demand in the recession period from the mid-seventies on, and because of a number of serious failures caused by stress corrosion cracking in some particular structures, such as storage tanks (Suzuki, Sakai, Moriyama, 1981). On the grounds of their superior mechanical properties they are now regularly used for the construction of pressure vessels, penstocks, bridges, earth-moving and general-purpose machinery, storage tanks etc. The extension of their application in line pipe production is expected in the near future (Rothwell, Cooke, 1981; Paxton, 1981).

Existing standards usually specify for quenched and tempered, as for other HSLA steels, only the basic mechanical properties, leaving the choice of alloying and production routes, as well as the specification of additional mechanical characteristics, important for specific application, to the customer and steel producer to agree on. This has resulted in a great variety of different steel compositions belonging to the same quenched and tempered quality class. In this situation, knowledge of the relationship between chemical composition and mechanical properties is essential for selection of the right steel in any particular application.

The object of this investigation is to assess the effect of different alloying, microalloying and cleanliness levels of weldable quenched and tempered steels of 700 MPa yield strength class, made by different producer and production routes, on the nonstandard mechanical features, with special reference to crack resistance properties.

MATERIALS

Three steels of different chemical composition, as shown in Table 1, but of the same strength class, quenched and tempered to a minimum yield strength of 700 MPa, are studied.

TABLE 1 Chemical Composition of the Studied Steels

Steel	C %	Alloying %					Microalloying %					Cleanliness ppm		
		Si	Mn	Cr	Ni	Mo	V	Zr	B	Al	N	P	S	O
A	0.16	0.61	0.94	0.77		0.34		0.05		0.043	0.0056	120	120	62
B	0.10	0.28	0.89	0.46	1.00	0.48	0.025		0.0016	0.066	0.0071	150	50	22
C	0.11	0.14	0.25	1.26	2.84	0.29	0.07			0.060	0.0110	100	160	110

The first of them, designated A, is supplied commercially, as a West German N-A-XTRA 70, 20 mm thick steel plate. This steel is the least defined material examined in this investigation because of the lack of precise data on the steelmaking route and thermomechanical treatment. Steel B is produced in Japan under the commercial name of SUMITEN 80P and is intended for penstock construction. The steel is made in BOF, vacuum degassed and cast with a shielded pouring stream into 25-ton ingots. After rolling, the plates are quenched and tempered on roller quenching equipment. Steel C is produced in Yugoslavia under the commercial name of NIONICRAL 70. It is made in an electric arc furnace, argon bubbled in a ladle and cast in an open pouring stream into 8-ton ingots. Rolling direction was changed 90^0 after roughing with the aim of minimizing plate anisotropy. Plate quenching is carried out in a quench.

MECHANICAL PROPERTIES

Basic mechanical properties of the steels are given in Table 2.

TABLE 2 Tensile Properties of the Studied Steels

Steel	Y.S. MPa	U.T.S. MPa	Elongation %	Contraction %	Y.S./U.T.S. ratio
A	739	784	25	58	0.94
B	785	834	23	70	0.94
C	735	785	21.5	72	0.94

TABLE 3 Charpy V Impact Energy, Ductile - Brittle and DW Nil - Ductility Transition Temperature

Steel	Plate thick-ness mm	Spe-cimen posi-tion	CVN (J) at temperature (^{0}C)									Ductile-brittle transition temperature (^{0}C)			
			20	0	-20	-40	-60	-80	-100	-120	-140	$vT_{3.6}$	vTE(½)	vTS	DW -NDT
A	20	L		112	111	85	57	41	26			-95	-65		-48
B	47	$\frac{t}{4}$ L		217	214	204	195	193	53	26		-108	-92	-94	-69
		T		137	137	132	76	56	39	31	16	-101	-66	-74	
C	45	$\frac{t}{3}$ L	112	107	107	102	98	70							
		T	99	100	97	91	83	61							
	18	L	129		128	125	119	116	109	95	59	<-140	-130		-112
		T	87		84	87	84	77							

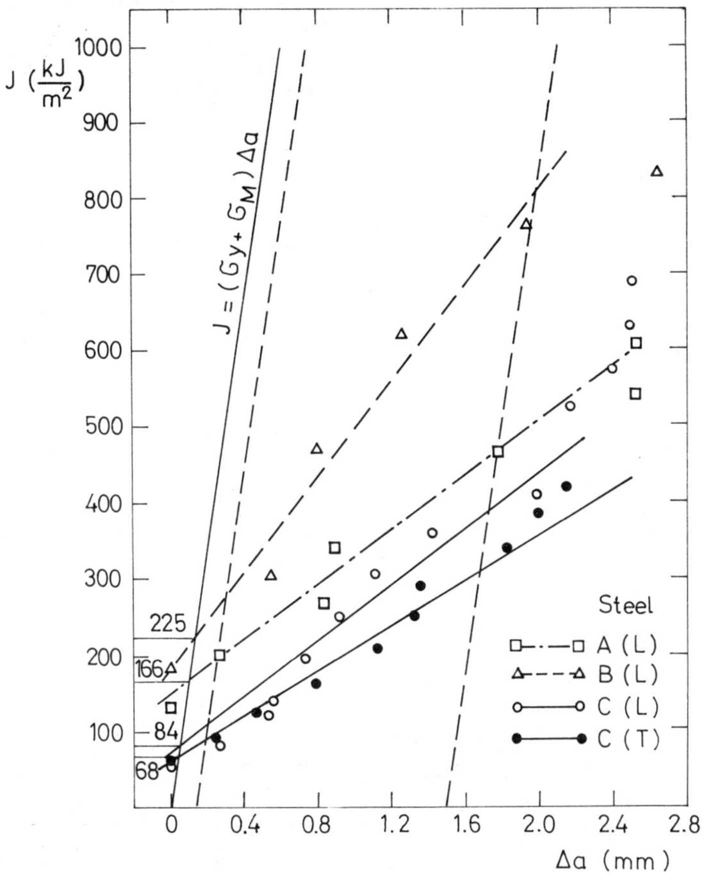

Fig. 1. J integral versus Δa plots for J_{Ic} evaluations

TABLE 4 Crack Initiation Parameters

Steel	Plate thickness	Specimen					Critical J integral value	Fracture toughness
		Orientation	Thickness	Width	Crack length	Crack ratio		
			B	W	a_0	a_0/W	J_{Ic}	K_{Ic}*
	mm		mm	mm	mm		kN/m	MPa $\sqrt{m}$
A	20	L	18.7	40.1	21.05	0.525	166	185
B	47	L	22.0	45.2	28.02	0.620	225	215
C	45	T	22.0	44.0	22.5	0.512	68	118
C	45	L	22.0	44.0	24.02	0.546	84	132

* Calculated as $K_{Ic}^2 = E \cdot J_{Ic}$

The Charpy V energies in longitudinal (L) and transverse (T) direction from 20°C to -140°C, together with the corresponding transition temperatures evaluated by various criteria and NDT temperature determined by the drop weight test are given in Table 3.

Crack resistance properties were studied on precracked three-point-bend specimens, tested under load-point displacement control on an Instron 1332 fatigue testing machine, using the partial unloading (30%) procedure. The results of J_{Ic} integral determinations in accordance with ASTM E 813 are shown in Fig. 1. Further testing details, together with calculated K_{Ic} values, are given in Table 4. In the same tests, crack growth was followed beyond 1.5 mm and this enabled the J_R curve to be drawn (Fig. 2), according to the Sumpter-Turner (1976) approach.

DISCUSSION

The results of mechanical property determination make it possible to assess performance of the steels studied by different criteria important for the structural design approach. The simplest of them is the traditional yield strength criterion. Steel B quenched and tempered to the highest yield strength without substantial loss in plasticity (Table 2) could be regarded as the best structural material for this design approach.

The transition temperature design approach relies on Charpy V energies and transition temperatures. The results in Table 3 show that steel B exhibits by far the highest notch impact toughness in the ductile, upper shelf energy range, while steel C has the lowest transition temperature and the highest notch impact toughness in the brittle fracture range. Accordingly steel B continues to be the best structural material in brittle-failure safe design, too, except at very low temperatures, at which steel C becomes superior. An additional feature of steel C significant for design purposes is the absence of Charpy V anisotropy in 45 mm thick plate.

The next more flaw-tolerant design approach is based on fracture toughness as a criterion for the determination of critical crack size. Assessment of the steels studied by this criterion can be based on the J integral values in Fig. 1 and 2 and Table 4. Putting aside the methodological problems concerned with the validity limitations of the applied procedure for J integral determination, the J_{Ic} values and J_R curves can be taken as a rough measure for the beginning of stable crack growth and for the crack-extension resistance resp. They can at the very least be taken as a relative measure for

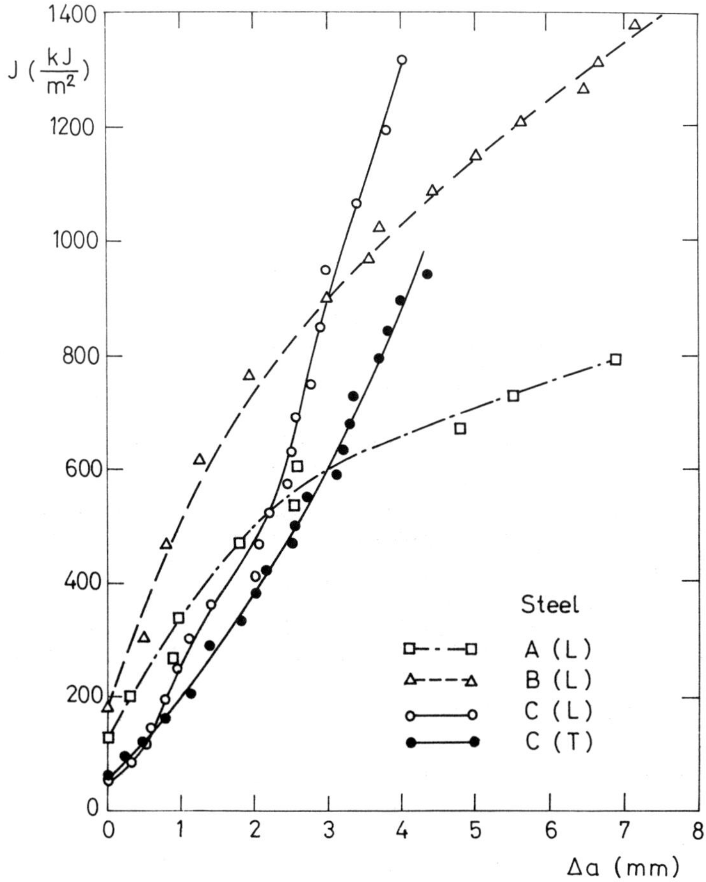

Fig. 2. J_R resistance curves

the distinction between the crack resistance behaviour of the studied steels.

According to J_{Ic} and the calculated K_{Ic} values, steel B is again rated as by far the best structural material, being able to tolerate the presence of the largest flaws in the structure. Due to the fact that fracture toughness correlates well with upper shelf Charpy V energy (Rolfe, Novak, 1970), this could already be anticipated from the high Charpy V upper shelf energies of steel B.

By contrast, the great difference between the J_{Ic} and K_{Ic} values of steel A and steel C is surprising, bearing in mind their similar Charpy V upper shelf energies. J_R curves reveal further difference in the crack resistance of steel C from both the other steels. After the early start of crack growth, at the lowest J_{Ic} values, the J_R curves of steel C continue to rise steeply (Fig. 2), intersecting first the J_R curve of steel A, and finally (one of them) even that of steel B. Thus increasing resistance to the crack extension of steel C counterbalances its initial low fracture toughness, suggesting at the same time its good crack-arresting properties.

An explanation for the observed discrepancies in the crack-resistance properties of the three steels should obviously be sought in their different metallurgical features resulting from different alloy design concepts.

For better elucidation of the alloy design concept of each steel, the elements in Table 1 are divided into alloying, microalloying and cleanliness groups, a separate place being given to carbon, and then based on this, the corresponding factors in Table 5 are calculated. The final result of microalloying greatly depends on the efficiency of thermomechanical treatment, and the cleanliness level on the type, amount and success of secondary refining.

TABLE 5 Chemical Composition Factors

Steel	Alloying*	Microalloying (V + Zr + B + Al + N) %	Cleanliness (S + O) ppm
A	12.7	0.0986	182
B	15.3	0.0997	76
C	20.9	0.141	270

* The factor is a sum of the concentrations of alloying elements (Table 1) multiplied by alloying values, specified by DIN 17006 (4 for Si, Mn, Cr and Ni, 10 for Mo).

The microalloying factor is obtained by adding up the concentrations of all microalloying additions, and cleanliness factor in view of the similar phosphorus content in all three steels, as the sum of their sulphur and oxygen content.

As can be seen from the data in Tables 1 and 5, the alloy design concept of steel A, obviously aimed at cheap alloying, relies on a higher carbon content, the lowest alloying, simple microalloying by Zr with some potential sulphide shape control and with no paricular emphasis on cleanliness level.

The alloy design concept of steel B is based on a low carbon content, moderate alloying additions and complex microalloying, clearly aimed at efficient thermomechanical and subsequent heat treatment. Special emphasis is laid on secondary refining, which results in an exceptionally high cleanliness level, 2.5-3 times higher than in the other two steels.

The alloy design concept of steel C is primarily based on low carbon content and higher alloying especially with an unusually high nickel addition, without much emphasis on cleanliness, but with good microalloying potential. The main reason for this alloy design approach seems to be to make the steel as insensitive as possible to the degree and efficiency of thermomechanical and ladle treatment.

The highest Charpy V upper shelf energy of steel B can now for the most part be explained by its exceptionally high cleanliness level. Similar upper shelf energies in steel A and C at similar cleanliness levels are further confirmation of this ratio. In agreement with other observations (Priest, 1977), nickel does not appreciably affect upper shelf Charpy V energy.

The best fracture toughness of steel B is a consequence not only of high cleanliness but presumably of highly effective thermomechanical and heat treatment. The lower fracture toughness of steel C, compared to steel A, might be accounted for by less effective thermomechanical treatment. On the other hand, the higher nickel concentration in steel C increased its resistance to crack extension and, as expected, its low temperature notch impact energy.

CONCLUSION

The results of notch impact and fracture toughness determination for three differently alloy designed, quenched and tempered HSLA steels to min. 700 MPa yield strength show that cleanliness and thermomechanical treatment, and not alloying, are decisive for good notch impact upper shelf energy and room temperature fracture toughness measured by J_{Ic} values. Alloying with nickel does not affect the Charpy V upper shelf energy, but increases resistance to the crack extension and low temperature notch impact energy.

REFERENCES

Paxton, H. V. (1981). The metallurgy of steels for large diameter linepipe. In R. Q. Barr (Ed.) Alloys for Eighties. Climax Molybdenum, Greenwich, Connecticut. pp. 185-213.

Priest, A. H. (1977). Influence of strain rate and temperature on the fracture toughness and tensile properties of several metallic materials. In M. G. Dawes (Ed.) Dynamic Fracture Toughness. The Welding Institute, Abington Hall, Cambridge. pp. 95-111.

Rolfe, S. T., and S. R. Novak (1970). Slow-bend K_{Ic} testing of medium-strength high toughness steels. In Review of Developments in Plain Strain Fracture Toughness Testing. ASTM STP 463. pp. 124-129.

Rothwell, A. B. and R. J. Cooke (1981). Linepipe requirement in the eighties. In R. Q. Barr (Ed.) Alloys for the Eighties, Climax Molybdenum, Greenwich, Connecticut. pp. 171-185.

Sumpter, J. D. G., C. E. Turner (1976). Method for laboratory determination of J_c. In Cracks and Fracture. ASTM STP 601. pp. 3-18.

Suzuki, H., T. Sakai and K. Moriyama (1981). The current general uses of QT steels and Nippon Steel Corp.´s HT steels. Trans. Iron Steel Inst. Japan, 21, pp. 742-755.

NEW APPROACHES IN FRACTURE MECHANICS

K. B. Broberg

Division of Solid Mechanics, Lund Institute of Technology, S-220 07 Lund, Sweden

ABSTRACT

Whereas linear fracture mechanics of monotone loading is easy to apply, this is not the case with non-linear fracture mechanics, stable crack growth being present. Currently used methods are surprisingly weakly founded, experimentally and theoretically, and their region of applicability is extremely small, since most structures are exposed to repeated, not to monotone loading. A simple procedure is suggested by which failure loads can be predicted for moderately low numbers of major load applications. It involves tests under cyclic loading, but is based on the J-integral at the first load application.

KEYWORDS

Fracture mechanics; Low-cycle fatigue; J-integral methods; Design criteria; Fracture tests.

INTRODUCTION: FRACTURE MECHANICS OF MONOTONE LOADING

The J-integral criterion is widely used for predicting the onset of crack growth, whereas the J-resistance curve seems to be the preferred instrument for prediction of fracture. The theoretical foundations for the methods are *near-tip autonomy* and *path-independence of the J-integral outside the region of autonomy*.

Autonomy implies that the same processes are going on in the *process region* (Broberg, 1968, 1971) in situations which are identical with respect to material, environmental conditions and fracture mode but different (though within certain limitations) with respect to body geometry, crack length and load distribution.

Path independence of the J-integral is generally very good before stable crack growth, but at an amount Δa of crack growth, approximate path-independence of the J-integral cannot be relied upon for paths closer than about Δa from the crack tip.

Figure 1 illustrates situations in which the J-integral is strictly path-independent, but near-tip autonomy is not prevalent. *Loss of autonomy implies that different local (near-tip) criteria for onset of crack growth give different results.* Thus in the example shown by Fig. 1, the J-integral criterion predicts onset of crack growth at the remote stress $\tau_\infty = 0.9$ k (Fig. 1: $a \approx 0.026\ a_s$) for a crack that is more than three times as long as predicted by the criterion of critical maximum strain at a given (small) distance from the crack tip (Fig. 1: $a \approx 0.008\ a_s$). A third result follows from the COD-criterion, etc.

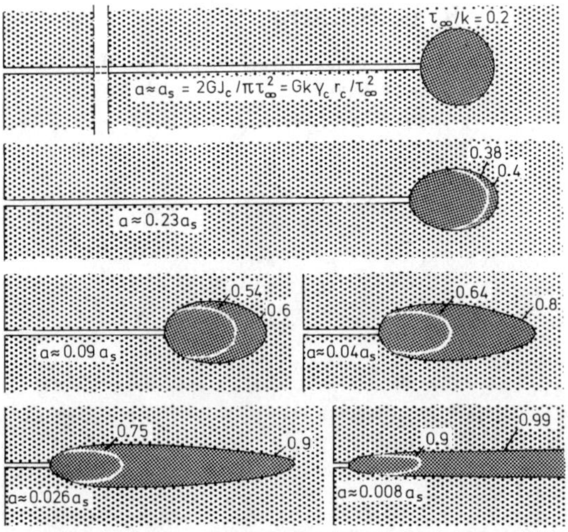

Fig. 1. Plastic regions at mode III edge cracks with different lengths but at the same value J_c of the J-integral (black contours) or the same value γ_c of the maximum strain at a specified distance r_c from the crack tip (white contours). Calculations of contours are based on a work by Rice (1966). Perfect plasticity, point-sized process region and no crack growth is assumed. G is the modulus of rigidity, k the yield stress in shear and τ_∞ the remote shear stress.

Figure 2 illustrates situations in which near-tip autonomy is prevalent, but the J-integral is not path-independent. Although it assumes a rigid-perfectly-plastic material it should also depict reasonably well cases of elastic-plastic material at net section yielding. *The almost linearly ascending J-resistance curves often reported from materials testing are most probably due to this particular consequence of poor path-independence of the J-integral, significant already before net section yielding is reached.*

Autonomy and path-independence of the J-integral can be found within any desired level of accuracy, if restriction is made to sufficiently small scale of yielding. If the scale of yielding is so small that the whole plastic region and its wake at stable crack growth lies inside the region of autonomy, then the J-integral is assuredly path-independent outside this region, as required.

In cases without noticeable stable crack growth a region of autonomy can

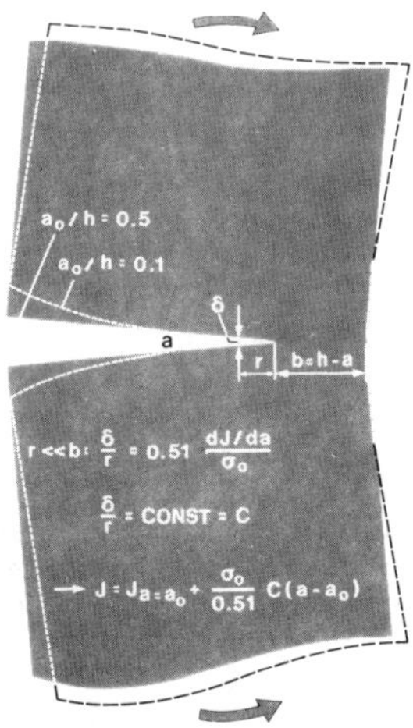

Fig. 2. Two cases with the same current near-tip state (δ/r is the same) but different history due to different original crack length. This implies different values of the J-integral for a remote path (Rice, Drugan and Sham, 1980). J increases linearly with the crack growth ($a-a_o$) if constant near-tip state is assumed. σ_o = yield stress.

exist inside the plastic region and then an extension to larger scale of yielding is justified. The necessary requirement can be expressed by the inequality

$$a - a_o \ll 0.04\ EJ_{c1}/\sigma_Y^2 \qquad \text{(mode I)} \qquad (1)$$

where the left member is the total extent of stable crack growth and the right member equals, approximately, the minimum extension of the plastic region in the crack direction at incipient crack growth, occurring at $J = J_{c1}$. However, inequality (1) is seldom satisfied. In most cases the total extent of stable crack growth is of the order of $0.1\ EJ_{c2}/\sigma_Y^2$. A typical example is shown in Fig. 3. J_{c2} is the J-integral at instability.

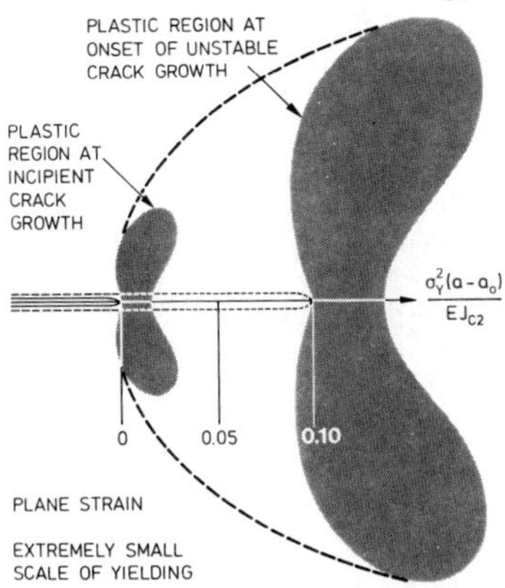

Fig. 3. A typical example of stable crack growth, which is assumed to start at about half the load at which unstable growth occurs. In many cases onset of crack growth occurs at a substantially lower load, but more seldom at a higher load.

Attempts to extend experiments to large scale of yielding often give rise to difficulties in interpretation. Shear lips and slanting crack surfaces very often influence - and distort - test results considerably. J-integral calculations become ambiguous and different specimen geomtries give different results.

In retrospect it seems that little progress has been made about fracture mechanics of monotone loading beyond small scale yielding. Use of complicated methods, as the one based on the J-resistance curve, is, therefore, questionable and a simpler method has been suggested (Broberg, 1977). In this method (after a slight modification) onset of unstable crack growth is assumed to occur when

$$J(Q,\ a_o + 0.1\ EJ_{c2}/\sigma_Y^2) = J_{c2} \tag{2}$$

where Q is a load parameter and J is calculated as

$$J(Q,a) = J_E(Q,a)\left[\frac{1+2(Q/Q_Y)^4}{1-(Q/Q_L)^4}\right]^{\frac{1}{2}} \tag{3}$$

where J_E is the J-integral calculated under the assumption of linear elasticity, Q_L is the limit load with respect to flow across a ligament from the crack tip, and Q_Y is such a limit load, also, but determined under the assumption that the material is elastic-perfectly plastic with yield stress σ_Y.

Some justification for a simple method of calculating J for non-linear cases came quite gratuitously from a "round-robin" (Larsson, 1981) in which several prominent experts took part, demonstrating wide scatter in results of numerical J-computations.

STATIC LOADING = MONOTONE LOADING?

Taken literally, this question may be easily answered: most engineering structures that are statically loaded experience several load applications during their lifetime and are thus not subject to monotone loading. Different materials exhibit differential sensitivity to load cycling. The present dominance of monotone loading fracture mechanics for analysing such structures seems therefore astonishing and might lead to serious mistakes.

IMPLICATIONS OF DESIGN PHILOSOPHY

In spite of all criticism regarding safety factors, no better choice seems available except in special cases. A safety factor can, however, be defined in many ways. Usually it compares the expected load with the load that causes failure if the structure possessed the expected strength. Suppose that a structure will be exposed to N load cycles, all of the same magnitude, Q. Then the "expected strength" could be expressed in two different ways:
1. The load magnitude Q_1 expected to cause failure at monotone loading after the structure has been exposed to N load cycles of magnitude Q.
2. The cyclic load magnitude Q_2 expected to cause failure after N load cycles.

The safety factor may be expressed as Q_1/Q or Q_2/Q. Some calculations, too lengthy to describe here, suggest that the same value of Q_1/Q may correspond to values of Q_2/Q ranging from those indicating a fair safety to those overshooting unity by only a few per cent. The conclusion is that Q_2/Q is the most suitable of the two candidates.

Like other physical disciplines with a strong foundation in engineering, fracture mechanics involves *theory*, *experiments* (including *tests*), and *practice*. Ideally all three aspects should be considered simultaneously but frequently focus rests on two aspects, overlooking the third. This might detract from relevance to practical situations and result in unduly emphasis on the behaviour of test pieces.

CRACKS IN ENGINEERING STRUCTURES AND IN TEST SPECIMENS

Cracks of engineering interest belong almost invariably to mode I. Under the general rubric of mode I, however, a large variety of cases can be found. Individual cases differ mainly with respect to load composition and to the significance of the crack length.

As an informative measure of the load composition one can take the constant stress at the crack tip in a linear elastic analysis (the term after the singular one in an eigenfunction expansion). For cracks of engineering interest - mainly edge-cracks, surface flaws and internal flaws - this stress is usually small and compressive. The significance of crack length can be described by the J-integral at constant load. For cracks of engineering interest the J-integral at constant load is approximately proportional to the crack length.

In contrast to cracks of engineering interest, CT test specimens exhibit large tensile stresses along the crack surfaces, whereas these stresses are small for 3-point bend specimens.

For most conventional test specimens the J-integral at constant load increases *much* more rapidly than in proportion to the crack length. However, the 3-point bend specimen shows approximate proportionality between J at constant load and the crack length a within a fairly wide range of crack lengths if the original (prefabricated) crack length a_o is chosen substantially smaller than half the specimen height h.

PROPOSED PROCEDURE

The procedure to be proposed concerns failure in a structure that is exposed to rather few - say a few thousand or less - major load applications during its life-time. It is based on the J-integral, which should be calculated according to (3). The material properties are found from tests with (preferably) 3-point bend specimens, the heights of which should equal four times their thicknesses, and the length a_o^T (T = Test) of the prefabricated crack should be about one quarter of the height.

In each test the load is cycled between zero (or a very small load) and some predetermined value until fracture occurs. The number N_T of cycles to failure and the *J-integral* J_T *at the first load application* are determined. By making a few different load choices a curve J_T versus N_T is obtained. An

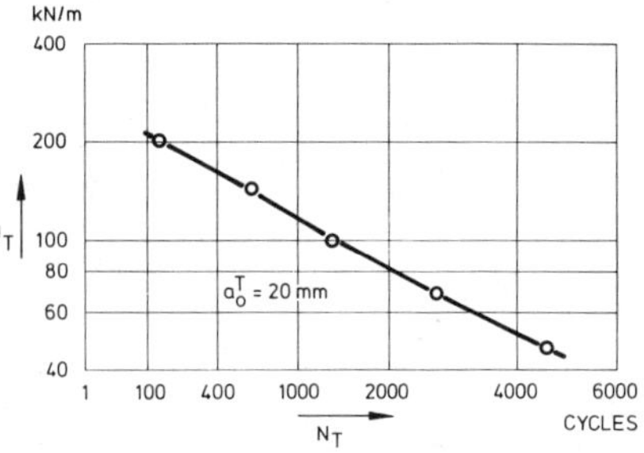

Fig. 4. Relation between J_T and N_T for a Swedish heavy wear steel OX812E (σ_Y=720 MPa, σ_u=770 MPa).

Courtesy Mr. A. Lundström, Lund Institute of Technology, Lund, Sweden example is shown in Fig. 4.

Even though the original crack length is chosen to only about one quarter of the specimen height, J increases faster with a than in most cases of engineering interest. In order to avoid too conservative results the crack length should not be allowed to exceed half the specimen height.

Assume now that the same kind of tests was performed on an engineering structure with a crack of original length a_o^S (S = Structure). Then a relation between J_S, the J-integral at the first load application, and N_S, the number of cycles leading to failure, would be obtained. Assuming the material to be the same as in the testpiece the J_S-N_S-curve would come close to the J_T-N_T-curve if $a_o^S = a_o^T$. However, as previously discussed different dependence of the J-integral on the crack length would in general imply that the J_S-N_S-curve would lie somewhat above the J_T-N_T-curve.

If the original crack length in the structure is $a_o^S \neq a_o^T$ a transformation to the same original crack length must be done. Since the J-integral at constant load is approximately proportional to the crack length, then $J \approx J_S a/a_o$ at each load peak. The general assumption that the crack growth rate depends on J gives

$$\Delta a/\Delta N = s(J) \approx f(a/a_o) \qquad (4)$$

where s and f are functions characteristic of the body. After changing $\Delta a/\Delta N$ to da/dN one obtains

$$N \approx \int_{a_o}^{a_c} [f(a/a_o)]^{-1} da = a_o \int_{1}^{a_c/a_o} [f(a/a_o)]^{-1} d(a/a_o) \qquad (5)$$

where a_c is the crack length at onset of unstable crack growth. a_c is obtained from the condition $f(a/a_o) \to \infty$, i.e. a_c/a_o is independent of a_o. Hence (5) shows that N is proportional to a_o i.e. the number N_S of load cycles for the structure corresponds to the number $N_T = (a_o^T/a_o^S)N_S$ of load cycles for the test piece.

In some cases there might be some confusion or ambiguity about the choice of proper crack length parameter a for comparison between J_S-N_S- and J_T-N_T-curves. A simple rule is that a crack tip advance equal to the original crack length parameter should double this parameter. For a surface crack the situation is less well defined, but if it is small compared to significant body dimensions, then a certain (material-dependent) shape - not very far from the semi-circular one - exists that would be preserved during crack growth. Then the crack length parameter could be taken as equal to the crack depth.

DEMONSTRATION EXAMPLE

Assume that tests are made on 3-point bend specimen with thickness b = 15 mm, height h = 60 mm, length L = 240 mm and original crack length $a_o^T \approx 15$ mm. Suppose further that it is known that the yield stress $\sigma_Y = 800$ MPa, the ultimate stress $\sigma_u = 1100$ MPa and Young's modulus E = 210 GPa. Three tests are performed. The loads are chosen to Q = 50,55 and 60 kN. Fracture occurs after $N_T = 1420$, 412 and 71 cycles, respectively. The corresponding original crack lengths turn out to be 15.0, 14.7 and 14.9 mm.

The limit loads (see Green and Hundy, 1956) are $Q_Y = 1.261\, \sigma_Y b(h-a_o^T)^2/2\sqrt{3}h \approx$

$\approx$ 147 kN and $Q_L \approx Q_Y\sigma_u/\sigma_Y \approx$ 200 kN.

Use of (3) and expressions for J_E for 3-point bend specimens gives

$$J_T = 25,\ 31 \text{ and } 37 \text{ kN/m, respectively.}$$

Plotting J_T against N_T gives the result shown in Fig. 5.

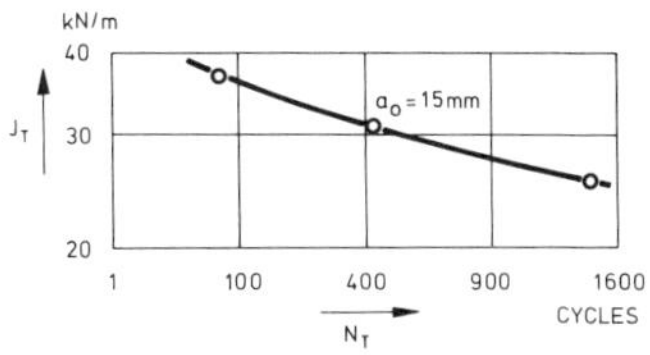

Fig. 5. Plot of three (hypothetical) test results.

The test results will now be applied to the case of a semi-elliptical surface crack with depth a_o^S = 3 mm and length 8 mm in a 15 mm thick plate, subject to tensile loading, stress = σ. The expected number of load cycles is N_S = 200. Thus the critical value of J for the surface crack should equal J_T at

$$N_T = (15/3) \times 200 = 1000 \text{ cycles}$$

i.e. $J_S \approx$ 26 kN/m.

By (conservatively) putting $Q/Q_L = \sigma/\sigma_u$ and $Q/Q_Y = \sqrt{3}\sigma/2\sigma_Y$ for the surface crack one obtains with $J_E \approx 0.65\pi\sigma^2 a_o/E$: σ = 755 MPa. This value is the one on which a safety factor should be applied.

MERITS OF THE PROCEDURE

Fig. 4 immediately demonstrates the hazard encountered when using monotone loading fracture mechanics (e.g. J-resistance curves) even when the expected number of load cycles appears to be very low.

The tests are very simple to perform. No measurement of crack growth is needed.

The test procedure comes very close to most common situations in engineering practice and the results obtained are on the conservative side with respect to design and risk evaluation. The practical application is very simple. The procedure is rather similar to linear fracture mechanics of monotone loading even though it covers both non-linearities and load cycling. Use of lower J-values than at monotone loading implies less non-linearity.

The presence of several load cycles before fracture implies much less stable crack growth at each load cycle than at monotone loading. It is believed that this in turn implies that the way of manufacturing the crack in the test specimen is less critical than at J-resistance curve determinations and thus contributes less to deviations from the practical reality.

ACKNOWLEDGMENT

The present study was carried out while I was a Visiting Professor at Brown University, Providence, Rhode Island. It was supported by the Materials

Research Laboratory at Brown University, funded by the U.S. National Science Foundation. Subsequently I have had several valuable discussions with Mr. A. Lundström at Lund Institute of Technology.

REFERENCES

Broberg, K.B. (1968). Critical review of some theories in fracture mechanics. *Int. J. Fract. Mech.*, 4, 11-18.

Broberg, K.B. (1971). Crack growth criteria and non-linear fracture mechanics. *J. Mech. Phys. Solids*, 19, 407-418.

Broberg, K.B. (1977). On the treatment of the fracture problem at large scale of yielding. In G.C. Sih and C.L. Chow (Eds.), *Proc. International Conference on Fracture Mechanics and Technology*, Hong Kong 1977, Vol. II. Sijthoff and Noordhoff International Publishers. pp 837-859.

Green, A.P., and Hundy, B.B. (1956). Initial plastic yielding in notch bend tests. *J. Mech. Phys. Solids*, 4, 128-145.

Larsson, L.H. (1981). A calculational round robin in elastic-plastic fracture mechanics. Extended version of a paper presented at *the 6th Int. Conference on Structural Mechanics in Reactor Technology*, Paris, 17-21 August, 1981.

Rice, J.R. (1966). Contained plastic deformation near cracks and notches under longitudinal shear. *Int. J. Fract. Mech.*, 2, 426-447.

Rice, J.R., Drugan, W.J., and Sham, T.L. (1980). Elastic-plastic analysis of growing cracks. In *Fracture Mechanics Twelfth Conference*, ASTM STP 700 Am. Soc. Testing and Mt'ls., Phil., Pa., 189-219.

ON CRACK PROBLEMS RELATED TO WELDING RESIDUAL STRESS FIELDS IN LINEAR ELASTIC FRACTURE MECHANICS

X. R. Wu and J. Carlsson

Department of Strength of Materials and Solid Mechanics, Royal Institute of Technology, S-100 44 Stockholm, Sweden

ABSTRACT

The present study is devoted to crack problems related to welding residual stress fields within the linear elastic fracture mechanics (LEFM) regime. The determination of residual stress intensity factors, K_{res}, using the generalized weight function method for 2-D cracks with different boundary conditions is discussed. Finite element K_{res}-results for half-elliptical surface cracks in butt-welded thin and thick plates are given for some typical residual stress fields. Experimental results showing the effect of welding residual stresses on brittle fracture are presented.

KEYWORDS

crack; fracture; stress intensity factor; residual stress; weldments; LEFM

INTRODUCTION

Cracks in welded structures are of great concern in engineering practice since such structures are susceptible to brittle fracture due to welding defects, material embrittlement and high residual stresses. In the LEFM regime, residual stresses may simply be assumed to be additive to the stresses caused by applied loads. Thus a LEFM assessment of cracks in welded structures becomes a matter of determining K_{res}, taking due account of whether surface traction and/or displacement control is at hand.

The determination of K_{res} with different boundary conditions is, however, not very clearly described in the literature. One question of concern is under which circumstances thermal or residual stresses behave as primary stresses (which are redistributed but not relieved by the introduction of a crack) or secondary stresses (which are partly relaxed by introducing a crack), (Chell and Ewing,1977). Another question is whether K_{res} is affected by the redistribution of residual stresses in the uncracked ligament due to the development of the crack (Rybicki, Stonesifer and Olson,1981). These questions are dealt with in the present paper and methods of calculating K_{res} are discussed.

MBM VOL2-M

RESIDUAL STRESSES

Due to incompatible elastic strains, caused mainly by inelastic deformation, and metallurgical transformations, residual stresses build up in the weld region. For structures welded under constraints, additional reaction stresses, often being transverse to the weld line, will arise. This type of reaction stresses are much simpler and will not be considered here.

Residual Stresses in Butt-welded Thin Plates

In thin plates (approximately with thickness < 25 mm), residual stresses are often assumed to be uniform in the plate thickness direction. Typical residual stress fields for thin plates are shown in Fig. 1a. The longitudinal stress σ_{zz} is commonly approximated by the formula (Masubuchi, 1980)

$$\sigma_{zz}(x) = \sigma_{zz}(0)\left[1-\left(\frac{x}{\ell}\right)^2\right] \cdot \exp\left[-\left(\frac{x}{\ell}\right)^2/2\right] \qquad (1)$$

where $\sigma_{zz}(0)$ is the maximum stress (at $x = 0$), which is of the order of the material uniaxial yield stress σ_y. The parameter ℓ is half the width of the tensile stress zone, which may be estimated from the material properties, the plate thickness and the welding data (Wu and Carlsson, 1983). The transverse stress σ_{xx} varies slowly in the central part of the plate, and the maximum σ_{xx} is typically lower then $0.2\ \sigma_y$ (Andersson and Karlsson, 1981).

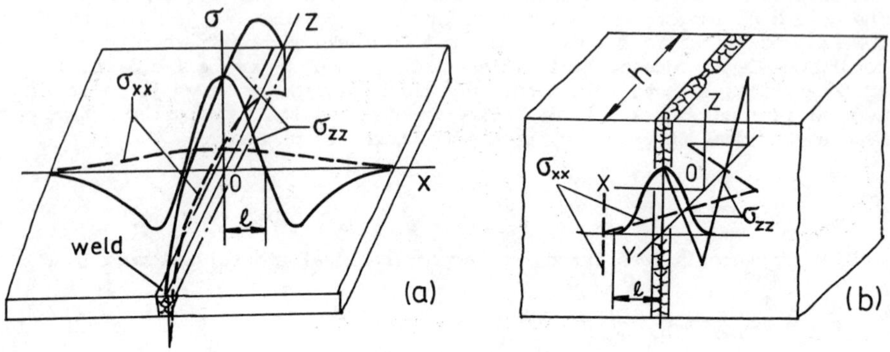

Fig. 1. Typical welding residual stress fields in (a), thin; (b), thick butt-welded plates.

Residual Stresses in Butt-welded Thick Plates

The distribution of residual stresses in thick plates (approximately with thickness > 100 mm) is quite different from that in thin plates, the main difference being that all the stress components vary considerably in the plate thickness direction. The stress state is tensile at the surfaces and compressive in the central part, the maxima of the tensile and compressive stresses being about the same. Typical variations of residual stresses, as shown in Fig. 1b, may be approximated by (Wu and Carlsson, 1983),

$$\sigma_{xx} = \sigma_{xxm}(4|y|/h - 1) \qquad (2)$$

for the transverse stress, and

$$\sigma_{zz} = \sigma_{zzm}(4|y|/h -1)\ (1 + \cos \frac{\pi x}{\ell})/2, \ |x| \leq \ell \tag{3}$$

for the longitudinal stress.(h denotes the plate thickness)

In eqn. (3) representative values of ℓ are in the range $(0.5 \sim 2)h$. The maximum stresses σ_{xxm} and σ_{zzm} are ordinarily of the order of σ_y. In particular situations, however, experimental measurements are recommended in order to determine the magnitude more accurately.

Residual stresses in plates of intermediate thickness show some transitional behaviour between the above two cases. As the thickness increases the thickness-variation of the stresses increases.

RESIDUAL STRESS INTENSITY FACTORS

Principle of Superposition

Consider a crack-free body containing residual stresses, σ_{res}, and with surface tractions T_i specified on part S_T of the boundary and displacements u_i specified on part S_u. Upon introducing a crack AB, Fig. 2a, the original residual stress field σ_{res} will be disturbed (the disturbed field is denoted as σ'_{res}). To determine K for the crack due to T_i, u_i and σ_{res}, one can apply the superposition principle of linear elasticity.

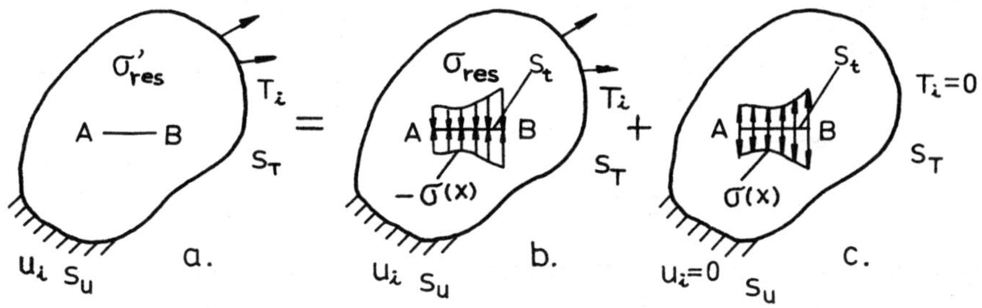

Fig. 2. Principle of superposition.

If tractions which are equal and opposite to the crack line stress $\sigma(x)$ defined as the stress at line AB in the crack-free body ($\sigma(x) = \sigma(T_i, u_i, \sigma_{res}, x)$),are applied to the crack surfaces S_t, Fig. 2b, the crack will be perfectly closed (no opening or overlapping) and the stress intensity factor K is zero. Superposing another case with $u_i = 0$ on S_u, $T_i = 0$ on S_T and a crack face pressure $p(x) = \sigma(x)$, Fig. 2c, we obtain the case for which the stress intensity factor is sought, Fig. 2a. Since $K_a = K_b(=0) + K_c$, we have $K_a = K_c$. Therefore in order to determine K, one may convert the original problem (Fig. 2a) to an equivalent problem in which the only non-vanishing loading is the crack line stress $\sigma(x)$ (Fig. 2c).(Note the separation of the boundary surface into S_T and S_u must be kept the same in the two problems). One can first compute the crack line stress $\sigma(x)$ in the crack-free body and then calculate K by applying $\sigma(x)$ on the crack surfaces,with T_i and u_i set to be zero. The computation of $\sigma(x)$ involves no fundamental difficulties since the problem is non-singular. In this way the K-determin^ation will be grately simplified, in particular when the generalized weight function method is used for 2-D crack problems.

The Generalized Weight Function Method for 2-D Crack Problems

A powerful method of determining K for 2-D cracks is the weight function method (Bueckner,1970; Rice,1972). For two identical geometries, if the the stress intensity factor $K^{(1)}$ and the crack face displacement $u^{(1)}$ are known for a load system (1), then $K^{(2)}$ for any load system (2) can be determined with this method. However the loadings in Rice's(1972) version of weight function method are of surface traction prescribed type only.

Using Betti's reciprocal theorem, Carlsson(1975) has derived a more general expression for the mixed boundary condition in which the applied loadings are functions of both position and crack length a.(Bowie and Freese(1981) recently obtained a same result):

$$K^{(2)}=\frac{E'}{2K^{(1)}}\left[\int_{S_T} T_i^{(2)}\frac{\partial u_i^{(1)}}{\partial a}ds-\int_{S_u} u_i^{(2)}\frac{\partial T_i^{(1)}}{\partial a}ds\right] \qquad (4)$$

where T_i and u_i are boundary tractions and displacements respectively,and S_T (S_u) denotes tractions(displacements) prescribed boundaries, and E'=E(plane stress) and $E'=E/(1-\nu^2)$(plane strain).

It is noted that eqn.(4) is not easy to use since $\partial u_i^{(1)}/\partial a$ and $\partial T_i^{(1)}/\partial a$ ar not readily determined on S_T and S_u respectively. However the corresponding relation for the equivalent problems are particularly simple. Since $T_i^{(2)}=0$ on S_T and $u_i^{(2)}=0$ on S_u, we have (write $u_i^{(1)}$ as $u^{(1)}$),

$$K^{(2)}=\frac{E'}{2K^{(1)}}\int_{S_t}\sigma^{(2)}(s)\frac{\partial u^{(1)}}{\partial a}ds=\frac{E'}{K^{(1)}}\int_0^a\sigma^{(2)}(x)\frac{\partial u^{(1)}}{\partial a}dx \qquad (5)$$

This result is, in its form,the same as Rice's. However the present formulation is more general since it has considered the mixed boundry condition. It should be noted that case (2) must have the same separation of S into S_T and S_u as case (1) from which $\partial u^{(1)}/\partial a$ is derived.

The determination of $\partial u^{(1)}/\partial a$ on S_t is relatively simple. Although exact values of $u^{(1)}$ are rarely available,especially for cracks in finite bodies, good approximations of $u^{(1)}$ are possible to obtain. The reader is referred to Wu(1983c),and Petroski and Achenbach(1978) for details.

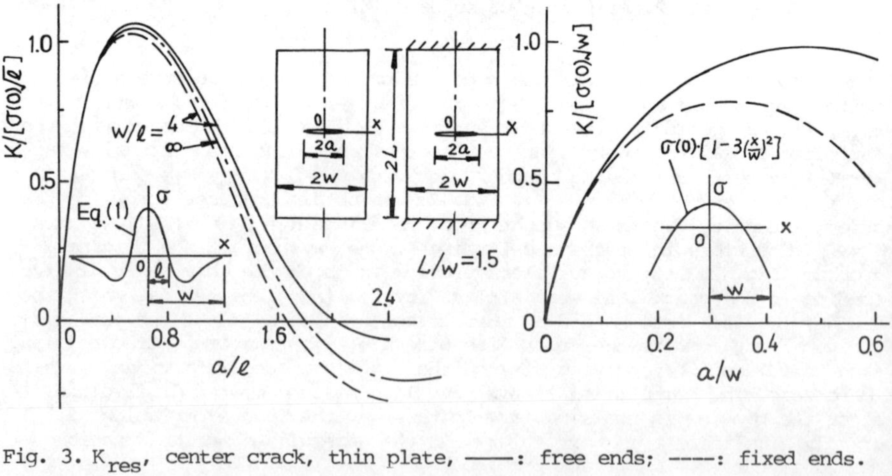

Fig. 3. K_{res}, center crack, thin plate, ——: free ends; ----: fixed ends.

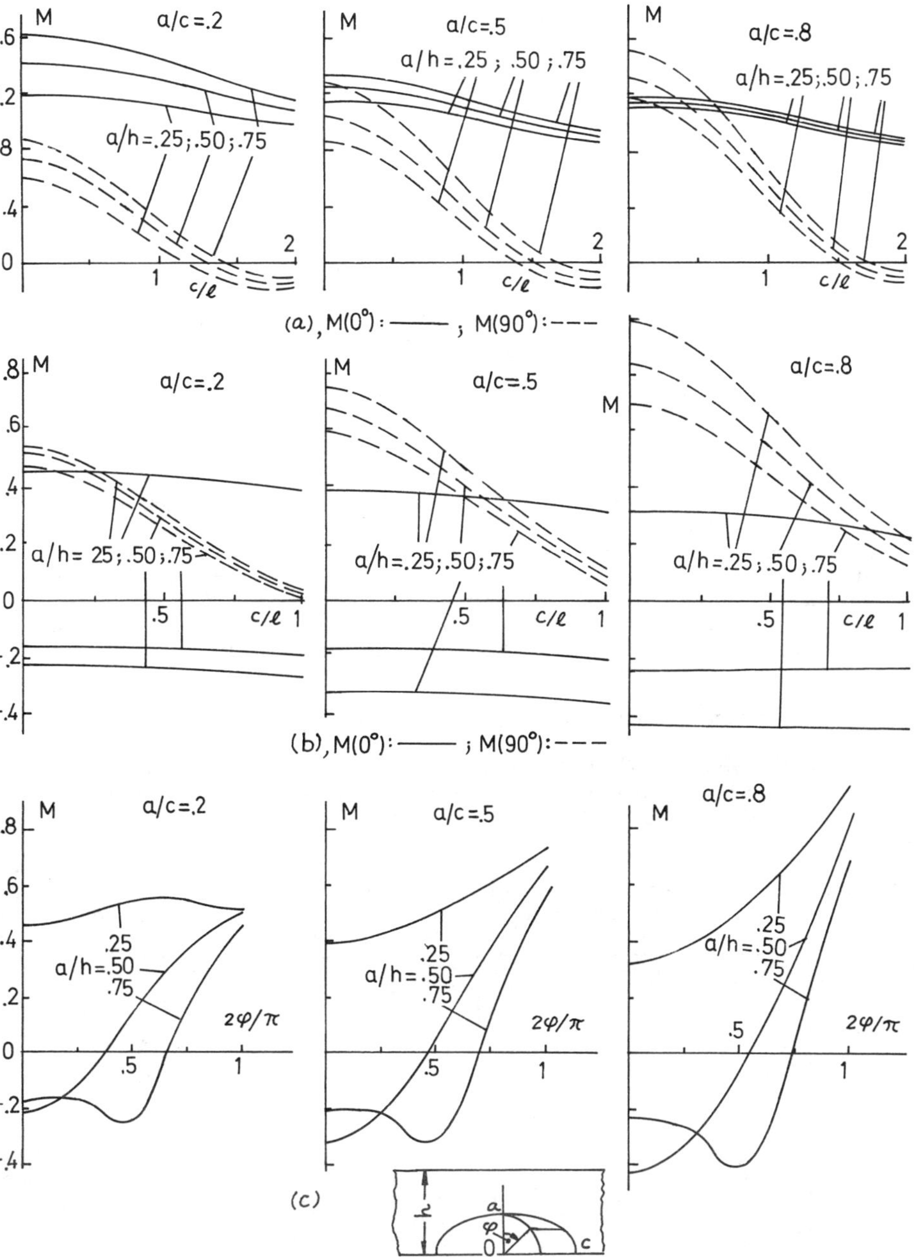

Fig. 4. M-factors for surface cracks in welding residual stress fields: (a), cracks perpendicular to the weld, thin plates, σ_{res}: eqn (1), (b), cracks perpendicular to the weld, thick plates, σ_{res}: eqn (3), (c), cracks along the weld, thick plates, σ_{res}: eqn (2).

Examples of K calculated with this method are shown in Fig. 3. Two cases of crack line stresses combined with two types of boundary conditions: free ends and fixed ends, were considered. The influence of the boundary condition depeneds on $\sigma^{(2)}(x)$, a/W, and the plate geometry (L/W).

Surface Cracks in Welding Residual Stress Fields

The weight function method is much more difficult to use in 3-D surface crack problems. For half-elliptical surface cracks in welding residual stress fields, K_{res} can be calculated using the finite element method. Engineering magnification factors M for half-elliptical surface cracks, defined by $M = K/(\sigma_o(\pi a)^{1/2}/\Phi)$, where $\Phi = (1+1.464\ (a/c)^{1.65})^{1/2}$, a is the crack depth, c is the crack half-length, and σ_o is the stress at the origin, Fig. 4, have been tabulated for a wide range of crack geometries and stress fields (Wu,1983a; Wu and Carlsson,1983). In these calculations the plates were assumed to be free from external loadings and constraints. The main results are shown in Fig.4.

Some features of K_{res} for surface cracks in welding residual stress fields are worth noting. First , K_{res} varies very much along the crack front and very irregularly with crack size. The variations are much more pronounced for cracks in thick plates. Secondly, due to the geometrical nature of surface cracks and the variation of residual stress fields, K_{res} for surface cracks are quite different from those for through-thickness center cracks having the same crack length. A simple 2-D through-thickness crack analysis may well underestimate K_{res} (Wu and Carlsson, 1983).

EFFECTS OF WELDING RESIDUAL STRESSES IN LEFM

In the LEFM regime, residual stresses are additive to the stresses due to external loads. Since the welding residual stresses are often of high magnitude, their effects in LEFM are significant.

The brittle fracture behaviour of welded structures is strongly influenced by residual stresses. Many examples of failure due to residual stresses in such structures containing through-thickness cracks have been analysed in

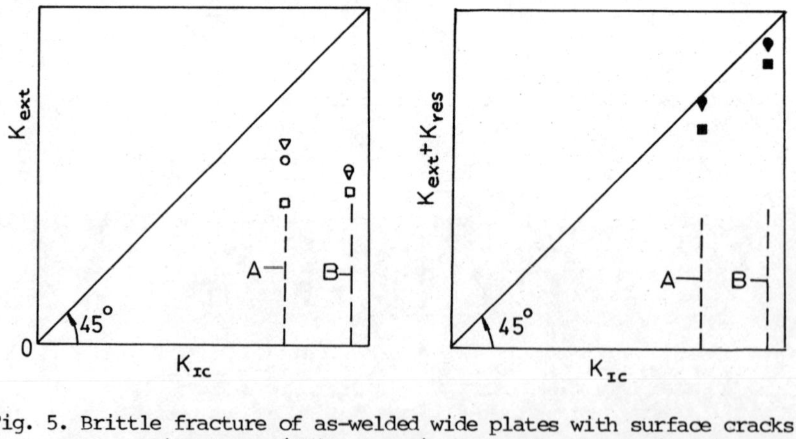

Fig. 5. Brittle fracture of as-welded wide plates with surface cracks, A: tensile tests (thin plates); B: bending tests (thick plates).

the literature. Recent investigations have shown the effect of residual stress on brittle fracture of wide plates with half-elliptical surface cracks(Wu, 1983b). The results of the experiments were explained well by a criterion: K_{ext} (external loads) + $K_{res} = K_{Ic}$, where K_{Ic} is the material fracture toughness. The effect of K_{res} is significant as shown in Fig. 5.

In connection with fatigue crack growth, residual stresses influence the R-ratio, affecting the growth rate, and especially the fatigue threshold value ΔK_{th}. The compressive residual stress in the middle of a thick plate can considerably retard the local crack growth or even arrest the crack.

Residual stresses rather than design stresses are often responsible for environment sensitive fracture such as stress corrosion cracking. High residual stress substantially increases the susceptibility to such fracture.

DISCUSSION AND CONCLUSIONS

Three topics have been discussed. These include welding residual stress fields in plates, determination of K_{res} for 2-D cracks using the generalized weight function method and finite element K_{res}-results for surface cracks, and finally effects of welding residual stresses in LEFM.

The central problem in treating crack problems related to residual stress fields in the LEFM regime is the determination of K_{res}. For 2-D cracks in particular, the generalized weight function method is simple, powerful and efficient whether displacement boundary conditions are involved or not. Whatever method is applied, K_{res} can be determined regarding the crack line σ_{res} as crack face pressure and taking the relevant external constraints into account when determining the weight function. In this connection, the concept of primary or secondary stress (Chell and Ewing, 1977) needs not to be introduced. It is also seen that K_{res} is not affected by the redistribution of σ_{res} in the uncracked ligament.

In LEFM residual stresses can be directly superimposed to the applied stresses and their effects are significant.

REFERENCES

Andersson, B. and Karlsson, L. (1981). J. of Thermal Stresses, 4:491-500.
Bowie, O.L. and Freese, C.E. (1981). Engng. Fracture Mech., 14, 519-526.
Bueckner, H.F. (1970). ZAMM, 50, 9, 529-546.
Carlsson, J. (1975). Report 2, Dept. Strength of Materials and Solid Mech. Royal Inst. of Technology, Stockholm
Chell, G.G. and Ewing, D.F.G. (1977). Int. J. Fracture, 13, 467-479.
Masubuchi, K. (1980). Analysis of welded structurtes, M.I.T., Pergam. Press
Petroski, H.J. and Achenbach, J.D. (1978). Engn. Fracture Mech 10, 257-266.
Rice, J.R. (1972), Int. J. Solids Structures, 8 751-758.
Rybicki, E.F., Stonesifer, R.B. and Olsson, R.J. (1981). J. Pressure Vessel Technology, 103, 66-75.
Wu, X.R. (1983a). Accepted for publication in Engineering Fracture Mechanics, 1983.
Wu, X.R. (1983b). Accepted for publication in Engineering Fracture Mechanics, 1983.
Wu, X.R. (1983c). (In preparation)
Wu, X.R. and Carlsson, J. (1983). Accepted for publication in Engineering Fracture Mechanics, 1983.

A DISTRIBUTED DAMAGE APPROACH TO COMBINED BENDING AND AXIAL LOADING OF ROCK BEAMS

F. Ouchterlony

Swedish Detonic Research Foundation, SveDeFo, P.O. Box 32058, S-12611 Stockholm, Sweden

ABSTRACT

The predictive capabilities of continuum damage mechanics (CDM) as to eccentric uniaxial loading and pure bending of strength test specimens of rock are shown. For eccentric loading much smaller strength reductions than usual are predicted, for bending rupture factors that depend properly on section shape. Experimental results support these predictions.

KEYWORDS

Continuum damage mechanics; non-linear material behavior; rock mechanics; tensile strength testing; eccentric loading; strain gauges; rupture factor.

INTRODUCTION

Eccentric loading makes the uniaxial testing of brittle materials a difficult tensile strength test to perform satisfactorily. The conventional beam theory approach to such loading as a case of combined bending and axial stresses (Timoshenko 1940) yields a reduced apparent strength

$$\sigma^f_{nom}/\sigma_t = 1/(1+ec/r_i^2). \tag{1}$$

Here σ^f_{nom} is the nominal section stress at failure and σ_t the uniaxial tensile strength. See Fig. 1 for notation. Eq. 1 is used mainly to demonstrate the necessity of avoiding eccentric loading which the uniaxial tensile testing practice for rock tries to do.

The tensile strength of rock is however mostly determined from indirect tests with resulting values that usually are much higher than σ_t. The apparent failure strength of a beam of brittle material in bending, σ_{tb}, is an example. In non-dimensional form

$$R = \sigma_{tb}/\sigma_t = [M_m/(I/c)]/\sigma_t, \tag{2}$$

it is called the rupture factor. M_m is the maximum bending moment. R depends on the beam material, the shape of the cross section, and the type of bending. Typical values for rock lie between 1.5 and 2.5 (Vutukuri and co-workers, 1974).

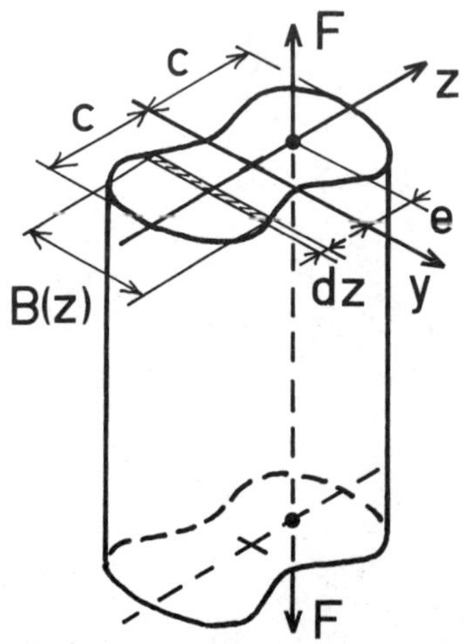

Notation	
load	F [N]
principal axes	x,y
bending plane	zF
breadth function	B(z) [m]
cross section area	$A = \int B dz$ [m^2]
moment of inertia	$I = I_y = \int Bz^2 dz$ [m^4]
radius of gyration	$r_i = \sqrt{(I/A)}$ [m]
extreme fiber position	c [m]
amount of eccentricity	e [m]
nominal section stress	$\sigma_{nom} = F/A$ [Pa]

Fig. 1 Uniaxial tensile testing of a specimen with doubly symmetric cross section of constant area, subjected to eccentric loading along one of the principal axes.

The failure criterion behind Eq. 1, that failure occurs when the tensile extreme fiber stress reaches σ_t, obviously can't explain why $R > 1$. A new simple approach which can predict both the effects of eccentric loading in uniaxial strength testing and the variation of R with section shape is continuum damage mechanics or CDM, as this contribution will show. See also Ouchterlony (1981).

CDM smears out the micro-cracking or damage and it results in non-linear material behavior, as is usually the case with rock. Other CDM work which is related to the tensile testing of brittle materials is that of Jansson (1978), Hult (1979), Krajcinovic (1979), Grady and Kipp (1981), and Krajcinovic and Fonseka (1981). A general review is given by Chaboche (1981).

CDM IN UNIAXIAL TENSION

The derivations rely on scalar CDM (Jansson, 1978). Consider the specimen in Fig. 1, subjected to uniaxial tension. Let A be the macroscopically observable section area and A_{eff} the effective load carrying one. Then

$$F = \sigma A = sA_{eff} \tag{3}$$

define the apparent stress σ and the net stress s. The amount of damage is defined by the scalar

$$\omega = (A-A_{eff})/A \quad \text{whence} \quad s = \sigma/(1-\omega). \tag{4}$$

Thus $\omega = 0$ describes intact material and $\omega = 1$ total inability to carry load. We assume that the material is linearly elastic and linearly damaging. Thus

$$s = E\varepsilon \text{ and} \tag{5}$$

$$\omega = \omega_o + s/D \text{ when } s > -\omega_o D \quad \text{and} \quad \omega = 0 \text{ when } s \le -\omega_o D. \tag{6}$$

Here ε denotes the engineering strain, E the Young's modulus, D the damage modulus, and ω the initial damage. For a non-damaging material $D \to \infty$. ω_o represents the frequently large number of pores and micro-cracks in rock.

These equations yield $F = s(1-\omega_o-s/D)A$ expressed in net stress. The specimen will fail when it looses its load carrying capacity, that is when $dF/d\varepsilon \propto$ $\propto dF/ds = 0$ and F(s) has a maximum. The resulting failure stresses become

$$s_t = \frac{1}{2}(1-\omega_o)D \quad \text{and} \quad \sigma_t = \frac{1}{4}(1-\omega_o)^2 D, \tag{7}$$

for the damaging material. For the non-damaging material F(s) is linear and a failure criterion $s = s_t$ has to be invoked directly.

CDM IN ECCENTRIC UNIAXIAL LOADING

Consider eccentric uniaxial loading of the specimen in Fig. 1, applied on a principal axis such that in-plane bending and axial loading results. We assume that plane cross sections remain plane and that the material is linearly elastic such that $\varepsilon(z)$ and $s(z)$ become linear. We may then write

$$s = \bar{s} + \Delta s \cdot z/c, \tag{8}$$

where $\bar{s}$ is the average net stress and $2\cdot\Delta s$ its variation over the cross section. Thus $\omega(z)$ will also be linear as long as $\omega \ge 0$. The apparent stress distribution $\sigma(z)$ for a damaging material becomes

$$\frac{\sigma}{\sigma_t} = 2\left(\frac{\bar{s}}{s_t} + \frac{\Delta s}{s_t} \cdot \frac{z}{c}\right)\left[1 - \frac{1}{2}\left(\frac{\bar{s}}{s_t} \div \frac{\Delta s}{s_t} \cdot \frac{z}{c}\right)\right]. \tag{9}$$

It is parabolic with max $\sigma = \sigma_t$ at $z/c = (s_t-\bar{s})/\Delta s$. For the non-damaging material $\sigma(z)/\sigma_t = s/s_t$ is clearly linear.

The force contribution from the strip $dA = Bdz$ in Fig. 1 is $dF = \sigma Bdz =$ $= s(1-\omega)Bdz$. The load induces a moment $M = F\cdot e$ and the equilibrium requirements are

$$F = \int_{-c}^{c} dF \quad \text{and} \quad M = F\cdot e = \int_{-c}^{c} z dF. \tag{10}$$

Expressing the load as the ratio of nominal section stress to uniaxial tensile strength and the eccentricity as $\rho = e/r_i$ they become

$$\left.\begin{aligned} \frac{\sigma_{nom}}{\sigma_t} &= \frac{F/A}{\sigma_t} = \frac{2}{\rho^2}\left(1-\frac{\bar{s}}{s_t}\right)\left\{\left[(1-\rho^2)\left(1-\frac{\bar{s}}{s_t}\right)^2+\rho^2\right]^{1/2}-\left(1-\frac{\bar{s}}{s_t}\right)\right\} \text{ and} \\ \frac{\Delta s}{s_t} &= \frac{c}{\rho r_i}\left\{\left[(1-\rho^2)\left(1-\frac{\bar{s}}{s_t}\right)^2+\rho^2\right]^{1/2}-\left(1-\frac{\bar{s}}{s_t}\right)\right\}. \end{aligned}\right\} \tag{11}$$

They give the load as a function of the complementary monotonic loading parameter, $F(\bar{s})$, and $\Delta s(\bar{s})$.

The specimen fails when $dF/d\bar{s} = 0$ at $\bar{s} = \bar{s}_1$ and $F(\bar{s}_1)$ is a maximum, that is when

$$\bar{s}_1/s_t = 1-\left[\frac{\rho}{2(1+\rho)}\right]^{1/2} \quad \text{and} \quad \Delta s_1/s_t = (c/r_i)\left[\frac{\rho}{2(1+\rho)}\right]^{1/2}. \tag{12}$$

Thus the reduced apparent strength follows as

$$\sigma^f_{nom}/\sigma_t = 1/(1+e/r_i). \tag{13}$$

For a doubly symmetric cross section $1 < c/r_i < \sqrt{12}$. If r_i is independent of bending axis then the load needn't be applied on a principal axis. Since $c/r_i > 1$ the apparent strength decreases less for a damaging material than for a non-damaging one, compare Eqs. 13 and 1. For a solid circular section $c/r_i = 2$ and the result is $1/(1+2e/c)$ for a damaging material as compared with $1/(1+4e/c)$ for a non-damaging one.

The maximum of $\sigma(z)$ in Eq. 9 at failure equals σ_t and lies at $z = r_i < c$. Thus the tensile extreme fiber stress $\sigma(c)$ is already decreasing, having been σ_t at some previous point in the load history $F(\bar{s})$. For a non-damaging material $\sigma(z)$ at failure is obviously linear with $\sigma(c) = \sigma_t$. The weakening of the damaging material redistributes the stresses over the section so that more of it sustains heavy loads. This explains why a damaging material is less sensitive to eccentric loading than a non-damaging one. The parallel to plastic flow lies close at hand.

Normally resistance strain gauges are used to measure the eccentricity e in a given test. Whether the material is damaging or not one can obtain $\bar{s}$ and Δs from suitably positioned gauges. Denote the ratio $\Delta s/\bar{s}$ at failure by η and regard it as a measured or known quantity. Then the predicted eccentricity for a damaging material will be

$$\rho = e/r_i = 2\eta^2/[(c/r_i)^2+2\eta c/r_i-\eta^2], \tag{14}$$

for a non-damaging one $\rho = \eta r_i/c$. The apparent strength predictions become

$$\sigma^f_{nom}/\sigma_t = 1-2\eta^2/(c/r_i+\eta)^2 \tag{15}$$

and $1/(1+\eta)$ respectively. The corresponding strength reductions are obviously $O(\eta^2)$ and $O(\eta)$ for small η, they differ by an order of magnitude!

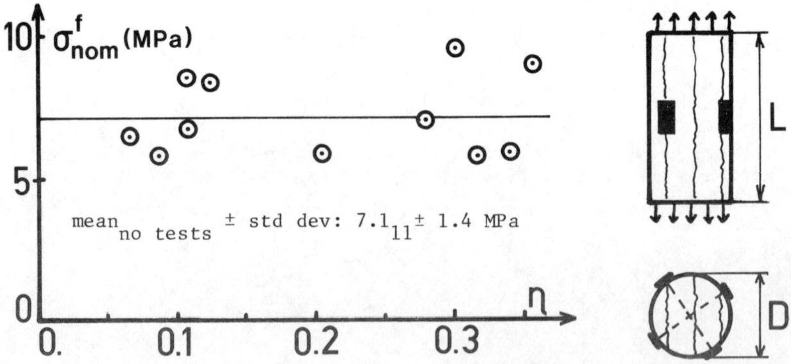

Fig. 2 Apparent uniaxial tensile strength of cylindrical core specimens of Klinthagen limestone. D = 38-41 mm and L/D > 2. Four symmetrically arranged strain gauges were mounted on each specimen.

Take the tests made by Mäki (1982) as an example, see Fig. 2. The largest strain ratio was $\eta \approx 0.35$ which for a non-damaging material would be interpreted as $e/c \approx 0.09$ and an apparent strength of $0.74\sigma_t$. For a damaging one there results $e/c \approx 0.02$ and $0.96\sigma_t$. Since the limited data shows no significant strength reduction they support the damaging approach.

CDM IN PURE BENDING

Consider the same specimen as before but subjected to pure bending, $M = F \cdot e$ with $F \to 0$ this time. $\varepsilon(z)$ and $s(z)$ will again be linear. The cross section will have to support both tensile and compressive stresses. Denote the neutral axis by $z = -c_n$ and the point where the initial damage closes up by $z = -c_o$. Geometrical proportionality requires that $\bar{s}/\Delta s = c_n/c$. Here we replace $\bar{s}$ by c_n and use $s(z) = \Delta s \cdot (z + c_n)/c$ instead of Eq. 8. See Fig. 3.

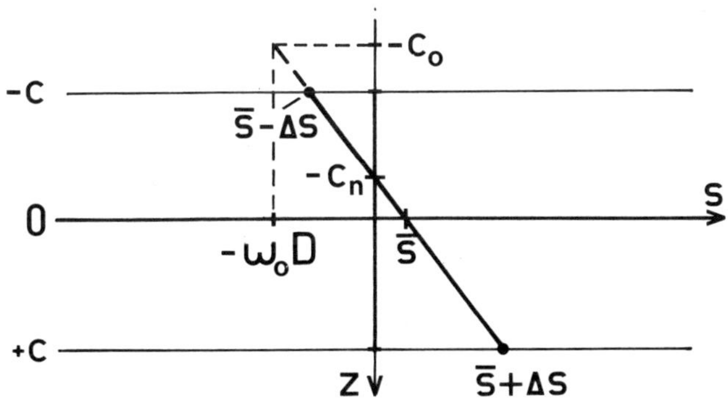

Fig. 3 Net stress distribution $s(z)$ over cross section $-c \leq z \leq c$ in pure bending, when damage closure is incomplete, that is when $\bar{s} - \Delta s > -\omega_o D$ or $-c_o < -c$.

For $\omega_o = 0$ complete damage closure will occur already on the neutral axis and for ω_o sufficiently large the compressive stresses fail to close the damage completely such that $c_o > c$. These two cases will be investigated separately.

We express the moment as the ratio of apparent linear elastic tensile extreme fiber stress to uniaxial tensile strength, σ_b/σ_t. As long as <u>complete damage closure doesn't occur</u> Eqs. 10 yield

$$\left.\begin{aligned} \frac{\sigma_b}{\sigma_t} &= \frac{M/(I/c)}{\sigma_t} = 2\,\frac{\Delta s}{s_t}\left[1-\left(\frac{r_i}{c}\right)^2\left(\frac{\Delta s}{s_t}\right)^2\right]^{1/2} \quad \text{and} \\ \frac{\bar{s}}{\Delta s} &= \frac{c_n}{c} = \frac{s_t}{\Delta s}\left\{1-\left[1-\left(\frac{r_i}{c}\right)^2\left(\frac{\Delta s}{s_t}\right)^2\right]^{1/2}\right\}, \end{aligned}\right\} \qquad (16)$$

with $F = 0$. This time the complementary monotonic loading parameter is Δs. The specimen looses its load carrying capacity when

$$\Delta s_1/s_t = (c/r_i)/\sqrt{2} \quad \text{and} \quad c_{n1}/c = (\sqrt{2}-1)(r_i/c). \qquad (17)$$

Eqs. 9 and 13 with $\rho \to \infty$ yield $\sigma(z)$ at failure. Again max $\sigma = \sigma_t$ at $z = r_i$. Thus the rupture factor in bending for a specimen of damaging material is

$$R = \sigma_{tb}/\sigma_t = c/r_i > 1, \tag{18}$$

when $c_o \geq c$. Eq. 18 predicts that $R = 2$ for a solid circular section and that $R = \sqrt{3} \approx 1.7$ for a rectangular one. Experimental values (Ouchterlony, 1981) for Bohus granite, 2.1 and 1.7 respectively, and Ekeberg marble, 2.3-2.4 and 2.0 respectively, aren't far off.

For a material with complete damage closure under compressive stresses the equilibrium integrals are separated into compressive and tensile parts. In this case $c_n = c_o$. Introduce the dimensionless variables $\zeta = z/c$, the breadth function $b(\zeta) = B(z)/B$ so that $b(0) = 1$, and $\zeta_o = c_o/c$. Then Eqs. 10 reduce to

$$\left.\begin{aligned} \frac{\sigma_b}{\sigma_t} &= \frac{4A}{Bc} \cdot \frac{\zeta_o}{J_2(\zeta_o)}\{1+\zeta_o^2\left(\frac{c}{r_i}\right)^2[1-\frac{J_3(\zeta_o)}{\zeta_o J_2(\zeta_o)}]\} \text{ and} \\ \frac{\Delta s}{s_t} &= \frac{2A}{Bc}\,\zeta_o/J_2(\zeta_o) \quad \text{where} \quad J_n(\zeta_o) = \int_{\zeta_o}^{1} (\zeta+\zeta_o)^n b(\zeta)\,d\zeta. \end{aligned}\right\} \tag{19}$$

A sufficient condition for R being the maximum of σ_b/σ_t at ζ_{o1} is that $\Delta s(\zeta_o)$ has an inverse in the relevant interval. This will obviously depend on the explicit forms of $J_n(\zeta_o)$.

Take the rectangular section first (Krajcinovic, 1979). For it $A = 2Bc$, $c/r_i = \sqrt{3}$, $b(\zeta) = 1$, and $J_n(\zeta_o) = (1+\zeta_o)^{n+1}/(n+1)$. Thus we obtain

$$\left.\begin{aligned} \sigma_b/\sigma_t &= 6(4\zeta_o - 9\zeta_o^2 + 35\zeta_o^3)/(1+\zeta_o)^3 \text{ and} \\ \Delta s/s_t &= 12\zeta_o/(1+\zeta_o)^3. \end{aligned}\right\} \tag{20}$$

Since $\zeta_{o1} = (13-\sqrt{97})/18 \approx 0.175$ and $\Delta s/s_t$ increases monotonically until $\zeta_o = 0$ then $R = 1.63$. This agrees with the result of Bert and Kumar (1980) who defin strength failure differently than Krajcinovic (1979).

The solid circular section is more important for rock. For it $A/Bc = \pi/2$, $c/r_i = 2$, and $b(\zeta) = \sqrt{(1-\zeta^2)}$. In this case $\zeta_{o1} = 0.149$ with $\Delta s/s_t$ still increasing monotonically. Hence the rupture factor becomes $R = 1.85$.

Unfortunately complete damage closure under compressive stresses lowers the predicted R-values, but not much. Thus an analysis of the intermediate cases $c_n < c_o < c$ wasn't made. The damage law $\omega(s)$ can probably be chosen so that theory and experiments agree but that choice should have a sounder physical basis.

The results for pure bending also apply to three point bending, a point loaded cantilever, and four point bending, all cases with statically determinate mome distributions. For them the damaging material will redistribute the stress ove the cross section but not from one section to another.

CONCLUSIONS

CDM permits a unified approach to both eccentric uniaxial loading and pure bending of strength test specimens of rock. The damaging implies a realistic deviation from linearly elastic material behavior which changes the conventional prediction formulas for apparent strength. The damage law used here is simple. Hence these CDM results should only be regarded as qualitative, yet they are quite reasonable when compared with experiments.

Several multiaxial CDM formulations exist (Chaboche, 1981). Thus a unified approach to all common strength tests for rock is possible, not only multiaxial tensile ones like the hydraulic tension test, the Brazilian disc test, and the point load test. CDM even connects the uniaxial compression test with σ_t (Krajcinovic and Fonseka, 1981).

REFERENCES

Bert, C. W., and M. Kumar (1980). Discussion of Krajcinovic (1979). *J. Appl. Mechs.*, 47, 449-450.

Chaboche, J.-L. (1981). Continous damage mechanics - A tool to describe phenomena before crack initiation. *Nucl. Engng. Des.*, 64, 233-247.

Grady, D. E. and M. E. Kipp (1980). Continuum modelling of explosive fracture in oil shale. *Int. J. Rock Mechs. Min. Sci.*, 17, 147-157.

Hult, J. (1979). CDM - Capabilities, limitations and promises. In proc. interdisciplinary conf. *Mechanisms of Deformation and Fracture*, ed. K. Easterling, Pergamon Press 233-247.

Jansson, J. (1978). *Crack in Material with Damage Formation - A Mechanical Analysis*. Diss. Chalmers Institute of Technology, Gothenburg, Sweden.

Krajcinovic, D. (1979). Distributed damage theory of beams in pure bending. *J. Appl. Mechs.*, 46, 592-596.

Krajcinovic, D. and G. U. Fonseka (1981). The continuous damage theory of brittle materials. In 2 parts. *J. Appl. Mechs.*, 48, 809-824.

Mäki, K. (1982). Personal communication.

Ouchterlony, F. (1981). A distributed damage approach to combined bending and axial loading of rock beams. Swedish Detonic Research Foundation report DS 1981:19, Stockholm, Sweden.

Timoshenko, S. (1940). *Strength of Materials Part I - Elementary Theory and Problems*, 2nd ed. Van Nostrand, Toronto. Chap. 51.

Vutukuri, V. S., R. D. Lama, and S. S. Saluja (1974). *Handbook on Mechanical Properties of Rock*, Vol. 1. Trans Tech Publications, Aedermansdorf, Switzerland. Chap. 3.

THREE METHODS FOR DETERMINING THE J-R CURVE DIRECTLY FROM A LOAD DISPLACEMENT RECORD

S. Kaiser

Department of Strength of Materials, The Royal Institute of Technology, S-100 44 Stockholm, Sweden

ABSTRACT

Two methods to experimentally determine the growth of two and three dimensional cracks with knowledge only about the load displacement curve of the specimen and the deformation hardening of the material, are described. Furthermore is a method to determine the material tearing modulus from the load and displacement at maximum load given.

KEYWORDS

Stable crack growth, tearing modulus, key curve, J-R curve, surface crack, crack length measurement.

INTRODUCTION

The J resistance (J-R) curve is today the most widely used parameter to characterize stable crack growth. To determine the J-R curve one has to determine the J integral, which is rather easy, and the crack extension, which is more difficult. Normally the latter is done by the multi specimen technique or the unloading compliance method. Both these methods have been found to give reliable results, if used with care. The multi specimen technique, however, could be rather expensive and tiresome in a large test program and the unloadings required for the compliance method can be undesireable or impossible to carry out. Examples are dynamic testing like instrumented charpy tests (Joyce, 1981) and creep testing. A third method, the so called key curve method, has been proposed by Ernst and co-workers (1979). With this method the crack propagation can be determined directly from a load displacement record. The key curve, however, has to be obtained from subsized specimens of the same geometry.

In this study two methods which both essentially are key curve method, but only require a tensile test in addition to the P-Δ curve, are described. This should be a considerable simplification since a tensile test is often done anyway. The tearing modulus approach was also used to determine the slope of the J-R curve directly from the maximum load of the specimen.

MATERIAL

To test the applicability of the methods proposed in this study, CT and bend specimens of two structural steels, OX 540 and OX 812, were loaded monotonically and the load and displacement were registered (ASTM E813). Properties of these materials are given in table 1. The stresses (true stresses) and the hardening exponent N were determined by conventional tensile tests, and the tearing modulus by the unloading compliance method.

Table 1.

	σ_y (MPa)	σ_u (MPa)	N	T_{mat}
OX 540	375	600	6.5	400
OX 812	750	850	13.8	170

The experimental results are only briefly discussed due to limitations in the length of the paper. The interested reader is therefore referred to Kaiser (1983b) for more information.

DIRECT SEPARATION OF HARDENING AND CRACK GROWTH

The most basic approach to determine crack extension from a load displacement record is to try to separate the effects of strain hardening and crack growth on the load. If one assumes that the load P is a function of displacement Δ and crack length a, the change in P becomes:

$$dP = \left(\frac{\partial P}{\partial a}\right)_\Delta da + \left(\frac{\partial P}{\partial \Delta}\right)_a d\Delta \qquad (1)$$

The deflection can be separated into two parts, one elastic and one plastic. If the stress in the plastic region is assumed to be power law strain hardening and if the strain is proportional to the deflection, one can write:

$$\Delta_{pl} \sim \varepsilon_{pl} \sim \sigma^N \sim P^N \quad \rightarrow \quad \Delta_{pl} = D\,P^N \qquad (2)$$

Since the proportionality constant D only depends on crack length and geometry, the derivative of eq. (2) becomes:

$$\left(\frac{\partial \Delta_{pl}}{\partial P}\right)_a = D\,N\,P^{N-1} = \frac{N\,\Delta_{pl}}{P} \qquad (3)$$

Eq. (3) and eq. (1) together with the elastic compliance C gives an expression for the change in crack length da, namely:

$$da = \left(\frac{\partial P}{\partial a}\right)_\Delta^{-1} \left[dP - \left(C + \frac{N\Delta_{pl}}{P}\right)^{-1} d\Delta\right] \qquad (4)$$

For a three point bend specimen the load can be assumed to be proportional to the square of the ligament length b (eq. (12)), and eq. (4) thus becomes:

$$da = \frac{b\,d\Delta}{2\,(PC + N\Delta_{pl})} - \frac{b\,dP}{2P} \qquad (5)$$

For the CT specimen Ernst and co-workers (1981) has given a crack length dependence according to eq. (13), and eq. (4) becomes:

$$da = \frac{d\Delta}{\beta\ (PC + N\Delta_{pl})} - \frac{dP}{\beta P} \qquad \text{with} \qquad \beta = \frac{2}{b} + \frac{0.522}{W} \tag{6}$$

Looking at eq. (5) and eq. (6) one sees that these equations contain load and displacement which both can be determined from a P-Δ record, the deformation hardening which can be obtained from a tensile test, and the ligament length which is known in the beginning of the test and then can be updated through da.

When applying eq. (5) and eq. (6) to actual test records the da determinations showed large fluctuations in the beginning of the test (small Δ_{pl}). For larger displacements, however, and especially past maximum load (dP negative), the results became quite acceptable and comparable in accuracy to those of the compliance technique and physical measurements. The resulting J-R curves (for large displacements) are given in table 2 as the material tearing modulus. The latter since the slope of the J-R curve was close to constant during the whole phase of J-controlled crack growth

Half Elliptical Surface Cracks

Starting from a half elliptical surface crack, one can still use eq.(4) if the crack length is substituted with the ligament area A. Denoting the crack depth by a and the width by 2c, A and the aspect ratio R become:

$$A = BW - \frac{\pi ac}{2} \qquad\qquad R = \frac{a}{2c} \tag{7}$$

If the plate is loaded in tension, the load is proportional to the ligament area and eq. (4) becomes:

$$dA = \frac{A\ dP}{P} - \frac{A\ d\Delta}{(PC + N\Delta_{pl})} \tag{8}$$

Assuming that the crack grows the increments da and dc respectively, which causes the change dR in R, then the change in the ligament can be obtained by differentiating eq. (7):

$$dA = \left(\frac{\pi a^2}{4R^2}\frac{dR}{da} - \frac{\pi a}{2R}\right) da = \pi c\ \left(c\ \frac{dR}{da} - 1\right)\ da \tag{9}$$

One can further assume that R changes linearly with a (this is of course a limitation, it is, however, less severe than e.g. assuming R constant during the propagation). After the test it is easy to determine the ininitial crack lengths a_o and c_o and the final ones a_f and c_f, from the broken specimen. The change in aspect ratio can then be expressed as:

$$dR = \frac{1}{(a_f - a_o)}\left(\frac{a_f}{2c_f} - \frac{a_o}{2c_o}\right) da = h\ da \tag{10}$$

Eq. (8), eq. (9), eq. (10) together gives:

$$da = \frac{A}{\pi c\ (1-ch)} \left[\frac{d\Delta}{(PC + N\Delta_{pl})} - \frac{dP}{P}\right] \quad (11)$$

Like eq. (5) and eq. (6) this equation can be used to incrementally determine the crack depth from a load displacement record. A and c can then be updated through eq. (7) and eq. (10). Test results of surface notched wide plates given by Garwood (1979) were evaluated and showed less good results than did the CT and bend specimens. The scatter in $\Delta a_{est}/\Delta a_{act}$ was in this case up to ± 20%. Reasons for this could both be the difficulty in taking experimental data from another paper and the more complicated geometry. The method could therefore still be said to be promising.

As a conclusion the method of eq. (5), eq. (6), eq. (11) is very simple to apply and can easily be extended to more complicated geometries. The crack length determination is rather exact except for small displacements where it can be somewhat dubious. The method seems therefore to be best suited as a complement to other methods.

THE KEY CURVE METHOD

The key curve, which is the P-Δ relation for a specimen with a non growing crack, was given for the bend specimen by Ernst et. al. (1979):

$$P = b^2\, k\, \left(\frac{\Delta_{pl}}{L}\right)^{1/N} \qquad k \sim \frac{B}{L} \quad (12)$$

and for the CT specimen (Ernst et al, 1981):

$$P = b^2\, e^{\frac{0.522\, b}{W}}\, k\, \left(\frac{\Delta_{pl}}{W}\right)^{1/N} \quad (13)$$

The material constants k and N were originally determined by testing subsized specimens (Ernst co-workers, 1979). The crack length at a given point on the load displacement curve can then be determined by inserting the corresponding P and Δ_{pl} in eq. (12) or eq. (13).

Following the reasoning in eq. (2) the hardening exponent N can be obtained by a tensile test. If N is known, k can be determined in the following manner: Keep the ligament length b equal to the initial value. Calculate k for some points on the P-Δ curve and plot these values as a function of plastic deflection. Finally make a linear extrapolation back to zero plastic displacement (zero crack growth) were a value of k can be read off. This means that all information, except the hardening exponent, can be obtained from the load displacement curve. Another advantage is that an individual k is determined for each specimen and thus the material scatter is taken care of to some extent. T_{mat} determined with this method is given is in table 2.

Since eq. (12) and eq. (13) gives eq. (5) and eq. (6) on differentiating, the two approaches should give the same results under ideal circumstances. In practice, however, the key curve method shows less problems for small displacements since one always loses accuracy while differentiating. Furthermore the errors in crack length for small Δ_{pl} doesn't affect the result for larger Δ_{pl} since one always has a reference to the absolute crack length. The crack growth for specimens loaded both monotonically and cyclically (Kaiser 1983a) was determined and the final crack extensions calculated were within ± 11% of the actual ones for both CT and bend specimens of two materials. This is close to the error allowed by ASTM E813 for a valid J_{IC} test. A conclusion is therefore that it is better to use eq. (12) and eq. (13) than eq. (5) and eq. (6) despite the extra work caused by determining the constant k.

The Key Curve for 3-D Cracks

The load for a half elliptical surface crack loaded in tension can be written as:

$$P = A\,k\,\left(\frac{\Delta_{pl}}{W}\right)^{1/N} \tag{14}$$

Where A is the ligament area given in eq. (8) and W is the width of the specimen. The constant k can be determined by keeping A constantly equal to the initial ligament area, as was described in the 2-D case. To determine the actual liagament during a test, P and Δ_{pl} are inserted into eq. (14). If the the aspect ratio changes during the crack growth, eq. (10) can be used. The liagament area as a function of crack depth then becomes:

$$A = BW - \pi a^2[2a_o/c_o + 4h(a-a_o)]^{-1} \tag{15}$$

Using eq. (14) and eq. (15) and solving for a gives:

$$a = \frac{2hS}{\pi} + \left[\left(\frac{2hS}{\pi}\right)^2 + \frac{4S}{\pi}(a_o/2c_o - ha_o)\right]^{1/2} \tag{16}$$

Where:

$$S = BW - \frac{P}{k}\left(\frac{\Delta_{pl}}{W}\right)^{-1/N} \tag{17}$$

Applying eq. (16)(17) to the results of Garwood (1979) the crack growth estimates were better (± 15%) than those of eq. (11) but still less accurate than in the 2-D case. A reason for the latter has been discussed earlier.

DIRECT DETERMINATION OF T_{MAT}

It was shown by Kaiser and Carlsson (1981) that the material T-modulus can be determined by introducing the slope of the P-Δ curve in the expression for T_{app}. An even simpler method could be devised by studying instability under load control (i.e. when $P = P_{max}$). Hutchinson and Paris (1979) have proposed the following expression as "exact" for the case of a bend specimen.

$$T_{app} = \frac{E}{\sigma_0^2}\left(\frac{4P^2}{Bb^2}\,\frac{1}{\frac{1}{C} + \left(\frac{dP}{d\Delta_c}\right)_a} - \frac{J}{b}\right) \tag{18}$$

Load control is equal to putting C = ∞. Further the derivative in eq. (18) was already determined in eq. (3). Eq. (18) can thus be expressed as:

$$T = \frac{E}{\sigma_0^2}\left(\frac{4P}{Bb^2}\left(N\Delta_{pl} + PC_c\right) - \frac{J}{b}\right) \tag{19}$$

Eq. (19) is the applied tearing modulus under load control, which means that it should be equal to T_{mat} when $P = P_{max}$. The strain hardening exponent N and the flow stress σ_o can be obtained from an ordinary tensile test. Δ_{pl}, P and at maximum load can easily be obtained from a load displacement diagram of a bend specimen. The ligament length b at maximum load is not known, however, but as a first approximation the initial ligament length could be used. An approximate value for dJ_R/da could then be calculated through eq. (19). Since J_{max} at maximum load is known (from the area under the P-Δ curve), the crack growth could be roughly estimated from:

$$\Delta a = \frac{J_{max} - J_{IC}}{dJ_R/da} \qquad J_{IC} \approx \frac{K_{IC}^2}{E} \tag{20}$$

The current ligament length at P_{max} can then be calculated and a corrected T be determined. The error in using K_{IC} instead of the real J_{IC} is small since it only gives a minor error in a correction term (100 % error in J_{IC} gives a 2 % error in T for these specimens)

For the CT-specimen the same procedure can be used. T_{app} for the CT-specimen can be found in Kaiser (1983) or Ernst and co-workers, (1981). In the case of load control the equation takes the form:

$$T = \frac{E}{\sigma_0^2} \left(\frac{\beta^2 P}{B} (N\Delta_{pl} + PC_c) + \frac{1}{B} \left(\frac{2}{b^2} - \beta^2\right) \int_o^{\Delta} P \, d\Delta \right) \tag{21}$$

With β from eq. (6). To test the validity of eq. (19) and (22) they were applied to bend and CT-specimens of standard size (W = 40 mm in both cases) of both material OX812 and OX540. The higher strain hardening in OX540 resulted in a larger deformation at maximum load. Therefore the difference between a T based on the original crack length and the corrected T based on the current crack length, was larger for OX540 than for OX812. The results are given in table 2 where T-values both uncorrected and corrected for crack growth are presented. The mean value for each specimen type is approximately the same as for the conventional T_{mat}, table 1, and the scatter is only slightly larger. The same scheme, as given above, can of course be used to determine the slope of the whole J-R curve, which is necessary if T_{mat} is not a constant. In this case one has to use eq. (18) instead, and put $C = C_{nc} - d\Delta/dP$, where $d\Delta/dP$ is the slope of the P-Δ curve (Kaiser & Carlsson, 1981). The correction for the actual crack length can, however, be used as discussed above

Table 2.

Material	Specimen type	Number of spec.	T eq.(5&6)	T eq.(12&13)	T eq.(19&21)	T_{corr}
OX 812	Bend	8	170	172	168	177
OX 812	CT	3	140	146	129	136
OX 540	Bend	6	422	440	343	382
OX 540	CT	3	365	367	332	382

Three Dimensional Cracks

The method discussed above can be applied to surface cracks. T_{app} for the half elliptical surface crack can be found in Kaiser (1983b):

$$\frac{\partial J_{app}}{\partial A} = - \frac{P^2}{A^2} \frac{1}{\frac{1}{C} + (\frac{\partial P}{\partial \Delta})_A} \quad (22)$$

With C = ∞ for load control, the change in ligament area and aspect ratio from eq. (9), eq. (10) and $(\partial P/\partial \Delta)_A$ from eq. (3), eq. (22) becomes:

$$T_{app} = \frac{E}{\sigma_0^2} \frac{\pi c(1-ch)\ P}{A^2} (N\Delta_{pl} + PC) \quad (23)$$

Starting from the initial values of a and c given by Garwood (1979) and updating them through the obtained approximate dJ/da as before, the results are as given in table 3. T is given both with an h factor (eq. (10)) calculated from the initial and final crack lengths and with h=0. The latter is equal to assuming propagation under constant aspect ratio. As a comparison Garwood (1979) reported $T_{mat} \approx 610$ when assuming constant aspect ratio and $T_{mat} \approx 220$ when assuming constant aspect ratio but using the actual values of a and c. σ_0 was taken as 530 MPa.

Table 3

Spec.	P_{max} (MN)	Δ_{pl} (mm)	a_o (mm)	c_o (mm)	h (1/μm)	T (h≠0)	T_{corr} (h≠0)	T (h=0)	T_{corr} (h=0)
1006	33.6	23.2	22.8	76	6.58	233	288	466	1265
1007	30	14.3	38.7	120	3.8	246	297	452	732
1046	24.1	10	56.2	164	2.8	242	298	448	613

ACKNOWLEDGEMENT

The financial support from the Swedish board for technical development project 80-3850 is greatfully acknowledged.

REFERENCES

Ernst, H.,Paris, P., Rossow, M. and Hutchinson, J. (1979) ASTM STP 677 p.581-599.

Ernst, H.,Paris,P. and Landes J. (1981) ASTM STP 743 pp. 467-502.

Garwood, S. (1979). Crack propagation toughness of structural steel at temperaturs above the brittle - ductile transition temperature. Final contract report, The Welding Institute. (Also in ASTM STP 700).

Joyce, J. (1981) Dynamic J-R curve testing NUREG/CR-2274 U.S. Nuclear Regulatory Commission.

Kaiser, S., Carlsson, J. (1981) Studies of different criteria for crack growth instability in ductile materials. ASTM 2:nd Int. Conf. on EPFM Phil. 1981 To be published.

Kaiser, S. (1983a) On the relation between stable crack growth and fatigue. Fatigue of Engineering Materials and Structures.

Kaiser, S. (1983b) Report 42. Dep. of Strength of Materials, Royal Institute of Technology Stockholm.

Hutchinson, J. and Paris, P. (1979). ASTM STP 668, pp. 37-64.

AN ELASTIC-PLASTIC ANALYSIS OF RELATIONSHIPS BETWEEN K_I, J AND CTOD

S. Pilcer* and N. G. Ohlson**

*Institute of Materials Science and Technical Mechanics,
Wroclaw Technical University, Wroclaw, Poland
**Materials Research Center, Royal Institute of Technology, Stockholm,
Sweden

ABSTRACT

Relation problems between fracture toughness parameters are discussed. It is found that generally known relations, valid in linear elasticity, can be extended into the elastic-plastic range through use of certain factors which take into account the strain hardening exponent. These modified relations can be used for the design of full-scale parts working under linear-elastic conditions, employing, as fracture parameters for the material, data from tests performed on small specimens.

KEYWORDS

NLFM, J-integral, CTOD, boundary integral equation method, strain-hardening material, unloading compliance technique.

INTRODUCTION

The fracture process has often been characterized by means of the stress intensity factor K_I, the J-integral, and the crack tip opening displacement (CTOD) δ. Closed solutions have been found for certain crack geometries in cases of linear-elastic and rigid-plastic models of a material. In other cases, taking into account strain hardening and the presence of a plastic zone larger than what is assumed in small scale yielding models, only approximate methods can be used. Certain elastic-plastic analyses have been verified by experiments or by comparison with solutions obtained in the case of small scale yielding. As a result of such considerations, relations have been suggested between K_I, J and δ in the form of the generally accepted formulae derived in linear elasticity, supplemented with numerical correction factors. The results obtained on the basis of these suggestions are often incompatible. This may either be due to errors in estimating K_I, J or δ by numerical means, or it may be due to difficulties in the evaluation or interpretation of experimental data.

Observations by the present authors as well as reports in various papers support the reliability of the numerical J-integral calculations due to its practically established path independence. Similar estimations of stress intensity factors, especially those based on the stress analysis near the crack tip, are more complicated. The problem consists of the proper calculation of the order of singularity and the choice of points where the function $K(r,\theta)$ has been established, r denoting the distance from the crack tip and θ the polar angle. The crack tip opening displacement, however, is the most controversial quantity in numerical calculations. There are many suggestions concerning how displacements δ should be defined, starting from the tangent definition, through the opening at an intercept on the profile of a chord at some particular angle from the tip, and finishing with displacements measured on the elastic-plastic interface just behind the crack tip. Therefore, numerical analyses of the relations between K_I, J and δ have been connected with conditions under which these parameters were calculated.

Purpose of Investigation.

The present work has two aims. The first is to express of relationships between K_I, J and δ as functions of the strain hardening exponent n.

$$\varepsilon = (\sigma/\sigma_o)^{1/n} \tag{1}$$

The second is to compare numerical results obtained for a particular value of this exponent with the results of experiments carried out on an aluminium alloy. At the same time, the type of stress state which is assumed to prevail in the specimens during the tests may be verified. The knowledge of the type of stress state is essential for the purpose of determining what relations should be used in fail-safe design based on the results of the tests.

Material Used in Experiment.

The material used was an aluminium alloy Polish standard PA 6N of the following composition, 94% Al, 4% Cu, 0.7 % Mg, 0.4%Mn. It was tested in the as-received condition.

Experimental and Numerical Procedures.

The experiments were made on three-point bend specimens according to British standard (BS 5762, 1979). Load versus crack opening displacement and load versus deflection under load were measured for each specimen. Four independent methods -electric potential drop (DD19, 1972), impedance (Öberg, 1975a), high frequency (Öberg, 1975b), and unloading compliance techniques (Paris & Herman, 1975)- were used to define the onset of crack initiation. These methods all indicate that crack initiation occurs at a normalized load $\sigma/\sigma_y \approx 0.5$, σ being the maximum net stress disregarding the effect of the notch and σ_y the yield stress of material.

Calculations of the stress intensity factor K_I were performed according to the formula,

$$K_I = \frac{P}{B\sqrt{W}} \cdot f(a/w), \tag{2}$$

P being the point force on the specimen, a crack length, W ligament and B width of the specimen.
The J-integral was calculated from the formula

$$J = \frac{2}{B(W-a)} \int_o^u P du, \tag{3}$$

see for example Clarke et al, employing the one-specimen method for determing J. Here, u denotes the displacement under P.
The crack tip opening displacement was calculated according to BS 5762, 1979.

The specimen subjected to pure bending has been considered in the numerical analysis. The boundary integral equation method was employed to solve the compatibility equation

$$\nabla^2(\nabla^2\phi) = h(x_i) \tag{4}$$

in plane strain and plane stress, respectively. Plastic strains were established by means of the Prandtl-Reuss rule and the Huber-Mises plasticity criterion. The method of successive elastic solutions (Mendelson, 1968) has been applied in the iteration process to find the solution of the compatibility equation. Numerical integration of equations

$$u_i = u_i + \int_L u_{i,j}\, dx_i \tag{5}$$

where

$$u_{i,j} = u_{i,j}^o + \int_L (\varepsilon_{ij,k} + \varepsilon_{ik,j} - \varepsilon_{jk,i}) dx_k \tag{6}$$

has been performed to calculate displacements from the established strain distributions.

The stress intensity factor K_I and the J-integral have been obtained from calculated stresses, strains and displacements by means of the formula

$$K_I = \sigma\sqrt{2\pi} \cdot r^n, \tag{7}$$

for $\theta = 0$, and the definition

$$J = \int_\Gamma (n_1 W - T_i u_{i,1})\, ds \tag{8}$$

where σ denotes the normal stress perpendicular to the crack line and n is the order of the singularity.

The crack tip opening displacement δ has been defined under the following two assumptions. Firstly, definition of δ is sensible only if there is a plastic zone. Secondly, δ has been defined as the opening displacement of a point on the elastic-plastic interface, which is the tangent point to the straight line intersecting the crack line at the distance "a" from the crack tip, that is to say at the point of the clip gage displacement measurement in the COD test, Fig. 1. The difference between the present definition and the tangent definition is negligible so long as the displacement line is straight.

RESULTS

1. The relationships

$$J = K_I^2/E \text{ and } J = (1-\nu^2)K_I^2/E \tag{9}$$

are valid for the linear elastic model only in plane stress and plane strain, respectively. Some numerical functions of the strain hardening exponent, $\kappa_1(n)$ and $\kappa_2(n)$, should be introduced in these formulae in the form of a factor. Fig. 2 shows J vs. σ, using suitably normalized quantities. Fig. 3 shows the function κ in the case of plane strain.

2. Similarly, functions $m_1(n)$ and $m_2(n)$ should be introduced in the relation

$$J = \sigma_Y \, \delta, \tag{10}$$

(Rice, 1968). Fig. 4 shows δ vs. σ (normalized) and Fig.5 shows the factor m in the case of plane strain.

3. The relationships

$$\delta = K_I^2/E\,\sigma_Y \quad \text{and} \quad \delta = (1-\nu^2)\,K_I^2/E\,\sigma_Y \tag{11}$$

may be modified in the same manner as eq. (9) and eq. (10), through use of the factors $\lambda_1(n) = \kappa_1/m_1$ and $\lambda_2(n) = \kappa_2/m_2$, respectively.

4. A good agreement has been observed between experimental and plane strain numerical values of the J-integral for n = 0.14. The strain hardening exponent was obtained from a tensile test.

5. The experimental relation between J and K_I (cf. Fig.3) supports the validity of the first suggestion.

6. Comparison of the measured and calculated plastic components of the clip gage displacements suggest a presence of plane strain during experiments, Fig. 6, even although some of the requirements of recommended testing procedures in COD testing are not fulfilled.

CONCLUSIONS

The numerical evaluations together with experimental investigations show that the relations between fracture toughness parameters, valid for a linearly elastic material, can be extended into the plastic range for a strain hardening material. It is suggested that this extension be made by means of factors which are functions of the strain hardening exponent. The exponent for the material in question is determined from an ordinary tensile test of a smooth specimen. This extension makes possible important enginering applications in fail-safe design, using fracture mechanics data obtained from tests of small specimens.

ACKNOWLEDGEMENTS

This paper summarizes some of the results achieved in a scientific collaboration established between the technical universities of Wroclaw and Stockholm. Financial suppport for this work from both universities is gratefully acknowledged. The authors also would like to thank all those who have assisted in the course of the project, through practical help as well as through advice. In particular, the skilful preparation of the typescript by Ingunn Wester is much appreciated.

REFERENCES

BS 5762. British standard methods for crack opening displacement testing (1972). British Standards Institution, London.

Clarke, G.A., Andrews, W.R., Paris, P.C. and Schmidt, D.W. Single specimen tests for J_{IC} determination. In Mechanics of Crack Growth, ASTM STP 590 pp. 27–42.

DD 19. Methods for crack opening displacement testing (1972). British Standard Institution.

Mendelson, A. (1968). Plasticity. Mac Millan, New York.

Paris, P.C. and Herman L. (1975). Improved compliance techniques for observing crack growth with reference to J testing. Presented at the 9th Nat. Symp. on Fr. Mech., Pittsburg.

Rice, J.R. (1968). A path-independent integral and the approximate analysis of strain concentration by notches and cracks. J. Appl.Mech. 35, 379–386.

Server, W.L., and Wullaert, R.A. The use of small specimen strength ratio for measuring fracture toughness. EFM 11, 477–486.

Öberg, H. (1975). An eddy current crack detector, report 15. Dept. Strength of Materials. Royal Inst. Techn., Stockholm.

Öberg, H. (1975). An electrical impedance method for determination of crack growth, report 14. Dept. Strength of Materials. Royal Inst. Techn., Stockholm.

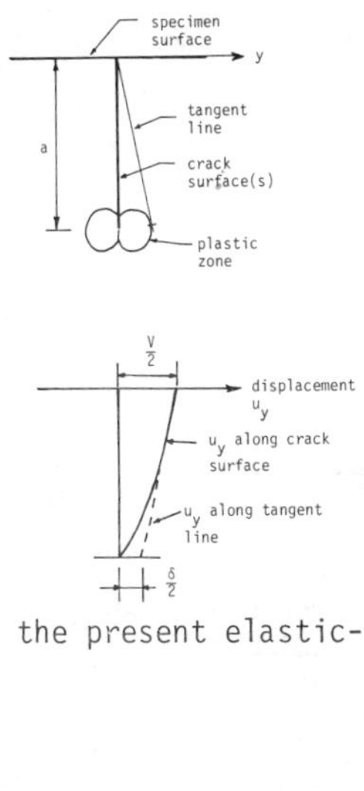

Fig. 1. Definition of CTOD in the present elastic-plastic analysis.

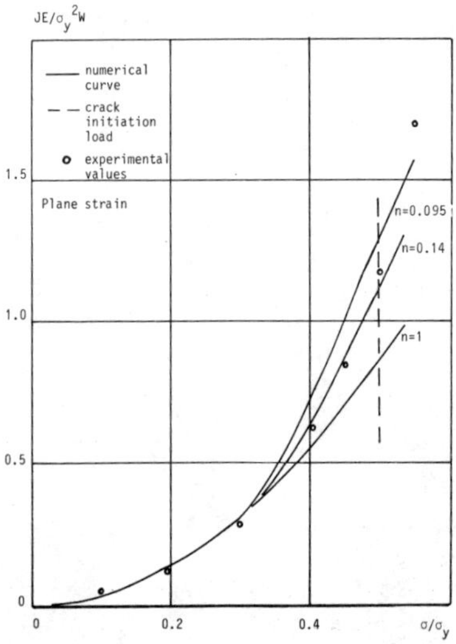

Fig. 2. Computed non-dimensional values of J integral vs. applied load in plane strain for two different values of n. Experimental values inserted for comparison.

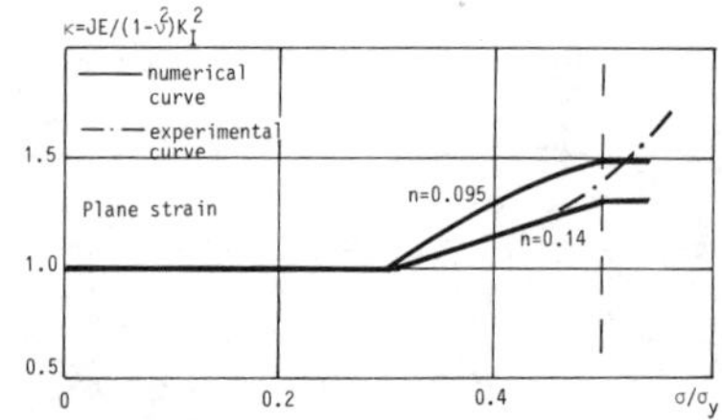

Fig. 3. Correction factor κ as a function of load for two values of n, together with experimental results.

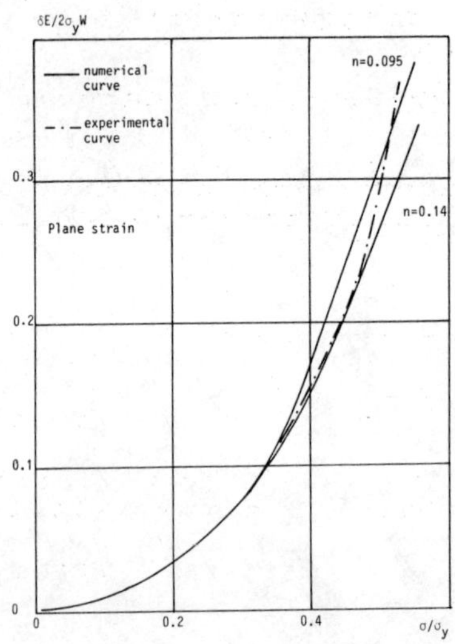

Fig. 4. Computed non-dimensional values of crack opening vs. applied load for two values of n and exp. results.

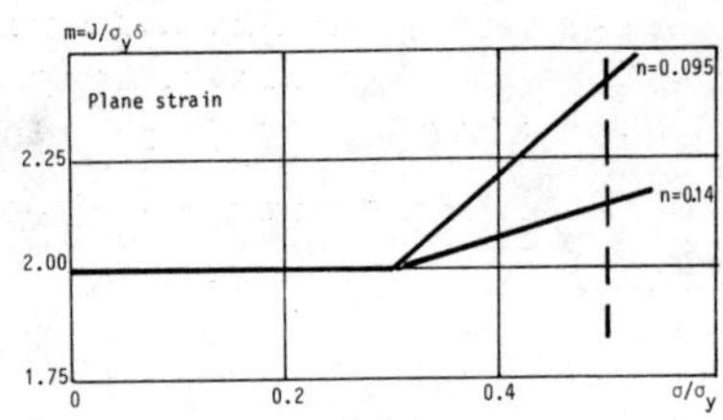

Fig. 5. Correction factor m as a function of load in plane strain.

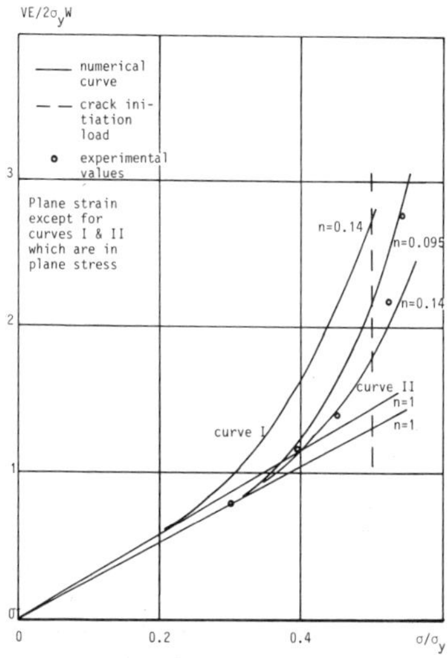

Fig. 6. Crack opening displacement V as measured by means of clip gage: computed values for plane strain and plane stress as well as experimentally obtained values as functions of applied load.

STRESS HOLD TIME, STRESS RATE AND FREQUENCY EFFECTS ON FRACTURE UNDER CORROSIVE ENVIRONMENT AND UNDER HIGH TEMPERATURE CREEP, FATIGUE AND CREEP-FATIGUE INTERACTION CONDITIONS

Takeo Yokobori*, A. Toshimitsu Yokobori, Jr and N. Takasu****

*Professor Emeritus of Mechanical Engineering, Tohoku University, Sendai, Japan

**Department of Mechanical Engineering II, Tohoku University, Sendai, Japan

ABSTRACT

In the present paper studies have been made on the effect of stress hold time, stress rate and frequency on fracture under corrosive environment, and the experimental formula for the crack propagation rate was obtained as a function of stress hold time, stress rise time and stress frequency. Also, the experimental formula for the crack propagation rate under high temperature creep, fatigue and creep-fatigue interaction conditions was obtained as a function of stress hold time. Some considerations were made and the results were discussed.

KEYWORDS

Fracture under corrosive environment; high temperature creep, fatigue, and creep-fatigue interaction; stress hold time effect; stress increasing rate effect; stress frequency effect.

INTRODUCTION

It is necessary to clarify the interaction of stress hold time, stress increasing rate, stress decreasing rate and stress frequency on time-dependent fracture, such as fracture under corrosive environment and fracture under high temperature creep, fatigue and creep-fatigue interaction conditions. There exist, however, few studies in which the effect of each factor was obtained in separate terms. Therefore, in the present paper we study each effect of stress hold time, stress increasing rate and frequency on time-dependent fracture of this type separately. Thus the experimental formula for the crack propagation rate under corrosive environment is obtained as a function of stress hold time, stress rise time (inversely proportional to stress increasing rate) and stress frequency. Also, the experimental formula for the crack propagation rate under high temperature creep, fatigue and creep-fatigue interaction conditions is obtained as a function of stress hold time.

EXPERIMENTAL RESULTS

The stress wave used is as shown in Fig. 1.

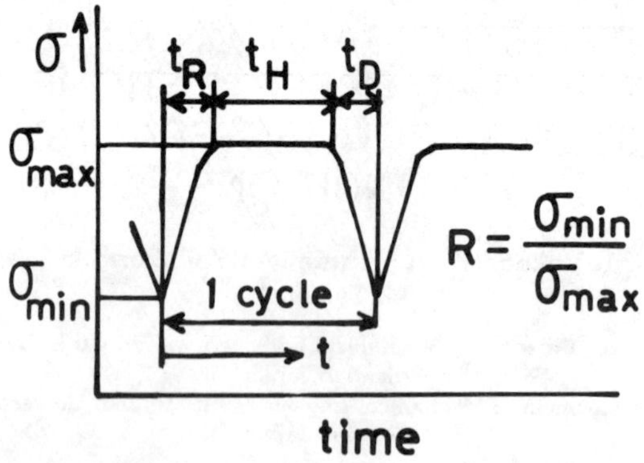

Fig. 1. Stress wave used.

For fracture under corrosive environment, stress hold time t_H and stress increasing rate were used as parameters. As it is reported (Barsom, 1972; Endo and Komai, 1965; Seilines and Pelloux, 1972) that the effect of stress decreasing rate is smaller as compared with that of stress increasing rate, the stress decreasing rate was kept constant in this study. As the measure of stress increasing rate, stress rise time t_R was used as inversely proportional to the stress increasing rate.

For fracture under high temperature creep, fatigue, and creep-fatigue interaction conditions, stress increasing rate and decreasing rate were kept equal.

Fracture under Corrosive Environment

Specimen, testing apparatus and testing method

The specimen, testing apparatus and the principle of the testing method is the same as reported in previous papers (Yokobori, Kuwano and Takizawa, 1981; A.T. Yokobori, Jr., T. Yokobori and N. Takasu, 1982). The specimen used is low alloy Cr-Mo steel, and the chemical composition is shown in Table 1.

TABLE 1 Chemical composition (wt%)

C	Si	Mn	P	S	Cr	Mo
0.38	0.34	0.84	0.018	0.004	1.07	0.17

It was machined as shown in Fig. 2 and oil quenched after heating for 10 min at 850°C in a salt bath and water quenched after heating for 60 min at 520°C.

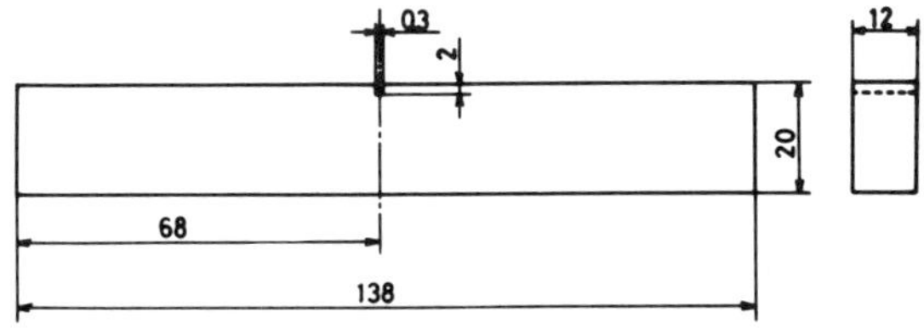

Fig. 2. Specimen shape and dimension for stress corrosion cracking and corrosion fatigue (in mm).

Static mechanical properties are shown in Table 2.

TABLE 2 Static mechanical properties

Yield stress N/mm^2	Ultimate tensile strength N/mm^2	Elongation %	Reduction of area %
1164.5	1270.3	12.53	42.49

After heat treatment a slit was provided by a crystal cut. For the stress corrosion cracking test and the corrosion fatigue test, a testing machine of cantilever type was used. 3.5% NaCl water as soluted in distillated one was used. The test specimen temperature and pH were kept at 18.0 ± 3.0°C and 6.3 ± 0.2, respectively, throughout testing.

In a previous paper (A.T. Yokobori, Jr., T. Yokobori and N. Takasu, 1982) a study has been carried out under the condition that the value of K_{max} increases from below the value of K_{ISCC} and with crack propagation it becomes larger than K_{ISCC}. The present article concerns the stress condition that the value of K_{max} was used at the starting stage of the test as larger than K_{ISCC}, that is, $K_{max} > K_{ISCC}$ throughout the crack propagation during the test (A.T. Yokobori, T. Yokobori and N. Takasu, 1983). The stress ratio $R = \sigma_{min}/\sigma_{max}$ was chosen as 0.16, where σ_{max} = maximum stress in the cycle and σ_{min} = minimum stress in the cycle. The range of t_H is $0 \sim \infty$ and the range of t_R is from 0.81 sec to 15.0 sec, respectively.

Effects of stress hold time t_H and stress rise time t_R on crack propagation rate

Figures 3 and 4 show the effect of stress hold time t_H on the plot of the logarithm of crack propagation rate per cycle da/dN against the logarithm of stress range intensity factor ΔK for each specific value of stress rise time t_R, respectively. From these figures, it can be seen that da/dN and the width of the plateau of da/dN independent of ΔK increases with increase of t_H. In Figs. 5 and 6 the logarithm of crack propagation rate da/dt is shown against the logarithm of maximum stress intensity factor K_{max} for each specific value of stress rise time t_R, respectively. From these figures, it can be seen that da/dt decreases with increase of t_H and becomes lower than the value of da/dt in stress corrosion cracking. Furthermore, Fig. 5 shows that when t_H becomes larger than some critical value, then da/dt will not increase more. In Fig. 7 log da/dN is plotted against log ΔK with both t_H and t_R as parameters. From Fig. 7 it can be seen that t_H at which the plateau appears becomes shorter with decrease of t_R, and da/dN becomes nearly independent of t_R when t_H is longer. Figure 7 also shows that the effect of t_R is large when t_H is small.

The effect stress frequency f on da/dN and da/dt

In Fig. 8 the value of log da/dN at $\Delta K = 150$ $kgmm^{-3/2}$ in the plateau of log da/dN versus log ΔK diagram was plotted against stress frequency f with the parameter t_R where f is defined as follows:

$$f = \frac{1}{t_H + 2t_R} \tag{1}$$

The solid line in Fig. 8 corresponds to the case of $t_H = 0$. The value of $(da/dN)_{scc} = (1/f)(da/dt)_{plat}$ is shown by dash-dot line in Fig. 8 as the crack propagation rate in the stress cracking corrosion where $(da/dt)_{plat}$ is the crack propagation rate at the plateau in log (da/dN) versus log ΔK diagram. From Fig. 8 it can be seen that in the range of higher frequency f, da/dN is much accelerated by t_R, whereas in the range of lower frequency f, da/dN is almost dependent of t_H only, but independent of t_R.

In Fig. 9 the value of log da/dt at $K_{max} = 180$ $kgmm^{-3/2}$ in the plateau of log da/dt versus log K_{max} diagram is plotted against stress frequency f with parameter of t_R. Figure 9 shows that da/dt decreases with increase of t_H, becomes lower than that in stress corrosion cracking, then reaches a minimum value and at last increases up to the value of crack propagation rate in stress corrosion cracking. In the shorter range of t_H, da/dt increases with increase of t_R. Whereas in the larger range of t_H, da/dt is almost determined by t_H only, but nearly independent of t_R.

According to the linear summation hypothesis (Wei and Landes, 1969) the crack propagation rate da/dN is given by the following formula:

$$\frac{da}{dN} = \left.\frac{da}{dN}\right|_E + \left.\frac{da}{dN}\right|_A \tag{2}$$

where $da/dN|_E$ = the contribution due to stress corrosion to da/dN, and $da/dN|_A$ = crack propagation rate unaffected by environment. Or in terms of time t (Gallagher, 1971), equation (2) is expressed as follows:

$$\frac{da}{dt} = \left.\frac{da}{dt}\right|_E + f\left.\frac{da}{dN}\right|_A \tag{3}$$

The solid straight line in Fig. 9 is the calculated value of $f\left.\frac{da}{dN}\right|_A$ from the data in air (Yokobori, Kuwano and Takizawa, 1981) and the dashed curve in Fig. 9 is the calculated value of da/dt by equation (3).

From Fig. 9 it can be seen that in the shorter range of t_H, da/dt will increase more than that given by the linear summation hypothesis and vice versa in the longer range of t_H.

Fracture under High Temperature Creep, Fatigue, and Creep-Fatigue Interaction Conditions

Specimen, testing apparatus and test method

The chemical composition and static mechanical properties at room temperature of 304 stainless steel used are shown in Tables 3 and 4, respectively. A double edge notched specimen was used as shown in Fig. 10. Tests were carried out in vacuum of 10^{-5} mmHg. The crack length was measured continuously during the tests without stopping. The minimum stress σ_{min} was zero and the gross section stress σ_g corresponding to the maximum value of

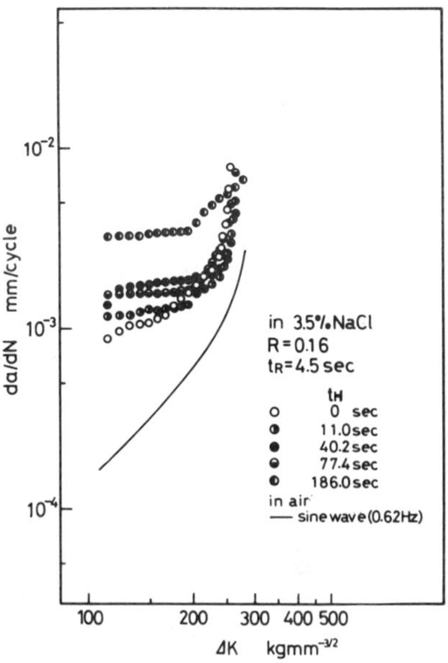

Fig. 3. da/dN versus ΔK (t_H = 0 ~ 186 sec, t_R = 4.5 sec)

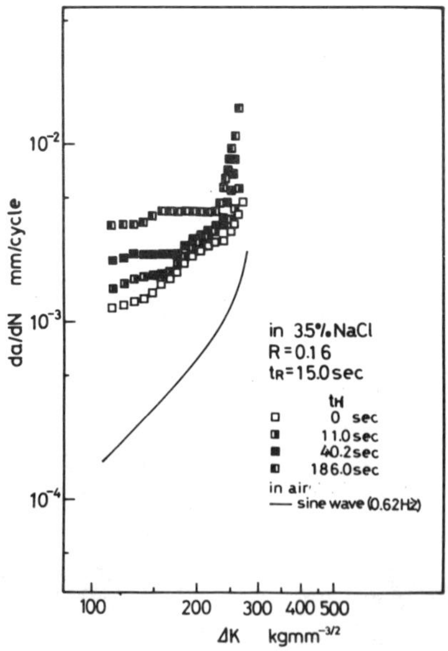

Fig. 4. da/dN versus ΔK (t_H = 0 ~ 186 sec, t_R = 15.0 sec)

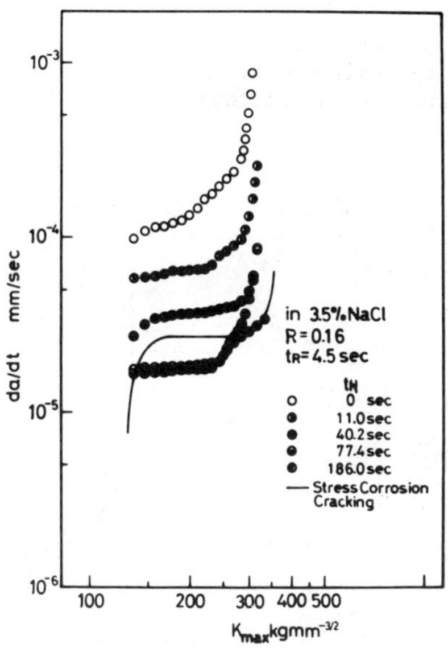

Fig. 5. da/dt versus K_{max} (t_H = 0 ~ ∞, t_R = 4.5 sec)

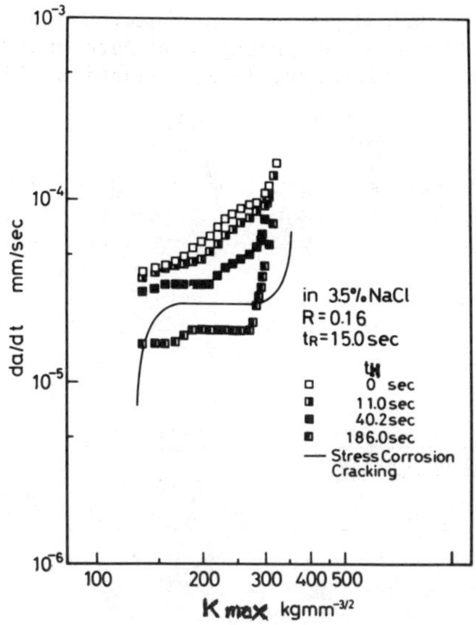

Fig. 6. da/dt versus K_{max} (t_H = 0 ~ ∞, t_R = 15.0 sec)

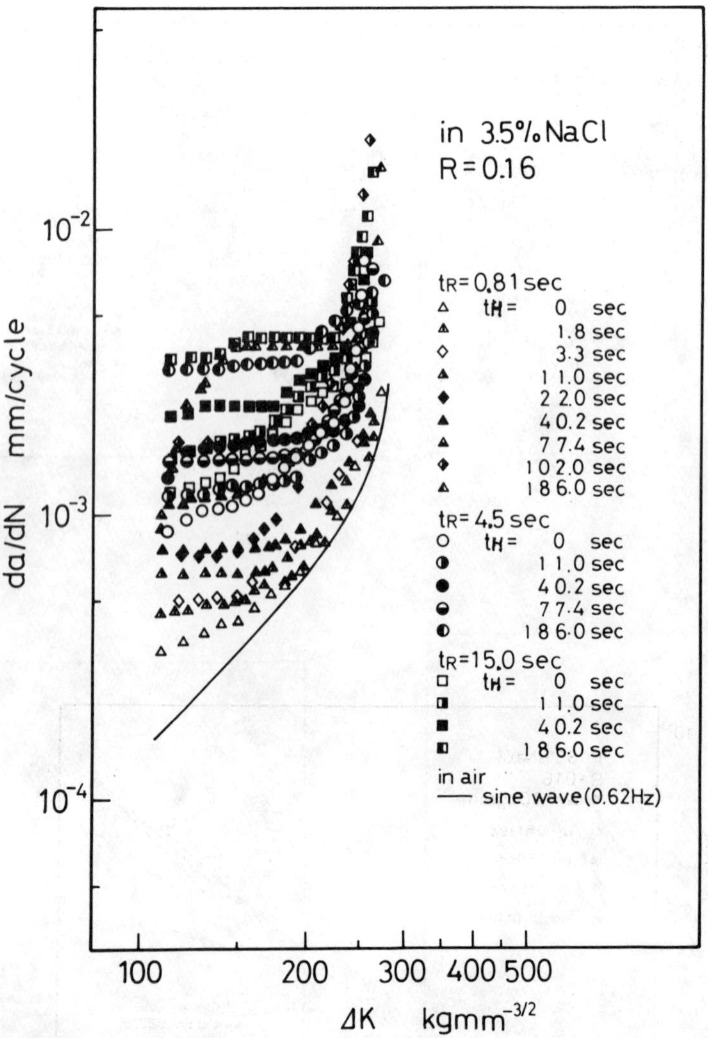

Fig. 7. da/dN versus ΔK with parameters, $t_H = 0 \sim 186$ sec and $t_R = 0.81 \sim 15.0$ sec, respectively.

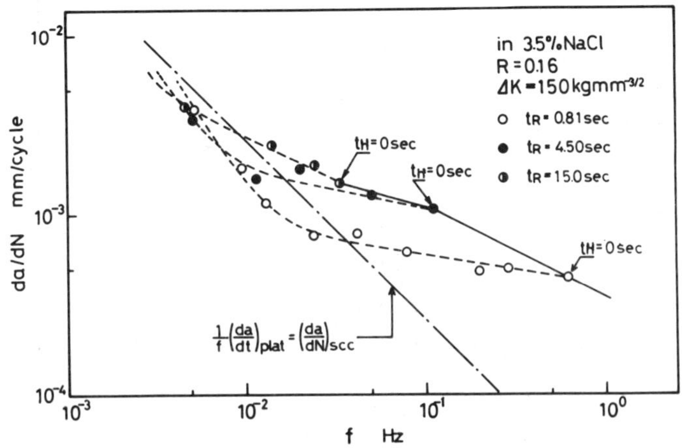

Fig. 8. The relation between da/dN and frequency f as affected by t_H and t_R.

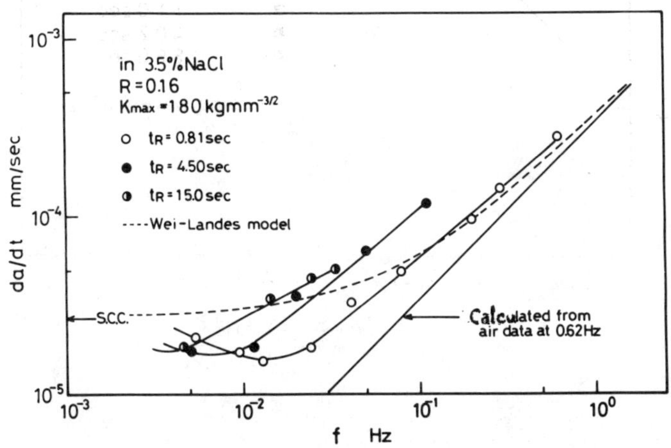

Fig. 9. The relation between da/dt and frequency f as affected by t_H and t_R.

tensile stress σ_{max} was 181, 195 and 209 N/mm^2 and test temperature was 600, 650 and 700°C, respectively (Yokobori and Sakata, 1979).

TABLE 3 Chemical composition (wt%)

C	Si	Mn	P	S	Cr	Ni
0.094	0.56	1.28	0.027	0.013	18.0	8.81

TABLE 4 Static mechanical properties at room temperature

Yield stress N/mm^2	Ultimate tensile strength N/mm^2	Elongation %
336	663	47

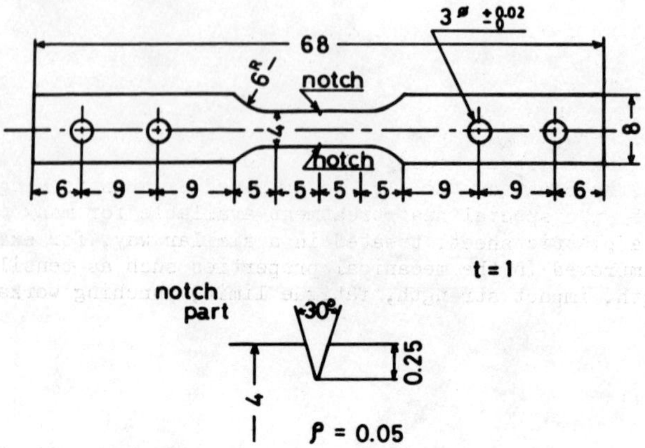

Fig. 10 Specimen shape and dimension for high temperature creep, fatigue, and creep-fatigue interaction tests (in mm).

Effects of stress hold time t_H on da/dt

In Figs. 11 and 12 log da/dt is plotted against the logarithm of frequency as affected by t_H with parameters of temperature and gross stress σ_g, respectively. It can be seen from Figs. 11 and 12 that da/dt is not simply a monotonical function of frequency f, although it is not so much affected by the frequency f.

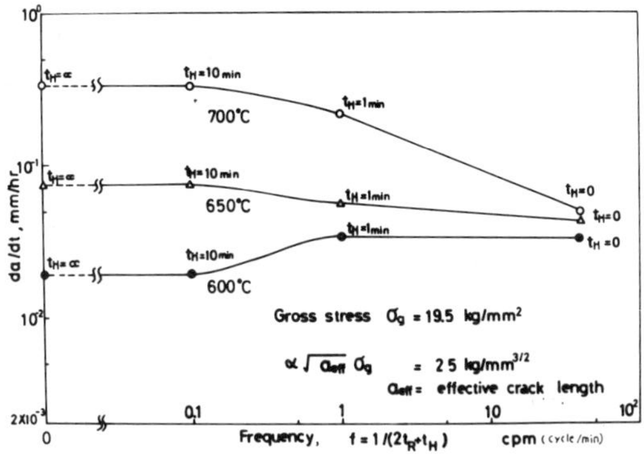

Fig. 11. The relation between da/dt and frequency as affected by hold time t_H in creep, fatigue and creep-fatigue interaction conditions with temperature parameter (304 stainless steel).

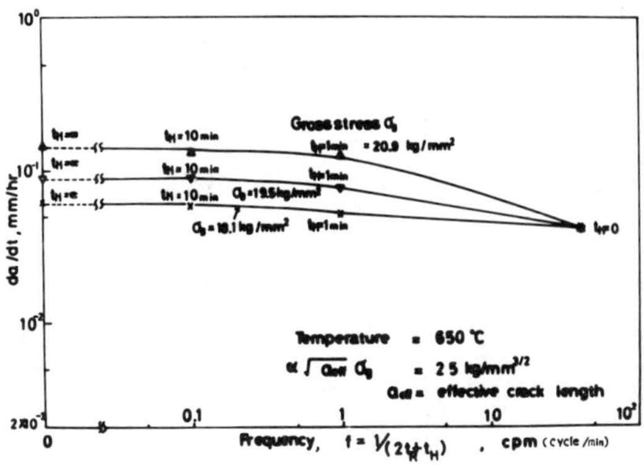

Fig. 12. The relation between da/dt and frequency as affected by hold time t_H in creep, fatigue and creep-fatigue interaction conditions with gross stress parameter (304 stainless steel).

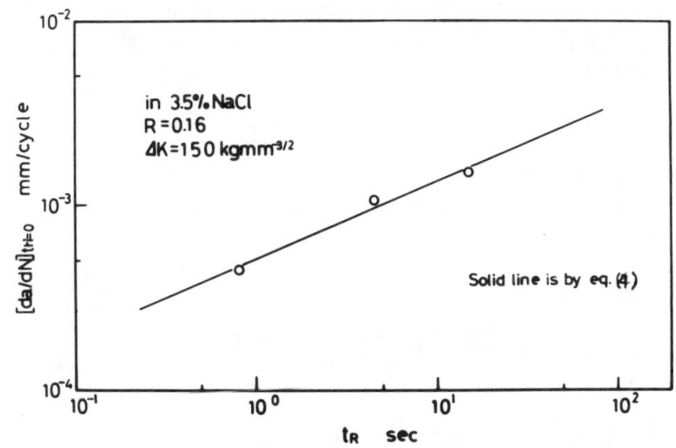

Fig. 13. The relation between da/dN and t_R (when t_H = 0).

ANALYSIS AND FORMULATION

Fracture under Corrosive Environment

Let us attempt to get the experimental formula of da/dN at the plateau in log da/dN versus log ΔK diagram under corrosive environment in terms of t_H and t_R. The crack propagation rate for the case of $t_H = 0$ is denoted by $(da/dN)_{t_H=0}$. Plotting $(da/dN)_{t_H=0}$ against log t_R we then get the linear relationship between them as shown in Fig. 13. It can be seen that the straight line is fairly well expressed by the following formula:

$$\left(\frac{da}{dN}\right)_{t_H=0} = 5.09 \times 10^{-4} t_R^{0.421} \tag{4}$$

or

$$\left(\frac{da}{dN}\right)_{t_H=0} = 3.80 \times 10^{-4} f^{-0.421} \tag{5}$$

Let us denote the crack propagation rate for the case with hold time t_H by $(da/dN)_{t_H}$. Plotting the logarithm of $[(da/dN)_{t_H} - (da/dN)_{t_H=0}]/(da/dN)_{t_H}$ against log t_H, we then get the linear relationship between them as shown in Fig. 14. It can be seen that this straight line can be fairly well expressed by the following equation:

$$\frac{\left(\frac{da}{dN}\right)_{t_H} - \left(\frac{da}{dN}\right)_{t_H=0}}{\left(\frac{da}{dN}\right)_{t_H}} = 6.34 \times 10^{-2} t_H^{0.482} \tag{6}$$

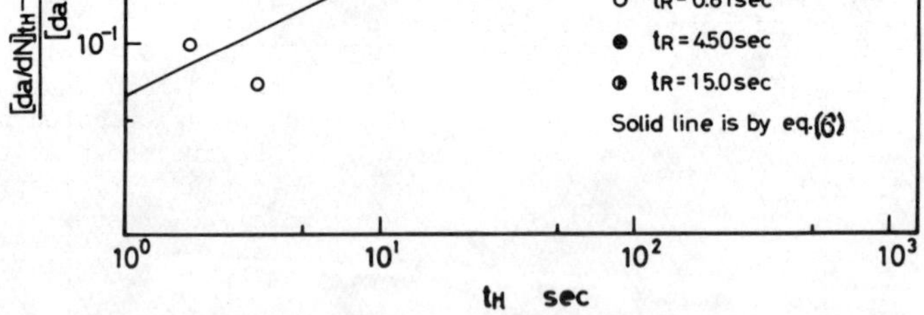

Fig. 14. The relation between da/dN and t_H with parameter of t_R.

From Eqs. (4) and (6) we get the following equation of crack propagation rate under corrosive environment in terms of t_H and t_R as parameters:

$$\left(\frac{da}{dN}\right)_{t_H} = \frac{5.09 \times 10^{-4} t_R^{0.41}}{1 - 6.34 \times 10^{-2} t_H^{0.482}} \tag{7}$$

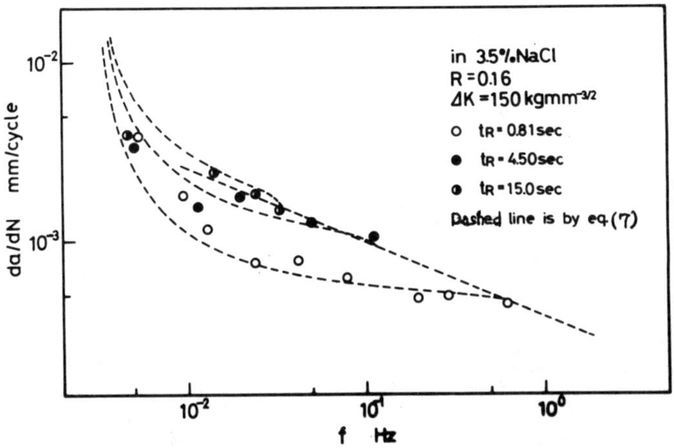

Fig. 15. The comparison of equation (7) with the experimental data.

In Fig. 15 the logarithm of the experimental value of da/dN is plotted against the logarithm of frequency f with parameter t_R and the calculated value by equation (7) is shown by dashed line in Fig. 15.

Fracture under High Temperature Creep, Fatigue and Creep-Fatigue Interaction Conditions

Based on the analysis (Yokobori et al, 1981) of experimental data (T. Yokobori and H. Sakata, 1979), high temperature crack propagation rate under creep, fatigue and creep-fatigue interaction of stainless steel can be commonly expressed by

$$\frac{da}{dt} = B_t \sigma_g{}^m (\alpha\sqrt{a_{eff}}\ \sigma_g)^{n_1} \exp[-\{\Delta f_1 - \Delta f_2 \ln(\alpha\sqrt{a_{eff}}\ \sigma_g/G\sqrt{b})\}/RT] \quad (8)$$

where Δf_1 is an apparent activation energy; Δf_1, Δf_2, m, n_1 and B_t are constants dependent of hold time t_H (for creep $t_H = \infty$; for fatigue $t_H = 0$) and a_{eff} = effective crack length, α = numerical constant, G = modulus of rigidity and b = Burger's vector. Values of these are shown in Table 5 against hold time. From Table 5, Δf_1, Δf_2, m, n_1 and B_t are plotted against t_H in Figs. 16 to 19. The values of Δf_1, Δf_2, m och B_t are shown as the ratio to $\Delta f_1{}^*$, $\Delta f_2{}^*$, m* and $B_t{}^*$ corresponding to the value for creep. From Figs. 16 to 19 Δf_1, Δf_2, m, n_1 and B_t are approximately expressed by the following formulas (T. Yokobori, A.T. Yokobori, Jr. and Sakata, 1983):

$$\Delta f_1 \simeq \Delta f_1{}^*[1 - 0.85 \exp(-0.45\ t_H)] \quad (9)$$

$$\Delta f_2 \simeq \Delta f_2{}^*[1 - 0.85 \exp(-0.45\ t_H)] \quad (10)$$

$$m \simeq m^*[1 - \exp(-2.1\ t_H)] \quad (11)$$

$$n_1 \simeq [181.1/(t_H + 2.2)^{2.5}] - 21.5 \exp(-1.48\ t_H) \quad (12)$$

$$B_t \simeq B_t{}^*[10^{35.5(10^{-0.544 t_H} - 1.2e^{-0.766 t_H})}] \quad (13)$$

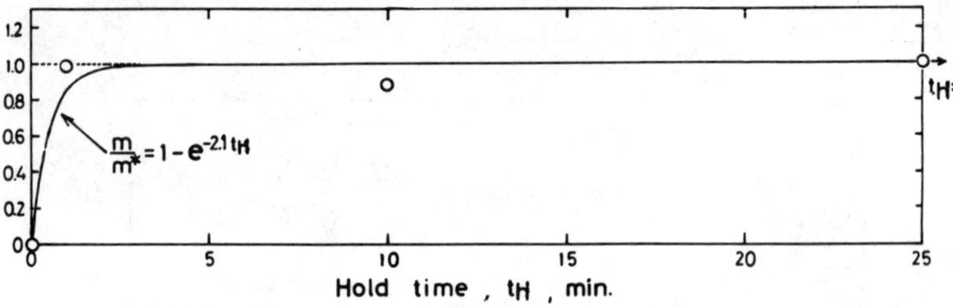

Fig. 16. The relation between m/m* and hold time t_H.

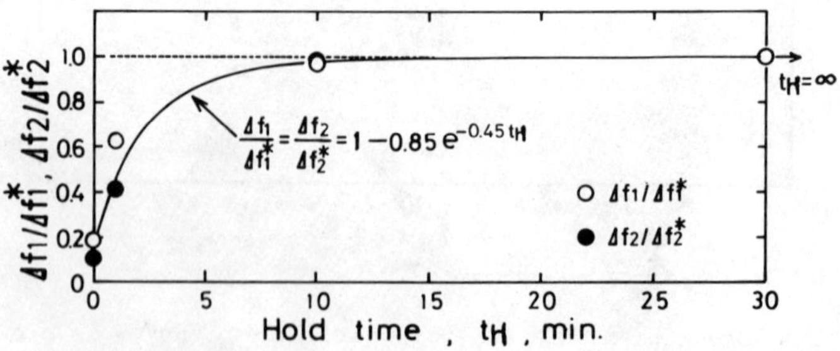

Fig. 17. The relation between $\Delta f_1/\Delta f_1^*$, $\Delta f_2/\Delta f_2^*$ and hold time t_H, respectively.

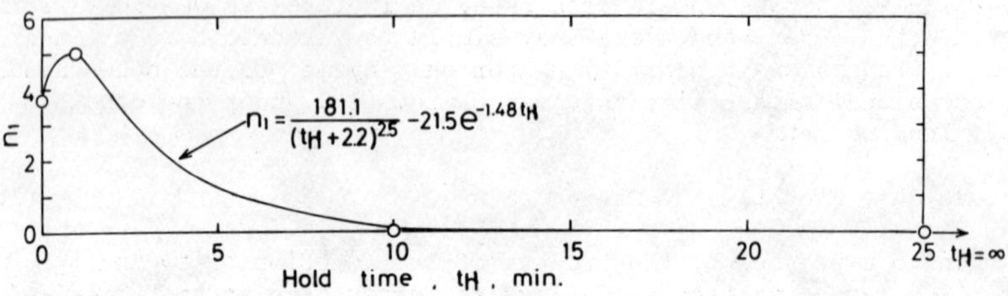

Fig. 18. The relation between n_1 and hold time t_H.

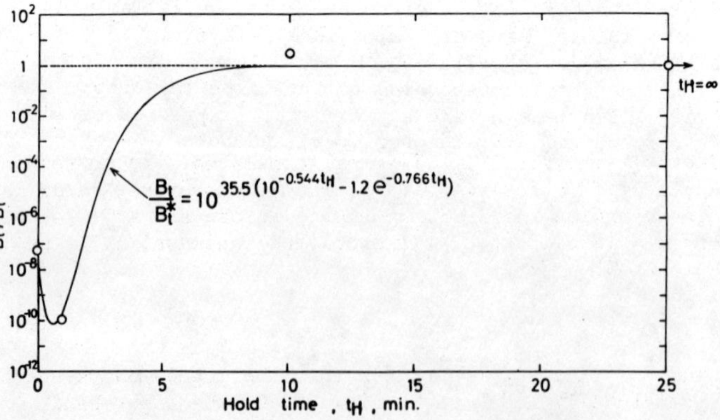

Fig. 19. The relation between B_t/B_t^* and hold time t_H.

TABLE 5 The Values of Constant in Equation (8)

	Δf_1 $K_{cal/mole}$	$\Delta f_1{}^*$ $K_{cal/mole}$
Creep ($t_H = \infty$)	78.2	85.6
t_H = 10 min	76.2	84.9
t_H = 1 min	49.2	51.4
Fatigue ($t_H = 0$)	14.8	17.0

CONSIDERATIONS AND DISCUSSION

Fracture under aggressive environments or at high temperatures can be treated as a thermal activation process. In the considerations a transition probability per unit time μ may be in general written as (Yokobori, 1974):

$$\mu = f(\sigma, t) \tag{14}$$

For the case where the stress changes with time as shown in Fig. 1, σ in Eq. (14) is:

$$\left.\begin{aligned} \sigma &= \dot{\sigma}t \simeq \frac{\sigma a}{t_R}\, t && \text{for } t < t_R \\ \sigma &= \sigma a && \text{for } t \geqq t_R \end{aligned}\right\} \tag{15}$$

where $\dot{\sigma}$ = applied stress application rate. σa is taken as $\sigma_{max}-\sigma_{min}$ for convenience. On the other hand, da/dN may simply be taken to have a linear relation to transition probability μ_o per unit cycle. On the other hand, μ_o is proportional to the time integral of Eq. (14) under the condition of Eq. (15). Thus we get

$$da/dN = F(\sigma, t_H, t_H + 2t_R) \tag{16}$$

When μ_o can be expressed by the product of the function of t_R and $t_H + 2t_R$, then Eq. (16) may be written as follows as far as time parameter is concerned:

$$da/dN = AF(t_R) \cdot F(t_H + 2t_R) \tag{17}$$

For the case $t_H \gg t_R$, Eq. (17) reduces to:

$$da/dN = AF_1(t_R) \cdot F_2(t_H) \tag{18}$$

where A is independent of time. The experimental formula Eq. (7) corresponds to Eq. (18). For the case of fatigue ($t_H = 0$) at room temperature, da/dN is given by a dislocation dynamics model (T. Yokobori, Konosu and A.T. Yokobori, Jr., 1977) or a vacancy diffusion model (Yokobori and Ichikawa, 1971) as follows:

$$\frac{da}{dN} = \frac{B}{f^{\lambda}}\,\Phi(\sigma) \tag{19}$$

Here B is constant and $0 < \lambda < 1$. Eq. (19) shows that the log da/dN versus log f relation is schematically as shown in Fig. 20. On the other hand, since da/dt = f da/dN, then

$$\frac{da}{dt} = \frac{B}{f^{\lambda-1}}\,\Phi(\sigma) \qquad (20)$$

which is schematically shown in Fig. 21. From Figs. 20 and 21 it can be seen that log da/dt versus log f characteristics show different behaviour as $\lambda \gtreqless 1$, whereas the log da/dN versus log f characteristics show only the change of the slope of the straight line as far as λ is positive. For fracture under corrosive environment, $\lambda < 1$ for the range of higher frequency, $\lambda > 1$ for the range of lower frequency and $\lambda \simeq 1$ for the interim range as shown in Fig. 9. For fracture under high temperature creep, fatigue and creep-fatigue interaction conditions, $\lambda \leqq 1$ as shown in Figs. 11 and 12.

Equation (7) holds at least in the range of $t_R = 0.81 \sim 15.0$ sec, but might not hold in a much shorter range of t_R. For the case $t_H = 0$, $(da/dN)_{plat} = 1.3\times10^{-4}\,f^{-0.8}$ is obtained for X-65 line-pipe steel at 25°C under 3.5 per-cent NaCl solution in distilled water (Vosikovsky, 1975), and $(da/dN)_{plat} = 1.87\times10^{-4}\,f^{-0.5974}$ for low carbon steel SA333-Gr6 at 288°C in water with oxygen content of 8ppm (Prater, Ford and Coffin, 1980). Equation (5) in our paper is about the same order although the material is different.

For instance, in the case of high temperature creep, fatigue and creep-fatigue interaction condition, such characteristics are reflected in Figs. 16 to 19. It can be seen from these figures that the value of n_1 and B_t shows the curve of two different trends, according to large range and small range of t_H. That is, in the small range of t_H, n_1 increases with increase of t_H, but in the large range of t_H vice versa. Also in the small range of t_H, B_t decreases with increase of t_H, but vice versa in the large range of t_H. As shown in Table 5 the value of activation energy Δf_1 obtained for creep is approximately equal to that of self-diffusion. On the other hand, the value of Δf_1 obtained for fatigue is the order of the activation energy for overcoming the resistance to the dislocation movement. The value of Δf_1 decreases with decrease of t_H. From these results it may be suggested that in fatigue the loading rate is effective, and, on the other hand, with increase of hold time, the hold time becomes more predominant.

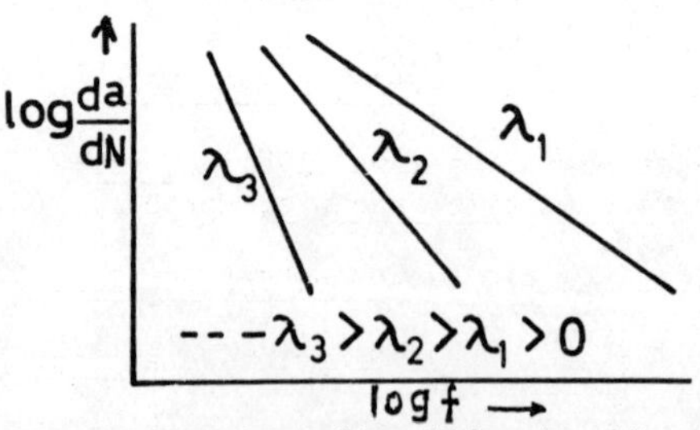

Fig. 20 Schematic illustration of log da/dN versus log f

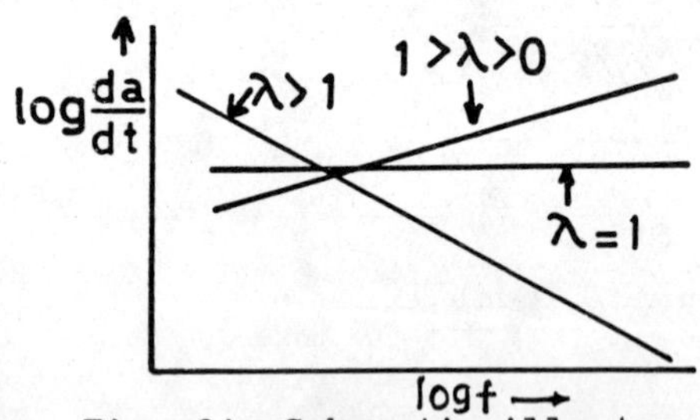

Fig. 21 Schematic illustration of log da/dt versus log f

CONCLUSIONS

(1) The crack propagation rate da/dN and da/dt under corrosive environment or at high temperature creep, fatigue and creep-fatigue interaction conditions shows different behaviour according to the range of stress hold time against stress frequency as affected by hold time. Such different behaviour is considered as corresponding to different mechanisms of crack propagation.

(2) An experimental formula for the crack propagation rate under corrosive fatigue of low alloy Cr-Mo steel under 3.5 percent NaCl solution was obtained in non-linear terms of stress hold time and stress increasing rate (stress rise time).

(3) A unified parametrical experimental formula of crack propagation rate for stage II of 304 stainless steel under creep, fatigue and creep-fatigue interaction conditions at high temperatures was given in terms of stress intensity factor, gross section stress, temperature and stress hold time.

ACKNOWLEDGEMENT

The authors would like to express their gratidude to Mrs. H. Kuwano, H. Takizawa and Dr. H. Sakata for their experimental cooperation.

REFERENCES

Barsom, J.M. (1972). Proc. Int. Conf. on Corrosion Fatigue, NACE-2, 424.
Endo, K. and K. Komai (1965). J. Soc. Mats. Sc. Japan, 14, 827.
Gallagher, J.P. (1971). J. Mater. 6, 941.
Prater, T.A., F.P. Ford and L.F. Coffin (1980). Metal Science 14, 424.
Seilines, R.J. and R.M. Pelloux (1972). Met. Trans. 3, 2525.
Vosikovsky, O. (1975). Trans. ASME, J. Engng. Mater. Techn. 97, 298.
Wei, R.P. and J.D. Landes (1969). Mats. Res. Standards 9, 25.
Yokobori, T. and M. Ichikawa (1970). Rep. Res. Inst. Strength & Fracture, Tohoku Univ., 6, 75.
Yokobori, T. (1974). Fundamentals and Methodologies of Fracture or Materials (in Japanese), Iwanami Shyoten, 172; Russian Edition (1978), Dumka Kiev, 189.
Yokobori, T., S. Konosu and A.T. Yokobori, Jr. (1977). Proc. 4th Int. Conf. Fracture, I, Pergamon Press, 665.
Yokobori, T. and H. Sakata (1979). Engng Fract. Mech. 13, 509.
Yokobori, T., H. Kuwano and H. Takizawa (1981). Trans. Japan Soc. Mech. Engrs. A 47 (in Japanese), 689.
Yokobori, T., A.T. Yokobori, Jr., H. Sakata and I. Maekawa (1981). Three-Dimensional Constitutive Relationships and Ductile Fracture, North Holland Publ. Co., 365.
Yokobori, T., A.T. Yokobori, Jr. and H. Sakata (1983). Proc. International Symposium on Fracture Mechanics, Beijing.
Yokobori, A.T., Jr., T. Yokobori and N. Takasu (1982). Preprint Japan Soc. Mech. Engrs. No. 820-2, 167.
Yokobori, A.T., Jr., T. Yokobori and N. Takasu (1983). Preprint Japan Soc. Mech. Engrs. No. 830-2, 65.

CREEP, FATIGUE AND ENVIRONMENT INTERACTIONS IN LOW ALLOY STEELS

A. Narumoto

Research Laboratories, Kawasaki Steel Corporation,
1 Kawasaki-cho, Chiba, Japan

ABSTRACT

The present work proposes a new method for predicting low cycle fatigue life at elevated temperatures on the bases of creep failure and environment assisted fatigue failure modes.

KEYWORDS

fatigue, high temperature fatigue, environmental effect, oxidation, creep, vacuum, low alloy steel, Cr-Mo steel.

INTRODUCTION

Low cycle fatigue at elevated temperatures has been studied with respect to the creep-fatigue interaction which is used to predict the fatigue life in ASME Boiler and Pressure Vessel Code Sec. III Case N-47. It has been noticed, however, that the environment plays an important role in low cycle fatigue at elevated temperatures. The present paper describes the effects of strain rate, temperature, wave form and environment on low cycle fatigue life of low alloy steels. It also proposes a new prediction method that takes into account the creep, fatigue and environment interactions.

EXPERIMENTAL PROCEDURE

The steels tested were four kinds of low alloy steels with various Cr and Mo contents. Their chemical compositions are shown in Table 1. The fatigue test specimens employed were smooth round bars with a gage section of 10 mm in diameter and 30 mm in length as shown in Fig. 1. Prior to the test, each specimen was polished in the longitudinal direction with #400 emery paper. The low cycle fatigue test was carried out by controlling the axial strain in air and vacuum at elevated temperatures using a servo-hydraulic testing machine with an induction heating equipment and a vacuum chamber which could reduce the pressure to less than $1x10^{-5}$ mmHg.

The axial strain was measured with an LVDT type extensometer whose gage length was 25 mm. The strain wave forms were even-even, fast-slow, slow-fast and strain hold triangular as illustrated in Fig. 6. The strain rate was between 1 %/s and 1×10^{-3} %/s. The fatigue life, N_f, was defined as the number of cycles required to reduce the load from a stabilized level by 25%.

EXPERIMENTAL RESULTS AND DISCUSSION

Creep, Fatigue and Environment Interactions

Figure 2 shows the effects of temperature and strain rate on the fatigue life under even-even wave forms for an ASTM A387-22 steel. The tests were performed at the total strain ranges, ε_{tR}, of 0.8% and 1.4%. The lives for the plastic strain ranges, ε_{PR} , of 0.5% and 1.0% were obtained by short range extrapolation or interpolation from the two experimental points. For both plastic strain ranges, no reduction of life was observed when $10^3/T$ was higher than 1.9 K^{-1} i.e. the temperature was below 250 °C. When the temperature was higher than 250 °C, however, the fatigue life decreased as the temperature increased and/or the strain rate decreased. The same behaviors of decrease in life could be observed in the other steels which are listed in Table 1. Figure 3 shows the pictures of subcracks after the fatigue test. When the strain rate was 2×10^{-1} %/s. and the temperature was 350 °C (a), the crack initiated along a slip band and propagated at an angle of 45° with respect to the specimen surface as observed in the specimen tested at room temperature. At a low strain rate (b), however, a crack propagated in the direction normal to the surface and the crack width increased due to the oxidation of the crack surface. The thickness of the oxide layer increased as the temperature increased and/or the strain rate decreased (c,d). It was almost uniform through the crack depth up to the crack tip. This fact indicates that a crack propagates accompanied by the oxidation at the crack tip. The growth of the oxide layer is coincident with the decrease in life at slow strain rate and/or high temperatures.

Figure 4 shows the effect of wave form on the life. In the test, one cycle period was about 280 s. and the total strain range was 1.4 %. The life under a fast-slow wave form coincided with that under an even-even wave form. On the other hand, the life obtained under a slow-fast wave form was also coincident with those under other wave forms at 450 °C ($10^3/T$=1.4). Above this temperature or in the so called creep temperature region, however, the life under a slow-fast wave form decreased as the temperature increased much more than in the other wave forms. Therefore, the fracture mode can be classified into two types. They are the environment assisted fatigue mode for even-even and fast slow wave forms and the creep damage mode for slow-fast wave form. This conclusion will be proved by the following observations. Figure 5 shows the relation between the compression time, τ_c , and N_f in air and vacuum at 650 °C. The tension time, τ_t , was 140 s. and ε_{tR} was 1.4 %. In vacuum, N_f increased with the increase in τ_c and reached the value obtained at room temperature. The fact that no difference existed between ε_{PR}-N_f curves obtained at room temperature and in vacuum at elevated temperatures for the even-even wave form has been reported by Coffin (1972). In air on the other hand, no environmental effect was observed when τ_c was shorter than a critical value. When τ_c exceeded the critical value a significant environmental effect on N_f was observed and the fatigue life began to decrease with the increase in τ_c due to the change in fracture mode. ASME B&PV Code Sec. III

Case N-47 assumes that both tensile and compressive creep is damaging. Figure 5 indicates, however, that the compressive creep recovers the tensile creep damage.

Figure 6 shows the relation between the life and the time for one cycle under various wave forms for an A387-22 steel at 650°C. The lives obtained under slow-fast wave forms are significantly shorter than those under the other wave forms. Though the strain hold period has been considered to be damaging, the damage should be discussed in comparison to the life under even-even wave form with equivalent cycle period. Figure 7 indicates that the strain hold cycle is less damaging than the even-even wave form with the equivalent cycle period. It is attributed to the oxidation behavior. Since the oxide layer is not cracked during the holding period, the environmental damage is less than for even-even wave form, in which the oxidation and cracking of oxide layer occur alternately.

Environment Assisted Fatigue Mode

Suppose that the environmental damage induced in one cycle, ϕ_e, is defined by the following equation:

$$\phi_e = 1/N_f - 1/N_o \qquad (1)$$

where N_f is the life at a given strain rate (or time for one cycle) and temperature, and N_o is the life at room temperature without environmental damage. The relations between ϕ_e and temperature for low alloy steels are shown in Figs. 7 and 8 for the plastic strain ranges of 1.0% and 0.5%, respectively. The value of ϕ_e was not significantly affected by Cr and Mo contents and is expressed by the following equation :

$$\phi_e = A\,(\dot{\varepsilon}_o \,/\, \dot{\varepsilon})^{0.27} \exp\left(-\frac{To}{T}\right) \qquad (2)$$

where $\dot{\varepsilon}_o$ is the reference strain rate equal to 1%/s, To is 2500K, $\dot{\varepsilon}$ is the strain rate in %/s, T is the test temperature in K and A is the plastic strain range dependent constant. The value of A was 8×10^{-3} for the plastic strain range of 1.0% and 4×10^{-3} for 0.5%. The environment assisted fatigue life for any combination of plastic strain range, strain rate and temperature can be predicted from the life obtained at room temperature using Eqs. 1 and 2 and the Manson-Coffin relationship. At thoroughly low strain rate and/or at thoroughly high temperature, this prediction method can be described as follows:

$$\phi_e \gg 1/N_o \qquad (3)$$

$$N_f \propto \phi_e^{-1} \qquad (4)$$

$$N_f \propto \nu^{0.27} \qquad (5)$$

where ν is the frequency. Equation 5 is equivalent to the frequency modified life concept which was proposed by Coffin (1970). The method proposed above is able to cover the transition region between room temperature and high temperature, and does not require the experimental constants which are used in the frequency modified strain method.

Creep Damage Mode

The creep damage mode was observed in the slow-fast wave forms. As τ_c increased, the tensile creep damage decreased. Figure 9 shows the comparison between the static creep and fatigue test results obtained under slow-fast wave forms at 550 °C and 650 °C using the Larson-Miller parameter which is given by the following equation :

$$P_{LM} = T \{ 20 + \log t_r \} \quad (6)$$

where t_r is the creep rupture time in h. In the present study, the creep stress and the rupture time for the fatigue test were defined by the maximum stress in a stabilized hysteresis loop and Eq. 7, respectively.

$$t_r = N_f (\tau_t - \tau_c) / 3600 \quad (7)$$

A considerably good agreement was observed between the static creep and fatigue test results. Thus the fatigue life for the creep damage mode can be predicted using Eq. 7 and the P_{LM} value which is obtained from the static creep test under a stress equal to the maximum cyclic stress.

Prediction of Life

The predicted lives for an environment assisted fatigue mode are shown in Fig. 10 in comparison with the experimental ones. The figure includes the data obtained under constant strain range at various strain rates and temperatures and under various strain ranges at a constant strain rate and temperature for several low alloy steels listed in Table 1. The predicted life agreed with the experimental one within a factor of 1.5. Figure 11 shows the experimental and predicted lives versus τ_c for the A387-22 steel tested at 650 °C. The values of τ_t were 28, 140 and 700 s. The experimental points fell on the prediction line for creep mode or that for environment assisted fatigue mode whichever gave the shortest life.

CONCLUSIONS

The conclusions obtained in the present study are as follows:

1) The fracture mode in low cycle fatigue at elevated temperatures for low alloy steels can be classified into two. They are the creep mode and the environment assisted fatigue mode. The fracture mode which gives the shortest life determines the life.
2) The compressive creep recovers the tensile creep damage.
3) The strain hold period is less damaging than the even-even wave form with an equivalent cycle period.
4) A life prediction method for two fracture modes is proposed.

REFERENCES

Coffin, L. F. Jr. (1970) Proc. Air Force Conf. on Fatigue and Fracture of Aircraft Structure and Materials, AFFDL TR7 0-144 pp301-312.

Coffin, L. F. Jr. (1972) Proc. Int. Conf. Fatigue, NACE-2, Houston, Tex. pp590-600.

TABLE 1 Chemical Compositions of Steels Tested.

	C	Si	Mn	P	S	Cu	Ni	Cr	Mo
A204B	0.25	0.24	0.73	0.010	0.008	0.01	0.03	0.04	0.49
A387-12	0.16	0.26	0.58	0.010	0.003	0.01	0.20	1.03	0.54
A387-22	0.13	0.08	0.51	0.009	0.008	0.12	0.12	2.41	1.01
A387-5	0.09	0.24	0.44	0.007	0.007	0.03	0.07	4.65	0.47

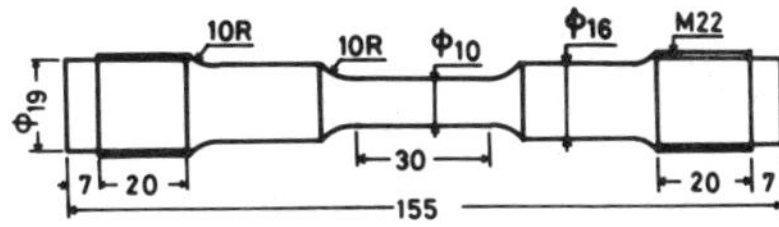

Fig. 1 Specimen dimensions in the unit of mm.

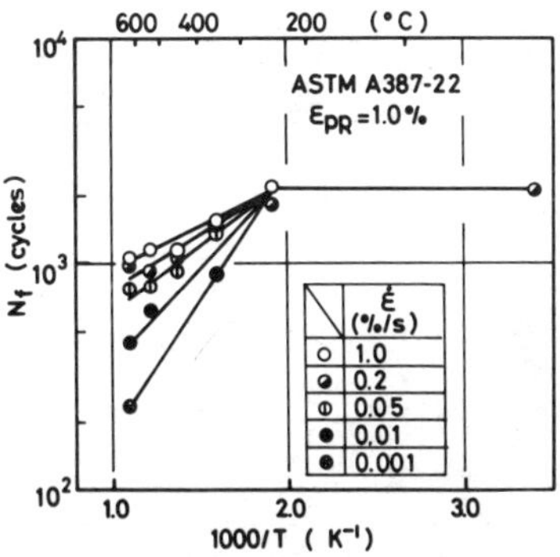

Fig. 2 Relation between N_f and 1000/T for an A387-22 at ε_{PR} of 1.0 %.

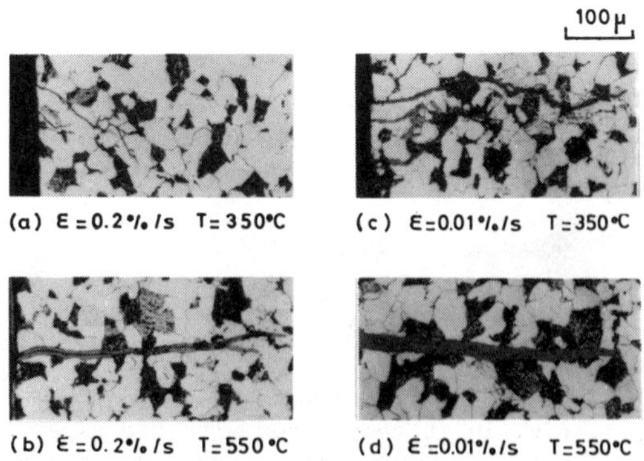

Fig. 3 Subcracks observed in the longitudinal cross sections of A387-22 specimens tested at the total strain range of 1.4 %.

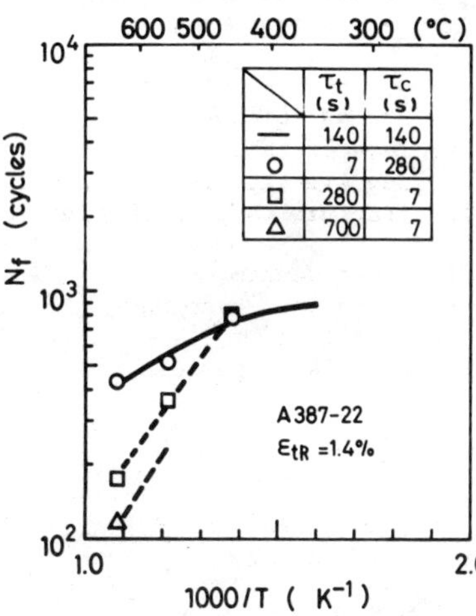

Fig. 4 Effect of wave form on the relation between fatigue life and temperature.

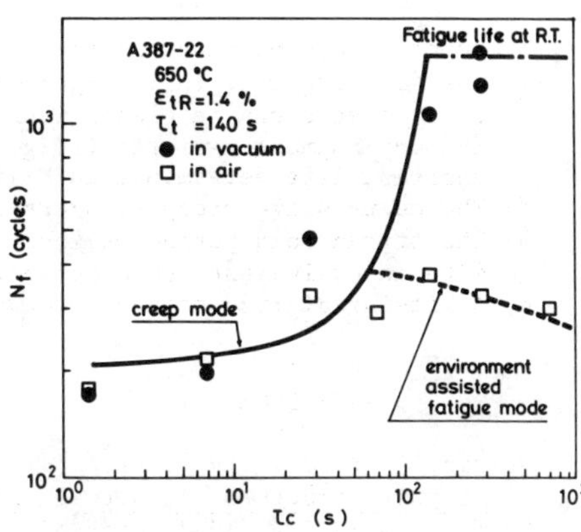

Fig. 5 Relation between τ_c and N_f obtained in air amd vacuum at 650°C and ε_{TR} of 1.4 %.

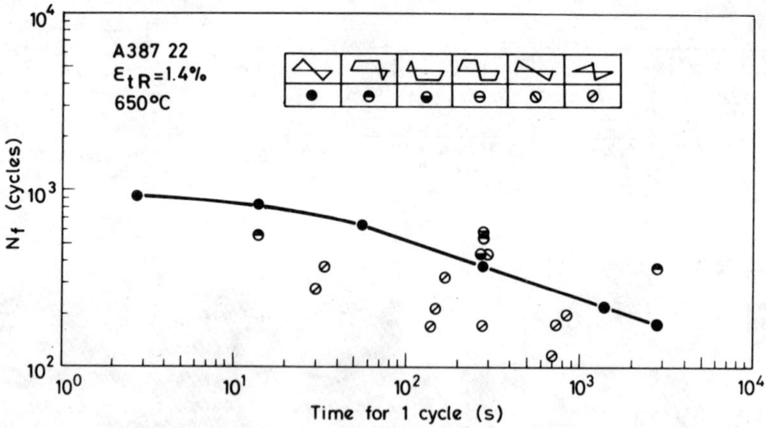

Fig. 6 Effect of wave form on the relation between N_f and time for one cycle.

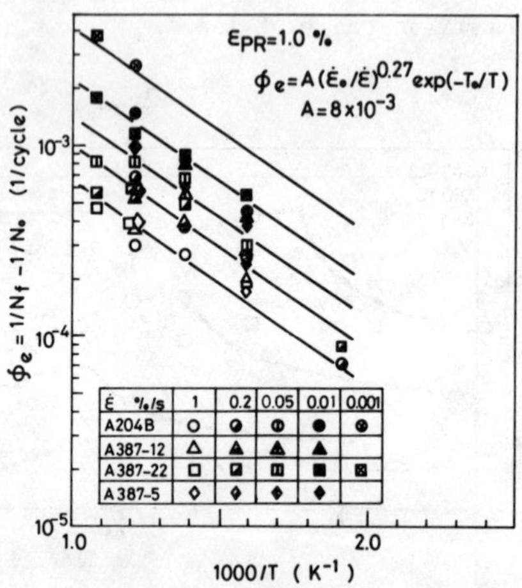

Fig. 7 Effects of temperature and strain rate on environmental damage in one cycle at ε_{PR} of 1.0% for several low alloy steels.

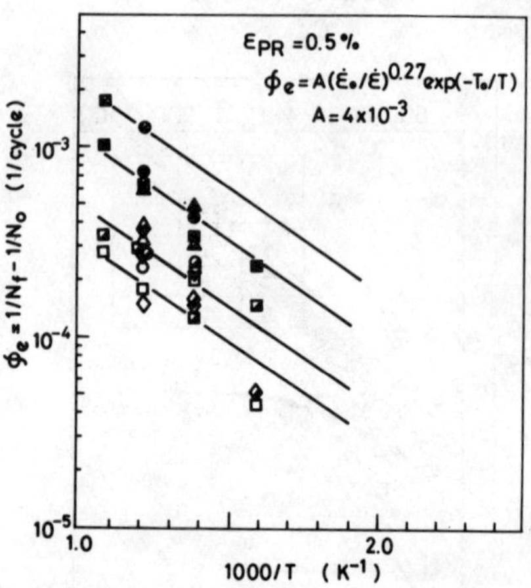

Fig. 8 Effects of temperature and strain rate on environmental damage in one cycle at ε_{PR} of 0.5% for several low alloy steels.

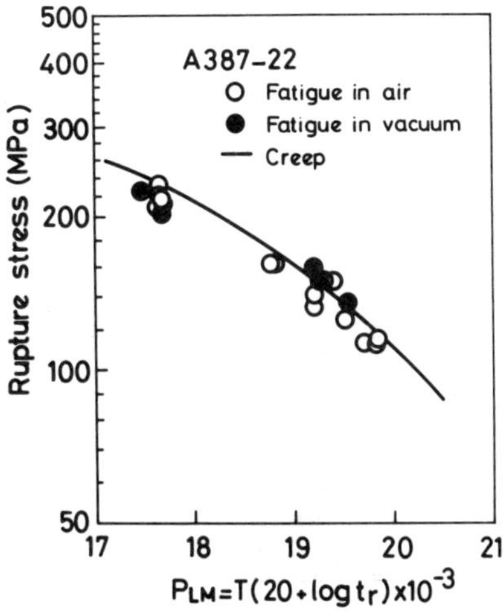

Fig. 9 A comparison of fatigue test results with monotonic creep test results for an A387-22 steel.

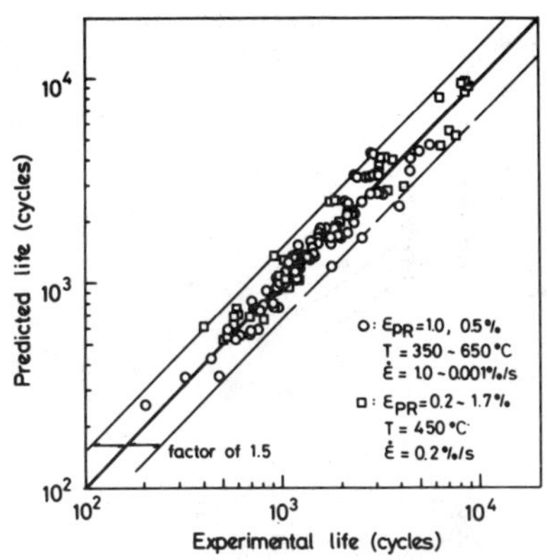

Fig. 10 A comparison of predicted life with experimental one obtained in environment assisted fatigue mode for several low alloy steels.

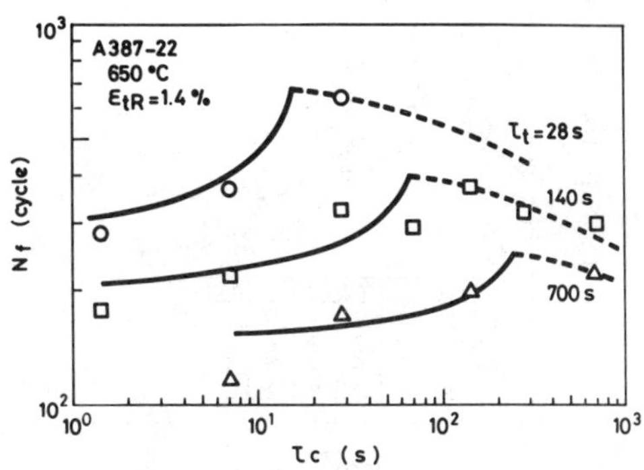

Fig. 11 A Comparison of predicted life and experimental one obtained in various sawtooth wave forms at 650°C and total strain range of 1.4 % for an A387-22 steel.

PHENOMENON AND APPLICATIONS OF THE RECOVERY-INDUCED THERMAL DEFORMATION OF THERMOPLASTICS

T. Machida, T. Okai and F. Hamayoshi

Department of Mechanical Engineering, Tamagawa University, 6-1-1, Tamagawagakuen, Machida-shi, Tokyo, Japan

ABSTRACT

A drastic deforming behaviour of thermoplastics and its applications are to be experimentally investigated. This phenomenon which is called "recovery-induced thermal deformation" seems to be caused by the shape memory effects. By utilizing this, a new simple process which is connected mainly with the local heating is here to be developed for cup forming and bending, the parts of which are confirmed to be acceptable in the dimensional stability. And furthermore, the deformation can also be obtained by transforming the cold working strain by the special heat-treatment available for many other purposes. Thus a plastic sheet, treated in a similar way, for example, can be remarkably improved in the mecanical properties such as tensile strength, shearing strength, impact strength, fatigue limit, punching workability and so on.

KEYWORDS

Thermoplastics; strain recovery, recovery-induced thermal deformation; sheet forming; dimensional stability; transforming of strain; special heat-treatment; thermo-responsive element; strengthening; mechanical property.

INTRODUCTION

It has for many years been believed that cold worked thermoplastics had a tendency to change their shapes, while the fabricated material in molten stage was most stable. However, it has been recently recognized that there is a strain recovery phenomenon of the melting worked thermoplastics and it is called "recovery-induced thermal deformation" by the authors (1979, 1980a). The principal reason behind these phenomena is a recovery of residual strains memorized by the thermoplastics or a shape memory effect of material.

Generally, the strain recovery in cold worked thermoplastics, which makes the shape of parts change, is a great trouble in processing plastics

(Maeda, 1978). But the authors (1980b, 1980c, 1981) also indicate that the recovery-induced thermal deformation, which does not appear in usual conditions, can be successfully utilized on the view points of two main purposes as follows: (a) applications of its own deformability to forming and thermo-responsive element etc., and (b) improvements of material properties by its memory mechanism. In the present paper, understanding quantitatively the phenomenon and the origin of the deformation is, first of all, mentioned. Its applications to sheet forming and useful attempts to make up the more desirable deformability and to improve the mechanical behaviour of material are on further considerations.

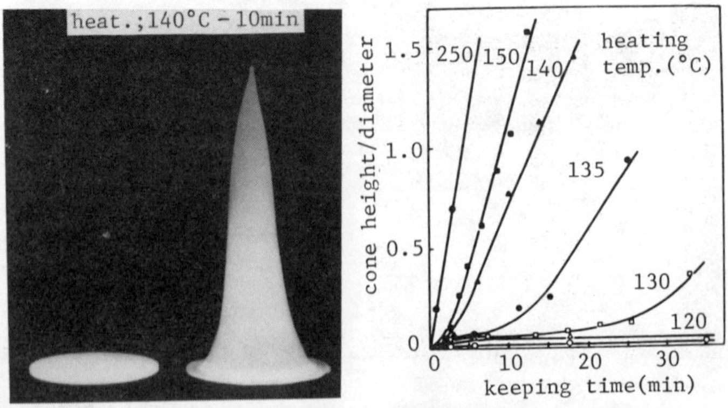

(a)disc and deformed cone (b)deformational behavior

Fig. 1. Recovery-induced thermal deformation of polyethylene bar.

RECOVERY-INDUCED THERMAL DEFORMATION

Phenomenon

Commercial thermoplastics bars of diameters 50 mm and 100 mm and sheets of thicknesses 1 mm, 3 mm, and 6 mm were mainly used for the study. In the case of bar, circular discs were sliced off by a saw and then the both sides of them were polished by abrasive papers. The polished discs were then kept at elevated temperatures for isothermal heating in an oil bath for a given time. As a result of only this heating, the discs change their shapes into cones due to inherent flow action. A typical example of the deformation, in case of using a screw extruded polyethylene, is photographically shown in Fig. 1(a). A great yielding or flow of 2 mm thick and 50 mm diameter disc has taken place to take a cone shape at an elevated temperature without application of external force. Fig. 1(b) shows the growth in height of the cone being formed with the increase of keeping time. It is clear from this result that at a low temperature the growth rate is very small, while at higher temperatures, especially above 140°C, the growth is very fast. Similar deformations were also observed in other thermoplastics, such as in polypropylene, polyacetal, polyamide, acrilic, polyvinyl chloride, polycarbonate and so on.

On the other hand, 50 mm square blanks were cut off from sheet of 3 mm thickness and then they were kept in a drying oven for a given time. The results of observation indicated distinct changes in length and thickness of the specimens as shown in photographs in Fig. 2(a). The length measured in the calendering direction was found to be shortened, while the thickness increased remarkably. This change is prominent in the case of polyethylene. The effect of temperature and keeping time on the deformation of the specimen, having 50 mm length in the calendering direction and 10 mm width, is illustrated in Fig. 2(b). The shrinkage starts after a lapse of small duration of heating which may be called latent period, and then it increases uniformly with the increase of keeping time until it reaches a saturation value of about 70%.

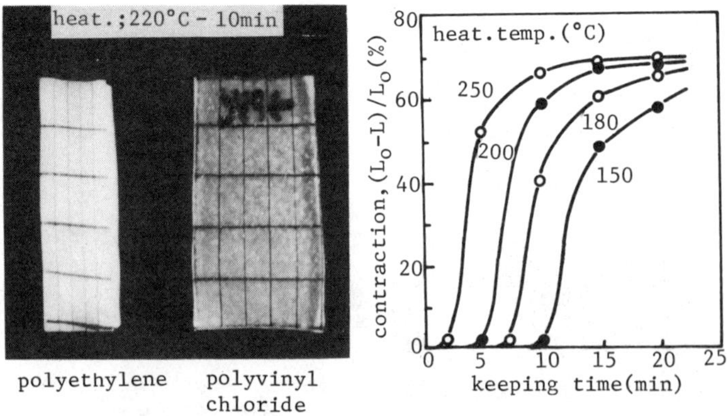

(a)shrinkage of blank (b)deformational behavior

Fig. 2. Recovery-induced thermal deformation of polyethylene sheet.

Application to Cup Forming and Bending

It is generally difficult to make cups with uniform distribution of thickness by conventional processes, such as blow forming, vacuum forming and others. The problem these processes face may, however, be solved by introducing the new forming process to be discussed now. To make a deep cup having thin wall and uniformly distributed thickness, the recovery-induced thermal deformation phenomena can successfully be used. Fig. 3(a) shows the logarithmic strain distributions in various directions of a cone formed of polyethylene. It is interesting to note that the circumferential strain values ε_θ are almost nil while the radial strain values ε_r are positive. Furthermore, the distribution of these strain values are remarkably uniform. As a result of this, the variation in the wall thickness of the cone is very small and this deep cone hence may be the most desirable preform for further blow forming.

With the experimental apparatus, a blank which is sliced off from a 100 mm diameter bar into 3 mm thickness is allowed to deform freely into a preform cone by the recovery-induced thermal deformation under a given temperature and time, and then it is formed into a desirably bottomed cup due to less stretching by an air blowing than usual methods. The result of the cup forming process is photographically shown in Fig. 4. The qualities of these

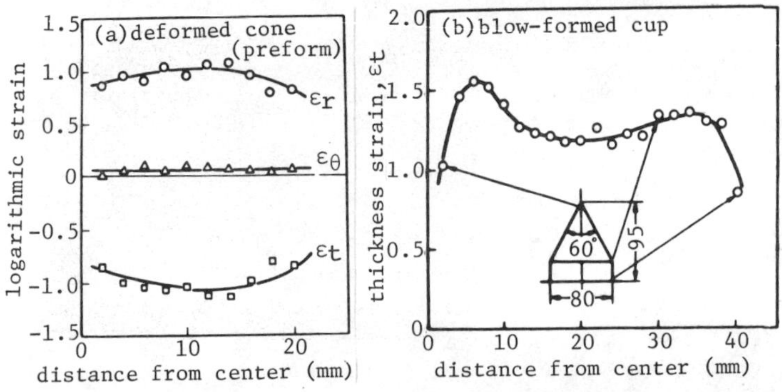

Fig. 3 Strain distribution of preform cone and formed cup of polyethylene.

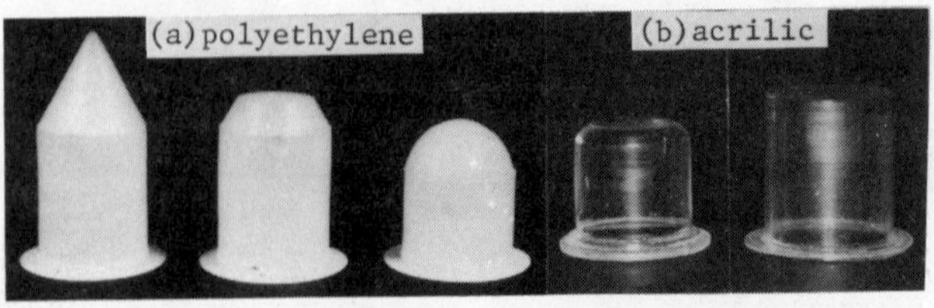

Fig. 4. Appearances of cup-formed parts.

cups are observed to be much higher than the products made by conventional processes from the point of view of thin but uniform wall thickness, thermal stability, sharp contours, dimensional accuracy etc. in the product. For example, since the cone shaped bottom of the cup is almost similar to the cone, the distribution of thickness strain is considerably uniform, as shown in Fig. 3(b). This is because of the fact that the forming by utilizing the recovery-induced thermal deformation is performed in hot condition, and it is based upon an unpinning mechanism which brings back every point in the blank to its original position before the displacement during extrusion.

On the other hand, the deforming behaviour of sheet can be utilized for bending without any die and external force. A blank sheet is placed on a flat metallic plate and heated locally by a spot ray from a radiator, which used an infrared lamp with a capacity of 600W in the present experiment, as shown in Fig. 5(a). The temperature in the surface of heated zone goes up rapidly to the molten state which can cause the material shrinkage in the calendering direction. Because the shrinkage induces a tensile stress in the direction about a bending line in only the upper layer, the blank is automatically bent. The sequence of bending steps of a polyethylene sheet of length 50 mm, width 10 mm and thickness 3 mm is shown by photographs in Fig. 5(b), in which the local heating is carried out by a lamp of 400W. In the initial stages (1 and 2), as the heating is affective only on the top of surface, thermal expansion occurs in it resulting in warping of the specimen. However, as the heat penetrates into the surface, the specimen regains its

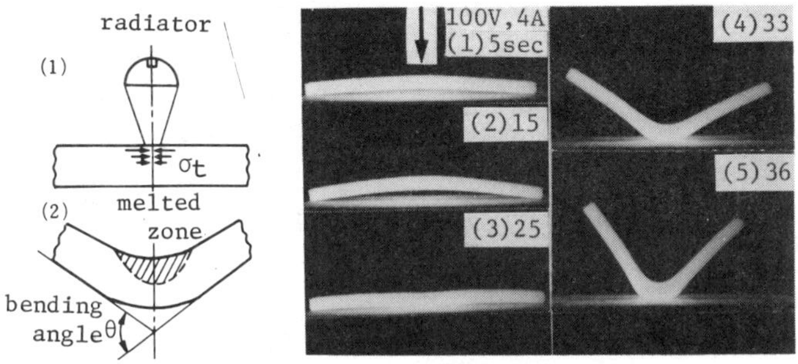

(a) bending method (b) deformation sequence

Fig. 5. Bending procedure of polyethylene sheet by local heating.

initial flat position (stage 3), and on further heating bending starts and continues by lifting of both ends about the zone being heated. This results into a sharp V-shape bend (stage 4 and 5). The bending may continue to contact of both ends, unless the deforming zone is cooled by switching off the lamp. Here, it is found that the heating condition has a prominent effect on the angle of bend. For example, the bending rate is accelerated with increasing lamp output, because in case of using a lower lamp output, a longer heating time is required to get the same bend angle. And also the distance of the lamp from the specimen affects clearly the bend radius, because the heated zone width depends on the width of the infrared ray. Hence, it is possible to control the opening angle and the radius of V-bend by a suitable choice of the heating condition. Thus, the process offers a die-less bending different from other conventional methods.

MECHANISM AND CONTROL OF THE DEFORMATION MEMORY

Cause of the Deformation

It has been recognized that the recovery-induced thermal deformation is caused by the long memory of shape, in other words, the release of the residual strains possessed by the product at the time of primary working such as extrusion and calendering of molten thermoplastics. Although thermoplastics bars and sheets are produced directly in a molten state, they possess a large amount of residual strain to be recovered, as if they were cold worked. Such residual strain cannot be released unless the product is heating to the working temperature. It is confirmed by the authors (1980d) in the deformation test of experimentally manufactured bar and sheet from molten polyethylene. Hence it should be abandoned that the melting working had for long time been believed to produce the plastics in a stabilized shape.

Trial Process for the Deformation Memory

A necessary condition for the recovery-induced thermal deformation to be utilized is not only thermo-responsiveness at an elevated temperature, but the deformation must be completely latent in the ordinary use of parts. On the other hand, the only possible method for memorizing it was to form or mold the molten plastics in the same way as commercial ones, since the as cold worked plastic is unstable in its shape, especially in long term use under warm environments. However, hot or melting worked material is more stable but has too little memory for us to use easily. And also, fabricating it is extremely hard because of many troubles such as heat loss, handling, unstable dimension and so on. On the further experiments, it is confirmed that the deformability can be more easily given and the amount and mode of deformation can successfully be controlled. The most promising process is essentially based on the rapid annealing of cold worked material under the complete restriction of its spring back. In this process, for example, cold rolled sheet is closely held in the holding device to prevent the strain recovery during heating and is rapidly annealed. As a result, the material, treated in such a way, shows the recovery-induced thermal deformation at a desirable elevated temperature, while it is almost perfectly fixed in shape at a lower temperature up to the testing temperature of about 100°C.

To transform the cold working strain into the recovery-induced thermal deformation, the rapid-annealing conditions such as annealing temperature, keeping time, rolling reduction etc. may be optimally selected on the annealing condition map. For example, a constitution of the variously cold rolled polyacetal is illustrated in Fig. 6, in which the boundary conditions between permanent shape and changeable shape in the annealed sheet at 100°C for 20 min. are represented with each curve. Hence, the techniques and the products become widely applied to various purposes such as sheet forming, mechanical combination system, thermo-sensor, thermo-sensitive element, especially, production of high strength plastics and so on.

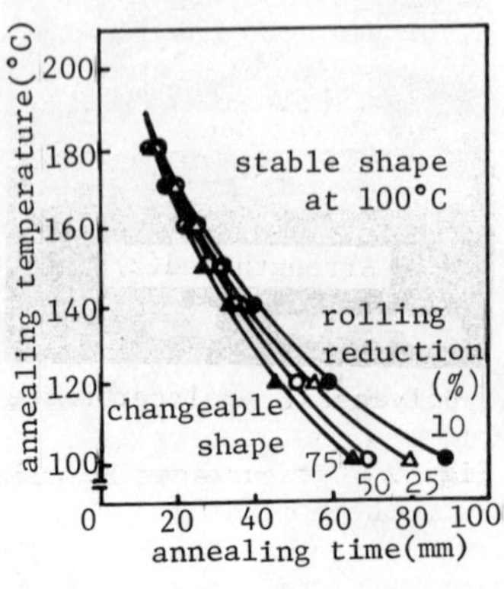

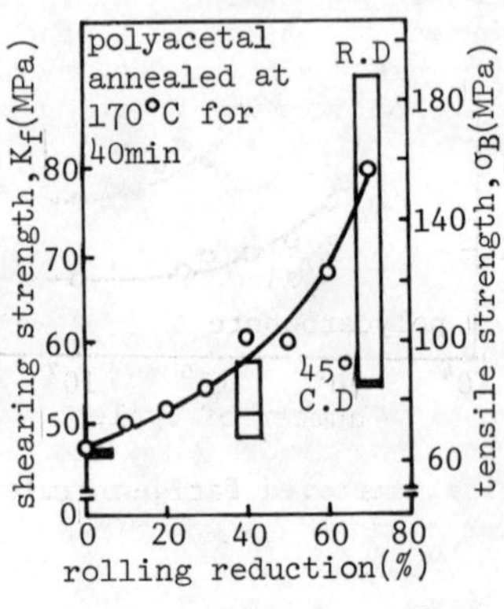

Fig. 6. Annealing condition. Fig. 7 Strengthening effects.

APPLICATION TO STRENGTHENING SHEET

It is clear that such heat-treated thermoplastics sheet can be remarkably improved in its strength due to the restriction of the strain recovery or the effects of residual stress, as shown in Fig. 7, in which effects of annealing condition on the shearing resistance and tensile strength of the treated polyacetal are given as results of punching test and tension test. For example, shearing stress increases itself uniformly with the increasing cold rolling reduction until the limit of cold rolling. Higher annealing temperature tends to make the material stronger. However, annealing time up to 40 min. affects little the strength. On the optimum condition, the most improved values go up about 60% for polyacetal and about 80% for polyethylene. And also it is clear from this figure that tensile strength of rolling direction is improved uniformly with the increasing cold rolling reduction and annealing temperature up to about 3 times for polyacetal and 4 times for polyethylene than the untreated sheet, while that of cross direction is small. Plastics sheet by the new process has not only extremely increased strength, but also many other improvements such as in fatigue, impact strength and punched surface. Fig. 8 shows the change in fatigue limit of plane bending. Compared with untreated plastics, the strengthened polycarbonate is remarkably improved in fatigue limit. The increasing seems to depend mainly on the strengthening.

By the way, the punching workability of plastics sheet is generally less than that of metal, mainly because plastics has too low elastic modulus and too high ductility to cause the early initiation of cracks and its linear propagation between punch point corner and die corner. To make plastics more rigid and brittle than the usual one seems to be, therefore, is important in order to make the desirable precise products. Surfaces cut by punching of variously treated polyacetal and polycarbonate with 20 mm diameter punch are photographically shown in Fig. 9, in which in the case of the untreated material the cut surface is observed to have too sharp and deep notch to use as it is, while in the case of strengthened sheets its contour is fortunately confirmed to be almost smooth and linear all over the surface.

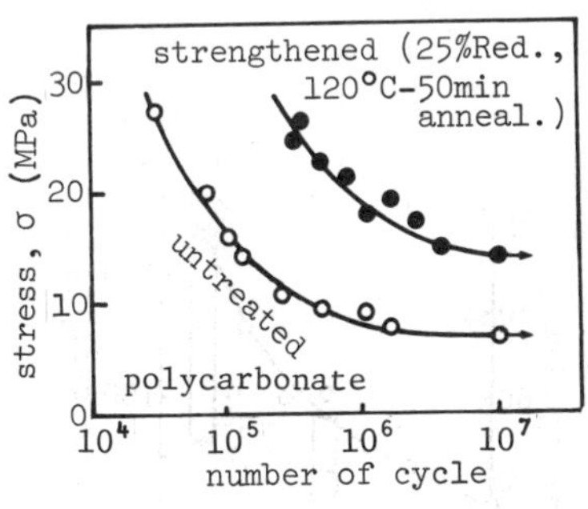

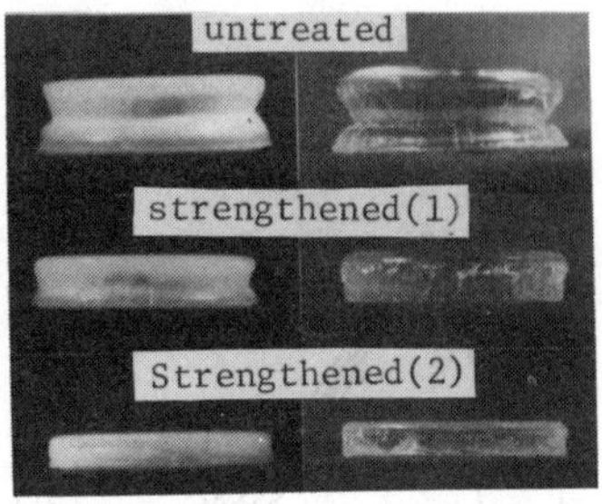

Fig. 8 Improved fatigue limit. Fig. 9 Cut surfaces by punching.

CONCLUSION

Phenomena and application of recovery-induced thermal deformation of thermoplastics are experimentally investigated. Results are as follows:

(1) Circular discs sliced off from extruded thermoplastics bar in molten state are found to deform into a cone shape by heating them to elevated temperatures without any application of external force. And also a usual thermoplastics sheet shrinks in the calendering direction along with an increase in the thickness at elevated temperatures.

(2) This thermal deformation induced as a result of the recovery of residual strain behaves itself as the memory of the material processing.

(3) By utilizing the recovery-induced thermal deformation new simple processes for cup forming and bending, which make best quality products by mainly using heating, are successfully developed.

(4) The special heat-treatment can transform from the cold working strain which can be easily recovered into the recovery-induced thermal deformation

which can be utilized on wide uses. The transformed deformation does not appear under ordinary use, but can function only at the desired elevated temperature.
(5) The transforming based on the restraint of free change in the shape given by cold working can remarkably improve the strength of plastics sheet up to about 3 to 4 times that of conventional material, along with other mechanical properties such as fatigue limit, impact strength, punching workability, etc.

ACKNOWLEDGEMENTS

We, authors, wish to thank Mr. Tetsuro Obara, President of Tamagawa Univ. for his encouragement in doing this work. We would also like to acknowledge the useful suggestions of Dr. T. Maeda, Emeritus Professor of Univ. of Tokyo, and Dr. T. Nakagawa, Professor of Univ. of Tokyo. We also extend our thanks to Mr. H. Komaki, Professor of Tamagawa Univ., for the preparation of the manuscript.

REFERENCES

Machida, T., T. Nakagawa and T. Okai (1979). J. Jap. Soc. Tech. Plasticity, 20, 610-617.
Maeda, T. (1978). J. Jap. Soc. Precis. Eng., 12, 476-480.
Machida, T. and T. Okai (1980a). J. Jap. Soc. Tech. Plasticity, 21, 152-159.
Machida, T. (1980b). J. Jap. Soc. Tech. Plasticity, 21, 1033-1038.
Machida, T., T. Okai and T. Nakagawa (1980c). Annals of the CIRP, 29, 179--184.
Machida, T. and T. Okai (1981). Kobunshi Ronbunshu, 38, 331-442.
Machida, T. and T. Okai (1980d). J. Jap. Soc. Tech. Plasticity, 21, 1102--1107.

THERMAL FATIGUE OF STRUCTURAL MATERIALS WITH REGARD TO STRUCTURAL STRESSES

Y. F. Balandin, A. F. Malyguin and Y. V. Yankovsky

Strength Problems Institute of the UkSSR Academy of Sciences, Kiev, USSR

ABSTRACT

Test data on thermal cycle damage of heterogeneous structural materials containing non-metallic inclusions, intermetallic compounds and artificially reinforcing fibres are discussed. It has been found that the thermal fatigue life of these materials decreases considerably due to heterogeneous phases. The increased liability of heterogeneous materials to damage under thermal cycling results from an additional cyclic structural stress. The test results which were obtained suggest relationships for the determination of the cyclic deformation range of heterogeneous materials with regard to structural stresses during thermal cycling.

KEYWORDS

Thermal fatigue; heterogeneous materials; thermal cycling; damage; structural stresses.

INTRODUCTION

Thermal fatigue of structurally inhomogeneous materials can be reduced by dozens of times compared to isothermal fatigue (Balandin, 1979; Malyguin, 1980). A quantative method is given to estimate the liability of heterogeneous materials to damage under thermal cycling with regard to loading conditions, physical and mechanical properties of structural components and specific phase distribution in metal volume.

EXPERIMENTAL RESULTS

The investigation of liability to damage during thermal cycling of heterogeneous materials containing nitrides, oxides, sulphides. silicates, graphite is based on test data for austenitic, perlitic and ferritic-martensitic steels, heat-resistant Ni-base alloys and cast iron. The appearance of cracks 0.3 mm in length was considered as the ctiterion of heterogeneous material damage. The number of cycles to fracture N (at a given cyclic deformation range and with non-metallic inclusions in the material) and N_o (with their absence) was determined from the 50% probability level specified on the basis of 20 cross-sectional specimen examinations. The cyclic deformation range $\Delta\varepsilon$ over the specimen surface and the cross-section under thermal shock was calculated with **regard** to temperature dependent physical and mechanical properties of the material and physical properties of the medium. The $\Delta\varepsilon$ value was 0.1-1.1%. It was found that the liability to damage N/N_o of a heterogeneous steel containing high-strength non-metallic inclusoins depends on their volume fraction ω ,the difference $\Delta\alpha$ between thermal expansion coefficients of inclusoin material and matrix metal, thermal cycling range Δt ,Fig.1a.The maximum cycletemperature t_{max} (350-900 C), active direction of main temperature stress components (tension, compression),as well as long-term creep duration from 0.I to 1 hour do not noticeably contribute to damage. The analysis of the effect of the above factors on thermal fatigue resistance of heterogeneous materials has shown that their increased liability to damage during thermal cycling results from additional cyclic structural stresses (Malyguin, 1979; Brooksbank,1968). This is illustrated in Fig.1a, which shows a sharp dependance of steel thermal fatigue damage from $\Delta\alpha$ and Δt ,and also freedom from such damage under cyclic isothermal loading (Δt =0). Thus, for example, $\Delta\alpha$ of a heterogeneous material with silicates, nitrides, oxides may be as great as $\Delta\alpha \approx$ (7-12) x 10^{-6} 1°C which is manifested in an up to 100-fold thermal fatigue resistance reduction (Fig.1a).

The higher the non-metallic inclusions content in the material, the more typical is their distribution in the form of chains and layers and less probable is their uniform distribution. When the distance r between inclusions significantly exceeds their typical dimension d($r > 3d$),there is no interaction between nearby local stresses, and resulting thermal fatigue cracks around inclusions are locked. Accordingly, when $r \gg d$ the liability to damage of a heterogeneous material is practically the same as that of matrix metal (Fig.1a).When non-metallic inclusions cluster in chains ($r \approx d$) and layers ($r < d$) the fatigue life of the heterogeneous material sharply decreases (Fig.1a).

Liability of a heterogeneous material to damage depends on the level of mechanical properties of the structural components. The higher the non-metallic inclusion elasticity modulus values E_o the lower are the cyclic life data. When a heterogeneous material contains low-strength inclusions (such as graphite in cast iron or slag and pores in weld metal) the matrix metal is not effected by inclusion force action in thermal cycling. The damage of such a heterogeneous material is controlled by local temperature stresses in the matrix metal due to different thermal conduction ($\lambda_0/\lambda \approx 3$) of the structural components, Fig.1a. However, a high level of local temperature gradients due to this factor can be attained only in sudden thermal shocks and when inclusions are at least 100 μm in size. The importance of the inclusion shape regarding higher geometrical stress concentrations increases with the inclusions size. In particular, thermal fatigue of spheroidal and flake graphite shows a 4-fold decrease and that of weld metal with pores and slag inclusions a 2-fold decrease (Malyguin,1979;Malyguin, 1978).

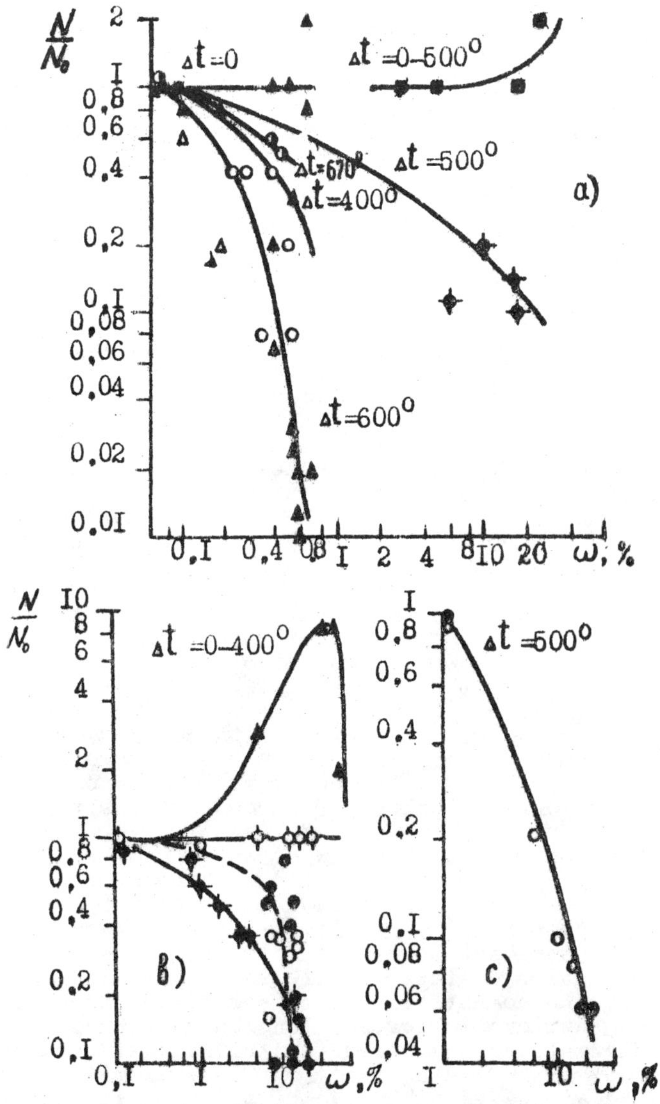

Fig. I. Thermal cycle damage of structural materials in relation to non-metallic inclusions (a), intermetallic compounds (b) and metal fibres (c) content;

a) ▲ -nitrides, ◐ -sulfides, ● -carbides in austenitic steel; O -oxides in perlitic steel; △ -nitrides in Ni-base alloy; ◆ -graphite in cast iron, pores in perlite steel

b) ◇,◆ -σ-phase, O, ● -Laves phase in smooth (◇,O) and notched (◆,●) specimens; ▲ -γ′-phase in Ni-base alloys

c) ● -steel fibres in Al-base alloys; O -W-fibres in Ni-base alloys.

Thermal fatigue of intermetallic compound hardened structural materials was investigated for Cr-Ni steels and Ni-base alloys which contained both original and thermal aging precipitating σ , Laves and γ'-phases (Malyguin,1982). Intermetallic particles possess physical properties similar to those of matrix metal and they have a uniform distribution within it (except for sigma phase). As a consequence, structural stresses induced by thermal cycling are not high. Reduced deformability of the matrix (Laves phase, sigma phase) due to the intermetallic particles and their increased strength (γ' -phase) have the major effect on thermal cycling damage of such structural materials (Fig.1b). Earlier damage of the sigma phase compared to the ductile matrix is the result of its lower strength ($\sigma_{UTS} \approx 700$ MPa). Notch sensitivity of such heterogeneous material encourages this process (Fig.1b). The above effects diminish when the cyclic loading temperature is increased to 850°C, because of coalescence of intermetallic compounds or their solution in the matrix.

The major features of thermal cycle damage of artifically hardened metallic materials were studied for Ni-base alloy with Hf dioxide particles, eutectic Ni-Al-Nb alloys with intermetallic lamellars, and also for Al- and Ni-base alloys reinforced by steel and tungsten fibre, respectively. Thermal cycle damage of these materials is to a great extent governed by their deformability (Malyguin,1979). Precipitation hardening alloys show the same tendency as for materials containing non-metallic inclusions (Fig.1a), whereas for eutectic alloys with delta phase in the form of lamellae the thermal fatigue damage is the same as specified for different base alloys with intermetallic hardening (Fig.1b). Alloys reinforced with 100-300 μm diameter metallic fibres showed a prominent effect of heterogeniety on thermal cycle damage (Fig.1c). The great difference in thermal expansion coefficients and also in thermal conductivity coefficients of structure components ($\Delta\alpha \approx (13-18) \times 10^{-6}$ 1/°C, $\lambda/\lambda 2$) was responsible for the initiation of structural cyclic stress of different origins in thermal cycling. Thermal fatigue damage of the reinforced materials occured at the boundary between the matrix and the reinforcing elements, and also between fibres in the matrix. At the same time W-wire reinforced Ni-base alloys revealed no reduction of fatigue life in isothermal fatigue testing compared to non-reinforced material (Morris,1970). This suggested that cyclic deformation of a heterogeneous material in thermal cycling can be determined using the followi[ng] relation

$$\Delta\varepsilon = (\Delta\varepsilon_0 + \Delta\alpha\Delta t\, E_c/E)\, K_c \qquad (1)$$

Here $\Delta\varepsilon_0$ is the material's cyclic deformation intensity range excluding heterogeneous phases and k_c is the effective structura[l] stress concentration factor presented as

$$K_c = (1+\omega)\{1 + f\sqrt{[1 + m \lg(\lambda_0/\lambda)]\,S}\} \qquad (2)$$

where ω is the volume fraction of the heterogeneous phase, $0 < \omega < 0.5$; f-inclusion shape or cluster characteristic, f =0.5-1 for single inclusions; f =1.5-2 for elongated inclusions or clusters; m-rate of temperature change in thermal cycling (m = 0 for slow heating and slow cooling, m = 0.5 for thermal shock within a semi-cycle and m = 1 for sudden

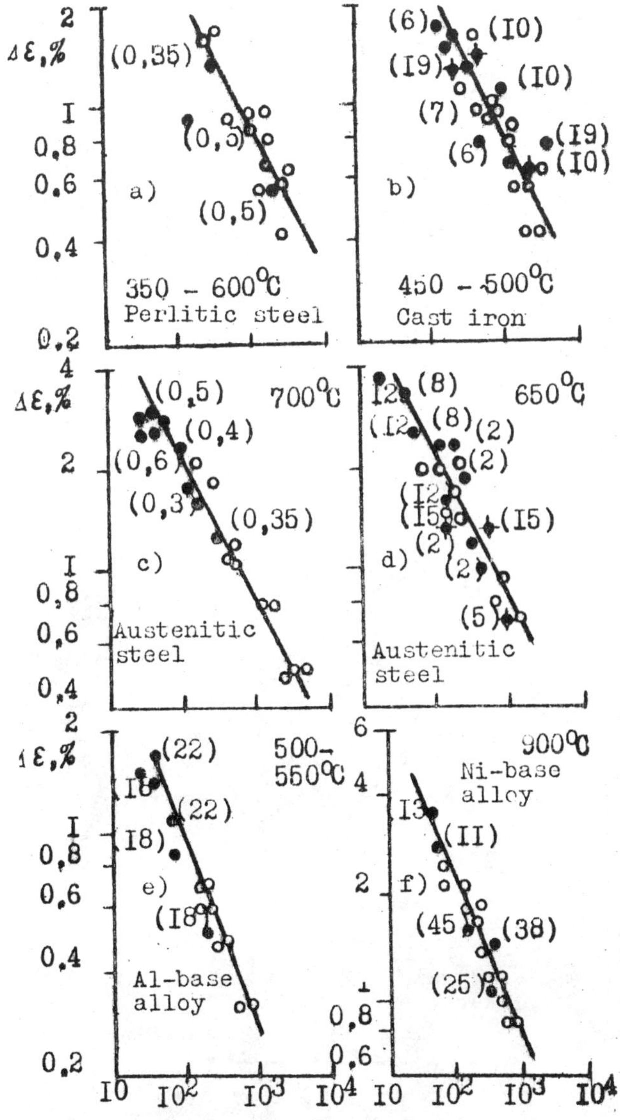

Fig. 2. Comparison of thermal fatigue strength of heterogeneous materials with regard to non-uniform structure and matrix metal (curves calculated from Malyguin's equ.)Values in brackets indicate volume fraction of inclusions in pct
a,c-non-metallic inclusions (●); b-spherical (●) and flake (♦) graphite; d-σ-phase (●) and Laves phase (♦); e- steel fibres (●); f -W-fibres (●); (o -martix metal).

heating and cooling); λ_0/λ - thermal conductivity coefficients ratio, $\lambda_0/\lambda > 1$; S - maximum inclusion size or cluster in mm. An estimation of the effect of matrix metal deformation behaviour and heterogeneous phases on matrix metal thermal fatigue strengths was performed using deformation fracture criterion developed by A.F. Malyguin. Figure 2 shows the comparison of thermal fatigue test results for different heterogeneous materials (including phase effect), corresponding matrix metal (separate points) and design curves. It is evident that the thermal fatigue strength of structural materials with different structures, physical and mechanical properties can be determined using a unified approach.

REFERENCES

Balandin, Y.F., Malyguin, A.F., Yankovsky, Y.V. Features of thermal cycle damage of metallic materials. Problems of strength and plastic behavior of solids. Leningrad, Nauka Press, 1979, 129-136.

Berezhko,B.I., Malyguin, A.F. and Yankovsky, Y.V. The effect of non-metallic clusters on structural materials damage under thermal cycling. Voprosy sudostroenia,seria Metallovedenie, 1980, v. 30, 10-21.

Malyguin, A.F. and Yankovsky Y.V. Features of fatigue behaviour of heterogeneous and composite materials. Problemy prochnosty 1979, 7, 20-27.

Brooksbank, D., Andrews, K.W. Thermal expansion of some inclusions found in steel and relation to stresses. J. Iron and Steel Institute, 1968,206, 595-599.

Malyguin A.F. and Ivanova T.I. Thermal cycle strength of welded joints in austenitic and perlitic steels with weld defects. Svarochnoe proizvodstvo, 1978, 24-27.

Malyguin A.F., Nikolaev Y.K. and Volovelsky D.E. Low-cycle fatigue of a Cr-Ni steel with regard to high temperature embrittlement. Voprosy sudostroenia seria Metallovedenie, 1982,v.35, 3-9.

Morris A.W.H.,Burwood-Smith A. Some properties of a fibre reinforced Ni-base alloy. Fibre science and technology, 1970, v.5, 1, 53-78.

Balandin Y.F. and Malyguin A.F. Thermal fatigue strength of power plant structural elements. Proceedings of the 3d All-union symposium on low-cycle fatigue of structural elements, Palanga, 1979, v. 2,12-19.

HIGH-TEMPERATURE BEHAVIOUR OF STEELS WITH PHASE TRANSFORMATION AND THE SIMULATION OF QUENCHING AND WELDING PROCESSES

T. Inoue and Z. G. Wang

Department of Mechanical Engineering, Kyoto University, Kyoto, Japan

ABSTRACT

General framework of coupling effect between stress, temperature and phase change is presented in the courses of quenching and welding processes, and discussions are made on the proper forms of inelastic constitutive relationship considering the phase transformation for both processes. Finite element simulations of the processes are made to obtain the induced distributions of metallic structures and stress fields.

KEYWORDS

Inelastic constitutive relation; phase transformation; quenching; welding.

INTRODUCTION

Stress analyses in the course of some engineering processes with phase transformation due to temperature variation have been treated by many researchers; especially in the cases of quenching [Ericsson, 1981; Inoue, 1978; Mura, 1957; Rammerstorfer, 1981] and welding [Andersson, 1978; Ueda, 1971]. Little attention is paid, however, in these papers to the coupling effect between temperature, phase change and stress, which sometimes plays a predominant role in the processes.

As is known in such cases, phase transformation affects the stress and strain fields in the body as the result of volumetric dilatation and the temperature redistribution due to latent heat generation. Mechanical work, especially inelastic dissipation, also has effects on the temperature field, and so on. Figure 1 represents the schematic feature of such coupling effects between temperature, stress and phase change as well as induced phenomena, which means that continuum thermodynamic treatment [Inoue, 1981, 1982] is inevitable to obtain proper results of analyses of the processes.

The purpose of the paper is to obtain the inelastic constitutive relationship of steels with phase transformation relevant for the analyses of quenching and welding as well as the equation of heat conduction and the

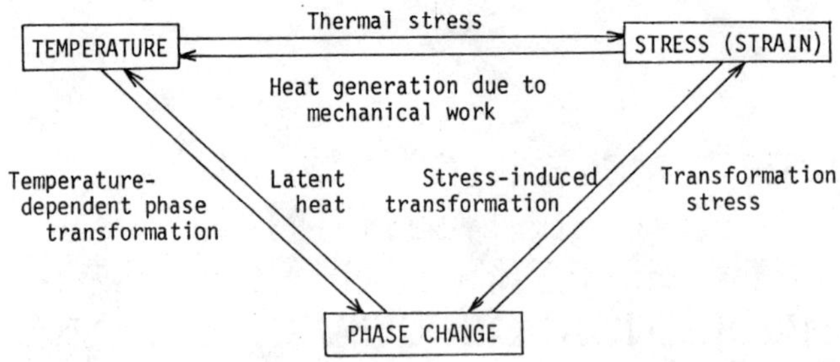

Fig. 1. Coupling between temperature, stress and phase change.

rate equation of each phase. Simulation of the processes is also made when considering the effect of coupling mentioned above.

INELASTIC CONSTITUTIVE RELATIONSHIPS OF MATERIAL

Two kinds of constitutive equation are employed: One is the time-independent elastic-plastic relation with solid-solid phase transformation, for example austenite-pearlite and -martensite, which is applied to the simulation of quenching of steel, and the other is the viscoplastic one applied to the welding process with solid-liquid transformation. The change of material properties under phase transformation is characterised by the concept of mixture [Bowen, 1976] without interaction that a material point is composed of N kinds of constituents with volume fraction ξ_I (I=1,2,...,N), and the material property X is assumed to be a linear combination of properties X_I of each constituent in the form:

$$X = \sum_{I=1}^{N} \xi_I X_I \equiv \Sigma \xi_I X_I \quad \text{with} \quad \Sigma \xi_I = 1 . \tag{1}$$

The Claudius-Duhem inequality and other ordinary ways of thermodynamics are applied to general form of constitutive equation including the volume fraction of constituents to obtain the elastic component of strain

$$\begin{aligned} \underset{\sim}{\varepsilon}^e &= \left(\Sigma \frac{1+\nu_I}{E_I} \xi_I\right)\underset{\sim}{\sigma} - \left(\Sigma \frac{\nu_I}{E_I} \xi_I\right)(\mathrm{tr}\underset{\sim}{\sigma})\underset{\sim}{1} + (\Sigma \alpha_I \xi_I)(T - T_o)\underset{\sim}{1} + (\Sigma \beta_I \xi_I)\underset{\sim}{1} \\ &\equiv \frac{1+\nu}{E}\underset{\sim}{\sigma} - \frac{\nu}{E}(\mathrm{tr}\underset{\sim}{\sigma})\underset{\sim}{1} + \alpha(T - T_o)\underset{\sim}{1} + \beta\underset{\sim}{1} , \end{aligned} \tag{2}$$

where E_I, ν_I, α_I, β_I denote the Young's modulus, Poisson's ratio, thermal expansion coefficient and dilatation due to phase transformation of the I-th constituent, respectively, and E, ν, α, β correspond to those of the global material.

Flow Law and Elastic-Plastic Constitutive Relation

Taking into account that the flow stress of the solid depends on the metallic structure, the yield function F is assumed to be determined by the constituent ξ_I as well as stress $\underset{\sim}{\sigma}$, temperature T, inelastic strain $\underset{\sim}{\varepsilon}^p$ and hardening parameter κ. Then the yield function of kinematic hardening is reduced to

$$F = F(\underset{\sim}{\sigma}, T, \underset{\sim}{\varepsilon}^p, \kappa, \xi_I) \equiv \sqrt{3J_2^*} - K(\kappa, T, \xi_I) \tag{3}$$

Here, $J_2{}^*$ stands for the second invariant of the deviatoric stresses of the form

$$\underset{\sim}{s}^* = [\underset{\sim}{\sigma} - 1/3(\mathrm{tr}\underset{\sim}{\sigma})] - [\underset{\sim}{a} - 1/3(\mathrm{tr}\underset{\sim}{a})] \,, \tag{4}$$

with the evolutional equation of back stress $\underset{\sim}{a}$

$$\dot{\underset{\sim}{a}} = A(\kappa,\, T,\, \xi_I)\dot{\underset{\sim}{\varepsilon}}^p \,. \tag{5}$$

The normality rule and the consistency relation of time-independent plasticity are applied to give the flow law

$$\dot{\underset{\sim}{\varepsilon}}^p = \Lambda \frac{\partial F}{\partial \underset{\sim}{\sigma}} = \hat{G}[\mathrm{tr}(\frac{\partial F}{\partial \underset{\sim}{\sigma}}\dot{\underset{\sim}{\sigma}}) + \frac{\partial F}{\partial T}\dot{T} + \Sigma \frac{\partial F}{\partial \xi_I}\dot{\xi}_I]\frac{\partial F}{\partial \underset{\sim}{\sigma}} \,, \tag{6}$$

with

$$\hat{G}^{-1} = -[\mathrm{tr}(\frac{\partial F}{\partial \underset{\sim}{\varepsilon}^p}\frac{\partial F}{\partial \underset{\sim}{\sigma}}) + A\mathrm{tr}(\frac{\partial F}{\partial \underset{\sim}{a}}\frac{\partial F}{\partial \underset{\sim}{\sigma}}) + \frac{2}{3}\frac{\partial F}{\partial \kappa}\,\mathrm{tr}\,(\frac{\partial F}{\partial \underset{\sim}{\sigma}}\frac{\partial F}{\partial \underset{\sim}{\sigma}})^{1/2}]. \tag{7}$$

The total strain rate relevant for the analysis of quenching is easily determined by summing up the elastic and plastic strain rate in Eqs. (2) and (6).

Elastic-Viscoplastic Constitutive Relation

The temperature range, on the other hand, is so wide during welding that the constitutive relation for the process should account for the following material properties: time-independent elastic-plastic characteristics in low temperature, which is already shown in the form of Eq. (6), viscoplastic behaviour in the high temperature range close to melting point, and some kind of fluid behaviour beyond that temperature.

A constitutive model of viscoplasticity, modified from that of Perzyna [1971], is used as

$$\dot{\underset{\sim}{\varepsilon}}^p = \frac{1}{2\mu}\,\psi(F)\,\frac{\partial F}{\partial \underset{\sim}{\sigma}} \,. \tag{8}$$

Here, $\mu = \Sigma\mu_I\,(\bar{\varepsilon}^p,\, T)\xi_I$ is the coefficient of viscosity, and $\psi(F)$ is a function of the yield function F of the same form as Eq. (3) under the condition that the material is elastic if $F < 0$, and viscoplastic if $F \geq 0$.

As a special case when $\kappa=0$, the total strain rate is reduced to a form of viscoelastic material as

$$\dot{\underset{\sim}{\varepsilon}} = \frac{1+\nu}{E}\dot{\underset{\sim}{\sigma}} - \frac{\nu}{E}(\mathrm{tr}\dot{\underset{\sim}{\sigma}})\underset{\sim}{1} + \frac{3}{2\mu}\underset{\sim}{s}^* \,, \tag{9}$$

if $\psi(F)=F$. This may be realistic when the temperature approaches the melting point. Above the liquidus temperature where the material changes to Newtonian fluid

$$\underset{\sim}{\sigma} = \frac{2}{3}\mu\dot{\underset{\sim}{\varepsilon}} - \frac{2}{3}\mu\,(\mathrm{tr}\dot{\underset{\sim}{\varepsilon}})\underset{\sim}{1} - p\underset{\sim}{1} \,, \quad \text{where} \quad p = -\frac{1}{3}\,\mathrm{tr}\underset{\sim}{\sigma} \,, \tag{10}$$

is obtainable by substituting the elastic shear component of deformation from Eq. (9). When the viscosity μ vanishes, and also F tends to zero in the decreasing temperature range, the value $\sqrt{3}\psi(F)/2\mu$ approaches some constant value γ in Eq. (8), and we have a time-independent plastic strain rate.

$$\dot{\underset{\sim}{\varepsilon}}^p = \frac{\underset{\sim}{s}^*}{\sqrt{J_2{}^*}} \,. \tag{11}$$

HEAT CONDUCTION EQUATION AND RATE EQUATION OF PHASE CHANGE

To complete the whole set of fundamental relations governing the coupled phenomena, the heat conduction equation and the rate form of constitutions are to be developed as well as the constitutive relationship discussed above.

The local form of energy balance leads to a heat conduction equation of the form

$$\rho c\dot{T} + T\mathrm{tr}\left(\frac{\partial \underset{\sim}{\varepsilon}^{e}}{\partial T}\dot{\underset{\sim}{\sigma}}\right) - \left[\mathrm{tr}(\underset{\sim}{\sigma}\dot{\underset{\sim}{\varepsilon}}^{p}) - \rho\,\mathrm{tr}\left(\frac{\partial H}{\partial \underset{\sim}{\varepsilon}^{p}}\dot{\underset{\sim}{\varepsilon}}^{p}\right) - \rho\,\mathrm{tr}\left(\frac{\partial H}{\partial \underset{\sim}{a}}\dot{\underset{\sim}{a}}\right)\right] + \rho\Sigma\ell_I\dot{\xi}_I = k\,\mathrm{div}\,(\mathrm{grad}\ T), \tag{12}$$

where H is the enthalpy density, and ρ, c and k denote the density, specific heat and heat conductivity which are also given in the form of Eq. (1). The second and third terms in the left hand side are heat generation due to mechanical work, and the fourth arises from phase transformation with latent heat ℓ_I.

The rate equation of the phase change is expressed generally as a function of stress as well as temperature as

$$\dot{\xi}_I = \dot{\xi}_I(\xi_I, T, \underset{\sim}{\sigma}) , \tag{13}$$

and the practical form of the relation is obtainable from experiments [Gautier, 1979; Bhattacharyya, 1955] as is discussed by the authors [Inoue, 1982].

RESULTS OF CALCULATION AND EXPERIMENT

Two kinds of steel were employed: 0.45%C plain carbon steel (S45C, equivalent to ASTM 1045) was used for quenching, and 0.08%C-1.38%Mn steel (SM41 or A131-75) for a model of welding. Material parameters of the constitutive relations (6) and (8) were obtained from tension tests under various strain rates in the range from room temperature to 1100°C. The isothermal transformation curve (T-T-T diagram) and dilatation-temperature diagram are quoted to determine the data of phase transformation.

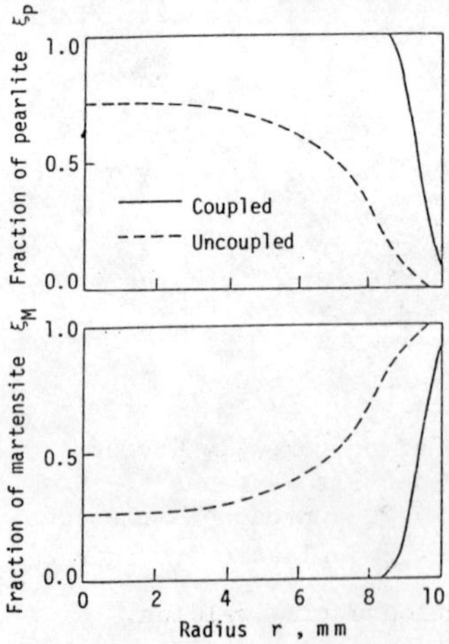

Fig. 2. Volume fractions of pearlite and martensite after quenching.

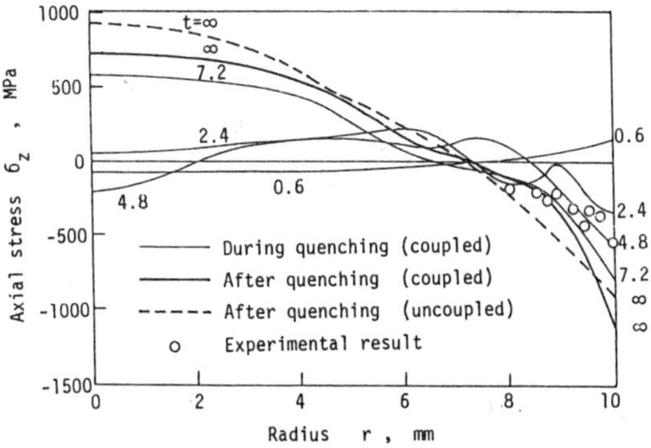

Fig. 3. Axial stress distribution during and after quenching.

Results of Quenching of Cylinder

A solid cylinder of 20mm diameter was water-cooled from 850°C as an example of quenching. Calculated distributions of pearlite and martensite transformed from austenite after quenching are shown in Fig. 2. Solid lines represent the results when considering the effect of stress on transformation kinetics, where the broken lines are those without coupling. The stress distribution is illustrated in Fig. 3 with experimentally measured residual stress.

Results of a Model of Welding

As a simplified model of welding, a circular plate of 300mm diameter and 8mm thickness with a 10mm-hole at the center was arc-welded in the hole. Temperature variations at several points against elapsed time is shown in Fig. 4, and Fig. 5 represents the residual stress distribution, which shows a reasonable simulation of welding processes.

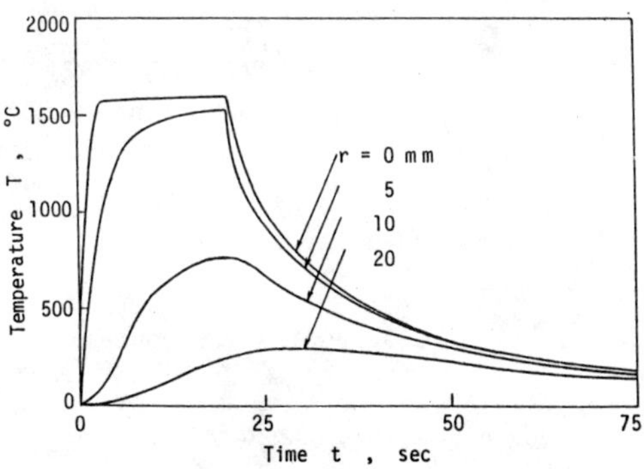

Fig. 4. Temperature variation during welding.

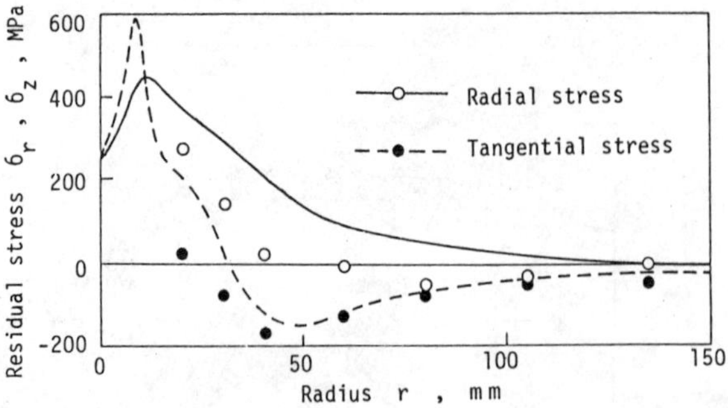

Fig. 5. Radial and tangential residual stresses of arc-welded plate.

CONCLUDING REMARKS

Thermoinelastic constitutive relations were presented in this paper, when considering phase transformation, and thermodynamic theory taking into account the coupling effect between temperature, stress and phase change, was developed. Processes of quenching and welding were simulated by use of the theory, which brought out reasonable results compared with experimental data.

ACKNOWLEDGEMENT

The authors wish to express their gratitude to Mr. M. Tanaka, graduate student of Kyoto University, who assisted in the experiments, and acknowledgements are also due to Nippon Kokan K.K. for providing testing material.

REFERENCES

Andersson, B.A.B. (1978). Thermal stresses in a submerged-arc welded joint considering phase transformation. Trans. ASME, EMT, 100, 356-362.

Bhattacharyya, S., and G. Kehl (1955). Isothermal transformation of austenite under externally applied tensile stress. Trans. ASM, 47, 351-379.

Bowen, R.M. (1976). Theory of mixture. Continuum Physics, Vol. 3, Ed. A.C. Eringen, Academic Press.

Ericsson, T., and B. Hildenwall (1981). Thermal and transformation stresses. Proc. 28th Sagamore Arm. Mat. Res. Conf. on Residual Stress and Stress Relaxation, 19-37.

Gautier, E., A. Simon, and G. Beck (1979). Deformation of eutectoid steel during pearlitic transformation under tensile stress. Proc. 5th Int. Conf. on Strength of Metals and Alloys, 2, 867-883.

Inoue, T., and S. Nagaki (1981). Description of transformation kinetics, heat conduction and elastic-plastic stress in the course of quenching and tempering of some steels. Ing.-Archiv., 50, 315-327.

Inoue, T., and B. Raniecki (1978). Determination of thermal-hardening stress in steels by use of thermoplasticity theory. J. Mech. Phys. Solids, 26, 187-212.

Inoue, T., and Z.G. Wang (1982). Finite element analysis of coupled thermo-inelastic problem with phase transformation. Proc. Int. Conf. on Numerical Method in Industrial Forming Processes, 391-400.

Mura, T. (1957). Residual stresses due to thermal treatments, Res. Rep. Fac. Eng., Meiji Univ., 10, 14-27.

Rammerstorfer, F.G., and others (1981). On thermo-elastic-plastic analysis of heat-treatment processes including creep and phase changes. Computers and Structures, 13, 771-779.

Perzyna, P. (1971). Thermodynamic theory of viscoplasticity. Advances in Applied Mechanics, Vol. 9, 315-354.

Ueda, Y., and T. Yamakawa (1971). Analysis of thermal elastic-plastic stress and strain during welding. IIW Doc. X-616-71.

PRODUCTION AND MECHANICAL PROPERTY OF CLAD-METALS LAMINATED WITH PARTICLE REINFORCED COMPOSITE METAL AND CERAMIC POWDER

Manabu Kiuchi and Sumio Sugiyama

Institute of Industrial Science, University of Tokyo, 7-22-1, Roppongi, Minato-ku, Tokyo, Japan 106

ABSTRACT

New metal working processes, such as mashy-state stirring and mixing process, mashy-state stirring and cooling process, mashy-state forging, mashy-state rolling and mashy-state extrusion, have been developed in order to produce particle reinforced composite metals and to work them into bars, wires, tubes, plates, sheets, machine parts and other required products. It has become clear that these new mashy-state working processes are very effective and useful to produce and work the composite metals which include reinforcing particles with high volumetric percentage. These processes can be applied in various ways for developments of new industrial materials composed with metals and ceramics. Furthermore, new types of particle reinforced clad-metals have been developed using masy-state working processes. These clad-metals have the particle reinforced composite laminae being bound on surfaces of the base metals or reinforced surface layers to which reinforcing particles are directly embedded in mashy-state. In order to produce these particle reinforced clad-metals, mashy-state forging and/or rolling have been used and it has become clear that these processes are very effective to attain very high binding strength between laminae and base metals.

KEY WORDS

Mashy-state, extrusion, forging, rolling, particle reinforced composite metal, particle reinforced clad-metal.

INTRODUCTION

The authors have been working on development of new metal working processes which are making use of characteristics of metals and their alloys in mashy-state. In mashy-state, metals (and their alloys) include the solid and liquid components in themselves and the liquid components exist at their grain boundaries. According to this, they have quite different characteristics comparing with metals in solid-state. They have distinctively low flow-stress and high deformability. They can be split into parts or connected together very easily. They can be stirred up like the sherbet in a cup and can be mixed up with other materials, such as powder, particles and short-cut fibers.

Making use of these characteristics, the author have developed various kinds of metal working processes, such as mashy-state extrusion, mashy-state forging and mashy-state rolling, in which a large amount of deformation can be attained easily with small working loads. Moreover new processes to produce many types of composite materials and to work them into required products such as bars, wires, tubes and machine parts have been developed.

At the first step of this paper, production of particle reinforced composite metals and their mechanical properties are being reported. Secondly, production and mechanical properties of new clad-type composite metals is going to be explained.

CHARACTERISTICS OF THE PROCESS

Schematic diagram of the process newly developed in this study in order to produce particle reinforced metals (P.R.M.) is shown in Fig. 1. Each step of the process is depending upon above mentioned characteristics of metals in mashy-state. The basic characteristics of the process are as follows.

(1) The matrix metal is heated up to the required temperature, that is corresponding to the aimed mashy-state, where the matrix metal includes solid and liquid components which are co-existing with appropriate ratio. The liquid component of the matrix metal usually exists at its grain boundaries and connecting forces among grains are weakened. Therefore, the matrix metal in mashy-state can be stirred up and mixed up with reinforcing particles. Here, in this study, aluminum alloys are used as matrix metals and particles and/or powder of Al_2O_3 (aluminum oxide), SiC (silicon carbide) and SiN (silicon nitride) with various dimensions are used as reinforcing particles.

(2) The particle reinforced composite metals made by above method can be worked into required bars, wires, tubes and various machine parts by newly developed mashy-state extrusion and mashy-state forging processes. In these processes, the particle reinforced composite metals, which are very hard and brittle and have little deformability in solid-state, have low flow stress and high deformability, therefore, they can be worked into required products by small loads.

(3) When the matrix metals are heated up to mashy-state and cooled gradually under stirring, they can be made into granule and/or powder. This is a new process to produce various powders of the matrix metals. It is named as the mashy-state stirring and cooling process. These powders can be mixed with reinforcing particles and thus the composite metals are obtained. The mixed powder can also be made into required products by mashy-state extrusion and mashy-state forging.

MECHANICAL PROPERTIES OF PRODUCTS

Fig. 2 (a) and (b) show effects of the volume fraction (V_p) on extrusion pressure (p) in **mahsy-state extrusion of composite** metals such as <(A-7075)+$A\ell_2O_3$>, <(A-5056)+$A\ell_2O_3$>, <(A-2011)+$A\ell_2O_3$> and <(A-5056)+Graphite>. Where the solid fraction (φ_c) of the billet is changed. In the figures λ shows the extrusion ratio.

From these figures, it can be easily known that when the solid fraction of the billet (φ_c) is high, the necessary extrusion pressure (p) becomes too high to extrude the products.

These results show the fact that the conventional hot extrusion process can not be applied to the extrusion of the particle reinforced composite metal with high V_p , but only the mashy-state extrusion with appropriate amount of liquid component, that is with low φ_c , can extrude such kind of composite metals with high V_p into required products.

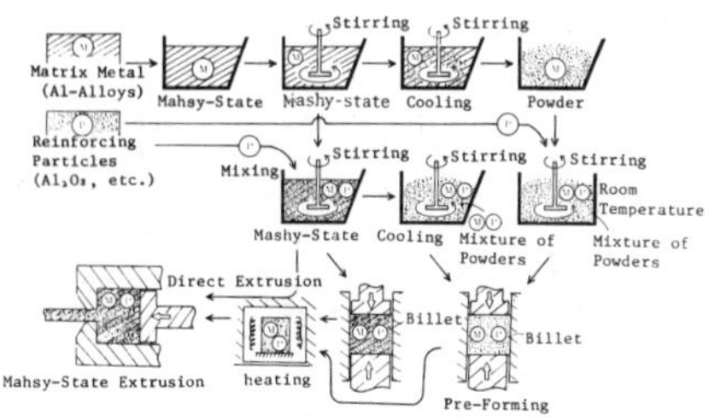

Fig. 1 Schematic diagram of mashy-state working processes of particle reinforced composite metals.

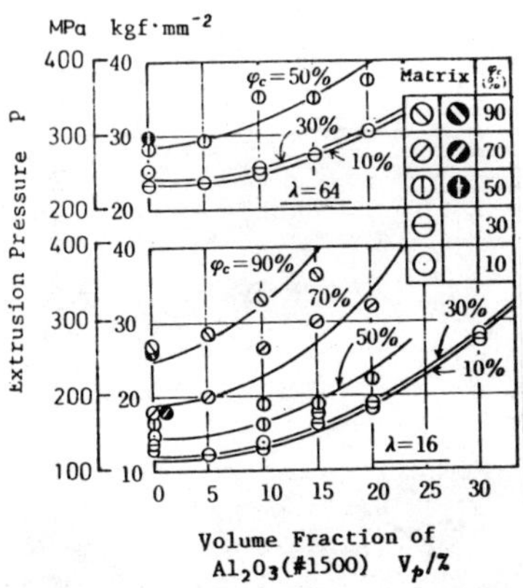

Fig. 2 (a) Effects of volume fraction of reinforcing particle (V_p) on extrusion pressure (p).

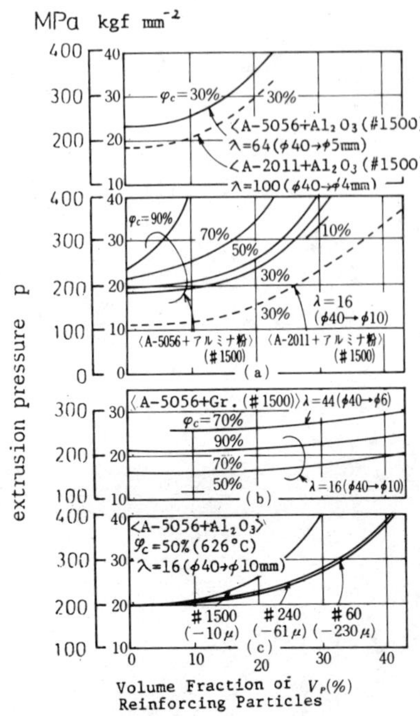

Fig. 2 (b) Effects of volume fraction of reinforcing particle (V_p) on extrusion pressure (p).

Fig. 3 shows effects of the volume fraction of reinforcing particle (V_p) on elongation ($E\ell$), tensile strength (σ_B) and hardness (H_V) of the extruded bar of <(A -7075)+$A\ell_2O_3$>. It can be noticed that as V_p increases, $E\ell$ decreases and H_V increases. Regarding the tensile strength (σ_B), it is difficult to evaluate effects of V_p on it, because of that as V_p increases, $E\ell$ decreases distinctively and the test pieces break abruptly in small strain range, therefore, it becomes difficult to measure the real tensile strength. Anyway, as V_p increases, the extruded products become more and more hard and less ductile. It can also be known that when the solid fraction of the billet (φ_c) is high, σ_B and H_V of the extruded bar is higher than those extruded from billets with low φ_c. This is caused by the fact that when φ_c of the billet is high, the matrix metal of the extruded composite metal becomes harder than that obtained from the billet with low φ_c.

Fig. 4 shows stress-strain curves of extruded bars of <A -7075+$A\ell_2O_3$> in compression tests. Fig. 5 shows relationships between flow stress of extruded composite metal <(A-7075)+$A\ell_2O_3$> and temperature. It should be noticed that the flow stress of the bar including 30% of $A\ell_2O_3$ is almost twice of that of the bar which includes no reinforcing particles. Fig. 6 shows similar relationships between flow stress of extruded composite metals, such as <(A-5056)+$A\ell_2O_3$>, <(A-2021)+$A\ell_2O_3$> and <(A-5056)+Graphite>, and temperature.

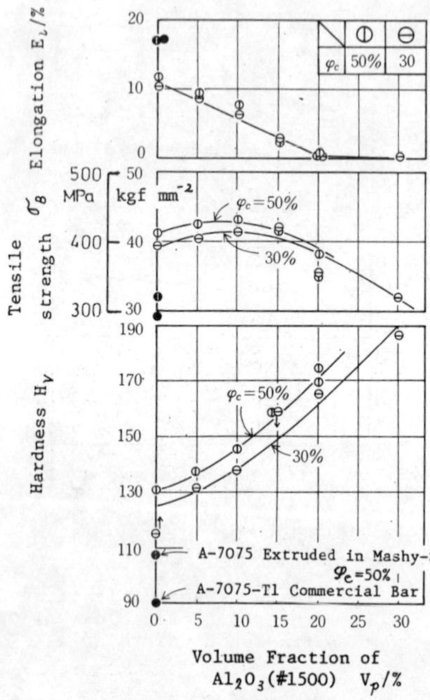

Fig. 3 Effects of volume fraction of reinforcing particles (V_p) on mechanical properties of bars extruded in mashy-state.

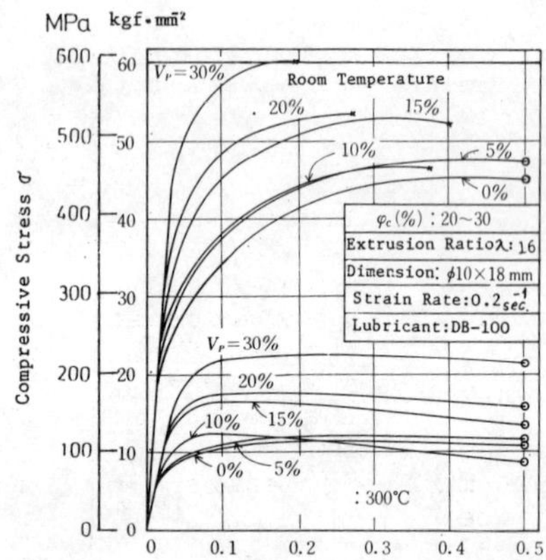

Fig. 4 Stress-strain curves of extruded composite metals.

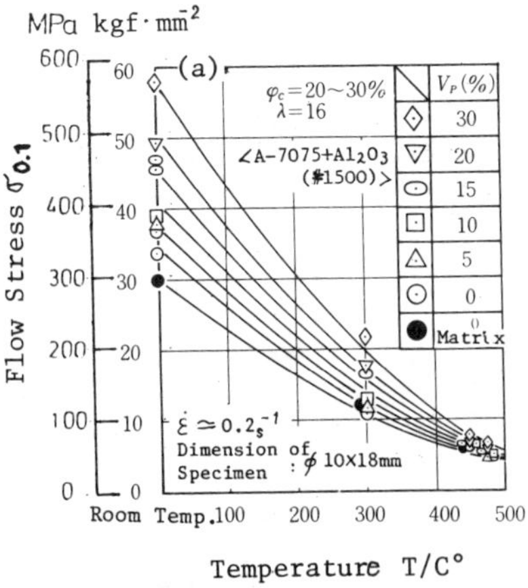

Fig. 5 Relationships between flow-stress of extruded composite metals and temperature.

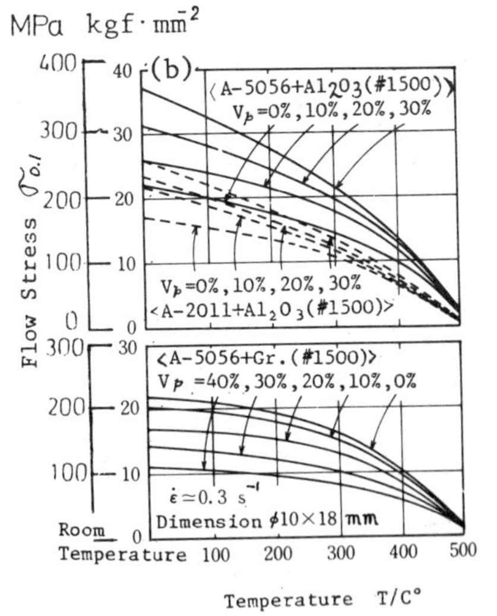

Fig. 6 Relationships btween flow-stress of extruded composite metals and temperature.

Fig. 7 shows effects of the volume fraction (V_p) on tensile strength (σ_B), hardness (H_V), elongation (E_ℓ), impact strength (E) and wear resistance (M) of the extruded composite metals such as $\angle$(A-5056)+A$\ell_2O_3$$>$, $\angle$(A-2011)+A$\ell_2O_3$$>$ and $<$(A-5056)+Graphite$>$. Here, the wear resistance is evaluated by the specific amount of wear per unit time. From these results, it can be known that as V_p increases, the hardness (H_V) increases almost proportionally, the elongation (E_ℓ) decreases extremely and it becomes almost zero when V_p reaches to 30%, and the impact strength also decreases intensely and it also becomes almost zero when V_p reaches to 30%. Moreover, it can be noticed that as V_p increases, the wear resistance increases distinctively, that is to say, the specific amount of wear decreases extremely. When V_p gets to 20%, the wear resistance of these composite metals based on aluminum alloys becomes almost equal to those of the stainless steel SUS 316 and the high carbon steel S45C. When the graphite particles are mixed into A-5056, the wear resistance of the composite metal decreases almost proportionally to the amount of graphite.

Fig. 8 shows some examples of extruded bars of the composite metals. Many kinds of bars, wires and tubes with sound surfaces and excellent inner structures have been successfully produced under appropriate mashy-state conditions.

Fig.9 shows the inner structures of extruded bars of $\angle$(A-5056)+A$\ell_2O_3$$>$ and $\angle$(A-5056)+Graphite$>$. The reinforcing particles embedded in the matrix metal can be observed clearly. These pictures have been obtained from the broken cross-section of the specimen in the impact test. The matrix metal shows clear ductility.

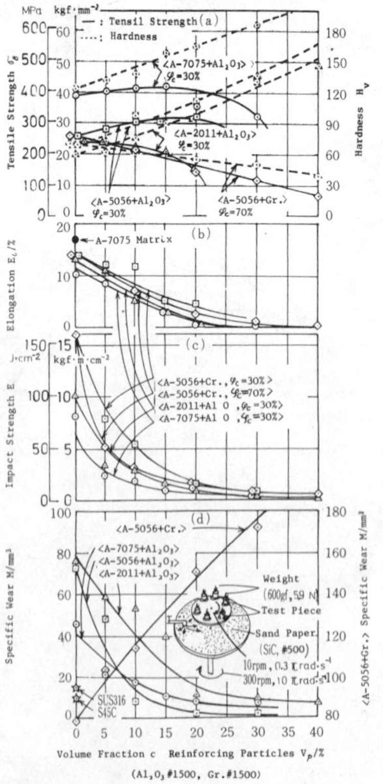

Fig. 7 Effects of volume fraction of reinforcing particles (V_p) on mechanical properties of extruded composite metals.

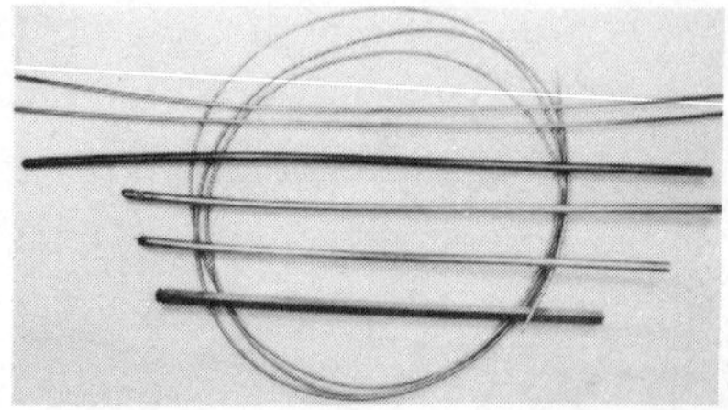

Fig. 8 Some examples of extruded bars and wires of composite metals.

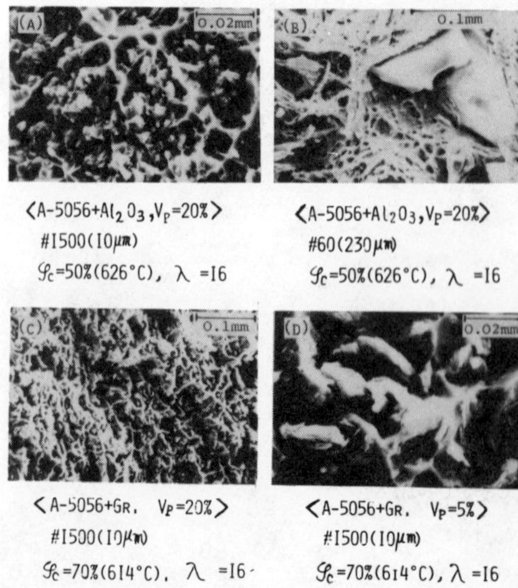

Fig. 9 Inner structures of extruded bars of composite metals.

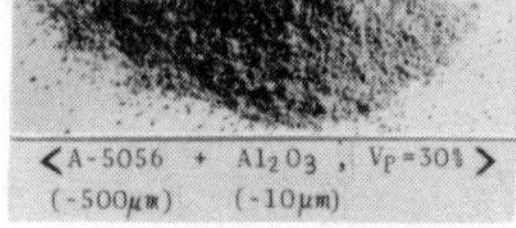

Fig. 11 Composite mixture of powders

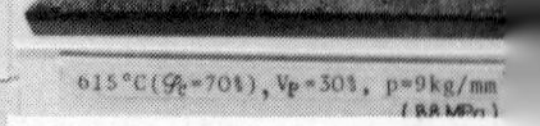

Fig. 12 External appearance and cross-section of a product.

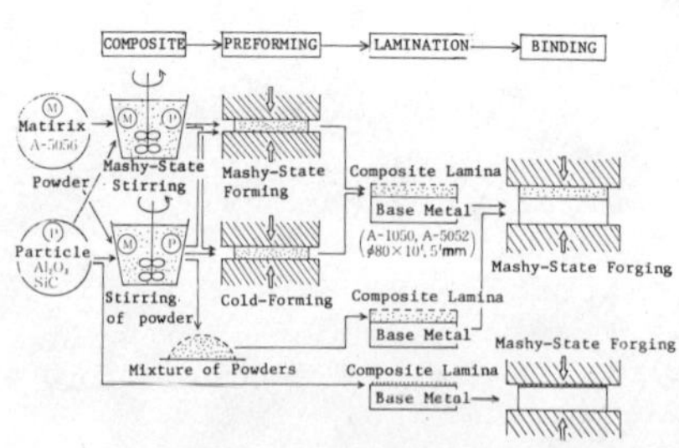

Fig. 10 Schematic diagram of mashy-state working processes of particle reinforced clad-metals.

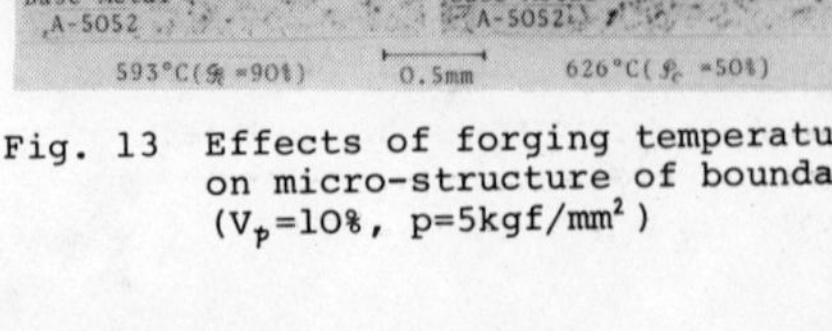

Fig. 13 Effects of forging temperature on micro-structure of boundary (V_p=10%, p=5kgf/mm^2)

PRODUCTION OF PARTICLE REINFORCED CLAD-METALS

At the second step of this paper, production of particle reinforced clad-metals and their mechanical properties are being explained.

Schematic diagram of the processes is shown in Fig. 10. The particle reinforced composite metals used as laminae are made by the processes explained in the previous sections. Fig. 11 shows the particle reinforced composite metal in powder state made by the mashy-state stirring, mixing and cooling process. In those processes shown in Fig. 10, the particle reinforced composite metals are usually preformed into plates and/or sheets with required thickness. After that, those plates and/or sheets are laminated on the base metals. In some cases, the composite mixture of the matrix metal and the reinforcing particle shown in Fig. 11 is directly laminated on the base metal without preforming. When the particle reinforced composite metal is laminated on the surface of the base metal by mashy-state forging and/or rolling process, the matrix metal of the composite and the base metal are both heated up to the required temperature, where they are in mashy-state and include appropriate amount of liquid component respectively. In the binding process performed by mashy-state forging and/or rolling, the liquid components included in both metals penetrate into each other and after being cooled and solidified they works as the effective binding agent. By using mashy-state forging and/or rolling like this, very high binding strength is obtained.

When the reinforcing particles, such as various ceramic particles, are embedded into surface layer of the base metal, the base metal is also heated up to the required temperature, where the base metal is in mashy-state. Each reinforcing particle is pressed into the liquid component of the base metal by a punch and/or a roll. After solidification of the liquid component, the reinforcing particles are fixed in the surface layer of the base metal. The surface layer of the base metal obtained by this process becomes very thin particle reinforced composite lamina which include the reinforcing particles by high ratio. This composite lamina shows very high hardness and wear resistance. Moreover, in this case, the base metal keeps its original ductility, therefore, the composite metal with particle reinforced lamina usually has enough deformability, and it can be made into various products by hot and/or cold plastic workings.

MECHANICAL PROPERTY OF PARTICLE REINFORCED CLAD-METALS

Fig. 12 shows an external appearance and a cross-section of a product made by mashy-state forging process.

Fig. 13 shows effects of the forging temperature on the micro-structure of the boundary between particle reinforced composite lamina (upper part) and base metal (lower part). From these pictures, it can be known that when the forging temperature is high enough, the liquid components of the matrix metal of the composite lamina and the base metal penetrate into each other and consequently the very firm and stable boundary is obtained.

Fig. 14 shows effects of the forging pressure (p) on the

micro-structure of boundary between composite lamina and base metal. From this result, it can be known that as the forging pressure (p) increases, the micro-stracture of the boundary becomes more and more improved, that is to say penetration of the liquid components is becoming more and more distinctive. When the forging pressure (p) is low or the forging temperature is not so high, the penetration of the liquid components hardly takes place and the boundary is left to be almost flat.

Fig. 15 shows effects of the base metal on the micro-structure of the boundary. In the left side case, A-1050 is used as the base metal and in the right side case, A-5052 is used, where in both cases, A-5056 is used as the matrix metal of the composite lamina. As it is well-known, A-1050 is a pure aluminum and it has no temperature zone of mashy-state. Therefore, in the forging process, the boundary of the left side case can not be bound in mashy-state, and consequently it is almost flat.

Fig. 16 shows relationships between volume fraction (V_p) of reinforcing particles of composite lamina and shear strength (τ_{max}) of boundary. Each curve is corresponding to different forging pressure (p). It is clear that the shear strength (τ_{max}) of the boundary increases according to the increase of the forging pressure (p). Moreover, τ_{max} increases according to the increase of V_p. However, if the liquid components of the matrix metal of the composite lamina and the base metal are not enough, the sound binding can not be performed. Consequently, the normal shear strength can not be obtained. The drop of τ_{max} shown with broken lines is caused by the lack of the liquid component in the binding process. When the composite lamina includes enough amount of liquid component and the enough forging pressure (p) is applied in the binding process, the shear strength of the boundary (τ_{max}) reaches to the shear strength of the base metal. In this figure, when φ_c is 70% and p is 13 kgf/mm , τ_{max} is almost equal to the shear strength of the base metal.

Fig. 17 and Fig. 18 show hardness (H_V) and wear resistance (specific amount of wear M) of the composite lamina made by above-mentioned processes. These results are almost same with the measured results of the particle reinforced composite metals discussed in the previous section.

Fig. 19 shows the micro structure of the surface layer of the composite metal to which the reinforcing particles are embedded directly by mashy-state forging and/or rolling. The reinforcing particles embedded into the surface layer of the base metal are observed clearly. They become a very hard composite lamina and its wear resistance is very high.

Fig. 20 shows some examples of products of cold-bending made with the particle reinforced clad-metals obtained by above mentioned processes. The left side example is a product which is bent from a flat clad-metal and has a particle reinforced composite lamina on its inner side. In this case, the bending has been performed successfuly The right side shows a product made from the composite metal surface layer of which is embedded with reinforcing particles. In this case, the outer surface layer include the reinforcing particles by high ratio. However, the bending has been performed successfully and the surface is sound and stable.

CONCLUSION

Through this study, new metal working processes, which use various characteristics of metals in their mashy-states, have been developed in order to produce the particle reinforced composite metals and to work them into bars, wires, tubes, plates, sheets and other required products.

Furtheremore, new types of particle reinforced clad-metals have been developed.

REFERENCE

[1] Mehrabian, R., Riek, R. G. and Flemings,M. C., (1974). Metallurgical Trans., 5, 1899. [2] Kiuchi, M., Sugiyama, S., and Arai, K., (1979). J. Japan Soc. Tech. Plasticity, 20-224, 826-833. [3] Kiuchi, M., and Sugiyama, S., (1981). Prod. Japan Spring Conf. Tech. Plasticity, 575-578. [4] Kiuchi, M., and Sugiyama, S., (1982). J. Japan Soc. Tech. Plasticity, 23-258, 700-707. [5] Kiuchi, M. and Sugiyama, S., (1982). J. Japan Soc. Tech. Plasticity, 23-260, 915-923.

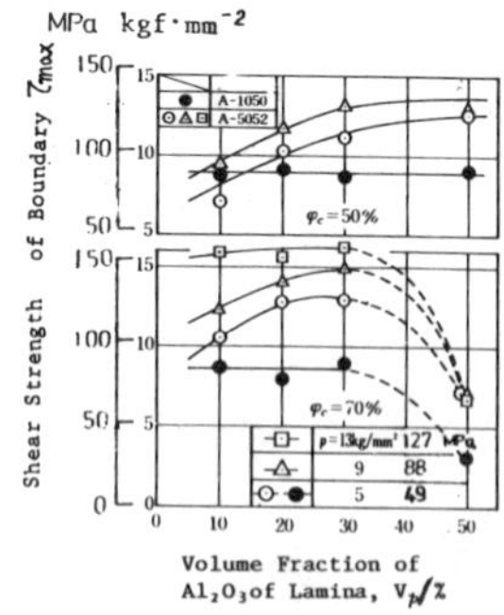

Fig. 16 Relationships between volume fraction of Al_2O_3 of lamina and shear strength of boundary.

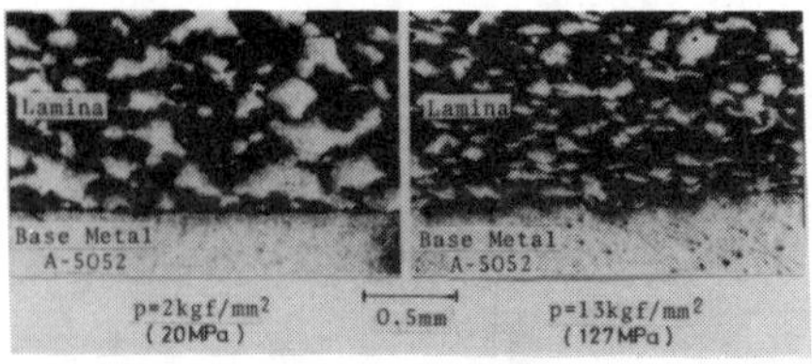

Fig. 14 Effects of forging pressure on micro-structure of boundary (V_p=30%, t=614°C ($\varphi_c \fallingdotseq$ 70%)).

Fig. 15 Effects of base metal on micro-structure of boundary (V_p=30%, p=5kgf/mm², φ_c=50%).

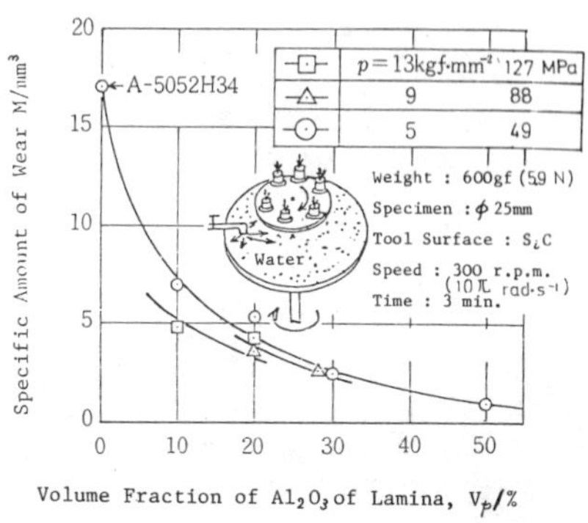

Fig. 18 Effects of volume fraction of Al_2O_3 of lamina on its wear resistance.

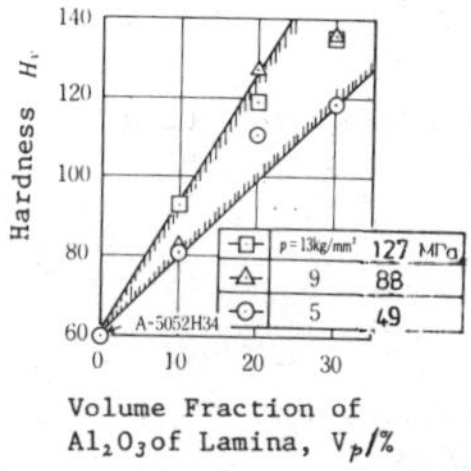

Fig. 17 Effects of volume fraction of Al_2O_3 of lamina on its hardness.

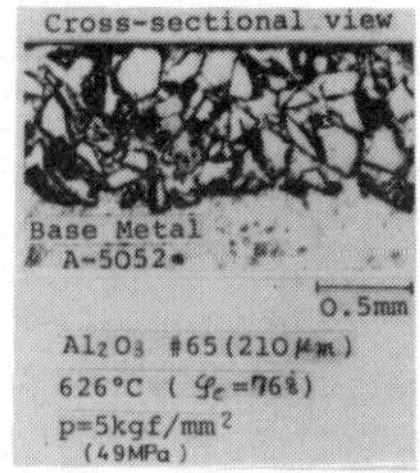

Fig. 19 Micro structure of surface layer reinforced with Al_2O_3 .

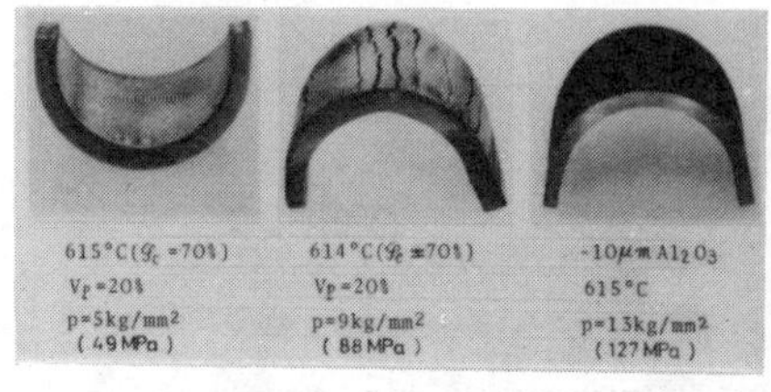

Fig. 20 Some examples of products of cold-bending.

FRACTURE BEHAVIOUR OF PRESSURE VESSEL SHELLS PLATED WITH ANTICORROSIVE CLADDING

V. A. Ignatov, B. T. Timofeyev, and V. A. Fedorova

Strength Problems Institute of Academy of Science UkSSR
252014, Kiev, Timirasevskaya 2, USSR

ABSTRACT

The fracture behaviour of pressure vessel heat resistant 15X2MΦA steel with anticorrosive austenitic cladding is discussed for static and cyclic loading in air and aqueous medium of boric acid. It is noted that a crack stops to propagate when it reaches the fusion line between cladding and base metal. The crack propagation rate in an aqueous medium is lower than in air under cyclic loading. The paper reports experimental data on fatigue crack propagation rate in the anticorrosive cladding metal and the base metal at different temperatures at various cycle ratios and loading frequencies. The effect of these parameters (taken separately) is demonstrated.

KEYWORDS

Anticorrosive cladding; static and cyclic loading; crack propagation resistance; aqueous medium; frequency; cycle ratio.

INTRODUCTION

In power plant engineering shells made of heat-resistant low-alloy steels are usually plated with a layer of anticorrosive metal to prevent material corrosion in the water environment. The anticorrosive layer of 9-12 mm thickness is applied to the shell inner surface by means of automatic plating with the use of austenitic strip electrodes, under protection of flux. The cladding is applied in three stages and then subjected to heat treatment by high-temperature tempering which results in residual tensile stresses at normal temperatures even in the absence of inner pressure. The level of these stresses is high enough to have an impact on development of crack-like production defects in the process of shell loading. The above factor makes it necessary to study fracture behaviour of the anticorrosive cladding metal and base metal in the transition zone directly adjacent to the austenitic layer in conditions of loading close to those in operation.

As the vessel inner surface in operation is in permanent contact with heat-transfer water of the first circuit, it is essential to determine the fracture resistance behaviour of the given materials in crack propagation stage under static and cyclic loading in these particular conditions, with appropriate asymmetry and frequency of the loading cycle. The paper reports the results of fracture resistance investigations of 15Х2МФА steel (widely used in power plant engineering) plated with two layers of anticorrosive cladding, the first of which is applied with Cb-07X25H13 strip electrode, the second with Cb-08X19H10Г2Б strip electrode. The mechanical properties of base metal and austenitic cladding near the fusion line were studied using indentation method (as specified by GOST 22762-77) and testing of 1.5 mm diameter microspecimens, after which the curves reflecting changes in strength and plasticity in the direction normal to the fusion line were plotted. Both methods indicated a similar trend in changes of the above properties depending on the distance from the fusion line.

The investigations established that the medium under study (aqueous solution of boric acid) has the highest effect on the corrosion resistance of heat-resistant steels not at the rated operational temperatures (250-300°C), but at rather lower temperatures (80-100°C) typical for the periods of power plant starting and stopping. The danger of corrosive effect at these periods is augmented by higher oxyden solubility in the medium in cases of free access of the oxygen. As the greatest corrosion damage to heat-resistant steel in operating power plants occurs in shutdown periods when the medium temperature is within 20-80°C, this temperature range was the main object of the present studies, though a number of experiments was carried out on the base metal at static and cyclic loading at temperature of 250°C to determine its crack propagation resistance in these conditions.

EXPERIMENTAL RESULTS

The material crack propagation sensitivity in aqueous medium under static loading is characterized by the critical value of stress intensity factor K_{1SCC} (Brown, 1971). To define this characteristic a non-linear fracture mechanics method was used based on the J-integral concept (Cherepanov, 1967; Rice, 1968). Under static loading of 16x18x180 mm cracked beam specimens during 1000 h in aqueous medium at 80°C the crack grows only when its tip is located in cladding material of Cb-07X25H13. It was observed that the crack propagated within the cladding only, just up to the line of fusion with base metal. At specified loading parameters the crack during the test had no propagation either in the base metal, or in the cladding layer directly contacting the aqueous medium. Besides, the tendency to subcritical crack propagation at long-time static loading is also noted in the heat-affected zone of this steel when subjected to automatic submerged-arc welding. It was also observed that none of the investigated materials was susceptible to fatigue crack propagation in aqueous solution of boric acid at 250°C at static loading not exceeding 1000 hours.
As pressure vessels usually operate under cyclic loading conditions, some tests were carried out to determine the crack propagation rate in anticorrosive cladding metal of type 07X25H13 susceptible to corrosion cracking, and in the base metal transition zone, both in the air and in a corrosive medium. The tests were conducted on 10x20x150 mm specimens where a fatigue crack was initiated from the notch; the specimens were subjected to pure bending in sinusoidal loading (with cycle ratios of R=0

and R=0.7), at loading frequencies of f=0.25 Hz and f=0.01 Hz, both in air and in aqueous medium with pH=8 (1% solution of boric acid in distilled water) at 80°C. As the boric acid at such temperatures dissociates in water losing some hydrogen ions, to maintain the required pH=8 (which corresponds to actual operating conditions) insignificant quantities of caustic potash were added.

Figure 1 contains the experimental data on fatigue crack propagation rate at pulsating loads with f=0.25 Hz in air at 20°C and in the aqueous medium at 80°C within the range of $4 \cdot 10^{-9} \div 10^{-6}$ m/cycle. The data indicate that in the anticorrosive cladding metal, at equal values of ΔK, the crack propagation rate in air is somewhat higher than in the aqueous medium (Fig. 1a).

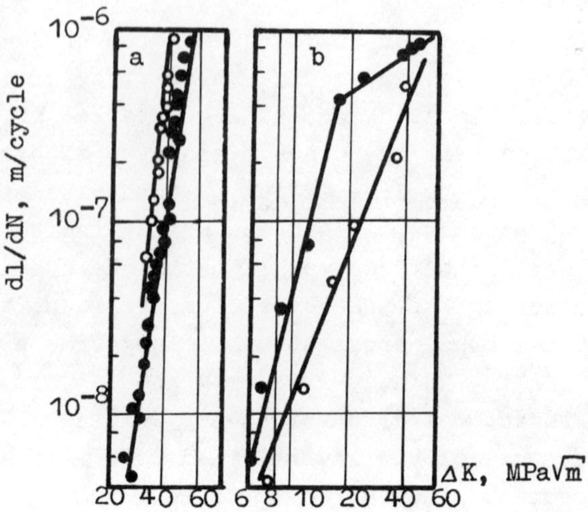

Fig. 1. Resistance to fatigue crack propagation in anticorrosive cladding metal (a) and base metal (b) in air (o) and in aqueous medium (•).

This result at first sight seems contradictory to common knowledge about corrosive action of the medium, for, as shown in Fig. 1b, the fatigue crack propagation resistance of pearlitic steels appears higher when the specimens are tested in air. Apparently, the lower crack propagation rate of anticorrosive cladding in water results from the crack tip getting blunt under the action of electrochemical processes accompanying the test. Even at the rather high loading frequency f=0.25 Hz the above corrosive medium can in certain conditions reduce the crack propagation rate in pearlitic steels by an order of magnitude as compared with tests in air. However, this significant difference in the rate can only be observed within a rather narrow range of ΔK values.

Comparing the crack propagation rates in steel and cladding metal, it should be noted that, starrting from a certain ΔK value, the crack propagation rate in cladding metal, when measured in air, is higher than in the base metal; in the aqueous medium, the crack propagation resistance in cladding metal is higher than in steel, over the whole range of stress intensity factor values. However, another point deserves consideration here: the crack propagation rate in cladding metal, both in air and in water, is appreciably higher than in pearlitic steel; and if, in the process of operation, the operating loads and residual stresses will bring forth the value of ΔK required for initiation of the crack, the latter will then grow quickly throughout the depth of cladding. The crack growth rate in cladding metal within the range of $10^{-8} \div 10^{-6}$ m/cycle at the above loading conditions can be described using the following exponential equations of Paris:

$$dl/dN = 6.44 \cdot 10^{-23} \cdot (\Delta K)^{9.92} \quad \text{in air,}$$

$$dl/dN = 1.06 \cdot 10^{-20} \cdot (\Delta K)^{8.13} \quad \text{in aqueous medium at } 80°C.$$

Since during actual operating conditions of frequency of loading of high pressure vessels with anticorrosive cladding is considerably lower and the cladding metal is subjected to cyclic loads with essential cycle asymmetry, similar specimens were tested by the authors in aqueous solution of boric acid at f=0.01 Hz and cycle ratio of R=0 and R=0.7. The crack propagation rate of $2\ 10^{-8}$ m/cycle corresponds to ΔK=12 MPa$\sqrt{m}$ at 0.01 Hz and R=0.7, to ΔK=18 MPa$\sqrt{m}$ at 0.01 Hz and R=0, and to ΔK= 36 MPa$\sqrt{m}$ at 0.25 Hz and R=0.7. Hence, the rate can be achieved by lowering the stress intensity factor 2-3 times. The results of these experiments are shown in Fig. 2. The data in

Fig. 2a indicate that crack propagation rate in water is influenced considerably by both the asymmetry and frequency of the loading cycle.

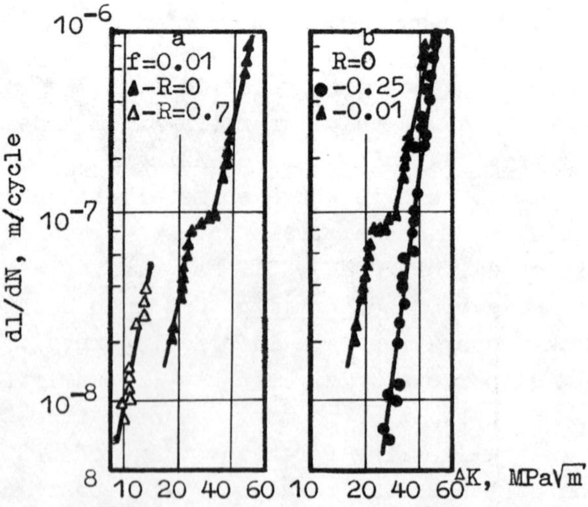

Fig. 2. Influence of cycle ratio (a) and loading frequency (b) on crack propagation rates in anticorrosive cladding.

The data in Fig. 2b indicate that at ΔK=30 MPa√m and with transition from frequency of 0.25 Hz to 0.01 Hz the crack propagation rate increases by more than an order of magnitude. Figure 3 shows primary experimental data reflecting the change in crack length with increase in the number of loading cycles (f=0.25 Hz)

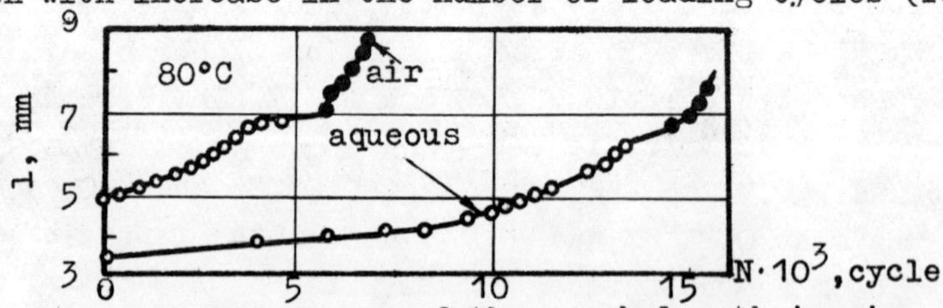

Fig. 3. The change of the crack length in air and in aqueous medium versus the number of cycles: o-cladding, ●-base metal

In the cladding metal a crack propagation at a certain speed (4000 cycles were required for a 2 mm length increase). Then in the zone of transition from the cladding to base metal it practically stops, then for 1000 cycles there is no crack penetration into the base metal and further crack propagation is going at a rate typical for base metal. Similar change in crack length with increase in the cycle number, but in a more modest degree, was observed at tests in the corrosive medium. The decrease of fatigue crack propagation rate in the zone of transition from cladding to base metal is made still more obvious by the relationships shown in Fig. 4.

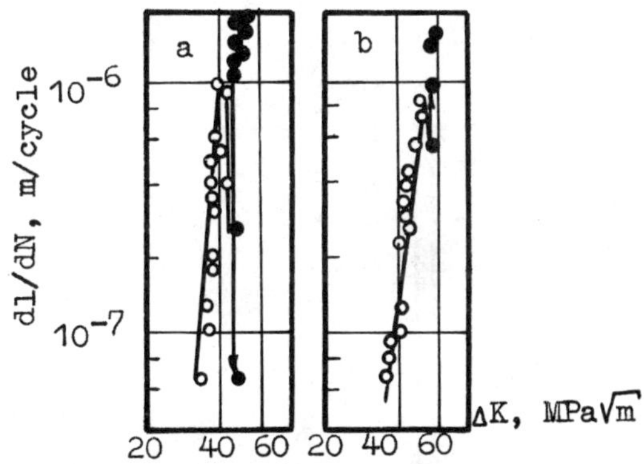

Fig. 4. Fatigue crack propagation in zone of transition from cladding (o) to base metal (•) in air (a) and in aqueous medium (b).

The attempts to explain the nature of this phenomenon have not been successful. It could be associated with the pattern of distribution of residual stresses in the vicinity of the line of fusion between the cladding and base metal, as it known (Kupka, 1977) that the above stresses, when passing through the interface, change their sign and become compressive stresses instead of tensile, thus slowing down the crack propagation in the transition zone. The phenomenon in question can also be attributed to increase in the level of strength characteristics in the event of crack tip passing from the austenitic cladding (U.T.S.=563 MPa and $\sigma_{0.2}$=308 MPa) into the base metal where, at a distance of 1 mm from the fusion line, U.T.S.=785 MPa and $\sigma_{0.2}$=597 MPa. As a result of such a change in strength characteristics, the plastic deformation zone at the crack tip remains mainly in the austenitic cladding metal and hardly ever spreads out into more strong pearlitic steel. Evaluation of the fatigue crack propagation rate in aqueous medium of average parameters (250°C and 10 MPa) for steel of 15X2MΦA grade was carried out using disk specimens of 4 mm thickness. The results of these experiments shown in Fig.5 indicated that the above material, at a higher temperature and f=0.01 Hz, displays a higher resistance to fatigue crack propagation than if tested in an aqueous medium with temperature at 80°C.

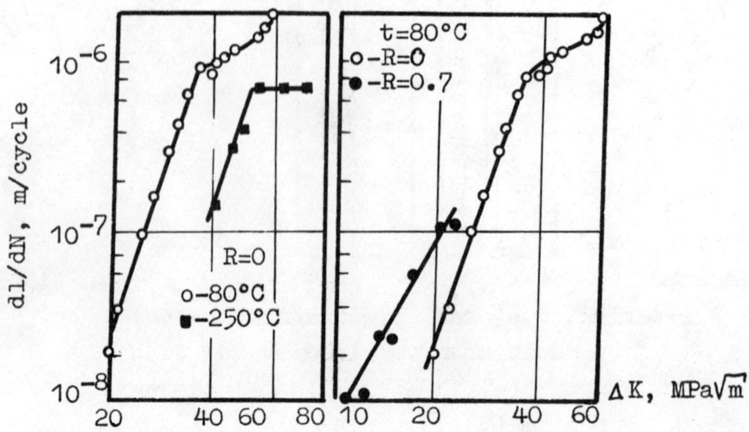

Fig. 5. Influence of aqueous medium at 250°C and 80°C on crack propagation rate in base metal.

A marked influence on the fatigue crack propagation rate in an aqueous medium is produced by the cycle ratio. The experiments described above demonstrate quite clearly the particulars of fracture of bimetallic parts made of 15X2MΦA steel with anticorrosive cladding at cyclic loading, the causes of which are to be found in joining of structurally-heterogenous materials - namely, austenitic cladding with cast coarse-dendrite structure and pearlitic steel with wrought structure.

REFERENCES

Brown, B. F. (1971). Stress corrosion cracking. The Theory of Stress corrosion cracking in alloys, NACE, Houston, 215-236.

Cherepanov, G. P. (1967). On crack propagation in solids. Appl. Math. and Mech., 31, 476-488.

Kupka, I., and P. Mrkous (1977). A contribution to the question of stress-corrosion cracking of austenitic stainless steel cladding in nuclear power plants. Proc. of Symposium IAEA-SM-218 Vienna, 317-329.

Rice, I.R. (1968). A path-independent integral and the approximate analysis of strain concentration by nothes and cracks. J. of Appl. Mech., 35, 287-298.

DELAYED FAILURE OF INELASTIC COMPOSITES

J. V. Suvorova

Mechanical Engineering Research Institute, Moscow, USSR

ABSTRACT

An approach is proposed which allows the description of nonmonotonous dependence of strength on loading rate and temperature. Two different ranges of these dependences are connected with a possible change of failure modes. The proposed model is confirmed by special experiments.

KEYWORDS

Composite material, failure mode, damage accumulation, delayed failure, hereditary theory.

INTRODUCTION

So far little attention has been given to the processes of delayed failure of composites, though failure processes for elastic composites are well known to depend considerably on loading history. Practically no mechanical models have been proposed being both general enough and taking into account the specific features of a composite. The present approach is based on macroscopic analysis, in which possible attempts are made to interpret the mutual components work. It allows to outline the applicability of the mathematical formulations.

The development of a delayed failure model is impossible without a preliminary analysis of the failure modes corresponding to the given loading conditions. Figure 1 shows photographs of carbon plastic samples which have failed at two different rates. The failure modes are seen to be different.

The possibility of change of failure mode under varying loading conditions leads to nonmonotonous dependence of the strength on the variable parameter. Figure 2 shows the data for two types of carbon plastics, where the rate of loading is the variable parameter. Figure 3 shows similar experimental results obtained by Sogolova and Demina [1972], the variable parameter being the temperature.

In both cases the first region represents damage accumulation failure, the second one macrocrack propagation.

To develop a delayed failure model all features of each region have to be taken into consideration. This may be done as follows.

Let us consider the region of increasing strength where failure is caused by damage accumulation. To develop a proper model it seems reasonable to regard the macroscopic failure as a consequence and a critical point of the process of deformation. For the general case of viscous and damage accumulating material it is necessary to describe both these properties in similar terms. The two mechanisms while different in nature both lead to increasing deformation under loading.

Fig. 1. Samples failed at two different rates of loading.
1 - $\dot{\sigma}=6\times10^{-1}$ kg/mm^{2}s, 2 - $\dot{\sigma}=8\times10^{2}$ kg/mm^{2}s.

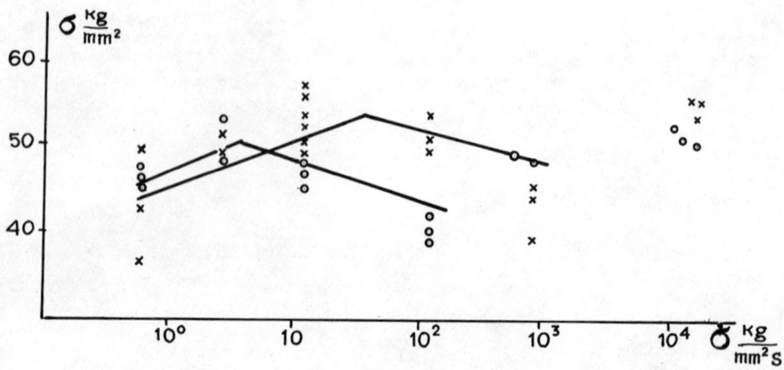

Fig. 2. Dependences of strength on loading rate for two types of carbon plastics with different adhesion properties. Solid lines are the results of calculations.

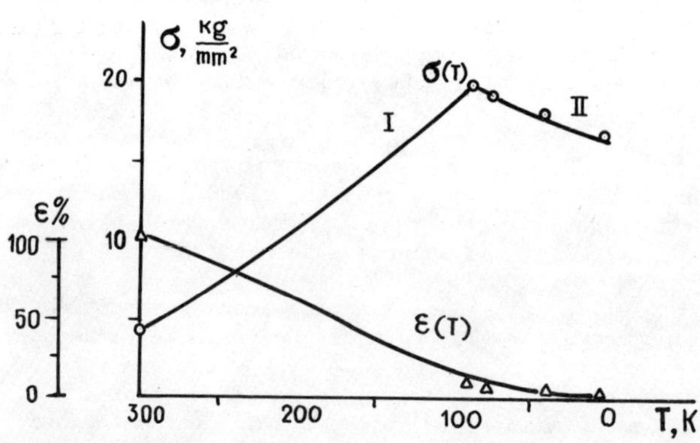

Fig. 3. Dependence of strength on temperature for polyarilate. Solid lines are the result of calculation.

In terms of the hereditary type theory the following constitutive equation may be written

$$\varphi(\varepsilon) = \sigma + \pounds^*\sigma + M^*\sigma \qquad 0 \leqslant t \leqslant t_* \tag{1}$$

$$\varphi(\varepsilon) = \sigma + \pounds^*\sigma + M^*\sigma|_{t_*} \qquad t > t_* \tag{2}$$

Here t_* is the moment of the beginning of unloading, $\varphi(\varepsilon)$ is the curve of the instantaneous deformation and the structure of the operators $\pounds^*$ and M^* is the following

$$\pounds^*\sigma = \int_0^t \pounds(t-\tau)\, f_l(T)\sigma(\tau)d\tau$$

$$M^*\sigma = \int_0^t M(t-\tau)\, f_m(T)\sigma(\tau)d\tau$$

The functions $f_l(T)$ and $f_m(T)$ describe the temperature influence.

It should be underlined that such an approach requires the independence of the instantaneous deformation curve $\varphi(\varepsilon)$ on temperature.

The equations (1) and (2) involve two operators to distinguish reversible and irreversible deformation. The former is reversible creep and described by the operator $\pounds^*$. The latter results from the damage accumulation process and is described by the operator M^*. Since failure deformation will not decrease under unloading its value in the equation (2) is fixed.

The loading equation (1) allows to formulate a strength criterion. Since failure is brought about by damage accumulation and is not directly connected with the reversible creep process the following failure criterion may be stated

$$\sigma + M^*\sigma = \sigma_o^* \tag{3}$$

Here σ_o^* is the limit point of the instantaneous deformation curve which can be found from some macro-experiment. It should be stressed that the parameters of the kernel $M(t-\tau)$ and the function $f_m(T)$ in equation (3) are the same as in equations (1) and (2).

If the material is not viscous as in the case of carbon or boro-plastic then failure will be caused by damage accumulation only, due to fiber breaking. In this case the constitutive equation involves only one operator describing this process, however the strength criterion will be the same.

The substitution of the criterion (3) into equation (1) gives the dependence of the failure strain on the parameters of the loading process:

$$\varphi(\varepsilon_*) = \sigma_o^* + \pounds^*\sigma|_{t=t_*} \tag{4}$$

The failure strain ε_* evidently does not vary if the material does not possess viscous properties. Figures 4 and 5 show the data of the dependences of the failure strain on stress rate for the carbon-plastic, which does not possess viscous properties, and for viscous organo-plastic.

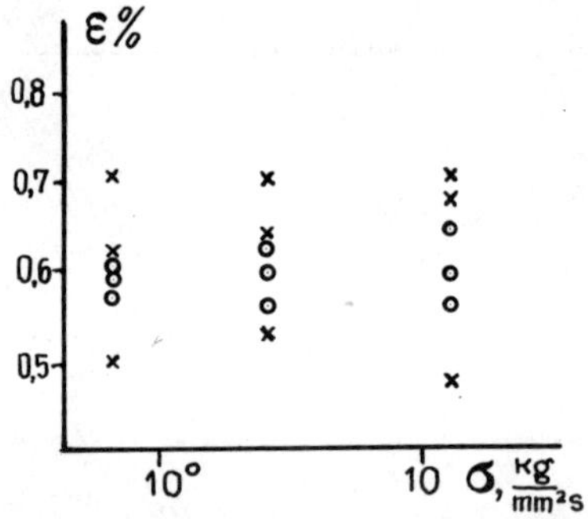

Fig. 4. Dependence of failure strain on loading rate for two types of carbon-plastic.

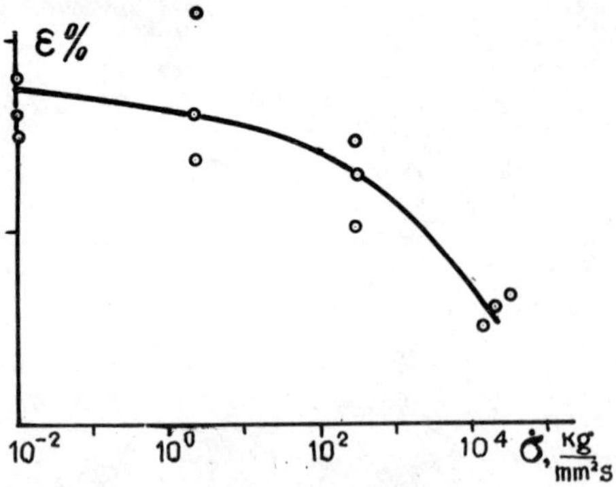

Fig. 5. Dependence of failure strain on loading rate for organo-plastic.

The latter failure strain is seen to decrease to a great extent. Equation (4) predicts this change rather well. The calculations in Fig. 5 were made supposing that the kernel can be taken in the form of Abel's type $\mathcal{L}(t) = l(1-\alpha)t^{-\alpha}$, the loading process being $\dot{\sigma}$ = const. Then equation (4) yields

$$\varphi(\varepsilon_*) = \sigma_o^* + \frac{l}{2-\alpha}\,\sigma_* t_*^{1-\alpha}$$

The values of the parameters in Fig. 5 being: $l = 1.3\ \mathrm{min}^{-(1-\alpha)}$, $\alpha = 0.97$, $\sigma_o^* = 107\ \mathrm{kg/mm^2}$.

Let us consider now the process of failure for the second region (see Fig. 2). An analysis of the ruptured surfaces of the samples allows to assume that the cause of rupture is macro-crack propagation. The possibility of macro-crack propagation in the material obviously depends to a great extent on its inner structure, the bond between fibers and resin being of great importance. Figure 2 shows the data for the two types of carbon-plastic differing in the adhesive properties only, that is, in the bond between fibers and resin. The weaker the bond the higher stress rates are required for the transition from the one type of failure to the other. This is due to the fact that a weak bond between fiber and resin leads to delamination preventing macro-crack propagation. For this reason a material with poor bonding appears to possess higher strength characteristics in a certain range of stress rates.

To formulate the failure criterion for the second region the following scheme may be proposed. Let the material possess some initial inner cracks or voids. Every of these defects clearly is a stress concentrator. The stress concentration σ' depends on the geometry of a void. If it is of ellipsoidal form with the axis sizes 2a and 2b, the axis "a" being the axis in the direction of loading, the following expression can be written:

$$\sigma' = \sigma[1 + f(\tfrac{b}{a})]$$

This relation, as shown by Neuber [1947], is almost linear in the range of the b/a ratio under consideration. Hence σ' may be taken as:

$$\sigma' = \sigma(1 + \delta'\tfrac{b}{a}), \qquad \delta' = \mathrm{const.}$$

Since σ' > σ, σ' will control the process of failure. But the void size "a" is elongated in the direction of loading. With the growth of "a" the magnitude of the concentration will decrease. When the value of "a" becomes fairly large the stress concentration will not be dangerous.

Along with the process of change of the void sizes the process of damage accumulation takes place. The latter leads to the decrease of the available section resisting the load and to the increase of the true stress. Thus failure can be due to a critical value of the true stress or the most dangerous void transforming into a macro-crack.

A macro-crack will appear when the value of the concentration becomes as large as the ultimate strength of material - σ_o^* :

$$\sigma' = \sigma(1 + \delta'\frac{b}{a}) = \sigma_o^* \tag{5}$$

The changing of "a" (the value of Δa) depends on the viscous properties of the material and can be expressed with the help of the corresponding operator $\pounds^*$. The initial "a" size of the void depends on the parameters of the damage accumulation process. So for homogeneous material the next formula can be written:

$$a = a_o\left(1 + \frac{\Delta a}{a_o}\right) = a_o\left(1 + \frac{\pounds^*\sigma}{\sigma + M^*\sigma}\right)$$

The criterion (5) takes the form

$$\sigma\left(1 + \delta\,\frac{\sigma + M^*\sigma}{\sigma + M^*\sigma + \pounds^*\sigma}\right) = \sigma_o^* \ , \quad \delta = \delta'\,\frac{b_o}{a_o} \tag{6}$$

The formula (6) shows that the more viscous effects are pronounced, the less dangerous stress concentration is. On the other hand, the higher loading rates and lower temperatures, the less is the value of denominator as well as σ. This accounts for the decrease of the stress in the second region.

The moment of transition from the first failure mode to the second one can be determined from the following condition

$$\sigma + M^*\sigma = \sigma\left(1 + \delta\,\frac{\sigma + M^*\sigma}{\sigma + M^*\sigma + \pounds^*\sigma}\right)$$

For nonhomogeneous materials such as composites, the analysis is a bit more complicated due to the specific composite structure. In composites it is the resin that possesses the viscous properties, the failure depending on the fiber strength. In this case, as it was shown by Suvorova (1982) the failure criterion takes the form:

$$\sigma_{of}^* = \sigma_f\left(1 + \delta\,\frac{(\sigma_m + M_m^*\sigma_m)/E_m}{(\sigma_m + \pounds_m^*\sigma_m + M_m^*\sigma_m)/E_m + M^*\sigma/E}\right) \tag{7}$$

Here the index f denotes the fiber and index m the matrix. Taking into account that $\varepsilon_m = \varepsilon = \sigma/E$, $\sigma_{of}^*/E_f = \sigma_o^*/E$ formula (7) can be rewritten in the following way:

$$\sigma_o^* = \sigma\left(1 + \delta\,\frac{\sigma}{\sigma + M^*\sigma}\right) \tag{8}$$

This formula determines the strength for the second mode of failure. It is obvious that the lower the temperature and the higher the stress rate the less the value of σ (both factors are accomodated in the operator M^*).

This approach has been used for calculation of the theoretical dependences strength - loading rate and strength - temperature. Figure 2 shows strength - stress rate dependence for the two types of carbon-plastics. Their parameters are:

for material I: $m = 0.29\ s^{-(1-\alpha)}$, $\alpha = 0.92$
$\sigma_o^* = 65\ kg/mm^2$, $\delta = 0.56$

for material II: $m = 0.946\ s^{-(1-\alpha)}$, $\alpha = 0.96$
$\sigma_o^* = 98.3\ kg/mm^2$, $\delta = 1.18$

For the calculations formulae (3), (8) were used.

Figure 3 shows the strength - temperature dependences for the homogeneous material - polyarilate. The calculations for the second region were carried out according to formula (6). If the loading rate is constant and the kernels $\mathcal{L}$ and M are chosen in the similar form: $\mathcal{L}(t) = l(1-\alpha)t^{-\alpha}$, $M(t) = m(1-\alpha)t^{-\alpha}$, the formula (6) yields

$$mf_m(T)T^{1-\alpha}\left(1 + mf_m(T)^{1-\alpha} + lf_l(T)t^{1-\alpha}\right) = \delta\left(1 + mf_m(T)t^{1-\alpha}\right)$$

In general the temperature functions f_l and f_m may be different, but for the results depicted in Fig. 3 it was sufficient to choose them as power functions distinguished only by a multiplier. The expression (9) becomes

$$m'T^\gamma(1 + m'T^\gamma + l'T^\gamma) = \delta(1 + m'T^\gamma)$$

where $m' = mt^{1-\alpha}$, $l' = lt^{1-\alpha}$

The parameters of the material have been found to be equal to $\sigma_o^* = 12.96\ kg/mm^2$, $m' = 3.622$, $l' = 97.48$, $\gamma = 3.0$, $\delta = 0.33$. The curves calculated with help of equations (3), (4) and (6) give a good approximation of the experimental data.

The possible change of the failure mode due to temperature and loading rate results in a significant change of the material strength. To predict failure the phenomenological models should be used very carefully, since their applicability is limited by the possible change of the failure mode.

REFERENCES

Sogolova, T.I. and M.I. Demina (1972). Temperature dependence of mechanical behaviour of polymers with different chemical structure in the temperature interval 4.2 ~ 300 K. Mech. of polymers, 3, 387-391.

Neuber, H. (1947). Stress concentration. OGIZ, Moscow-Leningrad.

Suvorova, J.V. (1982). Failure modes of nonelastic materials according to strain rate effects and temperature, and associated failure criterions. Mech. comp. mater., 5, 797-803.

FINITE ELEMENT ANALYSIS OF THERMAL RESIDUAL STRESSES IN ANTISYMMETRICALLY LAMINATED COMPOSITES

T. Hirai*, A. Inoue, T. Semba*** and M. Hirai******

*Department of Mechanical Engineering, Doshisha University, Japan
**Graduate Student of Doshisha University, Japan
***Institute of Industrial Science, University of Tokyo, Japan
****Osaka Prefectural Industrial Research Institute, Japan

ABSTRACT

This paper describes a theoretical approach to prediction of coupling effect on the out-of-plane displacement for antisymmetrically laminated composites under thermal stressing during the curing process. The distortion of laminate and the equivalent stress distribution in the inter-surface caused by above origin are predicated from a ply-by-ply hybrid finite element modelling with respect to the midplane of macroscopic adhesive surface. Experimental results are compared with the numerical results and there is good correlation between theory and experiment.

KEYWORDS

Antisymmetrically laminated composites; thermal residual stress; interlamina stress; out-of-plane displacement; dimensional effect; coupling effect.

INTRODUCTION

Unlike homogeneous materials, composites have a heterogeneous structure with distinct interfaces. These may be classified as microscopic surfaces within a lamina and macroscopic surfaces between laminae. An accurate evaluation of the interface stress distribution in the boundary layers should enable the failure processes of composites to be better understood. Knowledge of the macroscopic interface stress is particularly important since delamination of a macroscopic surface results in a severe drop in strength of FRP.

In relation to symmetrically laminated composites, attempts to estimate the peak stress in adhesive layers have been made by Pagano and Pipes(1970), Puppo and Evensen(1970) and Wang and Crossman (1977). They have solved the two dimensional interlamina stress distribution for a representative range of composites. The three-dimensional microscopic distribution has been studied by Hirai et al.(1982) using Finite Element Analysis. The results indicate that stress peaks occur near the traction-free edges and the clamped end, for a laminate of finite dimensions. For symmetrically layered laminates a distortional mode is not evident, but where symmetry is not achieved the stress peaks at the boundaries will affect the flexural mode. Detailed numerical studies of the distortion of composites due to the existence of edge stresses have not previously been made.

In this paper the thermal reşidual in-ply stress and accompanying deformations of antisymmetrically laminated composites are computed three-dimesionally by Finite Element Methods for a case where the behaviour is far removed from that of a symmetrical composite. The effects of thermal residual stress on out-of-plane flexural deformation of a laminate during the curing process following hot-press forming are discussed.

NUMERICAL PROCEDURE

Macroscopically, unidirectionally reinforced lamina show a typical orthotropy; so that the $[+\theta^\circ_n/-\theta^\circ_n]_T$ laminate is modelled as a set of two anisotropic in-plane shells and three-dimensional prismatic joint elements representing the adhesive layer between the lamina shown as Fig. 1. The variation of in-plane stress (such as σ_x, σ_y, τ_{xy}) along the normal direction in a prismatic element are supposed to be trivially small, since the thickness of the adhesive layer is negligible in practice. The ply orientation against Y-direction is difined as $+\theta$ on clockwise rotation.

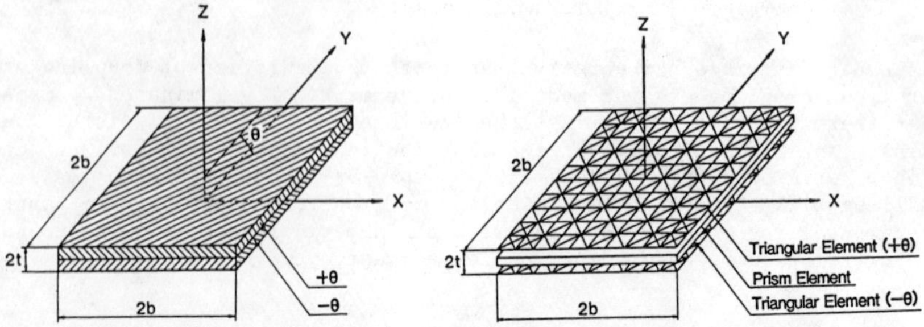

Fig.1 Illustration of modelling on composite laminate.

The principle of virtual displacement can be expressed for each constituent element at any equilibrium condition.

$$\int_v (\sigma_{ij} + \Delta\sigma_{ij})\delta(\Delta E_{ij})dv = \int_s (T_i + \Delta T_i)\delta(\Delta U_i)ds. \tag{1}$$

The strain increment E_{ij} is found by assuming that in-plane and out-of-plane components as following.

$$\Delta E_{ij} = \Delta\varepsilon_{ij}{}^t + \Delta\varepsilon_{ij}{}^b. \tag{2}$$

The constitutive equation could be assumed considering the thermal strain caused by thermal increment ΔT,

$$\sigma_{ij} = C_{ijkl}(\Delta E_{kl} - \alpha_{kl}\Delta T\delta_{kl}). \tag{3}$$

Introducing Eq.(2) and Eq.(3) into Eq.(1), the equation yields

$$\int_v \Delta\varepsilon_{kl}{}^t C_{ijkl}\delta(\Delta\varepsilon_{ij}{}^t)dv + \int_s \Delta\kappa_{kl}C'_{ijkl}\delta(\Delta\kappa_{ij})ds = \int_s \Delta T_i\delta(\Delta u_i)ds$$

$$+ \int_v (\alpha_{kl}\Delta T)C_{ijkl}\delta(\Delta\varepsilon_{ij})dv - \int_{C\sigma}\{Mxy\delta(\Delta w,x) + My\delta(\Delta w,y)\}dl. \tag{4}$$

The effect of the twisting moment can be duplicated by four self-equilibrating forces acting at the free corner shown as a example of Mxy,

$$Mxy = 2\int_0^{h/2} \tau_{xy}.zdz = \tau_{xy}^{0} h^2/6. \qquad (5)$$

Increments of linear strain and curvature tensor are found using that of in-plane and out-of-plane displacement.

$$\Delta\varepsilon_{ij} = (\Delta u_{i,j} + \Delta u_{j,i})/2,\ \Delta\kappa_{ij} = -\Delta w,ij\ (i=j),\ \Delta\kappa_{ij} = -2\Delta w,ij\ (i\neq j). \qquad (6)$$

Substituting Eq.(6) into Eq.(4) and expressing Δu_i and Δw by shape functions ϕ_{ij}, ψ_i and the nodal displacement $\Delta r_i^{(e)}$ as shown in Eq.(7), the element equilibrium equation can be derived as Eq.(8).

$$\Delta u_i = \phi_{ij}\Delta r_j^{(e)},\ \Delta w = \psi_j \Delta r_j^{(e)}. \qquad (7)$$

$$(K_{Eij}^{(e)} + K_{Bij}^{(e)})\Delta r_j^{(e)} = \Delta f_i^{(e)} + \Delta f_{Ti}^{(e)} + \Delta f_{Ci}^{(e)}, \qquad (8)$$

where:

$K_{Eij}^{(e)}$ is the linear component of element stiffness;

$K_{Bij}^{(e)}$ is the matrix of bending stiffness;

$\Delta f_i^{(e)}$ is the equivalent nodal force;

$\Delta f_{Ti}^{(e)}$ is the equivalent nodal force to thermal residual stress caused by that of strain;

$\Delta f_{Ci}^{(e)}$ is the equivalent nodal force to the distributed moment along the stress boundary.

Accordingly, the equilibrium equation can be derived by superposing the element equation into the whole constituent element.

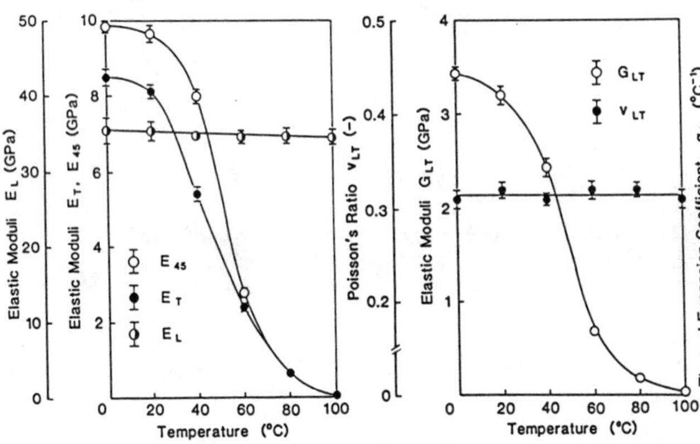

(a) (b)

Fig.2 Effect of temerature on the Elastic Moduli.

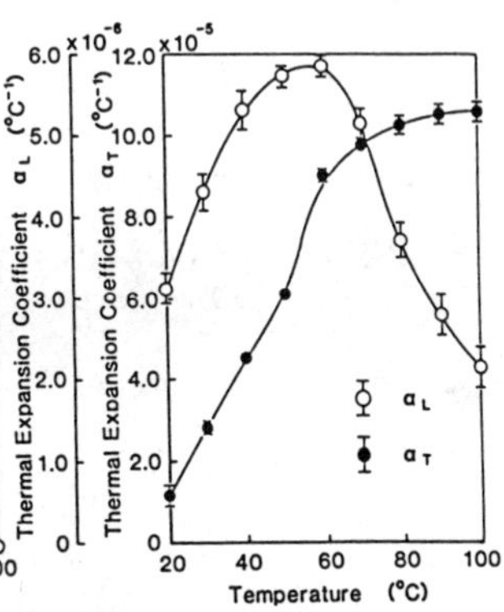

Fig.3 Effect of temperature on the Thermal Expansion Coefficient.

MATERIAL PROPERTIES DEPENDED ON THERMAL CONDITIONS

Fig. 2 shows the thermal dependance of anisotropic material constants such as E_L, E_T, G_{LT} and ν_{LT} obtained experimentally. The transverse Young's

Modulus E_T varies considerably with ambient temperature. The inflexion point on the curves correspond closely to the glass point for the epoxy resin matrix. Corresponding variations of anisotropic coefficients of thermal expansion are illustrated in Fig. 3. It is evident that the tangential expansion coefficient α_T is ten times larger than α_L.

ESTIMATION OF LAMINATE PROPERTIES

The Influence of Ply Orientation and Group Number

For the antisymmetrical $[+\theta^\circ_n/-\theta^\circ_n]_T$ laminate the out-of-plane flexural and torsional deformation might be predicted; the thermal residual stress and the distortional mode are estimated for the ply orientation θ=15°, 30° and 45° in the same ply group number. The value of the analysis is verified by comparing numerical and experimental results of distortion mode of the top face. Three-dimensional shape measurement is employed in the experiment by grid-irradiation type Moiré topography. The analysis is performed assuming that the laminate is uniformly cooled from the Tg point (55°C) to room temperature (20°C). Figs. 4 - 7 (a) show the distortion derived when out-of-plane displacement at the edges of the specimen is allowed. The calculated contour lines of thwe distortion are compared with the Moiré fringe pattern in Fig. 8 and there is good correlation between theory and experiment. The dotted points in the figures indicate the contact points on the standard grid pattern plate. Both numerical and experimental results illustrate the oblique symmetrical pattern against the diagonal lines except cross ply laminate showing a suddle pattern along the axes. Then the distortion figure shows a torsional mode.

The corresponding interface stress, that is the equivalent stress distributions in a prismatic joint element are illustrated in Figs. 4 - 7 (b). Measurement of interlamina stress is considered to be difficult in practice. All the figures show the similar tendency of stress distribution showing a uniformity around the inside domain but the stress concentration at the free edge. The value of the stress peak depends on the ply orientation, as the orientation $[+\theta^\circ_n/-\theta^\circ_n]_T$ increases the stress peak increases. Where a ±45° stacking angle is employed, the interlamina equivalent stress shows maximum value. The equivalent stress distribution on the cross ply laminate $[0^\circ_2/90^\circ_2]_T$ shows a pattern similar to that of the $[+45^\circ_2/-45^\circ_2]_T$ laminate.

The effect of ply group number on the interlamina stress peak caused by the coupling effect (Hyer, 1982) of antisymmetrical laminate are shown in the

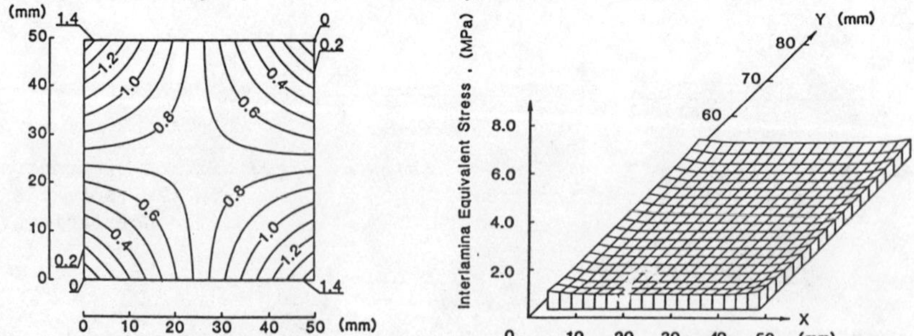

(a) Distortiom mode (b) Interface stress distribution

Fig.4 Numerical results on $[+15^\circ_2/-15^\circ_2]_T$ laminate for the case of free edge.

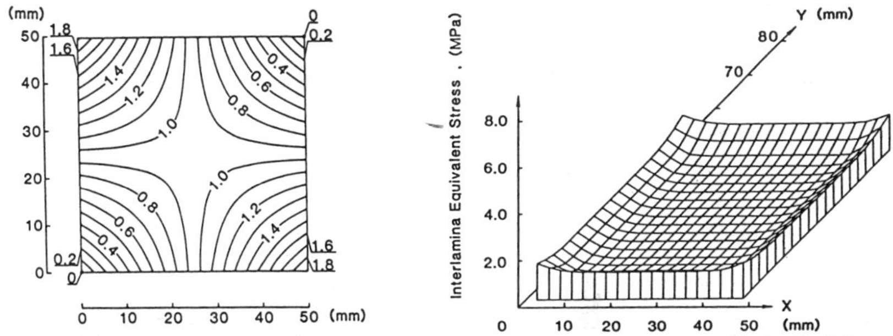

(a) Distortion mode (b) Interface stress distribution

Fig.5 Numerical results on $[+30^\circ_2/-30^\circ]_T$ laminate for the case of free edge.

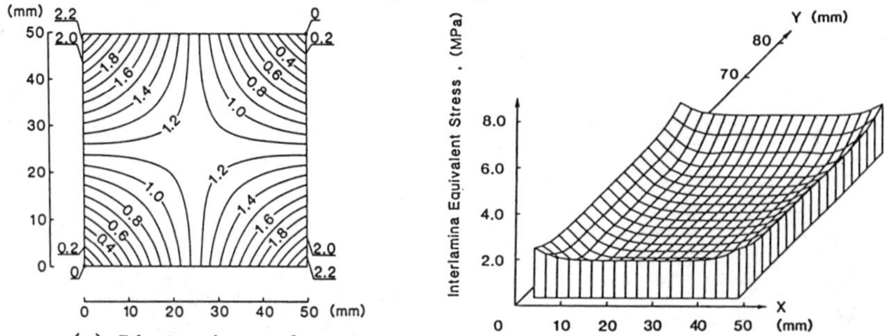

(a) Distortion mode (b) Interface stress distribution

Fig.6 Numerical results on $[+45^\circ_2/-45^\circ_2]_T$ laminate for the case of free edge.

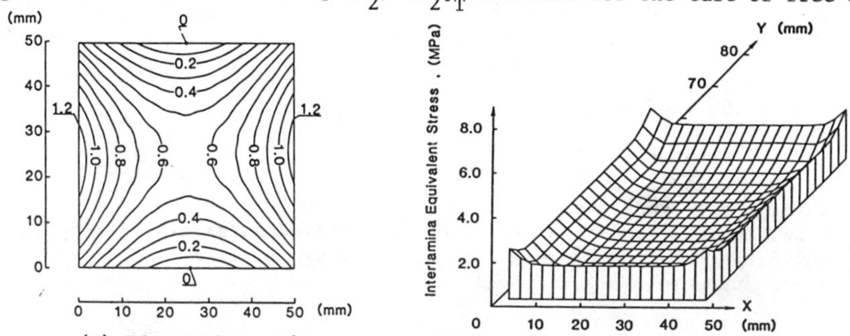

(a) Distortion mode (b) Interface stress distribution

Fig.7 Numerical results on $[0^\circ_2/90^\circ_2]_T$ cross ply laminate for the case of free edge.

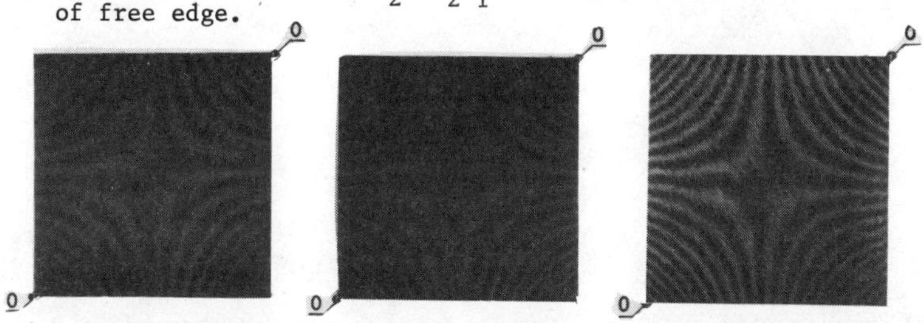

(a) $[+15^\circ_2/-15^\circ_2]_T$ (b) $[+30^\circ_2/-30^\circ_2]_T$ (c) $[+45^\circ_2/-45^\circ_2]_T$

Fig.8 Fringe pattern illustrating the distortion mode by Moiré topography.

figures as following. In selecting the ply orientation for the model, to intensify the expression of the effect, ±45° stacking angle is employed in considering the stress peak. The numerical results obtained in addition to $[+45^{\circ}_{2}/-45^{\circ}_{2}]_{T}$ laminates are shown in graphical form in Fig. 9 and Fig. 10 where n is 1 and 3. The results show that a progressively higher value of stress peak occurs at the corner of free edge for the higher values of ply group number. In order to produce a uniform disribution of the interlamina stress, it might be recommended to decrease the ply group number of the same orientation.

Effect of the Clamped Edge and Heterogeneous Nature in the Lamina

The analysis has been performed assuming that the laminate can have out-of-plane displacement at the free edges. For the case where the corresponding edge displacement is restricted unidirectionally, the different types of distribution can be derived, as shown in Fig. 11 (a)'. Fig. 11 (a) shows the stress distribution at a macroscopic interface. The peak stresses shown in Fig. 11 (a) are more pronounced than those in Fig. 6 (a). Edge and gripping effects on the interface stress can be seen, particularly at the corner of the laminate.

The state of stress distribution over the macroscopic interface will be influenced by non-uniformity of fibre density in the lamina. In order to emphasize the in-plane boundaries it is assumed that overlapping tapes provide the reinforcement. Fig. 11 (b)' shows the numerical results corresponding to the previously discribed boundary conditions and indicates the distortion and equivalent stress distrbution respectively. Heterogeneity of interface stress dependent on the non-uniformity of fibre density due to the tape overlap can be observed. The magnitude of the stress peak in the corner is reduced to some extent compared with uniformly laminated composites.

CONCLUSION

Summarising, the thermal residual stress and accompanying flexural deformations for antisymmetrically laminated composites are analysed on the microscopicscale. Considering the widespread use of laminates for mounting electronic circuits with high precision, the development of micromechanics for estimating the peak stresses and stress distribution could have considerable practical significance.

ACKNOWLEGMENT

We wish to acknowledge the support and encouragement of Dr. J. B. Hawkyard and Dr. P. D. Soden of the Mechanical Engineering Department, UMIST, U.K.. This work is supported by Grant-in-Aid for Scientific Research, Japan.

REFERENCES

Hirai, T., T. Semba and M. Tarui (1982). In T. Hayashi, K. Kawata and S. Umekawa (Ed.), Progress in Science and Engineering of Composites, ICCM-4, Vol. 1, 248-250.

Hyer, M. W. (1982). In T. Hayashi, K. Kawata and S. Umekawa (Ed.), Progress in Science and Engineering of Composites, ICCM-4, Vol. 1, 373-380.

Pipes, R. B. and N. J. Pagano (1970). J. Composite Materials, 4, 538-548.
Puppo, A. H. and H. A. Evenson (1970). J. Composite Materials, 4, 204-220.
Wang, A. S. D. and F. N. Crossman (1977). J. Composite Materials, 11, 92-106.

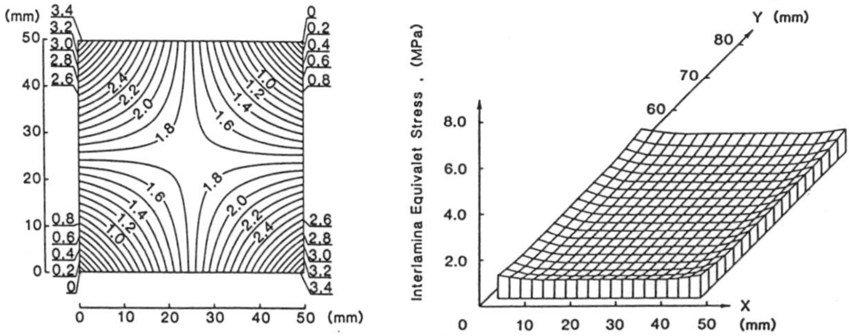

(a) Distortion mode (b) Interface stress distribution

Fig.9 Numerical results on $[+45°/-45°]_T$ laminate for the case of free edge.

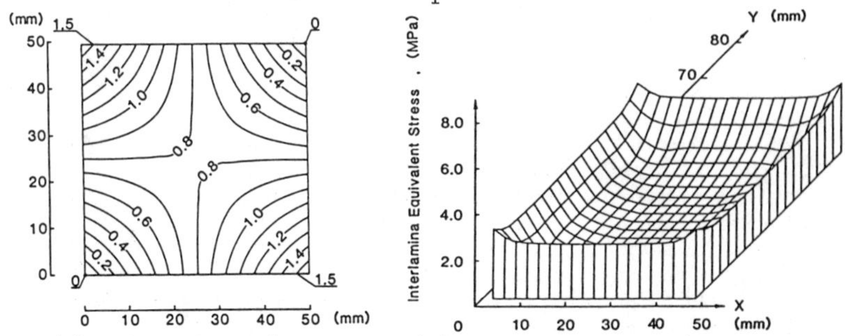

(a) Distortion mode (b) Interface stress distribution

Fig.10 Numerical results on $[+45^\circ_3/-45^\circ_3]_T$ laminate for the case of free edge.

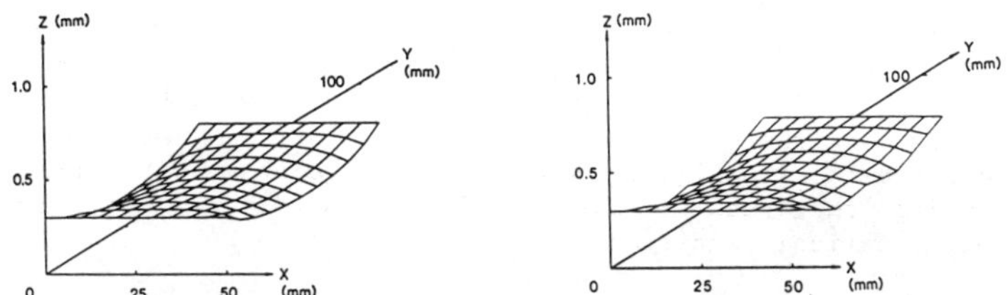

(a)' Distortion mode (b)' Distortion mode

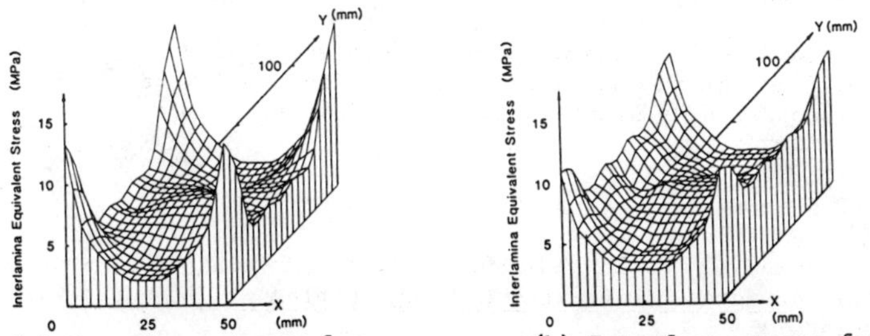

(a) Interface stress for homogeneous ply model (b) Interface stress for heterogeneous ply model

Fig.11 Numerical results on $[+45^\circ_2/-45^\circ_2]_T$ for the case of two side clamped.

A THREE-DIMENSIONAL DUCTILE DAMAGE MODEL APPLIED TO DEEP-DRAWING FORMING LIMITS

J. Lemaitre

Laboratoire de Mécanique et Technologie, ENSET, Univ. Paris 6, CNRS, 61, Avenue du Président Wilson, 94230 Cachan, France

ABSTRACT (Keywords)

The purpose of this paper is to present a new model for ductile fracture based on continuous damage mechanics and its applications to metal forming, particularly to deep-drawing of sheets. The damage is linear with equivalent strain and shows a large influence of triaxiality. This model is suitable either for analytical predictions of forming limits in the case of proportional loading or numerical calculation for any loading path.

1. INTRODUCTION

The prediction of metal-forming limits uses more and more continuum mechanics calculation in which the states of stress and strain are governed by equilibrium and kinematic equations and elasto-plastic constitutive equations. To predict rupture as the initiation of a macro-crack by the same technique, constitutive equations for evaluation of damage are needed.
The phenomenon of initiation and growth of cavities and micro-cracks induced by large deformations in metals and called "ductile plastic damage" has been extensively studied by means of micro-mechanics analysis [1] [2] , but the results are difficult to include in large scale structural calculations. Between these two scales there exists a macro-scale level (10^{-1} to 1 mm) which is the usual scale level of constitutive equations. The continuous damage mechanics approach deals within that macro-scale, defining a damage variable as an effective surface density of cracks or cavity intersections with any plane.

2. ELEMENTS OF DAMAGE MECHANICS

As many developments of the theory have already been published [3][4] , only the principal features are summarised here.

Damage variable

If S is the overall section area of $\vec{n}$ orientation, of a damaged body and $\tilde{S}$ the effective resisting area taking into account the areas of cavities and crack intersections, the micro-stress concentrations and the interactions between closed defects, then by definition

$$\boxed{D = \frac{S - \tilde{S}}{S}}$$

is the scalar measure of damage in the $\vec{n}$ direction.
If the damage is supposed to be isotropic, D does not depend on $\vec{n}$, and characterizes completely the state of damage.

$D = 0$ corresponds to the undamaged state,
$D = D_c$ with $D_c \simeq 0.2$ to 0.8, depending upon materials, characterizes rupture of the element into two parts.

Concept of effective stress [5]

If $\vec{F}$ is the load acting on the section S of the element considered, $\vec{T} = \frac{\vec{F}}{S}$ is the usual stress vector which leads to the CAUCHY stress tensor : σ , ($\sigma . \vec{n} = \vec{T}$).
The effective stress vector relative to the effective resisting area $\tilde{S}$ is $\tilde{\vec{T}} = \frac{\vec{F}}{\tilde{S}} = \frac{\vec{T}}{(1-D)S}$ and the effective stress tensor is :

$$\boxed{\tilde{\sigma} = \frac{\sigma}{1-D}}$$

Hypothesis of strain equivalence [3]

"The strain behavior of a damaged material is represented by the constitutive equations of the virgin material (D = 0) in which the stress is simply replaced by the effective stress".
Example of one dimensional linear elasticity of a damaged material : (E being YOUNG's modulus)

$$\varepsilon_e = \frac{\tilde{\sigma}}{E} = \frac{\sigma}{(1-D)E}$$

Thermodynamics

In order to model isothermal and isotropic elasticity, plasticity and damage, the following variables have to be introduced.

Observable variables	Internal variables	Associated variables
Elastic strain tensor ε^e	----------------------	stress tensor σ
	Accumulated plastic strain p ------------	Radius of yield surface $R_y + R$
	Damage D ------------	Damage strain energy release rate Y

Here p is define by $\dot{p} = \frac{dp}{dt} = (\frac{2}{3} \dot{\varepsilon}^p_{ij} \cdot \dot{\varepsilon}^p_{ij}$)

ε^p_{ij} being the components of the GREEN-LAGRANGE plastic strain tensor.

The variable Y associated with D is derived from the free energy potential Ψ quadratic in $\boldsymbol{\varepsilon}^e$ and linear in (1-D) to ensure the hypothesis of strain equivalence

$$Y = \rho\frac{\partial\Psi}{\partial D} = \frac{\partial(\frac{1}{2}\mathbb{a}:\boldsymbol{\varepsilon}^e:\boldsymbol{\varepsilon}^e[1-D]) + \Psi_p(p)}{\partial D} = -\frac{1}{2}\mathbb{a}:\boldsymbol{\varepsilon}^e:\boldsymbol{\varepsilon}^e$$

$\mathbb{a}$ being the fourth order tensor of elasticity and ρ the mass density. In the case of isotropic elasticity : ν being POISSON's ratio, $\sigma_m = \frac{1}{3}tr(\sigma)$ being the mean stress and $\sigma_{eq} = (\frac{3}{2}[\sigma_{ij} - \sigma_m\delta_{ij}][\sigma_{ij} - \sigma_m\delta_{ij}])^{1/2}$ being the VON MISES equivalent stress, Y reduces to :

$$-Y = \frac{\sigma_{eq}^2}{2E(1-D)^2}\left[\frac{2}{3}(1+\nu) + 3(1-2\nu)\left(\frac{\sigma_m}{\sigma_{eq}}\right)^2\right]$$

This quantity has the same meaning for damage as the strain energy release rate "G" for cracks in fracture mechanics [6] .

The constitutive equations for damage are derived from a potential of dissipation :

$$\varphi^*(\dot{\varepsilon}^p, \dot{p}, Y, q; \varepsilon^e, T, p, D)$$

$$\dot{D} = -\frac{\partial\varphi^*}{\partial Y}$$

3. FORMULATION OF A DUCTILE DAMAGE MODEL LINEAR IN STRAIN

A simple model thermodynamically admissible and in accordance with phenomenological observations is derived from the following potential :

$$\varphi^* = \frac{S_0}{2}\left(\frac{-Y}{S_0}\right)^2 \cdot \dot{p}$$

$$\dot{D} = -\frac{\partial\varphi^*}{\partial Y} = \frac{-Y}{S_0} \cdot \dot{p}$$

In the expression for Y , replace σ_{eq} by its value taken from the RAMBERG-OSGOOD hardening law coupled with damage and written for the threedimensional case :

$$\tilde{\sigma}_{eq} = \frac{\sigma_{eq}}{1-D} = K p^{1/M}$$

where K and M are material constants together with S_0

$$\boxed{\dot{D} = \frac{K^2}{2ES_0}\left[\frac{2}{3}(1+\nu) + 3(1-2\nu)\left(\frac{\sigma_m}{\sigma_{eq}}\right)^2\right] p^{2/M}\dot{p}}$$

This is the general constitutive equation for ductile plastic damage depending upon 3 material constants : $(K^2/2ES_0)$, ν , M . It has to be integrated in each particular case of loading history represented by p(t) , and $(\sigma_m/\sigma_{eq})(t)$: the ratio which gives the influence of the stress triaxiality.

In case of proportional loading the ratio σ_m/σ_{eq} can be considered as constant and the constitutive equation for damage may be integrated with initial condition for damage threshold p_D : $p < p_D \longrightarrow D = 0$

$$D = \frac{K^2}{2ES_0}\left[\frac{2}{3}(1+\nu) + 3(1-2\nu)\left(\frac{\sigma_m}{\sigma_{eq}}\right)^2\right]\frac{M}{2+M}\left\langle p^{\frac{2+M}{M}} - p_D^{\frac{2+M}{M}}\right\rangle$$

This expression can be written in a simpler fashion by considering a large strain range, when $M \rightarrow \infty$ which yields $(2+M)/M \simeq 1$ and by introducing the accumulated strain to rupture p_R defined by $D = D_C \rightarrow p = p_R$

$$D_C = \frac{K^2}{2ES_0}\left[\frac{2}{3}(1+\nu) + 3(1-2\nu)\left(\frac{\sigma_m}{\sigma_{eq}}\right)^2\right](p_R - p_D)$$

Dividing D by D_C and with $\frac{p_D}{p_R} = \frac{\varepsilon^p_D}{\varepsilon^p_R} \simeq \frac{\varepsilon_D}{\varepsilon_R}$ (ε_D and ε_R being the one dimensional strain damage threshold and strain to rupture for which $\sigma_m/\sigma_{eq} = 1/3$) :

$$\boxed{D = \frac{D_C}{\varepsilon_R - \varepsilon_D}\left(p\left[\frac{2}{3}(1+\nu) + 3(1-2\nu)\left(\frac{\sigma_m}{\sigma_{eq}}\right)^2\right] - \varepsilon_D\right)}$$

4. IDENTIFICATION OF THE MODEL

Identification of such a model consists in the quantitative evaluation of the three coefficients ε_D , ε_R and D_C which can be determined from uniaxial tests :

$$D = D_C \frac{\varepsilon - \varepsilon_D}{\varepsilon_R - \varepsilon_D}$$

Here ε_D and D_C need a measurement of damage which comes from the coupling between elasticity and damage [7] already written as an example in section 1. $\tilde{E} = (1-D)E$ being the elastic modulus of the damaged material measured as a function of strain ε .

$$\varepsilon_e = \frac{\sigma}{(1-D)E} = \frac{\sigma}{\tilde{E}} \qquad \text{or} \qquad \boxed{D = 1 - \frac{\tilde{E}}{E}}$$

An example is given figure 1 :

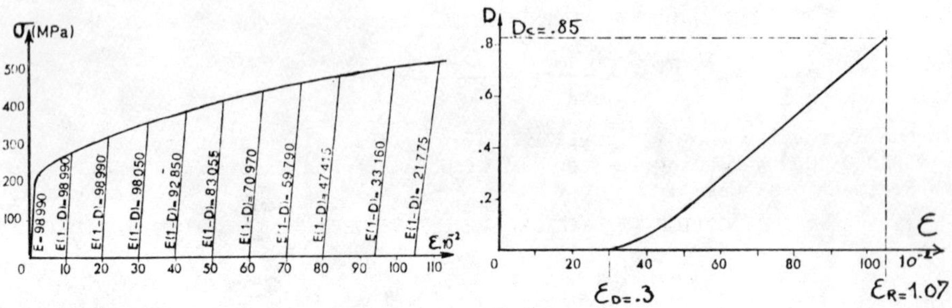

Fig. 1. Elasticity modulus change and damage evolution. 99,9 % copper - Room temperature.

This kind of experiment has been made on many materials such as steels, stainless steels, aluminium alloys, titanium alloys. In each case we found a linear dependence of D on ε , and this justifies the choice of the potential φ^* as a quadratic function of (-Y).

Another interesting feature of the model is the dependence of the damage upon the triaxiality through the expression $\left[\frac{2}{3}(1+\nu) + 3(1-2\nu)\left(\frac{\sigma_m}{\sigma_{eq}}\right)^2\right]$

The validity of this property can be checked on the accumulated strain to rupture $p = p_R$ which corresponds to $D = D_C$ in the integrated model :

$$p_R = \varepsilon_R\left[\frac{2}{3}(1+\nu) + 3(1-2\nu)\left(\frac{\sigma_m}{\sigma_{eq}}\right)^2\right]^{-1}$$

Figure 2 shows a comparison between values given by this expression and those calculated by the Mc CLINTOCK or RICE and TRACEY models and with the results of experiments taken from reference [8] . These results are also in accordance with those of reference [9] and [10].

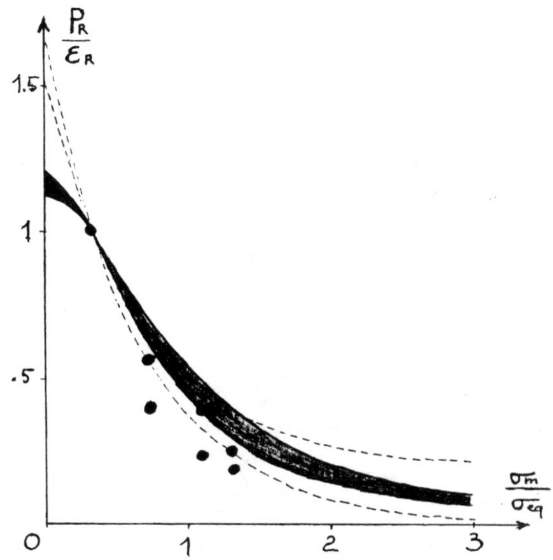

Fig. 2. Influence of triaxiality factor on ductility

▬ Model developed in this paper (for ν = 0.25 and ν = 0.33)

---- Mc CLINTOCK model (for two values of damage threshold $\varepsilon_D = 0$, $\varepsilon_D = 0.2\ \varepsilon_R$)

•••• Experiment on notched specimens at room temperature A 508 steel [8]

5. APPLICATION TO FRACTURE LIMITS OF DEEP-DRAWING FORMING PROCESS

The analysis of the deep-drawing process of sheets is usually made within the hypothesis of plane stress and the limits are drawn in the plane of principal strains (ε_1 , ε_2), with $\varepsilon_3 = -(\varepsilon_1 + \varepsilon_2)$.
The fracture limit here is defined as the state of local maximum strain (ε_1 , ε_2) for which a macro-crack initiates ; that is when the damage variable reaches its critical value D = D_C within the hypothesis of isotropic damage. This corresponds to the accumulated strain to rupture $p = p_R$.

$$p_R = \varepsilon_R \left[\frac{2}{3}(1+\nu) + 3(1-2\nu)\left(\frac{\sigma_m}{\sigma_{eq}}\right)^2 \right]^{-1}$$

For the case of proportional or radial loading.

$$p = \left(\frac{2}{3}\varepsilon_{ij}\ \varepsilon_{ij}\right)^{1/2}$$

or with ε_1 , ε_2 and $\varepsilon_3 = -(\varepsilon_1 + \varepsilon_2)$ as limiting values of the strains:

$$p_R = \frac{2}{\sqrt{3}}\left(\varepsilon_1^2 + \varepsilon_2^2 + \varepsilon_1\,\varepsilon_2\right)^{1/2}$$

The equation of the limiting curve is obtained by stating the equality of the two expressions of p_R :

$$\varepsilon_1^2 + \varepsilon_1 \varepsilon_2 + \varepsilon_2^2 - \frac{3}{4}\varepsilon_R^2\left[\frac{2}{3}(1+\nu) + 3(1-2\nu)\left(\frac{\sigma_m}{\sigma_{eq}}\right)^2\right]^{-2} = 0$$

This is an equation of degree 2 in ε_1 or ε_2 which can be solved for $\varepsilon_1(\varepsilon_2)$ (or $\varepsilon_2(\varepsilon_1)$) with the values of the triaxiality ratio σ_m/σ_{eq} calculated for each loading path :

- Assume several proportional loading paths defined by :

  $\varepsilon_2 = \alpha\,\varepsilon_1 \quad , \quad (-0.2 < \alpha < 0.3)$

- The HENCKY law of plasticity shows that $\frac{\varepsilon_2}{\varepsilon_1} = \frac{s_2}{s_1}$ (s_2 and s_1 being the components of the stress deviator) from which it is possible to calculate σ_2 as a function of σ_1 (σ_3 = 0 in plane stress hypothesis) and α . Then σ_m and σ_{eq} are calculated and finally :

$$\frac{\sigma_m}{\sigma_{eq}} = \frac{\alpha + 1}{\sqrt{3}\,(\alpha^2 + \alpha + 1)^{1/2}}$$

- Then ε_1 is calculated from the equation of the limiting curve in which ε_2 is replaced by $\alpha\,\varepsilon_1$:

$$\varepsilon_1 = \varepsilon_R\left(\left[\frac{1}{3}(2+\alpha)^2 + \alpha^2\right]^{1/2}\left[\frac{2}{3}(1+\nu) + 3(1-2\nu)\left(\frac{\sigma_m}{\sigma_{eq}}\right)^2\right]\right)^{-1}$$

  and ε_2 is deduced from $\varepsilon_2 = \alpha\,\varepsilon_1$.

This method has been applied to several materials of which an example is given in figure 3. The calculated curve (solid line) is compared with experiments taken from reference [11] (the experimental limit should be just in between the points and the crosses) and with a more sophisticated theory (dashed line) taking into account the evolutions of the plastic strain and damage all along the loading path [12] . The present theory gives a curve which is almost straight as observed by many investigators.

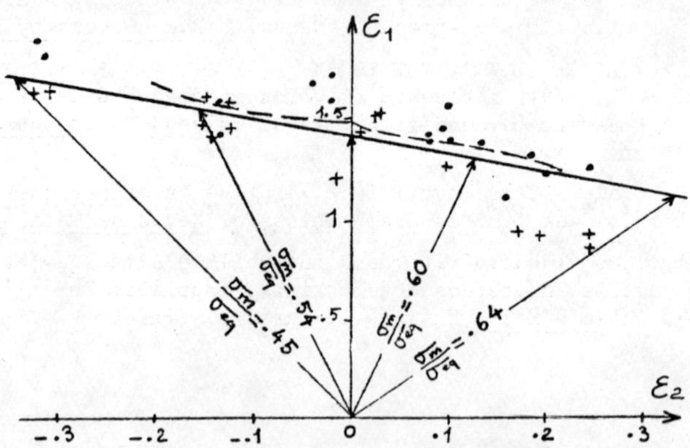

Fig. 3. Fracture limit curve of deep-drawing mild steel (ν = 0.27, ε_R = 2,16).

When the loading differs much from the proportional or radial loading a step by step numerical procedure is needed. It needs an elasto-plastic calculation using large deformation theory and step by step integration of the differential constitutive equation of ductile damage given in section 3 - [12] - As a consequence the fracture limit curve of deep-drawing is not unique. It characterizes the material but also the hisotries of loading.

6. REFERENCES

[1] Mc Clintock, F. (June 1968). A criterion for ductile fracture by the growth of holes. Journal of Applied Mechanics.

[2] Rice, J., D. Tracey (1969). On the ductile enlargement of voids in triaxial stress fields. Journal Mech. Phys. Solids, 17.

[3] Lemaitre, J., et J.L. Chaboche (1978). Aspect phénoménologique de la rupture par endommagement. Journal de Mécanique Appliquée, 2,3, Gauthier-Villars.

[4] Murakami, S. (1981). Effects of cavity distribution in constitutive equations of creep and creep damage. Euromech Colloquium, Cachan, France.

[5] Kachanov, L.M. (1958). Time of the rupture process under creep conditions. Izv Akad Nauk U.S.S.R Otd Teckh Nauk, 8-26.

[6] Chaboche, J.L. (1978). Description thermodynamique et phénoménologique de la visco-plasticité cyclique avec endommagement. Thèse Université PARIS 6, France.

[7] Lemaitre, J., J.P. Cordebois et J. Dufailly (1979). Sur le couplage endommagement-élasticité. Compte rendu à l'Académie des Sciences, B. 391. Paris.

[8] Mudry, F. (1982). Etude de la rupture ductile et de la rupture par clivage d'aciers faiblement alliés. Thèse Université de Technologie de Compiègne, France.

[9] Needleman, A. and N. Triantafyllidis (1977). Void growth and local necking in biaxially streched sheets. Brown University Report 16421/1.

[10] Tvergaard, V. (1979). Influence of voids on shear band instabilities under plane strain conditions. Tech. University of Denmark Report n°159.

[11] Grumbach, M. et G. Sanz (Avril 1982). Influence de quelques paramètres sur les couches limites d'emboutissage. Revue de Métallurgie

[12] Cordebois, J.P. (1983). Critères d'instabilité plastique et endommagement ductile en grandes déformations. (application à l'emboutissage). Thèse d'Etat, Université Paris 6, France

ON THE LAW OF MIXTURE IN DUAL PHASE STEELS

M. T. Ma, D. G. Wang and B. R. Wu

Central Iron and Steel Institute, Beijing, China

ABSTRACT

Based on the principle of discontinuous fibre composite and considering the effect of change of carbon content in martensite on the strength of martensite and the effect of martensite transformation on the strength of ferrite, an expression of the law of mixture in dual phase steel was deduced. The applicability of this expression for different grades of dual phase steels was studied and the significance of this expression to the study of sheet formability was discussed.

KEYWORDS

Law of mixture; dual phase steel; discontinuous fibre composite; strength of dual phase steel; stress distribution of two phases; sheet formability.

INTRODUCTION

Low carbon or HSLA steels which have been intercritically annealed or control-rolled to produce a mixed structure of ferrite and martensite are now known as dual phase steels. Because this kind of steel possesses excellent formability and high strength a lot of steel users have shown interest in these steels in the past ten years. Some investigators (Davies, 1978a, 1978b; Marder, 1981; Marder, 1982) found that the tensile strength of dual phase steel increases linearly with increasing volume fraction of martensite (f_m) and can be expressed by the law of mixture in composites (Davies, 1978a; 1978b). However, other authors (Koo, Young and Thomas, 1980; Lei and Shen, 1982; Speich and Miller, 1979) indicated that the relationship between tensile strength and f_m was not linear and deviated from the law of mixture. The purpose of this paper is to deduce an appropriate expression of the law of mixture in dual phase steel and examine the applicability of this expression to different grades of dual phase steels.

EXPRESSION OF THE LAW OF MIXTURE IN DUAL PHASE STEEL

The initial expression of the law of mixture was introduced by Kelly and Davies (1965), that is

$$\sigma_C = \sigma_f V_f + \sigma_M'(1-V_f) \qquad V_{crit} \leq V_f \leq 1 \tag{1}$$

in which σ_C is the tensile strength of composite, σ_f is the tensile strength of fibre, σ_M' is average **stress of martix when fibre is broken**, V_f is volume fraction of fibres, and V_{crit} is critical volume fraction of fibre which strengthens matrix.

Equation (1) is used for describing the mechanical behavior of composites containing continuous brittle fibres, assuming that the fibres and matrix are well bonded and the fibre strain, matrix strain, and composite strain are all equal.

Later, Mileiko (1969) deduced the expression of tensile strength of composites, σ_C, containing continuous ductile fibres, assuming also that the fibre, matrix and composite are in same strain condition. σ_C is given by

$$\sigma_C = \lambda_f V_f \sigma_f + \lambda_M \sigma_M (1-V_f) \tag{2}$$

in which λ_f and λ_M are constants that have relations to critical strain of composite, fibres and matrix. In general, they are approximately equal to 1. Therefore, equation (2) can be simplified into equation (1). **Where σ_M is tensile strength of matrix.**

According to Kelly and Davies (1965) the tensile strength of discontinuous fibre composites, (defined as composites containing fibres the length of which, L, is less than the critical length of fibre, Lc,) is given by

$$\sigma_C = \frac{\tau_M L}{2 r_0} V_f + \sigma_M (1-V_f) \tag{3}$$

where τ_M is shear yield strength of matrix and r_0 is radius of fibre. When L>Lc failure of composites results from breaking of fibres due to plastic flow of matrix. When L<Lc failure of composites results from voids initiated at the interface of fibre and matrix due to plastic flow of matrix.

In above equations, σ_M, σ_M', σ_f and τ_M do not change. However, in dual phase steels conditions are different. For instance, the strength of martensite, σ_m decreases with the increase of annealing temperature due to change of carbon content in austenite (or martensite). Moreover, the strength of ferrite, σ_α, and so τ_α changes when martensite transformation induces dislocations in the ferrite or when the f_m changes to cause variation of th the mean free path of ferrite.

After considering the factors above mentioned we used the expression of discontinuous fibre composites, i.e. equation (3) to describe the tensile strength of dual phase steel and rewrote i it as follows:

$$\sigma = \sigma_\alpha(1- f_m) + \frac{\tau_\alpha L}{2 r_0} f_m \tag{4}$$

According to Kelly and Davies (1965) $\tau_m = \frac{r_0 \sigma_f}{L_c}$. In dual phase steel it means that $\tau_\alpha = \frac{r_0 \sigma_m}{L_c}$. Substituting $\tau_\alpha = \frac{r_0 \sigma_m}{L_c}$ into equation (4), we got

$$\sigma = \sigma_\alpha(1- f_m) + \lambda \sigma_m f_m \tag{5}$$

in which $\lambda = L/2L_c$. By using Von Mises yield criterion $\sigma_y^\alpha = \sqrt{3}\,\tau_\alpha$ and substituting $\tau_\alpha = \sigma_m d/2L_c$ into it

$$L_c/d = \sqrt{3}\,\sigma_m / 2\,\sigma_y^\alpha \tag{5a}$$

In equation (4) to (5a), σ is tensile strength of dual phase steel, σ_α is tensile strength of ferrite, f_m is volume fraction of martensite, σ_m is tensile strength of martensite, λ is correction coefficient the value of which is less than 1, L and d are length and diameter of martensite island respectively, τ_α is shear yield strength of ferrite, σ_y^α is the yield strength of ferrite under uniaxial tensile condition, Lc is the critical length of martensite.

According to Koo, Young and Thomas (1980)

$$\sigma_m = \sigma_m^0 + K_2 C_m \tag{6}$$

From recent experimental results (Lei and Shen, 1982; Ma, Wang and Wu, 1983) it was assumed

$$\sigma_\alpha = \sigma_\alpha^0 + K_1 f_m \tag{7}$$

where σ_m^0 is tensile strength of martensite at zero carbon, σ_α^0 is tensile strength of ferrite at zero f_m, C_m is carbon content in martensite calculated assuming that it is equal to the carbon content of austenite at intercritical annealing temperature and σ_m^0, σ_α^0, K_1, K_2, are regressive constants.

Substituting equations (6) and (7) into equation (5):

$$\sigma = \sigma_\alpha^0 + [\,K_1 - \sigma_\alpha^0 - K_1 f_m + \lambda(\sigma_m^0 + K_2 C_m)\,]\, f_m \tag{8}$$

$$\lambda = \sigma_y^\alpha L/\sqrt{3}\,\sigma_m d \qquad \text{assuming} \qquad \sigma_y^\alpha = 0.6\,\sigma_\alpha$$

$$\lambda = \sqrt{3}(\sigma_\alpha^0 + K_1 f_m) L/5(\sigma_m^0 + K_2 C_m)\,d \tag{8a}$$

The method of calculating λ has been published elsewhere (Ma, Wang and Wu, 1983)

COMPARISON BETWEEN EXPERIMENTAL RESULT AND THAT CALCULATED FROM EQUATION(8)

The alloy used for comparison was vacuum melted, low carbon Mn-V steel (0.10%C, 0.44%Si, 1.31%Mn, 0.05%V). The tensile specimens were heated at 730-800°C for 1 hour then water quenched. After final machining the specimens were tested in an Instron

1251 testing machine. The f_m was determined by point counting. Microhardness was measured on Hanemann hardometer. Crack initiation and propagation in the sectional surface of tensile specimen was observed using SEM.

The relation between microhardness of martensite and carbon content in Mn-V dual phase steel is shown in Fig.1. That between

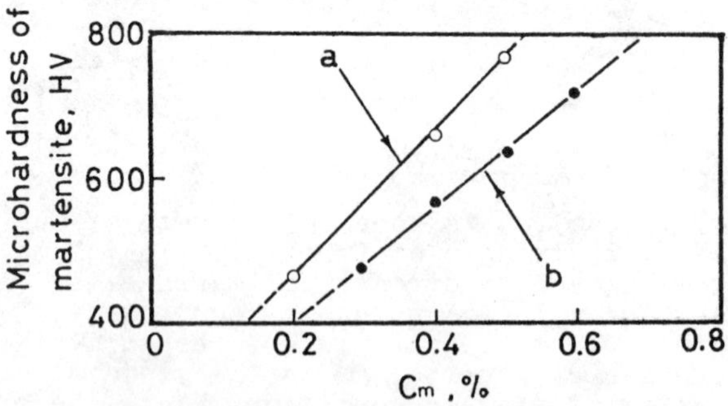

Fig.1. Microhardness of martensite vs C_m
a--100% martensite in iron-carbon alloy (Grange, Hribal, and Porter, 1977)
b--martensite in Mn-V dual phase steel

microhardness of ferrite and f_m is shown in Fig. 2. It can be seen that microhardness of ferrite increased with the increase of f_m The observations of the tensile specimen using SEM (Ma, Wang and Wu 1983) indicated that there were a few voids at interfaces of martensite and ferrite.

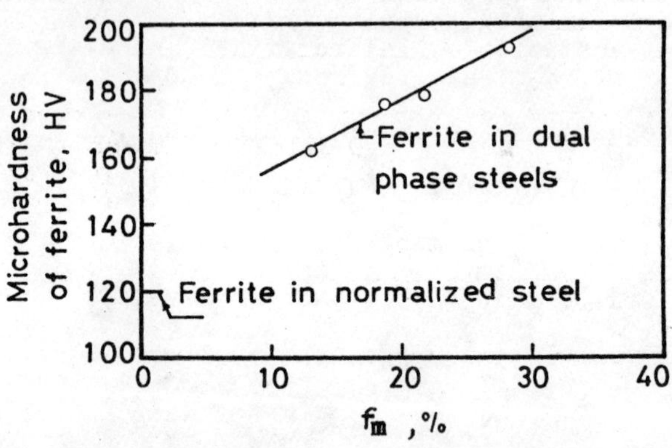

Fig.2. Microhardness of ferrite in dual phase steel vs f_m

After converting the measured microhardness of ferrite and martensite into tensile strength and using equation(6) and (7) to regress, two regressive equations were found

$$\sigma_m = 788 + 2970\ C_m \tag{9}$$

$$\sigma_\alpha = 482 + 634\ f_m \tag{10}$$

Substituting equations (9) and (10) into equation (8), then

$$\sigma = 482 + \left[\ 152 + \lambda(788 + 2970\ C_m) - 634\ f_m\right] f_m \tag{11}$$

The measured and calculated tensile strength of Mn-V dual phase steel are listed in Table 1.

Table 1 Measured and Calculated Tensile Strength of Mn-V Dual Phase Steels

Annealing temp. °C	f_m %	C_m %	λ	σ Calculated MPa	σ Measured MPa
730	13.0	0.6	0.50	669	685
750	18.5	0.5	0.60	759	761
780	21.4	0.4	0.71	779	778
800	28.0	0.3	0.90	895	901

APPLICABILITY OF EQUATION(11) FOR OTHER DUAL PHASE STEELS

Using equation (11) to calculate the data published by early investigators (Koo, Young and Thomas, 1980; Lei and Shen, 1982; Speich and Miller, 1979), the results in comparison with that calculated and measured by the original authors are shown in Table 2.

Table 2 Tensile Strength of Dual Phase Steels Calculated and measured by Different Investigators

Alloy	Annealing temp. °C	f_m %	C_m* %	λ	σ Calcd by Eq.(11), MPa	Measured σ MPa	σ Calcd by other authors MPa	Ref.
1010	775	30	.31	.40	676	683	716	(Koo and others, 1980)
	795	35	.27	.45	709	718	752	
	815	45	.21	.57	786	780	815	
1020	725	25	.74	.25	667	662	1014**	(Koo and others, 1980)
	740	40	.49	.39	788	842	1120	
	755	50	.39	.48	870	876	1165	
GB20	740	30	-	.40	678	680	1020	(Lei and Shen, 1982)
	770	40	-	.47	758	740	1120	
	800	48	-	.53	817	800	1310	
	830	80	-	.72	966	970	1410	
.12C	740	28	.42	.52	775	771	772	(Speich and Miller 1981)
.16C	740	39	.43	.57	889	872	900	
.20C	740	44	.46	.58	989	1040	1010	

*Calculated from the law of mass balance
**Calculated by authors using Koo and others(1980) Eq. (3)

DISCUSSION

(i) From the data listed in Table 1 and Table 2 it can be seen that the correlation between the measured values and that calculated from equation (11) is good. It indicates that this expression of the law of mixture in dual phase steel, which has taken account of the strengthening principle of discontinuous fibre composite, the effect of carbon content on the tensile strength of martensite and the effect of martensite transformation on the tensile strength of ferrite, shows the nature of strengthening mechanism and partition of stress between two phases in dual phase steels better than other.

(ii) If substituting the values of σ_m, Cm, fm at annealing temperature 730°C or 800°C into equation (8a) and assuming martensite island being spherical (i.e. d=L), λ will be equal to 0.076 or 0.136. These values will be lower than those used in Tables 1 and 2. These obvious differences resulted from that in calculating the former the deformation of martensite and the work-hardening of ferrite have not been taken into account. In fact, during deformation ferrite work-hardened so that σ_y^α increased and during necking martensite deformed slightly. These factors will cause the value of λ to increase. The lower the strength of martensite is, the more it deforms and the higher the λ value will be. The calculated values of tensile strength with λ as a constant and as a variant are shown in Fig.3. Obviously, when λ is constant the calculated value deviates from linear relationship.

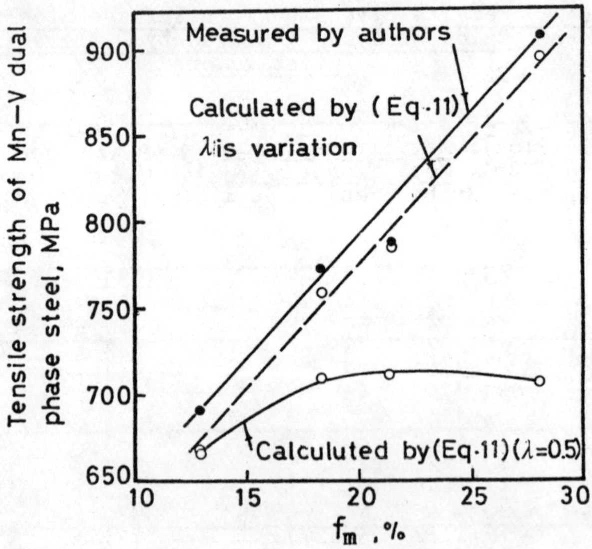

Fig. 3. Tensile Strength of Mn-V Dual Phase Steel vs f_m

(iii) The law of mixture is not only an expression of the strength of dual phase steel but also an equation of stress partition between two phases. If substituting the known constitutive equation of two phases into the expression of the law of mixture, the stress-strain equation of composite can be derived. According to the latter equation and the criteria of plastic instability, the work-hardening coefficient n can be predicted. Since n can be related to FLD₀ which is the intersection point of major strain and forming limit curve in forming limit diagram (Conrad, 1978), therefore, the law of mixture in dual phase steel is useful to assess the sheet metal formability.

CONCLUSION

The law of mixture in dual phase steel can be expressed as

$$\sigma = \sigma_\alpha^o + [K_1 - \sigma_\alpha^o - K_1 f_m + \lambda(\sigma_m^o + K_2 C_m)] f_m$$

using this equation to calculate the tensile strength of some dual phase steels, the correlation between calculated values and measured values is good. The value λ in this expression is a variant which is affected by the strength of ferrite and the strength, shape, and size of martensite.

The law of mixture is also significant to the formability of dual phase steel sheet.

ACKNOWLEDGEMENT

The authors would like to express their gratitude to Prof. Q.G. Cai for his valuable discussions.

REFERENCES

Conrad, H.(1978). J. of Mech. Working Technology, 2, 67-74.
Davies, R.G.(1978a). Met. Trans., 9A, 41-52.
Davies, R.G.(1978b). Met. Trans., 9A, 671-679.
Grange, R.A., C.R. Hribal, and L.F. Porter (1977). Met. Trans., 8A, 1775-1780.
Kelly, A., and G.J. Davies (1965). Met. Rev., 10, 1-79.
Koo, J.Y., M.J. Young, and G. Thomas (1980). Met. Trans., 11A, 852-854.
Lei, T.Q., and X.P. Shen (1982). Acta Metallurgica Sinica, 18, 75-84 (in Chinese).
Ma, M.T., D.G. Wang, and B.R. Wu (1983). CISRI Technical Bulletin, 3, (in Chinese), Accepted for publication.
Marder, A.R. (1981). Met. Trans., 12A, 1569-1579.
Marder, A.R. (1982). Met. Trans., 13A, 85-92.
Mileiko, S.T. (1969). J. Materials Science, 4, 974-977.
Speich, G.S., and R.L. Miller (1979). Mechanical properties of ferrite-martensite steels, In R.A. Kot, and I.W. Morris (Eds.), Structure and Properties of Dual Phase Steels, TMS/AIME, New York, pp. 183-201.

THERMOACTIVATION ANALYSIS OF FATIGUE CRACK GROWTH OF ALUMINIUM ALLOYS

V. S. Ivanova*, L. I. Maslov*, S. A. Kunavin* and T. Yokobori**

***A. A. Baikov Institute of Metallurgy of the USSR Academy of Sciences, Moscow, USSR**
****Research Institute for Strength and Fracture of Materials, Tohoku University, Sendai, Japan**

ABSTRACT

Stage II of the fatigue crack growth consists of two parts: 1 is quasi-elastic subcritical crack growth during which the striation spacing is slowly growing with the crack length; 2 is elastic-plastic subcritical crack growth during which the striation spacing is increasing with the crack length. In this paper thermoactivation and fractographic analyses are used for estimating threshold stress intensity factors K_{I*} controlling local fracture mechanism changes.

KEYWORDS

Crack growth mechanism; aluminium alloys; activation energy; striation; fatigue fracture; activation volume.

INTRODUCTION

As was shown by Yokobori (1978), it is important to combine the fracture mechanics approach with analysis of fracture micromechanism. These approaches take into account macroconcentration of stress from crack defects and microconcentration from dislocations, respectively. This part of mechanics called "Combination of macro- and micromechanism of fracture" [CMMMF] thus takes into consideration the plastic deformation process as well as the fracture process. The purpose of this paper is to find the threshold value ΔK_{I*} corresponding to the change of fracture mechanism using Paris-Erdogan crack growth law and making thermoactivation analysis of fatigue crack growth.

ANALYSIS

When fracture is thermally activated as in time-dependent fracture the fracture criterion will be given by the single formula (1)

$$U = A_1 \cdot \exp\left[-\Big(U_1 - \phi_1(\sigma_\ell)\Big)/(kT)\right] \qquad (1)$$

where U_1 is the activation energy, ϕ_1 is increase of local (including both microscopic and macroscopic) stress σ_ℓ caused by the applied stress, A_1 is a constant and k is Boltzmann's constant. A fatigue crack may grow under elastic-plastic conditions implying that the local plastic stress distribution plays the dominant role. This may also be concluded since the plastic deformation near the fatigue crack tip may play an important role whatever the physical model is. Thus, for the case of fatigue, even under small-scale yielding it is necessary to determine the local stress distribution near the crack tip. The result obtained is (Yokobori, 1981)

$$\sigma_\ell = f(\beta)\sigma_{cY}\left[\Delta K/(\sqrt{\varepsilon}\cdot\sigma_{cY})\right]^{\frac{2\beta}{1+\beta}} \qquad (2)$$

where ΔK is the range of the stress intensity factor, σ_{cY} is the initial yield stress in cyclic straining, β is the cyclic strain hardening exponent, $f(\beta)$ is some function of β, ε is the distance from the crack tip. At the same time the $da/dN - \Delta K$ dependence in Stage II is expressed as (Paris-Erdogan, 1963)

$$da/dN = C(\Delta K)^n \qquad (3)$$

where C and n are the constants. Also for the case of fatigue crack in Stage II eq. (1) may be expressed as (Yokobori, 1969)

$$da/dN = A\cdot\exp\left[-(U_o - a\ln\Delta K)/(kT)\right] \qquad (4)$$

where U_o is the activation energy of the fatigue fracture, A and a are constants. If we denote $A\cdot\exp[-U_o/kT] = C$ and $a/RT = n$ where R is the universal gas constant then eq. (4) can be transformed into eq. (3) (Jarema et al., 1977). In this case the effective activation energy $U_e = U_o - a\ln\Delta K_I$ can be expressed as

$$U_e = U_o - n\cdot RT\cdot\ln\Delta K_I \qquad (5)$$

From eq. (5) it follows that the effective fatigue fracture process activation energy in Stage II is a function of n, T and ΔK_I. It allows us to find the threshold value of $\Delta K = K_{IS}$ corresponding to the elastic-plastic transition during fatigue crack growth in Stage II.

EXPERIMENTAL TECHNIQUES

The materials investigated were aluminium alloys. Their chemical composition is listed in Table 1.

TABLE 1 Chemical Composition of the Al-alloys (%)

Alloy	Cu	Si	Fe	Mn	Mg	Cr	Zn
2024-T3	4.81	0.09	0.34	0.72	1.50	0.01	0.02
AMG6	0.10	0.25	0.30	0.65	6.1	-	0.15
D16AT	4.60	0.15	0.40	0.73	1.55	-	0.10
W95	1.62	0.50	0.50	0.45	2.4	0.18	0.02

For determination of the thermoactivation parameters disk type specimens (thickness 5 mm) were used for ensuring crack growth at ΔK = const (Yarema et al., 1977). Also plate (alloy 2024-T3) and cylindrical specimens (AMG6) with diameter 7.52 mm were used. Alloys D16AT and W95 were tested at +20°C and -70°C, alloy 2024-T3 at +24 to 300°C and alloy AMG6 at +20, +105 and +240°C. For obtaining kinetic da/dN - ΔK curves for cylindrical specimens the method proposed by Maslov et al. (Maslov et al., 1977) was used. The effective activation energy of the fatigue fracture was calculated using eq. (6)

$$U_e = \left\{R\left[\ln(da/dN)_{T=T_1} - \ln(da/dN)_{T=T_2}\right]\right\} / \left(\frac{1}{T_2} - \frac{1}{T_1}\right) \qquad (6)$$

The fractographic analysis was made to establish by using typical microfractographic features the change of the crack growth mechanism in Stage II.

RESULTS AND DISCUSSION

Figure 1 shows the $U_e - \Delta K$ dependence for AMG6 alloy.

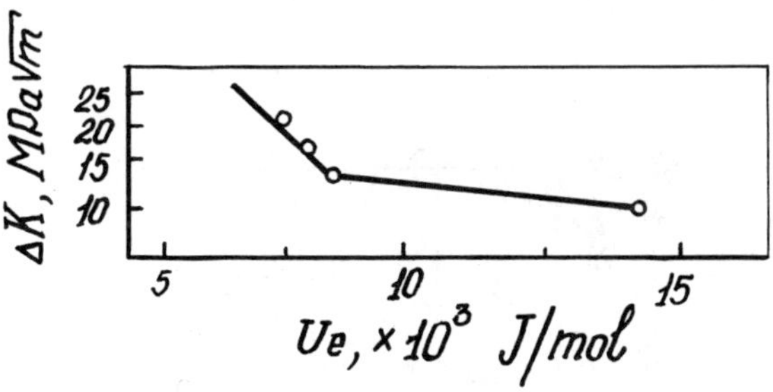

Fig. 1. ΔK versus U_e for AMG6 alloy.

The kink in the $U_e - \Delta K$ curve is connected to the elastic-plastic transition. Also kinks in the da/dN - ΔK dependence were observed (Fig. 2). Figure 3 shows the dependence between effective fracture activation energy and the range of the stress intensity factor ΔK for the 2024-T3 alloy.

For the 2024-T3 aluminium alloy two methods of determining the crack growth rate were used: the macroscopic method measuring the crack increment and the microscopic method measuring the size of the striations. As has been shown (Yokobori, Aizava, 1974) the fracture activation energy does not depend on the ΔK value if one determines the crack growth rate from measurements of the size of the striations (Fig. 3, symbol ▲).

However, the fracture activation energy depends on ΔK if determined from the macroscopic crack growth rate. As the fractographic analysis has shown the size of the striations increases with the crack length if $\Delta K > K_{IS}$, where $K_{IS} = 10$ MPa$\sqrt{m}$. This means that the elastic-plastic crack growth is realized in this case. For the AMG6 alloy the size of striations was constant at $\Delta K < K_{IS}$, where $K_{IS} = 13{,}0$ MPa$\sqrt{m}$. Table 2 presents the threshold values of the effective activation energy (U_e^*) and threshold values of the stress intensity factor range (K_I*) for aluminium alloys.

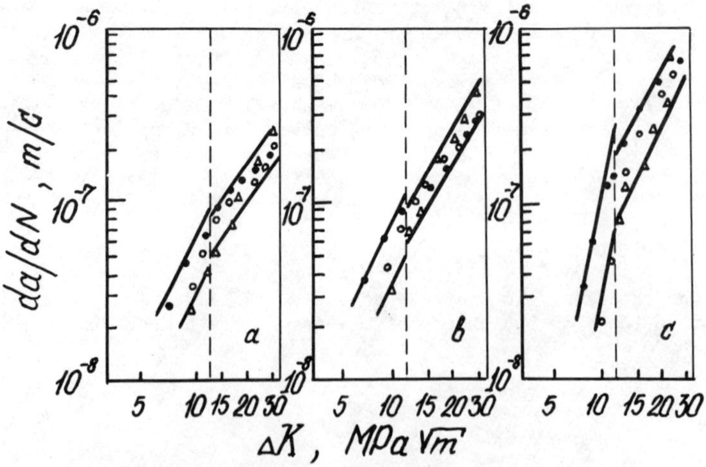

Fig. 2. da/dN - ΔK dependence for AMG6 alloy at different temperatures; (a) 20°C, (b) 110°C, (c) 240°C, Δ σ = ±200MPa, o σ = 200 MPa, • σ = 150 MPa.

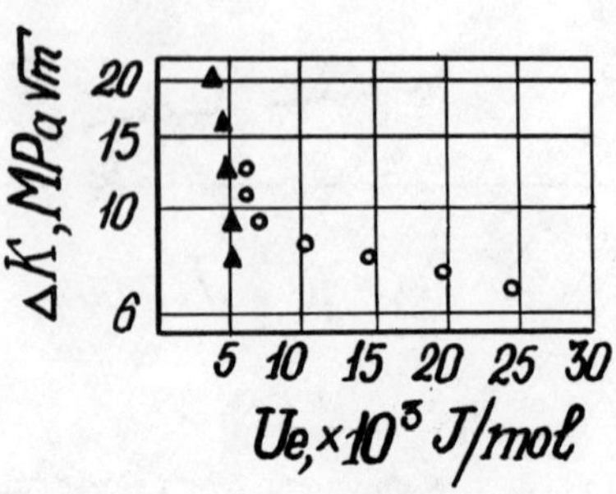

Fig. 3. ΔK versus U_e for 2024-T3 alloy.

TABLE 2 Thermoactivation parameters and threshold values of K_I* in Stage II fatigue crack growth for aluminium alloys

Alloys	$U_o \cdot 10^{-3}$ J/mole	$U_e^* \cdot 10^{-3}$ J/mole	K_I* MPa√m
AMG6	-	8.8	11.8
2024-T3	-	7.5	9.3
D16AT	43.2	5.0	20.5
		6.3	10.2
W95AT1	49.4	14.7	7.8
		4.2	18.6
		16.8	6.8
W95AT2	37.3	4.2	18.6
		10.5	5.6
W95AT3	29.7	2.1	19.2
		4.6	9.9
		10.5	5.6

Using these data the $U_e^* - \ln K_{I*}$ dependence was studied. According to eq. (5) this dependence should be linear. As Fig. 4 shows there are two straight lines - one is the line U_0 = 50 J/mole (upper, Fig. 4) and the other is the line with U_0 = 30 J/mole (lower, Fig. 4). For the 2024-T3 alloy the experimental data are in good agreement with the lower line and for the AMG6 alloy with the upper line. Thus it may be concluded that under quasielastic behaviour of the fatigue crack the activation energy U_0 in eq. (4) is equal to $U_0 = U_0^{max}$ = 50 J/mole and under the elastic-plastic fatigue crack growth U_0 is equal to $U_0 = U_0^{min}$ = 30 J/mole. This analysis allowed us to estimate the fundamental threshold value of the stress intensity factor to be $\Delta K = (K_{I*}^p)$ = 18.5 MPa$\sqrt{m}$ which is constant for aluminium alloys (this corresponds to the point U_e = 0 in Fig. 3). The K_{I*}^p value controls the upper boundary of elastic-plastic crack growth for aluminium alloys at minimum

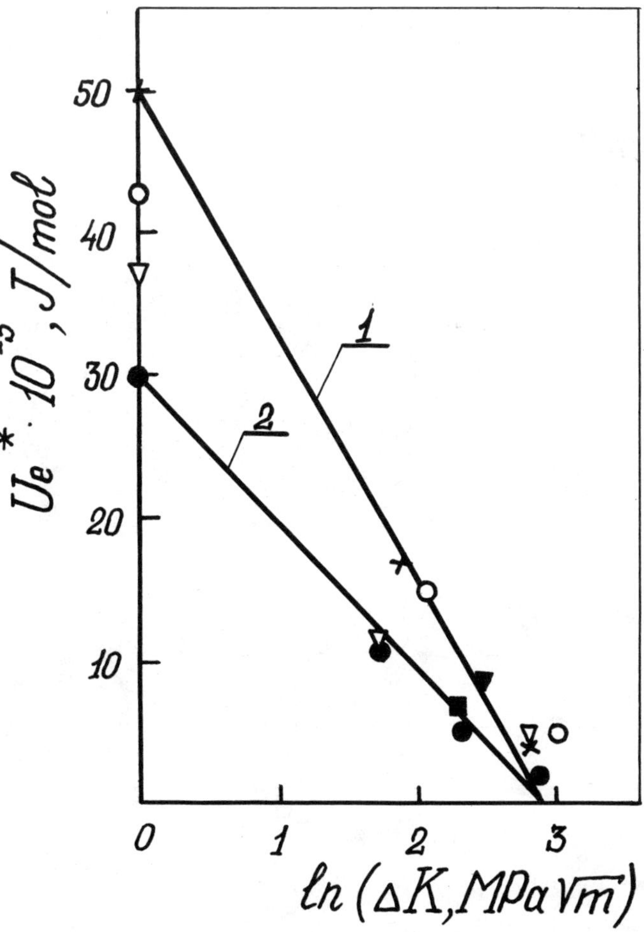

Fig. 4. $U_e^* - K_{I*}$ dependence for aluminium alloys.
1 - quasielastic crack growth; 2 - elastic-plastic crack growth; o D16AT; • W95AT3; ∇ W95AT2; × W95AT1; ▼ AMG6; ■ 2024-T3.

plastic constraint factor. This value is in good agreement with another fundamental threshold value of ΔK_p for aluminium alloys which is $(K^c_{I*})_{max}$ = = 40 MPa$\sqrt{m}$ and is connected to $(K^p_{I*})_{max}$ through eq. (7). One finds

$$[K^p_{I*}]_{max}/[K^c_{I*}]_{max} = \Delta^{1/2} \quad (7)$$

where Δ is the fracture constant which is equal to 0.22 for aluminium alloys (Ivanova, 1975). The $(K^c_{I*})_{max}$ value controls the upper boundary for quasi-elastic fatigue crack growth for aluminium alloys at maximum plastic constraint factor.

CONCLUSIONS

Fractographic studies and thermoactivation analysis of fatigue crack growth of aluminium alloys show that there is a linear $U^*_e - K_{I*}$ dependence with U^{max}_o = 50 J/mole for quasielastic crack growth and U^{min}_o = 30 J/mole for elastic-plastic crack growth.

REFERENCES

Ivanova, V.S. (1975). Spectra of the atomic reargement energy activation in metals. Docladi Academii Nauk, USSR, 220, 3, 579-582.

Ivanova, V.S. (1979). Fracture of metals. Metallurgia (Moscow), 351.

Maslov, L.I., M. Arita and A.Y. Begenov (1977). Fatigue crack growth at the steels Ti alloys and Ni alloys. Fiz.-Chim. Mech. Materialov, 13, 3, 20-26.

Paris, P. and F. Erdogan (1963). A critical analysis of crack propagation laws. Trans. ASME, S.D. 85, Basic Engng, 4, 528-534.

Jarema, S. Ja., O.P. Ostash, V.M. Belezkiy, V.N. Beljaev and A.A. Zboromirskiy (1977). On the crack growth change in from D16A and W95A alloys plates at decrease of temperature. Fiz.-Chim. Mech. Materialov, 13, 2, 5-10.

Yokobori, T. and T. Aizawa (1974). The influence of temperature and stress intensity factor upon the striation spacing and fatigue crack propagation rate of aluminium alloy. J. of the Japan Inst. of Metals, 38-8, 720-724.

Yokobori, T. (1978). Fundamentals of strength and fracture of materials, Naukova Dumka Inc., Kiev, 351.

Yokobori, T. (1981). On the critical problems in physico-structural foundations of fracture. Proc. 5th Int. Conf. on Fracture, Cannes, France, 3, 1145-1166.

TENSILE BEHAVIOR AND FRACTURE CHARACTERISTICS OF A DUAL-PHASE Fe/Mn/C STEEL

De-Zun Wang

Department of Metals and Technology, Harbin Institute of Technology, Harbin, People's Republic of China

ABSTRACT

A study was made on the tensile behavior of dual - phase (martensite plus ferrite) structure produced by quenching a Fe/Mn/C steel from the $\alpha+\beta$ range of temperature. The in-situ observation of deformation of the two phases indicate that the deformation is mainly concentrated in ferrite grains, and martensite islands practically deform only during the after-necking stage of tension. Microcracks initiate predominantly along the M/F interface, and they propagate at the ferrite side of the M/F interface. On the basis of a load transmission phenomenon an expression for the strength of dual-phase steels is suggested namely:

at 0.002 strain $\sigma_{0.2}=\sigma^f_{0.2}+\tau_{0.2}V_M$,

at tensile strength $\sigma_b=\sigma^f_b+\tau_b V_M$,

in which $\sigma^f_{0.2}$ and σ^f_b are regarded as ferrite strength, $\tau_{0.2}$ and τ_b are the shear stress caused by load transmission through ferrite to martensite.

KEYWORDS

In-situ observation; Microscopic deformation of phase; Load transmission through ferrite to martensite; Strength of dual-phase steel; Crack path.

INTRODUCTION

Martensite plus ferrite dual-phase steels due to their superior formability at the same strength level as commercial HSLA steels have sttracted much attention of investigators in the field of material engineering (Davies and Magee, 1978; Demeri, 1981). However, the law of strength of dual-phase steel is not entirely known, Martensite islands are similar neither to the particles in dispersion-strengthened alloys nor to the fiber in composite materials. Furthermore, microscopic deformation of the martensite and ferrite phase and the specific nature of crack initiation and propagation have not been studied in detail. Thus the aim of the present study is to clarify the deformation and fracture behavior of two phases in a dual-phase steel and to make an attempt in suggesting a simplified formula for evaluating the strength of such ateels.

EXPERIMENTAL PROCEDURES

The material used was a Fe/1.55%Mn/0.24%Si/0.15%C steel, normalized from 930°C. The microscopic deformation of the martensite and ferrite phase was in-situ studied in a SEM using miniature tensile specimens 2mm in width, 0.5mm in thickness and 3mm in gage length. Relative changes in the aspect ratio of ferrite grains and martensite islands were measured on micrographs taken from the central part of the specimens at various stages of tension. Tensile properties of the steel with dual-phase structure were examined on bulk specimens 8mm in diameter and 12mm in gage length at a cross-heat rate of tension of about 0.5mm/min.

Bulk specimens heated at 720, 730, 740 and 750 C for 15 min and miniature specimens heated at 720 C for 10 min were quenched into water and then tempered at 200 C for 2 h.

RESULTS AND DISCUSSION

Microscopic Deformation of the Two Phases

It was found that at an early stage in the process of plastic deformation by tension the slip lines appeared only in a few of the ferrite grains, and then the slip lines and the number of deforming ferrite grains gradually increased in the course of the tensile test. The plastic deformation of ferrite grains is strongl constrained by the adjacent martensite islands and thus the slip lines appear to be bent, giving rise to a build-up of strain concentration at the martensite/ferrite interface, as shown in Fig.1. The degree of deformation of ferrite grains and martensite islands as a function of the strain of a miniature speciman is shown in Fig.2. It is clear that the martensite constituent does not deform at all until strains well in excess of the maximum uniform strain are reached (Balliger and Gladman, 1981). It is implicate that the strain is concentrated in the ferrite. Therefore, partition of strain between the constituents does occur.

Crack Nucleation and Propagation

In the dual-phase structure strain concentration leads to formation of void nuclei at the martensite/ferrite interface. Internal necks are formed between the voids and final separation occurs along the knifeedges, to give a fracture surface composed of cusps and dimples. Plastic deformation is thus essential in ductile rupture. The strength is lower and the ductility is greater in ferrite than in martensite. Therefore the above proces developed at the ferrite side of the martensite/ferrite interface

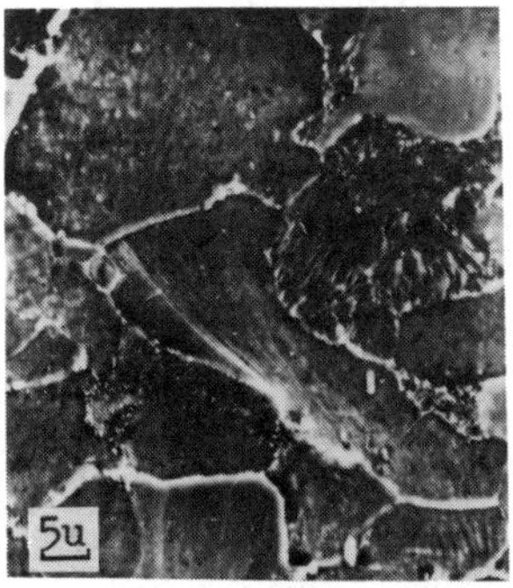

Fig. 1. SEM photomicrograph of a miniature tensile specimen at the uniform stage of deformation. The plastic deformation of ferrite grains is constrained by the adjacent martensite islands.

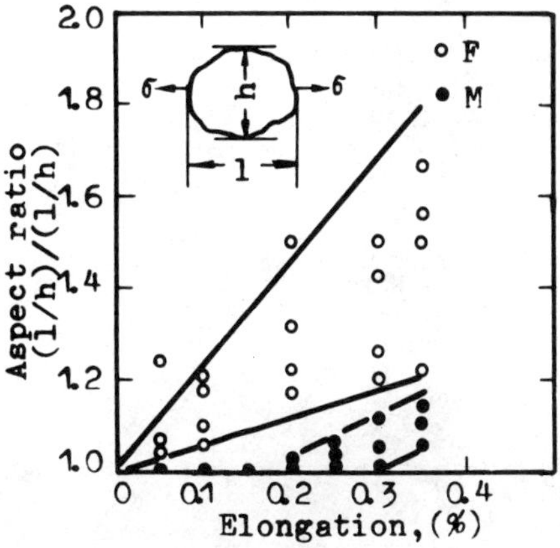

Fig. 2. Microscopic deformation of ferrite grains (F) and martensite islands (M). l_0 and h_0 are the original sizes of ferrite grain and martensite island, respectively.

(Fig.3a). A model, as shown in Fig.3b, is thus proposed, which illustrates the ductile rupture characterized by fracture surfaces composed of dimples along the martensite/ferrite interface.

It has been found that voids form ahead of the growing crack, and then an internal neck is formed between the nearest void and

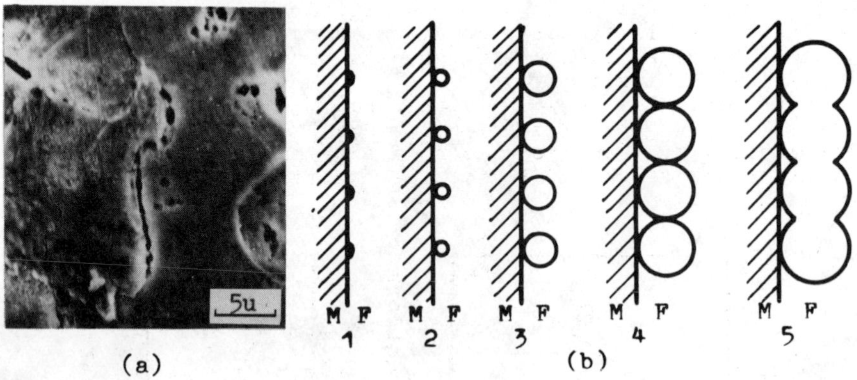

Fig. 3. Void nucleation and linkage along the M/F interface. SEM photomicrograph (a) and schematic model (b).

the crack tip. Most of the voids ahead of the crack tip are along the martensite/ferrite interfaces; the metal between the void and the crack tip could be ferrite or martensite, so that the crack propagates predominatly along the interface, but transgranular cracking has been found (Fig. 4).

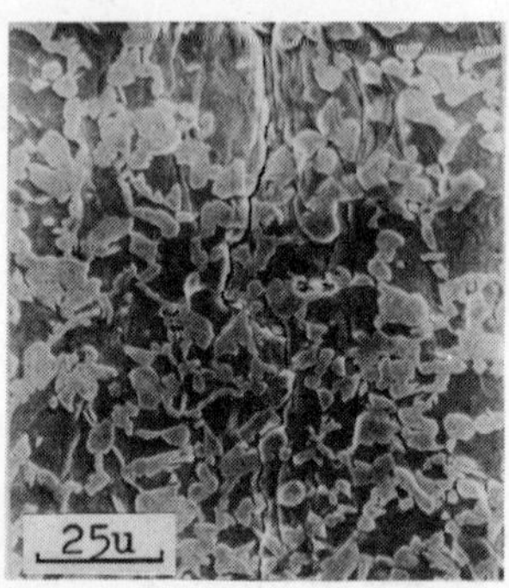

Fig. 4. SEM photomicrograph of crack propagation.

Variation of Strength with Martensite Content

There are different martensite contents in dual-phase structures quenched from different temperatures in which martensite islands are surround by ferrite. The flow stress as a function of volum fraction of martensite is shown in Fig.5a. It can be seen that the flow stress is linearly dependent upon the martensite content. The straight lines can be expressed in terms of percent martensite, V_M, as follows:

at 0.002 strain $\sigma_{0.2} = 95 + 10V_M$, (MPa) (1)

at tensile strength $\sigma_b = 350 + 16V_M$, (MPa) (2)

Uniform elongation and reduction of area as a function of percent martensite are shown in Fig.5b. The ductility decreases with the martensite content.

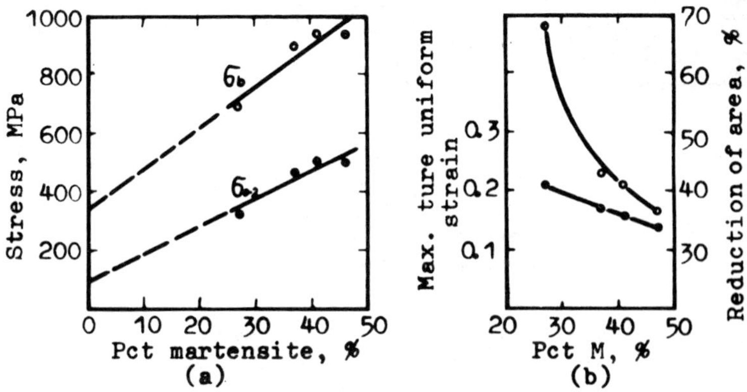

Fig. 5. 0.2 pct flow stress and tensile strength (a), and max. true uniform strain (∘) and reduction of area (•)(b) as a function of percent martensite.

When the martensite content is less than about 50% mechanical priperties of dual-phase steel can be characterized as follows (Davies, 1978): (1) The strength of the steel is linearly dependent on martensite content; (2) The strength does not depend upon the carbon content of martensite, and therefore upon the martensite strength; (3) The higher the strain, the more pronounced the dependence of the flow stress on martensite content. In the dual-phase structure the size and the spacing of martensite islands as a second phase, are much larger than those of dispersion-strengthened alloys. Thus the dual-phase structure could not be strengthened by martensite in the same way as for dispersion particles in the Orowan strengthening mechanism. The three factors that could be contributing to strengthening of dual-phase steel are: (1) ferrite is hardened due to strain from martensite transformation; (2) the plastic deformation of ferrite is constrained by adjacent martensite during tension; (3) ferrite serves to transmit the load to martensite, therefore the martensite carries the greater load, as in composite materials. The mechanism transmiting load is shown in Fig.6. At interfaces AA and BB ferrite can transmit a stress S_f which is equal to the ferrite internal sterss. There is shear stress τ_f at the interface AB, and its magnitude is dependent upon the strain difference between ferrite and martensite (Fig.6). The stress of martensite, S_M, is

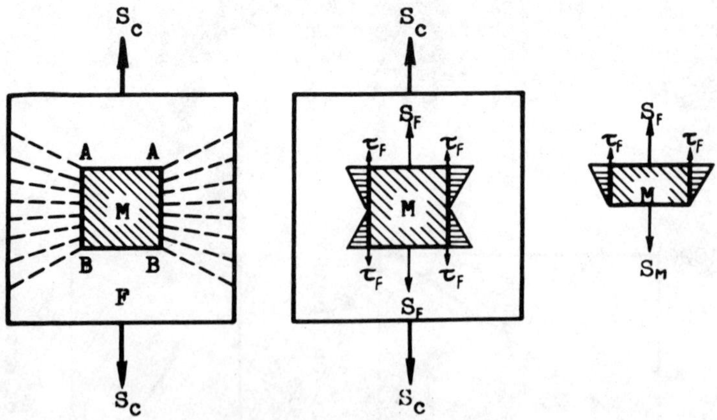

Fig. 6. Schematic model which shows ferrite transmits load to martensite.

given by

$$S_M = S_F + \tau_F \ . \qquad (3)$$

The stress of the dual-phase structure, S_c, is given by

$$S_c = S_F(1-V_M) + S_M V_M \ ,$$

or

$$S_c = S_F + \tau_F V_M \ . \qquad (4)$$

Where V_M is the volum fraction of martensite. So, the strength of dual-phase structure can be given by:

at 0.002 strain $\sigma_{0.2} = \sigma^f_{0.2} + \tau_{0.2} V_M$, (5)

at tensile strength $\sigma_b = \sigma^f_b + \tau_b \ V_M$, (6)

in which $\sigma^f_{0.2}$ is the ferrite stress when the strain of the dual-phase structure is 0.002 (ferrite yield strength); σ^f_b is the ferrite stress when the tensile strength of dual-phase structure is reached (ferrite tensile strength); $\tau_{0.2}$ and τ_b is the shear stress on interface AB when the strength of the dual-phase structure is strength at 0.002 strain and tensile strength, respectively. Their values can be determined by experiment. Comparing equations (1) and (2) with equations (5) and (6), the values of $\sigma^f_{0.2}$ and σ^f_b are equal to 95MPa and 350 MPa, respectively. They are the strength obtained by extrapolating to 100 pct ferrite. This result is expected. The shear stresses $\tau_{0.2}$ and τ_b are 10 MPa and 16 MPa, respectively. The stress of martensite which is about 10 pct more than that of ferrite make martensite islands not deform plastically at the uniform stage of tensile deformation.

CONCLUSION

1. During the uniform stage of tensile deformation of a dual-phase structure the martensite constituent can not deform. Therefore the strain is concentrated in the ferrite.

2. In the dual-phase steel void nuclei form at the martensite/ ferrite interface, and the crack propagates prodominatly along

the ferrite side of the interface.

3. The strength of a dual-phase structure can be shown to be given by:
at 0.002 strain $\sigma_{0.2} = \sigma^{F}_{0.2} + \tau_{0.2} V_M$,
at tensile strength $\sigma_b = \sigma^{F}_{b} + \tau_b V_M$.

ACKNOWLEDGEMENTS

The author is grateful to professor T. C. Lei for instruction and to associate professor M.Yao for many helpful discussions and to Miss B. Lu for typing the manuscript.

REFERENCES

Balliger, N. K., and T. Gladman (1981). Work Hardening of dual-phase Steels. Metal Science, March, 95-108.
Davies, R. G. (1978). Influence of Martensite Composition and Content on the Properties of Dual Phase Steels. Met. Trans., 9A, 671-679.
Davies, R. G., and C. L. Magee (1978). J. Metals, 30, 17.
Demeri, M. Y. (1981). The Formability of a Dual Phase Steel. Met. Trans., 12A, 1187-1196.

APPLICATION OF THE J INTEGRAL CONCEPT TO MICROSCOPIC FRACTURE MECHANICS

Masanori Kikuchi and Hiroshi Miyamoto

Department of Mechanical Engineering, Faculty of Science and Technology, Science University of Tokyo, Noda, Chiba (278), Japan

ABSTRACT

In this paper, the J integral concept is applied to microscopic fracture mechanics. Some modifications of the J integral are carried out. First, the J vector concept is introduced for the asymmetric problem in the microstructure of the material, and second the definition of this J vector is modified to the crack problem in multi-phase material.
Several examples are analyzed using the finite element method. They are:
(a) a crack existing in a plate which is composed of two phase materials,
(b) cracks existing in metal crystals, and (c) cracks uncleated at the phase boundary between base metal and inclusions.
The effect of a phase boundary on the J integral value is discussed for each case, and it is shown that the microcrack problems are well treated by the method proposed in this paper.

KEYWORDS

J vector; phase boundary; microcrack; grain boundary; inclusion; fracture mechanics.

INTRODUCTION

The analysis of micro problems such as the behavior of microcracks in the early stage of fatigue, the nucleation of voids from inclusions and so on are necessary to clarify the mechanism of fracture. Applications of fracture mechanics to these problems were rare, rather it has been thought to be impossible to apply it to the microscopic phenomenon. From the microscopic viewpoint, the material is composed of grains, grain boundaries, and some inhomogeneous inclusions. Then the material is neither isotropic nor homogeneous.
On the other hand, the J integral concept (Rice, 1968) can be interpreted as a generalized force acting at the singularities. The physical meaning of the J integral does not change even in the microscopic field, and it may be possible that this concept can be applied to microscopic fracture mechanics on the grain size level. For this purpose, the following two points are

considered in this paper.
Firstly, the J vector concept is used for the asymmetric problem (Kikuchi, 1979). Secondly, the definitions of these J integrals are modified to the multi-phase materials (Miyamoto, 1980). This modification is carried out by introduction of an additional term to the J integral definition. The new definition of the J integral preserves the path-independent characteristics as long as the material is elastic or non-linear elastic.
In the following, these modifications of the J integral are introduced briefly and several examples are analyzed.

J VECTOR IN MULTI-PHASE MATERIAL

The J_ℓ integral, the ℓ-th component of the J vector, is defined as follows:

$$J_\ell = \int_\Gamma (W\delta_{j\ell} - \sigma_{ij}u_{i,\ell})ds_j, \tag{1}$$

where Γ is the path from point A to point B surrounding the crack tip(Fig.1). J_ℓ is considered to be a component of the force acting on the crack tip.

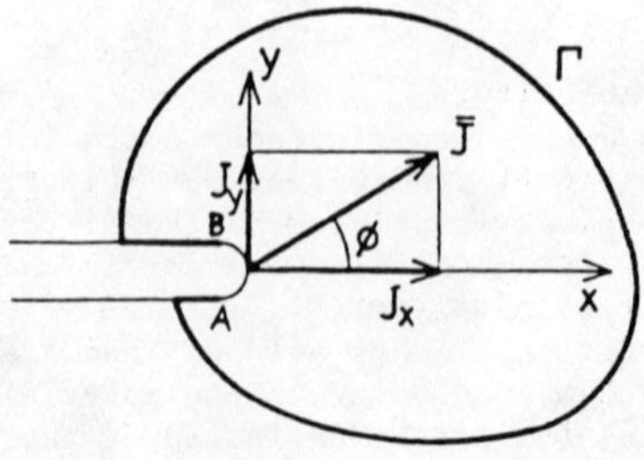

Fig.1 J_x and J_y of a crack.

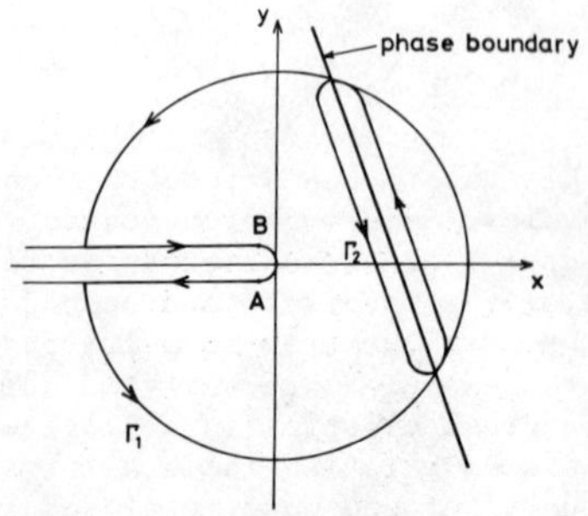

Fig.2 A crack and a phase boundary.

In Fig.2, a crack exists near the phase boundary. In this case, the J_ℓ integral is defined as follows:

$$J_\ell = \int_{\Gamma_1} (W\delta_{j\ell} - \sigma_{ij}u_{i,\ell})ds_j - \int_{\Gamma_2} (W\delta_{j\ell} - \sigma_{ij}u_{i,\ell})ds_j. \tag{2}$$

The physical meaning of eq.(2) is a generalized force acting on a crack tip, that is, the first term relating to Γ_1 gives the generalized force acting on both the crack tip and singularities on the phase boundary, and the second term, relating to Γ_2, gives the force only on the latter. By subtracting the latter from the former, the generalized force acting on the crack tip can be evaluated.
In Fig.3, a crack exists perpendicular to the phase boundary. The J_x value is evaluated by elastic analyses. Each phase is denoted by 1 and 2, as shown in Fig.3 and the corresponding material constants are also indicated with 1 or 2. E_1 is chosen to be 206 GPa, ν_1=0.3. Two cases, where E_2/E_1=0.5 and 2.0 are analyzed. In both cases ν_2 is considered to be equal to ν_1. Results are shown in Fig.4, where x indicates the J integral values where the second term of eq.(2) is neglected, and where it is considered. The results show that the J_x integral has path independent characteristics and it is necessary to consider the second term of eq.(2) to evaluate the J values of a crack in multi-phase materials.

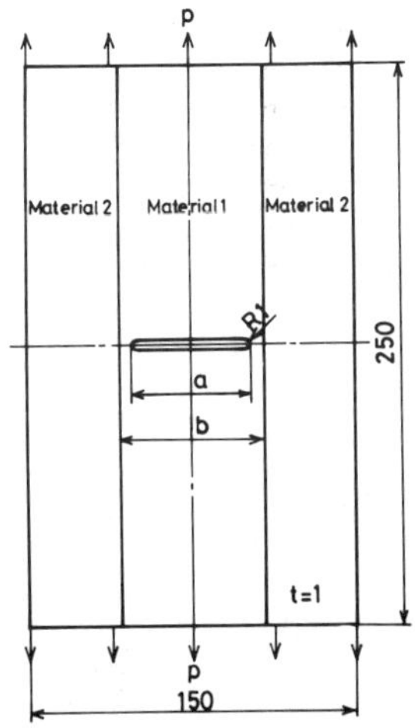

Fig.3. A crack in three storied lamella.

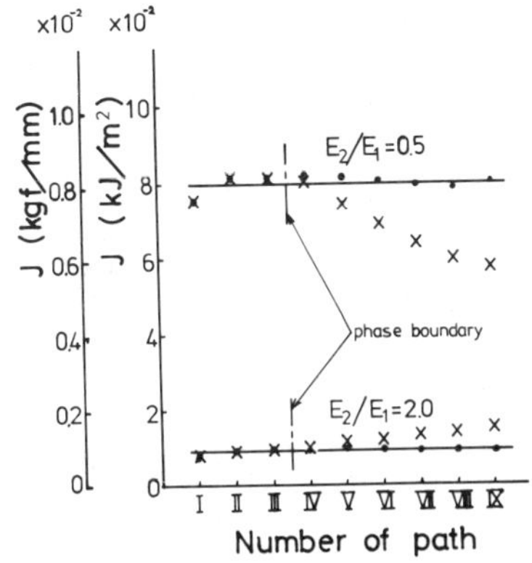

Fig.4. J_x value of Fig.3.

J VECTOR AND THE METALLURGICAL ANISOTROPY

From the microscopic viewpoint, the fracture of a metal structure is equal to the crack propagation in metal crystals. Frequently, a microcrack is nucleated on a grain boundary, and propagates along it. On this level, the boundary condition is asymmetric due to the anisotropy of the crystal, and the grain boundary behaves as a phase boundary. So the modified J concept, mentioned in the last section, is applicable.

Fig.5 (a)-(c) show the models analyzed in this section. They are: (a) a crack existing on the grain boundary, (b) a crack existing perpendicular to the grain boundary, and (c) a crack existing at the triple point. The materials are assumed to be Fe and Cu, which are cubic crystals, and the material constants are shown in Table 1. The finite element analyses are carried out by

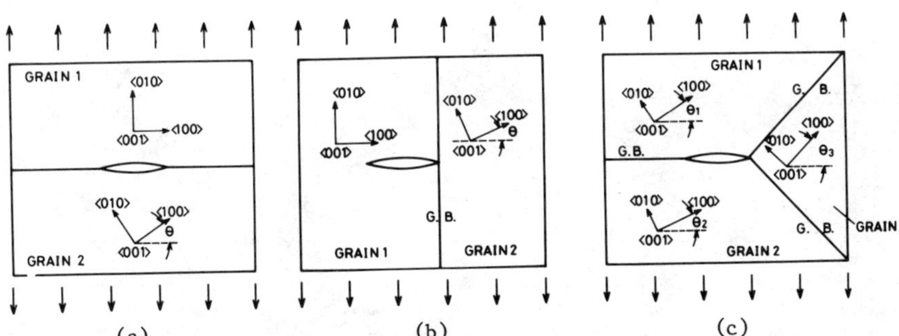

Fig.5. Models of crack in metal polycrystals.

TABLE 1 Material Constants

	A	A'	D
Fe	232.43	138.27	227.51
Cu	166.71	120.62	147.10

(A, A', D; GPa)

assuming two dimensional plane strain condition. This is accomplished by fixing the <001> axis of the cubic coordinates with the z axis of the macroscopic coordinates. The misorientation, θ, is expressed as the rotation around the z axis from 0° to 90°.

Fig. 6 shows the results of model (a), where the orientation of grain 1 is fixed and only that of grain 2 is changed from 0° to 90°. The J_x value varies strongly with the change of θ. For θ<50°, J_x is smaller than that of a single crystal (θ=0°), and for θ>50°, J_x is larger. In every case, J_y is sufficiently small in comparison with the J_x value.

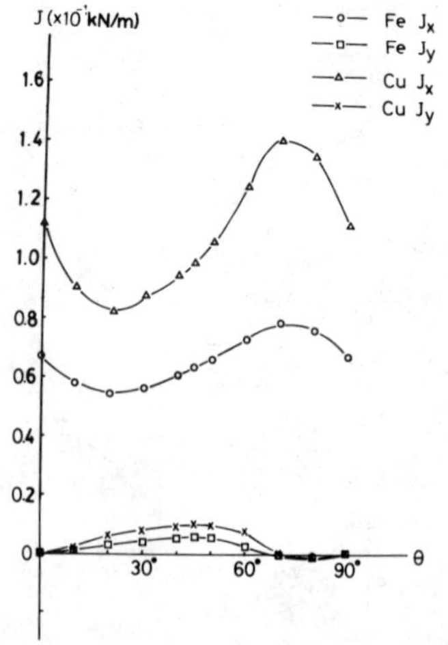

Fig. 6. J_x and J_y of model (a).

Fig. 7 shows the results of model (b) where the θ value of grain 1 is 0°. The J_x value is maximum when θ=0°, single crystal, and decreases as the misorientation increases. If the crack propagation is controlled by J values, it is concluded that the grain boundary acts as a barrier to the crack propagation. Then the orientation of grain 1 is changed to 45°, and the same analyses are carried out. The results are shown in Fig. 8. The tendency of the curves is the same as in Fig. 7, but in this case, the plate is a single crystal when θ=45°. So in this case, the J value becomes larger as the misorientation increases. That is, the grain boundary acts as an accelerator for the crack propagation. It is interesting that the role of the grain boundary on the crack propagation changes, depending on the orientation of each crystal.

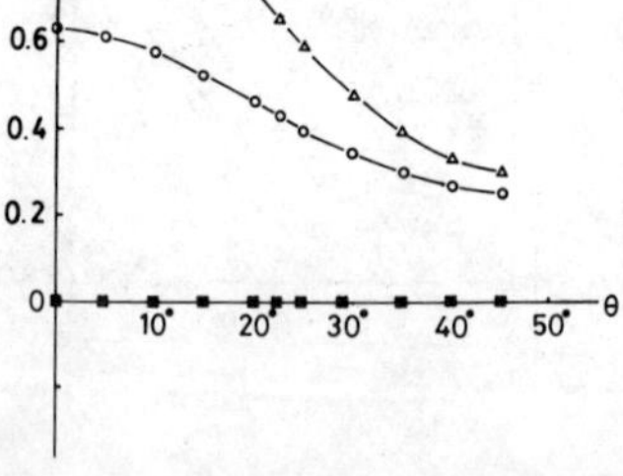

Fig. 7. J_x and J_y of model (b) (θ=0°; single crystal.)

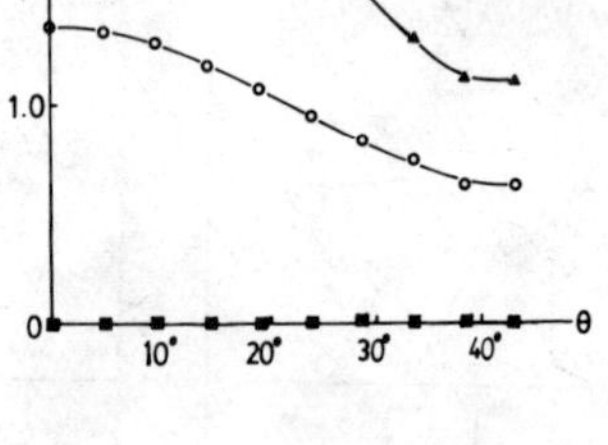

Fig. 8. J_x and J_y of model (b) (θ=45°; single crystal.)

Fig. 9 shows the results of model (c). The orientations of grain 1 and 2 are chosen to compose a Σ5 coincidence grain boundary. J_y values are not as small, and the direction of the J vector, ϕ, is large as shown in Fig. 10. By comparing with Fig. 7 and 8, it is noticed that when a crack exists of the grain boundary, J_y is not as small as when it exists in the grain.

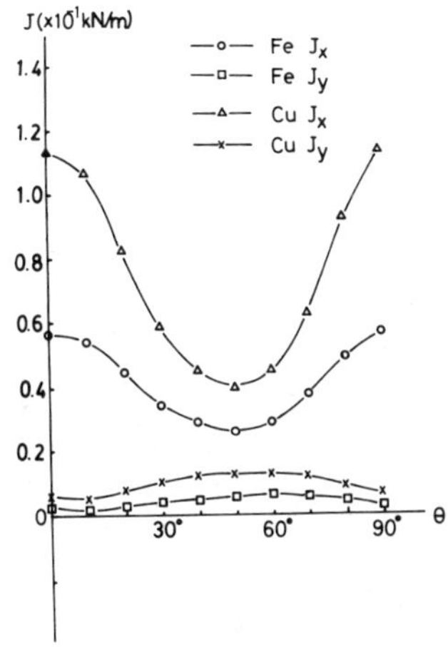

Fig. 9. J_x and J_y of model (c).

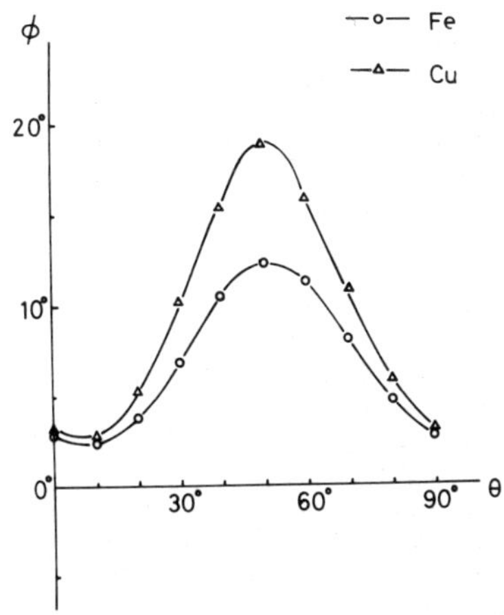

Fig. 10. The direction of J vector for model (c).

Fig. 11 shows the three dimensional model for model (b). In the two dimensional analyses, only the tilt boundary can be analyzed, but in this case, the twist and rotation boundaries can be treated as well as the tilt. Fig. 12 shows the results of the three cases. It is shown that when the grain boundary is of the twist type, the change of the J_x value with the change of θ is the largest, and for the rotation type, the change is the smallest.

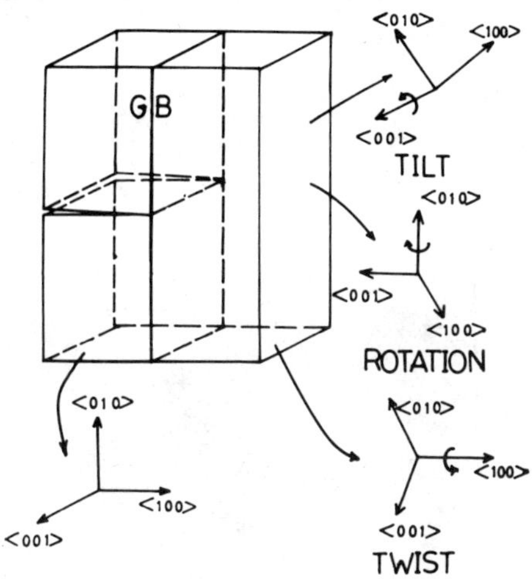

Fig. 11. Three dimensional model of model (b).

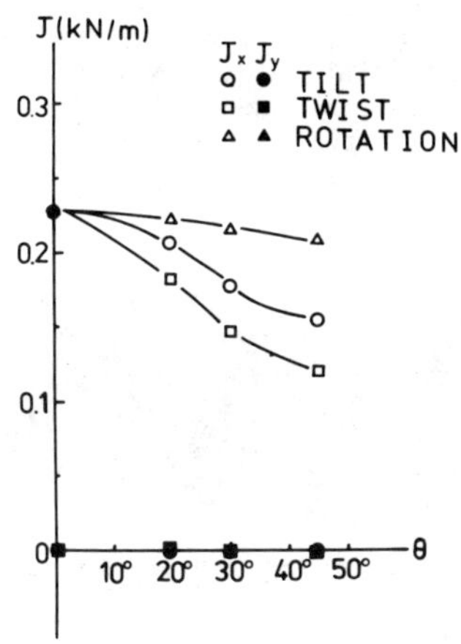

Fig. 12. J_x and J_y of three dimensional model.

J VECTOR OF A CRACK NUCLEATED FROM INCLUSIONS

Due to the stress concentration at the phase boundary of inclusions and matrix, micro cracks are nucleated at the phase boundary and propagates along it. The propagation of a crack is equal to separation of the inclusion from the matrix, and the condition for separation is the fracture criterion in this case.

Fig. 13 shows the model of a crack nucleated from an inclusion loaded along the y axis by a constant load p=9.8 MN/m^2. The matrix is assumed to be ferrite, and the inclusion is Al_2O_3. The material constants are shown in Table 2. The crack length, 2a in Fig.13, is changed and only elastic analyses are carried out.

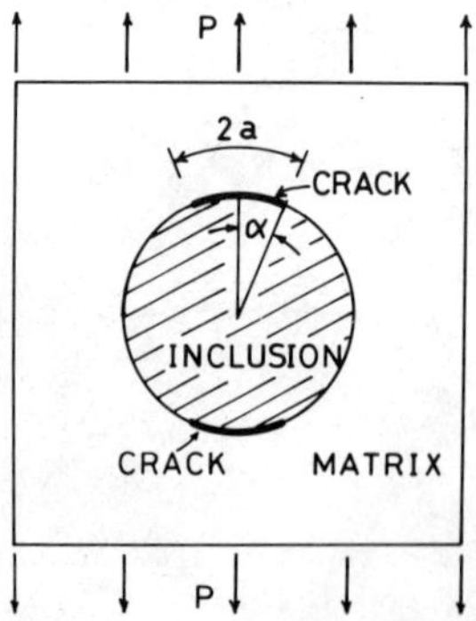

Fig. 13 Crack nucleated from an inclusion.

TABLE 2 Material Constants

	ferrite	Al_2O_3
Young's modulus	206 GPa	414 GPa
Poisson's ratio	0.3	0.3

Fig. 14 shows J_x and J_y, and the magnitude of the J vector, respectively. J_x and J_y increase with the increase of the α value, and then decrease gradually. Fig. 15 shows the $\bar{J}$ value for both cases, with and without inclusion. The $\bar{J}$ value is larger than for the case without inclusion. Fig. 16 shows the J vector for each case and each crack length. The J vector direction is toward the outside of the inclusion.

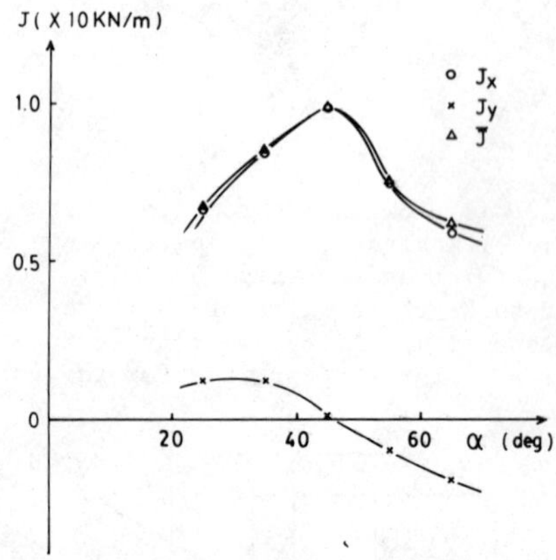

Fig. 14. J_x, J_y and J of the model.

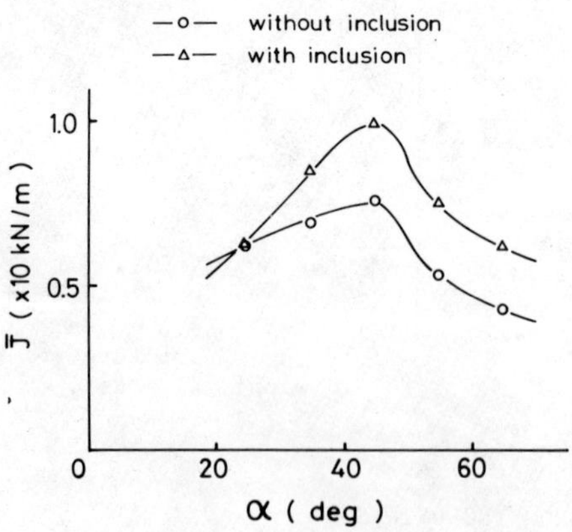

Fig. 15. J value for different crack length.

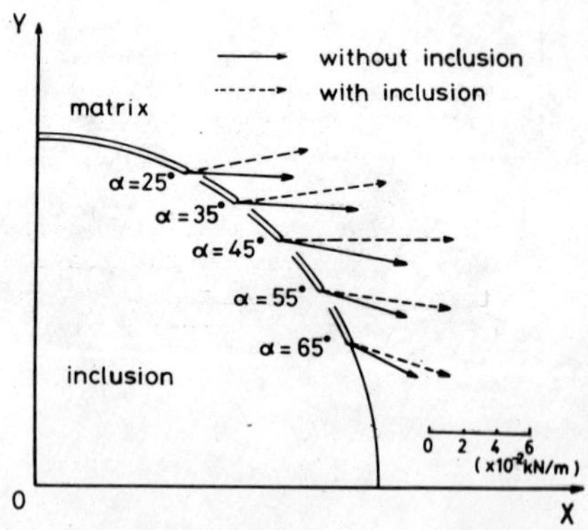

Fig. 16. J vector for different crack length.

The material constant of the inclusion is changed for several cases, and the results are shown in Fig. 17 where E_1 is the Young's modulus of the matrix and E_2 is that of the inclusion. As the inclusion hardness changes from soft to hard, the J_x and J_y values change considerably and seem to become saturated to a constant value. The directions of the J vector for each case are shown in Fig. 18. The magnitude of the J vector decreases and its direction tends outward as E_2 increases.

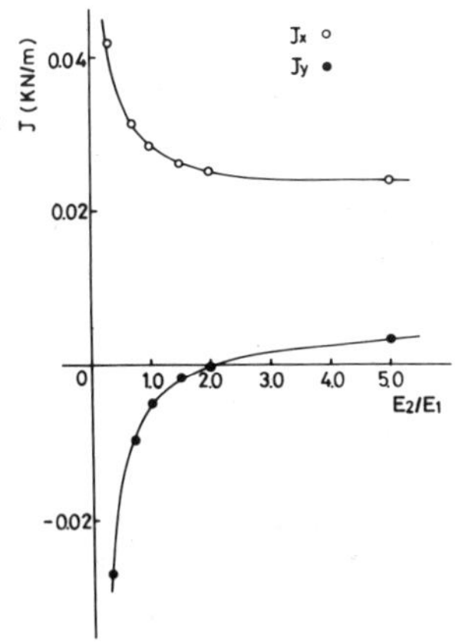

Fig. 17. J value for different Young's modulus of inclusion.

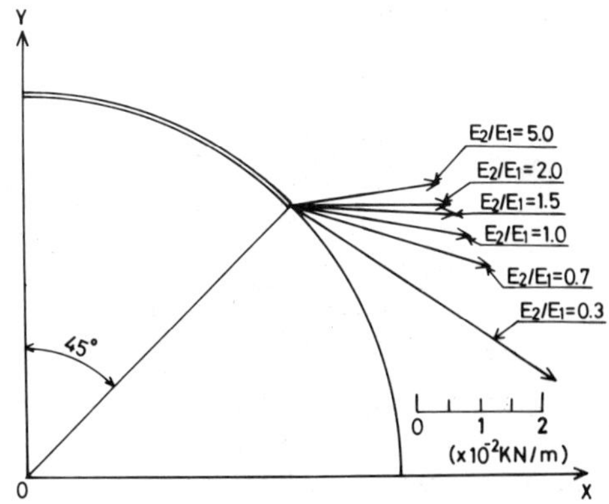

Fig. 18. The direction of the J vector

Then the matrix is assumed to be a single crystal, and the effect of the orientation of the matrix is studied. The inclusion is assumed to be Al_2O_3. Fig. 19 shows the results. The J_x value changes ±12% from the average value. The J_y value varies strongly but the absolute value is so small in comparison with J_x that the J vector does not greatly change.

SUMMARY

Several examples simulating the micro mechanisms of fracture are treated and the J integral concept is applied. Experimental comparisons are necessary for future study. Although many important factors are neglected to simplify the problem, for example, dislocation motion, instability of energy on the grain boundary, etc., the modified J integral is applicable to the microscopic fracture mechanics.

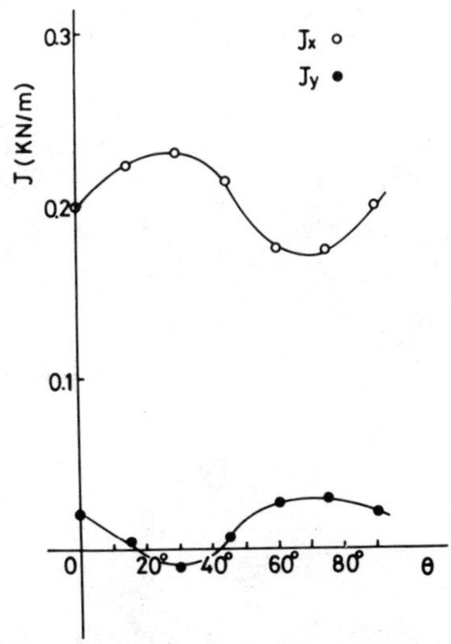

Fig. 19. J_x and J_y for different matrix orientation.

REFERECES

Rice, J.R. (1968). J. Appl. Mech., 35, 379-386

Kikuchi,M., and Miyamoto,H. (1979). Trans. of 5th SMiRT, G7/2

Miyamoto,H., and Kikuchi,M. (1980). Numerical Methods in Fracture Mechanics, Owen,D. R. J., and Luxmoore, A. R., 359-370

QUASI-STATIC AND IMPACT FRACTURE TOUGHNESS OF 5140 H STEEL AFTER DIFFERENT THERMAL TREATMENTS

J. R. Klepaczko* and A. Solecki**

***Institute of Fundamental Technological Research, Polish Academy of Sciences, 00-049 Warsaw, Poland**
****Łódź Polytechnics, Machine Building Faculty, 43-300 Bielsko Biała, Poland**

ABSTRACT

The effects of various thermal treatments, i.e. oil quench and different tempering conditions, on quasi-static and impact fracture toughness, stress-strain characteristics, hardness and Charpy energy of 5140 H steel were examined. During quasi-static and impact loading notched round tensile specimens were used with a prefatigued crack. A specially designed device involving a pendulum hammer and electronic measuring system was used for measurement of the opening mode fracture toughness at loading rates of up to $\dot{K}_I = 3\times10^6$ MPa$\sqrt{m}$/s. It has been found that within the region of the lower tempering temperatures, 500 K $\leqslant T^* \leqslant$ 650 K, the critical stress intensity factor K_{Ic} determined from impact testing is lower than that obtained during slow loading. Whereas at the higher tempering temperatures, 650 K $\leqslant T^* \leqslant$ 900 K, dynamic K_{Ic} values have a tendency to be higher than their quasistatic counterparts. This behavior was analysed quantitatively using the Hahn-Rosenfield model which relates tensile properties to fracture toughness. A good agreement was found between quasi-static experimental results and the model. The relation between Charpy energy K_V and the critical stress intensity factor K_{Ic} was also evaluated.

KEYWORDS

Fracture dynamics, alloy tempering, alloy steel 5140 H, Charpy V - energy of alloy steel.

INTRODUCTION

It is well known that alloy steel 5140 H is used after thermal treatment for highly to moderately stressed machine parts. Consequently, it is of great interest for practical applications to verify relations between such mechanical parameters as yield stress, tensile strength, reduction in area, hardness, etc., from one side, and fracture toughness from the other. All of these parameters should be referred to a specific thermal treatment such as quenching and subsequent tempering. The primary purpose of tempering is to impart plasticity or toughness to the steel, and the loss in strength is only incidental to this very important increase of toughness. Special interest is focussed on comparison of quasi-static and impact fracture toughness of 5140 H steel after quenching and different tempering treatments. The effect of thermal treatment on the mechanical properties together with quasi-static and impact fracture toughness of this steel has rarely been examined. The second purpose of this paper is to correlate parameters obtainable from ordinary tensile tests using round bar specimen with fracture toughness as defined by the critical stress intensity factor in the opening mode, i.e. K_{Ic}.

MATERIAL, SPECIMEN PREPARATION AND EXPERIMENTAL TECHNIQUES

All specimens were cut from the plate in a direction perpendicular to the rolling direction. The chemical composition of the specimen material is given in Table 1. This composition exactly fits the U.S. standard for the chromium steel 5140 H, ASTM A322-64 or AISI-SAE. The suffix H is used to distinguish specific hardenability requirements (SAE Standard J 407 d). This steel is designated according to the Polish Standards as 40 H. In Table 1 are also given limits for the alloying elements complying with the U.S. Standards for 5140 H steel. Three types of specimen were prepared: (i) ordinary tensile specimen with diameter D = 15.6 mm and gage l_o = 80 mm; (ii) notched round tensile specimen (NRT) in the quasi-static version with diameter D = 15.6 mm, length of cylindrical portion l_o = 80 mm, notch diameter d = 11.7 mm, the V-notch with $\alpha = 60^o$ and the radius of the notch tip $\rho = 0.1$ mm; and (iii) a reduced version of the notched round tensile specimen with the following dimensions: D = 12 mm, d = 9.6 mm, $\alpha = 45^o$, $\rho = 0.1$ mm. The change in size for (iii) was crucial in a substantial reduction of vibrating masses and consequently "ringing" during the tensile impact. In addition, a series of standard V-notch Charpy specimens was cut out from the plate with the notch and the specimen length perpendicular to the rolling direction. Thus, during bending impact the Charpy notch tip was loaded in a similar way as the notch tip of tensile specimens. In order to investigate the effect of tempering all specimens were normalized for 30 min. at 1133 K and oil-quenched. The tempering temperature T* was varied and five values of T* were applied: 573 K, 673 K, 773 K, 823 K, 873 K. A particular series of tempered specimens was held 60 min. at T* and then water cooled. After thermal treatment all notched round specimens underwent fatigue precracking. The fatiguing procedure resulted in almost exactly concentric cracks. The obvious advantages of the round-bar precracked specimen are that there are no edge effects after thermal treatment and an equal stress field around the crack can be created. A more exact discussion on the application of such a specimen in quasi-static fracture toughness determination is given by Brown and Srawley [1967] and Krafft [1971]. All quasi-static tests were performed on a closed-loop testing machine. Tensile tests were performed with the actuator velocity $V = 5\times10^{-2}$ mm/s which results in a strain rate $\dot{\varepsilon} = 6.25\times10^{-4}$ s^{-1}. The actuator velocity during fracture toughness testing was kept constant at $V = 1\times10^{-2}$ mm/s. Impact tests were performed on a standard Charpy hammer with pendulum providing impact velocity V = 5.4 m/s. The hammer was equipped with a special device to load the NRT specimen axially and with minimum bending. A schematic diagram of this device is shown in Fig. 1. The cross-bar CB is connected in series with the NRT specimen S and dynamometer D. One end of the dynamometer is attached to the hammer base B. The impact with velocity V is provided by a pendulum set to a proper energy, usually 1.479×10 J. The cross-bar undergoes the impact symmetrically at two points A, and the total tensile force P is monitored by the dynamometer with two short base 3 mm strain gages which eliminate all bending signals. To ensure axial impact the cross-bar CB is guided at both sides into slots (not shown in Fig. 1). Accurate measurement of the cross-bar displacement δ as a function of time is achieved by using an optical device based on frequency modulation coding. The cross-bar has a set of grids with the wave length $\lambda = 0.25$ mm that were produced mechanically on its bottom surface. The black screen S has one slot parallel to grids with the width $\lambda/2$. Since a beam of the parallel light is supplied from the bottom to the grids and part of the light is reflected through the screen to the photodiode F through bifurcated fiber-optics leads, the displacement δ is coded as a periodic voltage signal from the photodiode. It may be mentioned that the incisions of the grid are filled with a black cement. During impact the photodiode responds to changes in light intensity caused by fringes passing across the screen slot. The displacement resolution in this method is approximately $\lambda/8$, i.e. $\sim 3.1 \times 10^{-2}$ mm. This measurement technique has many advantages which makes it ideally suited to impact testing. Both electric signals, one from the dynamometer and the other from photodiode are monitored with a dual beam osilloscope and recorded as a function of time by Polaroid camera. It was found that displacement recordings are more important for less brittle materials in order to monitor COD. For a more brittle specimen, force as a function of time P(t) is adequate to determine impact fracture toughness. The tensile pulse which is applied to load the notched specimen with a circumferential fatigue crack is produced by the pendulum impact onto the cross-bar. Because of the masses involved the series cross-bar - specimen - dynamometer starts to vibrate, and these vibrations are superimposed on the main tensile pulse which breaks specimen. The vibrating system can be reduced to the system shown in the upper part of Fig. 1. The recorded signal $\overline{P}(t)$ from the dynamometer will differ slightly as compared to P(t). Signal P(t) is the input signal and $\overline{P}(t)$ is a recorded response of the system. The problem is that the response $\overline{P}(t)$ is known as a recorded oscillogram and force P(t) acting on the notch cross section is unknown. It is very frequently assumed that $\overline{P}(t) = P(t)$. But in the present case the whole system has been analyzed numerically using a desk-top computer and a plotter. Details of the analysis of the vibrating system are out of scope of this paper and are given elsewhere, Garbarski and Klepaczko [1980]; for a more general analysis see also Tobota, Klepaczko and Gronostajski [1980]. The experimental technique described here seems to be cheap as well as effective and is specially designed for screening tests. More exact experimental procedures for accurately establishing the dynamic fracture initiation properties of structural metals have also recently been proposed: Costin, Duffy and Freund [1977], Klepaczko [1979], [1982], but these are much more expensive and elaborate. However, the loading rates with the stress wave techniques, as mentioned above, can be still about one order of magnitude higher than for the present technique.

EXPERIMENTAL RESULTS

The influence of tempering temperature on the tensile properties of 5140 H steel is summarized in Fig. 2. Both the yield stress $\sigma_{0.2}$ and the ultimate stress σ_u show almost linear drop with tempering temperature T*. Plasticity on the other hand is substantially enhanced by an increase of tempering temperature. Uniform logarithmic deformation φ_u increases linearly with T* starting from brittle behavior at T* = 470 K. Such a behavior is frequently observed after tempering: Kuch [1978]. The logarithmic fracture strain $\varphi_f(T^*)$ has been estimated using measurements of reduction in area in the neck $r(T^*)$ and uniform deformation measurements $\varphi_u(T^*)$. Thus,

$$\varphi_f(T^*) = \varphi_u(T^*) + \varphi_r(T^*) \tag{1}$$

where

$$\varphi_r(T^*) = \ln[1-r(T^*)]^{-1} \tag{2}$$

By definition, the reduction of the area is

$$r = 1 - \left(\frac{D_n}{D_u}\right)^2 , \quad 0 \leqslant r \leqslant 1 \tag{3}$$

D_n and D_u are the specimen diameters after tensile fracture at the neck and along its uniform portion, respectively. The $\varphi_f(T^*)$ function shows a maximum at $T^* \approx 800$ K. A slight decrease in φ_f at higher tempering temperatures may be attributed to temper brittleness. This phenomenon is usually manifested as a loss of ductility observed after slow cooling from tempering temperatures of ~850 K or higher. It may also occur after tempering and slow cooling in the temperature range between approximately 650 K and 850 K. Thus, a steel that is susceptible to temper embrittlement may lose much of its fracture toughness. It is supposed that this effect is only slightly accentuated as a decrease of $\varphi_r(T^*)$, since after tempering all specimens were water cooled. Figure 4 shows changes of Rockwell-C hardness and Charpy V-notch energy as a function of tempering temperature T^*. After quenching hardness was ~48 HRC, and within the range of tempering temperatures from 575 K to 875 K hardness decreases almost linearly with temperature T^*. Charpy V-notch energy increases almost linearly as a function of T^*. All experimental points have been obtained as an arithmetic mean of five tests. Quasi-static force-displacement records $P(\delta)$ as well as oscillogram recordings of the tensile force versus time $P(t)$ for all prefatigued NRT specimens made it possible to determine quasi-static and impact critical stress intensity factors K_{Ic}. Values of K_{Ic} have been obtained for all five tempering temperatures. Five quasi-static tests were performed for each tempering temperature and three tests with the impact device described above. The results are shown in Fig. 3. The following formula was used to calculate K_{Ic}

$$K_{Ic} = P_Q d_f^{-3/2} f(\xi) \tag{4}$$

where P_Q is the critical tensile force, in the present case $P_Q = P_{max}$, d_f is the diameter of the remaining circular ligament after fatiguing, $f(\xi)$ is the stiffness or size function where $\xi = d_f/D$, D is the outer diameter of the NRT specimen. The following expression for $f(\xi)$ covers a wide range of ξ, Panasiuk et al [1977]:

$$f(\xi) = \frac{(1-\xi)^{1/2}}{2\pi} (1 - 0.5\xi - 0.125\xi^2 + 0.2757\xi^3 - 0.2082\xi^4 + $$

$$+ 0.0663\xi^5 + 0.0048\xi^6 - 0.013\xi^7)^{-1} \tag{5}$$

Other expressions for the size function in the case of radial symmetry are discussed by Brown et al [1967] and Tada [1973]. The test for K_{Ic} is valid when the following conditions are satisfied:

$$d_{fc} \geqslant 1.6 \left(\frac{K_{Ic}}{\sigma_{0.2}}\right)^2 \quad \text{and} \quad D_c \geqslant 2.3 \left(\frac{K_{Ic}}{\sigma_{0.2}}\right)^2 \tag{6}$$

All experimental values obtained during the course of fracture toughness testing are given in Table 2. In order to apply linear elastic fracture mechanics the conditions (6) must be satisfied. Thus, values of D_c were calculated and are given in Table 2. For the case of the two lowest tempering temperatures, $T^* = 573$ K and $T^* = 673$ K, the condition (6) is satisfied and the critical stress intensity factors have been obtained. For higher tempering temperatures T^* the conditions (6) are not satisfied and consequently the K_Q values have been obtained. The limit between K_{Ic} and K_Q is indicated by the vertical line in Fig. 3. It may be mentioned that values of D_c for the impact test are overestimated because the quasi-static values of $\sigma_{0.2}$ were used in these calculations. It is interesting to note that the fracture toughness for $T^* = 573$ K determined under impact loading is lower than the quasi-static value. The decrease is about -10%. This observation is in agreement with other literature sources, for example: Costin et al [1977], Klepaczko [1982], Kalthoff et al [1981], Krabiell and Dahl [1981]. After higher tempering temperatures T^* a substantial increase in the impact fracture toughness occurs (dark circles in Fig. 3), and K_Q determined from the impact tests are much higher than quasi-static K_Q values. This trend is specially observed for the two highest tempering temperatures $T^* = 823$ K and 873 K. The quasi-static values of K_Q (open circles in Fig. 3) clearly exhibit a maximum at $T^* = 773$ K. A substantial increase in the impact toughness of the 5140 H steel at higher T^* may be related directly to the strain rate sensitivity, i.e. to the increase of the whole stress-strain diagram at a high strain rate. The loading rates in terms of $\dot{K}_I$ are in the order of 10 MPa$\sqrt{m}$/s for the quasi-static tests and 2×10^6 MPa$\sqrt{m}$/s for the impact tests, where

$$\dot{K}_I = \frac{K_{Ic}}{t_c} \qquad \text{or} \qquad \dot{K}_I = \frac{K_Q}{t_c}$$

the time interval t_c is measured from the beginning of loading to the instant when the maximum force is reached. Scanning electron microscopy SEM can provide additional important data needed to correlate fracture appearance with fracture toughness. It has been reported in some papers that substantial differences were observed on SEM fracture appearance micrographs taken for steel after quasi-static and impact fracture, for example Marandet et al [1981], Rock et al [1979]. After each type of fracture toughness test (quasi-static and dynamic) SEM micrographs were taken and the observed fracture surfaces were quite consistent with $K_{Ic}(T^*)$ data. The fracture appearances after dynamic loading are characterized by a larger amount of intergranular cracking, in spite of the fact that the ligament diameters were smaller in comparison to the quasi-static specimens. It has been shown previously that the stretch-zone width can be correlated with fracture toughness of a material, including effects of test temperature and loading rate, Spitzig [1968], Krasovsky and Vainshtok [1981]. For the present case the observed stretch-zone shape depends on tempering temperature and loading rate. For the steel studied the stretch-zone width was almost reduced to zero for the lowest T^* and impact loading. For quasi-static loading the stretch zone increased more than for the impact cases when tempering temperature was higher. Generally, 5140 H steel behaves in a more brittle manner under impact loadings after lower tempering temperatures.

ANALYSIS

There have been a number of attempts to correlate the critical stress intensity factor K_{Ic} with properties measured during standard tensile test.

Since fundamental parameters like yield stress, ultimate stress (see Fig. 2), hardness (see Fig. 4), were measured after different tempering temperatures, an attempt has been made to correlate these quantities with K_{Ic}, Klepaczko and Solecki [1983]. This may explain general effects of heat treatment on K_{Ic}. A very convenient model for a simple application is that due to Hahn and Rosenfield [1968]. The model is based on the combination of the critical strain criterion and the effect of strain hardening. This model was successfully applied to predict K_{Ic} for a number of aluminum and titanium alloys as well as for steels, Fiarro and Roberti [1982]. A separate study on pure Al-Cu aluminum alloys confirmed the usefulness of this model, Garret and Knott [1976]. The same approach was utilized to describe the variation of the upper shelf fracture toughness with temperature and loading rate for two alloy steels (SA 533 B-1 and SA 302 B) which are used in the manufacture of nuclear pressure vessels,Ritchie et al [1979]. The results of calculations are shown in Fig. 3. The transition temperature $T^*_o = 473$ K for brittle behavior is indicated by the vertical dotted line. Values of K_{Ic} as a function of tempering temperature T^* were calculated according to the Hahn and Rosenfield model [1968]. The result is shown by the broken line in Fig. 3. There is an excellent agreement between calculated and experimental values of $K_{Ic}(T^*)$ within the range of valid K_{Ic}, i.e. for the lower tempering temperatures. The upper shelf of K_{Ic} predicted by the model is generally higher than experimental K_Q values, and reaches the level of $K_{Ic} \approx 85$ MPa$\sqrt{m}$. Since experimental K_Q values are generally lower than valid K_{Ic} values, the trend represented by the calculated values of $K_{Ic}(T^*)$ for the higher tempering temperatures is correct. For both quasi-static cases a maximum of the fracture toughness is observed. However, the model predicts the maximum at $T^*_m = 830$ K, the experimental value is $T^*_u = 730$ K. Within the range of tempering temperatures $T^*_o \leqslant T^* \leqslant 550$ K the model definitely predicts too low values of K_{Ic}. At this range a more brittle fracture dominates and a more appropriate approach should be represented by the critical fracture stress model, for example the RKR model, Ritchie et al [1973] as applied by Ritchie et al [1979]. The expanded region of the lower tempering temperatures is shown in Fig. 5. Experimental values of $K_{Ic}(T^*)$ are shown by solid lines, S means quasi-static and D means dynamic, the broken line denoted by HR is calculated according to the model of Hahn and Rosenfield. Since Charpy V-notch energy has been determined as a function of tempering temperature T^*, it was possible to estimate $K_{Ic}(T^*)$ using the correlation proposed by Barsom and Rolfe [1970]:

$$K_{Ic}(T^*) = 1.49\times10^{-2}\, E^{1/2}\, [K_v(T^*)]^{3/2} \qquad (7)$$

where K_V is in Joules and E in MPa; note that values of K_V in Fig. 4 must be multiplied by the ligament cross-section $A_l = 8\times10^{-5}$ m^2 to get the Charpy energy K_V(Joules) in eq. (7). The formula (7) predicts quite well the dynamic but not quasi-static values of K_{Ic} within the range of the lower tempering temperatures. For higher T^*, values of K_{Ic} calculated from eq. (7) are much larger than those predicted by the HR model or K_Q values determined experimentally.

DISCUSSION

Knowledge of effects of various thermal or thermomechanical treatments on fracture toughness determined under slow and impact loadings is of considerable importance for industrial applications. The improvement of alloy steels by thermal or thermomechanical treatments to find the best combination of strength and toughness has been subject of recent publications: Rock et al

[1979], Larson and Nunes [1961], Nunomura et al [1977], Briant and Banerji [1982], Jahn et al [1982]. A fracture toughness test clearly shows the quality of a steel and provides a useful measure of it as K_{Ic}. The heat treatment which results in a good K_{Ic} value, also assures the best performance of a machine component. However, the maxima of plain strain fracture toughness are observed when K_{Ic} is determined as a function of tempering temperature. In the present case of 5140 H steel such a maximum was observed at $T_m^* \approx 730$ K, but for a high-carbon chromium steel (ball bearing steel) the maximum was noted at $T_m^* \approx 490$ K, Nunomura [1977]. The influence of loading rate on K_{Ic} seems to be one of the important factors from two points of view: namely, theory of fracture and machine part service. It is frequently observed that K_{Ic} decreases at increasing loading rates $\dot{K}_I$. This effect was also observed for 5140 H steel after $T^* \approx 573$ K and the decrease of K_{Ic} under impact loading was about -10%. For very brittle alloy steels the loading rate effects are negligible. For the case of a more plastic response K_Q values are higher under impact than during quasi-static loading, as shown in Fig. 3, and the strain rate effects seem to dominate. The transition between more brittle and more plastic behavior at different loading rates and temperatures was demonstrated for three structural steels by Krabiell and Dahl [1981]. Fractography can provide an additional piece of information to rationalize quasi-static and impact fracture behavior. Without discussing the details it may be stated that at a particular tempering temperature the fracture appearances after impact loading indicate a generally more brittle behavior. The most promising is a possibility to predict quasi-static fracture toughness after different heat treatments by relating K_{Ic} with ordinary tensile characteristics. This would enable a link between existing experience on tensile characteristics with fracture toughness problems. It has been shown in this study that effects of different microstructures obtained through different thermal treatments may be interpreted using the framework of fracture toughness behavior.

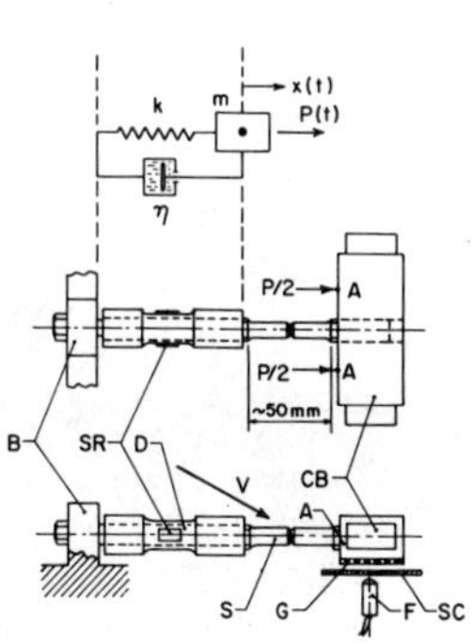

Fig. 1. Schematic diagram of device for impact fracture testing; S - NRT specimen; CR - cross bar with points A; D - dynamometer with strain gages SR; B - base; G - grid on the cross bar; SC - screen with a slot; F - photodiode; V - impact velocity.

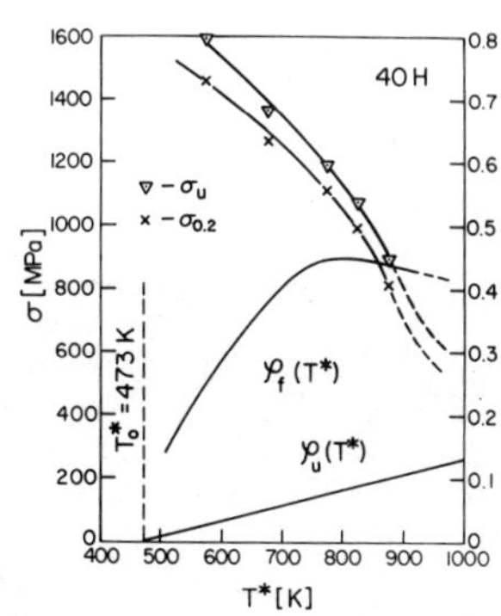

Fig. 2. Tensile properties of 5140 H steel at room temperature after different tempering temperature T^*, φ_u - logarithmic uniform deformation, φ_f - logarithmic deformation at fracture, T_o^* - temperature of brittle behavior.

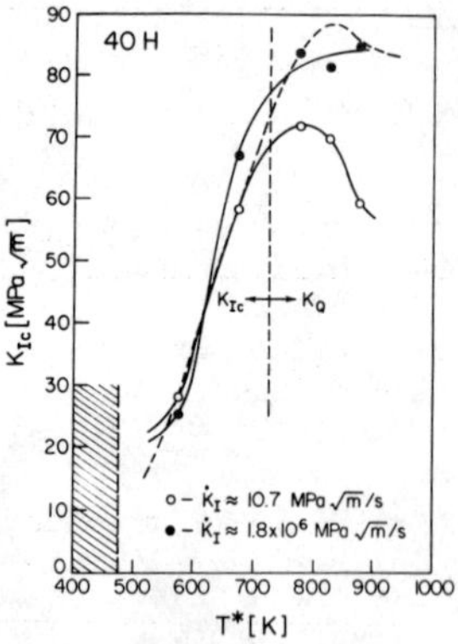

Fig. 3. Quasi-static and impact values of the critical stress intensity factor K_{Ic} for 5140 H steel as a function of tempering temperature T*; the shadow region shows brittle behavior; the broken line obtained according to eq. (6).

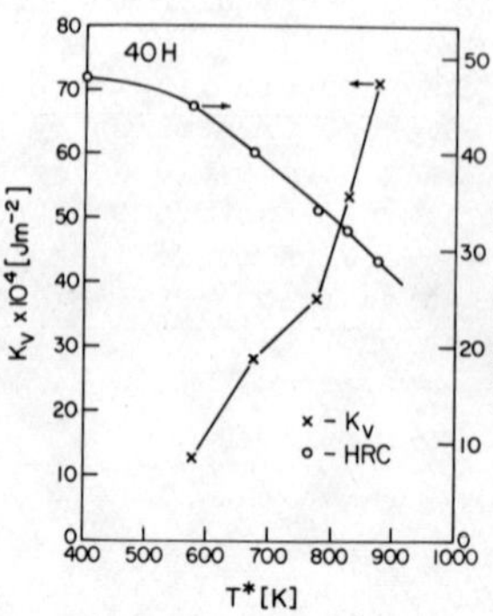

Fig. 4. Hardness Rockwell-C and Charpy V-notch impact energy K_V as functions of tempering temperature T*; note that K_V is referred to the specimen ligament cross section in m^2.

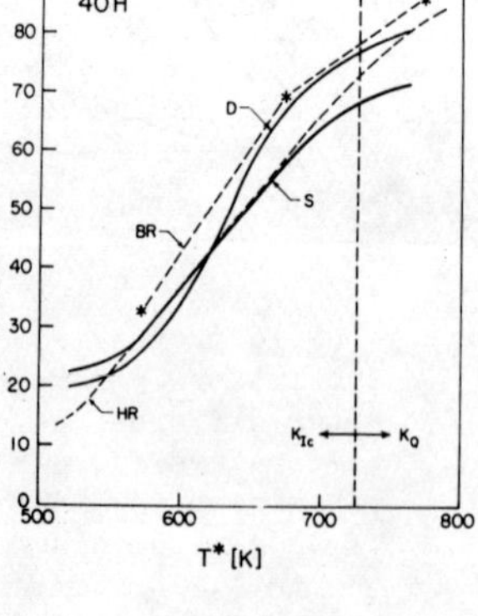

Fig. 5. Expanded region of lower tempering temperatures T* showing valid experimental K_{Ic} by solid lines; S - quasi-static; D - dynamic; HR - values of K_{Ic}(T*) predicted by Hahn and Rosenfield model, BR - values predicted according to Barsom and Rolfe based on K_V(T*) test data.

TABLE 1 5140 H steel, chemical composition

5140H 40H	Element	C	Mn	Cr	Si	Ni	Mo	S	P
	Weight %	0.44	0.61	0.90	0.27	0.10	0.065	0.014	0.025
	from	0.37	0.60	0.60	0.15				
	to	0.44	1.00	1.00	0.30			0.040	0.035

AISI-SAE or ASTM-A-322-64
Specific hardenability requirements: SAE J407d

TABLE 2 5140 H steel - experimental results for fracture toughness tests, quasti-static and impact

	T [K]	573	673	773	823	873
Static	K_{Ic} MPa$\sqrt{m}$	28.16	58.21	71.88	69.80	59.27
	$\dot{K}_I$ MPa$\sqrt{m}$/s	7.68	10.58	14.39	11.84	8.87
	D_c mm	0.8	4.0	9.0	11	12
Dynamic	K_{Ic} MPa$\sqrt{m}$	25.60	67.03	83.06	81.26	84.61
	$\dot{K}_I$ MPa$\sqrt{m}$/s $\times 10^6$	0.207	1.48	2.20	1.88	3.18
	D_c mm	0.7	6.4	12.8	15.4	25.2

Quasi-static values are arithmetic means of five tests.
Dynamic values are arithmetic means of three tests.

REFERENCES

Barsom, J.M. and S.T. Rolfe (1970). ASTM STP 466, Philadelphia PA, 281.

Briant, C.L. and S.K. Banerji (1982). Met. Trans., 13A, 827.

Brown, W.F. and Y.E. Srawley (1967). ASTM STP 410, ASTM.

Costin, L.S., J. Duffy and L.B. Freund (1977). ASTM STP 627, ASTM.

Fiarro, D. and R. Roberti (1982). Proc. 6th Int. Conf. on Strength of Metals and Alloys, Pergamon Press, 947.

Garbarski, J. and J.R. Klepaczko (1980). Theoretical and Applied Mechanics, 18, 265, (in Polish).

Garret, G.G. and J.F. Knott (1976). Proc. 2nd Int. Conf. on Mechanical Behavior of Materials, ASM 488.

Hahn, G.T. and A.R. Rosenfield (1968). ASTM STP 432, Philadelphia PA, 5.

Jahn, M.T., C.K. Huang and C.M. Wan (1982). Proc. 6th Int. Conf. on Strength of Metals and Alloys, Pergamon Press, 793.

Kalthoff, J.F., S. Winkler, W. Bohruc and W. Klemm (1981). Proc. 5th Int. Conf. on Fracture, Vol. 1, Pergamon Press, 363.

Klepaczko, J.R. (1979). Mechanical Properties at High Rates of Strain. Proc. Int. Phys. Conf. Ser. No. 47, Bristol.

Klepaczko, J.R. (1982). J. Eng. Mat. and Technology, 104, 29.

Klepaczko, J.R. and A. Solecki (1983). Submitted to Metallurgical Trans. s.A.

Krabiell, A. and W. Dahl (1981). Proc. 5th Int. Conf. on Fracture, Vol. 1, Pergamon Press, 393.

Krasowsky, A.J. and V. Vainshtok (1981). Int. J. of Fracture, 17, 579.

Krafft, J.M. (1971). Techniques of Metals Research, ed.
Bunshah, R.F. , Vol. V part 2, Interscience Publ.

Kuch, E.R. (1978). Metals Handbook, 9th ed. ASM, Metals Park, 455.
Larson, F.R. and J. Nunes (1961). Trans. ASM, 53, 663.
Marandet, B., G. Phelippcau and G. Sanz (1981). Proc. 5th Int. Conf. on Fracture, Vol. 1, Pergamon Press, 375.
Nunomura, S., T. Kashiramura, K. Machida and S. Sakui (1977). Fracture Mechanics and Technology. Sijthoff and Noordhoff, 553.
Panasiuk, V.V., A.E. Andrejkiv and S.E. Kovchik (1977). Evaluation methods for fracture toughness of structural materials. Science Publishers, Kiev, (in Russian).
Ritchie, R.O., J.F. Knott and J.R. Rice (1973). J. Mech. Phys. Solids, 21, 395.
Ritchie, R.O., W.L. Server and R.A. Wullaert (1979). Metallurgical Trans., 10A, 1557.
Rock, H.J., S. Beitscher and D. Kalish (1979). J. Eng. Mat. and Technology, 101, 91.
Spitzig, W.A. (1968). Trans. ASM, 61, 344.
Tada, H. (1973). The stress analysis of cracks handbook, Del. Research Corp. Hellertown.
Tobota, A., J.R. Klepaczko and J. Gronostajski (1980). Theoretical and Applied Mechanics, 18, 432 (in Polish).

THE INFLUENCE OF MICROSTRUCTURE AND TEMPERATURE ON STATIC AND DYNAMIC FRACTURE INITIATION IN HEAT RESISTANT STEEL

Z. Bílek, J. Buchar and M. Holzmann

Institute of Physical Metallurgy, Czechoslovak Academy of Sciences, 616 62 Brno, Czechoslovakia

ABSTRACT

The influence of initial microstructure of a low alloy weldable and heat-resistant steel having chemical composition 0.13% C, 0.61% Mn, 0.10% P, 0.28% Si, 0.007% S, 2.39% Cr, 0.22% Ni, 0.95% Mo, 0.05% Cu on static and dynamic fracture toughness is demonstrated. The results of extensive K_{Id} measurements at high loading rates produced by a Hopkinson split bar technique obtained within a wide temperature range are compared with the K_{Id} values from Charpy instrumented tests for seven different microstructures. The results of a dynamic elastic-plastic finite element stress analysis are briefly summarized. The calculation models relating the K_{Ic} and K_{Id} values to the mechanical properties describing a materials crack tip behaviour and microstructure are discussed.

KEYWORDS

Heat resistant steel, brittle fracture, fracture toughness, strain rate, Charpy instrumented test, Hopkinson split bar technique, finite element method.

INTRODUCTION

During welding of low alloy steels various types of microstructures are created around a weld in dependence on the welding conditions that determine cooling rates tempering effects. This usually leads to complex microstructures consisting of martensite and bainite or mixture of both. Welds are often metallurgically inferior and are frequently located in the most stressed parts of structures, and therefore they are of critical importance for structural reliability. The fracture properties of complex microstructures have been studied mainly at low strain rates produced by static loading [Naylor and Blondenau, 1976]. To design dynamically loaded structures it is necessary to determine the corresponding dynamic fracture data at high strain rates. One important material characteristic describing a resistance against brittle fracture is the dynamic fracture toughness K_{Id}. In this paper an attempt is made to study the influence of various microstructures on the temperature dependence of K_{Id}-values determined by a Charpy instrumented test ($\dot{K}_I = 10^5$ MPa $m^{1/2}s^{-1}$) and by a modified Hopkinson split bar technique which makes it possible to achieve the higher loading rate $\dot{K}_I = 10^6$ MPa $m^{1/2}$ s^{-1}. The latter procedure helps us to find out the minimum value of dynamic fracture toughness K_{Idm} on the $K_{Id}(\dot{K}_I)$ curves.

EXPERIMENTAL EQUIPMENT

The instrumented Charpy tests (ICHT) and static fracture toughness measurements were performed within a wide temperature range ($-196^{\circ}C$ + $25^{\circ}C$) in agreement with corresponding ASTM standards as described by Holzmann and co-workers [1980]. To produce high loading rates $\dot{K}_I$ we have modified Hopkinson split bar technique (HSBT) for dynamic fracture studies following the procedure proposed by Klepaczko [1979]. Our experimental set up and specimen geometry is shown in Fig. 1.

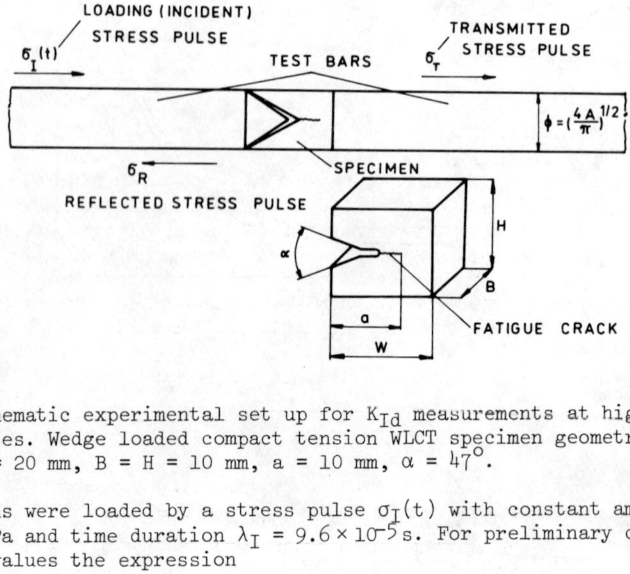

Fig. 1. Schematic experimental set up for K_{Id} measurements at high strain rates. Wedge loaded compact tension WLCT specimen geometry. W = 20 mm, B = H = 10 mm, a = 10 mm, $\alpha = 47^{\circ}$.

The specimens were loaded by a stress pulse $\sigma_I(t)$ with constant amplitude σ_{Im} = 500 MPa and time duration $\lambda_I = 9.6 \times 10^{-5}$ s. For preliminary calculations of the K_{Id}-values the expression

$$K_{Id} = \frac{A\sigma_T(t_c)\ f(a/W)}{2B\sqrt{x}\ \tan(\alpha/2 + \mu)} , \tag{1}$$

based on static stress equilibrium in the specimen, was used. The meaning of symbols A, B, W, a, α is apparent from Fig. 1 and f(a/W) denotes the compliance function, μ is the friction coefficient and $\sigma_T(t_c)$ represents the transmitted stress at the moment of fracture initiation t_c. It follows from the extensive experimental evidence given by Buchar and co-workers [1981] that $\sigma_T(t_c)$ equals to the amplitude of transmitted stress pulse $\sigma_T(t_c)$. Since comprehensive dynamic photoelasticity studies indicated the limited validity of eq. (1) we have computed the correct values of the dynamic stress intensity factor $K_I(t)$ by the dynamic finite element method [Buchar and co-workers, 1982]. Therefore, the results obtained from eq. (1) were corrected for stress wave effects. For the particular microstructure as received we also used stress pulses with rather short time duration λ_I = 2.8 and 5.2×10^{-5}s in order to reach high loading rates $\dot{K}_I$

$$\dot{K}_I = K_{Id}/t_c \qquad (2)$$

EXPERIMENTAL RESULTS, DISCUSSION

Analysing our dynamic fracture toughness data we observe two different kinds of behaviour. For initial microstructure, highly tempered martensite and bainite (ultimate tensile strength σ_{UTS} = 526 MPa + 664 MPa) we can see from Fig. 2 that the K_{Id} values obtained by means of ICHT are well above the HSBT data.

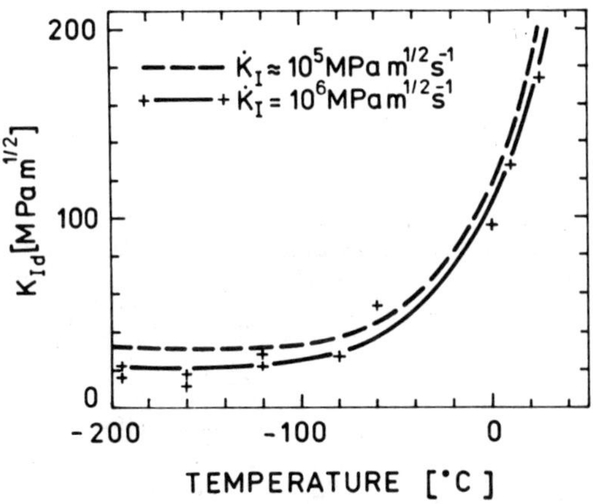

Fig. 2. Temperature dependence of K_{Id} for high tempering temperature microstructures. (Quenched from 940°C water, tempered at 550°C.)

However, this observation holds only at low testing temperatures for microstructures tempered at low temperatures (σ_{UTS} = 862 MPa ÷ 1342 MPa) - Fig. 3.

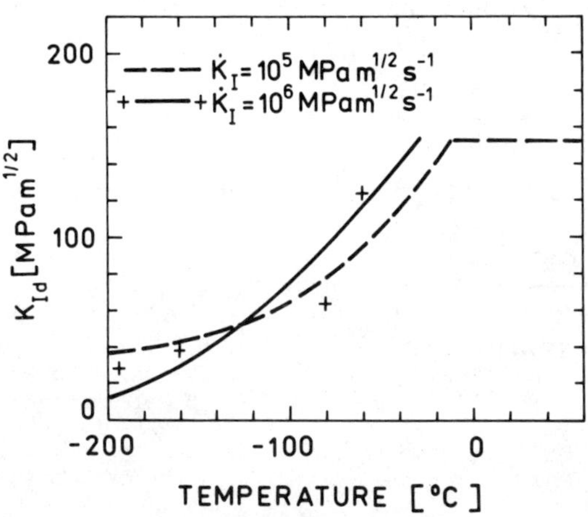

Fig. 3. Temperature dependence of K_{Id} typical for low tempering temperature microstructures. (Quenched from 940°C water.)

The sequence of microstructural stages from the point of K_{Id} magnitude is identical for both loading rates at low temperature region. With respect to the basic principle of HSBT [Andrzejewski and co-workers, 1981] it becomes obvious that further increase of $\dot{K}_I$ is possible by reducing the stress pulse time duration λ_I. The influence of λ_I on the temperature dependence of K_{Id} is apparent from Fig. 4. Shorter stress pulses lead to the higher dynamic fracture toughness most pronounced at low temperatures. Closer inspection of Fig. 4 reveals that the increase of K_{Id} due to shorter λ_I is more marked than the increase of the loading rate $\dot{K}_I$. The difference in the K_{Id} values to which corresponds a difference in the $\sigma_T(t)$ patterns in accordance with eq. (1), suggests a change of stress state in the specimen. Apparently the decrease of stress pulse duration produces more pronounced stress wave effects related to the rapid loading in our experiments. The correct evaluation of K_{Id} data requires careful dynamic finite element analysis. Figure 5 compares the $K_{Id}(\lambda_I)$ values calculated from eq. (1) with the K_{Id} values found out by numerical computations. It appears that the stress pulse duration $\lambda_I = 9.6 \times 10^{-5}$s gives the minimum value for which eq. (1) may be applied without greater error. In this connection it should be pointed out that Klepaczko [1982] works with $\lambda_I \gtrsim 2 \times 10^{-4}$s. For very short stress pulses $\lambda_I < 9.6 \times 10^{-5}$s we note a sharp increase of K_{Id} data and hence a higher crack stability. Similar observations were made by Kalthoff and Shockey [1977] based on different types of experiments. From their interpretation it follows the necessity to adjust fracture mechanics principles for short stress pulse loading. In addition, the defect stability increases not only with decreasing λ_I but also with increasing crack size which is a completely different situation than at low loading rates. Time dependent behaviour further complicates the development of dynamic fracture mechanics models relating microstructure to the K_{Id} values. The application of critical stress criterion for cleavage fracture initiation under stress pulse loading [Buchar, Bílek, Knésl, 1982] involves additional variable - incubation time to satisfy the above mentioned

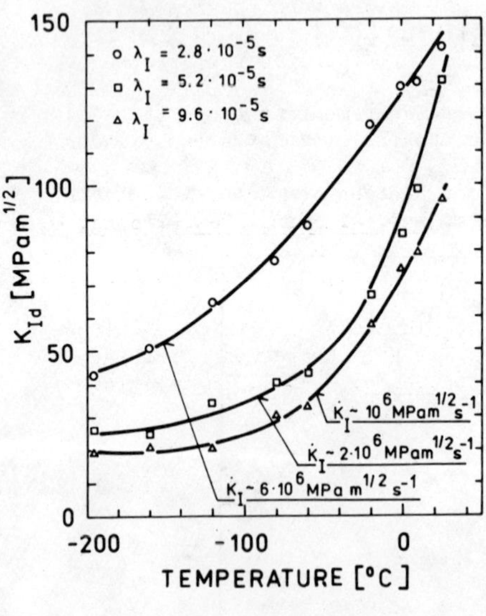

Fig. 4. The influence of λ_I on $K_{Id}(T)$.

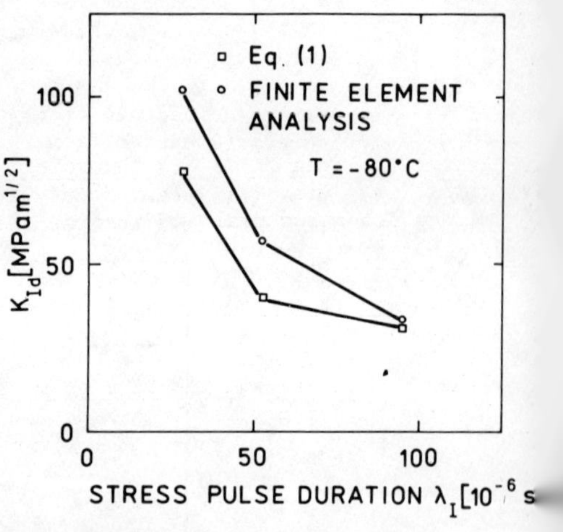

Fig. 5. Comparison of the $K_{Id}(\lambda_I)$ values determined by finite element analysis and from eq. (1). Bainite with σ_{UTS} = 613 MPa.

criterion. For example, this incubation time was estimated as 5×10^{-6}s for microstructure of untempered martensite at room temperature.

CONCLUSION

In summary, the increase of loading rate one order of magnitude above the current impact tests does not change substantially the dynamic fracture toughness of highly tempered martensite and bainite. Restricting ourselves to low testing temperatures we conclude that for all seven microstructures of C-Cr-Mo steel, Charpy instrumented test provides the minimum value of dynamic fracture toughness K_{Idm} important in engineering design. Loading by short stress pulses leads to marked increase of K_{Id} outside the classical fracture mechanics framework. With regard to Fig. 3 we propose that the influence of stress wave effects to a great extent does not depend only on the loading and geometry conditions but also on type of microstructure, especially at higher temperatures.

REFERENCES

Naylor, J.P., and R. Blondenau (1976). The respective roles of the packet size and lath width on toughness, Metallurgical Trans., 7A, 891-898.

Klepaczko, J.R. (1979). Application of the Split Hopkinson Pressure Bar to Fracture Dynamics, In J. Harding (Ed.) Mechanical Properties at High Rates of Strain. Inst. Phys. Conf. Ser. No 47, Bristol, pp. 201-214.

Buchar, J., Bílek, Z., Dušek, F., and Mazanec, K. (1981). Application of Hopkinson Split Bar Technique to dynamic fracture toughness evaluation. Metallic Materials, 19, 709-725.

Andrzejewski, A., Klepaczko, J.R., and G. Pluvinage (1981). Experimental determination of high loading rate effects on fracture toughness of aluminium alloys, In D. François (Ed.) Advances in fracture mechanics, Vol. 1. Pergamon Press, Oxford, New York, Toronto, Sydney, Paris, Frankfurt, pp. 385-392.

Klepaczko, J.R. (1982). Discussion of a New Experimental Method in Measuring Fracture Toughness Initiation at High Loading Rates by Stress Waves. J. Engng Matls and Technology, 104, pp. 29-35.

Holzmann, M., Man, J., and Z. Bílek (1980). Reference fracture toughness curves for structural and turbine steels. Int. J. Press. Ves. Piping, 8, pp. 451-459.

Kalthoff, J.F., and D.A. Shockey (1977). Instability of cracks under impulse loads, J. Appl. Phys., 48, pp. 986-993.

Buchar, J., Bílek, Z. and Z. Knésl (1982). The influence of steel microstructure on dynamic fracture toughness, In K.L. Maurer and F.E. Matzer (Eds.) Fracture and the role of microstructure, Vol. 1, Engineering Materials Advisory Services Ltd, London, pp. 280-287.

EFFECT OF SULFUR CONTENT ON FRACTURE BEHAVIOUR OF Ni-Cr-Mo-V STEEL

Y. Tanaka and T. Iwadate

Research Laboratory, Muroran Plant, The Japan Steel Works Ltd., Muroran, Hokkaido, Japan

ABSTRACT

In this study, the effect of overheating on the fracture toughness behaviour was investigated using Ni-Cr-Mo-V steels with various sulfur contents. Also, the effect of overheating on the ductility of the steels in the forging temperature range was studied. Overheating leads to a decrease of the ductility and fracture toughness of the materials, depending on sulfur content, heating temperature and cooling rate. Quenching and tempering heat treatment improves the toughness properties of the overheated steels.

KEY WORD

Ductility; impact energy; fracture toughness; tearing modulus; overheating; sulfur content; precipitate; forging.

INTRODUCTION

In the manufacturing process of large heavy section forgings, a forging process is one of the most important processes to obtain the sound quality of the products. The forging process of large products sometimes requires more than a month where many times heating to high temperature is repeated. The heating of steels up to extremely high temperatures during forging causes precipitation of sulfide inclusions at the grain boundaries which makes the grain boundaries weaken (Preece, 1946, 1950; Schulz, 1973; Glue, 1975; Mcleod, 1982). This phenomenon is called overheating.

The overheating of steel has been one of the most critical problems for steelmaking because it affects the mechanical and impact properties of the steel products. This phenomenon is sometimes revealed as a loss of ductility and toughness in mechanical and impact tests. In a special case, grain boundaries separation is found in the steel products.

In this paper, the effect of overheating on the fracture behaviour of steels was inveatigated using Ni-Cr-Mo-V steels with different sulfur content.

The fracture mechanics approach was used to evaluate the overheating phenomenon quantitatively.

MATERIALS

The materials used were nine heats of 3.7% Ni-Cr-Mo-V steels. Seven heats of them are laboratory heat steels using 50kg ingot and the other two heats are large forgings (solid marks in Figs. 1, 3 and 4). The ingots are forged and the soaking treatment was carried out subsequently. The chemical compositions of the steels are the same except for the sulfur content, that is, 0.25wt% C, 0.04wt% Si, 0.24wt% Mn, 3.70wt% Ni, 1.7wt% Cr, 0.4wt% Mo and 0.11wt% V. The sulfur content ranges from 0.0007wt% to 0.0099wt%.

EXPERIMENTAL PROCEDURES

As a first stage, the fracture behaviour of grain boundaries during hot working was investigated. Tensile test specimens having 14mm diameter were heated in vacuum at high temperature between 1373°K and 1593°K and then cooled to temperatures between 923°K and 1273°K, where the tensile tests were performed. The strain rate during the tensile tests was 50%/min. Some specimens were deformed to the prescribed strain and Charpy impact tests were conducted after the quenching and tempering (QT) heat treatment. The cross sections of the tensile specimens were investigated to examine the separation of overheated grain boundaries.

As a second stage, the fracture toughness J_{IC} test per ASTM E813-81 was carried out to investigate the fracture bahaviour of the overheated steels. The steels were heated at a temperature between 1373°K and 1623°K for 2 hrs and then cooled to 1103°K with a cooling rate of 0.5°K/min or 20°K/min followed by quenching. Some of the steels were directly quenched from the heating temperature. After quenching a tempering heat treatment followed. This treatment is referred to as heating hereinafter. For some materials, an additional quenching and tempering treatment was conducted, where the quenching temperature was 1103°K and tempering temperature was 933°K. The yield and tensile strength of these heat treated steels are approximately 650MPa and 800MPa, respectively.

The fracture toughness J_{IC} and tearing modulus T_J were measured by the unloading compliance method using a servo hydraulic testing machine connected with a computer. The specimen used was a 0.5 inch thick round compact tension (0.5T-RCT) specimen with 20% side groove. The fatigue crack length to specimen width ratio, a/W, of the specimens was about 0.6. The final precracking stress intensity factor, K_{max}, was kept less than $25MPa\sqrt{m}$. The fracture surfaces of the specimens were examined using a scanning electron microscope to study in detail the fracture behaviour of the overheated materials.

FRACTURE BEHAVIOUR AT HIGH TEMPERATURE

The overheating occurs during the heating for hot working. The subsequent hot working process may cause separation of the grain boundaries weakened by overheating. This phenomenon leads to a considerable degradation of the ductility and toughness of the steels.

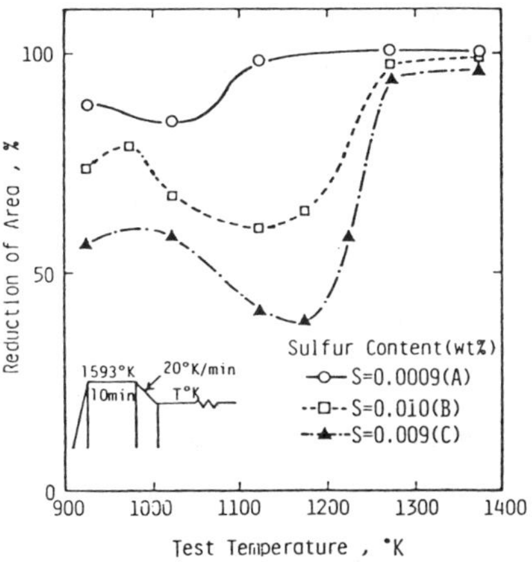

Fig. 1. Tensile test results in the forging temperature range.

Fig. 2. Appearance at the cross section of pre-strained specimens.

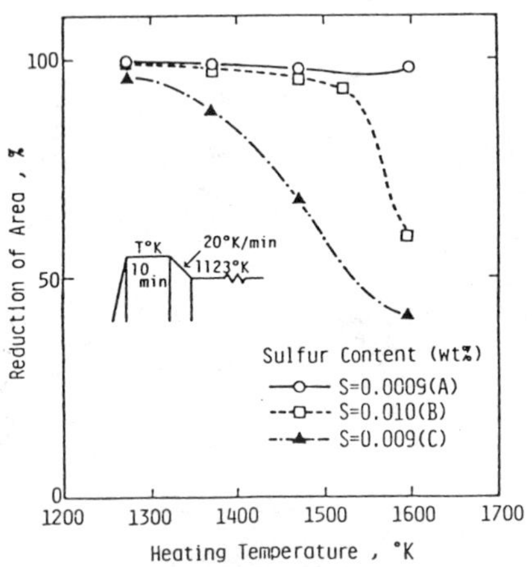

Fig. 3. Effect of the heating temperature on the ductility at 1123°K.

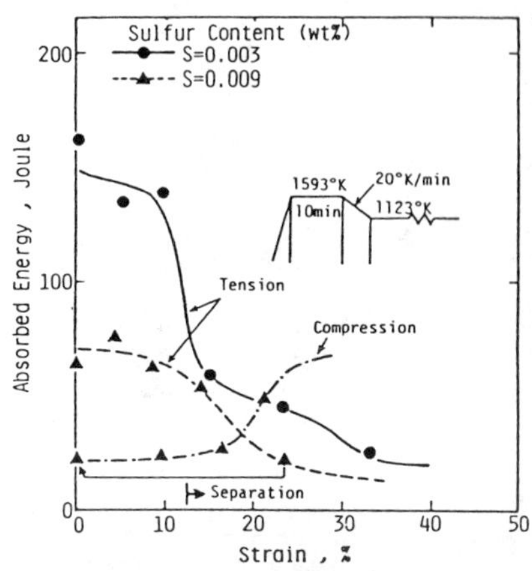

Fig. 4. Effect of pre-straining on the Charpy impact energy at 373°K

Fig. 1 shows the ductility of the 1593°K heated steel in the high temperature range. The high sulfur steel shows a remarkable decrease of ductility in the temperature range between 973°K and 1273°K, whereas the low sulfur steel shows no such tendency. Fig. 2 shows the cross section of the 15% pre-strained specimens tested at 1123°K for steel A and steel C. In the high sulfur steel C, the grain boundary separation was clearly observed. Fig. 3 shows the effect of heating temperature on the ductility of the specimens tested at 1123°K. Higher heating temperature above 1473°K have a significant effect on the ductility. The high sulfur steels are more sensitive to the degradation of the ductility during high temperature heating.

Fig. 4 presents the Charpy impact energy of the pre-strained material, where the pre-strain was done at 1123°K and the tests were conducted after the QT treatment. The test temperature was 373°K. The Charpy energy is severely affected by the pre-strain, especially at higher strain than 11% where the grain boundary separation occurs (Fig. 2). In Fig. 4, the Charpy energy of the materials given the compressive strain after 24% pre-straining is also plotted. A recovery of the toughness is clearly observed for the material strained more than 20% in compression.

The above mentioned phenomenon is enhanced by the precipitation of sulfide inclusions on the grain boundaries. The test results suggest that the precipitation is significantly affected by the sulfur content of the steel, heating temperature and cooling rate, and that the pre-strain makes the grain boundaries weakened resulting in much more decrease of ductility and toughness of the steel. Thus, the overheating has a detrimental effect on the soundness of the steel products.

FRACTURE TOUGHNESS BEHAVIOUR

To investigate the effect of overheating on the fracture toughness behaviour of the material, Charpy impact test and the fracture toughness J_{IC} tests were widely conducted. The specimens were prepared using the materials with 0.0007 to 0.0099% sulfur content where the materials were heated at high temperatures before the machining.

Fig. 5 shows the Charpy impact test results of the steels. A decrease in Charpy impact energy is observed for all steels heated at 1573°K or higher temperatures. In the material heated at 1473°K, the low sulfur content steels show a decrease in impact energy, whereas the high sulfur content steels show almost equivalent impact energy to the QT steel. The decrease of impact energy was accompanied by intergranular fracture.

Fig. 6 shows the relationship between the fracture mode observed in the 0.5T-RCT specimens having various sulfur content and the heating temperature. In the figure the effect of cooling rate is also shown. Fig. 7 shows the fracture surfaces of the 0.5°K/min cooled specimens shown in Fig. 6.

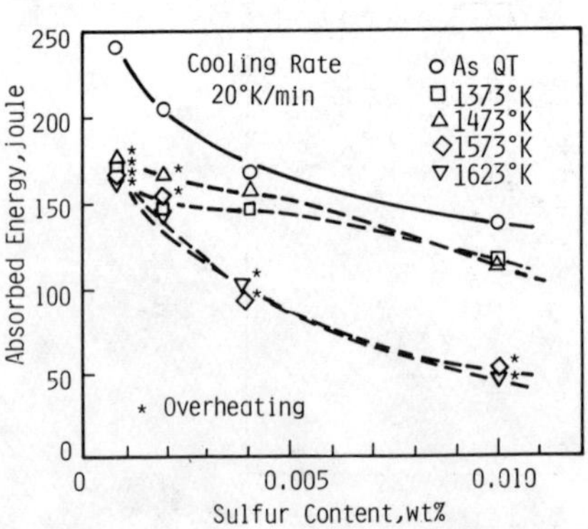

Fig. 5. Effect of heating temperature on the Charpy impact energy.

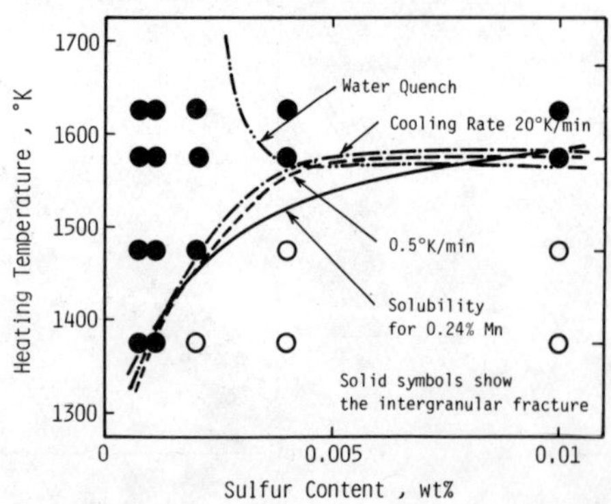

Fig. 6. Fracture mode observed in the 0.5T-RCT specimens.

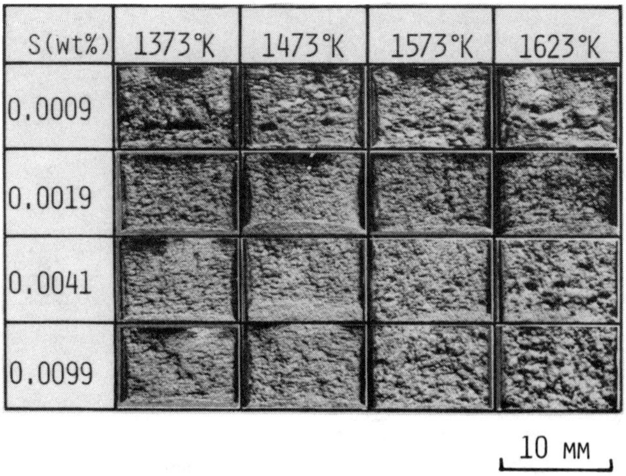

Fig. 7. Appearance of the fracture surfaces of the 0.5°K/min cooled specimens.

These figures indicate that the lower the sulfur content, the higher the susceptibility to overheating. The cooling rate does not affect significantly the intergranular fracture mode except for the very high cooling rate at water quenching. Fig. 6 also indicates that the boundary line between the intergranular and ductile fracture modes coincides well with the solubility of sulfur for the 0.24% Mn steel.

Fig. 8 presents the J versus stable crack extension R-curves for the 1573°K heated steels and the QT steels. The significant change of the R-curves due to sulfur content is demonstrated. Fig. 9 shows the relationship between the fracture toughness J_{IC} at the onset of stable crack growth and the

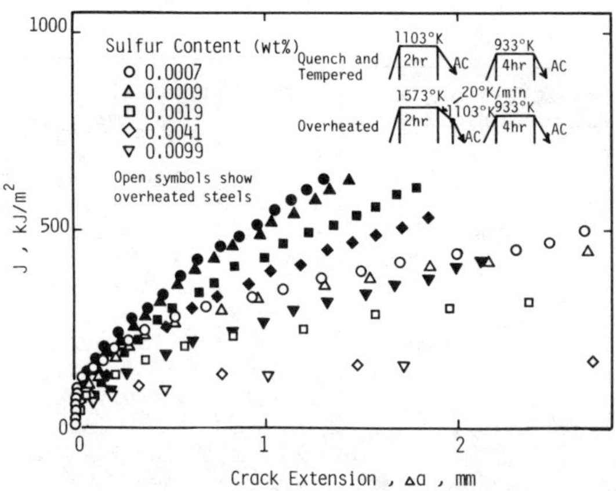

Fig. 8. J versus Δa R-curves for the QT'ed and overheated steels.

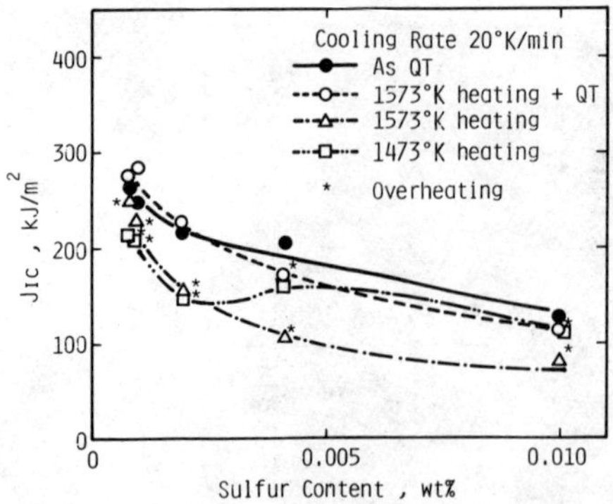

Fig. 9. Relationship between J_{IC} and sulfur content.

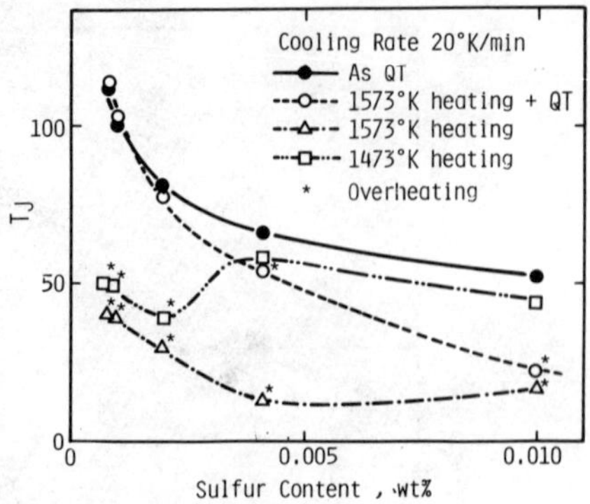

Fig.10. Relationship between T_J and sulfur content.

sulfur content. The figure indicates that the overheating decreases the fracture toughness J_{IC} conpared to that of standard quench and tempered steels. However, in the 1473°K heated steels, the degradation of J_{IC} values is noticeable in low sulfur steels. Note that the additional QT treatment after the heating is effective for the recovery of the fracture toughness, especially for the low sulfur steels.

Fig. 10 shows the tearing modulus T_J versus sulfur content calculated from the slope of the R-curves in Fig. 8 using the following equation.

$$T_J = \frac{E}{\sigma_y^2} \frac{dJ}{da} \tag{1}$$

where E is Young's modulus and σ_y is the yield strength of the steel. The effect of sulfur content and heat treatment on T_J values is more remarkable than that on J_{IC} values. The T_J value of the overheated steels decrease about a half of that of the QT condition. The T_J value is a measure of the resistance against stable crack growth of the material. The test results and the fracture appearance (Fig. 7) suggest that the amount of intergranular crack growth along the grain boundaries leads to a higher decrease of the T_J value.

The effect of cooling rate on the fracture toughness J_{IC} and T_J values was also examined. The cooling rate of around 20°C/min minimized these toughness values which is confirmed by many researchers. (Glue, 1975;Mcleod 1982)

In order to analyze the fracture toughness behaviour, the fracture surfaces of the specimens were examined using a scanning electron microscope (SEM). Fig. 11 shows the relationship between the critical stretch zone width (SZWc) and the J_{IC} value. The resistance against the development of the stretch zone (dJ/d(SZW)) gives a linear relationship for the tested specimens. However, the SZWc of low sulfur steels is larger than that of high sulfur steels, and the SZWc of overheated steels is smaller than that of QT steels. This fact means that in the specimens where the grain boundaries weakened due to overheating, the stable crack growth initiates along the grain boundary at the early stage of crack tip blunting, resulting in lower fracture toughness J_{IC}. SEM observations also showed that the lower cooling rate enhances the growth of precipitates and reduces the amount of precipitate on the grain boundary.

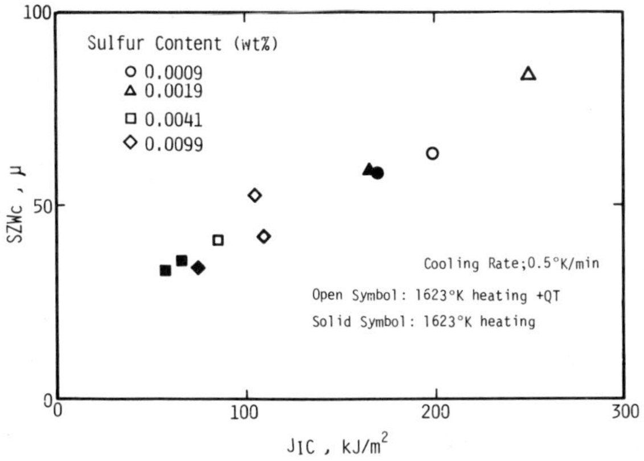

Fig. 11. Relationship between the critical stretch zone width (SZWc) and the J_{IC} value.

Thus, overheating leads to a decrease of fracture toughness and tearing modulus of the materials, depending on sulfur content, heating temperature and cooling rate. Test results also indicate that the quality heat treatment, that is quenching and tempering, improves the toughness of overheated steels up to those of sound steel.

CONCLUSIONS

(1) Overheating affects the ductility of steels in the forging temperature range. The loss of ductility becomes greater with increasing sulfur content.

(2) Tensile pre-straining of overheated steel in the forging temperature range leads to further embrittlement of the steel.

(3) The low sulfur steels have high susceptibility to overheating, although they have high toughness even under the overheated conditions.

(4) Overheating leads to a decrease of fracture toughness J_{IC} and tearing modulus T_J, which depends on sulfur content, heating temperature and cooling rate.

(5) Quenching and tempering heat treatment in the steel product improves the toughness properties of the overheated steels. The large effect is expected for low sulfur steels below 0.005wt%S.

REFERENCES

Preece, A., et al (1946). J. Iron and Steel Institute, 153, 237-254.
Preece, A., Nutting, J., and Hartley, A. (1950). J. Iron and Steel Institute, 164, 37-45.
Schulz, B. J. and McMahon Jr. C. J. (1973). Met. Trans., 4, 2485-2489.
Glue, D. R., Jones, C. H. and Lloyd, H. K. M (1975) Met. Tech., 2, 416-421.
McLeod, N. P. and Nurring. J. (1982). Met Tech., 9, 399-404.

MICROMECHANICS OF FRACTURE IN MICROALLOYED STEEL

A. N. Kumar and R. K. Pandey

Applied Mechanics Department, Indian Institute of Technology, Hauz Khas, New Delhi-16, India

ABSTRACT

A quantitative investigation of the micromechanics of fracture has been carried out in hot-rolled microalloyed steel. The inclusion size and inter-inclusion spacing distribution have been determined using statistical analysis. Tensile properties and toughness data have been obtained and used for determination of stress and strain distribution in the vicinity of crack tip. The fracture strain and dimple sizes for fracture are found to be highly sensitive to triaxiality factor. The process zone size has been determined from the critical strain level in the fracture toughness (FT) specimen and related to the dimple size. The effective inclusion size governing the ductile fracture process has been identified.

KEYWORDS

Inclusion distribution; dimple size; fracture stress and strain distribution; process zone; effective inclusion; microfracture size scale.

INTRODUCTION

The initiation of fracture in the microalloyed steel in the ambient condition, takes place by formation of voids and their coalescence. A number of microstructural parameters like size, shape and distribution of inclusions/second phase particles, grain size, matrix structures etc. are known to influence the mechanics of microcrack nucleation and its propagation leading to fracture. The present investigation aims at studying quantitatively the micromechanics of fracture in the microalloyed steel which is a high strength structural steel and finds extensive structural applications. The effective inclusions in the plastic zone are believed to control and determine the toughness of the material. The resulting size of the dimples are related to the size of effective inclusions and the inter-inclusion spacing. An attempt has been made in this investigation to determine the effective inclusion size and their relation to dimple dimensions. The effect of stress state and the triaxiality factor on critical fracture strain has been studied and their influence on the characteristics of dimples has been investigated.

MATERIAL PROPERTIES

The wt % composition and the tensile properties of the microalloyed steel are given below. A typical microstructure is shown in Fig. 1.

C - 0.12, Mn - 0.99, P - 0.022, S - 0.027, Si - 0.025, Nb - 0.044, V - 0.043. Grain size - 8-10 μm. Yield strength (σ_{ys}) = 404 MPa, T.S. = 516 MPa, %El. = 24, %RA = 73, true fracture strain, ε_f = 1.257, strain hardening exponent, n = 0.171.

The SEN FT specimens of dimensions 10×25×200 mm were machined with the longitudinal axis parallel to the rolling direction of the plate. The specimens were loaded in 3-point bend to determine the Crack Opening Displacement (COD) and the J-integral, J_c toughness values as reported below using the standard methods:

COD_c = 155 μm, J_c = 46 kJ m^{-2}, K_c (as estimated from J_c) = 102.5 MPa$\sqrt{m}$

MECHANICS OF FRACTURE

Crack Tip Stress Distribution

The crack tip stress distribution in the FT specimen was obtained on the basis of HRR theoretical analysis for elastic-plastic stress distribution, (Hutchinson, 1968; Rice and Rosengren, 1968). The normal stress at distance x directly ahead of the crack tip (σ_{yy}) is given by Hutchinson, 1968)

$$\frac{\sigma_{yy}}{\sigma_{ys}} = f(n)\left(\frac{J}{\sigma_{ys}\,\varepsilon_o\,Ix}\right)^{n/n+1} \tag{1}$$

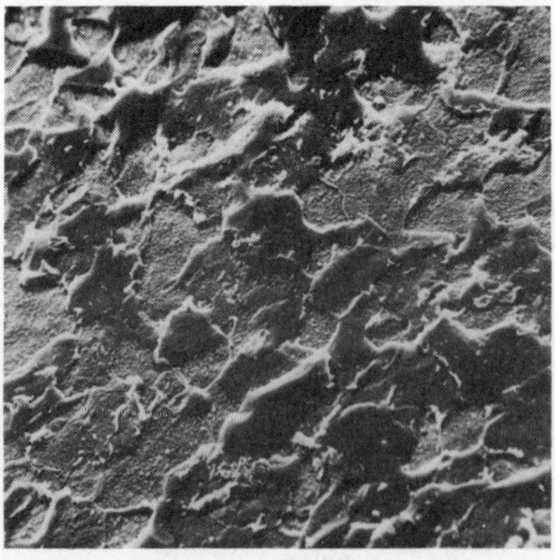

Fig.1 SEM micrograph

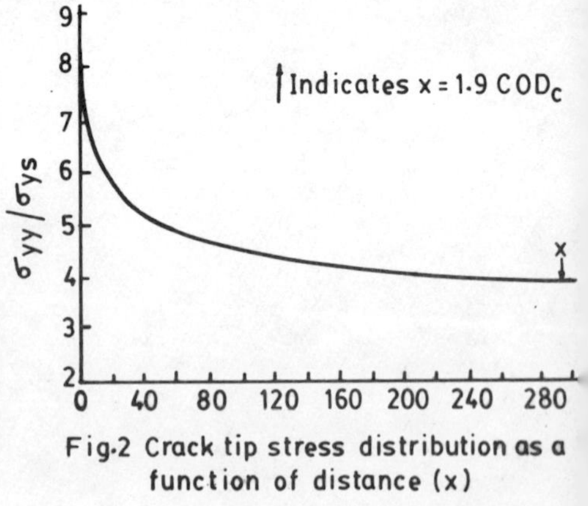

Fig.2 Crack tip stress distribution as a function of distance (x)

where ε_o is yield strain and f(n) and I are numerical constants dependent on n.

Equation (1) has been used to compute the stress distribution at $J = J_c$ (Fig. 2). However, due to blunting effects and large geometry changes, the actual stress distribution gets modified in the vicinity of the crack tip (Rice and Johnson, 1970). Also, the slipline fields focus intense plastic straining over these regions. The σ_{yy} level is expected to be maximum at a distance $X = 1.9\ COD_c$ from the crack tip.

Critical Fracture Strain

The crack tip fracture strain has been determined by

(a) McClintock's Void Coalescence Model (McClintock, 1968)

$$\varepsilon_{cf} = \frac{(1-n)\ \ln(l_o/2r_o)}{\sinh\left[(1-n)(\sigma_{yy} + \sigma_{zz})/\frac{2\bar{\sigma}}{\sqrt{3}}\right]} \qquad (2)$$

where l_o - initial void spacing, $\bar{\sigma}$ - effective stress, r_o - initial minor semi-axis of the pre-existing voids.

The $l_o/2r_o$ value in the above equation was determined for the tensile case of fracture using ε_f data. The $l_o/2r_o$ value is believed to be independent of stress state (Kumar, 1982) and found to be = 1.45. The ε_{cf} value is reported in Table I.

(b) The Experimental COD Data was also used to obtain the crack tip fracture strain (ε_{cf*}) which is given by (Robinson and Tetelman, 1975),

$$\varepsilon_{cf*} = COD/l \qquad \text{where } l \simeq 1.8\ n^2 \qquad (3)$$

The corresponding fracture strain values are reported in Table I.

TABLE I Critical Fracture Strain Values

ε_f	ε_{cf}	$\varepsilon_{cf}/\varepsilon_f$	ε_{cf*}	$\varepsilon_{cf*}/\varepsilon_f$
1.257	0.19	0.15	0.12	0.09

The above results evidence the high sensitivity of fracture strain to triaxiality factor $\sigma_m/\bar{\sigma}$ at the crack tip in the present study, where σ_m is the mean stress. The value of $\sigma_m/\bar{\sigma}$ was observed to be 0.58 which shows a reduction in crack tip fracture strain to a value of 10-15% of the tensile fracture strain.

The strain level at a characteristic distance, X (= 1.9 COD_c) where maximum triaxiality is expected to prevail in the plane strain specimen was also computed using eq. (2). The corresponding values of σ_{xx} and σ_{zz} were obtained from the relations (Sailors, 1973)

$$\sigma_{xx} = \sigma_{yy} - H\ \sigma_{ys} \quad , \quad \sigma_{zz} = 1/2(\sigma_{xx} + \sigma_{yy}) \qquad (4)$$

where H represents a correction factor for the flow stress. The strain level

(ε_{xf}) and the triaxiality factor were found to be 7.5×10^{-2} and 1.20, respectively. This further reveals that the fracture ductility is a very sensitive function of triaxiality factor. The fracture strain at the crack tip where biaxiality exists ($\sigma_m/\bar{\sigma}$ = 0.58) is thus around 2.5 times larger than the same at highly triaxial region ($\sigma_m/\bar{\sigma}$ = 1.20). Somewhat similar observation for lower yield strength materials have been reported elsewhere (Ritchie and Co-workers, 1979; Otsuka and Co-workers, 1979).

Inclusion Size and Spacing

The inclusion morphology and their distribution were studied using the quantitative microscopic techniques. Inclusions were found to be of almost spherical shape. The randomness of inclusions was checked using grid square technique (Ashby and Ebeling, 1966). From size distribution of inclusions based on volume fraction (Fig. 3), it may be noted that the two most probable size ranges of inclusions are 1.6 - 2.4 μm and 7.2 - 8.0 μm on WT plane. The interinclusion spacing was measured by a concentric circle technique. The statistical analysis was done by grouping the nearest neighbour distances into various ranges. The most probable interinclusion spacing was found to be between 2 - 4 μm on WT plane.

Statistical Distribution of Dimples

The fractographs (SEM) for tensile and FT specimens revealed the microvoid coalescence resulting in dimples. The areal distribution of dimples was obtained from the statistical analysis of size distribution of dimples as

$$f_{A(r)} = \frac{A(r)\ N(r)}{\sum\limits^{n} A(r)N(r)} \tag{5}$$

where A(r) is the area of dimple of size r, N(r) is the number of dimples/ /area of size r and n is the number of size groups.

Figure 4 presents typical histograms of areal fraction distribution analysis. The most probable dimple size range was found to be smaller in the case of tensile specimen, 3 - 4 μm, than the same for FT specimen, 10 - 14 μm.

DIFFERENT SIZE SCALE FOR FRACTURE

Ductile Fracture Zone (Process Zone for Fracture)

The size of ductile fracture zone which represents the distance over which the strain is equal to or greater than the critical fracture strain has been computed for the steel under investigation using the critical fracture strain values and crack tip strain distribution for elastic-plastic situation (Rice and Johnson, 1970). The values obtained are given below:

Process zone ($\bar{X}$) from ε_{cf} = 185 μm

Process zone ($\bar{X}^*$) from COD = 204 μm

Size of intense strain region (X) = 294 μm

The size of ductile fracture zone was found to lie within the intense strain region (X) which can be approximately related to the ductile fracture zone ($\bar{X}$) by

$$X \simeq 1.5\ \overline{X} \tag{6}$$

The PZS was also found to be around 20 times larger than the ferrite grain diameter. It is believed that the size of process zone is governed more by the size of inclusion and interinclusion spacing rather than the ferrite grain size. Therefore, the dimple size can be related to PZS for this steel. A comparison of effective dimple size ($\overline{D}$) with ductile fracture process zone shows that

$$PZS \simeq 15\ \overline{D} \tag{7}$$

Thus, it appears that formation of voids over a number of effective inclusions and their coalescence sets in the process for initiation of a critical crack (Fig. 5), rather than the coalescence of voids on just one or two inclusions (near the crack tip) as conceived earlier.

Effective Inclusion Size

The computation of effective inclusion size for fracture is based on the independent calculation of $l_o/2r_o$ (effective inclusion spacing to diameter) in eq. (2). From a knowledge of evaluated $l_o/2r_o$ value (= 1.45) and the experimentally determined interinclusion spacing, the size of effective inclusion was computed. $l_o = 1.45(2r_o) = 2.5$ to 3.6 µm, where $2r_o = 1.6$ to 2.5 µm (based on volume fraction analysis). It is quite interesting to note that value of l_o matches fairly well with the most probable dimple size (3 - 4 µm). Moreover, the most probable nearest neighbour spacing of inclusion, i.e. 2 - 4 µm, lies in the same range. Thus, the most probable inclusions, i.e. 1.6 µm - 2.5 µm, based on volume fraction analysis appear to be the effective inclusion for the initiation of fracture under uniaxial loading condition.

In plane strain fracture toughness specimens, relatively bigger size (10 - 14 µm) of the most probable dimples led to conclude that there is extensive expansion of the voids under triaxial stress field. The ratio of dimple to inclusion diameter lies in a range of 5 to 7 which agrees well with the reported data in literature.

CONCLUSIONS

1. The effective inclusion size for ductile fracture initiation may be given by the most probable size of inclusion based on volume fraction distribution in the microalloyed steel.

2. In plain tensile specimen, the effective dimple size is of the order of interinclusion spacing, however, in the presence of plane strain triaxiality, the effective dimple size is increased by a factor of 3 to 4.

3. The ductile fracture strain is very sensitive to plane strain triaxiality factor. A triaxiality factor of around 0.6 brings down the fracture strain at 10% of the tensile fracture strain.

4. The fracture process zone is fully enveloped by the intense strain region as predicted by slip line field. The initiation of ductile fracture is preceded by void-coalescence over a number of inclusions (around 15 or so) in the crack tip region.

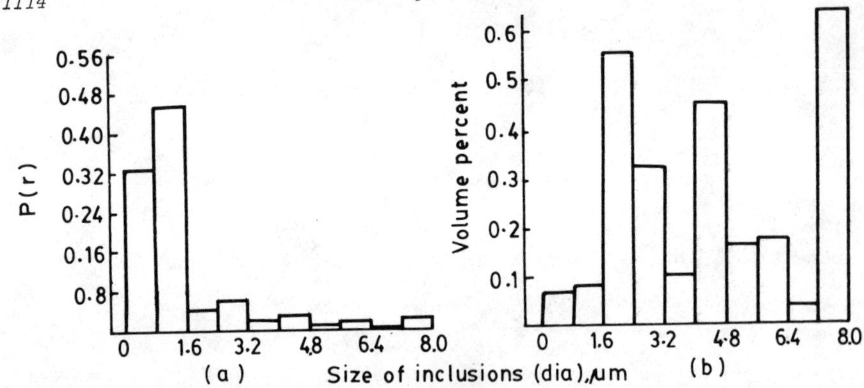

Fig.3 Size distribution of inclusion based on (a) population (b) volume fraction

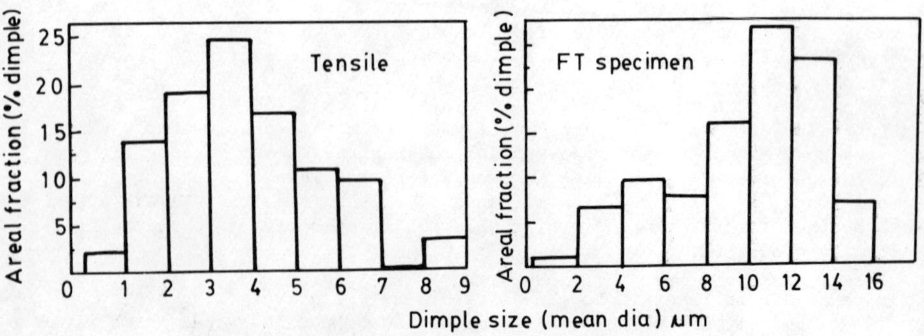

Fig.4 Areal fraction distribution of dimples

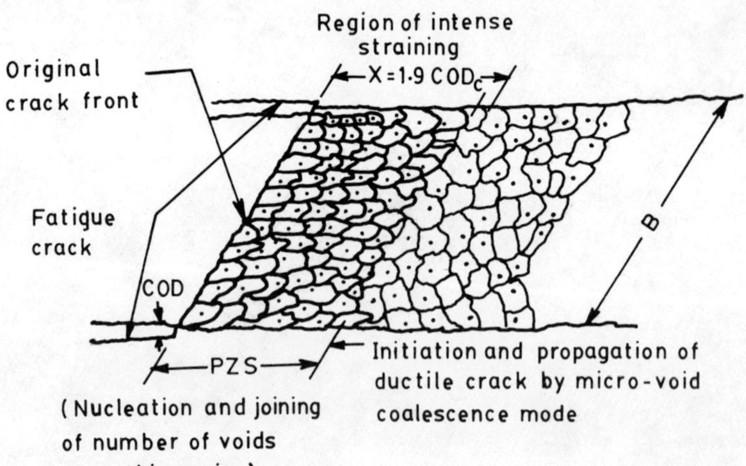

Fig. 5 Schematic illustration of the micro-fracture process in microalloyed steel

REFERENCES

Hutchinson, J.W. (1968). Plastic stress and strain fields at a crack tip. J. Mech. Phys. Solids, 16, 337-347.

Rice, J.R. and G.F. Rosengren (1968). Plane strain deformation near a crack tip in a power law hardening material. J. Mech. Phys. Solids, 16, 1-12.

Rice, J.R. and M.A. Johnson (1970). The role of large crack tip geometry changes in plane strain fracture. In M.F. Kanninen and Colleagues (Eds.) In-Elastic Behaviour of Solids, McGraw Hill, New York, pp. 641-672.

McClintock, F.A. (1968). A criterion for ductile fracture by the growth of holes. J. Appl. Mech., 35, 363-371.

Kumar, A.N. (1982). Ambient Fracture Behaviour and Environmental Crack Growth Kinetics in a Low Alloy and a Microalloyed Steel. Ph.D. Thesis, I.I.T., New Delhi, India.

Robinson, J.N. and A.S. Tetelman (1975). The relationship between crack tip opening displacement, local strain and specimen geometry. Int. J. Fracture, 11, 453-468.

Sailors, R.H. (1973). Relationship Between Tensile Properties and Microscopically Ductile Plane Strain Fracture Toughness, Ph.D. Thesis, Univ. of Illinois, Illinois.

Ritchie, R.O., W.L. Server and R.A. Wullaert (1979). Critical fracture stress and fracture strain models for the prediction of lower and upper shelf toughness in nuclear pressure vessel steels. Met. Trans. A., 10, 1557.

Otsuka, A., T. Miyata, S. Nishimura and M. Ohashi (1979). Effects of stress triaxiality on ductile fracture initiation and ductile brittle transition in prestrained steel. In G.C. Sih and S.R. Valuri (Eds.) Fracture Mechanics in Engineering Application, Bangalore, India, 111-122.

Ashby, H.F. and R. Ebeling (1966). On the determination of the number, size, spacing, and volume fraction of spherical second-phase particles from extraction replicas. Met. Soc. AIME, 236, 1396-1404.

AN INTERPRETATION OF SCATTER AND THICKNESS EFFECT OF FRACTURE TOUGHNESS IN THE TRANSITION REGION

Naotake Ohtsuka

Material and Welding Technology Department,
Chiyoda Chemical Engineering and Construction Co. Ltd.,
3-1-4 Ikegamishincho, Kawasaki, Japan

ABSTRACT

A new model is proposed to explain the scatter and the thickness effect of cleavage fracture in the ductile-brittle transition region. Cleavage fracture is triggered when a detonator encounters an igniter at the high stress field ahead of the crack tip. Experiments for 2-1/4Cr-1Mo steel are compared with the model. This model suggests that a modified J_{IC} test can evaluate K_{IC} using small specimens, if the initiation of microscopic ductile tearing along the crack tip is examined fractographically.

KEYWORDS

Fracture toughness; fracture mechanics; J_{IC}; K_{IC}; transition; cleavage; ductile tearing; crack; stress; size effect; scatter; probability.

INTRODUCTION

It is generally said that the smaller the thickness of a toughness specimen, the larger the scatter and the average value of the cleavage fracture toughness K_C in the ductile-brittle transition region (Sumpter, 1976). This observation is explained in terms of the loss of constraint model (Milne and Chell, 1979) and the weakest link model (Landes and Shaffer, 1980). It is usually considered that the toughness K_{in} which corresponds to the initiation of ductile tearing overestimates K_{IC}, which is the critical lowest value of K_C in the temperature region (Sumpter, 1976; Dawes, 1979). These experimental results are the reason why the J_{IC} test methods (JSME S001-1981 and ASTM E813-81) can not be applied in the transition region.

Recent fractographic examinations of loaded specimens, however, indicate that microscopic ductile tearing precedes cleavage fracture so that a modified J_{IC} test method to determine its inititation is considered to be applicable in the transition region (Ohtsuka, 1982, 1983). This observation can not be explained by the two previous concepts on the thickness effect.

In the paper, the initiating condition of cleavage fracture was discussed considering the deformation and stress state ahead of crack tip during loading. The scatter and thickness effects of K_C for 2-1/4Cr-1Mo steel were interpreted by a new model.

FRACTURE TOUGHNESS TEST OF 2-1/4Cr-1Mo STEEL

The test material was a quenched and tempered 90-mm thick ASTM A387 Gr.22 Cl.2 (2-1/4Cr-1Mo) steel. The compact tension (CT) type specimen was used for fracture toughness test. The specimen thickness, B, covers a range from 3 to 87.5mm. The widths are 175mm for the 87.5mm-thick specimen and 50mm for the others. The ratio of crack length to width, a_o/W, is 0.5. The fracture toughness test was performed in accordance with ASTM E399 except that the K_I value was converted by the following equation.

$$K_I(J) = \sqrt{EJ/(1-\nu^2)} \equiv \sqrt{E'J} \qquad (1)$$

where E and E' are Young's modulii for plane stress and plane strain, and ν is Poisson's ratio. J is a J-integral which is calculated in terms of area under load/load-line displacement cruve (JSME S001-1981 and ASTM E813-81).

The cleavage fracture toughness $K_C(J)$ obtained at 6 different temperatures is shown in Fig. 1. Solid symbols in the figure satisfy the valid K_{IC} condition of ASTM E399. Fig. 1 indicates that test temperatures below 10°C are in the transition region and that the range of variability increases as the thickness of specimen decreases at each temperature. Some of the CT specimens were unloaded prior to cleavage fracture and broken open by subsequent fatigue loading. The fracture surfaces were examined by scanning electron fractography. When a specimen is loaded over K_{IC}, localized ductile tearing by dimpled rupture can be observed along the tip of the stretched zone at the mid-thickness crack front as shown in Photo 1.

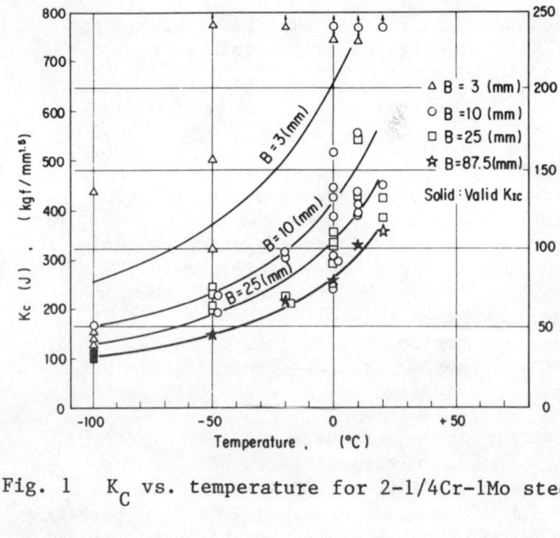

Fig. 1 K_C vs. temperature for 2-1/4Cr-1Mo steel.

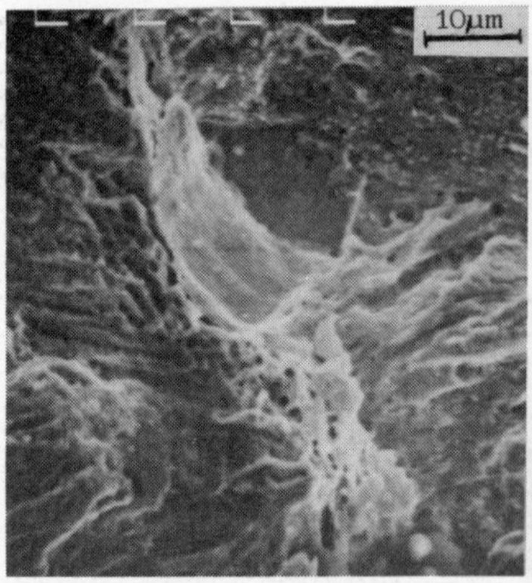

Photo 1 An example of ductile tearing of 2-1/4Cr-1Mo steel (-20°C).

DISCUSSION ON EXISTING MODELS FOR THICKNESS EFFECT ON K_C

The loss of constraint model: A loss of stress triaxiality ahead of the crack due to loss of through-thickness constraint causes higher measured toughness in the small specimen (Milne and Chell, 1979). This model can not sufficiently well explain the large scatter of K_C (even in the case of a side-grooved specimen) and is inconsistent with sufficient stress triaxiality at the crack tip at the mid-thickness of specimen.

The weakest link model: Fracture toughness being governed by the point or region of lowest toughness along crack front, smaller specimens which sample less of the variation in toughness exhibit higher toughness than larger (thicker) specimens (Landes and Shaffer, 1980). The variability of K_C supports a statistical behavior of cleavage fracture. However, whether the localized toughness has such a wide variation as expected by the model is questionable. Some experiments indicate that a Weibull plot of the cleavage toughness is not always a straight line. The model suggests the restriction that a very large number of small specimens should be used to characterize the cleavage toughness of a large specimen.

The two models do not consider the interacted effect of crack tip blunting and initiation of microscopic ductile tearing on cleavage fracture.

A NEW MODEL ON CLEAVAGE FRACTURE

Considering that cleavage fracture is governed by the maximum principal stress ahead of crack tip, a statistical hypothesis is proposed to explain the observed behavior. Cleavage fracture occurs when one of the detonators encounters an igniter at the triaxial stress field ahead of crack as illustrated in Fig. 2. The detonators which are distributed in a random manner within a specimen are assumed to have a nearly constant ignition level. This is contrary to the weakest link concept which is based on the assumption that the toughness itself has a large scatter. An effective crystal grain whose cleavage plane coincides with a crack plane or which has a large diameter or a specific angle against a slip plane in the adjoining grain is considered to act as the detonator. The igniter is assumed to originate when the stress intensity of the specimen exceeds K_{IC}. Ductile tearing or localized high normal stresses ahead of the crack tip can be effective igniters.

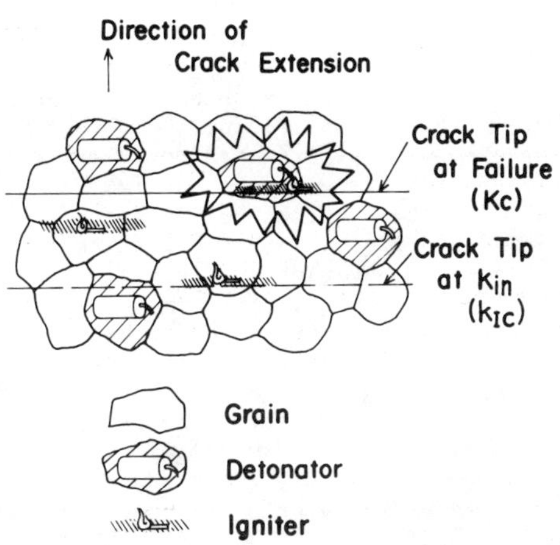

Fig. 2 Schematic fracture model.

A role of ductile tearing as an igniter of cleavage fracture is suggested in the APPENDIX by a brief discussion on the maximum principal stress at crack front. Crack tip blunting during loading lowers the stress concentration ahead of the crack. Consequently the possibility of the stress normal to the crack plane to exceed the cleavage stress is limited. Due to the formation of voids and their subsequent coalescence, ductile tearing initiates along the tip of the stretched zone at the mid-thickness where plane strain condition exists. Resharpening of the crack tip raises the normal stress which may cause cleavage fracture.

Failure probability is estimated as follows: The total number of effective igniters which appear in the course of crack growth Δa is

$$N = \int_o^{\Delta a} \gamma(J)B_e \, d(\Delta a) \tag{2}$$

where $\gamma(J)$ is the density or the number of igniters along the crack tip per unit area on the crack plane at a loading level of J. B_e is a length along the crack tip in plane strain state. Prior to the onset of ductile tearing, the crack extension is equal to the stretched zone width which is related to the J-integral by the equation (Kobayashi and co-workers, 1977, 1979)

$$\Delta a = mJ, \ (m = m_1/E \text{ or } 1/(m_2\sigma_Y)) \tag{3}$$

where, m_1 = 89, m_2 = 2 ∿ 4, E = Young's modulus and σ_Y = yield strength. If we assume that the effective igniter initiates after that the J level exceeds J_{IC}, $\gamma(J)$ can be written in the form

$$\gamma(J) = \gamma_o(J/J_{IC} - 1)^{\beta} \quad \text{for } J \geq J_{IC} \tag{4}$$

where γ_o and β are constants. Approximating B_e as half the thickness of the specimen, and inserting eqns (1), (3) and (4) into eqn (2), we obtain

$$N = \gamma_o B\Delta a_{IC}\{(K_I/K_{IC})^2 - 1\}^{(\beta+1)}/\{2(\beta+1)\} \tag{5}$$

where Δa_{IC} is Δa at $K_I = K_{IC}$. If the probability, P_c, of occurrence of a detonator at the crack tip is constant everywhere in a material, the probability that a given specimen has a fracture toughness less than K_c can be given by

$$P = 1 - (1-P_c)^{N(K_c)} \tag{6}$$

where $N(K_C)$ is the value of N obtained by eqn (5) for $K_I = K_C$.

Assigning appropriate values to the constants in eqns (5) and (6), the estimated probability for β = 0 is shown on a Weibull paper in Fig. 3 together with experiments for 2-1/4Cr-1Mo steel at 0°C. The thickness effect of K_C is given by putting P = 0.5 in eqn (6) and is illustrated in Fig. 4. The experimental data are replotted in the figure from Fig. 1. Fig. 4 and Fig. 5 indicate that the proposed model on cleavage fracture can explain the behavior sufficiently well.

A detailed study is needed to identify what the detonator and the igniter are. This model suggests that the small specimen toughness test can be used to evaluate the toughness of large specimens in the transition region, provided that the initiation of the igniter is experimentally determined. The validity of this model was discussed on a modified J_{IC} test (Ohtsuka, 1983). In the test, fractographic examination was made on the mid-thickness fracture surfaces of specimens made from C-1/2Mo steel and 2-1/4Cr-1Mo steel. The value of J or K_I at the initiation of microscopic ductile tearing at the crack front corresponded to K_{IC}.

CONCLUSIONS

1. The scatter and thickness effect of cleavage fracture toughness in the ductile-cleavage transition region can be explained by using a new model. The fracture of a specimen is triggered by an encounter of a detonator and an igniter ahead of the crack tip in the model.

2. This model suggests that a modified J_{IC} test can be used for evaluating the toughness of large specimen using small specimen results in the transition region, provided that the initiation of microscopic ductile tearing along the mid-thickness crack tip is fractographically examined.

REFERENCES

Dawes, M.G. (1979). ASTM STP, 668, 307-333.
Hutchinson, J.W. (1968). J. Mech. & Physics of Solids, 16, 13-31, 337-347.
Kobayashi, H., Hirano, K., Nakamura, H. and Nakazawa, H. (1977). Proc. ICF4. Vol.3, Pergamon Press, 583-592.
Kobayashi, H., Nakamura, H., and Nakazawa, H. (1979). Proc. ICM3. Vol.3, Pergamon Press, 529-538.
Landes, J. D. and Shaffer, D. H. (1980). ASTM STP, 700, 368-382.
Milne, I. and Chell, G.G. (1979). ASTM STP, 668, 358-377.
Neuber, H. (1961). Trans. ASME, Ser. E, 28-4, 544-550.
Ohtsuka, N. (1982). Proc. JSME Ann. Meeting, 820-12, 21, and 827-2, 20, in Japanese.
Ohtsuka, N. (1983). Development of J_{IC} Test in Transition Temperature Region. To be presented at 7th Int. Conf. on SMiRT, Chicago, USA.
Rice, J.R. (1968). Trans. ASME, Ser. E, 35-2, 379-386.

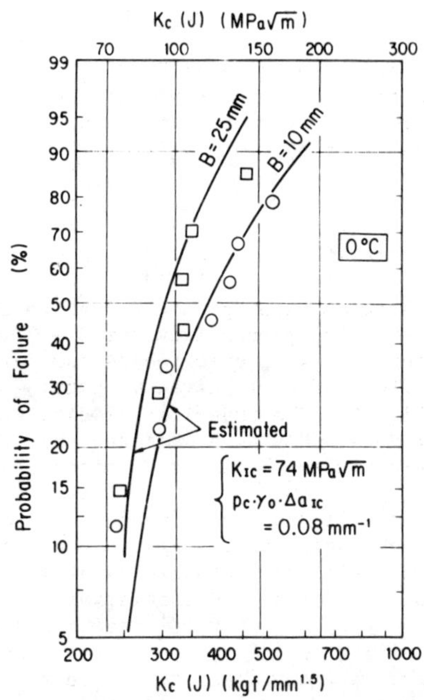

Fig. 3 Weibull plot of K_C for 2-1/4Cr-1Mo steel.

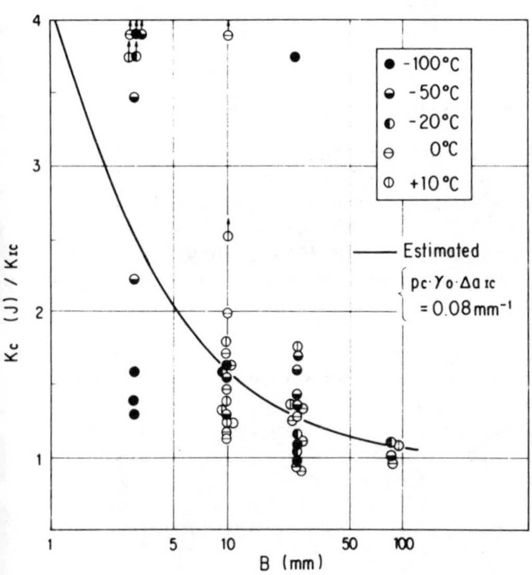

Fig. 4 K_C/K_{IC} vs. thickness for 2-1/4Cr-1Mo steel.

Rice, J.R. and Rosengren, G.F. (1968). J. Mech. and Physics of Solids, 16, 1-12.
Sumpter, J.D.G. (1976). Metal Science, 10, 354-356.

APPENDIX

Discussion on Maximum Principal Stress at Crack Front

Near the crack tip the stress and strain fields are given by the HRR singularity form for hardening materials (Hutchinson, 1968; Rice and Rosengren, 1968). However, this form can give neither the maximum value of normal stress ahead of crack due to its singularity nor the effect of crack tip blunting. No analytical expression of the maximum stress being available, the blunted crack tip is approximated as a sharp notch whose root radius, ρ, is governed by the previous eqn (3) where Δa is equated to ρ. The material behavior is modelled by

$$\sigma/\sigma_Y = (\varepsilon/\varepsilon_Y)^n \quad \text{for } \sigma \geq \sigma_Y \tag{7}$$

where σ and ε are the stress and the strain, σ_Y and ε_Y are the yield stress and the yield strain, and n is a hardening exponent ($0<n<1$). Neuber's equation (1961) on a sharp notch for antiplane deformation is used to estimate the maximum value of the normal stress ahead of the crack tip.

$$K_t^2 = (\sigma_{max}/\sigma_n)(\varepsilon_{max}/\varepsilon_n) \tag{8}$$

where σ_n and ε_n are the nominal stress and the nominal strain, and K_t is an elastic stress concentration factor. The object of discussion is for the sake of simplicity limited to a two-dimensional crack (whose crack length is 2a) in an infinite tensile body in a uniform remote stress field σ_n. The stress intensity factor, K_I, at the small-scale-yielding condition and the stress concentration factor, K_t, for the corresponding notch whose dimension is the same as the crack are given by

$$K_I = \sigma_n \sqrt{\pi a} \quad (9), \qquad K_t = 1 + 2\sqrt{a/\rho} \fallingdotseq 2\sqrt{a/\rho} \quad (10)$$

where ρ is the root radius of the notch. Substituting eqns (7), (9), (10) and (1) into eqn (8), the maximum stress is obtained by

$$\sigma_{max} = \{(4E'J/\pi\rho)^n \sigma_Y^{(1-n)}\}^{1/(n+1)} \tag{11}$$

Equating Δa in eqn (3) to ρ in eqn (11), the maximum stress in the course of crack tip blunting is given by

$$\sigma_{max,\ blunting} = (4E'/\pi m)^{n/(n+1)} \sigma_Y^{(1-n)/(n+1)} \tag{12}$$

A similar result is obtained using Rice's equation (1968) on a flat-surface notch in a two-dimensional deformation field.

$$\sigma_{max} = \sigma_Y \left\{ \frac{(n+3/2)(n+1/2)\Gamma(n+1/2)}{\sqrt{\pi}\ \Gamma\ (n+1)} \frac{J}{\sigma_Y \varepsilon_Y \rho} \right\}^{n/(n+1)} \tag{13}$$

$$\sigma_{max,\ blunting} = \sigma_Y \left\{ \frac{(n+3/2)(n+1/2)\Gamma(n+1/2)}{\sqrt{\pi}\ \Gamma\ (n+1)} \frac{E}{m\sigma_Y^2} \right\}^{n/(n+1)} \tag{14}$$

where Γ( ) is the gamma function. Eqns (12) and (14) indicate that the $\sigma_{max,blunting}$ is independent of J (or σ_n) and ρ (or Δa), and is not very high (for example; $\sigma_{max,blunting}$ < 3GPa for n = 0.2 and σ_Y < 1GPa).

When ductile tearing initiates at the crack front, the blunted crack tip is resharpened and the root radius is in the order of the radius of voids, ρ_v. In this case, the value of σ_{max} is estimated by substituting ρ_v into ρ in eqns (11) and (13). σ_{max} for σ_Y = 1GPa and E=210 GPa is illustrated in Fig. 5. The figure indicates that the difference between the results from Neuber's equation and Rice's equation is small, and that σ_{max} increases as J/ρ_v increases. Considering the interaction effect of double notches (i.e., blunted crack tip plus ductile tearing) in eqn (10) for a_o/a_v >> 0, σ_{max} in eqn (11) is approximately multiplied by

$$\alpha = \{1 + 8a_v/\rho_v + 4\sqrt{2a_v/\rho_v}\}^{n/(n+1)} \tag{15}$$

where a_o is half the crack length of the original crack and a_v is the length of ductile tearing. ρ_o and ρ_v are root radii of the blunted crack tip and of the voids. An example of the variation of σ_{max} during the loading of the cracked body is illustrated in Fig. 6 for n=0.2 and σ_Y=1GPa. Fig. 6 suggests that the maximum normal stress ahead of the crack tip which is nearly constant in the course of blunting increases rapidly at the onset of ductile tearing.

It should be noted that this analysis includes the following hypotheses and problems; (1) the influence of the stress histories is not taken into account, (2) formulii on the maximum stress at the notch tip is used instead of that below crack tip in plane strain state, and (3) as σ_{max} in Fig. 6 is yet insufficient for cleavage fracture, an additional discussion on dislocation theory is required.

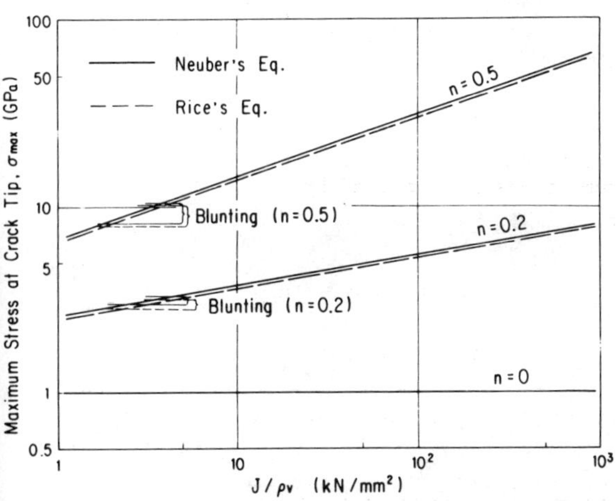

Fig. 5 σ_{max} vs. J/ρ_v (σ_Y=1GPa, E=210GPa).

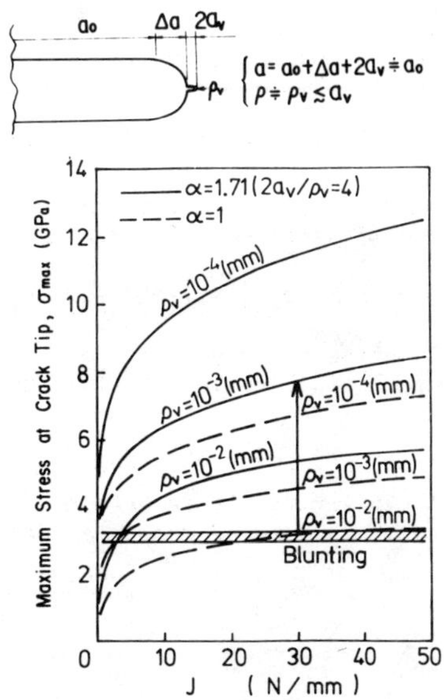

Fig. 6 Variation of σ_{max} for σ_Y=1GPa and n=0.2. The arrow indicates an example of rapid increase of σ_{max}.

THE DETERMINATION OF S-N CURVES WITH PRE-CRACKED SPECIMENS

J. T. Wang

Department of Mechanics, Zhejiang University, Hangzhou, China

ABSTRACT

The S-N curves of steels are very useful for design purposes. However, it is usually a very difficult job to obtain such a curve or a family of curves from experiments, if we use the conventional smooth specimens (i.e. specimens not pre-cracked). The experimental data are so scattered that one has to be satisfied by obtaining a low bound envelope of the multitude of experimental points as an S-N curve on the safe side.

The initiation life of materials are dependent upon the material constants such as C and m in Paris' equation:

$$da/dN = C(\Delta K)^m ,$$

and the sizes of micro-defects in the materials. The latter depends on the quality of steel, and is different for each specimen. They are unknown to the experimenter. For this reason it is impossible to determine the material constants from the data of the experiments with smooth specimens. Moreover, for each smooth specimen, the initiation life varies from around 10^5 cycles to 10^9 cycles according to the amplitude of the applied load. For each load level at least three to five specimens are needed so it would take very long time to carry out the tests, usually it would exhaust the life of the testing machines themselves.

KEYWORDS

Pre-cracked specimens, S-N curves, Threshold values, Fatigue limits, Material and time saving, Effective.

INTRODUCTION

By using pre-cracked specimens, we study the crack propagation lifes. Now the size of the initial defects will be known to the experimenter, because the natural micro-defects are much smaller and are negligible in comparison with the artificial cracks. By measuring the size of the cracks after a few hundred thousand cycles, we can obtain quite a number of data points from a single specimen. Since the size of the artificial crack is significantly larger than the natural micro-cracks, the measured propagation lifes are much shorter than the initiation lifes. We can calculate the equivalent size of initial defect of the smooth specimen from the tests of a few of them. For each equivalent size of initial defect, we can obtain without difficulty an S-N curve, as well as lower threshold value and fatigue limit. The time required for experiments on the pre-cracked specimens are much shorter than for the conventional methods; it consists of only a small fraction (usually less than one hundredth) of the latter.

EXPERIMENTS AND RESULTS

The experiments were carried out with pre-cracked specimens as shown in Fig. 1. For each piece of pre-cracked specimen, we can obtain more than 10 sets of data of da/dN and K. On the log-log coordinate paper, they appear as a broken line OABC as shown in Fig. 2. The line segment OA represents the state of initiation or the stage of very slow propagation of the crack. It might be thought of as the state of initiation or as the process of the propagation of the micro-defect to macro-size. In fact, the line segment OA is hardly a straight line. It has the tendency of bending slightly to being nearly vertical, where people defines the "Lower Threshold Limit" (ΔK_{th}). For better simulating the initial stage of the growth of the crack we preferably work with as small ΔK as possible. Obviously, the material constants thus obtained will place the resulting S-N curve on the safe side with regard to engineering practice.

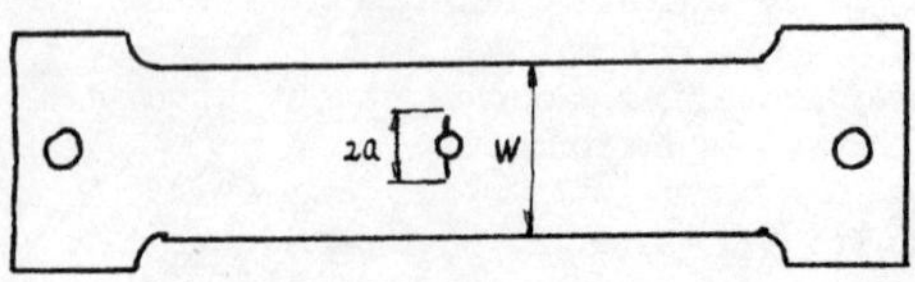

Fig. 1. Pre-cracked specimen

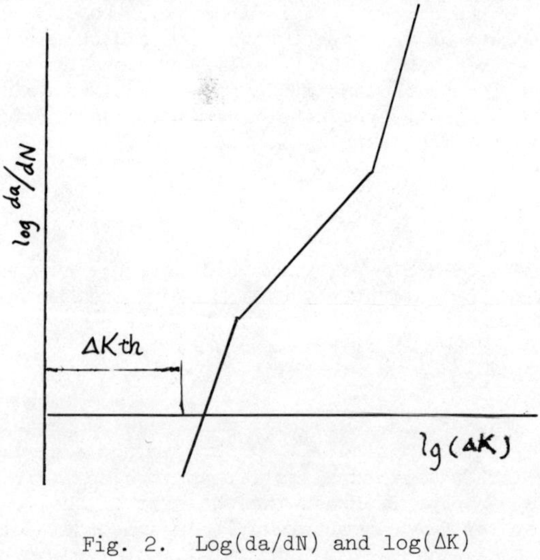

Fig. 2. Log(da/dN) and log(ΔK)

We shall illustrate our method with an example. Three specimens made of 50Cr steel offer seven sets of crack propagation data as shown in Table 1. The mechanical properties of 50Cr steel are: yield strength σ_Y = 97 kgf./mm^2, ultimate strength σ_u = 117 kgf./mm^2, elongation δ_5 = 23.5%, reduction in area ψ = 47.1%.

TABLE 1 da/dN and (ΔK)

Specimen	B mm	W mm	a mm	N cycles	da/dN mm/cycle	P kgf	K kgf/mm$^{3/2}$
1	9.7	40	5.565	185×10^3	0.0157	3600	40.11
			6.665	100 "	0.110	"	44.58
			8.195	90 "	0.170	"	50.83
2	10	40	5.608	501×10^3	0.0058	"	39.08
			6.79	100 "	0.1182	"	43.73
3	9.8	40	5.665	200×10^3	0.00335	"	40.11
			6.525	105 "	0.0819	"	43.57

By the method of least square, the material constants in the Paris' equation for the initiation period are C = 0.8123×10^{-30} and m = 15.08, so that the Paris' equation for the initiation is:

$$da/dN = 0.8123\times10^{-30}\ (\Delta K)^{15.08} \tag{1}$$

In the equation (1)

$$\Delta K = YS\sqrt{\pi a} \tag{2}$$

where Y = shape factor of the specimen,
S = amplitude of the applied stress.

By substituting (2) into (1), we have

$$da/dN = CS^m \pi^{m/2} a^{m/2}$$

by integrating and using the condition: a = a_o when N = 0, we obtain the formula for determining the initial size of the defect as follows:

$$\left(\frac{1}{a_o}\right)^{\frac{m-2}{2}} - \left(\frac{1}{a}\right)^{\frac{m-2}{2}} = \frac{m-2}{2}\ C^{\frac{m}{2}}\ S^m\ N = \frac{m-2}{2}\ CM^m\ N$$

where

$$M = S\sqrt{\pi} \tag{3}$$

Then we have

$$\frac{1}{a_o} = \left(\frac{m-2}{2}\ CM^m\ N\right)^{\frac{2}{m-2}} \tag{4}$$

For determining a_o, we used seven pieces of smooth specimens each with area of cross-section of about 100 mm^2.

TABLE 2 Equivalent sizes of initial defects

Specimen	σ_{max}	σ_{min}	$\Delta\sigma$	M	N	a_o
1	53.05	5.305	47.7	84.546	2.25×10^5	0.1638
2	51.08	5.108	45.97	81.48	2.44 "	0.1762
3	47.7	4.770	42.93	76.09	1.14×10^6	0.1631
4	43.48	4.348	39.13	69.356	1.35 "	0.1967
5	45.10	4.510	40.59	71.944	1.26 "	0.1826
6	41.74	4.174	37.57	66.59	$> 10^7$	< 0.159
7	42.7	4.27	38.43	68.115	"	"

From this table it is obvious that the sizes of initial defects in the seven pieces of smooth specimens are between 0.15 and 0.2 mm.

THRESHOLD, FATIGUE LIMIT AND S-N CURVES

The threshold value (ΔK_{th}) really depends upon the required number of cycles N_i before initiation occurs. If the size of initial defects a_o is known we can find the value of M from equation (4)

$$M = \left(\frac{2}{(m-2)CN_i}\right)^{\frac{1}{m}} \left(\frac{1}{a_o}\right)^{\frac{m-2}{2m}} \tag{5}$$

Then

$$M = \left(\frac{10^{30}}{5.31\ N_i}\right)^{\frac{1}{15.08}} \left(\frac{1}{a_o}\right)^{\frac{6.54}{15.08}} \tag{6}$$

We can define the fatigue limit S_{-1} as

$$S_{-1} = \frac{M}{\sqrt{\pi}} \tag{7}$$

Table 3 shows the different threshold values (ΔK_{th}) and S_{-1} defined by using specimens with different number of cycles N_i to initiation.

TABLE 3 Threshold values and fatigue limits

a_o mm	N_i cycle	M	(ΔK_{th}) kgf/mm$^{3/2}$	S_{-1} kgf/mm^2
0.15	10^9	50.33	19.49	28.4
	10^{10}	43.20	16.73	24.37
	10^{11}	37.08	14.36	20.92
0.20	10^9	44.42	19.86	25.06
	10^{10}	38.13	17.05	21.50
	10^{11}	32.73	14.64	18.46

From (3) and (4) we have

$$S^m \cdot N = \frac{2}{(m-2)C}\left(\frac{1}{a_o}\right)^{\frac{m-2}{2}}\left(\frac{1}{\pi}\right)^{\frac{m}{2}}$$

or

$$m \lg S + \lg N = -\lg\left(\frac{m-2}{2}C\right) - \frac{m-2}{2}\lg a_o - \frac{m}{2}\lg \pi \qquad (8)$$

For the present example of 50Cr steel

$$15.08 \lg S = 25.5264 - 6.54 \lg a_o - \lg N$$

Figure 3 shows the S-N curve with $a_o = 0.15$ and 0.20, respectively, on log-log paper.

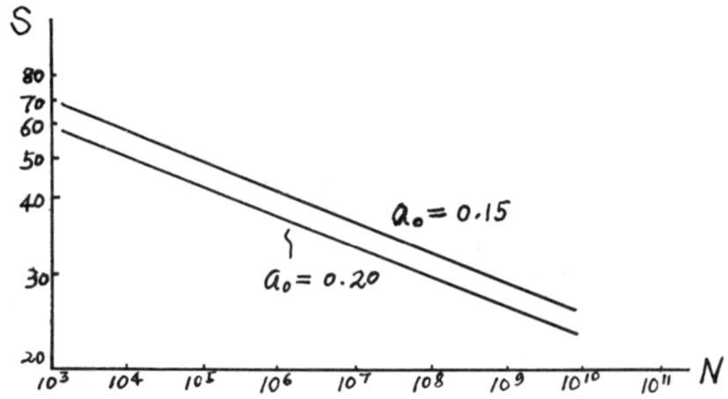

Fig. 3. S-N curve for a_o = 0.15 and 0.20 mm

CONCLUSIONS

1. The determination of the S-N curves (including the threshold value $(\Delta K)_{th}$ and the fatigue limit S_{-1}) needs only about 3 to 5 pieces of pre-cracked specimens with cross sectional area each about 400 mm^2 and in addition about 10 pieces of smooth specimens each with cross sectional area about 100 mm^2. The total number of cycles in the test amounts to less than 10^8 cycles. In comparison with the conventional method or P-S-N method, the saving of material and labour is more than 90%.

2. The determination of the material constants C and m at the propagation stage with $K = 2(\Delta K)_{th}$ ensures a resulting S-N curve on the safe side for engineering practice.

3. There exists the prospect of determination of the equivalent size of initial micro-defect a_o from the metallographic study of the material. Particularly the size and shape of the inclusions might be the dominating factors.

LITERATURE

May, R.A., A. Stuber and S.T. Rolfe. Effective Utilization of High-Yield Strength Steels in Fatigue, WRC. Bull. No. 243.

Wilson, A.D. (1979). The Influence of Inclusions on the Toughness and Fatigue Properties of A 516-70 Steel, J. Eng. Mat. & Tech., July 1979, Vol. 101, Trans. of ASME.

INFLUENCE OF CREEP VOID GROWTH ON MOTION OF A MACRO-CRACK

A. Neimitz

Technical University of Kielce, Al.Tysiaçlecia, 25-314 Kielce, Poland

ABSTRACT

The Dugdale model is used to describe the motion of the crack under constant loading and at high constant temperature. By postulating the model of the crack and the way of its motion it is possible to include the mechanisms of nucleation and growth of the creep voids into a macroscopic description of the crack motion. As a result the formula for the velocity of the trailing edge of the crack in secondary creep is obtained as a function of the external loads and macro- and microscopic parameters defining the properties of the metal.

KEYWORDS

Crack motion, creep void growth, nucleation, metal creep, Dugdale model.

MODEL OF THE CRACK AND MECHANISM OF ITS MOTION

In earlier papers [Neimitz, 1981, 1982, 1983] the model of the crack was proposed and the way of its motion under constant external loading and at high constant temperature was postulated. The Mode I of fracture was analysed in plane stress although the proposed model can be generalized to Mode II and III of fracture. In this paper the second period of the creep is discussed - the period of the stable motion of the crack. The jump-like motion of the trailing edge of the crack, which is an idealized model of the real situation is proposed (see Fig. 1). The model of the crack is simply a modified Dugdale model where the computational void, CV, (see Fig. 1) is inserted into the Dugdale zone henceforth referred to as the fracture zone, FZ. The CV changes its dimensions with time and models the nucleating and growing voids on the grain boundaries. The void nucleation and growth depends on the stress level, and thus it is assumed that in the FZ these processes are considerably intensified.

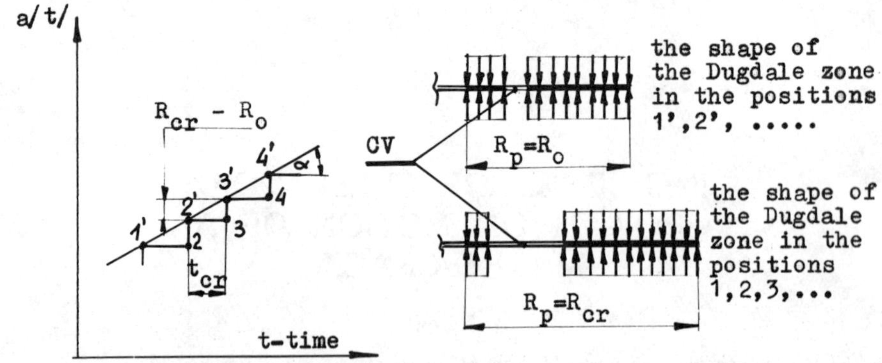

Fig. 1. Trajectory of the trailing edge of the crack and the shape of the FZ.

The growth of the total volume of all the voids in the FZ is modelled by the growth of the CV. Growing CV causes growth of the FZ length and opening displacement of the crack faces from the initial values R_o, δ_o to the critical R_{cr}, δ_{cr}. The motion (sudden jump) of the trailing edge of the crack starts when the opening of the crack faces and the FZ length reach critical values (points 2, 3, ... in Fig. 1). The jump is completed when the domain ahead of the trailing edge is in the similar condition (the number and the size of the voids) as it was in the points 1', 2', ... Thus according to Fig. 1 the mean crack-tip speed can be found from the formula:

$$\beta = (R_{cr} - R_o)/t_{cr} \qquad (1)$$

where R_{cr} is the critical length of the Dugdale zone that can be measured experimentally or defined theoretically, R_o is the FZ length at the points 1', 2', ... which is defined in the papers [Neimitz, 1982, 1983]; this value will not be used in the final formulas here, t_{cr} is the time interval between two jumps following each other; this magnitude needs a special procedure to define it, which is described in detail in the papers [Neimitz, 1982, 1983]. Only the main steps to get value of t_{cr} will be mentioned. The t_{cr} value is calculated from the assumption that the volume of the CV and the total volume of all the voids inside the FZ are equal to each other at the time t_{cr} The volume of the CV is calculated from the condition that the stress intensity factor is equal to zero at the moment t_{cr} (this condition leads to the critical diameter of the CV) postulating the height of the CV. The total volume of the growing voids on the grain boundaries is calculated by means of the void nucleation and growth laws. The sought equation has the following form:

$$2ww_m\left\{\left(R_{cr} - R_s(t_{cr})\right)R_{cr}\left(1 - P/P_{max}\right)^2 - 0.25R_{cr}^2\left(1 - P/P_{max}\right)^4\right\}^{1/2} = V_c \qquad (2)$$

where P is the applied external load, P_{max} is the critical load that applied to the specimen causes an abrupt fracture, w is the specimen width, w_m is the postulated height of the CV (it will be discussed later), R_s is the location of the CV center at the moment t_{cr}; this magnitude is found [Neimitz, 1982, 1983] by means of the void nucleation and growth laws, V_c is the total volume of all the voids inside the FZ and it can be found from the following expression:

$$V_c = V_o + \dot{N}\left\{\int_o^{t_k} J(\tau)d\tau - t_k J(t_o)\right\} \tag{3}$$

where

$$J(t) = \int \beta_v dt , \tag{4}$$

V_o is the total volume of all the voids inside the FZ of the minimal critical equilibrium diameter, $\dot{N}$ is the void nucleation rate,

$$t_k = t_{cr}\left(R_{cr}/(R_{cr} - R_o)\right) , \tag{5}$$

$$\beta_v = \beta_{vd} + \beta_{vp} , \tag{6}$$

β_{vd} is the volume growth rate of the individual void by the diffusional mechanism and β_{vp} is the volume growth rate of the individual void by plastic deformation of the surrounding material.

NUCLEATION AND GROWTH OF THE VOIDS

The variables $\dot{N}$, β_v, β_{vd}, β_{vp} were discussed in detail in the present author's papers [1982, 1983] where the following results were obtained:

$$\dot{N} = \frac{N_c}{t_k} = \frac{R_{cr} w w_z}{db^2}\,\overline{\alpha} \tag{7}$$

$$\beta_{vd} = \frac{2\pi\Omega D_B \delta}{kT}\,\frac{\sigma}{1.75} \tag{8}$$

$$\beta_{vp} = \pi b^2\left(\dot{\varepsilon} - \dot{\varepsilon}_{ss}\right)d \tag{9}$$ [1)]

These equations are valid for the shape of voids proposed by [Edward, Ashby, 1979] and

$$\beta_{vp} = 1.42\pi\dot{\varepsilon} r_c^3 \exp\left\{\left(1.06/h(\alpha)\right)t\dot{\varepsilon}\right\} \tag{10}$$

for voids of ellipsoidal shape in a visco-elastic material. The nomenclature used in the above formulas is the following: N_c is the number of the voids inside the FZ at the moment t_{cr}, w_z is the height of the FZ, d is the grain size, b is the distance between voids, $\overline{\alpha}$ is the numerical coefficient describing how the structure is deformed; $0 < \overline{\alpha} < 1$; and for the regular shape of the grains $\overline{\alpha} \cong 0.75$, Ω is the volume of the atom, D_B is the grain boundary self-diffusion coefficient, δ is the boundary thickness, k is the Boltzmann constant, T is the temperature, $\dot{\varepsilon}$ is the creep strain rate found from the Norton law, $\dot{\varepsilon}_{ss}$ is the strain rate according to the dislocation creep law, r_c is the critical value of the void radius when the surface of the void becomes free of stresses inside the FZ, $h(\alpha)$ is the shape function and σ is the stress at the grain boundary.

---

1) Eq. (9) is the corrected form of eq. (38) [Neimitz, 1982] where a mistake in calculation was made.

FINAL RESULTS, DISCUSSION

Inserting formulas 3 to 8 and 10 into eq. (2) the following equation can be obtained:

$$2\left\{\frac{(1-\lambda_o)}{t_{cr}}\frac{\left(\frac{1}{3}D_1\, t_{cr}^3\left(\frac{1}{1-\lambda_o}\right)^3+\frac{R_1}{R_2^2}\left\{t_{cr}\frac{1}{1-\lambda_o}\exp\left(R_2 t_{cr}\frac{1}{1-\lambda_o}\right)-\frac{1}{R_2}\left[\exp\left(R_2 t_{cr}\frac{1}{1-\lambda_o}\right)-1\right]\right\}\right)}{\left(\frac{1}{2}D_1\, t_{cr}^2\left(\frac{1}{1-\lambda_o}\right)^2+\frac{R_1}{R_2^2}\left[\exp\left(R_2 t_{cr}\frac{1}{1-\lambda_o}\right)-1\right]\right)}\cdot\right.$$

$$\left.\cdot\left(1-P/P_{max}\right)^2-0.25\left(1-P/P_{max}\right)^4\right\}^{1/2}=$$

$$=M\left\{\frac{1}{2}D_1\, t_{cr}\frac{1}{1-\lambda_o}+\frac{1}{t_{cr}}\left(1-\lambda_o\right)\frac{R_1}{R_2^2}\left[\exp\left(R_2 t_{cr}\frac{1}{1-\lambda_o}\right)-1\right]\right\} \tag{11}$$

Using eq. (9) instead of eq. (10) gives:

$$2\left\{\frac{(1-\lambda_o)}{t_{cr}}\frac{\left[\frac{1}{2}V_o t_{cr}\frac{1}{1-\lambda_o}+\frac{1}{3}\pi\Delta\dot{\varepsilon}\cdot b^2 d t_{cr}^2\left(\frac{1}{1-\lambda_o}\right)^2+\frac{1}{3}D_1\, t_k^2\right]}{\left[V_o+\frac{1}{2}\pi\Delta\dot{\varepsilon}\cdot b^2 d t_{cr}\left(\frac{1}{1-\lambda_o}\right)+\frac{1}{2}D_1\, t_k\right]}\left(1-P/P_{max}\right)^2-\right.$$

$$\left.-0.25\left(1-P/P_{max}\right)^4\right\}^{1/2}=M\left[V_o+\frac{1}{2}t_{cr}\frac{1}{1-\lambda_o}\left(D_1+\pi b^2\Delta\dot{\varepsilon}d\right)\right] \tag{12}$$

In the above equations $R_1 = 1.42\pi\dot{\varepsilon}r_c^3$; $R_2 = [1.06/h(\alpha)]\dot{\varepsilon}$; $\Delta\dot{\varepsilon} = \dot{\varepsilon} - \dot{\varepsilon}_{ss}$;

$D_1 = \frac{\pi\Omega D_B\delta\sigma}{kT1.75}$; $\lambda_o = \frac{R_o}{R_{cr}}$. The unit of the parameter M is $[1/m^3]$. M is hence a geometrical parameter. In the most general form it can be written:

$$M = \frac{w_z}{w_m}\frac{1}{db^2}\overline{\alpha} \tag{13}$$

The quotient w_z/w_m can be used as so-called "adjustable parameter" and can be established in experiments. This parameter can also be postulated on the basis of the rational reasoning, postulating the value w_m - "the model height of the FZ, e.g.

$$w_m = [\text{number of rows of voids} \times \text{concentration coefficient} \times$$
$$\times \text{height of the mean void inside the FZ}] = (w_z/d)\,\overline{\alpha}\, a_{mean} \tag{14}$$

According to the above assumption:

$$M = 1/a_{mean}b^2 \tag{15}$$

Equations 11 and 12 can in many cases be considerably simplified. A discussion is made for copper that has well known, easy accessible material constants. The strain rate is in many cases (for copper) of a very small magnitude (it was assumed that $\dot{\varepsilon}_{ss} = \dot{\varepsilon}$ in eq. (11) and $\dot{\varepsilon}$ was calculated for various temperature and stresses - the results are listed in Table 1).

TABLE 1 The values of $\dot{\varepsilon}$ [1/s] for various Y and T

Y [MN/m²] T [K]	10	40	60
$0.5T_m$	$0.8\ 10^{-10}$	$0.6\ 10^{-7}$	$0.1\ 10^{-5}$
$0.6T_m$	$0.75\ 10^{-7}$	$0.22\ 10^{-4}$	$0.16\ 10^{-3}$
$0.7T_m$	$0.21\ 10^{-5}$	$0.26\ 10^{-2}$	$0.10\ 10^{-1}$
$0.8T_m$	$0.55\ 10^{-4}$	$0.40\ 10^{-1}$	0.3

In Table 1 T_m is the melting point and Y is the stress greater than the yield stress. According to the primary calculations the values for t_{cr} under the most unfavorable conditions turn out to be of the order 10^2, the value R_2 is very small and therefore in eq. (11) the exponential terms can be developed in a Taylor series and all the terms of the smaller orders can be omitted. As a result eq. (11) is reduced to the form:

$$2\left\{\frac{\frac{1}{3}D_1\left(\frac{t_{cr}}{1-\lambda_o}\right)+\frac{R_1}{R_2}}{\frac{1}{2}D_1\left(\frac{t_{cr}}{1-\lambda_o}\right)+\frac{R_1}{R_2}}\left(1-P/P_{max}\right)^2-0.25\left(1-P/P_{max}\right)^4\right\}^{1/2} = M\left[\frac{1}{2}D_1\left(\frac{t_{cr}}{1-\lambda_o}\right)+\frac{R_1}{R_2}\right] \tag{16}$$

In the above formula, the quotient R_1/R_2 is the volume of the mean starting void inside the FZ. From microstructural observations it is known that the radius of the voids changes from 0.2 µm to 5 µm. For these values of the radii the quotient R_1/R_2 is much smaller than the term $D_1t_{cr}/(1-\lambda_o)$. The values of D_1 for copper are shown in Table 2.

TABLE 2 The values of D_1 [m³/s] at various Y and T

Y [MN/m²] T [K]	10	40	60
$0.5T_m$	$0.17\ 10^{-15}$	$0.68\ 10^{-15}$	$0.10\ 10^{-14}$
$0.6T_m$	$0.14\ 10^{-15}$	$0.56\ 10^{-15}$	$0.84\ 10^{-15}$
$0.8T_m$	$0.10\ 10^{-15}$	$0.40\ 10^{-15}$	$0.60\ 10^{-15}$

R_1/R_2 have the values as in Table 3.

TABLE 3 The volumes of the mean voids in the FZ

r_c [μm]	0.03	0.20	2	5	20
R_1/R_2 [m^3]	0.7 10^{-22}	0.2 10^{-19}	0.2 10^{-16}	0.67 10^{-16}	0.21 10^{-13}

A comparison of the values in Tables 5 and 6 enables us in many cases to omit the term R_1/R_2. Finally, eq. (11) can be further simplified to the form:

$$t_{cr} = \frac{4}{MD_1}(1-\lambda_o)\left[0.67(1 - P/P_{max})^2 - 0.25(1 - P/P_{max})^4\right]^{1/2} \qquad (17)$$

Inserting eq. (17) into eq. (1), the formula for crack-tip velocity in secondary creep can be written:

$$\beta = \frac{R_{cr}MD_1}{4\left[0.67(1 - P/P_{max})^2 - 0.25(1 - P/P_{max})^4\right]^{1/2}} \qquad (18)$$

Formula (18) can be plotted using nondimensional variables $\beta(4/R_{cr}MD_1) = f(P/P_{max})$.

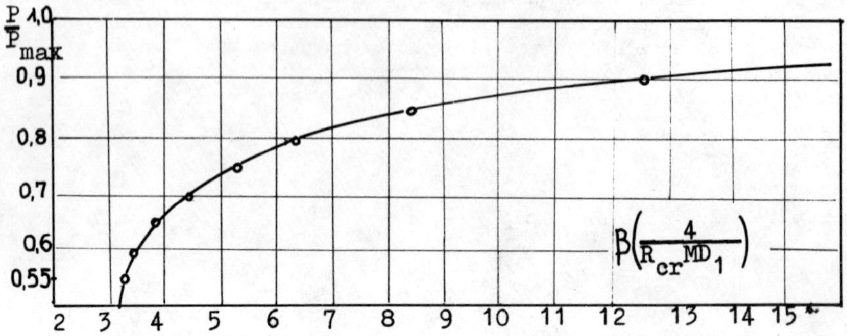

Fig. 2 The influence of the external load on the crack-tip velocity.

From eqs. (18) and (13) it can be seen that the crack-tip velocity is dependent essentially on the structural parameters. The distance between two voids has a very important influence on the crack-tip speed: $\beta \sim 1/b^2$. The influence of the shape and dimensions of the grains can be changed by the parameter w_z/w_m, which can be a function of the grain diameter (as was shown in eq. (15)). The uncontroversial influence of the void size will be visible for the large dimensions of the voids (Tables 2 and 3) or when w_z/w_m is postulated in a special way. The plastic properties of the metal at some particular temperatures are represented by R_{cr}. At very high temperatures (Table 1) eq. (11) cannot be simplified and the crack-tip velocity is a complex function of the structural parameters. For small r_c and for $T \leqslant 0.6T_m$ eq. (10) can also be simplified to the form of eq. (18).

REFERENCES

Edward, G.H. and Ashby, M.F. (1979). Acta Met., 27, 1505.
Neimitz, A. (1981). The mechanical-physical model of the crack in the creep condition, Symposium on Crack Formation and Propagation, IUTAM/CISM/PAS, Tuczno, Poland.
Neimitz, A. (1982). Motion of the crack under constant loading and at high constant temperature, Proceedings of the Second International Symposium on Defects, Fracture and Fatigue, Mont Gabriel, Canada.
Neimitz, A. (1983 in print). Crack motion in elastic-plastic bodies, Mechanika-Zesz. Nauk. Polit. Swięt. (in Polish)

MECHANICAL PROPERTIES OF SURFACE COATED MATERIALS

A. S. Argon

Massachusetts Institute of Technology, Cambridge, MA 02139, USA

ABSTRACT

The intrinsic mechanical properties of near-surface regions of homogeneous solids are briefly reviewed for both elastic brittle behavior and for plastic, ductile behavior. The many different ways in which surface coatings alter intrinsic behavior by suppressing certain critical processes, by triggering certain successions of auto-catalytic events, or by sealing the solid against the environment. The use of wear resistant coatings on cemented carbide cutting tools, and the procedure of chemical strengthening of glass articles by ion-exchange layers resulting in large residual compressive stresses are discussed in greater detail as two very successful examples of the use of surface coatings to dramatically improve mechanical performance.

KEYWORDS

Surface coatings, surface films, suppression of brittle fracture, suppression of fatigue fracture, wear resistant coatings, chemical strengthening.

INTRODUCTION

The surface of a solid has certain basic characteristics that are so familiar that they are often overlooked. Apart from being a discontinuity of mass and stiffness, a surface acts as the boundary through which the solid interacts with its surroundings. In most cases forces are transmitted into a body through its surface by contact with other objects. It has long been recognized that the pressures at such contact sites are so large as to produce local plastic flow or crushing. The former is the source of friction and the principal mechanism of wear, while the latter is responsible for the degradation of the strength of brittle solids as a result of mechanical handling.

The environment acts on the solid through its surfaces, producing localized chemical attack that even in the absence of stress can be in the form of pits. In the presence of stress or while the solid is plastically deforming, chemical attack can be very specific and can be the cause of severe embrittlement in the form of stress corrosion cracking or liquid metal embrittlement. Diffusive mass transfer can also occur across surfaces and can result in changed surface mechanical properties such as in carburization or decarburization of steels, surface plasticization of some glasses and polymers by absorbed water or solvents and the like.

Since surfaces offer little resistance to the deformation of the solid below, there are no compatibility restrictions at a surface to out-of-surface deformation, and many inhomogeneous modes of deformation of the near-surface material can readily develop. This has important consequences not only in the lowered levels of work hardening of the near surface material, but also in the development of shear bands and surface roughness in monotonic deformation, persistent slip bands in cyclic deformation, nucleation of fatigue cracks, and early ripening of intergranular creep damage, even in the absence of aggressive environments.

In addition to the different properties and sensitivities of surfaces and near-surface material of solids, many purely geometrical considerations in engineering components make surfaces the preferred sites for enhanced deformation and early fracture. Thus, in many instances, components are subjected to complex states of stress incorporating some bending or torsion where a strong stress gradient exists, and where the stress is highest on the surface. This is always the case in rotating shafts. Even in nominally uniformly stressed components the long range applied stress is highest at the surface due to the unavoidable presence of bending or twisting moments resulting from eccentricity. In addition, the stress level at surfaces is sensitive to surface topography. Many machining processes produce characteristic surface topographies that commonly have stress concentration factors of the order of 2 in tension. In normal machining operations, however, a thin plastically deformed surface layer usually results that is under a residual compressive stress.

It is well known in elasticity that heterogeneities including holes and cracks axi-symmetrically stressed in three dimensions in the interior of a solid, have lower interfacial stress concentrations than have corresponding discontinuities with the same cross sectional dimensions stressed in two dimensions in plane strain or plane stress near a surface. This results in higher stress concentrations for heterogeneities near a free surface than in the interior of a solid.

In view of these considerations there have been many attempts to coat surfaces with protective layers of higher strength, greater plastic resistance, improve resistance against environmental attack, or against dissolution in a surrounding medium. The mechanical effects of such coatings are often complex and out of proportion of their volume fraction. In the following sections we will review first the intrinsic mechanical properties of near-surface material, and then consider in some detail the mechanical effects of a number of different real or virtual coatings and films. We will follow this with two selected case studies of the successful use of surface coatings or layers to substantially improve the mechanical properties of parts.

MECHANICAL PROPERTIES OF NEAR-SURFACE MATERIAL

Fracture of Brittle Solids

Fracture in brittle solids is exclusively initiated from crack-like defects. In some instances such defects are produced upon the fracturing or separation of an internal weak heterogeneity or interface in a stressed solid, or upon generation of large local stress concentrations that result from the formation of intense shear bands along interfaces. In the majority of instances, however, the critical cracks occur on the surface due to environmental attack, due to mechanical contact damage produced by hard objects of high curvature, or due to machining processes. Thus, in oxide glasses strength limiting cracks are almost always on the free surfaces (Ernsberger, 1980). In glassy polymers, crazes that act as forerunners of cracks are nucleated from surface imperfections (Argon and Hanoosh 1977). In most ceramics the majority of the most dangerous flaws are the result of mechanical damage and machining (Evans, 1980).

The condition for the extension of such cracks in stressed brittle solids into final fracture have been stated by Griffith (1920, 1924) many years ago and have been verified in numerous studies since that time. Although the removal of such flaws by chemical means is possible (Ernsberger 1980), the use of real or virtual coatings to either prevent the formation of flaws, or at least to prevent their growth, is almost self evident and in the long run far more effective. We will discuss the methods of such strengthening by surface coatings in the next major section.

Plastic Flow in the Vicinity of Surfaces

Plastic flow in a crystalline solid is initiated when a critical magnitude of a shear stress is present on slip planes to overcome either the lattice resistance to the motion of dislocations or the collective action of the discrete slip plane obstacles pinning the dislocations. As in the case of brittle fracture the stress concentrations due to surface roughness or the accidental presence of higher stresses near the surface due to bending, etc., will result in the initiation of plastic flow near a free surface. In a polycrystal where the yielding of a large assembly of grains requires overcoming inter-crystalline constraints of geometrical nature, the absence of such constraints at the surface grains permits slip activity to begin there earlier. On a smaller scale still, line energies of dislocations near free surfaces are lower than those of dislocation in the interior of a solid so that they can be curved more acutely under the same level of applied stress and can overcome obstacles readily very near the surface (Fisher, 1952). In addition to the relatively easier generation of dislocations from a surface, and the easier overcoming of slip plane obstacles, the ready escape of dislocations through the surfaces that arrive there from the interior result in reduced statistical encounters of dislocations. This in turn results in less strain hardening and a positive gradient of plastic resistance from the surface into the interior. In single crystals this layer of reduced plastic resistance is relatively far reaching (Fourie, 1968, Mughrabi, 1970), having a mean penetration length equal to that of slip lines. In polycrystals the effect is probably localized to a dimension equal to either that of a slip line or a grain--whichever is smaller. Since the lattice resistance to the motion of dislocations in transition metals and inorganic compounds is strongly affected by the extension of the core of dislocations (Peierls 1940, Nabarro 1947), which in turn is directly linked to the ideal shear strength of a solid, any effect that reduces the latter increases very significantly the mobility of dislocations. There are many examples (Nabarro, 1977) of such increased mobility of surface dislocations in the presence of a cohesion impairing environment or agency. The above effects, which have been overly dramatized in the past, can only be important in parts with a large surface to volume ratio and are not likely to be of great importance in the uniform deformation of parts of substantial dimensions.

In many forms of machining of very hard solids where plastic flow is quite superficial, however, such effects can be of very considerable importance (Westwood and Mills, 1977; Westwood 1983).

The localization of deformation into necks in bars or into strips under monotonic uniaxial or multiaxial tension when the local strain hardening rate reaches some low level are well known mechanical instabilities (Argon, 1973). It is less well known that when the strain hardening rate becomes smaller than a certain value, perturbations to the flatness of the surface of a plastically deforming polycrystal can begin to grown into increasing roughness with a characteristic wavelength. Such surface instabilities have been investigated theoretically by Hutchinson and Tvergaard (1980), who have found that on the basis of continuum plasticity theory that surface roughening results primarily in response to corners that develop on yield surfaces in the direction of straining. Their analysis, which does not consider the stabilizing effect of surface energy, predicts that such instabilities should set in when the principal logarithmic strain ε in the surface reaches a magnitude

$$\varepsilon\ (1-\exp(-2\varepsilon)) = n \qquad (1)$$

where $n(<1)$ is the exponent on the plastic strain in the parabolic strain hardening law. Some experimental evidence (Onyewuenyi and Hirth, 1982; Hirth, 1983) exists that surface free energy does indeed have a stabilizing influence and is an important factor in these surface roughening instabilities in that its reduction predictably hastens the onset of roughening. Clearly, the discrete nature of crystal plasticity inside individual grains acts as an initial perturbation to trigger the roughening process. In cyclic deformation the very important role of the Bauschinger effect as a destabilizing influence to trigger surface roughening in structureless media was pointed out earlier by McClintock (1963) who also demonstrated the effect of surface roughening, dramatically in cyclically twisted and untwisted bars of silicone putty.

Clearly such instabilities producing surface roughening are no mere laboratory curiosities. They are linked to the hastening of ductile fracture inside shear bands (Hahn and Rosenfield, 1975) in plane strain deformation. They are, no doubt, also the principal cause of the formation of persistent slip bands in cyclically strained plastic solids where the geometrical constraints of crystal plasticity merely provide regularity of the features.

Grain boundaries in crystalline solids are planar conduits of enhanced matter transport, and by virtue of their random atomic structure, they are regions where the solubility of many trace components is much higher than in the perfect lattice. This results in enhanced chemical activity and embrittling phenomena. The regions of such particular activity are where grain boundaries intersect with free surfaces. Phenomena such as acute liquid metal embrittlement (Stoloff, 1983), stress corrosion cracking (Parkins, 1983), gas assisted intergranular cavitation in creeping alloys (Bricknell and Woodford, 1981) are all of this type. Even in a vacuum intergranular creep cavitation by decohesion of particles along grain boundaries at high temperature is initiated from the free surface (Chen and Argon, 1981), where trace amounts of residual oxygen could have played a role, but where most probably the reduced intercrystalline constraint to grain boundary sliding as the chief source

that fatigue crack initiation occurs and preferentially at grain boundaries by a grooving process resulting from combined sliding and migration of grain boundaries is aided by residual levels of oxygen in the system. In a real hard vacuum ($p < 10^{-10}$ torr) fatigue cracking at elevated temperatures once again reverts to a persistent slip band formation process inside individual grains (Coffin, 1972).

EFFECTS OF SURFACE COATINGS ON THE MECHANICAL BEHAVIOR

General Considerations

It is clear that if the effect of surface coatings were non-interactive with their substrate, their enhanced hardness or other beneficial property would be only in proportion to the volume fraction of the surface layer in the solid. This, in nearly all cases of thin surface coatings, would produce very small and non-dramatic effects. Generally, however, surface coatings are attractive because they are strongly interactive with their substrate and produce very important effects. Clearly, dramatic effects are only possible if the coating suppressed certain critical catastrophic events that initiate from surfaces; triggered certain successions of auto-catalytic events that changed the nature of routine processes; or sealed the interior to the environment that would otherwise adversely affect the mechanical behavior in monotonic or cyclic deformation. In principle, the interactions that we will enumerate below are those that alter significantly the several surface related mechanical phenomena that we presented in the preceding section.

Suppression of Brittle Fracture by Surface Compression

That the normal weakness of glass could be dramatically alleviated was demonstrated in the 17th century with the famous Prince Rupert drops produced by quenching molten droplets of glass into water. That the resulting strenghening effect is due to the introduction of large surface compressive stresses that counteract the weakness in tension had already been suggested by Robert Hooke in 1664 (Cooper, 1982). In spite of this ancient awareness, the large scale production of tempered glass is relatively recent and is based on the development of the air jet method of quenching glass plate (Cooper, 1982). In the usual tempering process a parabolic self equilibrated distribution of residual stress is set up across the glass that is compressive at the surface. These compressive stresses in the surface layers play a dual role of prestressing the existing surface cracks shut, and preventing new ones from forming by making the surface more resistant to mechanical damage. Clearly in this case no specific coating is developed on the glass that has different physical, chemical, or mechanical characteristics. Therefore, we term such surface layers under compressive stress that result from thermal tempering virtual coatings, which must be distinguished from true surface coatings that we are about to discuss. Similar procedures of thermal tempering have been implemented on polymers with similar results (So and Broutman, 1976).

In thermal tempering the entire glass aricle is subjected to a transient thermal cycle. The resulting stress distribution parallels the parabolic temperature distribution during cooling. The parabolic distribution of stress results in significant levels of tensile stress in the center plane of the article, which is undesirable. A more successful procedure that results in steeper stress gradients, i.e., both in higher levels of compressive stress in much thinner surface layers and therefore lower levels of tensile stress in the interior, relies on the production of volume expansion in thin surface layers. This so-called chemical strengthening is achieved by either an ion exchange process where in the surface layer some of the regular network modifying ions in the glass are exchanged with ions having larger atomic radius, or by a similar ion exchange process that results in a less stable glass that undergoes at least partial crystallization, with an accompanying volumetric expansion. In this case a true surface layer or coating is produced that has distinct physical, chemical, and mechanical characteristics, and as a result of these differences is under a high level of compressive stress in the altered surface layer (Bartholomew and Garfinkel, 1980). A complementary but less permanent procedure has been used to prevent the weakening by mechanical damage of freshly made glass surfaces. Such surfaces are exceptionally clean and have very high coefficients of friction in contacts with other hard inorganic objects where surface microcracking results. To prevent this it is now relatively routine practice to coat the surfaces of glass articles with chemically neutral lubricants that radically lower the coefficient of friction and thereby prevent the formation of surface damage (Scholes, 1971). This practice is now used regularly on high strength optical fiber before winding and handling.

Suppression of Fatigue Crack Initiation by Surface Compression

Although all phases of fatigue crack initiation from persistent slip bands are not yet fully understood (Mughrabi, 1979), it is well known that fatigue crack initiation is supressed on surfaces which are under large residual compressive stresses that prevent them from ever being subjected to tension during cycling. Thus, the presence of surface compressive stresses is usually beneficial in extending fatigue life in metallic components. To achieve this, shot peening has been used on metallic surfaces in which repeated and completely overlapping plastic indentations of the surface by quasi-spherical metallic shot produce a compressive residual stress that reaches its peak value somewhat underneath the surface. A combination of some case hardening followed by shot peening often gives the best results, where the beneficial effects of surface plastic flow by micro indentations reinforces the compressive residual stresses that are produced in the surface layer by the volumetric expansion accompanying martensitic transformations that produce the principal hardening effect (Wohlfahrt, 1982).

Surface Coatings Enhancing Work Hardening in Substrates

As stated above a hard surface film not interacting with its substrate synergistically contributes to the overall plastic resistance of the assembly only in proportion to its volume fraction that in many instances of very thin films and coatings should be negligible (Argon, 1977). There are, however, many documented situations where thin surface films have relatively very strong effects on the plastic resistance of the assembly (Kramer and Demer 1961, Nabarro 1977). These can only be the result of synergistic interaction of the film with the substrate, forcing the latter to behave in a qualitatively different manner. Although some of these effects have been reported to be present upon initial deformation, most of them are a result of enhanced strain hardening of the substrate material enforced by the surface coating.

When the hard surface layer is epitaxial with the substrate and has a lattice parameter substantially different from that of the substrate, coherency can not be maintained and the otherwise very large stresses that would be produced in the surface layer are relieved by a network of misfit dislocations in the interface between the two regions, that was discussed in great detail by Frank and van der Merve (1949). Such dislocations that are geometrically necessary can glide only in the interface, and displacing them even virtually should require relatively large stresses as that would substantially raise the energy of the system. In addition, these dislocations may in many cases have also impaired mobility in the interface. Although the line energies of such dislocations will be small if they are near to the surface, their spacing will also be quite small. Hence they can act as a very strong barrier to lattice dislocations of the substrate that must exit the deforming solid. When substrate dislocations are trapped by the interface, they will initiate additional slip activity on other slip systems of the substrate and catalyze the strain hardening process in the substrate in the vicinity of the surface layer. This can have far reaching effects and to a first approximation result in an elevated plastic resistance in a layer that is thicker than the actual surface layer by the ratio of the intrinsic initial plastic resistance of the surface layer to that of the substrate (Argon, 1977). The resulting total rise in overall plastic resistance can be typically anywhere between 30 – 100%, when compared with the plastic resistance of the unrestrained substrate (Argon, 1977). Although this is the predominant effect of surface layers where a substantial rise in plastic resistance of the assembly results, in at least one case surface layers have actually been known to lower the plastic resistance. Sethi and Gibala (1977) have reported that in their niobium single crystals at low temperatures when the easily mobile edge dislocations can be continually generated at an oxide interface, producing substantial increments of easy strain, hardening by the far less mobile screw dislocations is retarded. When the oxide film is removed, easy generation of edge dislocations is no longer possible and the hardening reverts to that which is observable in bulk.

When the surface layers are not epitaxial but are the result of any other mechanism such as superficial hardening, ion implantation, a much smaller surface grain size, etc., the processes of catalyzed accentuated hardening will still be present due to the jump in inelastic strain across the interface.

Hard surface layers in cyclic deformation below a critical amplitude that fractures the surface layer can be beneficial in suppressing the ubiquitous crystallographic or non-crystallographic surface roughening processes that are known to be the principal mechanism of fatigue crack initiation (Argon, 1972). In a most dramatic example of this, Alden and Backofen (1961) have demonstrated with aluminum single crystals that anodic Al_2O_3 films of about 1,000 Å thickness prevented the formation of persistent slip bands, and if cycled at strain amplitudes less than 0.2%, remained uncracked. In this manner the fatigue life of some crystals could be increased by a factor of 2. When the anodic film was broken in the cyclically hardened crystals or removed, however, persistent slip bands formed very rapidly and led to rapid fatigue fracture. Similar but less dramatic effects are known in steel machine elements with carburized surface coatings. When such elements undergo fatigue fracture and the fracture surfaces are examined, it is often found that the sources of the fatigue cracks are underneath the carbide coating and are due to some non-typical microstructural flaw.

Surface Coatings as Barriers against Environments

Most technologically important environments encountered in energy systems result in thermal oxidation, or they form other detrimental corrosion products that in most cases interact strongly with the mechanical behavior of the part, and more or less severely impair the performance by hastening a fracture process or "catalytically" accelerating fatigue damage as in corrosion fatigue. Parts of gas turbines operating at elevated temperatures in hot combustion products, parts of central station boilers, structural parts in fluidized beds of combustion, and high speed cutting tools for machining operations are some of the more prominent examples. Thus, if the performance of the part in the given thermal and loading schedule is otherwise acceptable, it is attractive to use barriers to aggressive environments by means of surface coatings.

The most successful oxidation preventing coatings for high temperature service of structural parts are those that develop naturally from the preferential oxidation of a key alloying element near the surface that then forms a diffusion barrier against further oxidation. The alloying of chromium into stainless steel or aluminum into the surface layer of nickel chromium superalloys are specific examples. In the latter case aluminum is diffused into the surface layers of super-alloys, which converts these layers into an intermetallic compound in the form of a metal-aluminide. Underneath this layer a reaction diffusion zone forms. Upon oxidation of the metal aluminide, a tenacious aluminum oxide film forms on the surface of the aluminide that then acts as the primary diffusion barrier to protect against further oxidation and against thermal fatigue processes usually accelerated by oxidation. Even if part of the oxide film is damaged or spalled by thermal fatigue cycling, the underlying aluminide oxidizes further to repair the protective film (Grisaffe, 1972).

In high speed machining operations of steel, the temperatures reached on the surfaces of cutting tools are high enough to break down tungston carbide, which then partially dissolves in the steel machining chips. The technology of metal cutting has benefited greatly in the last decade from the development of special carbide or oxide coatings on the tool surfaces acting as very effective diffusion barriers and radically changing the local tool wear mechanism. We will discuss some applications of this in the next section in considerable detail as a special case study of the use of surface coatings.

TWO EXAMPLES OF SURFACE COATINGS TO ENHANCE MECHANICAL BEHAVIOR

Surface Coatings of Cutting Tools to Inhibit Wear

Normal requirements of efficient machining demand the highest cutting speeds consistent with manufacturing controls and tool replacement procedures. As a result, metal cutting tools are routinely pushed to their limits of performance during manufacturing processes. Under these conditions, tool life is almost always governed by high temperature wear processes of the tool material by the part and the chips that are removed from it by the machining process.

Figure 1 shows an idealized chip formation process in an orthogonal machining operation by the advance of a cutting tool parallel to the surface at a certain depth of cut. Figure 1a shows the tool in profile advancing upward into a part and removing a chip toward the right, while Fig. 1b shows the rake face of the tool face-on. The chief wear process occurs in the region of plastic contact (AB) on the rake face of the tool where the highest temperatures generated by the rubbing of the chip on the tool face produces wear in the form of a crater as shown. A second area of intense wear is at the cutting flank, shown as the line A-A in Fig. 1b. The total force on a tool and its direction during cutting are generally well known, together with some knowledge about the pressure distributions along the faces and edges, and the temperature distribution on the rake face and in the chip (Boothroyd, 1963).

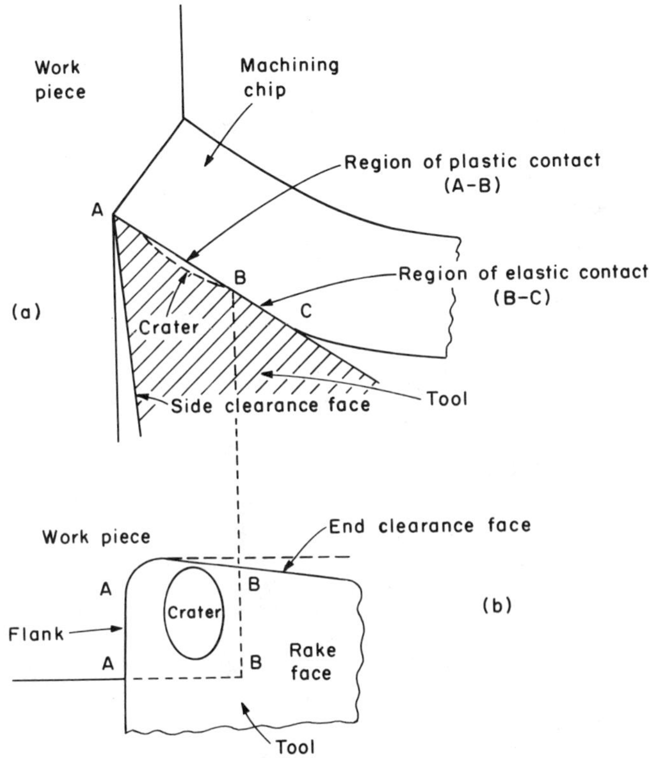

Fig. 1 Schematic representation of a cutting tool performing orthogonal machining, a) plan view, b) top view. The flank edge and the crater region of the rake face are two sites of rapid tool wear. (After Smidt et al (1982), courtesy of Naval Research Laboratory)

Empirical correlations have established that the wear rates of two rubbing solids is proportional to the square of the reciprocal ratio of the two hardnesses. On this basis it is desirable to choose the material with the highest hardness in the operating temperature range for the tool. Figure 2 shows the temperature dependence of the hardnesses of a number of prominant carbides. Clearly, in the temperature range of 300 - 900°C, WC has the highest hardness at a level of nearly 18 GPa (=1800 Kg/mm^2), that even at 900°C is above 11 GPa. Other carbides perform well also, but have lower high temperature hardnesses. A simple analysis of Suh (1977), considering a cutting tool as a Prandtl indenter, fully surrounded by the material to be cut, indicates that the cutting tool must at least be 4.5 times harder than the piece being machined to escape plastic distortion during the cutting operation. This indicates that WC tools should be able to machine parts with hardnesses up to 2.5 GPa at a temperature of 900°C. This gives a comfortable margin as it is above the hardnesses of most steels at that temperature. In spite of this, however, WC tools wear rapidly at such elevated temperatures because WC is not a very stable carbide and breaks down, losing its carbon to the part being machined by going into solution (Suh, 1977). In fact, crater wear rate measurements of Cook and Nayak (1969) of cemented WC and WC-TiC (both using Co as binder) tools in the cutting of AISI 1045 steel have established that these rates parallel those of the rate of diffusion of carbon through either α iron or WC, below and above the α-γ transition in iron respectively, as shown in Fig. 3. Thus, the wear mechanism of cemented carbide tools is not primarily one of adhesive or abrasive wear (with or without delamination of flakes or wear particles), where the rate of plastic deformation of the tool material governs the wear process, but one of dissolution, where loss of the hardening constituent by diffusion is followed by "wiping off" the surface layer of much reduced plastic resistance. If the constitutional degradation of the cemented carbide by leaching of the carbon could be prevented, the high hardness of the carbides should assure acceptably low rates of wear of the tool.

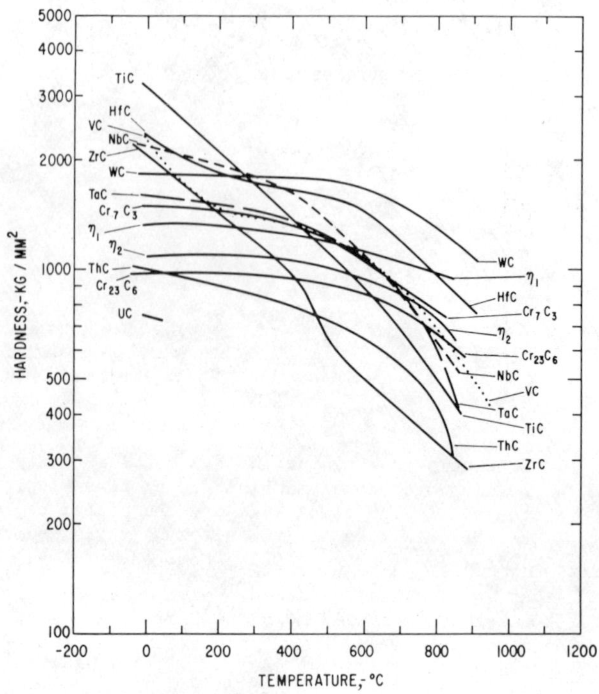

Fig. 2 Temperature dependence of microhardness of carbides (from Westbrook and Stover (1967), courtesy of J. Wiley & Sons)

This requirement has stimulated the development of one of the most dramatic improvements in tool life over the past decade by the introduction of diffusion-barrier coatings stable against dissolution into the machining chip or into the substrate cemented carbides.

Kramer and Suh (1980) have pointed out that the ideal coating material for cemented carbides is one that combines: a)high temperature stability, i.e.

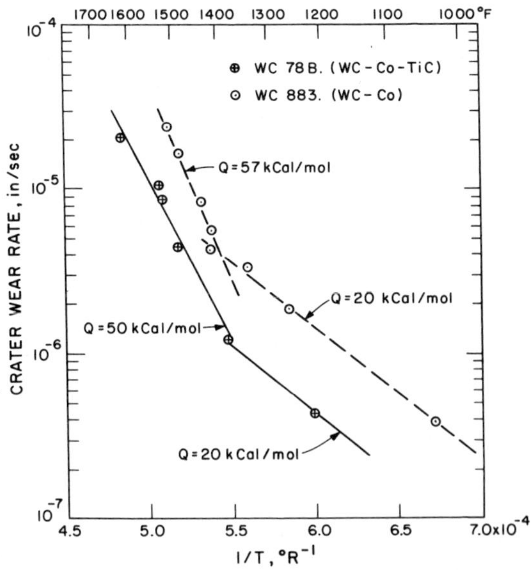

Fig. 3 Crater wear rates of two cemented tungsten carbides as a function of reciprocal temperature. The low temperature process parallels the diffusion of carbon in the alpha iron lattice, while the high temperature process parallels the diffusion of carbon in the tungsten carbide (from Cook and Nayak (1969), courtesy of AFML)

has the highest binding energy (lowest free energy at formation), and b)the lowest solubility in the chip material. In addition the coating must adhere strongly to the substrate cemented carbides over the entire operating temperature range (i.e. have good "wetting" characteristics, and similar coefficients of expansion so as not to generate large thermal stresses), and must have adequate high temperature hardness of its own. Figure 4 shows the free energy of formation of a family of carbides and oxides as a function of temperature that are candidates for usage as surface films on cemented carbides. The hardnesses of these carbides as a function of temperature is shown in Fig. 2, while those of some oxides are given in Table I. The equilibrium solubilities of a number of prominent candidate carbides and oxides in γ-Fe at 1600°K were estimated by Kramer and Suh (1980) on the basis of Henry's Law and are listed in Table II together with the energies of formation of these materials. The table also shows some experimentally measured solubilities of these materials at 1600°K, which are in reasonable agreement with the estimates. Thus, considering the crater wear rate of coated cemented carbide tool faces to be proportional to the solubility of the coating material in the tool chip material (γ-Fe) and to the relative velocity of rubbing during the cut, Kramer and Suh (1980) have estimated wear rates of several metal carbide coating materials in the 1300 - 1600K temperature range. These are shown in Table III in units of the wear rate of HfC, the most superior of the carbide coating materials on the basis of its resistance of going into sol-

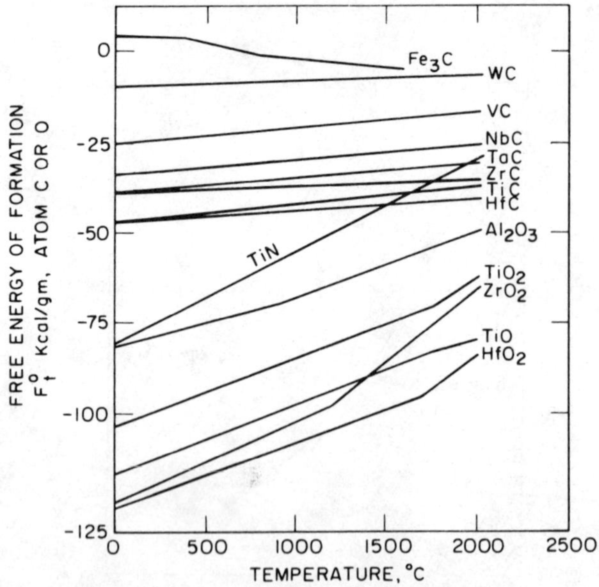

Fig. 4 Energies of formation of some prominent carbides and oxides as a function of temperature (from Suh (1980), courtesy of Elsevier S.A.)

ution in gamma iron.

Examination of the solubility results and the energies of formation of other potential coating materials indicates that oxides, and particularly Al_2O_3 and Ti_2O_3, should be far superior to HfC if the wear rate were governed purely by dissolution. Experience, however, indicates that this potential of the oxides

TABLE I. The Temperature Dependence of the Microhardness of some Oxides

Oxide Type	Hardness H_V (kgf mm^{-2})		
	400°C	600°C	800°C
TiO	1300	1000	650
FeO	350	210	50
MgO	320	220	130
NiO	200	140	100
MnO	120	60	45
Al_2O_3	1300	1000	650
SiO_2	700	2100	300
ZrO_2	650	400	350
TiO_2	380	250	160
$MgAl_2O_4$	1250	1200	1050
$ZrSiO_4$	400	290	140

Data from Suh (1980)

is not always realized. The reasons for this are several. As Table I and Fig. 2 show, the high temperature hardnesses of oxides are lower in comparison to those of carbides, so that conventional mechanisms of adhesive or abrasive wear remove the coating. In many instances it is also found that oxide coatings show poor adhesion to cemented carbides and quite often crack-off or flake-off readily from the substrate. This is consistent with the well known poor adhesion of metal oxides to metal matrixes that are not a constituent of the oxide. When the latter effects are successfully counteracted, however, that the oxide coatings can be more effective in reducing wear rate than carbide coatings has been demonstrated by Hale and Graham (1981) in connection with the machining of AISI 4340 steel utilizing Al_2O_3 coated cemented carbide tools.

TABLE II. Free Energies of Formation, Estimated, and Measured Equilibrium Solubilities of Some Tool Materials in Gamma Iron at 1600 K

Potential tool material	Free energy of formation (cal mol^{-1})	Estimated equilibrium concentration (solubility)	Experimental results extrapolated to 1600 K
ZrO_2	-190 300	3.60×10^{-8}	
Al_2O_3	-278 300	5.55×10^{-7}	
Ti_2O_3	-260 800	8.22×10^{-7}	
TiO_2	-156 300	1.52×10^{-6}	
TiO	- 91 020	1.40×10^{-5}	
HfN	- 52 604	1.53×10^{-4}	
HfC	- 49 122	1.97×10^{-4}	
ZrN	- 51 356	2.93×10^{-4}	
$TiC_{0.75}O_{0.25}$	- 52 395	5.42×10^{-4}	
ZrC	- 42 714	8.42×10^{-4}	
TiN	- 45 150	1.04×10^{-3}	
TaC	- 34 604	1.41×10^{-3}	2.1×10^{-3}
TiC (iron)	- 39 520	1.86×10^{-3}	6.1×10^{-3}
NbC	- 32 236	2.01×10^{-3}	6.8×10^{-3}
BN(graphitic)	- 26 100	9.65×10^{-3}	
WC	- 8 144	-	2.6×10^{-2}
VC	- 23 416	-	3.2×10^{-2}
TiC(nickel)	- 39 520	2.24×10^{-2}	6.3×10^{-2}
Diamond	-	9.30×10^{-2}	
Si_3N_4	- 51 850	9.50×10^{-2}	
β-SiC	- 14 548	4.30×10^{-1}	

Data from Kramer and Suh (1980)

A surprising aspect of the effectiveness of coatings in controlling crater wear rates is that these rates continue to remain relatively low even when the coating in the crater region of the rake face has worn through. Figure 5 shows that the crater wear resistance (reciprocal of the rate of wear) after the coating has worn through is still linearly proportional to the initial thickness of the coating, -- or the rate of material removed from the crater is inversely proportional to the coating thickness. This indeed should be the case if the wear crater enlarges geometrically similar to itself and if the overall wear rate is still dominated by the rate of wear of the tapered coating material framing the hole of the crater. A similar effectiveness of the coating material in reducing the rate of wear of the cutting flank is also found and can be explained similarly (Hale and Graham, 1981).

TABLE III. Theoretically Predicted Normalized Wear Rates (relative to HfC) based on Equilibrium Solubilities of Carbides in Gamma Iron. (Values in parentheses based on measured solubilities)

Carbide	Relative wear rate				
	1600 K	1500 K	1400 K	1300 K	Test results
HfC	1	1	1	1	1
TiC	7.65	8.82	10.6	12.8	2.75
$TiC_{0.75}O_{0.25}$	2.26	2.41	2.61	2.86	--
ZrC	4.44	4.87	5.47	6.20	5.51
TaC	6.33	7.19	8.39	9.98	10.6
	(9.43)	(10.7)	(12.5)	(14.9)	
NbC	9.13	10.6	12.7	15.6	23.7
	(31.0)	(36.0)	(43.1)	(52.8)	
WC	(107.0)	(153.0)	(215.0)	(332.0)	237.0

(from Kramer and Suh (1980)).

A final and surprising feature is that abrasive wear due to the hard phase particles in steels that are being machined is quite small in the wear of the coatings protecting the rake face. Most steels contain a multitude of carbides nitrides, and other intermetallic compounds making up their microstructures. When the steels are machined, these hard particles move past the crater region of the rake face and could result in abrasive wear of the coating. That this is not an important contribution to the crater wear results from the fact that most microstructural particles in steels that are incorporated to provide a source of hardness are on a micron or submicron scale and are often more or less equi-axed. Under the usual contact pressures of machining that are governed by the thickness and the plastic resistance of the machining chip, the ratio of the contact diameter of the particle on the coating to the particle size is most likely less than 0.1. Under these conditions for such small particle sizes the hard particles will at most produce shallow plastic indentations, but will not remove material and the abrasive wear rate should be negligibly low (Suh, Sin and Saka, 1980).

Chemical Strengthening of Glass Surfaces by Ion Exchange Layers

As discussed earlier above, chemical strenghening of glass surfaces by ion exchange to form thin surface layers under a high bi-axial compressive stress is very effective, both in inactivating pre-existing surface cracks by compressing them shut, and by making the surfaces far more resistant against the introduction of new mechanical damage resulting from contact with hard objects with small radius of curvature. In the so-called "stuffing" method this surface compression is achieved by exchanging a relatively small alkali ion for a larger one. Depending upon the transient nature of the ion exchange process, usually performed well below the glass transition temperature T_g of the glass, to prevent overall distortions of the glass article, the thickness of the ion exchanged surface layer and the gradient of composition into the surface can be controlled. Thus, in a typical case of a two dimensional large plate of thickness t, an ionic concentration C(x) of exchanged ions symmetricall distributed across the thickness will result in a bi-axial surface stress σ_{zz} that can be obtained readily by elementary techniques and is

$$\sigma_{zz}(x) = \frac{E\theta}{3(1-\nu)} \left[- C(x) + \frac{1}{t} \int_{-t/2}^{t/2} C(x)\, dx. \right] \tag{2}$$

where E and ν are the Young's modulus and Poisson's ratio perspectively, $\theta = (1/V)(dV/dC)$, the local specific dilatation of ion exchange per unit exchanged ionic concentration of large ions for smaller ions initially present in the glass, and where C(x) is the local exchanged concentration of ions. In an ion exchange process θ is not readily determinable from the published unconstrained ionic radii, since local relaxations or constraints at the exchange site are likely. On the other hand,

$$\frac{\theta}{tC_o} \int_{-t/2}^{t/2} C(x)\, dx = \frac{3\overset{\circ}{\varepsilon}_{yy}}{C_o} = \frac{(\overline{\Delta V/V})}{C_o} \tag{3}$$

the average total swelling divided by the average total concentration exchange is readily measurable by measuring the change in overall dimensions and overall mass. Thus, the bi-axial ion exchange stress σ_{zz} becomes

$$\sigma_{zz}(x) = AF_\sigma \tag{4}$$

where

$$A = \frac{E\theta}{3(1-\nu)} \left[- C_o + \frac{1}{t} \int_{-t/2}^{t/2} C(x)\, dx \right] \tag{5}$$

$$F_\sigma(x) = \frac{C(x)/C_o - (\overline{\Delta V/V})/C_o}{1 - (\overline{\Delta V/V})/C_o} \tag{6}$$

As Eqn (2) shows, the ion exchange stress parallels the ion exchange concentration gradient. The results for a typical case of distribution due to chemical strengthening is given in Fig. 6, showing the normalized stress and composition distribution in a glass of an initial composition by weight percent of 5 Li_2O; 26 Al_2O_3; 67 SiO_2; 1 TiO_2 in which some of the smaller Li ions were exchanged for larger Na ions by a 4 hr. soak in $NaNO_3$ salt at 400°C. The agreement between the computed values given by the solid curve and the measured values shown by circles is generally good. The typical levels of

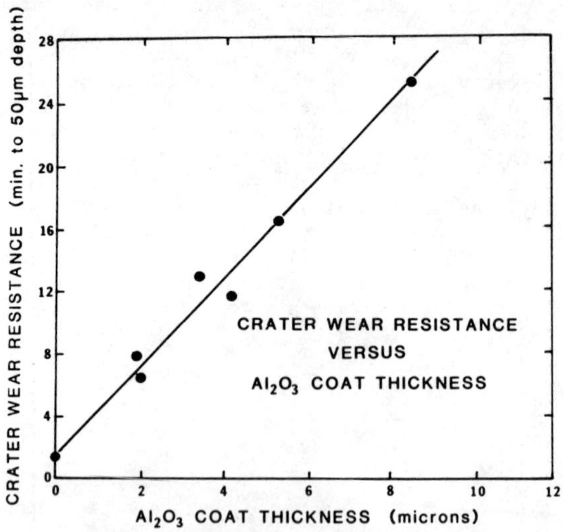

Fig. 5 Dependence of the crater wear resistance on the thickness of the initial thickness of the aluminum coating -- after the coating has been worn through (from Hale and Graham (1981), courtesy of the A.S.M.)

surface bi-axial stress are 300 MPa, while top values of around 500 MPa have been obtained in some cases. These are very substantial stresses and for relatively boxy distributions of ion exchange concentrations favoring thin surface layers only, they are accompanied with rather small levels of tensile stress in the interior of the plate. As Fig. 6 shows, the interior tensile stress need only be of the order of 10% of the surface stress and can be even lower in thicker plates where the ratio of the thickness of the ion exchanged surface layer to the thickness of the plate is smaller. Clearly in the presence of such large surface compressive stresses, inactivating all previous surface cracks and preventing the formation of new mechanical damage the local tensile stress induced by service requirements or by surface contact will have to reach at least comparable levels to the residual stress to produce fracture. This imparts to the glass a very substantial elastic range of performance.

The many variations on how to achieve chemical strengthening of glass are discussed in detail by Bartholomew and Garfinkel (1980).

DISCUSSION

We have presented in this overview of mechanical properties of surface coated materials the important intrinsic properties of surfaces and their role in affecting the strength and inelastic response of solids. We have then considered how in many instances these properties are synergistically affected by modifications of surfaces by both real and virtual surface layers and coatings. In this accounting we have emphasized the more dramatic effects in which

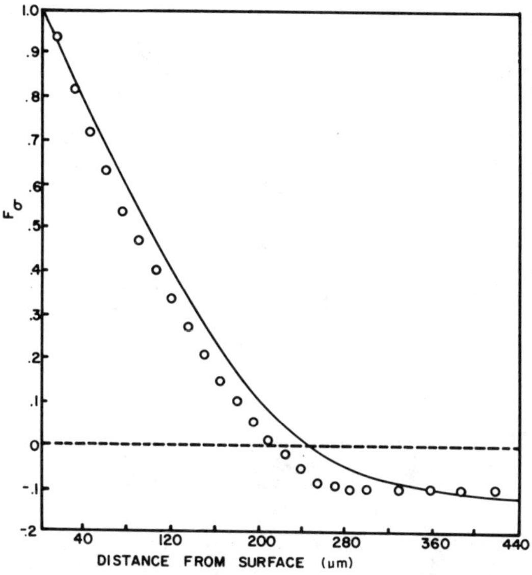

Fig. 6 Comparison of the calculated normalized residual stress distribution F_σ (solid line) to the experimentally measured distribution (circles) in a 0.05 Li_2O; 0.26 Al_2O_3; 0.67 SiO_2; 1 TiO_2 glass chemically strengthened by ion exchange in a 4 hour soak treatment in $NaNO_3$ (from Bartholomew and Garfinkel (1980), courtesy of Academic Press).

modifications performed to the surface have produced effects that were out of proportion to the volume fraction of the surface layer in any consideration based only on a rule of mixtures.

Clearly, there are numerous examples where relatively thicker surface coatings or sandwich structures have been employed where the surface coating performs almost as if it were an infinite solid and does not merely synergistically improve the behavior of the substrate. In these cases the substrate is merely a filler for the purpose of the surface performance. Even in such cases, however, there may be certain non-negligible interactions between surface coatings and the remainder of the part that can benefically affect the overall stiffness of the member or may act as a diffuser of contact pressures as a plate on elastic foundation. These applications are too numerous and the variations too many for us to cover them. A number of such ingeneous applications will be presented during this conference.

Furthermore, we have also not devoted much of this overview to a discussion of the many methods in forming or altering surface coatings ranging from the more exotic modern techniques of intercallation, ion implantation, ion exchange, laser glazing, electron beam glazing, formation of epitaxial coatings by surface chemical reactions, vapor deposition, to techniques of sandwich rolling, explosion bonding, to relatively old fashioned ways of soldering, brazing, glueing, or painting. Any one of these techniques has been dealt with extensively in review articles and even technical monographs that should be consulted.

ACKNOWLEDGEMENT

Discussions with Professors B. Kramer and N. Suh at M.I.T., and Dr. B. Rath of the Naval Research Laboratory were useful in the preparation of this overview.

REFERENCES

Alden, T.H., and W.A. Backofen (1961). The formation of fatigue cracks in aluminum single crystals. Acta Met., 9, 352-366.

Argon, A.S. (1972). Effects of surfaces on fatigue crack initiation. In Corrosion Fatigue, NACE, Houston, Texas, pp. 176-182.

Argon, A.S. (1973). Stability of Plastic Deformation. In The Inhomogeneity of Plastic Deformation, ASM, Metals Park, Ohio, pp. 161-189.

Argon, A.S. (1977). Inelastic deformation and fracture of crystalline solids. In Surface Effects in Crystal Plasticity, edited by R.M. Latanision and J.T. Fourie, Noordhoff, Leyden, pp. 383-430.

Argon, A.S., and J.H. Hannoosh (1977). Initiation of crazes in polystyrene Phil. Mag., 36, 1195-1216.

Bartholomew, R.F., and H.M. Garfinkel (1980). Chemical strengthening of glass. In Glass: Science and Technology, edited by D.R. Uhlmann and N.J. Kreidl, Academic Press, New York, pp. 217-270.

Boothroyd, G. (1963). Temperatures in orthogonal metal cutting. Proc. Inst. Mech. Eng., 177, 789-802.

Bricknell, R.H., and D.A. Woodford (1981). Cavitation in nickel during oxidation and creep. In Creep and Fracture of Engineering Materials and Structures, edited by B. Wilshire and D.R.J. Owen, Pineridge Press, Swansea, U.K. pp. 249-262.

Chen, I-W., and A.S. Argon (1981). Creep cavitation in 304 stainless steel. Acta Met., 29, 1321-1333.

Coffin, L. (1972). The effect of vacuum on the high temperature, low cycle fatigue behavior of structural metals. In Corrosion Fatigue, NACE, Houston Texas, pp. 590-600.

Cook, N.H., and P.N. Nayak (1969). Development of improved cutting tool materials. U.S. Air Force Technical Report, AFML-TR-69-185.

Cooper, A.R. (1982). Residual stress development in ceramics. In Residual Stre and Stress Relaxation, edited by E. Kula and V. Weiss, Plenum Press, New Yor pp. 439-465.

Ernsberger, F.M. (1980). Techniques of Strengthening Glasses. In Glass: Science and Technology, edited by D.R. Uhlmann N.J. Kreidl, Academic Press, New York. pp. 133-144.

Evans, A.G. (1980). Structural and microstructural design in brittle materials In Mechanical Behavior of Materials, edited by K.J. Miller, and R.F. Smith, Pergamon Press, Oxford, pp. 279-311.

Fourie, J.T. (1968). The flow stress gradient between the surface and centre o deformed copper single crystals. Phil. Mag., 17, 735-756.

Frank, F.C. and J.H. van der Merve (1949). One dimensional dislocations. II. Misfitting monolayers and oriented overgrowth. Proc. Roy. Soc., A198,

pp. 216-225.
Griffith, A.A. (1920). The phenomena of rupture and flow of solids. Phil. Trans. Roy. Soc., A221, pp. 163-198.
Griffith, A.A. (1924). The theory of rupture. Proc. First Intern. Conf. Appl. Mech., (Delft), pp. 55-63.
Grisaffe, S.J. (1972). Coatings and protection. In The Superalloys, edited by C.T. Sims, and W.C. Hagel, J. Wiley, New York, pp. 341-370.
Hahn, G.T., and A.R. Rosenfield (1975). Metallurgical Factors Affecting Fracture Toughness of Aluminum Alloys. Met Trans., 6A, 653-668.
Hale, T.E., and D.E. Graham (1981). The influence of coating thickness and composition upon metal-cutting performance. In Cutting Tool Materials, ASM, Metals Park, Ohio, pp. 175-191.
Hirth, J.P. (1983). Private communication.
Hutchinson, J.W., and V. Tvergaard (1980). Surface instabilities on statically strained plastic solids. Int. J. Mech. Sci., 22, 339-354.
Kramer, I., and L.J. Demer (1961). Effects of environment on mechanical properties of metals. In Progress in Materials Science, edited by B. Chalmers, Pergamon, Oxford, England, 9, pp. 131-199.
Kramer, B.M., and N.P. Suh (1980). Tool wear by solution: a quantiative understanding. J. Engineering for Industry, 102, pp. 303-308.
McClintock, F.A. (1963). On the plasticity of the growth of fatigue cracks. In Fracture in Solids (Met. Soc. AIME, Conf. 20), Interscience, New York, pp. 65-102.
Mughrabi, H. (1970). Investigations of plastically deformed copper single crystals in the stress applied state. Phys. Stat. Sol., 39, 317-327.
Mughrabi, H. (1980). Microscopic mechansisms of metal fatigue. In Strength of Metals and Alloys, edited by P. Haasen, V. Gerold, and G. Kostorz, Pergamon, Oxford. England, vol. 3, pp. 1615-1638.
Nabarro, F.R.N. (1947). Dislocations in a simple cubic lattice. Proc. Phys. Soc., 59, 256-272.
Nabarro, F.R.N. (1977). Surface effects in crystal plasticity--overview from the crystal plasticity standpoint. In Surface Effects in Crystal Plasticity, edited by R.M. Latanision and J.T. Fourie, Noordhoff, Leyden, pp. 49-125.
Onyewuenyi, O.A., and J.P. Hirth (1982). Plastic instability in U-notched bend specimens of spheroidized AISI 1090 steel. Met. Trans., 13A, 2209-2218.
Parkins, R.N. (1983). Stress corrosion cracking. In Atomistics of Fracture, edited by R.M. Latanision and J.R. Pickens, Plenum Press, New York, pp. 969-995.
Peierls, R. (1940). On the size of a dislocation. Proc. Phys. Soc., 52, 34-37.
Scholes, A.B. (1971). Surface treatment material and technique for container glass. Proc. 9th Intern. Glass Confr. Sect. A11.7, pp. 1155-1166.
Smidt, F.A., J.K. Hirvonen, and S. Ramalingam (1982). Preliminary evaluation of ion implantation as a surface treatment to reduce wear of tool bits. In The Use of Ion Implantation for Materials Processing, NRL Memorandum Report 4821, edited by F.A. Smidt, Naval Research Laboratory, Washington, D.C., pp. 70-125.
So, P., and L.J. Broutman (1976). Residual Stresses in polymers and their effect on mechanical behavior. Polymer Eng. and Science, 16, 785-791.
Stoloff, N.S. (1983). Liquid and solid metal enbrittlement. In Atomistics of Fracture, edited by R.M. Latanision and J.R. Pickens, Plenum Press, New York, pp. 921-953.
Suh, N.P. (1977). Coated carbides--past, present and future. Carbide J., 9, pp. 3-9.
Suh, N.P. (1980). New theories of wear and their implications for tool materials. Wear, 62, 1-20.

Suh, N.P., H-C. Sin, and N. Saka (1980). Fundamental aspects of abrasive wear. In Fundamentals of Tribology, edited by N.P. Suh, and N. Saka, M.I.T. Press, Cambridge, Mass., pp. 493-518.

Westbrook, J.H., and E.R. Stover (1967). Carbides for high temperature applications. In High Temperature Materials and Technology, edited by I.E. Campbell and E.M. Sherwood, J. Wiley, New York, pp. 312-348.

Westwood, A.R.C., and J.J. Mills (1977). Applications of chemomechanical effects to fracture-dependent industrial processes. In Surface Effects in Crystal Plasticity, edited by R.M. Latanision, and J.T. Fourie, Noordhoff, Leyden, pp. 835-862.

Westwood, A.R.C. (1983). Fracture -- applying the breaks. In Atomistics of Fracture, edited by R.M. Latanision and J.R. Pickens, Plenum Press, New York, pp. 65-91.

Wohlfahrt, H. (1982). Shot peening and residual stresses. In Residual Stress and Stress Relaxation, edited by E. Kula and V. Weiss, Plenum Press, New York, pp. 71-92.

PRESS FORMING PROBLEMS OF HSLA STEEL SHEETS

K. Miyauchi

Deformation Processing Laboratory, The Riken Institute, Wako-Shi, Saitama, Japan

ABSTRACT

The press forming problems of HSLA steel sheets in Japan are reviewed briefly and discussed taking into account a unique development of Japanese steel sheets and autobody forming. Also described are some trials for assessment of press performance in HSLA steel including the Yoshida Buckling Test, development of more suitable forming techniques, and forecast of HSLA steel sheets in the future automobiles.

KEYWORDS

HSLA steel sheets; press formability; forming accuracy; shape fixability; fittability; surface deflection; elastic behavior; buckling test

INTRODUCTION

The Japanese automotive industry was surrounded by the well-balanced neighbors of industrial suppliers when they started again to produce their own cars after World War II. As for the steel industry, it was matured in many respects, resulting partly from a long history of strong governmental backup since the beginning of the Meiji era. It has long been regarded as the main root of all other industries and has played such an important role very exactly for the past hundred years. The automotive industry was just like a child as compared with the steel industry, which is one of the special Japapanese features in the relationship of car and steel industries.

During the postwar period of innovation, the steel people had a strong sense of responsibility as the leading industry and were almost ready to help their younger brothers modernize the production system. There was necessary, however, a trigger for the start of cooperative research activities between the steel sheet and stamping engineers. In 1957, Fukui and Yoshida made a timely hit that they succeeded in organizing a small group of volunteers from the steel and car companies to investigate the Fukui conical cup test and establish their common standard for assessing the press formability of steel sheets. This group has now grown up to be the Japan Sheet Metal Forming Research Group (president: Yoshida) which is a member of the International Deep Drawing Research Group since 1960.

Although the group activities have not always been successful, it may be said that those have been a very great impact on the development of new materials and press forming techniques as well as a better mutual understanding and more smooth communication between the steel metallurgists and stamping engineers. As the press formability of steel sheets has been made clear in relation with the basic material properties through the cooperative investigations, the steel engineers have exactly understood the right direction to improve the material properties and have developed various steel sheets of higher formability, for example,

open coil annealed decarburized rimmed steel,
retarded-aging rimmed steel,
low yield point rimmed steel, and
Ti-killed steel.

Under these circumstances, the die engineers have had no need to be too much afraid of a risky fashionable design of autobody style and have been able to challenge a new trial for improving the productivity of their stamping lines. After they decided a die design of considerable severity taking into account the optimal productivity only, the steel engineers came to see it and select the most suitable material. When the assessment of forming severity was not appropriate and some failures occurred in the trial stamping, more formable steel sheets were supplied so immediately that the automotive engineers need not have increased the lead time before the start of production. A number of attempts have been made by the autobody engineers in the presence of such high formability materials, to achieve the present high productivity in the Japanese automotive industry.

The most formable material thus developed has:

as low yield point as possible,
as high n and r values as possible,
as high ultimate ductility as possible, and often consequently
lower ultimate tensile strength;

For example, the yield point is around 140 MPa, the tensile strength around 270 MPa, the n value higher than approx. 0.26, and the r value higher than approx. 2.0.

When HS steel sheets appeared before the footlights in response to a strong demand for autobody weight reduction, the Japanese autobody engineers felt the tensile strength of 350 MPa was high enough for HS steel sheets as compared with ordinary mild steel sheets. In Japan, the steel sheets of more than about 340 MPa in the tensile strength are classfied into HS steel sheets (cold rolled), which makes an important difference between Japan and some other countries like the United States of America. Such a discrepancy is expected to be corrected in the near future for better international exchange of technical information.

According to Yoshida's request, the Japanese group started to investigate the possibility of HS steel application to autobody panels in 1972, and have continued their efforts to try new HS steel sheets in new forming conditions through the cooperation of the steel and autobody engineers. They have so far made the investigations of actual panel forming in HS steel sheets, laboratory tests on press formability and galling, geometrical forming defects including the springback, surface deflection and sidewall curl, and the Yoshida buckling test not only for assessing the elastic and plastic behaviors relating to the shape fixability but also for developing the control technique of the material behaviors.

During these collaborative investigations, the steel companies have developed new HS steel sheets of better formability and/or higher strength as they have done so far. Consequently, the most of present formable HS steel sheets are of low alloy in order to make possible the HS steel application to autobody panels without any serious change in the present press forming system. In spite of the active cooperation of steel engineers, the autobody engineers are facing many difficult problems relating to the higher strength and lower formability of HS steel sheets while they are requested to minimize the lowering of productivity. As they are steadily solving such difficulties, they now forecast such a more concrete status of HS steel sheets in the future automobiles as the maximum HS steel application to autobody parts will be about 50% of white body weight and autobody panels will be made of HS steel sheets of at most 400 MPa in the tensile strength, as long as they do not employ a drastic change in the present system.

FORMABILITY PROBLEMS OF HS STEEL SHEETS

Mechanical Properties of HS Steel Sheets

Strength-elongation balance. Although the purpose of HS steel application is to increase the sheet material strength, it is very difficult to make such materials without some decrease in ductility. The tensile strength is therefore limited by the least required elongation which depends on the severity of forming. A relation between the tensile strength and elongation is very important for selection of HS steel sheets. An example of the so-called "strength-elongation balance" is Fig. 1 (Takechi, 1981). This balance is considerably affected by changing the addition of alloying elements and/or the heat treatment even in a group of dual phase steels. There will be a high possibility of remarkable improvement in such a balance in the near future.

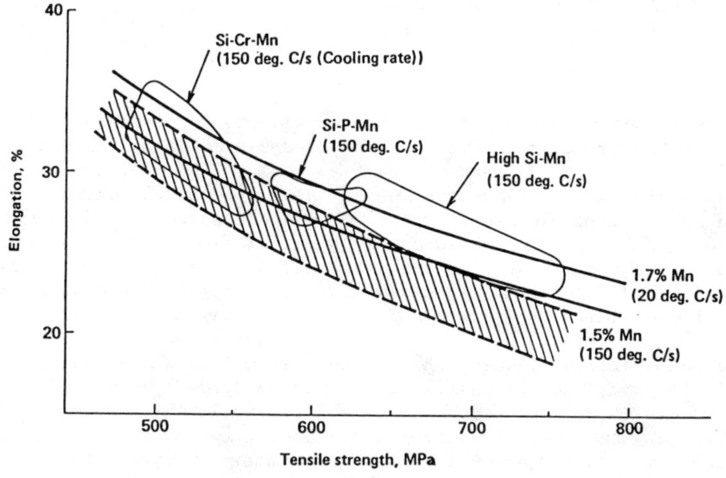

Fig. 1 Tensile strength-elongation relationship

Workhardenability and plastic anisotropy. It has been said that dual phase steels have high n value (that is, low yield ratio) but low r value, while rephosphorized HSLA steel can easily have high r value as compared with low yield ratio. As the common sense is apt to be old-fashioned, it may be worth noticing that the r value in low Mn dual phase steels can be increased to more than 1.5 by the following control techniques of steel production:

1) Make higher the coiling temperature to improve the texture of major ferrite phase (Kubotera, 1976; Nishimoto, 1980)
2) Control the minor martensite phase in such manners as
   a) decrease the hardness ratio of martensite to ferrite,
   b) decrease the volume fraction of martensite, and
   c) increase the mean distance between the nearest martensite grains (Kurihara, 1982).

As shown in Fig. 2, we can see that the effects of the hardness ratio c, martensite volume fraction f and intermartensite distance d on the r value are delicately affected by each other. This may be a sign of coming innovation in the production of HS steel sheets.

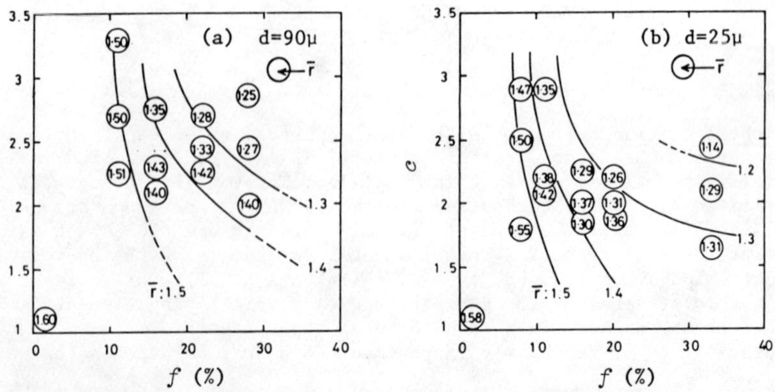

Fig. 2 Effect of hardness ratio, c, and second phase volume fraction, f, on the r value of dual phase steels for different mean distance, d, between second phase grains nearest to each other

Press Formability of HS Steel Sheets

Punch-stretchability and deep drawability. Nowadays we are readily supplied either of high n or high r material in HS steel sheets, depending on the type of press forming. In the forming of simple mode, there seems to be no serious problems, but we have to expect many difficult problems in press shops. For example, the relation between the n-value and the punch stretching limit in the range of a change in the n value of HS steel sheets is very complicatedly affected by the presence of friction especially in the forming of large-sized complex panels (Miyauchi, 1968). The effect of friction usually reverses or at least weakens the above relation, depending on the deformability in the range of diffused necking as well as the geometry of stretch deformation area. In addition, the fine geometrical defects like the surface deflection make more complicated the press performance in the forming of outer panels.

The expression of forming limit in deep drawing is changed from the conventional L.D.R. or limiting depth to the formable range by the blankholder force as shown in Fig; 3 (**Furubayashi, 1978; Takahashi, 1981; Yoshida, 1971**) because the scatter of material properties and fluctuation of forming conditions are inevitable where the stability of press performance is the most important. In other words, the forming limit by fracture is not convenient for practical use, but the possible change of forming conditions like the balnkholder force for a given depth of a panel is more available in press shops.

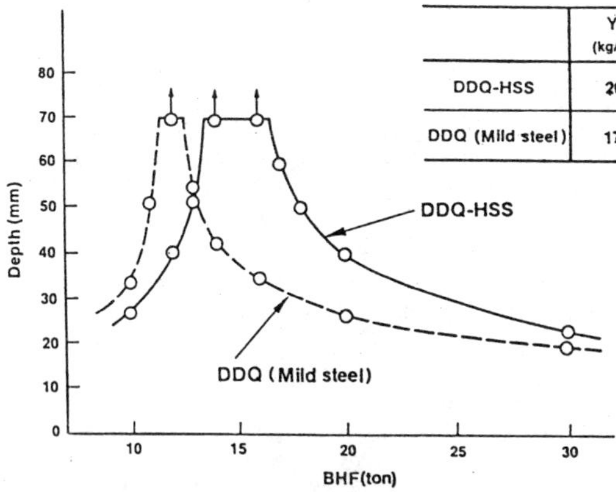

	YS (kg/mm²)	TS (kg/mm²)	El (%)	$\bar{r}$
DDQ-HSS	20.5	40.1	41	2.09
DDQ (Mild steel)	17.2	32.6	46	1.78

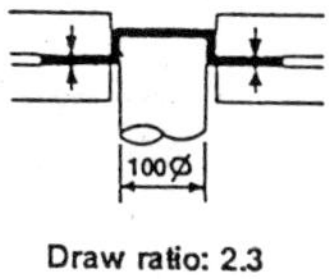

Fig. 3 Press formability of DDQ-HS steel sheets (Takahashi, 1981)

The required forming depth of large panels is not expected to be so large in the future but will be almost the same as or less than the present, as long as a strong demand continues for the improvement of productivity. Such a small depth, however, has brought a considerable difficulty of deformation control relating to the fitting behavior and shape fixation in complex panel forming. The good shape fixation reduces the manual operations and makes much easier the automation of subsequent production lines. The formable range for the required forming depth is now actually a measure of forming severity in press shops, in which the optimal blankholder force is selected by taking into account the best shape fixation and other requirements.

Fitting Behavior and Shape Fixation

Since Yoshida (1964) proposed the introduction of new criteria, fittability and shape fixability, into the category of press formability, these problems have repeatedly been investigated by the Japanese group. Their experiences are of benefit to solve new difficulties which are geometrical defects in the press forming of HS steel sheets. The main investigations (Furubayashi, 1978 ; Ishigaki, 1978) started to deal with the deforming behavior and its resulting geometrical surface defects in a large-sized flat panel. Recently there have been achieved many fruitful results relating to the mechanism of surface deflection, effect of material properties, improvement of forming techniques, and development of the Yoshida buckling test.

On the other hand, rather traditional problems of springback and sidewall curl in bending and/or unbending processes have not been investigated extensively, though they strongly govern the forming accuracy of joint portions in a panel and consequently the following body assembly process, probably because they have been regarded as a very old problem which is of simple deformation mechanism and can be solved by the past experiences and knowledge of bending operations. As the stamping engineers, however, have increased the application of HS steel sheets, they have found it necessary to recheck the material behavior in bending and unbending processes, to result in many interesting works (Hayashi, 1982; Hirata, 1981; Miyauchi, 1982; Umehara, 1980).

The fitting behavior and shape fixation may be divided into two problems now as described above. One is a slight surface deflection in the almost flat or lightly curved area of a large-sized panel which is caused by the non-uniform stress state during loading or unloading. Although the resulting elastic or plastic compression is small and does not harm the forming geometry of a considerably curved portion, it is large enough in many cases to make a deflection out of the reference surface which is nearly flat and has very low structural rigidity.

Another is a bending or unbending defect which is characterized by the non-uniform distribution of stress and strain in thickness direction. The springback is a typical example resulting from the elastic return deformation during unloading.

Surface deflection. In the press forming of flat panels, the punch bottom area is mainly subjected to biaxial tension, which non-uniformness causes such a compression as shown in Fig. 4. This type of compression is always present in complex panel forming and sometimes causes a surface deflection depending on the geometry of a panel, as in Fig. 5.

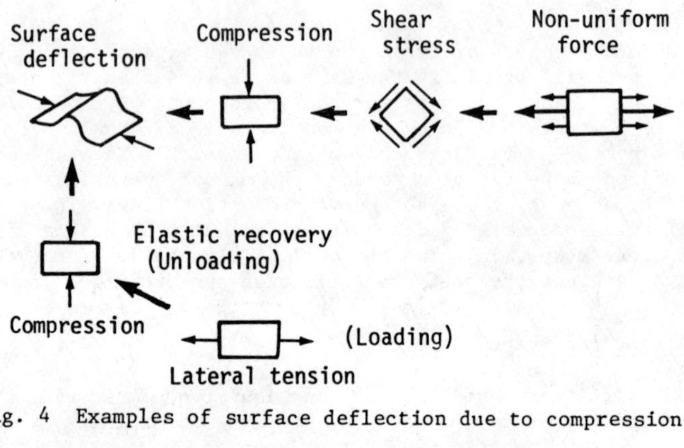

Fig. 4 Examples of surface deflection due to compression

Another type of surface deflection due to tension is also schematically shown in Fig. 5. The fitting behavior and shape fixation are in a very interesting contrast to the above ones in compression.

The surface deflection due to the elastic return deformation (elastic recovery) during loading has recently been found to take place in many cases (Umehara, 1980). As this surface deflection is strongly governed by the Young's modulus and the deformation strength for a given strain, it very often makes very difficult the application of HS steel sheets with respect to the geometrical accuracy of a panel. Even if the fitting behavior is satisfactory under loading, there are some combinations of forming geometry

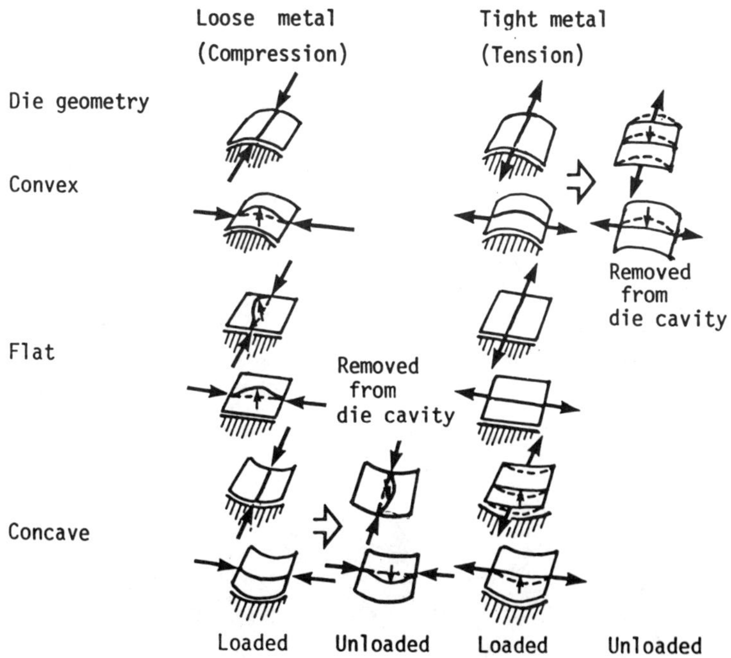

Fig. 5 Types of surface deflections for different geometry and loading

and stress state which are apt to cause loose or tight metal due to elastic recovery, as shown in Fig. 5. Naturally, in cases that such a compression or tension is very large during press forming, the fitting behavior is insufficient, to result mostly in a large surface deflection.

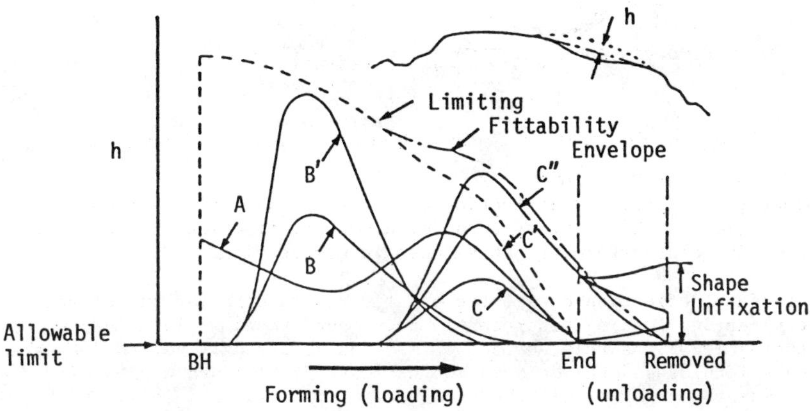

Fig. 6 Schematic diagram of typical fitting behavior curves

The surface deflection during press forming corresponds to the fitting behavior, which typical curves are shown in Fig. 6. In addition to these curves, as already described (Umehara, 1980), the shape fixation is considerably affected by the elastic recovery even when no surface deflection is present at the end of loading. A typical example is seen around a hollow for the door handle in Fig. 7.

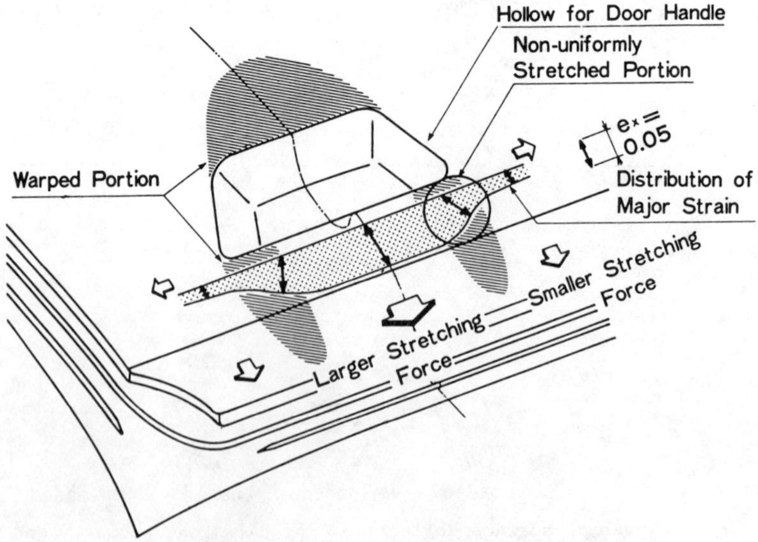

Fig. 7 Distribution of major strain and variation of stretching force at upper area of door handle hollow (Umehara, 1980)

The effect of mechanical properties on the surface deflection has also been investigated extensively in actual autobody panels. Some examples are shown in Figs. 8 and 9 (Sato, 1980).

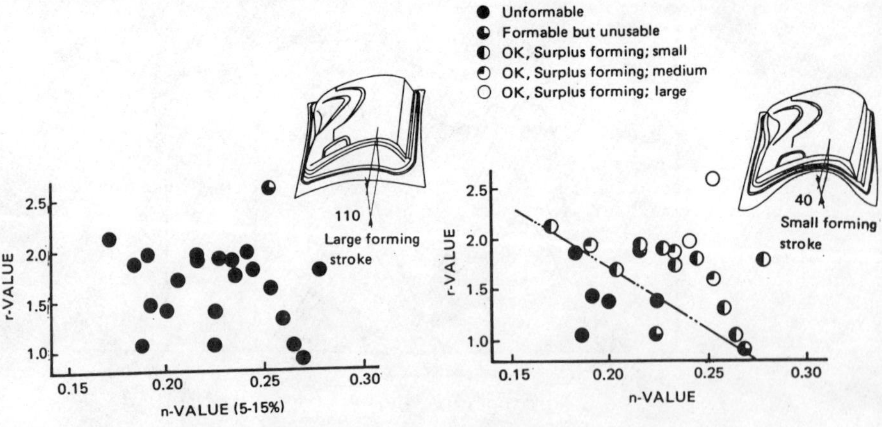

Fig. 8 Press performance in front fender panels of different forming stroke type (Sato, 1980)

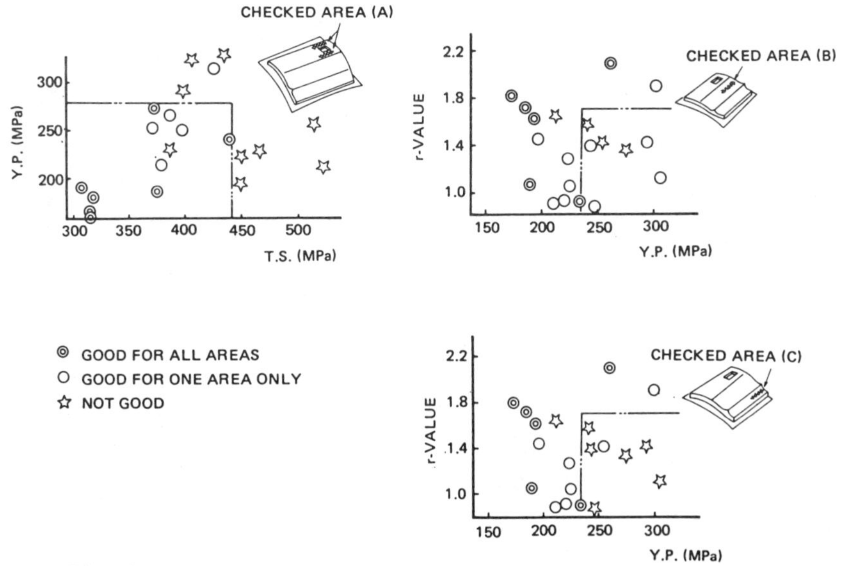

Fig. 9 Geometrical surface defects and material properties in press forming of door outer panels (Sato, 1980)

The correlation between the mechanical properties and surface deflection has not been made clear enough yet, though the n value or yield point and the r value are seen considerably correlated with the surface deflection in some panels. An increase in the n value or a decrease in the yield point makes more uniform the strain distribution, while an increase in the r value is regarded as effective for increasing the fracture and buckling resistances so that more severe forming can be employed if necessary to remove the deflection under stronger stretching. In order to find some unknown factors in the mechanism of surface deflection, the Yoshida buckling test was proposed as the diagonal tensile test of a square specimen to give a non-uniform stress field where the buckling behavior can easily occur (Jalinier, 1982; Yoshida, 1982).

Springback and sidewall curl. Another big problem is how to find the optimal bending condition for removing or minimizing the springback or sidewall curl in HS steel sheets. The conventional techniques for the control of springback in the press forming of hat-shaped channels are to make smaller the profile radius of bending dies, increase the blankholder pressure, and desrease the clearance. In addition, there has recently been reported an interesting phenomenon that the over-running behavior of a bent portion in the die cavity may be a strong factor to control the springback and sidewall curl in the range of smaller blankholder pressure and smaller clearance (Hayashi, 1982; Umehara, 1980). Such geometrical defects in the bend-unbend operations are inevitably larger in higher strength sheets, while they are affected by the Young's modulus and the Bauschinger effect.

The control of these defects is an emergent problem for a wider application of HS steel sheets especially in the forming of structural members. Some simple and economical techniques are found effective to reduce the sidewall curl as shown in Fig. 10 (Miyauchi, 1982). The adoption of elliptic profile has widely been common as avoids the shock line at the hold-down of a blank for outer panels and, at the same time, brings a favorable effect on the geometry of sidewalls.

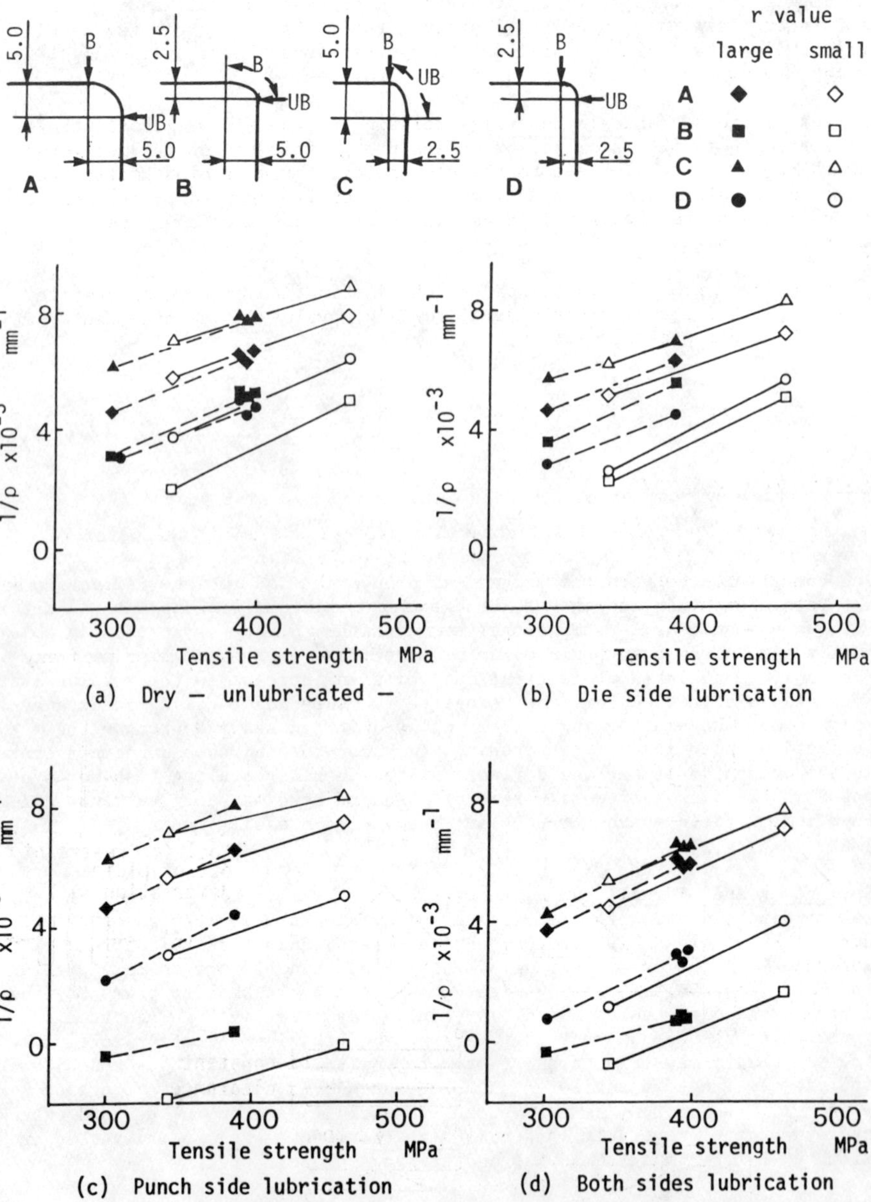

Fig. 10 Effects of bending profile geometry, lubrication and mechanical properties of steel sheets
sheet thickness: approx. 0.8 mm

The effect of lubrication on the radius of curvature of a sidewall is also very interesting in relation with the bending profile geometry and material properties. Although the punch side lubrication is often unfavorable from the drawability point of view, it gives the highest possibility to form a satisfactory sidewall in HS steel sheets of 500 MPa or more in the tensile strength for the bending profile of B type. The same lubrication, however, is not always the best for the different profiles.

As regards the mechanical properties relating to the sidewall curl, the tensile strength is the major factor unless the profile radius is very large and the backward tension due to the blankholder force is also small. The effect of r value on the sidewall curl is strongly governed by the geometry of bending die profiles. An increase in the r value is unfavorable in case of profile type B.

A special attention should be paid to the combination of bending profile geometry, lubrication, blankholder force and r value in the hat-shape forming of HS steel sheets.

ASSESSMENT OF FITTABILITY AND SHAPE FIXABILITY

Yoshida Buckling Test

Testing method. Figure 11 illustrates a typical one among the possible methos of this test. The test procedure is as follos:

(1) A specimen is stretched to λ_{75} = 1%, $\Lambda(\lambda_{101})$ = 2% or $\Lambda = \Lambda_B$ (up to fracture) and the buckle height is measured under loading: h_1

(2) The buckle height is measured after unloading: h_2
$\Delta h = h_1 - h_2$: A decrease in the buckle height due to elastic recovery

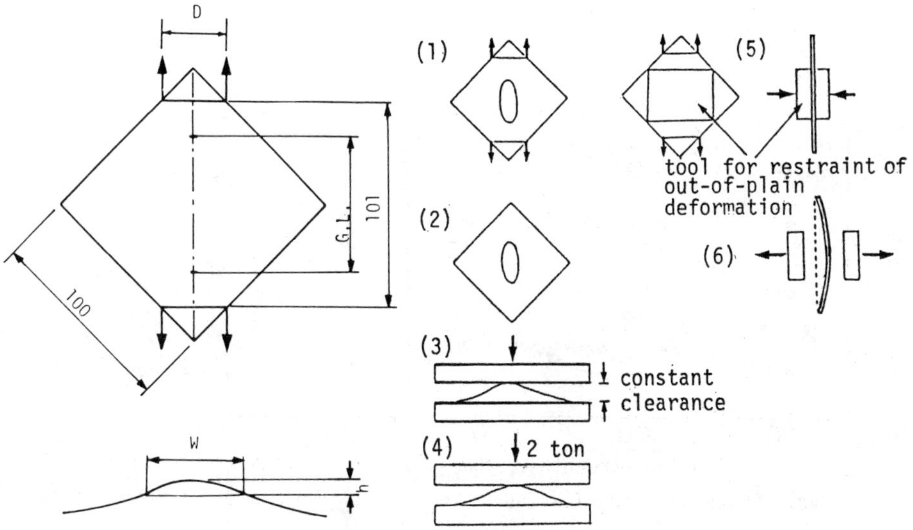

Fig. 11 Typical procedure and specimen geometry of the Yoshida buckling test
Standard size of a square specimen: L = 100mm, D = 40mm, G.L. = 75mm, W = 25mm

(3) The buckled specimen which was stretched to λ_{75} = 1% is compressed in thickness direction to a certain amount (clearance between tools is constant) and then compressive load is let off: h_3

(4) The buckled specimen which is given the same stretching as above is compressed by a certain load (2 ton) and the compressive load is let off: h_4

$\Delta h' = h_2 - h_3$: A decrease in the buckle height due to the compression

$\Delta h'' = h_3 - h_4$: The effect of the load on the decrease in the buckle height

(5) A specimen is stretched to λ_{75} = 1% under the restraint of out-of-plane deformation

(6) The loading and restraint are removed, then the specimen is buckled elastically due to the residual stresses: h_5 (w = 25), h_6 (w = 50)

Buckling behavior and material properties. The buckle height h_2 for λ_{75}=1% increases as the yield point increases and the r value decreases (Yoshida, 1980), which causes a decrease in the gradient of strain in and around the buckling region, resulting in the reduction of lateral compression. In comparison with λ_{75}, the elongation for a longer gage length λ_{101} includes all the deforming behaviors between the gripping devices and may be more similar to the deformation characteristics between both the ends of a blank in press forming. It means that λ_{101} may work more effectively for better understanding of actual complex relations among the surface deflection, material properties and forming conditions in such cases as the surface deflection is not simply or directly affected by changing the forming conditions, either.

An interesting relation between h_2 and the yield point at 2% λ_{101} can be rearranged by using $\sigma_y \cdot r/(1+r)$ instead of σ_y, as shown in Fig. 12 (Yoshida, 1982). The effect of the yield point and r value on the deforming and buckling behaviors is schematically explained in Figs. 13 and 14, where we can see a complex relation among the buckling behavior, material properties and changing mode of deformation, as seen in actual press performance.

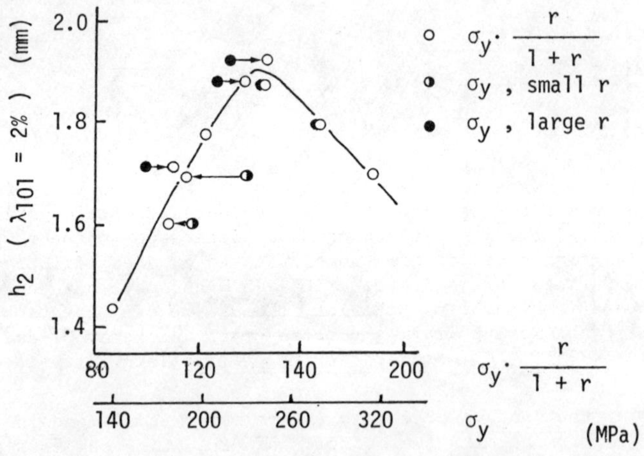

Fig. 12 Relation between mechanical properties and buckling height for λ_{101} = 2%

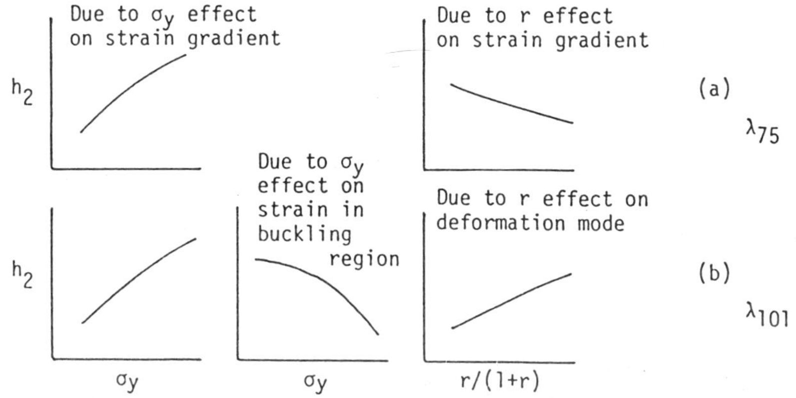

Fig. 13 Different effects of mechanical properties on the buckling height for different gage lengths

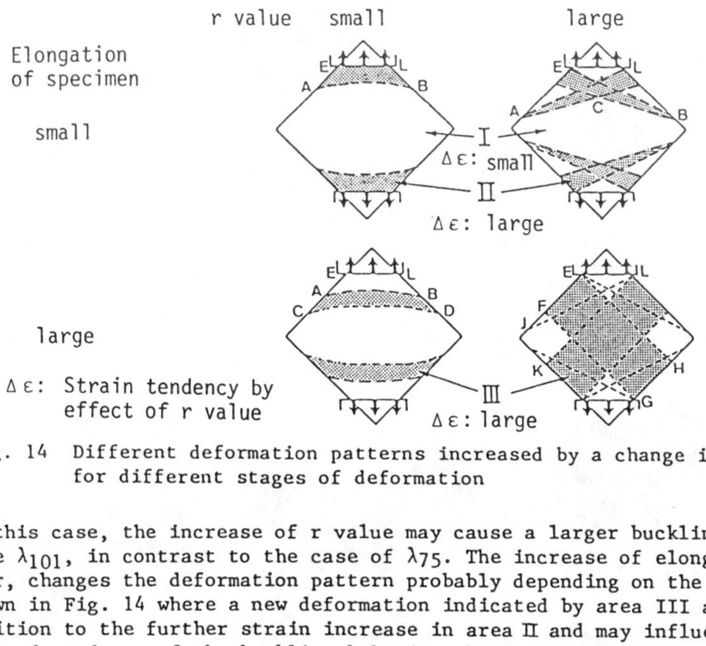

Fig. 14 Different deformation patterns increased by a change in r value for different stages of deformation

In this case, the increase of r value may cause a larger buckling for the same λ_{101}, in contrast to the case of λ_{75}. The increase of elongation, however, changes the deformation pattern probably depending on the r value, as shown in Fig. 14 where a new deformation indicated by area III appears in addition to the further strain increase in area II and may influence the r value dependence of the buckling behavior, because we can expect there will be a tendency to make more uniform the total deforming behavior due to the presence of area III type deformation in the case of larger r value.

It is interesting that a similar change in the deformation mode is often seen in the press forming of actual complex panels, which suggests a unique capability of the Yoshida buckling test for a wide variety of assessment of forming severity and material formability.

Elastic buckling due to unloading. This test is easily available for measuring the buckling height during loading as well as after unloading. It means that it is possible to find the unloading effect on the buckling behavior as shown by Δh in Fig. 15.

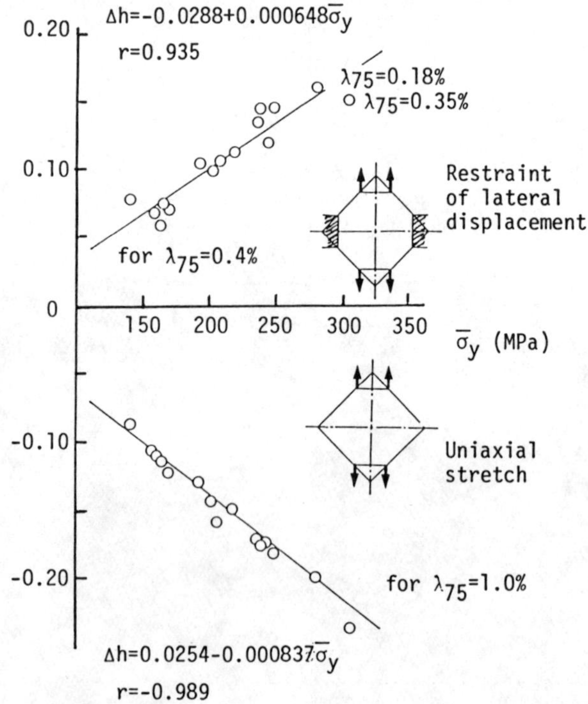

Fig. 15 Yield point and buckling height change due to unloading for different stretching conditions (Yoshida, 1983)

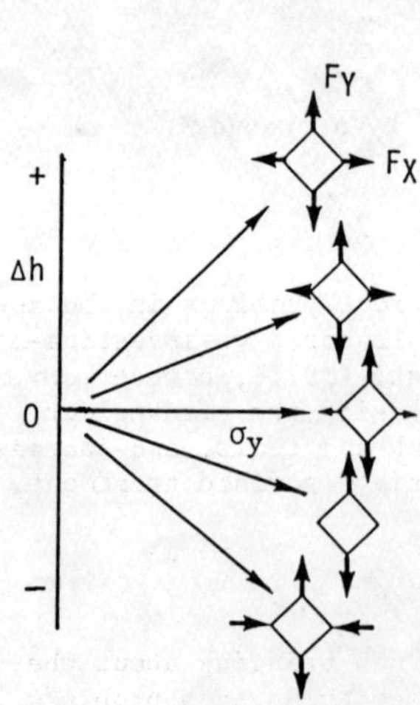

Fig. 16 Schematic relation between σ_y and Δh for different modes

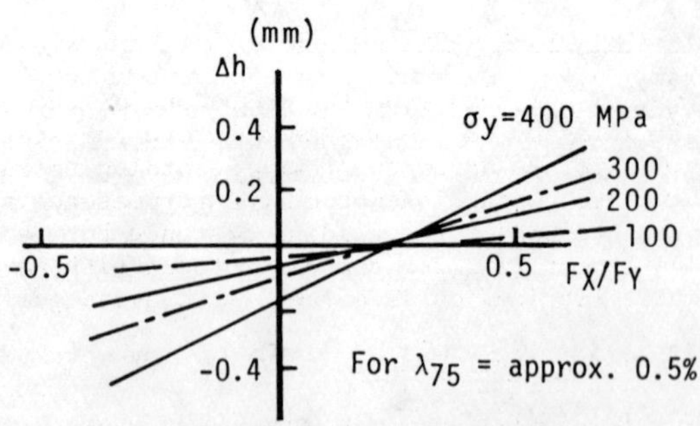

Fig. 17 An example of relations between Δh and F_x/F_y for different yield points

A change in the buckling height during unloading is strongly affected by the stress state of areas around the buckle as shown in Fig. 16. The effect of loading condition on Δh is approximately illustrated in Fig. 17. Although actual large-sized panels have more complex characters including the structural rigidity, the Yoshida buckling test is expected to tell us much information about a countermeasure suitable for avoiding possible surface defects before or after the tryout of forming.

The stress state in the surrounding areas is naturally varied with the progress of deformation and gives an interesting effect on Δh as in Fig. 18 (Yoshida, 1983). The control of stress state and strain level is very important among the techniques of high geometrical accuracy in press forming.

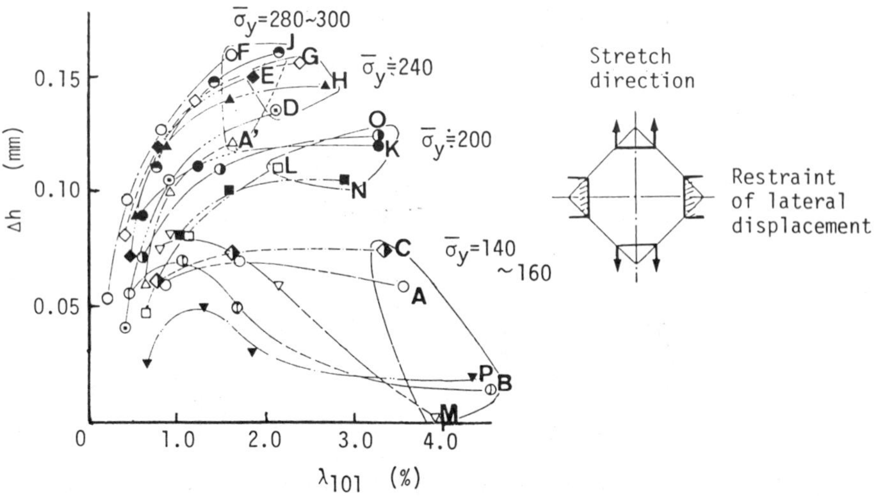

Fig. 18 Effect of given elongation on the buckling height change due to unloading (Yoshida, 1983)

SURFACE PROBELMS IN PRESS FORMING

Galling of HS Steel Sheets

Galling had been expected to be one of the most serious problems in the application of HS steel sheets before the start of collaborative investigations in the Japanese group. According to Furubayashi (1977), however, there have been established such countermeasures as decreasing the blankholder pressure and/or the sliding distance between a blank and tools, and increasing the galling resistance of tools with an appropriate surface treatment, while lubrication is effective to avoid galling.

Coated HS Steel Sheets

The increasing severity of corrosion regulation is now bringing about the development of a big variety of coated steel sheets with serious problems in press forming as well as spot welding operations. This problem consists mainly of:

(1) deformability and strength of coated layer itself

(2) bonding strength between coated layer and mother material
(3) anti-corrosion mechanism.

The present coated steels are partly satisfactory from the above-mentioned points of view. Every coated steel shows different characters in different forming conditions (Sudoo, 1980). Although new coated steels such as alloyed coating and double coating will probably work well in the near future, it is very important at the same time to establish the standardization of coated steel sheets for the systematized techniques of press forming.

POSSIBLE WAYS BEFORE HS STEEL SHEETS

Geometrical Accuracy Problems

Shear deformation. In a large complex panel forming, there are present many of shear deformation due to differential flow. A typical example is surface deflections around the door handle portion as shown in Fig. 19.

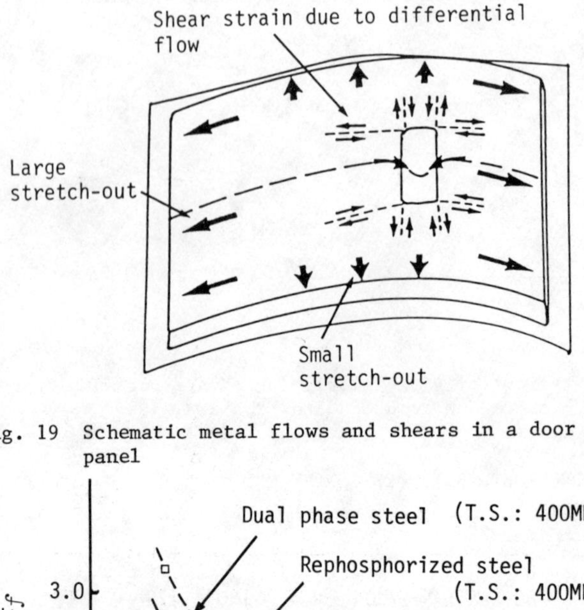

Fig. 19 Schematic metal flows and shears in a door outer panel

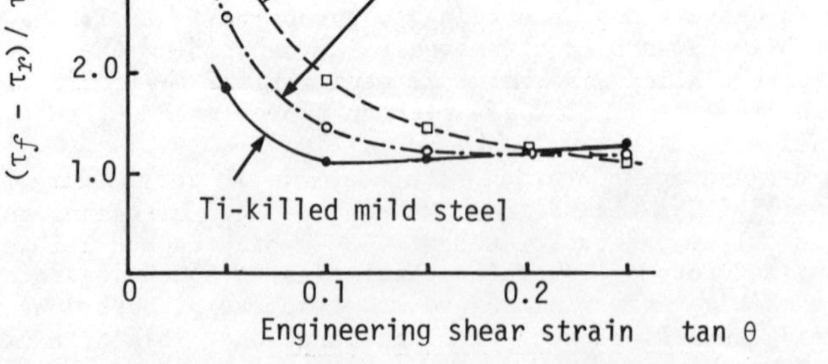

Fig. 20 Bauschinger effect in shear deformation for different steels (tan θ is given in reverse shear)

The stress-strain relation in shear deformation will be more important to explain the complex material behaviors relating to surface deflections as in a door outer panel. According to a newly developed planar shear test (Miyauchi, 1983), dual phase and rephosphorized HS steel sheets show a strong planar anisotropy in the stress-strain curve, while Ti-killed mild steel has almost the same curve in any direction. Recently, Amaike (1983) has found that the planar directional dependence of surface deflections around an embossed portion is present in the most of HS steel sheets but not in Ti-killed steel, which interestingly corresponds to the above-mentioned.

Bauschinger effect. This effect has long been left obscure in press forming, but we probably need to pay a more careful attention to the material behavior relating to this effect for achieving a higher geometrical accuracy of press forming. Figure 20 gives an example of the Bauschinger effect in planar shear deformation (Miyauchi, 1983a).

Bake Hardenability

This aging effect is strongly requested in order to apply HS steel sheets to the forming of higher severity, but it is considerably affected by the defor-

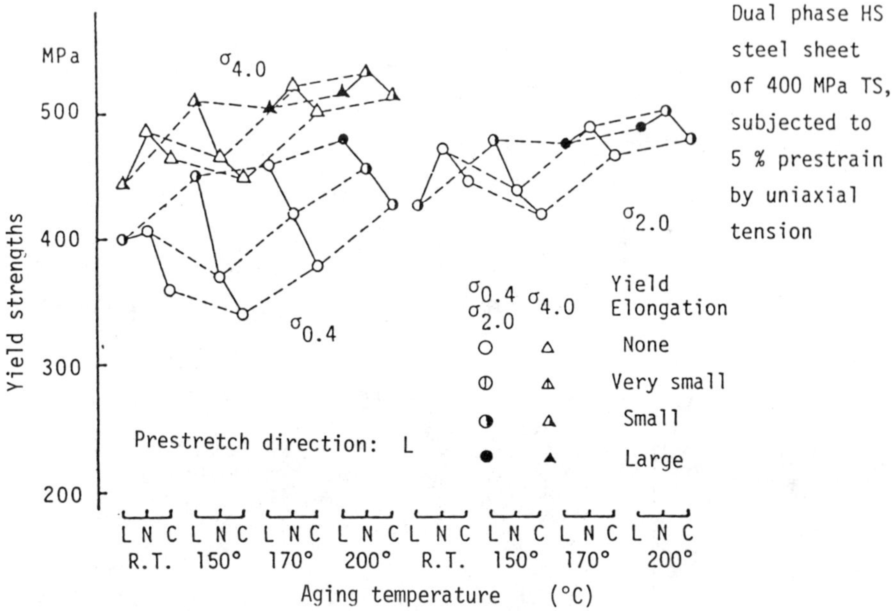

Fig. 21 Planar anisotropy of strain aging in a bake-hardenable dual phase steel

mation mode and loading condition after baking. An interesting behavior is shown in Fig. 21 (Miyauchi, 1983b). We can see aged softening in test directions different from the prestretch at 150°C which causes a lower bake hardenability at 170°C or more in such directions.

Fracture and Thermal Problem in Press Forming

The fracture limit is always a traditional important problem. As HS steel sheets suffer inevitably from lower ductility, the fracture problem is more serious. In addition, the accumulated heat in the tools due to the high strength of steel sheets may sometimes cause a sudden fracture after a certain number of stamping. From this point of view, it will be necessary to change, clean and repair the tools more frequently, which makes more important the speed of die change.

There are many other problems for a wider application of HS steel sheets to automobiles, but we may be optimistic, expecting remarkable developments of forming techniques and HS steel sheets in the near future.

REFERENCES

Amaike, T. and coworkers (1983). JDDRG Reports, No. 58-9.
Furubayashi, T. (1977). 60th Sym. Text of JSTP, 15-23.
Furubayashi, T., Shiokawa, M., and Yamazaki, K. (1978). 10th IDDRG Congress Text, 39-44.
Hayashi, Y., and Takagi, M. (1982). Jour. ISIJ, 68, 1236-1243.
Hirata, M., Abe, H., and Nakagawa, K. (1981). Proc. 1981 Jap. Spring Conf. Tech. Plasticity, 477-480.
Ishigaki, H., Umehara, Y., and Okamoto, I. (1978). 10th IDDRG Congress Text, 53-66.
Jalinier, J. M. and coworkers (1982). 12th IDDRG Congress Text, 201-220.
Kubotera, H. and coworkers (1976). Jour. ISIJ, 62, 846-855.
Kurihara, K., Araki, K., and Nakaoka, K. (1981). Jour. ISIJ, 67, 343-352.
Miyauchi, K., Komorida, H., and Yoshida, K. (1968). Jour. JSTP, 9-86, 139-150.
Miyauchi, K. (1982). 4th Joint Meeting of ADDRG and JDDRG, Warren.
Miyauchi, K., and Ishida, M. (1983a). Proc. 1983 Jap. Spring Conf. Tech. Plasticity, 5-8.
Miyauchi, K. (1983b). Proc. 34th Jap. Joint Conf. Tech. Plasticity, to be presented.
Nishimoto, A., Hosoya, Y., and Nakaoka, K. (1980). Jour. ISIJ, 66, A201, 81-84.
Sato, S. and coworkers. (1980). 11th IDDRG Congress Text (memoir. scientifiq. metallurgie, 77), 237-246.
Sudoo, C., Hayashi, Y., and Nishihara, M. (1980). 11th IDDRG Congress Text (memoir. scientifiq. metallurgie, 77), 353-362.
Takahashi, N. (1981). CAPL Symposium Text, Dusseldorf, III-1 - 35.
Takechi, H. (1981). CAPL Symposium Text, Dusseldorf, II-1 - 46.
Umehara, Y. (1980). 11th IDDRG Congress Text (memoir. scientifiq. metallurgie, 77), 247-256.
Yoshida, K. (1964). Jour. JSTP, 5-37, 56-64.
Yoshida, K., and Yoshii, K. (1971). Proc. ICSTIS (Trans. ISIJ), 856-861.
Yoshida, K. and coworkers. (1980). Proc. 31st Jap. Joint Conf. Tech. Plasticity, 109-116.
Yoshida, K., Hayashi, H., and Miyauchi, K. (1982). Report IDDRG WG III, "Yoshida buckling test".
Yoshida, K. (1983). Jour. JSTP, to be published.

MECHANICAL BEHAVIOR OF ENERGETIC POLYMERIC MATERIALS

S. Nicolaides, D. Wiegand and J. Pinto

US Army Armament Research and Development Center, US Army Armament, Munitions and Chemical Command, Dover, New Jersey 07801, USA

ABSTRACT

Understanding the mechanical behavior of energetic materials under the strain rate, temperature, and dormant storage environments of military applications is necessary for predicting their performance and safety tolerances and their useful life. This work focuses on the mechanical integrity of nitrocellulose base gun propellants and the correlation of brittleness with abnormal energetic performance. Compressive mechanical properties at a strain rate of 1000 percent-of-length per second as a function of temperatures descending to -60°C, and the failure appearances of the tested samples, are used to identify ductile-brittle transition temperatures and relate to propellant functioning. Preliminary data of mechanical property changes with time indicate that embrittlement takes place in these materials with age, shifting the ductile-brittle transition to higher temperatures or less severe loading conditions; glass transition temperatures, determined using thermo-mechanical and calorimetric methods, are suggested as promising materials parameters for indexing such time dependent changes.

KEY WORDS

Nitrocellulose-base propellant, Compressive failure, Closed bomb combustion, Ductile-Brittle and Glass Transitions, Aging, Embrittlement

INTRODUCTION

Nitrocellulose is the product of partial nitration of cellulose, which is a natural polymer of empirical formula $(C_6H_{10}O_5)_n$; it is the principal explosive ingredient in gun propellant. Containing three explosive ingredients: viz., nitrocellulose, nitroglycerin as the oxidant plasticizer, and either nitroguanidine or RDX, these so-called "triple-base" gun propellants are the energetic polymeric materials studied in this work. The objective was to determine how the mechanical properties of propellant grains could relate to the erratic and sometimes unsafe functioning of multiple-grain gun propellant charges, particularly at low temperatures.

Continuing efforts to understand the causes of abnormal propellant performance occurring during the ignition and combustion phases of the interior ballistics cycle, have implicated grain fracture as a suspect mechanism (Trask 1979; Russell & Goldstein 1961; Wachtell 1963; Horst, May, & Clarke 1977; Horst & Minor 1978). The prevailing hypothesis being that propellant grains in various stages of burning are accelerated and impacted against the confines of the gun chamber; this produces grain fracture that can result in sudden and significantly altered surfaces for burning and increased gas evolution or chamber pressurization rates. This provided the rationale for this work and accordingly localized the temperatures and strain rate used to those associated with the military use of propellants and further to those most contributing to grain breakup. The loading conditions employed were temperatures between room temperature and the lower limit of terrestial temperatures of military interest, approximately -60°C, and a rate, producing failure times of 2 to 10 milliseconds and consistent wtih rise times observed in the pressure-time curves of gun firing records (Hepner 1968).

The approach used was to identify loading conditions that could produce brittle propellant grain behavior, illustrate the effects of such brittleness on grain failure and on propellant burning and performance, and suggest parameters indicative of the brittleness and potential breakup of the propellant grains during gun firing.

EXPERIMENTAL APPROACH

Materials and Test Samples

The compositions of the propellants studied are given in Table 1; they are two-phase composites of a crystalline monopropellant of nitroguanidine, or RDX, dispersed in a plastic monopropellant binder of approximately a 1:1 ratio of nitroglycerin plasticizer and nitrocellulose polymer.

Table 1. Compositions in Weight Percent

	Propellant	
	Nitroguanidine	RDX
Nitrocellulose	27.40	29.3
(% Nitration)	(12.57)	(12.0)
Nitroglycerin	22.52	22.7
Nitroguanidine	45.77	5.0
RDX (Class E)	-	36.5
Ethyl Centralite	1.48	1.5
Dioctyl Phthalate	-	5.0
Potassium Nitrate	2.83	-
Graphite	0.12	-

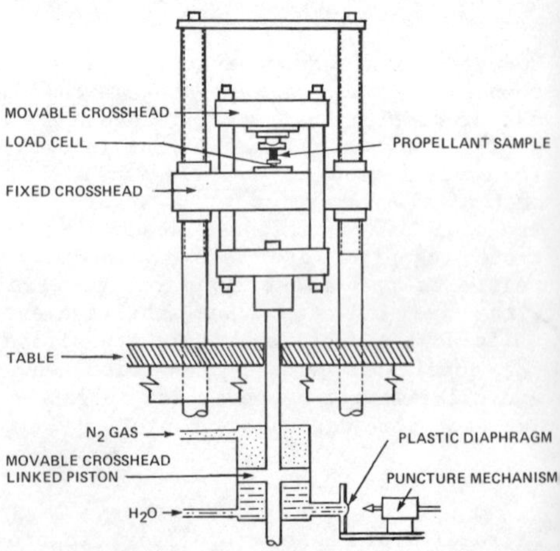

Fig. 1. Fast Rate Loading Apparatus.

The propellants were extruded into 1.27 cm diameter strands, cut and machined to 1.27 cm lengths for test samples having a length-to-diameter ratio of 1. The ends were as flat and parallel and as normal to the axis as possible. The nitroguanidine propellant was multi-perforated while the RDX propellant, which was an experimental formulation, was not; perforations are axisymmetric holes running the length of the grain, used for engineering the propellant burning by providing additional surfaces for burning. The ends of the test samples were coated with graphite at the start of the test. The sample was positioned within the load column, conditioned for approximately 1 hour at the test temperature, and tested in fast rate compression.

Mechanical Testing

Fast rate compression tests were performed in a Hesse-Eastern pneumatic/hydraulic loading apparatus that gave strain rates of approximately 10 $second^{-1}$ and loading times of typically less than 10 milliseconds. A sketch of the loading appratus is shown in Fig. 1. The system is pressurized through an equilibration of pressures above and below the movable crosshead-linked piston, the puncturing of a polyester diaphram suddenly vents the restraining liquid and allows the pressurized gas to expand and drive the movable crosshead toward the stationary crosshead; spacers are used to define the amount of travel.

A temperature chamber was designed to fit within the confines of the load column, Fig. 2a. Test temperatures were obtained using Optron temperature regulating equipment; the temperature variation during a 1 hour sample conditioning was ± 3°C. The load and the change in specimen length were measured using a load cell and a linear variable differential transformer (LVDT), respectively. The LVDT coils were mounted on the stationary crosshead and the LVDT core, via a calibration micrometer, on the movable crosshead. Load-time and displacement-time were recorded off a dual beam oscilloscope, Fig. 3.

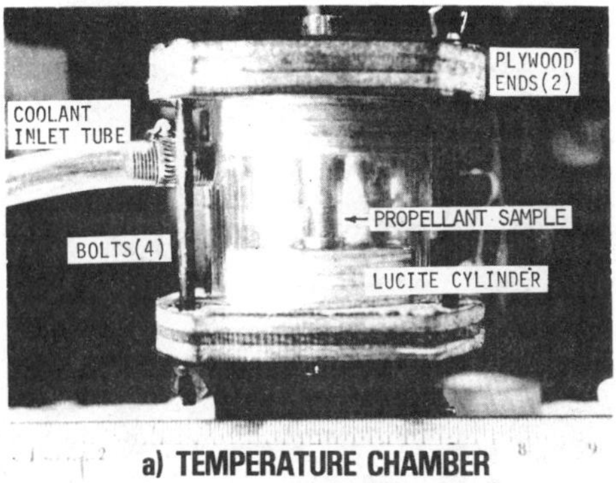

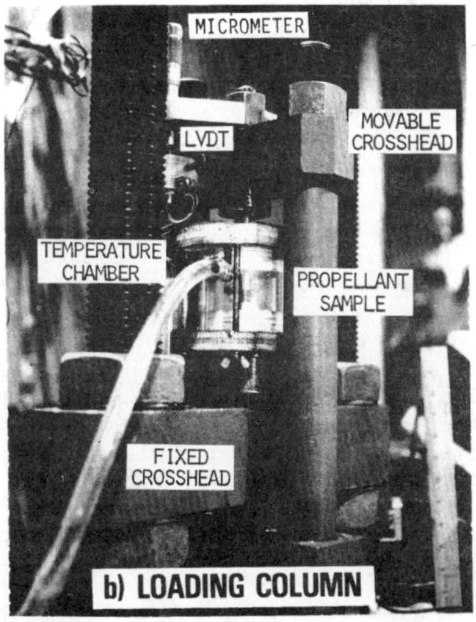

Fig. 2. Experimental Setup Showing the (a) Temperature Chamber and (b) Column Features.

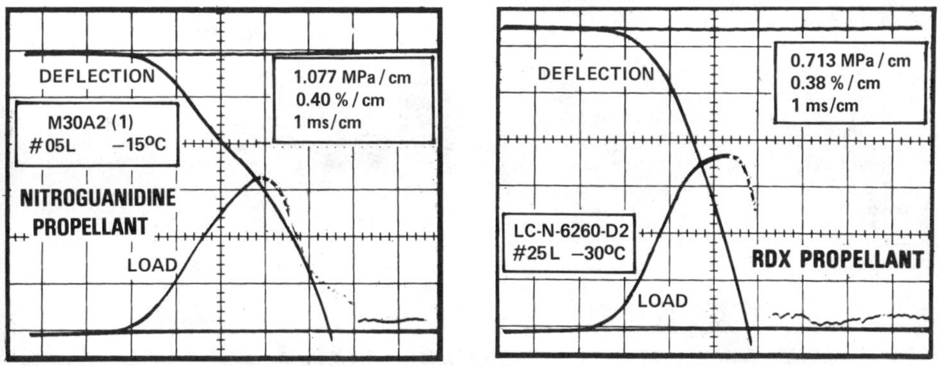

Fig. 3. Oscilloscope Records of Fast Rate Compression Tests on Gun Propellant. (a) Nitroguanidine, (b) RDX Propellant (lower traces-load, upper traces - displacement

Data reduction was performed in a film reader system that magnified the oscilloscope photographs, and used an x-y cross wire setup for reading and digitizing the traces. A computer program was used to produce stress-strain data and plots.

Closed Bomb Tests

Closed bomb tests are used to assure that the burning and pressure-time characteristics of successive propellant compare with an established standard of the same formulation (MIL-STD-286B, Method 801.1). The bomb is instrumented to record the rate of pressure buildup versus pressure. Comparison of the test propellant data with that of the standard provides relative performance information.

Thermal Analysis

Measurements of linear expansion coefficients and heat capacities for glass transition studies were made using the DuPont 900 Thermal Analysis System with the 940 Thermomechanical Analyzer (TMA) and Differential Scanning Calorimeter (DSC) cell base modules, respectively; the coolant used was liquid nitrogen. The TMA measurements were made on samples 0.635 cm long and 0.635 cm in diameter. The y-axis sensitivity was 6.5×10^{-4} cm probe displacement per cm of chart, and the heating rate used was 0.2°C per minute. The DSC measurements were made on chopped or granulated samples. The heat flow rate range was 5 millicalories per second, and the scan rate used was 1.25 K per minute.

RESULTS AND DISCUSSION

The effect of different ambient temperatures on the fast rate mechanical behavior of the propellants is seen in the corresponding changes in their stress-strain properties. In general, the elastic moduli and the compressive strengths increase as the temperature decreases and the strains to fracture decrease. The stress-strain curves for the nitroguanidine and RDX base propellants are shown in Fig. 4, with photographs of the deformed and fractured samples, in Fig. 5.

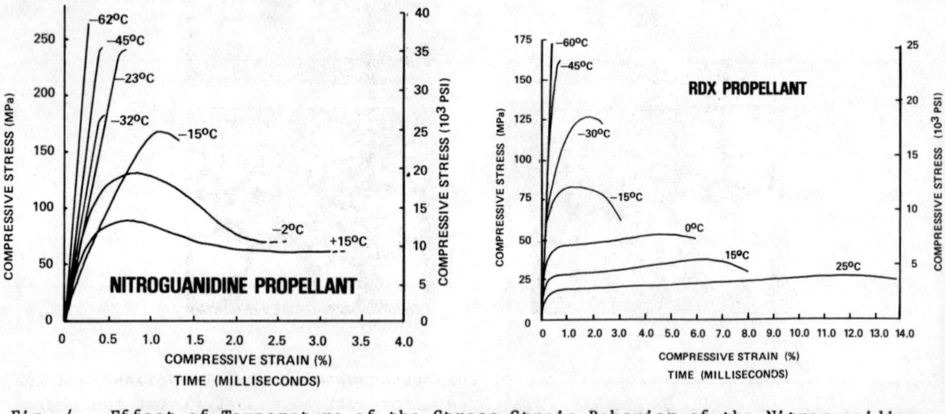

Fig. 4. Effect of Temperature of the Stress-Strain Behavior of the Nitroguanidine (MP) and the RDX Propellants, respectively.

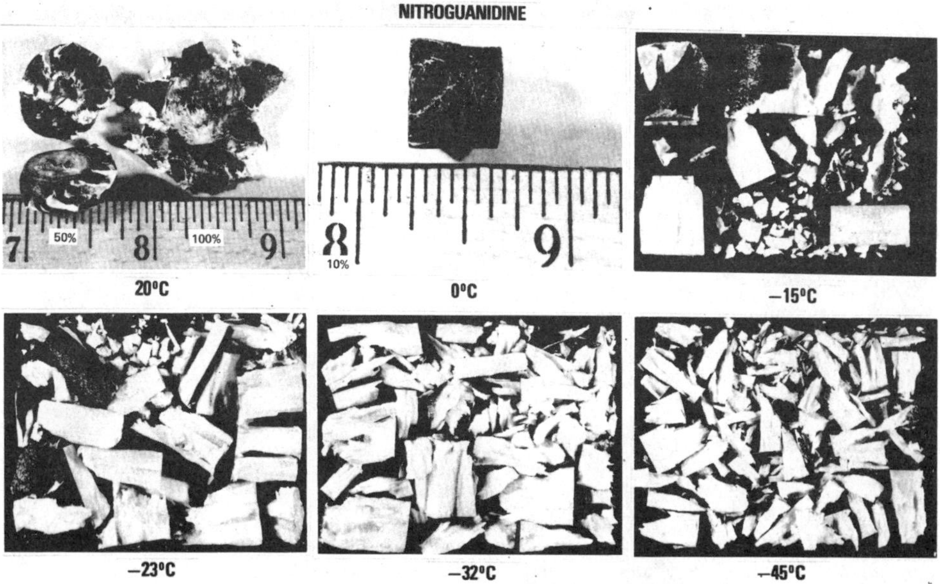

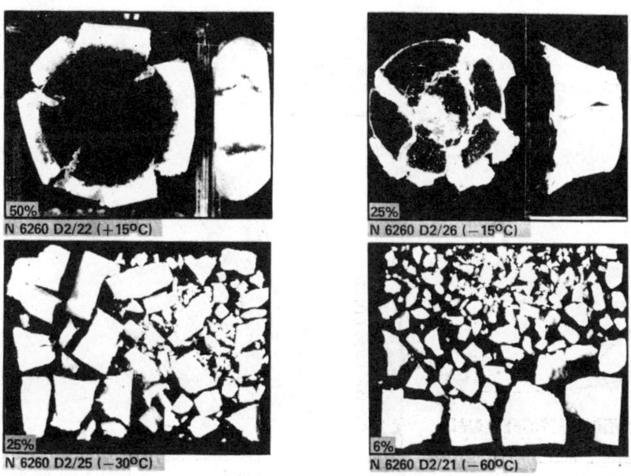

Fig. 5. Fast Rate Compression Failure of the Nitroguanidine (MP) Propellant and the RDX Propellant, respectively. (Note: For both cases, the first two photos represent 2 test samples each, the others are 1 test sample each.)

Deformation at room temperature shows obvious plastic, barrel-type failure; compression to well beyond stress-strain failure showed grain cohesion with large fissures and unpropagated surface cracks.

Stress-strain curves are identified with the mechanical failure exhibited by the propellant samples. Changes in the mechanical failure from nonfragmenting to fragmenting, occur at approximately -15°C for the nitroguanidine (MP) propellant and -30°C for the RDX propellant; corresponding strains at fracture are less than two percent. These data show that these propellants in thermal equilibrium at temperatures below approximately -30°C deformed at a rate of 10 second^{-1} (time to fracture: 4 milliseconds) will exhibit multiple-breakup and significantly affect the propellant burning. Faster strain rates at -30°C or lower temperatures at a strain rate of 10 second^{-1} will result in greater grain breakup can greater changes in the propellant burning.

The effect of the relative amounts of propellant breakup at temperatures of 20°C and at -45°C, on the burning characteristics of the propellant grains was examined using the instrumented closed bomb technique. Undeformed propellant grains were burned in the closed bomb to serve as standards for comparison and illustrate the rate of pressure buildup, dP/dt versus pressure, P curve typically expected. Closed bomb techniques have been employed successfully for screening triple base gun propellant lots for safe and unsafe ballistic use (Russel & Goldstein, 1961). Increases in relative quickness (vertical differences) greater than 15% above the standard were classified, and shown, to be unsafe. The closed bomb dP/dt-P curves for the nitroguanidine and the RDX propellants is shown in Figs. 6. These show increases in the relative quickness considerably greater than the 15% used for unsafe propellant lot classification, and changes in the pressure at which the maximum rate of pressure buildup occurs. Implicit here is that the mechanical behavior and grain failure produced at temperatures above -45°C would result in lesser changes in burning characteristics and that those produced below -45°C should result in greater changes in burning characteristics.

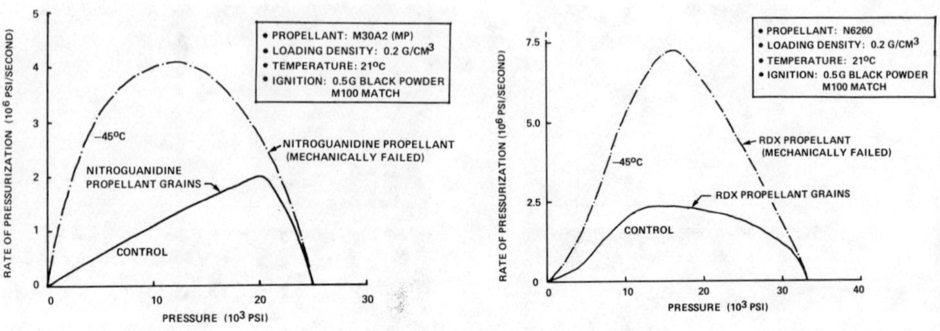

Fig. 6. Closed Bomb Combustion Traces of dP/dt vs. P for Undeformed and Mechan ically-Failed Nitroguanidine (MP) and RDX Propellants, respectively.

It is significant at this point to mention that aging may be an important factor affecting the propellant properties in such a way that brittle mechanical behavior and breakup occur under less severe loading conditions of temperature and strain rate. Table 2 lists increases in compressive strength and corresponding changes in the strain-at-fracture observed after a 1 year period in samples of the same propellant lots. It must be noted, however, that the samples were not stored under controlled conditions during this period, and even though this effect was also independently reported by another investigator (Zucker 1977) more systematic and careful studies are necessary to substantiate these observations.

Table 2. Aging effects noted in the fast rate mechanical properties of nitroguanidine propellant tested at the beginning and end of a 1 year period

Propellant	Temperature	Mechanical Properties	June 1977	June 1978
RAD 69711	Room temperature	Compressive strength (MPa)	53.8	86.9
		Strain-at-failure (%)	5.3	2.4
	-45°C	Compressive strength (MPa)	202.2	248.4
		Strain-at-failure (%)	6.6	0.5
RAD 69713	Room temperature	Compressive strength (MPa)	62.1	87.6
		Strain-at-failure (%)	4.9	2.0
	-45°C	Compressive strength (MPa)	211.8	218.7
		Strain-at failure (%)	4.4	0.9

The temperature at which the amorphousness of a polymer or a polymer system changes from a more flexible, rubbery state to a more rigid, glassy state is called the glass transition temperature. During the transition, significant changes in thermal expansion coefficient, viscosity, elastic modulus, etc., can be measured. The addition of a plasticizer to the polymer suppresses the temperature at which the transition takes place. The effect of a plasticizer content on the glass transition temperature of nitrocellulose is illustrated in Fig. 7.

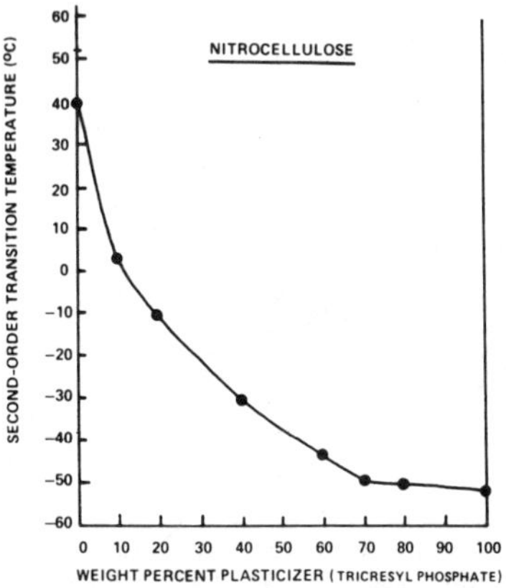

Fig. 7. Effect of Plasticizer Content on the Glass Transition

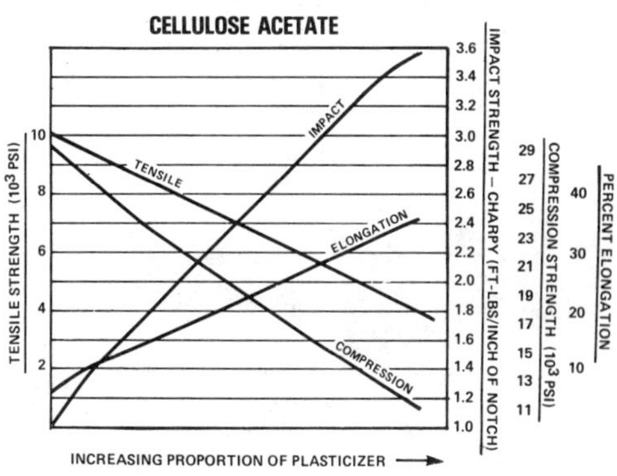

Fig. 8. Effect of Plasticizer (Tricresyl Phosphate) Content on Properties

If a plasticized polymer system, such as the nitroglycerin-nitrocellulose propellant system, is chemically and physically stable, and is insensitive to conditions that could alter the plasticizer concentration, then the glass transition temperature could be the most important characteristic parameter of the system. The mechanical properties for the system could then be evaluated in terms of plasticizer content as illustrated in Fig. 8, (Carswell & Nason 1944), and the generation of similar data at strain rates associated with the interior ballistics cycle would make the mechanical behavior of the propellant more predictable.

Thermal mechanical analysis was performed on both propellants to determine if this technique could be used to measure glass transition temperatures in these propellants. The glass transition temperature, T_g, might then be used as an index of the propellant's stability and provide a means of following aging and degradation processes. It could then be useful in predicting premature brittle failure under otherwise routinely safe loading conditions. A curve of the axial expansion as a function of temperature for the RDX propellant is presented in Fig. 9 showing the glass transition temperature for the RDX base propellant at approximately -57°C. A glass transition is suggested in the same temperature range for the nitroguanidine (MP) propellant, but was not as obvious. The use of differential scanning calorimetry to measure heat capacity and determine glass transition temperatures essentially confirmed the above, that is, a change in heat capacity for the RDX base propellant was found at approximately -57°C, but was not apparent for the nitroguanidine base propellant.

These glass transition temperatures, at -57°C, correlate with temperatures at which the fast rate mechanical breakups of these propellants exhibit advanced degrees of brittleness, Fig. 5, the failures occur within the Hookean portions of their stress-strain curves, Fig. 4. These results, although preliminary, suggest that glass transition temperatures in propellant characterization can be useful merit.

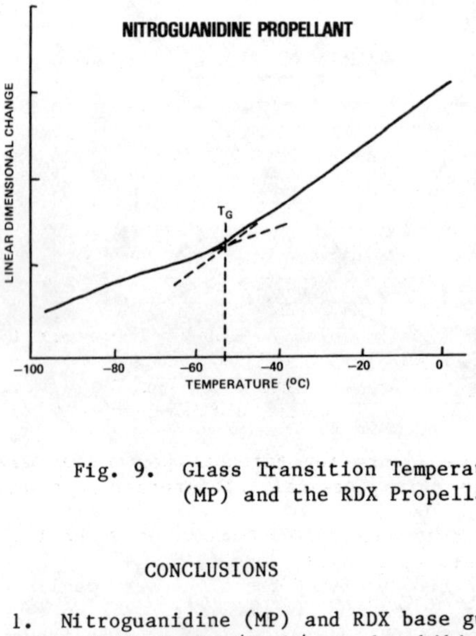

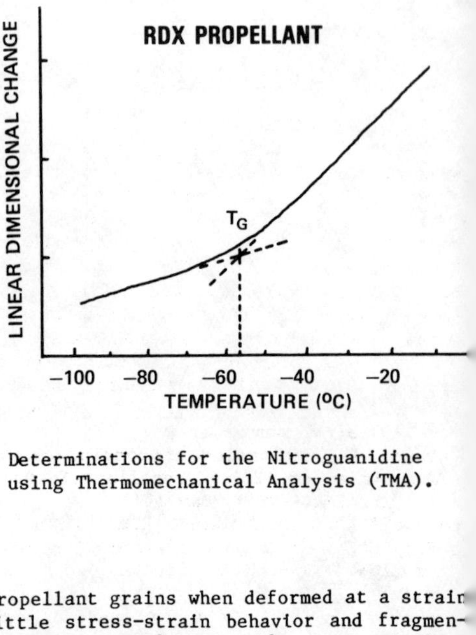

Fig. 9. Glass Transition Temperature Determinations for the Nitroguanidine (MP) and the RDX Propellants using Thermomechanical Analysis (TMA).

CONCLUSIONS

1. Nitroguanidine (MP) and RDX base gun propellant grains when deformed at a strain rate of 10 inches/inch/second exhibit brittle stress-strain behavior and fragmentation-type grain failure at ambient temperatures of -15°C and -30°C, respectively.
2. The effect of fragmentation-type grain failure on closed bomb, dP/dt versus P, curves show:
   a. Ballistically unsafe increases in relative quickness.
   b. Shifting of the maximum dP/dt to lower pressures, particularly for the multi-perforated grains where burning changing from progressive to degressive.

3. Changes in propellant mechanical properties, such as increases in strength and decreases in strain-to-failure, observed after a 1 year period, suggest a time-dependent embrittlement for the nitroguanidine propellant. Although the effect of aging requires systematic verification, the implication that resulting brittle propellant behavior and fragmentation-type grain failure would occur under less severe loading conditions or conditions initially determined to be normal and safe should be heeded.
4. The glass transition temperatures for the nitroguanidine and the RDX propellants tested were observed at approximately -57°C; these temperatures, coincidentally, compare with those producing Hookean, brittle, stress-strain behavior at the fast rate.
5. Glass transition temperatures may be useful for characterizing propellant behavior and grain failure, as well as for monitoring important time-dependent changes taking place within the propellant.

ACKNOWLEDGMENT

The authors wish to thank Messrs. R. Trask, E. Costa, B. Lehman and J. Zucker for many tutorial discussions, Mr. D. Anderson for his expertise in the thermal analysis, Messrs. C. Lenchitz, L. Shulman, and R. Young for their generous help with the closed bomb tests, and Ms. C. Novak for her timely performing of the word processing.

REFERENCES

Trask, R.L., ARRADCOM, Dover, NJ, Private communication of a 1954 paper presentation on the Processing Behavior of Nitroguanidine (M17) Propellant.

Russell, K. and Goldstein, W. "Investigation and Screening of M17 Propellant Production for Lots Subject to Poor Low Temperature Ballistic Performance," Technical Report PATR DR-TR: 7-61, Picatinny Arsenal, Dover, NJ, June 1961.

Wachtell, S. "Anomolous Ballistic Behavior of Propellants under Extreme Pressure Conditions," Tech Memo PATM 1228, Picatinny Arsenal, Dover, NJ, July 1963.

Horst, A.W., May, I.W., and Clarke, E.V., "The Missing Link Between Pressure Waves and Breechblows," 14th JANNAF Combustion Meeting, CPIA Publ. 292, Dec 1977.

Horst, A.W. and Minor, T.C., "Ignition-Induced Flow Dynamics in Bagged Charge Artillery," 4th International Symposium on Ballistics, 17-19 Oct 1978.

Hepner, L.D. "Final Report on Special Study of Setback and Spin for Artillery, Mortar, Recoilless Rifle and Tank Ammunition," Final Report USATECOM Project No. 9-7-0023-27, Jan 1968.

Zucker, J. "Evaluation of Bonding on Adhesive Agents for Effects Upon Mechanical Properties of Filled Propellant Systems," Picatinny Arsenal Laboratory Report No. 77, DRDAR-LCE-TF-1, May 1977.

Carswell, T.S. and Nason, H.K., "Effect of Environmental Conditions on Mechanical Properties of Organic Plastics," Modern Plastics, 21, 121, June 1944.

ANELASTIC CREEP BEHAVIOUR OF RR-58 ALUMINIUM ALLOY AT 180°C

L. O. Bueno

Departamento de Engenharia de Materiais, Universidade Federal de São Carlos, Caixa Postal 676, 13560 São Carlos, Brazil

ABSTRACT

The strain relaxation behaviour after full stress removals during creep of RR-58 aluminium alloy has been observed. Tensile creep tests were carried out at 180ºC, with stresses of 120, 170 and 230 MPa. Specimens were subjected to a series of unloading cycles after different creep time periods. In all tests strain relaxation was measured until its virtual saturation. At least two stages of relaxation were observed and attempts were made to interpret the results according to a model based on thermally activated dislocation kinectics.

KEYWORDS

Creep recovery; anelastic creep; anelastic strain; creep transients.

INTRODUCTION

Time dependent recovered strain following stress variations during creep has been studied by several authors (Lubahn (1953), Gibbs (1966), Alquist and Nix (1971), Davies and co-workers (1973), Threadgill and Mordike (1977), Morris (1978), Pahutova and Cadek (1979)). Different metals and alloys have been tested and attempts have been made to determine the physical mechanisms responsible for this behaviour. Gibbons and co-workers (1975) proposed a model for creep recovery based on the thermally activated unbowing movement of dislocation links. Lupinc and Gabrielli (1979) tested this model with success for γ' strengthened Ni-Cr base alloys and identified the occurrence of two stages of anelastic recovery in their experiments. Recently Gibeling and Nix (1981) also detected different stages of anelastic relaxation for both single and polycrystalline pure metals. This work describes a preliminary set of results regarding some aspects of creep anelasticity of RR-58, a precipitation-hardened aluminium alloy.

MATERIALS AND METHODS

The material was supplied by the Royal Aircraft Establishment (R.A.E.), Farnborough, in the form of sheets with 1.6mm thickness, grain size of 20 microns and the following composition (w%): 2.69Cu, 1.40Mg, 1.11Fe, 1.12Ni, 0.23Si. The alloy was solution-treated at 530ºC and aged at 190ºC for about 20 hours to reach its peak-hardness condition (150 HVN_{10}). Between the solution and ageing treatment some sheets were prestrained in tension and specimens were labelled accordingly by: A for 0%, B for 2%, C for 4% and D for 6% prestrain. In the present report, however, we are not concerned with the effect of prestrain on anelastic behaviour. Although differences were detected on the total amount of recovered strain for the various prestrain levels, the general trend of anelastic relaxation was the same for all specimens.

Four constant load tensile creep machines of high standard were also supplied by R.A.E. to be used in this programme. The equipment and the related extensometry, furnaces, grips, devices for specimen alignement, etc. were all described by Webb (1971, 1973, 1976) in detail. Changes in length of the specimens were monitored with capacitive transducers and an ASL Data Logger System. The maximum rate of data acquisition of this system was 1 reading /second and a series of lower rates up to 1 reading/10^4 seconds could be selected at any time during the tests.

All the experiments reported here were carried out at 180ºC and the selected creep stresses were 120, 170 and 230 MPa. Creep recovery was studied by fully removing the applied stress after different loading and reloading times, and following relaxation until the anelastic recovered rates decreased to very low values, usually between 10^{-6} to 10^{-7} h^{-1}. In all these tests a residual stress of 8 MPa was left in the system to mantain a certain degree of stiffness on the test assembly.

RESULTS AND DISCUSSION

Figure 1 shows schematically the loading histories of two specimens which were crept at 170 MPa for different time periods and left to relax after full stress removals. In the case of specimen B2, the secondary creep stage was fully reached on the 1st. loading cycle. In the 2nd. cycle, when the specimen was unloaded creep had just entered the secondary stage, but was still in primary in the other cycles. For specimen D1 the secondary creep stage was reached on all cycles.

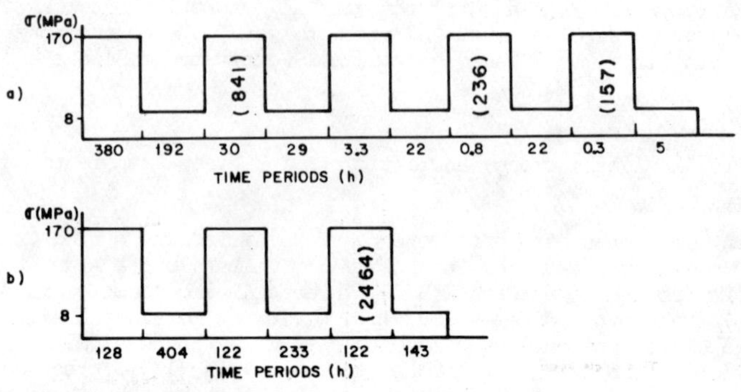

Fig. 1. Loading histories of: a) specimen B2; b) specimen D1. Values in brackets are creep strains introduced in each cycle in units of microstrain (10^{-6}).

Figure 2 shows similar diagrams for two other samples. Specimen D13 was subjected to creep stress cycles of 230 MPa and reached the secondary stage on the first two loading cycles. Specimen D12 was creep tested at 120 MPa and reached the onset of secondary creep stage only on the first loading cycle. From each relaxation experiment a number of 100 to 250 readings were recorded and plotted in the form of total strain recovered vs time. Figure 3 shows a typical strain relaxation curve which has been plotted in sections at different time scales. This graphical procedure was used for all relaxation curves to enable a more precise calculation of the anelastic strain rates during the test. Figure 4 shows mono-logarithmic plots of anelastic strain rate vs strain rate for the relaxation tests corresponding to Fig.1. This type of plotting has been suggested by Gibbons and co-workers (1975) for an analysis of anelastic

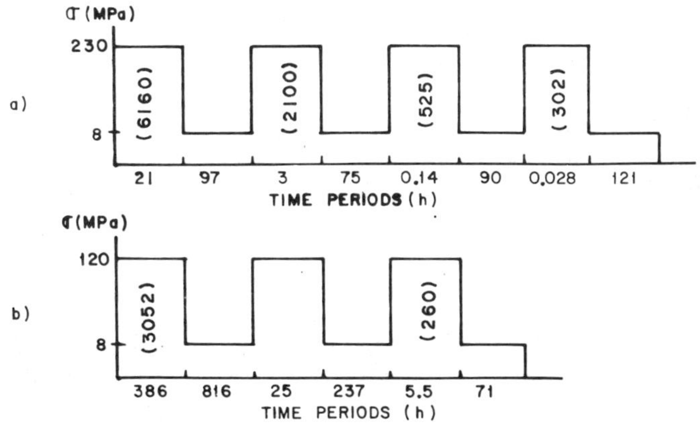

Fig.2. Loading histories of: a) specimen D13; b) specimen D-12

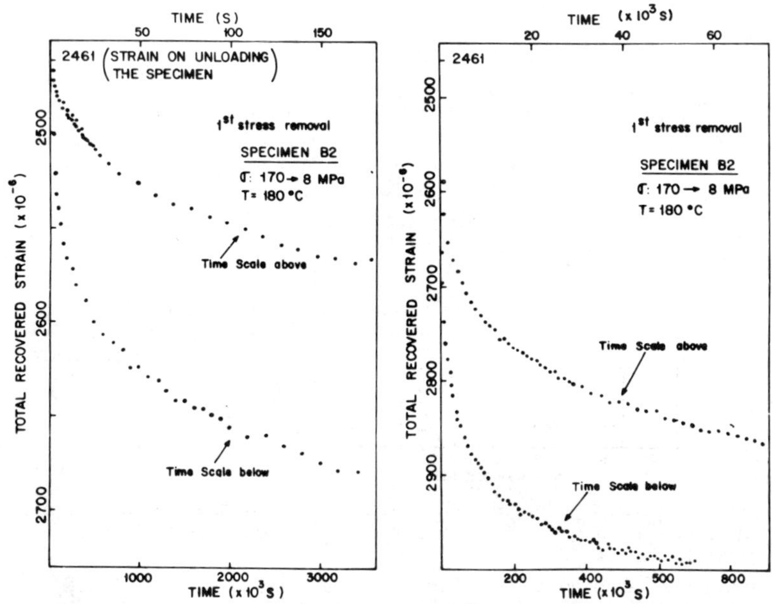

Fig.3. Data corresponding to the first unloading cycle of Fig.1.

creep based on the thermally activated unbowing of dislocations. This procedure was used as an attempt to analyse the present results. From Fig.4.a it is apparent that specimen B2 exhibited two relaxation regimes in the 1st. and 2nd. unloading cycles carried out during the secondary creep stage. The three other curves obtained from unloadings during primary creep showed only one regime of relaxation. Similar stages have been recently detected by Lupinc and Gabrielli (1979) and Gibeling and Nix(1981). Stage 1 corresponds to a fast relaxation regime which starts immediately after unloading the specimen,

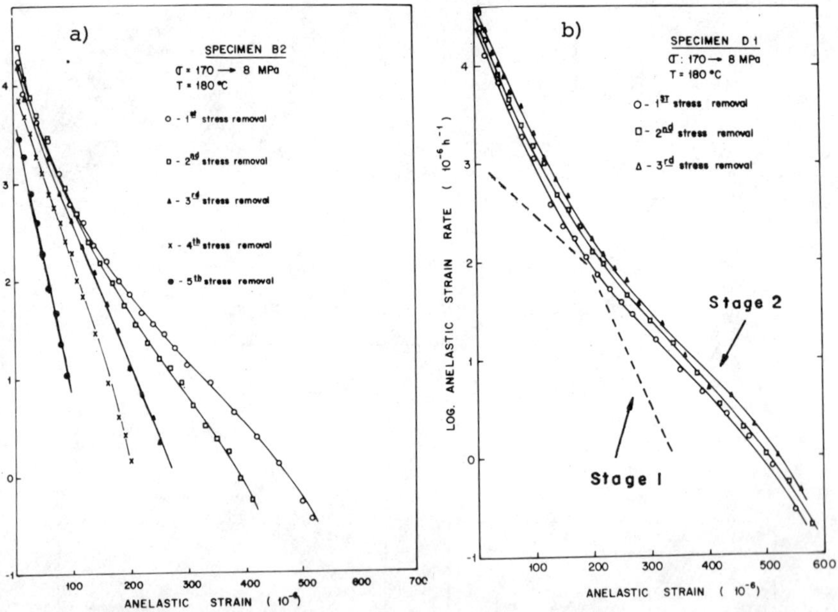

Fig.4. Log(anelastic strain rate) vs anelastic strain plots from: a) specimen B2; b) specimen D1. See Fig.1.

and Stage 2 corresponds to a slow relaxation regime which appears to operate in parallel with Stage 1. Fig.4.b. shows that two relaxation stages are also present on the three curves of specimen D1 unloaded during secondary creep. Figure 5 shows similar plots for the unloading experiments carried out with specimens D12 and D13. The presence of two stages is again clearly observed on the 1st. relaxation curve of specimen D12 unloaded from 120 MPa at the onset of secondary creep. Although a time period of 816 hours was allowed in this experiment, there was still no sign of a tendence to saturation of creep recovery. For the other two cycles there was a predominance of Stage 1 during relaxation. In the case os specimen D13 unloaded from the higher stress level of 230 MPa, the occurrence of Stage 1 is not evident. On the other hand these high stress tests present a clearer indication of the trend to saturation of the phenomenon. According to Fig.5.b. it seems that the saturation stage can persist for an appreciable part of the test since all curves present no positive sign of having reached a final level of anelastic strain.

The quantitative analysis proposed by Gibbons and co-workers (1975) considers that the anelastic strain rate, $\dot{\epsilon}_a$, can be expressed by the equation:

$$\dot{\epsilon}_a = A \exp(-Q/kT) \sinh\left[v\Delta\tau(1-\epsilon_a/\epsilon_a^s)/kT \right] \qquad ....(1)$$

where A is a structure sensitive constant, Q the activation energy, k Boltzman's constant, T absolute temperature, v the activation volume, $\Delta\tau$ the shear stress corresponding to the stress change $\Delta\sigma$, and ϵ_a^s the total anelastic strain after saturation. The term $\Delta\sigma(1-\epsilon_a/\epsilon_a^s)$ is considered to be a residual stress, σ_r, associated with the line tension which drives the bowing or unbowing movement of the dislocation links. Figure 6.a. shows that a good fit can be obtained with the hyperbolic sine equation on Stage 2 of the 1st.unloading

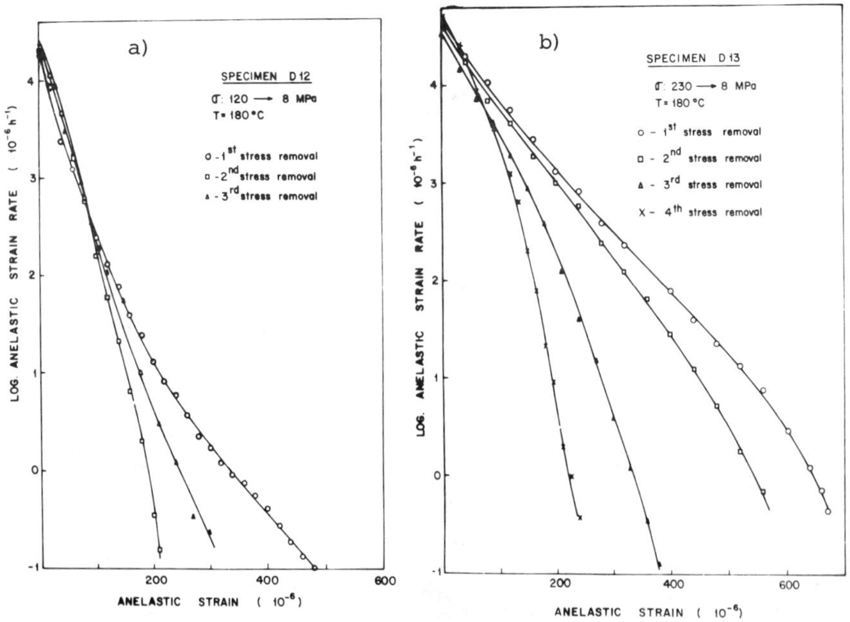

Fig.5. Log (anelastic strain rate) vs. anelastic strain plots from: a) specimen D12; b) specimen D13. See Fig.2.

cycle of specimen B2. In this plot the saturation strain was taken as 550 microstrain (1 microstrain = 10^{-6}). Stage 1 can be as well expressed by a sinh equation which coincides with a single exponential curve at the higher stress values of σ_r. In this way the parameters of Equation 1 can be calculated for the two Stages. In the case of Fig.6.a, for Stage 2, $v_2 = 0.58\ (nm)^3 \cong 25b^3$, and for Stage 1, $v_1 = 1.24\ (nm)^3 = 53b^3$. It should be pointed out, however, that these values correspond to the assumption of a unique value of ϵ_a^s for the relaxation test. If the mechanisms of each stage are operating in parallel, as seems to be the case, they will probably produce different saturation anelastic strains. in this case, by taking lower values of ϵ_a^s for stage 1, higher values of v_1 would be obtained.
Since sinh expressions with constant coefficients can be well fitted to each relaxation stage it seems that the parameters A, Q and v can be considered to remain constant throughout each relaxation process. This of course represents an oversimplification of the material behaviour and should not be taken as probable. Neverthless, Equation 1 can be easily integrated with respect to time if temperature, and the physical parameters Q, A and v are considered as constants during relaxation. The following expression is obtained:

$$\epsilon_a = \epsilon_a^s - \frac{1}{B}\ln\left[(1+\beta e^{-\alpha t}) / (1-\beta e^{-\alpha t})\right] \qquad ....(2)$$

where $B = v\,\Delta\sigma/kT.\epsilon_a^s$, $\beta = \mathrm{tgh}\,(B.\epsilon_a^s/2)$, $\alpha = A.B$

Equation 2 was tested with the same data corresponding to Fig.6.a. on Stage 2. The best values found for the coefficients were B = 0.015, β =1.0 and α =0.0115, with ϵ_a given in units of microstrain and t in hour. Fig.6.b. shows the original relaxation data fitted by Equation 2. It can be seen that the Equation does not fit the early part of the curve where Stage 1 predominates, but fits the data better and better as Stage 2 assumes control of the test. It should

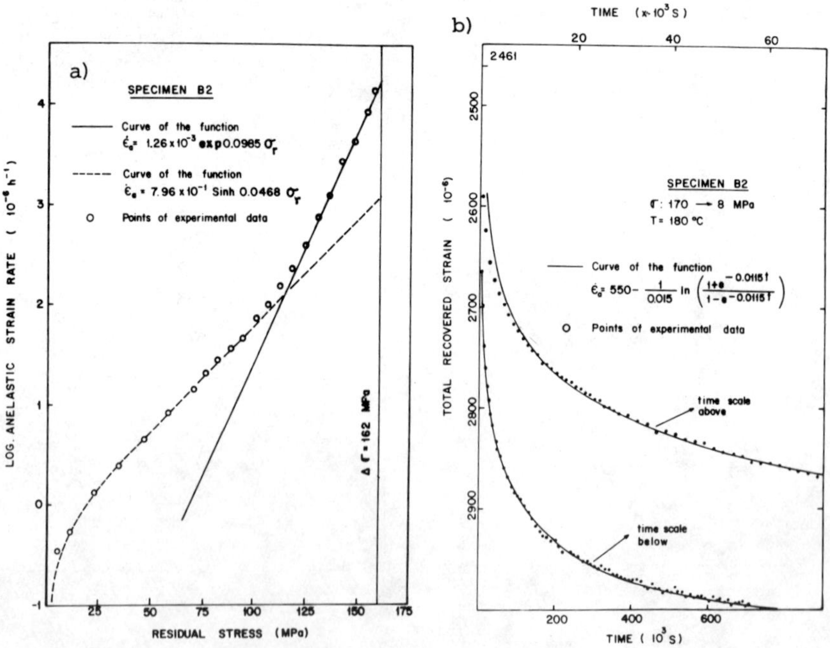

Fig.6. a) Log (anelastic strain rate) vs residual stress; b) Data fitted by the integrated rate equation for anelastic strain.

be noticed that this test was prolonged for 192 hours and that Stage 1 predominated over Stage 2 for about 1 hour only. The 2nd. relaxation curve of B2 and all the curves of D1 were analysed using both Equation 1 and 2 and the same kind of agreement was obtained. Although the analysis seems to give good results, it should be pointed out that the values of ϵ_a^s were always arbitrarily chosen. In fact, for all the relaxation tests reported here, there was never a positive indication of the attainement of definite values of ϵ_a^s. This uncertainty makes difficult a consistent check of the validity of the sinh rate equation. It is exactly on the portion of the relaxation curve where the saturation trend is manifested that the sinh equation has to be checked. At 170MPa, the curves of log $\dot{\epsilon}_a$ vs ϵ_a revealed only a limited number of points in this region. At 120 MPa, the situation was still worse since the saturation trend seems to have been displaced to lower strain rate levels. On the other hand at 230 MPa the curves reveal better this tendence to saturation. Tentatives to fit Equations 1 or 2 to the curves of Fig.5.b were not succesfull. Fig.7.a. shows for instance a plot of log $\dot{\epsilon}_a$ vs σ_r for the 3rd. relaxation test of D13. In this case relaxation was followed up to the lowest levels of strain rate resolution (10^{-7}/h). The value of ϵ_a^s was taken as 450 microstrain. It is clearly seen that a sinh equation that fits the initial straight line part of the curve fails to fit the data at the saturation stage. The sinh rate equation goes to saturation much more quickly than does the experimental data.
It should be mentioned also that the relaxation tests from the stress of 230MPa showed an appreciable difference between the instantaneous strain measured on unloading and the expected elastic strain corresponding to measured values of the Young Modulus at 180ºC. This difference was equal to about 150 microstrain for all the unloading cycles of specimen D13. At the other stresses this difference was lower: about 50 microstrain at 170 MPa and 20 microstrain at 120MPa.
Fig.7.b. shows a comparative plot of the 1st. relaxation curves of the specimens

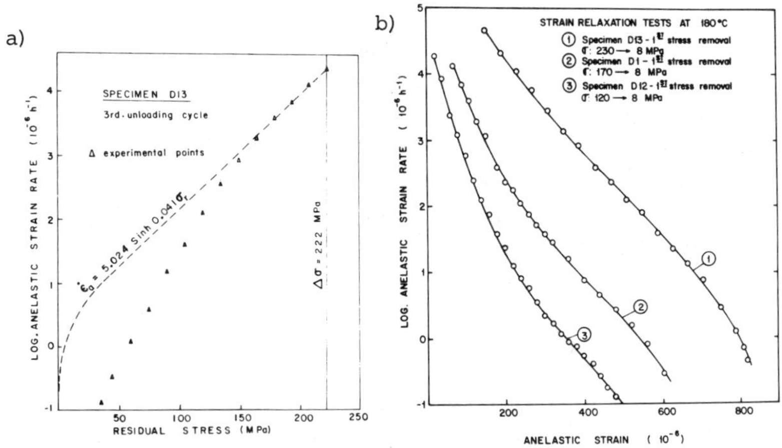

Fig. 7.a) Log $\dot{\epsilon}_a$ vs σ_r for the 3rd. relaxation data of specimen D13. b) Log $\dot{\epsilon}_a$ vs ϵ_a at different stresses.

tested at the different stress levels, with the incorporation of these differences in the anelastic strain readings. The absence of Stage 1 at the stress level of 230 MPa can be probably explained by the difficulty of the strain measuring apparatus to detect strain rate values higher than about 10^5 microstrain/h. At the lowest stress level of 120 MPa, the curve shows evidence of Stage 1 but fail to reveal the presence of the final stage leading to saturation. From a comparative analysis of Log $\dot{\epsilon}_a$ vs ϵ_a plots, it seems that the good fit obtained in Fig.6.a might be apparent. Care must be taken when trying to fit sinh equations to Stage 2 if only a small number of points are considered in the saturation range.

An extensive series of experiments was carried out with the same material at other temperatures and stresses and after different levels of forward creep strain including the tertiary stage (BUENO, 1980). These results will be discussed in more detail in a future publication.

ACKNOWLEDGEMENT

This work was carried out at the Engineering Materials Laboratories of The University of Southampton. The author would like to express his gratitude to Prof.R.L. Bell, Dr. D.J.Cartwright and Dr.R.C. Gifkins for the helpful discussions during this work. Thanks are also due to the following institutions: The Royal Aircraft Establishment, for the loan of the equipment and materials used in this research; The British Council and The Federal University of São Carlos, Brazil, for the financial support given to the author during his stay in England.

REFERENCES

ALQUIST, C.N. and W.D.NIX (1981), Acta Met., 19, 373-385.
BUENO, L.O. (1980), Ph.D. thesis-University of Southampton.
DAVIES, P.W., G.NELMES, K.R. WILLIAMS and B.WILSHIRE (1973), Met.Sc.J., 7, 87-92.
GIBBONS, T.B., V.LUPINC and D. McLEAN (1975), Met.Sc.J., 9, 437-444.
GIBBS, G.B.(1966), Phil.Mag., 13, 317-329.
GIBELING, J.C. and W.D. NIX, Acta Met., 29, 1769-1784.
LUBAHN, J.D. (1953), Trans. ASM, 45, 787-838.
LUPINC, V. and F. GABRIELLI (1979), Mat.Sc. & Eng., 37, 143-149.
MORRIS, D.G. (1978), J.Mat.Sc., 13, 1849-1854.
PAHUTOVA, M., J.CADEK and P.RYS (1979), Mat.Sc. & Eng., 39, 169-174.
THREADGILL, P.L. and B.L. MORDIKE (1977), Met.Sc.J., 11, 254-256.
WEBB, J.N. (1971), RAE Technical Report C.P. 1173.
WEBB, J.N. (1973), RAE Technical Report TR 73015.
WEBB, J.N. (1976), RAE Technical Report TR 76043.

BEHAVIOUR OF MATERIALS AT HIGH TEMPERATURES

D. R. Hayhurst* and F. A. Leckie**

***Department of Engineering, University of Leicester, Leicester LE1 7RH, UK**
****Department of Mechanical and Industrial Engineering, University of Illinois at Urbana-Champaign, Urbana, IL 61801, USA**

ABSTRACT

The paper presents the results of experiments carried out on materials subjected to sustained stress at high temperatures. The concept of continuum damage is introduced in the context of both material and component behaviour. It is shown how the results of experiments have been used to develop constitutive equations for deformation and rupture lifetimes by engineering and by materials scientists. The commonality of the approaches is shown for proportional loading. For non-proportional loading it is shown how the directional characteristics of the development of creep damage influences deformation and rupture behaviour. The consequences of the observations are discussed in terms of the design of structural components.

KEYWORDS

Cavity nucleation and growth; creep damage; continuum damage; multi-axial stress rupture; creep rupture; creep-fatigue.

INTRODUCTION

When metals operate for long periods, typically 10^5h , at temperatures in excess of $0.3\ T_m$, where T_m is the melting temperature of the base metal in degrees Kilvin, and at stresses below one third of the uni-axial yield stress consideration must be given to the possibility that the lifetimes of structural components may be limited either by excessive creep deformation or by failure due to creep rupture. In pressure vessels, or in bolted assemblies, where creep resistant materials are employed, lifetimes may be governed by creep rupture. For these types of components it is necessary to perform calculations which take account of the effects upon structural performance of two aspects of material behaviour. There are: the increased strain rates which occur during tertiary creep and the form of the multi-axial stress rupture criterion of the material. Before structural calculations can be performed it is first necessary to develop constitutive equations which accurately describe these phenomena.

In the development of constitutive equations two approaches have been used. The first is the so-called phenomenological method due to Kachanov (1960) and to Rabotnov (1969) which has been shown by Leckie and Hayhurst (1975) to be equivalent to a single state damage variable theory. The second is the approach of the materials scientist in which the physics of the microstructural processes are described either by experimental measurement (Dyson and McLean, 1977; Greenwood, 1973) or by the usage of suitable material models (Dyson, 1979; Cocks and Ashby, 1982). The effectiveness of both approaches has to be judged by their ability to describe the strain rates and lifetimes measured in long-term tests under both uni-axial and multi-axial stress.

The purpose of the paper is to highlight the importance of the growth of continuum damage in structural mechanics, to show how both engineering and materials scientists have developed constitutive equations to describe deformation and material degradation due to creep and to demonstrate the duality of the approaches for proportional loading. In addition, experimental results are presented which show that when the loading is non-proportional then it is necessary to take account of the directional characteristics of creep damage. The usage of such constitutive equations to determine the behaviour of components is briefly discussed for proportional and non-proportional loading.

CONTINUUM DAMAGE

At the lower stresses used in design applications, typically less than $0.3\,\sigma_y$ when the lifetimes are in excess of 10^5h, micro-structural changes take place. Measurements of the changes in volume which have been made during creep deformation by Boettner and Robertson (1961) indicate that defects begin to grow in primary creep. The growth of defects ultimately causes a monotonic increase in strain rate from a constant value during secondary creep, to an infinite value at failure. This part of the process, referred to as tertiary creep, can last for up to fifty percent of the lifetime and is therefore an important consideration in the design of structural components.

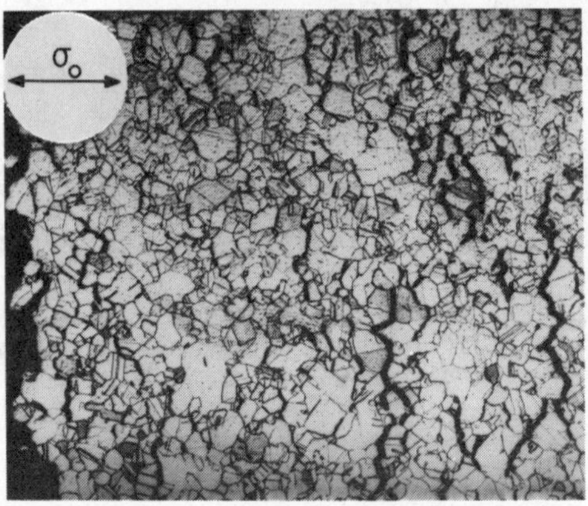

Fig. 1. Micrograph taken from a uni-axial tension specimen in copper, tested at 250°C; plane of failure is at the left-hand edge. (Mag. x 6)

An example of a uni-axial copper specimen which has undergone long term creep at 250°C under low stress is given in Fig. 1. The micrograph is of a diametral section taken after failure has occurred. Rupture has taken place due to the initiation and time-dependent growth of grain boundary cavities; the cavities grow and eventually link to form short grain boundary cracks, which in turn grow and link up to form macrocracks. The striking feature about Fig. 1 is that the density of cracking, or level of creep damage, is uniformly distributed throughout the homogeneously stressed region. Final separation of the specimen has taken place on a plane which is perpendicular to the direction of the applied stress. Similar features have been observed in tests carried out under pure shear by Hayhurst and Storåkers (1976),in tests carried out under equal bi-axial tension stress (Hayhurst, 1972) and in tests carried out under tri-axial tension stress in axi-symmetrically notched tension bars (Hayhurst, Leckie and Morrison, 1978).

The principal conclusion made from these tests is that when metals are subjected to sustained stress at high-temperatures they undergo material degradation due to the nucleation and the growth of grain boundary cavities; there is no evidence to support the notion that preferential growth of a single crack takes place; instead, it is clear that if the stresses are spatially uniform then creep damage forms during tertiary creep which is also spatially uniform.

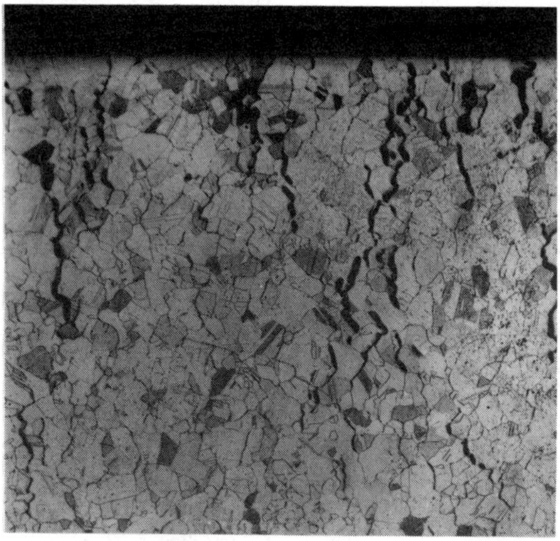

Fig. 2. Mid-thickness micrograph of a copper beam close to failure after creep due to a steady moment at 250°C. Black top edge denotes tensile surface.

The consequence of the stress field being spatially non-uniform may be examined from the results of tests carried out on beams subjected to uniform bending moments. On elastic loading the stress varies linearly through the depth of the beam from a tensile value at the top to a compressive value at the bottom. As creep and the growth of damage occur the stress distribution becomes non-linear but the stresses preserve their original signs. In Fig.2 it may be seen that grain boundary fissures and cracks have grown most rapidly nearer to the surface of the copper beam. The effect of the lower stresses can be observed some distance away from the surface of the beam, there is little evidence of the formation of damage close to its centre where the stresses are small. It is evident from these results that creep damage is distributed in a manner which has similar characteristics to the distribution of stress. Alternatively it may be said that creep damage possesses a field or continuum property similar to that of stress, hence the term CONTINUUM DAMAGE.

The effects of this property are most dramatically illustrated in the case of plane strain structures which contain sharp cracks which grow by creep. For these situations the stresses are non-uniform and the creep damage is confined to a narrow region ahead of the crack.

This is shown in the micrograph of a double edge-cracked tension specimen in

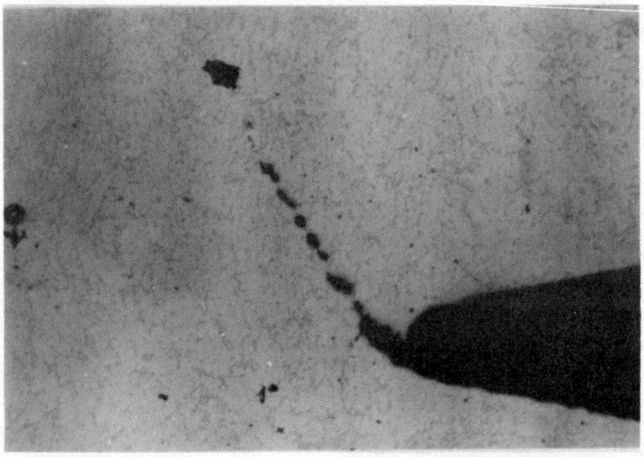

Fig. 3. Crack tip shear damage, close to failure, in a double edge cracked plane strain aluminium alloy specimen tested at 150°C under steady load (Mag. x 48).

Fig. 3 for an aluminium alloy tested at 150°C. Creep damage may be observed to be confined to a plane which is inclined at an angle of approximately 55° to the plane of the initial crack. The strong dependence of the mode of crack growth upon the characteristics of material behaviour is shown in the micrograph of Fig. 4 for the same specimen geometry as that shown in Fig. 3; the material is 316 stainless steel and the specimen was tested in tension at 550°C. Crack growth has initially taken place on the plane of the original crack but at 60% of the lifetime it can be seen that widespread damage has formed at the crack tip, and the direction of crack growth has begun to move away from the original crack plane.

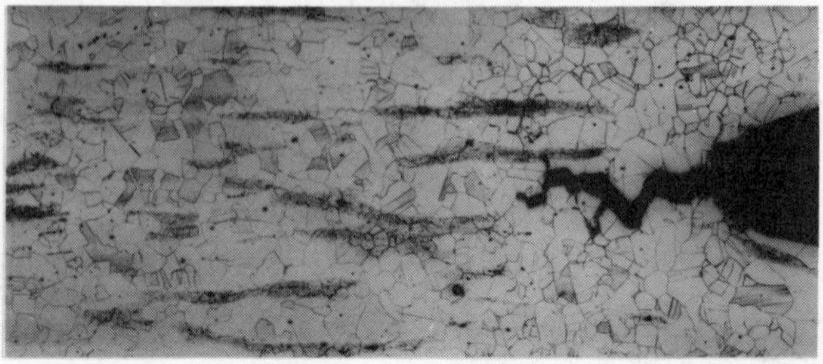

Fig. 4. Mid-thickness micrograph taken from the minimum section of a double edge cracked plane strain tension specimen in 316 stainless steel tested at 550°C at 60% of the lifetime.

The apparent different types of behaviour in the aluminium alloy and in the 316 stainless steel has been shown (Leckie and Hayhurst, 1980; Hayhurst, Brown and Morrison, 1983) to be due to the growth of continuum damage confined in the region of the crack, to the stress redistribution which results, and to the dependence of the rate of growth of creep damage upon the multi-axial stress state. Before the lifetimes of structural components, which contain either stress concentrations or cracks, can be predicted using the techniques available in structural mechanics, it is necessary to have available constitutive equations which accurately reflect the physical processes which control material degeneration. In the next section the development of such equations is discussed for steady loading conditions.

CONSTITUTIVE EQUATIONS FOR STEADY LOAD AND TEMPERATURE

Constitutive equations have been developed for primary and secondary creep deformation under both uni-axial and multi-axial stress conditions. The equations are well proven particularly for steady-load conditions and they will merely be referred to here. The majority of the section will be devoted to showing how uni-axial primary-secondary-tertiary creep behaviour can be described using a single state damage variable theory and how this theory can be generalised for steady multi-axial stress conditions.

Primary-Secondary Creep

The uni-axial creep strain rate is usually described by Norton's law,

$$\dot{v}/\dot{v}_o = (\sigma/\sigma_o)^n \; K(t) \; , \tag{1}$$

where n is a material constant, $\dot{v}_o$ is the uni-axial strain rate due to the stress σ_o and $K(t)$ is a function of time selected to describe the decrease in strain rate with time.

The generalisation of equation (1) to multi-axial stresses has been carried out by Odqvist (1974) and results in the following equation

$$\dot{v}_{ij}/\dot{v}_o = (3/2)(\sigma_e/\sigma_o)^{n-1}(S_{ij}/\sigma_o)K(t) \tag{2}$$

where $S_{ij} = \sigma_{ij} - \delta_{ij}\sigma_{kk}/3$. Both equations (1) and (2) have been verified by the test results of Johnson et al (1962).

State Variable Description for Uni-axial Continuum Damage

If, for convenience of discussion, the primary portion of the creep curve is neglected then the creep rate increases from the initial steady state value as damage causes an advance into tertiary creep. The steady-state creep rate can be expressed as a function of the applied stress σ alone. In order to account for the increase in strain rate during a constant stress test it is necessary to introduce a new variable into the strain equation. Since the increase in strain rate is the result of a damage process the variable ω introduced is referred to as the damage state variable. The strain rate equation then takes the form

$$dv/dt = f(\sigma,\omega) \; , \tag{3}$$

where f is a function which has yet to be defined. However, in the undamaged state when $\omega = 0$ the equation must reduce to the form observed for steady state conditions. To define the strain rate it is necessary that the value of ω as well as the stress σ be known and consequently an equation must be introduced which defines the growth of the damage state variable. The assumption that the damage rate depends both on the current state of stress and damage gives the equation

$$d\omega/dt = g(\sigma,\omega) \; . \tag{4}$$

The problem is how to find the functions f and g in a systematic manner. This may be achieved by following a testing procedure suggested by Leckie and Hayhurst (1975) which makes use of the results of tests with step changes in stress. For example, consider a creep test carried out at the stress σ , to the point b shown on the curve in Fig. 5, if, at this point the stress is increased to σ_2 for the remainder of the specimen lifetime a creep curve results which may be identified with the portion, bcde, of the curve for the constant stress σ_2 , shown in Fig. 5. In this way the points b on the curves for σ_1 and σ_2 may be shown to have the same value of the state variable ω_2 . By carrying out a series of step loading and unloading tests for the stresses σ_1, σ_2, σ_3 and σ_4 it is possible to construct the state

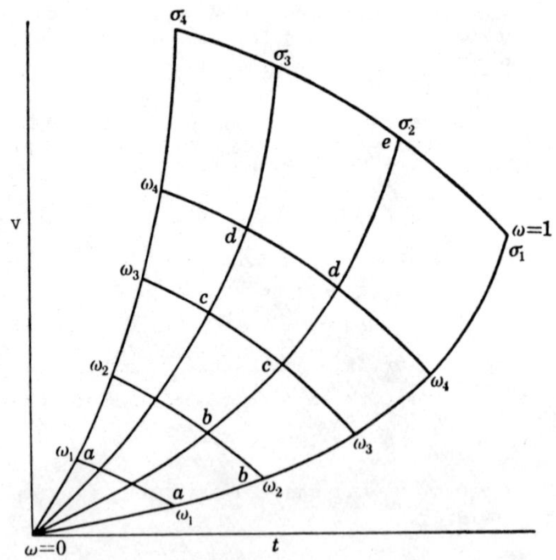

Fig. 5. Contours of constant damage

variable contours $\omega_1 < \omega_2 < \omega_3 < \omega_4 = 1$ shown schematically in Fig. 5. This information can then be used to predict the variation of strains and damage as a result of variable stress histories. The procedures described above are applicable provided that the state of damage can be described by a single damage parameter. If a single state parameter suffices it might still be necessary to modify the equations (3) and (4) for loading and unloading structures. This is certainly the case for creep deformations in the primary region when it is found that that the growth laws of dislocation density differ for loading and unloading. This point has been discussed by Leckie and Ponter (1974).

Rabotnov-Kachanov Equations or Effective Stress Approach to Uni-Axial Creep

Unfortunately it is difficult and time consuming to conduct the experiments described in the previous section and the variable stress experiments which have been performed are limited to the prediction of rupture life under cyclic conditions of loading. Faced with this difficulty Rabotnov (1969) proposed modifications of constitutive equations first suggested by Kachanov (1958) which described existing experimental results. A similar or effective stress approach has been used by Lemaitre and Chaboche (1978). For uni-axial stress tests the growth laws are assumed to have the simple form:

$$\dot{v}/\dot{v}_o = (\sigma/\sigma_o)^n \ K(t)/(1-\omega)^n \ , \quad \text{and} \tag{5}$$

$$\dot{\omega}/\dot{\omega}_o = (\sigma/\sigma_o)^\chi/(1+\phi)(1-\omega)^\phi \ , \tag{6}$$

where χ , ϕ and $\dot{\omega}_o$ are constants. In a constant stress test the value of ω monotonically increases from zero to a value of unity at failure. When $\omega = 0$ equation (5) reduces to Norton's law, given by equation (1), and when $\omega = 1$ the strain rate is infinite. Integration of equation (6) between the limits $\omega = 0$, at $t = 0$, and $\omega = 1$, at $t = t_f$, yields the experimentally observed expression for the rupture lifetime.

$$t_f = 1/\dot{\omega}_o \ (\sigma/\sigma_o)^\chi \ . \tag{7}$$

Equations (5), (6) and (7) have been shown (Hayhurst, 1973) to accurately describe uni-axial creep behaviour.

Multi-Axial Generalisation of State Variable Description

Generalised strain rates. The generalisation of equations (5) and (6) for multi-axial stresses has been achieved by making the assumption that the influence of continuum damage on the deformation rate processes is scalar in character. Equation (2) can then be written as:

$$\dot{v}_{ij}/\dot{v}_o = (3/2)(\sigma_e/\sigma_o)^{n-1}(S_{ij}/\sigma_o)K(t)/(1-\omega)^n \; . \qquad (8)$$

Equation (8) has been verified using the results of Johnson et al (1962) for tension-torsion tests. If in such a test, γ is the shear strain, v is the axial strain and γ_f and v_f denote their values at failure, then it may be shown from equation (8) that the ratio $(\gamma/\gamma_f)/(v/v_f)$ must remain constant throughout the test. The experimental verification of this may be seen in Fig. 6, where the results of the tests by Johnson et al on copper and aluminium have been plotted. The implication of this result is that the state variable ω has a scalar interpretation in the strain rate equation. In addition it is necessary to describe the effect of stress state on the rate of change of ω.

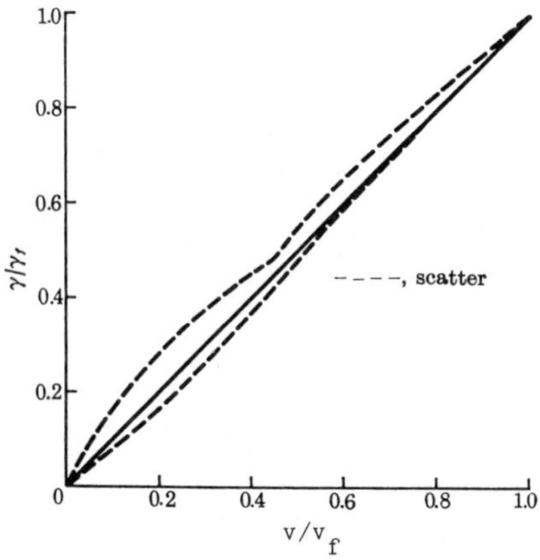

Fig. 6. Variation of axial and shear creep strains for copper and aluminium alloys

Isochronous rupture surface. In the generalisation of equation (6) the effects of stress-state must be reflected; this is achieved by the introduction of the homogeneous stress function $\Delta(\sigma_{ij}/\sigma_o)$ as follows:

$$\dot{\omega}/\dot{\omega}_o = \Delta^{\chi}(\sigma_{ij}/\sigma_o)/(1+\phi)(1-\omega)^{\phi} \quad (9)$$

Integration of equation (9) for the conditions $\omega = 0$, $t = 0$ and $\omega = 1$, $t = t_f$ yields the experimentally observed result $t_f = 1/\dot{\omega}_o \Delta^{\chi}(\sigma_{ij}/\sigma_o)$, which may be normalised to give

$$t_f/t_o = 1/\Delta^{\chi}(\sigma_{ij}/\sigma_o) \; ; \quad (10)$$

substitution of $t_f = t_o$ gives the equation of the isochronous surface $\Delta(\sigma_{ij}/\sigma_o) = 1$. The isochronous surface connects points in stress space for which a constant lifetime may be observed. Experimental observations, discussed by Hayhurst (1972), have shown that to describe material behaviour in principal plane stress space ($\sigma_1 > \sigma_2$, $\sigma_3 = 0$) it is necessary, in general, to use the maximum principal tension stress, the second deviatoric stress invariant J_2' $(=\sigma_{ij}'\sigma_{ij}'/2)$ and the first stress invariant $J_1(=\sigma_{ii})$; the corresponding expression for equation (10) is:

$$t = M\{(\alpha\sigma_1 + \beta J_1 + \gamma J_2'^{\frac{1}{2}}\}^{-\chi} \quad , \qquad (11)$$

where M is a constant independent of stress, α, β and γ are constants. Substitution of $t = t_f$ in equation (11) results in the isochronous loci shown in Fig. 7.

In practice the β-term is usually small and the behaviour of most materials can be represented using $\beta = 0$. Under these conditions $\alpha + \gamma = 1$ and equation (11) can be written as

$$t = M\{\alpha\sigma_1 + (1-\alpha)J'^{(\frac{1}{2})}\}^{-\chi} \quad ,$$

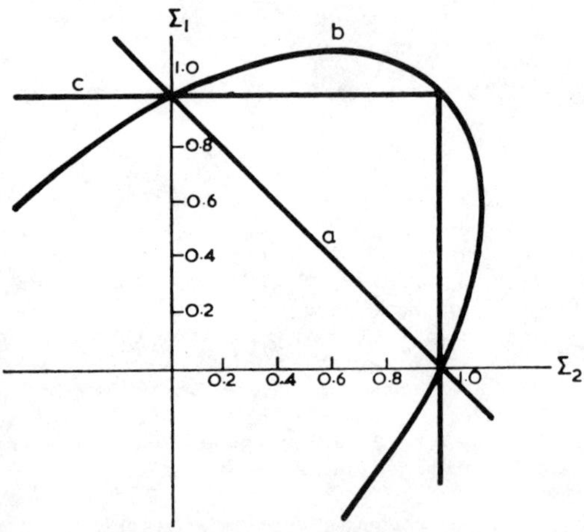

Fig. 7. Loci of constant rupture time for values of (α,β,γ): maximum, (a) hydrostatic stress (0,1,0), (b) octahedral stress (0,0,1), and (c) tensile stress (1,0,0).

or, to a close approximation by

$$t = M\{\alpha\sigma_1 + (1-\alpha)\sigma_e\}^{-\chi}. \quad (12)$$

The term in brackets corresponds to the function $\Delta(\sigma_{ij})$ introduced in equation (9). The corresponding damage rate equation is

$$\dot{\omega}=\{\alpha\sigma_1+(1-\alpha)\sigma_e\}^{\chi}/(M(1+\phi)(1-\omega)^{\phi}). \quad (13)$$

The material constants M, ϕ and χ may be determined from the results of uni-axial tests but the determination of the constant α requires the results of rupture tests to be obtained under two different states of stress. The tests usually carried out to determine α are uni-axial and pure shear tests ($\sigma_1 = -\sigma_2$, $\sigma_3 = 0$). The parameter α is zero for aluminium alloys (Hayhurst, 1972), 0.7 and 0.75 for copper and 316 for stainless steel respectively (Hayhurst, Dimmer and Morrison, 1983). Cane (1981) has shown that for ferritic steels the function $\Delta(\sigma_{ij}/\sigma_o)$ takes the form $\Delta(\sigma_{ij}/\sigma_o) = (\sigma_e^{\chi}\,\sigma_1^{\chi-\zeta})$, where χ and ζ are material constants.

The Approach of the Materials Scientist

In this section an attempt is made to present the approach of the materials scientist and to identify those features which are common with the single state damage variable approach described earlier. A general formulation for the growth of damage at grain boundaries has been given by Ashby and Raj (1975). The growth of damage is expressed in terms of two mechanisms. One mechanism, referred to as nucleation, gives a measure of the rate at which grain boundary voids are formed. The other mechanism, referred to as growth, gives a measure of the rate of growth of void size.

The nucleation rate is represented by dn/dt and is measured in the number of voids formed per unit time on unit area of grain boundary. Holes nucleate at points of strain concentration which can occur at inclusions contained in the grain boundary or at the extremities of slip bands within the crystal. The growth of holes can be the result of the diffusion of vacancies along the grain boundary or of concentration of creep strains at inclusions. Since voids are forming and growing continuously it is necessary to use the integral relation derived by Ashby and Raj to obtain a measure of the total damage. Suppose at time τ the nucleation rate is $\dot{n}(\tau)$ so that in time interval $d\tau$ the number of new voids formed is $\dot{n}(\tau)d\tau$. At a later time, t, these holes will be growing and the rate of increase of cross sectional area of the voids formed at time τ is $\dot{a}(t,\tau)$. During the time interval $(t-\tau)$ the holes formed at time τ will then have a cross sectional area

$$\dot{n}d\tau \int_{\tau}^{t} \dot{a}(t,\tau)dt \; .$$

The total area of the voids at time t_f is then

$$A(t_f) = \int_{\tau=o}^{\tau=t_f} \dot{n}(\tau)d\tau \int_{t=\tau}^{t=t_f} \dot{a}(t,\tau)dt \; . \qquad (14)$$

Instead of using A as a measure of damage, some authors prefer to use the volume of the voids. Using the same method, illustrated above, the total volume V of the voids is

$$V(t_f) = \int_{\tau=o}^{\tau=t_f} \dot{n}(\tau)d\tau \int_{t=\tau}^{t=t_f} \dot{\nu}(t,\tau)dt \; , \qquad (15)$$

where $\dot{\nu}$ is the volume growth rate of the voids.

The formulation is quite general but it implies that to calculate damage it is necessary to keep track of the size of each void as it is formed. This is equivalent to a multi-state variable theory in which the number of state variables corresponds to the number of voids and is increasing with time. It is unlikely that the integral can be calculated in closed form except for simple loading histories or for special forms of the rate equations. Various studies have been made with the objective of the formulation of equations which define the nucleation and void growth rates. Two particular examples are now discussed against the background of the state variable theories already discussed.

The Greenwood Equations

Greenwood (1973) has studied the growth of creep damage in copper at a temperature of 500°C. The tests were performed under conditions of constant uni-axial stress and measurements made of the number and size of voids. The visual observations indicate that the voids are all approximately the same size which implies that when a void is first formed it grows rapidly in size until it catches up with the voids formed earlier. It is possible to express this observation in a suitable mathematical form (Leckie and Hayhurst, 1977) which can be substituted in equation (14), but it is easy to see that the total cross-sectional area is simply $A = na$. No attempt was made to study the effect of damage on strain rate although it appears that the tertiary strains were little greater than those predicted by steady state theory. Suppose that a uni-axial test is conducted at constant stress σ_o. Let the rupture time be t_o when the strain is v_o , the hole density n_o and the average volume of the holes ν_o . With reference to these physical values the growth equations for a uni-axial stress σ are

$$\begin{aligned} d(n/n_o)/dt &= (\sigma/\sigma_o)^2 d(v/\dot{v}_o)/dt \; ; \\ d(\nu/\nu_o)/dt &= (\sigma/\sigma_o)/t_o \; ; \\ d(v/v_o)/dt &= (\sigma/\sigma_o)^5/t_o \; . \end{aligned} \qquad (16a\text{-}c)$$

In the first of these equations the nucleation rate is proportional to the strain rate but is, in addition, proportional to the square of the applied stress. In the second the volume rate of the voids is assumed to be controlled by a diffusion process which is proportional to the stress. The third equation illustrates that the strain rate is proportional to the fifth power of applied stress and the effect of damage on strain rate is neglected. The damage is defined as

$$\omega = n\nu^{2/3}/n_o\nu_o^{2/3} = A/A_o \;, \tag{17}$$

and the rupture condition as $\omega = 1$. This is the same condition as proposed by Greenwood (1973) who used the concept of a critical area fraction at failure. For multi-axial states of stress Leckie and Hayhurst (1977) suggested that equations (16) and (17) take the form

$$\begin{aligned} d(n/n_o)/dt &= (\sigma_1/\sigma_o)^2 d(v_e/v_o)/dt \;; \\ d(\nu/\nu_o)/dt &= (\sigma_1/\sigma_o)/t_o \;; \\ d(v_e/v_o)/dt &= (\sigma_e/\sigma_o)^5/t_o \;; \end{aligned} \tag{18a-c}$$

and the damage is $\omega = n\nu^{2/3}/n_o\nu_o^{2/3}$.

Integration of the above equations, for the rupture condition $\omega = 1$, gives the isochronous surface for rupture time t_o :

$$\Delta(\sigma_{ij}/\sigma_o) = (\sigma_1/\sigma_o)^{8/23}(\sigma_e/\sigma_o)^{15/23} = 1 \;. \tag{19}$$

The corresponding isochronous rupture surface is shown in Fig. 8. For constant stress, σ , the equations are easily re-expressed to give the following expression for damage rate:

$$d\omega/d(t/t_o) = (5\;\Delta^{23/5}(\sigma_{ij}/\sigma_o)/3)\omega^{2/5} \;. \tag{20}$$

If now the damage rate equation is written in the following particular Rabotnov-Kachanov form

$$d\omega/d(t/t_o) = 2\;\Delta^{23/5}\;(\sigma_{ij}/\sigma_o)/3(1-\omega)^{\frac{1}{2}} \;, \tag{21}$$

with the rupture conditions $\omega = 1$, it is found that this gives a damage-time graph which is closely represented by equation (20).

The corresponding strain rate expression to equation (21) is given by

$$d(v_e/v_o)/d(t/t_o) = (\sigma_e/\sigma_o)^5 \;. \tag{22}$$

These equations are a particular form of the Rabotnov-Kachanov equations in which the damage state variable ω is the normalised form of the damage expressed by equation (14).

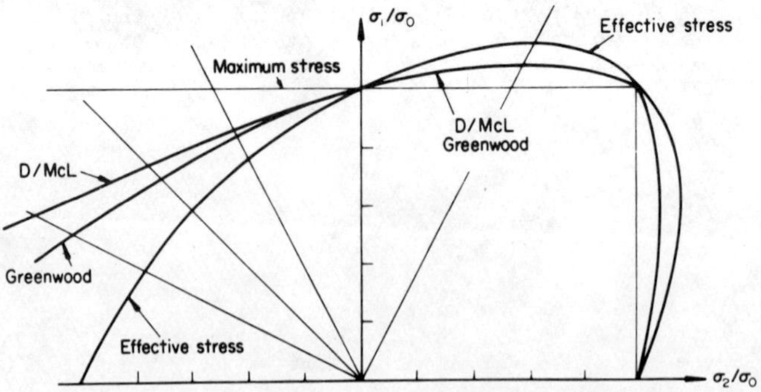

Fig. 8. Predictions of isochronous rupture loci

The Dyson-McLean Equations

Dyson and McLean (1977) have performed constant uni-axial tension and torsion tests on Nimonic 80 A at 750°C. In addition to measuring tertiary strains and times to rupture, measurements were also made of the number of voids per unit area of grain boundary and of the total volume of voids. For constant uni-axial stress tests carried out at stress σ_o , the rupture time is t_o at which time the void density is n_o and the total volume of the voids is V_o . The constitutive equations proposed by Dyson and McLean (1977) give expressions for the rate of void nucleation and of void growth. Damage is expressed as the total volume of voids and the effect of damage on the creep rate is also included. For multi-axial states of stress the proposed rate equations may be written in the form

$$\begin{aligned} d(v_e)v_o)/dt &= 3(\sigma_e/\sigma_o)^4 (V/V_o)^{4/9}/t_o \ ; \\ d(n/n_o)/dt &= \tfrac{1}{2}(\sigma_1/\sigma_e)^2 (v_o/v_e)^{1/2} d(v_e/v_o)/dt \ ; \\ d(\nu/\nu_o)/dt &= (\sigma_1/\sigma_e)^{0.7} d(v_e/v_o)/dt \ ; \end{aligned} \qquad (23a\text{-}c)$$

where $\nu_o = 3V_o/2n_o$.

The damage is defined in terms of the total volume of voids which are defined by the integral relation of equation (15). Then

$$\omega = \frac{V}{V_o} = \frac{3}{4}\int_{v_o=o}^{v_1=v} (\sigma_1/\sigma_e)^2 (v_o/v_1) d(v_1/v_o) \int_{v_1}^{v} (\sigma_1/\sigma_e)^{0.7} d(v/v_o). \qquad (24)$$

For proportional loading, the ratio (σ_1/σ_e) is constant and the general damage equation (24) can be integrated and the constitutive equations (23) reduced to the single state variable form:

$$\begin{aligned} d(v_e/v_o)/d(t/t_o) &= 3(\sigma_e/\sigma_o)^4 \omega^{4/9} \\ d\omega/d(t/t_o) &= 9\,\Delta^4(\sigma_{ij}/\sigma_o)\,\omega^{7/9}/2 \end{aligned} \qquad (25a\text{-}b)$$

where, $\Delta(\sigma_{ij}/\sigma_o) = [\sigma_1^{1.8}\,\sigma_e^{2.2}/\sigma_o^4]^{1/4}$,

and the rupture condition is

$$\omega = V/V_o = \sigma_o/\sigma_1 \ ; \qquad (26)$$

in addition, it may be shown, for constant stress, that:

$$\frac{V}{V_o} = \left(\frac{\sigma_1}{\sigma_e}\right)^{2.7} \left(\frac{v_e}{v_o}\right)^{3/2} \ . \qquad (27)$$

If now the damage rate equation is written in the Rabotnov-Kachanov form:

$$d\omega/d(t/t_o) = \Delta^4(\sigma_{ij}/\sigma_o)/5(1-\omega)^4 \ , \qquad (28)$$

and used with the rupture condition $\omega = 1$, it is found that this gives a damage-time graph which is closely represented by equation (25b). The corresponding strain rate expression to equation (25a) is given by

$$d(v_e/v_o)/d(t/t_o) = (\sigma_e/\sigma_o)^4/(1-\omega)^4 \ . \qquad (29)$$

When $\Delta(\sigma_{ij}/\sigma_o) = 1$ the isochronous surface is obtained for the rupture time t_o , Fig. 8.

The Duality of the Approaches of the Materials and Engineering Scientists

It would appear that for proportional stress or loading histories the phenomenological constitutive equations developed earlier using a single state damage variable satisfy the macroscopic observations and are also capable of physical interpretation. However, for non-proportional loading this approach indicates that it is necessary to provide rate equations for both nucleation and void growth and in these circumstances at least two state variables will be required. Fortunately, in many practical components, while stresses vary they do remain sensibly proportional. Nevertheless problems will arise in which the stress fields have large rotations when the sequence of loading is applied. Growth and nucleation will occur at different rates in particular directions and it is to be expected therefore that the description of damage will be tensorial in form. It will also be necessary to understand how and if damage introduces important anisotropic effects.

CONSTITUTIVE EQUATIONS FOR ISOTHERMAL UNSTEADY LOADING

Creep Under Unsteady Loading in Nimonic 80A

Equations (23a) and (27) may not be applied to non-proportional loading situations unless the nucleation, the growth and the fracture mechanisms all have scalar properties. Should any one of the mechanisms have directional properties then it is necessary to integrate equation (15) in a piecewise manner for each of the critical directions corresponding to each phase of the non-proportional loading. The integration is tedious and is best carried out numerically. Before such calculations may be performed it is necessary to assign a specific physical interpretation to nucleation, growth and fracture. To gain some insight into the physical significance of equations (23), consider a single reverse torsion test carried out on a thin tube. The role of V/V_o in the deformation process has been established by Leckie and Hayhurst (1974) to be scalar in nature and one would expect this to be the case for non-proportional loading. A number of possible interpretations of equations (23b) and (23c) can be made; these are;

1. ν/ν_o vector and n/n_o scalar with V/V_o being interpreted as either a scalar or a vector in the failure criterion.

2. ν/ν_o vector and n/n_o vector with V/V_o being interpreted as either a scalar or a vector in the failure criterion.

Although Leckie and Onat (1981) have shown that the growth of damage may be described by an even order tensor, for convenience, it will be described in this presentation as a vector.

The possibility of ν/ν_o being a scalar has been excluded because there is overwhelming microstructural evidence (Hayhurst; 1972) to confirm that in most materials damage grows on planes orthogonal to the maximum principal stress direction. To the authors' knowledge no conclusive microstructural evidence exists from which it is possible to determine the directional characteristics of the processes of nucleation. Hence, the possibility of n/n_o being either a scalar or a vector must be considered. The total volume fraction at failure V/V_o may be interpreted in two ways. Firstly, if the damage accumulated on two independent planes is V_1 and V_2 then for a scalar interpretation of damage at failure the rupture condition expressed by equation (26) becomes $|V_1/V_o|+|V_2/V_o|=\sigma_o/\sigma_1$. Secondly, if a vectorial interpretation of damage at failure is taken then the failure condition is $|V_1/V_2,V_2/V_o|=\sigma_o/\sigma_1$.

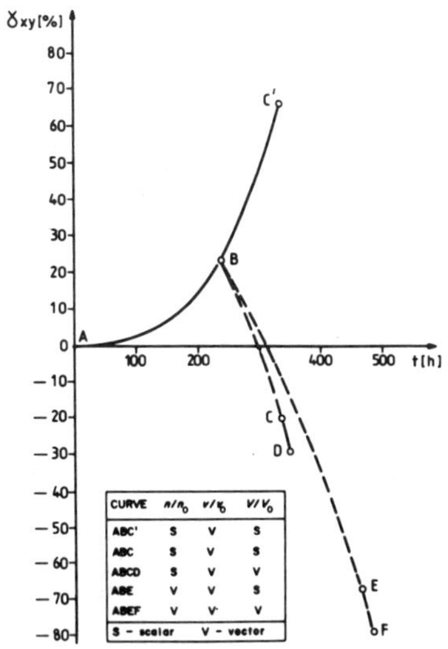

Fig. 9. Theoretical shear strain-time curves for a single reverse torsion test on virgin material (σ_e=297 MPa)

The form of the shear strain-time curve which ensues for single reverse torsion tests may be anticipated for the interpretations given above. The theoretical creep curves for a steady load and a single reverse load torsion tests are shown in Fig. 9 for σ_e = 297 MPa . The theoretical creep curve for steady torsion is represented by ABC'. This curve has been determined from equations (25) using the failure condition expressed by equation (26). Consider the torsion load being reversed at point B. If n/n_o is scalar ν/ν_o is vectorial and V/V_o is scalar then the curve BC will result; the curve BC is a mirror image of the steady load curve BC' about the horizontal axis through B. If n/n_o is scalar, ν/ν_o is vectorial and V/V_o is vectorial then the curve BCD will result. The additional curve CD is due to the vectorial fracture condition. The curve ABE will result if n/n_o is vectorial, ν/ν_o is vectorial and V/V_o is scalar. The curve BE must be to the right of the tangent to the curve BC at B since V/V_o influences the strain rates in a scalar manner. The curve ABEF will result if n/n_o is vectorial, ν/ν_o is vectorial and the fracture condition V/V_o is vectorial. By piecewise integration of equation (15) it may be shown that the time function which determines the shape of BEF if different from the $(t/t_o)^3$ function which determines the shape of ABC' and BCD . Similar arguments may be applied to multi-reverse torsion tests. It is clear that the results of single and multi-reverse torsion tests may be used to highlight the differences between different physical interpretations of the theory. In the following section of the paper results are presented for steady torsion and reverse torsion tests on thin tubes and are compared with theoretical predictions.

The Nimonic 80 A was prepared in the same manner as in an earlier investigation (Dyson and Rodgers; 1974). Thin-walled tubular torsion test pieces were machined with a gauge length of 16 mm, 0.8 mm wall thickness and outer diameter of 8.0 mm. The specimen temperature was maintained at 750 ± 1°C.

The results of a single reverse torsion test conducted at σ_e = 297 MPa and of a steady torsion test, carried out at the same effective stress, are presented in Fig. 10 and compared with theoretical results. The material constants used in the theoretical predictions are σ_o = 309 MPa, t_o = 93 h, v_o = 9.6%.

In the test result given in Fig.10, the torsion load has been reversed at γ_{xy} = 22.5%. The experimental curve for the reverse torsion test is to the left of the theoretical curve. The theoretical curve has been computed assuming that n/n_o is a scalar, ν/ν_o is a vector and V/V_o is scalar (point A) and n/n_o is a scalar, ν/ν_o is a vector and V/V_o is a vector (point B). Comparison of the results presented in Fig. 9 with those of Fig. 10 clearly show that n/n_o and ν/ν_o should be interpreted in equations (23b) and (23c) as a scalar and as a vector respectively and that the fracture

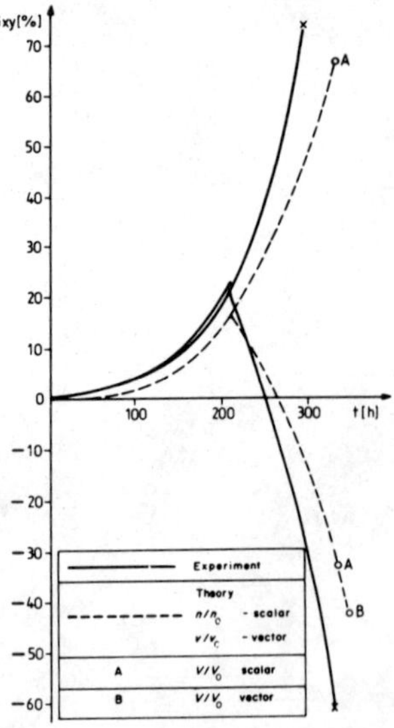

Fig. 10. Comparison of shear strain-time curves for a single reverse torsion test and constant torsion test with theoretical results for σ_e = 297 MPa.

condition, given by equation (26) is a vector.

More extensive studies on the creep of Nimonic 80 A have been carried out under non-proportional loading (Hayhurst, Trampczynski and Leckie, 1980). From the experimental results it is clear the cavity nucleation and growth laws, developed for steady load conditions, may be used to predict creep deformation and rupture under non-proportional loading, provided that the directional characteristics are correctly modelled.

In practice the rate equations for cavity nucleation and growth may not be available and to determine them is a major experimental task. In the absence of such information it is worth investigating if equations (8) and (13) can be used to describe material behaviour under non-proportional loading. This possibility is investigated in the next section.

Creep Under Non-Proportional Loading in Copper and Aluminium

Creep damage, characterised by the damage state variable ω, grows on grain boundaries perpendicular to the direction of maximum principal tension stress, even for aluminium alloys. The question then arises as to whether equations (8) and (13) are capable of describing the deformation and rupture behaviour of copper and aluminium alloys in situations where the stress field rotates relative to the material element during deformation. This type of loading is characteristic of a class of loading conditions found in practice. To test the validity of equations (8) and (13) under these conditions tests have been carried out on thin tubes where the magnitude of σ_e and σ_1 are maintained constant but where the direction of σ_1 rotates through approximately 34°. The loading conditions of the test are shown schematically in Fig. 11; loading proceeds from 0 to A, the load is maintained at A for a period, unloading to 0 then occurs and is followed by reloading to B, where the load is maintained constant until rupture. The magnitude of the stresses have been arranged to give equal strain rates in order to avoid large shear strains and associated buckling effects.

A steady load test (loading to OA only) was carried out first on copper tested at 250°C; the observed behaviour is shown in Fig.12, the lifetime was 230 h. A single reverse torsion test was then performed, Fig.13, on the same material at the same stress level; the stress reversal being carried out just prior to rupture. The measured lifetime was 530 h, which is slightly more than twice the steady load lifetime. This result has been discussed in detail by Trampczynski, Hayhurst and Leckie (1981) and metallographic studies have revealed the formation of two distinct sets of damage planes which do not interact despite their normals being separated by only 34°. The measured axial and shear strain rates are almost equal, as arranged using equation (9). After the reversal of the torsion load a slight reduction in strain rate occurred.

Similar tests were conducted on aluminium tubes. For these tests the life-

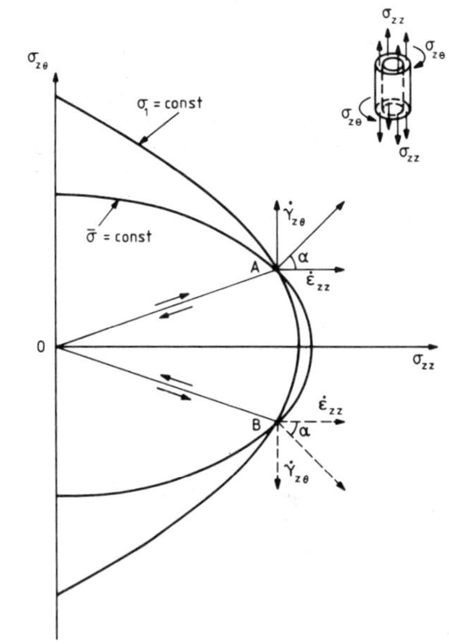

Fig. 11. Schematic representation of loading conditions for copper tube.

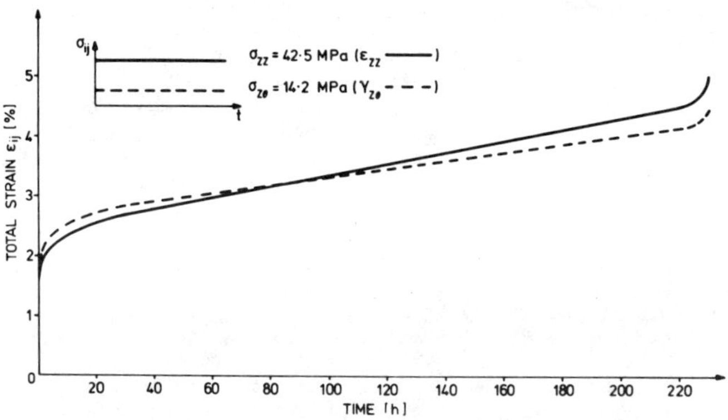

Fig. 12. Steady load tension-torsion test on copper tube.

times of the steady load tests and of the single reverse load tests were almost equal. This result is the one predicted by equations (8) and (13). Equations (8) and (13) can then be used to predict lifetimes of copper and aluminium alloys in which the principal stress directions rotate. For aluminium alloys the damage is included in both equations as a scalar and no special treatment is required. For copper the damage level on the most highly damaged plane determines the strain rate behaviour in equation (8) and the damage histories on all possible planes must be integrated, failure takes place on the plane where the failure criterion is first satisfied.

USE OF CONSTITUTIVE EQUATIONS IN COMPONENT DESIGN

Isothermal Steady Loading

The usage of equations (8) and (13), in studies of the behaviour of engineering components, has been widely accepted to describe the growth of continuum damage and the interaction between creep damage and the strain rate process. The single state variable theory implicit in these equations is also capable of describing the results obtained by materials scientists. Equations (8) and (13) have been used to predict the behaviour of a wide range of structural components. Plane stress plates subjected to in-plane loading have been studied by Hayhurst (1973) and by Hayhurst, Dimmer and Chernuka (1975) to predict component lifetimes and deformations. Also Hayhurst, Morrison and Leckie (1975) have shown that the effect of continuum damage is to cause considerable stress redistribution at stress concentrations and, as a consequence, to effectively blunt the geometrical discontinuity. The influence of the multi-axial stress rupture criterion, or the form of the isochronous rupture surface, has been shown by Leckie and Hayhurst (1974) and by

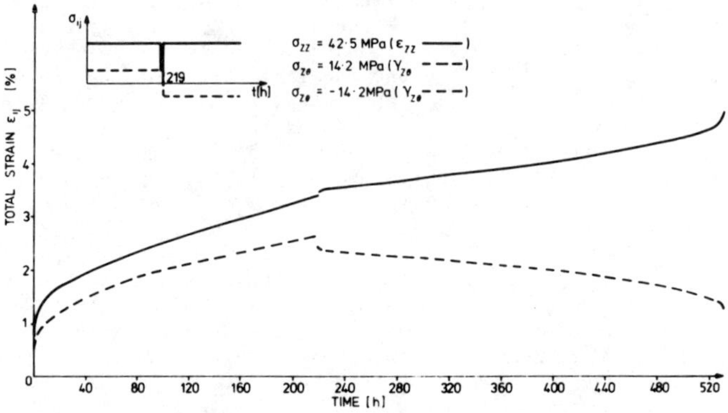

Fig. 13. Single reverse torsion, steady tension load test on a copper tube.

Hayhurst, Leckie and Morrison (1978) to be responsible for the phenomena of notch weakening and of notch strengthening. More recently finite element techniques have been used by Hayhurst, Dimmer and Morrison (1983) to predict the lifetimes of axi-symmetrically notched tension bars and by Hayhurst, (1983) to determine the lifetimes of components in which failure takes place by creep crack growth. In addition to the numerical investigations carried out on components simplified bounding techniques have been developed by Leckie and Hayhurst (1974) and by others.

Isothermal Unsteady Loading

The results of tests carried out under unsteady loading in Nimonic 80 A show the role of the processes of cavity nucleation and growth in the determination of lifetime and of strain rates. The same directional characteristics have been revealed in the tension/torsion tests reported on copper and on an aluminium alloy. These phenomena can be important in components subjected to non-proportional loading and in which the stress-state can vary over the loading cycle. Under these circumstances it is difficult to make broad generalisations and it may be necessary to integrate equations (8) and (13) over the entire loading cycles for all possible sets of damage planes throughout the body in order to isolate the most critical sets of planes.

CONCLUSIONS

The concept of continuum damage has been introduced and it has been shown that a single state variable theory may be used to describe the process of cavity nucleation, growth and failure. This approach has been shown to be equivalent to that used by the materials scientist for proportional loading. When materials are subjected to non-proportional loading it is necessary to consider the directional characteristics of the growth of damage and the dependence of its rate of growth upon stress state.

In the design of components for steady load operation equations (8) and (13) can be used but it is important to identify the shape of the isochronous surface particularly when the components are subjected to shear stress and to tri-axial states of stress. For the design of components which are subjected to non-proportional loading it is not easy to make accurate simplifications using equations (8) and (13), but none the less sufficient knowledge exists to enable the determination of realistic bounds on material behaviour.

REFERENCES

Ashby, M.F. and Raj, M, (1975). Proceedings of the conference on Mechanics and Physics of Metals, Cambridge: The Metals Society, Institute of Physics.

Boettner, R.C. and Robertson, W.D. (1961). Trans. Metall. Soc. A.I.M.E., 221, 613.

Cane, B.J. (1981). Acta. Metall., 29, 1581.

Cocks, A.C.F. and Ashby, M.F. (1982). Progress in Materials Science, 27, 3-4, 189.

Dyson, B.F. (1979). Canadian Met. Quart., 18, 31.

Dyson, B.F. and McLean, D. (1977). Met. Sci., 2, 37.

Dyson, B.F. and Rodgers, M.J. (1974). Metal Sci., 8, 261.

Greenwood, G. (1973). Int. Congress on Metals, Cambridge, 1973, Microstructure and the design of Alloys, 2, 91, 1973.

Hayhurst, D.R. (1972). J. Mech. Phys. Solids, 20, 381.

Hayhurst, D.R. (1973). Jnl. Appl. Mech., 1, 40, 244.

Hayhurst, D.R. (1983). Engineering Approaches to High Temperature Design, Ed. B. Wilshire and D.R.J. Owen, Pineridge Press, Swansea, pp. 85-176.

Hayhurst, D.R. and Storåkers, B. (1976). Proc. Roy. Soc. Lond. A., 349, 369.

Hayhurst, D.R., Brown, P.R. and Morrison, C.J. (1983). Leicester University Engineering Department, Report No. 83-4, February 1983.

Hayhurst, D.R., Dimmer, P.R., and Chernuka, M.W. (1975). J. Mech. Phys. Solids, 23, 335.

Hayhurst, D.R., Dimmer, P.R. and Morrison, C.J. (1983). Leicester University Engineering Department, Report No. 83-2, Jan. 1983.

Hayhurst, D.R., Leckie, F.A. and Morrison, C.J. (1978). Proc. Roy. Soc. Lond. A. 360, 243.

Hayhurst, D.R., Morrison, C.J. and Leckie, F.A. (1975). J. Appl. Mech., 42, 613.

Hayhurst, D.R., Trampzcynski, W.A. and Leckie, F.A. (1980). Acta. Met. 28, 1171.

Johnson, A.E., Henderson, J. and Khan, B. (1962). 'Complex-stress Creep, Relaxation and Fracture of Metallic Alloys', Ch. 4, H.M.S.O., London.

Kachanov, L.M. (1958). IZV Akad. Nauk. SSSR O.T.N., Teckh. Nauk. 8, 26.

Kachanov, L.M. (1960). 'The Theory of Creep', (English translation edited by Kennedy, A.J.) Chs. IX,X, National Lending Library, Boston Spa.

Leckie, F.A. and Hayhurst, D.R. (1974). Proc. R. Soc. London. A., 340, 323.

Leckie, F.A. and Hayhurst, D.R. (1975). Mech. Res. Communications, 2, 1.

Leckie, F.A. and Hayhurst, D.R. (1977). Acta. Met. 24, 1059.

Leckie, F.A. and Hayhurst, D.R. (1980). 'Experiments in high temperature fracture', February 1980, Departmental Report, Dept. Mech. & Ind. Eng., University of Illinois, Urbana-Champaign, U.S.A.

Leckie, F.A. and Ponter, A.R.S. (1974). Ing. Arch., 43, 158.

Leckie, F.A. and Onat, E.T. (1981). in Physical Non-linearities (edited by J. Hult and J. Lemaitre) 140-155, Springer-Verlag, Berlin.

Lemaitre, J. and Chaboche, J-L, (1978). Jnl. de Mécanique Appliquée, 2, 3, 317.

Odqvist, F.K.G. (1974). 'Mathematical theory of creep and creep rupture', 2nd edn. London. Oxford University Press.

Rabotnov, Y.N. (1969). 'Creep Problems in Structural Members', (English translation edited by Leckie, F.A.) Amsterdam: North Holland Publishing Company.

Trampczynski, W.A., Hayhurst, D.R. and Leckie, F.A. (1981). J. Mech. Phys. Solids, 29, 5/6, 353.

RELIABILITY-BASED FATIGUE-PROOF DESIGN OF COMPOSITE MATERIALS

H. Ishikawa*, A. Tsurui, T. Tanimoto*** and H. Kimura***

*Department of Information Science, Kagawa University, Takamatsu 760, Japan
**Department of Applied Mathematics and Physics, Kyoto University, Kyoto 606, Japan
***Department of Mechanical Engineering, Doshisha University, Kyoto 602, Japan

ABSTRACT

The reliability-based design safe lives of composite materials have been investigated through the statistical modelling of the failure probability under service loads from the viewpoint that the failure probability has to be evaluated only through the limited number of experimental data which would, of course, vary from sample to sample.

KEYWORDS

Reliability-based design; design safe life; failure probability; fatigue life distribution; damage; FRP; fatigue-proof design; log-normal distribution.

INTRODUCTION

Composite materials, especially glass fiber reinforced plastics (referred to as FRP), have been recently in more extensive use as machine parts and/or structural components. Accordingly, the appropriate estimation of fatigue life of FRP has come to be of great importance for those concerned with design by composite materials. Undoubtedly, fatigue life of the material would have an inherent scatter and, in addition, FRP possess less material uniformity in comparison to conventional metallic materials. Hence, a good understanding of the statistical aspect of fatigue properties of FRP is deemed essential to their successful application in the practical design.

In view of the above, the first phase of the present paper deals with the theoretical modelling of the statistical failure probability in the sense that the failure probability can be evaluated only through the limited number of experimental data which would, of course, vary from sample to sample. The distribution function for the statistical failure probability under service loads has been derived by making use of the linear damage accumulation hypothesis. Also, a simple but important fatigue-proof design principle is proposed. This design principle requires that a value smaller than a pre-specified allowable value for the failure probability be correct with a pre-specified reliability level.

In the second phase of the paper, the statistical fatigue life data of satin woven glass cloth FRP were obtained experimentally under rotating bending of constant stress amplitude with not less than twenty replication tests at each stress level. Then they were analyzed with the aid of a log-normal probability model to clarify the statistical fatigue properties of composite materials. Finally, the reliability-based design safe lives of composite materials have been investigated so as to meet the requirement in the practical design from the viewpoint of the proposed reliability-based fatigue-proof design principle.

THEORETICAL MODELLING OF FAILURE PROBABILITY UNDER SERVICE LOADS

Failure probability has to be treated as a statistic since it can be evaluted only through experimental data.

Let Z_n be a normalized damage accumulated during n cycles of stressing of not necessarily the same level. Z_n is assumed of the following form by Miner's rule:

$$Z_n = \sum_{i=1}^{n} (S_i/s_0)^b \tag{1}$$

where S_i = applied i-th stress level, s_0 = normalization stress level, b = a constant and S_i's are statistically independent, each having a distribution function $R(s) = P[S_i \le s]$.

Let Z_c be the critical value of Z_n. Introducing stress distribution R(s) of the form:

$$R(s) = \begin{cases} 0 & \text{for } s < s_0 \\ 1 & \text{otherwise} \end{cases} \tag{2}$$

we have, from Eq.(1),

$$Z_n = \sum_{i=1}^{n} (s_0/s_0)^b = \sum_{i=1}^{n} 1 = n$$

Therefore, Z_c is deemed to be the critical value of Z_n under constant amplitude of s_0. In other words, Z_c gives a measure associated with fatigue life under constant amplitude s_0.

Then, the distribution function $F_N(n)$ of fatigue life N under service loads with arbitrary distribution R(s) can be derived in the following way. For $n \gg 1$, the distribution of Z_n can be well approximated by a normal distribution by virtue of central limit theorem.

$$P[Z_n \le z_n] \cong \frac{1}{\sqrt{2\pi n}\nu} \int_{-\infty}^{z_n} \exp[-\frac{(\xi - n\lambda)^2}{2n\nu^2}]d\xi$$

$$= \Phi[(z_n - n\lambda)/\sqrt{n}\nu] \tag{3}$$

where

$$\Phi(u) = \frac{1}{\sqrt{2\pi}} \int_{-\infty}^{u} \exp[-\frac{u^2}{2}]du \tag{4}$$

$$\lambda = E[Z_n]/n = \int_0^{\infty} (s/s_0)^b dR(s) \tag{5}$$

$$\nu^2 = E[\{Z_n - E[Z_n]\}^2]/n = \int_0^{\infty} (s/s_0)^{2b} dR(s) - \{\int_0^{\infty} (s/s_0)^b dR(s)\}^2. \tag{6}$$

Consequently, $F_N(n)$ is given in the following form:

$$F_N(n) = P[N \le n] = P[Z_n > Z_c]$$
$$= \int_{-\infty}^{\infty} F_{Z_c}(z_c) \cdot \frac{1}{\sqrt{2\pi n\nu}} \exp[-\frac{(z_c - n\lambda)^2}{2n\nu^2}] dz_c \qquad (7)$$

where $F_{Z_c}(z_c)$ is the distribution function of Z_c. Making use of Taylor's series expansion of $F_{Z_c}(z_c)$ around $z_c = n\lambda$, we have

$$F_N(n) = F_{Z_c}(n\lambda) + (n\nu^2/2)F''_{Z_c}(n\lambda) + O(1/n^2). \qquad (8)$$

Generally, since the second and higher terms of the right-hand side of Eq.(8) is sufficiently small enough in comparison with the first term, $F_N(n)$ can be approximated by (Sogliero and Srinivasan, 1980; Tsurui, 1981)

$$F_N(n) \cong F_{Z_c}(n\lambda). \qquad (9)$$

Fatigue life distribution has often been reported to be well approximated by a log-normal probability model (Nakagawa and Nakabayashi, 1979; Ishikawa and others, 1980; Johno and others, 1981). Therefore, it is reasonable to assume that Z_c follows a log-normal distribution. With Z_c thus assumed, $\ln Z_c$ becomes a normal random variable with mean μ and variance σ^2 defined as

$$\begin{cases} \mu = E[\ln Z_c] \\ \sigma^2 = E[(\ln Z_c - \mu)^2] \end{cases} \qquad (10)$$

and consequently $(\ln Z_c - \mu)/\sigma$ follows a standard normal distribution (Kimura and Ishikawa, 1982). Hence,

$$F_N(n) \cong F_{Z_c}(n\lambda) = P[Z_c \le n\lambda] = P[(\ln Z_c - \mu)/\sigma \le \{\ln(n\lambda) - \mu\}/\sigma]$$
$$= \begin{cases} \Phi[\{\ln(n\lambda) - \mu\}/\sigma] & \text{for } n\lambda > 0 \\ 0 & \text{otherwise.} \end{cases} \qquad (11)$$

Fatigue life distribution can be given by the above equation. However, since the true value of parameters μ and σ are usually unknown, they need to be replaced, for instance, by their maximum likelihood estimates $\hat{\mu}$ and $\hat{\sigma}$ from a sample of size k, thus the failure probability

$$P = \Phi[\{\ln(n\lambda) - \mu\}/\sigma] \qquad (12)$$

by its estimate $\hat{P}$. Here, $\hat{\mu}$ and $\hat{\sigma}$ are considered as random variables in the sense that they vary from sample to sample, and so is $\hat{P}$. Putting

$$\ln(n\lambda) = q \qquad (13)$$

we have, as the statistical failure probability,

$$\hat{P} = \Phi[(q - \hat{\mu})/\hat{\sigma}] \qquad (14)$$

The joint distribution $Q(\xi, \eta)$ of maximum likelihood estimates $\hat{\mu}$ and $\hat{\sigma}$ from a sample of size k (assuming normal population) can be expressed in the following form with the fact that $(\hat{\mu}-\mu)/\sigma$ and $k\hat{\sigma}^2/\sigma^2$ are statistically independent and that they have, respectively, a normal distribution with mean zero and variance $1/k$ and χ^2 distribution of k-1 degrees of freedom (Mikami, 1969):

$$Q(\xi, \eta) = P[\hat{\mu} \le \xi, \hat{\sigma} \le \eta] = P[(\hat{\mu}-\mu)/\sigma \le (\xi-\mu)/\sigma, k\hat{\sigma}^2/\sigma^2 \le k\eta^2/\sigma^2]$$

$$= (\frac{k}{2\pi})^{\frac{1}{2}} \int_{-\infty}^{(\xi-\mu)/\sigma} \exp[-\frac{ku^2}{2}]du \cdot [2^{(k-1)/2} \Gamma(\frac{k-1}{2})]^{-1} \int_0^{(k/\sigma^2)\eta^2} v^{(k-3)/2} e^{-v/2} dv \quad (15)$$

From the above, the distribution function $H(p)$ of the statistical failure probability $\hat{P}$ can be derived:

$$H(p) = P[\hat{P} \le p] = P[\Phi[(q-\hat{\mu})/\hat{\sigma}] \le p] = P[q-\hat{\mu} \le \Phi^{-1}(p)\hat{\sigma}]$$

$$= \iint_{q-\xi \le \Phi^{-1}(p)\eta} dQ(\xi, \eta)$$

$$= 1 - \int_0^{\infty} (1/\Gamma(\frac{k-1}{2}))\Phi[\frac{\sqrt{k}}{\sigma}(q-\mu) - \Phi^{-1}(p)\sqrt{2w}]w^{(k-3)/2}e^{-w}dw \quad (16)$$

where $\Phi^{-1}(\cdot)$ is the inverse function of $\Phi(\cdot)$. Namely, if $\Phi(u) = p$, then $u = \Phi^{-1}(p)$.

RELIABILITY-BASED FATIGUE-PROOF DESIGN PRINCIPLE

The integrand in Eq.(16) is exemplified in Fig.1 as a function of w, by introducing a new parameter x defined as

$$x = (q - \mu)/\sigma. \quad (17)$$

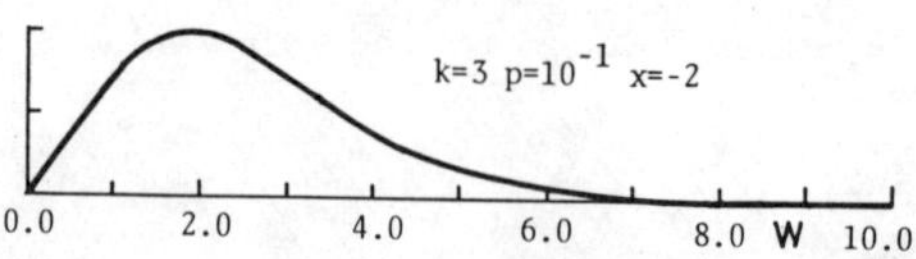

Fig. 1. Integrand in Eq.(16) as a function of w.

Examples of numerical integration of $H(p)$ in Eq.(16) are shown in Figs. 2(a)∿(e), with x as a parameter, for sample sizes (a) k=3, (b) k=5, (c) k=10, (d) k=20 and (e) k=50. $H(p)$ thus defined is now utilized for the reliability-based fatigue-proof design of composite materials, whose principle requires that the decision of the failure probability being less than a certain allowable value p_H during design life be correct with a pre-specified level of reliability H. In other words, fatigue-proof design has to be made so as to obtain as large value of $n\lambda$ as possible under the condition that sample size k, reliability level H and allowable value of failure probability p_H are all given. Together with the fact that λ can be evaluated through Eq.(5) when the distribution function $R(s)$ of applied stress is given, we are now in a position to determine the reliability-based design safe life.

RELIABILITY-BASED DESIGN SAFE LIVES OF COMPOSITE MATERIALS

Fatigue tests were carried out on the satin woven glass cloth FRP specified in Tables 1 to 3 under rotating bending of constant amplitude with not less than twenty replication tests at each stress level. Then, they were analyzed, with the aid of a log-normal probability model, to clarify the statistical fatigue properties of FRP as shown in Fig.3 where maximum likelihood estimates $\hat{\mu}$ and $\hat{\sigma}$ are also listed at each stress level.

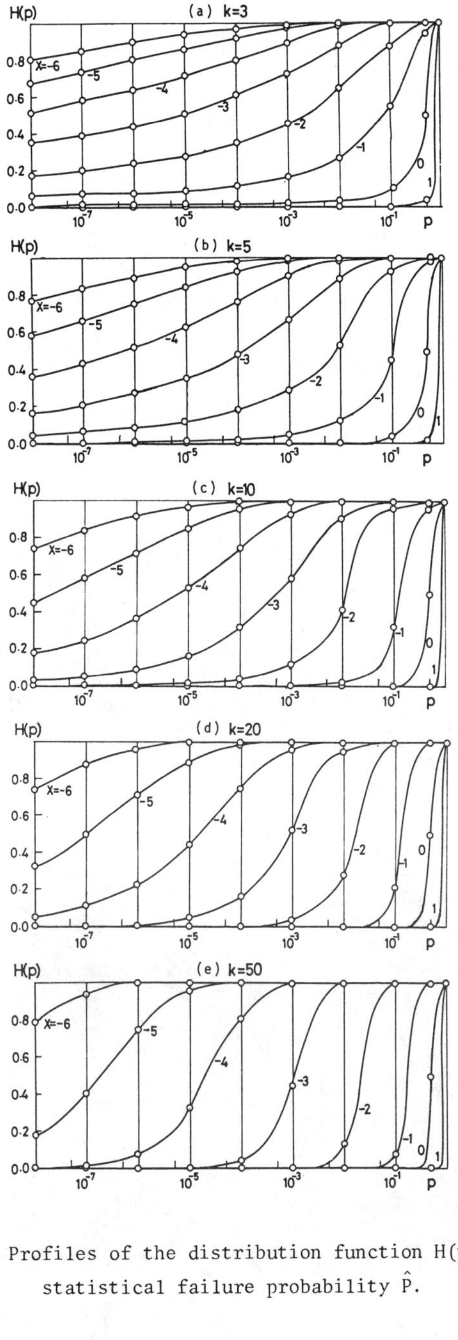

Fig. 2. Profiles of the distribution function H(p) of the statistical failure probability $\hat{P}$.

Table 1. Specification of the material.

Matrix	Polyester resin (polylite FG-284)
Reinforcement	Satin woven glass cloth (SLS-212)
Catalizer	Methyl ethyl ketone peroxide
Hardener	Cobalt naphthenate 0.6% solution

Table 2. Properties of the polyester resin.

Specific density	1210 kg/m^3
Tensile strength	58 MPa
Compressive strength	155 MPa
Young's modulus	3500 MPa

Table 3. Glass fiber content of the specimen and its mechanical properties.

Weight content W_g	54.47 %
Volume content V_g	35.78 %
Tensile strength σ_t	300 (MPa)
Bending strength σ_b	402 (MPa)

Now, the procedure of determining the reliability-based design safe lives on the basis of the proposed fatigue-proof design principle has been exemplified by using only the result at stress level of S = 110 MPa because of the limitation of pages. In this case the sample size k is equal to 20. Therefore, through Fig.2-(d) is first plotted Fig.4, where the variation of failure probability p_H is given as a function of x for a certain fixed value of reliability level H. Then the design safe life, the value of $n\lambda = E[\ Z_n]$, can be determined so as to make the failure probability less than a certain allowable value, say, $p_H = 5\times10^{-4}$ with reliability level of, for instance, H = 90%. As shown in the figure by dotted lines, we get x=-3.8 for $p_H=5\times10^{-4}$ and H=90%. On the otherhand, we get maximum likelihood estimates of $\hat{\mu}=10.842$ and $\hat{\sigma}=0.104$ from experimental data. Hence, $n\lambda=\exp[\hat{\mu}+x\hat{\sigma}]=\exp[10.842-3.8\times0.104]=3.44\times10^4$. By adopting this value of $n\lambda$, we can conclude, with reliability level of 90%, that the probability of fatigue failure during design lifetime would be less than 5×10^{-4}.

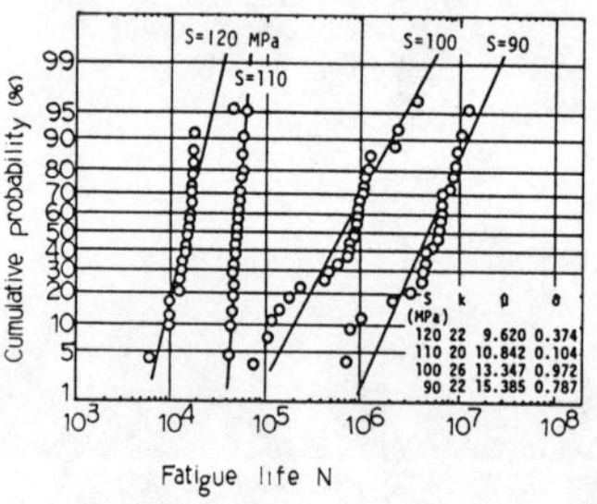

Fig. 3. Fatigue life distribution (log-normal probability model).

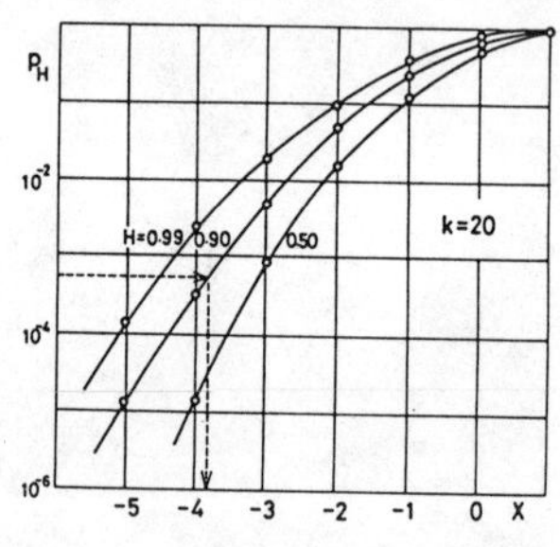

Fig. 4. The allowable failure probability p_H as a function of x for a fixed value of reliability level H.

CONCLUDING REMARKS

The failure probability is treated as a statistical quantity from the viewpoint that the probability has to be evaluated only through experimental data, and its distribution function is derived theoretically based on the assumption of linear damage accumulation and a log-normal type of critical damage distribution. Then a simple but important fatigue-proof design principle is proposed utilizing the above results. This design principle requires that a value smaller than a pre-specified allowable value for the failure probability be correct with a pre-specified level of reliability. Finally, reliability-based design safe lives of composite materials have been determined so as to meet the requirement in the practical design on the basis of the proposed reliability-based fatigue-proof design principle.

REFERENCES

Mikami, M. (1969). Statistical Inference. Chikuma Press, Tokyo. p. 189.
Nakagawa, T., and Y. Nakabayashi (1979). J. Jap. Soc. Mat. Sci., 28, 9-12.
Ishikawa, H., T. Tanimoto, and H. Kimura (1980). The Kagawa Univ. Economic Review, 52, 18-59.
Sogliero, G., and A. V. Srinivasan (1980). AIAA J., 18, 318-326.
Johno, M., T. Deishi, and M. Kikkawa (1981). J. Jap. Soc. Mat. Sci., 30, 8-14.
Tsurui, A. (1981). Int. J. Fatigue, 3, 125-127.
Kimura, H., and H. Ishikawa (1982). Introduction to Statistics. Jikkyo Press, Tokyo. Chap. 8, pp. 98-116.

MECHANICAL PROPERTIES OF MAGNETIC COMPOSITES

R. K. T. Hsieh

Department of Mechanics, Royal Institute of Technology, S-100 44 Stockholm, Sweden

ABSTRACT

This paper within the continuum approximation deals with the theoretical and experimental description of magneto-elastic composites. Cu-1% Co composites have been treated explicitely.

KEYWORDS

Young's modulus, magneto-elastic interactions, composites, vibrations resonant rod.

INTRODUCTION

Since electromagnetic fields have been found as pervading the universe electromagnetoelastic interaction applications have not ceased to develop. They gave rise to components implementing various functions: piezo-electric transformators, transducers, radar, matched filtering, spectral analysis, etc... It is the aim of this work to investigate some properties of a composite of ferromagnetic grains in an unpolar elastic matrix. As known in the presence of an external magnetic field B_i, a torque will be exerted on the grains. Besides these effects there are also other magneto-elastic interactions, see e.g. Hsieh (1979). However considering the ferromagnetic material as being composed of monodomain grains uni-formly distributed in a non-ferromagnetic elastic matrix with the size of particles such that inter-magnetic particle interactions are negligible, considering furthermore that crystal anisotropy has a dominant effect and the mode of rotation in unison is valid, the present situation is valid and in the continuum approximation, we have a medium with one more con-stant as compared with linear isotropic elastic materials. From the vibra-tions of a resonant bar, some theoretical expressions for this additional constant are given. Quantitative studies were performed on Cu-1% Co composites.

VIBRATIONS OF A MAGNETO-ELASTIC ROD

The vibrations of an elastic rod constitute a known problem. Let us therefore concentrate the study on the magnetic and the magneto-elastic contributions. On the micro-scale the grains have essentially saturated magnetization within a given domain and thus magnetization can only be changed by the rotation of the magnetic moment. This means that macroscopically the medium's temperature is below the Curie point. The medium is paramagnetic. Now small deformations of the medium as a continuum consist both in a displacement u_k of the mass elements and in a small rotation Φ_k of the magnetic moments. The equations of motion on the macro-scale then write (Hsieh & Vörös, 1981)

$$\sigma_{ij,i} - \rho a_j = 0 \tag{1}$$

$$\varepsilon_{kij}\sigma_{ij} + \rho \ell_k = 0 \tag{2}$$

where a_j is the acceleration vector, ρ is the density and ε_{kij} the permutation symbol. The constitutive law writes

$$\sigma_{ij} = \lambda\varepsilon_{kk}\,\delta_{ij} + (2\mu+\alpha)\varepsilon_{ij} + \alpha\varepsilon_{ijm}(r_m - \Phi_m) \tag{3}$$

and the body couple is as follows

$$\tilde{\ell} = (\tilde{M} + \tilde{\Phi} \times \tilde{M}) \times \tilde{B} \tag{4}$$

where $\tilde{r} = \frac{1}{2}\,\mathrm{curl}\,\tilde{u}$ is the infinitesimal rotation, σ_{ij} and ε_{ij} are respectively the stress and the strain tensors and α is the coupling modulus. It is easily seen that this system has 6 degrees of freedom and this material has one additional material modulus as compared to isotropic elastic solids. The constitutive law (3) is, of course, based on micro-considerations as are the equations of motion (1),(2). On that level, the inclusions are treated as being rigid and the matrix, elastic. It is also to be noted that the dynamic behaviour of the magnetization in the present case is fully described by the equations of motion (1), (2). More general situations can be and are considered, see e.g. Hsieh (1979, 1981, 1983). However in order to make the experimental part more manageable, they will not be considered here.

Let us now consider the vibrations of such a composite rod with its axis in equilibrium lying along an axis y_1 in a Cartesian coordinate system. By introducing a local coordinate system ξ_i and integrating the local equations (1), (2) for the volume of an element of length ds and limited by two cross sections $A_1 = A$ and $A_2 = A$ of the deformed rod, after some algebra, we obtain

$$\frac{d^2}{ds^2} \int_A (\varepsilon_{kmn}\,\xi_m\,\sigma_{jn}\,dA_j + \varepsilon_{kmn}\,t_m \rho A\ddot{Y}_n)\,ds + \frac{d}{ds}\int_{\Delta V} \rho\ell_k\,dV = 0 \tag{5}$$

where $Y_i\,(y_\ell, t)$ is the displacement at time t of any point y_ℓ of the elastic line of the rod from its equilibrium position, t_m is a unit vector tangential to the rod regarded as an elastic line and a superposed dot denotes a material time derivative. Recalling that the first integral on the r.h.s. of eq. (5) is the torque of the elastic stresses and using the approximation $(d/ds) \approx (d/dy_1)$, eq. (5) becomes

$$E\Theta Y^{(4)} - \frac{2\alpha\rho\, A\, \tilde{B}.\tilde{M}}{2\alpha + \rho\tilde{B}.\tilde{M}}\, Y^{(2)} + \rho A\ddot{Y} = 0 \tag{6}$$

where E is the Young's modulus of the composite material and Θ is the principal moment of inertia. Y stands for both Y_2 and Y_3 as it can be seen that no coupling exists between the two flexural vibrations. It also appears that the rotation Φ_i in eq. (4) can be formally eliminated through the displacement u_i. The rotational effects are still retained, for instance through the second term on the r.h.s. of eq. (6). In this treatment, we assumed that there are no macroscopic torsional strains and that the free surfaces are stresses free.

RESONANT ROD

The equation for lateral vibrations of the rod is given by (6). We assume that we have a periodic solution and that the rod is clamped at both ends. We then have

$$Y(y_1,t) = Y_0(y_1)\cos\omega t \tag{7}$$

and the space displacement $Y_0(y_1)$ fulfills the four conditions

$$Y_0(0) = Y_0'(0) = Y_0(L) = Y_0'(L) = 0 \tag{8}$$

where primes denote differentiation with respect to y_1.

The solution Y_0 has therefore the form

$$Y_0(y_1) = C_1\cos my_1 + C_2\sin my_1 + C_3\cosh ny_1 + C_4\sinh ny_1 \tag{9}$$

where

$$m = \left[\frac{\rho A\,\underset{\sim}{B}.\underset{\sim}{M}}{EI\,\kappa} - \left[\left[\frac{\rho A\,\underset{\sim}{B}.\underset{\sim}{M}}{EI\,\kappa}\right]^2 + \frac{4\omega^2\rho A}{EI}\right]^{\frac{1}{2}}\right]^{\frac{1}{2}},$$

$$n = \left[\frac{\rho A\,\underset{\sim}{B}.\underset{\sim}{M}}{EI\,\kappa} + \left[\left[\frac{\rho A\,\underset{\sim}{B}.\underset{\sim}{M}}{EI\,\kappa}\right]^2 + \frac{4\omega^2\rho A}{EI}\right]^{\frac{1}{2}}\right]^{\frac{1}{2}}$$

and we define the expression $\kappa = 1 + \rho\,\frac{\alpha}{2}^{-1}$

Substituting (9) into (8), we are left with a homogeneous system of equations for the constants C_1, C_2, C_3, C_4. In order to have a non trivial solution, its determinant has to vanish. This leads to

$$2\,m\,n\cosh nL.\cos mL + (m^2 + n^2)\sinh nL.\,\sin mL = 2\,mm \tag{10}$$

Expression (10) gives the polar constant as a function of the frequency.

In an attempt to obtain some quantitative results a resonant rod apparatus has been built (see fig. 1). A Cu-1%Co rod vibrates under the action of an alternative Lorentz force $\underset{\sim}{F}_d(\Omega) = I_0\underset{\sim}{L} \times \underset{\sim}{B}_d(\Omega)$. The magnetic field $\underset{\sim}{B}_d$ is produced by the magnetic circuit (a) and the constant current $I_0\underset{\sim}{L}$ is produced from the circuit (b). The magnetization of the Co-particles embedded into the rod under the action of the static body couple $\underset{\sim}{\ell}_S = \underset{\sim}{M} \times \underset{\sim}{B}_S$ are responsible for the elastic rotational effects. The static magnetic field $\underset{\sim}{B}_S$ is produced by the circuit (d). The frequencies of vibration are then collected through a recorder in circuit (c), amplified, shown on an oscillograph, read by a digital frequency counter and shown in an amplitude-time graph. At the resonance frequency $\omega_r = \Omega$, the amplitude of vibration has its maximum value and the coupling material constant α is obtained from eq. (10). Though we used strong magnetic fields and a material with high magnetic moment, no microeffects were detected. For Cu-1% Co the effects of the micro-rotation are thus negligible. This can also be seen by considering the 2nd term on the r.h.s. of eq. (6) as being balanced by an external traction force. An estimation of the order of magnitude of this traction force then shows that its magnitude is too small for present instrumentations. More sophisticated experimental apparatus are being considered (Hsieh and Vörös, 1983).

The material chosen was a Cu-1%Co, the length of the proof was L = 160 mm, its diameter Ø = 1 mm, the measured Young's modulus $E = 10^4$ kp/mm^2, the density $\rho = 10^{22}$ atom/mm^3, the magnetic moment of Co was $M = 2\ 10^{-20}$ CGS. The constant current varied from 0,1 to 0,5 A while the alternative magnetic field from 0,5 to 1 kG (In circuit (a), the number of turns of the magnetizing current was 300 while in circuit (d), 900. The path in vacuum was 0,15 cm. The frequency Ω varies from 0 to 600 Hz). The body couple wrote $\underset{\sim}{\ell} \sim \underset{\sim}{M} \times \underset{\sim}{B}_s$ with the static magnetic field varying from 1 kG to 5 kG.

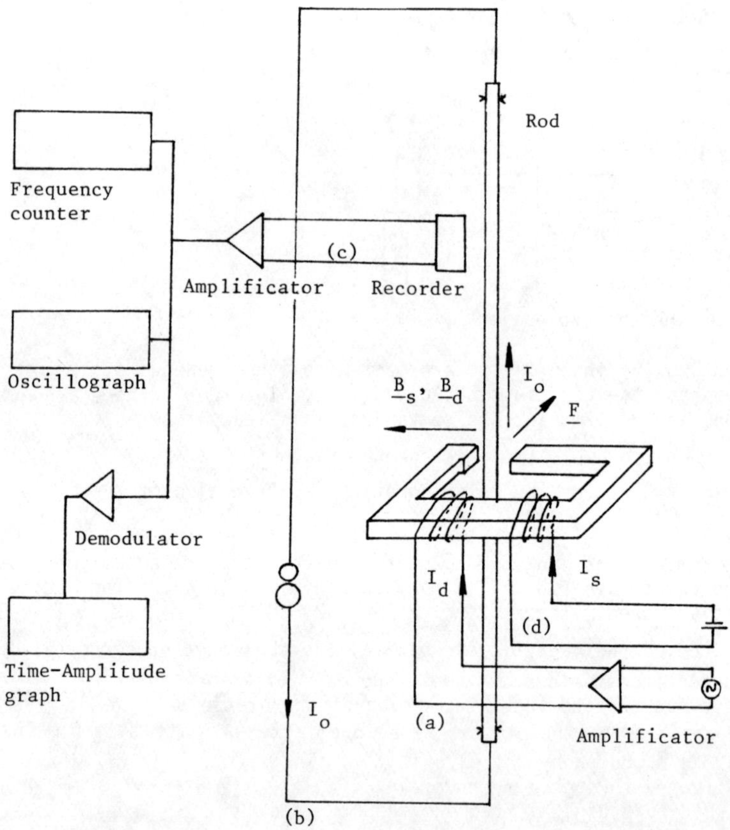

Fig.1 : Principle of the measurements.

ACKNOWLEDGEMENT

The author wishes to thank Prof. I. Kovács from Eötvös University for providing the sample and Dr. G. Vörös for the carriage of the measurements.

REFERENCES:

Hsieh, R.K.T. (1979). Micropolarized and magnetized media. In O. Brulin and R.K.T. Hsieh (Eds.), *Mechanics of Micropolar Media.* International Centre for Mechanical Sciences Courses and Lectures, Udine. pp. 187-266.

Hsieh, R.K.T. (1981). Dislocation in a ferromagnetic continuum. *Arch. of Mech., 33, No.6*, 957-960.

Hsieh, R.K.T. (1983) Multipolar lattice defects in electromagnetic continua. In G.A. Maugin (Ed.), *Proceedings Iutam-Iupap Symposium on the Mechanical Behaviour of Electromagnetic Solid Continua.* North Holland Pub. Co., Amsterdam.

Hsieh, R.K.T., and G. Vörös (1981). Microanelasticity of a vibrating rod containing particles with magnetic moment. *Int. J. Engng. Sci., 20, No.2*, 261-270.

Hsieh, R.K.T., and G. Vörös (1983). Internal friction of continuous magnetic composites. *Proceedings 5th International Symposium on Composite Metallic Materials*, Bratislava.

THE EFFECT OF NON-LINEAR STRAIN PATHS ON THE PLASTIC PROPERTIES OF STEELS

W. B. Hutchinson* and T. Davis**

***Swedish Institute for Metals Research, Drottning Kristinas väg 48, 114 28 Stockholm, Sweden**
****Department of Metallurgy, University of Aston in Birmingham, England**

ABSTRACT

The effects of changes in strain path on the plastic properties of steel sheet have been examined for plain carbon, rephosphorised and dual phase steels. All showed similar behaviour. A Bauschinger effect was observed in some cases but more commonly the pre-strain produced a latent hardening phenomenon. This increase in flow stress together with a reduction in work hardening rate leads to severely reduced ductility. The effect is attributed to changes in the dislocation substructure and is shown to be greatest if the two states of deformation are nearly orthogonal when described by vectors in strain-space.

KEYWORDS

Strain path change; steel; plasticity; anisotropy; ductility; formability.

INTRODUCTION

Variations in material properties associated with changes in strain path may have important implications for the formability of sheet subjected to complex forming operations and to the behaviour in service of the final components. A number of investigations in recent years have reported the effect of such strain path changes on subsequent flow stress, work hardening and plastic anisotropy (e.g. Hutchinson, Arthey and Malmström, 1976; Laukonis and Ghosh, 1978; Lloyd and Sang, 1979; Charpentier and Piehler, 1983; Yoshida, Miyauchi and Mizunuma, 1969). Although these investigations were generally of restricted scope they do show in all cases that the concept of a single equivalent stress-equivalent strain relationship for a metal is quite inadequate to describe its properties under non-linear deformation. The present work was carried out to investigate the phenomena in a more systematic manner and so to obtain a better general understanding of them. It may be noted also that forming limit curves are strongly influenced by changes in straining path (e.g. Kikuma and Nakazima, 1971; Müschenborn and Sonne, 1975) and that a diversity of theoretical models have been proposed in recent years to explain this behaviour. All of these models are inadequate, especially when biaxial stretching is followed by a drawing operation. The reason for their failure is probably a result of neglecting the effect of the change in strain path on the plastic flow characteristics of the material.

EXPERIMENTAL PROCEDURE

Three different sheet steels of 0.7 mm thickness have been used. These were an aluminium-killed deep-drawing quality steel (AK), a rephosphorised steel (RP), and a dual phase steel (DP). The latter two had comparable flow stresses but were strengthened by different mechanisms (solid solution hardening and dispersion strengthening respectively). Chemical compositions and some important mechanical property data are shown in Table 1.

TABLE 1 Properties of the Starting Materials

Steel	C	Mn	Cr	P	Al	0.2% Proof Stress	UTS	$\bar{r}$	Uniform Elong.	Total Elong.
AK	0.06	0.25	0.02	0.012	0.030	175 MPa	312 MPa	1.65	23.3%	41.2%
RP	0.08	0.33	0.02	0.087	-	266 MPa	419 MPa	1.74	18.2%	31.4%
DP	0.045	1.21	0.57	0.017	0.055	235 MPa	449 MPa	1.18	20.3%	34.3%

Samples of each steel were pre-strained in (a) simple tension using very large sheets (b) plane strain tension using box sections with an internal former to prevent lateral contraction, and (c) equi-biaxial stretching of clamped specimens using a hollow punch of circular cross-section. In all cases the sheet surfaces were free from contact with the tooling and at least six different pre-strain levels were used. For simple and plane strain tension the applied stress was along the transverse direction of the rolled sheet.

Testing was carried out using simple tension tests at different angles in the pre-strained sheet. Results presented here relate to testing along the rolling and transverse directions. Great care was taken to avoid ageing of the pre-strained materials during storage and machining of the specimens for testing. That these precautions were successful was shown when sheets pre-strained in simple tension were subsequently tested in the same direction. Flow stresses and ductility values were found to be unaffected by the interuption and interposed specimen preparation. During testing, the length and width strains were monitored continuously with sensitive extensometers so that r-values could be determined at any stage of the test. Length strains and ductility measurements were made using a gauge length of 50mm. All results shown below are averages of two or more tests; agreement between duplicate tests was always very close.

RESULTS AND DISCUSSION

Typical tensile test results can be considered by reference to Fig.1 which shows data for the steel AK pre-strained in simple tension along the transverse direction of the sheet followed by tensile testing along the rolling direction. Flow stress - strain curves are shown in Fig.1(a) where the dashed line is the curve for the virgin material also tested along the rolling direction. To avoid confusion, many curves have had to be truncated and these are shown ending with arrows. Crosses indicate the point of load instability where necking commenced and beyond which stresses and strains could no longer be evaluated.

There is a fundamental difficulty in assigning an appropriate level to the

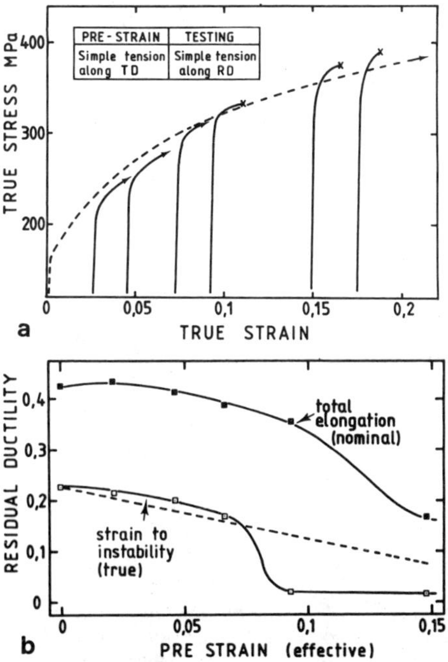

Fig.1 Tensile results for steel AK

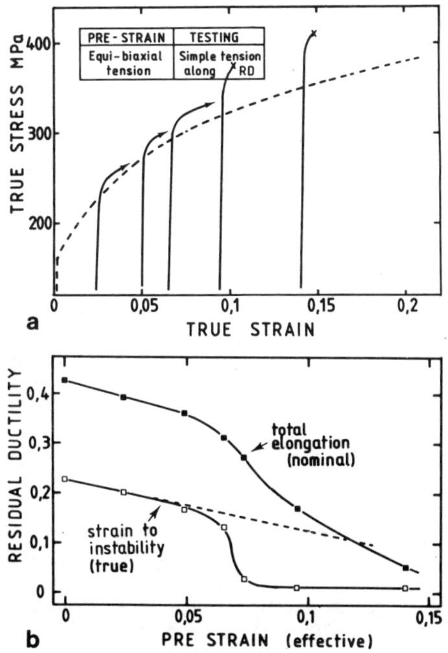

Fig.2 Tensile results for steel AK

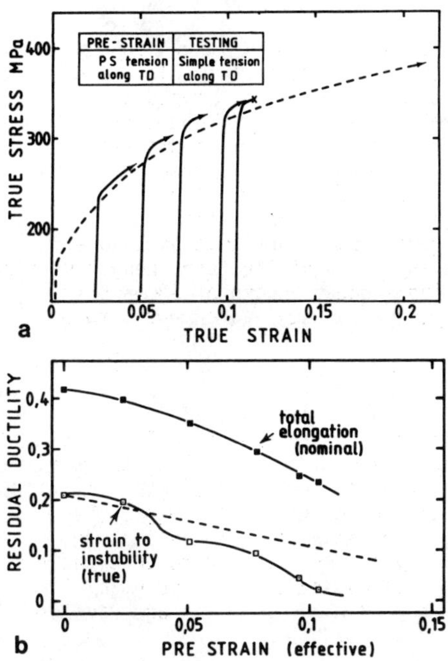

Fig.3 Tensile results for steel AK

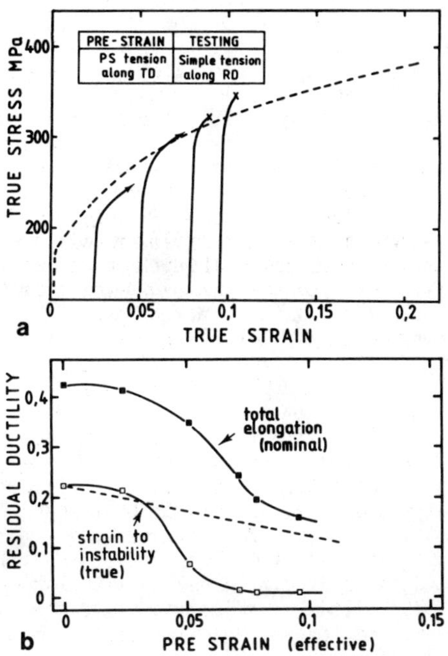

Fig.4 Tensile results for steel AK

pre-strain as the basis for the solid lines in Fig.1. A satisfactory measure of effective strain is debatable even for isotropic metals and the problem is still more acute when significant plastic anisotropy exists as in steels AK and RP. According to crystal slip calculations (Hosford and Kim, 1976), Hill's theory of anisotropic plasticity greatly over-estimates the range of effective strain in such steels. We have, accordingly, used the von Mises effective strain expression and neglected plastic anisotropy. This procedure cannot be strictly correct, but examination of the possible errors shows that none of the conclusions will be modified whatever the choice of expression for effective strain. The pre-strains involved here are also sufficiently small that changes in crystallographic texture are negligible.

For pre-strains of less than about 8% a Bauschinger effect can be seen where the subsequent flow stress is reduced while at higher pre-strains the flow stress increases above the corresponding value for the virgin material (we describe this effect as latent hardening). Figure 1(b) shows that the uniform elongation decreases gradually with pre-strain at first and then very rapidly at around 8% strain. Reasons for this behaviour, which confirms the earlier observations of Hutchinson, Arthey and Malmström (1976), will be discussed below. The dashed line in Fig.1(b) shows the expected behaviour assuming that the two stages of deformation are exactly equivalent to one another. The total elongation during testing also decreases with pre-strain level without showing an abrupt drop at any particular value.

Results for steel AK pre-strained equi-biaxially are shown in Fig.2. In this case the flow stress levels lie above those of the virgin material throughout and the latent hardening increases with pre-strain level. It is possible that the discrepancy at small pre-strains may be a result of neglecting anisotropy when evaluating effective pre-strain but this factor could not account for the higher strain behaviour. There is an abrupt fall in strain to instability around 7% pre-strain and a steady decrease in total elongation (Fig. 2(b)).

Pre-straining in plane strain tension followed by testing by simple tension in the same direction (steel AK) gave results shown in Fig.3. The level of flow stress is slightly increased and the ductility decreases more rapidly than would be expected on the basis of effective strain alone.

Plane strain tension followed by orthogonal simple tension produced a behaviour in steel AK which is shown in Fig.4. There is a Bauschinger effect initially which gives way to a latent hardening contribution as the level of pre-strain increases. This strain path change causes the most severe loss of ductility such that the uniform elongation falls very rapidly for a pre-strain of only 5%.

It was anticipated at the start of this work that the rephosphorised and dual phase steels might respond quite differently from each other and from the steel AK because of the different strength levels and types of strengthening mechanisms. However, a surprising result was that all three steels showed the same type of behaviour in almost every respect. Examples are given in Figs. 5 and 6 for samples deformed first in plane strain tension along the transverse direction and then tested in simple tension along the rolling direction. Comparison with the data for steel AK in Fig. 4 shows that the pattern of behaviour is virtually identical in each case. The phenomena in Figs. 1 to 4 seem to be quite general, at least so far as sheet-formable steels are concerned.

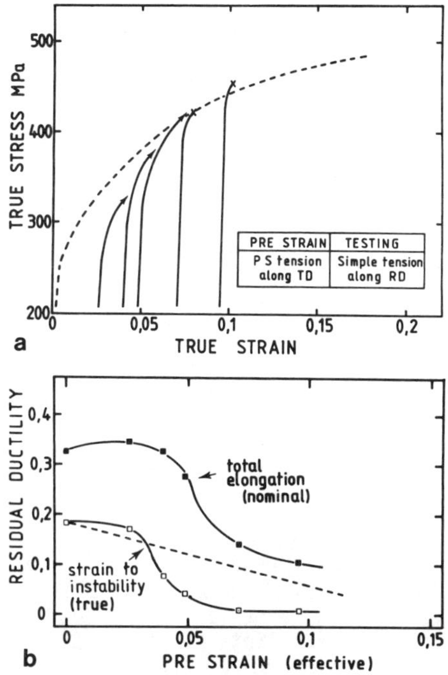

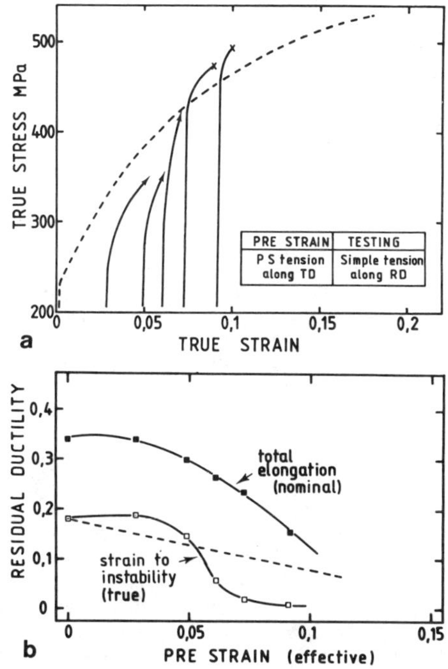

Fig.5 Results of testing for steel RP Fig.6 Results of testing for steel DP

Plastic strain ratios for the steels were found to vary according to the type and magnitude of the pre-strain and the extent of strain during testing. Some conventional r-values measurements are shown in Fig. 7 for steel AK.

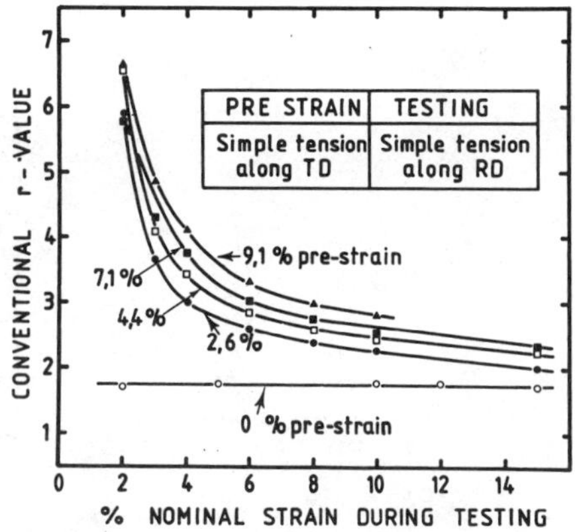

Fig.7 Measurements of conventional r-values during testing of steel AK

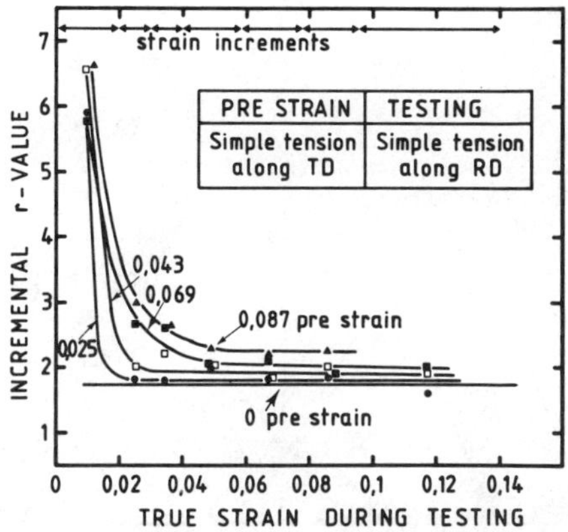

Fig.8 Same results as for Fig.7 plotted as incremental r-values for small strain increments

However, this measure of plastic anisotropy is based on the total accumulated strain during testing and not the current state of anisotropy which is a more fundamental property. When the same data are plotted in terms of incremental strains (Fig.8) it is seen that the change in anisotropy is a very transient phenomenon which is almost entirely eliminated within the first 3% strain during testing. For this reason we believe that the effect is associated with exhaustion of the mobile dislocations remaining from the pre-strain stage and should be considered separately from the other phenomena which persist to high strains.

We shall now consider reasons for the abrupt loss of ductility occasioned by changes in strain path, and present a means of rationalising the different observations. There are two factors which lead to premature instability and loss of ductility. Firstly, there is a loss of work hardening capacity as was demonstrated previously (Hutchinson, Arthey and Malmström, 1976). In addition, there is frequently an increase in flow stress which may be described as latent hardening. This flow stress effect has been emphasised by other workers (Laukonis and Ghosh, 1978). The effect of latent hardening can be estimated to a first approximation by translating the stress-strain curves horizontally until the stress levels become coincident. In most cases the residual ductility indicated from this manipulation is much greater than that actually observed suggesting that the major influence must be due to the loss of work hardening capacity. It seems reasonable to suppose that the dislocation structure produced in the first stage of deformation provides a latent hardening effect which is progressively eliminated during subsequent straining as the original structure is destroyed and replaced by a new one. The progressive decrease in this hardening component manifests itself as a loss of work hardening. Thus, the two effects which together contribute to instability and loss of ductility probably share a common origin in the changing dislocation structure.

In order to rationalise the different changes of strain path it is helpful to consider the deformation in terms of the coordinate system shown in Fig. 9.

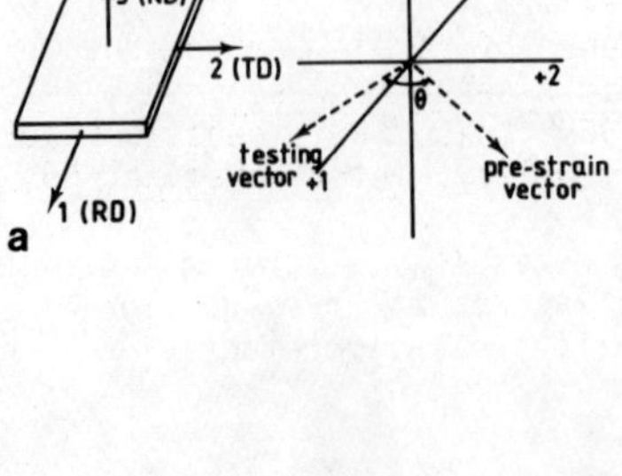

Fig.9 The coordinate system used to describe the strain vectors during pre-straining and testing

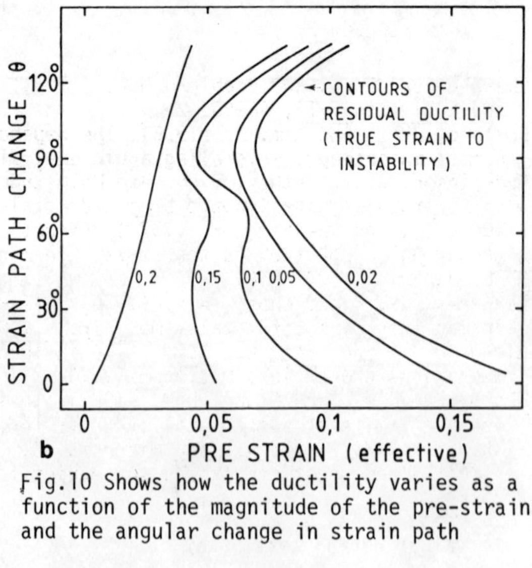

Fig.10 Shows how the ductility varies as a function of the magnitude of the pre-strain and the angular change in strain path

Since the principal axes are fixed in these experiments, both the pre-strains and the testing conditions may be described by vectors in a common strain-space. The change in strain state can then be associated with the angle between the two strain vectors which in the present work varies from 0° to 127°. As a consequence of the constancy of volume all the vectors are constrained to lie in the π-plane and so all angles are measured around a common axis. The final properties of the sheet should depend on the magnitude of the (effective) pre-strain and the angular change in strain path. Figure 10 shows a contour plot of the residual ductility as a function of these two parameters. It is very interesting that a clear minimum exists with strain path changes of approximately 90°. The material properties are therefore most strongly affected when the two strain states are nearly orthogonal. One may speculate that this condition corresponds to the greatest change in the active slip systems and the populations of mobile dislocations.

CONCLUSIONS

It is not possible to describe the plastic flow behaviour of steel by a single relationship between effective stress and effective strain when sharp changes in strain path occur during deformation. The levels of flow stress and work hardening are modified and there is a large but transient change in plastic anisotropy. Plain carbon, rephosphorised, and dual phase steels show similar behaviours.

Ductility is significantly reduced with non-linear straining because of the changes in flow stress and work hardening. The phenomenon is attributed to gradual removal of the prior dislocation substructure and its replacement by a new substructure. The effects are most severe if the two states of deformation are nearly orthogonal when described by vectors in strain-space.

ACKNOWLEDGEMENTS

The authors thank SERC and BL Cars for financial support and G.Davies, P.Bate and D.V.Wilson for helpful discussions.

REFERENCES

Charpentier, P.L., and Piehler, H.R. (1983). In-plane compression. J. Testing and Evaluation, 11, 46-55.

Hosford, W.F., and Kim, C. (1976). The dependence of deep-drawability on normal anisotropy; crystallographic analysis. Met.Trans, 7A, 468-472.

Hutchinson, W.B., Arthey, R.R., and Malmström, P. (1976). On anomalously low work hardening rates in pre-strained metals. Scripta Met., 10, 673-675.

Kikuma, T., and Nakazima, K. (1971). Effects of deforming conditions and mechanical properties on the stretch forming limits of steel sheets. Trans. ISI Japan, 11, 827-836.

Laukonis, J.V., and Ghosh, A.K. (1978). Effect of strain path changes on the formability of sheet metals. Met. Trans., 9A, 1846-1856.

Lloyd, D.J., and Sang, H. (1979). The influence of strain path on subsequent mechanical properties. Met.Trans., 10A, 1767-1772.

Müschenborn, W., and Sonne, H.M. (1975). Influence of strain path on the forming limits of sheet metal. Archiv. für das Eisenhüttenwesen, 46,597-602.

Yoshida, K., Miyauchi, K., and Mizunuma, S. (1969) Anisotropic behaviours of prestrained sheet steels and subsequent formability. Tetsu-toHagane, 55, 26-47.

RELATIONSHIP BETWEEN MICROSTRUCTURE AND HYDROGEN EMBRITTLEMENT IN FRACTURE MECHANICS ASPECT

J. Romański

Department of Mechanical Engineering, Technical University of Gdańsk, Poland

ABSTRACT

The relation between hydrogen embrittlement index and microstructure was given. Fracture mechanics approach as Bruecner's formula is used for studying the problem of hydrogen - induced cracks in steel construction. The correlation between hydrogen embrittlement and carbon equivalent was estimated.

KEYWORDS

Mechanical behaviour of metals, fracture mechanics, hydrogen embrittlement, microstructure, fractography.

INTRODUCTION

Problem of hydrogen - induced cracking of structural alloys has been a subject of intense study by a number of research groups in last years. It is well known that structural changes of low alloyed steels has the influence on the susceptibility to hydrogen induced cracking. It is possible to find some relation between stress and hydrogen level.In the present paper fracture mechanics is appliced to the problem of hydrogen - induced cracking. Much of this attention has been directed forward the behaviour of low alloyed steels with the level of hydrogen in the microstructure. A number of carbon equivalent equations have been tried as a one factor of structure intensity to hydrogen - induced cracking. One of the feature of hydrogen - induced cracking is that the fracture toughness parametr such as the critical size of tensile yield zone changes with the concentration of hydrogen at the crack tip. Information on the quantitative relationship between the fracture toughness parametr and hydrogen density in microstructure has never been obtained, because of laboratory difficulty. The damaging effects of hydrogen in microalloyed steels have been recognized and investigated using embrittlement index. To date, basically five major theories have been proposed to explain the embrittlement effects of hydrogen, but application of fracture mechanics gives new quantitative assesing this problem.

EXPERIMENTAL PROCEDURE AND MATERIALS

Materials used are C - Mn - Nb shipbuiding microalloyed steel, which have following carbon equiwalent from 0.31 to 0.45 %. Specimens with machined notch were tested for comparism. During research constant load delayed failure test were used according to IIW procedure.

RESULTS AND DISCUSSION

In the specimens, the cracking location is difficult to determine because of the complicated mechanism. No single mechanism can explain all the experimental result to date. Recent studies have suported a model which is connected with the influence of the hydrogen pressure, microstructure and triaxial stress in the notch area.

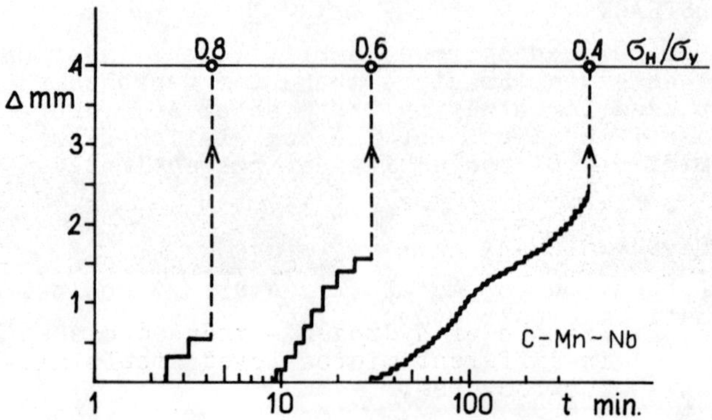

Fig. 1. Effect of applied constant stress concentration on hydrogen - induced crack growth.

Figure 1 shows the growth of crack for three levels of constant applied stress. The step curve or repetition of crack initiation and arrest is found during the growth of crack because of certain time, which is required for embrittling effects of hydrogen in the tip of the crack newely formed. This is the reason of diffirent fracture initiation time and the rate of propagation.

Figure 2 shows some examples of hydrogen induced cracking in different microalloyed steels.

Metallographical observations of microstructure shows diffirent mode of cracks. It is possible to observe hydrogen - induced cracking to be connected with the microstructure type and tiaxial stress level. The influence of hydrogen level is suposed to be more important for initiation than for propagation and final fracture of cracks. The fracture morphologies of specimens are shown in fig. 3.

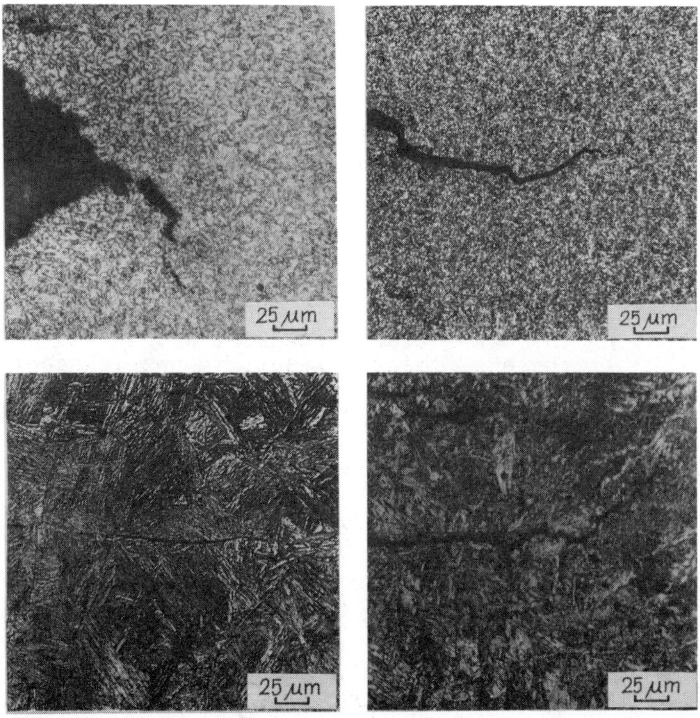

Fig. 2. Example of hydrogen - induced cracks in different microalloyed steels microstructures.

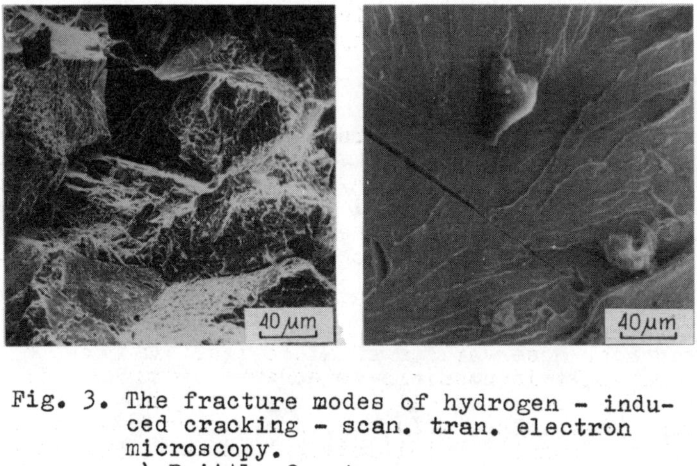

Fig. 3. The fracture modes of hydrogen - induced cracking - scan. tran. electron microscopy.
a) Brittle fracture.
b) Quasi - cleavage fracture.

The fracture mode of crack initiation was trangranular, quasi - cleavage with little contribution from intergranular failure. Sectoning of notched bar specimens showed that cracks has different shape and size, what was connected with influence of microstructure.
The present work shows that cracks is not so much effected by hydrogen. Bigger influence has stress and type of microstructure. Higher stress in notched specimens caused quasi - cleavage fracture with or without hydrogen. It is quite clear that hydrogen can strongly affect the yielding and flow characteristics of microalloyed steels.
The quantitative relation between hydrogen embrittlement and structure factor shows fig. 4.

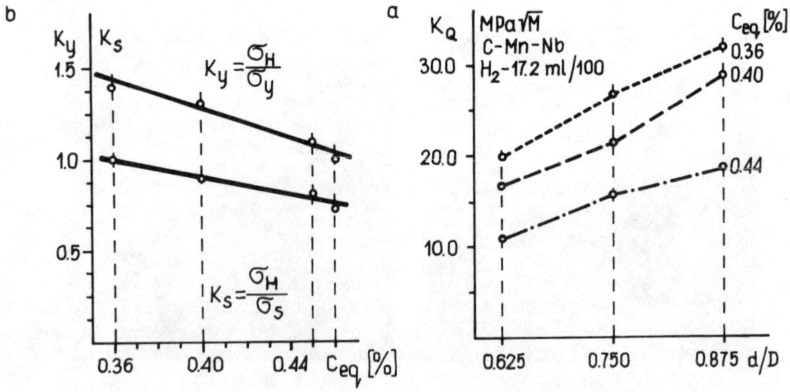

Fig. 4. Variation of K_Q in microalloyed steels with notch size and carbon equiwalent (a), relation between critical stress, ratio K_y, K_s and carbon equivalent (b).

The statistical analysis showed hydrogen - induced cracks susceptibility to be dependent on composition of the low carbon steels studied.
Figure 5 shows the relation between stress intensity factor and carbon equivalent.
The following formulae have been determined by the regression analysis:

$$I_H = 1.85 - 0.35\ C_{eq} \quad r^2 = 0.90$$

$$K_c = 37.4 - 41.5\ C_{eq} \quad r^2 = 0.89$$

$$K_y = 2.4 - 2.8\ C_{eq} \quad r^2 = 0.87$$

where r - correlation coefficient

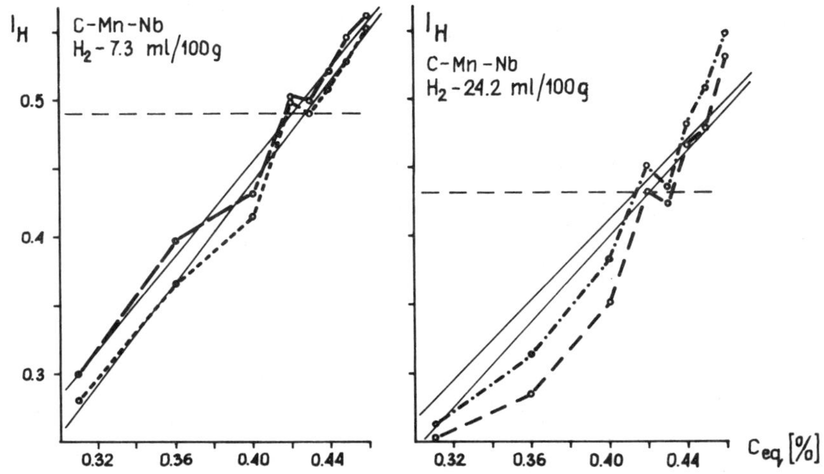

Fig. 5. Relationship between hydrogen embrittlement index and carbon equivalent for C - Mn - Nb steels.

It has been shown that certain microstructures are more prone to hydrogen induced cracking according to increase of carbon equivalent. This statistical correlation has been proposed to relate composition to cracking susceptibility.

CONCLUSION

From the above experimental results and discussions the following conclusions can be drawn:
The experimental results shows that carbon equivalent had greater embrittling effect than hydrogen level. The consideration of fracture mechanics may be used together with hydrogen embrittlement as a possibility for estimate criterion of the avoiding hydrogen induced cracks.
The correlation formulae can be adapted to the prediction of hydrogen induced cracks susceptibility data for avoiding this type of cracks in low alloyed steels.

ACKNOWLEDGMENTS

The autor expresses his sincere appreciation to Professor J. Carlsson, Division of Strenght of Materials and Solid Mechanics, Royal Institute of Technology for his support and help in research work. The autor also wishes to thank Dr N. G. Ohlson, Material Research Center, Royal Institute of Technology in Stockholm for his valuable discussions.

REFERENCES

Satoh, K., and M. Toyoda (1974). Proceedings of the Jap. U. S. Sem. S2, 71-81.
Erdmann, F., (1976) . Proceedings Mett. 4.243.
Pressouyre, G., Dollet, J., Baron, B., (1982). Rev. de Metallurgie, 161-176.
Vasudevan, R., Stout, R., Pense, A., (1981) . Welding J., 155-16
Takeda, Y., and Mc Mahon, C. J., (1981) . Met.Trans., 1255-1266.
Navara, E., and Easterling, K. E., (1982) . Scand. J. Mettalurgy, 169-172.
Carlsson, J., (1976) . Brottmekanik, Ingerijosforlaget, (Stockholm) .
Boniszewski, T., Watkinson, F., (1973) . Metal and Materials, 261-282.
Butnicki, S., (1979). Spawalność i kruchość stali. WNT, Warsaw.

AUTHOR INDEX

SUBJECT INDEX

X

Y

Z